# Principles of Organic Chemistry

**A SERIES OF BOOKS IN CHEMISTRY**

*Linus Pauling, Editor*

*Third Edition* / # Principles of Organic Chemistry

## T. A. GEISSMAN
*University of California, Los Angeles*

**W. H. FREEMAN AND COMPANY**
*San Francisco and London*

# Preface

A textbook on the principles of organic chemistry requires occasional revision because experience gained from its use discloses many ways in which it can be made more useful to the student. Continuing progress in the science and in the way the subject is presented in modern courses makes it necessary to remove some material that is no longer emphasized in the present-day classroom and to add new and illuminating examples of experimental fact and theory. Progress in the development of organic chemistry continues at a pace that is constantly increasing, and although a textbook cannot adequately record the advances that are being made on all fronts, it must present the subject in such a way as to provide the student with an understanding of its present state of development.

The basic purpose for which this book was originally intended has not been altered. It is to present organic chemistry as a blend of theory and experimental observation and to describe the nature and the reactions of organic compounds as expressions of the characteristic behavior of important functional groups. In an elementary course, too much emphasis cannot be placed upon fundamental general concepts, such as the formation and breaking of chemical bonds, the nature of acid-base reactions, the principles of stereochemical relationships, the recognition of the reactive sites in an organic molecule. It is for this reason that this book, too, emphasizes the nature of the important types of organic reactions rather than the descriptions of classes of compounds. It is hoped that, from the study of this book, the student will gain the ability to recognize why reactions take place as they do, to recognize the essential unity in classes of reactions that may appear to be quite various in kind, and to make an intelligent assessment of the ways in which a compound that may be quite new to his experience can be expected to behave.

Changes in the teaching of general chemistry, both in secondary school classes and in introductory college courses, have made it possible to enlarge upon certain topics and to de-emphasize others. The concept of the spatial arrangement of the atoms in organic molecules (stereochemistry) is introduced at the start of the book and is a recurring theme throughout. The description of the course of organic reactions as a series of events occurring in the course of the changes from reactants to products is discussed in terms of energy profiles, transition states, and stable and unstable intermediates. The consequences of resonance stabilization in determining both reaction rates and equilibria are given special attention.

The widespread use in the modern laboratory of physical methods for the study of organic compounds, and the accessibility of spectrometric instruments of several kinds in present-day undergraduate laboratories, require that the beginning student be introduced to the principles and application of ultraviolet, infrared, and nuclear magnetic resonance spectroscopy. These methods are described in separate chapters, and the spectral properties of organic compounds, and how they reflect the chemical behavior of these compounds, are frequently referred to. Mass spectrometry is described in a brief section.

The general plan of the second edition is retained. Throughout the third edition, emphasis is placed upon applications of the characteristic behavior of functional groups to structural diagnosis and to the recognition and characterization of compound classes. Even though the present-day organic chemist is more likely to use nuclear magnetic resonance or absorption spectrometry than to prepare a derivative or to perform a color test for the detection of functional groups, the instructional value of diagnostic chemical behavior in the recognition of chemical structures is so great that I have resisted the tendency to discard it in favor of the use of physical data alone.

The determination of structure by the use of physical data, chemical degradation, and synthesis is presented in a separate chapter and is also represented by additional examples in the text. Although many interesting examples of structure determination involve compounds of rather complex structures, it is in the study of such examples that the student comes to recognize that the characteristic behavior of the common structural units occurs in complex situations just as it does in the simpler cases in which such behavior was first introduced. Most of the examples of structure proof are drawn from the class of naturally occurring compounds, because it is among such compounds that one finds the widest diversity of compound classes, from the simplest structures to those of the most extreme complexity.

Certain features of the second edition have been retained and improved. The problems have been revised and new ones have been added; and an answer section is included.

This edition has not been greatly expanded, for it is not intended to be a reference work of encyclopedic scope. It is a textbook for the one-year college course in organic chemistry, and I hope that it does not encompass more than the student can deal with in this time. Certain classes of compounds are discussed only briefly, and others are not covered at all, because I feel that students should be urged to do additional reading—in more comprehensive books; in the smaller monographic paperbacks that are now available; in such works as *Organic Syntheses* and *Organic Reactions*; and in the periodical literature.

During the years in which the first and second editions of this book have been used, I have received many comments, criticisms, and suggestions. These have been of great help to me in the preparation of this edition.

T. A. GEISSMAN

*March, 1968*

# Contents

*Chapter one* / **Introduction.**

# Characteristics of Organic Compounds

The aim of this book is to present the chemistry of organic compounds in terms of the principles that govern their behavior and account for their properties. The number of known organic compounds is so large and the kinds of reactions they engage in are so numerous and varied that it would be nearly impossible—and certainly ill-advised—for one to attempt to master the subject merely by acquiring a collection of facts. Organic chemistry is not simply an accumulation of facts, but a logical and consistent body of interrelated ideas; it will be our goal to perceive the relationships not in terms of outward similarities but in terms of fundamental modes of behavior.

It is a highly satisfying intellectual experience to perceive a relationship that brings unity to a set of apparently unconnected observations, to discern the thread of design that binds the various parts of a complex tapestry into a harmonious pattern. But it is our purpose to go beyond recognition and explanation and, by adding imagination to understanding, to devise new experiments, prepare new substances, and apply past experience to the solution of new problems.

## 1.1 What is organic chemistry?

With the rapid growth of the science of chemistry in the past half-century has come its separation into a number of special areas: physical chemistry, analytical chemistry, biological chemistry, organic chemistry, and others. These areas are not mutually exclusive, for each permeates the others; but each is characterized by the emphasis that is placed upon it, by some special approach in the techniques utilized, or by the kinds of compounds that are dealt with. *Organic chemistry* deals with the chemical compounds of carbon, and principally with compounds in which carbon is combined with hydrogen, oxygen, nitrogen, sulfur, and the halogens. It is a special area of chemistry because of the enormous number of carbon compounds. About a million organic compounds are known, and the number is increasing at the rate of about five percent per year! More than ninety percent of these are synthetic substances, the remainder having been isolated from living organisms (animals, plants, fungi, microorganisms) and their fossil remains (coal, petroleum).

The naturally occurring organic compounds are of concern to the biochemist as well as to the organic chemist, and indeed it is in dealing with such substances that biochemistry and organic chemistry come together and lose their individual identity. The naturally occurring compounds include the proteins, fats, carbohydrates, vitamins, and hormones that make up the structure of living cells; many of the drugs that man uses to control disease and relieve suffering; the perfumes and colors of the plant world; and thousands of other substances that participate in the metabolic activities of living things.

The synthetic organic compounds are derived largely from natural sources of carbon—coal and petroleum—but are the products of man's voluntary ingenuity rather than of the involuntary activities of growing organisms. There is no limit to the number of organic compounds that can be made, and indeed the synthetic compounds that are known include many of the known naturally occurring compounds, which have been prepared synthetically as final confirmation of their structures. The only *kinds* of compounds that have so far resisted the attack of the synthetic organic chemist are the proteins, the complex carbohydrates, the nucleic acids, and the complex combinations of these that make up the essential stuff of the living cell. Although the complete synthesis of one of these substances, identical in structure and function with the natural material, has not yet been accomplished, this achievement will surely come in time. It is to the new generations of scientists who will reach these goals that this book is addressed, in the hope that it will help to impart that basic understanding of the scientific principles of organic chemistry that must underlie future progress.

## 1.2 The development of organic chemistry

The scientific revolution of the sixteenth and seventeenth centuries was characterized by the birth of a spirit of inquiry and skepticism that marked a new phase of the intellectual development of mankind. The influence of this spirit upon chemistry was felt at a time when alchemy had given way to the application of chemistry to medicine, and the dominance of an earlier mysticism was crumbling. The new approach to chemistry was the experimental attack, and, through its chief proponent, Robert Boyle (1626–1691), this new philosophy gave to chemistry the status of an independent science. The chemists of the seventeenth and eighteenth centuries soon began the task of systematizing the growing body of empirical facts that their experiments revealed and began to create generalizations into which numbers of facts could be grouped. Although these early theories and much of the work devoted to their study were of little importance to the development of organic chemistry, they did a great deal to define the area of intellectual activity that is the science of chemistry and thus to attract the attention of men of an inquiring turn of mind to these new studies.

Toward the end of the eighteenth century chemists began to turn to the examination of living organisms, and a great many compounds were isolated from plant and animal sources. The use of plants in medicine had been practiced since ancient times, and enlargement of interest in chemistry led to the study of plants of medicinal importance in the first years of the nineteenth century and to the isolation in crystalline form of such complex substances as strychnine ($C_{21}H_{22}O_2N_2$), quinine ($C_{20}H_{24}O_2N_2$), and morphine ($C_{17}H_{19}O_3N$), the structures of which were to remain unknown for another hundred years.

A bar to progress during these early stages of what was known as "organic" chemistry was the persistent belief that compounds formed in living organisms had properties and owed their formation to laws that set them apart from compounds of inanimate origins. This devotion to a doctrine of a vital force was the residue of an earlier mysticism, religious in origin, and was undoubtedly reinforced by the bewildering complexity of many of the known organic compounds, as well as by the fact that prior to 1800 the only organic substances known were produced by plants and animals and were not the products of man's invention.

The doctrine of vitalism persisted through the first quarter of the nineteenth century. It can be readily understood how a belief that was itself not rational could prevent the development of a rational interpretation of science. Such a belief could not, however, prevent the progress of experimental studies, and important developments took place, starting with Lavoisier's (1743–1794) recognition of the fact that organic compounds could be burned and that by weighing the products of the combustion accurately it was possible to determine the composition of a compound. Lavoisier had at his disposal only a limited number of compounds, most of them from vegetable sources, and his experiments led him to believe that organic compounds were composed of carbon, hydrogen, and oxygen. His contemporaries and followers soon discovered the presence of nitrogen, sulfur, and phosphorus in organic compounds of animal origin, and before long it was realized that there is no sharp distinction between organic substances that is based upon their origin. The importance of Lavoisier's contribution was that it introduced the *quantitative* element into the study of chemistry, and it so influenced further development of the subject that his work marks a real chemical revolution and can be regarded as the beginning of the modern science.

The residues of supersitition that were embodied in the doctrine of the vital force were soon to be removed. An important discovery was that of Wöhler

(1800–1882), who in 1828 found that urea, a typical organic substance that derives its name from urine, in which it is found, could be prepared by heating the "inorganic" compound ammonium cyanate:

$$NH_4NCO \rightleftharpoons NH_2CONH_2$$

ammonium          urea
cyanate

Wöhler's discovery did not immediately demolish vitalism, but the disintegration of that vague and enervating doctrine began to take place and continued from that time on, as similar instances of syntheses of other organic compounds from inorganic materials appeared in increasing numbers.

## 1.3   The introduction of analytical methods

Developed chiefly by Gay-Lussac (1778–1850) and Liebig (1803–1873), the quantitative analysis of organic compounds by combustion became a powerful tool for the discovery of many new facts. By 1830 the improvements in methods for determining the carbon and hydrogen content of organic compounds, as well as the analytical method for nitrogen (Dumas, 1800–1884), made it possible to ascertain the empirical composition of these substances with a high degree of accuracy. From then on, the development of organic chemistry proceeded with amazing rapidity. In less than fifty years there was developed a valid structural theory that survives to the present day; thousands of new compounds were discovered and synthesized; and an enormous chemical industry developed in Europe. It is a significant and perhaps fortunate turn of circumstances that the industrial revolution in England and Europe, with the consequent rise in production of the organic raw materials obtained from coal tar, took place during the period in which the theories that were to guide organic chemists in their exploitation of this rich source of organic chemicals were developed.

Thus we see how the scientific revolution of the seventeenth century set mankind on the road of free inquiry; how Lavoisier's chemical revolution a century later pointed out the importance of accurate, quantitative observations; and how, another century later, chemistry was at last established upon a firm base of fact and the beginnings of theory. The stage was now set for the twentieth-century discoveries on the nature of atoms and molecules and the forces that bind them together, and for the development of the modern theories of organic chemistry around which the discussions in this book are constructed.

## 1.4   Isomerism and the concept of structure

One of the most important discoveries that resulted from the development of accurate methods of quantitative analysis of organic compounds was the recognition that it is possible for two or more compounds to possess identical chemical *compositions*, yet to be quite different in their chemical and physical properties. Liebig, in 1823, found that both silver cyanate and silver fulminate, two quite distinct substances, have the composition AgCNO. Wöhler observed that both ammonium cyanate and urea were represented by the formula $CH_4N_2O$, and in 1828 he discovered that ammonium cyanate could be converted into urea by simple heating. Up to this time it was regarded as self-evident that substances of the same composition were identical. The observations of Liebig and Wöhler were soon followed by others of a similar nature, and Berzelius (1779–1848), a Swedish chemist, proposed the term *isomerism* to denote the relationship between two substances differing in properties but having identical compositions. We must also bring into this concept the molecular weights of the substances concerned, and define *isomers* as compounds having different properties but identical *molecular formulas*.

The recognition of isomerism was the first step toward the solution of the problem of the molecular structure of organic compounds. Berzelius recognized that isomerism could be explained only by supposing that the relative positions of the atoms in isomeric compounds must be different. The *fact* of difference—the experimental observations of composition and chemical behavior—leads to the question of the *reason* for the difference, and thus the *concept of isomerism has within it the concept of structure.*

## 1.5 Molecular and structural formulas

It will be seen that the molecular formula of a compound does not define either its chemical properties or the class to which it belongs. As simple a molecular formula as $C_2H_6O$ represents two quite different compounds: ethyl alcohol ($C_2H_5OH$) and dimethyl ether ($CH_3OCH_3$). Three quite different compounds are represented by $C_3H_8O$, and $C_4H_{10}O$ is the molecular formula for seven compounds.

It is evident that even if carbon, hydrogen, and oxygen atoms possess definite and invariable combining powers there can be more than one way in which a number of these atoms can be joined together; but there cannot be an unlimited number of ways. There exists, for example, only one compound $CH_4O$, only one $C_2H_6$, one $CH_2O$, one $C_2H_2O_2$, one $C_2H_2O_4$. But as the number of carbon atoms increases, the number of isomers increases rapidly, as the example of $C_4H_{10}O$ has shown.

The problem that confronted the chemists of Liebig's time was to express the constitution of organic compounds in terms of their internal nature and to develop a means of representing the spatial relationships of the atoms within molecules by means of formulas that not only clearly distinguished isomers from one another but expressed the chemical difference between them. One of the earliest observations contributing to the solution of this problem was made by Liebig and Wöhler, who showed (in 1832) that there existed certain groups of atoms that behaved as units in passing through a series of chemical transformations without change. The compound benzaldehyde, $C_7H_6O$, was converted to benzoic acid, $C_7H_6O_2$, and this in turn was transformed into benzoyl chloride, $C_7H_5OCl$, benzamide, $C_7H_7ON$, and methyl benzoate, $C_8H_8O_2$. All of these compounds can be regarded as consisting of the group of atoms, $C_7H_5O$, called the benzoyl *radical* (from *radix* = root), in combination with H, OH, Cl, $NH_2$, or $CH_3O$ in the compounds mentioned. Earlier, Gay-Lussac had observed that the cyanogen *radical* could appear in the compounds HCN, $(CN)_2$, BrCN, ClCN, and so on.

The discoveries and theories of the next twenty-five years culminated in 1858 in the publication by Kekulé (1829–1896), and independently in the publication by Couper (1831–1892), of a structure theory that for the first time represented the individual atoms in organic compounds and showed how they were joined. Kekulé proposed that the "combining power" of the elements was fixed, and that carbon had four, oxygen had two, and hydrogen had one combining unit (or, as we would say, a *valence* of four, two, or one). Thus, methane, $CH_4$, could be

represented by 

$$H-\overset{\overset{\displaystyle H}{|}}{\underset{\underset{\displaystyle H}{|}}{C}}-H,\quad \text{ethane by } H-\overset{\overset{\displaystyle H}{|}}{\underset{\underset{\displaystyle H}{|}}{C}}-\overset{\overset{\displaystyle H}{|}}{\underset{\underset{\displaystyle H}{|}}{C}}-H,\quad \text{and carbon } \textit{chains} \text{ could be}$$

constructed by linking together carbon atoms,

$$C-C-C-C-C \qquad \text{"straight" or unbranched chain}$$

$$C-C-\overset{\overset{\displaystyle C}{|}}{\underset{\underset{\displaystyle C}{|}}{C}}-C-C \qquad \text{branched chain}$$

and filling the remaining combining capacities of the carbon atoms with other atoms or groups.

We can now represent ethyl alcohol and dimethyl ether, both $C_2H_6O$, in a way that clearly shows the difference between them:

$$H-\overset{\overset{\displaystyle H}{|}}{\underset{\underset{\displaystyle H}{|}}{C}}-\overset{\overset{\displaystyle H}{|}}{\underset{\underset{\displaystyle H}{|}}{C}}-O-H \qquad H-\overset{\overset{\displaystyle H}{|}}{\underset{\underset{\displaystyle H}{|}}{C}}-O-\overset{\overset{\displaystyle H}{|}}{\underset{\underset{\displaystyle H}{|}}{C}}-H$$

$CH_3CH_2OH$  
ethanol (or ethyl alcohol)

$CH_3OCH_3$  
dimethyl ether

These formulas are *structural formulas*. Those in which all the bonds are shown are *graphic* structural formulas; the others (e.g., $CH_3CH_2OH$) are *condensed* structural formulas. Note that in both of the above structural formulas the

---

**Exercise 1.** Write graphic structural formulas for the three isomeric compounds $C_3H_8O$.

---

*group* of atoms $CH_3$— appears. This, the *methyl group*, is what in earlier years was called a *radical*. At the present time we prefer to speak of *groups* rather than radicals, since the latter term is used to denote actual chemical species with independent existence, which we shall refer to at a later time; the word *group* denotes a structural entity only: it is not a substance, but a component part of a substance.

The short lines that join the atoms together in the graphic formulas are symbols for the *bonds* between the atoms. We speak of the carbon-carbon bond, the carbon-hydrogen bond, the oxygen-hydrogen bond, and often abbreviate these by the terms C—C bond, C—H bond, O—H bond.

It is important to recognize that structural formulas are only conventional symbols for molecules. Ethanol is one compound, whether it be written

$$C_2H_5OH, \quad CH_3CH_2OH, \quad H-\overset{\displaystyle H}{\underset{\displaystyle H}{C}}-\overset{\displaystyle H}{\underset{\displaystyle H}{C}}-OH, \quad CH_3CH_2-O-H, \quad \text{or} \quad H-CH_2CH_2-OH$$

Different ways of writing the formula for a given compound may have particular advantages in describing one or another aspect of its behavior. For example, if we wish to depict the reaction of ethanol with sodium, we might write

$$C_2H_5OH + Na \longrightarrow C_2H_5ONa + \tfrac{1}{2}H_2$$

Its oxidation to acetic acid, in which the methyl group remains unchanged, could be written

$$CH_3CH_2OH \xrightarrow{\text{oxid.}} CH_3COOH$$

to show the kind of change that takes place in the oxidation. As we grow more familiar with ways of writing structural formulas we shall feel at ease even with the use of symbols such as Me for $CH_3—$ (*methyl*), *n*-Pr for $CH_3CH_2CH_2—$ (*normal-pr*opyl), and others. Thus, the structure

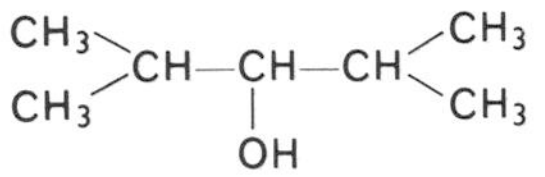

$$\begin{matrix} CH_3 \\ CH_3 \end{matrix} CH-\underset{\underset{\displaystyle OH}{|}}{CH}-CH \begin{matrix} CH_3 \\ CH_3 \end{matrix}$$

would be quite unnecessarily confusing if it were written out with each bond shown individually in a graphic formula. A simpler, yet adequate and comprehensible way of writing the structure would be

$$(CH_3)_2CHCHCH(CH_3)_2$$
$$\underset{\displaystyle OH}{|}$$

or even the more abbreviated form (where Me = methyl)

$$Me_2CHCHCHMe_2.$$
$$\underset{\displaystyle OH}{|}$$

In this book we shall use sometimes one, sometimes another, convention for writing the structure of a compound. The student should bear in mind that the formulas we write are symbols only; they are means of conveying ideas of structure and thus can be written in any way that permits the clearest communication of these ideas.

The student will probably find it easiest to use the graphic structural formulas, such as those on p. 6, until he gains confidence in his ability to observe the valence requirements of four for carbon, three for nitrogen, two for oxygen, and so on, but he will soon find it more convenient to use the condensed structural formulas and should accustom himself to these as soon as possible.

## 1.6 The nature of organic compounds

Organic chemistry is the chemistry of the compounds of carbon. The answer to the question of whether such substances as the metal carbonates, cyanides, and carbides are organic is an entirely arbitrary one, and is of no real concern because among the million organic compounds known, all but a small proportion contain hydrogen, and by far the greatest number are composed of carbon, hydrogen, and one or more other elements—usually oxygen or nitrogen or both.

Let us examine a list of 6,500 organic compounds that has been compiled for a well-known chemical handbook. Of these, only about 70 (1%) contain no hydrogen, and most of these are derived from hydrogen-containing compounds by the replacement of hydrogen by halogen atoms. Taking from this list of 6,500 compounds those that contain six carbon atoms only, we find those abundances listed in Table 1-1.

The list in Table 1-1 is an arbitrary sample that includes only a small percentage of the known six-carbon-atom compounds. It does show that the preponderance of compounds contain hydrogen as well as carbon, and that the bulk of these also contain oxygen or nitrogen.

Everyday experience has already made us familiar with other characteristics of many organic compounds. The burning of wood, paper, coal, and petroleum is evidence of the susceptibility of organic substances to oxidation. The end products

**Table 1-1.** *Some compounds containing six carbon atoms*

| Elements in $C_6$ Compounds | Number Listed | Percentage |
|---|---|---|
| C and O, N, X, or S only | 8 | 1.5 |
| C, H only | 24 | 4.6 |
| C, H, O only | 185 | 35.2 |
| C, H, O, N only | 128 | 24.3 |
| C, H, N, X only | 58 | 11.0 |
| C, H, X only | 56 | 10.5 |
| C, H, O, X only | 55 | 10.4 |
| C, H, O, S only | 13 | 2.5 |

of such oxidations are carbon dioxide (the ultimate oxidation state of carbon), water, and other elements (or their oxides) that may be present. The charring of paper, the caramelization of sugar, and the roasting of foodstuffs show us that heat can cause changes in and decomposition of organic substances. Extreme temperatures cause *pyrolysis*, with the result that extensive decomposition, often to simpler compounds that escape as vapors, takes place. The "coking" of coal, with the production of a distillate known as coal tar, and the (now obsolete) process of "destructive distillation" of wood, with the formation of a distillate containing methanol (wood alcohol), are examples of pyrolysis. Coal and wood, however, are not typical examples of organic compounds because of their complex nature and the high molecular weight of their component substances; and whereas most organic materials, simple or complex, are decomposed at high temperature, most of them survive the moderate elevation of temperature required to melt them if they are normally solids or to distill them if they are liquids.

*Solid organic compounds* usually melt to liquids when heated to temperatures which lie for the most part between about 30°C and 400°C. Let us have another look at the list of 6,500 compounds. Examining the physical properties that are recorded in our handbook, we find the data that are listed in Table 1-2. Of our arbitrary sample of 6,500 compounds, about 2,100 are liquids at ordinary temperature (those that are gases are relatively very few).

*Liquid organic compounds* are those that are liquids at "ordinary" temperatures. It is obvious that this description imposes quite arbitrary limits, since a *gas* is liquid at a sufficiently low temperature and a *solid* is liquid if heated above its melting point. Liquid compounds may be mobile, nonviscous, colorless substances, such as ethyl alcohol, carbon tetrachloride, or gasoline; some, such as glycerol (glycerin), are viscous and nonvolatile. The *boiling points* of liquids may vary over quite as wide a range as do the *melting points* of solids, but those with high boiling points (in the range of 300 to 400°C) often suffer partial or extensive decomposition when an attempt is made to distill them at ordinary (atmospheric)

**Table 1-2.** *Melting points of some organic compounds*

| Ranges of Melting Points | Number Listed |
|---|---|
| 31–100°C | 1580 |
| 101–150°C | 1020 |
| 151–200°C | 860 |
| 201–300°C | 840 |
| over 300°C | 100 |
| (over 360°C) | (16) |
| *Total* | 4400 |

pressure. For high-boiling liquids, distillation at reduced pressure (1 to 10 mm of mercury, or about 0.01 atmosphere) is usually practiced.

The melting point (or the boiling point) of an organic compound is a valuable and characteristic physical constant. It can usually be determined with ease, rapidity, and accuracy, and it is a value that is of considerable usefulness in the recognition of identity or the establishment of structure. Pure compounds usually melt sharply over a very narrow range of temperature; the presence of impurities usually causes a broadening of the melting point such that the impure compound may melt over a range of several degrees and at a temperature lower than that at which a pure specimen melts. Thus, the melting point serves as an index of purity as well as an aid in identification.

## 1.7 Relationship of melting point and boiling point to structure

The boiling point of a compound can often give a clue that aids in the establishment of its structure. Within a given class of compounds—for example, in a group of alcohols—the boiling point rises as the molecular weight increases (see Section 4.15). There is, however, no such simple relationship between boiling point and molecular weight when compounds of different types are compared. The influence of structural type upon boiling point will be discussed in Chapter 4.

There is no useful general relationship between structure and melting point or between molecular size and melting point. Within certain restricted groups of compounds that possess common structural features, a higher degree of molecular symmetry is usually associated with a higher melting point. This statement is a generalization to which exceptions exist, and, like many generalizations that come from the accumulation of experience, it is useful only to the person who recognizes its limitations.

## 1.8 The determination of structure

In this chapter we shall be able to describe only the first steps toward the determination of the structures of organic compounds. This problem is one to which organic chemists must apply all of their skill and knowledge. There is no single systematic way or set of rules for establishing the structures of all organic compounds. There are, however, certain systematic ways of proceeding with an attack upon a problem of structure determination and of gathering the information that must be possessed before the structure can be learned.

First, we must recognize that the chemist deals with *substances*, and uses theories and concepts of structure as a guide in his experiments and as a final means of generalizing the results of his experiments. Such abstractions as methods of writing structural formulas, theories of the nature of reactions, and concepts of structure and of the forces that bind atoms together into molecules are generalizations reached through *experimental observation*, as well as aids in the design and performance of new experiments.

Let us suppose that we are confronted with an organic compound of unknown composition and structure. It may be a solid or a liquid, or even a gas, but it is a *substance*, whose appearance may give us no clue to its chemical nature. Our ultimate objective is to be able to write a complete structural formula that will account for its chemical and physical properties and enable us to predict ways in which it might behave and be used.

The first steps in the study of an organic compound are as follows:

1. Its purification by crystallization, distillation, chromatography, or other means, and the determination of those physical properties that will serve to characterize it (melting point, boiling point).

2. The determination of the kinds of elements it contains; if the compound is known to be organic, the *qualitative analysis* is usually confined to tests for the presence of nitrogen, halogens, sulfur and, for salts, metals.

3. The *quantitative analysis* for the percentage composition of the compound in terms of carbon, hydrogen, oxygen, and any other elements it has been found to contain.

4. The calculation of an empirical formula and, if necessary, the determination of the molecular weight in order that the molecular formula can be obtained.

These four initial stages in the attack upon the problem of structure determination are discussed in the following sections.

## 1.9  Purification of organic compounds

It is usually unprofitable to attempt to determine the melting point, boiling point, or other physical property of a compound that is, or may be, impure. It is particularly inadvisable to place confidence in analytical values unless the purity of the sample analyzed is assured. Purification may be a process of removing small amounts of contaminating materials, or it may be a separation (or *isolation*) of a compound from a mixture in which the undesired substances are present in relatively large amounts.

Solid compounds are effectively purified by a process of *recrystallization*. The compound is dissolved in a solvent in which it is more soluble when hot than when cold. The compound separates as crystals (*crystallizes*) when the solution is cooled, and impurities, if present in small amounts, usually remain in solution. The compound is collected on a filter, washed free of the solution with fresh solvent, dried, and its melting point is taken. A second recrystallization is carried out, and the melting point is again determined. When the melting point no longer changes upon further recrystallization, the substance may be regarded—at least provisionally—as pure and suitable for analysis.

Liquid compounds that can be boiled without decomposition are usually purified by distillation. Impurities that are more volatile than the compound to be studied will distill at a lower temperature and thus will be removed and discarded before the desired compound reaches its boiling point. Less volatile or nonvolatile impurities will remain behind as a residue after the desired compound has been removed. With refined apparatus, separations by distillation can be made with ease and completeness.

### 1.10  Chromatographic separations. Column chromatography

A technique that is widely used by organic chemists in the separation of mixtures of organic compounds into the several pure constituents of the mixture is known as *chromatography*. The term comes from the word *chroma*, for color, because in its earliest applications the method was used for the separation of mixtures of colored compounds, and resolution could be followed by visual observation. In its simplest form, chromatography consists of passing a solution of a mixture of compounds through a column packed with a suitable adsorbent and collecting the effluent solution, or eluate, in a series of receivers. The constituents of the mixture pass through the column at different rates, with the result that a series of bands appear on the column and move down in sequence. The solution flowing out of the column may be collected in a series of receivers, and as each band moves down and out of the column it is collected separately in a number of fractions, which are combined and subsequently evaporated. Alternatively, the column packing may be extruded after the bands have developed and each band then removed by cutting the cylinder of adsorbent into sections. Each section is then extracted separately with a suitable solvent to give separate solutions of each component of the mixture.

When colorless compounds are separated by chromatography, the usual procedure is to collect numerous arbitrary fractions, locate those in which individual compounds are present (for example, by a color reaction or by observation of some physical property), combine the fractions into groups, each containing one component of the mixture, and evaporate each group of fractions separately.

In Figure 1-1 is shown a diagram of a typical chromatographic column. Many kinds of packing materials are used, depending upon the kinds of compounds being separated. The commonest, for organic compounds, are alumina and silica gel. The choice of the solvents used for development of the column and for elution of the separate components will depend upon the character of the compounds in the mixture. The following generalizations can be made: the lowest elution power is found in saturated petroleum fractions ("petroleum ether," pentane, hexane), followed by benzene, chloroform, ether, and methanol or ethanol; the ease of elution from the column increases in this order: saturated hydrocarbons (alkanes), unsaturated hydrocarbons, aromatic hydrocarbons, ethers, esters, ketones, alcohols, phenols, acids. The usual procedure is to allow the mixture to be separated to be absorbed on the top of the column, and then to

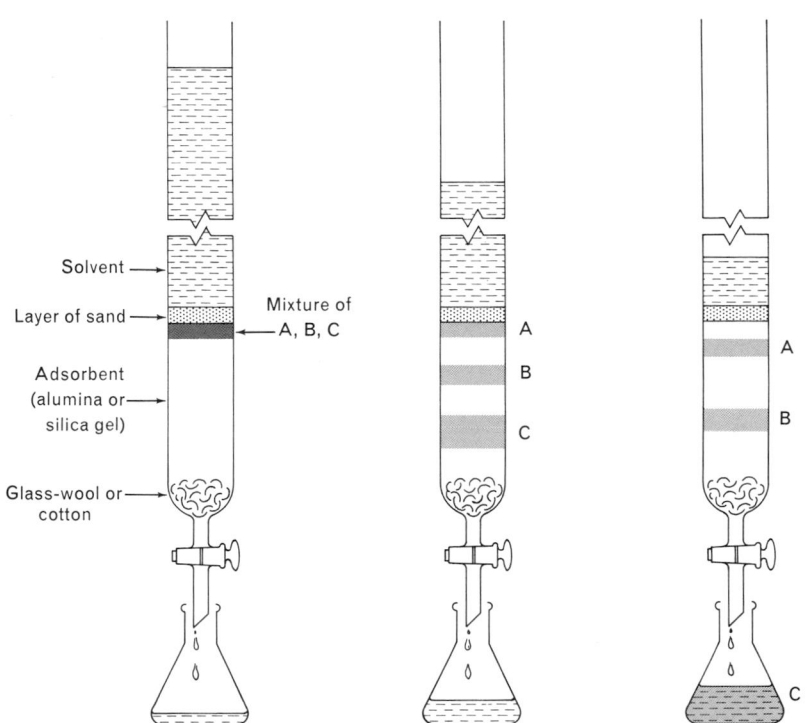

Solvent

Layer of sand

Mixture of
A, B, C

Adsorbent
(alumina or
silica gel)

Glass-wool or
cotton

A

B

C

A

B

C

**Figure 1–1.** Separation of organic compounds by column chromatography. *(a)* Start: mixture adsorbed on top of column. *(b)* Development of column proceeding. *(c)* Faster moving component has been eluted from column and collected in receiver.

elute, with the collection of fractions, successively with hexane, hexane-benzene mixtures of gradually increasing benzene content, benzene, benzene-ether mixtures with increasing ether content, and finally ether and ether-methanol mixtures. Many variations of this routine are possible.

## 1.11   Establishment of purity by chromatography

The use of a thin layer of an absorbent permits the separation of the components of a mixture on a sheet or plate, upon which they form individual bands or spots. These can be examined visually or under irradiation with ultraviolet light (which causes many compounds to fluoresce), or they can be subjected to the action of suitable reagents, with the production of colored products that can be observed visually. There are two common techniques for performing this kind of chromatography.

*Paper chromatography.*   The compound or mixture is placed at the bottom of a sheet of an absorbent paper (filter paper) as a small spot. The end of the sheet is dipped into a solvent and the solvent allowed to rise up the paper strip by capillarity. The separation is effected to some degree by differential absorption by the paper fibers, but to a greater degree by differential partition of the compound between the developing solvent and the water phase that is present in paper under ordinary laboratory conditions. This is called *paper partition chromatography*. If a single pure compound is applied to the paper, it will move from the origin, usually lagging behind the advancing solvent front, and form a single spot at some point on the paper. If a mixture is applied to the paper, a series of spots, corresponding to the components of the mixture, will usually result.

*Thin-layer chromatography.*   A thin (0.25-mm thickness) layer of a specially prepared silica gel is spread uniformly (as a slurry) on a glass plate, and, after drying, serves as the absorbent film. The compound or mixture is applied as a small spot at one end of the plate and the chromatogram is developed as in paper chromatography. The individual components may be seen as discrete spots, either

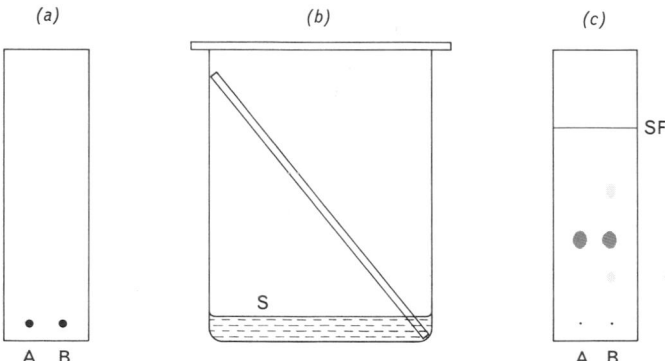

**Figure 1–2.** Thin-layer chromatography. *(a)* Compounds A (authentic specimen) and B (compound A containing small amounts of impurities), at the origin of a thin-layer plate. *(b)* Plate being developed by upward flow of solvent, S, in a tank with a close-fitting lid. *(c)* Plate after removal from tank, showing a single spot of pure A and minor contaminants in B (SF = solvent front), after development by a suitable spray reagent.

by visual examination or by the use of a reagent (applied as a fine spray) that will produce a colored spot (Figure 1-2).

Both paper chromatography and thin-layer chromatography are ordinarily used as analytical procedures, because the amounts of materials that are applied to such chromatograms are usually very small. Both methods can be adapted to separations on a preparative scale, however, by the use of thick paper or thicker (1- or 2-mm) layers of silica gel. These separation procedures are extraordinarily sensitive; amounts of materials as small as 1 microgram can often be detected with ease.

Identification of compounds on chromatograms is usually effected by applying a known compound to the chromatogram and noting the rate of travel of the known and unknown compounds on the same paper sheet or thin-layer plate. Impurities in a compound whose purity may be suspect can usually be detected by the appearance of one or more spots in addition to that of the principal component.

## 1.12 Vapor phase chromatography (VPC)

Compounds that are gases, or that can be vaporized at temperatures below their decomposition points, can be separated by chromatography in the vapor phase. In this method the mixture of compounds to be separated is passed in the vapor phase in a current of inert gas (nitrogen or helium) through a column containing a solid porous support that is impregnated with a nonvolatile liquid. The compounds are separated by differences in their partition between the stationary liquid phase and the moving gaseous phase; in many cases they move out of the column in successive discrete fractions of the separate, pure components. Their presence in the exit gases may be detected and recorded (*analytical VPC*), or they may be collected in suitable traps (*preparative VPC*). VPC is a very efficient method for separation of mixtures and is widely applied in organic chemical studies.

## 1.13 Melting points and boiling points

Organic compounds are commonly characterized by their melting points, if solids, or their boiling points, if liquids. *Boiling points* can be measured in the course of a purification by distillation by observing the temperature of distillation and the pressure. When the distillation is carried out at atmospheric pressure the barometer reading should be recorded along with the temperature of distillation. For example,

| | |
|---|---|
| *n*-butyric acid | b.p. 163.5°/757 mm |
| chloroacetic anhydride | b.p. 163°/116 mm |

Distillations at reduced pressure are often necessary, since ordinary distillation may require heating to a temperature high enough to cause some pyrolytic decomposition. The pressure is reduced by the use of a vacuum pump of a suitable kind and is measured with a manometer.

*Melting points* of solid compounds are commonly determined by placing the sample in a thin-walled glass capillary tube and inserting the tube into a bath heated by an external flame or an electrical heating coil. The sample is mounted close to the bulb of the thermometer used to measure the temperature. The melting point is recorded as a range of temperature (usually of one degree or less), since during the interval of time that is required for the sample to change from a crystalline solid to a liquid the temperature of the heating bath is slowly rising. The observed values are recorded as in the following examples:

> acetanilide　　m.p. 113–114°C
> glycine　　　　m.p. 232–236°C dec.

The term "dec." following a melting point means that the compound decomposes rather than melts, or that it decomposes during the process of melting. Melting points of this kind are subject to some uncertainty, since they often depend upon the rate at which the bath is heated. Most crystalline compounds, when *pure*, melt *sharply*, changing from a "dry" solid state to a clear liquid in an interval of about a degree (or less) at their melting points.

Melting points and boiling points are useful physical properties and are valuable aids in *identifying* organic compounds. The *identification* of an organic substance is a process of proving the compound to be identical with a compound that is already known and has been described; it is a process of comparison and recognition. *Structure proof* is quite a different process. It is the establishment for the first time of the structure of a compound. Physical properties such as melting points and boiling points are of some, but limited, value in the establishment of structure; their principal importance in structure proof is as an index of purity in the first steps of the process.

## 1.14　Qualitative analysis for the elements present in an organic compound

After the compound is purified and before the quantitative analysis is carried out, it is necessary to determine whether it contains any elements other than carbon, hydrogen, and oxygen. We may suppose that in most cases our observations of the compound during its purification have assured us that it is organic. If this is not certain, we can prove that it is organic by a few simple operations. If a sample of an organic substance is heated on a piece of platinum foil in the flame of a burner, it will burn or char, and if the latter occurs the charred mass will at length burn away. If the compound is a metallic salt (or an organic acid, for instance), the metal or its oxide will be left as a residue that can be identified by standard methods of inorganic analysis. Another way of testing for the presence of carbon and hydrogen is to mix the sample with dry, powdered copper oxide and heat the mixture in a test tube to redness. Water, formed from hydrogen in the compound, will condense as drops on the colder parts of the test tube; carbon dioxide, formed from the carbon in the compound, can be recognized by passing the gases formed into barium hydroxide solution.

Other elements are detected by means of the *sodium fusion test*, in which a sample of the compound to be analyzed is dropped onto metallic sodium in a test tube that has been heated to redness in the direct flame of a burner.* In the vigorous reaction that ensues the compound is decomposed, and the sodium combines with halogens, sulfur, and nitrogen that may be present to form sodium halides, sodium sulfide, and sodium cyanide. For example,

$$CH_3SO_2NHCH_2CH_2Br \xrightarrow[\text{heat}]{\text{Na}} NaBr, Na_2S, NaCN + \text{carbonaceous decomposition products}$$

The charred mass that results from the reaction is cooled, boiled with water, and the solution is filtered. The presence of *halogen* is recognized by the formation of the insoluble silver halide when silver nitrate solution is added to the acidified (with nitric acid) test solution. *Sulfur* is detected by the formation of black lead sulfide when lead acetate is added to another portion of the solution which has been acidified with acetic acid. *Nitrogen* is detected by a test for cyanide ion: the reaction between cyanide ion and ferrous hydroxide yields ferrocyanide ion, and this, upon the addition of a ferric salt, is transformed into the deeply colored, characteristic "Prussian blue," ferric ferrocyanide.

## 1.15   Quantitative analysis for elements other than carbon and hydrogen

If the qualitative analysis shows the presence of halogen, sulfur, or nitrogen, these must be determined by separate quantitative analyses. Analyses for halogen can be carried out by decomposing a sample of the compounds with nitric acid and silver nitrate at a high temperature (200°C) in a sealed tube (the Carius method) and collecting and weighing the silver halide that is formed. Sulfur and phosphorus can be determined by nitric acid oxidation in the same way, the resulting sulfate (or phosphate) being estimated by the usual methods.

An alternative procedure is to burn the sample in a Parr bomb with sodium peroxide, which oxidizes carbon and hydrogen to carbon dioxide and water and converts halogen, sulfur, and phosphorus into the alkali halide, sulfate, and phosphate.

*Example*

A 7.96 mg sample of an organic compound, known to contain bromine, was heated at 200°C in a sealed tube with 0.5 ml of fuming nitric acid and 1 g of silver nitrate. After cooling,

---

\* Other methods of carrying out this decomposition are also used. For example, the compound to be tested is allowed to react with a strongly heated mixture of powdered magnesium and potassium carbonate. The overall result is the same: the compound is decomposed with the formation of the inorganic ions noted above.

the tube was opened, water was added, and the contents were filtered. The silver bromide was dried and weighed, and amounted to 7.55 mg.

The percentage of bromine in the sample can now be calculated:

$$7.55 \text{ mg AgBr} \times \frac{80}{187.88} = 3.21 \text{ mg Br}$$

$$\% \text{ Br} = \frac{\text{wt. of bromine in sample}}{\text{wt. of sample}} \times 100 = \frac{3.21}{7.96} \times 100 = 40.3$$

*Nitrogen* is usually determined by the *Dumas* method. The Dumas method depends upon the fact that when a nitrogen-containing organic compound is heated strongly with copper oxide it is decomposed into carbon dioxide, water, and nitrogen and its oxides. In this procedure a weighed sample of the compound to be analyzed is mixed with powdered copper oxide and heated in a tube through which pure $CO_2$ can be passed. The exit end of the tube is attached to a gas-measuring burette (a nitrometer) filled with a strong solution of potassium hydroxide. The body of the tube is packed with copper and is heated in an electrical furnace; the copper reduces oxides of nitrogen to pure nitrogen. The tube is then swept with a stream of carbon dioxide, and the issuing gases are conducted into the nitrometer where the carbon dioxide is absorbed and the nitrogen collected. The amount of nitrogen in the sample is read directly as a volume on the calibrated burette, and is corrected to standard conditions of temperature and pressure in the usual way.

*Example*

3.36 mg of an organic compound gave 0.45 ml of nitrogen (corrected to 0°C, 760 mm). This volume of nitrogen is equal to $0.45/22.4 \times 28 = 0.56$ mg:

$$\% \text{ nitrogen} = \frac{0.56}{3.36} \times 100 = 16.75$$

## 1.16 Carbon-hydrogen analysis

The quantitative analysis of organic compounds is the most important of the initial stages in the determination of structure. The analysis tells us what elements the compound contains and the proportions in which they are present. With this information we can calculate the simplest, or *empirical*, formula. The true, or *molecular*, formula can then be written when the molecular weight is known.

Since all organic compounds contain carbon and nearly all of them hydrogen, the analysis for these two elements is the most universally applicable to organic substances. Carbon-hydrogen analysis is carried out as a routine operation on most organic compounds that have been prepared for the first time. Even if the compound is one that has been synthesized by methods that leave no doubt about its structure, a carbon-hydrogen analysis is reported as confirmatory evidence of its constitution, and most of the chemical journals of the world require that such confirmatory analytical data accompany a description of the work reported.

This method of analysis of organic compounds is essentially that developed by Liebig and Gay-Lussac from the procedures first used by Lavoisier. The modern refinements are principally a reduction in the size of the sample from the 0.5 to 1.0 g used in Liebig's day, to about 0.1 g used early in the twentieth century, to the present 3 to 5 mg. This refinement was made possible by improvements in the balance utilized. The modern microbalance is capable of weighing samples of a few milligrams with a precision of 0.002 mg.

If a sample of ordinary sugar, sucrose, is burned in a stream of oxygen, the reaction that occurs is

$$C_{12}H_{22}O_{11} + \text{excess } O_2 \longrightarrow 12CO_2 + 11H_2O$$

Ethanol would be burned according to the equation

$$C_2H_5OH + \text{excess } O_2 \longrightarrow 2CO_2 + 3H_2O$$

The combustion is carried out in a quartz tube packed with granular copper oxide. The tube is heated for about 8 inches of its length in a furnace kept at about 750 to 800°C. The sample, in a platinum boat, is heated with a flame or an auxiliary electric heating coil while a slow stream of carefully dried oxygen is passed through the tube. The combustion of the compound is completed by the heated copper oxide in the main body of the tube (and any copper formed is reoxidized by the oxygen), and the products of the combustion are swept by the oxygen stream into accurately weighed absorption tubes containing, first, anhydrous magnesium perchlorate (Anhydrone), and, second, a specially prepared material containing asbestos impregnated with sodium hydroxide (Ascarite). The absorption tubes are weighed on the microbalance, and the weights of water in the Anhydrone and carbon dioxide in the Ascarite are thus determined.

The *results* of a carbon-hydrogen analysis are expressed as percentages of carbon and hydrogen. For example, a 3.42 mg sample of a compound was found to yield 5.28 mg of $CO_2$ and 1.98 mg of water:

$$\text{wt. of carbon} = 5.28 \times \frac{12.01}{44.01} = 1.44 \text{ mg}$$

$$\% \text{ of carbon in sample} = \frac{1.44}{3.42} \times 100 = 42.11$$

$$\text{wt. of hydrogen} = 1.98 \times \frac{2.016}{18.016} = 0.220 \text{ mg}$$

$$\% \text{ of hydrogen in sample} = \frac{0.220}{3.42} \times 100 = 6.43$$

The unaccounted for 51.45% [100 − (42.11 + 6.43)] is taken as the oxygen content if it has been found that no halogen, nitrogen, or other elements are present in the compound. If one or more of these is present, separate analyses must be carried out by the methods already described.

## 1.17 Determination of the empirical formula

Let us carry out the combustion of a sample of 4.64 mg of an organic compound that we have shown by qualitative analysis to contain no elements other than C, H, and O. There is formed 11.70 mg of $CO_2$ and 1.46 mg of water:

$$\% \, C = \frac{11.70 \times 12/44}{4.64} \times 100 = 68.80$$

$$\% \, H = \frac{1.46 \times 2.016/18.016}{4.64} \times 100 = 3.51$$

$$\% \, O \text{ (by difference)} = 27.69$$

The relative *atomic* proportions of C, H, and O can be calculated by dividing each of the above values by the atomic weight of the element it represents (Table 1-3). The atomic ratios of $C:H:O$ are thus $5.74:3.49:1.73$, or $3.32:2.02:1.00$. That is, we could express the empirical formula as $C_{5.74}H_{3.49}O_{1.73}$ or $C_{3.32}H_{2.02}O_{1.00}$, but since a molecule cannot contain a fractional part of an atom, these represent mathematical relationships only. The empirical formula represents only the relative numbers of atoms, and so the formula $C_{3.32}H_{2.02}O$ contains the atoms in the same proportions as does the formula $C_{10}H_6O_3$, which is the *empirical* formula of our compound.

We can now confirm our result by calculating the percentage composition of a compound having the formula $C_{10}H_6O_3$:

carbon   $= 10 \times 12^* = 120$
hydrogen $= 6 \times 1.008 = 6$
oxygen   $= 3 \times 16 = 48$

$$\text{\textit{Total}} \quad \overline{174}$$

$\% \, C = 120/174 \times 100 = 68.95$ calc'd; 68.80 found,
$\% \, H = 6/174 \times 100 = 3.45$ calc'd; 3.51 found,
$\% \, O = 48/174 \times 100 = (27.60)$.

The agreement between the "calculated" and "found" values is within the usual limits of variation of the C-H analysis. For a compound having a content of

---

* The atomic weights 12.01 for carbon and 1.008 for hydrogen are ordinarily used by analysts in calculating the results of carbon-hydrogen analyses. We shall use the approximation 12 for carbon in the examples in this chapter.

**Table 1-3.** *Relative atomic proportions of carbon, hydrogen, and oxygen*

| Relative Weights | | Relative Atomic Proportions | Atoms, Relative to O=1 |
|---|---|---|---|
| C | 68.80 | 68.80/12=5.74 | 5.74/1.73=3.32 |
| H | 3.51 | 3.51/1.008=3.49 | 3.49/1.73=2.02 |
| O | 27.69 | 27.69/16=1.73 | 1.73/1.73=1.00 |

carbon of, say, 70.40%, "found" values of between 70.2 and 70.6 are regarded as generally acceptable.

The importance of knowing the molecular weight of the unknown compound can be appreciated when it is recognized that the same composition by weight is found for $C_{10}H_6O_3$, $C_{20}H_{12}O_6$, $C_{30}H_{18}O_9$. However, with reliable carbon-hydrogen analyses a relatively less accurate determination of molecular weight is usually adequate. For instance, the above three formulas correspond to molecular weights of 174, 348, and 522. Thus, if the molecular weight were found by an experiment to be 185, this result, even though not an accurate one, would clearly rule out all but the first of these.

Here is another example. A substance is found to contain nitrogen, and an analysis by the Dumas method gives the value 11.40% nitrogen. From 3.69 mg of the compound are obtained 9.26 mg of $CO_2$ and 2.36 mg of water:

$$\% \ C = \frac{9.26 \times 12/44}{3.69} \times 100 = 68.33,$$

$$\% \ H = \frac{2.36 \times 2.016/18.016}{3.69} \times 100 = 7.15,$$

$\% \ N = 11.40; \ \% \ O \ (\text{difference}) = 13.12.$

| | | |
|---|---|---|
| Atoms C = 68.33/12 = 5.69 | Atoms C/atom N = 5.69/0.814 = 6.9 |
| H = 7.15/1 = 7.15 | H/N = 8.8 |
| N = 11.40/14 = 0.814 | N/N = 1.0 |
| O = 13.12/16 = 0.820 | O/N = 1.0 |

Thus, the empirical formula appears to be $C_7H_9ON$:

Calculated for $C_7H_9ON$: C, 68.28; H, 7.33; N, 11.38.
Found: C, 68.33; H, 7.15; N, 11.40.

This is acceptable agreement, and the $C_7H_9ON$ formula may be taken as correct. A molecular weight determination will show whether the molecular formula is $C_7H_9ON$, $C_{14}H_{18}O_2N_2$, etc.

## 1.18   Preliminary deductions from the molecular formula

The results of an analysis and molecular weight determination are expressed as a molecular formula. This is the result of the first steps that are taken toward the determination of the *structure* of an organic compound. Establishment of the molecular formula is often a long way from the ultimate goal of being able to write a complete structure showing the arrangement of all the atoms of which the compound is constructed, but it is an essential stage in the study.

What information does the molecular formula give? When we recall that the

**Table 1-4a.**   *Some compounds composed of carbon, hydrogen, and oxygen*

| Structural Formula | | Name | Molecular Formula |
|---|---|---|---|
| Graphic | Condensed | | |
| (a) 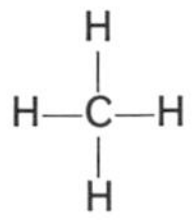 H—C—H (with H above and below) | $CH_4$ | methane | $CH_4$ |
| (b) 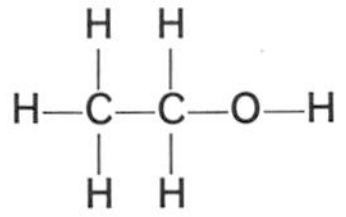 H—C—C—O—H (with H's) | $CH_3CH_2OH$ | ethanol | $C_2H_6O$ |
| (c) 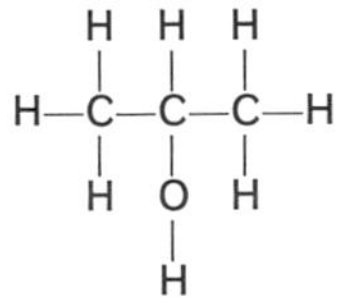 H—C—C—C—H (with OH) | $CH_3CHCH_3$ (OH) | 2-propanol | $C_3H_8O$ |
| (d) 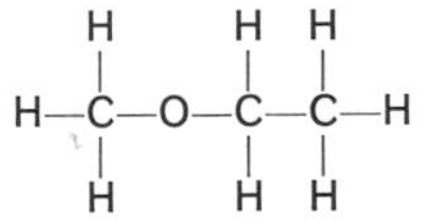 H—C—O—C—C—H | $CH_3OCH_2CH_3$ | methyl ethyl ether | $C_3H_8O$ |
| (e) 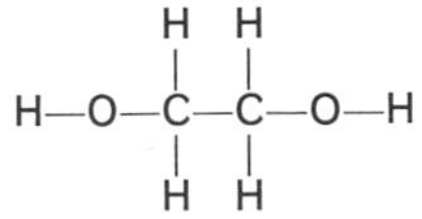 H—O—C—C—O—H | $HOCH_2CH_2OH$ | ethylene glycol 1,2-ethanediol | $C_2H_6O_2$ |
| (f) 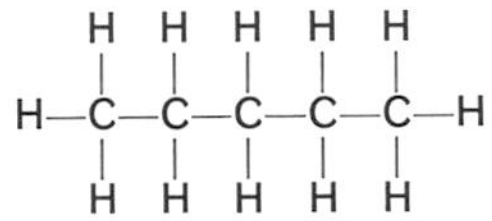 H—C—C—C—C—C—H | $CH_3CH_2CH_2CH_2CH_3$ | pentane | $C_5H_{12}$ |
| (g) 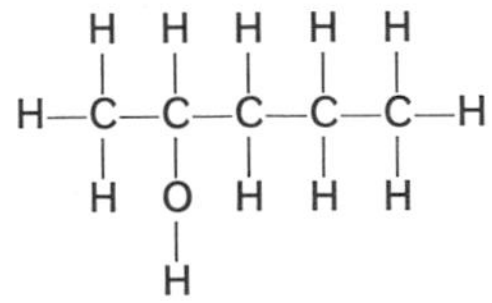 H—C—C—C—C—C—H (with OH) | $CH_3CHCH_2CH_2CH_3$ (OH) | 2-pentanol | $C_5H_{12}O$ |

formula $C_4H_{10}O$ is that of seven different compounds it becomes apparent that a molecular formula by itself gives us very little information about structure. Nevertheless, certain useful preliminary conclusions can be drawn. These conclusions are often only tentative and serve to do no more than suggest the course of further experiments, but they are often valuable.

Let us examine the structures shown in Table 1-4; they are those of some compounds composed of carbon, hydrogen, and oxygen. Note that in every compound in Table 1-4a, the normal valences of four for carbon, two for oxygen, and one for hydrogen are observed.

Note that in all of these compounds the ratio of carbon to hydrogen is $n:2n+2$. We find that in all compounds of carbon and hydrogen, or carbon, hydrogen, and oxygen, in which only *single bonds* are present and in which there are no rings (a structural feature that will be discussed further on), the carbon-hydrogen ratio is expressed by $C_nH_{2n+2}$.

**Table 1-4b.** *Some compounds composed of carbon, hydrogen, and oxygen*

| Structural Formula | | Name | Molecular Formula | C/H Ratio |
|---|---|---|---|---|
| Graphic | Condensed | | | |
| (h) | 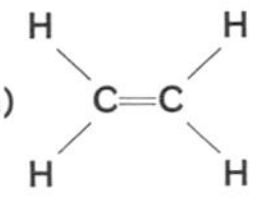 $CH_2{=}CH_2$ | ethylene (ethene) | $C_2H_4$ | $n:2n$ |
| (i) | 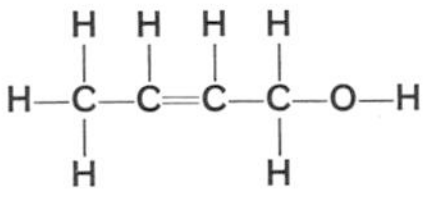 $CH_3CH{=}CHCH_2OH$ | crotyl alcohol (2-buten-1-ol) | $C_4H_8O$ | $n:2n$ |
| (j) | 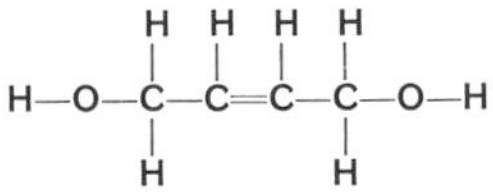 $HOCH_2CH{=}CHCH_2OH$ | 2-butene-1,4-diol | $C_4H_8O_2$ | $n:2n$ |
| (k) | 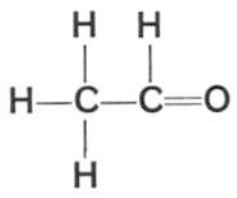 $CH_3CHO$ | acetaldehyde (ethanal) | $C_2H_4O$ | $n:2n$ |
| (l) | 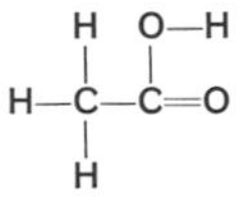 $CH_3COOH$ | acetic acid (ethanoic acid) | $C_2H_4O_2$ | $n:2n$ |
| (m) | 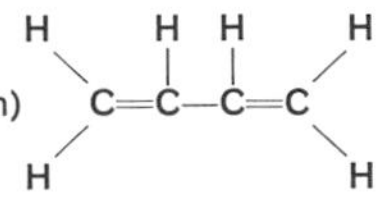 $CH_2{=}CH{-}CH{=}CH_2$ | 1,3-butadiene | $C_4H_6$ | $n:2n-2$ |
| (n) | 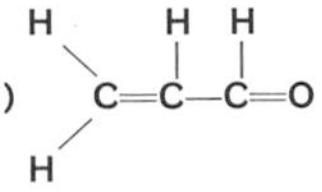 $CH_2{=}CH{-}CHO$ | acrolein (propenal) | $C_3H_4O$ | $n:2n-2$ |

Another way in which carbon can be bonded to carbon or oxygen is by a *double bond*. In Table 1-4b are representative examples of compounds that contain double bonds (again note that carbon has a valence of four; oxygen, two; and hydrogen, one).

Note that the compounds that contain *one* double bond ($C=C$ or $C=O$) have a carbon-hydrogen ratio of $C_nH_{2n}$, whereas those that contain *two* double bonds have a C:H ratio of $C_nH_{2n-2}$. It can be predicted (and the reader can check

**Table 1-4c.** *Some compounds composed of carbon, hydrogen, and oxygen*

| Structural Formula | Molecular Formula | C/H Ratio |
|---|---|---|
| (o)  —  141  591  — | $C_5H_{10}$ | $n:2n$ |
| (p)  —  141  592  — | $C_5H_{10}O$ | $n:2n$ |
| (q)  —  141  593  — | $C_6H_{10}$ | $n:2n-2$ |
| (r)  —  141  594  — | $C_6H_{10}O$ | $n:2n-2$ |
| (s)  —  141  595  — | $C_6H_8O$ | $n:2n-4$ |
| (t)  —  141  596  — | $C_6H_{10}O_2$ | $n:2n-2$ |
| (u)  —  141  597  — | $C_4H_8O_2$ | $n:2n$ |

this by constructing some appropriate formulas) that with *three* double bonds the C:H ratio will be $C_nH_{2n-4}$, with *four*, $C_nH_{2n-6}$, and so on.

Now let us examine one additional structural feature that is found in many organic compounds: a *cyclic* structure, or *ring*. Again, still observing the usual valence requirements, we can write a number of possible structural formulas (Table 1-4c).

Note that the ring is equivalent to a double bond in determining the C:H ratio: structures *(o)*, *(p)*, and *(u)* are $C_nH_{2n}$, as are *(h)*, *(i)*, *(j)*, *(k)*, and *(l)*. Compound *(r)*, with one ring and one double bond, is $C_nH_{2n-2}$, as are compounds *(m)*, *(n)*, *(q)*, and *(t)*.

It will now be apparent that the molecular formula, determined by elementary analysis, can give us some valuable preliminary information. Suppose a compound is found by analysis to have the composition $C_8H_{14}O_2$. A $C_8$-compound with no rings and no double bonds would have a carbon-hydrogen ratio of $C_8H_{18}$ [the oxygen can be neglected in this calculation of the C:H ratio: compare *(p)* and *(u)* with oxygen atoms in different kinds of combination]. Since our unknown compound has a C:H ratio expressed by $C_8H_{14}$, this is a difference of four hydrogen atoms. We conclude then that our compound has two double bonds, or two rings, or a double bond and a ring. The following possible structures, with these features included, confirm this.

$HOCH_2CH=CH-CH=CHCH_2CH_2CH_2OH$   $C_8H_{14}O_2$   (two C=C double bonds)

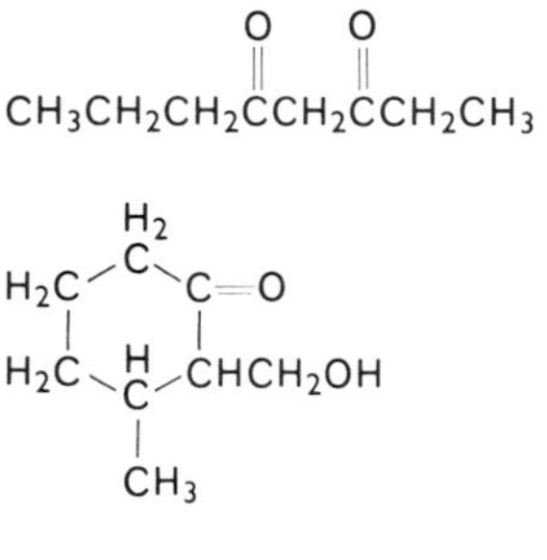

$CH_3CH_2CH_2\overset{O}{\overset{\|}{C}}CH_2\overset{O}{\overset{\|}{C}}CH_2CH_3$   $C_8H_{14}O_2$   (two C=O double bonds)

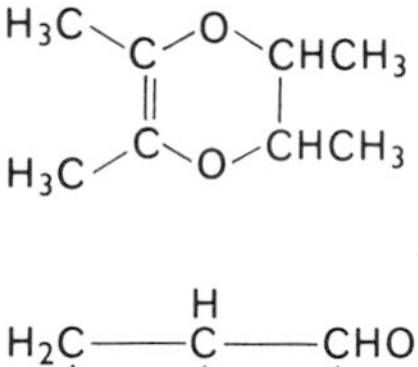

$C_8H_{14}O_2$   (one ring, one C=O)

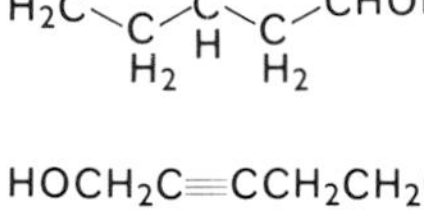

$C_8H_{14}O_2$   (one ring, one C=C)

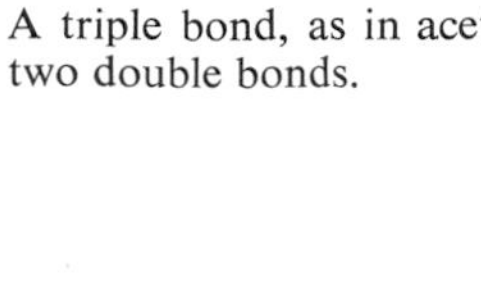

$C_8H_{14}O_2$   (two rings)

$HOCH_2C\equiv CCH_2CH_2CH_2CH_2CH_2OH$   $C_8H_{14}O_2$   (one C≡C)*

---

* A triple bond, as in acetylene ($HC\equiv CH$; $C_2H_2$), is equivalent in these calculations to two double bonds.

**Exercise 2.**    Given that nitrogen has a valence of 3, as in $NH_3$, $CH_3NH_2$, $CH_3NHCH_3$, and so on, determine, by calculations similar to those just discussed, the number of double bonds and/or rings in (a) $C_5H_5N$; (b) $C_6H_{14}N_2$; (c) $C_5H_9NO$. Hint: calculate the C:H ratios of a number of representative compounds, such as $CH_3CH_2NH_2$, $CH_2{=}CHCH_2NH_2$, and others that obey the valence rules, and use the C:H ratios in finding the number of rings and/or double bonds in the compounds given.

**Exercise 3.**    Count the total number of rings and double bonds in the following formulas and show how this number agrees with that predicted by the method just described.

(a)  $CH_2{=}CH{-}CH{=}CH{-}\overset{\displaystyle H}{\underset{}{C}}{=}O$

(d)

(b)  $HO{-}\overset{\displaystyle O}{\underset{}{C}}{-}\overset{\displaystyle O}{\underset{}{C}}{-}OH$

(e)

(c)

(f)

**Exercise 4.**    (a) A compound is found by analysis to have the composition $C_5H_8O_3$. Chemical evidence shows that it has one carbon-carbon double bond. Write four possible structures for the compound.
   (b) Do the same for $C_6H_{10}O_2$, which contains one carbon-oxygen double bond.

It should be recognized that to be able to write a number of structures that agree with a formula which is determined by an elemental analysis is only a way of laying a problem out for inspection and attack. We must now discover ways of recognizing the various structural features that we write in this way, and this recognition is achieved by chemical and physical experiments. The choice and interpretation of these experiments will depend upon the chemist's knowledge of the ways in which organic compounds react under various conditions and with various reagents. They will depend upon his understanding of the nature of the chemical bonds that link the atoms together—how these bonds can be formed and how they can be broken.

In succeeding chapters the structure of atoms and of organic molecules will be examined. In these chapters, we shall see what bonds *are*, because the nature and disposition of the bonds in space must be understood before we can attack the questions of how they are formed and broken in chemical reactions.

## Exercises

(Exercises 1–4 will be found within the text.)

**5** Devise and describe a method that could be used for the determination of silver in an organic compound, for example, a silver salt of an organic acid.

**6** It is possible to perform a direct analysis for *oxygen* in organic compounds. Why is it valuable to have an analytical figure for the percentage of oxygen in a compound?

**7** In the qualitative analysis for halogen by the sodium fusion method, when compounds containing nitrogen are tested it is necessary to boil the solution after acidification with nitric acid before adding silver nitrate. Why must this be done?

**8** Write the graphic (as in example *a* below) and the condensed (as in example *b*) formulas for all of the compounds of the composition $C_5H_{12}O$.

$$\text{(a) ethanol, } H-\overset{\overset{\displaystyle H}{|}}{\underset{\underset{\displaystyle H}{|}}{C}}-\overset{\overset{\displaystyle H}{|}}{\underset{\underset{\displaystyle H}{|}}{C}}-OH.$$

(b) diethyl ether, $CH_3CH_2OCH_2CH_3$.

**9** Write graphic and condensed structural formulas for all of the possible isomers of each of the following, remembering that carbon has a valence of four; oxygen, two; nitrogen, three; and halogens and hydrogen one. (a) $C_2H_5Br$; (b) $C_3H_7Cl$; (c) $C_2H_4Cl_2$; (d) $C_2H_7N$; (e) $C_3H_9N$.

**10** How much barium sulfate would be formed from 4.52 mg of $(CH_3)_2S$, after oxidation with nitric acid and precipitation with barium chloride?

**11** How large a sample of each of the following must be burned to give 4.40 mg of carbon dioxide? (a) $CBr_4$; (b) $C_3H_6$; (c) $CH_4O$; (d) $C_2H_2O_4$; (e) $C_6H_{12}O_6$.

**12** What is the percentage of bromine in each of the following? (a) $AgBr$; (b) $C_2H_5Br$; (c) $BrCH_2CH_2Br$; (d) $BrCH_2CONH_2$; (e) $CH_2ClBr$.

**13** What is the minimum possible molecular weight of a compound that contains 31.56 % bromine?

**14** A sample of 6.23 mg of a liquid organic compound gave on combustion 9.17 mg of $CO_2$ and 3.78 mg of water. In a molecular weight determination it was found that 357 ml of the vapor of the compound, measured at 27°C and 750 mm, weighed 0.890 g. Calculate the empirical and molecular formulas for the compound.

**15** A compound contained 69.95 % carbon and 11.60 % hydrogen (by analysis). A qualitative test showed that nitrogen was present. Calculate a possible empirical formula. If the molecular weight of the compound is less than 170, can a formula containing both nitrogen and oxygen be possible with the analytical figures given?

**16** Silver lactate has the composition C, 18.3 %; H, 2.5 %; O, 24.4 %; and Ag, 54.8 %. Calculate the molecular formula, assuming one silver atom per mole.

**17** Oxidation of lactic acid with potassium dichromate and sulfuric acid produces acetic acid, which has the molecular formula $C_2H_4O_2$, and contains a methyl group ($CH_3—$). Using this information, and that from problem 16, write a structural formula for lactic acid.

**18** Calculate the percentage composition for each of the compounds $C_{27}H_{46}O$ and $C_{28}H_{48}O$. What can you say about the practical limitations of the carbon-hydrogen analysis?

**19** A nitrogen-containing organic compound gave the following analytical figures: C, 75.45 %; H, 6.61 %; N, 8.40 %. Calculate an empirical formula and check it by comparing the percentage composition calculated from that formula with the analytical result.

**20** A compound analyzes as follows: C, 75.20%; H, 10.75%. The compound contains only carbon, hydrogen, and oxygen, and its molecular weight is found to be about 115.

Chemical tests prove that the structural element $-\overset{\displaystyle |}{C}=O$ is present. What else can be said about the structure?

**21** Suppose that the compound whose analysis is given in the preceding Exercise is found not to contain a carbonyl group (see p. 30). Write a possible structure for the compound.

**22** Write the molecular formula for each of the following:

  (a) $CH_2=CHCH_2CH_2CH=NCH_3$    (b) $CH_3C\equiv CCH_2CH_2CH_2NH_2$
  (c) $CH_3CH_2CH_2CH_2CH_2C\equiv N$.

**23** Write two structural formulas that could represent a compound whose molecular formula is $C_6H_9N$.

# Chapter two / Introduction to Some Important Classes of Organic Compounds. Sources of Carbon Compounds

We have now seen how carbon, by forming four bonds with other atoms, can become part of the structure of complex molecules. Oxygen and nitrogen can also be incorporated into organic molecules, usually by carbon-oxygen and carbon-nitrogen bonds. Hydrogen is nearly always present in organic compounds in C—H, O—H, and N—H bonds; and many organic compounds are known that contain halogen in C—F, C—I, C—Br, and C—Cl bonds.

The number of structural formulas that can be constructed by combinations of these few elements joined by single, double, and triple bonds according to these simple valence rules, is practically limitless. It is easy to understand why many more than a million organic compounds are known and why this number is growing at a rapid rate.

It will be a revelation to the beginning student of organic chemistry to learn that most of the structures he can construct from these first-period elements, simply by observing the few valence rules we have discussed, represent real substances that are actually known or could be prepared by a competent organic chemist. As our view of organic chemistry becomes more sophisticated we shall learn that certain structures that we might devise, which obey the simple valence rules, are nevertheless forbidden because the disposition of bonding orbitals in space requires that bond angles and bond lengths cannot vary too much from certain preferred values. Nevertheless, carbon, hydrogen, oxygen, and nitrogen can combine in many different ways and can occur in many different types of compounds.

Organic chemistry would be a quite bewildering study if it were necessary to deal with each one of more than a million substances as an individual compound that bore no systematic relationship to any other. Fortunately, this is not necessary. Many organic compounds can be grouped into *classes*, the members of which behave very similarly. Though each organic compound *is* an individual substance, the members of a given class usually have many chemical properties in common. Thus, the study of the properties of a class of compounds will enable us to recognize and predict, at least in a qualitative way, the properties of individual members of the class that we may be encountering for the first time.

Organic compounds are classified by the presence in the molecule of certain characteristic groups of atoms called the *functional groups*. As this term implies, the functional group represents that part of the molecule which confers a characteristic type of reactivity upon the compound.

## 2.1   The common functional groups

The commonest groups of atoms that lend characteristic behavior to organic compounds are the following.

***The hydroxyl group.***   The hydroxyl group, —*OH*, is present in two large classes of compounds: *alcohols*, and *phenols*. A typical alcohol is ethanol; a typical phenol is the compound that bears the name of the class, phenol itself:

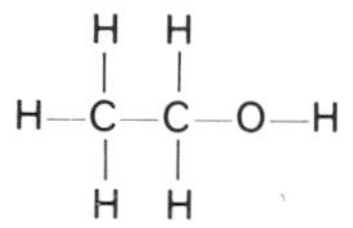

$CH_3CH_2OH$

ethanol

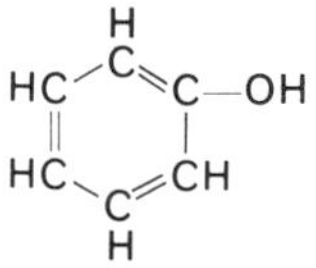

usually written

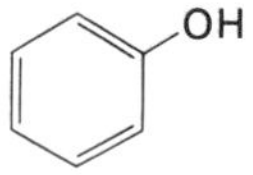

phenol

***The carbonyl group.***   The carbonyl group, —$C{=}O$, is the characteristic functional group of *aldehydes* and *ketones*. In aldehydes, one of the remaining two valences of carbon bears a hydrogen atom, as in acetaldehyde or benzaldehyde; in ketones, the carbonyl group is attached to two carbon atoms, as in acetone:

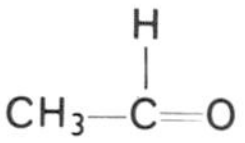

$CH_3{-}C{=}O$

acetaldehyde
(an aldehyde)

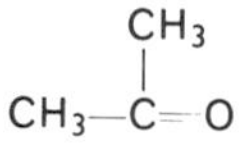

$CH_3{-}C{=}O$

acetone
(a ketone)

CHO

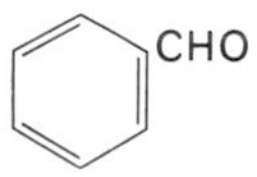

benzaldehyde
(an aldehyde)

In aldehydes, the whole grouping —$C{=}O$ is often referred to as the functional group, and is named the *formyl* group. Nevertheless, it is the carbonyl (—$C{=}O$)

grouping of both aldehydes and ketones that determines their characteristic, and similar, behavior.

***The carboxyl group.***   The carboxyl group, $-C\overset{O}{\underset{OH}{\diagup}}$ , in which the carbonyl and the hydroxyl group (hence the name) are combined in a single unit, is the characteristic functional group of the *carboxylic acids*. Because carboxylic acids have properties that are quite different from those of alcohols or aldehydes and ketones, they are not classed with these compounds. Their unique properties entitle them to separate classification. The carboxyl group can be found as a structural unit in compounds of many types. A few of the simplest carboxylic acids are the following:

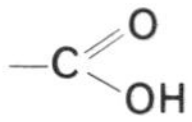

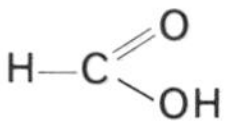

$$H-C\overset{O}{\underset{OH}{\diagup}} \qquad CH_3-C\overset{O}{\underset{OH}{\diagup}} \qquad \text{(benzene ring)}-C\overset{O}{\underset{OH}{\diagup}} \qquad \overset{O}{\underset{HO}{\diagdown}}C-C\overset{O}{\underset{OH}{\diagup}}$$

| formic acid | acetic acid | benzoic acid | oxalic acid |

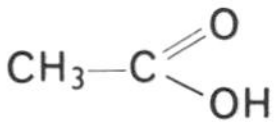

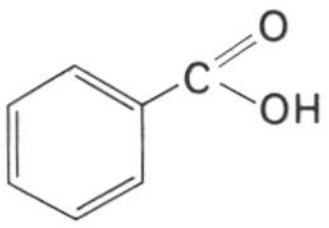

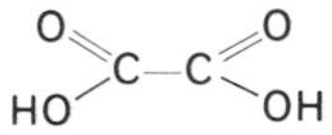

***The amino group.***   The amino group, $-NH_2$, is present in the primary amines. These compounds, substitution products of ammonia, are the organic bases. Like ammonia, they form stable salts with strong acids; indeed, the lower, more volatile members of the class have penetrating ammoniacal (often fishy) odors. Some typical primary amines are the following:

$$CH_3NH_2 \qquad CH_3CH_2NH_2 \qquad \text{(benzene ring)}-NH_2$$

| methylamine | ethylamine | aniline |

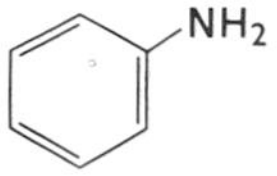

***Derivatives of alcohols and amines.***   Derivatives of alcohols and amines (in which the hydrogen atom of —OH, and one or both hydrogen atoms of —NH$_2$, are replaced by carbon) are *ethers* and *secondary* and *tertiary* amines:

$$CH_3CH_2OCH_2CH_3 \qquad CH_3-\overset{CH_3}{\underset{|}{N}}H \qquad CH_3CH_2-\overset{CH_2CH_3}{\underset{|}{N}}-CH_2CH_3$$

| diethyl ether | dimethylamine (a secondary amine) | triethylamine (a tertiary amine) |

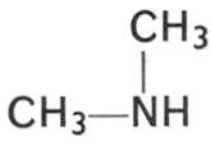

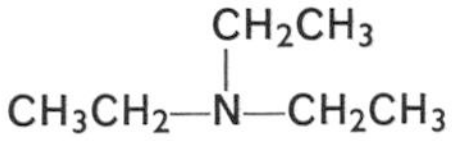

---

**Exercise 1.**   Note that ethers and alcohols of the same number of carbon atoms are isomeric. Write the structures of all of the possible isomers of the composition *(a)* $C_3H_8O$; *(b)* $C_5H_{12}O$.

A similar replacement of the hydrogen atom of the carboxyl group by a carbon substituent gives an *ester*:

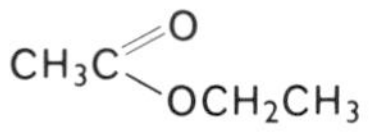

ethyl acetate
(an ester of acetic acid)

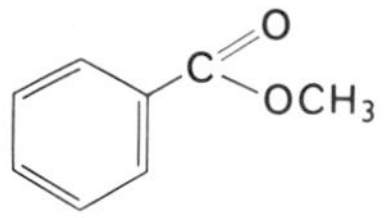

methyl benzoate
(an ester of benzoic acid)

***Acid derivatives.*** Esters are members of a large and diverse group of compounds called acid derivatives. These will be described in a later chapter, but are listed here without extended comment. The common acid derivatives include (besides esters) *acid halides* (acid chlorides, bromides, and so on):

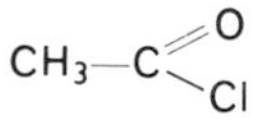

acetyl chloride
(a typical acid chloride)

*amides,* in which the —OH group of —COOH is replaced by —NH$_2$:

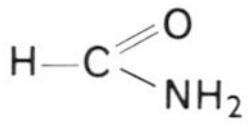

formamide (the amide
derived from formic acid)

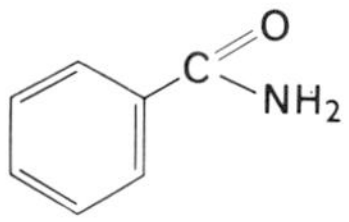

benzamide (the amide
derived from benzoic acid)

*acid anhydrides:*

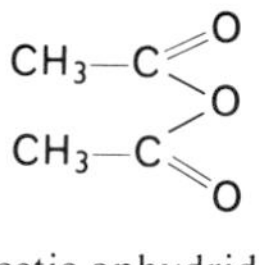

acetic anhydride

*Salts* of acids are not ordinarily regarded as acid derivatives because their relationship to the carboxylic acids depends simply upon a difference in $p$H. Nevertheless, they do constitute a class of organic compounds with definite composition and properties. An example is sodium acetate,

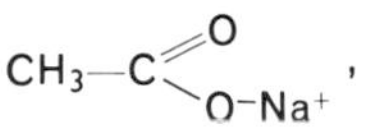

more commonly written simply as $CH_3COONa$ or $CH_3CO_2Na$.

***The carbon-carbon double bond.*** The carbon-carbon double bond, —$C$=$C$—, is a structural unit that is characteristic of a class of hydrocarbons (compounds

containing only carbon and hydrogen) that are known as *olefins*. A few simple olefins are the following (note that the tetravalence of carbon is observed in all
examples; the double bond accounts for two of the valence bonds on each carbon
atom):

$CH_2=CH_2$      $CH_3CH=CH_2$

$CH_2=CH-CH=CH_2$

   ethylene            propylene           cyclohexene          1,3-butadiene

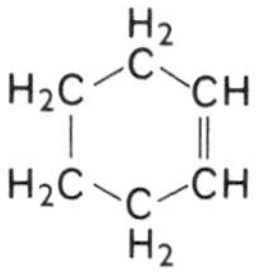

    The carbon-carbon double bond is a center of *unsaturation* in the molecule.
This term is used because unsaturated compounds contain fewer than the maximum possible number of hydrogen atoms. Carbon-carbon double bonds can be
present in molecules of all kinds: there are unsaturated alcohols, unsaturated
aldehydes and ketones, unsaturated acids, unsaturated amines, and so on. The
following are a few examples:

$CH_2=CHCH_2OH$         $CH_3CH=CHCHO$

   allyl alcohol             crotonaldehyde              benzalacetone

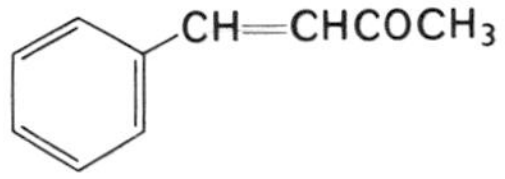

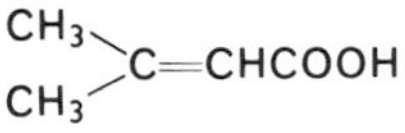

   β-methylcrotonic acid 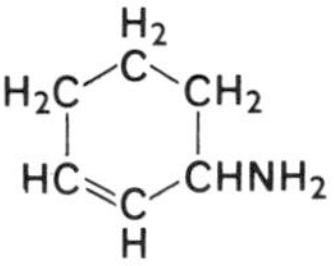        3-aminocyclohexene

***The carbon-carbon triple bond.*** The carbon-carbon triple bond, $-C{\equiv}C-$, is
the characteristic structural unit of the class of compounds known as *acetylenic*
compounds, from the name of the simplest member, acetylene. Acetylenic compounds are also unsaturated compounds. The following are typical examples:

$H-C{\equiv}C-H$

   acetylene            phenylacetylene 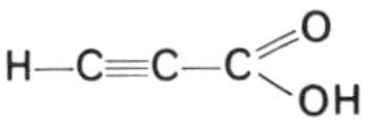            propiolic acid
                                          (propynoic acid)

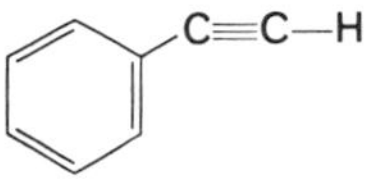

## 2.2 Polyfunctional compounds

Many organic compounds contain more than one of the typical functional groups
that have just been described, and often the presence of one functional group in a
molecule will alter the behavior of another from what might have been anticipated.

$$CH_3-\underset{\underset{OH}{|}}{\overset{\overset{H}{|}}{C}}-COOH$$

lactic acid
(an α-hydroxy acid)

$$\bigcirc\!\!\!\!\!\bigcirc CH_2-\underset{\underset{NH_2}{|}}{C}HCOOH$$

phenylalanine
(an α-amino acid)

$$HOCH_2CH_2OH$$

ethylene glycol
(a diol, or glycol)

$$CH_3-\underset{\underset{O}{\|}}{C}-COOH$$

pyruvic acid
(an α-keto acid)

$$CH_3-\underset{\underset{O}{\|}}{C}-CH_2COOH$$

acetoacetic acid
(a β-keto acid)

$$CH_3-\underset{\underset{O}{\|}}{C}-CH_2CH_2COOH$$

levulinic acid
(a γ-keto acid)

$$CH_3-\underset{\underset{O}{\|}}{C}-\underset{\underset{O}{\|}}{C}-CH_3$$

biacetyl
(an α-diketone)

$$CH_3-\underset{\underset{O}{\|}}{C}-CH_2-\underset{\underset{O}{\|}}{C}-CH_3$$

acetylacetone
(2,4-pentanedione or
pentane-2,4-dione)
(a β-diketone)

salicylaldehyde
(a phenolic aldehyde)

---

**Exercise 2.**   Write structural formulas for all of the isomers of the carboxylic acids of *(a)* the composition $C_6H_{10}O_3$ and *(b)* the composition $C_6H_{12}O_2$.

---

The best way to approach the problem of understanding, explaining, and predicting the properties of polyfunctional molecules is first to study the characteristic properties of the simple functional groups. After an understanding is reached it is possible, by rational extension of the simpler principles, to extend understanding to more complex situations.

## 2.3   The paraffin hydrocarbons

A large and industrially important class of organic compounds is characterized by the absence of functional groups of any kind and by a remarkable chemical inertness. These are the saturated, or paraffin, hydrocarbons (paraffin, from the Latin, means "little affinity"). They contain carbon and hydrogen only, and all of the carbon valences are utilized in bonding to carbon and hydrogen. There are no unshared electrons and none of the atoms possesses unsatisfied valence bonds; all of the carbon bonds are single bonds. The chemical inertness of the saturated hydrocarbons can be ascribed to this "saturation" of the carbon valences. In order for these compounds to react with other molecules, sufficient initial energy must be supplied to break one or more of the C—H or C—C bonds. This bond rupture exposes the carbon atom at which the break occurs to attack by reagents, which can thus form a new bond to replace the old.

**Table 2-1.** *Paraffin hydrocarbons*

| Structural Formula | Molecular Formula | Name |
|---|---|---|
| $CH_4$ | $CH_4$ | methane |
| $CH_3CH_3$ | $C_2H_6$ | ethane |
| $CH_3CH_2CH_3$ | $C_3H_8$ | propane |
| $CH_3CH_2CH_2CH_3$ | $C_4H_{10}$ | butane |
| $CH_3$<br> $\diagdown$<br> $\qquad CHCH_3$<br> $\diagup$<br> $CH_3$ | $C_4H_{10}$ | isobutane |
| $CH_3CH_2CH_2CH_2CH_3$ | $C_5H_{12}$ | pentane |
| $CH_3\!-\!\overset{\displaystyle CH_3}{\underset{\displaystyle CH_3}{\overset{\textstyle |}{\underset{\textstyle |}{C}}}}\!-\!CH_3$ | $C_5H_{12}$ | neopentane |
| $CH_3(CH_2)_{10}CH_3$ | $C_{12}H_{26}$ | dodecane |
| $CH_3(CH_2)_{28}CH_3$ | $C_{30}H_{62}$ | triacontane |

Typical saturated hydrocarbons are the group of compounds that start with the simplest member, methane, and proceed by increments of $CH_2$ up what is called a *homologous series* to large and complex compounds. The paraffin hydrocarbons may possess unbranched (i.e., "straight") or branched chains of carbon atoms, as in Table 2-1.

Note that the general molecular formula of the paraffin hydrocarbons is represented by $C_nH_{2n+2}$ (see also Chapter 1).

---

**Exercise 3.** Write the structures of all of the possible isomeric compounds of the composition: *(a)* $C_5H_{12}$; *(b)* $C_7H_{16}$

---

Although paraffin hydrocarbons are reluctant to engage in reactions, they can be induced to do so under proper conditions. Chemical transformations that the paraffin hydrocarbons can be made to undergo are the basis for many industrial processes that utilize the enormous natural sources of these substances, the most important of these being petroleum and natural gas.

*Thermal cracking*, or pyrolysis in the absence of air, is a process that uses heat energy to bring about extensive rupture of hydrocarbon molecules. At sufficiently high temperatures methane is decomposed into elementary carbon and molecular hydrogen:

$$CH_4 \longrightarrow C + 2H_2$$

Ethane can be "cracked" in a more useful manner to lose one molecule of hydrogen and yield ethylene:

$$H-\underset{\underset{H}{|}}{\overset{\overset{H}{|}}{C}}-\underset{\underset{H}{|}}{\overset{\overset{H}{|}}{C}}-H \longrightarrow \underset{H}{\overset{H}{>}}C=C\underset{H}{\overset{H}{<}} + H_2$$

When higher paraffin hydrocarbons are subjected to cracking, a number of processes can ensue because of the greater number of points at which bond breaking can occur. Carbon-carbon bonds may be broken at more than one place, and hydrogen may be lost from the fragments, with the production of a mixture of hydrocarbons of various numbers of carbon atoms, some saturated and some unsaturated.

Cracking thus leads to two important results: (1) large molecules are fragmented into smaller ones; (2) unsaturated hydrocarbons are produced. As we shall see, unsaturated hydrocarbons are reactive substances and so are fitted for use in synthetic processes from which a large number of useful products are obtained.

In modern petroleum technology cracking is carried out in the presence of catalysts. The petroleum vapors are passed over a bed of catalyst consisting of granules of an acid-treated clay or a synthetic aluminum oxide-silicon oxide complex, at temperatures of over 300°C. Catalytic cracking produces, in addition to lower molecular-weight paraffins and olefins, secondary products that arise from isomerization and aromatization, with the result that high quality motor fuels can be prepared at lower costs than if produced by purely thermal processes.

The fragmentation of large hydrocarbons into smaller ones plays an important part in petroleum refining. Petroleum is a naturally occurring mixture of carbon compounds, a large proportion of which consists of paraffin hydrocarbons. The low-molecular-weight members constitute the mixture known as gasoline. Gasolines differ in composition, but a typical automotive gasoline consists largely of a mixture of hydrocarbons ranging from about $C_5H_{12}$ to $C_{10}H_{22}$,* with a boiling-point range of about 30 to 200°C. Higher-boiling fractions comprise kerosene, fuel oils, and, finally, high-molecular-weight fractions that are viscous and oily and are used as lubricating oils. The chemical inertness of the paraffin hydrocarbons is of obvious value in preserving a lubricant that, in use, is exposed to oxygen at elevated temperatures.

Natural gas is usually composed largely of methane (a typical natural gas may be 80% methane), along with smaller amounts of ethane and propane. The chief use of natural gas is as a domestic and industrial fuel, but large quantities are also consumed by the chemical industry as the starting material for the synthesis of numerous organic chemicals.

---

* Present-day high-"anti-knock" gasolines are prepared by more complex processes than those described, and contain aromatic and unsaturated hydrocarbons as well as certain special additives, in addition to their chief content of saturated hydrocarbons.

## 2.4 Aromatic hydrocarbons

The earth, which yields petroleum and natural gas, also provides an almost limit-less supply of coal, another major source of energy and of organic compounds. The carbon compounds derived from coal are obtained chiefly from the produc-tion of coke by destructive distillation. As coal is converted into coke (which is largely a mixture of amorphous carbon and the mineral constituents of coal) there is produced a gas (coal gas) and a distillate which is condensed to yield a viscous black liquid known as *coal tar*. The redistillation of coal tar yields a number of fractions of which the lowest-boiling contains a mixture of hydrocarbons. Addi-tional amounts of these compounds can be obtained by cooling and scrubbing the coal gas.

The most important constituent of coal tar, both industrially and scientific-ally, is benzene, a cyclic hydrocarbon of the composition $C_6H_6$. We shall have more to say about its structure in a later chapter (Chapter 19), but for the present we can represent it as

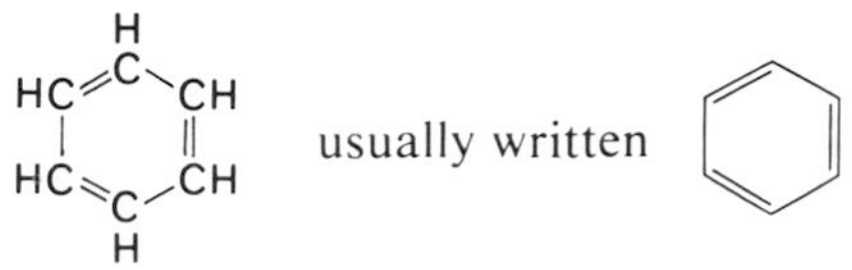

usually written

Accompanying benzene in the more volatile fractions of coal tar are the simple homologous hydrocarbons toluene and the three xylenes:

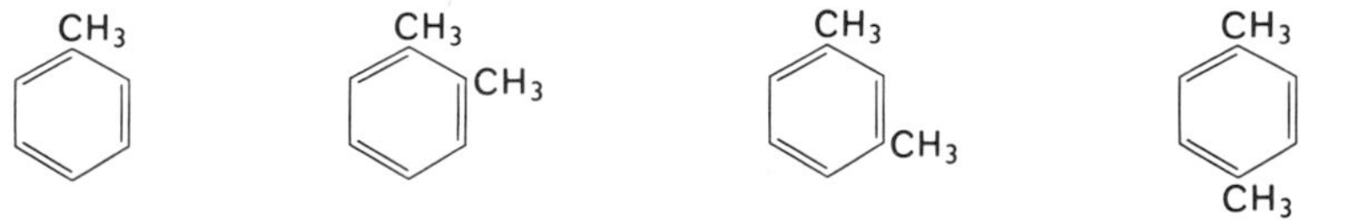

These compounds, which are characterized by the "benzenoid" six-membered ring, have unique properties, distinct from those of olefins, and are called "aro-matic" compounds. By continued substitution, on the ring or on the substituents, of saturated carbon atoms, a group of compounds as large and various as the group of derivatives of the paraffin hydrocarbons is possible.

It will be recalled that several other aromatic compounds have already been mentioned earlier in this chapter (phenol, benzoic acid, aniline, salicylaldehyde).

Benzene and its homologues are produced in enormous quantities as raw materials for the production of synthetic organic chemicals, dyes, drugs, insecti-cides, perfumes, flavoring materials, fibers, and a host of other useful products. The historical fact that benzene was first derived solely from coal tar has led to the widespread use of such terms as "coal-tar derivatives" to designate compounds prepared from benzene. At the present time, aromatic compounds are also pro-duced in large amounts from the products derived from petroleum.

## 2.5  Cyclic compounds other than "aromatic" compounds

We have seen that the angles between carbon-carbon bonds permit carbon atoms to be joined in rings with the formation of cyclic structures. (We shall discuss this at greater length in Chapter 3.) When five- and six-membered rings are formed, the angles between the carbon-carbon bonds are close to the "normal" tetrahedral angle of 109.5°; such rings are thus essentially without strain and show extraordinary stability. For example, cyclopentane and cyclohexane exhibit a chemical inertness comparable to that of the noncyclic paraffin hydrocarbons.

$$H_2C \overset{\overset{H_2}{C}}{\diagdown} CH_2 \\ | \qquad | \\ H_2C \text{———} CH_2$$

cyclopentane

$$H_2C \overset{\overset{H_2}{C}}{\diagdown} CH_2 \\ | \qquad | \\ H_2C \underset{\underset{H_2}{C}}{\diagup} CH_2$$

cyclohexane

It is possible to form rings of as few as three carbon atoms, as well as rings containing many more than six carbon atoms. The smaller rings (cyclopropane and cyclobutane) are under strain, owing to the considerable distortion of the carbon-carbon bond angles from the normal 109.5°. For this reason compounds containing 3- and 4-membered rings show a considerably greater reactivity than their

$$\overset{\overset{H_2}{C}}{\diagup \diagdown} \\ H_2C \text{————} CH_2$$

cyclopropane
(C—C—C angle 60 )

$$H_2C \text{—} CH_2 \\ |\quad | \\ H_2C \text{—} CH_2$$

cyclobutane
(C—C—C angle 90 )

open-chain counterparts. Their reactions commonly involve an opening of the ring to give compounds that are less highly strained.

Rings of greater than six members are not strained by the distortion of bond angles, but some of them are difficult to prepare because of other factors. Further discussion of this will be found in Chapter 17.

## 2.6  Limitations on structural possibilities

Owing to the requirements imposed by the normal bond lengths and bond angles of covalently bound carbon, certain restrictions are found to limit the number of permissible structures that can be drawn by simply obeying the usual valence rules. For example, in the compound allene, $CH_2\text{=}C\text{=}CH_2$, the three carbon atoms are known (by experimental measurements) to be arranged linearly; the same is true of the four carbon atoms of dimethylacetylene, $CH_3\text{—}C\text{≡}C\text{—}CH_3$. For this reason no such compounds as cyclopentyne or 1,2-cyclohexadiene have yet been prepared, owing to the great distortion that would be required to form them.

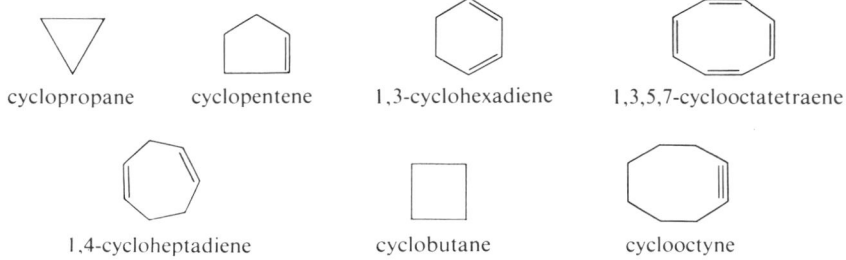

The endings used for the above names are to be noted: *-ane* (no double bonds); *-ene* (one double bond); *-diene* (two double bonds); *-yne* (one triple bond); *-tetraene* (four double bonds). How large must a ring be to accommodate the linear C—C≡C—C or the linear C=C=C unit as a part of the cycle? No final answer to this can yet be given; at the present time, cyclohexyne is not known, but it has been suggested that it is formed as a transitory and extremely reactive intermediate in certain reactions. Cyclo-octyne, however, is a known compound. Here, the joining of the ends of the linear —CH₂C≡CCH₂— unit by four —CH₂— groups can be accomplished, not *without* strain, but with energetically *permissible* strain.

## 2.7   The functional group

All of the types of compounds that have been introduced in these brief descriptions present various aspects of behavior that depend less upon their individual identity than upon the fact that the compounds contain one of the characteristic structural features that identifies the class to which they belong. The saturated hydrocarbons are exceptional in this regard: they are relatively inert and react slowly or not at all with most reagents under ordinary conditions. In the alkenes (olefins) the characteristic point of attack by reagents is the carbon-carbon double bond. The behavior of alcohols depends largely upon the hydroxyl group and the C—OH bond.

Aldehydes and ketones are characterized by the —C=O group, and most of their

**Table 2-2.** *General ways of representing some classes of compounds*

| Class of Compound | General Symbol | Example |
|---|---|---|
| paraffin hydrocarbons | RH | ethane ($CH_3CH_2$—H) |
| olefins | $RCH{=}CHR'$ | $CH_3CH{=}CHCH_2CH_3$ |
| alcohols | | |
| primary | $RCH_2OH$ | $CH_3CH_2OH$ |
| secondary | $R_2CHOH$ | $(CH_3CH_2)_2CHOH$ |
| tertiary | $R_3COH$ | $(CH_3)_3COH$ |
| aldehydes | RCHO | $CH_3CHO$ |
| ketones | $\overset{\overset{\text{O}}{\|\|}}{R}CR'$ | $\overset{\overset{\text{O}}{\|\|}}{CH_3}CCH_2CH_3$ |
| acids | $RC{\overset{\displaystyle /\!/O}{\underset{\textstyle \backslash OH}{}}}$ | $CH_3CH_2CH_2C{\overset{\displaystyle /\!/O}{\underset{\textstyle \backslash OH}{}}}$ |
| amines | | |
| primary | $RNH_2$ | $CH_3NH_2$ |
| secondary | $R_2NH$ | $CH_3CH_2NHCH_2CH_3$ |
| tertiary | $R_3N$ | $(CH_3)_3N$ |

reactions depend upon their possession of this function. Carboxylic acids possess the $-C{\overset{\displaystyle /\!/O}{\underset{\textstyle \backslash OH}{}}}$ group, and many of the types of reactions that acids undergo depend merely upon the presence of this grouping rather than upon what is attached to it.

Lest it appear that all alcohols behave alike and that all aldehydes behave alike, it should be emphasized here that differences (always in detail, and often in more general ways) between the behavior of the compounds of any given class will be observed, depending upon the particular structures of the individual compounds. Nevertheless, a great deal of the basic chemistry of a class of compounds can be described in general terms. In order to do this conveniently, organic chemists have long used a device for representing types of compounds in which a nonspecific symbol is used for the part of the molecule that is attached to the functional group. For example, Table 2-2 shows some general ways of representing some of the classes of compounds we have been describing.

## 2.8   The use of the symbol "R"

Suppose we wish to indicate that it is a general property of a carboxylic acid to react with a base, such as hydroxide ion, to form the anion of the acid and water. We can write the general expression

$$RC{\overset{\displaystyle O}{\underset{\displaystyle OH}{\Big\backslash}}} + OH^- \quad \rightleftarrows \quad RC{\overset{\displaystyle O}{\underset{\displaystyle O}{\Big\backslash}}} + H_2O$$

This equation does not mean, for example, that the equilibrium constant for the reactions of all acids with hydroxide ion is the same, regardless of the nature of R. It does mean that the expression as written represents a *characteristic* type of behavior of carboxylic acids.

What is gained by adopting a symbolism of this kind, since we have already said that individual compounds do not react in identical ways? We gain a great deal of economy and convenience in the expression of certain aspects of chemical behavior that may differ in degree but not in kind. We gain a succinct method of categorizing *types* of behavior and of outlining a general area of reactivity that can then be examined in a search for details in which individual compounds differ, or in which comparisons between individual compounds can be made.

## 2.9 Sources and uses of organic compounds

The importance of organic chemistry and the products of the organic chemical industry in the day-to-day life of modern man is too well recognized to need emphasis. We rely upon the organic chemist to provide us with plastics and synthetic textiles, drugs and pharmaceutical preparations, coatings and finishes, dyes, perfumes, flavoring substances, explosives, fuels. Our crops are protected by organic insecticides; our soils are treated with organic soil conditioners and fertilizers; and our fields are kept free of weeds by the use of synthetic herbicides. There is no aspect of life that is not touched by the hand of the organic chemist.

But organic chemists are not committed solely to the end of producing compounds that provide us with material comfort and convenience. The chemist is curious about the nature of the physical world; he may, for example, have a very esoteric interest in a minute detail of a chemical reaction or in the way in which a living organism transforms one substance into another. Whatever his interests are, he develops and pursues them by experimental attack; to perform experiments he requires the apparatus in which to carry out chemical reactions, instruments with which to measure and study the properties of the substances he prepares, and, above all, chemical compounds to work with.

The research laboratories and the manufacturing facilities of the organic chemical industry provide us with a great number of pure organic compounds, some of which are final products that can be put to immediate practical use, but most of which are starting materials used by the research chemist for the preparation of new compounds. The chemical industry draws upon nature for the carbon it needs for the preparation of its products. Carbon seldom occurs in nature in pure elemental form: even coal—a complex mixture of carbon compounds—contains, besides carbon, a small proportion of hydrogen, oxygen, nitrogen, and sulfur.

The utilization of elemental carbon in the form of coke (Section 2.4), for the preparation of organic compounds, depends first of all upon its transformation into a reduced form. There are two important ways in which this is done: (1) through the conversion of coke into calcium carbide by the high-temperature reaction of lime and carbon, followed by the generation of acetylene by the reaction of calcium carbide and water,

$$CaO + 3C \longrightarrow CaC_2 + CO$$
$$CaC_2 + 2H_2O \longrightarrow Ca(OH)_2 + HC\equiv CH$$

and (2) through the high-temperature reaction of coke with steam to form a mixture of carbon monoxide and hydrogen and through the subsequent reduction of the carbon monoxide by the hydrogen to produce methanol or (depending upon how the process is conducted) compounds of higher molecular weight:

$$C + H_2O \longrightarrow CO + H_2$$
$$CO + 2H_2 \longrightarrow CH_3OH$$

Acetylene and methanol are simple, abundant, inexpensive, and valuable starting materials from which many organic compounds can be prepared.

Some of the uses of natural gas and petroleum as organic raw materials have already been described. These materials, along with coal, constitute the bulk of the world's supply of carbon in a form that makes it readily available for transformation into organic compounds. The supplies of alcohols, ethers, aldehydes, ketones, acids, esters, and other organic chemicals on the market are for the most part derived from these raw materials. A "fine chemical" industry also exists; its role is to transform the simple compounds available cheaply and in large quantities into more complex substances.

Thus, the organic chemist has at his service a market in which a great many compounds can be obtained for use in his research  From this research comes the development of new ways of transforming carbon into useful substances, the discovery of new compounds useful to man, and the further understanding of nature.

## 2.10   Natural sources of carbon compounds

Nature provides additional sources of carbon compounds, many of them in large amount. The forests and the cotton plant provide cellulose, a complex compound of carbon, hydrogen, and oxygen; sugar cane and sugar beets provide enormous tonnages of a pure organic compound, sucrose; the fragrant exudates and essences of trees and flowers provide oils and perfumes; and, until the rise of the synthetic dye industry, plants were once important sources of dyestuffs. Most of the compounds of these classes are used without being subjected to extensive transformation: cellulose is used in the form of paper, textiles, and films; fragrant oils, as flavoring materials and perfumes; sugar, as a food; drugs (such as morphine and cocaine), as medicinals. The role of the organic chemist in dealing with compounds derived from living things is chiefly in discovering and isolating the compound, determining its structure, synthesizing it from simpler compounds, and, often, inquiring into the physiological processes by which it is formed in the living cell, or asking how it is transformed in the body when used as a food or as a drug.

In order to achieve these ends the organic chemist must be able to bring about and interpret the chemical transformations that take place when one organic compound is changed into another. The study of organic chemistry provides him with the understanding that he needs to manipulate organic compounds in predictable and understandable ways and that he needs to apply the knowledge he gains in this research to the recognition and solution of new problems.

### Exercises

(Exercises 1–3 will be found in the text.)

4 Write an acceptable structural formula for each of the following:
   (a) an ester of the composition $C_8H_{14}O_2$
   (b) a cyclic ketone of the composition $C_5H_6O$
   (c) a phenolic carboxylic acid
   (d) a carboxylic acid of the composition $C_5H_8O_2$
   (e) a cycloheptenone
   (f) a β-diketone of the composition $C_5H_6O_2$
   (g) the amide of the acid of (d)
   (h) 1,2-cyclo-octadiene.

5 What is the origin of the following words: (a) petroleum; (b) allyl; (c) formic; (d) oxalic; (e) methyl; (f) olefin; (g) sebacic; (h) suberic?

# Chapter three / Atomic and Molecular Structure and Chemical Bonds

## 3.1 Regularity of structure. Molecular architecture

In order for organic compounds to exist as unique substances having characteristic properties and showing definite behavior, the atoms of carbon, hydrogen, and oxygen (and other elements) that make them up must be held together in definite, stable arrangements. For example, the fifteen atoms of carbon, hydrogen, and oxygen that constitute a molecule of $n$-butyl alcohol form a constellation of atoms having a definite relationship to one another, joined by bonds of definite lengths, and at fixed angular relationships. The fifteen atoms of the isomeric compound diethyl ether are fixed in quite a different arrangement; yet both of these compounds are represented by the formula $C_4H_{10}O$. The differences between these and other isomers are to be found in the manner in which the constituent atoms are joined together; i.e., in the *structures* of the molecules With the aid of small spheres fitted with means for connecting them together (molecular models), it is possible to construct a model of a compound—such as $n$-butyl alcohol or diethyl ether—that is a good large-scale approximation of the actual molecule. The spheres will represent the atoms (carbon, hydrogen, oxygen) of which the compound is composed, and the connections between them will represent carbon-oxygen, carbon-carbon, carbon-hydrogen, or hydrogen-oxygen *bonds*.

## 3.2 Valence, or combining power

From general chemistry, and from the preceding chapters, we are familiar with the concept of valence, or the combining power of an element. Hydrogen is typically monovalent, and the valence of another element may be determined by noting the number of hydrogen atoms with which it combines Thus, in methane, $CH_4$, carbon is tetravalent; in ammonia, $NH_3$, nitrogen is trivalent; in water, $H_2O$, oxygen is divalent; in sodium hydride, NaH, sodium is monovalent. Although the concept of valence is an important one, it does not carry within it the concept of structure, and it is with structure that we are most concerned. For example, there are six compounds with the composition $C_4H_8$; in all of them carbon is tetravalent and hydrogen is monovalent. We must look beyond the idea of valence to discover the differences between these six compounds, and we can do this best by examining the nature of the bonds by which the atoms are joined: their length, direction, and relative disposition in space.

## 3.3 The electronic structures of the atoms

An atom of any element is composed of a dense, compact nucleus and one or more electrons. The nucleus bears a positive charge that is numerically equal to the number of electrons that surround it. This number, the *atomic number*, ranges from 1 for hydrogen, the lightest element, to 103 for lawrencium, the heaviest of the known elements. The mass of the atom is concentrated in the nucleus, and the *mass number* of an atom is equal to the total number of protons and neutrons in the nucleus. The atomic number is the number of protons, which are positively charged particles. Since the atom is electrically neutral, the number of protons, or positive charges, in the nucleus is equal to the number of electrons that surround the nucleus. Hydrogen, with atomic number 1, consists of a nucleus of one proton, and one extranuclear electron. The mass number of hydrogen is thus also 1. In deuterium, the nucleus contains a proton and a neutron, and thus the mass number is 2; but since the neutron carries no charge, there is one extranuclear electron and the atomic number is 1. Tritium with mass number 3 and atomic number 1, is a third *isotope* of hydrogen.

As more protons are added to the nucleus, with a corresponding increase in the number of extranuclear electrons, the elements that make up the periodic table are formed. Our chief concern will be with those of the *first long period*, or row, since in this group are found carbon, nitrogen, and oxygen. We can discuss the nature of most organic compounds quite adequately by dealing with hydrogen, carbon, nitrogen, oxygen, and the halogens (of which fluorine will serve as the type for our initial discussions). Organic compounds containing elements of the *second long period* (sodium, magnesium, aluminum, silicon, phosphorus, sulfur, and chlorine) are also well known and important, both practically and theoretically, and will be dealt with when discussion of them is appropriate.

## 3.4 Electron orbitals

The extranuclear electrons are constrained to occupy certain regions in space surrounding the nucleus. The region in which an electron is most likely to be found is known as an *orbital*, the characteristics of which are defined mathematically in terms of its location with respect to the nucleus. The electron orbitals are grouped in *shells*, or *principal energy levels*, which are defined by the letters $K$, $L$, $M$, and so on, or by the *principal quantum numbers* 1, 2, 3, and so on. The $K$ shell consists of a single orbital, the $1s$ orbital; the $L$ shell of four orbitals, the $2s$, $2p_x$, $2p_y$ and $2p_z$; and the $M$ shell of nine orbitals, one $3s$, three $3p$, and five $3d$ orbitals. It will be noticed that the shells consist of subshells equal in number to the principal quantum number ($M$ shell, principal quantum number 3, subshells $3s$, $3p$, $3d$).

The state of any one electron is described by four quantum numbers, and the *Pauli exclusion principle* states that *no two electrons in an atom can have the same four quantum numbers*. The result of this, expressed in terms of our present purpose, is that no more than two electrons can occupy a single orbital and these must have opposite spins.

The single extranuclear electron in the hydrogen atom occupies the $1s$ orbital of the $K$ shell. The $1s$ orbital is spherically symmetrical, and this electron may be regarded as moving randomly within this spherical volume. In the early treatment of the hydrogen atom by Bohr, the electron was placed in a planetary orbit at a distance of 0.53 Å.

The helium atom consists of a nucleus with a positive charge of 2 and two extra-nuclear electrons. These electrons are both found in the $1s$ orbital, but differ in their spin quantum number; i.e., they have opposite spins. The $1s$ orbital is now "filled", and in lithium, atomic number 3, the third electron occupies the orbital with the next higher principal quantum number, the $2s$ orbital. This

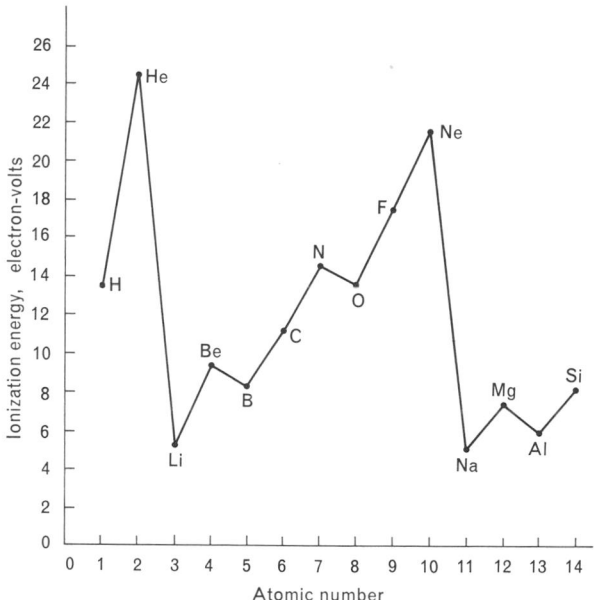

**Figure 3–1.** Ionization energies of the first fourteen elements of the periodic table.

orbital can contain two electrons with opposite spins, and the beryllium atom contains two $1s$ electrons and two $2s$ electrons.

With increasing atomic number, the additional electrons are found in successively higher atomic orbitals. The three $2p$ orbitals are occupied successively by single electrons (Hund's rule states that the several orbitals of a subshell are each occupied by a single electron before pairing of electrons begins) up to nitrogen, which may be represented by $1s^2 2s^2 2p_x{}^1 2p_y{}^1 p 2_z{}^1$. Pairing of the electrons in the $2p$ orbitals begins in oxygen, in which the electron distribution is $1s^2 2s^2 2p_x{}^2 2p_y{}^1 2p_z{}^1$.

The picture that has been given of the structures of the light atoms is supported by experimental measurement of the energy required to remove an electron from an outlying orbital. It is to be expected that an outer, loosely held electron will be more easily removed than an electron closer to the positively charged nucleus, by which it is thus more strongly held. In Figure 3-1 are shown the experimentally determined *ionization energies* for the elements from hydrogen (atomic number 1) to silicon (atomic number 14).

We shall not analyze all of these data in detail, but the following comments will indicate some of the conclusions that can be reached by their study. It will be noted that hydrogen is converted into the hydrogen ion ($H^+$) at an ionization energy* of 13.6 e.v.

Helium, however, is ionized only at the much higher energy of 24.58 e.v. Since the single electron of hydrogen and the two electrons of helium are in the $1s$ orbital and thus at the same relative energy level, it is apparent that the increased attraction exerted by the doubly positive helium nucleus increases the difficulty of removing one of the two helium electrons. Lithium, however, despite the $+3$ charge on the nucleus, has the low ionization energy of 5.39 e.v. This can only mean that the third electron of lithium is farther from the nucleus, and thus not as strongly held by it. As we have seen, the lithium atom has two electrons in the $1s$ orbital of the $K$ shell and a third in the $2s$ orbital of the $L$ shell. The $L$ shell surrounds the $K$ shell and lies, on the average, farther from the nucleus; thus, the single $2s$ electron is removed at a relatively low energy. When a second electron is present in the $2s$ shell (beryllium), the increased nuclear charge ($+4$) exerts a greater attraction upon the $2s$ electrons and as a consequence the ionization energy of beryllium is greater than that of lithium. It is to be noted that the alkali metals (Li, Na, K) have ionization energies that are at minima in the graph shown in Figure 3-1. In each of these, the electron removed is a single electron that occupies an orbital lying outside of a completed lower energy subshell.

## 3.5 Electronic structure and symbolism

Figure 3-2 summarizes the electronic structure of the lighter elements of the periodic table in terms of the disposition of the extranuclear electrons in the atomic orbitals just described. It will be noted that pairs of electrons occupying a given

---

* The unit in which ionization energy is expressed is the electron volt: 1 e.v. $= 23.06$ kilocalories per mole.

| | 1s | 2s | 2p | 3s | Ionization energy, e.v. | |
|---|---|---|---|---|---|---|
| H | ↑ | | | | 13.60 | H |
| He | ↑↓ | | | | 24.58 | He |
| Li | ↑↓ | ↑ | ☐ ☐ ☐ | | 5.39 | Li |
| Be | ↑↓ | ↑↓ | ☐ ☐ ☐ | | 9.32 | Be |
| B | ↑↓ | ↑↓ | ↑ ☐ ☐ | | 8.30 | B |
| C | ↑↓ | ↑↓ | ↑ ↑ ☐ | | 11.26 | C |
| N | ↑↓ | ↑↓ | ↑ ↑ ↑ | | 14.54 | N |
| O | ↑↓ | ↑↓ | ↑↓ ↑ ↑ | | 13.61 | O |
| F | ↑↓ | ↑↓ | ↑↓ ↑↓ ↑ | | 17.42 | F |
| Ne | ↑↓ | ↑↓ | ↑↓ ↑↓ ↑↓ | | 21.56 | Ne |
| Na | ↑↓ | ↑↓ | ↑↓ ↑↓ ↑↓ | ↑ | 5.14 | Na |
| Mg | ↑↓ | ↑↓ | ↑↓ ↑↓ ↑↓ | ↑↓ | 7.64 | Mg |

**Figure 3–2.** The electronic atomic structures of the first twelve elements of the periodic table.

orbital are distinguished by opposite spins. This is indicated by the vertical arrows in each orbital.

It is convenient to adopt a symbolism for the electronic structures of atoms that is less awkward than that in which each orbital is individually written. A satisfactory device for the first-period elements is to consider the nucleus and the $1s$ electrons as a unit, called the *kernel*, and to show the valence electrons of the $L$ shell as dots surrounding the usual symbol of the atom. Thus, we write

$$\text{Li·} \quad \text{·Be·} \quad \text{·B·} \quad \text{·C·} \quad \text{·N·} \quad \text{·O·} \quad \text{:F·} \quad \text{:Ne:}$$

Similarly, sodium is conveniently written as Na·, since only the single $3s$ electron is involved in the usual reactions of sodium:

$$\text{Na·} \longrightarrow \text{Na}^+ + e$$

The unspecified ten electrons of the filled $1s$, $2s$ and $2p$ shells are not involved in reactions that will concern us, and they are understood to make up the electronic complement of the sodium ion when it is written Na⁺. Chlorine and bromine will be encountered later in the text as the molecule, the atom, the cation, and the anion:

$$:\overset{..}{\underset{..}{Cl}}:\overset{..}{\underset{..}{Cl}}: \qquad :\overset{..}{\underset{..}{Cl}}\cdot \qquad :\overset{..}{\underset{..}{Cl}}+ \qquad :\overset{..}{\underset{..}{Cl}}:^-$$

In chlorine, the external shell, or *valence shell*, is represented (in the atom) by the seven electrons of the $3s$ (two), $3p$ (five ) subshells. The bromine atom is similarly represented as $:\overset{..}{Br}\cdot$ in which the seven valence-shell electrons are those of the $N$ shell ($4s^2$, $4p^5$). The complete electronic complement of the bromine atom can be shown by the symbol $1s^2 2s^2 2p^6 3s^2 3p^6 3d^{10} 4s^2 4p^5$.

---

**Exercise 1.** Using the notation just described for bromine, write the symbols for the electronic structures of *(a)* nitrogen; *(b)* aluminum; *(c)* carbon; *(d)* silicon; *(e)* phosphorus; *(f)* boron; *(g)* oxygen; *(h)* sulphur. Compare *(a)* with *(e)*; *(b)* with *(f)*; *(g)* with *(h)*; *(c)* with *(d)*. What similarities do you find?

---

## 3.6 The noble gas structure: the "octet"

An electronic interpretation of chemical bond formation was proposed in 1916 by G. N. Lewis (1875–1946). Lewis pointed out that the chemical inertness of the noble gases indicated a high degree of stability of the electronic complements of these elements: helium, with a shell of two electrons, neon with shells of two and eight, and argon with shells of two, eight, and eight. The first consequence to be noted is that elements near the ends of a period tend to lose or gain electrons in such a way as to acquire a noble gas structure. For instance,

$$\text{lithium} \left( \overset{1s}{\underline{\underset{..}{\phantom{x}}}} \; \overset{2s}{\underline{\underset{.}{\phantom{x}}}} \right) \xrightarrow{\;-e\;} \text{Li}^+ \left( \overset{1s}{\underline{\underset{..}{\phantom{x}}}} \right) \text{ helium structure}$$

$$\text{fluorine} \; \overset{1s}{\underline{\underset{..}{\phantom{x}}}} \; \overset{2s}{\underline{\underset{..}{\phantom{x}}}} \; \overset{2p}{\underline{\underset{..}{\phantom{x}}}} \; \underline{\underset{.}{\phantom{x}}} \; \underline{\underset{..}{\phantom{x}}} \xrightarrow{\;+e\;} \text{F}^- \left( \overset{1s}{\underline{\underset{..}{\phantom{x}}}} \; \overset{2s}{\underline{\underset{..}{\phantom{x}}}} \; \underline{\underset{..}{\phantom{x}}} \; \overset{2p}{\underline{\underset{..}{\phantom{x}}}} \; \underline{\underset{..}{\phantom{x}}} \right) \text{neon structure}$$

Since a chemical reaction in which lithium is converted to lithium ion involves a transfer of the electron to some other atom, the two changes shown in these equations can be written as a single reaction:

$$\text{Li}\cdot + :\overset{..}{\underset{..}{F}}\cdot \longrightarrow \text{Li}^+ + :\overset{..}{\underset{..}{F}}:^-$$

Lithium, by the loss of the valence electron in the $L$ shell, has now attained the helium structure; and fluorine, by the gain of an electron to acquire the full complement of eight electrons in the $L$ shell, has attained the neon structure.

The complete eight-electron grouping of four completely filled (one $s$, three $p$) orbitals was called by Lewis the *octet*. Helium is of course the unique exception, since it possesses only the $K$ shell. The completion of a pair of electrons in the $K$ shell by gain (hydrogen) or loss (lithium, beryllium) is the counterpart of the completion of the octet in the higher atoms.

Sodium, like lithium, is an alkali metal that readily loses an electron and is transformed into a positive ion. Chlorine, like fluorine, is a halogen, and readily gains an electron to become chloride ion.

$$\begin{cases} \text{Na } (1s^2 \;\; 2s^2 2p^6 \;\; 3s^1) & \xrightarrow{\;-e\;} & \text{Na}^+ (1s^2 \;\; 2s^2 2p^6) & \text{neon structure} \\ \text{Cl } (1s^2 \;\; 2s^2 2p^6 \;\; 3s^2 3p^5) & \xrightarrow{\;e\;} & \text{Cl}^- (1s^2 \;\; 2s^2 2p^6 \;\; 3s^2 3p^6) & \text{argon structure} \end{cases}$$

## 3.7   The ionic bond: electrovalence; organic salts

The bond that holds sodium ions and chloride ions in the definite relationship that we recognize as present in sodium chloride is an electrostatic attraction between the oppositely charged ions. A crystal of sodium chloride consists of a closely packed array of these two ions in which we cannot discern any pairs that might be called sodium chloride molecules. The attractive forces between these ions have no fixed direction in space, and each ion is surrounded by ions of the opposite kind. When a salt such as sodium chloride is dissolved in water the sodium ions and chloride ions become largely independent; each exerts its electrostatic attraction upon the surrounding water molecules. They are *hydrated ions*.

The ionic bond in organic chemistry is found chiefly in salts of organic acids and bases; in this respect it needs no special discussion. Organic salts are usually water-soluble for the same reasons that the more familiar inorganic salts are soluble: the ions, bound in a regular lattice in the crystal, dissociate and solvate when put into water and yield a solution of the separate ions. The organic portion retains its structural integrity; for example, sodium acetate dissolves in water to give a solution of sodium ions and *acetate* ions:

$$\text{CH}_3\text{COO}^-\text{Na}^+ + \text{H}_2\text{O} \longrightarrow \text{CH}_3\text{COO}^- + \text{Na}^+ \quad \text{(both solvated)}$$

Salts of amines consist of equal numbers of substituted ammonium cations and organic or inorganic anions, which, in the solid state, form a crystal lattice by virtue of the electrostatic attraction of the oppositely charged ions.

| $\text{NH}_4^+$ | $\text{Cl}^-$ | $\text{CH}_3\text{NH}_3^+$ | $\text{Br}^-$ |
|---|---|---|---|
| ammonium | chloride | methylammonium | bromide |

| $\text{CH}_3\text{CH}_2\text{NH}_3^+$ | $\text{CH}_3\text{COO}^-$ | $((\text{CH}_3)_3\text{NH})_2^+$ | $\text{SO}_4^=$ |
|---|---|---|---|
| ethylammonium | acetate | trimethylammonium | sulfate |

Most salts are solids, often with high melting points because the ionic interactions preserve the integrity of a crystal lattice so that vigorous thermal agitation, brought about by elevation of the temperature, is necessary to destroy the ordered arrangement of the ions. In this respect, ionic organic compounds (usually salts of the kind described) are not much different from salts consisting of inorganic ions. They differ chiefly in the melting points, those of many organic salts being much lower than those of the familiar inorganic salts. The reason for this is that, because of the more complex nature of the organic ions (compare the acetate ion, $CH_3COO^-$, with the chloride or nitrate ion; and the trimethylammonium ion with $Na^+$ or $K^+$), the crystal lattice formed is less compact and orderly than that of, say, sodium chloride, and consequently less energy is required to bring it to a state of disorder (melt it).

## 3.8 The covalent bond

The tendency for the elements near the ends of the periodic table to gain or lose electrons and thus attain the electronic structures of the rare gases is not found in the elements near the center of a period. The energy required to add successive electrons to nitrogen or carbon to form $N^{3-}$ or $C^{4-}$, or the energy required to remove successive electrons from boron or carbon to form $B^{3+}$ or $C^{4+}$, is prohibitive for simple electrostatic reasons. These atoms may, however, acquire a complete octet of electrons by *sharing* electrons with other atoms. This sharing of electrons joins the atoms together by a *covalent bond*, for the electrons are not transferred from an orbital of one atom to that of another. Rather, they spend part of the time in the orbitals of both atoms in order that each atom will have a filled shell.

## 3.9 The hydrogen molecule

The simplest and purest example of the covalent bond is found in the hydrogen molecule. When two atoms of hydrogen collide, the single electron of each can be accommodated into the $1s$ orbital of the other (for an orbital can contain a pair of electrons). The great gain in stability brought about by this combination is evidenced by the large quantity of energy liberated (for example, as heat) when the hydrogen molecule is formed:

$$H\cdot + H\cdot \longrightarrow H_2 + 104 \text{ kcal}$$

The covalent bond is thus a strong bond, for it would require 104 kcal to dissociate a mole of hydrogen into its constituent atoms.

What is the reason for this great stability of the covalent bond? The two positively charged nuclei are held together by the attractive force of two electrons, which are now to be found largely in the region between the nuclei.

In order for two hydrogen atoms to form a hydrogen molecule, the two atoms must clearly approach one another sufficiently closely to permit the $1s$ atomic orbitals to overlap, permitting both electrons to occupy either orbital and

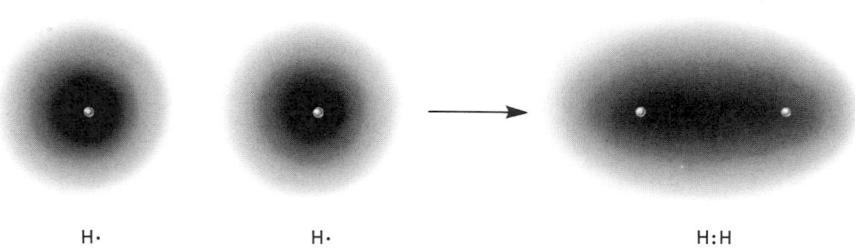

H·                    H·                                    H:H

**Figure 3–3.** The formation of the hydrogen molecule by combination of hydrogen atoms.

to find their way from one to the other without leaving the attractive influence of a nucleus (Figure 3-3).

But the atoms cannot approach closer than a certain distance, because the repulsive force of the two positively charged nuclei will increase sharply as the distance between them diminishes. In the normal hydrogen molecule, then, the atoms will be separated by a fixed distance, known as the *bond length*. Figure 3-4 is a graphical representation of the energy of the H—H system as a function of the internuclear distance.

### 3.10  The covalent bond between unlike atoms

Only a few elements form simple diatomic molecules. The halogens, by sharing the single unpaired electron in the outer ($2p$ in fluorine, $3p$ in chlorine) valence orbital, can form a covalent bond. As in hydrogen, the paired electrons occupy a *molecular*

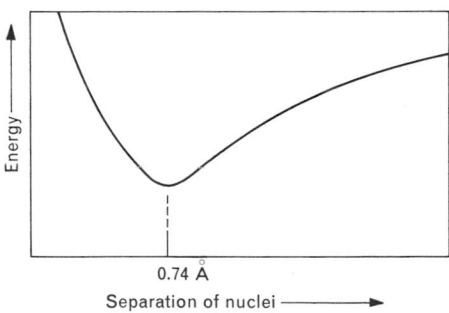

0.74 Å

Separation of nuclei ⟶

**Figure 3–4.** Energy diagram for the hydrogen molecule. The distance 0.74 Å is the H—H bond length.

*orbital* which includes the two atomic nuclei. The formation of the molecular orbital, with the completion of the electronic complement of all four of the *p* orbitals of each atom (the "octet"), represents a condition of increased stability over that of the separate atoms. This is shown by the energy of combination of the isolated atoms. The union of two fluorine atoms to form the diatomic fluorine molecule is accompanied by the liberation of 36.6 kcal/mole. Conversely, the rupture of the F—F bond to produce two separated fluorine atoms requires energy in the amount of 36.6 kcal/mole.

$$:\ddot{\text{F}}\cdot + :\ddot{\text{F}}\cdot \longrightarrow :\ddot{\text{F}}:\ddot{\text{F}}: + 36.6 \text{ kcal}$$

It is clear that the fluorine molecule does not consist of an equal number of $F^+$ and $F^-$ ions, joined by electrostatic attraction, from the following energy considerations: the ionization energy of fluorine is 17.42 e.v., or 403 kcal/mole

$$\text{F}\cdot \longrightarrow \text{F}^+ + e - 403 \text{ kcal}$$

and the electron affinity of fluorine is such that the formation of fluoride ion from the atom is exothermic, liberating 80.3 kcal/mole

$$\text{F}\cdot + e \longrightarrow \text{F}^- + 80.3 \text{ kcal}$$

The sum of these equations

$$\text{F}\cdot + \text{F}\cdot \longrightarrow \text{F}^+ + \text{F}^- - 322.7 \text{ kcal}$$

shows that the formation of a fluorine molecule that consisted of the two oppositely charged ions would require energy; and experiment shows that the combination of fluorine atoms to produce the molecule is exothermic.

The formation of the $Cl_2$ and $Br_2$ molecules follows the same pattern:

$$:\ddot{\text{C}}\text{l}\cdot + :\ddot{\text{C}}\text{l}\cdot \longrightarrow :\ddot{\text{C}}\text{l}:\ddot{\text{C}}\text{l}: + 58.0 \text{ kcal/mole}$$

$$:\ddot{\text{B}}\text{r}\cdot + :\ddot{\text{B}}\text{r}\cdot \longrightarrow :\ddot{\text{B}}\text{r}:\ddot{\text{B}}\text{r}: + 46.1 \text{ kcal/mole}$$

The majority of known covalent compounds are constructed of bonds between unlike atoms. The same process of electron sharing is involved in such bonds, but for elements near the center of a period, two, three, or four electrons are needed to complete the octet, with the result that more than one covalent bond can be formed.

Oxygen, with six electrons in its valence shell, can complete its octet by forming two covalent bonds with other atoms, as in the compounds water, methanol, chlorine oxide, dimethyl ether:

|  |  |  |  |
|---|---|---|---|
| H:Ö:H | H:C̈:Ö:H with H above and below | :C̈l:Ö:C̈l: | H:C:O:C:H with H's above and below |
| H—O—H, or $H_2O$ | $CH_3OH$ | $Cl_2O$ | $CH_3OCH_3$ |
| water | methanol | chlorine oxide | dimethyl ether |

Nitrogen, with five *L*-shell electrons, can acquire a complete octet by forming three covalent bonds:

$$:N{\equiv}N: \qquad H-\overset{\cdot\cdot}{\underset{H}{N}}-H \qquad H-\overset{H}{\underset{H}{C}}-\overset{\cdot\cdot}{N}{\underset{H}{\overset{H}{<}}} \qquad Cl-\overset{\cdot\cdot}{\underset{Cl}{N}}-Cl$$

Carbon, with four *L*-shell electrons, can form four covalent bonds. The ability of carbon to form covalent bonds, not only with atoms of other kinds but also with other carbon atoms, gives rise to the enormous number of organic compounds. The pairing of the four electrons in bond formation gives carbon its characteristic valence of four:

$$H{:}\overset{H}{\underset{H}{\overset{\cdot\cdot}{C}}}{:}H \quad \text{or} \quad H-\overset{H}{\underset{H}{C}}-H \qquad Cl-\overset{Cl}{\underset{Cl}{C}}-Cl \qquad H-\overset{H}{\underset{H}{C}}-\overset{H}{\underset{H}{C}}-\overset{H}{\underset{H}{C}}-H$$

Boron, however, can form only three bonds by a simple process of electron-pair formation with single electrons of other atoms:

$$\overset{\cdot\cdot}{\underset{\cdot\cdot}{:Cl:}} \atop :\overset{\cdot\cdot}{Cl}{:}B{:}\overset{\cdot\cdot}{Cl}{:} \qquad\qquad H-\overset{\underset{H}{\overset{H\ \ H}{\backslash\ |\ /}}C}{\underset{H}{|}}-B-\overset{\underset{H}{/}}{\underset{}{\overset{H\quad}{C}}}-H \quad \text{or} \quad B(CH_3)_3$$

Note that in $BCl_3$ and $B(CH_3)_3$ the boron atom has not completed its octet and thus has the capacity for accepting another pair of electrons. This property of boron (and its second-row counterpart, aluminum) has important chemical consequences which will be dealt with later in this chapter.

---

**Exercise 2.** Write complete electronic structures for the following compounds. (a) $CH_3NH_2$; (b) $CH_2Cl_2$; (c) $CH_3CH_2Br$; (d) $CH_3OCH_3$; (e) $(CH_3)_2CHOH$; (f) $HOCl$; (g) $H_2NOH$.
Show all covalent bonds and all unshared pairs of electrons. As an example,

$$(a) \quad H{:}\overset{H}{\underset{H\ H}{\overset{\cdot\cdot}{C}}}{:}\overset{\cdot\cdot}{N}{:}H$$

---

The nature of the covalent bond is independent of the way in which it is formed. For example, methane is the same compound, whether it is formed by the combination of carbon and hydrogen atoms,

$$\cdot\overset{\cdot\cdot}{C}\cdot + 4H\cdot \quad\longrightarrow\quad H{:}\overset{H}{\underset{H}{\overset{\cdot\cdot}{C}}}{:}H \quad (\text{or } CH_4)$$

by the combination of a methyl radical and a hydrogen atom,

$$
\begin{array}{ccc}
\text{H} & & \text{H} \\
\text{H:}\overset{..}{\text{C}}\cdot +\text{H}\cdot & \longrightarrow & \text{H:}\overset{..}{\text{C}}\text{:H} \\
\text{H} & & \text{H}
\end{array}
$$

or by the loss of carbon dioxide from acetic acid:

$$
\text{CH}_3\text{COOH} \xrightarrow[\text{soda-lime}]{\text{heat with}} \text{CH}_4 + \text{CO}_2
$$

## 3.11 Unequal sharing of electrons in the covalent bond

In the H—H bond of the hydrogen molecule the electrons are equally shared between the two atoms; that is, the molecular orbital is symmetrical with respect to the two nuclei. In diatomic molecules consisting of two *different* atoms joined by a covalent bond, a greater share of the pair of electrons is held by one of the atoms—the molecular orbital is unsymmetrical, and one end of the covalent bond is negative with respect to the other. This dissymmetry is reflected in the physical and chemical properties of the compounds in which such covalent bonds are present.

A conspicuous property of diatomic compounds containing dissymmetric covalent bonds is their tendency to orient themselves in an electrostatic field. The measure of this tendency is expressed numerically as the *dipole moment* of the molecule; it is the product of the difference in charge and the distance between the positive and negative centers. In Table 3-1 are given a few representative examples of the dipole moments of simple dipolar molecules.

The presence of a dipole moment in a molecule confers upon it certain properties that are the result of the electrical dissymmetry. Dipolar molecules tend to orient themselves toward each other so that opposite charges attract and like charges repel. The result of this is that what in a nonpolar substance would be randomly oriented population of molecules has, in the dipolar substance, a degree of structural organization in the liquid or solid state. The additional energy required to overcome these intermolecular attractions is reflected in higher boiling points for dipolar liquids and high melting points for dipolar solids—that is, higher than the boiling points and melting points of nonpolar substances of comparable molecular size and weight.

Dipolar character in a bond results in greater bond strengths than would be expected were it not a factor. The reason for this is that the energy required to break the bond (to produce the two atoms) must include the energy necessary to overcome the electrostatic attraction of the two oppositely charged atoms joined by the bond. For example, the H—H, Cl—Cl, and H—Cl bond energies are as follows:

$$
\begin{array}{l}
\text{H—H} \longrightarrow \text{H}\cdot + \text{H}\cdot - 104 \text{ kcal/mole} \\
\text{Cl—Cl} \longrightarrow \text{Cl}\cdot + \text{Cl}\cdot - 58 \text{ kcal/mole} \\
\text{H—Cl} \longrightarrow \text{Cl}\cdot + \text{H}\cdot - 102 \text{ kcal/mole}
\end{array} \left.\rule{0pt}{3.5em}\right\} \begin{array}{c} \text{experimental} \\ \text{values} \end{array}
$$

| Table 3-1. *Dipole moments of some simple molecules* | |
|---|---|
| Compound | Dipole Moment, D* |
| HF | 1.9 |
| HCl | 1.1 |
| HBr | 0.8 |
| HI | 0.4 |
| FCl | 0.9 |
| HCN | 3.0 |
| $NH_3$ | 1.5 |

*Dipole moments are recorded in Debye units, D, named after Peter Debye, who made the first experimental studies of this property. One Debye unit is the product of a charge of about $10^{-10}$ unit electrostatic (e.s.u.) at a distance of about one Ångstrom unit (Å), or $10^{-8}$ cm, and is therefore about $10^{-18}$ e.s.u. × cm.

If the contribution to the H—Cl bond energy were the sum of the contributions of one-half each of the H—H and Cl—Cl bond energies, the H—Cl bond energy should be $104/2 + 58/2 = 81$ kcal/mole. This bond is in fact much stronger (by 21 kcal/mole) than this, because of the unequal sharing of the electrons between H and Cl. The dipole moment of HCl (1.10 D) would then be represented as

$$\overset{+\longrightarrow}{\text{H—Cl}}$$

The direction of the arrow (indicating the greater share of electrons on chlorine) is as shown because of the much larger nuclear charge on chlorine. Thus, chlorine is more *electronegative* than hydrogen.

### 3.12   Electronegativity

The increasing ionization energy (Figure 3-1) of the elements in the row from lithium to fluorine is a measure of the attraction of the positively charged nucleus for the peripheral electron that is removed in the process $X \rightarrow X^+ + e$. The actual value of the energy is modified by the character and position of the orbital in which the electron is found, but an inspection of the values given shows that the ionization energy, and thus the nuclear attraction of the valence electrons, rises in going from left to right in the row. By the use of calculations that will not be given here, Pauling has assigned numerical values to this electron-attracting power and called this set of values an electronegativity scale. A portion of the scale is given in Table 3-2.

**Table 3.2.** *Electronegativity values for some of the common elements found in organic compounds* $(H=2.1)$

| | | | Period | | | |
|---|---|---|---|---|---|---|
| I | II | III | IV | V | VI | VII |
| Li | Be | B | C | N | O | F |
| 1.0 | 1.5 | 2.0 | 2.5 | 3.0 | 3.5 | 4.0 |
| Na | Mg | Al | Si | P | S | Cl |
| 0.9 | 1.2 | 1.5 | 1.8 | 2.1 | 2.5 | 3.0 |
| | | | | | | Br |
| | | | | | | 2.8 |
| | | | | | | I |
| | | | | | | 2.4 |

Although the actual nuclear charge in the halogens increases from fluorine to iodine, the effective nuclear charge, which is that of the *kernel* left after removal of the outer valence electrons, is $+7$ in each. The decreasing electronegativity of the halogens as the atomic number increases is the result of the greater distance of the electrons in the outer shell from the positively charged nucleus.

The greatest value of the concept of electronegativity to the organic chemist is in its qualitative applications. Only rarely are the numbers given in the table used in calculations. These values do, however, permit us to recognize the nature and, roughly, the extent of polarization in a chemical bond. For example, in the

molecule of methyl chloride, $H—\overset{\displaystyle H}{\underset{\displaystyle H}{C}}—Cl$, whose dipole moment is 1.9 D, the

greatest difference in electronegativity is found in the C—Cl bond, and chlorine, which is the more electronegative, will be the negative end of the dipole.

---

**Exercise 3.** The single-bond energies of the hydrogen molecule, the halogen molecules, and the hydrogen halide molecules are as follows (approximate values in kcal/mole):

| Bond energy | H—H | F—F | Cl—Cl | Br—Br | I—I |
|---|---|---|---|---|---|
| | 104 | 37 | 57 | 45 | 35 |

| Bond energy | H—F | H—Cl | H—Br | H—I |
|---|---|---|---|---|
| | 135 | 102 | 86 | 70 |

A calculated value for the H—F bond would be $104/2 + 37/2 = 70.5$. Calculate the other HX bond energies, and compare them with the experimental values given in the above table. How can you account for the differences between the calculated and observed values, taking into account *(a)* the electronegativities of the halogens, and *(b)* the effect of unequal electron distribution in the HX bond upon the bond energy?

---

Our chief concern in what follows will be with the charge distribution in single bonds between carbon and other elements of the first long row (Li to F). The nature of C—X (X = halogen), C—N, and C—O bonds will be of primary importance. It is easy to see that in each of these examples the *carbon will be positive with respect to the other atom*, and thus will be the center of attack by reagents that provide electrons to the formation of new bonds.

### 3.13 The extension of covalent bonds in space. The geometry of organic molecules

*Atomic orbitals.*    The atomic orbitals ($1s$, $2s$, and the three $2p$ orbitals of the first row elements) occupy definite regions in space. The extranuclear electrons are found distributed around the nucleus in certain well-defined ways.

The $1s$ orbital is spherically symmetrical; this is true also of the $2s$ orbital, which lies around and is concentric with the $1s$ orbital (Figure 3-5). The electrons in these orbitals are found somewhere in a diffuse spherical volume, and their location can be defined only as a probability. In the hydrogen atom, the most probable distance of the $1s$ electron from the nucleus is 0.53 Å; in heavier atoms, with greater nuclear charge, the $K$ shell becomes smaller with increasing atomic number.

The electron density in the hydrogen atom, as would be expected, is greatest in the vicinity of the nucleus, because of the electrostatic attraction. But in any spherical volume element with a small radius, the total charge would be small because of the small volume of such an element. At a great distance from the nucleus, a spherical volume element would be large, but the electron density low. Calculation shows that the effects of the decreasing probability of finding the electron at a given *point* (which decreases with distance from the nucleus), and the larger number of points in a larger spherical volume element (which grows larger

**Figure 3–5.** Spatial disposition of the $s$ atomic orbital.

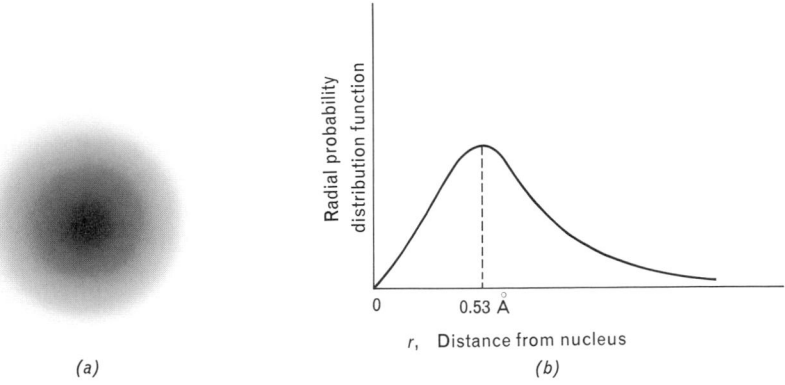

*(a)*        *(b)*

**Figure 3–6.** *(a)* Electron distribution in the hydrogen atom. *(b)* Probability of finding the 1s electron at a distance *r* from the nucleus.

with distance from the nucleus) reach a balance at a distance of 0.53 Å from the nucleus. Figure 3-6*a* shows the electron density about the hydrogen nucleus and Figure 3-6*b* shows the probability of finding the electron at a distance from the nucleus.

The most probable distance for the electrons in the 2*s* orbital will depend upon the magnitude of the nuclear charge, but again we find that the distribution is spherically symmetrical (Figure 3-5). The three 2*p* orbitals are not spherically symmetrical, but extend at right angles to each other along the *x*, *y*, and *z* axes (Figure 3-7).

*Molecular orbitals.*    When two hydrogen atoms combine to form the hydrogen molecule, the 1*s* orbitals overlap so that the resulting molecular orbital, containing the two electrons, can encompass both nuclei (Section 3.9). It is clear that overlap is a necessary condition for the formation of the molecular orbital; since electron density drops off sharply with increasing distance from the nucleus, the two hydrogen atoms must approach each other closely to permit both electrons to come under the attractive influence of both nuclei. This can be expressed in another way by saying that for the electron pair to have access to both atomic nuclei, the

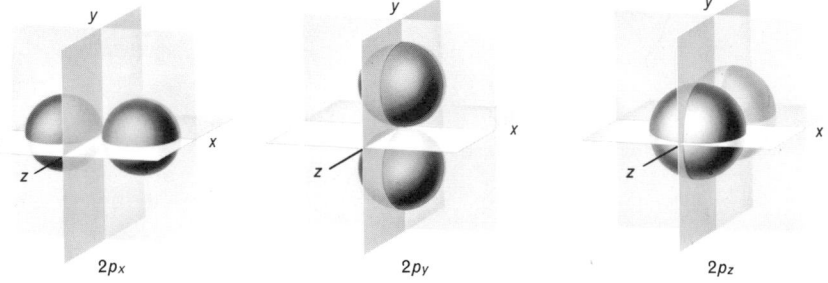

2*p*x       2*p*y       2*p*z

**Figure 3–7.** Spatial disposition of the *p* atomic orbitals.

orbitals must come together closely enough for the electrons to come under the influence of both nuclei without having to cross a high energy barrier (i.e., without having to leave the influence of one nucleus before coming under the influence of the other).

Although the atomic $s$ orbitals are spherical, and thus can engage in molecular orbital formation from any direction, the atomic $p$ orbitals have extension in space along the three axes at right angles to each other. Thus, if two atomic $p$ orbitals were to be used to form bonds with two other atoms, the two new bonds (molecular orbitals) so formed might be expected to be at right angles to each other.

This situation can be seen in the combination of two hydrogen atoms and an oxygen atom to form a molecule of water. Oxygen, with six electrons in the $L$ shell, can complete the full complement of eight electrons (four filled orbitals) by forming two covalent bonds with two hydrogen atoms:

$$O\left\{\begin{array}{cccc}1s & 2s & & 2p \\ \cdot\cdot & \cdot\cdot & \cdot\cdot & \cdot\ \ \cdot\end{array}\right\}+2H\left\{\begin{array}{c}1s \\ \cdot\end{array}\right\} \longrightarrow \left\{\begin{array}{cccc}1s & 2s & & 2p \\ \cdot\cdot & \cdot\cdot & \cdot\cdot & \overset{\cdot\cdot}{H}\ \ \overset{\cdot\cdot}{H}\end{array}\right\} \quad (H_2O)$$

What is the shape of the water molecule? Our first assumption might be that since two $p$ orbitals of the oxygen atom have been used, and these are at right angles to each other, the H—O—H bond angle would be 90°. The experimentally determined bond angle is in fact 104.5°. Figure 3-8 is a representation of the molecule: $(a)$ is a schematic drawing in which the molecular orbitals are indicated; $(b)$ is a molecular model.

It must be recognized, however, that when oxygen combines with hydrogen to form the water molecule, the H—O bonds found in the completed water molecule constitute molecular orbitals, and thus need not have the geometry of the atomic $p$ orbitals of the oxygen atom. The eight electrons around oxygen in the water molecule, found in the two H—O bonds and in the two unshared pairs of electrons on oxygen, can be expected to be disposed in such a way as to compose an arrangement of minimum energy. They are no longer to be described in the same way as the atomic $2s$ and $2p$ orbitals. For one thing, the two molecular orbitals of

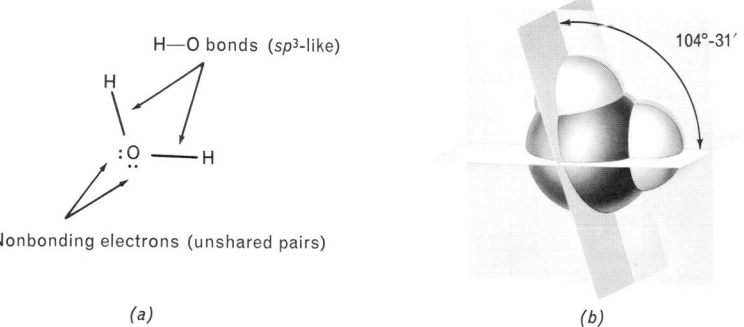

H—O bonds ($sp^3$-like)

H

:O —H

Nonbonding electrons (unshared pairs)

104°-31′

$(a)$          $(b)$

**Figure 3–8.** $(a)$ Disposition of the bonding and nonbonding electrons in the water molecule. $(b)$ A model of the water molecule.

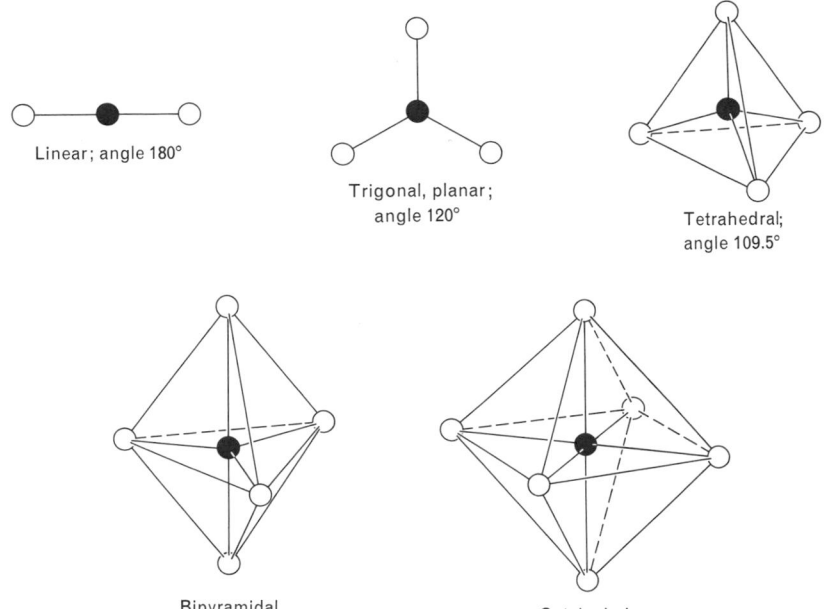

Linear; angle 180°

Trigonal, planar; angle 120°

Tetrahedral; angle 109.5°

Bipyramidal

Octahedral

**Figure 3–9.** Angular distribution of bonds.

the H—O bonds are now extended from the oxygen atom toward the hydrogen atoms; atomic orbitals, on the other hand, are symmetrical with respect to the nucleus and possess a nodal plane of zero density probability at their centers. If these two molecular orbitals are altered in character, it is to be expected that the remaining two orbitals on oxygen, each of which contains an unshared ("lone") pair of electrons, will also be different from the atomic $2s$ and $2p$ orbitals.

A useful guiding principle for assessing the probability distribution of electrons (or bonds) about an atomic nucleus is the simple rule that the electrostatic repulsion between the electrons will cause them to be arranged in such a way as to minimize these repulsive forces. Imagine the electrons as situated on the surface of a sphere. Two like charges or electron pairs on the spherical surface would tend to move to opposite sides and to arrive at a position 180° from each other. Three pairs of electrons would arrange themselves at 120° apart on a plane; four pairs at the corners of a regular tetrahedron. Similar considerations for penta-, hexa-, and octacovalent molecules lead to the distributions shown in Figure 3-9.*

The geometry of the water molecule can now be pictured as resulting from the distribution of four electron pairs about the oxygen kernel: two pairs constitute the two H—O bonds, two pairs are "lone". Were the tetrahedron that has been

---

* An excellent classroom demonstration of the distribution of bonds by mutual repulsion can be performed with the use of cylindrical (sausage-shaped) balloons. A description of this demonstration is given by H. R. Jones and B. R. Bentley, *Proceedings of the Chemical Society* (London), November, 1961, p. 438. The elongated balloons required are readily obtainable in model or toy shops.

formed by these four orbitals a regular one, the bond angles would be 109.5°. But the unshared pairs are closer to the oxygen nucleus than are those shared by the hydrogen atoms, and thus would be expected to exert greater repulsive force upon each other than upon the shared pairs. The result of this is that the angle between the lone-pair orbitals is slightly greater than 109.5°, and that between the H—O bond slightly less. This qualitative approach to the problem gives us a picture of molecular geometry that is close to reality, although the precise values for the bond angles can only be found by other ways.

In ammonia, in which nitrogen is bonded to three hydrogen atoms, the simplest description, in which the three atomic $p$ orbitals are utilized in bond formation, would lead to a molecule with three $s$-$p$ bonds at right angles to each other, and a $2s$ orbital possessing an unshared electron pair. This would present a picture of the ammonia molecule like that in Figure 3-10$a$.

The experimental facts are that the N—H bond angles in ammonia are 107°, and that the fourth pair of (unshared) electrons is not symmetrically distributed about the nucleus but has its major extension from the nucleus in a direction opposite to the three N—H bonds. The molecule thus has a distorted tetrahedral configuration, with N—H bonds to three corners of the tetrahedron and the lone pair extending toward the fourth corner (Figure 3-10$b$).

The larger atoms in the sixth period—sulfur, selenium, and tellurium—form hydrides in which the bond angles between hydrogen and the central atoms are nearly 90°. However, since these large atoms make use of orbitals above the $L$ shell, the electronic distribution in the hydrides is more complex, and simple qualitative arguments are not conclusive.

### 3.14 The tetrahedral carbon atom. The $sp^3$ bond and orbital hybridization

The formation of four bonds to carbon ($1s^2 2s^2 2p^2$) is one of the central facts of organic chemistry. Several alternative *a priori* pictures of bonding to carbon might be made: *(a)* the two $2p$ electrons could be removed to give an ionic compound

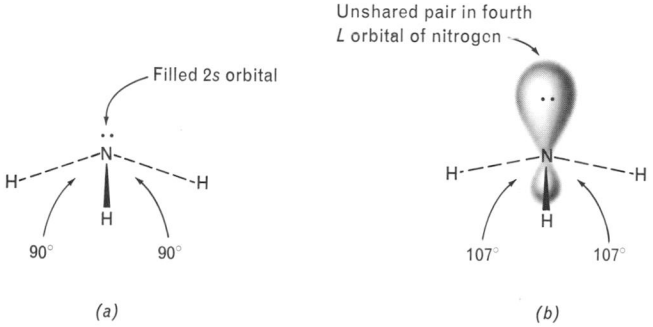

(a) (b)

**Figure 3–10.** *(a)* First approximation of $NH_3$ molecule with bonds disposed as in the atomic orbitals of nitrogen. *(b)* Actual structure of the $NH_3$ molecule.

containing $C^{++}$; *(b)* the two $2p$ electrons could engage in molecular orbital formation to give two covalent bonds; *(c)* the three $p$ orbitals might combine with, for example, two chlorine atoms and one chloride ion to give $:CCl_3^-$:

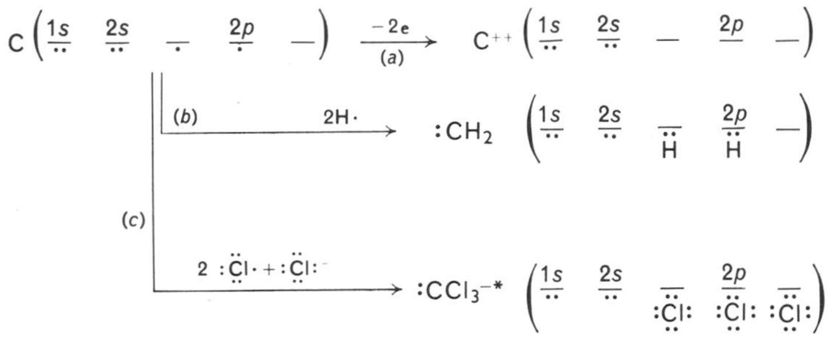

In fact, the products shown in *(b)* and *(c)* are known, but the species $:CH_2$ and $:CCl_3^-$ are unstable and readily enter into reactions, the products of which contain tetracovalent carbon. The conclusion is that carbon uses both the $2s$ electrons and the two $2p$ electrons to form four covalent bonds:

The disposition of the four C—H bonds in methane and in other compounds of the type $CX_4$ is *tetrahedral*. The bonds are all alike; no one can be distinguished from the other three. Were the four L-shell orbitals used in bond formation with retention of the atomic configuration, we should expect to find three C—H bonds mutually disposed at 90° and a fourth, involving the spherical $2s$ orbital, with no fixed direction. This is contrary to the facts, and in any case lacks *a priori* justification as a primary assumption.

That the configuration of the carbon atom carrying four substituents, *Cabde*, is tetrahedral was postulated on quite empirical grounds by van't Hoff and Le Bel in 1874, and has been the central postulate upon which the stereochemical aspects of organic chemistry have been based. The calculations of quantum mechanics lead to the same conclusion: that the most stable configuration for the four orbitals in a tetracovalently bound carbon atom is that in which the bonds are tetrahedrally arranged (Figure 3.11).

Indeed, we could have arrived at the same conclusion by the simple reasoning that is based upon considerations of electrostatic repulsions (Section 3-13). How can we distribute four bonds (pairs of electrons) around a central atom such that they would be positioned as far as possible from one another in order that repulsive forces between the electrons would be minimal? The answer is the simple

---

\* In these formulas, the $1s$ electrons on carbon are to be regarded as part of the symbol "C", which represents the "kernel".

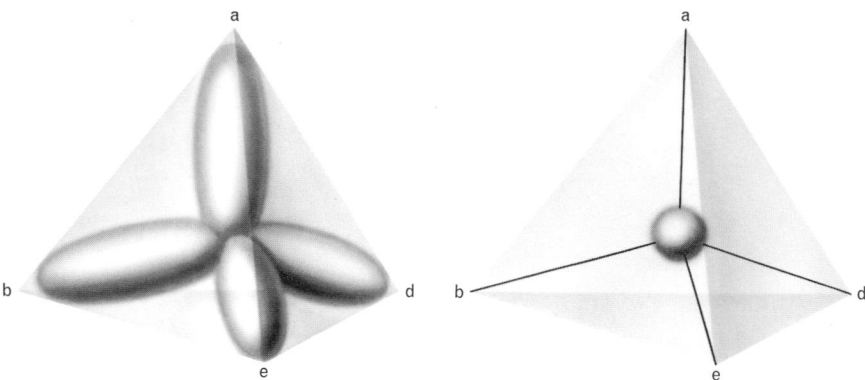

**Figure 3–11.** The tetrahedral disposition of the four ($sp^3$) bonds to the saturated carbon atom.

geometrical one: four bonds are uniformly disposed at the greatest mutual separation when they are arranged tetrahedrally.

Four equivalent, symmetrically disposed bonds (as in a molecule like methane, $CH_4$, or carbon tetrachloride, $CCl_4$) are, of course, indistinguishable from one another. To identify one as representing an atomic $2s$ orbital and the other three as atomic $2p$ orbitals would be meaningless. They are all alike.

The formation of four *equivalent* orbitals from one $2s$ and three $2p$ orbitals is called *hybridization* of the four orbitals, since the new molecular orbitals have neither the same angular relationship nor the same kind of symmetry with respect to the $x$, $y$, and $z$ axes as do the four original atomic orbitals. The new orbitals are called $sp^3$ orbitals, and the four bonds to a tetracovalent carbon atom like the one in methane are called $sp^3$ *bonds* ("$sp$-three" *bonds*).

### 3.15   Orbital hybridization in other first-row elements

Boron (atomic number 5; $1s^2\ 2s^2\ 2p$) forms compounds that are typically $BX_3$, e.g., $BF_3$, $BCl_3$, $B(CH_3)_3$. Thus, it utilizes the three electrons of the $L$ shell to form three covalent bonds by pairing these in two-electron molecular orbitals:,

$$\cdot \ddot{B}\cdot + 3:\ddot{F}\cdot \longrightarrow \quad :\ddot{F}:\overset{\displaystyle :\ddot{F}:}{\underset{\phantom{.}}{B}}:\ddot{F}:$$

Boron trifluoride is a gas at ordinary temperatures, and thus has none of the characteristics of ionic compounds. The three B—F bonds are covalent bonds and $BF_3$ is a covalent molecule. Moreover, it has been found by physical measurements that *the three B—F bonds are identical, and lie in a plane with equal* (120°) *angles between the bonds.* Now since $s$ and $p$ orbitals are quite different both in energy and geometry, it is apparent that the three B—F bonds, which are all alike, cannot have been formed with the use of atomic $s$ and $p$ orbitals of boron with the retention of the atomic configuration; and since the three $BF_3$ bonds are in a single plane and 120° apart, it is equally clear that they have not been formed with the use of the three $p$ orbitals of boron. Thus the three B—F bonds represent a new kind of molecular orbital. This is called a hybrid orbital, and, in the terminology already used for carbon, the bonds are called $sp^2$ ("$sp$-two") bonds.

Again, the electrostatic analogy gives a picture in agreement with the experimental facts: three pairs of electrons will occupy positions of minimum energy—i.e., will show minimum repulsive interactions—when they are arranged in a plane about the nucleus at mutual angles of 120° (a "trigonal" arrangement).

From the above discussion, and that in Section 3.14., we can draw some general conclusions about the geometry of a number of simple molecules and ions. For example, the carbonate ion, in which three oxygen atoms are attached to a central carbon atom, would be expected to be planar and trigonal. The methyl cation $CH_3^+$, and the sulfur trioxide molecule ($SO_3$ in the vapor state),* would also be expected to be planar and trigonal.

The tetrahedral configuration is possessed by such tetracovalent compounds as the ammonium ion, $NH_4^+$, the perchlorate ion, $ClO_4^-$, the fluoborate ion, $BF_4^-$, and the silicon tetrachloride molecule, $SiCl_4$. Despite the wide range in the character of the central atoms and the peripheral atoms in these substances, all of them have a property in common: each of them consists of a central atom surrounded by four bonds to identical substituents, and each bond consists of a shared pair of electrons; each possesses bonds formed with the use of $sp^3$ hybrid orbitals of the central atom.

Since very few organic compounds consist only of a carbon atom surrounded by four *identical* substituents, the simple regular tetrahedron, with all four bond angles equal to exactly 109.5° is seldom found. Nevertheless, a carbon atom carrying four substituents is tetrahedral in form, but when the four substituents are not the same there will be small deviations from the "regular" 109.5° bond angle.

### 3.16   Other kinds of hybridization of orbitals

Another kind of orbital hybridization involves the use of an $s$ and a $p$ orbital to form two colinear bonds. Examples are found in such compounds as $HgCl_2$ (Cl—Hg—Cl) and $Ag(NH_3)_2^+$. The bonds to mercury and silver in these molecules are called $sp$ bonds. When higher orbitals are called into play, more complex arrangements of atoms result. There is little application in organic chemistry of such bonding, in which orbitals higher than $3s$ and $3p$ are used, but one can be

---

* It is occasionally necessary to specify that a molecule whose structure is being described is in the vapor state since association and the formation of complex molecular aggregates in the solid or liquid state may cause alteration of the atomic arrangements.

mentioned here as an example. When one $s$, three $p$, and two $d$ orbitals of one of the heavier elements are used, the hybridization of these six orbitals leads to what are called $d^2sp^3$ bonds, which are disposed octahedrally (Figure 3-9).

### 3.17    The stereochemistry of the tetracovalent carbon atom

*Stereochemistry* (*stereo*, from the Greek word for "solid", and having the usual connotation of three-dimensional, or spatial) is concerned with the arrangement of atoms within a molecule. Most of the chemical properties of organic compounds bear a definite relationship to the shape of the molecule and the relative positions of its constituent atoms. Nearly all considerations of the nature of organic substances must include explicit reference to the stereochemical aspects of their structures, and the establishment of the stereochemistry (that is, the configuration) of an organic compound is an integral part of the description of its structure. The stereochemical features of the molecules will be a part of the description of most of the organic compounds to be discussed in this text. A later chapter will be devoted to an inquiry into many particular aspects of stereochemistry, but the fundamental aspects of the stereochemical properties of organic compounds will be dealt with throughout the chapters to follow.

There are a number of simple consequences of the tetrahedral arrangement of the four bonds to the carbon atom. For one thing, certain stereochemical effects are at once apparent:

1. The four hydrogen atoms in methane are equivalent, and the replacement of any one of them by another atom or group gives but one compound. *There is only one compound $CH_3Cl$, one $CH_3Br$, one $CH_3OH$, and so forth*:

four different drawings of one compound

If the four bonds were, for example, formed with the use of three $p$ orbitals, disposed at right angles to one another, and an $s$ orbital, with no fixed direction, there could be at least two compounds $CH_3Cl$, in one of which the Cl—C bond used a $2p$ orbital, in the other, used the $2s$ orbital.

2. Similarly, there is only one $CH_2X_2$, one $CHX_3$. Suppose the four bonds to carbon were in one plane. If so there could be two compounds $CH_2Cl_2$:

*No such isomerism is known*; the bonds are not in a plane.

3. A compound *Cabde* can exist in *two* forms. With a tetrahedral arrangement of the bonds to carbon, these two isomers of such a compound can be represented by*

A moment's study of these structures (best of all, of the actual models) will disclose that they are indeed different; they are not simply two ways of representing the same compound. The spatial relationships in this kind of carbon compound, represented in the most general terms as *Cabde*, and in a particular case by $(CH_3)CHBrCOOH$, is customarily represented as

It is asymmetric; it has no element of symmetry.†

---

**Exercise 4.** Examine the three-dimensional models (or drawings) of the following compounds, and find the plane about which they are symmetrical:

(a) $CH_3Cl$          (b) $CH_2Cl_2$          (c) $CH_2ClBr$     (d) $CH_3CHClCH_3$ (with Cl substituent)
(e) $CH_3—O—CH_3$     (f) $CH_3CH_2—O—CH_3$   (g) $H_2O$
(h) cyclopropane      (i) ammonia            (j) $NH_2OH$

Remember that a plane of symmetry of a molecule may cut *through* an atom or group, if that atom or group is itself symmetrical.

---

Moreover, a molecule $Ca_2bd$ is symmetrical—if b and d are themselves symmetrical—with a plane of symmetry represented as in Figure 3-12. Note that in part 3 of Figure 3-12 if the a-a substituents are a-c, the top one is not a reflection of the bottom, and the plane that is shown is no longer a plane of symmetry.

---

* This kind of representation of three-dimensional structures will be used throughout this book when it is desired to point out stereochemical details. The reader should study these drawings and compare them with the models they represent. A useful device for constructing such models is to insert toothpicks in tetrahedral disposition into a sphere of some material, such as cork (or a small potato), and to place on the outer ends of the toothpicks small colored spheres (e.g., candy jellybeans).

† The dictionary definition of symmetry is as follows: "An exact correspondence between the opposite halves of a figure, form, line, pattern, etc., on either side of an axis or center; the condition whereby half of something is the mirror image of the other half." This is a simple and commonplace concept in everyday life, and its application to the stereochemistry of organic compounds is quite straightforward.

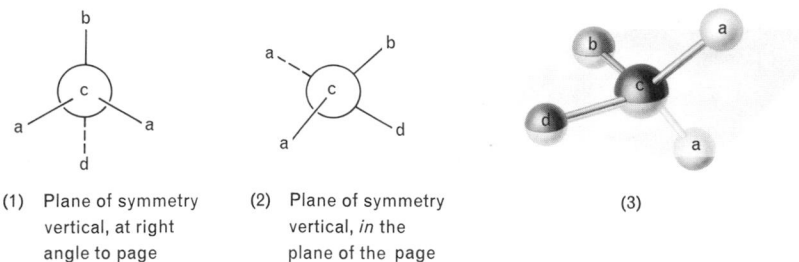

(1)  Plane of symmetry
     vertical, at right
     angle to page

(2)  Plane of symmetry
     vertical, *in* the
     plane of the page

(3)

**Figure 3–12.**   The molecule Ca$_2$bd, showing the plane of symmetry.

4. When carbon compounds are represented in the ordinary way, in writing or in type, as, for example, $CH_3CH_2CH_2CH_3$, these are but conventional expressions of nonlinear molecules. Butane, for example, has the configuration shown in Figure 3-13.

Carbon chains may exist in other conformations than the regular zig-zag shown in Figure 3-13; moreover, thermal agitation may bring about molecular motions that cause the chain to alter its shape by twisting or, in the case of long chains, coiling. One of the various possible conformations of an organic molecule is usually more stable than others and will represent the most probable shape of the molecule, and the population of less stable conformations will be dependent upon the temperature. What is to be emphasized here is that organic compounds, even though represented in two-dimensional ways on paper, are structures with extension in three dimensions in space, and many of them do not possess rigid, fixed conformations.

The tetrahedral nature of the four bonds to carbon can lead to the formation of stable cyclic structures, in which the ends of a chain of carbon atoms are joined to form a ring. Several cyclic compounds were illustrated in Chapter 2. The three-

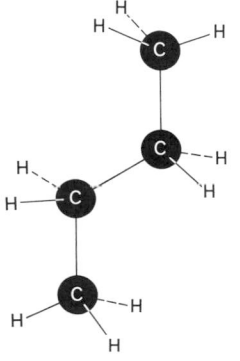

**Figure 3–13.**   Three-dimensional structure of butane, $CH_3CH_2CH_2CH_3$.

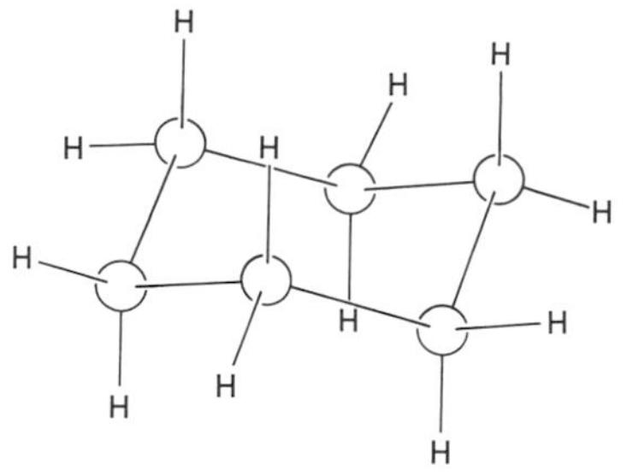

Figure showing three-dimensional representation of cyclohexane with hydrogen atoms labeled.

**Figure 3–14.** Three-dimensional representation of cyclohexane.

dimensional structure can be shown for a representative cpclic compound, cyclohexane.

It is to be noted that in butane (Figure 3-13) and cyclohexane (Figure 3-14) the arrangement of the four bonds around each carbon atom is tetrahedral.

### 3.18 Elemental carbon. Diamond and graphite

Carbon occurs in nature in pure form in diamond and graphite, and in impure form—contaminated with inorganic materials and miscellaneous other organic compounds—in coal and the products of combustion of organic substances such as wood.

Diamond and graphite have definite structures: diamond consists of a repeating pattern of tetrahedrally bound carbon atoms, and the graphite "molecule" is a planar array of carbon atoms joined by trigonal bonds in a repeating pattern of hexagons (Figures 3-15 and 3-16). The symmetrical and compact matrix of fully saturated carbon atoms in diamond forms a rigid network that accounts for its hardness and chemical inertness. Since all of the orbitals of every carbon atom are fully occupied with electron pairs that are confined to the axis joining the carbon

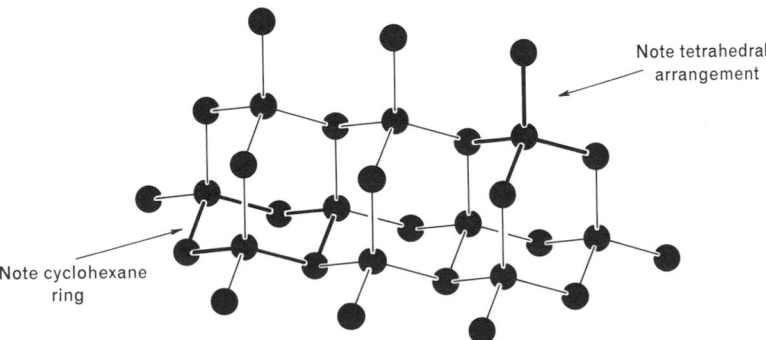

Note tetrahedral arrangement

Note cyclohexane ring

**Figure 3–15.** A portion of the diamond structure. The crystal lattice is composed of a repeating three-dimensional network of tetrahedrally bound carbon atoms.

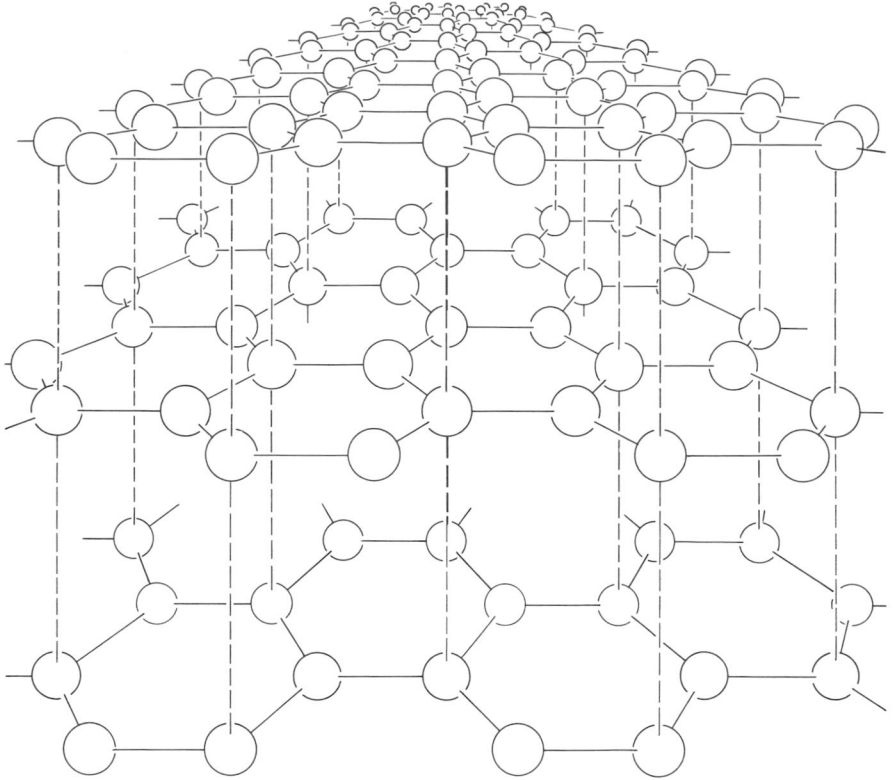

**Figure 3–16.**   A portion of the graphite structure. The crystal lattice is composed of parallel sheets of carbon atoms present in a repeating pattern of planar six-membered rings.

nuclei, no capacity exists for reaction with an external reagent. The only way in which diamond can undergo chemical reaction (for example, burning in oxygen) is by an initial rupture of a carbon-carbon bond, as by high temperature, followed by reaction at the carbon atoms so exposed to attack. Thus, diamond can be burned, but cannot be easily ignited.

Graphite consists of a planar "sheet" of hexagonally bonded carbon atoms extending indefinitely over a two-dimensional plane (Figure 3-16). The graphite crystal consists of a stack of these planar arrays. The C—C bond distance between the carbon atoms *in* the plane is 1.42 Å, and the distance *between* layers is 3.4 Å. Thus, there are no covalent or other strong bonds between the layers, and these can move with respect to each other. The usefulness of graphite as a lubricant depends upon this ability of the layers to slide over each other.

Since only three of the four *L*-shell electrons are used in forming the C—C single bonds between the carbon atoms in graphite, there is one electron for each carbon atom still to be accounted for. These extra electrons confer some double-bond character upon each of the C—C single bonds (making them shorter than the 1.54 Å C—C bond length found in diamond; see Section 3.19 for further discussion of bond lengths); but perhaps the most useful description of the graphite molecule is that of a nonlocalized electron cloud above and below the plane. The electrical conductivity of graphite can be accounted for with this description, because the flow of electrons through this non-localized electron atmosphere can lead to transfer of charge from one point in the polyhexagonal plane to another.

The foregoing discussion of diamond and graphite is important principally because these two forms of carbon provide excellent examples of some of the principles of carbon-carbon bond formation, and form the prototypes for carbon-carbon bonds present in the kinds of compounds to be discussed as we proceed.

## 3.19 Carbon-carbon multiple bonds

*The carbon-carbon double bond.* In ethylene, $C_2H_4$, the two carbon atoms are joined together, and two hydrogen atoms are attached to each carbon atom:

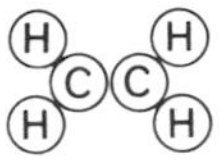

These facts are known from chemical evidence, because we know that ethylene can be converted to ethane ($CH_3CH_3$), and it can be oxidized to two molecules of $H_2C{=}O$. To write the structural formula of ethylene using conventional valence-bond notation, and assigning a valence of four to carbon, we write $CH_2{=}CH_2$ or, in a graphic formula,

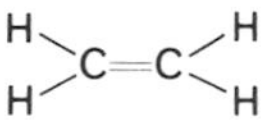

$$\text{H}{>}\text{C}{=}\text{C}{<}\text{H} \quad , \text{ or, in a perspective formula,} \quad$$

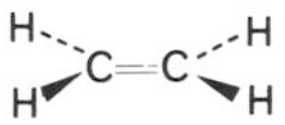

What kinds of bonds are involved in the structure of ethylene? Three physical facts must first be mentioned: (1) the carbon-carbon bond distance in ethylene is 1.34 Å, which is considerably less than that in ethane (1.54 Å); (2) the angle between the H atoms (that is, the H—C—H angle) is nearer to 120° than to the tetrahedral angle of 109°28'; (3) the carbon and hydrogen atoms are in one plane.

From these physical and chemical data, we can construct a reasonable picture of the ethylene molecule (Figure 3-17). We see, first of all, that the C—H bonds of each carbon atom and the carbon-carbon bond can most reasonably be arranged such that in each $\text{H}{>}\text{C}{-}$ the three bonds are symmetrical and planar. In this way the electron pairs are disposed for maximum stability by separation from each other, just as they are known to be in $BCl_3$, which is a symmetrical, planar, trigonal molecule.

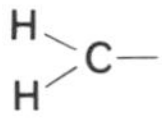

The planar, trigonal disposition of the three bonds to the carbon atoms in ethylene indicates that the bonds are formed with the use of $sp^2$-hybrid orbitals of carbon. We shall see that there are other consequences of this structure; for one, the

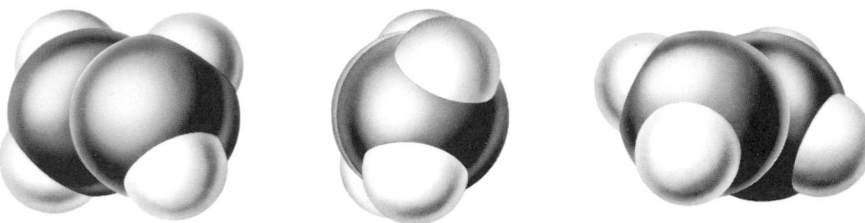

**Figure 3–17.** Molecular model of ethylene; several views.

C—H bond length in ethylene is 1.085 Å, compared with 1.10 Å for the C—H bond length in ethane. This is in accord with the view that $sp^2$ hybrid orbitals are used in bond formation, for the greater degree of $s$ character of such an orbital means that the orbital lies closer to the nucleus than does an $sp^3$ orbital, with a consequent shortening of the bond to the hydrogen atom.

But what can be done with the extra electron pair in the C=C bond? We can expect to find that the total of four electrons of this double bond will be disposed for minimum mutual interaction; this can best be done if the two electron pairs are located between the carbon atoms, but in two regions of space that are separated from one another, as shown in Figure 3-18.

The summary of these considerations is this: in ethylene each carbon is approximately trigonal, and since there must be minimal mutual interaction between the C—H bonding electrons and the C—C bonding electrons, the molecule as a whole is planar, with regions of electron density extending above and below the plane. It can be seen from Figure 3-18 that all of the electron pairs are arranged in such a way as to produce minimal mutual interaction. The two carbon nuclei are under the mutual attraction of two pairs of electrons. The carbon atoms are drawn closer together than are those in the C—C single bond. The electrons between the carbon atoms are distributed in such a way as to extend above and below the plane of the molecule and, furthermore, to extend into regions outside the carbon-carbon bond axis. Thus, these electrons are readily accessible to attack by electron-seeking reagents, and the carbon-carbon double bond is a center at which chemical reactions can occur readily. The nature of the reactions of the carbon-carbon double bond and further comments about its electronic structure will be found in Chapter 9.

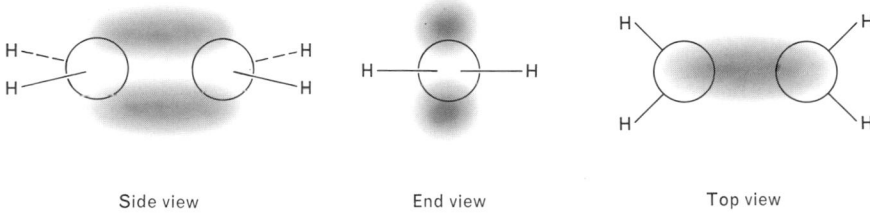

      Side view             End view             Top view

**Figure 3–18.** Several views showing electron disposition in the ethylene double bond.

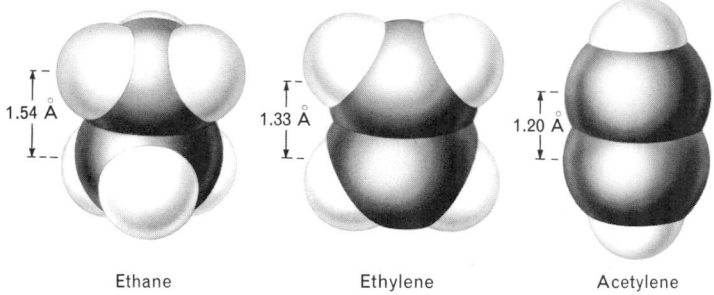

**Figure 3–19.** Molecular models of ethane, ethylene, and acetylene.

*The carbon-carbon triple bond.* The carbon-carbon triple bond, found in acetylene, $C_2H_2$, can be described in terms comparable to those used in the foregoing discussion of the double bond. Acetylene is a linear molecule, with a C—C bond distance of 1.20 Å and a C—H bond distance of 1.06 Å. The electrons that form the triple bond are disposed between the carbon atoms in such a way as to form a cylindrically symmetrical shell of electron density about the carbon-carbon bond axis. The increased *s* character of the *sp* carbon orbitals is reflected not only in the shorter bond lengths but in the increased acidity of acetylene. This will be discussed in Chapter 9; acetylene is a stronger acid than ethylene, which in turn is more acidic than ethane (that is, $HC{\equiv}C:^-$ is a weaker base than $CH_2{=}CH:^-$, which is a weaker base than $CH_3CH_2:^-$.

In Figure 3-19 are shown the structures of ethane, ethylene, and acetylene.

The conventional representation of double and triple bonds is shown by the following examples. The multiple bonds are shown as short lines and in this form no indication is given of the shapes of the molecules. A discussion of the stereochemistry of multiple bonds will be found in Chapters 9 and 16.

| | | |
|---|---|---|
| $CH_2{=}CH_2$ | $CH_3CH{=}CH_2$ | $CH_3CH{=}CH{-}CH{=}CH_2$ |
| ethylene | propylene | 1,3-pentadiene |
| $HC{\equiv}CH$ | $CH_3C{\equiv}CCH_3$ | $CH_3CH_2C{\equiv}CCH_2CH_3$ |
| acetylene | dimethylacetylene or 2-butyne | 3-hexyne |

## 3.20 The semipolar bond

An aspect of covalent bond formation that deserves special mention is encountered in those cases in which the bond is formed by the donation of *both* electrons of the binding pair by one atom to fill a vacant orbital of another. For example, boron, in its trivalent compounds such as boron trifluoride and trimethyl boron, has but six electrons in its valence orbitals and thus possesses a vacant orbital which is capable of forming a fourth covalence. Ammonia, on the other hand, is a compound in which the nitrogen atom possesses an *unshared pair* of electrons and which is capable of forming a fourth bond to nitrogen. It is found that ammonia reacts with boron trifluoride in the following way:

$$
\begin{array}{ccc}
\underset{\underset{H}{|}}{\overset{\overset{H}{|}}{H-N:}}+\underset{\underset{F}{|}}{\overset{\overset{F}{|}}{B-F}} & \longrightarrow & \underset{H}{\overset{H}{\diagdown}}N:B\underset{F}{\overset{F}{\diagup}} \quad \text{or} \quad \underset{H}{\overset{H}{\diagdown}}N{\rightarrow}B\underset{F}{\overset{F}{\diagup}}
\end{array}
$$

The product, $H_3N:BF_3$, is a neutral molecule, but an inspection of the electronic complements of the nitrogen and boron atoms discloses that the boron atom shares four electron pairs and thus *possesses* half of these, and thus has a formal negative charge of one; and the nitrogen, which possesses half of the eight electrons it shares, has one electron less than the normal complement of five (in the neutral nitrogen atom), and thus has a formal positive charge of one. The ammonia-boron trifluoride complex should thus be written $H_3\overset{+}{N}:B\overset{-}{F}_3$.

Similarly, the unshared electron pair on the nitrogen atom of a tertiary amine such as $(CH_3)_3N:$ can be donated to complete the octet of an oxygen atom, with the formation of a compound known as an amine oxide:

$$
\underset{\underset{CH_3}{|}}{\overset{\overset{CH_3}{|}}{CH_3-N:}}+\ddot{\overset{..}{O}}: \quad \longrightarrow \quad \underset{\underset{CH_3}{|}}{\overset{\overset{CH_3}{|}}{CH_3-\overset{+}{N}:}}\ddot{\overset{..}{O}}:^-
$$

|  trimethylamine  | trimethylamine oxide |
|---|---|

Dipolar covalent bonds such as the N—B and N—O bonds of the kinds described above are found in many organic compounds. The common grouping $-\overset{\overset{+}{}}{N}\overset{\diagup\overset{O}{}}{\diagdown}:\overset{..}{\underset{..}{O}}:$ , called the *nitro group*, possesses a nitrogen-oxygen bond of this kind. It is customary for organic chemists to refer to this kind of covalent bond as the *semipolar bond*, and to designate it by the convenient symbols shown in the following examples. It is a matter of individual preference whether the formal charges are written or the arrow-headed bond is used; both mean the same thing.

$$
\underset{\underset{CH_3}{|}}{\overset{\overset{CH_3}{|}}{CH_3-\overset{+}{N}-O^-}} \quad \text{or} \quad \underset{\underset{CH_3}{|}}{\overset{\overset{CH_3}{|}}{CH_3-N\rightarrow O}} \qquad CH_3-\overset{+}{N}\diagdown\overset{\diagup O}{O^-} \quad \text{or} \quad CH_3-N\diagdown\overset{\diagup O}{O}
$$

|  trimethylamine oxide  | nitromethane |
|---|---|

The formation of the complex fluoborate ion, $BF_4^-$, by the reaction of boron trifluoride with fluoride ion would appear to be an example of the formation of a

---

**Exercise 5.** Inspect the electronic distribution in the isomeric compounds N,N,O-trimethylhydroxylamine, $(CH_3)_2NOCH_3$, and trimethylamine oxide, $(CH_3)_3\overset{+}{N}-\overset{-}{O}$, and explain why no formal charges are written in the former.

---

semipolar bond. The process by which the B—F bond is formed is indeed the same as that in which the N—B bond is formed in the example described above, but there is no semipolar bond in the $BF_4^-$ ion, since, although the ion carries a negative charge of one, the four fluorine atoms are completely equivalent and no one of them carries a formal charge. In the combination of ammonia with a proton, $H^+$, also, the bond is formed by the donation of the unshared pair of nitrogen to be shared by the proton. But $NH_4^+$ does not have a semipolar bond: all four hydrogen atoms are equivalent and the positive charge is symmetrically disposed about the ion as a whole.

The presence of a semipolar bond in a compound confers upon the compound physical properties that are often quite different from those of its isomers which do not possess semipolar bonds. The electrical forces that exist as a result of the presence of the formal charges on the atoms joined by one bond are molecular orienting influences and often result in abnormally high melting points and boiling points (see Section 3.11).

---

**Exercise 6.** Look up the following compounds in a good handbook, and compare their physical properties: methyl nitrite, $CH_3ONO$; nitromethane, $CH_3NO_2$. Write the electronic structures of these compounds.

---

## 3.21 The representation of organic structures in written formulas

The representation of the structures of organic compounds in written formulas may take any of a number of forms. This subject has been discussed briefly in Section 1.5, but further comment will be useful here. The selection of one or another representation will depend upon the purpose for which it is intended, and one or another feature of the complete structure may often be omitted in a formula. For example, ethanol may be written as $C_2H_5OH$ or $CH_3CH_2OH$ if all that is desired is to show its constitution and that it contains a hydroxyl group. To write it as $C_2H_6O$ would not be useful, because methyl ether, $CH_3OCH_3$, also has this composition, and this molecular formula gives a minimum of information. If the possession of unshared electrons on the oxygen atom is relevant to the discussion, ethanol may be written $CH_3CH_2\overset{..}{\underset{..}{O}}H$. If the arrangement of the atoms in space needs to be represented, one can use such conventional figures as

$$CH_3CH_2-\overset{H}{\overset{|}{O}}: \quad \text{or} \quad H-\overset{H}{\underset{H}{\overset{|}{C}}}-\overset{OH}{\underset{H}{\overset{|}{C}}}-H$$

in which the heavy bonds (►) are projecting toward the reader, the dashed bonds (---) are behind the plane of the page, projecting away from the reader, and the light single bonds (—) are in the plane of the page. For certain purposes it is useful to draw a modified picture of a model of the molecule, in which the perspective is somewhat more apparent:

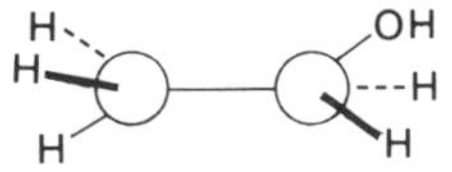

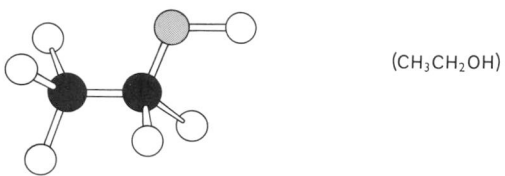

$(CH_3CH_2OH)$

Drawings of complete molecular models may also be used. These are usually of two kinds: (1) ball-and-stick models:

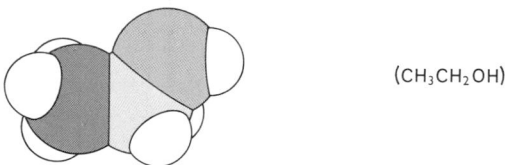

$(CH_3CH_2OH)$

(2) molecular models which are made to exact scale and show the relative sizes and arrangements of all of the atoms:

$(CH_3CH_2OH)$

The purpose of writing a chemical formula is to provide a visual aid to a discussion or a description of the molecule's properties, and to point out molecular features that are pertinent to what is being said. Many simplifying conventions are employed, partly for economy and partly to focus attention upon the particular structural feature that is being examined and to clarify it. For example, there are three compounds of the composition $C_2H_2Br_2$. These can first of all be written in a form that distinguishes one of them clearly from the other two:

$Br_2C{=}CH_2$    and    $BrCH{=}CHBr$

The second of these formulas represents two compounds, distinguished by the stereochemical disposition of the hydrogen and bromine atoms. A ball-and-stick drawing of these is the following:

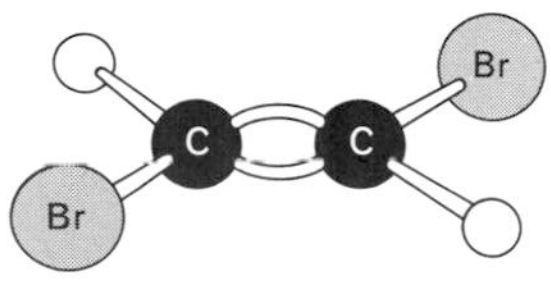

*trans*-1,2-dibromethylene 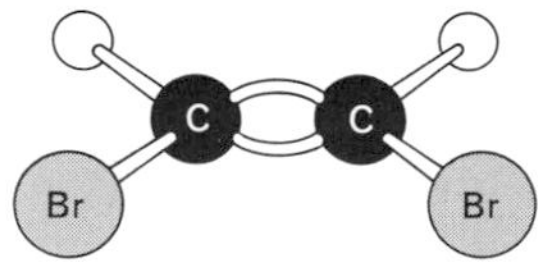                     *cis*-1,2-dibromoethylene

Because it would be awkward to have to make such drawings whenever one wrote about these compounds, various simple conventional substitutes for the drawings are commonly used, such as the following representations for the *cis* isomer:

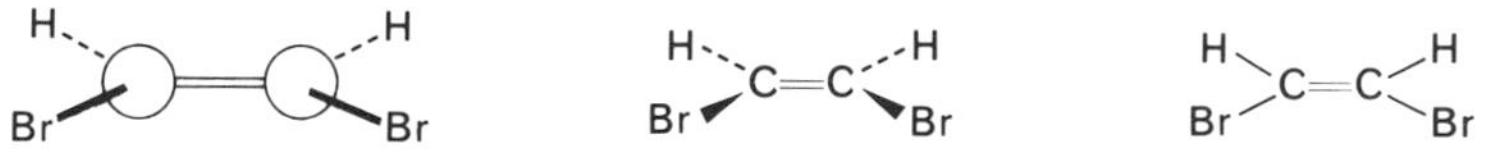

*cis*-1,2-dibromoethylene

In a subsequent chapter we shall discuss the chemical behavior of compounds of this class. One of their characteristic reactions is addition to the double bond, as in the following:

$$H_2 + CH_3CH\!=\!CHCH_3 \xrightarrow{\text{catalyst}} CH_3CH_2CH_2CH_3$$
$$\text{2-butene}$$

The *same compound*, butane, is obtained from both the *cis*- and the *trans*-2-butenes, and thus the equation need not be specific about which of these two isomers is shown. On the other hand, the compounds obtained by the addition of bromine to the two 2-butenes are different for the *cis* and *trans* isomers. The formulation of this reaction in an equation should show the stereochemical details:

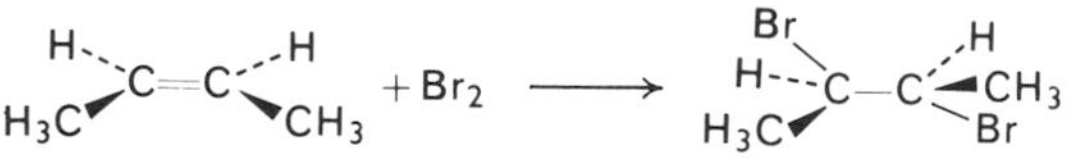

*cis*-2-butene

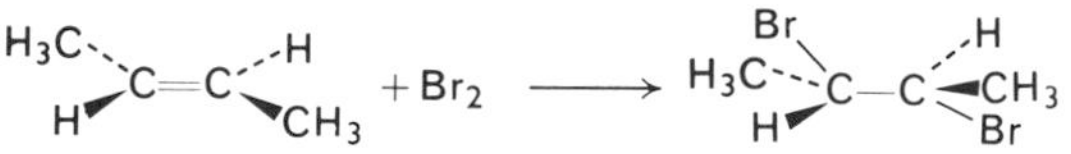

*trans*-2-butene

Certain other conventions that simplify representation of organic structures will be introduced from time to time as the reader gains facility and familiarity with the subject. One that will be useful from the start is the method usually adopted in writing cyclic structures. Cyclopentane, cyclopentene, cyclopentanol, and cyclohexanone can be written in a simple fashion that is adequate for most purposes:

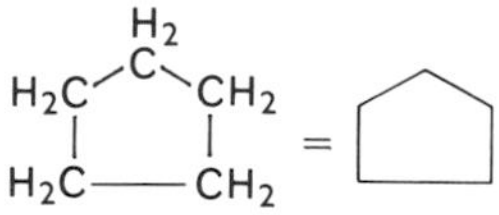

cyclopentane

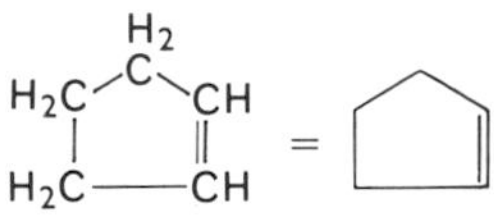

cyclopentene

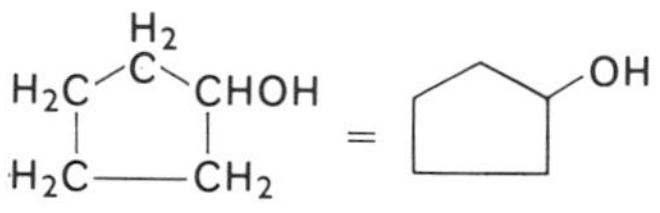

cyclopentanol

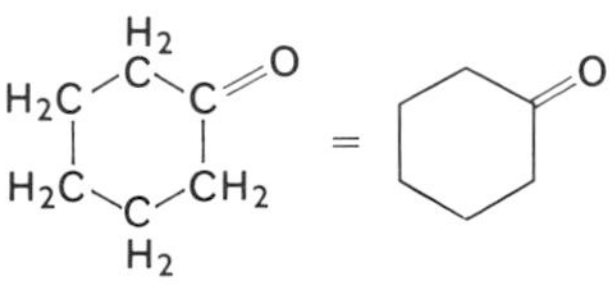

cyclohexanone

Similarly:

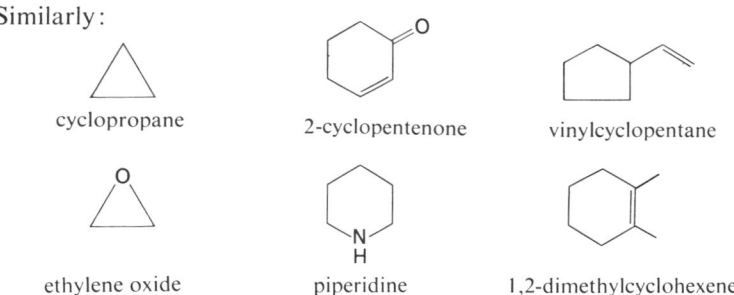

cyclopropane          2-cyclopentenone          vinylcyclopentane

ethylene oxide          piperidine          1,2-dimethylcyclohexene

## 3.22   Formal charges

Compounds in which carbon contains four covalent bonds, nitrogen three, oxygen two, and hydrogen one are the most frequently encountered in organic chemistry. If we examine these we see that electrons are utilized in covalent bond formation in such a way that a maximum of eight electrons (four pairs) is found in the valence shell of each atom (except hydrogen, which requires only two electrons for the completion of its $K$ shell).

For carbon this means that four single bonds, one double and two single bonds, one triple and one single bond, or two double bonds represent possible covalent states:

$$
\begin{array}{cccc}
\text{H} & \text{H} \\
| & | \\
\text{H—C—H} & \text{H—C=O} & \text{H}_2\text{C=C}\text{H}_2 & \text{H—C}\equiv\text{N} \\
| \\
\text{H}
\end{array}
$$

methane          formaldehyde          ethylene          hydrogen cyanide

$$
\text{H—C}\equiv\text{C—H} \qquad \text{O=C=O} \qquad \text{H}_2\text{C=C=O}
$$

acetylene          carbon dioxide          ketene

Carbon cannot form stable compounds with more than four covalent bonds, because only four stable orbitals (the $2s$ and $2p$ orbitals) are available for bond formation.

However, unusual valence states of carbon are known to exist in which less than four covalencies are present. The methyl "radical," $\cdot\text{CH}_3$, can be produced experimentally. It is, however, very unstable and reactive and seeks to complete its complement of eight 2-level electrons by reaction with other molecules. The "carbonium" ion is also recognized as an intermediate in many chemical reactions. The following is an example:

$$
\begin{array}{c}
\text{CH}_3 \\
| \\
\text{CH}_3\text{—C}^+ \\
| \\
\text{CH}_3
\end{array}
$$

As we shall see in later discussions, the ability of the carbon atom to exist in the radical or ionic condition depends very greatly upon the kinds of groups or atoms to which it is attached. For the present we shall note, for the purposes of inventory only, that carbon "possesses" one of the electrons in the pair that forms each covalent bond. Thus, we see that in a radical, such as the methyl radical,

$$\text{H} \\ \text{H:\overset{..}{C}\cdot} \\ \text{H}$$

the carbon kernel "possesses" its normal complement of four electrons; thus the radical $\cdot CH_3$ is neutral in charge. In the $(CH_3)_3C^+$ ion, on the other hand, the central carbon atom "possesses" but three electrons (one of each of the three carbon-carbon bonds); since this is one less than carbon's normal four, the ion is positively charged. In a carbon *anion*, such as

$$\text{Cl} \\ | \\ \text{Cl—C:}^- \\ | \\ \text{Cl}$$

carbon (although it has eight electrons in its valence shell) "possesses" five (one in each of the three covalent C—Cl bonds and two in the unshared pair). The ion is therefore negatively charged.

---

**Exercise 7.** Assign the proper charge (if any) to the carbon atom in each of the following:

(a) $:C{\equiv}N:$        (c) $:\overset{..}{C}:\overset{..}{\underset{..}{C}l}:$        (e) $:C{\equiv}C:H$
                    $:\underset{..}{\overset{..}{C}l}:$

(b) $H—C{=}\overset{..}{\underset{..}{O}}:$      (d) $:\overset{..}{\underset{..}{O}}{=}C{=}\overset{..}{\underset{..}{O}}:$      (f) $H—\overset{..}{\underset{..}{O}}—\overset{\overset{\text{H}}{|}}{\underset{\underset{\text{H}}{|}}{C}}$

Note: Here, as in the discussion in the text, the symbol for an element represents the atomic kernel; electrons and bonds that are shown involve external (valence) orbitals.

---

A neutral nitrogen atom has five valence electrons in its outer shell. If three of its electrons engage in covalent-bond formation, an "unshared" pair remains. For example, in the uncharged ammonia molecule, nitrogen "possesses"

$$\text{H:\overset{..}{N}:H} \\ \text{H}$$

a total of two (in the unshared pair) plus three (in the three covalent bonds) valence electrons, which is its normal complement of five.

In the ammonium ion,

$$\overset{\displaystyle H}{\underset{\displaystyle H}{H:N:H}}{}^{+}$$

the pair of electrons that was unshared in ammonia is now shared with hydrogen. Nitrogen now "possesses" only *one* of the electrons of this pair and thus owns a total of four. This is one less than the five possessed by the neutral nitrogen atom; the ammonium ion thus has a positive charge.

In the amide ion,

$$\overset{\displaystyle H:\ddot{N}:^{-}}{\underset{\displaystyle H}{}}$$

nitrogen possesses two unshared pairs of electrons plus one in each of two covalent bonds, or a total of six. The amide ion therefore has a negative charge.

With respect to oxygen, the three well-known combinations with hydrogen are water, the hydronium ion, and the hydroxide ion. These are respectively neutral, positively charged, and negatively charged:

$$H:\ddot{O}:H \qquad \overset{\displaystyle H:\ddot{O}:H}{\underset{\displaystyle H}{}}{}^{+} \qquad H:\ddot{O}:^{-}$$

The explanations of these charge states can easily be worked out from the knowledge that the normal electronic complement of the valence shell (i.e., external to the $1s$ orbital) of the neutral oxygen atom is six.

In summary, we can see that when carbon has four, nitrogen three, and oxygen two covalent bonds, no excess or deficit of electrons is involved.

*Tetracovalent nitrogen*, however, is positively charged; *tricovalent oxygen* is also positively charged. We can see at once, then, without counting electrons, that the following are all positively charged ions:

$$(CH_3)_3O{:}^{+} \qquad (CH_3)_4N^{+} \qquad CH_2{=}\overset{+}{N}\overset{\textstyle CH_3}{\underset{\textstyle CH_3}{}} \qquad \overset{\textstyle H_3C}{\underset{\textstyle H_3C}{}}\overset{+}{N}\overset{\textstyle CH_3}{\underset{\textstyle CH_3}{}}$$

$$\overset{\textstyle H_3C}{\underset{\textstyle H_3C}{}}\overset{+}{N}{=}C\overset{\textstyle CH_3}{\underset{\textstyle CH_3}{}} \qquad \overset{\textstyle H_3C}{\underset{\textstyle H_3C}{}}\overset{+}{N}\overset{\textstyle CH_3}{\underset{\textstyle \ddot{O}CH_3}{}} \qquad H\ddot{O}{-}C\overset{\textstyle CH{-}CH}{\underset{\textstyle CH{=}CH}{}}O{:}^{+}$$

The following are negatively charged:

$$CH_3{-}\ddot{O}{:}^{-} \qquad \overset{\displaystyle CH_3{-}\ddot{N}:^{-}}{\underset{\displaystyle CH_3}{}} \qquad H{-}C\overset{\textstyle \ddot{O}\cdot}{\underset{\textstyle \ddot{O}{:}^{-}}{}} \qquad BF_4^{-} \qquad H{-}\ddot{O}{-}\ddot{O}{:}^{-}$$

These should be worked out in detail by counting the electrons on the relevant atoms and noting how the charge types are arrived at. It can be pointed out here that the methyl group, $-CH_3$, as a substituent in a formula, can be dealt with as a unit and can be taken as a monovalent group, equivalent to a hydrogen or a halogen atom.

---

**Exercise 8.**   Assign the proper charges to the following:

(a) :Ö—CH$_3$         (c) H$_3$C—NH$_3$      (e) BeCl$_4$

(b) :NCl$_3$          (d) BCl$_4$         (f) CH$_2$=NH$_2$

---

### 3.23   Maximum covalency

We have seen that the tetravalency of carbon is due to the presence of four valence orbitals, which can form four single bonds, as in methane; two single and one double bond, as in ethylene or formaldehyde ($H_2C$=O); two double bonds, as in carbon dioxide or allene ($CH_2$=C=$CH_2$); or one single bond and one triple bond, as in acetylene. No more than four bonds can be attached to carbon because there are no more orbitals in the *L* shell, and the *M* orbitals are at an energy level so high they cannot be used by carbon. Experience has shown that *none of the atoms of the first row forms more than four covalent bonds.* All of them lack stable orbitals above the *L* shell, and when all four of the *L* orbitals are utilized, stable bond formation is at an end. This accounts for the nonexistence of compounds such as $NH_5$, in which the five electrons of nitrogen form electron-pair bonds with five hydrogen atoms.

An equivalent statement regarding the covalency maximum of the first period elements is that these atoms cannot have more than eight electrons in their valence shell.

The utilization of all four of the nitrogen orbitals in bond formation is observed in the ammonium ion, $NH_4{}^+$. Why, then, cannot oxygen form $H_4O^{++}$ and fluorine $H_2F^+$ and $H_3F^{++}$? The answer to this question is that although in $H_3O$:$^+$ (which is known) there remains a fourth unshared pair of electrons on oxygen,

$$
\begin{array}{c}
\text{H} \\
| \\
\text{H—O:}^+ \\
| \\
\text{H}
\end{array}
$$

this pair of electrons is so firmly bound to the oxygen by reason of the positive charge on the nucleus that the orbital lacks sufficient extension from the nucleus to form a stable bond. In HF, the high electronegativity of fluorine exerts a comparable restraint on its unshared electrons, and thus effective sharing of these electrons with other atoms does not occur.

In the ion :$CH_3{}^-$, isoelectronic with $H_3O^+$, the unshared pair would be so readily accessible for bond formation because of the smaller nuclear charge on carbon that the ion would be exceedingly avid for another atom's vacant orbital. Thus, carbon nearly always displays its maximum covalency.

Elements of the second period (sodium to argon) and of higher periods utilize orbitals in the $M$ shell, in which nine orbitals (in all) are available. Thus, covalencies of greater than four are not prohibited. Some examples are $PCl_5$, $SF_6$, $IF_5$, and $Fe(CN)_6^{3-}$. We shall have little occasion to deal with elements of higher periods and need not discuss them at length at this time.

## Exercises

(Exercises 1–8 will be found within the text.)

**9** What is the atomic number of *(a)* carbon; *(b)* nitrogen; *(c)* sodium; *(d)* aluminum; and *(e)* chlorine? Write electronic structures of these atoms, as in the example of fluorine, $1s^2 2s^2 2p^5$.

**10** How might dimethylberyllium, $(CH_3)_2Be$, react with ammonia?

**11** The ion $C^{4+}$, in which the helium structure is found, would be formed by the loss of four electrons from the neutral carbon atom. This ion is unknown. Why?

**12** Suggest a plausible reason for the fact that while the hydride ion, $H{:}^-$, is well known, the sodium ion, $Na{:}^-$, is not.

**13** Write electronic structures (using the $1s2s2p$ notation) for *(a)* beryllium; *(b)* the beryllium ion, $Be^{++}$; *(c)* the ion, $F^-$; *(d)* the sodium ion, $Na^+$; *(e)* the aluminum ion, $Al^{3+}$; *(f)* argon.

**14** The electronic representation of carbon tetrachloride is $:\!\ddot{C}l\!:\!\overset{\displaystyle :\ddot{C}l:}{\underset{\displaystyle :\ddot{C}l:}{C}}\!:\!\ddot{C}l\!:$. Using this notation, write electronic structures for *(a)* methanol; *(b)* methyl chloride (chloromethane); *(c)* ammonia; *(d)* magnesium chloride; *(e)* nitrogen trichloride; *(f)* hydrogen peroxide; *(g)* ozone; *(h)* ammonium chloride; *(i)* diethyl ether; *(j)* boron trifluoride.

**15** Write the equation for the reaction of boron trichloride with diethyl ether.

**16** Trimethylamine, $(CH_3)_3N$, can be oxidized to trimethylamine oxide, $(CH_3)_3\overset{+}{N}{-}\overset{-}{O}$. Why does not trimethylboron, $(CH_3)_3B$, form a corresponding oxide?

**17** The H—S—H bond angle in $H_2S$ is nearer 90° than is the H—O—H angle in water. What would you predict for the H—Se—H angle in hydrogen selenide?

**18** Calculate the formal charge on the nitrogen atom in each of the following compounds: *(a)* $CH_3NH_2$; *(b)* $CH_3NH_3Br$; *(c)* $HN_3$; *(d)* $(CH_3CH_2)_3NO$; *(e)* $(CH_3)_2NOCH_3$; *(f)* $NH_2NH_2$ (both nitrogen atoms); *(g)* $KNH_2$.

**19** What change would take place in the carbon-boron bond angles when trimethylboron reacts with trimethylamine to form $(CH_3)_3N{-}B(CH_3)_3$? What C—C bond angles would you expect to find in the trimethylcarbonium ion, $(CH_3)_3C^+$?

**20** What would you expect the configuration of the bonds to be around the following:
*(a)* nitrogen in the $NH_4^+$ ion
*(b)* boron in the $BF_4^-$ ion
*(c)* boron in the $NH_3 \cdot BF_3$ complex
*(d)* tellurium in dimethyl telluride, $(CH_3)_2Te$
*(e)* oxygen in $OF_2$
*(f)* beryllium in dimethylberyllium, $(CH_3)_2Be$
*(g)* oxygen in the hydronium ion $H_3O^+$.

**21** What can you suggest as *two* of the factors that would operate against the formation of a complex (of the $B^-$—$N^+$ type) between trimethylboron and tri-*tertiary*-butylamine $[(CH_3)_3C]_3N$? Suggestion: Make sketches of the configuration of the amine and of the amine-trimethylboron complex. What are the respective C—N bond dispositions? What influence would the large and bulky *tert*-butyl groups have?

**22** The protons of methane, $CH_4$, have no demonstrable acidic character. Can you suggest a reason why the H—C bond in chloroform, $CHCl_3$, has some small but demonstrable degree of acidic character?

**23** Why does nitrogen not form electron-pair bonds with the use of all five of its 2-shell electrons and thus form such compounds as $NH_5$ or $NCl_5$?

# Chapter four / Acids and Bases. Electrophilic and Nucleophilic Reagents

The chief concerns of the organic chemist are with the processes of bond formation and bond breaking. These processes are loosely designated by the terms synthesis and degradation, although those terms have broader connotations. Our attention will be directed first to the process of bond making and breaking without immediate reference to the utility of the consequences of the process. We shall first of all examine the manner in which two atoms become joined by an electron-pair bond, and the reverse of this.

## 4.1 Covalent bond formation

Let us consider the most general expression for the covalent bond between two atoms or groups, *A* and *B*,

$$A{:}B$$

There are three ways in which we can conceive of the formation of this bond:

$$A{:} + B \longrightarrow A{:}B \quad (1)$$

$$A + {:}B \longrightarrow A{:}B \quad (2)$$

$$A\cdot + \cdot B \longrightarrow A{:}B \quad (3)$$

Cases (1) and (2) are identical in type, but are given separately to show that a given atomic species can react either as *A*: or *A*. For example, chlorine may engage in some reactions as chloride (anion), $:\overset{..}{\underset{..}{Cl}}:^-$; in others, as the chlorine cation, $:\overset{..}{\underset{..}{Cl}}{}^+$.

Case (3) represents what is known as a free-radical, or simply *radical*, reaction. Using chlorine again as an example, the reactive species for a reaction of the third type would be $:\overset{..}{\underset{..}{Cl}}\cdot$. For the present we shall direct our attention to reactions in which both electrons of the pair that constitute the covalent bond are furnished by a single atom.

## 4.2   Acid-base reactions

First of all, it is necessary to restate the reaction of the first or second type by pointing out that the commonest form in which we encounter this covalent bond-making process is that in which species *B* [case (1)] is not free, but in which it relinquishes one bond as it forms another:

$$A:+B:C \longrightarrow A:B+:C$$

The most familiar example of this reaction is that between a base *(A:)* and an acid:

$$H:\overset{..}{\underset{..}{O}}:^-+H:\overset{..}{\underset{..}{Cl}}: \longrightarrow H:\overset{..}{\underset{..}{O}}:H+:\overset{..}{\underset{..}{Cl}}:^-$$

$$A:+B:C \longrightarrow A:B+:C$$

The simplified form of this corresponds to (1):

$$H:\overset{..}{\underset{..}{O}}:^-+H^+ \longrightarrow H:\overset{..}{\underset{..}{O}}:H$$

Although this equation is convenient (and, indeed, is much used) it should be recognized that free, uncombined protons ($H^+$) are not involved in acid-base reactions in solution.

The naked proton does not exist in solution to any but a very small extent, since the tendency for solvation (i.e., the acceptance of a pair of electrons into the $1s$ orbital of $H^+$) is so great that reaction of the proton with water,*

$$H_2O:+H^+ \longrightarrow H_2O:H^+ \quad (H_3O^+)$$

goes essentially to completion. Evidence for the nonexistence of free protons is found in the observation that hydrogen acids in the crystalline state do not form ionic lattices containing $H^+$. The crystalline perchloric acid monohydrate forms an ionic lattice that is isomorphous with ammonium perchlorate, and X-ray analysis has disclosed that it contains a lattice composed of $H_3O^+$ and $ClO_4^-$ ions. Moreover, the three hydrogen atoms of the $H_3O^+$ ion are indistinguishable: the

---

\* It is usual to designate explicitly only the relevant electron pair, and this only for the purpose of emphasizing the feature of reaction to which attention is being directed. Thus, we shall often write $Cl:^-$ instead of $:\overset{..}{\underset{..}{Cl}}:^-$, $H_2O:$ instead of $H_2\overset{..}{\underset{..}{O}}:$, and so on.

three H—O bonds are identical in type. Conductance measurements show that pure (liquid) HCl and HBr are not ionized, nor is gaseous HCl. The lack of solubility of HCl in such nonaqueous solvents as concentrated sulfuric acid and paraffin hydrocarbons indicates that the covalent H—Cl bond has no tendency to separate unless the solvent can participate in the process by *solvating* the proton (and, undoubtedly, the $Cl^-$ ion as well). That the solvent plays an important part in the dissociation process becomes apparent when we consider the remarkably different behavior of HCl in a variety of solvents such as water, alcohols, ethers, benzene, sulfuric acid, liquid ammonia, hexane, acetic acid.

The role of water in the ionization of hydrogen acids in aqueous solution can be expressed by writing the general equation

$$H_2O + H{:}A \rightleftharpoons H_3O^+ + {:}A^-$$

where H:A is an expression for an acid such as HCl, HBr, $HNO_3$, $H_2SO_4$, $HC_2H_3O_2$ (acetic), $HC_2O_4H$ (oxalic), and so on.

## 4.3 Dissociation of acids

The ionization of an acid H:A in aqueous solution is thus not a simple dissociation, but a *reaction with water*. The role of the water is twofold: to hydrate the ions; and, because of its high dielectric constant, to permit the separation of the charged particles. The hydration of ions may involve a known, definite number of water molecules: $Be(H_2O)_4^{++}$, $Al(H_2O)_6^{+++}$, $Mg(H_2O)_6^{++}$, are examples of hydrated ions. The hydronium ion, $H_3O^+$ [or $H(H_2O)^+$] may be hydrated further by water molecules attracted by electrostatic attraction between the (negative) oxygen of

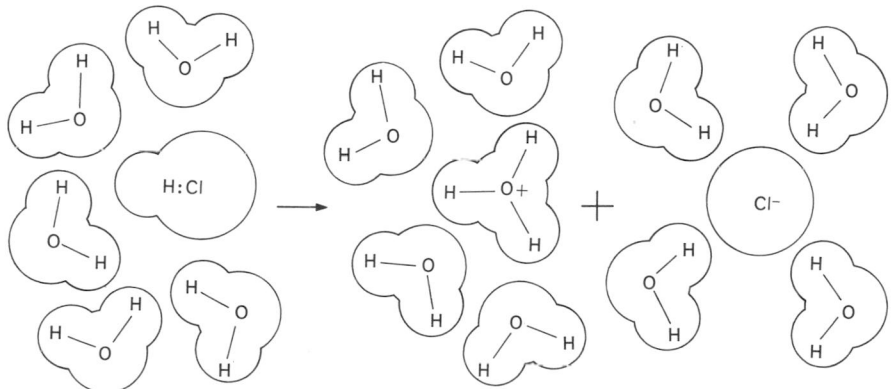

$$HCl + H_2O \rightarrow (H_3O^+)aq. + (Cl^-)aq.$$

**Figure 4–1.** Dissociation of HCl in water; solvation of ions as a factor in the reaction.

$H_2O$ and the positively charged $H_3O^+$. Thus, a proper expression for the ionization of HCl in water is

$$HCl + nH_2O \rightleftharpoons H_3O^+(aq.) + Cl^-(aq.)$$

where the "aq." indicates solvation of the ions by water (Figure 4-1). It is not customary to indicate the participation of solvent in "solvation" in so explicit a manner, but it should always be kept in mind that the role of the solvent in most reactions is not a passive one.

## 4.4 Dissociation as a displacement reaction

The dissociation of HCl into a hydronium ion and a chloride ion may be looked upon as an attack of the water molecule upon HCl, with the expulsion, or *displacement*, of the chloride ion:

$$H-\overset{\cdot\cdot}{\underset{|}{\text{O}}}: + H:Cl \longrightarrow H-\overset{\cdot\cdot}{\underset{|}{\text{O}}}-H + :Cl^- \tag{4}$$
$$\quad\; H \qquad\qquad\qquad\quad H$$

This simple reaction represents a prototype of a kind of reaction encountered frequently in organic chemistry, and we shall employ a symbolism in dealing with it that is exemplified as follows:

$$H-\overset{\cdot\cdot}{\underset{|}{\text{O}}}: \curvearrowright H:Cl \longrightarrow H-\overset{\cdot\cdot}{\underset{|}{\text{O}}}-H + :Cl^- \tag{5}$$
$$\quad\; H \qquad\qquad\qquad\quad H$$

The attack of the water molecule is shown by the curved arrow on the far left; the arrow on the right, embracing the chlorine atom *and its binding pair of electrons*, shows the expulsion of the chloride ion.

Since the displacement of chloride with its electron pair* is coupled with the formation of a new bond to the hydrogen atom, it is clear that the ability of the attacking molecule to provide the electron pair needed for the formation of the new bond to the hydrogen atom is an essential requirement for the dissociation of the HCl molecule. Indeed, we can generalize the reaction by writing in place of water a general species B:† whose essential feature is the possession of an unshared pair of valence electrons:

$$B: + H:Cl \longrightarrow B:H^+ + :Cl^- \tag{6}$$

---

\* We speak of *an* electron pair, although chlorine in HCl possesses a complete octet. This is because our concern is usually only with the electrons involved in bond breaking and bond making, and thus with the electron pair provided by oxygen and that which the chlorine atom retains when it leaves.

† B: is an abbreviation for "base" and is not to be confused with the symbol B for boron.

Of the elements in the first period, only nitrogen, oxygen, and fluorine can furnish examples of **B:** in their covalent compounds:

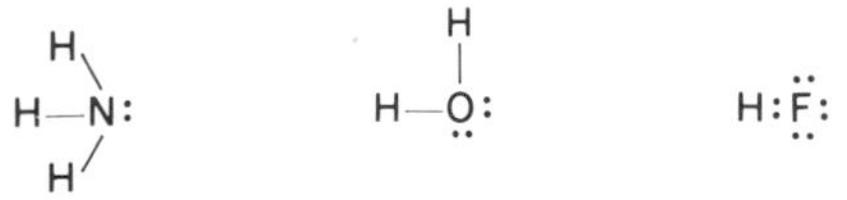

$$H—\overset{H}{\underset{H}{\big|}}N: \qquad H—\overset{H}{\underset{..}{\big|}}\underset{..}{O}: \qquad H:\overset{..}{\underset{..}{F}}:$$

Thus, the reaction between gaseous ammonia and hydrogen chloride is analogous to the "dissociation" of HCl in water:

$$NH_3 + HCl \longrightarrow NH_4^+ + Cl^-$$

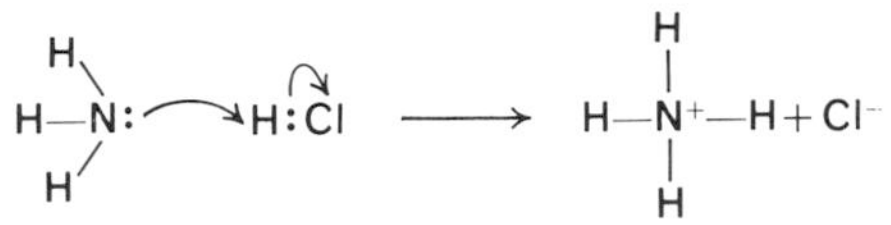

$$H—\overset{H}{\underset{H}{\big|}}N: \quad H:Cl \longrightarrow H—\overset{H}{\underset{H}{\overset{|}{N^+}}}—H + Cl^- \tag{7}$$

The reaction between ammonia and HCl is a typical acid-base reaction. It is clear that water is also a base, for it accepts the proton of HCl just as ammonia does.

## 4.5  Conjugate acids and bases

We know that reactions of the kind we are discussing are reversible; that is, in the ionization of HCl in water, there can be written the two separate reactions

$$H_2O + HCl \longrightarrow H_3O^+ + Cl^-$$

and

$$H_3O^+ + Cl^- \longrightarrow H_2O + HCl$$

or,

$$H_2O + HCl \rightleftharpoons H_3O^+ + Cl^-$$

It can be seen that chloride ion is a base and that it accepts a proton from the acid $H_3O^+$. There is a relationship between chloride ion and HCl, ammonia and ammonium ion, water and hydronium ion that is expressed by the statement that each such pair is a *conjugate pair* of an acid and a base.

$$\underset{\text{acid}_1 \ \ \text{base}_2}{HCl + H_2O} \rightleftharpoons \underset{\text{acid}_2 \ \ \text{base}_1}{H_3O^+ + Cl^-}$$

HCl and $Cl^-$ are a conjugate pair, as are $H_2O$ and $H_3O^+$. HCl is the conjugate acid of the base $Cl^-$; $H_3O^+$ is the conjugate acid of the base $H_2O$. In general terms:

$$H:A + :B \rightleftharpoons H:B + :A$$

conjugate pair

conjugate pair

It can be seen that a given compound can be both an acid and a base:

$H_2O$ {conjugate acid of $OH^-$
       {conjugate base of $H_3O^+$

$H_2PO_4^-$ {conjugate acid of $HPO_4^=$
            {conjugate base of $H_3PO_4$

## 4.6 Proton transfer

Another term that is convenient and often used is to speak of the process of *protonation*. The conjugate acid of a base is the *protonated base*. Thus, protonation of water gives protonated water, or the hydronium ion; protonation of ammonia gives the ammonium ion. We shall see, as we proceed, that the transfer of protons from one ion or molecule to another is a very common feature of a great many reactions. It is for this reason that the term protonation is so useful and convenient; it does not require the explicit identification of the source of a proton, but directs attention to the compound that accepts the proton rather than to the protonating agent.

The term *protonated* is a useful and convenient one in organic chemistry, because many organic reactions are catalyzed by acids, the function of which is to provide a proton to an organic molecule and thus to produce the reactive species. Protonation is a reaction that is widely encountered in organic chemistry, and most organic compounds that contain unshared electron pairs can be protonated (see Section 4.10).

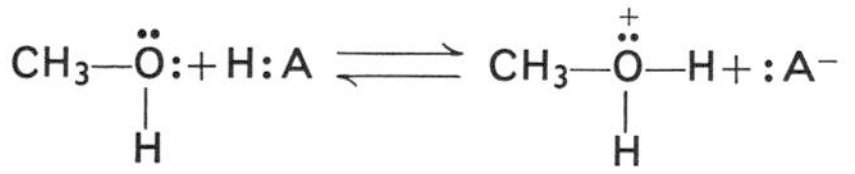

$$CH_3—\overset{..}{\underset{H}{O}}:+H:A \rightleftharpoons CH_3—\overset{+}{\underset{H}{O}}—H+:A^- \tag{8}$$

protonated methanol

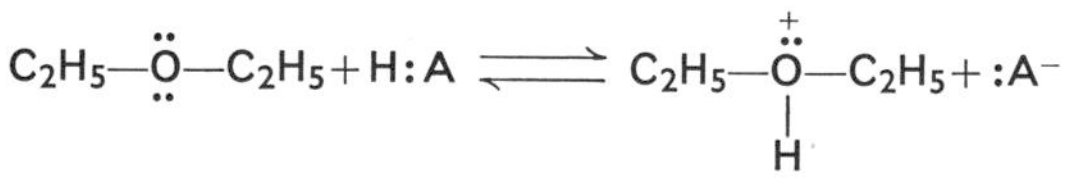

$$C_2H_5—\overset{..}{\underset{..}{O}}—C_2H_5+H:A \rightleftharpoons C_2H_5—\overset{+}{\underset{H}{O}}—C_2H_5+:A^- \tag{9}$$

protonated ethyl
ether

## 4.7 Equilibrium and the dissociation constant

The transfer of a proton from an acid to a base is a reversible reaction and the extent to which the proton is shared between the two conjugate bases $A:$ and $B:$ in the reaction

$$H:A+:B \rightleftharpoons :A+H:B.$$

can be expressed by the *equilibrium constant* of the reaction. For the ionization of an acid $H:A$ in water,

$$H:A+H_2O \rightleftharpoons A:^-+H_3O^+$$

the equilibrium constant is $K$ in the expression

$$K = \frac{[A:^-][H_3O^+]}{[H:A:][H_2O]} \tag{10}$$

Since in dilute solutions the concentration of water is very large compared with the concentrations of the other reacting species it is included in the constant $K_a$, which is the *ionization constant* of the acid $H:A$.

$$K_a = \frac{[A:^-][H_3O^+]}{[H:A]} = \frac{[base][H_3O^+]}{[acid]} \tag{11}$$

Ionization constants can be expressed as $K_a$ or as $pK_a$, the negative logarithm of $K_a$:

$$pK_a = -\log K_a$$

For acetic acid, $K_a = 1.75 \times 10^{-5}$:

$$pK_a = 5 - \log 1.75 = 4.75$$

The dissociation constant, or ionization constant, of a base in aqueous solution can be calculated from the expression (12) for the equilibrium*

$$A:^- + H_2O \rightleftharpoons A:H + OH^-$$

$$K_b = \frac{[A:H][OH^-]}{[A:^-]} = \frac{[acid][OH^-]}{[base]} \tag{12}$$

and

$$pK_b = -\log K_b.$$

A useful relationship between $pK_b$ and $pK_a$ of the conjugate acid can be derived by multiplying the expression for $K_b$ by $H_3O^+/H_3O^+$:

$$K_b = \frac{[A:H][OH^-][H_3O^+]}{[A:^-][H_3O^+]} = \frac{[A:H][K_w]}{[A:^-][H_3O^+]} = \frac{K_w}{K_a} = \frac{10^{-14}}{K_a} \tag{13}$$

$$pK_a = 14 - pK_b \tag{14}$$

The equilibrium constant will be discussed further in Chapter 5, where it will be seen that it is a measure of the extent to which a reaction proceeds and that its magnitude is determined by the relative free energies of the reactants and products of the reaction. The expression relating the free energy change with the equilibrium constant is the following:

$$-\Delta F = RT \log_e K$$

It is apparent that when the overall free energy change ($\Delta F$) is negative (reactants possess higher free energy than products) the equilibrium constant will be the logarithm of a number that is greater than zero, and the equilibrium constant will be greater than one.

---

* A base may be neutral or positively or negatively charged. In this equation the conjugate base, $A:^-$, of the acid $H:A$ is chosen for the example. Another equally valid expression is

$$B:+H_2O \rightleftharpoons B:H^+ + OH^-$$

Acid dissociation constants ($K_a$ values) thus reflect differences in free energy between the undissociated acid and the anion formed by loss of the proton, or, in other words, between the acid and its conjugate base. In assessing the relative strengths of acids, structural factors which alter the *relative* free energies of acid-conjugate base pairs can be examined for the purpose of discerning how they alter the energy of the acid and of its conjugate base. In Section 4.13 we shall see how this analysis of acid strength is applied to the study of the effects of structural factors upon acidity.

---

**Exercise 1.** *(a)* Calculate $pK_a$ for the conjugate acids of the following bases:

$CH_3NH_2$, methylamine, $pK_b = 3.37$

$(CH_3)_2NH$, dimethylamine, $pK_b = 3.22$

$(CH_3)_3N$, trimethylamine, $pK_b = 4.20$

*(b)* Two bases, 1 and 2, have $pK_a$ (of their conjugate acids) of 5.0 (1) and 10.0 (2). Which is the stronger base?

---

## 4.8 The Brønsted-Lowry concept of acidity and basicity

The discussion of acids and bases in the foregoing sections is an expression of a concept of acids and bases first proposed by Brønsted and Lowry. This view of the nature of acids and bases can be stated as follows:

*Bases* are substances that possess an unshared pair of electrons that can be shared with a proton.

*Acids* are substances that can transfer a proton to a base.

It is apparent that this concept is a very general one and is not constrained by considerations of the strengths of the bases or acids. Thus, both water and nitric acid are acids in the Brønsted-Lowry sense,

$$H_2O + NH_3 \rightleftharpoons HO^- + NH_4^+ \tag{15}$$

$$HNO_3 + NH_3 \rightleftharpoons NO_3^- + NH_4^+ \tag{16}$$

and both ammonia and diethyl ether (Section 4.6) are bases.

## 4.9 Strong and weak acids

The term *strong* applied (without qualification) to an acid refers explicitly to the degree to which it is dissociated in *water solution*. A strong acid is one whose conjugate base is a very weak base compared with water. Bases such as $Cl^-$, $Br^-$, $NO_3^-$, $HSO_4^-$, $H_2PO_4^-$ are conjugate to the strong acids $HCl$, $HBr$, $HNO_3$, $H_2SO_4$, $H_3PO_4$. The basicity of these anions is so low that they cannot compete successfully for the proton of $H_3O^+$.

"Weak" acids, on the other hand, are those whose conjugate bases compare to or exceed in strength the basicity of water. Acetic acid, which ionizes incompletely in water, is thus the conjugate acid of a moderately strong base, acetate ion. The ionization constant of acetic acid $(K_A)$ is $1.75 \times 10^{-5}$; a 0.1 $N$ solution of acetic acid is ionized to the extent of 1.3%. Now if acetate ion, in the form of a salt such as sodium acetate, is added to water, the latter, as the conjugate acid of the base $OH^-$, will provide a proton to the base, acetate ion

$$CH_3COO^- + HOH \rightleftharpoons CH_3COOH + OH^- \tag{17}$$

$$(B:^- + H:A \rightleftharpoons B:H + :A^-)$$

The reaction described in equation (17) is called the "hydrolysis" of the salt; it is clear that the reaction is simply an acid-base reaction of the general type we have been considering.

The hydrolysis of acetate ion—that is, the *protonation* of acetate ion by water—is incomplete, since in this case the proton must be shared between the base acetate ion and the very strong base, hydroxide ion. Now if acetate ion competes with a very weak base for a proton, the reaction will be complete:

$$\underset{B:^-}{CH_3COO^-} + \underset{H:A}{HCl} \longrightarrow \underset{B:H}{CH_3COOH} + \underset{:A^-}{Cl^-} \tag{18}$$

An important conclusion emerges from the comparison of equations (17) and (18): the degree to which acetate ion is protonated depends upon the strength of the acid with which it reacts. This conclusion can be restated in the more general ways:

1. the extent of ionization of an acid is related to the strength of the base with which it reacts, a weak acid being completely ionized by a sufficiently strong base;

2. the degree to which a base is protonated (i.e., transformed into its conjugate acid) depends upon the strength of the acid from which the proton is derived.

In other words, acid and base strengths bear reciprocal relationships to one another; an acid that is weak in respect to one base can be strong in respect to another.

Although it is a characteristic of a base that it possesses an unshared pair of electrons in its valence shell, it will be recalled (Chapter 3) that the *availability* of such an electron pair is determined by the nature of the atom in which it is found. Thus $H\overset{..}{O}:^-$, $H_2\overset{..}{O}:$, and $H_3O:^+$ all possess at least one unshared pair of electrons, but $OH^-$ is a strong base, $H_2O$ is a weak base, and $H_3O^+$ is not basic by any experimental criteria. Similarly, $Cl^-$ is a weak base, but HCl shows no tendency to form $H_2Cl^+$. Nevertheless, basic properties *can* be displayed by substances that we ordinarily class as acids. For example, nitric acid, $HONO_2$, which is a typical Brønsted acid of the type $H:A$ ($H:ONO_2$),

$$HONO_2 + :B \rightleftharpoons H:B^+ + NO_3^-$$

$$(B:= H_2O,\ OH^-,\ NH_3)$$

possesses unshared electrons on oxygen

$$H:\overset{..}{\underset{..}{O}}:NO_2$$

and thus has the essential characteristic of a base.

Indeed, toward a sufficiently strong acid—for example, concentrated $H_2SO_4$ —nitric acid does behave as a base, accepting a proton on oxygen:

$$H_2SO_4 + HNO_3 \rightleftharpoons HSO_4^- + H_2NO_3^+ \tag{19}$$

In this expression $HNO_3$ is the conjugate base of the acid $H_2NO_3^+$. Structurally, this reaction appears as,

$$\underset{O_2N}{\overset{H}{\diagdown}}\overset{..}{O}: \quad \longrightarrow H:\overset{\frown}{O}SO_3H \rightleftharpoons \overset{H\overset{+}{\underset{NO_2}{\diagup}}H}{\overset{..}{O}} + :OSO_3H^- \tag{20}$$

to which can be compared the reaction of sulfuric acid with water.

$$\underset{H}{\overset{H}{\diagdown}}\overset{..}{O}: \quad \longrightarrow H:\overset{\frown}{O}SO_3H \rightleftharpoons \overset{H\overset{+}{\underset{H}{\diagup}}H}{\overset{..}{O}} + :OSO_3H^- \tag{21}$$

Since, as we would expect, $HNO_3$ is an exceedingly weak base, it requires a very strong acid to evoke its basic character. This can be understood when we recall that the nitrate ion $:ONO_2^-$ is formed readily in the presence of the weak base water, an indication that the tendency for $ONO_2^-$ to attract a proton is not great. It follows that the tendency for $ONO_2^-$ to attract *two* protons will be very small indeed. Nevertheless, a strong acid such as sulfuric acid can provide the second proton to form $H_2ONO_2^+$.

## 4.10 Organic compounds as acids and bases

We have seen that water, ammonia, acetate ion, nitric acid, chloride ion, can act as bases and accept a proton, and that the distinctive feature which all of these compounds have in common is their possession of at least one unshared pair of electrons.

Organic compounds can be protonated—that is, can act as bases—when they possess an unshared pair of electrons. In most cases the basicity of organic compounds is associated with the presence of either oxygen or nitrogen. Thus, *alcohols and ethers*, which it is evident bear a close structural relationship to water, can be protonated to form *oxonium* salts (Section 4.6):

$$\underset{\substack{\text{methanol} \\ \text{(an} \\ \text{alcohol)}}}{CH_3OH} + HBr \rightleftharpoons \underset{\substack{\text{methyloxonium} \\ \text{bromide}}}{CH_3OH_2^+\}Br^-} \tag{22}$$

$$\underset{\text{diethyl ether}}{CH_3CH_2OCH_2CH_3} + HClO_4 \rightleftharpoons \underset{\underset{\text{diethyloxonium perchlorate}}{CH_3CH_2\overset{+}{\underset{H}{O}}CH_2CH_3}}{} \Big\} ClO_4^- \tag{23}$$

*Ketones and aldehydes* possess an oxygen atom with unshared electrons in its valence shell and are thus capable of accepting a proton from a strong acid:

$$\begin{array}{c}H_3C\\ \diagdown\\ H_3C\diagup\end{array}C=\ddot{O}:+H:Cl \quad \rightleftharpoons \quad \begin{array}{c}H_3C\\ \diagdown\\ H_3C\diagup\end{array}C=\ddot{O}:H^+\Big\}:Cl^- \tag{24}$$

acetone
(a ketone)

Comparable protonations of sulfur analogues of alcohols, ethers, aldehydes, and ketones are represented by the following example:

$$\begin{array}{c}H_3C\\ \diagdown\\ H_3C\diagup\end{array}\ddot{S}:+H:Br \quad \rightleftharpoons \quad \begin{array}{c}H_3C\\ \diagdown\\ H_3C\diagup\end{array}\ddot{S}:H^+\Big\}Br^- \tag{25}$$

dimethyl
sulfide

dimethylsulfonium
bromide

Alcohols, ethers, ketones, and aldehydes are very weak bases and can be protonated to an appreciable extent only by a very strong acid. Indeed, their strength as bases is so small that when they are in the pure state, or in solution in water, or in dilute aqueous acid, the amount of ionized species that is present is negligible. For this reason, they are usually classed as "neutral" compounds.

The very low degree of *intrinsic basicity* of oxygen is shown by the fact that water is not a strong base, just as the high degree of intrinsic basicity of nitrogen is shown by the fact that ammonia is distinctly basic.* In contrast to alcohols and ethers, amines exhibit basic properties of the "ordinary" kind; they are bases by the criteria applicable to the water system. Just as ammonia is protonated in water to an appreciable degree, so are alkyl-substituted derivatives of ammonia (amines) bases of comparable strengths:

$$CH_3\ddot{N}H_2+H_2O \rightleftharpoons CH_3NH_3^++OH^- \tag{26}$$
methylamine

$$(CH_3)_3N:+H_2O \rightleftharpoons (CH_3)_3NH^++OH^- \tag{27}$$
trimethylamine

We shall return later to further discussion of these organic bases.

---

* It is well here to digress and consider our use and interpretation of the terms "acid" and "base." The term *acid* is commonly applied to a substance that turns the color of indicators, tastes sour, can be neutralized with alkali, and reacts with metals or their oxides or carbonates to form salts. These criteria are highly arbitrary, and really apply only to acids that are appreciably stronger than water (as an acid). In the same kind of terminology— a kind of scientific vernacular—a *base* is a substance that tastes brackish, turns red litmus blue, or phenolphthalein red. These trivial definitions embrace acids and bases that are relatively strong, and even among such acids and bases there are wide variations in strength. The important fact to bear in mind is that a substance may be an acid and fail to have acidic properties that are marked enough to be recognizable according to these criteria.

#### 4.11 Solutions in sulfuric acid: freezing point lowering

The dependence of the degree of protonation of a base upon the strength of the acid with which it reacts can be well illustrated by the example of acetic acid in its various states of ionization. If sodium acetate is dissolved in water, hydrolysis of the acetate ion leads to a rise in the $pH$ of the solution above 7 as a result of the formation of $OH^-$ ions. If the salt is added to dilute aqueous acid, a more nearly complete protonation of the ion occurs:

$$CH_3COO^- + H_3O^+ \rightleftharpoons CH_3COOH + H_2O$$

If sodium acetate is dissolved in concentrated sulfuric acid, not only the acetate ion *but the acetic acid itself is protonated:*

$$CH_3COO^- + H_2SO_4 \rightleftharpoons CH_3COOH + HSO_4^-$$

$$CH_3COOH + H_2SO_4 \rightleftharpoons CH_3COOH_2^+ + HSO_4^-$$

*Acetic acid is thus a base* with respect to the strongly acid solvent $H_2SO_4$.

The protonation of acetic acid can be recognized by measurement of the freezing points of solutions of acetic acid in 100% $H_2SO_4$. It is well known that the freezing point of a solvent is lowered by the presence of dissolved *solutes*, and it is possible to define for a given solvent a *freezing point constant*. This constant is the number of degrees the freezing point is lowered (from that of the pure solvent) by the addition of 1 gram-mole of solute to 1000 g of the solvent. This freezing point lowering is conveniently referred to as *i*. Thus, a solution of *n* moles of undissociated solute in 1000 g of solvent shows an *i* value equal to *n*. For instance, a solution of 1 mole of perchloric acid ($HClO_4$) in 1000 g of pure sulfuric acid gives a solution that freezes at 4.35°C, whereas pure $H_2SO_4$ freezes at 10.5°C; this is a freezing point lowering of 6.15°C.

A solution of 1 mole (60 g) of acetic acid in 1000 g of sulfuric acid is found to freeze at −1.8°C; that is, the depression in freezing point is 12.3°C, or the observed *i* value is 2. This must mean that 1 mole of acetic acid has furnished 2 moles of solute particles. This observation is readily accounted for, since the protonation of one acetic acid molecule produces two new ions:

$$CH_3COOH + H_2SO_4 \longrightarrow CH_3COOH_2^+ + HSO_4^- \qquad (28)$$

| 1 mole | large excess | 1 mole | 1 mole |

in a large excess of $H_2SO_4$

In the same way, water shows an *i* value of 2, since 1 mole of water produces 2 moles of ions:*

$$H_2O + H_2SO_4 \longrightarrow H_3O^+ + HSO_4^-$$

Many organic compounds show *i* values of 2 or more because sulfuric acid is sufficiently strong to protonate most organically bound oxygen, nitrogen, and

---

* In practice such concentrated solutions as 1 mole/1000 g would not be used, since they would depart widely from ideality. The use of, say, 0.1 mole/1000 g, coupled with sensitive methods of temperature measurement, would give more reliable values.

sulfur atoms. The $i$ value may exceed 2 when the products of the first reaction can undergo further reaction to produce additional substances in the solution:

$$CH_3OH + H_2SO_4 \rightleftharpoons CH_3OH_2^+ + HSO_4^-$$

$$CH_3OH_2^+ + HSO_4^- \rightleftharpoons CH_3OSO_3H + H_2O \qquad (29)$$

$$H_2O + H_2SO_4 \rightleftharpoons H_3O^+ + HSO_4^-$$

Thus, methanol shows an $i$ factor of 3 in sulfuric acid, since 1 mole of the alcohol produces 1 mole each of methyl hydrogen sulfate ($CH_3OSO_3H$), hydronium ion ($H_3O^+$), and bisulfate ion ($HSO_4^-$).

It is clear that a great deal of useful information can be obtained by studying the behavior of substances in sulfuric acid solution; in this solvent studies can be carried out of very weak bases, bases that are neutral substances with reference to the water system.

It is likewise possible to study the strengths of acids in solvent systems other than water. For example, with the use of acetic acid as a solvent, it is possible to measure differences between strong acids, such as $H_2SO_4$ and HCl, both of which are so extensively ionized in water that differences between them are small in this solvent. In the solvent acetic acid, however, the fraction of $CH_3COOH_2^+$ ions would not be the same for the two "strong" acids.

The reciprocal relationship between acidity and basicity is well exemplified by the behavior of the chloro derivatives of acetic acid in sulfuric acid solution. The replacement of the hydrogen atoms of the $-CH_3$ group of acetic acid by chlorine atoms increases the acid strength. As the acids increase in strength *as acids* they become weaker *as bases*. The data in Table 4-1 show these relationships.

It should be added that it is not practicable in all cases to measure the base strength of an organic compound by freezing point measurements in sulfuric acid, since decomposition or oxidation may occur and make measurements unreliable or meaningless.

**Table 4-1.** *Strengths of chloroacetic acids*

| Acid | $K_A$ | % Ionized in 0.3 $M$ aqueous solution | $i$ Value for f.p. depression in $H_2SO_4$ |
|---|---|---|---|
| $CH_3COOH$ | $1.75 \times 10^{-5}$ | 2.4 | 2 |
| $ClCH_2COOH$ | $1.4 \times 10^{-3}$ | 22.5 | about 2 |
| $Cl_2CHCOOH$ | $5.0 \times 10^{-2}$ | 70.0 | >1 but <2 |
| $Cl_3CCOOH$ | $1.3 \times 10^{-1}$ | 89.5 | 1 |

## 4.12 Lewis acids

An extension of the concept of acids, as it is found in the Brønsted-Lowry theory, was offered by G. N. Lewis, who recognized the acidic character of many substances that possess no hydrogen atoms. It has long been known that certain compounds, such as $BF_3$, $SnCl_4$, $AlCl_3$, $ZnCl_2$, $FeCl_3$, have many properties in common with Brønsted acids. They combine with bases, such as amines, to form salt-like products; they affect indicators as acids; and they catalyze many reactions which are also catalyzed by strong hydrogen-acids. Lewis' extension of the Brønsted concepts of acids and bases was to define *bases* as substances having electron pairs that can be donated to form a (shared) covalent bond; and to define *acids* as substances, one of whose constituent atoms can receive into its valence shell an electron pair belonging to another atom.

It can be recognized that a base is regarded in the same way in both the Brønsted and the Lewis concepts. An acid, however, is a *proton donator* in the Brønsted theory but an *electron acceptor* in the Lewis theory. The Lewis theory is neither a substitute for, nor a modification of, the Brønsted theory. Rather, it is an extension, one which brings into consideration a number of compounds that possess a property also possessed by the proton: namely, the ability to form a bond with another atom by accepting into its valence shell a pair of electrons possessed by that atom.

Let us examine a typical Lewis acid, $BF_3$. In this compound, boron has utilized its three $L$-shell electrons in covalent bond formation. There remains a fourth orbital, unoccupied in $BF_3$, but capable of occupancy by a pair of electrons, and thus capable of forming a covalent bond with an atom that possesses an unshared electron pair. Trivalent boron compounds, then, should be able to react with bases to form compounds in which the boron atom has completed its octet. The following equations represent reactions in which $BF_3$ is reacting with bases of several types, all of which furnish an electron pair to complete the octet of the boron atom:

$$:\ddot{F}:^- + BF_3 \longrightarrow BF_4^- \tag{30}$$
fluoride ion

$$(CH_3)_2\ddot{O}: + BF_3 \longrightarrow (CH_3)_2\overset{+}{\ddot{O}} - \overset{-}{B}F_3 \tag{31}$$
dimethyl ether

$$(CH_3)_3N: + BF_3 \longrightarrow (CH_3)_3N^+ - BF_3^- \tag{32}$$
trimethylamine

Aluminum halides form analogous complexes with bases:

$$\begin{matrix} C_6H_5 \\ \phantom{x} \\ C_6H_5 \end{matrix}\!\!\!>\!C\!=\!\ddot{O}: + AlCl_3 \longrightarrow \begin{matrix} C_6H_5 \\ \phantom{x} \\ C_6H_5 \end{matrix}\!\!\!>\!C\!=\!\overset{+}{\ddot{O}}-\overset{-}{A}lCl_3 \tag{33}$$
benzophenone

$$N: + AlBr_3 \longrightarrow \overset{+}{N} - \overset{-}{A}lBr_3 \tag{34}$$
pyridine

In future discussion we shall use the term *acid* for a substance that is capable of furnishing a proton to a base; acids of the Lewis type will be explicitly referred to as *Lewis acids*.

Lewis acids are of considerable importance in organic chemistry, because such compounds as $AlCl_3$, $ZnCl_2$, $FeBr_3$, $SnCl_4$, and certain other covalent compounds of heavy metals are widely used, principally in the carrying out of acid-catalyzed reactions. It is urged that when, in this book, one of these compounds is encountered, the student examine its role in the reaction in which it takes part, with a view to discerning how its character as a Lewis acid is involved in its action.

### 4.13   Strengths of acids and bases. Inductive and structural effects

What factors govern the degree of ionization of a given acid with a series of bases? If the acid, or proton donor, is water, the question becomes, "What factors govern the degree of dissociation of a series of bases in aqueous solution?"

The basic dissociation constants (as $pK_b$ values) of several closely related aliphatic amines are given in Table 4-2, where it will be seen that the alkylamines are stronger bases than ammonia by about $1.5\,pK$ units in most of the examples given.

The effect of the alkyl groups upon the base strengths of the alkylamines is to increase electron density on nitrogen and thus to decrease the energy change that occurs upon protonation. This relationship between the free energy change in a reaction and the equilibrium constant has been described briefly in Section 4.7 and will be discussed further in Chapter 5. The effect of the alkyl substituents is a reflection of the unequal sharing of electrons in the C—N bond, and has the result

---

**Table 4-2.**   *Basic strength of some simple aliphatic amines*

| Amine | Formula | $pK_b$ | Relative Base Strength vs. Hydrogen Acids |
|-------|---------|--------|--------------------------------------------|
| ammonia | $NH_3$ | 4.75 | 6 |
| methylamine | $CH_3NH_2$ | 3.36 | 5 |
| dimethylamine | $(CH_3)_2NH$ | 3.29 | 4 |
| trimethylamine | $(CH_3)_3N$ | 3.28 | 3 |
| ethylamine | $C_2H_5NH_2$ | 3.25 | 2 |
| diethylamine | $(C_2H_5)_2NH$ | 2.90 | 1 |
| triethylamine | $(C_2H_5)_3N$ | 3.25 | 2 |

of causing nitrogen to be more electron-rich in methylamine than in ammonia. It is called an *inductive effect*.

The figures in Table 4-2 show that the dialkylamines are stronger bases than the corresponding monoalkylamines; but, surprisingly, the introduction of a third alkyl group, as in trimethylamine and triethylamine, causes a diminution rather than an increase in base strength. The reason for this is largely a steric one. The three alkyl groups in triethylamine are disposed in a pyramidal structure in which the bond angles are those that provide for minimal interference between these rather bulky substituents. In the protonated amine, $(C_2H_5)_3NH^+$, however, the four substituents on nitrogen are tetrahedrally disposed and are thus constrained to a conformation in which increased repulsive forces exist between the alkyl groups. These repulsions represent an energy increase upon protonation that is more than what would have resulted were they not present, with a consequent decrease in the basic dissociation constant that is more than what would have been anticipated from consideration of inductive effects only.

**Inductive effects.** It would be predicted that since methylamine is a stronger base than ammonia, methanol should be a stronger base than water. This is indeed true. Furthermore, acetic acid, $CH_3COOH$, should be a weaker acid than formic acid, $HCOOH$; or, in other words, acetate ion should be a stronger base than formate ion. This is also found to be true.

On the other hand, we have seen (Table 4-1) that chloroacetic acid is a stronger acid ($K_a = 1.4 \times 10^{-3}$) than acetic acid ($K_a = 1.75 \times 10^{-5}$). This is the result of an inductive effect of the opposite kind than that shown by the methyl group. The strongly dipolar character of the Cl—C bond, with chlorine at the negative end of the dipole, creates a positive center adjacent to the —COOH group (in the acid) and to the —COO⁻ group (in the anion). Now because of the electronegativity difference between carbon and oxygen, the carboxyl group is also dipolar in nature, the carbon-oxygen dipole creating a positive center at the carbon atom of —COOH. In chloroacetic acid, therefore, the two opposing dipoles create electrostatic repulsion between the positive centers at —CH₂— and —COOH, and this repulsive interaction raises the energy of the acid more than it would be raised were no such dipolar repulsion to exist. In the anion, on the other hand, the negative charge is stabilized by the electrostatic attraction of the Cl—C dipole, thus stabilizing the anion by lowering its energy more than it would be lowered if the dipolar attraction were absent. The result is therefore to *decrease the energy difference* between $ClCH_2COOH$ and $ClCH_2COO^-$ as compared with that between acetic acid and its anion. Since the free energy change in the dissociation reaction is reflected in the magnitude of $K_a$, the equilibrium constant for the dissociation reaction, the smaller the free energy increase in going from RCOOH to RCOO⁻, the greater the $K_a$ for the reaction and the stronger the acid RCOOH.*

---

* Since $K_a$ values for the aliphatic acids that are used for these examples are very small, $\Delta F$ for the dissociation reaction is positive. The smaller the magnitude of $\Delta F$, the larger $K_a$ becomes until, when $\Delta F$ becomes negative ($-\Delta F$ positive), $K_a$ becomes greater than 1.

**Table 4-3.** *Effect of structure of chlorobutyric acids upon dissociation constant*

| Acid | pK$_A$* |
| --- | --- |
| CH$_3$CH$_2$CH$_2$COOH | 4.8 |
| CH$_3$CH$_2$CHCOOH<br>        $\mid$<br>        Cl | 2.9 |
| CH$_3$CHCH$_2$COOH<br>   $\mid$<br>   Cl | 4.1 |
| CH$_2$CH$_2$CH$_2$COOH<br>$\mid$<br>Cl | 4.5 |

* To the nearest tenth.

The property of chlorine that is manifested in its effect upon the acidity of the chloro-substituted aliphatic acids, by creating an electron-deficiency at the atom to which it is attached, is alluded to as a *negative inductive effect*. In general, atoms and groups at the negative end of a dipole such as that in the Cl—C bond are said to have negative inductive effects.

The inductive effect drops off rapidly with distance. As the chlorine atom is moved farther from the atom from which the proton is removed, its effect diminishes, sharply at first, less markedly thereafter (Table 4-3).

We see from the foregoing discussion that the inductive effect has direction: halogen atoms draw electrons closer to their nuclei and decrease electron density on the atom to which they are attached; alkyl (e.g., CH$_3$—) groups act in the opposite sense. These shifts involve only displacements *within the covalent linkages,* not displacements out of one orbital and into another.

**Exercise 2.** In the following pairs, the stronger acid is placed first in *(a)* to *(d)*. For *(e)* to *(h)*, select the stronger acid of each pair:

    *(a)* HOCl, H$_2$O

    *(b)* BrCH$_2$COOH, CH$_3$CH$_2$COOH

    *(c)* NH$_4^+$, CH$_3$NH$_3^+$

    *(d)* CF$_3$CH$_2$OH, CH$_3$CH$_2$OH

    *(e)* CH$_3$CHBrCOOH, BrCH$_2$CH$_2$COOH

    *(f)* NH$_3$, (CH$_3$)$_3$NH$^+$

    *(g)* CH$_3$OH, BrCH$_2$CH$_2$OH

    *(h)* CH$_3$COOH, (CH$_3$)$_3$CCOOH

### 4.14 Nucleophilic and electrophilic reagents

Since the characteristic behavior of a base is its combination with a proton or a Lewis acid, a base is often described as a *nucleophilic* (nucleus-seeking) substance. The term *nucleophile* is commonly used for substances that possess the feature that we have so far taken as characteristic of a base: an unshared pair of valence electrons. The following are representative nucleophilic reagents; it is seen that all of them are capable of accepting a proton:

| | | | | | |
|---|---|---|---|---|---|
| $CH_3OH$ | $CH_3O^-$ | $H_2O$ | $HO^-$ | $CH_3S^-$ | $(CH_3)_2S$ |
| $CH_3NH_2$ | $(CH_3)_3N$ | $CH_3COO^-$ | | $HCOO^-$ | $CN^-$ |

---

**Exercise 3.**   Write the formulas for the conjugate acids of these bases.

---

The behavior of nucleophilic reagents in reactions with organic compounds will be discussed at length in Chapter 7.

*Electrophilic* reagents are those which act by acquiring a share in a pair of electrons belonging to another atom or molecule. The proton is the prototype of an electrophilic reagent, but acids in general are electrophilic reagents, since they furnish the electrophilic proton. Lewis acids are, of course, electrophiles, and other atoms that are not electron deficient but can accept a pair of electrons belonging to another atom, while simultaneously relinquishing a pair already in their possession, are electrophilic.

---

**Exercise 4.**   Identify the electrophile and the nucleophile in the following reactions:

$$Ag^+ + 2NH_3 \rightleftharpoons Ag(NH_3)_2^+$$
$$ZnCl_2 + 2Cl^- \rightleftharpoons ZnCl_4^=$$
$$(CH_3)_3C^+ + Br^- \rightleftharpoons (CH_3)_3CBr$$
$$NO_2^+ + H_2O \rightleftharpoons H_2ONO_2^+$$
$$CH_2{=}CH_2 + Ag^+ \rightleftharpoons (CH_2CH_2Ag)^+$$

---

## 4.15   Hydrogen bonding and association

When we examine a homologous series of compounds, such as the alcohols, the amines, and the hydrocarbons, we note that the boiling points of the individual compounds increase as their molecular weights increase (Table 4-4).

**Table 4-4.**

| Alcohols | b.p. | Amines | b.p. | Carboxylic Acids | b.p. |
|---|---|---|---|---|---|
| $CH_3OH$ | 56°C | $CH_3NH_2$ | −6.7°C | $CH_3COOH$ | 118°C |
| $CH_3CH_2OH$ | 78 | $CH_3CH_2NH_2$ | 16.6 | $EtCOOH$ | 141 |
| $CH_3CH_2CH_2OH$ | 98 | $CH_3CH_2CH_2NH_2$ | 50 | $PrCOOH$ | 164 |
| $CH_3CH_2CH_2CH_2OH$ | 117 | $CH_3CH_2CH_2CH_2NH_2$ | 78 | $BuCOOH$ | 186 |
| $CH_3(CH_2)_4OH$ | 138 | $CH_3(CH_2)_4NH_2$ | 104 | $AmCOOH$ | 205 |

**Table 4-5.**

| Compounds | Formula | Mol.Wt. | b.p. |
|---|---|---|---|
| *n*-butyric acid | $CH_3CH_2CH_2COOH$ | 88 | 164°C |
| *n*-amylamine | $CH_3CH_2CH_2CH_2CH_2NH_2$ | 87 | 104 |
| *n*-amyl alcohol | $CH_3CH_2CH_2CH_2CH_2OH$ | 88 | 138 |
| *n*-hexane | $CH_3CH_2CH_2CH_2CH_2CH_3$ | 86 | 69 |

It can be seen that the increase of molecular weight of 14 (addition of —$CH_2$—) causes the boiling point, within a given class, to rise by about 20 to 25°C. It is also to be noted that boiling point and molecular size bear no regular relationship if we compare alcohols with amines or with acids (Table 4-5).

In compounds such as water, alcohols, amines, carboxylic acids, and others that possess both (1) a hydrogen attached to oxygen or nitrogen and (2) a donor atom, such as oxygen or nitrogen, intermolecular proton exchange can occur. The hydrogen atom possesses a certain positive character, relative to the other atom, which may be described by the symbol

$$\overset{\delta^-}{R—O}—\overset{\delta^+}{H}$$

The result of this is that a certain degree of organization exists in a medium consisting of such molecules, brought about by a weak binding, largely electrostatic in nature, caused by the attraction of positive and negative ends of the dipoles. The lining up of the molecules in such liquids creates loose polymolecular aggregates, with the result that the *effective* molecular weight, *compared with what it would be in the absence of such intermolecular forces,* is on the average greater than that calculated for the simple molecules. It should be pointed out that seldom can a definite polymolecular species be identified in such a liquid, but the relationship between boiling point and chemical nature can be accounted for in qualitative terms on the basis of this concept.

It is probable that *hydrogen bonds* involve a certain degree of orbital overlap between the hydrogen atom and the donor atom, and are not to be attributed solely to dipolar interactions; were dipolar interactions the predominant factor, certain compounds with greater dipole moments, but without O—H or N—H bonds, would be associated to a greater degree than in fact they are.

In water we can picture a "water substance," which is impossible to define explicitly but can be represented by a picture such as the following,

$$
\begin{array}{c}
\quad\quad\quad H \\
\quad\quad\quad | \\
\cdots H—O\cdots H—O \\
\quad\quad | \\
\quad\quad H \\
\quad\quad \vdots \\
\quad\quad O—H\cdots O—H\cdots \\
\quad\quad | \quad\quad | \\
\quad\quad H \quad\quad H
\end{array}
$$

in which the dotted bonds indicate the dipolar attractions we have described. These hydrogen bonds are very weak, and are broken with rising temperature, until at the boiling point individual water molecules escape as vapor. But the temperature at which vaporization occurs is higher than it would be were no such organization present.

---

**Exercise 5.**   Alcohols are capable of forming hydrogen-bonded complexes similar to that shown for water. Draw a portion of such an aggregate for methanol. Would you expect methanol to be more highly organized in such an aggregate than is water, or less so? Why? Would you expect the boiling point of methanol to be more than or less than 100°C?

---

In certain cases, aggregates of definite size are known. Measurements of the molecular weight of benzoic acid in benzene solution show that the acid exists as a *dimer*; it appears that two molecules of a carboxylic acid can associate to form a definable aggregate through the formation of hydrogen bonds in the following way:

$$\text{R}-\text{C} \underset{\text{O}-\text{H}\cdots\text{O}}{\overset{\text{O}\cdots\text{H}-\text{O}}{\big<}} \overset{\big>}{\text{C}}-\text{R}$$

In most cases, however, the association of the molecules in a liquid or solution is more random and less clearly defined than this.

It can be seen that these properties are consonant with the acidic and basic properties of the compounds we are discussing. The greater base strength of amines—or, what is equivalent, their lower degree of acidity as hydrogen acids—indicates that the electron pair between N— and H— is less strongly attracted to the N— than it is to the O— in the alcohols. Consequently, the proton is more strongly bonded to nitrogen and is less able to engage in intermolecular association with electron-donor atoms. That *some* such association occurs, however, is shown by the boiling points of amines, compared with ethers and hydrocarbons of the same molecular weight.

An appreciation of the influence of structural type upon boiling point can be gained by an inspection of Table 4-6 in which compounds of various classes with molecular weights in the range 58–62 are compared.

When a semipolar bond is present, the effect of this very strong dipole upon the boiling point is marked. In a molecule such as trimethylamine oxide (Chapter 3) strong orienting forces result from the dipolar bond, $(CH_3)_3\overset{+}{N}\!-\!O^-$, with the result that the molecules in the aggregate have a high degree of organization. Trimethylamine oxide is quite involatile, compared with substances of comparable molecular weight but with no semipolar bonds, and sublimes from the solid state at about 180°C. It appears to be almost salt-like in nature, the strongly dipolar character of the molecule causing it to form a crystal lattice resembling that composed of the positive and negative ions of a salt like NaCl.

**Table 4-6.**

| Class | Name | Formula | Mol.Wt. | b.p. |
|---|---|---|---|---|
| hydrocarbon | *n*-butane | $CH_3CH_2CH_2CH_3$ | 58 | 0°C |
| tertiary amine | trimethylamine | $(CH_3)_3N$ | 59 | 4 |
| ether | methyl ether ether | $CH_3OC_2H_5$ | 60 | 6 |
| ester | methyl formate | $HCOOCH_3$ | 60 | 32 |
| secondary amine | methylethylamine | $CH_3NHC_2H_5$ | 59 | 36 |
| primary amine | *n*-propylamine | $CH_3CH_2CH_2NH_2$ | 59 | 50 |
| aldehyde | propionaldehyde | $CH_3CH_2CHO$ | 58 | 60 |
| alcohol | *n*-propyl alcohol | $CH_3CH_2CH_2OH$ | 60 | 98 |
| acid | acetic acid | $CH_3COOH$ | 60 | 118 |
| amino alcohol | ethanolamine | $HOCH_2CH_2NH_2$ | 61 | 171 |
| glycol | ethylene glycol | $HOCH_2CH_2OH$ | 62 | 197 |

**Exercise 6.** With the aid of a handbook, find the boiling points of all of the isomers of $C_4H_{10}O$. Can you suggest how the boiling points of these compounds can be reconciled with their structures?

## 4.16   Solubilities of organic compounds

A solid or liquid substance is an aggregate of individual molecules. If the inter-molecular forces are weak, such that the positions of the molecules are not fixed in definite orientations relative to one another, the physical state of the substance is that of a liquid; if the intermolecular forces are very weak, the substance is a gas, with the molecules in motion over appreciable distances. With increasing inter-molecular associating forces—which may be weak chemical bonds or electrostatic attractive forces—greater degrees of organization exist until at length the *crystalline* state is reached. In the crystalline state the molecules are packed into definite arrangements (lattices) in which random molecular movements do not occur (except for vibrational motion about an average position) and which retain an organized arrangement until molecular motion increases to the point at which intermolecular forces are overcome. When this happens, the regular lattice breaks down, the structure becomes randomized, and the substance changes from a crystalline solid into a liquid. This process is called *melting*. The heat that is supplied to raise the crystal to its melting point represents energy that is required to disrupt the intermolecular forces that hold the molecules in the regular arrange-ment of the crystal lattice.

The solution of an organic compound in a solvent is a process in which the intermolecular forces in the pure substance are replaced by forces that act between the molecules of the solute and those of the solvent. It is to be expected, therefore, that unless the balance between the molecule-molecule forces in the solute and the molecule-solvent forces in the solution is in favor of the latter, the compound will be insoluble.

Some organic compounds are readily soluble in water, but the great majority are not. Most organic compounds are soluble in other organic compounds—the so-called organic solvents. Typical organic solvents are the simple alcohols (methanol, ethanol), ethers (diethyl ether), halogenated hydrocarbons (chloroform), ketones (acetone), esters (ethyl acetate), and hydrocarbons (benzene, light petroleum fractions).

What structural features determine whether an organic compound will be soluble or insoluble in water? Water solubility would be expected to involve binding forces between the molecules of water and those of the organic solute. Since the association of water molecules with each other in liquid water involves hydrogen bonding between the hydroxyl groups of the associated molecules, we would expect that for an organic compound to be soluble in water, it would have to possess structural features that would permit it to fit into the "water structure" and to play a role similar to that played by the water molecules. For this reason, organic compounds that contain several hydroxyl groups (such as the sugars) or several carboxylic acid (—COOH) groups, or combinations of these, tend to be water-soluble. On the other hand, compounds of these kinds are usually insoluble in hydrocarbon solvents such as benzene or hexane. These solvents are lacking in functional groups that would permit the formation of associative bonds with hydroxyl or carboxylic acid groups. To put it in another way, the *intermolecular* forces within the crystal lattice of a hydroxyl-containing organic compound, such as a sugar, can be overcome only if acceptable substitute forces between the sugar molecules and the solvent molecules can take their place. In a solvent such as benzene or hexane this cannot happen, thus the sugar is insoluble; in water this can happen, and the sugar is soluble.

We are familiar with the solubility of inorganic salts in water. The "solvation" of the individual ions and the high dielectric constant of water contribute to disrupting the electrostatic forces in the ionic crystal lattice and dispersing the ions throughout the solvent as discrete ions, each within a shell of solvent molecules.

In the same way, organic compounds that can be ionized by aqueous solvents will be soluble in such solvents. For example, aqueous acids (e.g., dilute HCl) will dissolve most amines by protonating the amine molecules and dispersing the resulting ammonium ions as solvates. Aqueous bases (e.g., dilute NaOH) will dissolve organic acids by converting them into the corresponding anions by removal of the protons.

Very weak bases, such as alcohols, ethers, and many nitrogen-containing compounds other than amines, are insoluble in dilute aqueous acids but are usually soluble in more concentrated acids. For this reason, concentrated sulfuric acid will dissolve most organic compounds that contain oxygen, nitrogen, or sulfur; the ability of the concentrated acid to protonate even a very weak base permits the formation of the cation which can then be dispersed (i.e., dissolved) in the solvent acid.

Finally, it should be noted that low-molecular-weight organic compounds (containing less than five carbon atoms) that contain oxygen or nitrogen in any manner of combination, are usually water-soluble. The small size of these organic molecules does not provide opportunity for strong intermolecular binding forces between them; thus, even weak binding forces between the solvent water molecules and the oxygen or nitrogen atoms of the solute molecules are sufficient to bring the organic compound into solution. Low-molecular-weight hydrocarbons, such as pentane, hexane, and so forth, are insoluble in water, for reasons described in an earlier paragraph.

The solubility behavior of an organic compound provides valuable structural information. The examination of the solubility behavior of an unknown organic compound (by testing its solubility in water, dilute HCl, dilute NaOH, concentrated sulfuric acid, and in an organic solvent such as ether) is another stage in the step-by-step accumulation of information that provides the chemist with the means for the eventual establishment of its structure.

Table 4-7 is a summary of a few generalizations concerning solubility relationships. This information is presented only as an introductory guide to the solubility concepts discussed in this section. It must be recognized that solubility is not an all-or-none phenomenon. Chemists usually relate solubility or insolubility to some arbitrary ratio of solute to solvent. These details will not be discussed here; they are chiefly of importance in cases in which "borderline" solubility behavior is encountered.

**Table 4-7.**  *Some simple solubility characteristics*

| Soluble in: | Class of Compound | Typical Examples |
|---|---|---|
| water and in most organic solvents | low-molecular-weight alcohols, ketones, esters, acids, amines | ethanol, acetic acid, acetone |
| water, but not in most organic solvents | polyhydroxy compounds, salts of organic acids and bases | sugars, sodium acetate, amine hydrochlorides |
| most organic solvents | most organic compounds of many classes | — |
| dilute acid (1 $N$ HCl) | organic bases (amines) | aniline, triethylamine |
| dilute alkali (1 $N$ NaOH) | organic acids (carboxylic acids and phenols) | benzoic acid, phenol |
| concentrated sulfuric acid | most oxygen- and nitrogen-containing organic compounds | — |
| usually insoluble in sulfuric acid, but soluble in organic solvents | paraffin hydrocarbons, halogenated hydrocarbons | $n$-hexane, chloroform, dichlorobenzene |

## Exercises

(Exercises 1–6 will be found within the text.)

**7** Write the equations showing the reaction of HCl with *(a)* water; *(b)* hydroxide ion; *(c)* bicarbonate ion; *(d)* hydrazine; $NH_2NH_2$; *(e)* ammonia; *(f)* chloroamine, $NH_2Cl$; *(g)* acetate ion; *(h)* dimethyl ether.

**8** What would be a more descriptive name than perchloric acid monohydrate for $H_3O^+ClO_4^-$?

**9** What are the acids in the following solutions: *(a)* a solution of HBr in water; *(b)* A solution of HCl in liquid ammonia; *(c)* a solution of acetic acid in water; *(d)* a solution of acetic acid in *n*-hexane; *(e)* a solution of ammonium chloride in water; *(f)* a solution of acetic acid in concentrated sulfuric acid?

**10** Write the formula for the conjugate base of each of the following acids: *(a)* $H_2O$; *(b)* $H_3O^+$; *(c)* $NH_3$; *(d)* $CH_3OH$; *(e)* $NH_4^+$; *(f)* $CH_3COOH$; *(g)* $H_2CO_3$; *(h)* $CH_3NH_2$; *(i)* HBr; *(j)* $NH_2NH_3^+$; *(k)* $HSO_4^-$.

**11** Write the equation for the reaction of acetate ion, $CH_3COO^-$, with each of the following acids; *(a)* HCl; *(b)* $H_2O$; *(c)* $NH_4^+$; *(d)* $CH_3OH_2^+$; *(e)* $HSO_4^-$; *(f)* $CH_3OH$; *(g)* $H_3PO_4$.

**12** Write electronic structures for the following, showing only the unshared electron pairs; for each, write the protonated form (the conjugate acid): *(a)* $CH_3CH_2OH$; *(b)* $CH_3OCH_3$, *(c)* $H_2S$; *(d)* $(CH_3)_2S$; *(e)* $HO^-$.

**13** The following *i* values are found for solutions of each of the following substances in sulfuric acid. Account for these by writing the equations for the reactions between solute and solvent: *(a)* $NH_3$, $i=2$; *(b)* $CH_3OCH_3$, $i=2$; *(c)* $NH_2NH_3Br$, $i=3$; *(d)* $CH_3OH$, $i=3$.

**14** Arrange the following in order of decreasing strength as acids: *(a)* $H_2O$; *(b)* $HSO_4^-$; *(c)* $NH_4^+$; *(d)* $CH_3COOH$; *(e)* $CH_3COOH_2^+$; *(f)* $HClO_4$; *(g)* $OH^-$; *(h)* $NH_3$; *(i)* $H_2ONO_2^+$; *(j)* $HCO_3^-$; *(k)* $NH_2^-$.

**15** What is the pH of a 0.1 *M* solution of sodium acetate in water? (The $K_a$ of acetic acid is $1.75 \times 10^{-5}$.)

**16** Write the reaction of dimethyl ether, $CH_3OCH_3$, with each of the following Lewis acids: *(a)* Boron trichloride; *(b)* Trimethylboron; *(c)* $AlBr_3$; *(d)* $SnCl_4$; *(e)* $MgBr_2$, *(f)* $CH_3MgCl$; *(g)* $CH_3^+$.

**17** Would $BrCH_2CHCOOH$ be a stronger or weaker acid than $Br_2CHCH_2COOH$?
$\quad\quad\quad\quad\quad\quad$ |
$\quad\quad\quad\quad\quad\quad$ Br

Why?

**18** Look up and compare the boiling points of *n*-hexane, 1-pentanol, 1-aminopentane (*n*-amylamine), and *n*-butyric acid. Calculate the molecular weight of each compound, and discuss briefly the relationship between structure and boiling point as shown by these data.

**19** Is HOCl a stronger or a weaker acid than $H_2O$? Why?

**20** Water is nearly insoluble in liquid sulfur dioxide. Hydrobromic acid is soluble in liquid $SO_2$, but the solution is not a conductor of electricity. When water is added to a solution of HBr in $SO_2$, an amount of water *equivalent to the amount of HBr present* will dissolve, and the resulting solution is found to conduct electricity. Explain these observations.

*Chapter five* / **Equilibrium and Reaction Rate**

## 5.1 Reaction types

Chemical reactions most often encountered are those that may occur between

*(a)* two ions:

$$Ag^+ + Cl^- \rightleftharpoons AgCl \tag{1}$$
(in solution)   (insoluble salt)

*(b)* an ion and a neutral molecule:

$$NH_4^+ + H_2O \rightleftharpoons NH_3 + H_3O \tag{2}$$
$$OH^- + HCl \rightleftharpoons H_2O + Cl^- \tag{3}$$
$$OH^- + CH_3Br \rightleftharpoons HOCH_3 + Br^- \tag{4}$$

*(c)* two neutral molecules:

$$NH_3 + CH_3Br \rightleftharpoons CH_3NH_3^+ + Br^- \tag{5}$$
$$CH_3COOH + H_2O \rightleftharpoons CH_3COO^- + H_3O^+ \tag{6}$$

*(d)* a free radical and a neutral molecule:

$$Cl\cdot + CH_4 \rightleftharpoons HCl + CH_3\cdot \tag{7}$$

*(e)* two free radicals:

$$\cdot H + \cdot H \rightleftharpoons H_2 \tag{8}$$
$$CH_3\cdot + \cdot Cl \rightleftharpoons CH_3Cl \tag{9}$$

These reactions are written as equilibrium reactions, although in some of them the position of equilibrium is such that they are essentially complete in one direction. For instance, the association of two hydrogen atoms to give a molecule of hydrogen is accompanied by the liberation of a large amount of energy (104 kcal/mole), so that at ordinary temperatures the reverse reaction, dissociation of hydrogen into atoms, which would require this amount of energy to occur, would be unobservable. Similarly, the reaction of an acid such as HCl with a strong base such as $OH^-$ would be essentially complete.

There are two characteristics of a reaction, such as the generalized one

$$A+B \rightleftharpoons C+D \tag{10}$$

that are of principal interest: one characteristic is the *extent* to which the reactants A and B are transformed into the products C and D; the second is the *rate* at which the reaction proceeds.

## 5.2  Equilibrium

The familiar types of equilibrium reactions encountered in the case of acids and bases have been discussed in Chapter 4. Many of the reactions of organic compounds, which involve the making and breaking of covalent bonds [e.g., (4), (5), and (9) in Section 5.1], are equilibrium reactions, and most of them proceed, or can be made to proceed, essentially to completion.

The equilibrium constant of reaction (10) can be defined by the expression

$$K = \frac{[C][D]}{[A][B]} \tag{11}$$

and the equilibrium constant is related to the free energy change in the reaction by the equation

$$-\Delta F = RT \log_e K \tag{12}$$

Thus, in a reaction in which energy is liberated (an *exergonic* reaction), $\Delta F$ is negative (or $-\Delta F$ is positive), and $K$ is greater than one. When $\Delta F$ is positive (in an *endergonic* reaction), $K$ is less than one. A graphical representation of the energy changes in reactions of several types is shown in Figure 5-1.

The equilibrium constant, $K$, varies with temperature; and although $K$ is a constant at a given temperature, the *extent* to which a reaction proceeds can be manipulated by alteration of temperature or concentrations of the reagents. An increase in the concentration of A [Equation (10)] will cause a greater proportion of B to be consumed in the reaction; conversely, an increase in the concentration of D will reduce the concentration of C that is present at equilibrium.

A familiar example of an equilibrium reaction in organic chemistry is the reaction of an alcohol and a carboxylic acid to give an ester; in the particular case of ethyl acetate the reaction is as follows:

$$\underset{\text{ethanol}}{CH_3CH_2OH} + \underset{\text{acetic acid}}{CH_3COOH} \rightleftharpoons \underset{\text{ethyl acetate}}{CH_3COOC_2H_5} + H_2O \tag{13}$$

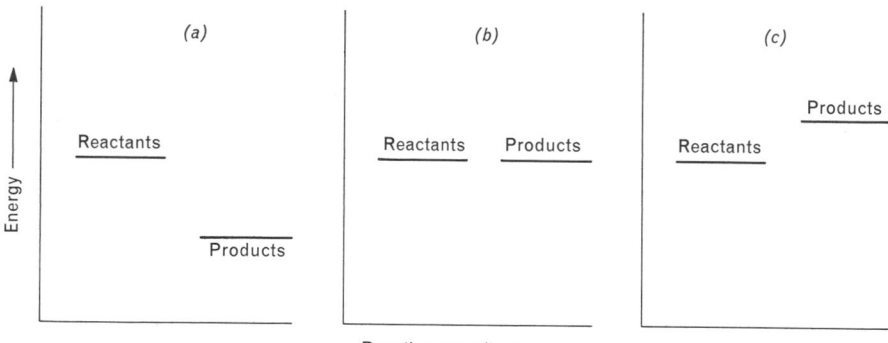

**Figure 5–1.** Energy change in reaction. *(a)* Exergonic, $K>1$; *(b)* $K=1$; *(c)* Endergonic, $K<1$.

The equilibrium constant for this reaction is expressed by the equation

$$K = \frac{[\text{ethyl acetate}][\text{H}_2\text{O}]}{[\text{ethanol}][\text{acetic acid}]} \tag{14}$$

If a mixture of $a$ moles of ethyl alcohol and $b$ moles of acetic acid is allowed to react, it is found that when equilibrium is attained there will be $x$ moles of the ester, $x$ moles of water, $a-x$ moles of ethanol, and $b-x$ moles of acetic acid present in the final mixture:

$$K = \frac{x^2}{(a-x)(b-x)} \tag{15}$$

A study of this reaction by Berthelot and St. Gilles showed that $K=4$. Thus, if 1 mole each of the alcohol and acid react, the final mixture will contain 0.667 mole of ester ($x$, $a-x$, and so on are concentrations, and so a constant volume is assumed when moles are specified). Table 5-1 shows the results of a series of

**Table 5-1.** *Equilibrium in ethanol-acetic acid reaction, at 25°C. Starting concentrations: Ethanol=x, Acetic Acid=1, Ester=0*

| Start; Acetic Acid=1; Alcohol, x | At Equilibrium | |
|---|---|---|
| | Ester (Obs.) | Ester (Calc.) |
| 0.10 | 0.05 | 0.05 |
| 0.50 | 0.41 | 0.42 |
| 1.00 | 0.66 | 0.67 |
| 2.00 | 0.86 | 0.85 |
| 8.00 | 0.97 | 0.97 |

experiments with varying concentrations of the alcohol.* It is clear from these data that if we wish to convert acetic acid into ethyl acetate, we can do this substantially quantitatively by using a large excess of ethanol. Conversely, if a large excess of acetic acid were used, practically all of the ethanol would be esterified. In general, when the organic chemist wishes to prepare an ester by this method he will use a large excess of the cheaper or less valuable reagent; if his primary concern is the conversion of an acid into its ester, he will use the alcohol as the solvent and thus in large excess. In this way, the esterification of the acid will be, for all practical purposes, complete.

Since the equilibrium constant varies with temperature (in the esterification reaction, the variation is small); and since volume changes often accompany chemical reactions, it is sometimes possible to shift an equilibrium in one or the other direction by changing the temperature or pressure. In the reaction

$$2NH_3 \rightleftharpoons N_2 + 3H_2 - heat \text{ (endothermic)} \tag{16}$$

an increase in pressure at constant temperature will diminish the degree of dissociation of ammonia; and an increase in temperature at constant pressure will increase the degree of dissociation.

The measurement of equilibrium constants in organic reactions is often an excellent way of studying the effect of structure upon the course of a reaction. For example, in the general case of addition to a carbonyl compound,

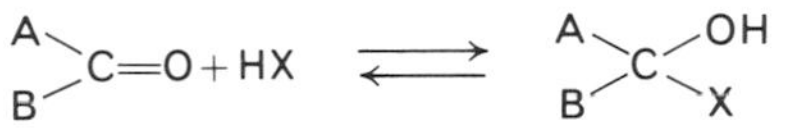

$$\begin{matrix} A \\ \diagdown \\ B \diagup \end{matrix} C{=}O + HX \rightleftharpoons \begin{matrix} A \\ \diagdown \\ B \diagup \end{matrix} C \begin{matrix} \diagup OH \\ \diagdown X \end{matrix} \tag{17}$$

the addition of HX to the trigonal carbonyl group alters the configuration around the carbon atom from trigonal (planar, 120° bond angles) to tetrahedral (bond angles 109°). The equilibrium constants of this reaction, carried out with a series of compounds containing different groups A and B, can be taken as an index of the relative stability of the two configurational states, and thus as a measure of the effect of the substituents A and B upon the relative stability of the two states. Some specific examples of this and of other uses of equilibrium data in studying chemical reactions will be given in later chapters.

## 5.3   Rates of reactions

The mathematical expressions for the equilibrium constant of a reaction do not contain terms involving time, and thus do not give information about how fast a reaction will proceed. Indeed, the rate of a given reaction can be varied over an enormous range by very small differences in reaction conditions. Hydrogen and oxygen combine to form water in a reaction for which the equilibrium constant is

---

* The course of a reaction such as the esterification reaction may be followed experimentally by withdrawing aliquot samples from time to time and determining the amount of acetic acid remaining by titration with standard alkali.

so large that the reaction may be said to go to completion. Yet a mixture of hydrogen and oxygen may remain unchanged for an indefinite period of time; but if a small amount of a catalyst (platinum) is added, or an electric spark is passed through the gases, an explosive reaction will occur. Similarly, the reaction of chlorine with methane

$$Cl_2 + CH_4 \longrightarrow CH_3Cl + HCl \tag{18}$$

is so slow as to be nearly undetectable if a mixture of the dry gases is kept in the dark; but upon exposure to sunlight immediate reaction occurs.

Reactions between ions, and proton transfers between acids and bases are usually very fast. Proton transfers are fast because the approach of the atom bearing an unshared electron pair is unimpeded because of the small size of the hydrogen atom and is aided by the fact that the hydrogen nucleus is not shielded by an external array of electrons. Free radical reactions, such as the combination of chlorine atoms to give molecular chlorine, are usually very fast because the bond that results is formed by the overlap of external orbitals, and all that is required for reaction to occur is that the atoms or radicals collide. In the case of complex radicals (such as $CH_3CHCH_3$, for example), a suitable orientation at collision is necessary, but this is a probability factor that is seldom important enough to slow the overall reaction to a marked degree.

When covalent molecules react with each other or with ions, reaction rates are often slow enough to be measured in the laboratory. The reaction of ethanol and acetic acid, discussed in Section 5.2, is an example of a reaction that proceeds at a rate that is easily measured, and that can be altered by the reaction conditions. Pure acetic acid and pure ethanol react quite slowly at room temperature, and the reaction requires many days to reach equilibrium. If a drop of sulfuric acid is added to the reaction mixture the reaction proceeds much more rapidly, and equilibrium may be reached in hours instead of days. This is discussed further in Section 8.16.

## 5.4   Activation energy and the transition state

A typical reaction in which an organic compound undergoes rupture of one bond and formation of a new one is the following:

$$HO^- + CH_3Br \rightleftharpoons HOCH_3 + Br^- \tag{19}$$

This reaction is a simple example of a large class of organic reactions which are important both because of their wide application in practical ways and because of the great amount of theoretical and experimental study that has been devoted to them. This is a typical example of a *nucleophilic displacement reaction*; it is so called because the attacking agent, hydroxide ion, is a nucleophile, and the reaction involves the formation of an oxygen-carbon bond with displacement of the bromide ion. In general terms, this reaction can be represented by

$$B: + CH_3:X \rightleftharpoons B:CH_3 + :X \tag{20}$$

where the symbol B: is used to represent a nucleophile (a base) and :X, also a nucleophile, is called the *leaving group* (or sometimes the *departing group*).

The reaction of methyl bromide with hydroxide ion proceeds at a measurable rate, and one can follow it experimentally by periodically withdrawing an aliquot portion of the solution and performing a suitable analysis—for instance, acidimetric titration to determine the OH⁻ concentration, or argentimetric titration for bromide ion concentration.

An examination of experimental data obtained in this way for reaction (19) permits two conclusions to be drawn:

1. The rate of the reaction (the rate of disappearance of OH⁻ ion or the rate of appearance of Br⁻ ion) is proportional to the concentration of both OH⁻ and $CH_3Br$. Doubling the concentration of either doubles the rate of the reaction

$$\text{rate} = k[OH^-][CH_3Br] \tag{21}$$

where $k$ is the *rate constant* for the reaction.

2. At given concentrations of the reactants, the rate of the reaction increases with increasing temperature.

The nature of this reaction and a description of the stages through which it proceeds are embodied in a general theory called the *transition state theory*. This theory presents a description of chemical reactions in which the reacting molecules collide to form an unstable combination called the *activated complex* or the *transition state*, after which this unstable complex decomposes in such a way as to yield either the components from which it was formed, or in another way to yield new compounds, the products of reaction.

The activated complex is a high energy state of the system, and is formed only when molecules with sufficient energy collide. A graphical representation of the course of a reaction that proceeds through the high-energy transition state is shown in Figure 5-2.

What is activation energy, and what is the activated complex? Why is the activated complex formed in some collisions and not in others? Why is the activated complex an unstable state of the system? In the following discussion we shall

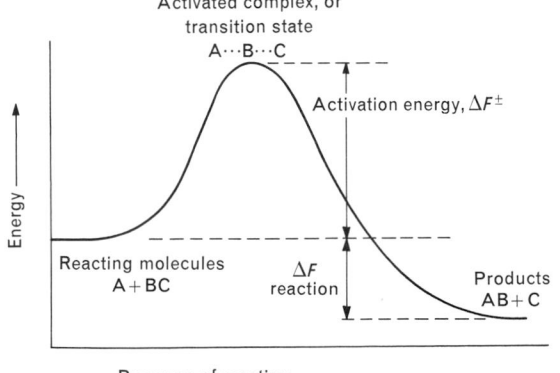

**Figure 5–2.** Energy diagram for reaction $A + BC \rightarrow AB + C$.

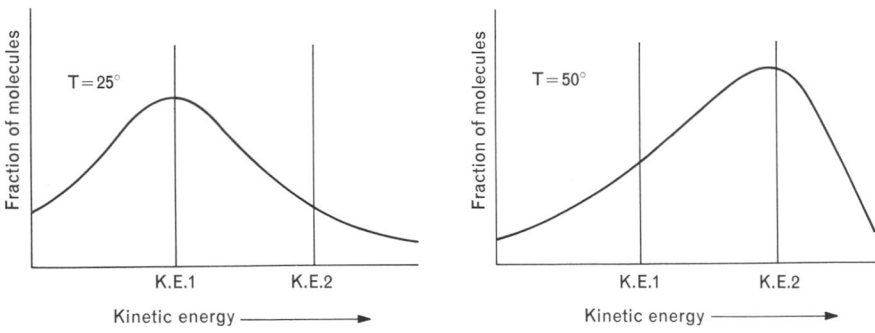

**Figure 5–3.** Distribution of energy among molecules and its temperature dependence.

trace the course of events that take place in the formation of methanol and bromide ion from the reaction of methyl bromide and hydroxide ion (19), and try to give answers to the above questions.

First of all, let us consider a system containing the reacting species, methyl bromide and hydroxide ion. The molecules of these substances are in motion; they move from point to point, collide, and bounce from one another, from the walls of the vessel, and from solvent molecules. By the operation of the laws of chance, they move at various speeds. Some have low kinetic energy, some high. The differences in energies can be described by the Boltzmann distribution law, and the proportion of molecules of low and high energies varies with temperature. As it is reasonable to assume, and, as is shown by the Boltzmann distribution function, the higher the temperature, the greater the number of molecules with high kinetic energy. A graph of a Boltzmann distribution function at two temperatures is shown in Figure 5-3; it can be seen that at the higher temperature the fraction of molecules with high kinetic energies is larger.

The collisions that occur between the hydroxide ions and the methyl bromide molecules may or may not lead to a chemical reaction. Collision must occur in such a way that the orbital overlap necessary for bond formation can occur, and thus collision must take place with proper orientation; and it must take place with sufficient force to permit sufficient interpenetration of the two substances to bring them within overlap distance. Reaction can occur when an $OH^-$ ion with sufficient energy collides with a $CH_3Br$ molecule in such a way that a molecular orbital can begin to form between the carbon atom and the unshared pair of electrons on oxygen. Various possible modes of collision are pictured in Figure 5-4; of these, that marked (*c*) places the two molecules in a position to enter into a reaction.

But the four carbon orbitals in methyl bromide are filled, the octet is complete, and there appears to be no way in which a new bond can form. How can the hydroxide ion begin to engage in bonding with the "saturated" carbon atom? It will be recalled that in our earlier discussion of covalent bonds it was pointed out that bonds between unlike atoms are polar in character, with the electron pair closer to the more electronegative atom (Section 3.12). The molecular orbital that

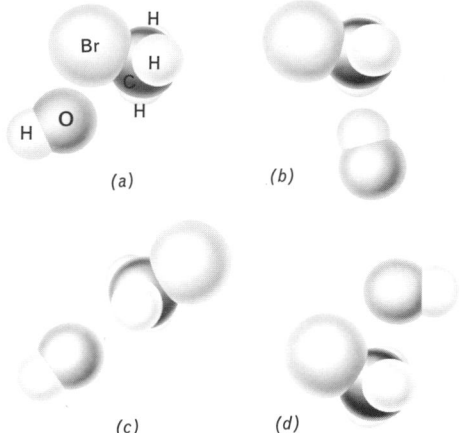

(a)

(b)

(c)

(d)

**Figure 5-4.** Various possible aspects of collision between OH⁻ and CH₃Br molecules.

constitutes the carbon-bromide bond is not only distorted so that there is a greater share of the electron density at the bromine atom, but the carbon atom, because of its partial positive character, has some bonding capacity of its own. This can be represented by sketching the shape of the carbon orbitals as in Figure 5-5, where a lobe of the $sp^3$ orbital of the C—Br bond extends in the direction away from this bond. The sequence of events that leads to reaction is pictured in Figure 5-5: the high-energy hydroxide ions strike the methyl bromide molecule (or the latter may be the energetic missile) with sufficient velocity to overcome the repulsive forces of the hydrogen nuclei: the orbital on oxygen that contains an unshared pair of electrons overlaps with the residual lobe of the carbon-bromine bond orbital; a very unstable (high energy) complex, in which OH⁻ has preempted somewhat more of the carbon orbital and bromine has relinquished some, forms; and finally either the OH⁻ ion recoils, or the bromine is expelled as the anion.

The decrease in energy in the overall process ($\Delta F$ reaction, Figure 5-1) is such that the equilibrium in the reaction (19) will lie far to the right. This means that the transition state will decompose into $CH_3OH$ and $Br^-$ more often than it will revert to OH⁻ and $CH_3Br$.

It is now apparent why an increase in concentration of OH⁻ or $CH_3Br$ will increase the rate of the formation of $CH_3OH$: at higher concentrations of the reactants there is a higher probability of fruitful collisions to form the activated complex. The effect of temperature is also understandable, since at higher temperatures there are relatively more molecules with sufficient energy to collide effectively to form the activated complex.

Most organic reactions proceed through a transition state, the properties of which are seldom known in detail, because the activated complex is a transitory species that cannot be isolated. Nevertheless, it is usually possible to write a structure that is a reasonable approximation of what the transition state is like, and to use this hypothetical model as a basis for a consideration of its properties. The structures of transition states that will be written in later discussions in this text

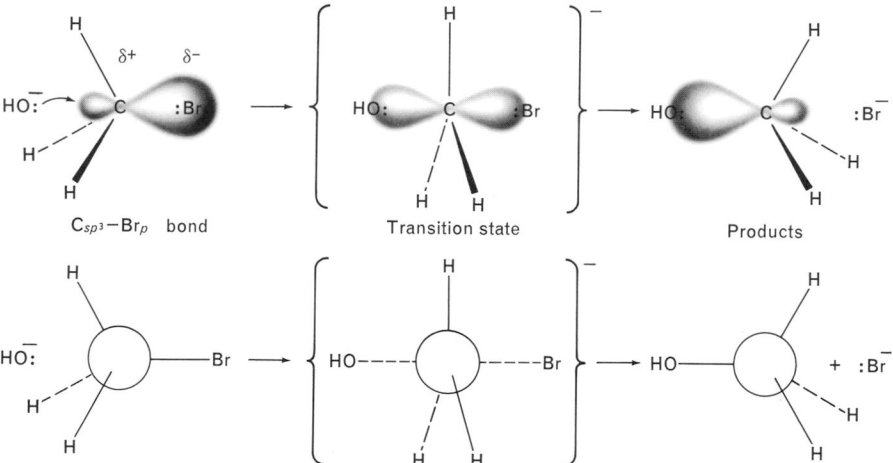

**Figure 5–5.** Stages in the reaction of methyl bromide with hydroxide ion. Note that in the transition state pictured, the three C—H orbitals are $sp2$ in type. The fourth orbital, a $p$-orbital of carbon, cannot contain more than two electrons; thus, the HO—C bond and the C—Br bonds cannot be full bonds. This can be expressed by saying that the oxygen and bromine atoms are partially bonded to carbon in the transition state.

will be of this kind: reasonable structural assignments whose value will be in their usefulness in helping one to reach an understanding of a reaction.

### 5.5  Competing reaction paths

It is frequently possible to devise, *a priori*, two reasonable and possible paths by which a reaction might proceed to yield a given product; and in some cases it is possible for a given pair of reactants to react in two different ways to lead to two different products. Often a consideration of the probable transition states for the possible reactions permits a decision as to the preferred course for the reaction. If the products of two competing reaction courses are of comparable structure and energy content, the preferred reaction will be that whose transition state is at the lower energy level—that is, whose transition state is the most stable (Figure 5-6).

The reason for the preference for the lower energy transition state is found in a consideration of the expression for the dependence of reaction rate upon the energy of the transition state.

$$\text{rate constant, } k = Ce^{-(A/RT)} \tag{22}$$

where $C$=a constant relating to effective collision frequency; $A$=activation energy; $R$=gas constant; $T$=temperature. Since the activation energy is exponential in this expression, small changes in its value have profound effects upon the value of $k$; and, more important to our present purpose, an increase in the value of

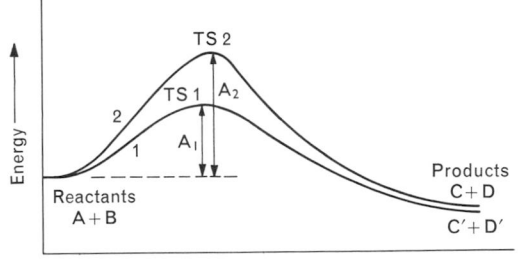

**Figure 5–6.** Energy profile of the reaction of A+B to yield different products, by way of transition states, 1 and 2, with activation energies $A_1$ and $A_2$.

$A$ causes a decrease in $k$. A simple numerical calculation will illustrate this point. Consider two reactions with activation energies of (1) 20,000 cal/mole and (2) 22,000 cal/mole.*

$$\frac{k_A}{k_B} = \frac{e^{-20,000/600}}{e^{-22,000/600}} = \frac{e^{-33.3}}{e^{-36.7}}$$

$$\frac{k_A}{k_B} = e^{3.4} = 30$$

Thus, the reaction rate is increased by a factor of 30 by a reduction in activation energy of only about ten percent.

In the hypothetical case for which the energy profiles are shown in Figure 5-6, the reaction that proceeds through transition state TS 1 to products $C'$ and $D'$ would proceed much faster than the alternative reaction through TS 2 to products C and D. If $A_2$ were 22,000 cal/mole and $A_1$ were 20,000 cal/mole, the ratio of products would be $C'+D'/C+D=30$.

## 5.6 Multistage reactions

The displacement reaction (19) proceeds through a single transition state and thence to the products. Although a great many organic reactions are of this kind, many others proceed through a sequence of intermediate stages, with the formation of one or more discrete intermediates which can usually be assigned specific structures. Intermediates in a reaction may be of two kinds: *unstable* (that is, high-energy) *intermediates* whose existence is transitory and can only be inferred from indirect experimental evidence; *stable intermediates*, which can often be isolated from the reaction mixture. Energy profiles for these various reaction courses are shown in Figures 5-7 and 5-8. In Figure 5-7a the reactants combine to form the first transition state (TS 1), which then proceeds to the intermediate product whose

---

\* In the units used in this calculation, $RT$ is approximately 600 cal/mole.

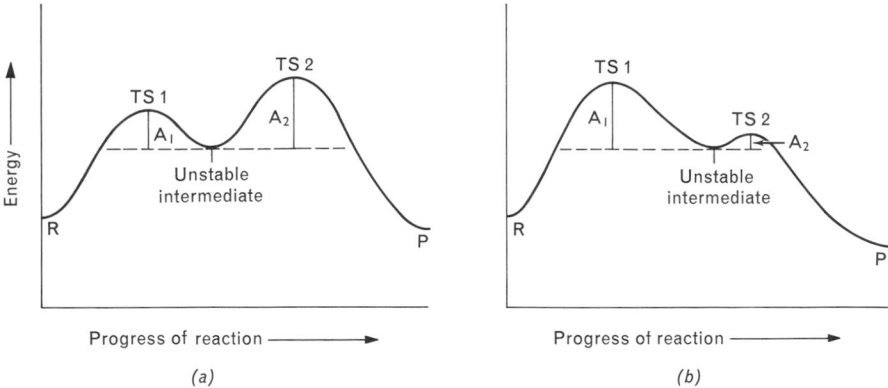

(a)                                        (b)

**Figure 5–7.** Energy profiles for reactions involving two transition states and an unstable intermediate. R=reactants; P=products: A=activation energies. *(a)* Second step is rate determining, $A_2 > A_1$. *(b)* First step is rate determining, $A_1 > A_2$.

energy state is defined by the minimum in the curve. This intermediate then takes part in a second reaction, distinct from the first, leading through a second transition state (TS 2) to the final products of the reaction. Since transition state (TS 1) is at a lower energy than TS 2, the reaction rate for the process R (reactants) → P (products) will be governed by the (slower) rate of the second step, and this step is called the *rate-determining step* of the overall reaction.

Figure 5.7b shows the energy profile of a two-step reaction in which the first step is rate determining.

Figure 5-8 shows the energy profile for a two-stage reaction in which a stable intermediate is formed. The energy difference between the reactants (R) and the intermediate (I) shows that the equilibrium R ⇌ I lies on the side of the intermediate. The step, I ⇌ products, is a separate reaction, proceeding through a second transition state, and with its own equilibrium constant. Whether or not the intermediate accumulates in the reaction mixture will depend upon the rate of its conversion into the final products and thus upon the relative activation energies of the two steps of the reaction.

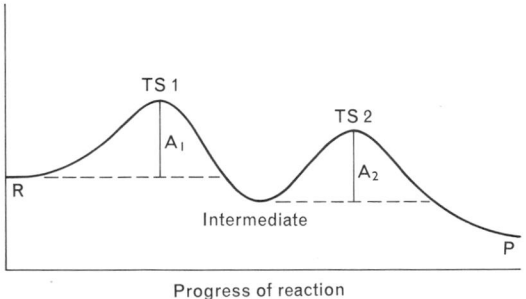

Progress of reaction

**Figure 5–8.** Energy profile for a two-step reaction in which a stable intermediate is formed.

### 5.7 Alternative reaction paths

The reaction between methyl bromide and hydroxide ion can be conceived of as proceeding in another way than that described in the foregoing discussion. We might formulate it as follows:

$$\text{Step 1:} \quad CH_3Br \longrightarrow CH_3^+ + Br^- \tag{23}$$

$$\text{Step 2:} \quad CH_3^+ + OH^- \longrightarrow CH_3OH \tag{24}$$

Unless there were some stabilizing influence acting to promote the formation of the $CH_3^+$ ion and to stabilize it, step 1 would be expected to require a very high energy indeed, for the following reasons. The $CH_3$—Br bond energy is 68 kcal/ mole, which is the energy required for process (25)—*homolytic* scission:

$$CH_3Br \longrightarrow CH_3 \cdot + Br \cdot \tag{25}$$

The energy required for step 1 (23)—*heterolytic* scission—which involves not only C—Br bond breaking but also the separation of two opposite charges, would be expected to be greater than 68 kcal/mole. The observed activation energies for reactions of the type given in (19) are, however, in the region of 20 kcal/mole. In Figure 5.9 are shown two energy profiles for a reaction, formulated in the general terms

$$A: + CH_3:X \rightleftharpoons A:CH_3 + :X, \tag{26}$$

for two reaction pathways: the dissociation-recombination path described by Equations (23) and (24); the second-order displacement reaction via the bimolecular transition state HO---$CH_3$---Br.

It will be obvious from this energy diagram that the reaction, with an experimental activation energy of around 20 kcal/mole, cannot proceed by the first pathway (Figure 5-9), which would involve an energy of activation (required for the heterolytic bond cleavage) of more than 70 kcal/mole.

The general displacement reaction of the type

$$B:^- + R:X \longrightarrow B:R + :X^-, \tag{27}$$

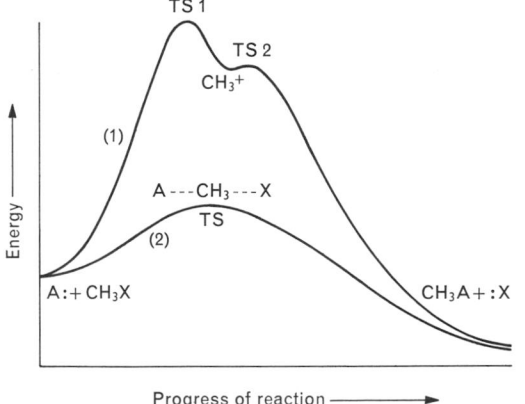

**Figure 5–9.** Energy profile for two hypothetical paths for the reaction

$$A: + CH_3:X \longrightarrow A:CH_3 + :X$$

(1) Heterolytic dissociation-recombination. (2) Bimolecular displacement.

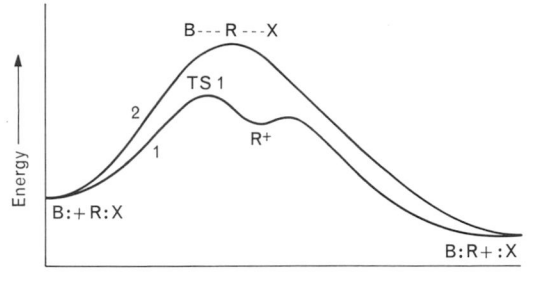

**Figure 5–10.** Energy profile for two hypothetical paths for the reaction $B: + R:X \to B:R+ :X$, in which structural factors stabilize $R^+$ (path 1) and increase the energy of the bimolecular transition state (path 2).

where B: is a nucleophile, R is an alkyl group, and $:X^-$ is a leaving group such as a halide ion, will be discussed more in Chapter 7. It will be seen there that in some cases, where R is other than $CH_3$, the reaction does indeed proceed by the route exemplified by path (1) in Figure 5-9. When these conditions obtain, the transition state for course (1) may be at a lower energy than that for course (2). Under such circumstances, a change in mechanism, caused by structural factors, occurs, although the overall reaction remains the same. An energy profile for a displacement that does in fact proceed through an ionization step to give $R^+$ as an unstable intermediate, is shown in Figure 5-10.

It may be concluded from the foregoing discussion that it is quite possible for structural effects to increase the rate of a reaction by one path and to decrease the rate of reaction by another so that under certain circumstances the reaction proceeds by both pathways simultaneously or perhaps by a pathway that cannot be described accurately by either of the two. This is quite true, and many organic reactions proceed either to give a given product by more than one mechanistic pathway, to give more than one product by two or more pathways, or to give two or more products by a pathway that leads through a single rate-determining step and then diverges into more than one second pathway. Examples of these various reaction courses will be examined in the chapters to follow.

# Chapter six / The Naming of Organic Compounds

Clarity and precision in the use of language are essential to effective communication of ideas in any field of human activity, and an important part of the language in any well-defined discipline is the terminology that is used to identify the materials and concepts that are dealt with by its practitioners. In chemistry, and particularly in organic chemistry, the communication of information is largely carried out with use of the formulas of chemical substances. These can be written in chemical symbols or they can be translated into names that are clear equivalents of the structural formulas. The proper and comprehensible naming of organic compounds is an important preliminary to a discussion of their chemical behavior. Names should first of all be unambiguous: a name should not be used unless it refers to one specific substance, and no other. Names should be simple; but when there is, as is often the case, a choice between two or more names for a certain compound, the selection of one of these—though not always the simplest name—can sometimes aid a discussion by evoking the clearest mental image of the substance.

The chemist, often faced with the necessity for choosing among several possible names, all correct and unambiguous, for a single compound, must be familiar with a variety of methods for arriving at a proper name. He must also develop the ability and judgment to select that name which best suits the purpose of his discussion and the requirements for clarity and conformity to accepted practice.

There are three general ways of naming: organic compounds may be given *trivial, derived,** or *systematic* names.

---

\* *Derived* names are largely *substitution* names, and are often so called.

## 6.1   Trivial names

*Trivial*, or common, names are usually arrived at by reference to a source from which a compound is derived or obtained, or they may refer to a characteristic property of the compound. Trivial names are usually of long standing, frequently having been adopted before systematic rules for naming were in general use. Trivial names are assigned to many compounds isolated from natural sources, as a means of referring to them before their structures have been determined, and they often persist and come to occupy a permanent place in the literature. Finally, trivial names are used for many compounds whose structures are so complex that any systematic name would be unwieldy and not readily comprehensible (e.g., quinine, morphine, glucose, cortisone).

Such simple compounds as oxalic acid (*Oxalis*), allyl alcohol (*Allium*), menthane (*Mentha*), tiglic acid (*tiglium*, the specific part of the name *Croton tiglium*), and malic acid (*Pyrus malus*, the apple) have names that are botanical in origin (the terms given in parentheses). Other compounds bear names that refer to other sources: uric acid (urine); hippuric acid (horse + urine); squalene (*Squalus*, shark); testosterone (testis); civetone (civet, a glandular secretion of the civet cat); sebacic acid (*sebum*, the fatty secretion in skin) are some examples. Some names refer to distinctive properties that characterize the compounds: acrolein (Latin *acris*, pungent); chrysene (Greek *chrysos*, golden); glucose (Greek *glucose*, sweet) refer to attributes of odor, color, and taste.

Trivial names have a firm place in the literature and language of organic chemistry. They cannot be learned systematically, but become familiar to the chemist and are added to his vocabulary through continued association.

An important class of names that are in part trivial and in part derived are those formed by alteration of the names of related compounds. For example, the pyrolysis of barium and certain other heavy metal salts of acids is a general method of preparing ketones:

$$(CH_3COO)_2Ba \xrightarrow{\Delta} CH_3COCH_3 + BaCO_3$$
$$\text{barium acetate} \qquad\qquad \text{acetone}$$

The name *acetone* is derived from *acetic* acid, the ending *-one* being generally used to designate the grouping $-\overset{|}{C}{=}O$ in *ketones*.

There is an increasing tendency among modern-day chemists to discontinue the use of trivial names other than those of very commonly encountered or very complex substances. For example, whereas the name acetone is still, and will continue to be, applied to $CH_3COCH_3$, the name *n*-butyrone would seldom be used for $CH_3CH_2CH_2COCH_2CH_2CH_3$; this is preferably called 4-heptanone (a systematic name, as discussed later); and isobutyrone, $(CH_3)_2CHCOCH(CH_3)_2$, is called 2,4-dimethyl-3-pentanone, or diisopropyl ketone.

A type of trivial name that will always be useful is exemplified by such names as methyl bromide ($CH_3Br$), isopropyl alcohol [$(CH_3)_2CHOH$], and ethylene dibromide ($BrCH_2CH_2Br$). In names of this kind it is often difficult to draw the line between a systematic and a trivial name. For example, isobutyl alcohol and allyl alcohol are both names of the same type,

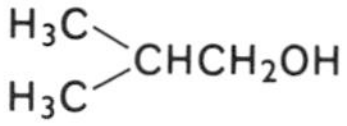

$$H_3C$$
$$\diagdown CHCH_2OH \qquad\qquad CH_2{=}CHCH_2OH$$
$$H_3C\diagup$$

isobutyl alcohol                       allyl alcohol

but the first is recognizable at once because it includes a systematic name for the group $(CH_3)_2CHCH_2$—, whereas the second requires a knowledge of the meaning of a term (allyl) that has no systematic relationship to any organized system of naming.

## 6.2  Substitution names

*Substitution* names are derived in a logical and systematic manner and are readily understood because they depend upon a recognition of a relatively few familiar common names. Substitution names of complex compounds are formed by regarding the substance to be named as having been derived from a simpler compound by the replacement of hydrogen atoms by other groups or atoms. As illustrations, two systems—the *methane* and the *carbinol* systems—will be considered in some detail.

Before we can do this, however, we must have a stockpile of group names to draw from. It is necessary to know the names of a few of the smaller alkyl groups, but the list is not a long one. This is so because, rather than employing complex trivial designations, resort is had to systematic names. Thus, we find the following *alkyl groups* an adequate starting point from which to form many names.

| | | |
|---|---|---|
| methyl | $CH_3$— | from methane ($CH_4$) |
| ethyl | $CH_3CH_2$— | from ethane ($C_2H_6$) |
| *n*-propyl | $CH_3CH_2CH_2$— | |
| isopropyl | $CH_3$—$\underset{\underset{CH_3}{\vert}}{CH}$— | from propane ($CH_3CH_2CH_3$) |
| *n*-butyl | $CH_3CH_2CH_2CH_2$— | |
| isobutyl | $\overset{CH_3\diagdown}{\underset{CH_3\diagup}{}}CHCH_2$— | |
| *sec*-butyl | $\underset{CH_3-CH_2}{\overset{CH_3-CH-}{\vert}}$ | from *n*-butane ($CH_3CH_2CH_2CH_3$) and isobutane ($CH_3$—$\underset{\underset{CH_3}{\vert}}{CH}$—$CH_3$) |
| *tert*-butyl | $CH_3-\underset{\underset{CH_3}{\vert}}{\overset{\overset{CH_3}{\vert}}{C}}$— | |

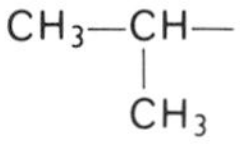

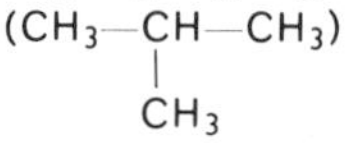

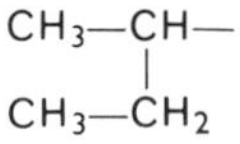

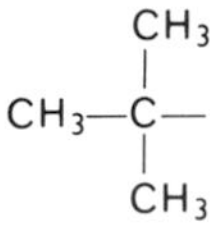

In general, an unbranched chain that carries $-CH\begin{smallmatrix} CH_3 \\ CH_3 \end{smallmatrix}$ at the end is an *iso* group.

$-CH_2CH_2CH_2CH\begin{smallmatrix} CH_3 \\ CH_3 \end{smallmatrix}$

isohexyl

$-CH_2CH_2CH_2CH_2CH\begin{smallmatrix} CH_3 \\ CH_3 \end{smallmatrix}$

isoheptyl

In *iso* groups above butyl, the name is derived from the Greek or Latin root (iso-*pent*yl, 5; iso*hept*yl, 7; iso*non*yl, 9, and so forth) corresponding to the total number of carbon atoms in the group.

In addition to these names for the lower alkyl groups, there are a number of other, often trivially derived, group names that are frequently used as part of substitution names. A few of the more common of these are given in the following list:

| | | | | |
|---|---|---|---|---|
| $CH_2=CH-$ | vinyl | as in | $CH_2=CHCl$ | vinyl chloride |
| $CH_2=CHCH_2-$ | allyl | as in | $CH_2=CHCH_2CH_2COOH$ | allylacetic acid |
| ⬡— | phenyl | as in | ⬡$-CH_2COCH_3$ | phenylacetone |
| $R_2N-$ | dialkylamino | as in | $(CH_3)_2NCH_2CH_2OH$ | 2-dimethyl-aminoethanol |

### The methane system of naming.

In the *methane system* of naming, the compound to be named is regarded as a derivative of methane:

$$H-\overset{\overset{\displaystyle H}{|}}{\underset{\underset{\displaystyle H}{|}}{C}}-H \qquad H-\overset{\overset{\displaystyle H}{|}}{\underset{\underset{\displaystyle H}{|}}{C}}-CH_3 \qquad CH_3-\overset{\overset{\displaystyle H}{|}}{\underset{\underset{\displaystyle H}{|}}{C}}-CH_3 \qquad CH_3-CH_2-CH\begin{smallmatrix} CH_3 \\ CH_3 \end{smallmatrix}$$

methane   *methylmethane*   *dimethylmethane*   *methyl isopropylmethane*

$(CH_3)_2C(CH_2CH_3)_2$
*dimethyldiethylmethane*

$(BrCH_2)_4C$
tetra(bromomethyl)methane

$[(CH_3)_2CH]_3CCH_3$
methyltriisopropylmethane

$(C_6H_5)_3CH$
triphenylmethane

In the above examples all of the names are correct because they are unambiguous and properly derived, but those in italics would seldom be used; other, usually simpler, names are preferable for these compounds, as will be shown in the discussion to follow.

### The carbinol system of naming.*

The *carbinol system* for naming alcohols is entirely comparable to the methane system for hydrocarbons. In this system, alcohols are regarded as substitution products of the simplest member of the series,

---

* In *Chemical Abstracts* the carbinol names are indexed under "methanol." Thus, "triphenylcarbinol" is "triphenylmethanol."

$CH_3OH$. This alcohol is called carbinol for the purposes of the carbinol method of naming, although the name always used for it, as a substance, is methanol or methyl alcohol.

Ethanol, $CH_3$—$CH_2OH$, is thus *methylcarbinol*; isopropyl alcohol, $(CH_3)_2CHOH$, is *dimethylcarbinol*. Further examples are shown in the following:

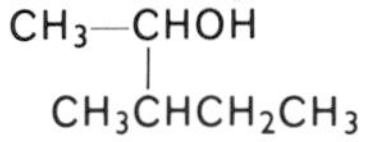

methyl-*sec*-butylcarbinol

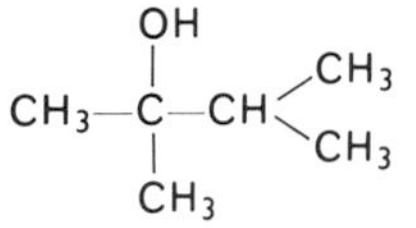

dimethylisopropylcarbinol

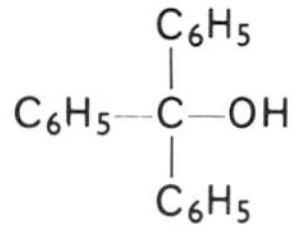

triphenylcarbinol

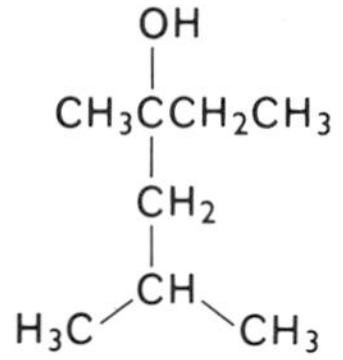

methylethylisobutylcarbinol

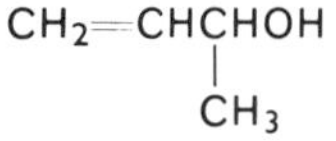

methylvinylcarbinol

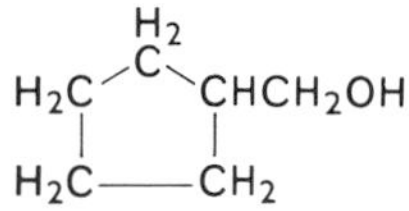

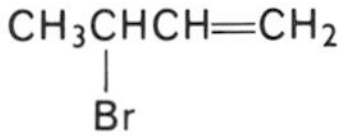

cyclopentylcarbinol

Carbinol names are not usually applied to alcohols for which common names are familiar or for which systematic names are not unwieldy:

$CH_2$=$CHCH_2OH$

allyl alcohol
(*vinylcarbinol* not used)

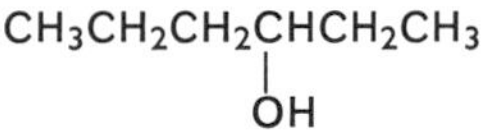

$CH_3CH_2CH_2CHCH_2CH_3$ with $OH$

3-hexanol
(*ethyl-n-propylcarbinol* not used)

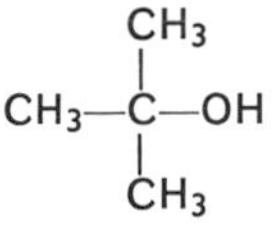

tertiary-butyl alcohol
(*trimethylcarbinol* not used)

The group corresponding to the alcohol for which a carbinol name is used is often called by the corresponding name ending in *carbinyl*:

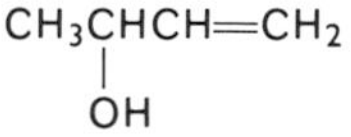

$CH_3CHCH$=$CH_2$ with $OH$

methylvinylcarbinol

$CH_3CHCH$=$CH_2$ with $Br$

methylvinylcarbinyl bromide

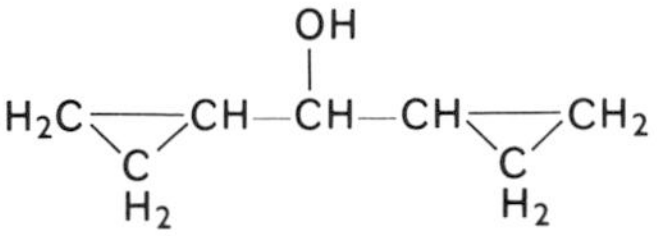

dicyclopropylcarbinol

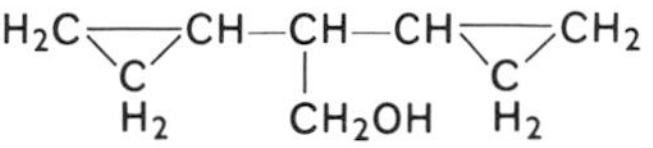

dicyclopropylcarbinylcarbinol

---

**Exercise 1.**   Write structural formulas for cyclopropylcarbinol, methylvinylcarbinyl chloride, diallylcarbinol.

***Other substitution names.***   A substitution name can be derived from the name of any simpler compound, provided that it is possible to number or otherwise designate the substituted carbon atom in an unambiguous way. Many compounds that are halogen derivatives are named in this way. For example, $ClCH_2COOH$ is chloroacetic acid, $BrCH_2CH_2COOH$ is $\beta$-bromopropionic acid, $CH_3CHCOOH$
            |
            Br

is $\alpha$-bromopropionic acid. The use of Greek letters is a convenient and generally useful way of designating the position of substitution. For example, in *n*-valeric acid the carbon atoms are designated as follows:

$$\overset{\delta}{C}H_3\overset{\gamma}{C}H_2\overset{\beta}{C}H_2\overset{\alpha}{C}H_2COOH$$

It is to be noted that the —COOH group is not given a letter, since no substitution can occur on that carbon atom (there is no replaceable hydrogen atom attached to carbon). Thus, the following names are proper:

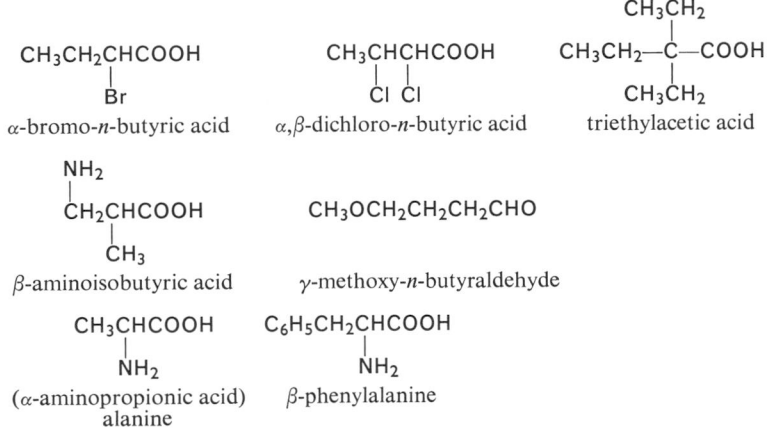

$CH_3CH_2CHCOOH$
        |
        Br
$\alpha$-bromo-*n*-butyric acid

$CH_3CHCHCOOH$
       |   |
       Cl  Cl
$\alpha,\beta$-dichloro-*n*-butyric acid

$CH_3CH_2$
         |
$CH_3CH_2$—C—COOH
         |
$CH_3CH_2$
triethylacetic acid

$NH_2$
  |
$CH_2CHCOOH$
       |
       $CH_3$
$\beta$-aminoisobutyric acid

$CH_3OCH_2CH_2CH_2CHO$

$\gamma$-methoxy-*n*-butyraldehyde

$CH_3CHCOOH$
      |
      $NH_2$
($\alpha$-aminopropionic acid)
alanine

$C_6H_5CH_2CHCOOH$
            |
            $NH_2$
$\beta$-phenylalanine

---

**Exercise 2.**   Write formulas for the following: $\alpha,\gamma$-dibromo-*n*-butyric acid; trimethylacetic acid; $\alpha$-phenylalanine; $\beta$-chloroisobutyric acid.

---

It should be said at this point that the examples given have been chosen primarily for their value in illustrating how names are formed. It will be recognized that for many compounds there are at least two names possible. The choice of the best name is a matter that must be decided on the basis of judgment that comes only with experience and familiarity. For any rule that might be given, exceptions could be cited. The best way to learn how to name organic compounds properly is to learn the *methods* of naming and to regard rules as guides but not always as rigid requirements. The strict observance of general rules is most important in the system of nomenclautre next to be discussed: the *systematic* methods of naming organic compounds.

## 6.3 Systematic names. The International Union of Chemistry (I.U.C.) system

In 1892 the rapid growth of organic chemistry had brought the number of known compounds to a very large figure, and more and more new substances were being reported year after year in the chemical literature. In order to stem what could shortly have become a monumental confusion in which trivial names could multiply beyond man's ability to comprehend them, and duplication of names for the same compound could cause costly waste of effort, a move was taken to bring some order into chemical nomenclature. The International Chemical Congress, meeting at Geneva in 1892, established a system of nomenclature, based upon simple and widely applicable principles, which was adopted internationally and became the accepted basis for naming organic compounds. The system was modified and improved in subsequent meetings of the International Union of

**Table 6-1.** *The normal paraffin hydrocarbons (normal alkanes) and group names used in the I.U.C. system*

| Hydrocarbon | | Formula | Alkyl Group | Formula |
|---|---|---|---|---|
| $CH_4$ | methane | $CH_4$ | methyl | $CH_3-$ |
| $C_2H_6$ | ethane | $CH_3CH_3$ | ethyl | $CH_3CH_2-$ |
| $C_3H_8$ | propane | $CH_2CH_2CH_3$ | *n*-propyl | $CH_3CH_2CH_2-$ |
| | | | isopropyl | $CH_3CHCH_3$ |
| | | | *n*-butyl | $CH_3CH_2CH_2CH_2-$ |
| | | | isobutyl | $CH_3 \backslash CHCH_2-$ / $CH_3$ |
| $C_4H_{10}$ | butane | $CH_3CH_2CH_2CH_3$ | *sec*-butyl | $CH_3CH_2CHCH_3$ |
| | | | *tert*-butyl | $CH_3 \overset{CH_3}{\underset{CH_3}{-C-}}$ |
| $C_5H_{12}$ | pentane | $CH_3CH_2CH_2CH_2CH_3$* | | |
| $C_6H_{14}$ | hexane | $CH_3(CH_2)_4CH_3$ | | |
| $C_7H_{16}$ | heptane | $CH_3(CH_2)_5CH_3$ | | |
| $C_8H_{18}$ | octane | $CH_3(CH_2)_6CH_3$ | | |
| $C_9H_{20}$ | nonane | $CH_3(CH_2)_7CH_3$ | | |
| $C_{10}H_{22}$ | decane | $CH_3(CH_2)_8CH_3$ | | |
| $C_{11}H_{24}$ | undecane | $CH_3(CH_2)_9CH_3$ | | |
| $C_{12}H_{26}$ | dodecane | $CH_3(CH_2)_{10}CH_3$ | | |
| $C_{20}H_{42}$ | eicosane | $CH_3(CH_2)_{18}CH_3$ | | |

* It is not often necessary to use the names of alkyl groups larger than butyl, since the best name will usually be formed without the necessity for doing so.

Chemistry (whence the term I.U.C. system). We shall refer to it as the *I.U.C. system*, and shall call names that follow its rules *I.U.C. names.*\*

*The basis for I.U.C. names is the series of paraffin hydrocarbons that begins with methane and proceeds by increments of —CH₂ to longer chains.* Table 6-1 presents the lower members of this series and a few representatives of higher members.

The saturated hydrocarbons are called *alkanes*, and have the general formula $C_nH_{2n+2}$. The groups corresponding to them are *alkyl* groups. When the group is a straight chain and the position of substitution is at the end, it is a normal (*n*) alkyl group, but in the I.U.C. system the prefix *n* is not used, it being understood that *in the absence of a prefix the normal group is meant.* The use of the prefix *iso* has already been described.

## 6.4  Naming of saturated hydrocarbons

The application of the I.U.C. system is exemplified by the naming of alkanes. Branched-chain hydrocarbons are regarded as derivatives of the *normal* hydrocarbons. Thus, the I.U.C. names are a special kind of substitution name. *The names of alkanes are derived by selecting the longest straight chain of carbon atoms, numbering it from one end to the other and naming the groups attached to this chain, designating them by the numbers of the carbon atoms to which they are attached.* This rule is the essence of the I.U.C. system, and should be fixed firmly in mind; it forms the basis for the naming of compounds that are not alkanes, as will be seen in the following discussion. The following examples should be analyzed with care.

*Example 1*

As shown in Table 6-1, the basic hydrocarbons are the *normal* alkanes, $CH_4$ (methane) $CH_3CH_3$ (ethane); $CH_3(CH_2)_nCH_3$ ($n=1$, propane; $n=2$, butane; $n=8$, decane, and so forth) The names butane, hexane, nonane, and so on are the *I.U.C. names* for the *normal* hydrocarbons.

*Example 2*

$CH_3CHCH_3$ is 2-methylpropane. The longest chain is C—C—C, so the name is derived
        |
      $CH_3$

from the name of the corresponding *n*-alkane, propane. The chain is numbered $\overset{1}{C}-\overset{2}{C}-\overset{3}{C}$; thus 2-methylpropane. This compound is *a butane*, but this is a description of its class, not a name.

*Example 3*

            $CH_3$
            |
$CH_2CH_2CCH_3$ is 2,2-dimethylbutane. When a carbon atom carries two like substituents,
            |
            $CH_3$

the number is repeated (2,2-dimethylbutane, not 2-dimethylbutane).

---

\* Later revisions and modifications of the system of organic nomenclature have been made, and continue to be made when necessary, by the International Union of Pure and Applied Chemistry. For this reason, the modern system of chemical nomenclature is often referred to as the IUPAC system.

*Example 4*

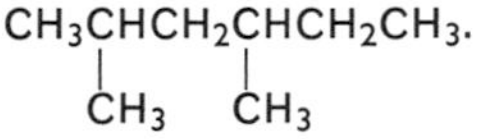

$CH_3CHCH_2CHCH_2CH_3$. The longest chain has six carbon atoms, thus the compound
    |      |
   $CH_3$   $CH_3$

is derived from hexane. The chain is numbered so that the substituents (the two methyl groups) have the lowest possible numbers:

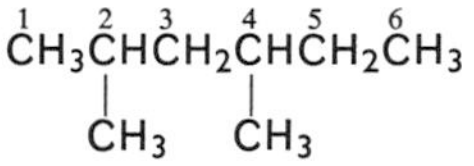

$\overset{1}{C}H_3\overset{2}{C}H\overset{3}{C}H_2\overset{4}{C}H\overset{5}{C}H_2\overset{6}{C}H_3$       *not*       $\overset{6}{C}H_3\overset{5}{C}H\overset{4}{C}H_2\overset{3}{C}H\overset{2}{C}H_2\overset{1}{C}H_3$
   |     |                               |     |
  $CH_3$   $CH_3$                        $CH_3$   $CH_3$
  2,4-dimethylhexane               3,5-dimethylhexane

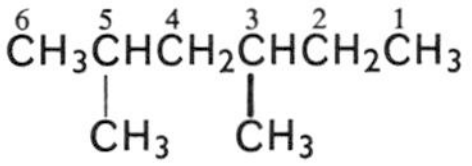

*Example 5*

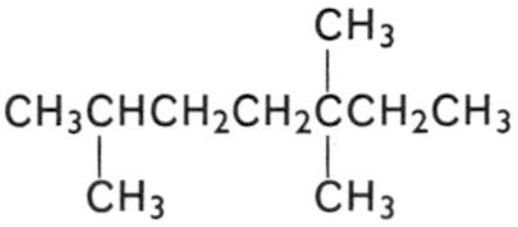

                $CH_3$
                |
$CH_3CHCH_2CH_2CCH_2CH_3$   is   2,5,5-trimethylheptane   (not  3,3,6-trimethylheptane).
    |        |
  $CH_3$     $CH_3$

Although the sum of 2,5,5- and 3,3,6- is the same, 2,5,5- is chosen because it contains the smaller number (2).

*Example 6*

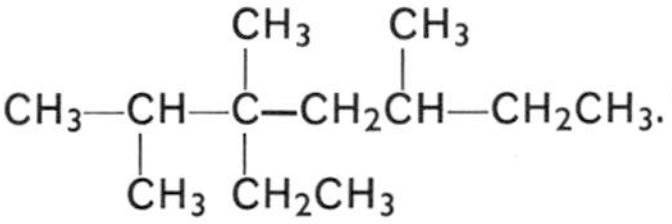

           $CH_3$    $CH_3$
           |       |
$CH_3$—$CH$—$C$—$CH_2CH$—$CH_2CH_3$. Two names, both of which are based on heptane, are
         |     |
        $CH_3$  $CH_2CH_3$

possible:

    3-ethyl-2,3,5-trimethylheptane
    3-isopropyl-3,5-dimethylheptane

The first of these is preferred, since it makes use of less complex substituents (ethyl instead of isopropyl). Although isopropyl, isobutyl, *sec*-butyl, *tert*-butyl are now allowed as group names in the I.U.C. system, they are ordinarily not used when a simpler alternative is available. See Example 7.

*Example 7*

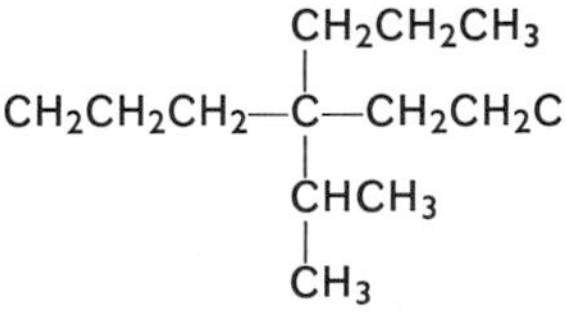

              $CH_2CH_2CH_3$
              |
$CH_2CH_2CH_2$—$C$—$CH_2CH_2CH_3$ is 4-propyl-4-isopropylheptane. An attempt to avoid the
              |
             $CHCH_3$
              |
             $CH_3$

use of "isopropyl" would require selecting *hexane* as the basic hydrocarbon; this is not acceptable; thus, the name given is preferable.

**General observation.** The selection of the longest chain does not depend upon how the formula is written. Thus the longest chain is *six* for each of the following:

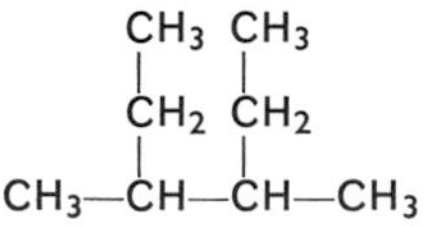

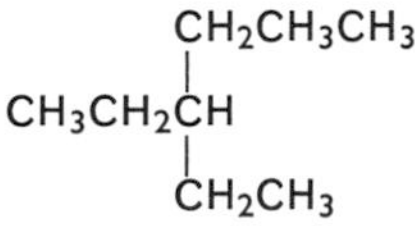

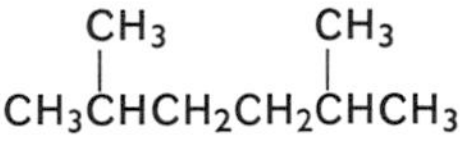

    $CH_3$  $CH_3$                $CH_2CH_3CH_3$          $CH_3$      $CH_3$
    |    |                      |                  |         |
    $CH_2$  $CH_2$        $CH_3CH_2CH$      $CH_3CHCH_2CH_2CHCH_3$
    |    |                    |
$CH_3$—$CH$—$CH$—$CH_3$            $CH_2CH_3$

***Naming of other substituents.***   Halogen-substituted alkanes are named according to the same basic rules as those outlined in the previous section. Some examples are the following:

$CH_3CH_2CH_2CH_2Br$                    1-bromobutane

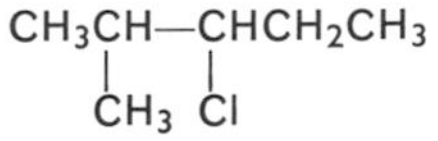

$CH_3CH—CHCH_2CH_3$                    2-methyl-3-chloropentane 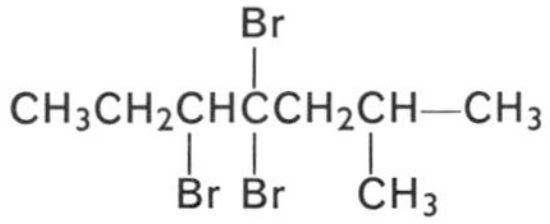
  |         |
 $CH_3$   $Cl$

            Br
            |
$CH_3CH_2CHCCH_2CH—CH_3$              2-methyl-4,4,5-tribromoheptane
       |  |         |
      Br  Br      $CH_3$

but

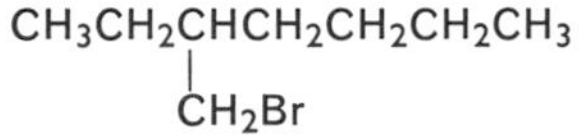

$CH_3CH_2CHCH_2CH_2CH_2CH_3$          3-(bromomethyl)-heptane
       |
     $CH_2Br$

rather than 1-bromo-2-ethylhexane, which uses a shorter basic chain.

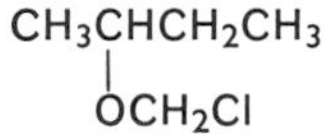

$CH_3CHCH_2CH_3$          2-chloromethoxybutane
     |
   $OCH_2Cl$

Note *(a)* that the longest chain of *carbon atoms* is chosen, not the chain of five atoms of which oxygen is one; and *(b)* that since the number 2 is the only number used, *chloro* and *methoxy* cannot refer to *two* substituents. See the example that follows:

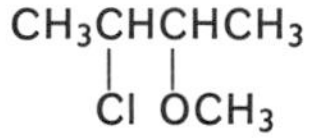

$CH_3CHCHCH_3$          2-chloro-3-methoxybutane
     |  |
    $Cl$ $OCH_3$

In general, groups such as halogens (bromo-, chloro-, iodo, fluoro-), nitro ($—NO_2$), nitroso ($—NO$), azido ($—N_3$), alkoxy (methoxy, $—OCH_3$; ethoxy, $—OCH_2CH_3$) are named as substituents. Hydroxyl groups are usually not named as substituents because there is a special way of dealing with alcohols, as will be described. Detailed discussions of the naming of alcohols, ketones, acids, and other classes of compounds will be reserved until their chemistry is dealt with. However, *the main principles of systematic naming have been covered already*; all that is necessary now is to describe how this basic structure of a system of nomenclature is altered for new classes of compounds.

## 6.5   The functional group as a basis for naming organic compounds

A *functional group* is one that confers characteristic properties upon a compound. Indeed, the way in which classes of organic compounds are distinguished is by reference to their functional groups. For example, one speaks of alcohols, ketones, aldehydes, acids, amines, in each of which a characteristic group ($—OH$, $—CO—$,

—CHO, —COOH, —NH$_2$) is present (Chapter 2). Table 6-2 summarizes the more common functional groups and the class of compounds that each characterizes.

It will be noted in the examples shown in the table that the I.U.C. names are based upon straight chains of carbon atoms in which (as in acids, aldehydes) the carbon atom of the functional group is counted as one of the members of the chain.

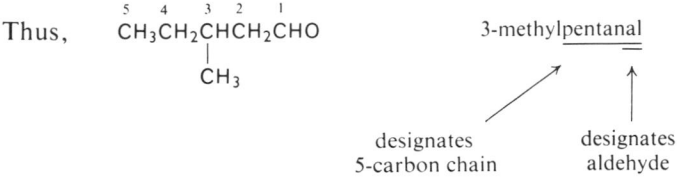

For the present, the only important modification of the basic rule for selecting the longest chain is this: when there is one functional group the fundamental chain is selected to contain this group:

$$\overset{6}{C}H_3\overset{5}{C}H_2\overset{4}{C}H-\overset{3}{C}HOH \qquad \text{4-methyl-3-hexanol}$$
$$\underset{\underset{2}{C}H_3}{|}\ \underset{\underset{1}{C}H_2CH_3}{|}$$

**Table 6-2.** *Common functional groups*

| Class of Compound | Example | Name* | Ending | Func-tional Group |
|---|---|---|---|---|
| olefins alkenes (G) | $CH_3CH{=}CH_2$ | propylene propene (G) | -ene | $C{=}C$ |
| acetylenes alkynes (G) | $CH_3C{\equiv}CCH_3$ | dimethylacetylene 2-butyne (G) | -yne | $C{\equiv}C$ |
| alcohols alkanols (G) | $CH_3CHCH_3$ $\quad\ \|$ $\quad\ OH$ | isopropyl alcohol 2-propanol (G) | -ol | OH |
| aldehydes alkanals (G) | $CH_3CHCHO$ $\quad\ \|$ $\quad\ CH_3$ | isobutyraldehyde 2-methylpropanal (G) | -al | —CHO |
| ketones alkanones (G) | $CH_3CH_2COCH_2CH_3$ | diethyl ketone 3-pentanone (G) | -one | —CO— |
| acids alkanoic acids (G) | $CH_3CH_2CH_2COOH$ | $n$–butyric acid butanoic acid (G) | -oic acid | —COOH |
| amines | $CH_3CH_2NH_2$ | ethylamine (no special I.U.C. suffix) | — | —NH$_2$ |

* The symbol (G) means the name is the I.U.C. name; this is not always the one ordinarily used.

This is in accord with the basic rule; but

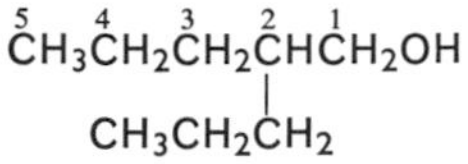

$$\overset{5}{CH_3}\overset{4}{CH_2}\overset{3}{CH_2}\overset{2}{CH}\overset{1}{CH_2}OH \qquad \text{2-propyl-1-pentanol}$$
$$| \\ CH_3CH_2CH_2$$

The longest chain of seven carbon atoms does not contain the —OH group, so it is not used as the basis for the name.

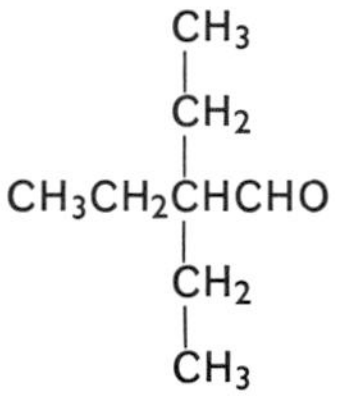

$$CH_3 \\ | \\ CH_2 \\ | \\ CH_3CH_2CHCHO \qquad \text{2,2-diethylbutanal} \\ | \\ CH_2 \\ | \\ CH_3$$

The longest chain (five) is not used, since the —CHO group is a part of no chain longer than four carbon atoms.

The manner of naming of compounds of the classes shown in Table 6-2 can now be summarized:

1. The numbering of the fundamental chain is done in such a way that the functional group is given the lowest number.

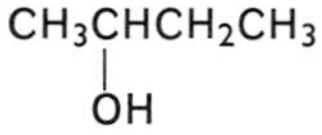

$$CH_3CHCH_2CH_3 \qquad \text{2-butanol; } not \text{ 3-butanol} \\ | \\ OH$$

$$CH_3 \\ | \\ CH_3C—CH_2CHCH_3 \qquad \text{4,4-dimethyl-2-pentanol;} \\ | \qquad | \qquad\qquad\qquad not \text{ 2,2-dimethyl-4-pentanol} \\ CH_3 \quad OH$$

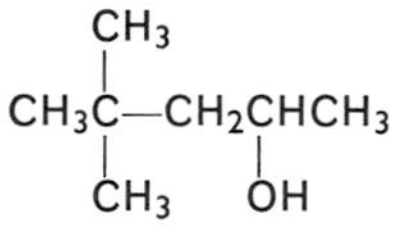

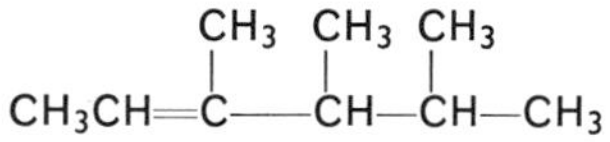

$$CH_3 \ CH_3 \ CH_3 \\ | \qquad | \qquad | \\ CH_3CH{=}C——CH—CH—CH_3 \qquad \text{3,4,5-trimethyl-2-hexene;} \\ \qquad\qquad\qquad\qquad\qquad\qquad not \text{ 2,3,4-trimethyl-4-hexene}$$

2. The fundamental chain is chosen to include the functional group, even though this may not be the longest chain. Examples have been given above.

3. Aldehydes and acids are numbered, starting with the carbon atom of the functional group as 1. (It should be recalled that when Greek letters are used in substitution names, the $\alpha$-carbon atom is the one *adjacent* to the functional group.)

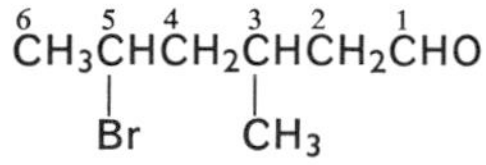

$$\overset{6}{CH_3}\overset{5}{CH}\overset{4}{CH_2}\overset{3}{CH}\overset{2}{CH_2}\overset{1}{CHO} \qquad \text{3-methyl-5-bromohexanal} \\ | \qquad | \\ Br \quad CH_3$$

but

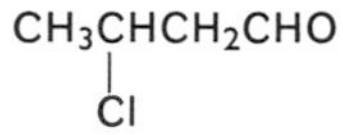

$$CH_3CHCH_2CHO \qquad \beta\text{-chloro-}n\text{-butyraldehyde; } not \ \gamma\text{-chloro-}n\text{-butyralde-} \\ | \qquad\qquad\qquad\qquad\qquad \text{hyde (not an I.U.C. name)} \\ Cl$$

4. When there is a choice between two chains of the same length, that chain which allows the most convenient numbering and naming of the substituents is chosen:

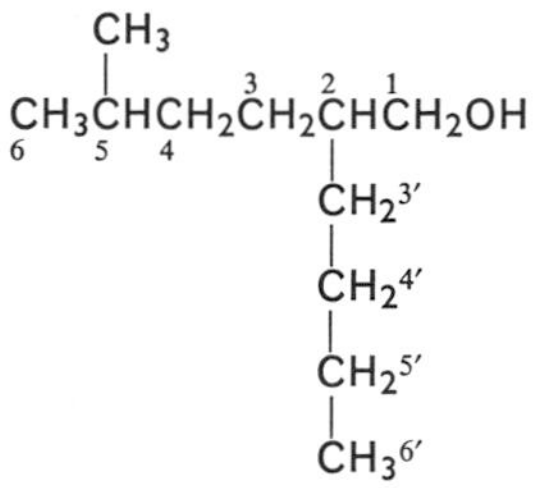

$$CH_3CHCH_2CH_2CHCH_2OH$$

with $CH_3$ branch on carbon 5, and the chain
$$\overset{6}{C}H_3\underset{5}{C}H\underset{4}{C}H_2\underset{3}{C}H_2\underset{2}{C}H\underset{1}{C}H_2OH$$
$$CH_2^{3'}$$
$$CH_2^{4'}$$
$$CH_2^{5'}$$
$$CH_3^{6'}$$

If chain 1,2,3',4',5',6' is chosen, the compound would be called 2-isopentyl-1-hexanol. This is less acceptable than the name based upon chain 1,2,3,4,5,6, which is 2-butyl-5-methyl-1-hexanol. (Note that *isopentyl* must be used, but butyl is not called *n*-butyl).

5. If more than one of a given functional group is present, the parts *di-*, *tri-*, *tetra-*, and so on are used.

| | |
|---|---|
| $CH_2=CHCH_2CHCH=CH_2$ | 1,5-hexadiene |
| $CH_2=CHCH=CHCH=CH_2$ | 1,3,5-hexatriene |
| $HOCH_2CH_2OH$ | 1,2-ethanediol or ethane-1,2-diol |
| $CH_3COCH_2CH_2COCH_3$ | 2,5-hexanedione |
| $HC\equiv C-C\equiv C-CH_3$ | 1,3-pentadiyne |
| $HOOCCH_2CH_2CH_2COOH$ | pentanedioic acid |

## 6.6   A note on the writing of names

The proper writing of a name requires correct punctuation, the use of correct endings, and a knowledge of when a name should be written as one word or as several words. *When the parts of a name are substituents on a basic structure, the name is written as one word*; but when a name is derived from the designation of the *class* of compounds to which the substance to be named belongs, the name is composed of separate words:

| | |
|---|---|
| $(CH_3)_3C-OH$ | *tert*-butyl alcohol<br>trimethylcarbinol<br>2-methyl-2-propanol |
| $C_2H_5OC_2H_5$ | diethyl ether |
| $CH_3OCH(CH_3)_2$ | methyl isopropyl ether<br>2-methoxypropane |
| 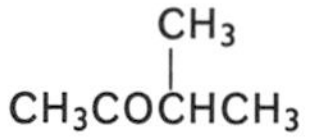 $CH_3COCHCH_3$ with $CH_3$ | methyl isopropyl ketone<br>3-methylbutanone |
| $CH_3CH_2CH_2CH_2Br$ | *n*-butyl bromide<br>1-bromobutane |

**Exercises**

(Exercises 1 and 2 will be found within the text.)

3 In a good unabridged dictionary find the derivation of the name of each of the following compounds: *(a)* ethyl alcohol; *(b)* glycerol; *(c)* glycine; *(d)* capric acid; *(e)* squalene; *(f)* carotene; *(g)* limonene; *(h)* benzene; *(i)* phytol; *(j)* aniline; *(k)* stearic acid; *(l)* sebacic acid; *(m)* chlorophyll; *(n)* crotonic acid.

4 Match the names in one column with the formulas in the other

| | |
|---|---|
| *(a)* $(CH_3)_2CHOH$ | *(a)* carbinol |
| *(b)* $ClCH_2CH_2OH$ | *(b)* vinylacetic acid |
| *(c)* $CH_3OH$ | *(c)* β-chloroethanol |
| *(d)* $CH_3CHClCOOH$ | *(d)* α-chloropropionic acid |
| *(e)* $(CH_3)_4C$ | *(e)* tetravinylmethane |
| *(f)* $(CH_2{=}CH)_4C$ | *(f)* dimethylcarbinol |
| *(g)* $CH_2{=}CHCH_2COOH$ | *(g)* tetramethylmethane |

5 Name the following compounds as derivatives of acetic acid, $CH_3COOH$.

*(a)* $BrCH_2COOH$

*(c)* $CH_3CH_2CHCH_2CH_3$
                 |
                COOH

*(b)* $Cl_2CHCOOH$

*(d)* $(CH_3)_3CCOOH$

6 Name the following compounds as derivatives of *n*-butyric acid, $CH_3CH_2CH_2COOH$.

*(a)* $CH_3CHCH_2COOH$        *(c)* $Cl_2CHCH_2CHCOOH$        *(e)* $F_3CCF_2CF_2COOH$
           |                                              |
           Cl                                            Cl

*(b)* $BrCH_2CHCH_2COOH$      *(d)* $CH_3CH_2CHCOOH$         *(f)* $F_3CCH_2CHCOOH$
            |                                      |                               |
            Br                                     F                               F

Answer to *(c)*: α,γ,γ-trichloro-*n*-butyric acid. Answer to *(e)*: Since there is only one possible acid in which all of the hydrogen atoms have been replaced by fluorine, it is unnecessary to designate the positions of substitution. Thus, the compound is heptafluoro-*n*-butyric acid. It is also called perfluoro-*n*-butyric acid.

7 Write the structures of the following compounds: *(a)* tetraisopropylmethane; *(b)* trimethylcarbinol; *(c)* divinylcarbinol; *(d)* allyl chloride; *(e)* di-*tert*-butylmethane; *(f)* tetrafluoromethane; *(g)* allylcarbinol; *(h)* methyl-*sec*-butylmethane; *(i)* triethylcarbinol.

8 Write the structures of the following compounds.

*(a)* 2,2-dimethylpropane
*(b)* 3-ethyl-2,3,5,5-tetramethylheptane
*(c)* 2,2,5,5-tetramethylhexane
*(d)* 1,2,3-propanetriol
*(e)* hexabromoethane

9 Give I.U.C. names to the following: *(a)* diisopropylmethane; *(b)* methylisobutylmethane; *(c)* tetramethylmethane; *(d)* *sec*-butylmethane; *(e)* dimethyl-*n*-propylmethane; *(f)* isoheptane.

10 Write the structures of all of the possible compounds of the formula $C_5H_{12}$ and name each by an I.U.C. name and by a non-I.U.C. substitution name.

11 Name the following compounds in two ways, using only substitution names of the methane system:

*(a)* $CH_3CH_2CH{\Large\langle}^{CH_3}_{CH_3}$

*(b)* $CH_3CH_2\underset{\underset{CH_3}{|}}{\overset{\overset{CH_3}{|}}{C}}CH_3$

*(c)* $CH_3\underset{\underset{CH_3}{|}}{CH}CH_2\underset{\underset{CH_3}{|}}{CH}CH_3$

# Chapter seven / The Nucleophilic Displacement Reaction. Substitution at a Saturated Carbon Atom

One of the most general reactions in organic chemistry is that in which a group possessing an unshared pair of electrons in its valence shell attacks and forms a bond to a carbon (or other) atom that is deficient in electrons. This electron deficiency may be complete (the carbon atom may bear a positive charge); or it may be partial (the carbon atom may be electron-deficient because of unequal sharing of an electron pair that binds it to another, more electronegative atom).

In its broadest aspects, this kind of reaction is central to most of the bond-forming processes that are encountered in organic chemistry, and it can be recognized as a feature of substitution reactions, addition reactions, elimination reactions, and rearrangement reactions. In this chapter we shall examine one of these: the replacement of a group, along with its bonding electron pair, by another group which provides the electron pair for the formation of the new bond.

## 7.1 General characteristics of displacement reactions

The reaction of methyl bromide with alkali, described in Chapter 5, is commonly referred to as the hydrolysis of the alkyl halide, and represents the simplest example of one of the most general reactions in organic chemistry. To recapitulate briefly, the reaction involves the attack of a *nucleophilic* reagent, Y:, upon the carbon atom of a general substance RX, with *displacement* of the group or atom :X, with its binding electron pair. The group :X is called the *leaving group*. In Chapter 5 it was pointed out that one route that can be followed in this reaction is that in which the attack of Y: leads to its partial bonding to carbon and formation of a molecular complex called the transition state or activated complex. This complex is in equilibrium with the components from which it is formed. as well as with the products into which it decomposes. The *extent* to which the initial reactants are converted into the final products is governed by the free energy change, and thus the equilibrium constant, for the overall process. The *rate* at which the reaction proceeds depends primarily upon the activation energy, or the height (in energy units) of the barrier over which the reaction must pass.

The reaction can be formulated in the following general way:

$$Y: \quad + R:X \rightleftharpoons \text{transition state} \rightleftharpoons Y:R+ \quad :X \tag{1}$$

nucleophile                                                                leaving group

This reaction has been studied extensively and in depth for many years, and the various factors that influence its course can be described with respect to the following variables:

*(a)* The mechanism of the reaction. It is known that there are two distinct mechanisms, which differ in respect to the rate-determining step, for reactions of the general type (1). These two mechanisms are influenced by:

*(b)* The structure of R in RX. We shall first of all take account of a few representative types of RX, where R may be:

  i. A primary alkyl group, as in ethyl bromide, $C_2H_5Br$, and longer-chain primarily halides, $RCH_2X$.

  ii. A secondary alkyl group, as in isopropyl halides, $(CH_3)_2CHX$ and *sec*-butyl halides $CH_3CHCH_2CH_3$.

  X

  iii. A tertiary alkyl group, as in *t*-butyl halides, $(CH_3)_3CX$.

Other kinds of R groups will be encountered in later chapters.

*(c)* The nucleophilic character of Y:.

*(d)* The nature of the leaving group, :X. As the reversibility of the reaction (1) implies, :X is also a nucleophile; but since our concern will ordinarily be with carrying out the reaction as it is written from left to right; our view of :X will be from the standpoint of its displaceability.

*(e)* The medium in which the reaction is carried out. Since most organic reactions are carried out in solution, we shall consider the way in which solvents of several kinds aid, impede, or otherwise alter the course of the reaction.

*(f)* Side reactions or alternative courses for the reaction.

*(g)* The stereochemistry of the reaction.

These variables in the reaction will be discussed in turn in the following sections.

## 7.2 The mechanism of the displacement reaction

***The $S_N2$ mechanism.*** The reaction [Section 5.4, Equation (19)] of methyl bromide with hydroxide ion is a bimolecular reaction; that is, its rate is dependent upon the concentrations both of $CH_3Br$ and of $OH^-$.

$$\text{rate} = k[CH_3Br][OH^-] \tag{2}$$

This expression means that both reactants are involved in the rate-determining step of the reaction; an increase in the concentration of either one increases the concentration of the activated complex and thus the rate at which the final products are formed. The energy profile for the reaction has been described (Figure 5-1, Chapter 5).

This reaction course, or mechanism, is called by the symbol $S_N2$, for substitution *(S)*, nucleophilic *(N)*, bimolecular *(2)*. The designation "second order" for the "*2*" is also used, but since the term $S_N2$ properly refers to the number of species that are undergoing covalency change at the *rate-determining step* (that is, to form the highest energy transition state), the word "bimolecular" is preferable.

The characteristics of the $S_N2$ displacement reaction are the following: the nucleophilic agent attacks the carbon atom at the opposite side to that at which the leaving group is attached; the energy necessary to break the R—X bond [Equation (1)] is partly provided by the energy of formation of the Y—R bond; the configuration of the groups attached to the carbon atom is *inverted* in the process of displacement. This very important stereochemical result will be discussed in more detail in Section 7-8.

***The $S_N1$ mechanism.*** When *t*-butyl bromide is treated with dilute alkali, the product of the reaction is again the alcohol, and the process may be represented by the equation

$$(CH_3)_3CBr + OH^- \rightleftarrows (CH_3)_3COH + Br^- \tag{3}$$

The kinetics of this reaction, however, are quite different from those of the hydrolysis of methyl bromide: it is found that the rate is dependent only upon the concentration of the alkyl halide and is independent of the concentration of the nucleophile, $OH^-$.

$$\text{rate} = k[(CH_3)_3CBr] \tag{4}$$

The *rate-determining step* of this reaction, then, must involve only the alkyl halide, and it is the conclusion drawn from much experimental study that this step is the *initial ionization* of the *t*-butyl halide into a carbonium ion and a halide ion:

$$(CH_3)_3CBr \xrightarrow{\text{slow}} (CH_3)_3C^+ + Br^- \tag{5}$$

The subsequent combination of the carbonium ion and the hydroxide ion is a very fast reaction and is not reflected in the kinetic measurements:

$$(CH_3)_3C^+ + OH^- \xrightarrow{\text{fast}} (CH_3)_3COH \tag{6}$$

An energy diagram for a reaction of this kind is given in Figure 7-1.

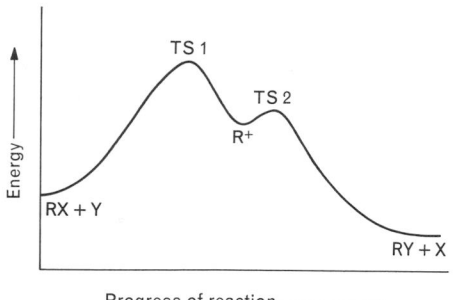

**Figure 7–1.** Energy profile for the nucleophilic displacement reaction in which ionization of the alkyl halide is the rate-determining step *(S$_N$1)*.

It is clear that the *t*-butyl alcohol cannot be formed faster than $(CH_3)_3C^+$ is generated by the ionization of the *t*-butyl halide; thus, process 5 is rate-controlling.

The symbol $S_N1$ is used to describe the overall reaction (3), the "*1*" in this case indicating that only one species, the alkyl halide, is involved in the rate-determining step, which is thus "unimolecular."

## 7.3   The nucleophilic agent

The characteristic feature of the agent Y: in Equation (1) is its possession of an unshared pair of electrons. It is therefore a base in the general meaning of this term as it was discussed in Chapter 4. Basicity is a measure of nucleophilicity toward the proton, and is assessed in terms of *equilibria* between proton acceptors and proton donors. Nucleophilicity towards carbon, with which we shall be primarily concerned, is a property whose value is assessed by measuring the *rates* of a series of displacement reactions. Nevertheless, the parallel between basicity and nucleophilicity toward carbon, while not exact, can be taken as a general guide, and strong bases are usually better nucleophiles than weak bases. The departures from this generalization will be considered later in this section.

In Table 7-1 are found a list of common nucleophiles and some typical displacement reactions in which they take part. It will be noted that no single charge type characterizes the nucleophile; both neutral molecules and anions may be nucleophilic, and both classes of reagents are used in preparative applications of the displacement reaction.

The nucleophiles in this list include the strong bases $OH^-$ and $CH_3O^-$; weak bases such as the acetate ion, $CH_3COO^-$, and the cyanide ion, $CN^-$; and the very weakly basic substances methanol, $CH_3OH$, and iodide ion, $I^-$. All of the nucleophiles, Y:, in this table are the conjugate bases of acids, Y:H. Iodide ion is an unusually effective nucleophile. This, and the nucleophilicity of $CH_3S^-$, which is much greater than that of $CH_3O^-$, represent marked departures from the parallelism between basicity and nucleophilic effectiveness towards carbon. It is observed that nucleophilic reactivity of the atoms within a single group of the periodic table increases with the size of the atom. Thus, the orders $I^- > Br^- >$

**Table 7-1.** *Some typical nucleophilic displacement reactions*

| Y:<br>(nucleophile) | + R:X | → Y:R | :X<br>+ (leaving<br>group) |
|---|---|---|---|
| 7. Br:$^-$ | + $CH_3OH_2^+$ | → Br:$CH_3$ | + $H_2O$ |
| 8. HO:$^-$ | + $CH_3Br$ | → HO:$CH_3$ | + Br$^-$ |
| 9. $H_3N$: | + $CH_3I$ | → $H_3\overset{+}{N}$:$CH_3$ | + I$^-$ |
| 10. $CH_3O$:$^-$ | + $CH_3Cl$ | → $CH_3O$:$CH_3$ | + Cl$^-$ |
| 11. $CH_3COO$:$^-$ | + $CH_3CH_2Br$ | → $CH_3COO$:$CH_2CH_3$ | + Br$^-$ |
| 12. I:$^-$ | + $CH_3Br$ | → I:$CH_3$ | + Br$^-$ |
| 13. $(CH_3)_2S$: | + $CH_3I$ | → $(CH_3)_2\overset{+}{S}$:$CH_3$ | + I$^-$ |
| 14. $CH_3S$:$^-$ | + $CH_3CH_2Br$ | → $CH_3S$:$CH_2CH_3$ | + Br$^-$ |
| 15. CN$^-$ | + $CH_3CH_2I$ | → NC:$CH_2CH_3$ | + I$^-$ |
| 16. $N_3^-$ | + $CH_3Br$ | → $N_3$:$CH_3$ | + Br$^-$ |
| 17. $CH_3\overset{\overset{\textstyle H}{\displaystyle |}}{—}O$: | + $CH_3OH_2^+$ | → $CH_3\overset{\overset{\textstyle H}{\displaystyle |}}{O}$:$CH_3$<br>$\quad\overset{+}{}$ | + $H_2O$ |

Cl$^-$ > F$^-$; $CH_3S^-$ > $CH_3O^-$; and HS$^-$ > HO$^-$ are observed. The reason for the unexpectedly high nucleophilicity of the larger atoms is that with the increase in the number of electrons and the size of the outer shells, two factors begin to assume increasing importance: (1) the deformability of the outer orbitals, or the *polarizability* of the electrons that are involved in bonding to the carbon atom; (2) the degree to which the nucleophilic anion is solvated, smaller ions being more extensively solvated than larger ones. Thus, nucleophilicity, as reflected in relative rates of reaction of a number of nucleophiles with a given compound R:X, is a property affected by the strength of the nucleophile as a base, the size of the nucleophilic atom, and the solvent in which the displacement reaction is carried out.

The following order of nucleophilicity is observed for displacement reactions that take place in such solvents as water or alcohols, or mixtures of these (*protic* solvents): HS$^-$, $CH_3S^-$ > CN$^-$ > I$^-$ > $NH_3$ > HO$^-$ > $N_3^-$ > Br$^-$ > Cl$^-$ > F$^-$ > $H_2O$. This order should not be regarded as absolute, for relative nucleophilicity is strongly influenced by the nature of the solvent. It is given in order to place some of the common nucleophilic agents in an order of reactivity that holds for reaction conditions that are often employed.

The high degree of nucleophilicity of the iodide ion is sometimes used to practical advantage in carrying out displacement reactions upon alkyl chlorides. As we shall see in the next section, the chloride ion is much more slowly displaced than is iodide; and iodide is a better nucleophile than the hydroxide ion. Thus, the addition of some iodide (which need not be in stoichiometric amount) to a reaction mixture containing an alkyl chloride and alkali causes a faster replacement of chlorine by the hydroxyl group:

$$\left.\begin{array}{l} RCl+I^- \longrightarrow RI+Cl^- \\ RI+OH^- \longrightarrow ROH+I^- \end{array}\right\} \text{faster than } RCl+OH^- \longrightarrow ROH+Cl^- \tag{18}$$

In general, the anions (conjugated bases) of strong acids, such as $SO_4^=$, $NO_3^-$, $Cl^-$, and $H_2O$ (the conjugate base of the strong acid $H_3O^+$), are poor nucleophiles. This is in accord with the generalization, subject to the exceptions described above, that nucleophilicity generally parallels basicity.

## 7.4 The leaving group

The group [:X in Equation (1)] that is expelled from RX in the nucleophilic displacement reaction can be any of a wide variety of structural types, but all of the readily displaced groups have one property in common: *they are very weak bases, usually the conjugate bases of strong acids*. Since the equilibrium (1) shows that the reaction is one in which two nucleophiles, Y: and :X, are competing to form the bond to the carbon atom, it would be expected that for the reaction to proceed to the formation of Y:R, the nucleophilicity of :X would ordinarily be much lower than that of :Y. The position of the final equilibrium expressed in Equation (1) is, however, determined by the overall energy change in the reaction. This does not necessarily parallel the height of the activation energy curve for the reaction, and thus it is possible for a stronger nucleophile to be displaced by a weaker one. An example has been given in Equation (18).

The commonest leaving groups in displacement reactions are the anions or conjugate bases of the strong inorganic acids. In Table 7-2 are listed a number of examples of leaving groups :X, along with their conjugate acids, H:X.

Certain other groups that can be displaced under some conditions but that are generally less reactive and not often used are $F^-$, $RCOO^-$, and $OSO_2O^=$.

An important leaving group, which is more prone to undergo ready displacement than are those listed in Table 7-2, is the diazonium group. Compounds containing this group are called diazonium salts, and can be prepared by the reaction of a primary amine with nitrous acid:

$$RNH_2 \xrightarrow[\text{HCl}]{\text{NaNO}_2} \underset{\text{a diazonium salt}}{RN_2^+Cl^- + 2H_2O} \tag{19}$$

The displacement of $-N_2^+$ by the attack of a nucleophile leads to its expulsion as molecular nitrogen:

$$Y:^- + RN_2^+ \longrightarrow Y:R + N_2 \tag{20}$$

The formation of this very stable molecule is a strong driving force for the breaking of the C—N bond in $RN_2^+$, and accounts for the great instability of diazonium salts and their high reactivity in displacement reactions. Examples of their use will be found frequently in later chapters.

**Table 7-2.** *Some typical leaving groups* : *X in the displacement reaction*
$$Y: + R:X \rightarrow Y:R + :X$$

| :X | Conjugate Acid, H:X |
|---|---|
| $Cl^-$ | HCl |
| $Br^-$ | HBr |
| $I^-$ | HI |
| $NR_3$ | $HNR_3^+$ |
| $SR_2$ | $HSR_2^+$ |
| $OH_2$ | $H_3O^+$ |
| $ROSO_2O^-$ | $ROSO_2OH$ |
| $RSO_2O^-$ | $RSO_2OH$ |
| HOPO—O—PO—O⁻* $\quad$ | HO—PO—O—PO—OH |
| $\quad$ | $\quad$ OH $\quad$ OH $\qquad$ OH $\quad$ OH |

\* This is the pyrophosphate group, of great importance in biological systems in reactions involving nucleophilic displacement.

The order of ease of displacement of the leaving groups most frequently used is the following:

$$X \text{ in } RX = N_2^+ > OSO_2R > I \sim Br > NO_3 \sim Cl > OH_2^+ \sim S(CH_3)_2^+ > NR_3^+ > OR$$

It will be noted that the most readily displaceable group, $N_2$, in this series is the conjugate base of the hypothetical very strong acid $HN_2^+$; and the least readily displaced group, $OR^-$, is a strong base, conjugate to the very weak acid ROH.

The position of iodine in this series may appear to be anomalous, for it was remarked in Section 7.3 that iodide ion possesses a high degree of nucleophilicity, the result of the ready polarizability of the large ion. Its ease of displaceability can also be accounted for by a high polarizability, for the ready deformation of the orbital that forms the C—I bond aids in the lengthening of this bond, which is an essential part of the process of forming the transition state complex. These properties of iodine as the nucleophilic anion and as the leaving group account for the catalytic effect of iodide described in the previous section.

## 7.5 Displacement reactions in synthesis

Many synthetic procedures of wide usefulness in organic chemistry have as their central process the nucleophilic displacement reaction. It will be useful here, before discussing some of the other factors that influence this reaction, to summarize the roles of nucleophilic reagents and leaving groups by describing some synthetic methods for the preparation of a number of important classes of compounds by the use of the $S_N2$ displacement reaction. A variety of types of groups :X in RX are chosen in these examples; and in each it is to be noted that the group R— is primary (i.e., $RCH_2$). Specific examples are used, but each may be regarded as a prototype of a general procedure.

*(a)* Preparation of alkyl bromides by the reaction of an alcohol with an inorganic bromide in strongly acidic solution. This is a general synthetic method for the preparation of *n*-alkyl halides (chlorides and iodides can be prepared in an analogous way), but is often attended by molecular rearrangements when R is other than a primary alkyl group:

$$Br^- + RCH_2OH_2^+ \longrightarrow RCH_2Br + H_2O \tag{21}$$

*(b)* Preparation of ethers:*

$$(CH_3)_2CHO^- + CH_3I \longrightarrow (CH_3)_2CHOCH_3 + I^- \tag{22}$$

*(c)* Preparation of esters:

$$CH_3COO^- + \underset{\text{benzyl methanesulfonate}}{\underset{}{\bigodot}CH_2OSO_2CH_3} \longrightarrow$$

$$\underset{\text{benzyl acetate}}{\bigodot CH_2OCOCH_3} + CH_3SO_2O^- \tag{23}$$

*(d)* Preparation of alcohols (hydrolysis of alkyl halides):

$$\underset{\text{1-bromobutane}}{CH_3CH_2CH_2CH_2Br} + OH^- \longrightarrow \underset{\text{1-butanol}}{CH_3CH_2CH_2CH_2OH} + Br^- \tag{24}$$

*(e)* Synthesis of nitriles (alkyl cyanides):

$$\underset{\text{1-bromopentane}}{CH_3CH_2CH_2CH_2CH_2Br} + CN^- \longrightarrow \underset{\text{1-cyanopentane}}{CH_3CH_2CH_2CH_2CH_2CN} + Br^- \tag{25}$$

*(f)* Preparation of dialkyl sulfides:

$$CH_3S^- + CH_3CH_2Br \longrightarrow \underset{\text{methyl ethyl sulfide}}{CH_3SCH_2CH_3} + Br^- \tag{26}$$

*(g)* Preparation of sodium alkanesulfonates:

$$\underset{\substack{\text{sulfite (as}\\\text{sodium sulfite)}}}{SO_3^=} + CH_3I \longrightarrow \underset{\substack{\text{methanesulfonic}\\\text{acid (anion)}}}{CH_3SO_2O^-} + I^- \tag{27}$$

---

* Reagents such as alkoxide ions, $RO^-$, are associated with some inorganic cation. Thus $(CH_3)_2CHO^-$ would be provided by a solution of $(CH_3)_2CHO^-Na^+$ in isopropyl alcohol. Similarly, such anionic reagents as $RCOO^-$, $CN^-$, $RS^-$ are the nucleophilic species in solutions of their salts; e.g., potassium cyanide, sodium acetate, and so on. Since the cation is unchanged in the reaction, it seldom needs to be made a part of the equation.

*(h)* Preparation of amines:

$$NH_3 + CH_3CH_2I \longrightarrow CH_3CH_2NH_3^+ + I^- \tag{28}$$
ethylamine<br>(protonated)

In the presence of an excess of ammonia, free ethylamine will be formed by the simple proton exchange

$$CH_3CH_2NH_3^+ + NH_3 \rightleftharpoons CH_3CH_2NH_2 + NH_4^+ \tag{29}$$

and the ethylamine, also a reactive nucleophile, can then react further with the ethyl iodide that is present

$$CH_3CH_3NH_2 + CH_3CH_2I \longrightarrow CH_3CH_2\overset{+}{N}H_2CH_2CH_3 + I^-. \tag{30}$$

Further proton exchange and further reaction with ethyl iodide leads to further alkylation of the amine to give as the final result a mixture of $EtNH_2$, $Et_3N$, and $Et_4N^+ I^-$ ($Et = -CH_2CH_3$). Thus, this method of preparing amines has obvious practical disadvantages. Other methods will be described later in the text.

*(i)* Preparation of isothioureas.

Thiourea, $S{=}C{\overset{NH_2}{\underset{NH_2}{\diagdown}}}$, is an excellent nucleophile, the sulfur atom acting as the nucleophilic center. It reacts with alkyl halides and sulfonic esters as follows: the nucleophilicity of the sulfur atom is aided by the ability of the $-NH_2$ group to provide electrons in the manner shown:

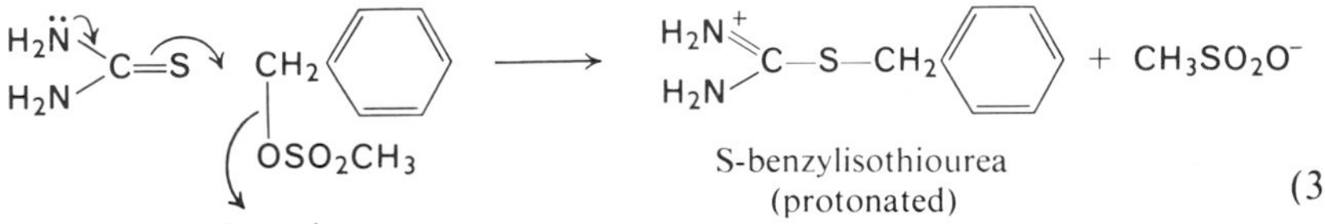

benzyl
methanesulfonate

S-benzylisothiourea
(protonated)

$$\tag{31}$$

## 7.6 The role of the solvent in the nucleophilic displacement reaction

The nature of the medium in which the displacement reaction occurs has a profound influence upon several of the factors that affect the reaction. A change in the solvent can

    *(a)* alter the relative nucleophilicity of a series of nucleophilic reagents;

    *(b)* alter the rate at which the reaction proceeds;

    *(c)* Change the mechanism of the reaction, so that a reaction that proceeds by the $S_N2$ mechanism in one solvent can become an $S_N1$ reaction in another.

Three properties of organic solvents are important in their effect upon the reaction: their dielectric constant; their ability to engage in hydrogen bonding; and their ability to solvate ions either by hydrogen bonding or by electrostatic

**Table 7-3.** *Dielectric constants of some common organic solvents*

| Solvents | Dielectric Constants (e.s.u.) |
|---|---|
| **Protic Solvents** | |
| ethanol | 26 |
| methanol | 32 |
| water | 81 |
| formic acid | 58 |
| acetic acid | 6.2 |
| **Aprotic Solvents** | |
| cyclohexane | 2.0 |
| benzene | 2.3 |
| ethyl ether | 4.3 |
| chloroform | 4.6 |
| nitrobenzene | 28 |
| acetone | 21 |
| acetonitrile | 38 |
| nitromethane | 101 |
| dimethyl sulfoxide | 45 |

association (dipole interactions). Solvents may be classed in two main groups: those that contain hydroxyl or amino groups, and thus can engage in hydrogen bonding (*protic* solvents); and those that do not contain such groups and cannot engage in hydrogen bonding (*aprotic* solvents). Both groups may contain solvents with high and with low dielectric constants (Table 7-3).

The effects of solvent composition upon the rate and mechanism of nucleophilic displacement reactions are many and often subtle, and will not be discussed here in detail. A suggestion of the many variable factors that may come into play can be had by a consideration of the following facts: the nucleophile may be neutral (Y:) or charged (Y:$^-$); the substrate may be RX or RX$^+$; and thus the leaving group may appear as :X$^-$ or :X. For these reasons, the transition state may be charged or neutral, and if it is charged, the charge may be negative or positive and dispersed to a greater or lesser degree over the activated complex. Consequently, whether a solvent is polar (high dielectric constant, protic) or nonpolar (low dielectric constant, aprotic) will have a profound effect upon the reaction because of the various degrees to which nucleophile, substrate, transition state, and leaving group can be solvated. We shall confine our attention here to the effect of the solvent in altering the rate and the mechanism of the displacement reaction.

The reaction of *t*-butyl chloride with a nucleophile has been described (section 7-2) as proceeding through a preliminary ionization into $(CH_3)_3C^+$ and $Cl^-$ ions, followed by a fast reaction of the carbonium ion with the nucleophilic reagent. It has not yet been emphasized that the solvent plays an important role in the reactions we have been considering; indeed, it is usually a direct participant in the reaction by virtue of its capacity for solvating the ions. Equation (5) is properly written as follows.

$$(CH_3)_3CCl + (solvent) \rightleftharpoons (CH_3)_3C^+ (solvated) + Cl^- (solvated) \tag{32}$$

The energy required to bring about the separation of the free, oppositely charged $(CH_3)_3C^+$ and $Cl^-$ ions would be expected to be much higher than the activation energy actually observed for the reaction carried out in, say, aqueous-alcoholic solution; it is, however, reduced by the amount of the energy gained by the solvation of the ions.

Water (or solutions containing water) promotes ionization by virtue of its high dielectric constant (which aids the separation of the charges) and by its ability to solvate the ions by hydrogen bonding and dipolar interactions. The process of ionization of an alkyl halide can be pictured as in Figure 7-2 (compare the diagram for the ionization of HCl in water, Figure 4-1).

The events pictured in Figure 7-2 are only approximations to reality because we do not know how many solvent molecules are involved in the "solvation shell" at the various stages along the path to final separation of the ions. Nevertheless,

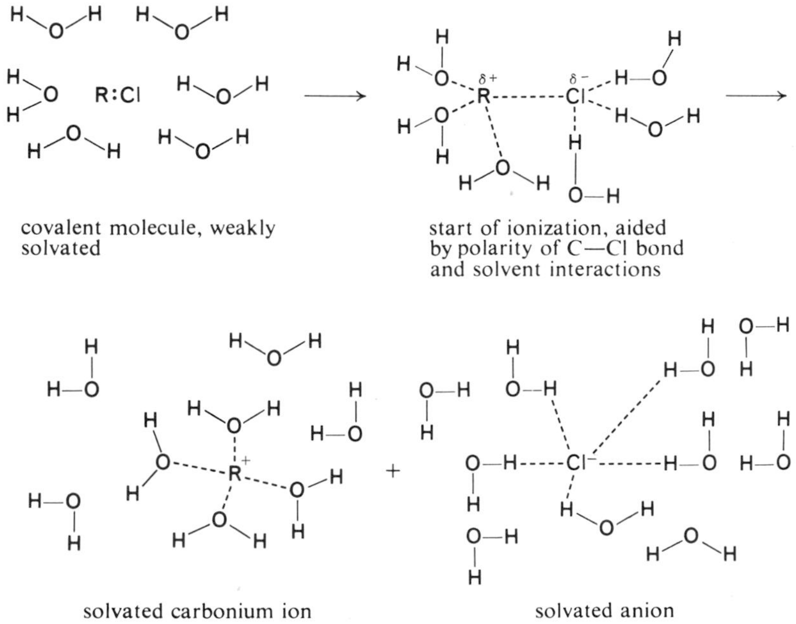

covalent molecule, weakly solvated

start of ionization, aided by polarity of C—Cl bond and solvent interactions

solvated carbonium ion

solvated anion

**Figure 7–2.** The role of solvent (water) in the ionization of an alkyl halide.

this concept is in accord with what is known about solvent effects and provides a basis for further understanding of the nucleophilic displacement reaction.

A good ionizing solvent would be expected to favor reaction by the $S_N1$ mechanism, and the hydrolysis of $t$-butyl chloride in aqueous ethanol (50%) is in fact of the order of $10^4$ times faster than in pure ethanol; and in anhydrous acetone, $t$-butyl chloride reacts (for example, with iodide ion) at a very slow rate.

In the discussion to follow, the solvents that will be chosen for use in the examples will be the poorly ionizing solvent, anhydrous acetone; the solvent of intermediate ionizing power, anhydrous ethanol; and the solvents of high solvating and ionizing ability, aqueous ethanol and aqueous acetone.

If a given displacement reaction of the general class shown in Equation (1) can, for structural reasons, proceed either by the $S_N1$ or the $S_N2$ mechanism, the $S_N1$ process will be favored in the good ionizing medium. A reaction that, for structural reasons, is $S_N2$ in type and cannot proceed by the $S_N1$ mechanism, will show relatively less effect of change of solvent upon rate of reaction, and the direction of change is not always easy to predict.

### 7.7   The effect of the structure of R in RX in displacement reactions

*The $S_N2$ reaction.*   The overall rate of a nucleophilic displacement reaction is not an index of the mechanism ($S_N1$ or $S_N2$) by which it proceeds. Methyl bromide and $t$-butyl bromide both react very rapidly in dilute aqueous-alcoholic alkali; yet the first reaction proceeds by the $S_N2$ route, the second by the $S_N1$ route. Ethyl bromide and isopropyl bromide react at intermediate rates, and a careful kinetic analysis of these hydrolyses shows that ethyl bromide reacts principally by the $S_N2$ mechanism but the isopropyl bromide reaction shows "mixed" kinetics. In Figure 7-3 is shown a diagram in which the rates of alkaline hydrolysis for these four alkyl bromides are plotted; the points are connected by lines which indicate the degree to which one of the other of the two mechanisms obtains.

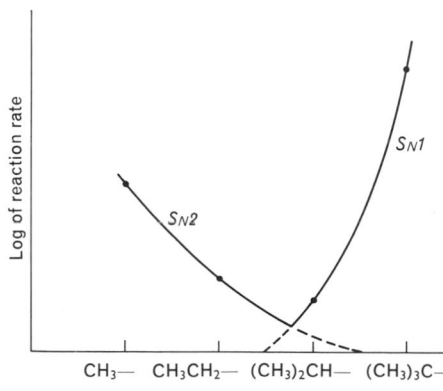

**Figure 7–3.**   Plot of the rates of hydrolysis of alkyl bromides in dilute aqueous-alcoholic alkali. Relative specific rates in 0.01 $N$ NaOH in a solution of 80% ethanol and 20% water, at 55°. Me=74; Et=17; $iso$Pr=(1); $t$-Bu=3500.

**Table 7-4.** *Relative rates of reaction for* $RBr + I \rightarrow RI + Br$ *under $S_N2$ conditions*

| | R | Relative Rate |
|---|---|---|
| methyl | $CH_3$— | 200,000 |
| ethyl | $CH_3CH_2$— | 1,000 |
| isopropyl | $(CH_3)_2CH$— | 12 |
| $t$-butyl | $(CH_3)_3C$— | (1) |

When the reaction

$$R\!:\!Br + I^- \longrightarrow R\!:\!I + :Br^- \tag{33}$$

is carried out in anhydrous acetone, the carbonium ion pathway ($S_N1$) is suppressed, and the reaction proceeds by the second-order ($S_N2$) process. Table 7-4 shows the effect of increasing substitution on the carbon atom at which reaction occurs.

It is evident that the reaction is strongly retarded when the hydrogen atoms of the methyl group are replaced by methyl groups. An important reason for this is readily understood when the nature of the $S_N2$ reaction is recalled:

$$Y\!:^- + \; \underset{R}{\overset{R''}{\underset{|}{\overset{|}{\underset{R'}{\diagdown}}}}}\!\!C\!-\!Br \; \rightleftharpoons \; \left\{ Y\cdots\underset{\underset{R}{|}}{\overset{R'\;\;R''}{\overset{\diagup}{C}}}\cdots Br \right\} \; \rightleftharpoons \; Y\!-\!\underset{R}{\overset{R''}{\underset{\diagdown}{\overset{|}{\diagup}}}}\!\!C\!\!-\!\!R' + :Br^- \tag{34}$$

The approach of Y: to within bonding distance of the carbon atom suffers the minimum of interference when R, R′, and R″ are all hydrogen atoms. As substitution progresses to ethyl (R=$CH_3$, R′=R″=H), isopropyl (R=R′=$CH_3$, R″=H), and $t$-butyl (R=R′=R″=$CH_3$), greater activation energy is required (Figure 7-4) for the attacking nucleophile to penetrate to within bonding distance, and thus the reaction rate (at a given temperature) decreases in the order shown in Table 7-4.

Primary alkyl halides, $RCH_2CH_2X$, undergo $S_N2$ displacement reactions at comparable rates; increasing the chain length of the R group has little effect on the rate, showing that in the change from $CH_3X$ to $RCH_2X$, where R is a $n$-alkyl group or methyl, the principal effect is produced by the substitution *at* the reaction center (Table 7-5).

The structural effects shown in the data of Tables 7-4 and 7-5 are exerted primarily at the reaction center, and represent hindrance to the approach of the nucleophile to the carbon atom at which reaction occurs. Very important structural effects upon reaction rate are observed when substitution occurs at the

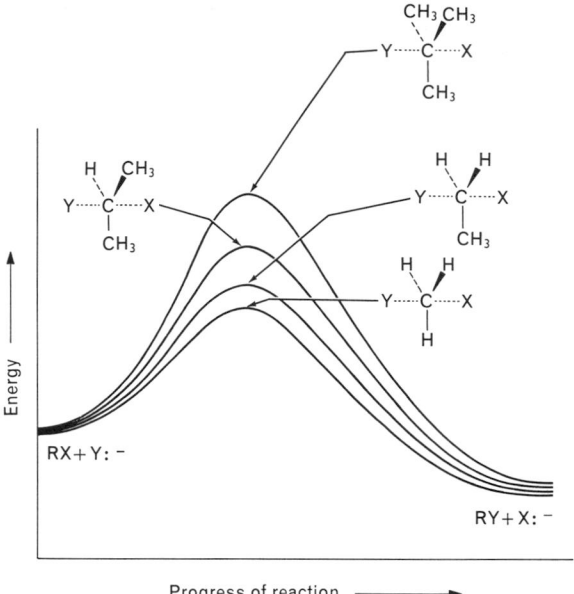

**Figure 7–4.** Energy profiles for $S_N2$ displacement reactions in the series $CH_3Br$, $CH_3CH_2Br$, $(CH_3)_2CHBr$, $(CH_3)_3CBr$. (Overall free energy change is arbitrary.)

**Table 7-5.** *Relative rates of $S_N2$ displacement reactions of n-alkyl bromides : (a) $RI + N(C_2H_5)_3 \rightarrow$*

$\overset{+}{R}N(C_2H_5)_3 + I^-$ *(in acetone) ; (b)* $RBr + C_2H_5O^- \rightarrow ROC_2H_5 + Br^-$ *(in absolute ethanol)*

| R | Relative Rate | |
|---|---|---|
| | *(a)* | *(b)* |
| $CH_3—$ | (100) | (100) |
| $CH_3CH_2—$ | 8.8 | 5.7 |
| $CH_3CH_2CH_2—$ | 1.7 | 0.17 |
| $CH_3CH_2CH_2CH_2—$ | 1.2 | 0.13 |
| $CH_3CH_2CH_2CH_2CH_2—$ | — | 0.12 |
| $CH_3(CH_2)_5CH_2—$ | 0.9 | — |
| (isopropyl-) | 0.18 | — |

**Table 7-6.** *Relative rates of displacement reactions of primary alkyl bromides in the reaction*
$$RBr + C_2H_5O^- \rightarrow ROC_2H_5 + Br^-$$

| R | Relative Rate |
| --- | --- |
| $CH_3CH_2-$ | 500,000 |
| $CH_3CH_2CH_2-$ | 28,000 |
| $(CH_3)_2CHCH_2-$ | 4,000 |
| $(CH_3)_3CCH_2-$ | (1) |

$\beta$-position to the reaction center.* In the data shown in Table 7-6, increasing $\beta$-substitution is present in the series of alkyl halides $CH_3CH_2X$, $RCH_2CH_2X$, $R_2CHCH_2X$, and $R_3CCH_2X$. These are all primary halides, but it is evident that they display an enormous range of reactivity in the $S_N2$ displacement reaction.

The remarkably low reactivity of neopentyl halides $[(CH_3)_3CCH_2X]$ in displacement reactions shows very dramatically the effect of steric hindrance to the attack of the nucleophile. The formation of the bimolecular transition state would require the penetration of the nucleophile past the protecting wall of the methyl groups; indeed, a model shows that considerable bond distortion would have to occur to allow the nucleophilic reagent to come within bond-forming distance of the carbon atom at which reaction occurs. Figure 7-5 is a representation of a molecular model of neopentyl bromide. It can be seen from this diagram that a nucleophile cannot readily approach the reaction center (designated by the heavily

* The designations $\alpha$, $\beta$, and so on, are commonly used when it is desired to indicate the position of a substituent with respect to a functional group: $\overset{\gamma}{C}-\overset{\beta}{C}-\overset{\alpha}{C}-X$. For example, $\beta$-chloroethyl acetate, $ClCH_2CH_2OCOCH_3$; $\beta,\beta'$-dichlorodiethyl ether, $ClCH_2CH_2OCH_2CH_2Cl$.

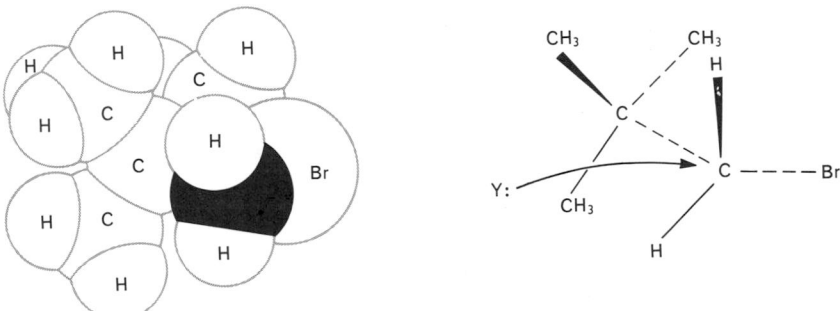

**Figure 7-5.** Molecular model of neopentyl bromide, showing sterically hindered environment of reaction center, $-CH_2Br$.

shaded carbon atom). The nature of the steric interference can be more readily seen by examination of a molecular model of neopentyl bromide.

***The $S_N1$ reaction.*** The data of Figure 7-3 show that a change in the mechanistic course of the displacement reaction takes place as the alkyl substituent in RX changes from a secondary to a tertiary group. Further, Figure 7-3 and the data in Table 7-4 show that although the *t*-butyl halide hydrolyzes at a very rapid rate, its reactivity by the $S_N2$ mechanism is very low. It is a reasonable conclusion from these observations that the *t*-butyl carbonium ion, $(CH_3)_3C^+$, is very readily formed, and thus that there must be some special property of this trisubstituted carbonium ion that lends it a high degree of stability. On the other hand, methyl halides, $CH_3X$, show little tendency to react by the $S_N1$ mechanism. Why is $(CH_3)_3C^+$ so much "better" a carbonium ion than $CH_3^+$?

The methyl carbonium ion would be expected to be quite unstable: it has an unfilled valence shell and a positive charge, and would be expected to have a strong attraction for an electron pair to complete its octet of valence electrons. The tendency, then, for the $CH_3$—X bond to ionize, even with solvent participation, would be very small. The experimental facts are in agreement with these expectations.

The *t*-butyl carbonium ion forms readily, although it, too, possesses a carbon atom with a positive charge and but six electrons in its valence orbitals. We are led to conclude that the presence of the three methyl groups provides a stabilizing influence that is lacking in the $CH_3^+$ ion.

In Chapter 4 (Section 4.13) it was stated that the methyl group exerts an inductive effect that is electron releasing: methylamine is a stronger base than ammonia; acetic acid is a weaker acid than formic acid (HCOOH) but stronger than propionic acid. This property of the methyl group affects the electron density on the carbonium carbon atom in $(CH_3)_3C^+$ in a way that can be represented as follows:

$$\overset{\displaystyle CH_3}{\underset{\displaystyle CH_3}{H_3C \rightarrow \overset{\downarrow}{\underset{\uparrow}{C}} \; +}} \tag{35}$$

Another (not unrelated) reason for the stabilizing effect of the methyl groups lies in the ability of the electrons of the H—C bond orbitals to contribute to, and thus partially to satisfy the electron-deficiency on the central carbon atom:

$$\tag{36}$$

This contribution, if carried to the extreme, would be represented formally as follows:

$$
\begin{array}{c}
\text{H} \quad \text{H}^+ \quad \text{CH}_3 \\
\text{C}=\text{C} \\
\text{H} \qquad \text{CH}_3
\end{array}
\tag{37}
$$

No such structure as (37) has actual existence; but it is apparent that the interaction represented in (36) has nine chances, and thus a high probability of taking place (because there are three $-CH_3$ groups), with the result that there is a high degree of *delocalization* of the positive charge. This statement can be paraphrased by saying that each of the nine hydrogen atoms of the methyl groups carries a small fractional positive charge, and thus the charge on the ion $(CH_3)_3C^+$ is dispersed over the whole molecule. The conclusion to be drawn from this is that what was at first regarded as a condition of instability—the electron-deficient central carbon atom—does not in fact exist: the *actual* structure of the *t*-butyl carbonium ion is not literally $(CH_3)_3C^+$ (although it is always written in this way). A true representation of the structure would be very difficult to write and a reasonable approximation would be (37), with the provision that eight other such formulas would have to be written, all of them being considered together to represent the single species.

An important corollary to this description of the *t*-butyl carbonium ion is that for the most effective electron delocalization of the kind pictured in (36) the ion must be symmetrically trigonal and planar. The three C C bonds are $sp^2$ hybrid bonds, and the $(CH_3)_3C^+$ ion would be expected to resemble the trimethylboron molecule, which is known to be planar and symmetrically trigonal (38):

$$
\begin{array}{cc}
\text{H}_3\text{C} \quad + \quad \text{CH}_3 & \text{H}_3\text{C} \quad \text{CH}_3 \\
\text{C} & \text{B} \\
\text{CH}_3 & \text{CH}_3
\end{array}
$$

both planar; bonds at 120°.

What if the carbonium carbon atom *cannot* assume a planar configuration? From the foregoing discussion, it would be concluded that the ability of the ion to become planar is a necessary condition for its formation. That this is indeed true is very elegantly shown by the behavior of the following compound:

$$
\begin{array}{ccc}
\begin{array}{c}
\text{Cl} \\
\text{C} \\
\text{H}_2\text{C} \qquad \text{CH}_2 \\
\text{H}_3\text{C}-\text{C}-\text{CH}_3 \\
\text{H}_2\text{C} \qquad \text{CH}_2 \\
\text{C} \\
\text{H}
\end{array}
&
\text{or}
&
\begin{array}{c}
\text{H}_3\text{C} \quad \text{CH}_3 \\
\text{C} \\
\text{HC} \qquad \text{C} \quad \text{Cl} \\
\text{CH}_2-\text{CH}_2 \\
\text{H}_2\text{C} \qquad \text{CH}_2
\end{array}
\end{array}
$$

$$(38)$$
$$(39)$$

A drawing of the molecular model of this compound is shown in Figure 7-6. This compound is remarkably inert to the displacement of the halogen: it is un-

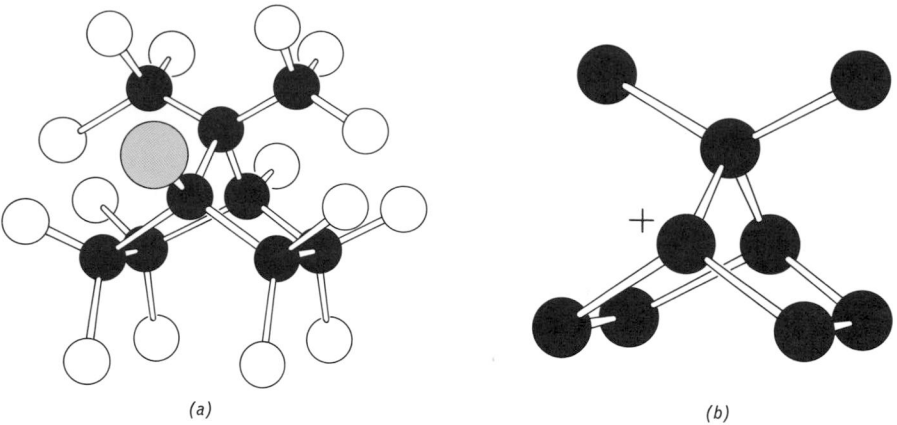

**Figure 7–6.** Ball-and-stick models of cyclic compounds (39) and (40):

In *(b)* the hydrogen atoms have been omitted in order that the carbon skeleton can be seen more clearly.

affected by heating with 30 % KOH for 21 hours, and does not react with boiling alcoholic silver nitrate (no silver chloride is formed). It is clear from the geometry of the molecule that the carbonium ion (40) that would be formed by ionization of the halogen cannot assume a planar configuration because of the rigid cage-like structure of the molecule.

(40)

The inability of the carbonium ion to adopt a planar configuration destroys the opportunity for stabilization by effective charge delocalization, and thus raises the energy of the transition state that leads to its formation. Furthermore, this cyclic, cage-like molecule *cannot* undergo $S_N2$ displacement, for it is manifestly impossible for a nucleophilic reagent to approach the carbon atom at the side opposite to the point of attachment of the chlorine atom. In consequence, the compound is inert to nucleophilic displacement by either mechanism.

*Other structural factors in displacement reactions.*    Structural interference with the formation of the transition state complex in $S_N2$ reactions can be of a number of kinds other than those described for the *t*-alkyl and neopentyl halides. The structure of the nucleophile may be such as to cause steric interference and to decrease the rate of the reaction. An example of this is found in the behavior of certain tertiary amines in the displacement reaction (41):

$$R_3N: + R'I \longrightarrow R_3\overset{+}{N}:R' + :I^- \tag{41}$$

Let us consider two tertiary amines, $R_3N$, similar in their general structure but differing markedly in their stereochemical features. These are triethylamine (42) and the cyclic compound quinuclidine (43):

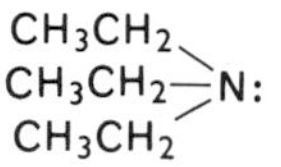

$$\tag{42}$$

triethylamine

or

$$\tag{43}$$

Quinuclidine

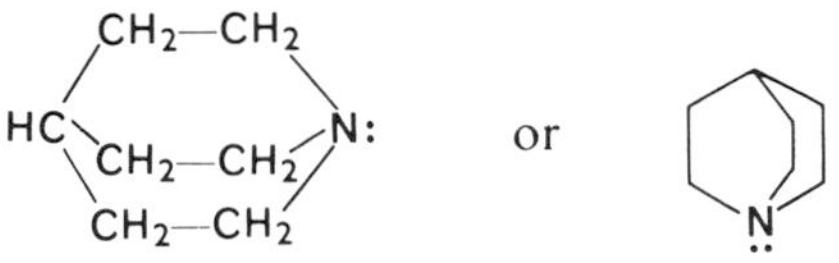

Quinuclidine has the same essential structural (i.e., electronic) characteristics of triethylamine, but the three groups attached to nitrogen are "tied back" and so present less steric interference to a molecule approaching the nitrogen atom. The rate constants for the reactions of these two amines in reaction (41) are shown for three alkyl halides in Table 7-7.

Quinuclidine reacts more rapidly than triethylamine in each case, but the *relative* reactivity of quinuclidine rises sharply with increasing substitution at the reaction center. It can be seen from these data that the steric interference between the ethyl groups of triethylamine and the methyl groups of isopropyl iodide assumes great importance:

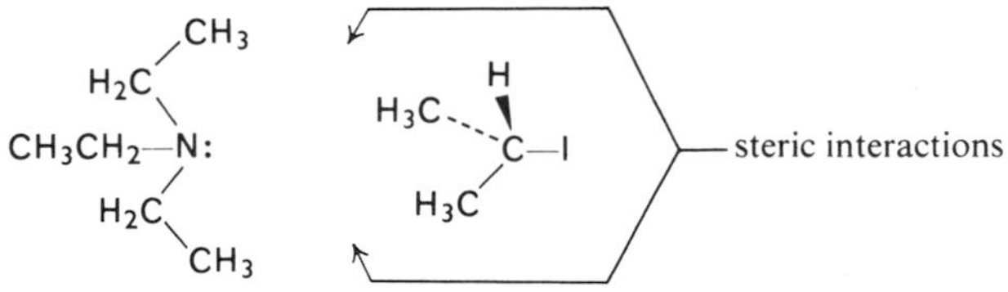

steric interactions

$$\tag{44}$$

Quinuclidine, on the other hand, in which the "alkyl" substituents on nitrogen are held away from the substituents on the alkyl halide, is less affected by increasing complexity of R' in R'I; thus, the relative rate of the reaction of the alkyl iodide with triethylamine falls markedly as R' becomes more highly substituted.

An important factor in the ionization of a tertiary alkyl halide in $S_N1$ reactions is the change in bond angles that takes place as ionization occurs. In *t*-butyl

**Table 7-7.** *Ratios of rate constants: k(quinuclidine)/k(triethylamine), in the reaction* $R'I + R_3N \rightarrow R'\overset{+}{N}R_3 + I^-$

| R′ in R′I | k(quinuclidine)/k(triethylamine) |
|---|---|
| $CH_3$ | 57 |
| $CH_3CH_2$ | 252 |
| $(CH_3)_2CH$ | 706 |

chloride, the central carbon atom is tetrahedral ($sp^3$), with bond angles between the substituents of about 110°. In the *t*-butyl carbonium ion, the bond angles are 120° ($sp^2$); thus, the formation of the ion results in a spreading of the bond angles with a diminution of the intergroup repulsive forces between the methyl groups. It would be expected that this relief of the "crowding" of the substituents would be more pronounced, the larger the substituents on the carbonium carbon atom. The data given in Table 7-8 show that this is indeed true. When the R groups in $R_3CCl$ are all *t*-butyl, the $S_N1$ hydrolysis (in 80% aqueous ethanol) of $(t\text{-Butyl})_3CCl$ is 600 times faster than that of $(CH_3)_3CCl$.

Cyclopropyl halides or sulfonic esters of cyclopropanol are almost inert to displacement by either the $S_N1$ or $S_N2$ mechanisms:

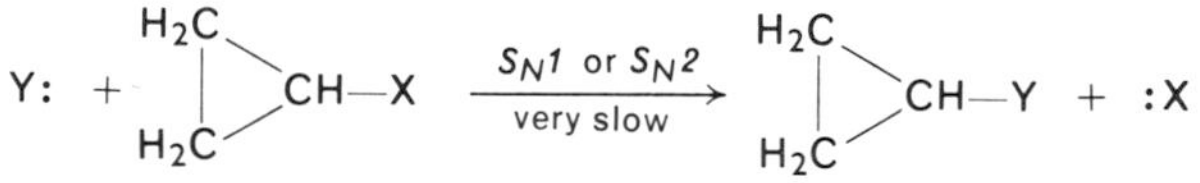
Y: +
H2C
H2C
CH—X
SN1 or SN2
very slow
H2C
H2C
CH—Y + :X

(45)

(X = halogen or $-OSO_2R$)

This is an example of the effect of the change in bond angle in the formation of the transition state. By either mechanism, the carbon atom at which reaction occurs tends to become planar in the transition state:*

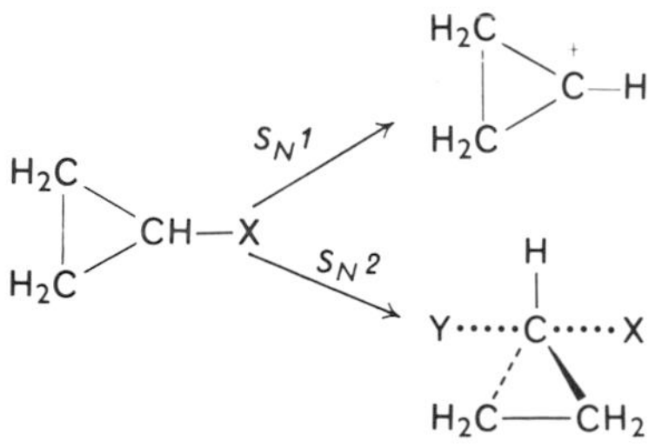
H2C
H2C
CH—X
SN1
SN2
H2C
H2C
C—H
H
Y·····C·····X
H2C
CH2

(46)

* The $S_N1$ *transition state* is not represented exactly by the structure shown, but must approximate this structure closely enough to allow the structure of the carbonium ion *intermediate* to be examined in connection with the arguments that are presented.

**Table 7-8.** *Relief of steric interaction in hydrolysis of t-alkyl chlorides (relative rates in 80% ethanol)*

| R in $R_3CCl$ | $Me_3$ | $Me_2(iPr)$ | $Et_2(t\text{-}Bu)$ | $(t\text{-}Bu)_3$ |
|---|---|---|---|---|
| Relative Rate | (1) | 14 | 48 | 600 |

$Me = CH_3$; $iPr = CH(CH_3)_2$; $Et = CH_2CH_3$; $t\text{-}Bu = C(CH_3)_3$.

For neither transition state can planarity of the carbon atom be accompanied by normal $sp^2$ hybridization (with $120°$ bond angles), because the C—C—C bond angle is necessarily $60°$. Since $sp^3$ hybridization, with $109.5°$ bond angles, is the most stable arrangement of the four bonds to carbon, a departure from this would impose an energy requirement. The change from

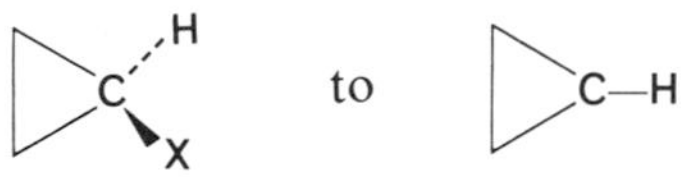

involves an increase in the degree of departure from the tetrahedral bond angle, and requires additional energy. This energy, which must be supplied to form the transition state, represents a rise in the energy of the barrier over which the displacement reaction must pass. The result is that reaction is very slow. The experimental fact is that the displacement reaction

$$\tag{47}$$

is not experimentally detectable, although the comparable reaction with *t*-butyl bromide, although slow in a nonionizing solvent such as dry acetone, proceeds at a measurable rate.

## 7.8 Stabilization of the carbonium ion intermediate

Stabilization of the $(CH_3)_3C^+$ ion by delocalization (dispersion) of the positive charge over the peripheral hydrogen atoms was discussed in Section 7-7, page 151. If the substituents attached to the carbon atom of C—X are capable of delocalizing (or in a manner of speaking, accommodating) the positive charge as it develops on the carbon atom as the transition state is approached, stabilization of the transition state and enhancement of the reaction rate will be the result. There are numerous ways in which this can happen, but most of these will be discussed later in this text. One example can be discussed to indicate the manner in which this stabilization occurs.

Ethers containing a halogen atom on the carbon atom adjacent ($\alpha$-) to the oxygen atoms are very reactive by both nucleophilic displacement mechanisms. An

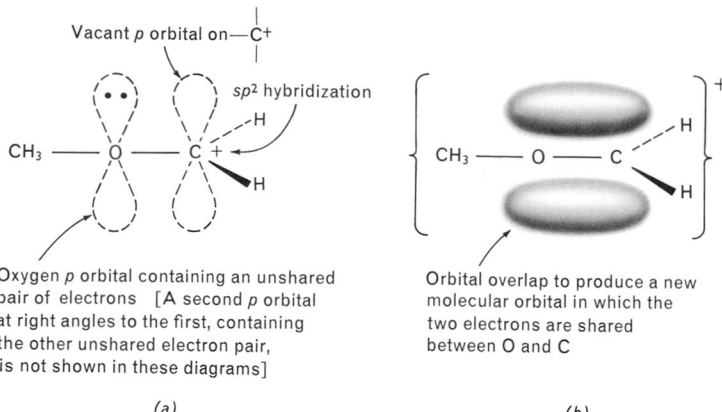

Oxygen *p* orbital containing an unshared pair of electrons [A second *p* orbital at right angles to the first, containing the other unshared electron pair, is not shown in these diagrams]

Orbital overlap to produce a new molecular orbital in which the two electrons are shared between O and C

(a)                                      (b)

**Figure 7–7.** Stabilization of the carbonium ion $CH_3OCH_2^+$ by charge delocalization. *(a)* The hypothetical "free" carbonium ion. *(b)* The actual ion with charge delocalized by orbital overlap.

example is chloromethyl ether, $ClCH_2OCH_3$, which undergoes hydrolysis at a very rapid rate. The reactivity of this compound under conditions favoring the bimolecular displacement reaction need not be dwelt upon, because it is clear that this halide resembles 1-chloropropane in its overall geometry, and thus that nucleophilic attack upon the $—CH_2Cl$ group is not subject to unusual steric interference.

In contrast to 1-chloropropane, whose reaction by way of a carbonium ion intermediate is slow (the primary carbonium ion $CH_3CH_2CH_2^+$ has no marked stabilizing features), chloromethyl ether and $\alpha$-haloethers in general are very reactive by the $S_N1$ route. This suggests that the transition state leading to the carbonium ion $CH_3OCH_2^+$ is unusually stable.

The reason for this stability can be seen by an examination of Figure 7-7, which presents a perspective view of the ion. The hypothetical unstabilized form shown in Figure 7-7a would not be expected to be more stable than $CH_3CH_2CH_2^+$; indeed, it would be less so, because of the inductive effect of oxygen, which would decrease the electron density on the carbonium carbon atom. The actual, stabilized ion shown in Figure 7-7b can be represented by the formal structure (48);*

$$\{\ CH_3O\overset{+}{—}CH_2\ \longleftrightarrow\ CH_3\overset{+}{O}{=}CH_2\ \} \tag{48}$$

The double-headed arrow used in (48) is a special symbol that is used to represent the kind of charge delocalization shown in Figure 7-7. Its meaning will be discussed further in Chapter 10.

---

\* Again it is to be noted that, although it is the stabilization of the transition state that is the important factor in determining the reaction rate, it is likely that the structural factors that stabilize this intermediate are those that affect the transition state as well.

**Exercise 1.** Show by an appropriate counting of the electrons on $CH_3\overset{+}{\ddot{O}}{=}CH_2$ why the oxygen carries a positive charge in this formula.

**Exercise 2.** Would you expect $CH_3CH{=}CHCH_2Cl$ to show high or low reactivity in the $S_N1$ displacement reaction? Why? Note: write the structure of the carbonium ion that would result from the loss of $Cl^-$ and examine it for stabilizing or destabilizing features.

## 7.9 The stereochemistry of the nucleophilic displacement reaction

An important feature of the bimolecular displacement reaction is shown in the following formulation (49):

$$
Y: \quad \overset{R''}{\underset{R}{\overset{R'}{\diagdown}}}C{-}X \quad \longrightarrow \quad Y{\cdots}\underset{R}{\overset{R' \quad R''}{\underset{|}{C}}}{\cdots}X \quad \longrightarrow \quad Y{-}\underset{R}{\overset{R''}{\diagup}}C\overset{R'}{\diagdown} \quad + \quad :X \qquad (49)
$$

The configuration of the carbon atom has been *inverted* in the course of the displacement of X; the carbon atom has been turned inside out, as an umbrella in a high wind.*

If the four substituents in the compound $R_3CX$ are different (e.g., as in 2-bromobutane), these can be arranged in two ways, and thus there can exist a pair of compounds that differ only in one respect; one of them rotates the plane of a beam of polarized light in a clockwise direction; the other in a counterclockwise direction. These two compounds are *optical isomers* (Chapter 3); they are non-identical mirror images (Section 3.17). The use of optically isomeric compounds in the study of the displacement reaction has shown that the picture of the reaction given in (49) is a correct one: every substitution reaction that follows the mechanistic course shown in (49) is accompanied by inversion of the configuration of the carbon atom upon which substitution takes place.

The experimental demonstration of inversion has been accomplished in the way shown in (50) [the letters *(a)*, *(b)*, and *(c)* refer to explanatory notes given below].†

*(a)* The conversion of the alcohol to the sulfonic ester does not involve breaking of the C—O bond; thus, there is no alteration of the configuration around the carbon atom.

---

\* The discussion of the stereochemistry of the tetrahedral carbon atom in Section 3.17 should be reviewed in conjunction with the study of this section.
† The generalized structures are shown in order to simplify the equations. In the actual experiment that was carried out, $R{=}C_5H_5CH_2$—(benzyl), $R'{=}CH_3$, and $R''{-}CH_3C_6H_5$ (*p*-tolyl).

$$(50)$$

*(b)* This is the displacement reaction, involving attack of the nucleophilic acetate ion and displacement of $R''SO_2O^-$ as the leaving group.

*(c)* The alkaline hydrolysis of the ester does not affect the C—O bond, and no change in configuration occurs in this step.

The conversion of the alcohol with an optical rotation of $+33°$ into its optical enantiomer (Gr. *enantios*, opposite) with rotation $-33°$ shows that *inversion of configuration has taken place at some point in this series of reactions*. Since it is known from much evidence that the formation of the sulfonic ester *(a)* and the hydrolysis of the acetate *(c)* do not involve inversion, the inversion must have occurred in the step involving the displacement of $R''SO_2O^-$ by attack of the acetate ion.

The stereochemical course of the unimolecular ($S_N1$) displacement reaction is neither so clear cut nor so predictable. The solvated carbonium ion may combine with the nucleophile at either side, with the result that the configuration of the product may be partially inverted with respect to that of the starting material, largely inverted, or inverted to a very small extent only (*retention* of configuration). The reason for this variability in the stereochemical result is that the carbonium ion may be symmetrically solvated on both sides, in which case the entering group may attack with equal probability at either side; or the nucleophile may attack at a stage in which solvation is heavier on one side than the other, in which case a predominance of one enantiomer may result; or the nucleophile may attack at a stage at which the leaving group is held in a "solvent cage" and has not left the vicinity of the carbon atom, in which case it will impede attack of the nucleophile at the "front" face, and thus promote predominant attack at the rear face, with inversion. These variable factors will be affected by experimental conditions: the nature of the solvent, the nature of the nucleophile and of the leaving group, and the conditions under which the reaction is performed. The usual stereochemical result in the $S_N1$ reaction is that some inversion occurs but the product is a mixture of both enantiomers.

### 7.10  Internal nucleophilic displacement. Ring formation

If the attacking nucleophilic center and the carbon atom that carries the leaving group are a part of the same molecule, the displacement reaction will lead to the formation of a cyclic product. An example of this is found in the formation of an ethylene oxide from a $\beta$-halo alcohol (51):

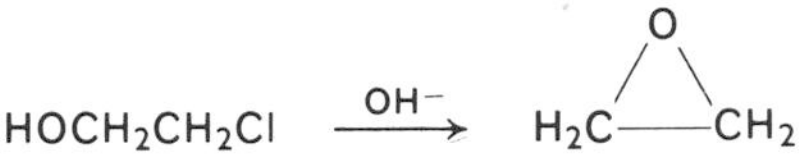

$$HOCH_2CH_2Cl \xrightarrow{OH^-} H_2C\overset{O}{\underset{\diagdown\diagup}{\text{——}}}CH_2 \tag{51}$$

The steps in this reaction are the following:

1. The proton exchange between the hydroxide ion and the alcoholic hydroxyl group leads to the acid-base equilibrium (52):

$$HO^- + HOCH_2CH_2Cl \rightleftharpoons {}^-O\text{—}CH_2CH_2Cl + H_2O \tag{52}$$

2. The conversion of —OH into —O$^-$ greatly enhances the nucleophilic character of the oxygen atom, with the result that an *internal* nucleophilic displacement can occur (53):

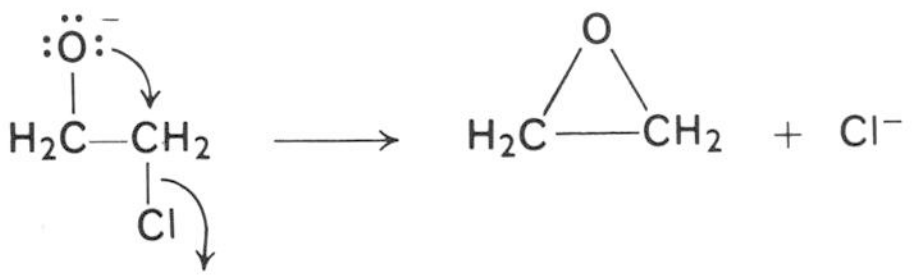

$$\underset{\underset{Cl}{|}}{\overset{\overset{\ddot{O}:{}^-}{|}}{H_2C\text{—}CH_2}} \longrightarrow H_2C\overset{O}{\underset{\diagdown\diagup}{\text{——}}}CH_2 \ +\ Cl^- \tag{53}$$

The reaction of a dihalide with ammonia or an amine can lead to ring formation by a comparable route (54):

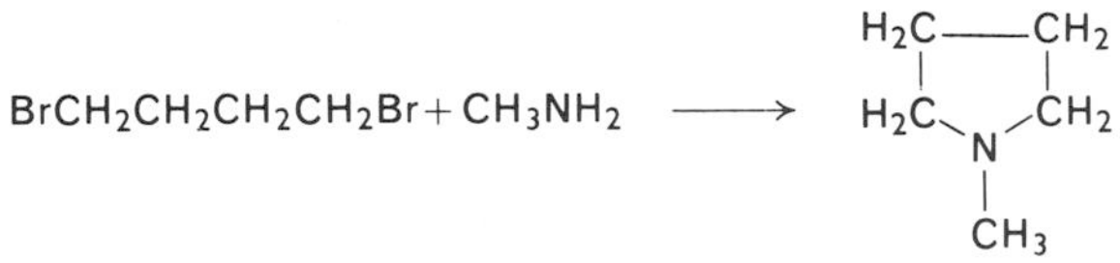

$$BrCH_2CH_2CH_2CH_2Br + CH_3NH_2 \longrightarrow \begin{array}{c} H_2C\text{——}CH_2 \\ |\qquad\quad| \\ H_2C\diagdown_{\underset{|}{N}}\diagup CH_2 \\ CH_3 \end{array} \tag{54}$$

This reaction takes place in a series of steps:
1. The initial displacement reaction is the quite unexceptional step (55):

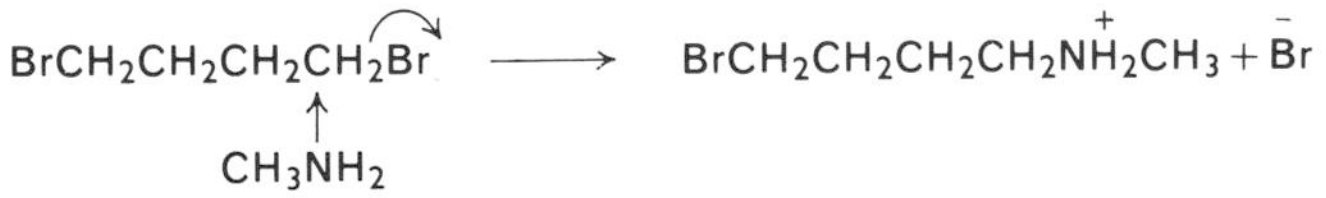

$$\underset{\underset{CH_3NH_2}{\uparrow}}{BrCH_2CH_2CH_2CH_2Br} \longrightarrow BrCH_2CH_2CH_2CH_2\overset{+}{N}H_2CH_3 + \overset{-}{Br} \tag{55}$$

2. Proton exchange between the product of reaction (54) and excess $CH_3NH_2$ generates the free amine (56):

$$BrCH_2CH_2CH_2CH_2\overset{+}{N}H_2CH_3 + CH_3NH_2 \rightleftharpoons BrCH_2CH_2CH_2CH_2NHCH_3 + CH_3\overset{+}{N}H_3 \tag{56}$$

3. Internal nucleophilic displacement in the bromo amine leads to ring formation (57):

$$\underset{\substack{\text{Br} \\ \text{CH}_3}}{\overset{\substack{\text{H}_2\text{C} - \text{CH}_2 \\ | \qquad | \\ \text{H}_2\text{C} \quad \text{CH}_2 \\ \diagup \quad \text{NH}}}{}} \longrightarrow \underset{\substack{\text{H} \quad \text{CH}_3}}{\overset{\substack{\text{H}_2\text{C} - \text{CH}_2 \\ | \qquad | \\ \text{H}_2\text{C} \underset{+}{\text{N}} \text{CH}_2}}{}} + \text{Br}^- \tag{57}$$

N-methylpyrrolidine (protonated)

The cyclic ammonium salt is but the protonated form of the cyclic amine, into which it can be transformed by any strong base. In practice, the reaction mixture would be made alkaline with sodium hydroxide and the cyclic amine isolated by extraction with some suitable organic solvent.

The synthetic method illustrated in Equations (53) to (56) is a general and valuable procedure for the preparation of cyclic amines. It is most widely applied in the preparation of five- and six-membered rings. Obvious variants are possible; for example, the treatment of the following bromo-substituted ammonium salt with aqueous alkali leads to ring closure to form N-methylpiperidine:

$$\text{BrCH}_2\text{CH}_2\text{CH}_2\text{CH}_2\text{CH}_2\overset{+}{\text{N}}\text{H}_2\text{CH}_3 \xrightarrow{\text{NaOH}} \text{N-methylpiperidine} \tag{58}$$

---

**Exercise 3.**   Formulate the steps in the reaction summarized in (58), and write the structure of N-methylpiperidine. Why will ring closure not occur before alkali is added?

---

## Exercises

(Exercises 1–3 will be found within the text.)

**4** Show all the details in the preparation of 1-bromopropane by the reaction of 1-propanol with a mixture of NaBr, $H_2O$, and $H_2SO_4$.

**5** Explain why $ClCH_2OCH_2CH_3$ is hydrolyzed more rapidly than $CH_3OCH_2CH_2Cl$.

**6** Allyl halides ($CH_2{=}CHCH_2X$) are very reactive by the $S_N1$ displacement reaction. Suggest a reason for stabilization of the $CH_2{=}CHCH_2^+$ ion.

**7** Write equations for the reactions between each of the following pairs of compounds (assume that the reactions are carried out in alcoholic or aqueous-alcoholic solutions):
   (a) ethyl bromide and potassium acetate
   (b) allyl chloride and trimethylamine
   (c) 1-bromopentane and potassium cyanide

   (d) methyl iodide and potassium t-butoxide ($\overset{+}{\text{K}}\overset{-}{\text{O}}\text{C(CH}_3)_3$) in t-butyl alcoholic solution
   (e) benzyl bromide and sodium hydroxide
   (f) ethyl methanesulfonate ($CH_3SO_2OC_2H_5$) and diethylamine.

**8** Draw estimated energy profiles for the alkaline hydrolysis (dilute aqueous alcoholic sodium hydroxide) of 2-bromopropane, showing the two chief mechanistic pathways (refer to Figure 7-3).

# Chapter eight / Alcohols and Ethers. The Hydroxyl Group

## 8.1 Alcohols

The name "alcohol" is applied to compounds that contain the hydroxyl group, —OH, attached to carbon in the following ways:

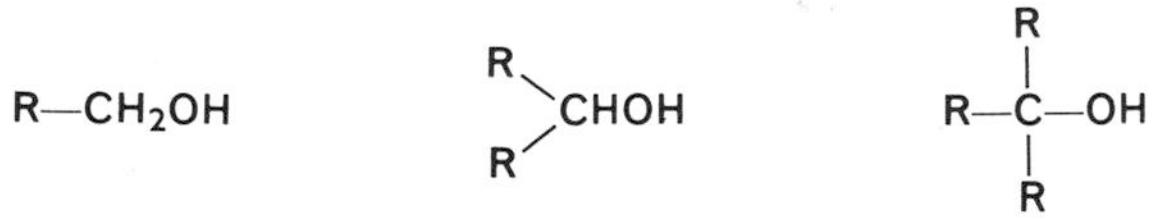

$$R—CH_2OH \qquad \begin{matrix} R \\ \diagdown \\ \diagup \\ R \end{matrix} CHOH \qquad R—\overset{\displaystyle R}{\underset{\displaystyle R}{\overset{|}{\underset{|}{C}}}}—OH$$

The groups R in these generalized formulas can include a wide variety of kinds of groupings, from simple alkyl groups to very complex structures. The simple alcohols (in which the R groups are the lower alkyl groups) include the well-known compounds methanol (methyl alcohol), ethanol (ethyl alcohol), isopropyl alcohol (2-propanol), and others in which, as in these, a hydrogen atom of an alkane is replaced by a hydroxyl group.

*Ethyl alcohol*, or ethanol (G), is the compound that is often referred to simply as "alcohol"; the phrase "an alcoholic solution" ordinarily refers to a solution in ethanol. Ethanol is one of the oldest known organic compounds because it is formed in the fermentation of fruit juices by the action of natural yeasts, with the formation of wines and ciders. The use of alcoholic beverages has been practiced by mankind since the dawn of history. Indeed, the fermentation of sugar, either as such (molasses) or as it is formed by the action of enzymes from starch (corn, potatoes, rice), is still an important industrial process for the manufacture of alcohol. When starch is the raw material it must first be converted into sugar (glucose) by enzymatic action. The subsequent conversion of sugar into alcohol can be represented by the overall reaction

$$C_6H_{12}O_6 \xrightarrow{\text{enzymes}} 2C_2H_5OH + 2CO_2 \tag{1}$$
glucose

the stages of which are described in a later chapter. The immediate product of the fermentation of an aqueous sugar solution is an aqueous solution containing up to about 15% of ethanol. The separation of the ethanol is accomplished by distillation. Beverage alcohols may contain traces of flavor elements derived from the source from which they are obtained (brandy from grapes, whiskies from grains), or may be essentially free of such distinctive substances (vodka). In general, they are used in the form of about 40% aqueous solutions ("spirits"). The undistilled solutions of fermented fruits—wines and beer—contain from about 3% to 12% of alcohol. Wines containing up to about 20% alcohol (e.g., sherry) are "fortified," usually with grape brandy.

Industrial alcohol is made by fermentation (usually of molasses, the residue from sugar refining), or from ethylene (Section 8.21). Ordinary alcohol is about 95% ethanol and 5% water (by volume) because this is the composition of the mixture that boils at a constant temperature (78.13°) below that of pure ethanol (78.3°). "Absolute" alcohol can be made by removing the water from "95%" alcohol by chemical or other means.

Besides the wide use of ethyl alcohol for beverage purposes (the American public spends over 10 billion dollars a year for alcoholic beverages), it is a very important industrial chemical. As it is innocuous when ingested in small amounts it finds wide use as a vehicle for drugs and in the preparation of tinctures, fluid extracts, elixirs, and other medicines. Ethanol finds wide use in industry as a solvent and as a primary raw material for the preparation of other industrial chemicals.

*Methyl alcohol* [methanol (G)] is also an important industrial chemical. Since methanol was formerly made by the destructive distillation of wood, it came to be called "wood alcohol"; but this method of manufacture has been largely superseded by a synthetic process. In this process a mixture of carbon monoxide and hydrogen is passed over a special catalyst at temperatures of 300 to 400°C and under high pressures. The reaction that occurs

$$CO + 2H_2 \longrightarrow CH_3OH \tag{2}$$

can be carried out with nearly quantitative yields.

Methanol is used to some extent as an "anti-freeze," as a denaturing agent for ethanol, for the manufacture of formaldehyde, as a solvent, and in many other ways. Methanol is a toxic substance, and when ingested can cause death or blindness. Small amounts of it (about 5%) may be added to ethanol to produce "denatured alcohol," which is suitable for most purposes for which ethanol is used but which must not be taken internally.

*Higher alcohols*, such as *n*-propyl alcohol, isopropyl alcohol, and the butyl alcohols, are made in a variety of ways and from various raw materials. A number of alcohols are made from the olefins that are by-products of petroleum refining. For example, in the commercial production of 2-propanol, propylene is "hydrated" by the addition of sulfuric acid and subsequent hydrolysis (Section 8.21), or by the direct addition of water at high temperature and pressure in the presence of a catalyst:

$$CH_3CH{=}CH_2 + H_2SO_4 \longrightarrow CH_3\overset{\displaystyle OSO_3H}{\underset{|}{C}}HCH_3 \xrightarrow{H_2O} CH_3\overset{\displaystyle OH}{\underset{|}{C}}HCH_3 + H_2SO_4 \qquad (3)$$

## 8.2 Properties of the simple alcohols

The lower-boiling members of the simple alkanols (methanol, ethanol, the propyl alcohols) are colorless, mobile liquids, completely miscible with water. As the number of carbon atoms increases, the water solubility drops; although *t*-butyl alcohol is miscible with water, the other butyl alcohols and alcohols of five or more carbon atoms are not. (Question: Why is *t*-butyl alcohol more soluble in water than 1-butanol? Refer to the discussion in Section 4.16.)

The boiling points of alcohols are higher than those of ethers and paraffin hydrocarbons of comparable molecular weights (Section 4.15) because of their association in the liquid state, as a result of hydrogen bonding. In compounds that contain more than one hydroxyl group, such as ethylene glycol ($HOCH_2CH_2OH$), the increased opportunity for association through hydrogen bonding results in markedly increased boiling points. For instance, 1-propanol (mol. wt. 60) boils at 98°C; ethylene glycol (mol. wt. 62) boils at 197°C. Glycerol, or 1,2,3-propanetriol, with three hydroxyl groups, boils at 290°C. The increased opportunity for hydrogen bonding in glycols and triols is reflected in their water solubility: glycerol (mol. wt. 92) is completely miscible with water, whereas the monohydroxy compound of comparable molecular weight (1-pentanol, mol. wt 88) is only slightly soluble in water.

Long-chain and high-molecular-weight alcohols [stearyl alcohol, $CH_3(CH_2)_{16}CH_2OH$, and cholesterol (Section 32.6) are examples] are often waxy solids, soluble in hydrocarbon solvents such as hexane and benzene but quite insoluble in water.

When economic considerations have permitted, ethanol and methanol have been used as fuels. However, since alcohols are in a partially oxidized state, compared with paraffin hydrocarbons, their heats of combustion are lower. The heat of combustion of ethanol is 328 kcal/mole; that of hexane, the hydrocarbon of comparable boiling point, is 991 kcal/mole, that of propane, the hydrocarbon of comparable molecular weight, is 526 kcal/mole.

## 8.3 Naming of alcohols

The saturated alcohols are hydroxyalkanes, or *alkanols*, and bear I.U.C. names that are formed according to the rules described in Chapter 6. Many alcohols are commonly called by trivial names; and substitution names and names based on

**Table 8-1.** *Names of some representative compounds*

| Compound | Name |
|---|---|
| $CH_3OH$ | methanol (G)*<br>methyl alcohol<br>carbinol† |
| $CH_3CH_2OH$ | ethanol (G)<br>ethyl alcohol |
| $CH_3CH_2CH_2CH_2OH$ | 1-butanol (G)<br>n-butyl alcohol‡ |
| $CH_3CH_2CHCH_2CH_3$<br>      $\vert$<br>      OH | 3-pentanol (G)<br>diethylcarbinol |
| $CH_3$<br>$\vert$<br>$CH_3-C-OH$<br>$\vert$<br>$CH_3$ | t-butyl alcohol<br>trimethylcarbinol<br>2-methyl-2-propanol (G) |
| $CH_3-CH-CH_2-CHOH$<br>    $\vert$       $\vert$<br>   $CH_3$     $CH_3$ | 4-methyl-2-pentanol (G)<br>methylisobutylcarbinol |
| $CH_2{=}CHCH_2OH$ | allyl alcohol<br>vinylcarbinol<br>2-propen-1-ol (G) |
| $CH_2{=}CH-CH-CH{=}CH_2$<br>           $\vert$<br>         OH | divinylcarbinol<br>1,4-pentadien-3-ol (G) |
| 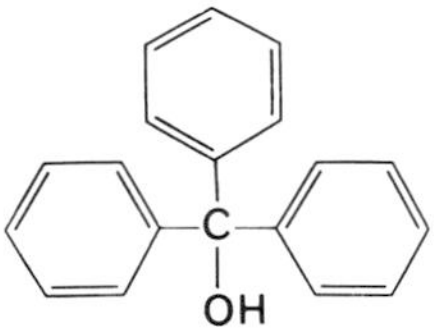 (triphenylcarbinol structure) | triphenylcarbinol<br>triphenylmethanol |
| $CH_3$<br>$\vert$<br>$CH_3-C-CH_2OH$<br>$\vert$<br>$CH_3$ | neopentyl alcohol<br>t-butylcarbinol<br>2,2-dimethyl-1-propanol (G) |
| $CH_3CH_2CH_2CH_2CH_2CH_2OH$ | 1-hexanol (G)<br>n-hexyl alcohol |
|    Cl   Cl<br>   $\vert$   $\vert$<br>$CH_3CH-C-CH_2OH$<br>       $\vert$<br>       Cl | 2,2,3-trichloro-1-butanol (G) |
| $CH_3$<br>     $\diagdown$<br>       CHOH<br>     $\diagup$<br>$CH_3$ | isopropyl alcohol§<br>2-propanol (G) |
| $BrCH_2CH_2OH$ | β-bromoethyl alcohol<br>2-bromoethanol (G) |

Footnotes for this table at foot of facing page.

the carbinol system are also employed. In Table 8-1 are found examples of some representative compounds. When more than one name is given, the first is the one that is likely to be used by most organic chemists, and the others, which are correct names, would probably be encountered less frequently or not at all. Whenever the choice of a suitable name is in question, the I.U.C. name is always acceptable.

The alcohols that are named in the above list include three different classes: those that contain the grouping —$CH_2OH$; those that contain —CH—; and those

$$\overset{|}{OH}$$

that contain —$\overset{|}{\underset{|}{C}}$—OH (the remaining bonds in each case being attached to carbon). These are described by the terms *primary*, *secondary*, and *tertiary*:

| $RCH_2OH$ | $R_2CHOH$ | $R_3COH$ |
|---|---|---|
| primary alcohol, | secondary alcohol | tertiary alcohol |

It should be noted that *sec*-butylcarbinol is a primary alcohol, *sec*-butyl alcohol is a secondary alcohol, di-*t*-butylcarbinol is a secondary alcohol, and tri-isopropylcarbinol is a tertiary alcohol:

$$CH_3CH_2\overset{\overset{\displaystyle CH_3}{|}}{CH}CH_2OH \qquad CH_3CH_2\overset{\overset{\displaystyle CH_3}{|}}{CH}OH \qquad (CH_3)_3CC\overset{\overset{\displaystyle OH}{|}}{H}C(CH_3)_3$$

sec-butylcarbinol $\qquad$ *sec*-butyl alcohol $\qquad$ di-*t*-butylcarbinol

$$(CH_3)_2CH\overset{\overset{\displaystyle CH(CH_3)_2}{|}}{\underset{\underset{\displaystyle CH(CH_3)_2}{|}}{C}}OH$$

tri-isopropylcarbinol

## 8.4 The hydroxyl group

The important chemistry of alcohols is essentially that of the hydroxyl group and the carbon-oxygen bond. But we cannot describe the chemical behavior of alcohols simply by making general statements about a general alcohol R—OH. Differences are found between the chemical behavior of compounds of this general type: when R is a primary, secondary, or tertiary alkyl group; when R is saturated or unsaturated; and when R is unsaturated, depending upon the location of —OH with respect to the double bond. However, a great deal can be said about the chemistry of alcohols by considering the properties of the hydroxyl function. Many of the

---

\* The symbol (G) means the name is the I.U.C. name.

† The name carbinol is not used to refer to the compound itself, but forms the basis for the carbinol system of naming. Nevertheless, "carbinol" *is* a name for $CH_3OH$.

‡ The prefixes *n* (normal), *i-* (iso), *t-* or *tert-* (tertiary), and *sec-* (secondary) are italicized; they are usually used as the abbreviations, although such names as "isopropyl" are quite common.

§ It is improper to use names that combine parts of two systems of nomenclature. Such names as isopropanol, *t*-butanol, *n*-hexanol, β-bromoethanol, should not be used.

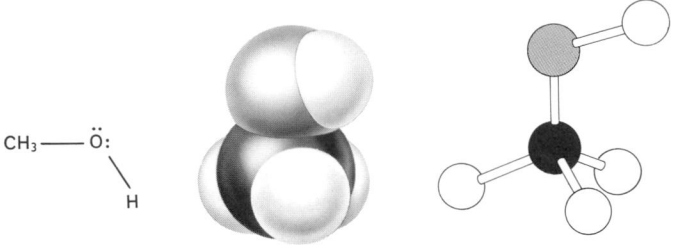

$CH_3 \overset{..}{\underset{\diagdown}{O}}:$
$\qquad H$

**Figure 8–1.** Conventional formula, molecular model, and ball-and-stick model of methanol, $CH_3OH$.

reactions of alcohols of all types proceed by way of initial stages that are common to all of them, diverging in a quantitative sense, and often not in a qualitative one, in subsequent stages of the process. That is to say, the nature of R affects a given *kind* of reaction chiefly in a matter of degree, and *chiefly in the effect of the nature of R upon the characteristics of the carbon-oxygen bond.*

Electronically, the hydroxyl group consists of an oxygen atom bonded to hydrogen and to a carbon atom, thus utilizing two of the six oxygen valence electrons and two of the four available orbitals (Figure 8-1). The C—O—H angle is in the neighborhood of (but smaller than) 109°.* Ordinarily, this angular nature of the bond is not made explicit in the formula, ROH usually being used. Whether the unshared electron pairs on oxygen are shown $(R\overset{..}{\underset{..}{O}}H)$ may depend upon whether it is desired to direct attention to their involvement in the reaction being discussed.

## 8.5 Alcohols as weak organic bases

Alcohols are weak bases, for the oxygen atom, with its two unshared pairs of valence electrons, has the properties of a *nucleophile* and can accept a proton.

In general, the initial protonation of the alcohol can be written:

$$ROH + HA \;\rightleftharpoons\; \left\{ R-O\!\!\begin{array}{c} {}^{H} \\ {}_{\diagdown H} \end{array} \right\}^{+} + \; A^{-} \tag{4}$$

alkyl oxonium ion

The degree to which the protonation occurs is dependent upon the nature of R and the strength of the acid HA. The basicity of alcohols is of the order of

---

* The C—O—H bond angle in methanol is similar to, and can be discussed in the same terms as in the discussion of the H—O—H bond angle in water (Chapter 3). The four orbitals of oxygen in the alcohol are hybrid orbitals, approximating *sp*³ orbitals in character.

magnitude of that of water. That is, strong acids will ionize (dissociate) in alcohols according to expression (4), and such solutions will contain $ROH_2^+$ (alkyl oxonium ions).

The protonation step is the first stage of most reactions of alcohols with acids, but the reactions of greatest interest to us are the subsequent events that follow protonation.

## 8.6 Effect of protonation upon the C—O bond

It will be helpful to further discussion to examine the nature of the carbon-oxygen bond in an alcohol and the way in which its properties are altered by protonation. Let us take methanol, the simplest primary alcohol, as an example. In methanol, the electronic structure of which is

$$H_3C\diagdown\ddot{\underset{H}{\overset{}{O}}}:$$

carbon is joined to the more (than carbon) electronegative element oxygen. The carbon-oxygen bond thus has partial ionic character, with the oxygen having the somewhat greater share of the binding pair of electrons. The oxygen-hydrogen bond is similarly polarized, and so the methanol molecule is a dipolar substance with oxygen relatively more negative than the alkyl group and the hydrogen atom. The dipole moment (1.7 D) of methanol is in accord with this concept (Section 3.11).

Suppose now that methanol is added to a solution of concentrated aqueous HBr. The initial equilibria will be the simple proton exchanges

$$H_2O + HBr \rightleftharpoons H_3O^+ + Br^- \tag{5}$$

$$H_3O^+ + CH_3OH \rightleftharpoons H_2O + CH_3OH_2^+ \tag{6}$$

and the solution will contain $CH_3OH_2^+$, $H_3O^+$, and $Br^-$ ions, as well as the undissociated species $CH_3OH$ and $H_2O$.

The methyl oxonium ion

$$H_3C\diagdown\overset{+}{\underset{\underset{H}{|}}{\ddot{O}}}\diagup H$$

*will contain a greatly weakened C—O bond*, since the positive charge on oxygen will attract the electrons of the carbon-oxygen bond toward the oxygen nucleus. The important result of this withdrawal of electrons from the carbon atom is to render the carbon atom *electron-deficient*. The consequence of this electron deficiency of the carbon atom is to permit it to accept electrons from an available nucleophile, $X:^-$,

$$X:^- \frown CH_3 — OH_2^+ \tag{7}$$

The ensuing process, through the transition state,

$$X^- \cdots CH_3 \cdots OH_2^+ \tag{8}$$

proceeds as described in Chapter 7; the leaving group $H_2O$ is displaced by the attack of $X:^-$. In a solution in which the concentration of bromide ion is high, a major part of the reaction will proceed to yield $CH_3Br$. An important by-product of a reaction of this kind results from the presence of unprotonated methanol in the reaction mixture. Methanol is, of course, a nucleophile, and can also take part in the reaction as in Equation (9):

$$\underset{H}{\overset{H_3C}{>}}\ddot{O}:\ \overset{+}{CH_3-OH_2} \longrightarrow \underset{H}{\overset{H_3C}{>}}\overset{+}{\underset{|}{\ddot{O}}}\overset{CH_3}{<} + H_2O \tag{9}$$

The product of this reaction is simply protonated dimethyl ether, or the dimethyl oxonium ion. Thus, dimethyl ether is one of the *products* of the reaction.

Throughout this text we shall often find that the *product* of a reaction makes its first appearance in an ionized form. For example, when the synthesis of an acid is carried out in alkaline solution, the anion ($RCOO^-$) instead of the free acid ($RCOOH$) is present until the solution is acidified. In reaction (9) the protonated form of dimethyl ether and not the ether itself is the immediate product. If a solution of dimethyl ether in a strong acid were poured into a large excess of water, the ether would be liberated because the base water, in large excess, would possess the major share of the protons:

$$\underset{H}{\overset{H_3C}{>}}\overset{+}{\underset{|}{O}}\overset{CH_3}{<} + H_2O \ \rightleftharpoons \ CH_3OCH_3 + H_3O^+ \tag{10}$$

large excess of $H_2O$ drives
equilibrium to the right

Consequently, we shall often ignore the niceties of the state of ionization, since the addition and removal of protons is simply a matter of $pH$ adjustment. For instance, we shall often write such summary equations as

$$CH_3CN + 2H_2O \xrightarrow{NaOH} CH_3COOH + NH_3 \tag{11}$$

$$2CH_3CH_2OH \xrightarrow{H_2SO_4} CH_3CH_2OCH_2CH_3 + H_2O \tag{12}$$

because they describe the result that is obtained after acidifying an alkaline reaction mixture, making basic an acidic one, or pouring a strongly acidic solution into a large volume of water.

## 8.7    The reaction of ethanol with sulfuric acid

Let us now examine in detail a series of transformations that take place when a typical primary alcohol, ethanol, is allowed to react with the strong acid $H_2SO_4$.

In a solution of concentrated sulfuric acid in ethanol, we find the equilibria shown below. This example is given here to show that what appears to be a very complex set of reactions is in reality much less complicated than it seems. The student will find it advantageous to recognize that what appears to be an involved situation is often no more than a group of individual examples of one or two simple principles.

$$CH_3CH_2OH + H_2SO_4 \rightleftharpoons CH_3CH_2OH_2^+ + HSO_4^-* \tag{13}$$

$$HSO_4^- + EtOH_2^+ \rightleftharpoons EtOSO_3H + H_2O \tag{14}$$

$$H_2O + H_2SO_4 \rightleftharpoons H_3O^+ + HSO_4^- \tag{15}$$

$$EtOH + EtOSO_3H \rightleftharpoons EtOH_2^+ + EtOSO_3^- \tag{16}$$

$$EtOSO_3^- + EtOH_2^+ \rightleftharpoons EtOSO_2OEt + H_2O \tag{17}$$

$$EtOH + EtOH_2^+ \rightleftharpoons Et_2OH^+ + H_2O \tag{18}$$

$$EtOH + EtOSO_3H \rightleftharpoons Et_2OH^+ + HSO_4^- \tag{19}$$

$$Et_2OH^+ + EtOH \rightleftharpoons Et_2O + EtOH_2^+ \tag{20}$$

This (at first sight, appalling) set of equations is much less formidable than it seems because it contains only two kinds of reactions: (13), (15), (16), and (20) are simple proton exchanges from an acid to a base; and the remaining equations are displacement reactions on carbon. We shall analyze these in detail because of the insight that can be gained into these universally important kinds of reactions.

Equations (13), (15), (16), and (20) are proton exchanges of the type $H:A + :B \rightleftharpoons H:B + :A$.

| H:A | :B | H:B | :A | |
|---|---|---|---|---|
| $H_2SO_4$ | EtOH | $EtOH_2^+$ | $HSO_4^-$ | (13) |
| $H_2SO_4$ | $H_2O$ | $H_3O^+$ | $HSO_4^-$ | (15) |
| $EtOSO_3H$ | EtOH | $EtOH_2^+$ | $EtOSO_3^-$ | (16) |
| $Et_2OH^+$ | EtOH | $EtOH_2^+$ | $Et_2O$ | (20) |

The remaining reactions are nucleophilic displacements on carbon of the type $Y: + A:X \rightarrow Y:A + :X$.

| Y: | A:X | Y:A | :X | |
|---|---|---|---|---|
| $HSO_4^-$ | $Et:OH_2^+$ | $HSO_4Et$ | $H_2O$ | (14) |
| $EtOSO_3^-$ | $EtOH_2^+$ | $EtOSO_2OEt$ | $H_2O$ | (17) |
| EtOH | $EtOH_2^+$ | Et—O—Et$^+$<br>    &#124;<br>    H | $H_2O$ | (18) |
| EtOH | $EtOSO_3H$ | $Et_2OH^+$ | $OSO_3H^-$ | (19) |

---

* In subsequent steps we shall use Et as a symbol for the ethyl group; thus, ethanol $C_2H_5OH$, is EtOH.

---

**Exercise 1.**   Expand the list of equations (14)–(20) by adding others of the two types described that utilize the species found in the solution. One added example is the displacement $Et_2OH^+ + HSO_4^- \rightarrow EtOSO_3H + EtOH$.

---

Nothing has yet been said of the constitution of the reaction mixture or of the products that can actually be *isolated*, because the answers to these questions depend upon the proportions of ethanol and sulfuric acid and upon the temperature at which the reaction is conducted.

## 8.8   Diethyl sulfate

At low temperatures, and with excess sulfuric acid, the chief constituent of the mixture is ethyl hydrogen sulfate, $EtOSO_3H$. If the mixture is now heated under reduced pressure to a temperature at which diethyl sulfate, $EtOSO_2OEt$, distils, the removal of the diethyl sulfate shifts the equilibrium $EtOSO_3H + EtOSO_3^- \rightleftharpoons EtOSO_2OEt + HSO_4^-$ to the right, with the result that the *overall* equation

$$2EtOH + H_2SO_4 \longrightarrow Et_2SO_4 + 2H_2O \tag{21}$$

represents the result of the process as a preparative operation.

## 8.9   Diethyl ether

If the temperature of the ethanol-sulfuric acid mixture is raised to about 140°C, and additional ethanol is introduced into the heated mixture, diethyl ether [e.g., (20)] distils out of the mixture. In this case the *overall* result of the reaction is represented by

$$2EtOH \longrightarrow Et_2O + H_2O \tag{22}$$

This process can be made continuous by adding alcohol as the ether is removed by distillation, but the reaction is eventually brought to a halt by accumulation of the water in the reaction mixture, and by the gradual accumulation of minor by-products (for example, oxidation products). Nevertheless, it can be seen that a small amount of sulfuric acid can catalyze the conversion of a relatively large quantity of ethanol into diethyl ether, since the acid is not consumed in the process.

## 8.10   Ethylene

At high temperatures (about 170°C) a new reaction, which we have not yet described, begins to assume importance. This reaction results in the formation of *ethylene*, by the overall process

$$H-CH_2-CH_2-X \longrightarrow H-X + CH_2{=}CH_2 \tag{23}$$

The increased formation of ethylene with higher temperatures illustrates a reaction principle that is often an important factor in the performance of an organic synthetic operation.

The rate of a reaction, expressed by the rate constant $k$, increases with temperature for reasons that were discussed in Chapter 5. This can be expressed in the equation relating the specific reaction rate constant to the activation energy as a function of the temperature:

$$\frac{\log_e k}{dT} = -\frac{A}{RT^2} \tag{24}$$

or

$$\log_e k = \text{a constant,} \; -\frac{A}{RT} \tag{25}$$

Thus, if $\log k$ is plotted against $1/T$, the resulting curve should be a straight line whose slope is $-A/R$. This is an important relationship, for it permits the calculation of $A$ from the reaction rate measured at two (or more) temperatures.

Now let us suppose we have two reactions occurring simultaneously, with different activation energies. The plot of Figure 8-2 shows a plot of the rate constants *versus* temperature; it is evident that of the two lines, that for the reaction with the higher activation energy (reaction 1) will have the greatest slope ($A/R$, negative slope).

At the lower temperature, $T_L$, reaction 2 is proceeding faster than reaction 1; but at the higher temperature $T_H$, reaction 1 is much the faster of the two. That is, the relative importance of reaction 1 increases with increasing temperature. If reaction 1 is the undesirable reaction, leading to unwanted by-products, it will be favored at higher temperature and so it would be advantageous to impose careful temperature control in carrying out the synthetic operation.

The formation of ethylene from ethanol is an example of a general type of reaction described in the most general terms by equation (23). In the formation of olefins from alcohols the reaction is spoken of as "dehydration," for the elements

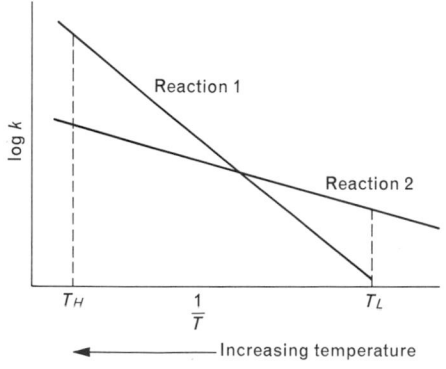

**Figure 8-2.** Plots of specific reaction rate versus reciprocal of absolute temperature for two reactions: activation energy 1 > activation energy 2.

of H—X in equation (23) are H—OH. It represents an example of an *elimination reaction*.

The mechanism of the dehydration of alcohols is not described in Equation (23), which is simply a statement of the overall process. A discussion of the course of elimination reactions and of alcohol dehydration is given in the following section.

## 8.11 Dehydration of alcohols. Elimination reactions

The elimination of the elements of H—X from adjacent carbon atoms, as in Equation (23), can take place by one of two mechanisms which are analogous to the $S_N1$ and $S_N2$ displacement reactions. These are called *E2* eliminations, which show second-order kinetics; and *E1* eliminations, which are first order in the compound undergoing the reaction. Like $S_N2$ displacement reactions, *E2* elimination reactions probably proceed through a single transition state in which the proton situated $\beta$- to the leaving group X is attacked by a proton acceptor (a base):

$$\text{B:} \longrightarrow \text{H} \quad CH_2 \text{—} CH_2 \text{—} X \longrightarrow CH_2 \text{=} CH_2 + :X^- \qquad (26)$$
$$\text{B:} H^+$$

The transition state can be represented by

$$
\begin{array}{c}
R' \qquad X \\
R\cdots C\cdots C\cdots R' \\
H \qquad R \\
B:
\end{array}
\qquad (27)
$$

in which the B—H bond, the H—C bond, the C—C bond, and the C—X bond are in the process of forming or breaking. In the *E1* elimination, the $\beta$-hydrogen atom is also removed by an attack of a base, but in this case the removal of the proton is a fast reaction, the slow (rate-determining) step being the formation of the carbonium ion:

$$H\text{—}\overset{|}{C}\text{—}\overset{|}{C}\text{—}X \underset{}{\overset{slow}{\rightleftharpoons}} H\text{—}\overset{|}{C}\text{—}\overset{|}{C}{}^+ + X^- \qquad (28)$$

$$\text{B:}^- \longrightarrow H\text{—}\overset{|}{C}\text{—}\overset{|}{C}{}^+ \underset{}{\overset{fast}{\rightleftharpoons}} \text{B:}H + \overset{|}{C}\text{=}\overset{|}{C} \qquad (29)$$

It is apparent from these equations that the elimination reactions involve the same reactants as the nucleophilic substitution reactions; the nucleophile can attack carbon (displacement of :X) or hydrogen (elimination). Indeed, many nucleophilic displacement reactions that involve the use of strong bases as nucleophiles give both substitution and elimination products in proportions that depend upon a number of factors; these will be considered at some length later on.

The dehydration of alcohols is usually carried out with use of strong acids (either proton donors, such as $H_2SO_4$ or $H_3PO_4$, or Lewis acids, such as alumina). The order of ease of dehydration of the alkanols is

tertiary > secondary > primary

which suggests that the reactions are principally *EI* in type.

Since the $\beta$-hydrogen atom [as in Equation (29)] is removed by attack of a base, and in strong acid solutions the available proton acceptors are either the unprotonated alcohol or the anion of the catalytic acid, neither of which is a strong base, it is evident that the removal of the $\beta$-hydrogen atom must be greatly facilitated by the positive charge on the adjacent carbon atom. This is to be expected, for the $\beta$—C—H bond would be strongly affected by the powerfully electronegative (electron-attracting) carbonium carbon atom.

The structures of the olefins derived by the dehydration of alcohols that can lose water in either of two ways is determined by the stability of the transition state that is formed in the proton removal step. Since the transition state is stabilized by structural factors that resemble those of the product into which it is converted, we can predict which of two pathways will be followed by inspecting the structures of the olefins that will be formed. In general, *the most highly substituted olefin is the more stable.* Let us examine an example of a dehydration that can lead to two different olefins:

$$CH_3-\underset{\underset{CH_3}{|}}{\overset{\overset{OH}{|}}{C}}-CH_2CH_3 \xrightarrow{H_2SO_4} CH_3-\underset{\underset{CH_3}{|}}{\overset{+}{C}}-CH_2CH_3+H_3O^+ \qquad (30)$$

The carbonium ion can lose a proton in direction *(a)* or direction *(b)*:

$$ \qquad (31)$$

The experimental facts are that route *(b)* predominates, and more than 80% of the resulting olefin mixture consists of 2-methyl-2-butene. An energy profile for a reaction of this kind is pictured in Figure 8-3.

## 8.12   Molecular rearrangements in elimination reactions

The dehydration of many alcohols follows a course that involves a *molecular rearrangement.* For example, the dehydration of 1-butanol with sulfuric acid (conditions similar to those used for the conversion of ethanol into ethylene; Section

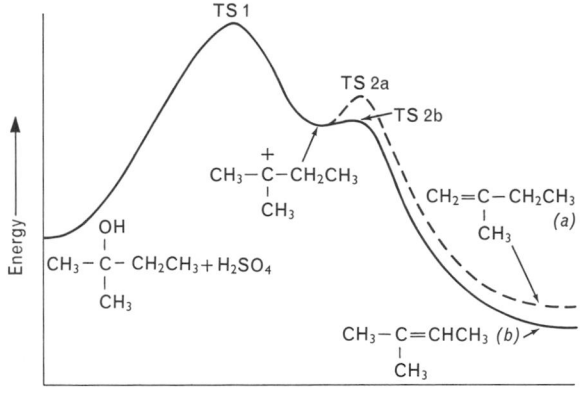

**Figure 8–3.** Energy profile for the dehydration of 2-methyl-2-butanol to give the olefins 2-methyl-1-butene *(a)* and 2-methyl-2-butene *(b)* [Equation (31)].

8.10) leads to the formation of 2-butene as the major product; 1-butene is formed in small amount:

$$CH_3CH_2CH_2CH_2OH \xrightarrow[170°]{H_2SO_4} \underset{\text{major product}}{CH_3CH=CHCH_3} + \underset{\text{minor product}}{CH_3CH_2CH=CH_2} \tag{32}$$

An even more striking example of rearrangement in alcohol dehydration is found in the acid-catalyzed dehydration of 3,3-dimethyl-2-butanol. The product is almost exclusively 2,3-dimethyl-2-butene; only minor amounts of other isomers are formed:

$$\underset{\substack{\text{3,3-dimethyl-2-butanol}}}{\overset{\displaystyle CH_3}{\underset{\displaystyle \underset{CH_3}{|}\ \underset{OH}{|}}{CH_3-C-\!\!\!-CH-CH_3}}} \xrightarrow{H_2SO_4} \underset{\substack{\text{major product}}}{\overset{\displaystyle}{CH_3-C=\!\!\!=C-CH_3}}_{\underset{CH_3\ CH_3}{|\ \ |}} \tag{33}$$

---

**Exercise 2.**   What other olefins could be formed in the dehydration of 3,3-dimethyl-2-butanol?

---

The explanation for rearrangements of the kinds illustrated in Equations (32) and (33) is based upon the *E1* reaction mechanism, in which a carbonium ion is an intermediate. Let us examine the reaction involving 3,3-dimethyl-2-butanol:

$$\underset{\substack{\underset{CH_3}{|}\ \underset{OH}{|}}}{CH_3-C-\!\!\!-CH-CH_3} \overset{HA}{\rightleftharpoons} \underset{\substack{\underset{CH_3}{|}\ \underset{OH_2^+}{|}}}{CH_3-C-\!\!\!-CH-CH_3} \rightleftharpoons \underset{\substack{\underset{CH_3}{|}}}{CH_3-\overset{+}{C}-CHCH_3} + H_2O \tag{34}$$

The secondary carbonium ion, by a migration of one of the methyl groups, can be transformed into the *more stable tertiary carbonium ion*; this reaction can be seen to involve the overlap of the vacant $p$ orbital on $-\overset{+}{C}H-$ with the $sp^3$-$s$ molecular orbital of the C—CH$_3$ bond. We have seen (Chapter 7) that carbonium ion stability follows the order tertiary > secondary > primary, and thus the rearrangement of the secondary into the tertiary carbonium ion represents a change to a condition of lower energy, or greater stability:

$$CH_3-\underset{\underset{+}{CH_3}}{\overset{CH_3}{\underset{|}{\overset{|}{C}}}}-CHCH_3 \longrightarrow CH_3-\underset{\underset{CH_3}{|}}{\overset{CH_3}{\underset{|}{\overset{|}{\overset{+}{C}}}}}-CHCH_3 \qquad (35)$$

Loss of the proton from the tertiary carbonium ion now leads to 2,3-dimethyl-2-butene. An energy profile for these reactions is shown in Figure 8-4.

The acid-catalyzed dehydration of alcohols is consequently often subject to uncertainty as to the structure of the product, because rearrangements may occur to give a product other than the "normal" one, or, more often, to give a mixture of two or more compounds.

---

**Exercise 3.**  The acid-catalyzed dehydration of 2,2-dimethylcyclohexanol,

gives mainly 1,2-dimethylcyclohexene:

. Explain.

---

In the practical preparation of olefins from alcohols* several methods are commonly used. The alcohol may be heated with a small amount of sulfuric or

---

\* The simple lower olefins are seldom prepared by the methods described in this chapter, for they are produced in very large amounts as by-products of petroleum refining.

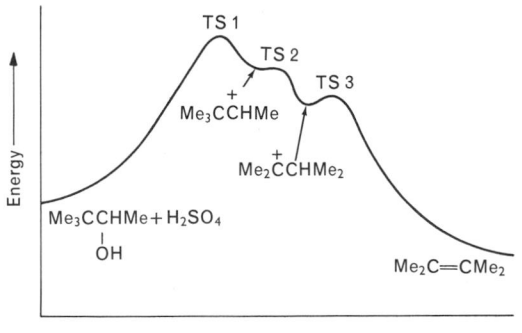

**Figure 8–4.** Energy profile of the dehydration of 3,3-dimethyl-2-butanol, showing the carbonium ion intermediates.

**Table 8-2.**   *Dehydration of 1-butanol*

| Conditions | Percent of Olefin Produced | |
|---|---|---|
| | $CH_3CH{=}CHCH_3$ | $CH_3CH_2CH{=}CH_2$ |
| $H_3PO_4/Al_2O_3$, 250° (vapor phase) | 55 | 45 |
| $H_2SO_4$, 160° (liquid phase) | 90 | 10 |
| $Al_2O_3$, 250° (vapor phase) | 10 | 90 |

phosphoric acid, and the olefin, which is of lower boiling point than the alcohol, is distilled out of the reaction mixture. Potassium bisulfate ($KHSO_4$) is also used as the dehydration catalyst. The lower, more volatile alcohols are conveniently dehydrated by passing them in the vapor state through a heated tube packed with pumice moistened with phosphoric acid  or with granules of aluminum oxide. It should be noted that the proportions of the olefinic products vary greatly with the conditions used for the dehydration. As an example  the results of the dehydration of 1-butanol under different reaction conditions are shown in Table 8-2.

One of the reasons for these variations in product composition is that olefins themselves can rearrange under the influence of acids (Chapter 9); consequently, the primary product of a dehydration may be the "normal" olefin (Chapter 9, 1-butene from 1-butanol), which then rearranges to a greater or lesser degree depending upon the time and conditions of subsequent contact with the strong acid. This is one reason why dehydration in the vapor state often gives quite different results from those obtained from dehydration in solution: the olefins produced in the reaction zone may be swept out of the tube before subsequent changes can occur.

### 8.13   Ether formation

Ether formation often occurs as a side-reaction in reactions of alcohols under acidic conditions. An example of this is given in Section 8.6, in which the alcohol acts as the nucleophilic reagent, attacking the protonated alcohol with displacement of $H_2O$.

The practical *preparation* of ethers is ordinarily carried out by a reaction which, although mechanistically allied to that just alluded to, takes place under different experimental conditions. In this method, usually referred to as the *Williamson synthesis*, the alcohol reacts in the form of its strongly nucleophilic conjugate base, the alkoxide anion.

Alcohols are weak acids with replaceable (ionizable) protons. Alcohols react with alkali metals, magnesium, and aluminum to form metal alkoxides: the reaction [(37) to (40)] can be seen to be comparable to the well-known formation of sodium hydroxide in the reaction of sodium with water (36):

$$Na+H_2O \longrightarrow \overset{+}{Na}OH^-+\tfrac{1}{2}H_2 \qquad (36)$$

$$K+(CH_3)_3COH \longrightarrow (CH_3)_3CO^-K^++\tfrac{1}{2}H_2 \qquad (37)$$
$$\text{\textit{t}-butyl alcohol} \qquad \text{potassium \textit{t}-butoxide}$$

$$Na+CH_3CH_2OH \longrightarrow CH_3CH_2O^-Na^++\tfrac{1}{2}H_2 \qquad (38)$$
$$\text{ethanol} \qquad \text{sodium ethoxide}$$

$$3(CH_3)_2CHOH+Al \longrightarrow (CH_3)_2CHO)_3Al+1\tfrac{1}{2}H_2 \qquad (39)$$
$$\text{aluminum isopropoxide}$$

$$2CH_3OH+Mg \longrightarrow (CH_3O)_2Mg+H_2 \qquad (40)$$
$$\text{magnesium}$$
$$\text{methoxide}$$

The alkali metal alkoxides are strong bases and effective nucleophiles. They undergo nucleophilic substitution reactions with primary alkyl halides or with the sulfate or sulfonate esters of primary alcohols to give ethers:

$$CH_3O^-+CH_3CH_2Br \longrightarrow CH_3OCH_2CH_3+Br^- \qquad (41)$$
$$(CH_3)_2CHO^-+CH_3I \longrightarrow (CH_3)_2CHOCH_3+I^- \qquad (42)$$
$$(CH_3)_3CO^-+CH_3OSO_2OCH_3 \longrightarrow (CH_3)_3COCH_3+CH_3OSO_2O^- \qquad (43)$$
$$\text{dimethyl sulfate}$$

The reaction of alkoxides with secondary alkyl halides gives poor yields of ethers because of the competing reaction of elimination:

$$CH_3CH_2O^-+(CH_3)_2CHBr \longrightarrow (CH_3)_2CHOCH_2CH_3+CH_3CH{=}CH_2 \qquad (44)$$
$$\text{little} \qquad\qquad \text{largely}$$

and tertiary alkyl halides do not yield ethers in this reaction, giving only the olefin by the elimination of HX:

$$(CH_3)_3CBr+CH_3O^- \nearrow \begin{array}{l} (CH_3)_3COCH_3+Br^- \\ \text{not obtained} \end{array}$$
$$\searrow CH_2{=}C\!\!\begin{array}{l} CH_3 \\ CH_3 \end{array} + CH_3OH+Br^- \qquad (45)$$

Since the step in which the proton is removed from the alkyl halide *(E2)* or the carbonium ion *(E1)* is a nucleophilic displacement on hydrogen, the effectiveness of a nucleophile in causing elimination (or in increasing the ratio elimination/substitution) increases with its strength as a base. In nucleophilic substititution on carbon the nucleophilic effectiveness is not directly related to the strength of the nucleophile as a base, as was pointed out in Section 7.3. It is sometimes possible to make use of this information in carrying out the substitution of —OH for —Br in an alkyl halide when direct hydrolysis with sodium hydroxide leads to an undesirable amount of elimination. Acetate ion is an effective nucleophile towards carbon, but is a weak base. Thus in the reaction series

$$\overset{\text{Br}}{\underset{|}{RCH_2CHR'}}+CH_3COO^- \longrightarrow \overset{\text{OCOCH}_3}{\underset{|}{RCH_2CHR'}}+Br^- \qquad (46)$$

$$\underset{RCH_2\overset{\overset{\displaystyle OCOCH_3}{|}}{CHR'}+H_2O/OH^-}{} \longrightarrow \underset{RCH_2\overset{\overset{\displaystyle OH}{|}}{CHR'}+CH_3COO^-}{} \qquad (47)$$

the use of acetate ion for replacement of the bromine (46) is attended with less elimination than would occur were the strong base, hydroxide ion, to be used. The subsequent hydrolysis (saponification) of the ester to yield the desired alcohol (47) does not affect the $-\overset{|}{CH}-O$ bond, and the result is that the final product is obtained in better overall yield than if the alkali were used to effect replacement directly.

Elimination reactions of other kinds will be considered elsewhere in the text.

## 8.14   Properties and reactions of ethers

Ethers are in general rather unreactive compounds and find important uses in organic chemistry by virtue of their relative inertness and their excellent solvent properties. Ethyl ether, $CH_3CH_2OCH_2CH_3$ is a readily available industrial chemical and is widely used as a solvent: it is immiscible with water and finds much use as a means for extracting compounds from aqueous solutions; it is very volatile (b.p. 35°) and thus is easily removed by distillation at moderate temperatures; it has excellent solvent powers for organic compounds, and in most of its applications as a solvent it is quite unreactive and serves as an inert reaction medium.

Ethers are usually very stable to reaction media that are neutral or alkaline. They can, however, undergo reaction under strongly acidic conditions, for they do contain an oxygen atom with two unshared electron pairs, and thus they can be protonated, or they can combine with Lewis acids. Although ethers composed of primary alkyl groups (such as ethyl ether, or methyl $n$-alkyl ethers) react only slowly with even strong acids (except for protonation, which is reversible and need not lead to further changes), ethers containing secondary and tertiary alkyl groups are susceptible to reaction with acids because of the ease with which they can give rise to the corresponding carbonium ions. In general, the ether $R-O-R'$ will be readily susceptible to cleavage by acids if either of the groups R or $R'$ is capable of forming a highly stable carbonium ion. For example, a $t$-butyl alkyl ether reacts with a strong acid in the following way:

$$(CH_3)_3COR+HA \rightleftharpoons (CH_3)_3C-\overset{\overset{\displaystyle +}{|}}{\underset{\underset{\displaystyle H}{|}}{O}}-R$$

$$(48)$$

$$\text{olefin} \longleftarrow (CH_3)_3C^+ + ROH$$

(or $(CH_3)_3COH$, if solvent is aqueous).

In short, the reaction of the ether $(CH_3)_3C$—OR in a strongly acidic medium would be essentially the same as that of *t*-butyl alcohol. Ethyl ether, on, the other hand, dissolves in cold concentrated HCl but is recovered unchanged when the solution is diluted with a large excess of water:

$$CH_3CH_2OCH_2CH_3 + HCl \rightleftharpoons \left. CH_3CH_2{-}\overset{+}{\underset{\underset{H}{|}}{O}}{-}CH_2CH_3 \right) Cl^- \qquad (49)$$

$$CH_3CH_2{-}\overset{+}{\underset{\underset{H}{|}}{O}}{-}CH_2CH_3 + H_2O \rightleftharpoons CH_3CH_2OCH_2CH_3 + H_3O^+ \qquad (50)$$

---

**Exercise 4.** What would be the final (qualitative) composition of the reaction mixture if $(CH_3)_3COCH_3$ were dissolved in cold concentrated aqueous HBr?

---

Even primary alkyl ethers can be cleaved by treatment with strong acids under more drastic conditions. A very effective acid for their cleavage is HI, since the iodide ion, a very effective nucleophile, can attack the protonated ether and bring about a nucleophilic displacement reaction:

$$RCH_2OCH_2R' + HI \rightleftharpoons RCH_2{-}\overset{+}{\underset{\underset{H}{|}}{O}}{-}CH_2R' + I^- \qquad (51)$$

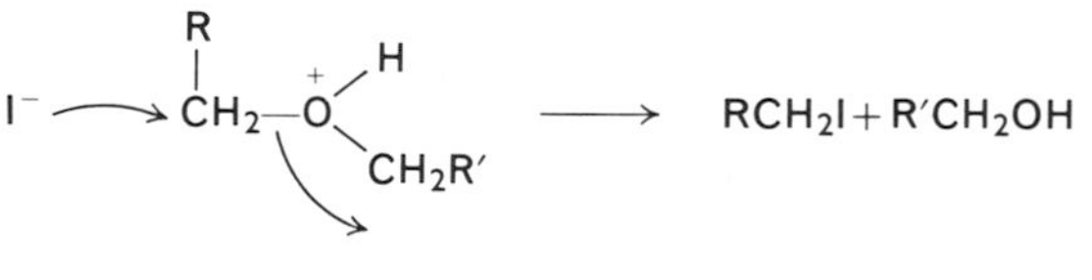

$$I^- \longrightarrow \overset{\overset{R}{|}}{CH_2}{-}\overset{+}{O}\overset{H}{\diagdown}_{CH_2R'} \longrightarrow RCH_2I + R'CH_2OH \qquad (52)$$

If R and R′ are different, the attack shown in (52) will take place upon the alkyl group most reactive in the $S_N2$ displacement reaction. Thus, if $RCH_2$— is $CH_3$—, the chief reaction will be that shown in (52). Indeed, the commonest use of the cleavage of ethers with HI is in the demethylation of aromatic methyl ethers such as anisole (53):

$$\underset{}{\text{(ring)}}{-}OCH_3 + HI \longrightarrow \underset{}{\text{(ring)}}{-}OH + CH_3I \qquad (53)$$

Additional examples of the acid-catalyzed cleavage of ethers (and of a related class of ether-like compounds called acetals) will be found in discussion to follow.

An important use of ethyl ether (which is used because it is the most readily available and least costly ether) is as a solvent in the preparation of the important *Grignard reagents*. These are described in Section 8.22.

### 8.15   Alcohols as nucleophilic reagents

We have already considered two expressions of the nucleophilic character of alcohols: (1) the protonation of the —OH group (Chapter 7) and (2) the formation of ethers (9).

The unshared pair of electrons of the alcoholic hydroxyl group can attack other electrophilic centers. Some reactions of this kind will be outlined here but discussed in greater detail in another chapter; they include the reactions of alcohols with carbonyl compounds: aldehydes, ketones, acids, and acid derivatives. All of these contain as the characteristic functional grouping the carbonyl group

—C=O, in which the carbon atom is electrophilic because of the greater electron-attracting power of the oxygen atom:

$$-\overset{|}{C}=\overset{..}{O}: \quad \longleftrightarrow \quad -\overset{|}{C}^{+}-\overset{..}{\underset{..}{O}}:^{-}$$

The electron-deficient carbonyl-carbon atom can accept the electron pair of the alcoholic —OH group to start a reaction, the consequences of which vary according to the nature of the groups attached to —C=O.

The *type* of reaction that takes place can be summarized in the partial terms (54), showing the initial step:

$$(54)$$

The process pictured in (54) is an example of one of the most general and widely encountered reactions we shall have to deal with. It should be examined with care and attention to its essential features. *It is a nucleophilic attack upon the carbon atom of the carbonyl group.* It can also be classed as a kind of displacement reaction, since the nucleophile displaces not a group or atom but a binding pair of electrons of the —C=O double bond. Compare the "typical" nucleophilic displacement

$$Y:^{-} \longrightarrow \overset{\diagdown}{\underset{\diagup}{C}} - Cl \quad \longrightarrow \quad Y:\overset{\diagdown}{\underset{\diagup}{C}} + :Cl^{-} \tag{55}$$

with

$$Y:^{-} \longrightarrow \overset{\diagdown}{\underset{\diagup}{C}} = O \quad \longrightarrow \quad Y:\overset{\diagdown}{\underset{\diagup}{C}} - O^{-} \tag{56}$$

In (56) the reaction has been written with a general nucleophile, Y: ; thus, (54) is simply a special case of (56), in which Y: is $\overset{R}{\underset{H}{\diagup}}\overset{..}{O}:$ .

## 8.16  Reaction of alcohols with carboxylic acids: ester formation

*Esters*, which have the general structure (57), and are typified by the following examples,

ethyl acetate                    isopropyl propionate

(57)

---

**Exercise 5.**   Write the structures of ethyl propionate and isopropyl acetate.

---

are formed by the reaction of alcohols with carboxylic acids in the presence of a strong (mineral) acid as a catalyst:

$$CH_3COOH + CH_3CH_2OH \xrightarrow{\text{H}_2\text{SO}_4} CH_3C\overset{O}{\underset{OCH_2CH_3}{}} + H_2O \qquad (58)$$

This reaction has been discussed briefly in Chapter 5 in relation to the question of equilibrium, but can now be considered in detail.

The reaction between acetic acid and ethanol appears to resemble an ordinary acid-base neutralization, but it is quite different. For one thing, it is slow, requiring a matter of days for the formation of the maximum yield of ester; it is an *equilibrium* reaction; and in the water that is produced in the reaction, the oxygen atom is derived from the —CO*OH* group and not from the *OH* group of the alcohol. This has been established by using alcohol containing an excess of $O^{18}$ in the hydroxyl group. The ester that is formed contains $O^{18}$ in the —O-alkyl group; the water that is formed does not contain excess $O^{18}$.

The equilibrium (58) has an equilibrium constant of approximately 4; that is,

$$K = \frac{[\text{ethyl acetate}][\text{H}_2\text{O}]}{[\text{acetic acid}][\text{ethanol}]} \cong 4$$

For other acids and alcohols, other equilibrium constants are found. The position of the equilibrium is not affected by the presence of the sulfuric acid that is used as the catalyst, but the *speed* with which equilibrium is established is greatly affected by the catalyst. A mixture of ethanol and acetic acid alone will form ethyl acetate very slowly, even at the boiling point of the mixture. The addition of a few percent of sulfuric or other strong acid will allow the reaction to be completed (reach equilibrium) in a very much shorter time.

What is the role of the catalyst? To answer this question we must examine the manner in which the reaction proceeds. When sulfuric acid is added to a mixture of the acid (RCOOH) and alcohol (R′OH), there are first of all established the usual acid-base equilibria in which the proton is shared between all of the various basic species in the solution, with the formation of the protonated species $R'OH_2^+$, $RCOOH_2^+$ in equilibria with the bases $R'OH$, $RCOOH$, $HSO_4^-$ (and after the reaction has proceeded with the formation of water, $H_3O^+$ and $H_2O$ will participate).

Since it is the acid and the alcohol that we are concerned with, the state in which these are present is of first importance. The acid can be protonated in two ways:

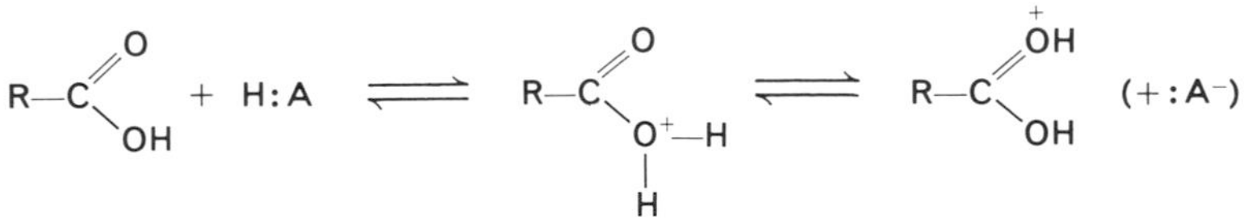

(59)                    (60)

The important feature to note is this: in (59) and (60) the positive charge on oxygen will further withdraw electrons toward oxygen and will create a greater dissymmetry in the C—O bond. Thus, *protonation of the carboxyl group has increased its ability to accept nucleophilic attack by the alcohol.*

The reaction that leads to ester is, then,

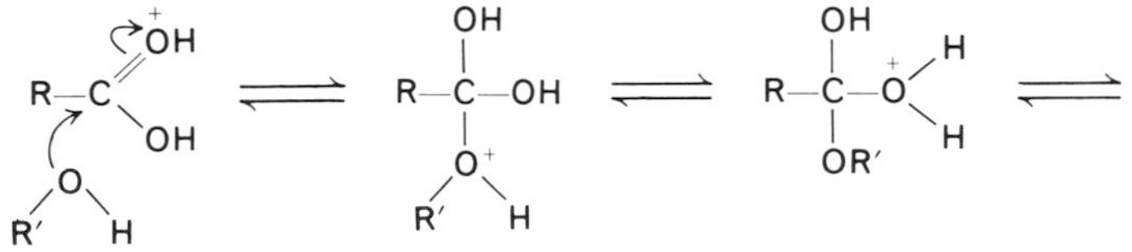

(61)              (62)              (63)

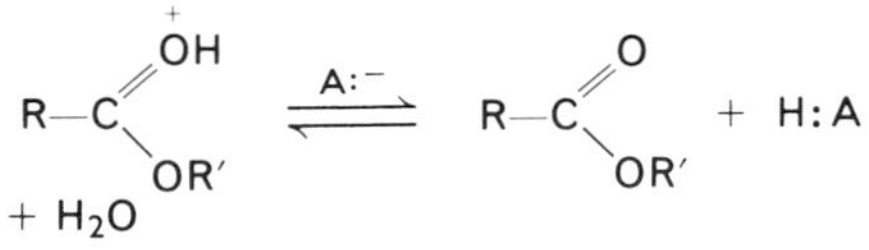

(64)                    (65)

The effect of protonation—the catalytic effect of the acid—can be understood by a consideration of the step (61) → (62), compared with the equivalent step (61a) → (62a) in which the unprotonated acid is attacked by the alcohol:

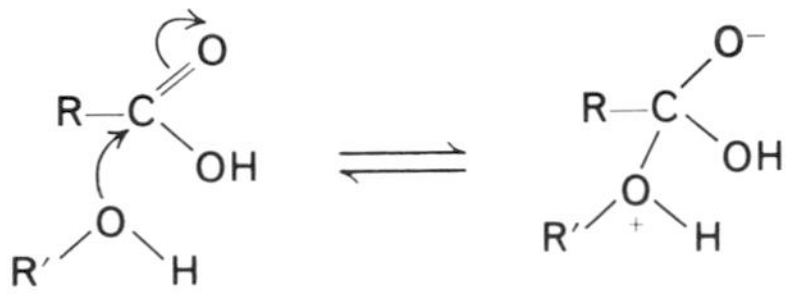

(61a)                   (62a)

In the latter case [(61a) → (62a)] the initial product of the reaction is formed in a step that involves a separation of charge; in the former case [(61) → (62)] no charge

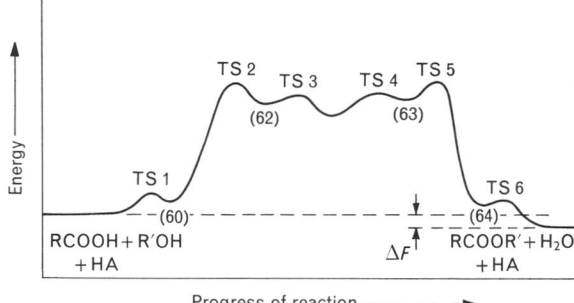

**Figure 8–5.**   An energy diagram for an acid-catalyzed esterification reaction,

$$RCOOH + R'OH \xrightleftharpoons{HA} RCOOR' + H_2O$$

separation, but a dispersal of the positive charge, is involved. Since charge separation is a process that requires energy (raises the energy of the system), the transition state leading to (61a) will be at a higher energy level than that leading to (61). The latter reaction, proceeding through a transition state of lower energy, will be faster.

After the initial attack (61), the succeeding steps (62), (63), and (65) are simply proton exchange reactions in which the proton is shared between the various proton acceptors (C=O, —OH, A⁻) that are present. Step (64) → (63) is the exact analogy of the step (61) → (62), but in (64) → (63) the nucleophile water attacks the protonated ester. Since all of these reactions are equilibria, it can be seen that *this series of equations represents acid-catalyzed hydrolysis of an ester* as well as acid-catalyzed esterification of an acid.

The steps in this sequence of equilibria* represent separate reactions, each with its own transition state; and the substances (60), (62), (63), and (64) are the unstable intermediates that can be recognized by the several minima in the energy profile for the reaction. In Figure 8-5 is shown an energy diagram for an esterification reaction. The small difference in energy of the initial and final states (a low value of $-\Delta F$) would indicate that the equilibrium for the generalized reaction for Figure 8-5 lies to the right, but that the equilibrium constant is not large.

---

**Exercise 6.**   Identify the intermediate at the unlabeled minimum in the curve of Figure 8-5. See footnote below.

---

* A step intermediate between (62) and (63), representing the loss of a proton from (62) and the reprotonation to give (64), is omitted from the series (61)–(65), but is shown on the energy profile.

## 8.17 Acid-catalyzed ester exchange

We can pursue these concepts further. Suppose we treat an ester with an alcohol containing an alkyl group different from R′ [in (65)], in the presence of the acid catalyst. In this situation, we can write a series of equations exactly paralleling (61) → (65), but with the ester RCOOR′ in place of the acid RCOOH:

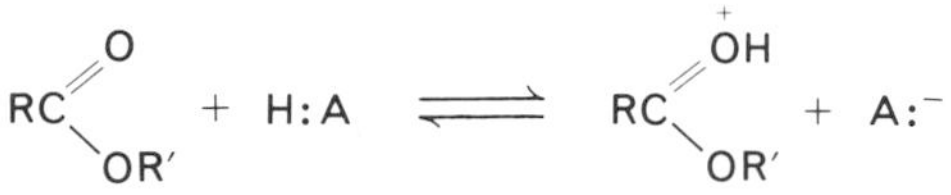

$$
RC\overset{O}{\underset{OR'}{\diagdown}} + H{:}A \;\rightleftharpoons\; RC\overset{\overset{+}{O}H}{\underset{OR'}{\diagdown}} + A{:}^-
$$

(65)

$$
RC\overset{\overset{+}{O}H}{\underset{OR'}{\diagdown}} + R''OH \;\rightleftharpoons\; RC\overset{OH}{\underset{\underset{R''\;\;\;H}{O^+}}{\overset{|}{\underset{|}{-}}OR'}} \;\rightleftharpoons\; R\overset{OH}{\underset{OR''}{\overset{|}{\underset{|}{-}C-}O}}\overset{H}{\diagdown}_{R'}
$$

$$
\big\Updownarrow
$$

$$
H{:}A + RC\overset{O}{\underset{OR''}{\diagdown}} \;\underset{A{:}^-}{\rightleftharpoons}\; RC\overset{\overset{+}{O}H}{\underset{OR''}{\diagdown}} + R'OH
$$

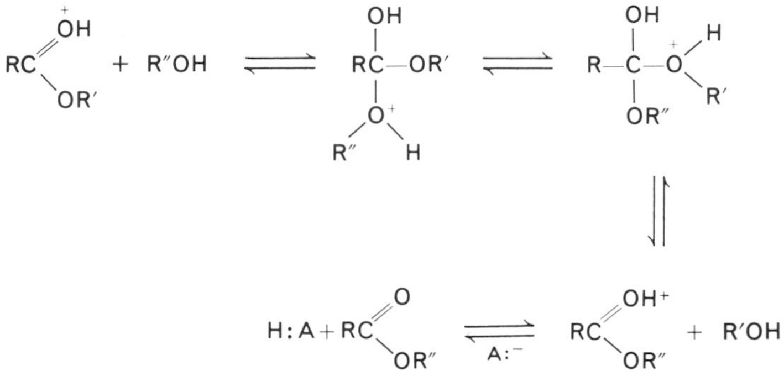

The result is the exchange of —OR′ in the ester for —OR″, to give a new ester. This, called "ester interchange," is a practical and useful method for converting one ester into another. If the "new" alcohol, R″OH in the above example, is present in large excess (for instance, is used as the solvent for the reaction), the equilibrium is shifted sufficiently far to the right to give essentially complete conversion of RCOOR′ into RCOOR″. Another device for shifting the equilibrium to the right is to remove R′OH by distillation if the boiling point differences of the various substances in the mixture permit.

## 8.18 The reaction of alcohols with acid halides and anhydrides

Esters are also prepared (indeed, most commonly) by the reaction of alcohols with other acid derivatives, especially acid halides and acid anhydrides (Chapter 14); for example,

$$
CH_3CH_2C\overset{O}{\underset{Cl}{\diagdown}} + CH_3OH \;\longrightarrow\; CH_3CH_2C\overset{O}{\underset{OCH_3}{\diagdown}} + HCl \tag{66}
$$

propionyl                          methyl
chloride                          propionate

$$CH_3C\overset{O}{\diagup}O-C\overset{O}{\diagdown}CH_3 + \overset{H_3C}{\underset{H_3C}{\diagup}}CHCH_2OH \longrightarrow CH_3C\overset{O}{\diagdown}_{OCH_2CH}\overset{CH_3}{\diagdown}_{CH_3} + CH_3COOH \qquad (67)$$

acetic anhydride          isobutyl acetate

In general terms,

$$R-C\overset{O}{\diagdown}_{X} + R'OH \longrightarrow RC\overset{O}{\diagdown}_{OR'} + HX \qquad (68)$$

Again, the reaction is initiated by the nucleophilic attack of $\overset{R}{\underset{H}{\diagup}}O:$ on the carbon atom of $-\overset{|}{C}=O$ in the acid derivative. When X is a halogen (e.g., Cl), the reaction is much more rapid than is that of the alcohol with the acid itself. This is because the halogen attached to the carbonyl group exerts a strong inductive withdrawal of the electrons of the C—Cl bond and thus creates a deficiency of electrons on the carbon that causes it to have enhanced electrophilic properties. The attack of the nucleophilic alcohol is thus strongly aided, and the reaction proceeds readily.

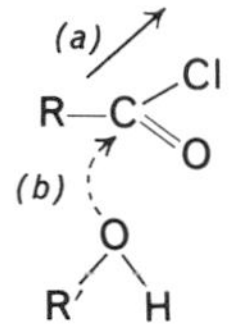

$$\begin{array}{c} (a)\nearrow \quad Cl \\ R-C\overset{\diagup}{\underset{\nearrow}{\diagdown}}O \\ (b) \vdots \\ O \\ R' \quad H \end{array}$$

Inductive effect *(a)* creates electron deficiency on carbon, aiding nucleophilic attack *(b)*.

The intermediate formed by this attack [compare this with (54)] can now expel either R'OH (a course of no consequence, since it gives back the starting reactants), or Cl⁻, in which case the ester results:

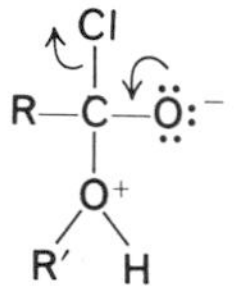

$$\begin{array}{c} Cl \\ R-C-\overset{..}{O}:^- \\ | \\ O^+ \\ R' \quad H \end{array} \longrightarrow R-C\overset{O}{\diagdown}_{OR'} + H^+ + Cl^- \qquad (69)$$

It will be clear now that the route shown in (69), the elimination of Cl⁻, will be greatly favored over the alternative, the elimination of R'OH, because R'OH is much more nucleophilic than Cl⁻. Thus, the equilibrium in a reaction such as (68), where X is a strongly electronegative atom or group, is far to the side of the products shown and the reaction will be substantially "complete".

The reaction of the anhydride (67) can be formulated in a comparable manner.

---

**Exercise 7.** Write the equations for the following reactions in the form shown in (68):
  (a) methanol + acetyl bromide ⟶
  (b) isopropyl alcohol + acetic anhydride ⟶
  (c) ethanol + propionyl chloride ⟶
  (d) 2-butanol + propionic anhydride ⟶

**Exercise 8.** Tertiary alcohols, when allowed to react with acid chlorides, give esters only when a base (such as tertiary amine) is present. The reaction of *t*-butyl alcohol with acetyl chloride gives chiefly *t*-butyl chloride. Can you account for this?

---

### 8.19   Esters of inorganic acids

The reaction of alcohols with inorganic acids and certain inorganic acid derivatives leads to the formation of compounds in which the —OH group of the inorganic acid is converted to —OR. These are esters of the inorganic acids, and many of them are important and useful substances.

Certain inorganic esters are formed simply by the reaction of the alcohol with the acid. When long-chain alcohols are esterified with sulfuric acid* in this way, the resulting alkyl hydrogen sulfates form sodium salts that are valuable detergents (cleansing agents; Section 14.5) and emulsifying agents. A common "synthetic" detergent of this type is lauryl hydrogen sulfate (as the sodium salt):

$$CH_3(CH_2)_{10}CH_2OSO_2O^-Na^+$$

Other synthetic detergents are the sodium salts of the acid esters made by the addition of sulfuric acid to the double bonds of high molecular weight olefins (see Section 8.21).

Esters of nitrous acid form readily by the direct reaction of alcohols with the acid, usually by the addition of sulfuric acid to a mixture of the alcohol and sodium nitrite; for example, isobutyl nitrite is formed when a solution of isobutyl alcohol and sodium nitrite is acidified:

$$\begin{array}{c} H_3C \\ H_3C \end{array}\!\!>\!\!CHCH_2OH + HONO \longrightarrow \begin{array}{c} H_3C \\ H_3C \end{array}\!\!>\!\!CHCH_2ONO + H_2O \qquad (70)$$

The alkyl nitrites are useful in many synthetic operations because in the presence of mineral acids they can hydrolyze and thus serve as an *in situ* source of nitrous acid. Their usefulness in this way is enhanced by their solubility in organic solvents. Isoamyl nitrite has profound physiological effects; it lowers blood pressure by relaxing smooth muscle, and for this reason is a valuable drug for anginal attacks, which it relieves (when inhaled) by producing coronary vasodilatation. Other drugs that are used clinically for the same purpose are the nitric acid esters of glycerol and pentaerythritol:

$$\begin{array}{ll}
CH_2ONO_2 & CH_2ONO_2 \\
| & | \\
CHONO_2 & O_2NOCH_2-C-CH_2ONO_2 \\
| & | \\
CH_2ONO_2 & CH_2ONO_2 \\
\text{glyceryl trinitrate} & \text{pentaerythritol tetranitrate} \\
\text{("nitroglycerin")} &
\end{array}$$

---

\* Commercially it is more common to use chlorosulfonic acid, $ClSO_3H : ROH + ClSO_3H \longrightarrow ROSO_3H + HCl$; the conversion of alcohol to alkyl sulfate is more nearly complete than when $H_2SO_4$ is used.

Nitrates of the tri- and tetrahydroxy alcohols, such as those of glycerol and pentaerythritol, are also powerful explosives. They are very sensitive to shock; but when "nitroglycerin" is mixed with an inert material, such as wood pulp or diatomaceous earth, the resulting material, called "dynamite," is relatively safe to handle. The discovery of dynamite by Alfred Nobel laid the foundation for the fortune that provides the Nobel Prizes.

Esters of phosphorus acids can be prepared by the reaction of alcohols with phosphorus halides and oxyhalides. For example, ethanol reacts with phosphorus oxychloride to form triethyl phosphate:

$$3EtOH + POCl_3 \longrightarrow O{=}P(OEt)_3 + 3HCl$$

When less than the above proportion of alcohol is used, alkyl hydrogen phosphates can be prepared.

Certain esters of phosphorus acids are of considerable economic importance as insecticides. Tetraethyl pyrophosphate (TEPP) and di-isopropyl phosphorofluoridate (DFP) are representative:

$$\begin{array}{cc} \underset{\text{tetraethyl pyrophosphate}}{\underset{\text{(TEPP)}}{\ce{EtO\underset{EtO}{\overset{O}{\|}}P-O-\overset{O}{\|}P\overset{OEt}{\underset{OEt}{}}}}} & \underset{\substack{\text{diisopropyl phosphorofluoridate}\\ \text{(DFP)}}}{\ce{H_3C\underset{H_3C}{}CH-O-\overset{O}{\underset{F}{\|}}P-O-CH\underset{CH_3}{\overset{CH_3}{}}}} \end{array}$$

These compounds are highly toxic to animals as well as to insects; they must be used with great care, and the worker must be adequately protected. They are potent inhibitors of the enzyme acetylcholinesterase, an enzyme that is an important participant in the process of conduction and transmission of nerve impulses. The toxic manifestations of the insecticides of the phosphate ester group, such as TEPP and DFP (and a large number of others of related structures that are extensively used both domestically and in agriculture), are severe derangements of nervous function that can result in serious illness or death. They can enter the body by way of the lungs, the mouth, or by absorption through the skin.

## 8.20 Reaction of alcohols with other carbonyl compounds

In aldehydes and ketones, the carbonyl group is similarly polarized in such a way as to possess electrophilic character on the carbon atom:

$$\underset{\text{aldehydes}}{\ce{\overset{R}{\underset{H}{}}C{=}O \longleftrightarrow \overset{R}{\underset{H}{}}\overset{+}{C}-O^-}} \qquad \underset{\text{ketones}}{\ce{\overset{R}{\underset{R}{}}C{=}O \longleftrightarrow \overset{R}{\underset{R}{}}\overset{+}{C}-O^-}}$$

This electrophilic character of the carbonyl carbon atom is greatly enhanced by protonation of the oxygen atom:

$$\ce{\overset{R}{\underset{H}{}}C{=}\ddot{O}{:} + H{:}A <=> \overset{R}{\underset{H}{}}C{=}\overset{+}{\ddot{O}}H + {:}A^-} \tag{71}$$

and in the presence of acid catalysts many aldehydes will react with alcohols in the following manner:

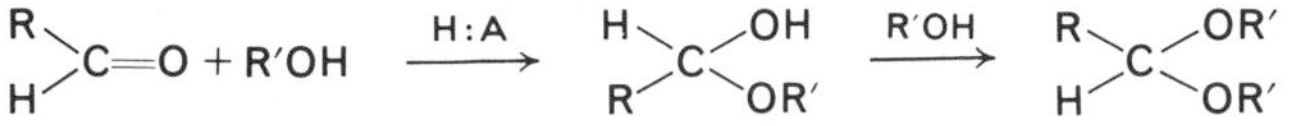

$$\underset{H}{\overset{R}{>}}C=O + R'OH \xrightarrow{\ H:A\ } \underset{R}{\overset{H}{>}}C\underset{OR'}{\overset{OH}{<}} \xrightarrow{\ R'OH\ } \underset{H}{\overset{R}{>}}C\underset{OR'}{\overset{OR'}{<}} \tag{72}$$

That this is again the result of a nucleophilic attack on the carbonyl carbon atom is evident from the result: a bond is formed between the carbon of $-\overset{|}{C}=O$ and the oxygen of R'OH, indicating that a process (73), comparable to the general process (54), occurs:

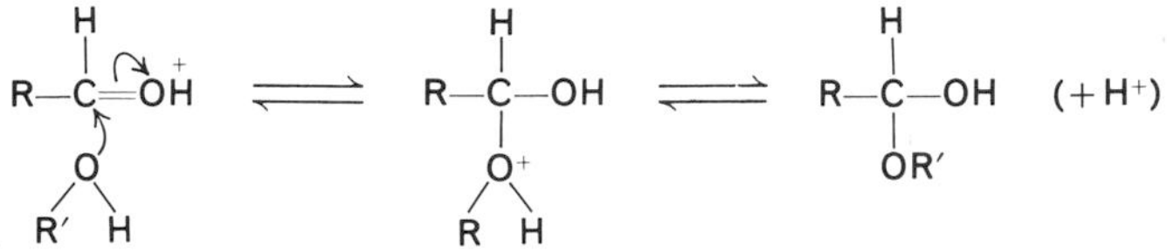

$$\tag{73}$$

The product $RCH\underset{OR'}{\overset{OH}{<}}$ is called a *hemiacetal*. If the solution is free of water at the start, and a trace of strong acid is present, the reaction proceeds further to the formation of a compound known as an *acetal*. In the reaction of acetaldehyde and methanol, the overall process can be written as

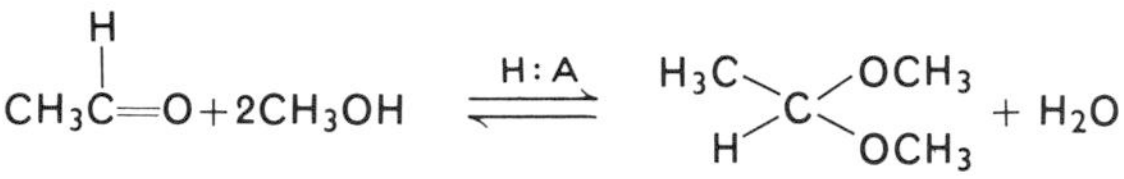

$$CH_3\overset{H}{\underset{|}{C}}=O + 2CH_3OH \xrightleftharpoons{\ H:A\ } \underset{H}{\overset{H_3C}{>}}C\underset{OCH_3}{\overset{OCH_3}{<}} + H_2O \tag{74}$$

acetaldehyde dimethyl<br>acetal

The dissection of this reaction into its separate steps serves as an excellent illustration of the principles of acid-base reactions we have been discussing:

1. The first stage is the nucleophilic attack of methanol upon the aldehyde carbonyl group to form the hemiacetal.

2. In the presence of a strong acid, the hemiacetal will be protonated (on either oxygen atom) to give (74) and (75):

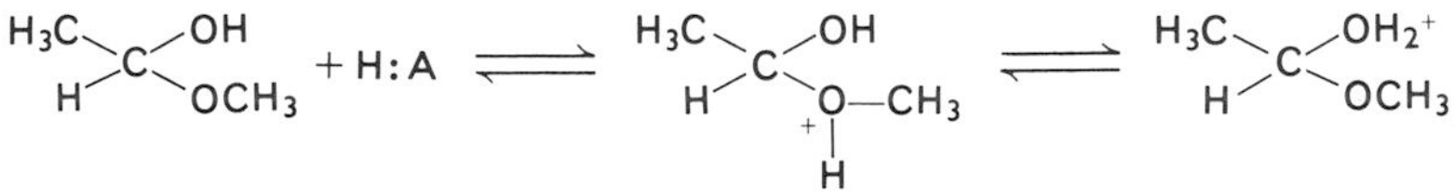

$$\underset{H}{\overset{H_3C}{>}}C\underset{OCH_3}{\overset{OH}{<}} + H:A \rightleftharpoons \underset{H}{\overset{H_3C}{>}}C\underset{\overset{+}{O}-CH_3}{\overset{OH}{<}} \rightleftharpoons \underset{H}{\overset{H_3C}{>}}C\underset{OCH_3}{\overset{OH_2^+}{<}}$$

(74)      (75)

3. The dissociation of (75) into a water molecule and the carbonium ion

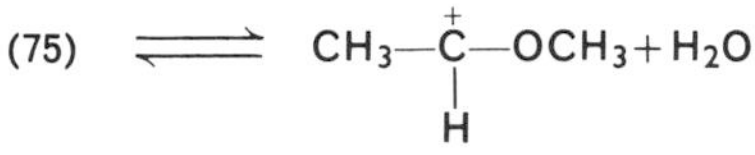

$$(75) \rightleftharpoons CH_3-\overset{+}{\underset{\underset{H}{|}}{C}}-OCH_3 + H_2O \tag{76}$$

resembles the ready dissociation of chloromethyl ether (Figure 7-7), the discussion of which applies directly to the present case.

4. Coordination of a second alcohol molecule with the carbonium ion:

$$CH_3-\overset{+}{\underset{H}{C}}-OCH_3 + CH_3OH \rightleftharpoons CH_3-\overset{H}{\underset{H}{C}}\overset{\overset{+}{O}-CH_3}{\underset{OCH_3}{}} \tag{77}$$

5. Loss of a proton to any available proton acceptor, to give the acetal:

$$CH_3-\overset{H}{\underset{H}{C}}\overset{\overset{+}{O}-CH_3}{\underset{OCH_3}{}} + :A^- \rightleftharpoons CH_3-\overset{OCH_3}{\underset{H}{C}}\overset{OCH_3}{\underset{OCH_3}{}} + H:A \tag{78}$$

the acetal

This series of transformations should be studied with care until each step is thoroughly understood, for it embodies principles found in much of the discussion throughout the previous sections of this book.

Most ketones do not form acetals with the same ease as do aldehydes, since the reactivity of the carbonyl group of ketones to addition reactions of this kind is considerably less than that of aldehydes.*

Because of a diminished degree of protonation of *thioacetals* and *thioketals*, with a consequently favorable effect upon the equilibrium of the reaction, these carbonyl derivatives are formed readily when mercaptans, the sulfur analogues of the alcohols (e.g., methyl mercaptan, $CH_3SH$), are used instead of alcohols; for example,

$$CH_3CH_2C\overset{O}{\underset{H}{\diagdown}} + 2CH_3CH_2SH \longrightarrow CH_3CH_2-\overset{SCH_2CH_3}{\underset{H}{\underset{|}{C}}}-SCH_2CH_3 + H_2O \tag{79}$$

propionaldehyde    ethyl mercaptan        propionaldehyde diethyl
           [ethanthiol (G)]              thioacetal

$$CH_3-\overset{}{\underset{CH_3CH_2}{\underset{|}{C}}}=O + 2CH_3SH \longrightarrow \overset{H_3C}{\underset{CH_3CH_2}{}}\overset{}{C}\overset{SCH_3}{\underset{SCH_3}{}} + H_2O \tag{80}$$

butanone                         butanone dimethyl
                                thioketal

Thioacetals and thioketals undergo a reaction that is very useful in synthetic work: when treated with Raney nickel (a special finely divided nickel used as a

---

\* One of the reasons for the difficulty of preparing ketals is the unfavorable equilibrium in the reaction [corresponding to (74)]. When the $sp^2$ configuration of the bonds in the ketone carbonyl group is transformed to the $sp^3$ configuration in the ketal, the two alkyl groups on the carbonyl group are necessarily brought closer together, creating new repulsive forces that raise the energy of the system.

catalyst) they undergo loss of the sulfur, with replacement of —SR by —H. For example, butanone dimethyl thioketal is converted into butane:

$$\underset{\substack{CH_3CH_2}}{\overset{\substack{H_3C}}{\diagdown}}C\underset{\substack{SCH_3}}{\overset{\substack{SCH_3}}{\diagup}} \quad \xrightarrow[\substack{(Raney)}]{\substack{H_2/Ni}} \quad \underset{\substack{CH_3CH_2}}{\overset{\substack{H_3C}}{\diagdown}}CH_2 \tag{81}$$

The hydrogen required in this reaction is present in the adsorbed state in the catalyst.

*Cyclic* ketals, however, can be prepared successfully with the use of a glycol instead of an alcohol. In this case, the two hydroxyl groups of, for example, ethylene glycol, fulfill the role of the two molecules of alcohol with the result that a cyclic acetal (with an aldehyde) or ketal (with a ketone) is formed. The acid catalyst used may be a strong mineral acid, a sulfonic acid, or a Lewis acid ($BF_3$):

$$\text{(cyclohexanone)} \quad + \quad \underset{\substack{CH_2OH}}{\overset{\substack{CH_2OH}}{|}} \quad \xrightarrow{\substack{BF_3}} \quad \text{(cyclic ketal)} \tag{82}$$

Since acetals and ketals are stable under neutral or alkaline reaction conditions, but are easily decomposed by acidic hydrolysis to regenerate the original carbonyl compound, they serve a valuable purpose in "protecting" a carbonyl group while transformations are performed at some other point in the molecule. An example is given under "Reduction of Carbonyl Compounds," p. 197.

## 8.21 Methods for the preparation of alcohols

*Hydration of olefins.* The process of dehydrating an alcohol to give an olefin can be reversed, and olefins can be *hydrated* to produce alcohols. Whereas some olefins can be hydrated by the direct addition of water, the process is ordinarily indirect, the immediate product being the result of the addition of sulfuric acid to the olefin to form a sulfuric ester of the alcohol. Hydrolysis of this ester yields the alcohol:

$$CH_3CH{=}CH_2 + H_2SO_4 \;\rightleftharpoons\; \underset{\substack{OSO_3H}}{CH_3CHCH_3} \;\xrightarrow{\substack{H_2O}}\; \underset{\substack{OH}}{CH_3CHCH_3} + H_2SO_4 \tag{83}$$

$$\text{propylene}$$

In the hydration of isobutylene it is unlikely that a discrete sulfuric ester is formed.

$$\underset{\substack{H_3C}}{\overset{\substack{H_3C}}{\diagdown}}C{=}CH_2 + H_2SO_4 \;\rightleftharpoons\; \left\{ \underset{\substack{CH_3}}{\overset{\substack{H_3C}}{\diagdown}}\overset{+}{C}\overset{\substack{CH_3}}{\diagup} \right\} OSO_3H^- \tag{84}$$

$$\text{isobutylene}$$

$$\downarrow H_2O$$

$$\underset{\substack{CH_3}}{\overset{\substack{CH_3}}{CH_3{-}C{-}OH}} + H_2SO_4$$

**Table 8-3.** *Relative rate of absorption in 80% $H_2SO_4$*

| | |
|---|---|
| isobutene | 158 |
| 2-butene | 2.6 |
| 1-butene | 1.0 |
| propene | 1.0 |
| ethylene | 0.003 |

These reactions are the reverse of those in which olefins are formed from alcohols, and show the importance of the experimental conditions—in this case temperature and acid concentration—in determining the course of reactions.

The rate of addition of sulfuric acid is markedly influenced by the structure of the olefin (Table 8-3).

The acid-catalyzed hydration of olefins is an example of a more general process: the addition of acids to olefins. The addition of sulfuric acid to propylene is an example of the reaction that may be formulated in the following general way:

$$CH_3CH=CH_2+HA \longrightarrow CH_3-\overset{\overset{\displaystyle A}{|}}{C}H-CH_3$$

HA = $H_2SO_4$;  A = $HSO_4$  ($\longrightarrow$ isopropyl hydrogen sulfate)
HA = HBr;  A = Br  ($\longrightarrow$ isopropyl bromide)
HA = HCl;  A = Cl  ($\longrightarrow$ isopropyl chloride)

The course of this reaction will be discussed in detail in Chapter 9.

*Hydroboration of olefins.* The acid-catalyzed hydration of olefins yields alcohols in which the hydroxyl group is attached to the most highly substituted carbon atom: isopropyl alcohol from propylene; *t*-butyl alcohol from isobutylene; 2-butanol from 1-butene, and so on. Another method for the addition of the elements of water to the olefinic double bond, but in which addition takes place in the opposite sense (so that propylene yields 1-propanol, and so on) is by a process involving two steps: (1) *hydroboration* of the double bond, followed by (2) oxidation of the resulting alkylboron compound. The hydroboration step is carried out by the addition of boron hydride, $B_2H_6$, to the double bond.

1. Preparation of $B_2H_6$:

$$3NaBH_4+4BF_3 \longrightarrow 2B_2H_6+3NaBF_4 \tag{85}$$

or $6NaBH_4+2AlCl_3 \longrightarrow 3B_2H_6+2AlH_3+6NaCl \tag{86}$

2. Addition of $B_2H_6$ to the carbon-carbon double bond:

$$6RCH=CH_2+B_2H_6 \longrightarrow 2(RCH_2CH_2)_3B \tag{87}$$

3. Oxidation of the alkylborane:

$$(RCH_2CH_2)_3B+3H_2O_2+3OH^- \longrightarrow 3RCH_2CH_2OH+BO_3{}^{3-}+3H_2O \tag{88}$$

The overall addition of the H and OH occurs on the same side of the double bond (*cis* addition):

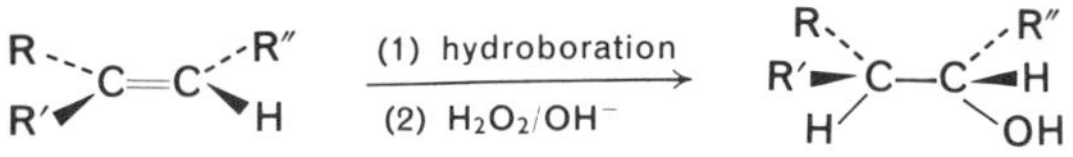

$$
\begin{array}{c}
R \cdots \quad \cdots R'' \\
\phantom{xx}C{=}C \\
R' \quad \quad H
\end{array}
\xrightarrow[\text{(2) } H_2O_2/OH^-]{\text{(1) hydroboration}}
\begin{array}{c}
R \cdots \quad \quad R'' \\
R' {-}C{-}C{\blacktriangleleft} H \\
H \quad \quad OH
\end{array}
\tag{89}
$$

Reactions that lead to products of predictable configuration are known as *stereospecific* reactions, and are of considerable importance in synthetic organic chemistry. Other examples of stereospecific reactions will be encountered throughout the text.

The mechanism of the addition may involve a four-center transition state, in which a B—H bond of the boron hydride is in the process of breaking while the H—C and C—B bonds are forming; as an example, the addition of boron hydride to 1,2-dimethylcyclohexene will be used to illustrate the course of the reaction (only one B—H bond of the $B_2H_6$ is shown):

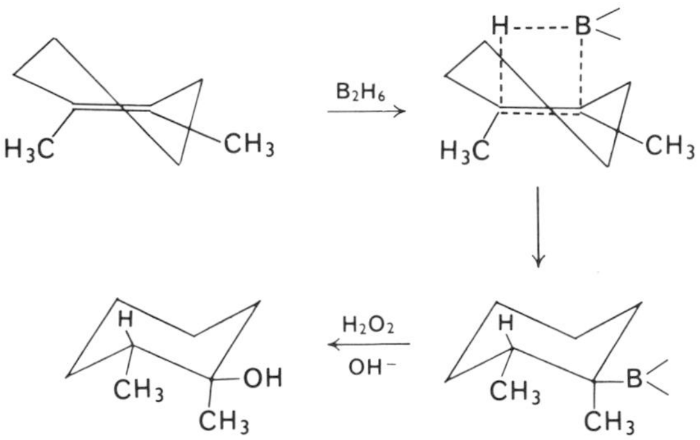

$$\tag{90}$$

***Hydrolysis of alkyl halides.*** Alkyl halides react with aqueous alkali to undergo displacement of the halogen by —OH, with the formation of alcohols:

$$RBr + OH^- \longrightarrow ROH + Br^-$$

This, one of the simplest examples of the displacement reaction, has been described earlier.

The generality of this reaction is diminished by the tendency for certain alkyl halides—in particular, tertiary alkyl halides—to react partly or extensively by a second course to give not the alcohol but an olefin. A treatment of this "elimination" has already been presented in connection with alcohol dehydration and the Williamson synthesis of ethers. For the present we shall note that the elimination (dehydrohalogenation) reaction is a side reaction in the hydrolysis of alkyl halides to alcohols, and is favored by hot, strong alkalis and branching of the alkyl group at the halogen-bearing carbon atom.

Tertiary alkyl halides do not require the presence of strong alkali for hydrolysis. A weak base such as sodium bicarbonate may be used to neutralize the acid formed, thus preventing reattack of the acid upon the alcohol:

$$CH_3-\underset{\underset{CH_2CH_3}{|}}{\overset{\overset{CH_3}{|}}{C}}-Cl+H_2O \xrightarrow{(NaHCO_3)} CH_3-\underset{\underset{CH_2CH_3}{|}}{\overset{\overset{CH_3}{|}}{C}}-OH+(HCl) \tag{91}$$

The usefulness of this means for the preparation of alcohols is lessened by the fact that alkyl halides are not obtainable by convenient general methods. Alkyl halides are usually prepared from alcohols rather than the reverse. Nevertheless, some alcohols are prepared commercially by hydrolysis of the corresponding halides when these are conveniently accessible.

*Allyl alcohol* is prepared by the hydrolysis of allyl chloride. The latter can be made by the high-temperature chlorination of propylene:

$$CH_2{=}CHCH_3 \xrightarrow[500°C]{Cl_2} CH_2{=}CHCH_2Cl \xrightarrow[NaOH]{H_2O} CH_2{=}CHCH_2OH \tag{92}$$

The chlorination of a "pentane" fraction obtained by the fractional distillation of petroleum yields a mixture of chlorinated pentanes from which the compounds of composition $C_5H_{11}Cl$ can be separated by distillation from di-, tri-, and polychlorinated products. Hydrolysis of the $C_5H_{11}Cl$ mixture yields a mixture of five-carbon alcohols. Since the original hydrocarbon fraction is a mixture of isomers, and since halogenation of paraffin hydrocarbons (alkanes) is not selective, the final alcohol mixture contains numerous isomeric compounds, all of them $C_5H_{11}OH$. This mixture (called "Pentasol") is used, without separation into individual isomers, as a commercial solvent.

**Reduction of carbonyl compounds.** The reduction of carbonyl compounds leads to hydroxyl compounds:

$$\underset{}{>}C{=}O \xrightarrow{2[H]} \underset{}{>}CHOH \tag{93}$$

In this equation the reduction is designated by 2[H], a noncommittal way of expressing that in the overall process two hydrogen atoms attach themselves to the ends of the $-\overset{|}{C}{=}O$ double bond. The reaction may be carried out in a number of ways; for example:

1. by the addition of hydrogen under the influence of a catalyst;
2. by the use of metal-solvent combinations such as sodium or potassium in alcohol;
3. by the use of metal hydrides;
4. by the use of other alcohols with the aid of metal alkoxides.

**Catalytic reduction.** Catalytic reduction (see also Chapter 9) is a general reaction applicable to all types of unsaturation. It consists of shaking a solution of the compound to be reduced with the finely divided catalyst in an atmosphere of gaseous hydrogen. The hydrogen is adsorbed onto the surface of the catalyst where it is "activated," probably by the formation of hydrogen atoms combined

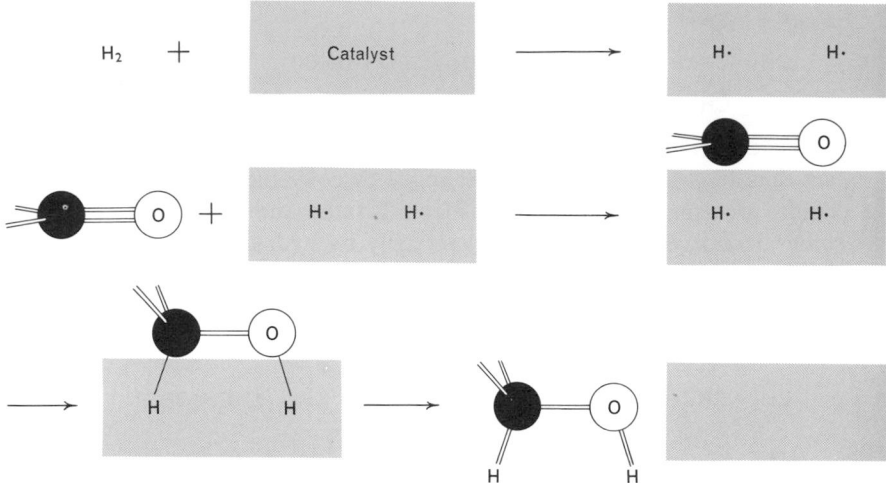

**Figure 8–6.** Catalytic hydrogenation of the carbonyl group:

$$\text{>C=O} \xrightarrow{\text{2H}} \text{>CHOH}$$

with the catalyst; the adsorption of the unsaturated compound to the catalyst is followed by attachment of two hydrogen atoms to the ends of the double bond; and finally the reduced compound is desorbed from the catalyst surface (see Figure 8-6).

Catalysts used for this reaction are most commonly platinum or palladium; they may be in the form of the oxides, which are reduced to the free metal by the hydrogen used. Other catalysts—nickel and complex heavy metal oxides—are sometimes used. The reaction is usually carried out under a positive pressure of hydrogen (1–200 atmospheres).

***Reduction by dissolving metals.*** Reduction by dissolving metals is not reduction by the molecular hydrogen that is liberated, but is brought about by the *donation of electrons from the metal* and subsequent neutralization of the resulting negative charges by proton donation by the solvent. Combinations commonly used are sodium amalgam, zinc and alkali, sodium and alcohol, and others.

Zinc (dust) and acetic acid (bimolecular reduction):

$$(C_6H_5)_2CO \xrightarrow[\text{HOAc}]{\text{Zn}} (C_6H_5)_2\overset{\displaystyle OH}{\underset{\displaystyle |}{C}}\text{—}\overset{\displaystyle OH}{\underset{\displaystyle |}{C}}(C_6H_5)_2 \tag{94}$$

benzophenone        benzpinacol

Zinc dust and sodium hydroxide:

$$(C_6H_5)_2C\text{=}O \xrightarrow[\text{NaOH}]{\text{Zn}} (C_6H_5)_2CHOH \tag{95}$$

benzophenone        benzohydrol

Sodium and alcohol (Bouveault-Blanc) reduction:

$$CH_3CH_2CH_2CH_2CH_2COOC_2H_5 \xrightarrow[\text{EtOH}]{\text{Na}} CH_3CH_2CH_2CH_2CH_2CH_2OH \tag{96}$$

ethyl hexanoate $\qquad\qquad$ *n*-hexyl alcohol

$$CH_3COCH_2CH_2CH_2CH_2CH_3 \xrightarrow[\text{EtOH}]{\text{Na}} CH_3CHOHCH_2CH_2CH_2CH_2CH_3 \tag{97}$$

2-heptanone $\qquad\qquad\qquad$ 2-heptanol

Aliphatic ketones are reduced by metallic magnesium with the formation of bimolecular reduction products. These are generically called "pinacols" from the name of the product derived from acetone. This reaction is described in Chapter 12.

*Metal hydrides.* One of the most elegant methods of reduction of carbonyl compounds is with the use of lithium aluminum hydride, LiAlH$_4$. This reagent, prepared by the reaction of lithium hydride with aluminum chloride,

$$4LiH + AlCl_3 \longrightarrow LiAlH_4 + 3LiCl \tag{98}$$

is soluble in diethyl ether, the solvent in which it is ordinarily used. It reduces aldehydes and ketones smoothly and quantitatively to the primary and secondary alcohols, respectively:

$$4RCHO + LiAlH_4 \longrightarrow (RCH_2O)_4AlLi \xrightarrow{\text{HA}} RCH_2OH + AlA_3 + LiA \tag{99}$$

$$4\,\underset{R}{\overset{R}{>}}C=O + LiAlH_4 \longrightarrow \left(\underset{R}{\overset{R}{>}}CHO\right)_4 AlLi \xrightarrow{\text{HA}} \underset{R}{\overset{R}{>}}CHOH + AlA_3 + LiA \tag{100}$$

An important use of LiAlH$_4$ lies in its ability to reduce carboxylic acids and their esters—reductions that are either difficult to accomplish, require somewhat involved manipulations, or cannot be accomplished cleanly and in good yield by other means:

$$4RCOOEt + 2LiAlH_4 \longrightarrow (RCH_2O)_4AlLi + (EtO)_3AlLiOEt$$
$$\xrightarrow[\text{H}_2\text{O}]{\text{HA}} RCH_2OH \;(+\,\text{Li and Al salts}) \tag{101}$$

Reductions with LiAlH$_4$ proceed readily, give excellent yields, and permit ready isolation of products. Other uses of this reagent will be dealt with elsewhere in the text.

Other complex metal hydrides are also used; they differ from one another chiefly in their capacity for reduction of carbonyl compounds of various kinds. Sodium borohydride, NaBH$_4$, is a commonly used reducing agent for aldehydes and ketones. It is incapable of reducing carboxylic acids and their esters and consequently can be used for selective reduction. An example of this selectivity is shown in the following; lithium aluminum hydride reduces both the ketone and the ester groupings; sodium borohydride reduces the ketone grouping only:

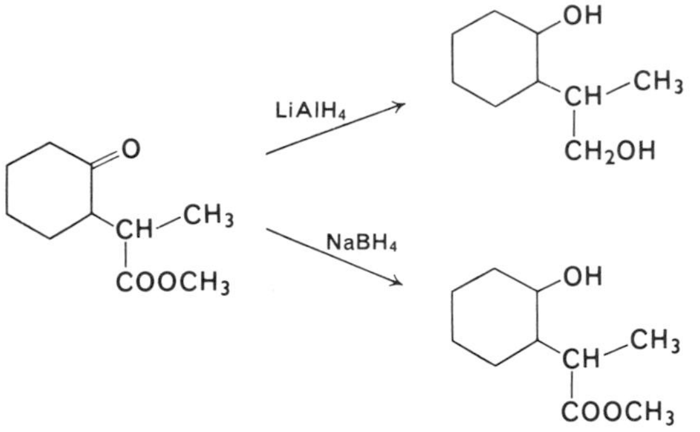

(102)

Selective reduction of the ester group only can be accomplished if the ketone carbonyl groups is first "protected" by conversion into a ketal (Section 8.20) After reduction of the ester grouping, hydrolysis of the ketal yields the hydroxy ester:

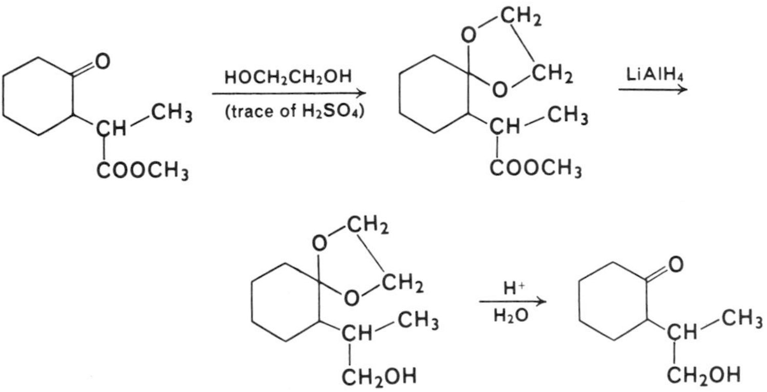

(103)

---

**Exercise 9.** Write the individual steps in the hydrolysis with aqueous acid of the acetal $CH_3CH_2CH(OCH_3)_2$, starting with the premise that the initial stage is the protonation of one of the oxygen atoms of the acetal.

---

*Reduction of aldehydes and ketones by metal alkoxides.* The reduction of aldehydes and ketones by alcohols can be represented by the general expression

$$\underset{R'}{\overset{R}{>}}C{=}O \ + \ \overset{*}{>}CHOH \ \rightleftharpoons \ \underset{R'}{\overset{R}{>}}CHOH \ + \ >C{=}O \tag{104}$$

The reducing agent (marked by the asterisk) is usually a readily available alcohol which is used in large excess: isopropyl alcohol is commonly employed  The reaction is catalyzed by an alkoxide of aluminum or magnesium, as in the following two examples:*

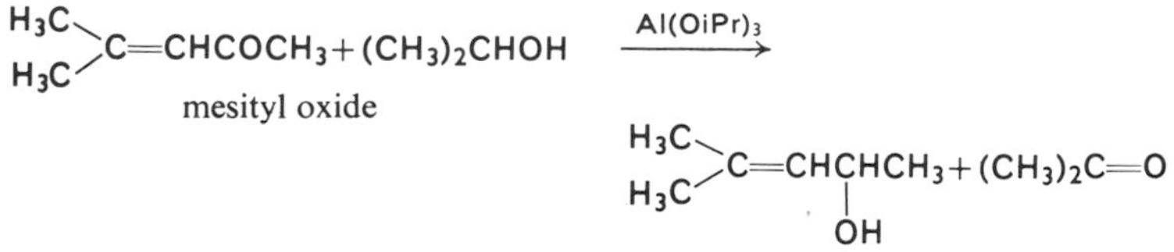

$$\underset{\text{mesityl oxide}}{\overset{\displaystyle H_3C}{\underset{\displaystyle H_3C}{>}}C=CHCOCH_3} + (CH_3)_2CHOH \xrightarrow{\text{Al(OiPr)}_3}$$

$$\overset{\displaystyle H_3C}{\underset{\displaystyle H_3C}{>}}C=CHCHCH_3 + (CH_3)_2C=O \quad (105)$$
$$\underset{\displaystyle OH}{|}$$

$$\text{menthone} \quad + \quad CH_3CH_2OH \quad \xrightarrow{\text{Mg(OEt)}_2} \quad \text{menthol} \quad + \quad CH_3CHO \quad (106)$$

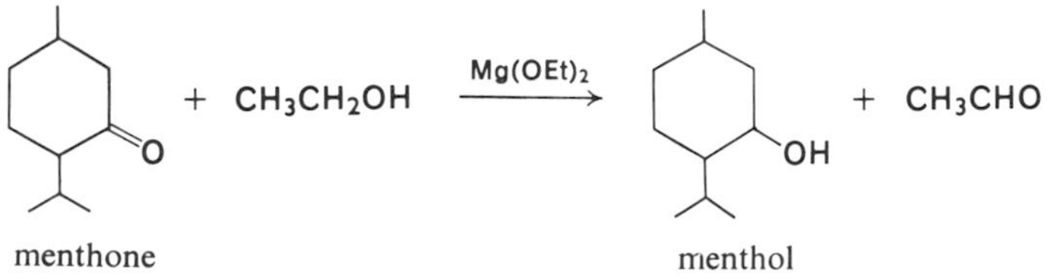

This reduction method is described in greater detail in Chapter 12.

## 8.22  The preparation of alcohols with Grignard reagents

The most general method of synthesizing alcohols of all types is by the use of the Grignard reagent. This reagent, named after its discoverer, Victor Grignard (1871–1935), is prepared by allowing an alkyl halide to react with metallic magnesium (as shavings or turnings) in ether. The magnesium dissolves as the exothermic reaction proceeds, and at length there results an ether solution of an alkylmagnesium halide, known as a "Grignard reagent":

$$CH_3CH_2Br \xrightarrow[\text{ether}]{\text{Mg}} \underset{\substack{\text{ethylmagnesium} \\ \text{bromide}}}{CH_3CH_2MgBr} \quad (107)$$

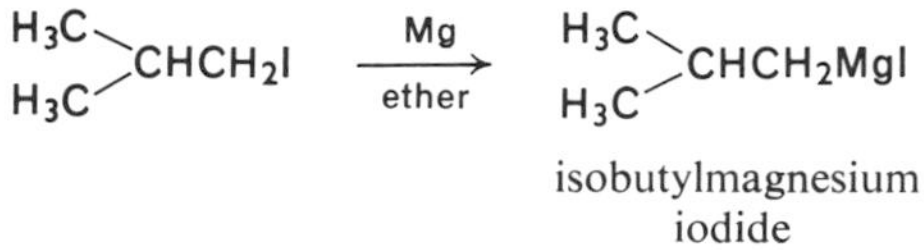

$$\overset{\displaystyle H_3C}{\underset{\displaystyle H_3C}{>}}CHCH_2I \xrightarrow[\text{ether}]{\text{Mg}} \overset{\displaystyle H_3C}{\underset{\displaystyle H_3C}{>}}\underset{\substack{\text{isobutylmagnesium} \\ \text{iodide}}}{CHCH_2MgI} \quad (108)$$

$$CH_2=CHCH_2Cl \xrightarrow[\text{ether}]{\text{Mg}} \underset{\substack{\text{allylmagnesium} \\ \text{chloride}}}{CH_2=CHCH_2MgCl} \quad (109)$$

Grignard reagents are very versatile in that they engage in a wide variety of reactions. The most valuable of these in synthesis is the reaction with carbonyl

---

* The symbols Al(OiPr)₃, for aluminum isopropoxide, and Al(OtBu)₃, for aluminum *t*-butoxide, will be used for simplicity and convenience.

compounds: the reagent adds to the carbon-oxygen double bond with the formation of a new carbon-carbon bond, the —MgX portion of the reagent becoming attached to the oxygen atom:

$$>C=O + RMgX \longrightarrow >\underset{R}{C}-OMgX \tag{110}$$

The immediate product of the addition is a halomagnesium alkoxide, which by the action of water (or dilute acids, to dissolve the basic magnesium salt that is formed on hydrolysis) is converted to the alcohol:

$$>\underset{R}{C}-OMgX + HX(H_2O) \longrightarrow >\underset{R}{C}-OH + MgX_2(H_2O) \tag{111}$$

The partial formulation (110) of the essential step in the reaction can be expanded by assigning substituents to the carbonyl group:*

$$\underset{H}{\overset{H}{>}}C=O + RMgX \longrightarrow \underset{H}{\overset{H}{>}}\underset{R}{\overset{OH}{C}} \tag{112}$$

a primary alcohol

$$\underset{H}{\overset{R'}{>}}C=O + RMgX \longrightarrow \underset{H}{\overset{R'}{>}}\underset{R}{\overset{OH}{C}} \tag{113}$$

a secondary alcohol

$$\underset{R'}{\overset{R''}{>}}C=O + RMgX \longrightarrow \underset{R'}{\overset{R''}{>}}\underset{R}{\overset{OH}{C}} \tag{114}$$

a tertiary alcohol

$$\underset{RO}{\overset{R'}{>}}C=O + RMgX \longrightarrow \left\{ \underset{O}{\overset{R'}{\underset{\parallel}{C}}}\overset{R} \right\} \longrightarrow \underset{R}{\overset{R'}{>}}\underset{R}{\overset{OH}{C}} \tag{115}$$

an ester                an intermediate
                        ketone, which reacts
                        further with RMgX

Some specific examples of these reactions are shown in the following equations.

Secondary and tertiary alkylmagnesium halides react with formaldehyde to give primary alcohols having one more carbon atom:

$$\underset{H_3C}{\overset{H_3C}{>}}CHMgBr + HCHO \longrightarrow \underset{H_3C}{\overset{H_3C}{>}}CHCH_2OH* \tag{116}$$

---

* The final step of hydrolysis of the —$\underset{|}{\overset{|}{C}}$—OMgX compound first formed (111) is omitted in this series of equations. It is not usually shown in writing the equation for a synthesis involving a Grignard reagent.

$$CH_3-\underset{\underset{CH_3}{|}}{\overset{\overset{CH_3}{|}}{C}}\cdot MgCl + HCHO \longrightarrow CH_3-\underset{\underset{CH_3}{|}}{\overset{\overset{CH_3}{|}}{C}}-CH_2OH \qquad (117)$$

neopentyl alcohol

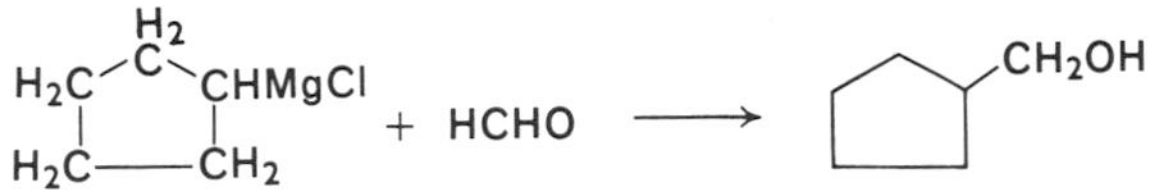

$$\qquad (118)$$

cyclopentylmagnesium
chloride

cyclopentylcarbinol

The formation of secondary alcohols from aldehydes, and of tertiary alcohols from ketones, is illustrated as follows:

$$CH_3CHO + CH_3CH_2CH_2CH_2MgBr \longrightarrow CH_3\overset{\overset{OH}{|}}{C}HCH_2CH_2CH_2CH_3 \qquad (119)$$

acetaldehyde      *n*-butylmagnesium         2-hexanol
bromide

$$CH_3CH_2CHO + (CH_3)_3CCH_2MgCl \longrightarrow CH_3CH_2\overset{\overset{OH}{|}}{C}HCH_2\overset{\overset{CH_3}{|}}{\underset{\underset{CH_3}{|}}{C}}-CH_3 \qquad (120)$$

propionaldehyde   neopentylmagnesium
chloride

5,5-dimethyl-3-hexanol

$$(CH_3)_2CHCOCH(CH_3)_2 + CH_3MgBr \longrightarrow (CH_3)_2CH\overset{\overset{CH_3}{|}}{\underset{\underset{OH}{|}}{C}}CH(CH_3)_2 \qquad (121)$$

di-isopropylketone    methylmagnesium
bromide

2,3,4-trimethyl-<br>3-pentanol

---

**Exercise 10.** Write the reaction between *n*-propylmagnesium bromide and *(a)* formaldehyde; *(b)* cyclopentanone; *(c)* diethylketone.

---

The reaction of esters (115) with Grignard reagents is another means of preparing tertiary alcohols. In this case, a ketone is the intermediate; but since ketones react with Grignard reagents more rapidly than do esters, it does not survive, but reacts further to yield the expected tertiary alcohol:

$$RCOOEt + R'MgBr \longrightarrow \{RCOR'\} \overset{R'MgBr}{\longrightarrow} R-\underset{\underset{R'}{|}}{\overset{\overset{R'}{|}}{C}}-OH \qquad (122)$$

Thus, the tertiary alcohol that is formed contains *two* of the groups furnished by the Grignard reagent; for example:

$$CH_3CH_2COOEt + CH_3CH_2MgBr \longrightarrow CH_3CH_2\overset{\overset{\displaystyle OH}{|}}{\underset{\underset{\displaystyle CH_2CH_3}{|}}{C}}{-}CH_2CH_3 \qquad (123)$$

ethyl propionate

triethylcarbinol

It can be observed, then, that tertiary alcohols in which at least two of the alkyl groups are alike can be prepared from esters or symmetrical ketones.

---

**Exercise 11.**   Show two ways of preparing, by means of the Grignard reaction, the following alcohols: *(a)* 2-butanol; *(b)* 2,3,4-trimethyl-3-pentanol.

---

There are some limitations of this synthetic method. In brief, poor yields of the desired alcohols are often obtained when both the ketone and the Grignard reagent contain alkyl groups that are highly substituted on the carbon atom attached to the carbonyl group or the magnesium atom.

## 8.23   Oxidation of alcohols: general

A primary alcohol may be regarded as the first stage of oxidation in the series

$$RCH_3 \xrightarrow{+O} RCH_2OH \xrightarrow{-2H} RCHO \xrightarrow{+O} RCOOH \qquad (124)$$

hydrocarbon   primary alcohol    aldehyde       acid

The loss of two hydrogen atoms $(-2H)$ or the gain of one oxygen atom $(+O)$ are purely formal designations here; they represent two-electron oxidation steps, and they may be brought about experimentally by a number of means. The reagents that are used and the methods of carrying out the oxidations are described in the sections to follow.

Our concern here is with the step alcohol $\rightarrow$ carbonyl compound. The oxidation of a primary alcohol yields the aldehyde; the oxidation of a secondary alcohol yields a ketone; and a tertiary alcohol cannot yield a carbonyl compound with the same carbon skeleton:*

$$\overset{R}{\underset{R}{\diagdown}}CHOH \xrightarrow{-2H} \overset{R}{\underset{R}{\diagdown}}C{=}O \longrightarrow \text{oxidized further only with rupture of the carbon chain} \qquad (125)$$

$$R{-}\overset{\overset{\displaystyle R}{|}}{\underset{\underset{\displaystyle R}{|}}{C}}{-}OH \longrightarrow \text{oxidized only with alteration of carbon skeleton or rupture of carbon chain} \qquad (126)$$

---

\* Further oxidations of other parts of the molecule can, of course, occur, depending upon the nature of the R groups in these generalized equations; but our attention is directed here to the changes at the functional group. Organic compounds are susceptible to oxidations of many kinds, but discussion of these will be deferred.

### 8.24 Oxidation of alcohols. Reagents and methods

*Chromic acid.* The oxidation of primary and secondary alcohols to aldehydes and ketones can be brought about by a variety of oxidizing agents, most of them inorganic compounds of metals that are capable of undergoing a change to a lower valence state.

Among the most widely used oxidizing agents are found compounds of chromium (VI), which is reduced in the process to chromium (III). Chromium trioxide (chromic acid) and potassium dichromate are the usual forms that are employed. The oxidation of a secondary alcohol to a ketone is represented by the half-reactions:

$$Cr^{VI}O_3 + 3e + 6H^+ \longrightarrow Cr^{III} + 3H_2O \qquad (127)$$

$$R_2CHOH \longrightarrow R_2C{=}O + 2H^+ + 2e \qquad (128)$$

$$\text{sum: } 6H^+ + 2Cr^{VI}O_3 + 3R_2CHOH \longrightarrow 2Cr^{III} + 3R_2C{=}O + 6H_2O \qquad (129)$$

These oxidations are usually carried out in acid solution (acetic acid or aqueous sulfuric acid); in the latter case the total balanced equation becomes

$$2CrO_3 + 3H_2SO_4 + 3R_2CHOH \longrightarrow 3R_2C{=}O + Cr_2(SO_4)_3 + 6H_2O \qquad (130)$$

---

**Exercise 12.** Write the complete balanced equation for the oxidation of ethanol to acetaldehyde ($CH_3CHO$) with sodium dichromate ($Na_2Cr_2O_7$) in aqueous sulfuric acid.

---

The oxidation of primary alcohols to aldehydes is difficult to carry out successfully by this method, for the aldehyde is very readily oxidized further to the corresponding carboxylic acid. It is necessary to interrupt the oxidation at the aldehyde stage, and this can often be done by removing the aldehyde as it is formed—by distilling it out of the reaction mixture if it is sufficiently volatile, or by extracting it into a nonmiscible organic solvent and thus removing it from contact with the oxidant. Secondary alcohols can be oxidized to ketones very effectively by the use of chromic acid, for ketones are sufficiently resistant to further oxidation to escape destruction.

The mechanism of the chromic acid oxidation of secondary alcohols to ketones is an interesting example of an elimination reaction of a kind that has general application to oxidations of several types. It can be formulated in general terms:

$$\overset{|}{\underset{B:\overset{\curvearrowright}{H}}{-\underset{|}{C}}}-O\overset{\curvearrowright}{-}X \longrightarrow -\overset{|}{C}{=}O + :X^- \qquad (131)$$
$$B:H^+$$

When chromic acid is the oxidizing agent, the course of the reaction is formulated in the following way:

1. Formation of a chromate ester

$$R_2CHOH + CrO_3 + H^+ \longrightarrow R_2CH{-}O{-}CrO_3H_2^+ \qquad (132)$$

2. Elimination, by attack of a nucleophile on the proton (of $\rangle$CH—) and loss of $CrO_3H_2$ as a leaving group

$$R_2C—O—CrO_3H_2^+ \longrightarrow R_2C{=}O + H_2CrO_3 \qquad (133)$$
$$+ B{:}H^+$$

This is the rate-determining step of the oxidation. The chromium compound formed in the displacement is $H_2CrO_3$, equivalent to $CrO_2$, in which the chromium is in the IV oxidation state.

3. Disproportionation of $Cr^{IV}$ to $Cr^{III}$ and $Cr^{V}$. The latter can then oxidize another mole of the alcohol to produce $R_2C{=}O$ and $Cr^{III}$

Since the —O—Cr bond in the ester is formed by nucleophilic attack of the OH group on the chromic acid, its two electrons are both provided by the alcohol. Since in the loss of $H_2CrO_3$ in Step 2 the two electrons of the O—Cr bond leave with the departing group, the overall consequence of the above reaction sequence is the loss of two electrons by the alcohol, or a two-electron oxidation.

***Other metals as oxidizing agents.*** Alcohols can be oxidized by a variety of inorganic reagents, among which are potassium permanganate and certain salts of cerium (IV) and cobalt; but these are less general and are used in special cases.

Certain oxidizing agents have a high degree of specificity for particular kinds of oxidations. An example is manganese dioxide, which is capable of oxidizing allylic alcohols to the corresponding carbonyl compounds under very mild conditions and without side reactions of other kinds:

$$RCH{=}CHCH_2OH \xrightarrow{MnO_2} RCH{=}CHCH{=}O \qquad (134)$$

$$RCH{=}CHCHOH \xrightarrow{MnO_2} RCH{=}CHC{=}O \qquad (135)$$
$$\qquad\quad | \qquad\qquad\qquad\qquad\quad |$$
$$\qquad\quad R \qquad\qquad\qquad\qquad\quad R$$

The reaction is useful in dealing with substances that contain additional functional groupings that would be susceptible to attack by other oxidizing agents, such as chromic acid or potassium permanganate. An interesting application of $MnO_2$ is in the oxidation of vitamin $A_1$ to retinene, a component of the photosensitive substance "rhodopsin" that is found in the retina of the eye:

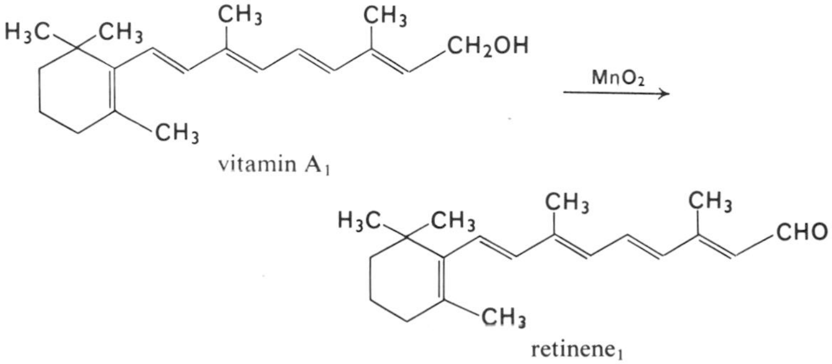

vitamin $A_1$

$$\qquad (136)$$

retinene$_1$

---

**Exercise 13.**   Write out the complete structures for vitamin $A_1$ and retinene$_1$, showing all carbon and hydrogen atoms.

---

The use of manganese dioxide for the oxidation of allylic alcohols provides another diagnostic test that is useful in structure determination. The oxidation of an unknown hydroxyl-containing compound to an aldehyde or ketone by means of manganese dioxide under mild conditions, coupled with recognition by other means of the presence of the C=C—C=O grouping in the product of the reaction, is an indication that the original compound contains the structural unit

C=C—CHOH. Such a result thus defines the relative placing of a group of several atoms in the molecule and may be a major step toward the final structure proof.

It should be added that manganese dioxide is not inert toward alcohols of other kinds; but if it is used under properly controlled, mild reaction conditions, the behavior described above is characteristic.

*Dimethyl sulfoxide.*   An interesting example of an oxidation reaction that is described by the general equation (131) is that which occurs when a sulfonic ester is treated with dimethyl sulfoxide. This reaction occurs in two stages: (1) (Equation 137) an $S_N2$ displacement of the sulfonate ion by nucleophilic attack of $(CH_3)_2S^+$—$O^-$ [dimethyl sulfoxide, prepared by oxidation of dimethyl sulfide, $(CH_3)_2S$]; followed by (2) an elimination reaction (Equation 138):

$$RCH_2OSO_2R' + (CH_3)_2\overset{+}{S}-\overset{-}{O} \longrightarrow RCH_2O-\overset{+}{S}(CH_3)_2 + R'SO_2O^- \tag{137}$$

$$RCH_2-O-\overset{+}{S}(CH_3)_2 \longrightarrow B:H + RCH=O + (CH_3)_2S: \tag{138}$$

The internal displacement of $(CH_3)_2S$ is favored by the strong polarization of the —O—S bond by the positive charge on the sulfur atom.

Since the first step of this reaction can proceed with other leaving groups than $R'SO_2O^-$, the reaction can be also used for the conversion of alkyl halides into aldehydes. As a method for oxidizing alcohols, it depends upon an initial conversion of the alcohol into the sulfonic ester:

$$RCH_2OH + R'SO_2Cl \longrightarrow ROSO_2R' + HCl \tag{139}$$

As in the case of the chromic acid oxidation, the oxidation step (138) involves the loss of the group attached to oxygen with the bonding electron pair.

---

**Exercise 14.**   Under certain conditions, bromine in alkaline solution can bring about the oxidation $R_2CHOH \rightarrow R_2C=O$. Suggest a course for this oxidation, showing the steps in a plausible mechanism.

---

*Ketones and metal alkoxides.* The oxidation of secondary alcohols to ketones can be accomplished by transferring the two hydrogen atoms of the $>$CHOH group to some "acceptor" molecule, usually a ketone. This is accomplished through the agency of a metal alkoxide, but the essential reaction can be expressed by the equilibrium (140):

$$\overset{*}{>}\!CHOH \;+\; >\!C\!=\!O \;\rightleftharpoons\; \overset{*}{>}\!C\!=\!O \;+\; >\!CHOH \qquad (140)$$

| alcohol being oxidized | acceptor | oxidized alcohol | reduced acceptor |

As the equilibrium (140) indicates, the reaction can also be used to reduce a ketone to the secondary alcohol. For these reactions to be of preparative value, the equilibrium must be caused to lie largely to the right or to the left. This can can be accomplished by applying the principle of mass action: in the case of (140) the alcohol can be substantially completely oxidized if *(a)* the "acceptor" carbonyl compound is used in large excess, or if *(b)* either the product (the "oxidized alcohol") or the reduced acceptor can be removed from the reaction mixture as it is formed.

A detailed discussion of the mechanism of this oxidation reaction (called the Oppenauer oxidation) will be given when its application to reduction (called the Meerwein-Ponndorf-Verley reduction), which has been alluded to briefly in Section 8.21, is described in a later chapter.

*Other methods.* A very simple process that is used industrially for the oxidation of the lower alcohols consists simply in the removal of hydrogen by the catalytic action of special copper-containing alloys at high temperatures:

$$RCH_2OH \underset{300°C}{\overset{Cu}{\rightleftharpoons}} RCH\!=\!O + H_2 \qquad (141)$$

Since, as we have seen earlier (Section 8.21), the carbonyl group can be reduced by the use of hydrogen, this dehydrogenation reaction proceeds to an equilibrium. The yield of aldehyde can be improved if air is present, in which case the hydrogen is removed by oxidation.

Catalytic dehydrogenation and controlled oxidation are both used to prepare such important industrial aldehydes as formaldehyde, acetaldehyde, and the three- and four-carbon aldehydes, but these methods are rarely used in laboratory syntheses.

### Exercises

(Exercises 1–14 will be found within the text.)

**15** Show the two possible courses for the reaction of sodium ethoxide with *(a)* ethyl bromide; *(b)* isopropyl iodide; *(c)* 2-bromo-2-methylbutane.

**16** Write the reaction showing the predominant result of allowing sodium ethoxide to react with each of the following. *(a)* $CH_3Br$; *(b)* $(CH_3)_3CCl$; *(c)* $CH_3CH_2CBr(CH_3)_2$; *(d)* $CH_3OSO_2OCH_3$; *(e)* $(CH_3)_2CHCHCH_3$.

**17** Using as starting materials any compounds of three carbon atoms or less, show how the following alcohols may be synthesized. *(a)* 3-pentanol; *(b)* isopentyl alcohol; *(c)* triethylcarbinol; *(d)* 2,3-dimethyl-2-butanol.

**18** Write the reactions involved in the acid-catalyzed ester interchange between ethyl acetate and 1-propanol.

**19** Write the reactions involved in the base-catalyzed ester interchange between ethyl acetate and 2-butanol. (A trace of sodium 2-butoxide may be regarded as the catalyst used.)

**20** Write the structures of the olefins that would be required to produce the following alcohols by hydration. *(a)* 2-propanol; *(b)* *tert*-butyl alcohol; *(c)* 2-butanol; *(d)* 2-methyl-2-butanol; *(e)* 1-methyl-1-cyclohexanol.

**21** Write the equations showing the reduction of the following compounds by means of lithium aluminum hydride. *(a)* acetaldehyde $(CH_3CHO)$; *(b)* isopropyl propionate $(CH_3CH_2COOCH(CH_3)_2)$; *(c)* $CH_3COCH_2CH_3$; *(d)* methyl acetate $(CH_3COOCH_3)$.

**22** Write the reactions for the following transformations. *(a)* 1-propanol into *n*-propyl isopropyl ether, $CH_3CH_2CH_2OCH(CH_3)_2$; *(b)* ethanol to 2-butene; *(c)* methyl ethyl ketone to 3,4-dimethyl-3-hexanol; *(d)* acetone and methyl iodide into isobutylene (2-methylpropene).

**23** What would be the result of dissolving acetaldehyde dimethyl acetal in an excess of 1-propanol, in the presence of a trace (a catalytic amount) of HCl?

**24** Acetals are stable to alkali, but hydrolyze readily in aqueous acid. Write the equilibria that would obtain in the hydrolysis of acetaldehyde dimethyl acetal in aqueous acid.

**25** Formulate the reduction of a ketone $R_2C{=}O$ to the secondary alcohol, $R_2CHOH$, by means of aluminum isopropoxide and isopropyl alcohol. This is described in detail in Chapter 10; attempt to work it out before referring to it there.

**26** Write the reactions between *n*-butylmagnesium bromide and *(a)* acetaldehyde; *(b)* ethyl acetate; *(c)* formaldehyde; *(d)* diethyl ketone; *(e)* menthone.

## Chapter nine

# The Carbon-Carbon Multiple Bond. Alkenes and Alkynes

## 9.1  Unsaturated compounds

Organic compounds that contain carbon-carbon double bonds ($C{=}C$) or triple bonds ($C{\equiv}C$) are *unsaturated* compounds, so called because they are able to undergo reactions in which *addition* to the double or triple bond takes place. Saturated compounds, which include such hydrocarbons as ethane and the homologous alkanes, as well as cyclic compounds that contain only single bonds, such as cyclohexane, do not undergo addition reactions. The "saturation" of a double bond is ordinarily construed to mean the addition of hydrogen; but addition reactions of many other kinds are known.

The simplest compounds containing the carbon-carbon double bond are hydrocarbons known as *olefins*, or by the I.U.C. term *alkenes*. Acetylenes, or *alkynes*, contain the carbon-carbon triple bond. Compounds containing two or more double or triple bonds are named in ways that are illustrated in Section 9.2.

The reactivity of olefins is due chiefly to the chemical properties of the double bond. The lower members of the series, produced in large quantities as products of the petroleum industry, are valuable raw materials for the industrial synthesis of alcohols, ethers, aldehydes, ketones, polymers and plastics, and organic halogen compounds. The double bond, which is a structural feature very commonly encountered in organic compounds of many kinds, is a point of attack upon the molecule; it permits the controlled degradation of the compound by oxidizing agents and the alteration of the structure of the compound by chemical manipulations which depend upon the reactivity of the double bond.

**Table 9-1.** *Comparison of physical properties of some olefins and their saturated analogs*

| Olefin | b.p. (°C) | Related Nonolefin | b.p. (°C) |
|---|---|---|---|
| $CH_2{=}CH_2$ | $-103.9$ | $CH_3CH_3$ | $-89.0$ |
| $CH_3CH_2CH{=}CH_2$ | $-6.3$ | $CH_3CH_2CH_2CH_3$ | 0.6 |
| $CH_3CH_2CH_2CH_2CH_2CH{=}CH_2$ | 93 | $CH_3(CH_2)_5CH_3$ | 98 |
| 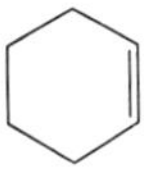 (cyclohexene) | 83 | 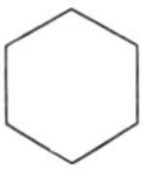 (cyclohexane) | 81 |
| 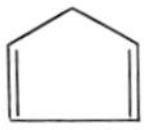 (cyclopentadiene) | 41 |  (cyclopentane) | 51 |
| 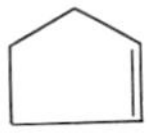 (cyclopentene) | 45 | — | — |
| $CH_2{=}CHCH_2Br$ | 71 | $CH_3CH_2CH_2Br$ | 71 |
| $CH_2{=}CHCOOH$ | 142 | $CH_3CH_2COOH$ | 141 |
| $CH_3CH{=}CHCH_2OH$ | 121 | $CH_3CH_2CH_2CH_2OH$ | 118 |

The physical properties of olefinic compounds resemble those of the corresponding saturated compounds  The carbon-carbon double bond has no marked effect upon boiling point, although it does increase water solubility to some degree, probably because the greater electron density at the double bond permits some degree of association with the dipolar water molecules  In Table 9-1 are given the boiling points of a few representative olefinic compounds, along with those of the corresponding saturated hydrocarbons. It is apparent that the double bond has no profound effect upon this property.

## 9.2  Nomenclature of unsaturated compounds

Most of the simpler olefins and acetylenes bear trivial names. The most suitable names for unsaturated compounds are the I.U.C. names, in which the endings -*ene*, for the double bond, and -*yne* for the triple bond, are used. The I.U.C. name is based upon the longest chain that bears the multiple bond. Substitution names are also used (Nos. 11, 13, 15; Table 9-2).

It is to be noted that when *cis* and *trans* isomers can exist, the designation should be added to the name. Thus, No. 4 is a nonexplicit formula for *cis*-2-butene or *trans*-2-butene.

In the writing of the structural formulas for unsaturated compounds, a number of conventions can be used. These have been described earlier (Section

**Table 9-2.** *Names of some unsaturated compounds*

| Compound | Name |
|---|---|
| 1. $CH_2\!\!=\!\!CH_2$ | ethylene <br> ethene (G) |
| 2. $CH_3CH\!\!=\!\!CH_2$ | propylene <br> propene (G) |
| 3. $CH_3CH_2CH\!\!=\!\!CH_2$ | 1-butene (G) |
| 4. $CH_3CH\!\!=\!\!CHCH_3$ | 2-butene (G) |
| 5. $(CH_3)_2C\!\!=\!\!CH_2$ | isobutylene <br> methylpropene (G) |
| 6. $CH_3CH\!\!=\!\!CHCH_2CH_2CH_3$ | 2-hexene (G) |
| 7. $CH_2\!\!=\!\!CCH_2CH_2CHCH_3$ <br> $\qquad\;\; |\qquad\quad |$ <br> $\qquad CH_2CH_3\;\; CH_3$ | 2-ethyl-5-methyl-1-hexene (G) |
| 8. $CH_2\!\!=\!\!CHBr$ | vinyl bromide <br> bromoethene (G) |
| 9. $CH_2\!\!=\!\!CHOCH_3$ | methyl vinyl ether |
| 10. $CH_2\!\!=\!\!CHCH_2Cl$ | allyl chloride <br> 3-chloro-1-propene (G) |
| 11. $(CH_3)_2C\!\!=\!\!C(CH_3)_2$ | tetramethylethylene |
| 12. $HC\!\!\equiv\!\!CH$ | acetylene <br> ethyne (G) |
| 13. $CH_3C\!\!\equiv\!\!CH$ | methylacetylene <br> propyne (G) |
| 14. $CH_2\!\!=\!\!CH\!\!-\!\!CH\!\!=\!\!CH_2$ | 1,3-butadiene (G) |
| 15. $CH_2\!\!=\!\!CH\!\!-\!\!C\!\!\equiv\!\!CH$ | vinylacetylene <br> 1-buten-3-yne (G) <br> (*or* but-1-en-3-yne) |
| 16. $CH_3CH\!\!=\!\!CH\!\!-\!\!CH\!\!=\!\!CH\!\!-\!\!CH\!\!=\!\!CH_2$ | 1,3,5-heptatriene (G) |

3-21) and several of them will be used, as convenience requires, in the discussion to follow.

## 9.3 The structure of the carbon-carbon double bond

The structure of ethylene has been discussed in Chapter 3. The experimental facts are that in ethylene the four hydrogen and two carbon atoms lie in a single plane, with a C=C—H angle of about 122° and a H—C—H angle of about 116°. These values are close to those that would be expected if the two carbon atoms each had three bonds hybridized to $sp^2$ configuration (compare with boron trihalides). The extra pair of electrons can be assigned to an orbital symmetrically disposed above and below the plane of the molecule in such a way that the electrons may be found above or below the plane of the hydrogen and carbon atoms. This picture of the ethylene molecule is called the sigma-pi model: the carbon-carbon bond of

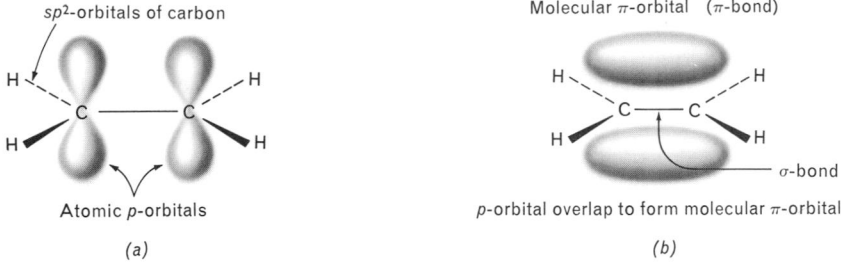

*sp²-orbitals of carbon*

Atomic *p*-orbitals

*(a)*

Molecular π-orbital (π-bond)

σ-bond

*p*-orbital overlap to form molecular π-orbital

*(b)*

**Figure 9–1.** Electron distribution for sigma-pi model of ethylene. *(a)* Disposition of orbitals without overlap. *(b)* Ethylene molecule as constituted by formation of π-orbital.

the $sp^2$ bonds being the sigma bond, the outlying orbital, the pi bond (since it is formed by overlap of the remaining $p$ orbital of each atom) (Figure 9-1).

There is another picture that can be constructed of the ethylene molecule that provides an equally good model of the molecule. Suppose each carbon atom were $sp^3$ in bond type. This would mean that each would possess four orbitals in tetrahedral arrangement: two of these are C—H bonds; the other two would, by overlap, form the C—C double bond (Figure 9-2).

The C=C—H angle for the $sp^3$ ("bent bond") model would be about 125°, and the H—C—H angle about 109.5°. Now ethylene does not exactly fit either the σ-π picture nor the "bent" tetrahedral bond picture, but the bond angles are closer to the former. However, ethylene seems to be more the exception than the rule. In $(CH_3)_2C=C(CH_3)_2$ the $CH_3$—C—$CH_3$ bond angle is 111.5°—close to the tetrahedral value. Propylene, $CH_3CH=CH_2$, has a C—C=C angle of about 125°; $Cl_2C=CH_2$ has a Cl—C=C angle of about 123°.

It is evident that neither of these models is an ideal representation of the carbon-carbon double bond. There is, however, no need to insist upon a decision in the matter, since for our purposes the consequences of either picture are exactly the same. It will be apparent that in both models the carbon and hydrogen atoms are in a plane and that there are regions of electron density lying above and below

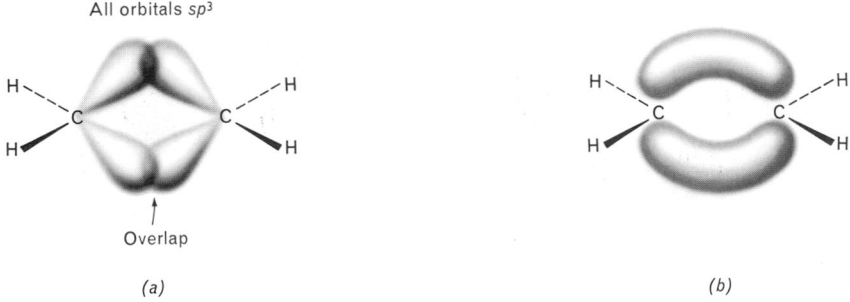

All orbitals *sp³*

Overlap

*(a)*

*(b)*

**Figure 9–2.** Electron distribution for $sp^3$ model of ethylene. *(a)* Disposition of orbitals showing nature of overlap. *(b)* Ethylene molecule as constituted by overlap.

this plane. To rotate one of the $CH_2$ groups with respect to the other would require "breaking" a bond; i.e., destroying the overlap of $p$ orbitals in the one case, of the $sp^3$ orbitals in the other. Such a bond rupture would require energy; and although energy can be supplied to bring about such rotation by suitable experimental means (heat or light energy), rotation about the double bond does not occur spontaneously (freely) under ordinary conditions. Thus, the double bond is a point of structural rigidity in the molecule. These conclusions are summarized in the next section.

## 9.4 The nature of the carbon-carbon double bond: *cis-trans* isomerism

Since the electronic distribution between the carbon atoms is not cylindrically symmetrical, the carbon atoms are not free to rotate with respect to one another. Thus, if the carbon atoms joined by the double bond hold groups other than hydrogen, as in 2-butene, there are two possible ways in which these substituents can be arranged. These are shown for two representative olefinic compounds in Figure 9-3.

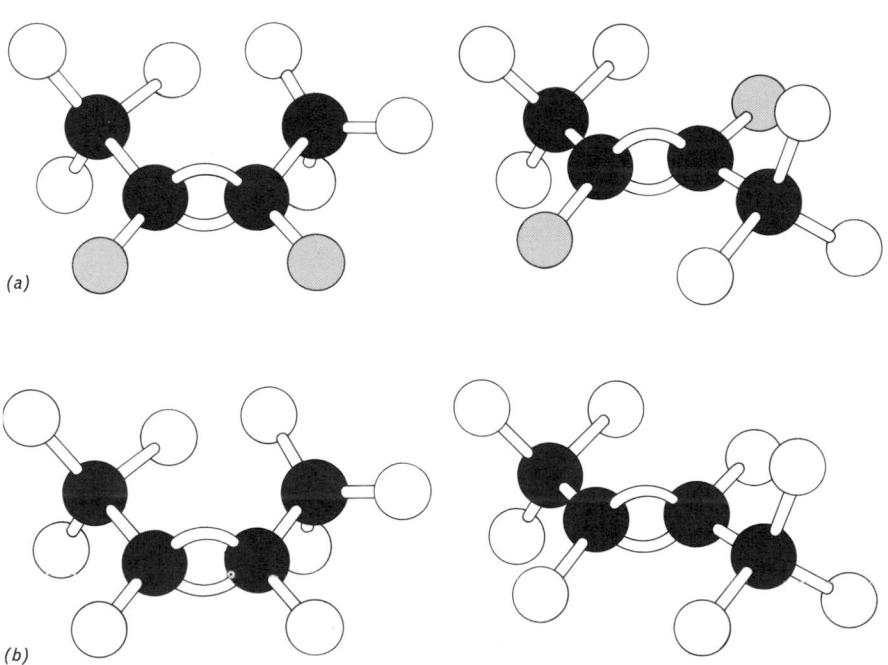

(a)

(b)

**Figure 9–3.** Ball-and-stick models of *(a) cis-* and *trans* 2,3 dibromo-2-butene ($CH_3CBr$=$CBrCH_3$), and *(b) cis-* and *trans*-2-butene ($CH_3CH$=$CHCH_3$).

Such *cis-trans isomers* are distinct individuals. They have different chemical properties, and different physical properties, such as boiling points, densities, and refractive indices. In order to convert one into the other, sufficient energy must be supplied to the molecule to break one of the bonds of the double bond (i.e., to destroy the overlap of one pair of orbitals). It is found that the more stable of the two 2-butenes is the *trans* compound, for in this configuration the two methyl groups are sufficiently far apart that their mutual interference is minimal.

*Cis-trans* isomerism is commonly encountered among unsaturated compounds; both isomers are known for many such compounds.

Since the electrons of the double bond are not confined to the region close to the axis joining the two carbon atoms, the molecule is accessible to attack by electron-seeking reagents. Thus, one of the two pairs of electrons of the double bond is available for combination with the reagent, leaving the two carbon atoms joined by the remaining electron pair, which forms the carbon-carbon single bond in the final addition product. We shall discuss this further in Section 9.5.

## 9.5 Reactivity of olefins

One of the characteristic properties of olefins is their reaction with strong acids ($HCl$, $H_2SO_4$, $HBr$) and with halogens ($Cl_2$, $Br_2$) to give products in which *addition to the double bond* has occurred. The following are examples of these reactions:

$$CH_3CH{=}CH_2 + HBr \longrightarrow CH_3\overset{\overset{\displaystyle Br}{|}}{C}HCH_3 \qquad (1)$$

propylene       *iso*propyl bromide
[propene (G)]

$$CH_2{=}CH_2 + Br_2 \longrightarrow BrCH_2CH_2Br \qquad (2)$$

ethylene      1,2-dibromoethane (G)

$$(CH_3)_2C{=}CH_2 + HCl \longrightarrow (CH_3)_2\overset{\overset{\displaystyle }{}}{C}CH_3 \qquad (3)$$

*iso*butylene
[methylpropene (G)]
$\overset{|}{Cl}$
*tert*-butyl chloride

$$CH_2{=}CHCH_2CH_3 + HOCl \longrightarrow ClCH_2\overset{\overset{\displaystyle }{}}{C}HCH_2CH_3 \qquad (4)$$

1-butene (G)
$\overset{|}{OH}$
1-chloro-2-butanol (G)

Let us examine these reactions in detail, selecting the addition of HBr to propylene for first consideration. Since the overall result is the addition of HBr as (H) and (Br) to the double bond, we might start by assuming that HBr is acting in its usual fashion and is providing a proton ($H^+$) and a bromide ion ($Br^-$). There is excellent evidence that this is indeed the case, and that the reaction is an electrophilic attack of HBr upon the olefin, in which the latter supplies an electron pair with which the proton coordinates in the first stage of the reaction:*

$$
\begin{array}{c} CH_3CH \\ \| \\ CH_2 \quad H—Br \end{array}
\quad \longrightarrow \quad
\left\{ \begin{array}{c} CH_3CH \\ | \\ CH_3 \end{array} \right\}^{+} Br^{-}
\tag{5}
$$

The second step is the addition of the bromide ion to the positively charged (carbonium) carbon atom:

$$
\left\{ \begin{array}{c} CH_3CH \\ | \\ CH_3 \end{array} \right\}^{+} Br^{-}
\quad \longrightarrow \quad
\begin{array}{c} CH_3—CH—Br \\ | \\ CH_3 \end{array}
\tag{6}
$$

The behavior of the olefin is thus that of a *nucleophilic, or basic, substance* since it supplies the electron pair to which the proton is furnished by the acid. As the picture of the olefinic double bond (Figure 9-1) shows, this pair of electrons is outside the region between the two carbon nuclei and is thus accessible to attack by an electrophilic reagent.

### 9.6 Experimental evidence in support of the hypothesis of stepwise addition

Evidence for the nucleophilic behavior of the olefinic double bond is found in the addition of bromine. Ethylene reacts with bromine to form 1,2-dibromoethane (G) (ethylene dibromide):

$$
CH_2{=}CH_2 + Br_2 \quad \longrightarrow \quad BrCH_2CH_2Br
\tag{7}
$$

ethylene dibromide

The reaction can be written as a nucleophilic displacement on bromine, with expulsion of bromide ion,

$$
\begin{array}{c} CH_2 \\ \| \\ CH_2 \quad Br—Br \end{array}
\quad \longrightarrow \quad
\left[ \begin{array}{c} CH_2 \\ | \\ CH_2Br \end{array} \right]^{+} Br^{-}
\tag{8}
$$

---

* This way of representing the course of a reaction is a convenient way of summarizing graphically the essential nature of the process. The arrow $\begin{array}{c} CH_2 \\ \| \\ CH_2 \quad H—Br \end{array}$ indicates that one of the electron pairs of the double bond is being supplied to the attacking reagent. The arrow $H—Br$ indicates that a bromide ion is being expelled (displaced) from the hydrogen bromide molecule while the electron pair of olefin is being accepted. This is a concerted process, at some stage of which the C—H bond is in the process of forming and the H—Br bond is in the process of breaking. It is to be emphasized that these are conventional representations, and are only devices used to illustrate a sequence of events. The arrows have no general significance beyond that which is ascribed to them in this explanation. It will be recalled that similar methods of representing the course of displacement reactions of other kinds have been used in earlier chapters.

followed by combination of the bromide ion with the electron-deficient carbon atom:

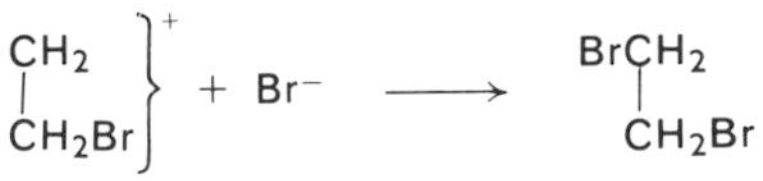

$$\left\{ \begin{matrix} CH_2 \\ | \\ CH_2Br \end{matrix} \right\}^+ + Br^- \longrightarrow \begin{matrix} BrCH_2 \\ | \\ CH_2Br \end{matrix} \tag{9}$$

The two-step nature of this process has been demonstrated by carrying out the addition of bromine to ethylene in the presence of "foreign" nucleophilic agents. For example, if ethylene is brominated in an aqueous solution of sodium chloride, the intermediate $\{\overset{+}{CH_2CH_2Br}\}$ accepts a chloride ion as well as a bromide ion:

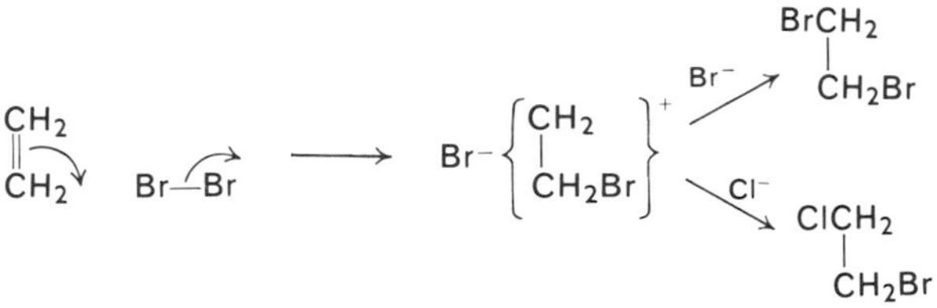

$$\begin{matrix} CH_2 \\ \| \\ CH_2 \end{matrix} \quad Br\!-\!Br \longrightarrow Br^- \left\{ \begin{matrix} CH_2 \\ | \\ CH_2Br \end{matrix} \right\}^+ \begin{matrix} \overset{Br^-}{\nearrow} \begin{matrix} BrCH_2 \\ | \\ CH_2Br \end{matrix} \\ \underset{Cl^-}{\searrow} \begin{matrix} ClCH_2 \\ | \\ CH_2Br \end{matrix} \end{matrix} \tag{10}$$

Bromination of ethylene in aqueous sodium nitrate solution produces both 1,2-dibromoethane and 2-bromoethyl nitrate, $BrCH_2CH_2ONO_2$. This and other experimental evidence shows clearly that in the addition reactions of olefins with strong acids and halogens there is an initial stage in which one electron pair of the double bond coordinates with the attacking reagent, followed by a subsequent step in which the positively charged intermediate ion accepts an anion to complete the process.

The stability of the intermediate, positively charged species varies with the nature of the olefin to which addition is taking place. In the case of ethylene, the ion $(CH_2CH_2Br)^+$ can be represented as a very unstable intermediate at a high-energy minimum in the energy diagram for the reaction (Figure 9-4). Examples are known, however, in which the intermediate is sufficiently stable to be isolated. A specific example of this is the following. The bromination of 1,1-dianisylethylene can be carried out in such a way as to yield the crystalline salt shown in the following equation:

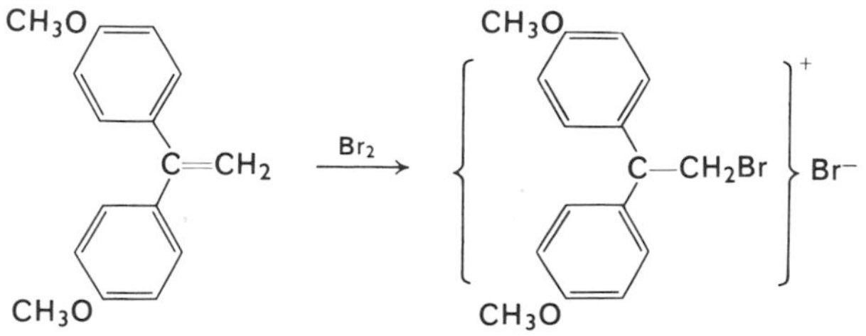

This, although an unusual example, shows that when the intermediate is stabilized by structural factors (for reasons that will become clear in the discussions of

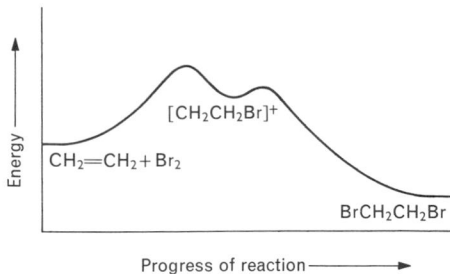

**Figure 9-4.** Energy diagram for the addition of bromine to ethylene:

$$CH_2{=}CH_2 + Br_2 \longrightarrow BrCH_2CH_2Br$$

Chapter 10), it can be isolated as a discrete intermediate. Further reaction of such an ionic intermediate, either *(a)* to lose a proton or *(b)* to add $Br^-$, can and often does occur as a subsequent, separate step.

Relative reactivities of olefins in addition reactions vary greatly with the nature of the substituents on the double bond. The energy diagrams for addition reactions with more reactive olefins will resemble that in Figure 9-4, but will show transition state maxima at lower energies, the faster the reaction. The relationship between structure and reactivity of olefins is discussed in the following section.

## 9.7  Effect of substituents upon the reactivity of olefins

Since the olefin furnishes the electron pair for the first stage of the addition reaction, the presence in the olefin molecule of substituents that increase the availability of electrons will aid the reaction. We have seen that alkyl groups are electron-repelling, and that halogen atoms are electron-attracting (Chapter 4):

$$\overrightarrow{CH_3CH{=}CH_2} \qquad \overleftarrow{BrCH{=}CH_2}$$

**Table 9-3.** *Rates of addition of bromine to olefins*

| Olefin | Relative Rate of Bromine Addition (in $CH_2Cl_2$ solution, $-78°C$) (ethylene$=1$) |
|---|---|
| $(CH_3)_2C{=}C(CH_3)_2$ | 14.0 |
| $(CH_3)_2C{=}CH_2$ | 5.5 |
| $CH_3CH{=}CH_2$ | 2.0 |
| $CH_2{=}CH_2$ | (1.0) |
| $CH_3CH{=}CHCOOH$ | 0.3 |
| $BrCH{=}CH_2$ | very slow |

In propylene, the methyl group increases electron density in the double bond; in vinyl bromide the halogen atom decreases it. It is found that propylene adds bromine faster than does ethylene, and vinyl bromide slower. Several alkyl groups increase the reactivity, as the figures in Table 9-3 show.

It will be noted in Table 9-3 that the acid $CH_3CH=CHCOOH$ (crotonic acid) undergoes addition of bromine more slowly than does ethylene. The inductive effect of the carboxyl group tends to draw the electrons of the double bond closer to the carbon nuclei and thus reduces their accessibility to the attacking bromine molecule. The effect is similar to that shown above for vinyl bromide.

## 9.8   Addition of unsymmetrical reagents

The addition of HCl to propylene yields isopropyl chloride [2-chloropropane (G)]; no appreciable amount of *n*-propyl chloride is formed:

$$CH_3CH=CH_2 + HCl \longrightarrow CH_3\overset{\underset{\displaystyle |}{Cl}}{C}HCH_3, \text{ not } CH_3CH_2CH_2Cl \qquad (11)$$

If the addition proceeds by way of the protonated olefin as an intermediate stage, it is clear from this result that $CH_3\overset{+}{C}HCH_3$, not $CH_3CH_2\overset{+}{C}H_2$, is the structure of the protonated olefin to which chloride ion coordinates in the last stage of the reaction. In general, addition of halogen acids to unsymmetrical olefins gives products in which the halogen atom is found on the most highly substituted carbon atom (the carbon atom having the lowest number of hydrogen atoms):

$$(CH_3)_2C=CH_2 + HCl \longrightarrow (CH_3)_2\overset{\underset{\displaystyle |}{Cl}}{C}-CH_3 \qquad (12)$$

$$CH_3CH_2CH=CH_2 + HBr \longrightarrow CH_3CH_2CHBr \qquad (13)$$
$$\underset{\displaystyle |}{\quad}\;CH_3$$

$$CH_3CH_2\overset{\underset{\displaystyle }{CH_3}}{C}=CHCH_3 + HBr \longrightarrow CH_3CH_2\overset{\underset{\displaystyle Br}{CH_3}}{C}-CH_2CH_3 \qquad (14)$$

---

**Exercise 1.**   Name the reactants and products in the above equations by the I.U.C. system.

---

This mode of addition of unsymmetrical reagents to olefins has been known for a very long time, and the Russian chemist Markovnikov proposed a "rule" that is exemplified by the above addition reactions and which states that ionic addition of unsymmetrical reagents proceeds in such a way as to place the negative part of the addend on the more highly substituted carbon atom of the double bond. Now in reagents such as HBr, HCl, HI, $H_2SO_4$ it is clear that the respective "negative" addends are $Br^-$, $Cl^-$, $I^-$, and $HSO_4^-$; but it is not at once apparent whether HOCl will add as HO and Cl or as H and OCl. It is found experimentally that HOCl adds to propylene as follows:

$$CH_3CH{=}CH_2 + HOCl \longrightarrow CH_3\overset{\underset{\displaystyle OH}{|}}{C}HCH_2Cl \tag{15}$$

1-chloro-2-propanol

This result indicates that HOCl supplies $Cl^+$ to the olefin, the reaction proceeding as follows:

$$CH_3CH{=}CH_2 + HOCl \longrightarrow \{CH_3\overset{+}{C}HCH_2Cl\}OH^- \tag{16}$$

$$CH_3\overset{\underset{\displaystyle OH}{|}}{C}HCH_2Cl$$

It is likely that this is oversimplified, and that in an acidic medium it is *protonated* HOCl that provides the $Cl^+$ fragment:

$$CH_3CH{=}\overset{+}{C}H_2 \quad Cl{-}\overset{+}{O}H_2 \longrightarrow CH_3\overset{+}{C}HCH_2Cl\}H_2O \tag{17}$$

$$CH_3\overset{\underset{\displaystyle +OH_2}{|}}{C}HCH_2Cl$$

The final product shown is simply the protonated form of the chloro alcohol, which by loss of the proton to any proton acceptor gives 1-chloro-2-propanol.

Markovnikov's rule is often ambiguous, and cannot be used to predict the outcome of some olefin addition reactions. For example, in $CH_3CH_2CH{=}CHCH_3$, both carbon atoms of the double bond possess one hydrogen atom and one alkyl substituent. The observed result of the addition of HI to 2-pentene is as follows:

$$CH_3CH_2CH{=}CHCH_3 + HI \longrightarrow CH_3CH_2CH_2\overset{\underset{\displaystyle |}{|}}{C}HCH_3 \quad \text{(largely)} \tag{18}$$

The best statement that can be made regarding this result is to say that of the two carbonium ions $CH_3CH_2\overset{+}{C}HCH_2CH_3$ and $CH_3CH_2CH_2\overset{+}{C}HCH_3$, the second is preferred (is more stable).

The addition of HBr to vinyl bromide ($CH_2{=}CHBr$) yields 1,1-dibromoethane, and "obeys" Markovnikov's rule. The reason for this is that of the two protonated species

$$Br\overset{+}{C}H{-}CH_3 \quad \text{and} \quad BrCH_2{-}\overset{+}{C}H_2$$

the former is the more stable and is formed through a transition state of lower energy in a step preceding subsequent reaction with $Br^-$ to give the final addition product. This course for the reaction will be discussed in the following paragraphs.

When Markovnikov's rule was first stated it was an expression of empirical results. While the "rule" is still used as a convenient and concise summary of the course of addition to olefins, its theoretical basis can now be described in terms of our knowledge of the details of the addition reaction. Another way of stating the rule is to say that the addition reaction proceeds through a transition state that leads to the more stable of the two possible ionic intermediates. In the case of the simple olefins the decision as to which of the two possible carbonium ions is the more stable is easy to make. In earlier chapters we have seen the evidence for the conclusion that stability is greatest for tertiary and least for primary alkyl carbonium ions:

$$(CH_3)_3C^+ > (CH_3)_2CH^+ > CH_3CH_2^+ > CH_3^+$$
$$\text{decreasing stability} \longrightarrow$$

The addition of HBr to isobutylene can be examined from this viewpoint. The initial protonation could take place in either of the two ways:

$$
\begin{array}{c}
H_3C \\
H_3C
\end{array}
C = CH_2 + HBr \quad \text{or}
\qquad
\left\{
\begin{array}{l}
H_3C \\
H_3C
\end{array}
\overset{+}{C} - CH_3 \right\} Br^- \\
\left\{
\begin{array}{l}
H_3C \\
H_3C
\end{array}
CH - CH_2^+ \right\} Br^-
\tag{19}
$$

It is clear that the protonation will lead to the tertiary carbonium ion rather than the far less stable primary carbonium ion. These two courses can be described in an energy diagram (Figure 9-5) in which the transition state leading to the more stable $(CH_3)_3C^+$ is at a lower energy level than that leading to the primary ion.

It is clear from this that the reason that *t*-butyl bromide is formed as the product of the addition is that the reaction leading to this product proceeds through the lower energy transition state and thus is much the faster of the two. It can be concluded from this that the primary halide is also a product of the reaction in a proportion that depends upon the relative rates of the two rate-determining steps. If, as is the case, the route via the *t*-butyl carbonium ion is very much faster than

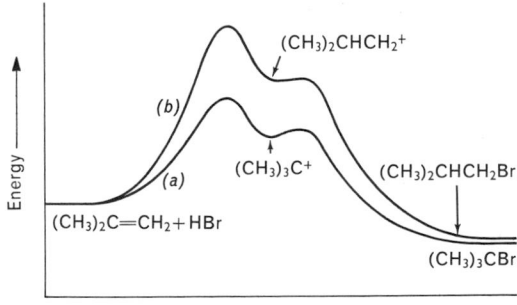

**Figure 9–5.** Energy diagram for the addition of HBr to isobutylene, $(CH_3)_2C=CH_2$. Route (a) with lower activation energy is much the faster of the two possible reactions.

the alternative route, the relative amount of *t*-butyl bromide is so great as to make it, for practical purposes, the sole product.

The addition of sulfuric acid to the double bond of an olefin with the formation of an ester of sulfuric acid:

$$CH_3CH{=}CH_2 + H_2SO_4 \longrightarrow CH_3CHCH_3 \qquad\qquad (20)$$
$$\underset{OSO_2OH}{|}$$

is clearly analogous to the addition of HCl or HBr, and proceeds in the same way:

$$CH_3CH{=}CH_2 + H{-}HSO_4 \longrightarrow CH_3\overset{+}{C}HCH_3\}HSO_4^- \longrightarrow CH_3CHCH_3 \qquad (21)$$
$$\underset{OSO_2OH}{|}$$

The alkyl hydrogen sulfate (in the above example, isopropyl hydrogen sulfate) is hydrolyzed by the addition of water:

$$CH_3CHCH_3 + xH_2O \longrightarrow CH_3CHCH_3 + H_2SO_4 \text{ (aq.)} \qquad (22)$$
$$\underset{OSO_2OH}{|} \qquad\qquad\qquad \underset{OH}{|}$$

with the formation of *iso*propyl alcohol. The preparation of alcohols by the hydration of olefins has been discussed earlier.

Among the byproducts often formed in the hydration of olefins are the dialkyl sulfates corresponding to the alkyl hydrogen sulfates:

$$CH_2{=}CH_2 + H_2SO_4 \longrightarrow CH_3CH_2OSO_2OH + CH_3CH_2OSO_2OCH_2CH_3 \qquad (23)$$

The formation of the dialkyl sulfate is readily accounted for, since the alkyl hydrogen sulfate is a strong acid and can add to the olefin in exactly the same way that sulfuric acid adds:

$$CH_2{=}CH_2 + C_2H_5OSO_2OH \longrightarrow CH_3CH_2OSO_2OC_2H_5 \qquad (24)$$

The dialkyl sulfate can also be hydrolyzed to the alcohol (2 moles) and sulfuric acid.

## 9.9   The stereochemistry of addition to the carbon-carbon double bond

The addition of bromine to propylene gives 1,2-dibromopropane. Since 1,2-dibormopropane contains an asymmetric carbon atom, two possible forms can exist. These differ *only* in their optical properties; one of them rotates the plane of polarized light to the right, the other to the left. They are designated as (+)-1,2-dibromopropane and (−)-1,2-dibromopropane. *Both are formed in equal amounts in the addition of bromine to propylene*, and the 1,2-dibromopropane that is the product of the reaction is optically inactive: it has no effect upon plane polarized light. This optically inactive product is designated as (±)-1,2-dibromopropane.

The reason for the formation of both the (+)- and the (−)-compounds is that bromine can approach the propylene molecule with equal probability from above as below the plane of the olefin molecule:

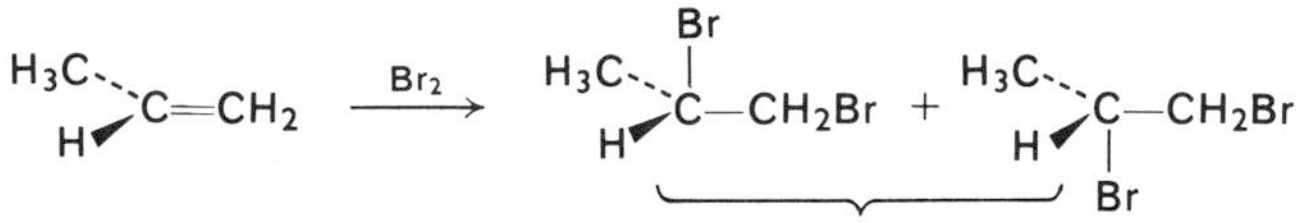

$$H_3C \diagdown C=CH_2 \xrightarrow{Br_2} H_3C \diagdown C-CH_2Br + H_3C \diagdown C-CH_2Br \qquad (25)$$

$$(\pm)\text{-}1,2\text{-dibromopropane}$$

The two dibromides can be redrawn as follows

to show their mirror-image relationship.

---

**Exercise 2.** What is the stereochemical result of the addition of bromine to *(a)* isobutylene; *(b)* 1-butene.

---

The products of reactions in which optically inactive reagents are used are themselves optically inactive; consequently, even though the products may possess asymmetric carbon atoms, this feature is ordinarily not made explicit, and the designation $(\pm)$- is usually omitted from the name. Further consideration of optical activity and molecular asymmetry will be found in Chapter 16.

The addition of bromine to 2-butene presents a new problem. First of all, there are two 2-butenes, the *cis-* and the *trans-*. The addition of bromine to *cis*-2-butene gives a different dibromide from that obtained from *trans*-2-butene. The experimental fact is that the 2,3-dibromobutanes obtained from the two 2-butenes have the configurations shown in the following equations:

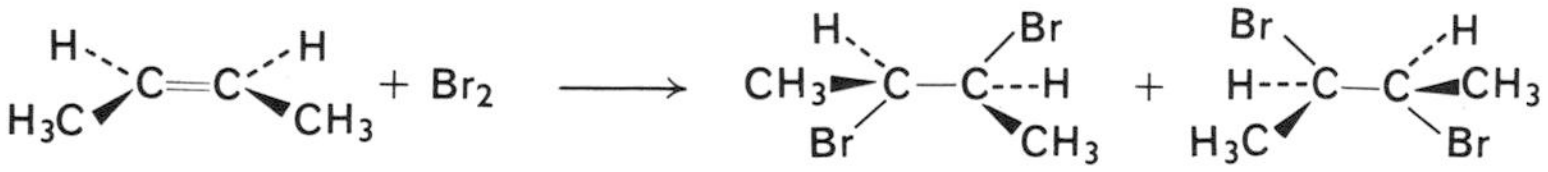

$$H \diagdown C=C \diagup H + Br_2 \longrightarrow CH_3 \diagup C-C \diagdown H + H \diagdown C-C \diagup CH_3 \qquad (26)$$

*cis*-2-butene $\qquad$ I $\qquad$ II

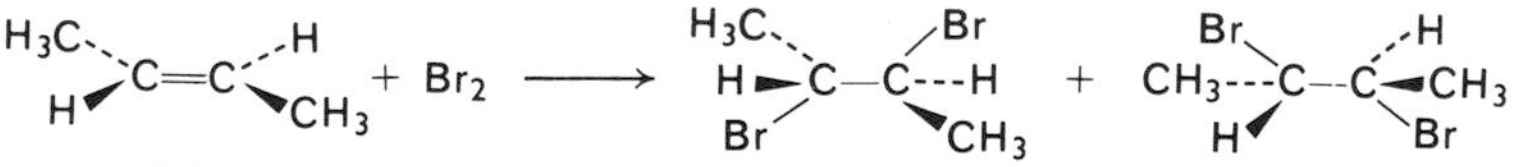

$$H_3C \diagdown C=C \diagup H + Br_2 \longrightarrow H \diagup C-C \diagdown H + CH_3 \diagdown C-C \diagup CH_3 \qquad (27)$$

*trans*-2-butene $\qquad$ III $\qquad$ IV

The following representation of equation (26) makes the three-dimensional aspects clear:

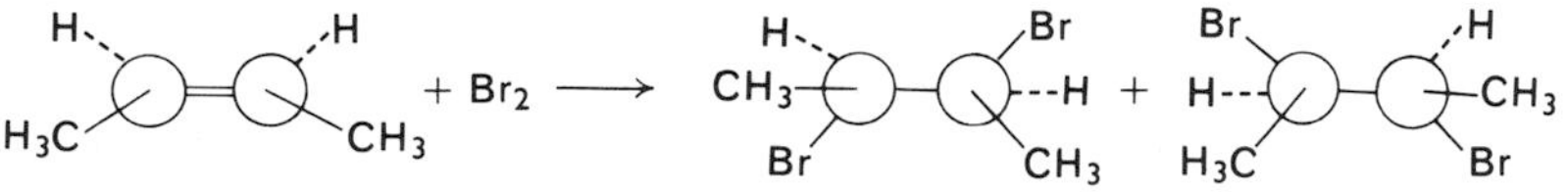

$$H \diagdown C=C \diagup H + Br_2 \longrightarrow CH_3 \diagup C-C \diagdown H + H \diagdown C-C \diagup CH_3 \qquad (28)$$

*cis*-2-butene $\qquad$ I $\qquad$ II

An inspection of the structures I, II, III, and IV discloses the following:

*(a)* I and II are nonidentical mirror images; they are $(+)$- and $(-)$-forms of 2,3-dibromobutane.

*(b)* III and IV are *identical*, and so represent only *one* compound; this is called *meso*-2,3-dibromobutane. *Meso*-2,3-dibromobutane has a plane of symmetry.

Analysis of those reactions in terms of the mechanism of addition that has been described in Sections 9.5 and 9.6 reveals the way in which the products are formed and explains the stereochemical results. Let us examine the reaction of bromine with *trans*-2-butene.

The first stage is the electrophilic attack of bromine upon the double bond (*or the nucleophilic attack of the olefin upon bromine*):

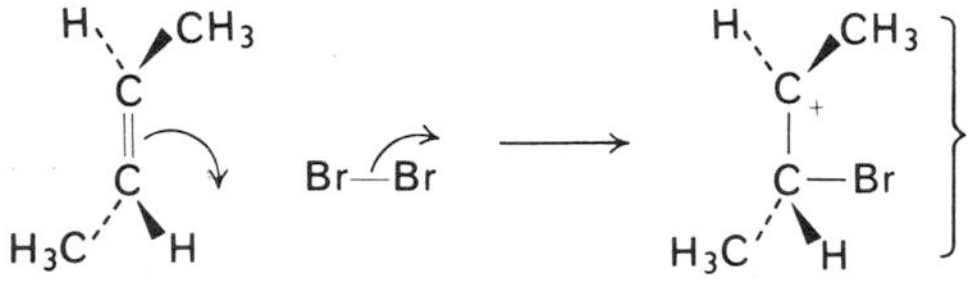

$$\tag{29}$$

As this step is written here, it is oversimplified; in the formation of the transition state leading to the ionic intermediate, it is probable that the developing carbonium ion remains under the influence of the bromine atom, the unshared valence electrons of which form a bond that utilizes the orbital on carbon. The resulting ion is pictured as in Equation (30) and is called a cyclic bromonium ion. The delocalization of the positive charge contributes to the stabilization of the ion. Another way of describing the initial step is the following:

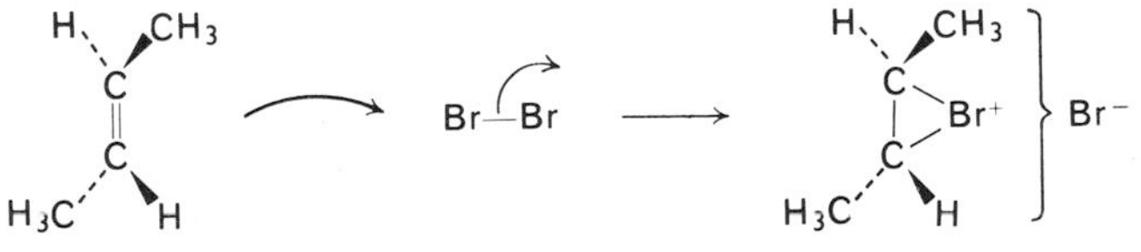

$$\tag{30}$$

It can now be seen that the intermediate shown in Figure 9-4 should be written in the explicit form $\overset{\overset{+}{Br}}{\underset{CH_2-CH_2}{\diagup\diagdown}}$.

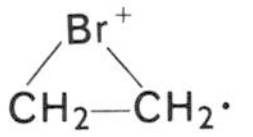

The second step of the addition reaction is that between the intermediate "-onium" ion and bromide ion to give the final product. It is clear that the attack of $Br^-$ upon the intermediate is *stereochemically confined to a single course:* the bromide ion displaces the C—Br bond and thus must attack the molecule from the side *opposite* to that occupied by the first bromine atom. The attack of $Br^-$ upon the cyclic bromonium ion can obviously take place at either carbon atom:

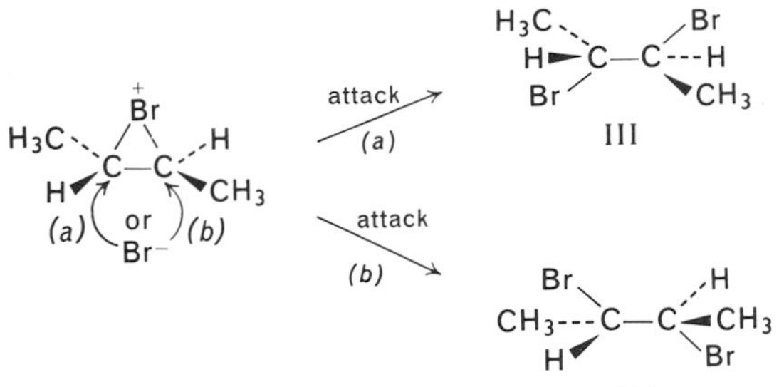

$$(31)$$

The products of this step are formulated as III and IV, but, as was pointed out above, these are identical and represent only the one compound, *meso*-2,3-dibromobutane.*

The addition of bromine to the carbon-carbon double bond is thus a *trans* addition: the two bromine atoms become attached to opposite sides of the molecule. It is a *stereospecific* reaction. Some additional examples are the following:

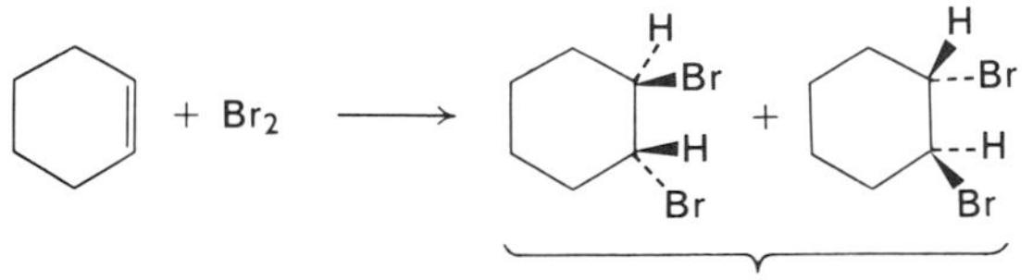

$$(32)$$

cyclohexene          ($\pm$)-*trans*-1,2-dibromocyclohexane

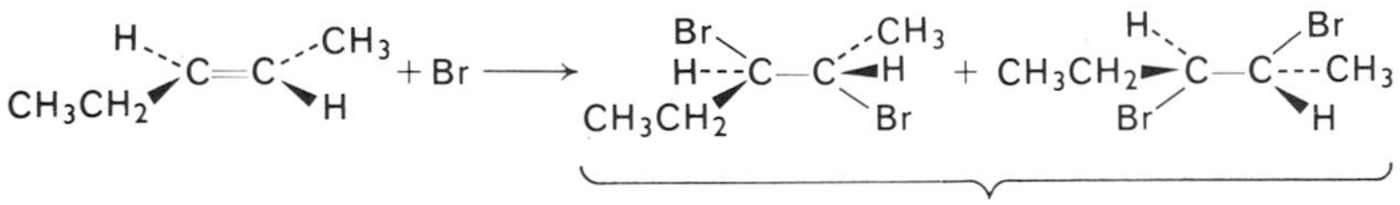

*trans*-2-pentene         ($\pm$)-*erythro*-2,3-dibromopentane      $(33)$

---

**Exercise 3.** Formulate the stereochemical course of *(a)* the addition of bromine to *cis*-2-pentene, and *(b)* the addition of hypochlorous acid to *cis*-2-butene (see Section 9.8). Note that there is no *meso* form of 2,3-dibromopentane; there are an *erythro*- [Equation (33)] and a *threo*-2,3-dibromopentane.

---

The addition of hypochlorous acid, HOCl, to the olefinic double bond is also a *trans* addition (Section 9.8).

---

* A thorough understanding of the spatial relationships of I, II, and III ($=$IV) should be reached by the student by study of the three-dimensional drawings or, better, by a study of actual models. If a commercial ball-and-stick model set is not accessible, adequate models can be constructed with small styrofoam spheres (found in hobby shops) and colored toothpicks; or toothpicks and colored jelly-beans; or colored styrofoam balls of various colors.

### 9.10   The 1,2-dibromoalkanes (olefin dibromides)

Alkenes can be regenerated from their dibromides, and the removal of bromine is stereospecific: the dibromide formed by the addition of bromine to a *trans* olefin can be debrominated to regenerate the *trans* olefin. This result shows that the elimination of bromine takes place in a *trans* manner:

$$CH_3CH{=}CHCH_3 \xrightarrow{Br_2} CH_3CHBrCHBrCH_3 \xrightarrow[EtOH]{Zn} cis\text{-}2\text{-}butene + ZnBr_2 \qquad (34)$$

*cis*-2-butene                 (±)-2,3-dibromobutane

The debromination can be carried out with the use of metallic zinc, as shown in the above equation. Since the reaction is stereospecific, it must proceed by way of a transition state in which the two bromine atoms and the two carbon atoms lie in a plane; in this way, the developing orbitals on carbon can combine to reconstitute the double bond while the two C—Br bonds are in the process of breaking:*

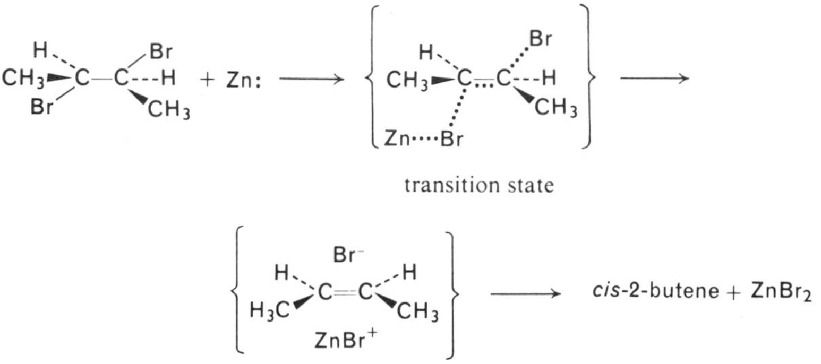

transition state

$$\longrightarrow cis\text{-}2\text{-}butene + ZnBr_2 \qquad (35)$$

Although this reaction has only limited use in practical organic chemistry, it is a good example of the stereospecificity of the elimination reaction. It does have some useful applications, however. Since most dibromides are high-boiling liquids or solids, and the simple olefins are gases or volatile liquids, an olefin may be "stored" as the dibromide and generated when needed by this debromination reaction. Another application can be illustrated by a simple example—a mixture of ethylene and ethane could easily be separated by the following procedure:

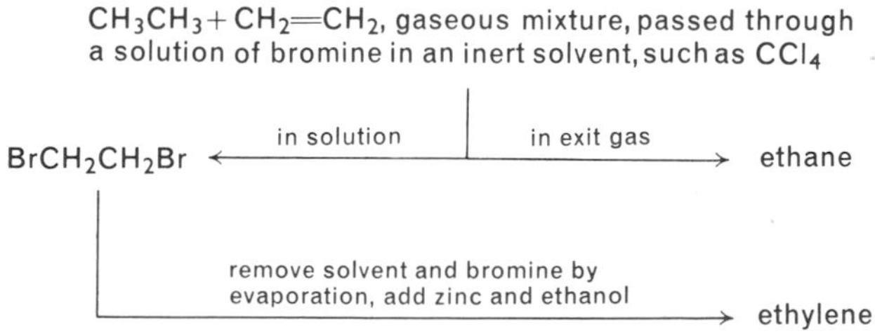

CH₃CH₃ + CH₂=CH₂, gaseous mixture, passed through a solution of bromine in an inert solvent, such as CCl₄

BrCH₂CH₂Br  ⟵ in solution | in exit gas ⟶ ethane

remove solvent and bromine by evaporation, add zinc and ethanol ⟶ ethylene

---

* Although only one [the (+) or (−)] form of the 2,3-dibromobutane is shown in this equation, it is ordinarily understood that the optically inactive (±) mixture is being used unless explicit reference is made to an optically active compound. Both (+) and (−) forms have identical reactivity in this reaction.

### 9.11 *Cis* addition: hydrogenation of multiple bonds

When an olefin, usually in an inert solvent such as ethanol, is shaken with molecular hydrogen in the presence of a finely divided catalyst, hydrogen adds to the double bond:

$$\ce{>C=C< + H2 ->[\text{catalyst}] >CH-CH<} \tag{36}$$

This process, called *catalytic hydrogenation*, is carried out with the use of catalysts that are in most cases finely divided forms of nickel, platinum, or palladium. A special form of nickel, called "Raney nickel," is prepared by the action of sodium hydroxide solution on a nickel-aluminum alloy. The aluminum dissolves (with the formation of sodium aluminate and the evolution of hydrogen), leaving the nickel as a black powder, pyrophoric when dry. Platinum and palladium are used as the oxides (which are reduced in the course of the hydrogenation to the elementary metals), or as the metals supported on charcoal, calcium carbonate, or other inert carriers.

In the hydrogenation of an olefin, the compound is dissolved in a solvent (e.g., alcohol, acetic acid, ethyl acetate, water), the catalyst is added, and a positive pressure of hydrogen is applied (after flushing the system to free it of air). The flask is shaken mechanically, and the course of the reaction is followed by measurement of the drop in pressure (or volume) of the hydrogen. Most double bonds ($\ce{C=O}$, $\ce{C=N}$, $\ce{C#N}$, as well as $\ce{C=C}$) are reduced readily, and the reaction is one of the most general and valuable tools of the organic chemist. A simple but useful apparatus for conducting quantitative hydrogenations is shown in Figure 9-6.

The addition of hydrogen to the double bond is a *cis* addition, in contrast to the *trans* additions of halogens and halogen acids.

The course of the addition can be illustrated by the example of the hydrogenation of 1,2-dimethylcyclohexene:

$$
\begin{array}{ccc}
\overset{\text{H}_2}{\underset{\text{H}_2}{
\begin{array}{c}
\ce{H2C^{C}\!C-CH3} \\
\ce{H2C_{C}\!C-CH3}
\end{array}}}
& \xrightarrow{\ce{H2/Pt}} &
\overset{\text{H}_2}{\underset{\text{H}_2}{
\begin{array}{c}
\ce{H2C^{C}CHCH3} \\
\ce{H2C_{C}CHCH3}
\end{array}}} \\
\text{1,2-dimethylcyclohexene} & & \text{1,2-dimethylcyclohexane}
\end{array}
\tag{37}
$$

The 1,2-dimethylcyclohexane formed in this reaction is the *cis* compound. This shows that both hydrogen atoms were added to the same side of the olefin molecule (see Figure 9-7).

Since molecular hydrogen does not add to olefins in the absence of a catalyst, it is believed that the function of the metallic catalyst is to adsorb the hydrogen and, by providing electrons to form metal-hydrogen bonds, to dissociate the molecular hydrogen into atoms. It is known that those metals that are effective hydrogenation catalysts have the capacity for absorbing large amounts of hydrogen. The actual hydrogenation process then consists in the absorption of the unsaturated compound on the surface of the catalyst and the addition of hydrogen

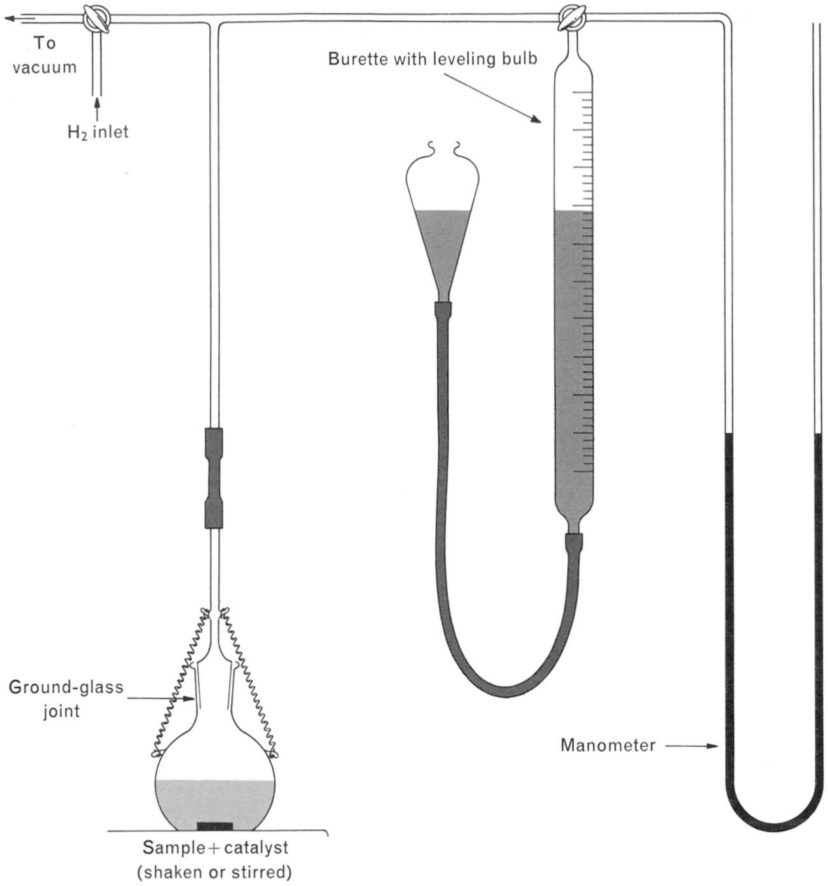

To
vacuum

H₂ inlet

Burette with leveling bulb

Ground-glass
joint

Manometer

Sample + catalyst
(shaken or stirred)

**Figure 9–6.** A simple apparatus for quantitative catalytic hydrogenation. Uptake by hydrogen by the sample measured with a calibrated buret.

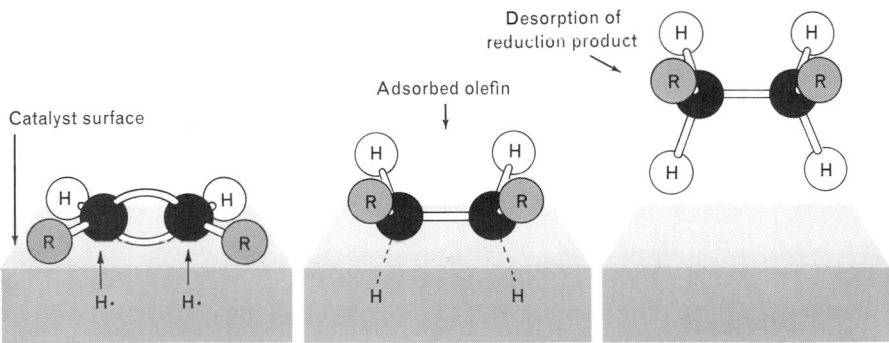

Desorption of
reduction product

Adsorbed olefin

Catalyst surface

H·   H·

H   H

**Figure 9–7.** Catalytic hydrogenation of an olefin. Note how stereospecificity results from adsorption on the catalytic surface.

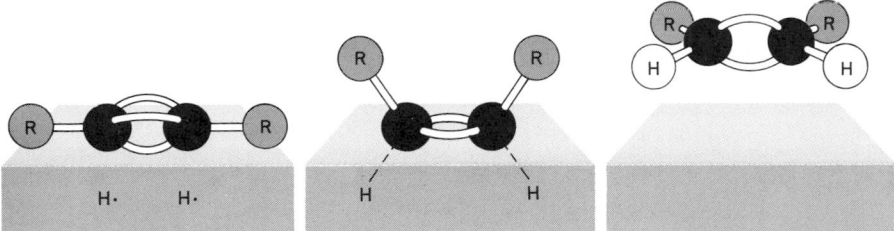

**Figure 9–8.** Catalytic hydrogenation of an acetylene to an olefin. Note how stereospecificity, with formation of the *cis*-olefin, results from adsorption on the catalytic surface.

to the double or triple bond. It is likely that the transition state for hydrogen addition involves the formation of hydrogen-carbon bonds as the substrate collides with the catalyst surface, followed by completion of the process to yield the hydrogenated product in an exothermic (energy yielding) final phase.

In Figures 9-7 and 9-8 are shown illustrations of models of the reaction. Figure 9-8 illustrates the addition of one mole of hydrogen to the carbon-carbon triple bond. The addition of two moles of hydrogen to an alkyne proceeds readily to yield the alkane; but by the use of special catalysts and by the interruption of the reaction after one mole of hydrogen has been added, an alkyne can be reduced to an alkene. As Figure 9.8 shows, the *cis* olefin is produced.

The stereochemical specificity of the catalytic hydrogenation of acetylenic compounds is of great value in the synthesis of organic compounds because it is thus possible to prepare olefins of *known cis* configuration starting from an acetylene. For example, oleic acid, a naturally occurring fatty acid, is known to have the *cis* configuration of the olefinic double bond because it is the product obtained by the catalytic hydrogenation of the following acetylenic acid:

$$CH_3(CH_2)_7C\!\equiv\!C(CH_2)_7COOH \xrightarrow{H_2/Pd} \underset{H}{\overset{CH_3(CH_2)_7}{\diagdown}}C\!=\!C\underset{H}{\overset{(CH_2)_7COOH}{\diagup}} \qquad (38)$$

octadec-9-ynoic acid                              oleic acid
                                    (*cis*-octadec-9-enoic acid)

## 9.12 Analytical application of catalytic hydrogenation

Catalytic hydrogenation of olefins is of practical importance in many ways. It is possible to determine the number of double bonds in a polyolefin (a polyene) by accurate measurement of the amount of hydrogen taken up. It is for this reason that the small-scale hydrogenation apparatus (Figure 9-6) includes a calibrated burette. On larger-scale hydrogenation apparatus a pressure gauge may be calibrated to measure the amount of hydrogen taken up in the course of hydrogenation.

For example, the group of highly unsaturated compounds known as carotenoids, of which carotene and vitamin A are important members, contain long chains of alternating double and single bonds. The first accurate estimate of the number of double bonds (eleven) in $\beta$-carotene was made by quantitative catalytic hydrogenation:

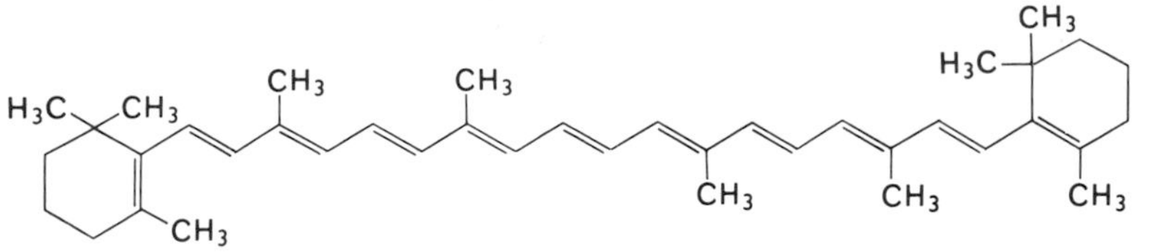

$\beta$-carotene

Pt/H$_2$ (11 moles)

$C_{40}H_{78}$ perhydro-$\beta$-carotene

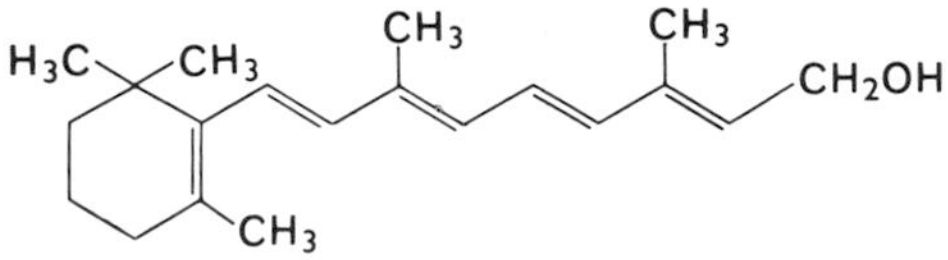

(39)

vitamin A$_1$

Pt | H$_2$ (5 moles)

(40)

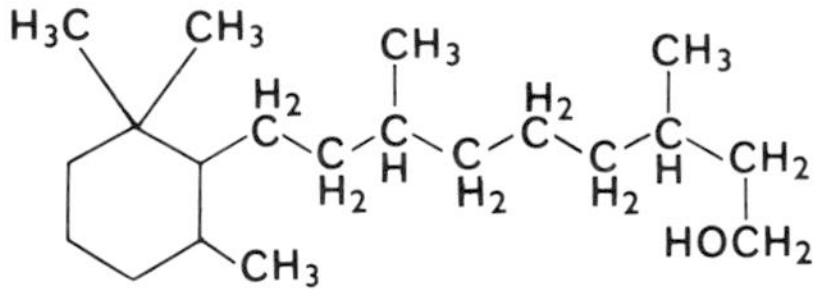

perhydrovitamin A$_1$

## 9.13   Other applications of catalytic hydrogenation

Industrial applications of catalytic hydrogenation are of considerable economic importance. Many vegetable oils are liquids or low-melting solids containing glycerol esters of unsaturated fatty acids, such as oleic, linoleic, linolenic acids. By bubbling hydrogen through the liquid fat in which finely divided nickel is suspended, the double bonds are hydrogenated. The ultimate product is tristearin:

$CH_3(CH_2)_7CH{=}CH(CH_2)_7COOH$  (oleic acid)
$CH_3(CH_2)_4CH{=}CHCH_2CH{=}CH(CH_2)_7COOH$  (linoleic acid)
$CH_3CH_2CH{=}CHCH_2CH{=}CHCH_2CH{=}CH(CH_2)_7COOH$  (linolenic acid)

(as esters of glycerol)

$\downarrow$ H$_2$/Ni

$CH_3(CH_2)_{16}COOH$   [stearic acid, as the glycerol ester (tristearin)]

Tristearin is a hard waxy solid; but by stopping the hydrogenation short of completion, intermediate, softer consistencies may be obtained. Hydrogenated oils, such as "Crisco" and "Spry," are widely used as cooking fats. Hydrogenated fats are more stable against the development of rancid flavor than are the more highly unsaturated oils, and are preferable for soap making as well as for some food uses.

## 9.14 Chemical reduction of olefinic double bonds

***Boron hydride.*** The preparation and oxidation of alkylboranes for the synthesis of alcohols was described in Section 8.21. If the intermediate boron compound is treated with acid, it decomposes in the following way:

$$R_3B + HA \longrightarrow 3RH + BA_3 \tag{41}$$

An organic acid such as acetic or propionic acid can be used. The addition of hydrogen in the overall reaction is *cis-*:

$$RCH{=}CHR' \longrightarrow (RCH_2CHR')_3B \longrightarrow RCH_2CH_2R' \tag{42}$$

***Diimide.*** Diimide, $HN{=}NH$, is an unstable compound that can be generated *in situ* by the action of an oxidizing agent on hydrazine ($H_2NNH_2$), or by an elimination reaction:

$$CH_3O^- + H_2NNHSO_2R^* \longrightarrow CH_3OH + HN{=}NH + RSO_2^- \tag{43}$$

Diimide transfers its two hydrogen atoms to the carbon-carbon double bond with the formation of nitrogen. The addition is *cis-* and probably proceeds through a transition state in which the diimide and the olefin react in the following way:

$$\tag{44}$$

One of the advantages of reductions with diimide, besides its stereospecificity, is that the reaction is quite selective for carbon-carbon double bonds. Catalytic hydrogenation, on the other hand, is not selective except under certain conditions in which specially prepared catalysts are used, and can lead to the hydrogenation of other reducible groups as well as of the olefinic double bonds.

***Other methods.*** The reduction of carbon-carbon double bonds that are not "isolated"—that is, that are adjacent to other carbon-carbon double bonds or to carbonyl groups—can be effected by certain metallic reducing agents. The description of these reductions will be deferred until later in the text.

---

* The derivative commonly used is the benzenesulfonyl compound, $H_2NNHSO_2C_6H_5$, prepared by the reaction between hydrazine and benzenesulfonyl chloride, $C_6H_5SO_2Cl$.

### 9.15 Oxidation of carbon-carbon double bonds

In its simplest form, the oxidation of the olefinic double bond consists simply in the addition of oxygen. This straightforward reaction can be brought about in more than one way. An important example is the industrial preparation of oxirane, or ethylene oxide, itself, by the oxidation of ethylene with air:

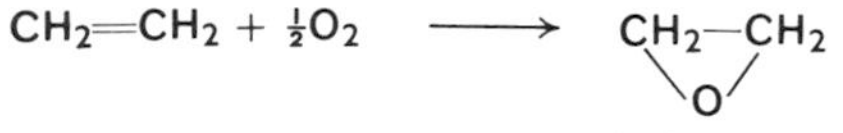

$$CH_2{=}CH_2 + \tfrac{1}{2}O_2 \longrightarrow \underset{\underset{\displaystyle O}{\diagdown \diagup}}{CH_2{-}CH_2} \tag{45}$$

ethylene oxide
[oxirane (G)]

However, this method is not of general usefulness for the epoxidation of the carbon-carbon double bonds of more complex compounds.

Another oxidation reaction involving the C=C double bond is the addition of the elements of $H_2O_2$ (as two —OH groups) to the ends of the double bond:

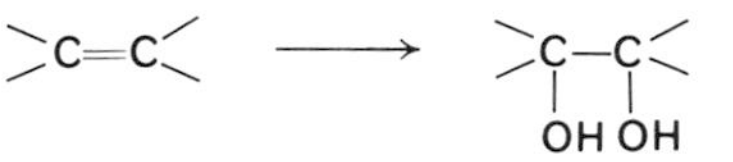

$$\underset{}{>}C{=}C\underset{}{<} \longrightarrow \underset{\underset{\displaystyle OH\ OH}{|\ \ |}}{>C{-}C<} \qquad \text{(a ``glycol'')} \tag{46}$$

There are two possible 1,2-diols that can be prepared by the hydroxylation of an olefin such as 2-butene; these are the *meso-* and ($\pm$)-2,3-diols:

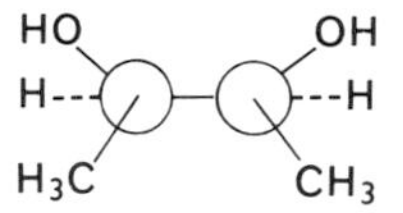

*meso-*2,3-butanediol

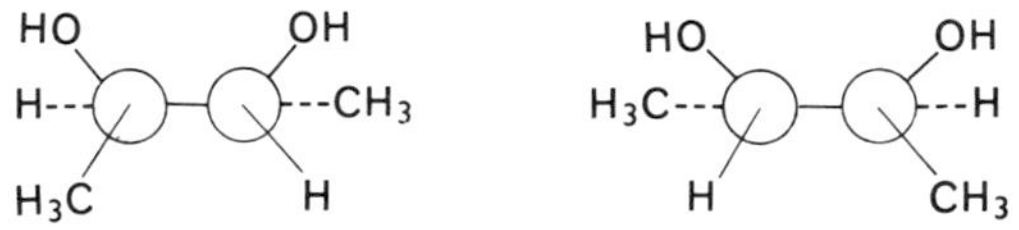

($+$)- and ($-$)-2,3-butanediol

**Table 9-4.**  *Possible stereochemical modes of hydroxylation of 2-butene*

| 2-Butene | Overall Mode of Addition of —OH Groups | 2,3-Butanediol Formed |
|---|---|---|
| *cis* | *cis* | *meso* |
| *cis* | *trans* | *dl* |
| *trans* | *cis* | *dl* |
| *trans* | *trans* | *meso* |

Now if the two —OH groups are added (by whatever mechanism) in the *cis* manner to *cis*-2-butene, the *meso*-diol will result; the other possibilities are shown in Table 9-4.

---

**Exercise 4.** Work these out with three-dimensional formulas, as was done in an earlier part of this chapter for the case of the addition of bromine to the two 2-butenes.

---

## 9.16 *Cis* hydroxylation of the double bond

*Cis* hydroxylation of the double bond can be accomplished by careful oxidation in the cold with dilute potassium permanganate solution. For example, cyclohexene is oxidized to *cis*-1,2-cyclohexanediol:

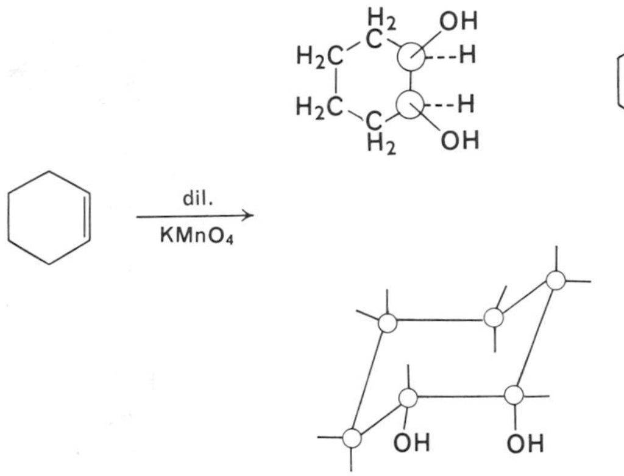

$$(47)$$

three ways of representing
*cis*-1,2-cyclohexanediol

*Cis*-2-butene gives, as stated above, *meso*-2,3-butanediol; and *trans*-2-butene gives the (±)-diol. *Cis* hydroxylation can also be accomplished by the use of osmium tetroxide, $OsO_4$. For example, cholesterol is readily converted by either $OsO_4$ or potassium permanganate into a triol, in which the two —OH groups introduced by the oxidation are *cis*-:

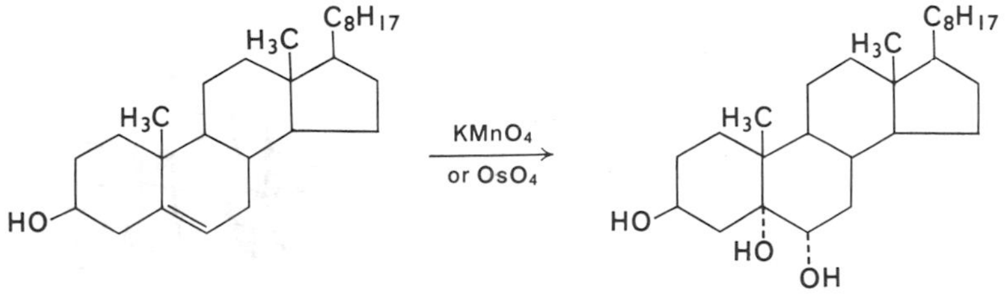

$$(48)$$

cholesterol
[cholest-5-en-3β-ol (G)]

cholestane-3β-5α-6α-triol (G)

Both osmium tetroxide and potassium permanganate form intermediate cyclic esters by *cis* addition to the double bond:

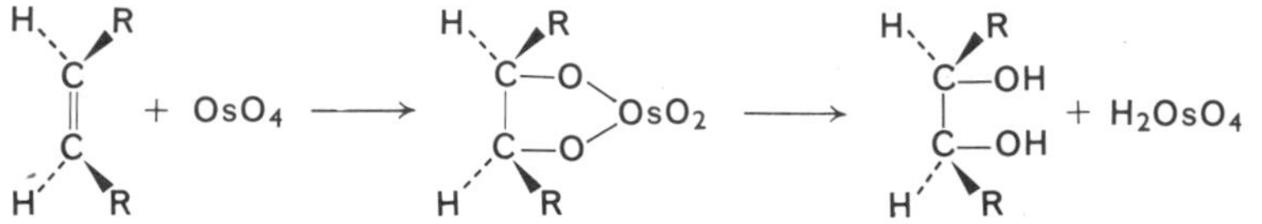

$$\text{(49)}$$

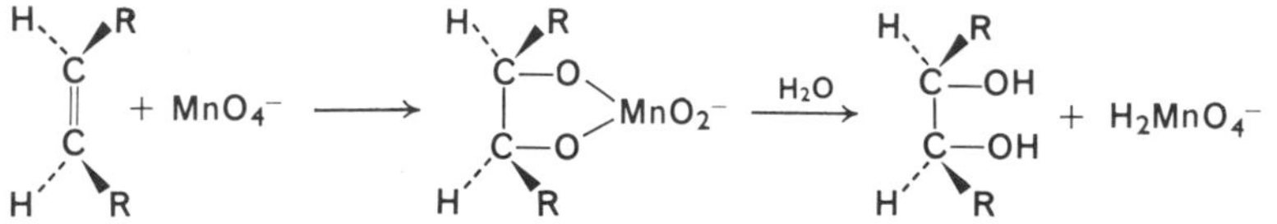

$$\text{(50)}$$

Hydrolysis of the cyclic esters proceeds with cleavage of the O-metal bonds, thus preserving the stereochemistry of the C—O bonds. In some cases the intermediate osmic ester can be isolated, and then hydrolyzed to the glycol in a separate step. The manganese (V) ester is hydrolyzed to the *cis* glycol in the reaction medium, which is maintained at a high $p$H in carrying out this hydroxylation. In neutral or acidic solution potassium permanganate reacts with carbon-carbon double bonds to give products that are the result of more extensive oxidation. Osmium tetroxide hydroxylation, although a convenient reaction, easy to carry out with good yields, has the disadvantage that the reagent is costly, and is a toxic compound that must be handled with great care.

### 9.17   *Trans* hydroxylation of the double bond

*Trans* hydroxylation is effected by reagents that appear to form an ethylene oxide (an epoxide) as a first step and then to open the oxide ring by a displacement reaction. The formation of the *trans* glycol by oxide ring opening is the normal consequence of the displacement reaction, which occurs by attack of the nucleophile (in the present case, the solvent) at the rear side of the carbon atom in the oxide ring. It is instructive to note the close similarity between the oxide ring opening and the opening of the cyclic bromonium ion intermediate in the addition of bromine to the double bond (Section 9.9):

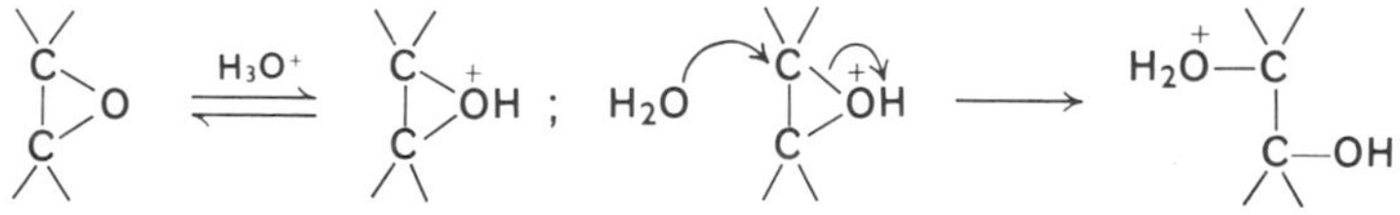

$$\text{(51)}$$

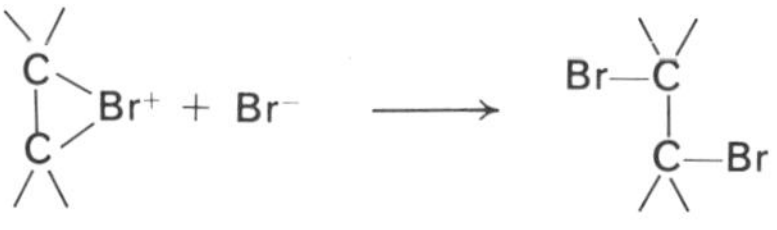

$$\text{(52)}$$

The *trans* hydroxylating agents most commonly used are *peroxy* acids, which are derivatives of hydrogen peroxide; or mixtures of concentrated hydrogen peroxide itself and organic acids. The reaction will be illustrated with the hydroxylation of *cis*-2-butene with a solution of hydrogen peroxide in acetic acid. The peroxy acid is formed in the equilibrium:

$$CH_3COOH + H_2O_2 \rightleftharpoons CH_3CO_3H + H_2O \tag{53}$$

and reacts with the olefinic double bond as follows:

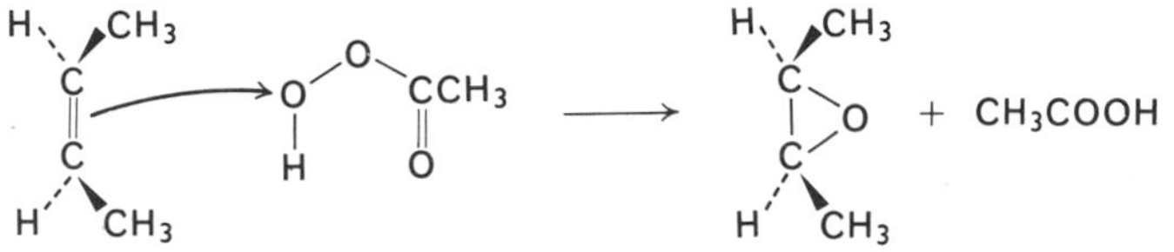

$$+ \; CH_3COOH \tag{54}$$

*cis*-2,3-epoxybutane

The ensuing step is that shown in Equation (56), with acetate acting as the nucleophile:

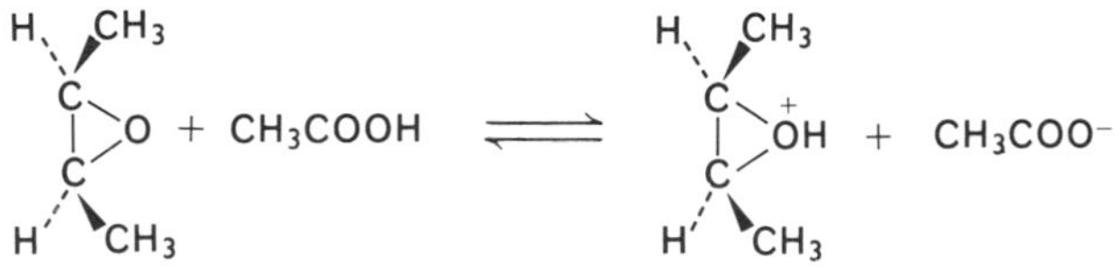

$$+ \; CH_3COO^- \tag{55}$$

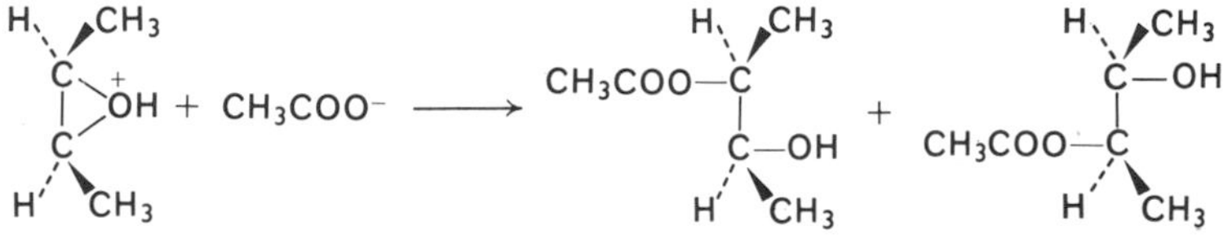

$$\tag{56}$$

The two products are monoesters of the dihydroxy compounds, and are converted into the glycols by a final hydrolysis of the ester. The two glycols obtained in this way from *cis*-2-butene have the following structures and configurations:

These are nonidentical mirror images; they constitute a ($\pm$) pair [compare Equations (26) and (28), Section 9.9].

The hydroxylation of *trans*-2-butene by the same method gives one compound, the symmetrical *meso* glycol,

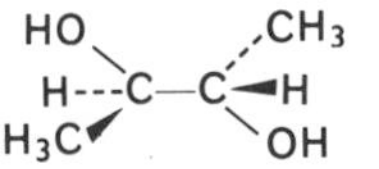

which, redrawn, is

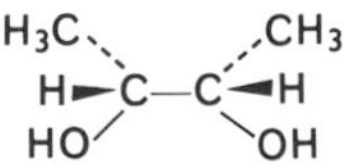

which, it can be seen, is symmetrical about a plane perpendicular to the page, passing between the two central carbon atoms.

---

**Exercise 5.**  The principles of molecular symmetry and asymmetry that were dwelt upon earlier in this chapter (Section 9.9) are illustrated again by the 2,3-butanediols. Make molecular models of these and study them until it is quite clear that the (+) and (−) forms are mirror images but not identical, and that the mirror image of the *meso* form is the same *meso* compound.

---

If the reaction of a peroxy acid with an olefin is carried out in a solvent such as ethyl ether or chloroform, the epoxide does not undergo further transformation and can be isolated as a pure substance. The peroxy acids used for the preparation of epoxides are prepared in the inert solvent (usually chloroform), and can be preserved for some time if stored at refrigerator temperatures.

Epoxide formation is accompanied by a drop in the peroxidizing titer of the peroxy acid used, and a useful analytical procedure can be performed to determine the number of olefinic double bonds in an unknown compound. A measured quantity of unsaturated compound is allowed to react with a measured (by titration for peroxide oxygen) amount of peroxy acid, and serial titrations on aliquot portions are performed. When there is no further drop in the peroxidizing titer, the amount of peroxy acid that has been consumed is calculated; each mole of peroxy acid used corresponds to one double bond. The method has some limitations, for some double bonds may react slowly, and if the reaction requires a long time some secondary reactions may occur to consume the oxidizing agent.

## 9.18  Epoxides from halohydrins

Epoxides can be prepared from chlorohydrins and bromohydrins by an internal nucleophilic displacement reaction which is an intramolecular version of the Williamson ether synthesis (Sections 7.9 and 8.13):

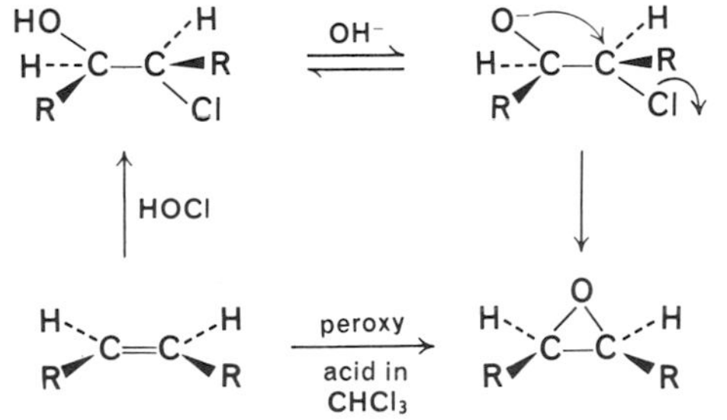

(57)

(58)

It will be noted that the same oxide is formed by direct (*cis*) epoxidation as by the two-step process involving (1) *trans* addition of HOCl and (2) a displacement reaction with inversion of configuration.

The stereochemical requirements for epoxide formation from a halohydrin are to be noted. In order for the nucleophilic attack to take place at the back side of the carbon atom holding the halogen, the OH and Cl must be *trans* disposed. This is clearly shown in cyclic halohydrins; for example, *cis*- and *trans*-2-chlorocyclohexanol:

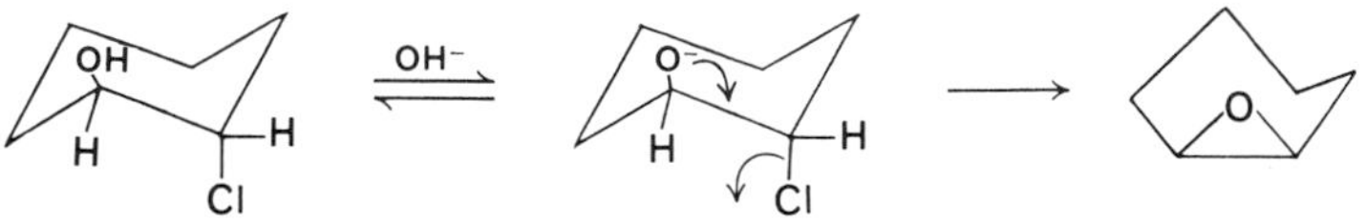

(59)

*trans*-2-chlorocyclohexanol

cyclohexene
oxide

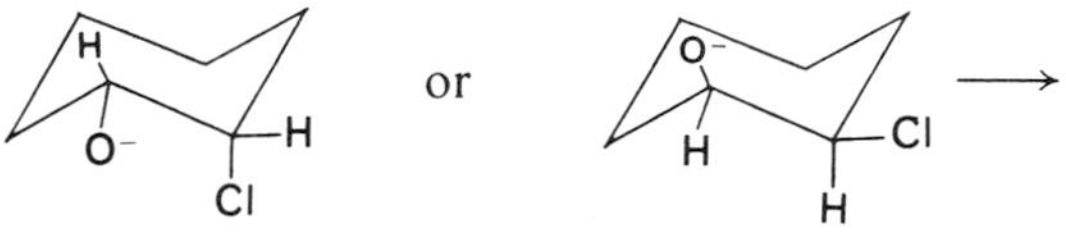

*cis*-2-chlorocyclohexanol

does not give the epoxide

The *cis*-chlorohydrin does not form the oxide because the anionic oxygen cannot reach a position at which it can attack C—Cl at the side opposite to and colinear with the carbon-chloride bond. The difference in the reactivity of these two chlorohydrins is very marked: the *trans* compound consumes hydroxide ion in an almost instantaneous reaction,* whereas the *cis* compound is scarcely affected under the same conditions.

## 9.19 Steric effects in epoxide ring closure

The following two chlorohydrins are those formed from *cis*- and *trans*-2-butene by the addition of HOCl:

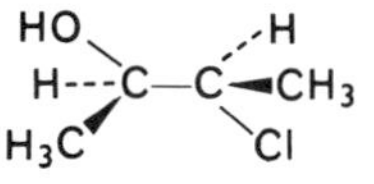

from *cis*-2-butene 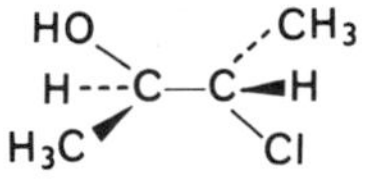  from *trans*-2-butene

An inspection of these structures discloses that both of them have the capability of undergoing epoxide formation by intramolecular reaction:

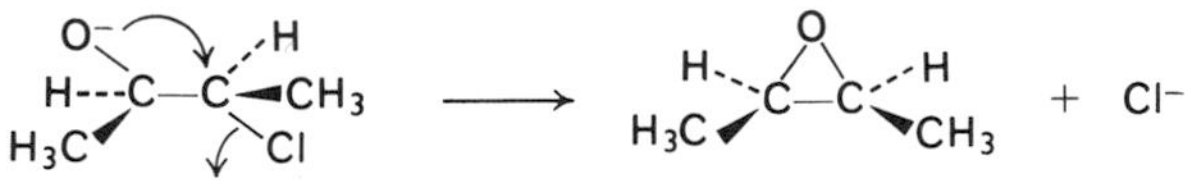

(60)

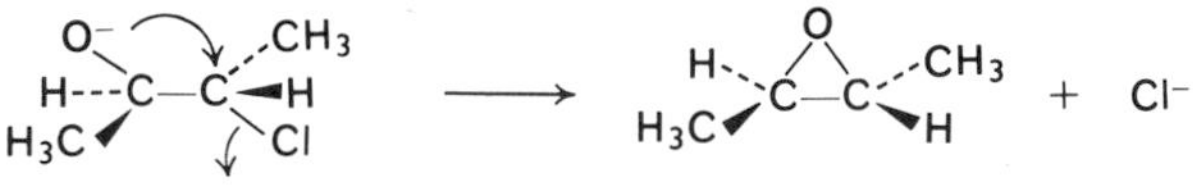

(61)

---

* It is to be noted that the "consumption" of alkali is the disappearance of OH⁻ and the appearance of Cl⁻ in the reaction mixture. This change involves a drop in $pH$ and can be followed by the use of an acid-base indicator such as phenolphthalein.

Experiment shows that the chlorohydrin derived from *trans*-2-butene reacts *more rapidly* than the other isomer. This means that the rate-determining step for the slower reaction shown in Equation (60) involves a higher-energy transition state than that for Equation (61), the faster reaction. First of all, both reactions of the chlorohydrins with alkali proceed to give the intermediate ionic species shown in (60) and (61); this proton-exchange step is fast, as is typical of acid-base reactions of this type. The second rate-determining step is the displacement reaction shown by the curved arrows in (60) and (61). This must proceed through a transition state in which the anionic oxygen has assumed a position in which the new O—C bond formation is beginning to occur. In the transition state for the reaction of Equation (60) the two methyl groups are on the same side and repelling each other by the steric interference caused by their proximity. The other, faster reaction proceeds through a transition state in which the two methyl groups are *trans* disposed : each methyl group is adjacent only to a hydrogen atom. The lesser degree of crowding in this transition state is favorable to its formation and is reflected in its greater stability, or lower energy. The situation is pictured in the energy diagrams in Figure 9-9 in which are shown *(a)* the transition states of the step leading to the formation of the intermediate anions, and *(b)* the transition states of the rate-determining intramolecular displacement steps.

Because the initial reactants, the chlorohydrins, are free to assume their most

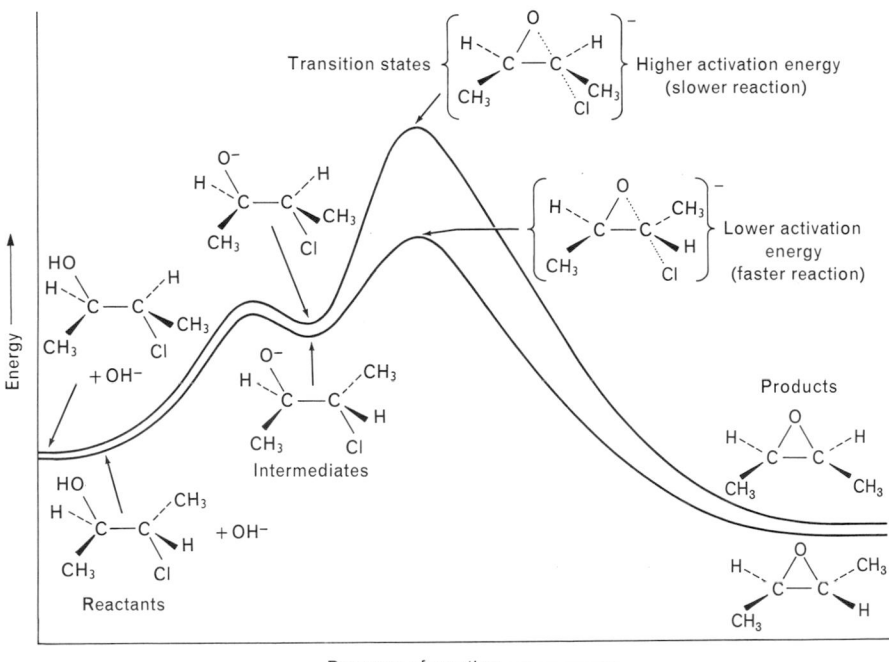

**Figure 9–9.** Energy diagrams for the epoxide ring closure of the isomeric chlorohydrins derived from *cis*- and *trans*-2-butenes. (Ground-state energies of reactants assumed to be equal.)

stable configurations by rotation about the C—C single bond, their energies would differ but little, and are assumed here to be about the same. In the transition states, however, the configuration is *necessarily* that shown in Figure 9-9; consequently, the steric repulsions play a role in affecting the activation energies and thus have a pronounced effect upon the rates of the ring-closing steps of the two reactions.

The effects of sterically imposed interactions upon the rates of reactions are often very important factors in the selection of alternative pathways for reaction and in determining the relative rates of two reactions that may appear superficially to be identical. The selectivity that is thus imposed often has the result that a single product is obtained from a reaction that could give two possible products.

An example of steric selectivity is found in the hydroxylation of cholesterol (48), and is shown in more detail in the permanganate hydroxylation of the double bond of the following compound:

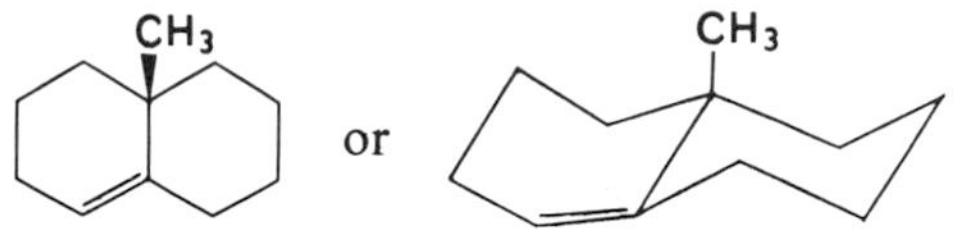

The three-dimensional drawing of this structure shows the steric disposition of the methyl group with respect to the double bond. The approach of the large $MnO_4^-$ ion to the double bond could take place from either side of the general plane of the molecule:

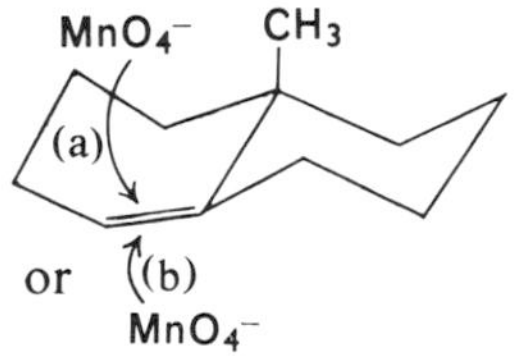

(62)

but it is apparent that the approach from above (a) would be impeded by the proximity of the methyl group, and the activation energy for the reaction would be increased by this additional requirement for overcoming steric repulsion. The approach from below (b) is sterically relatively less impeded. The product of the reaction is in fact the following glycol:

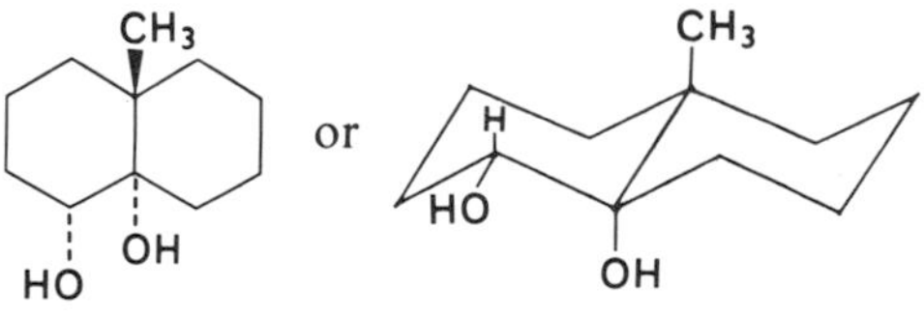

## 9.20  Ozonolysis of the carbon-carbon double bond

One of the most general reactions of the carbon-carbon double bond is the addition of ozone and the subsequent transformations of the *ozonide* that is so formed.

Ozone, $O_3$, is a highly active form of oxygen that is produced by the action of a silent discharge between electrodes at a high potential upon molecular oxygen. Ozone has a structure best represented by the two equivalent contributing structures:

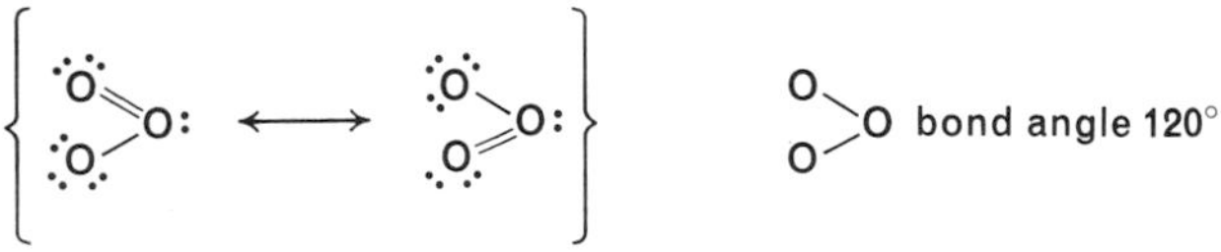

O bond angle 120°

The formal charges on the ozone molecule may be represented as $\overset{+}{O}=O-\overset{-}{O}$, and thus the attack of ozone upon the olefinic double bond may be regarded as an electrophilic attack upon nucleophilic $C=C$:

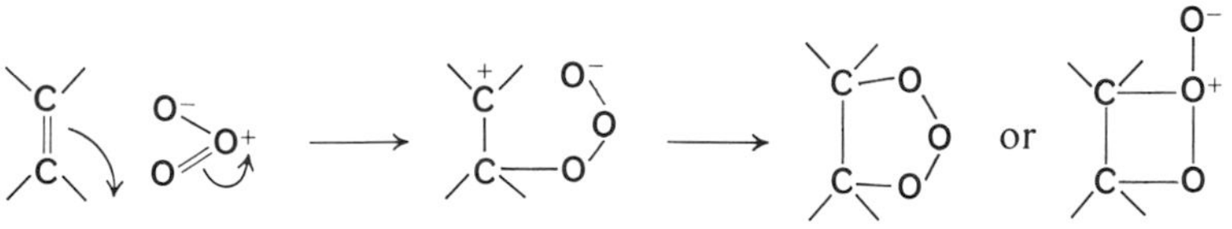

(63)

"primary ozonide"
(molozonide)

The structure of the molozonide is still not known with certainty; the reaction proceeds at once to give further products. The instability of the $-O^+-O-$ bond leads to the following probable sequence of changes:

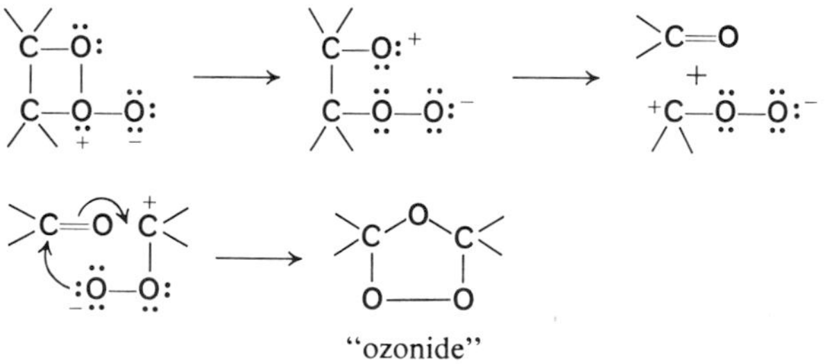

(64)

"ozonide"

The first isolable product is often the "ozonide"; but in most cases the so-called ozonides are of unknown, or at least unproved, structures, and are simply the uncrystallizable residues that remain after removal of the solvent in which the reaction with ozone has been carried out. Ozonides are usually highly unstable, and tend to decompose, often explosively. They are seldom isolated, but are treated further without attempts to purify them. The important thing to notice about ozonides is that *the carbon-carbon bond has been broken.*

Ozonides are treated further in one of several ways. They may be hydrolyzed directly by boiling with water:

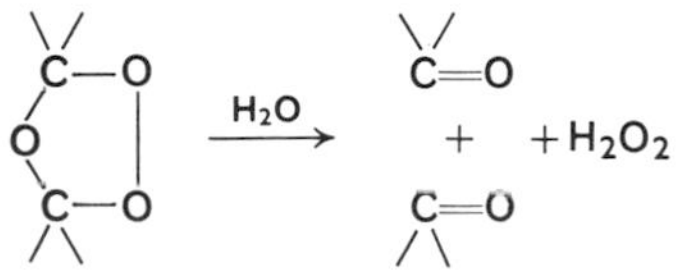

(65)

It is more common to carry out the hydrolysis under conditions such that the hydrogen peroxide (or peroxidic intermediates) is reduced in the process. This is done by adding powdered zinc before the hydrolysis is carried out. Another method is to reduce the ozonide catalytically with hydrogen in the presence of platinum:

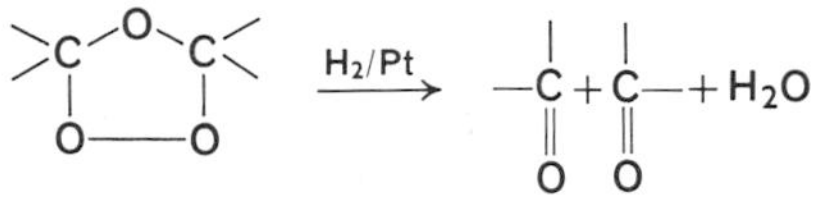

$$\underset{O-O}{\overset{O}{\underset{|}{C}}}\overset{O}{\underset{|}{C}} \xrightarrow{H_2/Pt} \quad -\overset{||}{\underset{O}{C}} + \overset{||}{\underset{O}{C}} - + H_2O \tag{66}$$

## 9.21 The use of ozonolysis in the proof of structure of olefinic compounds

The products of the ozonolysis of compounds containing C=C double bonds are carbonyl compounds, the nature of which depends upon the particular structure of the olefins. The following are examples:

$$CH_2{=}CH_2 \quad \longrightarrow \quad 2H_2C{=}O \tag{67}$$

$$CH_3CH{=}CH_2 \quad \longrightarrow \quad CH_3CHO + H_2CO \tag{68}$$

$$(CH_3)_2C{=}CHCH_3 \quad \longrightarrow \quad (CH_3)_2C{=}O + CH_3CHO \tag{69}$$

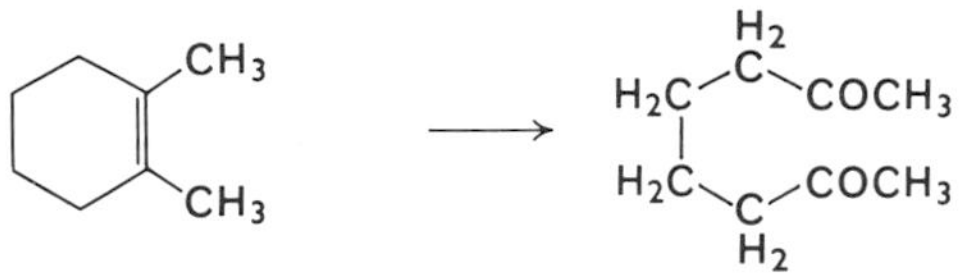

$$\tag{70}$$

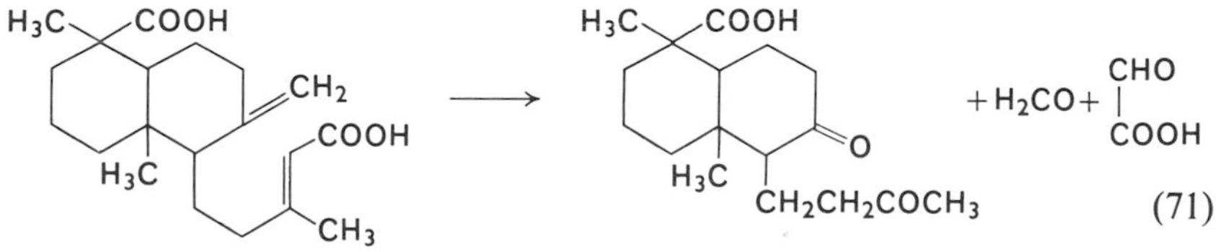

$$\tag{71}$$

agathene dicarboxylic acid

The chief utility of ozonolysis is in the proof of structure. By identifying the products of the ozonolysis, the structure of the olefin can be deduced.

$$\underset{R}{\overset{R'}{>}}C{=}C\underset{R'''}{\overset{R''}{<}} \xrightarrow{O_3} \underset{R}{\overset{R'}{>}}C{=}O + O{=}C\underset{R'''}{\overset{R''}{<}} \tag{72}$$

*Example 1*

An alcohol, $C_6H_{14}O$, was readily dehydrated to an olefin, $C_6H_{12}$. Ozonolysis of the olefin gave two products: acetone and a three-carbon aldehyde. Acetone is $CH_3COCH_3$, showing that the olefin has the grouping $\underset{H_3C}{\overset{H_3C}{>}}C{=}$. The three-carbon aldehyde can be only propion-aldehyde, $CH_3CH_2CHO$. Thus, the olefin has the structure $\underset{H_3C}{\overset{H_3C}{>}}C{=}CHCH_2CH_3$. Further evidence of other kinds would have to be secured to distinguish between two possibilities for the alcohol.

---

**Exercise 6.** What are the two possible structures for the alcohol $C_6H_{14}O$, and how could one distinguish between them?

---

*Example 2*

An olefin $C_7H_{12}$ was ozonized. Decomposition of the ozonide gave formaldehyde, $H_2C=O$, and a ketone, $C_6H_{10}O$. The ketone did not react with ozone, and so contained no carbon-carbon double bond; thus, it must be a cyclic ketone because an open-chain ketone (e.g., $CH_3COCH_2CH_2CH_2CH_3$) of six carbon atoms would have the formula $C_6H_{12}O$. One possible structure for the original olefin is

---

**Exercise 7.** Write two other possible structures for the olefin in Example 2.

**Exercise 8.** Ozonolysis of a compound yields three products: acetaldehyde; 2,5-hexane-dione (G); and *iso*butyraldehyde [2-methylpropanal (G)]. Write the structure of the original compound.

---

## 9.22 Acid-catalyzed reactions of olefinic double bonds

The product of the initial protonation of the carbon-carbon double bond—the protonated olefin, or carbonium ion—can undergo any of several subsequent reactions, depending upon its structure and the reaction conditions.

*(a) Rearrangement.* If the carbonium ion is less stable than one into which it can be transformed by a simple 1,2-shift, rearrangement to the more stable species can ensue. Usually this means that if the initial carbonium ion is a primary one, it may rearrange into a secondary or tertiary carbonium ion. Since protonation is a reversible reaction, the loss of the proton from the rearranged carbonium ion will produce a new olefin. The initial carbonium ion may also lose a proton in a different way to give a new olefin. After these equilibria have been established, the olefin present in greatest amount in the reaction mixture will be that with the greatest stability. Some examples of these reactions are the following.

Treatment of 1-butene with a strong acid causes it to be transformed into a mixture of 1- and 2-butenes:

$$CH_3CH_2CH=CH_2 + HA \rightleftharpoons CH_3CH_2\overset{+}{C}HCH_3 \rightleftharpoons CH_3CH=CHCH_3 + HA \quad (73)$$
$$+ A^-$$

In this reaction the position of the double bond "rearranges" but there is no altera-tion in the carbon skeleton.

Treatment of 3,3-dimethyl-2-butene with a strong acid (e.g., 25% aqueous sulfuric acid) causes a rearrangement to occur; the final equilibrium mixture con-tains as its chief constituent 2,3-dimethyl-2-butene (tetramethylethylene):

$$
\begin{array}{cc}
\text{CH}_3 & \text{CH}_3 \\
| & | \\
\text{CH}_3-\text{C}-\text{CH}=\text{CH}_2 + \text{HA} \rightleftharpoons & \text{CH}_3-\text{C}-\overset{+}{\text{CH}}-\text{CH}_3 \\
| & | \\
\text{CH}_3 & \text{CH}_3
\end{array} \Bigg\} \text{A}^-
$$

rearrange to tertiary
carbonium ion by 1,2-
shift of (CH₃:) $\downarrow$

(74)

$$
\begin{array}{cc}
\text{CH}_3 & \text{CH}_3 \\
| & | \\
\text{CH}_3-\text{C}=\text{C}-\text{CH}_3 \underset{}{\overset{-\text{H}^+}{\rightleftharpoons}} & \text{CH}_3-\overset{+}{\text{C}}-\text{CH}-\text{CH}_3 \\
| & | \\
\text{CH}_3 & \text{CH}_3
\end{array}
$$

The more highly substituted olefin is the product; the alternative deprotonation of the tertiary carbonium ion to give 2,3-dimethyl-1-butene occurs to a minor extent, and this olefin is formed in a yield of only 2 to 3%.

---

**Exercise 9.** What new olefins could be formed by the acid-catalyzed isomerization of 3-methyl-1-hexene?

---

Whenever carbonium ions are formed, whether by loss of $H_2O$ from a protonated alcohol, by protonation of an olefin, or in other ways, the possibility of a rearrange-ment must be kept in mind.

*(b) Polymerization.* The treatment of olefins with strong acids often gives, in addition to the products of simple addition of the acid to the double bond, poly-merization products consisting of two, three, or more molecules of the olefin. The case of *iso*butylene may be taken as typical, since the polymerization of this olefin to di*iso*butylene, tri*iso*butylene, and so forth, is well known and of industrial im-portance. The polymerization is best brought about by Lewis acid catalysts: boron trifluoride can be used, or solid catalysts such as aluminum chloride (on an inert support, such as pumice). Let us examine the process with the use of a generalized symbol, HA, for an acid.

The first reaction between the acid and the nucleophilic olefin is the coordina-tion of one of the electron pairs of the double bond with the acid:

$$
\begin{array}{c}
\text{H}_3\text{C} \\
\diagdown \\
\diagup \text{C}=\text{CH}_2 \quad \text{H}-\text{A} \rightleftharpoons \\
\text{H}_3\text{C}
\end{array}
\left\{
\begin{array}{c}
\text{H}_3\text{C} \\
\diagdown \\
\diagup \overset{+}{\text{C}}-\text{CH}_3 \\
\text{H}_3\text{C}
\end{array}
\right\} \text{A}^-
$$

(75)

The *tert*-butyl carbonium ion may now do several things: (1) it may lose the proton and revert to the olefin; (2) it may coordinate with the nucleophilic $A^-$; (3) it may coordinate with another molecule of the nucleophilic olefin to give a new carbonium ion:

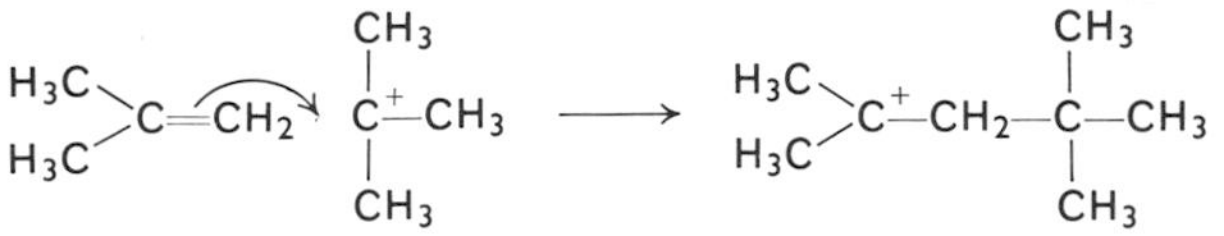

$$\begin{array}{ccc} & CH_3 & CH_3 \\ H_3C\!\!-\!\!C\!\!=\!\!CH_2 \quad C^+\!\!-\!\!CH_3 & \longrightarrow & H_3C\!\!-\!\!C^+\!\!-\!\!CH_2\!\!-\!\!C\!\!-\!\!CH_3 \\ H_3C & CH_3 & H_3C \qquad\quad CH_3 \end{array} \qquad (76)$$

The new carbonium ion may now follow one of the three courses mentioned above.

1.  Loss of a proton to some nucleophile to form an olefin (di*iso*butylene):

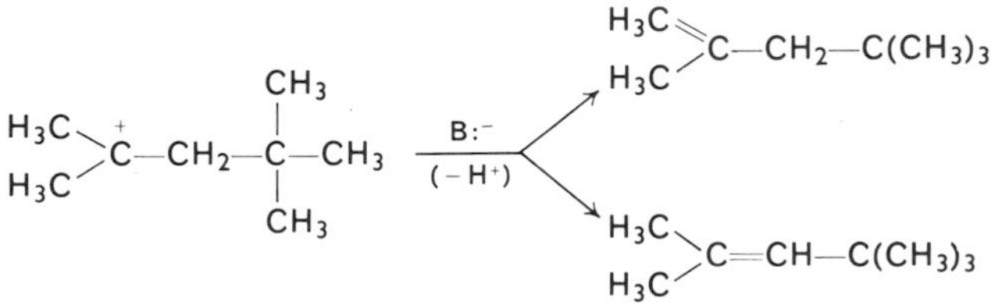

$$\qquad (77)$$

2.  Coordination of a nucleophile with the carbonium carbon atom.
3.  Reaction with another olefin molecule:

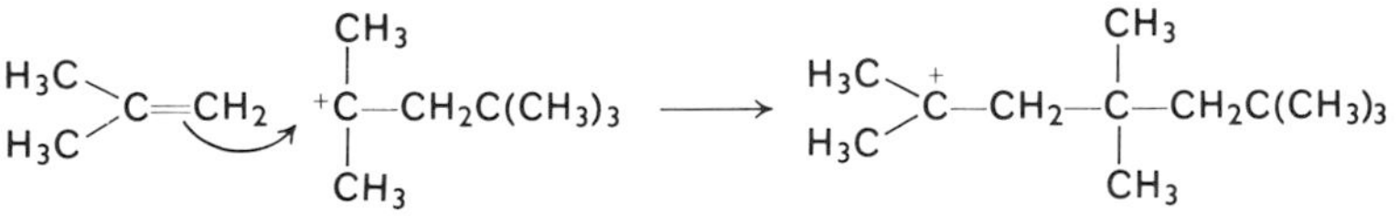

$$\qquad (78)$$

"Di*iso*butylene" is the term applied to the mixture of olefins formed in Equation (77); "tri*iso*butylene" can be derived in the same way from the product of Equation (78).

It can be seen that a continuation of the process

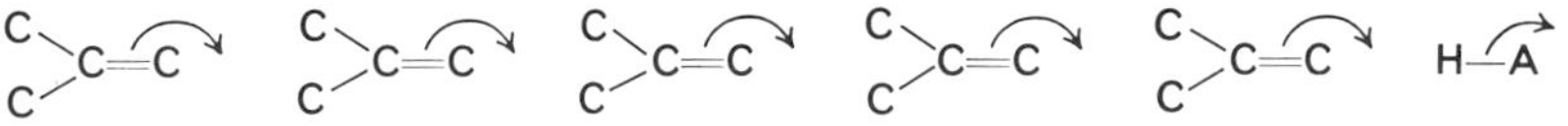

could give rise to large *polymeric* molecules containing numerous *monomer* units. When boron trifluoride (plus a trace of water) is used as the catalyst, viscous, high-molecular-weight polymers of the approximate structure $(\text{---}CH_2\text{---}C(CH_3)_2\text{---})_x$ are formed (Chapter 34).

Polymerization is a well known and industrially important process. However, most industrial polymerization reactions do not proceed by ionic mechanisms of the sort described for isobutylene, but are "free radical" reactions. The acid-catalyzed polymerization of *iso*butylene is discussed here because it is a good example of the application to a complex process of the principle involved in the simple addition reaction of the double bond, and shows that the fundamental processes involved in the reaction of olefins with acids start with the protonation of the double bond and the generation of the Lewis-acid-like carbonium ion.

*(c) Acid-catalyzed ring formation.* In reactions (75) through (78), the polymerization is the result of the addition of a carbonium ion to an olefin molecule. If the carbonium ion and the double bond to which it adds are a part of the same molecule, their interaction can lead to ring formation. If the ring so formed is five- or six-membered, reactions of this kind occur with great ease. A typical example is the following:

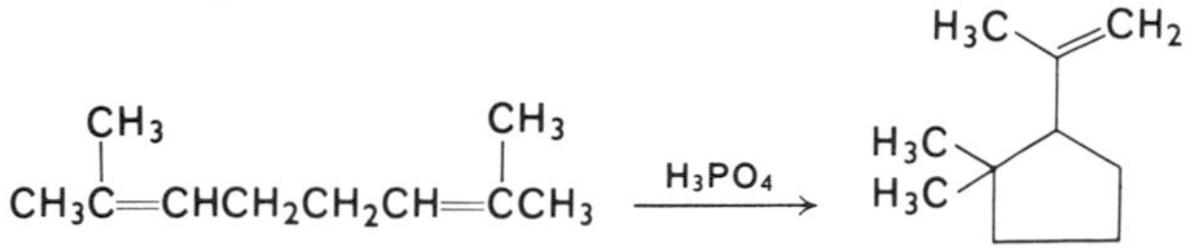

$$\underset{\substack{\displaystyle |\\ CH_3}}{CH_3C}=CHCH_2CH_2CH=\underset{\substack{\displaystyle |\\ CH_3}}{CCH_3} \quad \xrightarrow{H_3PO_4} \quad \text{[ring structure]} \tag{79}$$

The course of this cyclization becomes clear when we show the individual stages:*

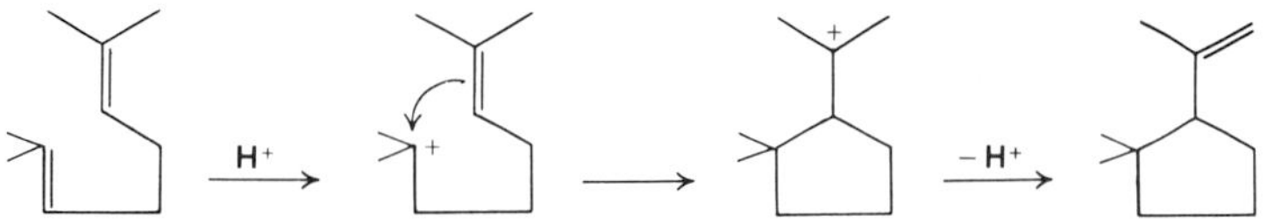

$$\text{[structures]} \quad \xrightarrow{H^+} \quad \text{[structures]} \quad \longrightarrow \quad \text{[structures]} \quad \xrightarrow{-H^+} \quad \text{[structures]} \tag{80}$$

The more commonly encountered examples of ring closure by intramolecular reaction of a carbonium ion and a double bond are those in which the carbonium ion is generated from an alcohol; for example:

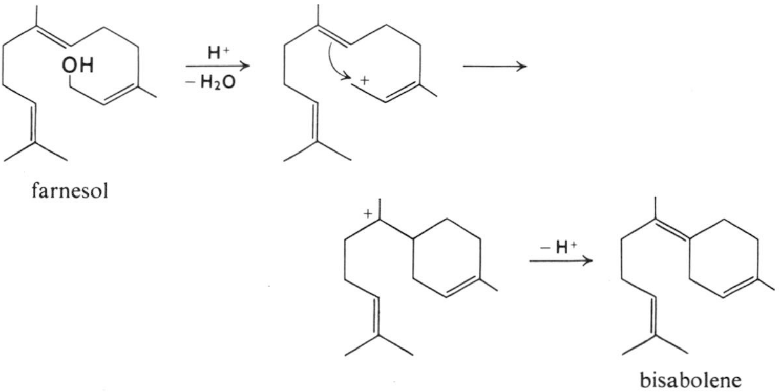

$$\text{[farnesol structure]} \quad \xrightarrow[-H_2O]{H^+} \quad \text{[structure]} \quad \longrightarrow$$

farnesol

$$\text{[structure]} \quad \xrightarrow{-H^+} \quad \text{[bisabolene structure]} \tag{81}$$

bisabolene

Farnesol and bisabolene are members of a large class of compounds, occurring in nature, that are called *terpenoid* compounds. Ring closure reactions of the kind shown above are of great importance in the chemistry of compounds of this group and will be dealt with again in a later chapter.

---

* This short and conventional notation is very convenient and is customarily used by organic chemists: the projecting lines are methyl groups; each "corner" represents a —$CH_2$— of the chain or ring. Thus, methylcyclohexane is [structure] ; isopropyl alcohol is [structure] —OH , etc.

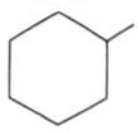

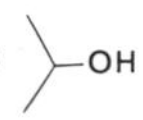

### 9.23  The carbon-carbon triple bond. Acetylenes

The alkynes form a homologous series of which acetylene, $HC \equiv CH$, is the first member. They are unsaturated compounds, and many of their reactions are similar to those that are characteristic of olefins. They add hydrogen in the presence of a catalyst to give olefins (Figure 9-8) or saturated hydrocarbons:

$$RC \equiv CR \xrightarrow[\text{catalyst}]{H_2} \underset{cis}{RCH = CHR} \xrightarrow[\text{catalyst}]{H_2} RCH_2CH_2R \tag{82}$$

Alkynes add halogens; for example, they add bromine to give tetrabromo compounds. In general, electrophilic addition reactions of alkynes are somewhat slower than the corresponding reactions of alkenes.

Alkynes can be prepared by a number of methods. Some of the most general are the following:

1. Double elimination of HX from a 1,1- or 1,2-dihalide:

$$\underset{\overset{|}{Br}\quad\overset{|}{Br}}{R-CH-CH-R} + (\text{alcoholic}) KOH \longrightarrow RC \equiv CR + 2KBr + H_2O \tag{83}$$

2. Elimination of HX from a vinyl halide:

$$\underset{\overset{|}{Br}}{RCH = CR} + \text{alcoholic } KOH \longrightarrow RC \equiv CR + KBr + H_2O \tag{84}$$

3. Elimination of halogen from a 1,1,2,2-tetrahalide:

$$\underset{\overset{|}{Br}\ \overset{|}{Br}}{\overset{\overset{|}{Br}\ \overset{|}{Br}}{R-C-C-R}} + Zn \text{ (in alcohol)} \longrightarrow RC \equiv CR + 2ZnBr \tag{85}$$

4. Alkylation of acetylenes with alkyl halides. This depends upon the ease with which the nucleophilic acetylide anion ($RC \equiv C:^-$) can be formed (Section 9.24):

$$RC \equiv CH + \text{strong base} \rightleftharpoons RC \equiv C:^- \tag{86}$$

The formation of the acetylide anion in this equilibrium can be accomplished by strong bases of several kinds. A useful reagent for this purpose is sodamide ($NaNH_2$), in which the anion $NH_2^-$ is the base. Potassium $t$-butoxide is also used. The strongly basic nucleophilic anion $RC \equiv C:^-$ reacts in the expected way with alkyl halides and alkyl sulfonates to give alkylated acetylenes:

$$RC \equiv C:^- + R'Br \longrightarrow RC \equiv CR' + Br^- \tag{87}$$

$$RC \equiv C:^- + R'OSO_2R'' \longrightarrow RC \equiv CR' + R''SO_2O^- \tag{88}$$

Secondary halides give poor yields in this alkylation reaction, and tertiary halides cannot be used. The strongly basic character of $RC \equiv C:^-$ promotes the dehydrohalogenation (elimination) reaction of the secondary and tertiary halides, and the chief or only products formed are the corresponding olefins and $RC \equiv CH$.

## 9.24  Acidity of acetylenic hydrogen

Although the hydrogen atoms on saturated hydrocarbons (e.g., ethane) and olefins (e.g., ethylene) have no demonstrable acidity, the $\equiv$CH hydrogen atom of
acetylenes is definitely, though weakly, acidic. The anion, $RC\equiv C:^-$ can be
formed by strong bases (Section 9.23), and acetylenes form characteristic silver
and cuprous salts:

$$RC\equiv CH + Ag(NH_3)_2^+ + OH^- \longrightarrow RC\equiv C:Ag + 2NH_3 + H_2O \qquad (89)$$

silver ammonia        a silver acetylide<br>complex

Since the silver and cuprous acetylides are insoluble compounds, this reaction
serves as a useful diagnostic test for the presence of the grouping $RC\equiv CH$.
Disubstituted acetylenes, $RC\equiv CR'$, which have no replaceable hydrogen atom,
do not form such salts.

Acetylenes also react with Grignard reagents to form halomagnesium derivatives:

$$RC\equiv CH + C_2H_5MgBr \longrightarrow RC\equiv CMgBr + C_2H_6 \qquad (90)$$

ethylmagnesium<br>bromide

This, too, is an indication of the acidity of the $\equiv$CH bond, for Grignard reagents
react with acids (proton donors) of many kinds in an analogous way:

$$RMgX + HX \longrightarrow RH + MgX_2 \qquad (91)$$

(where HX can be HCl, $H_2O$, ROH, RCOOH, $RNH_2$).

The peculiar character of the C—H bond in acetylenes is in sharp contrast to
that of the olefins and alkanes. It can be accounted for by examining the character
of the bonding to carbon in these three classes of compounds. The C—H bond in
acetylene is formed with the use of a molecular orbital of the $sp$ type, that in olefins
with one of the $sp^2$ type, and that in alkanes with one of the $sp^3$ type. The greater
degree of $s$ character in the acetylenic C—H bonding orbital requires that the
electrons in this bond are in a lower energy state than those in orbitals with increasing degrees of $p$ character; consequently, the relative stability of the $RC\equiv C:^-$ ion
is greater than that of $R_2C\equiv CH:^-$, and that of the latter is greater than that of
$R_3C:^-$. The equilibrium in which a proton is added to the carbon anion is a
measure of the relative energies of the anion and the protonated compound (i.e.,
$RC\equiv C:^-$ and $RC\equiv CH$), and because of the greater stability of the electron pair
in the acetylene anion the free energy difference between the acetylene (the acid)
and its anion (the conjugate base) is smaller than that between the corresponding
acid-conjugate pairs of alkenes and alkanes (see Section 4.13). Acetylenes,
$RC\equiv CH$, have demonstrable acidity (although they are exceedingly weak acids);
simple alkenes and alkanes have not.

The character of the C—H bonds in acetylenes, alkenes, and alkanes is also
to be seen in the carbon-hydrogen bond lengths in acetylene (1.063 Å), ethylene
(1.086 Å), and ethane (1.102 Å). These figures reflect the character of the bonding
orbitals; that of the acetylenic C—H bond, formed with the use of a bonding

**Table 9-5.** *A comparison of the effects of bond type upon basicity and upon acidity of conjugate acid in some nitrogen compounds*

| Compound | Bond Type | $pK_a$* | Corresponding (Isoelectronic) Carbon Compound |
|---|---|---|---|
| $CH_3\ddot{N}H_2$ | $sp^3$ | 10.6 | $CH_3$:$^-$ |
| 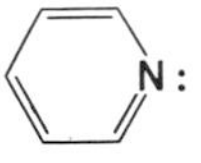 (pyridine) $N$: | $sp^2$ | 5.3 | $CH_2{=}CH$:$^-$ |
| $CH_3C{\equiv}N$: | $sp$ | $-10.1$ | $HC{\equiv}C$:$^-$ |

\* $pK_a$ refers to the dissociation constant of the conjugate acid, $-N$:$H^+$, of the nitrogen containing-compound.

orbital of carbon of the *sp* type, possesses an electron pair that is closer to the carbon nucleus. This reflects a greater electronegativity in carbon atoms with *sp* hybridization, and leads to the expectation that the proton would have enhanced acidity.

A parallel series in which can be seen the effect of bond type upon basicity (and acidity of conjugate acid) is found in nitrogen compounds in which the nitrogen atom possesses *sp*-, *sp²*-, and *sp³*-hybridized orbitals. The comparisons are made in Table 9-5.

The strongest acids in these two series are, respectively, $CH_3C{\equiv}N$:$H^+$ and $HC{\equiv}C$:$H$; the weakest, $CH_3NH_3{}^+$ and $CH_4$.

The synthetic uses of alkynes, both acetylene itself and acetylenes of the type $RC{\equiv}CH$, are many and important. Most of their applications depend upon the formation of a metal acetylide and the subsequent reaction of this with a compound having an electrophilic carbon atom. Some reactions of acetylenes with carbonyl compounds will be described in later chapters.

## 9.25   Acetylene

Acetylene itself is an important industrial chemical, and serves as the starting material for the preparation of many commercially valuable products. Acetylene is prepared by several processes, the most widely used of which is the hydrolysis of calcium carbide:

$$CaO + 3C\ (cokc) \xrightarrow{\text{electric furnace}} CaC_2 \mid CO \tag{92}$$

$$CaC_2 + 2H_2O \longrightarrow C_2H_2 + Ca(OH)_2 \tag{93}$$

Large amounts of acetylene are used for welding and cutting of metals. Its combustion with oxygen produces an intensely hot flame, because of the large amount of energy liberated in the reaction and because of the small (compared with, say, ethane) number of molecules of products formed:

$$HC{\equiv}CH + 2\tfrac{1}{2}O_2 \longrightarrow 2CO_2 + H_2O + 317 \text{ kcal/mole} \tag{94}$$

The acid-catalyzed addition of water to acetylene, a reaction analogous to the hydration of olefins to yield alcohols, yields acetaldehyde by way of an unstable intermediate compound:

$$HC{\equiv}CH + H_2O \xrightarrow[\text{HgSO}_4]{\text{H}_2\text{SO}_4} \left\{ \begin{array}{c} HC{=}CH \\ | \quad | \\ H \quad OH \end{array} \right\} \longrightarrow CH_3C{\diagup}_H^{\diagup O} \tag{95}$$

unstable enol<br>form of acetaldehyde

The mercuric salt that is used as a catalyst probably forms a mercury-acetylene complex which is the actual species to which water is added.*

More than one billion pounds of acetylene is made annually in the United States, a large proportion of which is used as a raw material for synthesis.

The following summary shows some of the compounds that are produced, and indicates briefly the method of their preparation. Most of the processes involve some kind of an addition to the triple bond as shown in the formulas at the top of page 246. Further transformations of these products give a wide range of useful materials, among which synthetic fibers, rubbers, and plastics are prominent.

## 9.26  Some synthetic applications

The chemistry of alcohols and unsaturated compounds that has been discussed to this point has included a great many individual reaction types. In this section we shall apply some of these to actual synthetic procedures.

The best general approach to developing the details of a synthesis, which may consist of a series of separate reactions, is to work it out in reverse order by selecting the *immediate* precursor in each step and working from the last step toward the first until the starting materials are reached.

---

* Substituted acetylenes react in an analogous way, addition taking place according to Markovnikov's rule:

$$RC{\equiv}CH + H_2O \longrightarrow \left\{ \begin{array}{c} RC{=}CH_2 \\ | \\ OH \end{array} \right\} \longrightarrow R{-}\underset{\underset{O}{\parallel}}{C}{-}CH_3$$

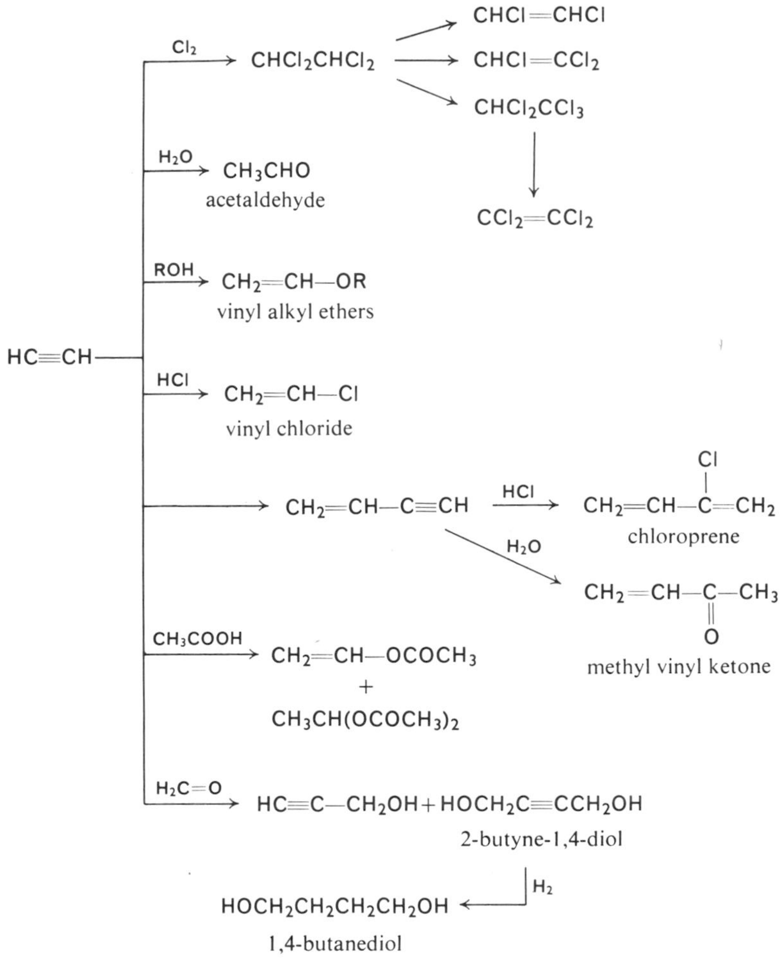

1. Prepare 2,3-dibromo-2-methylpentane, using as the starting material 1-propanol.

(a) The final step is the addition of bromine to 2-methyl-2-pentene:

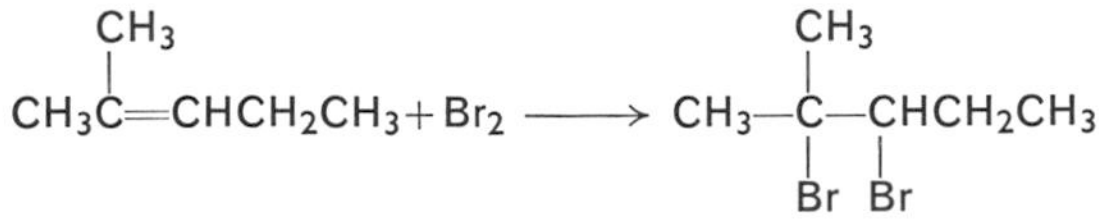

$$CH_3\overset{\overset{\textstyle CH_3}{|}}{C}=CHCH_2CH_3+Br_2 \longrightarrow CH_3-\overset{\overset{\textstyle CH_3}{|}}{C}-\overset{}{C}HCH_2CH_3$$
$$\qquad\qquad\qquad\qquad\qquad\qquad\quad \underset{Br\ \ Br}{}$$

(b) 2-methyl-2-pentene can be prepared by dehydration of 2-methyl-2-pentanol:

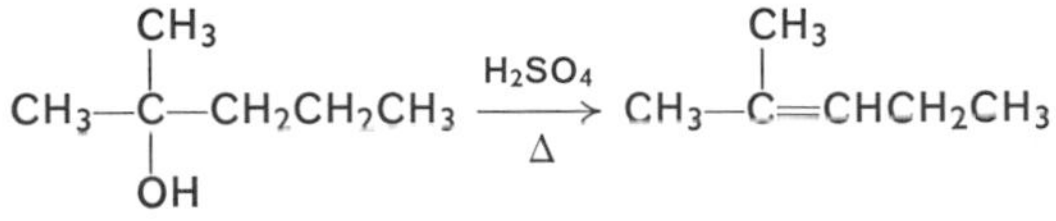

$$CH_3-\overset{\overset{\textstyle CH_3}{|}}{\underset{\underset{\textstyle OH}{|}}{C}}-CH_2CH_2CH_3 \xrightarrow[\Delta]{H_2SO_4} CH_3-\overset{\overset{\textstyle CH_3}{|}}{C}=CHCH_2CH_3$$

*(c)* 2-methyl-2-pentanol can be prepared by the Grignard reaction:

$$\underset{\underset{CH_3}{|}}{CH_3-\overset{\overset{CH_3}{|}}{C}}{=}O + CH_3CH_2CH_2MgBr \longrightarrow CH_3-\underset{\underset{OH}{|}}{\overset{\overset{CH_3}{|}}{C}}CH_2CH_2CH_3$$

*(d)* *n*-Propylmagnesium bromide is prepared from 1-bromopropane, which is prepared from 1-propanol by means of HBr. Acetone can be prepared by the oxidation of 2-propanol, and 2-propanol can be prepared from 1-propanol as follows:

$$\text{1-propanol} \xrightarrow[\Delta]{H_2SO_4} \text{propylene} \xrightarrow[H_2O]{H_2SO_4} \text{2-propanol}$$

2. Prepare 1,3-dichloro-2-propanone, starting with allyl alcohol:

*(a)*

$$\underset{\underset{OH}{|}}{ClCH_2CHCH_2Cl} \xrightarrow{CrO_3} \underset{\overset{||}{O}}{ClCH_2CCH_2Cl}$$

*(b)*

$$CH_2{=}CHCH_2Cl \xrightarrow{HOCl} \underset{\underset{OH}{|}}{ClCH_2CHCH_2Cl}$$

*(c)*

$$CH_2{=}CHCH_2OH \xrightarrow{\text{conc. HCl}} CH_2{=}CHCH_2Cl$$

3. Prepare 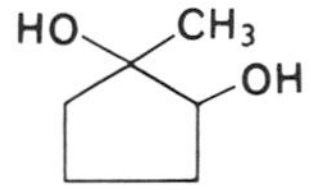 starting with cyclopentanone:

*(a)*

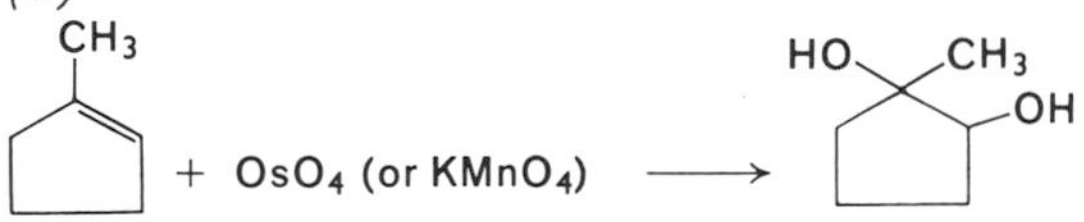

+ OsO₄ (or KMnO₄) $\longrightarrow$

*(b)*

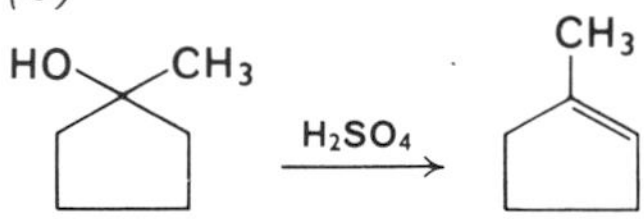

H₂SO₄

*(c)*

+ CH₃MgI $\longrightarrow$

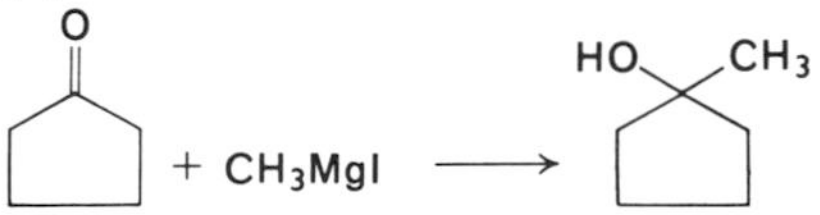

4. Prepare *cis*-9-octadecene [CH₃(CH₂)₇CH=CH(CH₂)₇CH₃]. The obvious starting material is oleic acid, CH₃(CH₂)₇CH=CH(CH₂)₇COOH, the double

bond of which is already in the *cis* configuration. The synthetic problem is to convert —COOH into —CH$_3$. One way of doing this is to make use of the reaction of Grignard reagents with acids [Equation (91), Section 9.24]:

$$RCOOH \xrightarrow{LiAlH_4} RCH_2OH \xrightarrow{PBr_3} RCH_2Br \xrightarrow[\text{Ether}]{Mg} RCH_2MgBr \xrightarrow[\text{H}_2\text{O}]{HCl} RCH_3$$

[where R=CH$_3$(CH$_2$)$_7$CH=CH(CH$_2$)$_7$—].

5. Prepare 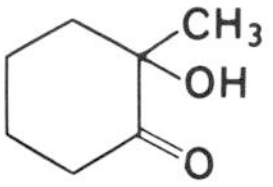 from cyclohexanone:

(*a*)

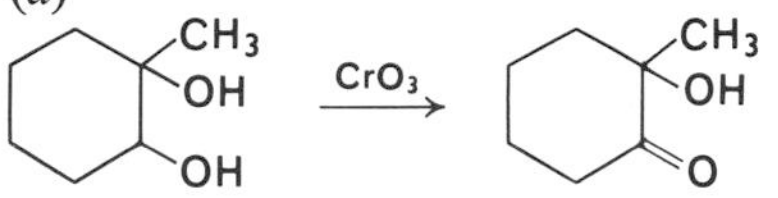

(tertiary alcoholic group not oxidized)

(*b*)

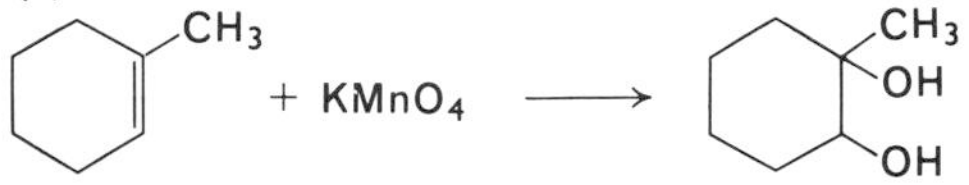

(*c*)

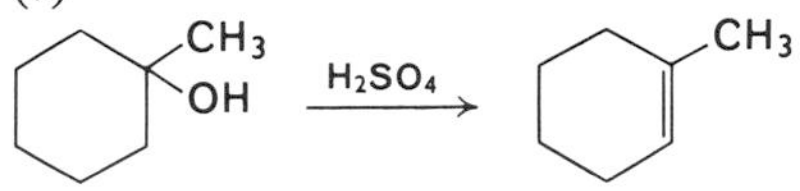

(*d*)

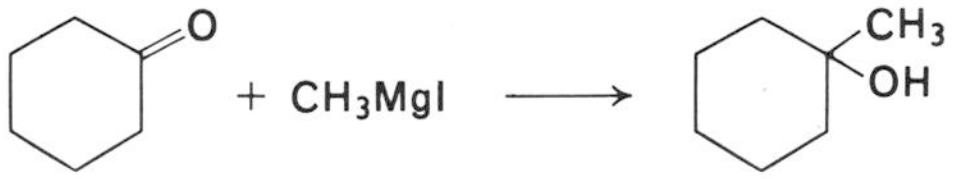

6. Prepare 2-ethoxybutane (ethyl *sec*-butyl ether) from ethanol:

(*a*)

$$CH_3CH_2\underset{\underset{CH_3}{|}}{C}HO^-K^+ + CH_3CH_2Br \longrightarrow CH_3CH_2\underset{\underset{CH_3}{|}}{C}HOCH_2CH_3 + KBr$$

(*b*)

$$CH_3CH_2\underset{\underset{CH_3}{|}}{C}HOH + K \longrightarrow CH_3CH_2\underset{\underset{CH_3}{|}}{C}HO^-K^+ + \tfrac{1}{2}H_2$$

(*c*)

$$CH_3CH_2MgBr + CH_3CHO \longrightarrow CH_3CH_2\underset{\underset{CH_3}{|}}{C}HOH$$

(*d*) Ethyl bromide and acetaldehyde can be prepared from ethanol by methods that need no comment.

## Exercises

(Exercises 1–9 will be found within the text.)

**10** Name the following olefins by the I.U.C. system.

(a) $CH_2\!=\!CH_2$

(b) $CH_3CH\!=\!CH_2$

(c) $CH_3CH_2CH\!=\!CHCH_2CH_3$

(d) $(CH_3)_2C\!=\!CH_2$

(e) $BrCH_2CH_2CH\!=\!CH_2$

(f) $CH_2\!=\!CH\!-\!CH\!=\!CH\!-\!CH_3$

(g) $CH_2\!=\!C\!=\!C\!=\!CHCH_2CH_3$

(h) $(CH_3)_2C\!=\!C(CH_3)_2$

(i) $(CH_3)_3CCH\!=\!CHC(CH_3)_3$

**11** Formulate the addition of one mole of HBr to each of the olefins, except *(g)*, in Exercise 10.

**12** Write the structures of the alcohols that would be obtained by the hydration (addition of sulfuric acid, followed by hydrolysis) of the following olefins.

(a) $CH_2\!=\!CH_2$

(b) $(CH_3)_2C\!=\!CH_2$

(c) $(CH_3)_3CCH_2CH_2CH\!=\!CH_2$

(d) $CH_3CH_2CH\!=\!C(CH_3)_2$

(e) $CH_3CH_2C\!=\!CHCH_3$
  $\qquad\qquad\ \ \overset{\displaystyle |}{CH_3}$

(f) $CH_3CH_2CH\!=\!CHCH_2CH_3$

**13** By the use of appropriate projection formulas write the structures of all of the products that would be obtained in the following reactions:

(a) *cis*-2-butene + HBr

(b) *cis*-2-pentene + $Br_2$

(c) *trans*-2-hexene + $Cl_2$

(d) cyclohexene + $Br_2$

(e) 1,2-dimethylcyclohexene + $H_2$ (catalytic reduction)

(f) 1,3-butadiene + $2H_2$ (catalytic reduction)

(g) propene + $Br_2$

(h) isobutylene + HBr

**14** Show how the application of a simple chemical test or analytical procedure will enable one to distinguish between:

(a) propylene and cyclopropane

(b) 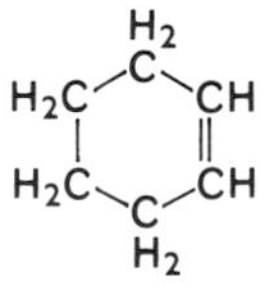 and

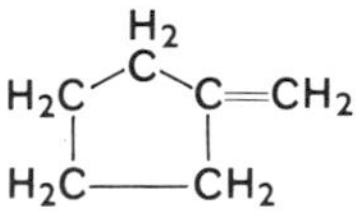

(c) 1-octene and 4-octene,

(d) 2,4-heptadiene and

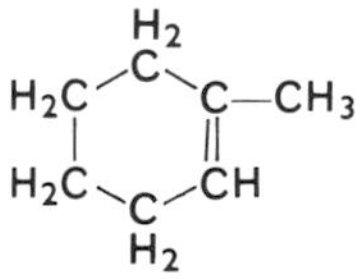

(e) dicyclohexyl and 1,11-dodecadiene

**15** Write the structures of the olefins from which the following products are obtained on ozonization: *(a)* formaldehyde only; *(b)* acetaldehyde only; *(c)* acetone and formaldehyde; *(d)* methyl ethyl ketone and acetaldehyde; *(e)* 2,6-heptanedione only; *(f)* cyclopentanone and acetone; *(g)* $CO_2$ and acetaldehyde; *(h* $CH_3COCHO$, $CH_3CHO$, and $(CH_3)_2CO$.

**16** Formulate the following reaction series, assigning structures to the compounds designated by letters. A compound *(A)*, $C_5H_{10}O$, does not react with cold, aqueous $KMnO_4$ and does not add bromine. When *A* is reduced, *B*, $C_5H_{12}O$, is formed. Treatment of *B* with hot 20 % sulfuric acid yields *C*, $C_5H_{10}$. Compound *C* instantly adds bromine and decolorizes aqueous $KMnO_4$. When *C* is dissolved in cold 70 % sulfuric acid and the resulting solution diluted with water, compound *D*, $C_5H_{12}O$, is formed. Although *B* can be reoxidized to *A*, a similar oxidation of *D* to a five-carbon-atom compound cannot be accomplished.

**17** Starting with isobutyl alcohol as the only organic compound, show how you could prepare: *(a)* *tert*-butyl alcohol; *(b)* isopropyl alcohol; *(c)* 1,2-dibromopropane; *(d)* 2-methylpropane (isobutane).

**18** Formulate the transformation of 1-butanol into 2-butanol.

**19** How could pure ethylene be prepared from a mixture containing ethylene and ethane?

**20** How could cyclohexene be removed from a sample of cyclohexane that contained some of the olefin as an impurity?

**21** Vitamin A is a primary alcohol of the composition $C_{20}H_{30}O$. Catalytic hydrogenation affords the fully saturated "perhydro" vitamin A, $C_{20}H_{40}O$. How many rings does vitamin A contain? How many double bonds (assuming no triple bonds) are present?

**22** Write the equations showing the course of the acid-catalyzed polymerization of 2-methyl-1-butene (to the stage of the trimer).

**23** A sample of 19.6 mg of an unsaturated compound absorbed 4.48 ml of hydrogen (measured at 0°C and 760 mm) upon catalytic hydrogenation. What is the minimum molecular weight of the compound?

**24** The addition of HBr to the olefin of part *(i)*, Exercise 10, could give rise to a product other than that formed by simple addition to the double bond shown. Show the structure of this other product, and write a mechanism for its formation.

**25** Show how the following compounds could be synthesized; use the starting material indicated and any other necessary organic or inorganic reagents:
*(a)* *cis*-2-hexene from propyne
*(b)* 2-chlorocyclopentanone from cyclopentene
*(c)* $CH_3COCH_2CH_2CH_2OH$ from $CH_3COCH_2CH_2COOEt$

# Chapter ten / The Concept of Resonance and Its Application to Organic Chemistry

The concept of resonance deals with the structures of molecules in terms of their energy content, or stability. To most organic chemists it is principally another means of further refining the representations of molecular structures, and of arriving at an understanding of the properties of organic compounds, many of which are not adequately or clearly expressed by the usual valence-bond structures.

Our interest in resonance will be primarily in its application to the representation of structure and to the meaning, in terms of chemical behavior, of the structural symbols that we use. An understanding of the meaning of structure is essential to the understanding of chemical reactivity if this is to be arrived at in a rational way.

## 10.1 The meaning of "stability"

We shall find, in the discussion to follow, that an organic compound is often a more stable substance (as determined by *experimental* means) than it would be expected to be (as determined by *mathematical* means) if its structure were as we ordinarily write it. The conclusion might be that there is something wrong either with our mathematics or with the structure we have written. We shall see that it is most often in our structural representations that the inadequacy lies.

It is well at this point to say something about the term "stability," a much-used word in organic chemistry. This term, and the corollary words "stable," "unstable," and so forth are used in more than one way. There is a popular, or colloquial, meaning of stability that is equivalent to inertness, or lack of reactivity. In this sense, methane and the homologous paraffin hydrocarbons are stable substances: they react slowly or not at all with many reagents at ordinary temperatures, and they may be stored for long periods of time without undergoing appreciable change. Other compounds may be stable to a given reagent, meaning that there is no reaction when the compound and reagent are brought together. Such uses of the term "stable" are not precise, although they usually have recognized meanings within given contexts. But these meanings are based upon the general agreement that comes from common usage, and do not relate to any absolute state of reference.

The meaning of the word stability that will be of greatest importance in our discussion of molecular structure carries with it an implicit allusion to some definite state of reference. Since we are usually concerned with the *difference* in the energies of two states, the reference state can be arbitrary. However, it is convenient to adopt as the reference the energy of the system in which all of the atoms are separated and in the gaseous state.

### 10.2 Bond energy and energy content

The measurement of the energy content of a substance is an experimental process. It consists of comparing the energy content with that of some reference state by determining how much energy (usually expressed in calories or kilocalories) is required to place the compound in that state. The reference state is often that of the separate, gaseous atoms. Thus, the energy of the compound is that amount which must be expended to break the bonds and separate the individual atoms; it is thus a measure of the energies of the bonds that join the atoms.

The energy of the hydrogen molecule is the H—H bond energy. It is the amount of energy required to break this bond and form two hydrogen atoms, according to Equation (1), and amounts to 104 kcal/mole:

$$H_2 \longrightarrow 2H\cdot \tag{1}$$

This also means that when hydrogen atoms combine to give molecular hydrogen, 104 kcal/mole is liberated—the heat of formation of molecular hydrogen.

Bond energies between atoms are usually calculated from thermochemical data. Since it is seldom possible to prepare organic compounds by the direct combination of the elements, heats of formation and bond energies are calculated indirectly from experimental data obtained by measuring the heat evolved when the compound is burned in oxygen.

One way to do this is to burn a known quantity of the compound (e.g., a hydrocarbon) to carbon dioxide and water and to measure the heat evolved in the process by carrying out the combustion in an instrument called a calorimeter. With the aid of the well-known values (determined in other ways) of the heats of formation of carbon dioxide and water from their constituent gaseous atoms, it is then possible to calculate the energy of the compound from the heat of combustion. This is the *experimentally determined* energy of the substance, and the measurements are not dependent upon a knowledge of its structure.

Now if we write a structure for the compound, an energy content can be *calculated* for this structure. This calculation is based upon the energies of the various bonds contained in the structure we have written; these bond energies are known from measurements made on many other compounds of known structure. For example, we can accumulate data that will give empirical values for the energies of C—H and C—C bonds by carrying out the combustion of a series of pure paraffin hydrocarbons; of C=C bonds by the combustion of a number of olefins, and so on. We can then, for example, calculate the heat of formation of a hydrocarbon such as propane, $CH_3CH_2CH_3$, by adding the bond energies of eight C—H bonds and two C—C bonds. Isobutylene, $(CH_3)_2C=CH_2$, contains eight C—H bonds, two C—C bonds, and one C=C bond. By reference to a table of bond contributions to heats of combustion we find that the heat of combustion of isobutylene is expected to be 646 kcal/mole; the experimental value is 646 kcal/mole.

It will be recalled that if the heat of combustion can be determined, the heat of formation can be calculated; thus, the most extensive tables of data available are the empirical heats of combustion obtained directly from calorimetric measurements.

## 10.3 Calculation of bond-energy values

To illustrate the method of calculating bond energies, let us determine the energy of the H—O bond, assuming this to be one-half the value of the energy of the following reaction (note: all reactants are in the gaseous state, and values are rounded off):

$$2H + O \longrightarrow H_2O + energy$$

The energy derived in this reaction is taken to be twice the H—O bond energy, since two H—O bonds are formed. Since we cannot carry out the calorimetric experiment in which hydrogen and oxygen *atoms* combine, we use the readily determined value for the reaction of molecular hydrogen with molecular oxygen:*

$$H_2 + \tfrac{1}{2}O_2 \longrightarrow H_2O(g) \qquad \Delta H = -58 \text{ kcal}$$

---

* The manner of writing thermochemical equations is to be noted. We can write

$$2H(g) \longrightarrow H_2(g) + 104 \text{ kcal}$$

or

$$2H(g) \longrightarrow H_2(g) \qquad \Delta H = -104 \text{ kcal}$$

In the first example, the 104 kcal is expressed as a product of the reaction. The second manner of writing the equation is more usual.

We have available, from other experimental studies, the values of the heats of the following reactions:

$$2H \longrightarrow H_2 \qquad \Delta H = -104 \text{ kcal}$$
$$O \longrightarrow \tfrac{1}{2}O_2 \qquad \Delta H = -59 \text{ kcal}$$

We now add the reactions shown:

$$H_2 + \tfrac{1}{2}O_2 \longrightarrow H_2O \qquad \Delta H = -58 \text{ kcal}$$
$$2H \longrightarrow H_2 \qquad \Delta H = -104 \text{ kcal}$$
$$O \longrightarrow \tfrac{1}{2}O_2 \qquad \Delta H = -59 \text{ kcal}$$

$$\overline{2H + O \longrightarrow H_2O \qquad \Delta H = -220 \text{ kcal}}$$

Note that $H_2$ and $\tfrac{1}{2}O_2$, on opposite sides of the equations, cancel out. Thus, the H—O bond energy is $220/2 = 110$ kcal.

---

**Exercise 1.** Given the following experimental heats of reaction (in kilocalories):

$$CH_4(g) + 2O_2(g) \longrightarrow CO_2(g) + 2H_2O(g) \qquad \Delta H = -192 \text{ kcal}$$
$$2H_2(g) + O_2(g) \longrightarrow 2H_2O(g) \qquad \Delta H = -116 \text{ kcal}$$
$$C(s) + O_2(g) \longrightarrow CO_2(g) \qquad \Delta H = -94 \text{ kcal}$$
$$2H(g) \longrightarrow H_2(g) \qquad \Delta H = -104 \text{ kcal}$$
$$C(g) \longrightarrow C(s) \qquad \Delta H = -172 \text{ kcal}$$

Calculate the C—H bond energy, assuming this value to be one-fourth the heat of formation of $CH_4$ from gaseous carbon and hydrogen atoms.

---

## 10.4 Resonance energy

Benzene is a hydrocarbon of the composition $C_6H_6$. This important compound will be discussed at length later in this book, but for our present purposes we can record that it is a cyclic hydrocarbon for which the conventional Lewis structure, (2), accounts for the six carbon and six hydrogen atoms and all of the valence electrons:

usually written as                    (2)

If we calculate the heat of formation of benzene by adding the values obtained empirically for the C—H bonds (six), C—C bonds (three) and C=C bonds (three), we obtain the value of 1287 kcal/mole as the heat evolved in the formation of benzene from the gaseous atoms. The calorimetrically *measured* energy is 1323 kcal/mole. The difference, 36 kcal/mole, is the extent to which benzene is actually more stable than it would be if it had structure (2). This is the extent to which benzene is said to be *stabilized by resonance.*

It should be noted here that if we compare the calculated and experimental heats of combustion of 1,4-cyclohexadiene, (3), there is very good agreement, and so this valence-bond structure appears to be an acceptable representation for this

1,4-cyclohexadiene

(3)

compound. Thus, the wide disparity in the calculated value [based on formula (2)] and the experimental value for benzene is not a result simply of the fact that the compound is cyclic or that the double bonds are in a ring, because if this were the case we might expect to find a similar difference between the calculated and experimental values for (3) as well, but we do not.

We have reason to conclude, then, that (2) is not a proper representation of the structure of benzene; neither is (4), because this also contains three carbon-carbon double bonds, three carbon-carbon single bonds, and six carbon-hydrogen bonds.

(2)                    (4)

Nevertheless, experimental evidence obtained from physical measurements tells us that benzene does consist of a six-membered ring. Furthermore, all of the carbon-carbon bonds are alike (there is only one C—C bond distance in the molecule), and the ring is perfectly symmetrical, with C—C—C and C—C—H bond angles of 120°. Both (2) and (4) would be expected to have two different carbon-carbon bond distances (one for C=C and one for C—C), and thus the ring would not have perfect hexagonal symmetry. Added to this, we have found that benzene is actually 36 kcal more stable than it would be if it had either structure (2) or structure (4).

What, then, *is* the structure of benzene? The physical evidence tells us that the six carbon and six hydrogen atoms are arranged in a six-membered ring, as formula (2) shows. But the thermochemical evidence tells us that benzene does not have the kinds of bonds shown in this structure. If we simply omit the double bonds the result is not acceptable, since it leaves six electrons unaccounted for.

The answer to the dilemma is simply that we cannot write a satisfactory single structure for benzene, using conventional valence-bond notation. However, we can arrive at a practical solution by introducing a new symbol. This is the double-headed arrow, shown in formula (5). This method of writing a structure is a concession to the established use of classical valence-bond formulas for the depiction of organic structures, and avoids the necessity for inventing a totally new kind of notation. From time to time, attempts have been made to introduce new kinds of symbolism in order to escape the inconvenience or ambiguity of writing more than one formula to represent only one substance. For example, benzene is sometimes written as (6) or (7). *All of these mean the same thing*; it must be kept clearly in mind that these are only representations; they are symbols for the compound benzene:

(5)          (6)          (7)

In the everyday use of structural formulas, chemists often simplify their problems of representing compounds by being arbitrary. Thus, benzene is commonly written as (2) *or* (4); the compromise of using formula (6) or (7) is not generally adopted, although it is becoming more common. Nevertheless, the use of formulas such as (6) or (7) does emphasize the fact that *benzene is a single substance with one structure only*. The representation (5) must *not* be taken to suggest that benzene has more than one structure or that the molecule oscillates between the two structures.

## 10.5  Hybrid bonds

The formulas (5), (6), or (7) tell us that the carbon-carbon bonds in benzene are a new kind of bond. This is a "hybrid" bond, neither a double bond nor a single bond. The six "extra" electrons that must be accounted for are thus distributed uniformly around the ring in molecular orbitals lying above and below the plane of the ring. This freedom from constraint of the electrons in defined orbitals, such as those of (2), is a more stable condition, and the greater stability is expressed as the *resonance energy* of benzene.

Another term that may be used is *delocalization* of the electrons. In a carbon-carbon single or double bond the electrons are localized—and can be defined by writing a Lewis structure—somewhere in the region between the carbon kernels. In benzene the six "extra" valence electrons are delocalized, and thus cannot be represented by a bond of the usual kind.

The concept of resonance is more than mere symbolism, and it is important to combine a description of its meaning, which will be given in the following pages, with a description of how we use it in writing the structures of molecules. We can do this best by making a number of statements regarding resonance and illustrating these with actual examples.

## 10.6   The structural conditions for resonance

Since a given compound is *only one substance, with one structure,* it consists of a group of atoms occupying definite regions in space,* joined by bonds of definite length and angular distribution. The electrons that constitute these bonds are distributed between the atoms in a way that defines these bond lengths and angles and confers a characteristic chemical behavior upon the compound.

The resonance hybrid can be represented only by structures in which the atoms of the molecule occupy the same, or nearly the same, relative positions. The structures that we write of the forms that contribute to the resonance hybrid differ only in the arrangement of the electrons in the bonds between the atoms. This statement may be regarded as a rule (*Rule 1*) that is to be observed when we assess the possibilities for resonance stabilization in an organic compound.

Now if we write a conventional valence-bond formula (which we shall call a Lewis formula or structure) for the compound, the simple rules of valence require that we take account of all valence electrons in writing our formula and assign to them definite locations in bonds or on atoms.

Let us examine a simple substance, the carbonate ion, $CO_3^=$. One Lewis structure for this would be (8); but it is apparent that (9) and (10) are no less likely and are equally valid:

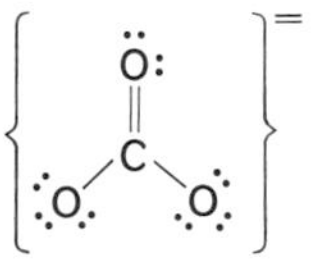

(8) 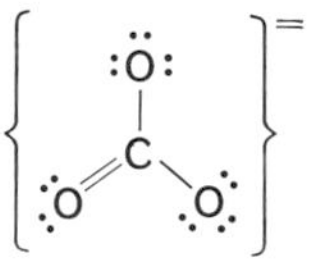          (9) 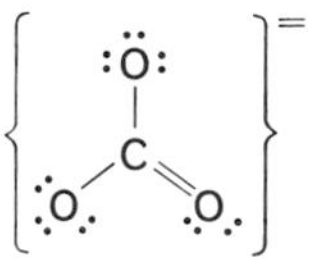          (10)

An additional structure,

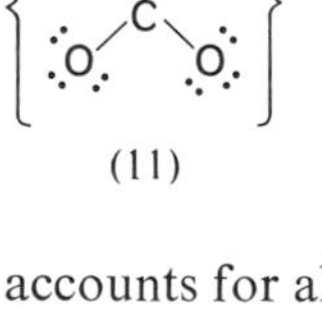

(11)

also accounts for all valence electrons, but has the disadvantage that it possesses one less covalent bond on carbon. Structures of this kind are relatively less important. We can thus state the following general rule, *Rule 2:* Of the various structures that can be written for a substance, those with the greatest number of covalent bonds are the more stable; i.e., they represent more nearly the actual structure.

---

* There are, of course, small vibrational motions that may be regarded as causing the individual atoms to take some *average* position, but this may be viewed as fixing them in a definite location in space.

Thus, structure (11), being less stable than structures (8) through (10), is a less likely representation for the carbonate ion than are the others. This does not mean that (11) is "wrong" any more than (8), (9), or (10) is "correct," but only that the carbonate ion is a resonance hybrid whose structure is *best* represented by (8) ↔ (9) ↔ (10). We can say that structures (8), (9), and (10) *contribute* equally (since they are identical) to the actual structure of the carbonate ion. If we use a notation similar to that of formula (6) for benzene, we might represent the carbonate ion as

(12)

This clearly implies [as does the resonance notation (8) ↔ (9) ↔ (10)] that the carbonate ion does *not* contain one carbon-oxygen double bond and two carbon-oxygen single bonds. It does lead to the prediction that there is only one kind of carbon-oxygen bond in the ion, and that this is neither a single nor a double bond. Measurements of bond lengths in calcite ($CaCO_3$) confirm this; the C—O bond length is found to be 1.30 Å. This is longer than the carbon-oxygen double bond and shorter than the C—O single bond. In short, it is a hybrid bond.

Consider the compound formaldehyde, $H_2C{=}O$. The usual (Lewis) structure for this is (13); another distribution of the electrons is shown in (14):

(13)              (14)

In view of rule 2, we would say that formaldehyde is *more nearly* represented by (13). The contribution of (14) to the structure would make the C—O bond length somewhat longer than the "true" C—O double bond, but the greater importance of (13) would make the C—O bond considerably shorter than the usual C—O single bond. If we take the carbon-oxygen distance in $CO_2$ as a reasonable approximation to the C—O double bond distance (1.16 Å), the measured C—O bond length of 1.21 Å in formaldehyde is seen to be greater than this; but it is much less than the C—O single-bond length in methanol (1.44 Å) or diethyl ether (1.43 Å).

The C=O bond in formaldehyde (and in aldehydes and ketones in general) is said to have "partial ionic character." This ionic character results in a charge distribution in the bond such that the oxygen is the negative end, the carbon the positive end, of the C=O dipole. This idea can be expressed as *Rule 3:* In a structure in which partial ionic character exists in a bond, the polarization will be such that the more electronegative element has the negative charge.

The concept of electronegativity was discussed in Chapter 3, where it was shown that nitrogen, oxygen, and the halogens are more electronegative than carbon. Thus a structure (15) does not make a significant contribution to

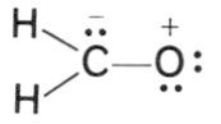

(15)

the structure of the compound formaldehyde.

For the compound nitromethane, two Lewis structures, identical in bond types, can be written, (16) and (17):

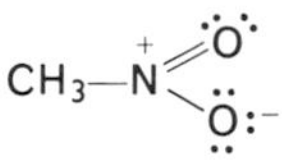

(16) 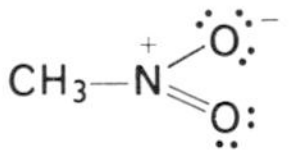                    (17)

The compound is actually a resonance hybrid, (16) ↔ (17), and has but one N—O bond length (1.21 Å). Another structure,

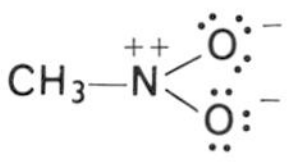

(18)

can be taken into consideration, but since the electronegativity difference between N and O is not large, the contribution of this would not be great. But structure (19) need not be considered at all, since oxygen is more electronegative than nitrogen:

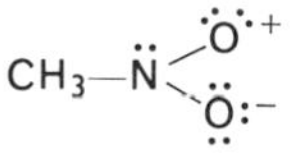

(19)

It is implicit in what we have already learned (Chapter 3) about maximum co-valency of the first period elements that structure (20) cannot contribute to the structure of nitromethane, for in this structure nitrogen has ten electrons in its valence shell:

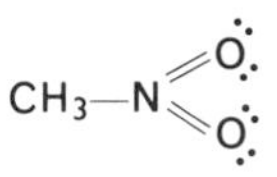

(20)

## 10.7 Resonance and isomerism

We cannot speak of (16) and (17) as isomers of nitromethane, since neither struc-ture alone represents a substance. Rather, they are *contributing structures* to the hybrid, which *is* nitromethane. In other words, (16) and (17) are symbolic approxi-mations, each of which is, in a sense, a "correction" of the other. Since they involve

---

**Exercise 2.** Do the "bookkeeping" for the electrons in nitromethane, and show why the formal positive and negative charges that are shown must be a part of the structures (16), (17), and (18).

---

identical kinds of bonds, the true nature of the nitro ($-NO_2$) group is that of a perfectly symmetrical group with two identical N—O bonds.

Now we can write another structure which has the same empirical composition as nitromethane:

$CH_3-\ddot{\underset{..}{O}}-\ddot{N}=\ddot{\underset{..}{O}}:$

    (21)

This is *another compound,* methyl nitrite (the methyl ester of nitrous acid). It is clear that nitromethane and methyl nitrite differ in more than just the way the electrons are distributed. In nitromethane, carbon is attached to nitrogen; in methyl nitrite, there is a carbon-oxygen bond. These two compounds have quite different atomic arrangements (Rule 1), and thus are *isomers.*

---

**Exercise 3.** Write another contributing structure for methyl nitrite.

    (Answer: $CH_3-\overset{+}{\ddot{O}}=\ddot{N}-\ddot{\underset{..}{O}}:^-$; account for the charges shown.)

---

### 10.8   Further examples

The question of how to decide whether the various contributing structures we can write are important or negligible remains to be discussed; but first let us examine some examples that show how the formulas can be written as Lewis structures.

The compound diazomethane has the composition $CH_2N_2$. It is known to be a linear molecule, and thus is not to be represented by the structure (22).

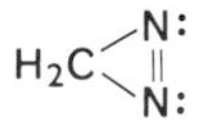

$H_2C\underset{\diagdown \ddot{N}:}{\overset{\diagup N:}{\big\|}}$

    (22)

Moreover, in many of its chemical reactions its ready decomposition to give nitrogen shows us that it has a nitrogen-nitrogen bond; other evidence demonstrates the presence of a carbon-nitrogen bond. The sum of this evidence leads us to conclude that diazomethane can be represented by

$CH_2=N=\ddot{N}:$                                 (23)

in which all of the electrons are accounted for and in which each atom possesses a complete octet.

It will be noted, however, that the quadricovalent nitrogen atom should bear a positive charge (Section 3.22), and the terminal nitrogen, with six fully-owned electrons, is negatively charged. Thus, structure (23) is incomplete and should properly be written as:

$$CH_2{=}\overset{+}{N}{=}\overset{..}{N}{:}^{-} \qquad\qquad (24)$$

Now there is another equally satisfactory way to write a structure for diazomethane that accounts for all of the primary requirements of valence-bond conventions and provides for complete occupancy of all available orbitals:

$$:\overset{-}{C}H_2{-}\overset{+}{N}{\equiv}N: \qquad\qquad (25)$$

Is the structure of diazomethane (24) or (25)? The answer is that it is neither of these: diazomethane is a resonance hybrid that can best be represented by the contributing structures (24) and (25) in the following way:

$$\{:\overset{-}{C}H_2{-}\overset{+}{N}{\equiv}N: \longleftrightarrow CH_2{=}\overset{+}{N}{=}\overset{..}{N}{:}^{-}\} \qquad\qquad (26)$$

Another common substance, boron trichloride, has the Lewis structure

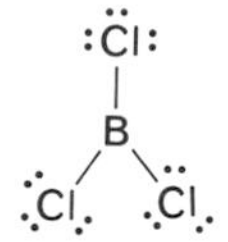

(27)

Since boron has the capacity for accommodating eight electrons in its valence orbitals, and since the chlorine atoms possess unshared electron pairs in their valence orbitals, each of the boron-chlorine bonds can assume some double bond character, as shown by the contributing structures of the resonance hybrid

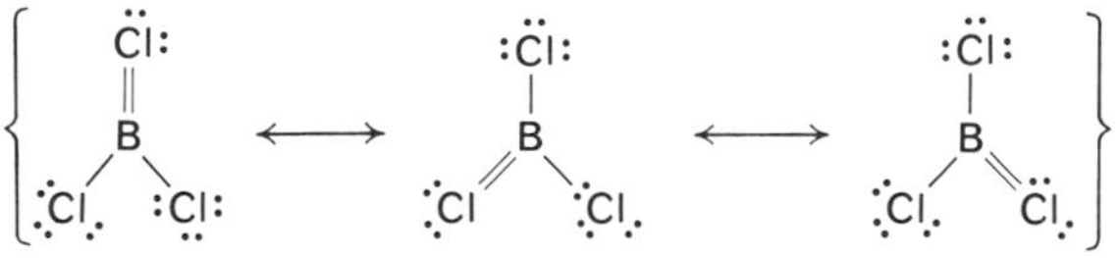

(28)

The formal resemblance between (27) and the resonance hybrid (28), and the structures written for the carbonate ion, is to be noted.

## 10.9 Resonance stabilization

We might ask why the actual structure of a compound is that of the resonance hybrid, when any one of the contributing structures adequately accounts for the bonding requirements. The answer must be that the resonance hybrid corresponds to a more stable structure than any of the others. This is because the compound is in its most stable state; if it were in an unstable condition it would tend to lose energy and assume the more stable condition.

Since an actual compound—the resonance hybrid—is more stable than it would be if it possessed any one of the contributing structures, we speak of *resonance stabilization* of a compound. In what we have said so far, resonance stabilization refers to a difference in energy between a real substance and an imaginary substance (one of the contributing structures). But the chief utility of the concept lies in its application to situations involving differences between real substances.

Consider the compounds 1,3-pentadiene (29) and 1,4-pentadiene (30):

$$CH_2\!=\!CH\!-\!CH\!=\!CH\!-\!CH_3 \qquad\qquad CH_2\!=\!CH\!-\!CH_2\!-\!CH\!=\!CH_2$$
$$\text{(29)} \qquad\qquad\qquad\qquad\qquad \text{(30)}$$

If we *calculate* the heats of formation of these two compounds, they turn out to be the same, for each has eight C—H, two C=C, and two C—C bonds. However, the *experimental* heats of combustion for these two compounds are different; 1,3-pentadiene is actually the more stable compound by about 3 kcal/mole. Evidently when the two double bonds are in adjacent positions, as in (29), added stability is conferred on the compound. Double bonds in such a relationship are "conjugated" double bonds, and the delocalization of the electrons is a stabilizing factor. As we shall see in the discussion to follow, the structure (29) is actually very close to the true representation for the compound, since the total resonance stabilization is quite small (only 3 kcal; compare the 36 kcal in the case of benzene).

In 1,4-pentadiene the heat of combustion data indicate that formula (30) is a completely adequate representation of the structure. Similar conclusions are reached by inspecting the heat of combustion data for most compounds in which there are no double bonds, or in which carbon-carbon double bonds are isolated from each other by intervening saturated carbon atoms. Thus, compounds such as propane, dimethyl ether, ethylamine, and isobutyl alcohol are adequately and properly represented by the structures (31) through (34); they are not resonance hybrids:

$$CH_3CH_2CH_3 \qquad CH_3OCH_3 \qquad CH_3CH_2NH_2 \qquad (CH_3)_2CHCH_2OH$$
$$\text{(31)} \qquad\qquad \text{(32)} \qquad\qquad \text{(33)} \qquad\qquad \text{(34)}$$

Indeed, it is by the measurement of the heats of combustion of such compounds as (30) through (34), and others of comparable structures, that we derive the values for bond energies that we have used in the calculations discussed earlier.

## 10.10   Rules for recognizing the existence of resonance

We have stated three rules under which resonance will be a factor to be considered in writing the structure of a compound. Most of the cases we have dealt with in which resonance must be taken into account have been compounds in which double- and triple-bonded atoms are present, and it will be recalled that it is in the ways in which the electrons are distributed in such multiple-bonded structures that the various contributing structures are derived. We can now state another general rule, *Rule 4:* The greater the number of important contributing structures, the greater the resonance stabilization.

How can we tell when a contributing structure is "important"? We must be able to answer this question in order to decide how many possible structures can be written. We can anticipate that one approach to this problem is to assume that structures of high energy—i.e., unstable structures—will not make important contributions.

In the case of methane, for example, there are obviously no other arrangements of bonds between the tetravalent carbon atom and four hydrogen atoms. The only other structures we can write for methane involve an actual separation of one or more of the carbon-hydrogen bonds, as in (35) or (36):

$$\overset{\overset{\displaystyle H}{|}}{\underset{\underset{\displaystyle H}{|}}{\bar{H}:\overset{+}{C}-H}} \qquad\qquad \overset{\overset{\displaystyle H}{|}}{\underset{\underset{\displaystyle H}{|}}{\overset{+}{H}\ :\overset{-}{C}-H}}$$

$$(35) \qquad\qquad\qquad (36)$$

That such structures cannot represent energy states comparable to that of the Lewis structure for methane is clear: in both of these structures one atom has a stable orbital that is vacant; both of them involve the separation of unlike charges; and both of them have but three covalent bonds instead of the four of the Lewis structure. Indeed, these structures appear to be quite unreasonable from the viewpoint of conventional theory. We can assume, as a rule of thumb, that any structures that are "unreasonable" on such grounds can usually be ignored. Thus, methane has only one important contributing structure (the Lewis structure), and it is not a resonance hybrid.

Other factors that lead to instability can be noted in a set of rules that may be regarded as subsidiary to Rule 4.

*Rule 4.1.*   Structures in which adjacent atoms bear like charges are of high energy and do not make important contributions to the structure of the resonance hybrid. This rule follows from simple electrostatic considerations. Work is required to bring like charges together and to separate unlike charges. Thus, structures of the kind to which Rule 4.1 applies have an additional energy (i.e., are more unstable) corresponding to the work required to sustain such an unstable charge distribution.

Let us consider the compound methyl azide, $CH_3N_3$. This linear molecule can best be represented as the hybrid of the following contributing structures:

$$\{CH_3-\overset{..}{\overset{+}{N}}=N=\overset{..}{N}:^- \longleftrightarrow CH_3-\overset{\overset{\displaystyle -}{..}}{N}-N\equiv N:\}$$   (37)

Another structure that might be considered is (38):

$$CH_3-\overset{+}{N}=\overset{+}{N}-\overset{..}{\overset{..}{N}}:^=$$   (38)

Although (38) has the same number of covalent bonds as (37), and no atom lacks a complete octet, this structure must be regarded as quite unstable because of the adjacent positive charges on the central two nitrogen atoms. We can thus regard (37) as an adequate representation of the structure of methyl azide and leave (38) out of consideration. The following rule should appear to be a logical consequence of what has just been said.

*Rule 4.2.* Among contributing structures of unequal stabilities, those of lowest energy will contribute most to the actual structure of the molecule. The demonstration of the validity of this rule is found in several of the examples that have been discussed. In the case of methane only *one* of the possible structures is regarded as significant. Of the two structures (37) shown for methyl azide there are no obvious differences that would make for a large disparity in their energy, but structure (38) is discarded for the reason stated. Thus (37) is accepted as an adequate representation for the structure of the hybrid.

When the contributing structures are identical, neither can be "closest" to the actual structure of the hybrid. This is the case for benzene, in which (2) and (4) are identical. In a situation of this kind the delocalization of the electrons is maximal, and thus the resonance energy is large. Thus we can state the following rule.

*Rule 4.3.* The more nearly the contributing structures are equal in stability, the greater the resonance energy. The case of methane can again be mentioned. Since the only structure we were able to write are the conventional Lewis structure and the high-energy structures (35) and (36), the difference in stability of the structures is so great that the resonance energy is negligible.

## 10.11   Resonance as a factor in chemical reactions

In Chapter 9 the addition of halogen acids to olefins was dealt with, and Markovnikov's rule was discussed. The addition of HBr to propylene by the ionic mechanism described in Chapter 9 leads to 2-bromopropane, a result that can be accounted for by the two-stage nature of the addition reaction. The protonation of propylene as the first stage to lead to (38) rather than (39) can be accounted for by noting that the secondary carbonium ion (38) can be stabilized by the contributions of six additional structures, whereas the primary carbonium ion (39) has only the two additional contributing structures:

$$CH_3\overset{+}{C}HCH_3 \longleftrightarrow CH_3CH=CH_2\overset{+}{H} \longleftrightarrow 5\ more \tag{38}$$
$$(38a)$$

$$CH_3CH_2\overset{+}{C}H_2 \longleftrightarrow CH_3CH\overset{H^+}{=}CH_2 \longleftrightarrow 1\ more \tag{39}$$
$$(39a)$$

No one of the structures (38*a*) or (39*a*) can be regarded as a significant stabilizing contribution, but since (38*a*) and (39*a*) are of the same kind, and since there are more structures like (38*a*), we can conclude that (38) is a "better" carbonium ion than (39). Thus, the 2-bromo compound is formed.

As was pointed out in Chapter 9 (Section 9.8), stabilizing effects of this kind exert their effect upon reaction rate in the way they influence the stability of the transition state, and thus the activation energy. An energy diagram for a comparable case to that just described is shown in Figure 9-5. It must be emphasized that although the discussion there, and in this section, has been concerned with the structure and stability of the *intermediate* carbonium ions and not with that of the transition state itself, the stabilization of the latter will be subject to the same structural factors that affect the stability of the intermediate to which it leads. Consequently, most descriptions of relative stabilities of transition states are based upon the structures of the closely related intermediates. This is generally valid, and leads to the correct conclusions.

It is very interesting to note that the addition of HBr to $CH_2{=}CHCF_3$ takes place in the opposite direction to that which Markovnikov's rule would predict (40):

$$CH_2{=}CHCF_3 \xrightarrow{\text{HBr}} BrCH_2CH_2CF_3 \quad \text{not} \quad CH_3\overset{\overset{\text{Br}}{|}}{C}HCF_3 \tag{40}$$

On the basis of the same arguments we have used to account for the "Markovnikov addition" of HBr to propylene, we would conclude from this result that the ion (41) must be a more stable product of the protonation of $CH_2{=}CHCF_3$ than is the ion (42):

$$\overset{+}{C}H_2CH_2CF_3 \qquad CH_3\overset{+}{C}HCF_3$$
$$(41) \qquad\qquad (42)$$

Why is (41) "better" (more stable) than (42)? If we recall that fluorine has a powerful inductive effect, which decreases the electron density on the carbon atom of $-CF_3$, we can see that this carbon atom has a good deal of positive character. Thus, (42), in which the carbonium-carbon atom is adjacent to the electron-deficient carbon atom of the $-CF_3$ group, is less stable than (41) (Rule 4.1). Thus, Markovnikov's rule is "violated," but for a readily discernible reason.

---

**Exercise 4.**   How would you expect HI to add to the ion $(CH_3)_3\overset{+}{N}CH{=}CH_2$?

Would $(CH_3)_3\overset{+}{N}CH_2CH_2I$ or $(CH_3)_3\overset{+}{N}\overset{\overset{|}{}}{C}H{-}CH_3$ be formed?

---

## 10.12   Resonance in the strengths of acids and bases

The concept of resonance can help explain differences between the acidity or basicity of compounds by allowing us to assess the stabilizing effects that accompany ionization.

Why, for example, is an alcohol an exceedingly weak acid whereas carboxylic acids are a great deal stronger? If we write the reactions involved in ionization, (43) and (44),

$$CH_3CH_2—\ddot{O}H+:B^- \;\rightleftharpoons\; CH_3CH_2\ddot{O}:^- +H:B \tag{43}$$

$$CH_3C\overset{O}{\underset{\phantom{O}}{\|}}—\ddot{O}H+:B^- \;\rightleftharpoons\; CH_3C\overset{O}{\underset{\phantom{O}}{\|}}—\ddot{O}:^- +H:B \tag{44}$$

we can see that in the formation of the ethoxide ion from ethanol there is no gain in resonance stabilization. For neither $CH_3CH_2OH$ nor $CH_3CH_2O^-$ can we write any other structures that need to be considered; the simple Lewis structures are adequate representations of the true structures, and neither ethanol nor the ethoxide ion is resonance-stabilized.

In the case of acetic acid, however, quite a different situation exists. The acetate ion is a resonance hybrid of two identical structures,

$$\left\{CH_3—C{\overset{\ddot{\;}\ddot{O}:}{\underset{\ddot{O}:}{\diagdown}}} \;\longleftrightarrow\; CH_3—C{\overset{\ddot{O}:}{\underset{\ddot{O}.}{\diagdown}}}\right\}^- \;\; or \;\; \left\{CH_3—C{\overset{\ddot{O}:}{\underset{\ddot{O}:}{\diagdown}}}\right\}^- \tag{45}$$

and has a high degree of resonance stabilization.

Now it must not be overlooked that the acid itself is also a resonance hybrid (46):

$$\left\{CH_3—C{\overset{\ddot{O}:}{\underset{\ddot{O}H}{\diagdown}}} \;\longleftrightarrow\; CH_3—C{\overset{\ddot{O}:^-}{\underset{\overset{+}{O}H}{\diagdown}}}\right\}$$

$$(46)$$

Where, then, is the "gain" in resonance energy in the ionization of the acid? It is found in the fact that the resonance hybrid (45) affords more stabilization than is found in (46), for in (45) there is no charge separation, and the contributing structures are identical. Thus, the ionization of acetic acid, shown in (44), leads to a gain in resonance energy. Hence, the ionization of the acid is energetically favored by resonance; that of the alcohol is not.

An example of the effects of resonance stabilization upon basic strength is encountered in amides [such as acetamide (47)] and amidines [acetamidine (48)]. Acetamidine is a much stronger base than acetamide; that is, the protonation of the amidine is energetically the more favorable:

$$CH_3C{\overset{\ddot{O}:}{\underset{NH_2}{\diagdown}}} +H:A \;\rightleftharpoons\; CH_3C{\overset{\ddot{O}H^+}{\underset{NH_2}{\diagdown}}} + :A^- \tag{47}$$

acetamide

$$CH_3—C{\overset{\ddot{N}H}{\underset{NH_2}{\diagdown}}} +H:A \;\rightleftharpoons\; CH_3—C{\overset{NH_2^+}{\underset{NH_2}{\diagdown}}} + :A^- \tag{48}$$

acetamidine

The protonated amide is a resonance hybrid of the two chief contributing structures (47a),

$$\left\{ CH_3-C\overset{\overset{+}{\ddot{O}H}}{\underset{\ddot{N}H_2}{\diagdown}} \quad\longleftrightarrow\quad CH_3-C\overset{\ddot{O}H}{\underset{\overset{+}{N}H_2}{\diagdown}} \right\} \qquad (47a)$$

whereas the amidinium ion is (48a).

$$\left\{ CH_3-C\overset{\overset{+}{N}H_2}{\underset{\ddot{N}H_2}{\diagdown}} \quad\longleftrightarrow\quad CH_3-C\overset{\ddot{N}H_2}{\underset{\overset{+}{N}H_2}{\diagdown}} \right\} \qquad (48a)$$

The reason for the great resonance stabilization of the protonated acetamidine (the acetamidinium ion) is seen in the symmetry of the ion: the two contributing structures are identical and thus the delocalization energy (stabilization) is larger than in the protonated amide, in which the two structures are not identical. In the latter, the actual structure of the ion must be closer to one of the contributing structures than to the other, with the result that the difference in energy between the actual ion and the most stable of the contributing forms is less. The resonance energy, or stabilization energy, then, is less than in the case of the amidine.

In guanidine we find a structure which upon protonation gives a perfectly symmetrical ion to which three identical structures contribute equally (49):

$$\overset{\overset{NH}{\|}}{NH_2-C-NH_2} + H:A \;\rightleftharpoons$$

guanidine

$$\left\{ \overset{NH_2}{\underset{NH_2}{\diagup}}\overset{+}{\underset{}{N}H_2} \quad\longleftrightarrow\quad \overset{\overset{+}{N}H_2}{\underset{NH_2}{\diagup}}\overset{}{\underset{}{N}H_2} \quad\longleftrightarrow\quad \overset{NH_2}{\underset{^+NH_2}{\diagup}}\overset{}{\underset{}{N}H_2} \right\} + :A^- \qquad (49)$$

The result of this is to make guanidine a very strong base; it is comparable in strength to the alkali metal hydroxides.

---

**Exercise 5.**  Write the structure of carbonic acid, $H_2CO_3$, and the equation for its protonation to $H_3CO_3^+$, using electronic Lewis structures. Why might carbonic acid be expected to be a strong base, as guanidine is? Can you suggest why it is not?

---

It is important to note that these effects upon the strengths of acids and bases are effects upon equilibrium, not upon rate. Stabilization of the ions produced in the ionization reactions affect the overall energy change in the reaction, and thus the value of $K$, the equilibrium constant.

## 10.13    Conjugate addition to dienes

The addition of halogens to mono-olefins was seen, in Chapter 9, to be a process involving the formation of an ionic intermediate followed by the combination of this with a halide ion. Disregarding stereochemical considerations for the present discussion, we can summarize the addition of chlorine to 2-butene by the following expression:

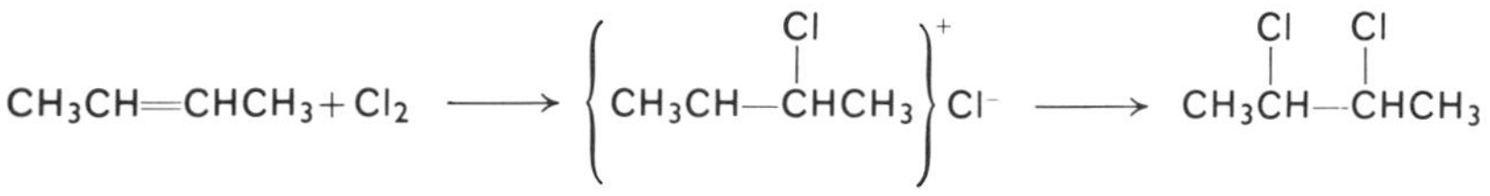

$$CH_3CH{=}CHCH_3 + Cl_2 \longrightarrow \left( CH_3CH{-}\underset{Cl}{CHCH_3} \right)^+ Cl^- \longrightarrow CH_3\underset{Cl}{CH}{-}\underset{Cl}{CHCH_3}$$

When one mole of chlorine is added to 1,3-butadiene, a mixture of two dichlorobutenes is formed. These are the result of 1,2-addition (50) and 1,4-addition (51):

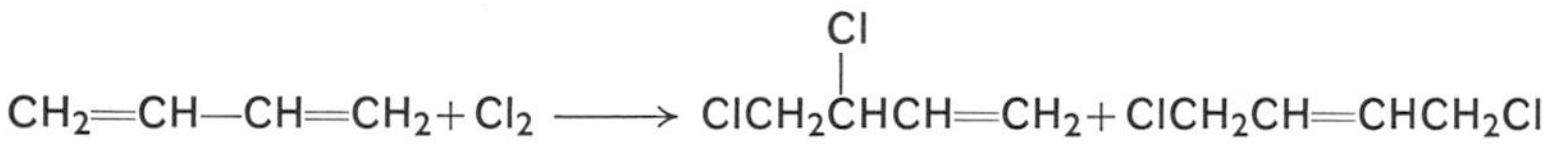

$$CH_2{=}CH{-}CH{=}CH_2 + Cl_2 \longrightarrow ClCH_2\underset{Cl}{CHCH}{=}CH_2 + ClCH_2CH{=}CHCH_2Cl$$

$$\qquad\qquad\qquad\qquad\qquad\qquad (50) \qquad\qquad\qquad\qquad (51)$$

This result can be accounted for by invoking considerations of resonance along with the recognized mechanism of halogen addition. The initial attack of chlorine on the diene would lead to the intermedite (52):

$$CH_2{=}CH{-}CH{=}CH_2 + Cl_2 \longrightarrow \{ClCH_2{-}\overset{+}{C}H{-}CH{=}CH_2\}Cl^-$$

$$\qquad\qquad\qquad\qquad\qquad\qquad\qquad (52)$$

---

**Exercise 6.**    Why would (52) be formed rather than the alternative

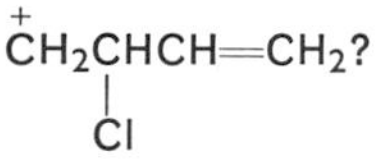

$$\overset{+}{C}H_2\underset{Cl}{CHCH}{=}CH_2 ?$$

---

The ion (52) can be seen to be a resonance hybrid, more properly represented by the contributing structures (53):

$$\{ClCH_2{-}\overset{+}{C}H{-}CH{=}CH_2 \longleftrightarrow ClCH_2{-}CH{=}CH{-}\overset{+}{C}H_2\} \qquad\qquad (53)$$

Whatever the relative importance of the two structures of (53), it is clear that the positive charge will be delocalized over the three-carbon system C⚊⚊C⚊⚊C, and thus a chloride ion can attack either at carbon 2 or at carbon 4:

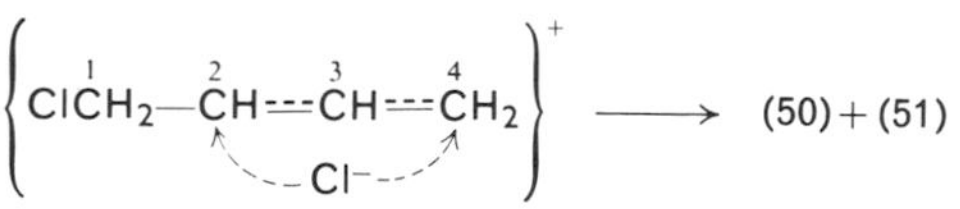

$$\left( Cl\overset{1}{C}H_2{-}\overset{2}{C}H{-}{-}{-}\overset{3}{C}H{-}{-}{-}\overset{4}{C}H_2 \right)^+ \longrightarrow (50) + (51)$$

The fact that (50) and (51) are produced in the proportion of about 2:1 is a result of subtle factors that we shall not enter into here. The fact that *both* of these products are formed, however, is readily explained in terms of the resonance concept.

---

**Exercise 7.** When HCl is added to 2-methyl-1,3-butadiene (isoprene) the only 1,4-addition product obtained is 1-chloro-3-methyl-2-butene. Can you explain why no 1-chloro-2-methyl-2-butene is formed? Hint: examine the structures of the protonated intermediates that would give rise to these two compounds.

---

## 10.14 Carbonium ion rearrangements

When HCl is added to *t*-butylethylene, the product is, surprisingly, a mixture of 2-chloro-2,3-dimethylbutane (54), and the compound that would normally be expected, namely, 2-chloro-3,3-dimethylbutane (55):

$$CH_3-\overset{\overset{\displaystyle CH_3}{|}}{\underset{\underset{\displaystyle CH_3}{|}}{C}}-CH=CH_2 + HCl \longrightarrow CH_3-\overset{\overset{\displaystyle Cl}{|}}{\underset{\underset{\displaystyle CH_3}{|}}{C}}-\overset{\overset{\displaystyle CH_3}{|}}{C}H-CH_3 + CH_3-\overset{\overset{\displaystyle CH_3}{|}}{\underset{\underset{\displaystyle CH_3}{|}}{C}}-\overset{\overset{\displaystyle Cl}{|}}{C}H-CH_3$$

$$(54) \qquad\qquad (55)$$

It is evident that in the formation of (54) a rearrangement of the carbon skeleton has occurred; (54) has the carbon skeleton (54*a*) and (55) has the skeleton (55*a*):

$$C-\overset{\overset{\displaystyle C}{|}}{C}-\overset{\overset{\displaystyle C}{|}}{C}-C \qquad\qquad C-\overset{\overset{\displaystyle C}{|}}{\underset{\underset{\displaystyle C}{|}}{C}}-C-C$$

$$(54a) \qquad\qquad (55a)$$

We may write the first step in the reaction, as usual, as the protonation of the olefin. This would be expected to produce the ion (56), which is the immediate precursor of (55):

$$CH_3-\overset{\overset{\displaystyle CH_3}{|}}{\underset{\underset{\displaystyle CH_3}{|}}{C}}-CH=CH_2 + HCl \longrightarrow \left. CH_3-\overset{\overset{\displaystyle CH_3}{|}}{\underset{\underset{\displaystyle CH_3}{|}}{C}}-\overset{+}{C}H-CH_3 \right\} Cl^- \longrightarrow \qquad (55)$$

$$(56)$$

The secondary carbonium ion (56) is, however, less stable than the ion (57), which can be produced by a migration of one of the —CH₃ groups to the open sextet adjacent to it:

$$CH_3-\overset{\overset{\displaystyle CH_3}{|}}{\underset{\underset{\displaystyle \overset{+}{C}H_3}{|}}{C}}-CH-CH_3 \xrightarrow{\text{rearrange}} CH_3-\overset{\overset{\displaystyle CH_3}{|}}{\underset{\underset{\displaystyle CH_3}{|}}{\overset{+}{C}}}-CH-CH_3 \xrightarrow{Cl^-} \qquad (54)$$

$$(57)$$

The *tertiary* carbonium ion (57) has gained resonance stabilization by charge delocalization among the seven hydrogen atoms adjacent to the carbonium carbon atom; for this reason, it is more stable than the secondary carbonium ion (56). Thus, the gain in resonance energy is a driving force for the rearrangement.

A related rearrangement is observed when an attempt is made to dehydrate the alcohol *t*-butylmethylcarbinol (58). The "normal" dehydration to (59) occurs to a very minor extent (about 3 %); the major product is the rearranged olefin (60):

$$CH_3-\underset{\underset{CH_3}{|}}{\overset{\overset{CH_3}{|}}{C}}-\underset{\underset{}{|}}{\overset{\overset{OH}{|}}{CH}}-CH_3 \xrightarrow{-H_2O} CH_3-\underset{\underset{CH_3}{|}}{\overset{\overset{CH_3}{|}}{C}}-CH=CH_2(\text{little}) + CH_3-\underset{}{\overset{\overset{CH_3}{|}}{C}}=\underset{}{\overset{\overset{CH_3}{|}}{C}}-CH_3(\text{most})$$

$$(58) \qquad\qquad (59) \qquad\qquad (60)$$

The explanation for this rearrangement is the same as that just discussed for the addition of HCl to *t*-butylethylene. In the acid-catalyzed dehydration of (58), the first steps would involve the protonation of the alcohol and formation of the carbonium ion (56):

$$CH_3-\underset{\underset{CH_3}{|}}{\overset{\overset{CH_3}{|}}{C}}-\underset{}{\overset{\overset{OH}{|}}{CH}}-CH_3 + H:A \rightleftharpoons CH_3-\underset{\underset{CH_3}{|}}{\overset{\overset{CH_3}{|}}{C}}-\underset{}{\overset{\overset{OH_2^+}{|}}{CH}}-CH_3 + :A^-$$

$$\downarrow{-H_2O}$$

$$CH_3-\underset{\underset{CH_3}{|}}{\overset{\overset{CH_3}{|}}{C}}-\overset{+}{CH}-CH_3$$

$$(56)$$

The loss of a proton from (56) will give (59), without rearrangement. But the secondary carbonium ion will, as we have seen in the previous case, rearrange to give the more stable tertiary carbonium ion (57). Loss of a proton from (57) will lead to the rearrangement product, the olefin (60).

---

**Exercise 8.** What isomeric olefin, of the same carbon skeleton as (60), would you expect to find along with (59) and (60)? This third olefin is actually found as another (minor) product of the acid-catalyzed dehydration of (58).

---

## 10.15 Stability of carbonium ions

In Chapter 7, before resonance had been considered in detail, some discussion was devoted to the stability of carbonium ions as this related to the $S_N1$ displacement reaction. This discussion should be studied again in the light of our more penetrating analysis of the nature of resonance. Let us reexamine one of the cases treated there. The $S_N1$ reactivity of $\alpha$-chloro ethers in nucleophilic displacement reactions is very high, and can be accounted for by the readiness with which the dissociation (solvolysis) occurs:

$$CH_3OCH_2Cl \rightleftharpoons \{CH_3OCH_2\}^+Cl^-$$

This dissociation is aided by resonance stabilization of the carbonium ion, by participation of the unshared electron pairs on oxygen in the delocalization of the positive charge.

$$\left\{ CH_3-O-\overset{+}{C}H_2 \quad \longleftrightarrow \quad CH_3-\overset{+}{O}=CH_2 \right\}$$

It is significant to note that $\alpha$-halogen-substituted ketones, such as bromo-acetone (61), react very slowly by $S_N1$ displacement mechanisms (although they are highly reactive by the $S_N2$ route). We conclude from this that it is energetically *disadvantageous* for the ionization (62) to occur:

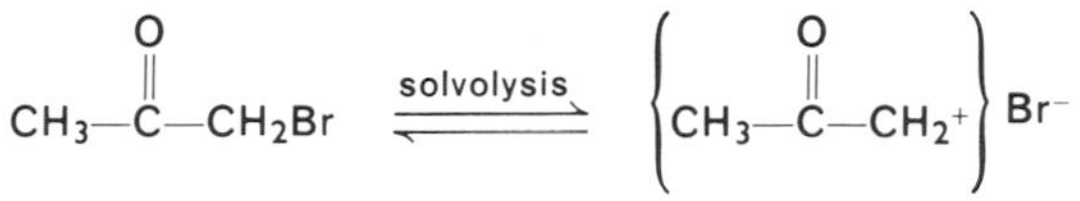

$$CH_3-\overset{\overset{\textstyle O}{\|}}{C}-CH_2Br \quad \underset{\text{solvolysis}}{\rightleftharpoons} \quad \left\{ CH_3-\overset{\overset{\textstyle O}{\|}}{C}-CH_2^+ \right\} Br^- \tag{62}$$

$$(61) \qquad\qquad\qquad\qquad (63)$$

Why is the ion (63) a "poor" carbonium ion? If we recall that the normal state of the carbonyl group (C=O) is that of a resonance hybrid in which one of the contributing structures bears a positive charge on carbon [(13) $\leftrightarrow$ (14), Section 10.6], we can see that the ion (63) contains like charges on adjacent atoms. Thus, the normal resonance stabilization of the carbonyl group in (61) would be diminished in (63), and, consequently, there would be a *loss* of resonance stabilization in the ionization reaction (62). The situation cannot be improved by postulating a contribution such as (64).

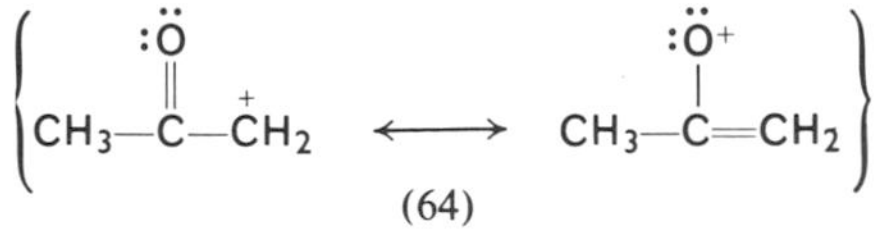

$$\left\{ CH_3-\overset{\overset{\textstyle :\ddot{O}}{\|}}{C}-\overset{+}{C}H_2 \quad \longleftrightarrow \quad CH_3-\overset{\overset{\textstyle :\ddot{O}^+}{|}}{C}=CH_2 \right\}$$

$$(64)$$

for this would be counter to the electronegativity difference between oxygen and carbon and would not be a stabilizing factor (Rule 3).

In general, reactions that would require passage through an intermediate state in which carbonium (positive) ionic character is created on a carbon atom adjacent to a carbonyl group are slow reactions, not aided by resonance.

## 10.16  A graphical representation of resonance stabilization

A graphical illustration of the energy relationships involved in resonance stabilization is given in the following diagrams. In these, $E_1$ and $E_2$ will represent the energies that would be possessed by compounds of the structures shown;* $E_h$ is the energy of the resonance hybrid—i.e., the *experimentally determined* energy of the actual compound. It will be noted that the closer together are $E_1$ and $E_2$, the farther $E_h$ is from the lower of these: i.e., the greater is the stabilization due to resonance. Particular examples are shown, but the energy values are entirely arbitrary (and are not related between the separate diagrams).

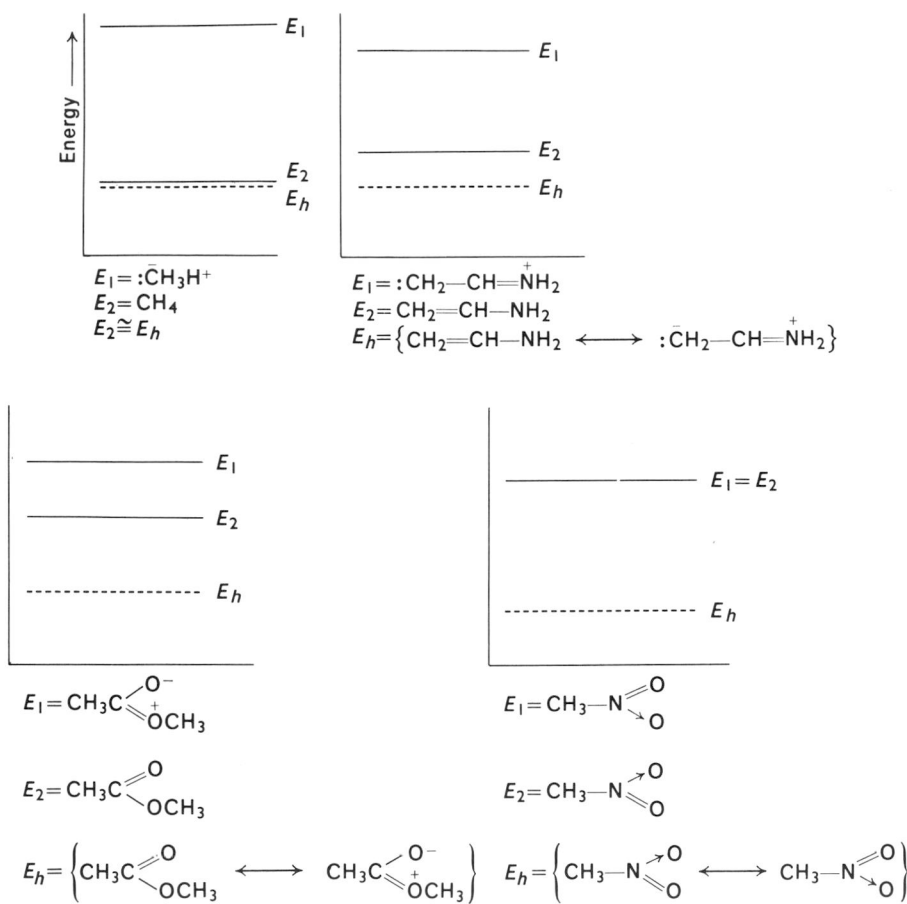

## 10.17  Resonance and bond lengths

If a bond in a compound that is a resonance hybrid of two contributing structures is neither a single nor a double bond, but a hybrid bond, it would be expected to have characteristics different from those of a true single or a true double bond.

---

\* The individual contributing structures $E_1$ and $E_2$ are hypothetical, and should be regarded as the structures of substances which, if they could somehow be prepared, would return by loss of energy to the lower energy level and structure of the resonance hybrid.

**Table 10-1.**   *Carbon-carbon single-bond lengths*

| Substance | Measured C—C Bond Length (Å) |
|---|---|
| ethane | 1.54 |
| ethanol | 1.55 |
| ethyl mercaptan | 1.54 |
| isopropyl bromide | 1.54 |
| glycerol | 1.54 |
| hexane | 1.54 |
| cyclohexane | 1.54 |

One readily measurable characteristic of a bond between two atoms is its length. Measurements of bond lengths can be made by such physical means as X-ray or electron diffraction, and accurate values are available.

Many measurements on saturated molecules containing carbon-carbon single bonds have shown that the C—C single bond is very close to 1.54 Å in length (Table 10-1).

Similar measurements on substances containing other kinds of bonds have given values as shown in Table 10-2.

These values are representative of those that have been obtained by measurements on a great many compounds. In all but acetone, the bonds in this table are present in compounds for each of which a single valence bond formula is an adequate representation of its actual structure.

**Table 10-2.**   *Typical bond lengths in organic compounds*

| Substance | Bond Type | Bond Length (Å) |
|---|---|---|
| ethylene | C=C | $1.34 \pm 0.02$ |
| acetone | C=O | $1.22 \pm 0.03$ |
| dimethyl ether | C—O | $1.43 \pm 0.03$ |
| ethanol | C—O | $1.43 \pm 0.02$ |
| methyl chloride | C—Cl | $1.77 \pm 0.02$ |
| ethylene dichloride | C—Cl | $1.78 \pm 0.01$ |
| methyl bromide | C—Br | $1.91 \pm 0.06$ |
| ethyl bromide | C—Br | $1.91 \pm 0.02$ |
| acetylene | C≡C | $1.20 \pm 0.01$ |
| methylacetylene | C≡C | $1.21 \pm 0.03$ |

For substances that are resonance hybrids, the measured length of a given bond is usually different from that predicted on the basis of any one of the contributing structures. Chloroethylene is found by measurement to have a carbon-chlorine distance of 1.69 Å. This is shorter than the C—Cl bond in such compounds as methyl chloride and ethylene dichloride (1.77 Å), an indication that in chloroethylene the C—Cl bond has some *double-bond character*. The representation of chloroethylene as the resonance hybrid accounts for this:

$$\{CH_2{=}CH{-}\overset{..}{\underset{..}{Cl}}: \longleftrightarrow \overset{..}{CH_2}{-}CH{=}\overset{+}{\underset{..}{Cl}}:\}$$

Benzene is found to have one kind of carbon-carbon bond, of 1.39 Å. intermediate between the C—C single bond (1.54 Å) and the C=C double bond (1.34 Å).

Carbon dioxide possesses a carbon-oxygen bond length of 1.15 Å, indicating the structure

$$\{O{=}C{=}O \longleftrightarrow \overset{+}{O}{\equiv}C{-}\overset{-}{O} \longleftrightarrow \overset{-}{O}{-}C{\equiv}\overset{+}{O}$$

the partial triple-bond character of the C—O bond causing an effective shortening of the bond from the expected (for C=O) 1.22 Å.

---

**Exercise 9.** In cyclo-octatetraene, two kinds of carbon-carbon bonds are found: one with a length of 1.50 Å, the other with a length of 1.35 Å. What information does this finding give us about the structure of this compound, in comparison with benzene? The molecule is not planar.

**Exercise 10.** Biacetyl (2,3-butanedione) is found to have a carbon-oxygen bond length of 1.20 Å, a $CH_3$—C length of 1.54 Å, and a C—C length (carbon-carbon) of 1.47 Å. From these

$$\qquad\qquad \underset{O}{\overset{\|}{\,}}\;\underset{O}{\overset{\|}{\,}}$$

values, and with the aid of Tables 10-2 and 10-3, write a probable important contributing structure.

---

The $CH_3$—C bond distances shown in Table 10-3 are given for a number of alkynes.

| **Table 10-3.** *Carbon-carbon single-bond lengths in alkynes* | | |
|---|---|---|
| Alkyne | Bond Type | Bond Length (Å) |
| $CH_3C{\equiv}CH$ | $CH_3$—C | 1.46 |
| $CH_3C{\equiv}CCH_3$ | $CH_3$—C | 1.47 |
| $CH_3C{\equiv}CCH_2CH_3$ | $CH_3$—C | 1.47, 1.54 |
| $CH_3C{\equiv}C{-}C{\equiv}CCH_3$ | $CH_3$—C | 1.47 |

The shortening of the bond distance from 1.54 Å (ordinary single bond) to the observed 1.47 Å indicates that the $CH_3$—C bond has some double-bond character, and thus that structures such as

$$\{\overset{+}{H}CH_2=C=\overset{..}{\overset{-}{C}}H \longleftrightarrow CH_3-C\equiv CH\}$$

contribute significantly. This kind of participation of the $\sigma$-bond electrons of the C—H bond in overlap with the $\pi$-orbitals of the triple bond is known as hyperconjugation. Hyperconjugation is also encountered in other situations in which $\sigma$-bond electrons can be utilized by an adjacent electron-deficient atom. The resonance stabilization of the *tert*-butyl carbonium ion is a case in point.

## Exercises

(Exercises 1–10 will be found within the text.)

**11** Write two permissible electronic structures for each of the following: *(a)* methyl vinyl ether (example: $CH_2=CH-O-CH_3 \leftrightarrow \overset{-}{C}H_2-CH=\overset{+}{O}-CH_3$); *(b)* methyl formate; *(c)* boron tribromide; *(d)* acetate ion; *(e)* acetamide; *(f)* acrolein (propenal); *(g)* tert-butyl carbonium ion; *(h)* nitrate ion; *(i)* nitrous acid; *(j)* acetaldehyde.

**12** Guanidine is a strong base, in part because of the high degree of resonance stabilization in the symmetrical guanidinium ion, $\overset{+}{N}H_2=C(NH_2)NH_2$. Write the contributing forms to the resonance hybrid, showing all of the unshared electron pairs in the structures.

**13** Given the data in Table 10-2 for interatomic distances (bond lengths), account for observed bond lengths shown for the following compounds.

*(a)* $HC\equiv\overset{a}{C}-\overset{b}{C}\equiv C-CH_3$    C—C (a) 1.38 Å;    C—C (b) 1.46 Å

*(b)* $CH_3-\overset{O}{\overset{\|}{C}}-\overset{a}{O}-\overset{b}{C}H_3$    C—O (a) 1.36 Å;    O—C (b) 1.46 Å

*(c)* $CH_3CN$    C—C 1.49 Å

*(d)* $CH_2=\overset{a}{C}H-\overset{b}{C}HO$    C—C (a) 1.36 Å;    C—C (b) 1.46 Å

*(e)* $CH_2=\overset{a}{C}H-CH=CH_2$    C—C (a) 1.46 Å

**14** Can you account for the greater acidity of the hydroxyl group of $CH_3-\overset{\overset{\displaystyle OH}{|}}{C}=CHCOOEt$ than of that of $CH_3\overset{\overset{\displaystyle OH}{|}}{C}HCH_2COOEt$? (Note: examine these two structures carefully so that the difference between them is clear. The first is the enol form of ethyl acetoacetate, the second is ethyl $\beta$-hydroxybutyrate.)

**15** Nitromethane dissolves in alkali with the formation of a sodium salt, $NaCH_2NO_2$. Write the electronic structure of this salt, indicating the resonance in the anion.

**16** State in *words* what is meant by the terms *(a)* resonance hybrid; *(b)* resonance stabilization; *(c)* contributing structures; *(d)* resonance energy.

**17** Make a sketch showing the approximate spatial relationship between the C, O, H, and N atoms in acetamide, $CH_3C{\displaystyle \begin{smallmatrix} \nearrow O \\ \searrow NH_2 \end{smallmatrix}}$ , given the information that the C—N bond distance is about 0.15 Å less than the C—N bond distance in methylamine.

**18** The hydrogen atoms of the methyl group in acetaldehyde, $CH_3C{\displaystyle \begin{smallmatrix} \nearrow O \\ \searrow H \end{smallmatrix}}$ , have a readily recognizable degree of acidic character. This can be attributed to the fact that the ion $^-:CH_2CHO$ derived from acetaldehyde is resonance-stabilized. Show this resonance by drawing the electronic structures involved. What prediction can you make regarding the acidity of the methyl-group hydrogen atoms in crotonaldehyde, $CH_3CH{=}CHCHO$? Explain.

**19** Explain the difference between the effect of resonance stabilization upon reaction rates and equilibrium.

## Chapter eleven

/ **Absorption Spectra**

**of Organic Compounds**

## 11.1 Energy absorption by organic molecules

*Electronic energy levels.* Molecules have available many levels of energy, into which they can be placed by the absorption and emission of radiant energy. Under ordinary conditions of temperature and in the absence of exciting influences, molecules are in their *ground state*. In this condition the electrons in the bonding orbitals and in the nonbonding orbitals (such as those occupied by unshared electrons in the valence shells) are in their most stable condition, distributed about the respective atomic nuclei according to the relative attraction of the nuclei for the electrons and the repulsive forces between the electrons themselves. The electrons are in ground state orbitals. Electrons can be caused to occupy other available molecular orbitals by the absorption of energy by the molecule. These orbitals are less stable than ground state orbitals; they are at higher energy levels. A molecule that is in this unstable condition as a result of the absorption of energy is in an *excited state*.

It is one of the principles of quantum mechanics that atomic and molecular energy levels are quantized; that is, they have certain discrete values, and the absorption of energy by a molecule occurs only when the magnitude of the incident energy—for example, in a beam of light—corresponds to the difference between two discrete energy levels. This absorption of light energy can be recognized experimentally by observing the diminution in the intensity of the light when it passes through the sample of the compound, and the measurement of this diminution of intensity can be carried out with the use of an instrument called a *spectrometer*.

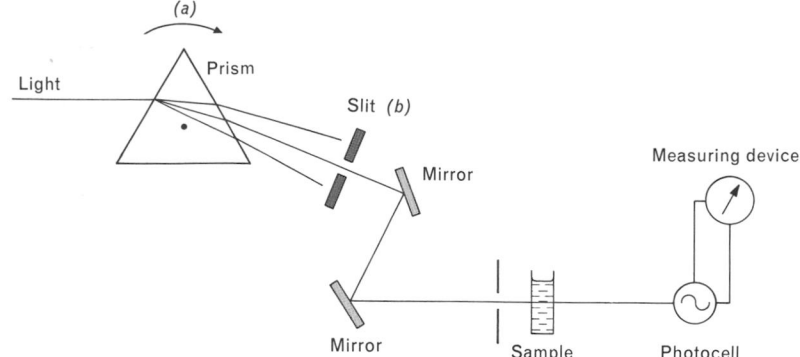

**Figure 11–1.** A schematic diagram of the essential elements of a spectrophotometer for the measurement of absorption spectra.

    The diagram in Figure 11-1 represents a simplified experimental system for measuring energy absorption at various wavelengths of light.

    When the measurement of the light intensity is made with a photocell, the instrument is called a *spectrophotometer*. Rotation of the prism ($A$), which is coupled with a calibrated wavelength dial, causes a continuous change in the frequency of the light that passes through the sample. When the frequency of the light selected by the slit *(B)* is such that its energy corresponds to a transition of the molecule from the ground state to a higher energy level, the light will be absorbed by the compound and the photocell reading will fall. Let us suppose that the compound used absorbs light of wavelength 3,000 Å, which corresponds to an energy of 96 kcal/mole. The plot of light absorbed versus the wavelength of light incident upon the sample will be like that of Figure 11-2a. This graph is an *absorption spectrum*. It is highly idealized, for ordinary molecules are complex and exist in

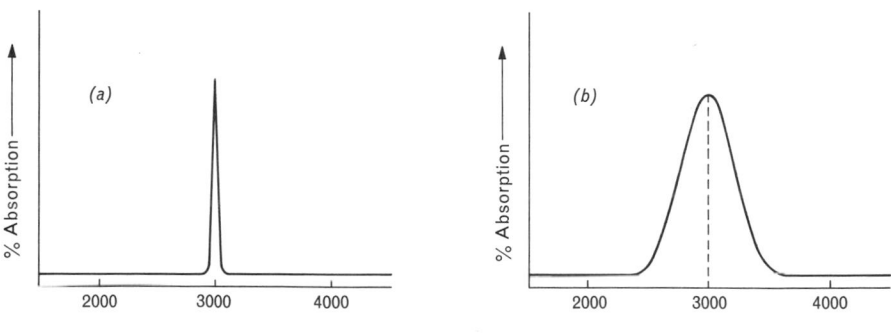

Wavelength, Ångstrom units

**Figure 11–2.** Absorption spectrum of a compound with an absorption maximum at 3000 Å: *(a)* Idealized spectrum. *(b)* What an actual spectrum would be like.

various vibrational and configurational states that permit the electronic transitions to occur over a small range of frequency. An actual absorption spectrum of a molecule with an *absorption maximum* at 3,000 Å would ordinarily look more like the one shown in Figure 11-2b.

An absorption spectrum in the spectral region chosen for this example is an *ultra-violet absorption spectrum.*

*Vibrational energy levels.* The atoms in molecules occupy average positions with respect to one another. They do undergo small motions in which their relative positions change; these motions have their origin in the repulsive and attractive forces between atomic nuclei and the electrons involved in the bonds between the atoms. These modes of vibration have periods that are characteristic of the kinds of bonds and the kinds of atoms joined by the bonds. It would be expected, and it is found, that each kind of bond—a paraffinic C—H bond, a C—C bond, a C=O bond, an olefinic C—H bond, and so on—possesses a vibrational mode that is approximately the same in one molecule as in another.

The vibrational modes that we shall be most concerned with are *(a)* stretching of bonds, a back-and-forth motion of the atoms joined by the bond, and *(b)* bending of bonds, a wagging of the atoms with respect to one another. The following are some examples:

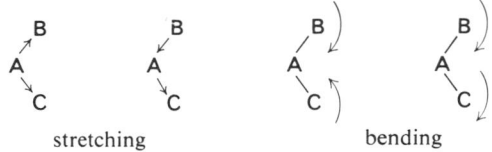

stretching          bending

Since the vibrational modes arise by interplay of forces between specific atoms joined by specific kinds of bonds, they possess definite frequencies. They can therefore absorb light of definite frequencies, the absorption of light energy resulting in increased oscillation of the atoms about their average positions. The absorption of light of definite frequencies gives rise to an absorption spectrum, which is a plot of absorption (or transmission) of light versus wavelength.

Vibrational transitions require much less energy than electronic transitions, and are caused to occur by light of longer wavelength than is required for the latter (see the following discussion on ultraviolet and infrared spectral regions). The characteristic wavelengths lie in a region of the electromagnetic spectrum that is called the infrared region, and the absorption spectra are *infrared absorption spectra.*

*Ultraviolet (UV) and infrared (IR) spectral regions.* The electronic transitions of greatest use in organic chemistry involve energy changes of the order of 30 to 150 kcal/mole. Vibrational transitions require much smaller increments of energy,

**Table 11-1.** *Some values of the magnitude of energy values associated with radiation of various wavelengths*

| Wavelength | $\Delta E$, kcal/mole |
|---|---|
| UV { 2000 Å | 143 |
| Visible { 4000 Å | 72 |
| 8000 Å | 36 |
| IR { 5 $\mu$ | 6 |
| 15 $\mu$ | 2 |

and take place with the absorption of the order of 5 to 20 kcal/mole. These energies can be expressed in terms of the wavelength of light by means of the relationship

$$\Delta E = h\nu = \frac{hc}{\lambda}$$

$\Delta E$ = energy transition on light absorption

$h$ = Planck's constant

$\nu$ = frequency, cycles/sec

$c$ = velocity of light = $3 \times 10^{10}$ cm/sec

$\lambda$ = wavelength, cm

Some values of the magnitude of the energy associated with radiation of various wavelengths are shown in Table 11-1.

## 11.2 The electromagnetic spectrum

It is fortunate that the energies required for most electronic and vibrational transitions are provided by light of wavelengths that lie within limits that are experimentally easily accessible. These wavelengths are between about 2,000 Å ($=200$ millimicrons$=2 \times 10^{-5}$ cm) and about 20 $\mu$ (20 microns). That part of the spectrum lying between about 400 and 750 m$\mu$ is the visible region; light of wavelengths between these limits is colored, and compounds that absorb light within this region are colored compounds. The relationship between the wavelength of light absorbed by a compound and the visible color of the compound is roughly as shown in Table 11-2.

The observation and measurement of absorption spectra are not limited by the relatively narrow spectral region to which the human eye responds, for instruments are available that permit the accurate and automatic determination of absorption spectra from the far UV (wavelengths below 200 m$\mu$) to the far IR, and,

**Table 11-2.** *The relationship between the wavelength of light absorbed by a compound and the visible color of the compound*

| Absorption (m$\mu$) at | Light Absorbed | Visible Color |
| --- | --- | --- |
| 400 | violet | yellow |
| 450 | blue | orange |
| 500 | blue-green | red |
| 530 | yellow-green | violet |
| 550 | yellow | blue |
| 600 | orange-red | green-blue |
| 700 | red | green |

indeed, into regions of electromagnetic radiation that far exceed the limits of what is ordinarily regarded as "light". The commonly used spectrophotometers are of two kinds: UV-visible spectrophotometers, and IR spectrophotometers. Since the components of optical systems operating in the UV range are quite different from those operating in the IR, it is customary to use separate instruments for the two ranges. UV instruments use quartz in their optical components; instruments suitable for the visible region only can use glass (although in the UV-visible instruments quartz is used for the whole range); and IR spectrophotometers use clear sodium chloride prisms and cells (or certain other salts). Diffraction gratings are commonly used instead of prisms for both spectral ranges.

## 11.3 Quantitative expression of spectral measurements

The fraction by which the intensity of a beam of incident light is reduced in passing through an absorbing substance depends upon a number of factors; one factor is the molecular structure of the absorbing compound, and one is the number of molecules of the absorbent in the light path. The number of molecules depends upon the concentration of the compound, $c$, and the length of the light path (the thickness of the sample cell), $l$. The relationship known as the Beer-Lambert Law is the following

$$\log(I_0/I) = kcl$$

where

$I_0$ = intensity of incident light at a given wavelength

$I$ = intensity of transmitted light

$c$ = concentration

$l$ = path length

$k$ = an absorption coefficient, characteristic of the absorbing substance

These quantities are ordinarily expressed in the units: $c = $ g moles/liter, and $l = $ cm, in which case $k$ is the *molar extinction coefficient*, designated as $\epsilon$. Thus, the usual expression is $\log(I_0/I) = $ optical density $= \epsilon$ $cl$. Since *optical density* is the form in which the absorption of light by the sample is provided by the measuring system of most instruments, the molar extinction coefficient is easily calculated by the equation

$$\epsilon = \text{O.D.}/cl$$

For various reasons beyond the scope of this discussion, extinction coefficients are not easily measured with IR instruments, and IR spectra are usually recorded as a graph of transmission versus wavelength.

UV spectra are conveniently recorded in the form of a plot of $\epsilon$ against wavelength (in $m\mu$ or Å). Since $\epsilon$ values range in practice from as low as 10 to as high as 100,000, it is convenient to use $\log \epsilon$ as the abscissa of the UV plot.

Most spectra (UV, visible and IR) are measured with the use of solutions in a solvent that is selected because of its transparency over the range of interest. For UV spectra, ethanol, methanol, hexane, and water are commonly used. IR spectra may be determined with the use of solutions in chloroform, carbon disulfide, or carbon tetrachloride; thin films of the pure compound; or uniform suspensions of the finely ground sample in an inert medium such as solid potassium bromide or mineral oil.

### 11.4   UV absorption and molecular structure

*Nonconjugated systems.*    The transition of electrons from the ground state to an excited state requires increments of energy that vary over a wide range, depending upon molecular structure. Electrons in sigma ($\sigma$) bonds are tightly bound between the atomic nuclei and can be excited to higher ($\sigma^*$) energy states ($\sigma \rightarrow \sigma^*$ transitions) only by very short wavelength ("far") UV radiation. Methane and ethane show $\sigma \rightarrow \sigma^*$ transition bands at about 130 $m\mu$. Light of this wavelength is very strongly absorbed by air; consequently, instruments suitable for measurements in this region have optical systems that operate in vacuum, and because of these special instrumental requirements the region below about 190 $m\mu$ is not ordinarily used.

The $\pi$ molecular orbital is excited by light of lower energy than that required for the $\sigma \rightarrow \sigma^*$ transitions, and many $\pi \rightarrow \pi^*$ transitions occur in regions of the spectrum that are accessible with some, but not all, instruments. Ethylene (vapor) shows an absorption maximum at 175 $m\mu$, and tetraalkylethylenes may show absorption up to about 205 $m\mu$. The energy of the $\pi \rightarrow \pi^*$ transition is not greatly different for C=C, C≡C and C=O. Acetone absorbs at 189 $m\mu$, 2-octyne at 178 $m\mu$. In general, the spectra of compounds containing simple carbon-carbon multiple bonds are not very useful and because they occur near the limit of the practical range of most instruments they are seldom measured.

Acetone and other simple aliphatic aldehydes and ketones show a second, less intense, absorption maximum at about 280 m$\mu$. This is attributed to an excitation of the nonbonding (unshared) electrons on the carbonyl oxygen atom to a $\pi^*$ orbital, and is called an $n \rightarrow \pi^*$ transition ($n$ refers to "nonbonding electron"). The extinction coefficients of $n \rightarrow \pi^*$ transitions are usually very low (10 to 100, compared with about 10,000 for the $\pi \rightarrow \pi^*$ bands), but since they are in a readily accessible region of the spectral range of the usual instruments, they are of considerable diagnostic value.

***Conjugated systems.***   When double bonds are conjugated, as in C=C—C=C and C=C—C=O, $\pi \rightarrow \pi^*$ transitions occur at lower energies and thus at longer wavelengths. Even the simplest systems, 1,3-butadiene and acrolein (CH$_2$=CH—CHO), have absorption maxima at 217 and 208 m$\mu$, respectively, and with increasing substitution in the molecule these wavelengths are raised to the region of 220 to 250 m$\mu$, which is readily accessible to the usual instruments.

There is no way in which conventional valence-bond (Lewis) structures can be written for excited states of complex organic molecules that have undergone one or another of the transitions described in the foregoing discussion, for the exact electronic structures of the excited molecules are not always known with certainty. There is, however, one generalization that will become apparent as we study the relationship between structure and light absorption: the greater the degree of conjugated unsaturation in the molecule, the smaller the energy difference between the ground states and excited states, and the lower the energy (the longer the wavelength of light) required for excitation. The data in Table 11-3 show that as conjugation increases, the wavelength of the absorption maximum increases.†

It can be seen from Table 11-3 that there are large differences between the compounds containing none, one, two, three, and more, conjugated double bonds, so that the UV absorption spectrum is a valuable diagnostic method for providing a clue to the nature of the chromophore†† present in an unknown compound.

It is not our intention to dwell at length upon the many aspects of UV spectroscopy. The discussion in this section is intended only to give the reader an introduction to its physical and chemical basis and to show how it can be applied to the solution of structural problems. It is seldom that a UV absorption spectrum can do more than indicate the structural unit, or chromophore, of a molecule that gives rise to the characteristic absorption peaks; only by showing the identity, by direct

---

† These figures refer to $\pi \rightarrow \pi^*$ transitions. The carbonyl compounds, which possess nonbonding electrons capable of excitation in a process that is described as an $n \rightarrow \pi^*$ transition, show additional absorption at longer wavelengths, but of low intensity. An example of the UV absorption spectrum of an $\alpha,\beta$-unsaturated ketone in which the long-wavelength absorption can be seen, is found in Figure 35-1.

†† Chromophore, meaning "color bearing," is the term used for the grouping that gives rise to the characteristic light absorption of a compound. Even though crotonaldehyde, for instance, is colorless, the system C=C—C=O is its "chromophore."

**Table 11-3.** *Principal absorption maxima for some conjugated and unconjugated carbonyl compounds and olefinic compounds*

| Compound | Name | $\lambda$ (Max.), m$\mu$ | $\epsilon$ (Max.) |
|---|---|---|---|
| **Carbonyl** | | | |
| $(CH_3)_2C=O$ | acetone | 189 | 900 |
| $CH_3CH=CHCHO$ | crotonaldehyde | 217 | 15,000 |
| $CH_3CH=CH-CH=CH-CHO$ | 2,4-hexadienal | 270 | 27,000 |
| $CH_3(CH=CH)_3CHO$ | 2,4,6-octatrienal | 312 | 40,000 |
| **Olefinic** | | | |
| $CH_2=CH_2$ | ethylene | 175 | 15,000 |
| $CH_2=CH-CH=CH_2$ | 1,3-butadiene | 217 | 21,000 |
| $CH_2=CH-CH=CH-CH=CH_2$ | 1,3,5-hexatriene | 258 | 35,000 |
| $CH_2=CH-CH=CH-CH=CH-CH=CH-CH=CH_2$ | 1,3,5,7,9-deca-pentaene | 334 | 125,000 |
| $\beta$-carotene (p. 226) | (11 conjugated double bonds) | 465 | 125,000 |

matching, of the UV spectra of an unknown with that of a known compound can one obtain evidence for a complete structure. It should be apparent from the foregoing discussion, however, that a valuable use of UV spectroscopy is to provide answers to specific questions about structure. For instance, if the structure of a compound could, from other evidence, be *(a)*, *(b)*, or *(c)*, the UV spectrum would

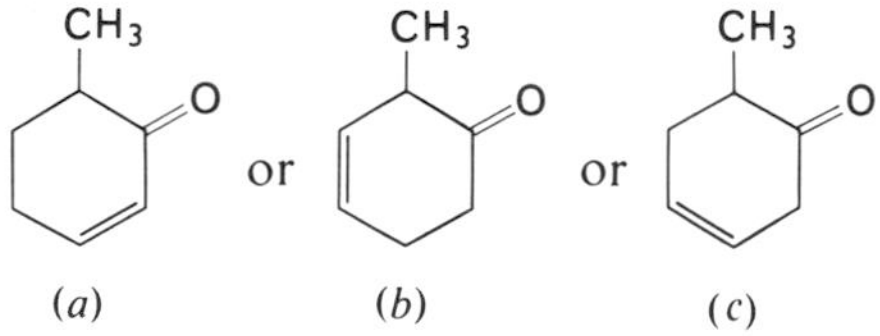

*(a)*       *(b)*       *(c)*

give an unequivocal answer if it showed an intense ($\epsilon \sim 10,000$) absorption maximum at about 225 m$\mu$. Only the $\alpha,\beta$-unsaturated ketone *(a)* could have this spectrum. If, on the other hand, no such maximum appeared in the UV spectrum, the compound would not be *(a)*; but this would leave unresolved the question of whether the compound had structure *(b)* or *(c)*.

Specific examples of UV spectra will be given in later chapters when the information they contain serves to illuminate the discussion.

### 11.5 IR absorption

***General.*** The IR spectra of organic compounds can yield a great deal of information on details of their structures, for nearly all of the important functional groups, alone and in combination (as by conjugation), show specific and highly characteristic absorption in the IR region.

IR spectroscopy, as it is applied to the solution of structural problems in organic chemistry, is quite empirical. By this is meant that the interpretation of IR spectra depends largely upon one's recognition of the characteristic absorption peaks of the various functional groups. Effective use of IR spectra depends greatly upon experience and practice, and familiarity with the spectra of compounds of a variety of structural types. For this reason we shall present only an introduction to the interpretation of IR spectra, and a brief description of how this technique can aid in analyzing organic structures. It will be seen also that the IR spectral characteristics of certain groups can add greatly to an understanding of their structures and can reinforce interpretations of their behavior that were arrived at in quite other ways.

IR spectra of organic molecules are usually quite complex. Since the kinds of possible vibrational transitions are numerous, and in a polyatomic molecule there are many individual vibrational modes, an IR spectrum ordinarily contains a great many individual peaks.* For example, in the spectrum of the relatively simple compound, $n$-butyl vinyl ether, $CH_3CH_2CH_2CH_2OCH{=}CH_2$, measured with a commercial IR spectrophotometer, there can be discerned more than 25 separate absorption peaks in the range 3 to 15 $\mu$.

*An IR spectrum is unique to a given compound*; it would be highly unlikely that two different compounds would have IR spectra identical in all details. For this reason, one of the important uses of IR spectroscopy is the identification of organic compounds by matching the spectrum of the known compound with that of the compound under study. If the compounds are the same (and equally pure), the spectra will, of course, be identical in every detail. If the compounds are not the same the noncorrespondence in at least some details of their spectra will reveal their nonidentity.

The wavelength scale of IR spectra can be in units of *wavelength*, in microns ($\mu$), or in *wave number*, which is the reciprocal of the wavelength in centimeters and thus has the units cm$^{-1}$.

wave number $= 1/\lambda$

for $\lambda = 5\,\mu = 5 \times 10^{-4}$ cm, wave number $= 2000$ cm$^{-1}$.

Both units are used, and the charts used for recording IR spectra normally carry both scales. We shall be concerned principally with the range 2.5 $\mu$ (4000 cm$^{-1}$) to 15 $\mu$ (667 cm$^{-1}$).

***Characteristic regions of the spectrum.*** In Figure 11-3 is shown the IR spectral range on which are noted some of the regions at which characteristic peaks are found. Only a few of the fundamental bond types found in compounds containing only C, H, and O are shown; there are many additional characteristic stretching and bending frequencies for N—H, C≡N, N=O, S—H, and other structural elements.

---

\* The actual number will, of course, depend upon the capacity of the instrument used to resolve absorption bands lying close together into a group of individual peaks.

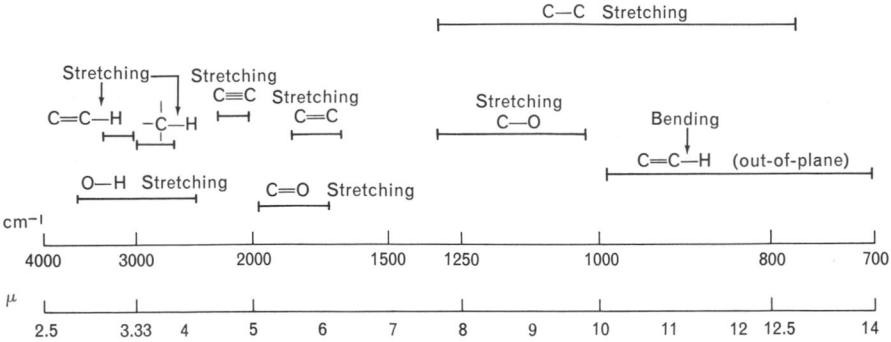

**Figure 11–3.** Spectral regions characteristic of various stretching and bonding vibrational nodes of some common structural units.

Several points are to be noted in this chart. The stretching frequency of C≡C is higher than that of C=C; that of C=O is higher than C—O. The shorter, stronger bonds require higher energy for vibrational (stretching) transitions. The same conclusion is illustrated by the C=O stretching frequency of saturated and unsaturated (conjugated) aldehydes (Table 11-4). The effects of conjugation upon the C=O stretching frequency can be described in terms of the resonance concept. In all of the carbonyl groups in the group of aldehydes in Table 11-4, the contribution

$$\text{>C=O} \longleftrightarrow \text{>}\overset{+}{\text{C}}\text{—}\overset{-}{\text{O}}$$

describes the charge delocalization in the double bond. When carbon-carbon double bonds are conjugated with C=O, the contribution of additional forms

$$\text{C=C—C=O} \longleftrightarrow \text{C=C—}\overset{+}{\text{C}}\text{—O}^- \longleftrightarrow \overset{+}{\text{C}}\text{—C=C—O}^-$$

provides for further lengthening of the C=O bond by augmenting the relative contribution of C—O forms to the hybrid. Thus, the C=O bond in the conjugated

**Table 11-4.** *The infrared C=O stretching frequency\* of some aldehydes*

| Compound | C=O Stretching Frequency ($cm^{-1}$) |
|---|---|
| acetaldehyde | 1733 (in $CCl_4$) |
| acrolein | 1704 (in $CCl_4$) |
| 2,4-hexadienal | 1677 (in $CHCl_3$) |
| 2,4,6-octatrienal | 1664 (in $CHCl_3$) |

\* Higher frequencies correspond to higher wave numbers.

aldehydes has more *single-bond character* than in acetaldehyde; and the single-bond character is enhanced with increasing conjugation. The result is that the stretching frequency decreases as conjugation increases; the wavelength of absorption increases.

Another illustration of the effect on the infrared absorption frequency by resonance-caused bond lengthening is found in the pair of compounds

$CH_2$=CH—$COCH_3$          C=O stretching 1689 cm⁻¹ (CCl₄)

$Et_2N$—CH=CH—$COCH_2CH_3$     C=O stretching 1664 cm⁻¹ (CCl₄)

In the diethylamino compound, the ability of the $R_2N$— grouping, with its un-shared electrons on nitrogen, to contribute in the following way

$$\left\{ Et_2\overset{\cdot\cdot}{N}-CH=CH-\overset{R}{\underset{|}{C}}=O \longleftrightarrow Et_2\overset{+}{N}=CH-\overset{R}{\underset{|}{C}}=C-O^- \right\}$$

causes an appreciable increase of the C=O bond length, and a marked decrease in the carbon-oxygen stretching frequency.

Additional examples of these effects upon the carbonyl absorption bands in the IR spectra of aromatic ketones (benzene derivatives) will be described in a later chapter.

***Interpretation of IR spectra.*** In Figures 11-4 to 11-8 are shown some representative IR spectra of a number of organic compounds. The purpose of these spectra is to show how certain functional groups can be recognized and how such information can be used to arrive at, or to confirm, structure. The compounds selected are rather complex, but this selection has been made quite deliberately, for it is important to recognize that *characteristic molecular groupings* can be recognized in complex spectra, even though a great many of the absorption peaks cannot be usefully or confidently interpreted. The multiplicity of separate peaks in the region between about 1500 and 750 cm⁻¹ (7 to 13 $\mu$) can be regarded as a pattern that is

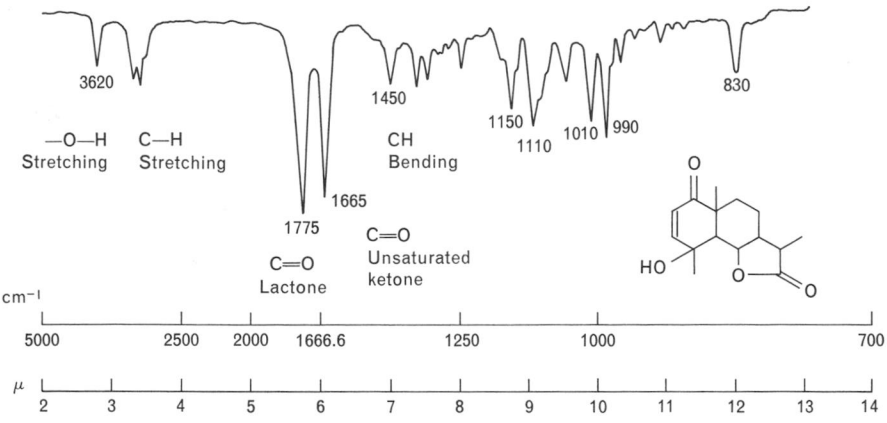

**Figure 11–4a.**

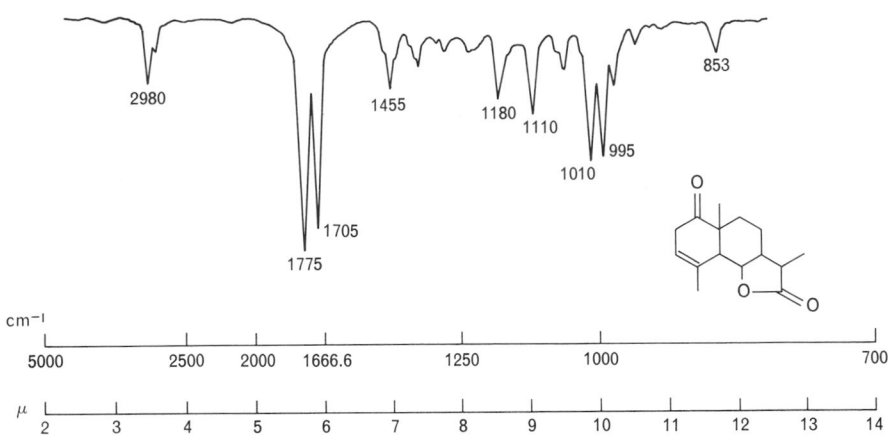

**Figure 11–4b.**

unique for a given compound, as a fingerprint identifies a person. This region is often called the "fingerprint" region, and its greatest value is for establishing identity by direct matching of spectra but without further detailed interpretation.

The following description of the significant features of these spectra will show how IR spectral data can be used in the interpretation of structure.

The spectra in Figure 11-4 are those of the compounds whose structures are shown on the figures. Compound 4a contains a hydroxyl group (O—H stretching at 3620 cm⁻¹), a conjugated ketone carbonyl group (C=O stretching at 1665 cm⁻¹), and a five-membered ring containing a carbonyl group in an ester grouping (a cyclic ester of this kind is called a *lactone*) (C=O stretching at 1775 cm⁻¹). The group of peaks near 1400 cm⁻¹ (7 μ) represent C—H bending in the various

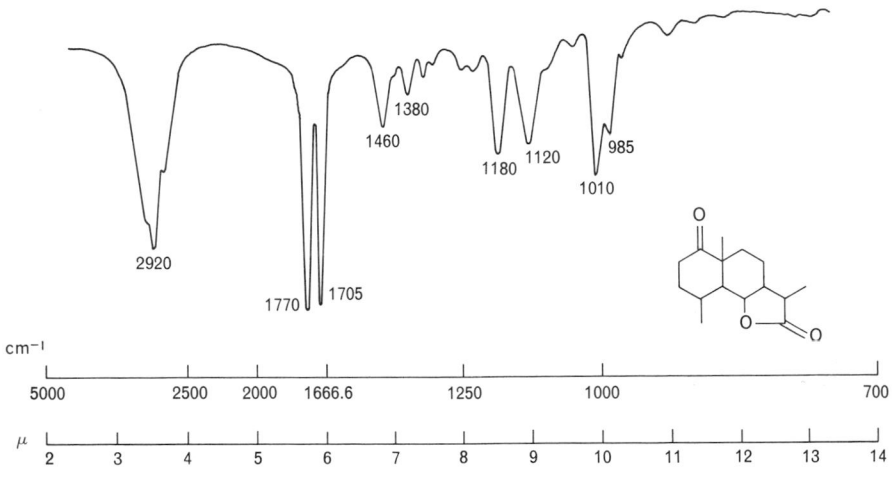

**Figure 11–5a.**

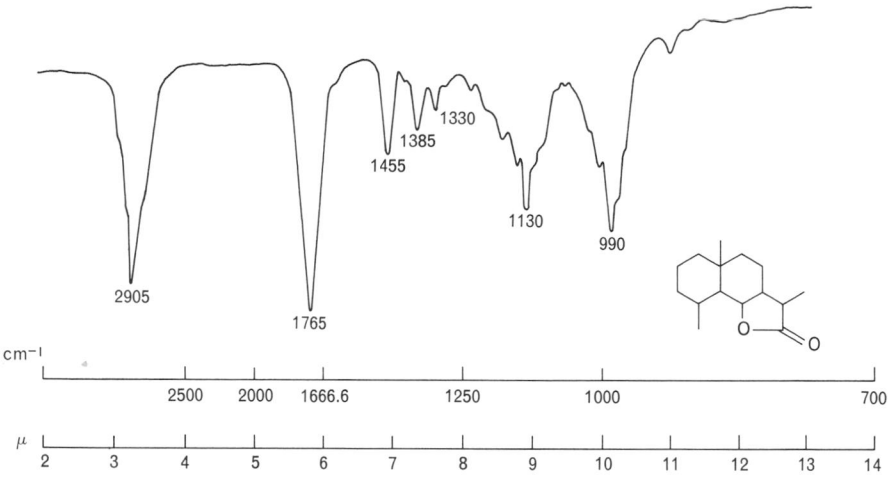

**Figure 11–5***b.*

—CH$_3$ and —CH$_2$— groups of the molecule. The fingerprint region contains peaks for various other vibrational transitions, but will not be discussed in detail.

Compound 4*b* can be seen to possess many of the same structural features as 4*a*, but the carbonyl group and carbon-carbon double bond *are not conjugated*, and the ketone C=O stetching frequency is 1705 cm$^{-1}$. The lactone C=O is seen at the same position as in 4*a*; clearly, the alteration in the structure of the left-hand ring has no appreciable effect upon the rest of the molecule. The C—H bending peaks are again seen near 7 $\mu$, and the fingerprint region shows some differences from that of 4*a*. The small peak at about 850 cm$^{-1}$ in 4*b* is due to out-of-plane bending of the =C—H hydrogen atom.

Figure 11-5 shows the spectra of two further modifications of the structures of Figure 11-4. Compound 5*a*, it will be seen, resembles 4*b* very closely in the positions of the C=O stretching absorptions. This shows that the nonconjugated double bond in 4*b* has no appreciable effect upon the absorption frequency of the C=O group, for this is 1705 cm$^{-1}$ in both 4*b* and 5*a*. The C—H bending vibration at 853 cm$^{-1}$ in 4*b* is missing in 5*a*, which contains no carbon-carbon double bond.

Compound 5*b* lacks the ketone carbonyl group, but the lactone carbonyl group is seen in the peak at 1765 cm$^{-1}$. This frequency is somewhat lower than might be expected (it is at 1770 to 1775 cm$^{-1}$ in 4*a*, 4*b* and 5*a*), but such small differences are sometimes observed. The remainder of the spectrum of 5*b* contains no other noteworthy features.

The three spectra shown in Figure 11-6 are those of the sterol, cholesterol, and two of its derivatives. The spectrum of 6*a*, cholesterol itself, shows the hydroxyl group absorption at about 3500 cm$^{-1}$ and a peak near 840 cm$^{-1}$, due to the out-of-plane bending vibration of the hydrogen attached to the C=C grouping. The characteristic C—H bending frequencies near 7 $\mu$ are again seen (compare Figures 11-4 and 11-5).

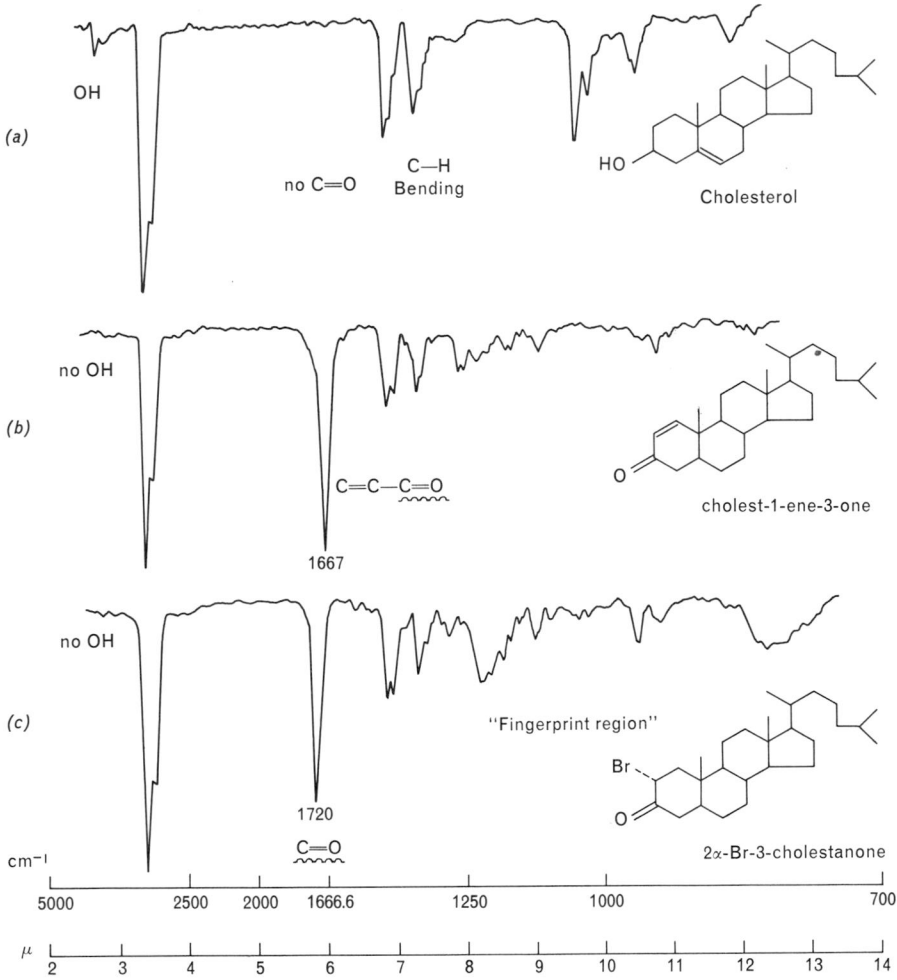

**Figure 11–6.**

Compound *6b* contains the system C=C—C=O, and shows the C=O stretching at 1667 cm⁻¹ (compare Figure 4a). The OH peak is absent, and other features of the spectrum resemble (but are not identical with) those of *6a*.

Compound *6c* shows the ketone carbonyl peak at 1720 cm⁻¹, an indication that it is not in the system C=C—C=O.* The fact that this frequency is higher than that of the ketone carbonyl group in *5a* (1705 cm⁻¹) is due to the effect of the

---

* Since carbonyl groups in five-membered rings (cyclopentanones) show their stretching frequency at about 1745 cm⁻¹, the introduction of a conjugated double bond will lower this to the region of 1710 to 1720 cm⁻¹. Thus, a peak at 1720 cm⁻¹ is not unequivocally diagnostic of the system of compound *6c*; it could represent an α,β-unsaturated five-membered-ring ketone.

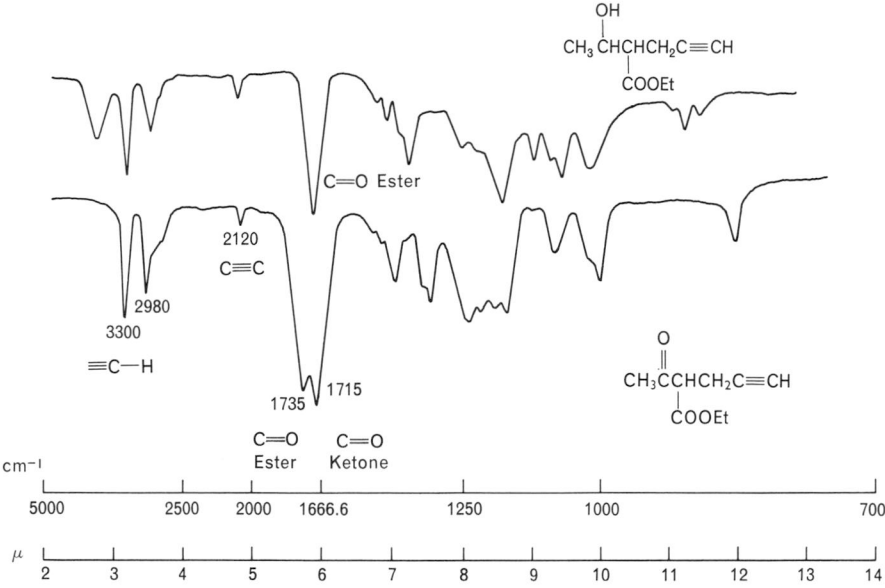

**Figure 11–7.**

bromine atom in the adjacent ($\alpha$-) position. Again, spectra of 6*b* and 6*c* are similar (but not identical) in the fingerprint region.

The spectrum of 7*a* shows two carbonyl stretching frequencies: that at 1715 cm$^{-1}$ is due to the carbonyl group in $CH_3CO$; that at 1735 cm$^{-1}$ is due to the carbonyl group of the ester grouping. Compound 7*b* is that of the compound obtained by reducing the C=O group of 7*a* to CHOH. Only the ester group is now seen in the 1700 cm$^{-1}$ region, and an OH peak at about 3600 cm$^{-1}$ is now present.

Two features of the spectra of 7*a* and 7*b* are the sharp peak at 3300 cm$^{-1}$ and the weak absorption at 2120 cm$^{-1}$. These are important diagnostic features: the

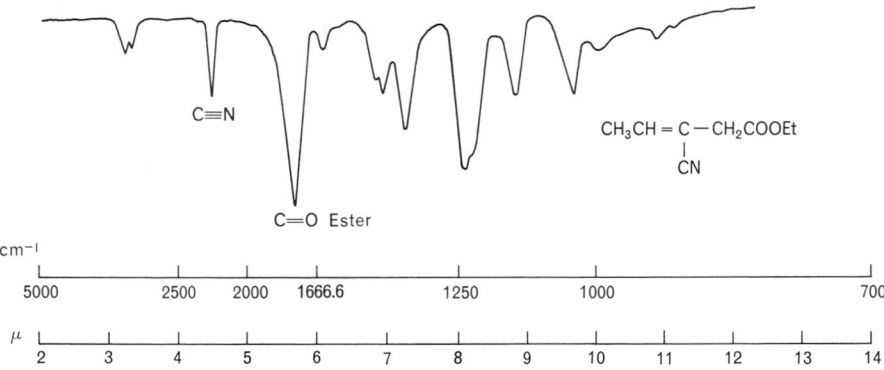

**Figure 11–8.**

absorption at 3300 cm$^{-1}$ is due to the stretching of the acetylenic C—H bond; and that at 2120 cm$^{-1}$ is due to the stretching of the C≡C bond. Another prominent absorption in both of these spectra occurs at around 1200 cm$^{-1}$. This is characteristic of the stretching of the C—O (single) bond in the ester grouping. Again, both of these spectra show general absorption in the 7 $\mu$ region due to C—H bending (of —CH$_3$ and —CH$_2$—).

The compound whose spectrum is shown in Figure 11-8 shows the ester carbonyl group (not conjugated with the carbon-carbon double bond, and thus at about 1740 cm$^{-1}$). The sharp peak at 2220 cm$^{-1}$ is due to the C≡N stretching vibration. This is to be compared with the C≡C stretching peak seen in the spectra of Figure 11-7 at 2120 cm$^{-1}$. It is to be noted that this stretching vibration occurs at approximately the same frequency for C≡N and C≡C, the difference between them being chiefly due to difference in mass between carbon and nitrogen.

### 11.6   Auxochromic groups. Effects on UV and IR absorption

The term "auxochrome" was first used in the field of chemistry of dyes and dyeing. An *auxochrome* is a group that deepens color; that is, that shifts the absorption maximum in the visible or UV spectrum to longer wavelengths and increases the intensity of absorption. In terms of visual observation, a "deeper" color is one that results from a higher $\lambda_{\text{maximum}}$. Thus, red is a "deeper" color than yellow, blue "deeper" than red, green "deeper" than blue (Section 11.2). Compounds whose absorption maxima lie below about 380 m$\mu$ are colorless. Nevertheless, auxochromic groups cause shifts in absorption at wavelengths below the visible range; for instance, an auxochromic group having its effect in the UV range might cause a

**Table 11-5.** *The effect of the auxochromic amino group upon the ultraviolet absorption of the C=C—C=O system*

| Compound | $\lambda_{\text{max}}$ | log $\epsilon$ |
|---|---|---|
| CH$_2$=CH—CO—CH$_3$ | 219 | 3.6 |
| CH$_3$CH$_2$CH$_2$CH=CH—CO—CH$_3$ | 228 | 4.0 |
| Et$_2$N—CH=CH—CO—CH$_2$CH$_2$CH$_3$ | 307 | 4.4 |
| CH$_3$—$\overset{\displaystyle \overset{NH_2}{\mid}}{C}$=CH—CO—CH$_3$ | 295 | 4.0 |
| CH$_3$—$\overset{\displaystyle \overset{CH_3}{\mid}}{C}$=CH—COOEt | 217 | 4.1 |
| CH$_3$$\overset{\displaystyle \overset{NMe_2}{\mid}}{C}$=CH—COOEt | 284 | 4.5 |

shift from, say, 230 m$\mu$ to 280 m$\mu$ in the principal absorption maximum. Two other terms used to describe spectral shifts are *bathochromic*—causing a shift of absorption to longer wavelengths; and *hypsochromic*—causing a shift of absorption to shorter wavelengths.

The most conspicuous property of auxochromes is their ability to provide additional opportunity for charge delocalization. A typical auxochrome is the amino (or substituted amino) group, —NH$_2$, —NHR, or —NR$_2$. An $\alpha,\beta$-unsaturated ketone is a resonance hybrid; the IR stretching frequency of the C=O bond has been discussed in Section 11.5, page 286. The effect of the auxochromic amino group upon the UV absorption of the C=C—C=O system is illustrated by the data in Table 11-5.

It will be apparent that the charge delocalization indicated by the forms

$$C=C-C=O \longleftrightarrow \overset{+}{C}-C=C-O^-$$

will be greatly augmented by the presence of an electron-donating —NH$_2$ (or —NR$_2$) group on the system:

$$\left\{ \overset{\cdot\cdot}{>N}-C=C-C=O \longleftrightarrow \overset{\cdot\cdot}{>N}-\overset{+}{C}-C=C-O^- \longleftrightarrow >\overset{+}{N}=C-C=C-O^- \right\}$$

The added opportunities for resonance stabilization of the $\pi \rightarrow \pi^*$ excited state provided by the amino group cause a marked lowering in the requirement for excitation energy, and thus a decreased frequency (longer wavelength) of absorption.

---

**Exercise 1.** Which compound of each of the following pairs would have the longer wavelength $\lambda_{max}$? Why?

(a) (CH$_3$)$_2$NCH=CH—COCH$_3$ or CH$_2$=CH—CO—CH$_2$N(CH$_3$)$_2$

(b) H$_2$N—CH=CH—CO—CH$_3$ or H$_3\overset{+}{N}$—CH=CH—CO—CH$_3$

(c) CH$_3$OCH=CH—CO—CH$_3$ or CH$_3$OCH$_2$CH=CH—CO—CH$_3$

---

A comparison of the frequency for the C=O absorption in the IR spectra of CH$_2$=CH—CO—CH$_3$ and Et$_2$N—CH=CH—CO—CH$_2$CH$_3$ (Section 11.5) reveals that the auxochromic amino group affects the IR absorption as well as the UV absorption. Indeed, the two effects are due to the same fundamental cause: the increased stabilization by charge delocalization (effect on UV absorption) causes the increase in the single-bond character of the C=O group (effect on IR absorption).

Experimental conditions and structural modifications that affect the nature of the auxochrome have the expected effect on the spectral observations. The data is Table 11-6 demonstrate some of these effects.

Protonation of —NH$_2$ changes it to —$\overset{+}{N}$H$_3$; this group no longer has an unshared electron pair to participate in charge delocalization. Alteration of

**Table 11-6.** *Carbonyl stretching frequencies (infrared) for some substituted acetophenones*

| Compound | C$=$O (IR) | $\lambda_{max}$ (UV)* |
|---|---|---|
| 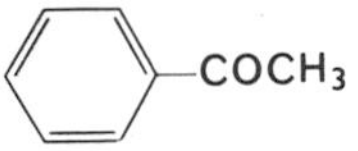 ⬡—COCH$_3$ | 1692 cm$^{-1}$ | 240 m$\mu$ |
| 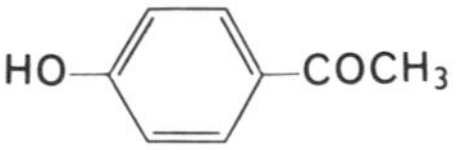 HO—⬡—COCH$_3$ | 1686 | 276 |
| 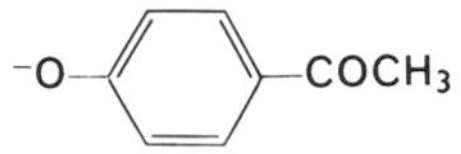 $^-$O—⬡—COCH$_3$ | — | 325 |
| 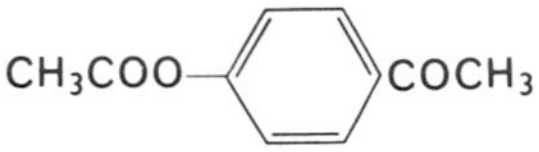 CH$_3$COO—⬡—COCH$_3$ | 1691 | — |
| 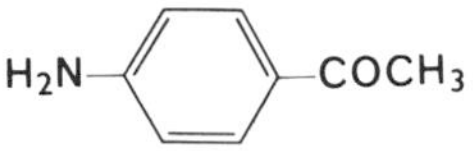 H$_2$N—⬡—COCH$_3$ | 1677 | 317 |
| 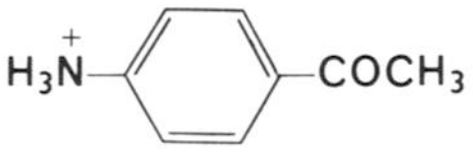 $\overset{+}{H_3N}$—⬡—COCH$_3$ | — | 240 |
| 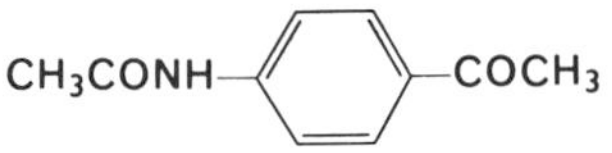 CH$_3$CONH—⬡—COCH$_3$ | 1686 | — |

* All of the $\epsilon$ values are in the neighborhood of 10,000–15,000; because their differences are not critical to the present discussion, they are omitted from this table.

—OH to the ion, —O$^-$, provides further opportunity for participation of unshared electrons on oxygen in charge delocalization. Thus, the change H $\rightarrow$ NH$_2$ is bathochromic; NH$_2$ $\rightarrow$ NH$_3{}^+$ is hypsochromic; OH $\rightarrow$ O$^-$ is bathochromic; and both of the changes, OH $\rightarrow$ OCOCH$_3$ and NH$_2$ $\rightarrow$ NHCOCH$_3$ (acetylation), are hypsochromic.

**Exercises**

(Exercise 1 will be found within the text.)

2 *(a)* Explain why the change NH$_2$ $\rightarrow$ NHCOCH$_3$ is hypsochromic. *(b)* NH$_2$ is a better auxochrome than OH. Can you relate this effect to another general property of NH$_2$ and OH groups? (Hint: Compare ammonia and water.)

3 Look up the meaning of the prefixes "hypso-" and "batho-" in a good unabridged dictionary.

## Chapter twelve

# The Carbonyl Group.
# Aldehydes and Ketones, and Addition
# Reactions of Carbonyl Compounds

We have seen in Chapter 3 that the sharing of four electrons between two carbon atoms constitutes the carbon-carbon double bond, $C{=}C$. Carbon may form double bonds to other atoms as well, as the following examples show:

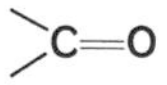

$$\text{>C=O} \qquad \text{>C=N—} \qquad \text{>C=S}$$

carbonyl 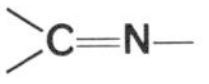            imino 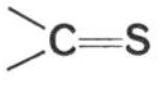            thiocarbonyl

By far the most widely encountered of these is the carbon-oxygen double bond, which constitutes the structural unit known as the *carbonyl group*. The carbonyl group is the characteristic functional group of aldehydes, ketones, and carboxylic acid derivatives. In aldehydes the carbonyl group bears one hydrogen atom; the resulting group is the *formyl* group,

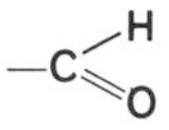

$$-C\overset{H}{\underset{O}{\lessgtr}}$$

usually written —CHO.

The simplest aldehyde, formaldehyde, is unique in that it has the structure

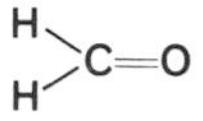

$$\overset{H}{\underset{H}{>}}C{=}O$$

and is the only possible compound with more than one hydrogen atom attached to the carbonyl group. Aldehydes have the general structure

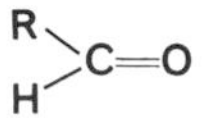

$$\overset{R}{\underset{H}{>}}C{=}O$$

in which R is H or a carbon group (alkyl or aryl). When both of the groups attached to the carbonyl group are alkyl or aryl groups, the compounds are *ketones:*

$$\begin{array}{c} R \\ \diagdown \\ \phantom{R}C{=}O \\ \diagup \\ R \end{array}$$

The nature of R in these compounds may vary widely; some representative aldehydes and ketones are the following:

| $CH_3CHO$ | $CH_3CH{=}CHCHO$ | $CH_3CHCH_2CHO$<br>$\phantom{CH_3}|$<br>$\phantom{CH_3}OH$ | $CHO$<br>$|$<br>$CHO$ |
|---|---|---|---|
| acetaldehyde | crotonaldehyde | 3-hydroxybutanal | glyoxal |

$$\begin{array}{c} H_3C \\ \diagdown \\ \phantom{H_3}C{=}O \\ \diagup \\ H_3C \end{array}$$

$CH_3COCH{=}CH_2$

(ring with N) $COCHCH_2CH_3$ / $CH_3$

acetone        methyl vinyl ketone        *sec*-butyl 3-pyridyl ketone

## 12.1 Aldehydes and ketones

*Aldehydes* are so called because they can be regarded as the products of *al*cohol *dehyd*rogenation. Some aldehydes can be prepared in this way (Section 8.21), although the reaction is in practice limited to the preparation of the simpler members of the class. The word *ketone* is an old modification of the name of the simplest member of the class, acetone, which is in turn derived from the fact that *acet*one can be formed from *acet*ic acid.

Although aldehydes and ketones are usually considered together, because most of their chemical properties derive from the carbonyl group found in both, no close relationship exists between the occurrence in nature, preparation, and practical uses of these two classes.

The most widely used aldehyde is *formaldehyde*, the most important commercial source of which is the oxidation of methanol by air in the presence of a catalyst. Large quantities of formaldehyde are produced industrially to supply demands for its use in the manufacture of plastics [Bakelite, urea-formaldehyde plastics, melamine resins (see Chapter 34)], preservation of biological specimens, and the production of pentaerythritol and products derived from it.

Formaldehyde is a gas (b.p. $-21°$), but is not usually handled in this form. It is commonly produced as an aqueous solution (formalin) or as a polymeric or trimeric solid. Paraformaldehyde, the polymer, and trioxymethylene, the trimer, regenerate monomeric formaldehyde on heating:

$HO(CH_2OCH_2OCH_2OCH_2O)_xH$

paraformaldehyde

$$\begin{array}{c} \phantom{xx}O \\ H_2C{\diagup}{\diagdown}CH_2 \\ |\phantom{xxxx}| \\ O{\diagdown}{\phantom{x}}{\diagup}O \\ \phantom{x}C \\ \phantom{x}H_2 \end{array}$$

trioxymethylene

*Acetaldehyde*, a low-boiling, pungent liquid, is an important industrial chemical which finds its chief uses as a raw material for the synthesis of other compounds. It can be prepared by the hydration of acetylene (compare the hydration of olefins) or by the air oxidation of ethanol:

$$HC\equiv CH + H_2O \xrightarrow[\text{HgSO}_4]{\text{H}_2\text{SO}_4} \left( \begin{array}{c} CH_2=CH \\ | \\ OH \end{array} \right) \longrightarrow CH_3CH=O$$

unstable intermediate
enol form

The hydration of terminally substituted acetylenes proceeds according to the following equation:

$$RC\equiv CH + H_2O \xrightarrow[\text{HgSO}_4]{\text{H}_2\text{SO}_4} \left( \begin{array}{c} R-C=CH_2 \\ | \\ OH \end{array} \right) \longrightarrow RCOCH_3$$

It is seen that the hydroxyl group of the added water goes to the alkyl-bearing carbon atom, a course that corresponds to the hydration of olefins according to Markovnikov's rule.

*Acetone*, the simplest ketone, is produced industrially by the dehydrogenation of isopropyl alcohol and, to a smaller extent, by a special fermentation process starting with starch or molasses. Acetone is a very important industrial chemical. It is a valuable solvent in the preparation of lacquers and other coatings, and is the starting material for the synthesis of other chemicals. Other ketones that are produced commercially in large quantities are *methyl ethyl ketone* (by oxidation of 2-butanol) and *cyclohexanone* (by oxidation or dehydrogenation of cyclohexanol).

The carbonyl group is one of the most important and most commonly encountered functional groups in organic chemistry; a large body of the material of organic chemistry is concerned with the behavior of carbonyl-containing compounds. Important carbonyl compounds occur in nature: retinene has already been mentioned (Section 8.24); the important substance camphor is a complex cyclic ketone; and numerous natural oils, perfumes, and coloring matters contain carbonyl groups. The chemistry of carbonyl compounds can be discussed with respect to reactions that occur at the $-\overset{|}{C}=O$ group itself, and to the influence of the carbonyl group upon the substituents attached to it.

## 12.2 Nomenclature

In the I.U.C. system, the carbonyl group is designated by the ending *-one* in ketones, *-al* in aldehydes. The name is based upon the longest chain of carbon atoms in which the functional group is found, and ketones are numbered in such a way that the carbonyl group is given preference for the lowest number. In aldehydes, the —CHO group is necessarily terminal and bears the number 1:

$CH_3CH_2CH_2CH_2CHO$

pentanal

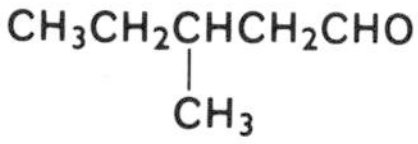

$\overset{6}{B}r\overset{5}{C}H_2\overset{4}{C}H_2\overset{3}{C}H_2\overset{2}{C}H\overset{1}{C}HO$
$\quad\quad\quad\quad\quad\quad|$
$\quad\quad\quad\quad\quad\quad OCH_3$

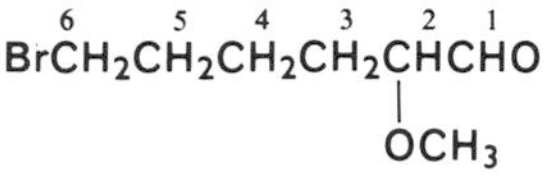

2-methoxy-6-bromohexanal

$\begin{array}{c} H_3C \\ \diagdown \\ \quad CHCOCH_2CH_2CH \\ \diagup \\ H_3C \end{array} \begin{array}{c} CH_3 \\ \diagup \\ \diagdown \\ CH_3 \end{array}$

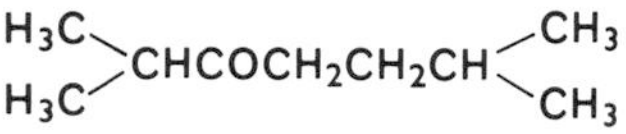

2,6-dimethyl-3-heptanone

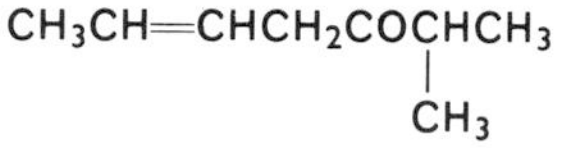

$CH_3CHCH_2COCH_3$
$\quad\quad|$
$\quad\quad OH$

4-hydroxy-2-pentanone

$CH_3CH_2CHCH_2CHO$
$\quad\quad\quad|$
$\quad\quad\quad CH_3$

3-methylpentanal

$CH_3CH_2CH{=}CH{-}CH{=}CH{-}CHO$

2,4-heptadienal

$CH_3CH{=}CHCH_2COCHCH_3$
$\quad\quad\quad\quad\quad\quad\quad|$
$\quad\quad\quad\quad\quad\quad\quad CH_3$

2-methyl-5-hepten-3-one

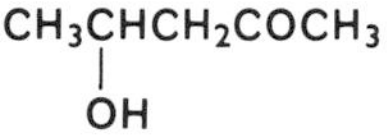

$CH_3CH_2COCH_2CH\begin{array}{c} \overset{H_2}{C} \\ \diagup\diagdown \\ | \quad\quad | \\ H_2C{-}{-}{-}CH_2 \end{array}CH_2$

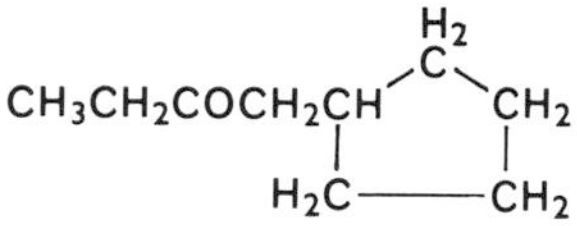

1-cyclopentyl-2-butanone

Notes to Preceding Examples: Conventions governing the order in which a name is written differ somewhat in other countries. In Great Britain, for example, the name *2-methyl-5-hepten-3-one* is written *2-methylhept-5-en-3-one*. British usage would require that the name for 4-pentyn-2-one ($CH_3COCH_2C{\equiv}CH$) be pent-4-yn-2-one, and 3,5-octadien-2,7-dione ($CH_3{-}CO{-}CH{=}CH{-}CH{=}CH{-}CO{-}CH_3$) be written octa-3,5-dien-2,7-dione.

It is also to be noted that a compound containing a carbonyl and a hydroxyl group is called a "... hydroxy ... one ..." and not by a name containing "... onol" or "... olone." This is in accord with the general rule that the "chief function" is expressed in the main part of the name, the chief function being specified by an arbitrary order of preference. Thus, 4-hydroxy-2-pentanone, and not 4-oxo-2-pentanol, since the ketone function precedes the alcohol function in order of preference.

A summary of the rules regarding these principles of nomenclature is to be found in the introduction to the 1952 Subject Index of *Chemical Abstracts*.

It is occasionally convenient to regard the formyl group, —CHO, as a substituent. In this case, naming follows the usual rules:

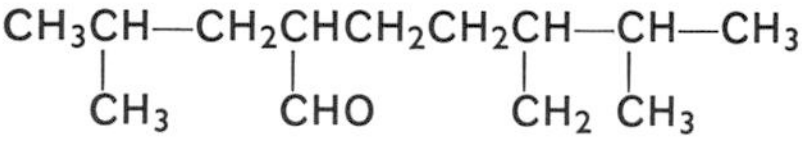

$CH_3CH{-}CH_2CHCH_2CH_2CH{-}CH{-}CH_3$
$\quad\quad|\quad\quad\quad\quad|\quad\quad\quad\quad\quad|\quad\quad|$
$\quad\quad CH_3\quad\quad CHO\quad\quad\quad CH_2\; CH_3$
2,3,8-trimethyl-6-formylnonane

*Trivial names* are used for many aldehydes and ketones:

| $CH_3COCH_3$ | $CH_2{=}CHCHO$ | $(CH_3)_2C{=}CHCOCH_3$ | $(CH_3)_3CCOCH_3$ |
|:---:|:---:|:---:|:---:|
| acetone | acrolein | mesityl oxide | pinacolone |

A systematic method of naming ketones consists in naming the two groups attached to the carbonyl group, following these by the word "ketone":

| $CH_3CH_2CH_2CH_2COCH_3$ | $CH_2{=}CHCOCH_2CH_3$ | $(CH_3)_2CHCH_2COCH_2CH(CH_3)_2$ |
|:---:|:---:|:---:|
| methyl *n*-butyl ketone | ethyl vinyl ketone | diisobutyl ketone |

Finally, the use of substitution names is still widely practiced. When this method of naming is used, the carbon atoms of the group attached to the formyl or keto group are designated by Greek letters, starting with the carbon atom attached to the carbonyl group:

$$\overset{\delta}{C}-\overset{\gamma}{C}-\overset{\beta}{C}-\overset{\alpha}{C}-CO-\overset{\alpha'}{C}-\overset{\beta'}{C}-\overset{\gamma'}{C}-\overset{\delta'}{C} \qquad \overset{\epsilon}{C}-\overset{\delta}{C}-\overset{\gamma}{C}-\overset{\beta}{C}-\overset{\alpha}{C}-CHO$$

$CH_3CH_2\underset{|}{CH}CHO$
  $\,\,\,Br$

α-bromo-*n*-butyraldehyde

$(CH_3)_2\underset{|}{C}CHO$
  $\,\,\,OCH_3$

α-methoxyisobutyraldehyde

$CH_3\underset{|}{CH}\underset{|}{CH}CH_2CHO$
  $\,\,\,Br\,\,\,Br$

β,γ-dibromo-*n*-valeraldehyde

$CH_3CO\underset{|}{CH}CH_3$
   $\,\,\,Br$

α-bromoethyl methyl ketone

## 12.3   The nature of the carbonyl group

The carbonyl group consists of a carbon and an oxygen atom linked by a double bond, and thus bears a formal resemblance to the olefinic double bond. The angle between the bonds in several simple carbonyl compounds (H—C—H, C—C—O, and C—C—O in formaldehyde, acetaldehyde, and acetone, respectively) is found to be close to 120°. The three bonds to the carbon atom of the carbonyl group are thus trigonal, and we can thus represent a carbonyl compound, such as formaldehyde (Figure 12-1),

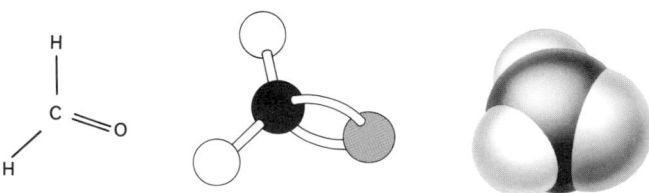

**Figure 12–1.** Conventional formula, ball-and-stick model, and molecular model of formaldehyde, $H_2CO$.

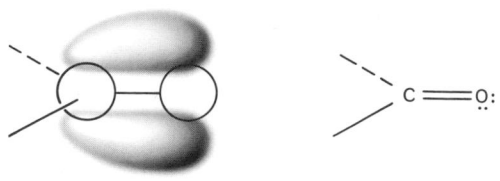

**Figure 12–2.** A representation of electron distribution in the carbon-oxygen double bond, showing electron-deficiency on carbon resulting from greater electronegativity of oxygen. (Unshared electrons on oxygen not shown on drawing of model.)

with the added provisions that (1) an additional pair of electrons is shared between carbon and oxygen, and (2) the oxygen atom possesses two additional pairs of unshared electrons. One of the two pairs of electrons that form the carbon-oxygen double bond is analogous to what we have called the $\pi$-electrons of the carbon-carbon double bond, but there is a marked difference between the C=O bond and the C=C bond, due to the greater electronegativity (electron-attraction) of oxygen as compared with carbon. This causes a distribution of the electron density in the bond between carbon and oxygen in such a way that the electrons are not equally distributed between the two nuclei. A convenient concept is that one electron pair occupies the $p$-orbital of oxygen with a greater probability than it occupies the region midway between carbon and oxygen. This concept can be illustrated by several symbolic representations. To depict the electron clouds, we can draw the carbonyl group as in Figure 12-2, or in conventional notation, the charge distribution can be represented as follows:

Thus, the carbon-oxygen bond is polarized to a marked degree, and has a partial ionic character, resulting in *electron deficiency* on the carbon atom. This is shown by the dipole moment, which is a measure of the tendency for a molecule to orient itself in an electrostatic field, and thus of the charge separation within the molecule. Acetaldehyde has a dipole moment of 2.65 Debye units, whereas propylene has the much lower moment of 0.35 D:

$$CH_3CH=O \qquad\qquad CH_3CH=CH_2$$
dipole moment: 2.65 D       dipole moment: 0.35 D

These figures indicate that the contribution of the structure $CH_3\overset{+}{C}H—O^-$ to the structure of acetaldehyde is greater than the contribution of $CH_3\overset{+}{C}H—CH_2^-$ to the structure of propylene.

## 12.4  Basic properties of the carbonyl group

The basicity of the carbonyl group that is conferred upon it by the unshared oxygen electrons is manifested by its ready protonation by strong acids and its coordination with Lewis acids:

$$CH_3CH=O+H_2SO_4 \rightleftharpoons CH_3CH=\overset{+}{O}H+HSO_4^-$$

The basicity of a carbonyl compound can be greatly increased if the positive charge of the resulting conjugate acid is delocalized by resonance; i.e., if a number of contributing structures of comparable energy can be written. For example, the compound 2,6-dimethyl-γ-pyrone (1),

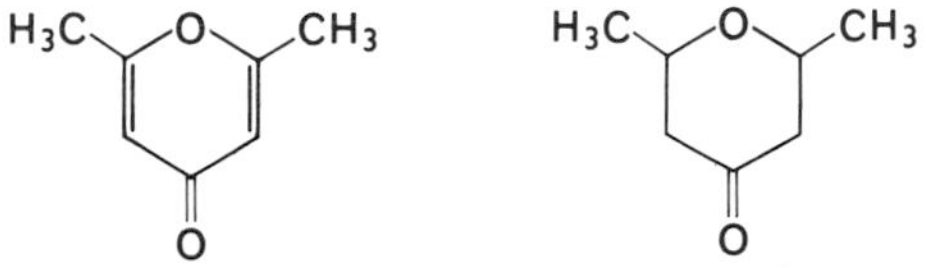

2,6-dimethyltetrahydro-γ-pyrone

(1)                                   (2)

is a much stronger base than (2), since the conjugate acid of 2,6-dimethyl-γ-pyrone is a resonance hybrid of several important contributing structures:

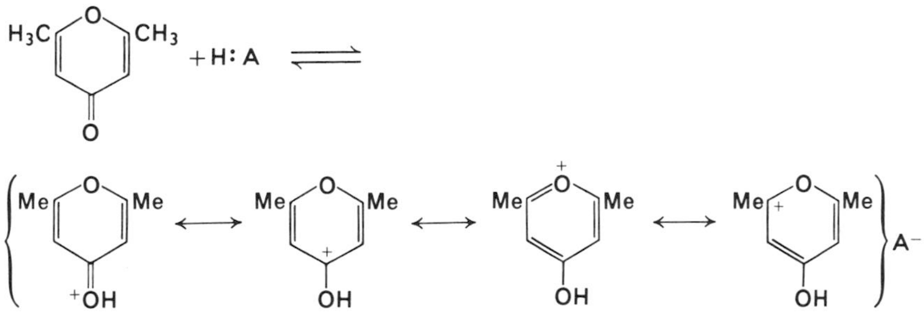

No comparable contributions are possible in the case of the conjugate acid of the saturated analogue (2).

---

**Exercise 1.** Write the reaction for the protonation of (2). Is more than one protonated form (conjugate acid) of (2) possible?

---

## 12.5 Spectral properties of carbonyl compounds

The characteristic ultraviolet and infrared absorptions of carbonyl groups and carbonyl compounds, both saturated and containing conjugated carbon-carbon double bonds, have been discussed in Chapter 11. Ultraviolet spectra are of

**Table 12-1.** *Ultraviolet absorption maxima ($\pi \to \pi^*$ transition bands) for some selected $\alpha,\beta$-unsaturated carbonyl compounds. (Values are for ethanol solutions)*

| Compound | $\lambda_{max}$, m$\mu$* |
|---|---|
| $CH_2{=}CH{-}CHO$ | 208 |
| $CH_2{=}CH{-}CO{-}CH_3$ | 219 |
| $CH_2{=}C{-}CO{-}CH_3$<br>    $\mid$<br>    $CH_3$ | 220 |
| $CH_3{-}CH{=}C{-}CO{-}CH_3$<br>           $\mid$<br>           $CH_3$ | 230 |
| $CH_3{-}C{=}CH{-}CO{-}CH_3$<br>    $\mid$<br>    $CH_3$ | 237 |
| $CH_3{-}C{=}C{-}CO{-}CH_3$<br>    $\mid$   $\mid$<br>  $H_3C$  $CH_3$ | 247 |

* Extinction coefficients are omitted from this list; for the absorption maximum cited they are mostly in the range of $\epsilon =$ about 10,000 (log $\epsilon =$ about 4). Data from *Constants of Organic Compounds*, M. Kotake, Editor. Asakura Publishing Company, Tokyo, 1963.

limited usefulness in the structural characterization of aldehydes and ketones in which the carbonyl group is not conjugated with carbon-carbon double bonds, but they provide valuable information about the structure of unsaturated carbonyl compounds. In Table 12-1 are given the ultraviolet absorption maxima for a number of unsaturated aldehydes and ketones. It will be noted that the wavelength of the absorption maximum rises with increasing number of substituents on the system $C{=}C{-}C{=}O$: acrolein has $\lambda_{max}$ 208 m$\mu$, while 3,4-dimethyl-3-penten-2-one has $\lambda_{max}$ 247 m$\mu$.

A distinguishing characteristic of the ultraviolet spectra of $\alpha,\beta$-unsaturated ketones is the appearance of a second, low-intensity absorption band at longer wavelengths than that of the intense absorption band in the 220 to 250 m$\mu$ region. This long-wavelength absorption is the result of $n \to \pi^*$ transition of the $C{=}O$ group, and corresponds with the similarly located band of saturated aldehydes and ketones. In Figure 12-3 are given the ultraviolet absorption spectra of two typical $\alpha,\beta$-unsaturated ketones in which the intense band at low wavelength and the weak band at longer wavelength are shown.

Infrared spectra give useful information about carbonyl compounds of all kinds, saturated and unsaturated. Simple carbonyl groups, such as those in acetaldehyde, acetone, cyclohexanone and so on, show strong and characteristic

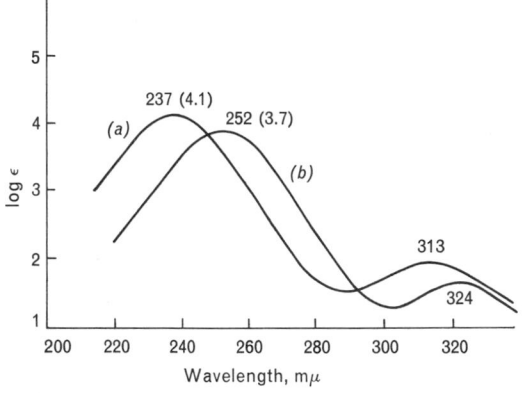

**Figure 12–3.** Ultraviolet spectra of α,β-unsaturated ketones (in ethanol).
(a) Mesityl oxide,
$(CH_3)_2C$=$CHCOCH_3$.

(b) Pulegone,

absorptions at about 1710 cm$^{-1}$, and conjugated carbonyl compounds show their characteristic C=O stretching absorption at lower frequencies (about 1670 cm$^{-1}$ for simple C=C—C=O compounds, and at lower frequencies for compounds with extended conjugated systems). Some additional examples are to be found in Chapter 11.

Ultraviolet and infrared spectra are complementary in the study of carbonyl compounds. Saturated ketones show only the weak absorption in the 280 mμ region, and their ultraviolet spectra are not nearly so informative as their infrared spectra. The infrared spectra, by the position of the carbonyl stretching band in the region of 1700 cm$^{-1}$, can often be used to determine whether the carbonyl group is present in a four-, five-, six-, or larger-membered ring; whether an electronegative substituent (halogen, hydroxy, acyloxy) is present in the *alpha* position; and sometimes the stereochemistry of such an *alpha*-substituent. When the infrared spectrum indicates the presence of an α,β-unsaturated carbonyl group, the highly characteristic ultraviolet spectrum provides corroborative evidence that places the conclusion beyond doubt.

## 12.6   Attack by nucleophilic reagents upon the carbonyl group

The positive character of the carbon atom causes it to behave as an *electrophilic* center, and thus to be accessible to attack by nucleophilic reagents (as represented by the generalized base, B:):

$$\text{>C=O:} \quad \rightleftharpoons \quad \text{>C—O:}^{-} \quad {}^{*} \tag{3}$$

B:

a nucleophile

---

* It is again to be noted that the nucleophile may be neutral or negatively charged. It if is negatively charged, of course, the positive charge shown on B in the product would not be present.

The process indicated in (3) represents the essential features of a *nucleophilic attack upon the carbonyl group*, and will find specific expression in many ways as we proceed. It should always be borne in mind that of the numerous particular forms which the reactions of carbonyl groups will be found to take, *the electrophilic character of the carbonyl carbon atom plays a dominant role*.

## 12.7    Addition of hydrogen cyanide

The typical addition reactions of carbonyl compounds are those in which the attacking species is a nucleophilic molecule or ion, in a medium whose $pH$ is adjusted such that the protonation of the first addition product can occur to complete the reaction and give a stable final product. A well-studied case is that in which hydrogen cyanide adds to an aldehyde to give a *cyanohydrin*. Since this reaction is quite general for alkanals, the general formulation, RCHO, will be used as an example:

$$RCHO + HCN \rightleftharpoons RCH\begin{smallmatrix} \diagup OH \\ \diagdown CN \end{smallmatrix}$$

a cyanohydrin

The addition of HCN to an aldehyde is catalyzed by bases and is very slow in the presence of excess acid. It is ordinarily carried out with the use of sodium or potassium cyanide, with the addition of mineral acid in such amount as to keep the reaction mixture buffered in an intermediate $pH$ range. A study of this reaction has disclosed that the actual attacking species is the cyanide ion, $:CN^-$, and that the rate of formation of the cyanohydrin is determined by the rate of the initial step

$$R-C\begin{smallmatrix} \diagup \ddot{O}: \\ \diagdown H \end{smallmatrix} \quad \rightleftharpoons \quad \begin{smallmatrix} R \\ NC \end{smallmatrix} \diagdown C \diagup \begin{smallmatrix} \ddot{O}: ^- \\ H \end{smallmatrix}$$
$$:CN^-$$

(4)

The intermediate ion (4) is a base, and can accept a proton from an acid in the medium to give the final product, the cyanohydrin (5):

$$RCH\begin{smallmatrix} \diagup O^- \\ \diagdown CN \end{smallmatrix} + HA \quad \rightleftharpoons \quad RCH\begin{smallmatrix} \diagup OH \\ \diagdown CN \end{smallmatrix} + A^-$$

(4)                                        (5)

If this reaction is carried out in the presence of excess strong acid (i.e., at low $pH$) the dissociation of HCN will be suppressed and the concentration of free $CN^-$ will be very small. At high $pH$ the protonation of (4) will be suppressed and the cyanohydrin will not be formed in appreciable amount. In a buffered medium at intermediate $pH$, $CN^-$ will be present, and acidic species (e.g., $HA = HCN$, $H_3O^+$) will also be available to perform the step (4) $\rightarrow$ (5).

It may be thought that at low $pH$, the protonation of the aldehyde according to

$$RCH{=}O + HA \rightleftharpoons RCH{=}\overset{+}{O}H + A^-$$

will increase the electrophilic character of the carbonyl carbon atom and thus favorably influence the addition. Although it is true that the protonated aldehyde indeed possesses a more electrophilic carbonyl carbon atom than does the free aldehyde, the addition of HCN cannot proceed because at the low $pH$ the concentration of the necessary nucleophile, $CN^-$, is low.

The addition of hydrogen cyanide to ketones is a less general reaction than the addition to aldehydes, since it is a slower reaction, frequently with a less favorable equilibrium, and is influenced more by structural factors. However, simple ketones, such as acetone, methyl ethyl ketone, and cyclohexanone, add HCN readily to form the corresponding cyanohydrins:

$$CH_3COCH_3 + HCN \longrightarrow \overset{H_3C}{\underset{H_3C}{>}}C\overset{OH}{\underset{CN}{<}} \tag{6}$$

α-hydroxyiso-<br>butyronitrile

$$CH_3COCH_2CH_3 + HCN \longrightarrow \overset{H_3C}{\underset{CH_3CH_2}{>}}C\overset{OH}{\underset{CN}{<}} \tag{7}$$

α-hydroxy-α-methyl-<br>*n*-butyronitrile

$+$ HCN $\longrightarrow$ (cyclohexane ring with OH and CN) $\tag{8}$

1-cyanocyclohexanol

Cyanohydrins are useful synthetic intermediates since they can be hydrolyzed (usually with $HCl/H_2O$) to α-hydroxy acids.

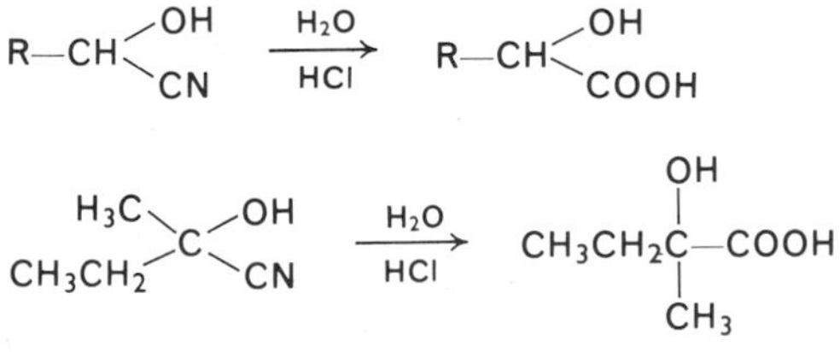

$$R{-}CH\overset{OH}{\underset{CN}{<}} \xrightarrow[\text{HCl}]{H_2O} R{-}CH\overset{OH}{\underset{COOH}{<}} \tag{9}$$

$$\overset{H_3C}{\underset{CH_3CH_2}{>}}C\overset{OH}{\underset{CN}{<}} \xrightarrow[\text{HCl}]{H_2O} CH_3CH_2\overset{OH}{\underset{CH_3}{\overset{|}{\underset{|}{C}}}}{-}COOH$$

α-hydroxy-α-methyl-<br>*n*-butyric acid

Since the cyano group can be reduced by means of lithium aluminum hydride, the reduction of cyanohydrins provides a route to amino alcohols:

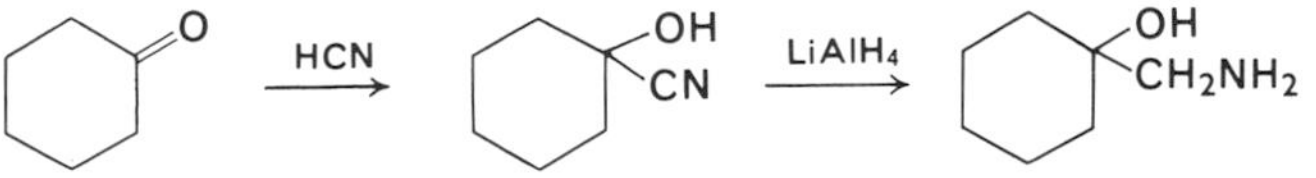

(cyclohexanone) $\xrightarrow{\text{HCN}}$ (cyclohexane ring with OH and CN) $\xrightarrow{\text{LiAlH}_4}$ (cyclohexane ring with OH and $CH_2NH_2$) $\tag{10}$

1-aminomethylcyclohexanol

## 12.8    Bisulfite addition compounds

The addition of sodium bisulfite to aldehydes and to the low-molecular-weight ketones is another characteristic carbonyl-addition reaction:

$$CH_3CH{=}\overset{+}{O} + NaHSO_3^- \rightleftharpoons CH_3CH\overset{OH}{\underset{SO_3^-Na^+}{\diagup}} \tag{11}$$

The product of this reaction is commonly called a "bisulfite addition compound." It is a salt, and if an attempt is made to convert it into the corresponding acid by treatment with a mineral acid, the aldehyde is regenerated and $SO_2$ is evolved. This result illustrates the reversible character of the addition reaction.

The bisulfite reaction follows the basic pattern of the addition reaction generalized in (3). The bisulfite ion

$$\left.\begin{array}{c} :\ddot{O}: \\ | \\ :S{-}\ddot{O}H \\ | \\ :\ddot{O}: \end{array}\right\}^-$$

bisulfite ion

has a plethora of available electron pairs, and it would be difficult to decide *a priori* whether the attack proceeded to form an oxygen-carbon or a sulfur-carbon bond.

$$\overset{>}{\underset{:OSO_2H^-}{C{=}\ddot{O}:}} \qquad or \qquad \overset{>}{\underset{:SO_3H^-}{C{=}O}}$$

Indeed, it was found necessary to carry out a detailed study of the constitution of the bisulfite addition compounds to establish that *it is actually a carbon-sulfur bond that is formed*, and that the reaction proceeds (with acetone in the following example) by attack of the unshared electron pair of the sulfur atom:*

$$\underset{:SO_3H^-~Na^+}{\overset{H_3C}{\underset{H_3C}{>}}C{=}O~H{:}A'} \rightleftharpoons \left.\underset{H_3C}{\overset{H_3C}{>}}C\overset{OH}{\underset{SO_3H}{\diagup}}\right\} A^- \rightleftharpoons \underset{Na^+}{\overset{H_3C}{\underset{H_3C}{>}}C\overset{OH}{\underset{SO_3^-}{\diagup}}} + HA$$

The addition of sodium bisulfite is a very useful means of isolating aldehydes from mixtures. The bisulfite addition compound (a sodium salt) frequently separates as a crystalline solid, especially when a concentrated (40%) sodium bisulfite solution is used, and can be separated by filtration and washed with

---

\* From this point on we shall frequently condense a series of acid-base exchanges involving the addition and abstraction of protons by a device in which the protonation of the carbonyl oxygen atom is written in the same step as the attack on the carbonyl carbon atom. Thus, the addition of HCN to RCHO might be condensed as follows.

$$\underset{A:{\to}H{-}CN}{R{-}C\overset{O}{\underset{H}{\diagdown}}} \quad H{-}A \rightleftharpoons \underset{A:H}{R{-}CH\overset{OH}{\underset{CN}{\diagdown}}} \quad A{:}^-$$

ether to remove contaminating substances. Treatment of the purified addition compound with dilute acid or alkali regenerates the aldehyde, which can then be separated by extraction or distillation.

---

**Exercise 2.** Write the reactions for the formation of the bisulfite addition compounds from cyclohexanone, isobutyraldehyde and 3-hexenal.

---

## 12.9 Addition of ammonia derivatives

Ammonia and amines, which are typical organic bases and effective nucleophilic compounds, would be expected to add to the carbonyl groups of aldehydes and ketones according to (3). Ammonia does add to aldehydes to yield unstable addition products such as the following,

$$CH_3CH{=}O + NH_3 \;\rightleftharpoons\; CH_3CH{<}^{OH}_{NH_2} \tag{12}$$

but the simple "aldehyde ammonias" are seldom isolable substances, condensing further to yield polymeric products. In the addition of ammonia to formaldehyde, the final product has the constitution $(CH_2)_6N_4$, and is called *hexamethylenetetramine*, or *urotropine:*

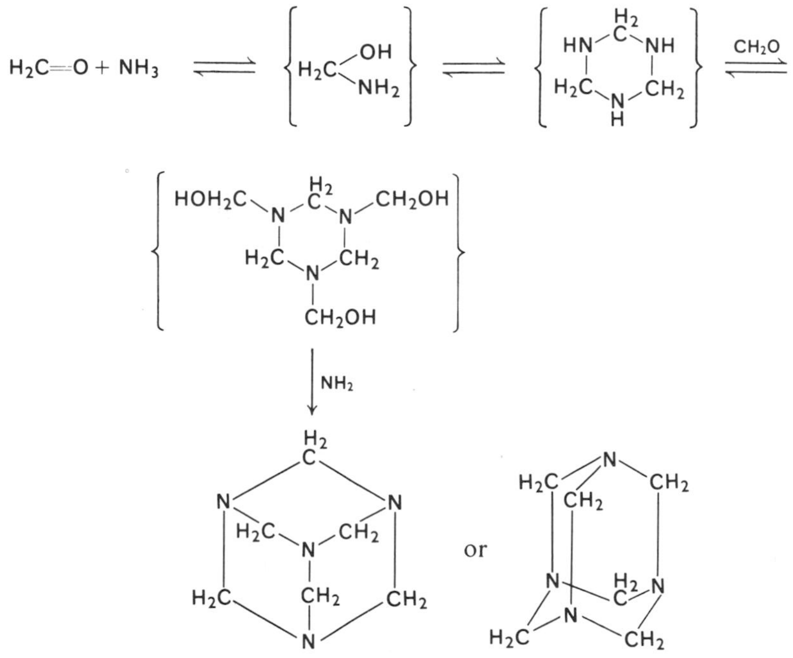

$$\tag{13}$$

hexamethylenetetramine

Aldehyde ammonias derived from other aldehydes are often of ill-defined structure.

*Amines* react with aldehydes and ketones to give addition products, and those formed from primary amines ($RNH_2$) usually lose the elements of water in a second stage of the reaction to give compounds known as *imines* (*aldimines* from aldehydes, *ketimines* from ketones).

$$RCH{=}O + R'NH_2 \rightleftharpoons RCH\begin{smallmatrix}OH\\NHR'\end{smallmatrix} \longrightarrow RCH{=}NR' + H_2O \qquad (14)$$
$$\text{an aldimine}$$

The aldimines are better known by the name *Schiff's bases*. Schiff's bases formed from aliphatic aldehydes and primary aliphatic amines are usually unstable, and polymerize or react further with more amine to give more complex products. When aromatic aldehydes, such as benzaldehyde, are used, the Schiff's bases are usually stable and readily isolated:

$$C_6H_5CH{=}O + CH_3NH_2 \longrightarrow C_6H_5CH{=}N{-}CH_3 \qquad (15)$$
$$\text{benzaldehyde}$$

$$C_6H_5CH{=}O + NH_2NH_2 \longrightarrow C_6H_5CH{=}NNH_2$$
$$\text{hydrazine} \qquad\qquad \text{benzalhydrazone}$$

$$\Big\downarrow C_6H_5CHO \qquad\qquad (16)$$

$$C_6H_5CH{=}N{-}N{=}CHC_6H_5$$
$$\text{benzalazine}$$

The most important of the reactions of carbonyl compounds with derivatives of ammonia are those in which the amino compound is one of the following (the reactive —$NH_2$ group is underlined):

$$C_6H_5NH\underline{NH_2} \qquad HO\underline{NH_2} \qquad NH_2CONH\underline{NH_2}$$
phenylhydrazine      hydroxylamine      semicarbazide

**Table 12-2.** *Aldehyde addition reactions and products*

| X in *(17)* | Product RCH=NX | Characteristics of Product RCH=NX |
|---|---|---|
| —H | aldehyde ammonia | often unstable, polymerize |
| —R (alkyl, aryl) | Schiff's base | often unstable; sometimes stable and isolable |
| —OH | oxime | stable and isolable |
| —NH₂ | hydrazone | stable, but can react again at —$NH_2$ to form an azine, RCH=N—N=CHR |
| —NHCONH₂ | semicarbazone | stable and isolable |
| —NHC₆H₅ | phenylhydrazone | stable and isolable |

The reaction of these compounds with a typical ketone [or aldehyde, in (17), where $R' = H$] may be summarized in the following equation. The formula $X—NH_2$ will be used in representing the reaction in the most general way; the reader should rewrite each equation (17), substituting for X the groups shown in Table 12-2:

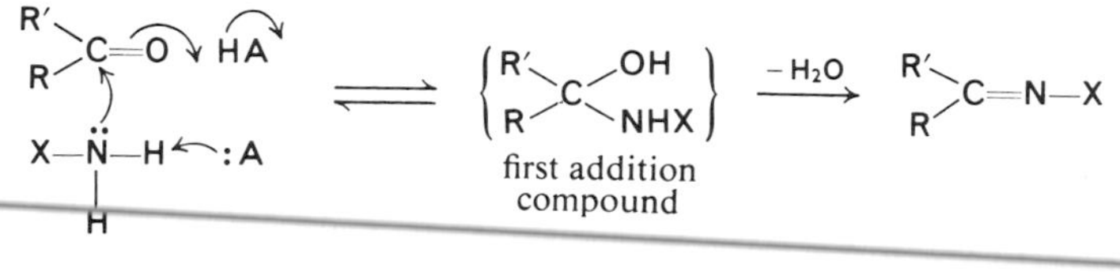

$$\begin{matrix} R' \\ R \end{matrix} C{=}O \quad HA \qquad \rightleftharpoons \qquad \left\{ \begin{matrix} R' \\ R \end{matrix} C \begin{matrix} OH \\ NHX \end{matrix} \right\} \xrightarrow{-H_2O} \begin{matrix} R' \\ R \end{matrix} C{=}N{-}X \qquad (17)$$

$$X—\overset{..}{N}—H \quad :A$$
$$\underset{H}{|}$$

first addition compound

---

**Exercise 3.** Formulate the reactions listed in Table 12-2 using a specific aldehyde (e.g. acetaldehyde) and a specific ketone (e.g. acetone) as in the following equation which shows the formation of the oxime of diethyl ketone.

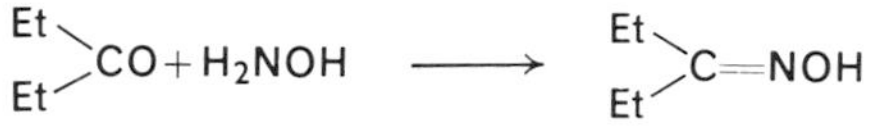

$$\begin{matrix} Et \\ Et \end{matrix} CO + H_2NOH \longrightarrow \begin{matrix} Et \\ Et \end{matrix} C{=}NOH$$

diethyl ketoxime

---

The formation of compounds such as Schiff's bases, oximes, semicarbazones, and others is reversible, and the aldehyde or ketone may be regenerated by acid hydrolysis:

$$RCH{=}NR' + H_2O(H^+) \longrightarrow RCHO + R'NH_2 \text{ (as the salt, } R'NH_3{}^+)$$

The hydrolysis is a nucleophilic attack by $H_2O$ upon the protonated imine*

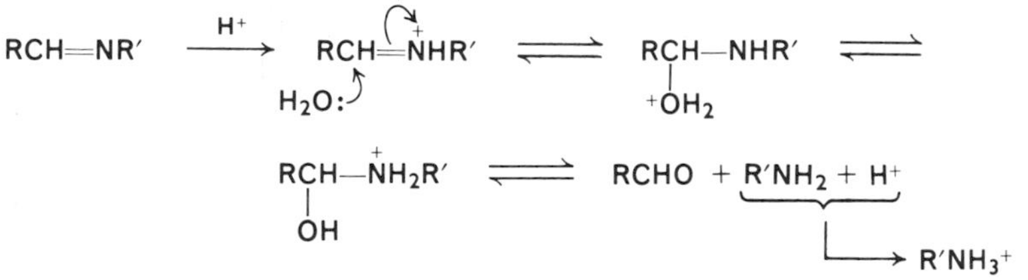

$$RCH{=}NR' \xrightarrow{H^+} RCH{\overset{+}{=}}NHR' \rightleftharpoons \underset{\overset{|}{{}^+OH_2}}{RCH{-}NHR'} \rightleftharpoons$$

$$\underset{\overset{|}{OH}}{RCH{-}\overset{+}{N}H_2R'} \rightleftharpoons RCHO + \underbrace{R'NH_2 + H^+}_{} \qquad (18)$$
$$\qquad\qquad\qquad\qquad\qquad \longrightarrow R'NH_3{}^+$$

and thus resembles in its essential details the carbonyl addition reaction [compare (3)].

---

\* Another simplification can be introduced into our writing of equations; that is, to write proton transfers simply as

$$B: + H^+ \longrightarrow B:H^+$$

instead of

$$B: + H:A \longrightarrow B:H^+ + :A^-$$

with the understanding that the former is an abbreviation of the latter.

## 12.10   The isolation and purification of aldehydes and ketones by means of derivative formation

The ready hydrolysis of the carbonyl derivatives of Table 12-2 permits them to be used for the isolation of carbonyl compounds from mixtures. A mixture containing an aldehyde or ketone can be treated with semicarbazide, the semicarbazone separated (for example, by crystallization) from noncarbonyl compounds that may be present, purified, and hydrolyzed to the pure aldehyde or ketone. Another useful reagent for this purpose is the compound called "Girard's reagent." This is a compound of the following structure:

$$Cl^-(CH_3)_3\overset{+}{N}\cdot CH_2C\overset{O}{\diagup}{-}NHNH_2$$
Girard's reagent

This structure should be compared with that of semicarbazide. Derivatives of aldehydes and ketones are formed with Girard's reagent as follows:

$$RCHO + Cl^-(CH_3)_3\overset{+}{N}CH_2CONHNH_2 \longrightarrow Cl^-(CH_3)_3\overset{+}{N}CH_2C\overset{O}{\diagup}{-}NH{-}N{=}CHR$$

The particular advantage of Girard's reagent is that the derivatives formed from it are water-soluble and ether-insoluble because of the presence of the $(Cl^-(CH_3)_3\overset{+}{N}{-})$ grouping, and thus can be separated from ether-soluble, water-insoluble materials that do not contain carbonyl groups. After separation of the derivative from ether-soluble noncarbonyl compounds, hydrolysis of the derivative regenerates the carbonyl compound, which can then be extracted with ether and obtained in pure form:

$$Cl^-(CH_3)_3\overset{+}{N}CH_2C\overset{O}{\diagup}{-}NH{-}N{=}CHR \xrightarrow{\text{H}_2\text{O/HCl}}$$

$$Cl^-(CH_3)_3\overset{+}{N}CH_2COOH + NH_2NH_2 + RCHO \qquad (19)$$

Girard's reagent has found extensive use in biological chemistry in studies on steroid metabolism. Many of the steroidal sex hormones are ketones, and can be separated from nonketonic constituents of urine by means of Girard's reagent prior to their purification by other means.

## 12.11   The use of carbonyl derivatives in identification

The most important uses of oximes, semicarbazones, and substituted hydrazones are in the *characterization* and *identification* of aldehydes and ketones. The *formation* of the derivative indicates the *presence* of the carbonyl function, whereas the *properties* of the derivative (melting point, crystalline form, color, and so on) serve to *identify* the carbonyl compound as a particular substance.

The importance of derivatives in the identification of organic compounds lies (1) in the ease of manipulating, purifying and measuring the physical constants of small amounts of a crystalline solid when the original compound may be a liquid; (2) in the fact that the ability of a given unknown compound to form the derivative is a confirmation of its chemical nature; and (3), most of all, in the fact that a single compound can be converted into several derivatives.

## 12.12 Addition of water and alcohols to carbonyl compounds

The following equilibrium probably exists in aqueous solutions of aldehydes,

$$RCH{=}O+H_2O \;\rightleftharpoons\; RCH\!\!\begin{smallmatrix}\diagup OH\\\diagdown OH\end{smallmatrix}$$

an "aldehyde hydrate"

but aldehyde hydrates are usually not stable substances and are seldom isolable. However, when the carbonyl group is attached to one or two strongly electron-attracting groups, the hydrate is often stable:

$$CCl_3CHO+H_2O \;\longrightarrow\; CCl_3CH\!\!\begin{smallmatrix}\diagup OH\\\diagdown OH\end{smallmatrix} \tag{20}$$

chloral         chloral hydrate

$$C_6H_5COCHO+H_2O \;\longrightarrow\; C_6H_5COCH\!\!\begin{smallmatrix}\diagup OH\\\diagdown OH\end{smallmatrix} \tag{21}$$

phenylglyoxal        phenylglyoxal hydrate

(22)

1,2,3-triketohydrindene        1,2,3-triketohydrindene hydrate

The hydrates shown in the above equations can be converted into the free carbonyl compounds by removal of the water; this can be accomplished with the use of a good dehydrating agent such as phosphorus pentoxide.

The effect of the electron-attracting substituents (such as the chlorine atoms in chloral) in stabilizing the hydrate is to be attributed to the effect of such substituents upon the character of the carbonyl group. The formation of the hydrate necessarily removes the possibility of charge delocalization (resonance) which characterizes the carbonyl group. The strong inductive effect of three chlorine atoms reduces electron density on the carbonyl carbon atom and reduces the importance of the contributing structure

$$Cl_3C\overset{\,+}{-}\!\!\underset{\underset{H}{|}}{C}\!-O^- \quad \text{in the hybrid} \left\{ Cl_3C\!-\!\!\underset{\underset{H}{|}}{C}\!\!=\!O \;\longleftrightarrow\; Cl_3C\overset{\,+}{-}\!\!\underset{\underset{H}{|}}{C}\!-O^- \right\}$$

(a)                          (b)

The reduced contribution of form *(a)* to the structure of the hybrid *(b)* should reduce the degree of single-bond character in the carbonyl group, and this is confirmed by the experimental evidence: the C=O bond length in chloral is 1.15 Å, compared with 1.22 Å in acetaldehyde. The consequence of these facts is that the loss in resonance energy in forming the hydrate of chloral is *less* than the loss in resonance energy in forming the hydrate of acetaldehyde. This difference is reflected in the overall energy change in the reaction, and thus in the equilibrium constant for hydrate formation.

### 12.13  Instability of 1,1-dihydroxy compounds

With the exception of the kinds shown in Section 12.12, it is generally true that structures such as

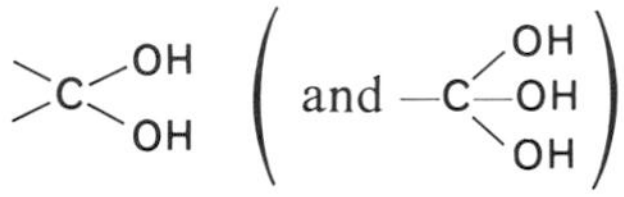

$$\underset{}{>}C\underset{OH}{\overset{OH}{<}} \quad \left( and \; -C\underset{OH}{\overset{OH}{<}}\; \right)$$

are not stable except as they exist in small equilibrium concentration in solution. For example, the compound $C_2H_6O_2$ could have the structure $HOCH_2CH_2OH$ but not $CH_3CH(OH)_2$; and $C_3H_8O_2$ could be

$CH_3CHCH_2OH$, $HOCH_2CH_2CH_2OH$, but not $CH_3CCH_3$. Moreover, if a
  |                                                          |
  OH                                                        OH

synthesis that should lead to a compound having two hydroxyl groups on a single carbon atom is performed, the *product* isolated will be the corresponding carbonyl compound:

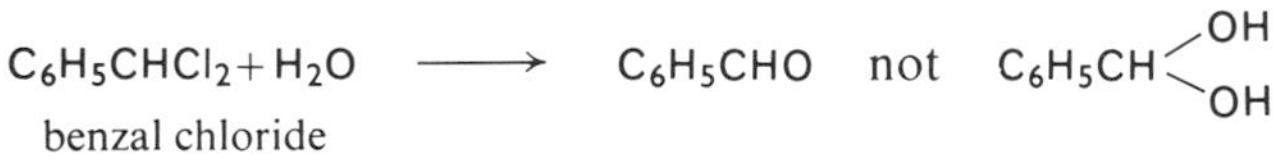

$$C_6H_5CHCl_2 + H_2O \longrightarrow C_6H_5CHO \;\; not \;\; C_6H_5CH\underset{OH}{\overset{OH}{<}} \tag{23}$$
benzal chloride

$$CH_3\underset{Cl}{\overset{Cl}{C}}CH_2CH_3 \xrightarrow[\text{(OH$^-$)}]{H_2O} CH_3COCH_2CH_3 \;\; not \;\; CH_3\underset{OH}{\overset{OH}{C}}CH_2CH_3 \tag{24}$$

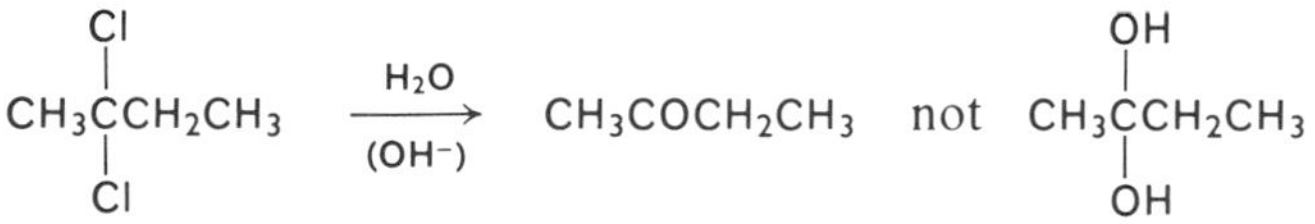

### 12.14  Hemiacetals and acetals

Alcohols add to aldehydes and ketones in the same manner as described for the addition of water:

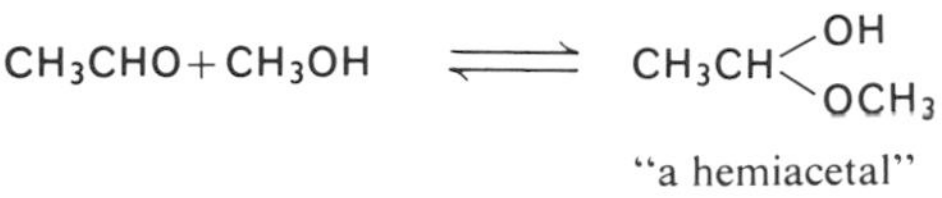

$$CH_3CHO + CH_3OH \;\; \rightleftharpoons \;\; CH_3CH\underset{OCH_3}{\overset{OH}{<}} \tag{25}$$
"a hemiacetal"

Hemiacetals are usually unstable, being comparable in this respect to hydrates, except that when the hydroxyl group is a part of the molecule that contains the carbonyl group, a *cyclic* hemiacetal may form.

When the intramolecular addition of —OH to the carbonyl group leads to a cyclic hemiacetal in which the ring is five- or six-membered, the cyclic structure is frequently the normal form in which the compound exists. For example, 4-hydroxybutanal and 5-hydroxypentanal would exist largely in the cyclic forms

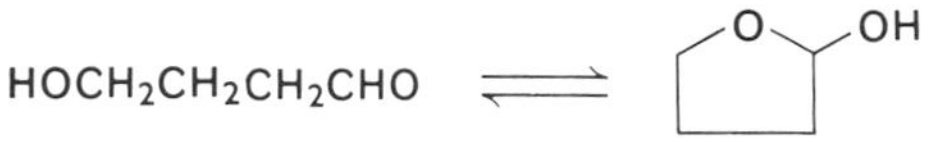

$$HOCH_2CH_2CH_2CHO \quad \rightleftharpoons \quad \text{(cyclic structure)} \tag{26}$$

$$HOCH_2CH_2CH_2CH_2CHO \quad \rightleftharpoons \quad \text{(cyclic structure)} \tag{27}$$

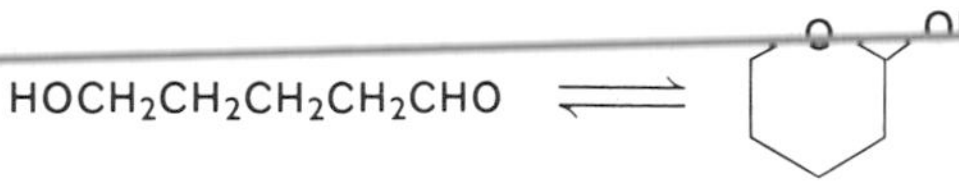

This situation is met in the sugars, as illustrated in the following example (this will be discussed fully in Chapter 18).

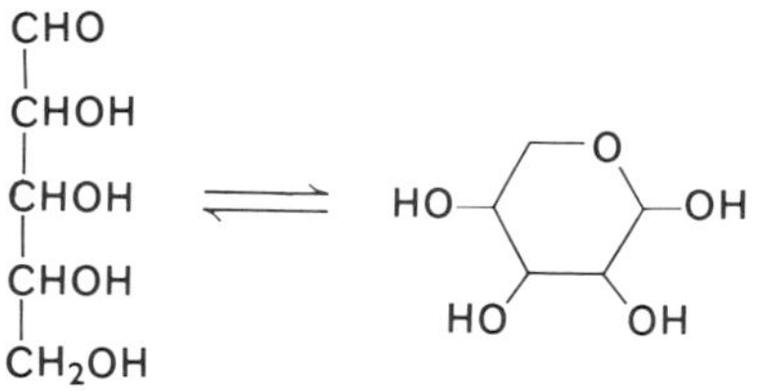

$$\tag{28}$$

D-ribose

D-ribopyranose
(cyclic hemiacetal form)

The formation of *acetals* as a stage subsequent to hemiacetal formation has been discussed in detail in Chapter 8 and should be restudied at this time. The ready formation of *cyclic* ketals and acetals, with the use of glycols, was also described in Chapter 8.

## 12.15 Effect of structure upon reactivity in additions to carbonyl compounds

The principle of the carbonyl addition reaction has so far been illustrated with the use of general structures RCHO and $R_2CO$, or with rather simple examples of the aldehydes and ketones involved. It should be noted now that the nature of the substituents R and R′ in R—CO—R′ has a profound influence in determining the rate at which addition to the C=O group will take place and in determining the position of the final equilibrium.

As the R groups in $R_2CO$ become more complex, as in the series H, $CH_3$, $CH_3CH_2$, $(CH_3)_2CH$, $(CH_3)_3C$, the rate of addition to the carbonyl group decreases, and the position of the equilibrium tends to lie more on the side of reactants than of products. Since both steric and electronic factors play a part

**Table 12-3.** *Addition of bisulfite to some carbonyl compounds*

| Carbonyl Compound | % Yield of $R_2C\begin{smallmatrix}OH\\SO_3K\end{smallmatrix}$ |
|---|---|
| $CH_3CHO$ | 80 |
| $CH_3COCH_3$ | 22 |
| $CH_3COCH_2CH_3$ | 14 |
| $CH_3COCH(CH_3)_2$ | 3 |
| $CH_3CH_2COCH_2CH_3$ | 2 |

in these reactions, this rule is not invariably valid for substituents of all kinds, but it is generally useful for comparing substituents of the same type though of different degrees of complexity, such as a series of alkyl groups.

In general, aldehydes and methyl ketones (RCOR', in which one R is —CH$_3$) react more rapidly and completely in addition reactions than do ketones in which both R groups are higher alkyl groups. As an example, the yields of the bisulfite addition compounds from a series of aliphatic carbonyl compounds ($KHSO_3$ was used, and the experimental conditions were the same in each case) are given in Table 12-3.

Why does increasing substitution at the carbonyl carbon atom reduce the capacity for the carbonyl group to undergo addition reactions? If we compare examples of a given reaction (addition of HCN or of NaHSO$_3$), we can ignore small differences in bond energies as a factor. Two other prominent factors remain: (1) loss or gain in resonance stabilization in reactants compared with products; (2) steric effects. Since in a series of alkyl ketones and aldehydes the resonance effects will be expected to be nearly the same from one case to another, it appears that the steric factors are the most important in determining the course of the addition reaction.

Steric factors can influence these reactions in two ways: they can affect the energy required for the formation of the transition state (the activation energy) and thus influence the *rate* of the reaction; and they can affect the stability of the final product (and the overall energy change in the reaction) and thus influence the equilibrium constant of the reaction.

The data of Table 12-3 reflect the equilibrium constant of the bisulfite addition reaction. These results can be accommodated to the steric factors involved in the following way. In the overall (general) reaction

$$\begin{smallmatrix}R\\R\end{smallmatrix}C{=}O + :A^- \rightleftharpoons \begin{smallmatrix}R\\R\end{smallmatrix}C\begin{smallmatrix}O^-\\A\end{smallmatrix} \xrightarrow{H^+} \begin{smallmatrix}R\\R\end{smallmatrix}C\begin{smallmatrix}OH\\A\end{smallmatrix} \tag{29}$$

the configuration of the carbonyl carbon atom changes from the trigonal arrangement (bond angles 120°) in the aldehyde or ketone to the tetrahedral arrangement in the addition product. Thus, the course of the reaction involves a "crowding" of the R groups by forcing them into closer proximity. These repulsive forces are reflected in the overall energy change in the reaction, and it is clear that the larger and more complex the R groups, the greater the repulsive forces and the less stable the final configuration.

Kinetic factors also involve steric effects. The attack of $:A^-$ upon the carbonyl carbon atom, leading to a transition state that probably resembles the initial addition product, will be progressively more blocked or hindered by increasingly complex R groups. Consequently, the activation energy required for the formation of the initial intermediate will increase with increasing hindrance to the initial attack. The steric effect here will be upon the *rate* of the reaction. It will be apparent however, that a given steric factor may influence *both* rate and equilibrium.

Another factor that affects the addition reaction is related to the fact that in the reaction (both in the attack that leads to the formation of the transition state complex, and in the overall reaction to the final product) the carbon atom of the resonance-stabilized carbonyl group is altered to the tetracovalent condition in which $C{=}O$ resonance no longer exists.

In a most illuminating experiment, it has been shown that acetaldehyde reacts with semicarbazide (Section 12.9) 100 times faster than does benzaldehyde; yet the equilibrium constant for the formation of benzaldehyde semicarbazone is seven times as great as that for the formation of acetaldehyde semicarbazone. In this case the initial attack upon the acetaldehyde carbonyl group is not impeded by unusual steric factors, but does involve some sacrifice of resonance stabilization in the carbonyl group, whereas the initial attack upon the benzaldehyde carbonyl group leads to a transition state in which conjugation of the carbonyl group with the benzene ring has disappeared. The final products in both cases contain the grouping $>C{=}N{-}NH{-}CO{-}NH_2$; thus, the final delocalization (resonance) energy provided by additional conjugation of the $C{=}N$ double bond with the benzene ring favors the equilibrium leading to the benzaldehyde derivative.

It is interesting to note that the three-membered ring ketone, cyclopropanone, is exceedingly difficult to prepare in a pure state, because in the presence of water or alcohols it forms a very stable hydrate or hemiketal. The following equilibrium lies completely on the side of the addition product:

$$
\underset{H_2C{-}CH_2}{\overset{CO}{\triangle}} + H_2O \ \rightleftharpoons \ \underset{H_2C{-}CH_2}{\overset{HO{\diagdown}C{\diagup}OH}{}} \tag{30}
$$

The $C{-}CO{-}C$ angle in cyclopropanone is $60°$, and thus the "distortion" from the normal $120°$ angle (as in acetone) is (for the two angles) $60°$. In the hydrate the "distortion" of the $60°$ $C{-}C{-}C$ angle is the difference between $60°$ and the normal tetrahedral angle of $109°$, or only $49°$. Thus the formation of the hydrate involves a large decrease in the strain in the bond angles. This is reflected in the completeness of the hydration equilibrium.

Cyclohexanone undergoes addition reactions much more readily than diethyl ketone, to which it is similar in the degree of substitution of the carbonyl carbon atom. This greater reactivity of cyclohexanone than of diethyl ketone has been explained in the following way: *cyclohexane* (all carbon atoms tetrahedral) is known to be a nearly strainless compound, and so any departure of the bond angles from 109.5° would be expected to lead to strain, and thus to lower stability. In *cyclohexanone* one of the carbon atoms (the —C=O) is nearly trigonal (C—C=O angle 120°), and thus the formation of an addition compound, such as the cyanohydrin, would tend to bring this carbon atom back to the preferred tetrahedral configuration. Furthermore, since the carbon atoms are "tied back" in the ring, there would be no increase in the crowding of these atoms, as there would be in the case of diethyl ketone, when addition occurred. In summary,

$$\underset{Et}{\overset{Et}{>}}C=O \xrightarrow[HX]{add} \underset{Et}{\overset{Et}{>}}C\underset{X}{\overset{OH}{<}}$$

$>$C=O trigonal          tetrahedral, "crowding" of Et groups

$$\begin{array}{c} H_2C—CH_2 \\ H_2C \qquad C=O \\ H_2C—CH_2 \end{array} \xrightarrow[HX]{add}$$

$>$C=O trigonal          tetrahedral, more favorable as part of six-membered ring

### 12.16 Tests for aldehydes and ketones and differentiation between them

The addition of bisulfite and HCN is not a property of aldehydes only, but of certain kinds of ketones (e.g., R—COCH₃) as well. Thus, the classification of an unknown carbonyl compound as an aldehyde cannot safely be based solely upon the observation that it forms a cyanohydrin or a bisulfite addition compound. On the other hand, a carbonyl compound that does not react in either of these ways is probably not an aldehyde. The clearest distinction between aldehydes and ketones lies in the ease with which aldehydes can be oxidized to acids with certain reagents, and the stability of ketones to these reagents. Two oxidizing agents that are commonly used for this purpose are *Tollens' reagent* and *Fehling's solution* (or the very similar Benedict's reagent). Tollens' reagent is an ammoniacal solution of the silver ion-ammonia complex; it is readily reduced by easily oxidized compounds with the formation of metallic silver (as a black precipitate or a silver mirror).

$$Ag^+(NH_3)_2 + RCHO \longrightarrow RCOOH + \underline{\underline{Ag}} \tag{31}$$

The criterion of a positive test is the appearance of metallic silver; the organic products of the oxidation are usually not demonstrated as the test is ordinarily performed.

Fehling's solution and Benedict's reagent are both cupric complexes, and a readily oxidized compound will reduce cupric ion to the cuprous state, with the formation of an orange to red precipitate of cuprous oxide ($Cu_2O$).

*Neither Tollens' reagent nor Fehling's solution is specific for aldehydes*, since other easily oxidized compounds will also give positive tests. But if the question is one of distinguishing between simple aldehydes and ketones, only aldehydes will give positive tests, ketones will not. In Chapter 18 it will be noted that the sugar fructose will reduce Fehling's solution and Tollens' reagent. Fructose is a hydroxy ketone, containing the grouping —CO—$\overset{|}{C}$H—OH; and it is found that Fehling's solution will also oxidize other compounds with this structure. For example, acetoin, $CH_3COCHOHCH_3$, will give a positive Fehling's test. Simple dialkyl ketones, however, will not.

Aldehydes can also be recognized by their ability to react with *Schiff's reagent*. This reagent is formed by adding sulfur dioxide to a solution of a magenta dye called "fuchsin," with the decolorization of the dye as a result of the addition to it of $SO_2$. Aldehydes react with the solution of the decolorized dye in a complex reaction that results in the reappearance of a magenta color (which is not identical with that of the original dye). Ketones do not restore the color.

---

**Exercise 4.**   Prepare a summary in tabular form of the behavior of aldehydes and ketones with the diagnostic and derivative-forming reagents that have been discussed.

---

## 12.17   The haloform reaction

Ketones that contain the grouping —$COCH_3$, such as acetone, methyl ethyl ketone, and methyl isobutyl ketone, react with alkaline solutions of bromine, chlorine, and iodine (i.e., with sodium hypobromite, hypochlorite, and hypo-iodite) to furnish the corresponding haloform (bromoform, $CHBr_3$, and so on) and the salt of the acid RCOOH (from $RCOCH_3$):

$$CH_3COCH_3 + 3NaOBr^* \longrightarrow CH_3CO\overset{-}{O}\overset{+}{Na} + CHBr_3 \qquad (32)$$

$$CH_3COCH_2CH\begin{subarray}{l} CH_3 \\ \diagdown \\ CH_3 \end{subarray} + 3NaOCl \longrightarrow (CH_3)_2CHCH_2CO\overset{-}{O}\overset{+}{Na} + CHCl_3 \qquad (33)$$

$$\begin{subarray}{l} H_3C \diagdown \\ \phantom{H_3}C=CHCOCH_3 + 3NaOI \\ H_3C \diagup \end{subarray} \longrightarrow \begin{subarray}{l} H_3C \diagdown \\ \phantom{H_3}C=CHCO\overset{-}{O}\overset{+}{Na} + CHI_3 \\ H_3C \diagup \end{subarray} \qquad (34)$$

---

\* Formation of hypobromite:

$$HO^- + Br-Br \rightleftharpoons HOBr + Br^-; \qquad HOBr + HO^- \rightleftharpoons H_2O + OBr^-$$

This reaction, called the "*haloform* "*reaction*, proceeds in two stages: the first (using NaOCl as the example) is the halogenation of the —COCH$_3$ group to —COCCl$_3$; the second is the cleavage of the resulting RCOCCl$_3$ by the hydroxide present in the alkaline solution.

$$CH_3CH_2COCH_3 + NaOCl \longrightarrow CH_3CH_2C{\overset{O}{\underset{CCl_3}{}}}$$

$$CH_3CH_2C{\overset{O}{\underset{CCl_3}{}}} \; \rightleftharpoons \; CH_3CH_2C{\overset{O}{\underset{CCl_3}{}}} \; \rightleftharpoons \quad (35)$$
$$HO: \qquad\qquad\qquad\qquad HO$$

$$\left\{ CH_3CH_2C{\overset{O}{\underset{OH}{}}} + :CCl_3 \right\} \longrightarrow CH_3CH_2C{\overset{O}{\underset{O^-}{}}} + CHCl_3$$

Reaction (35), it can be seen, is another example of the general reaction (3). The nucleophilic attack of OH$^-$ on the carbonyl carbon atom is aided by the strong inductive withdrawal of electrons by the three chlorine atoms of the —CCl$_3$ group. This inductive reduction of electron density on the carbon atom of —CCl$_3$ also permits its dissociation as the negative (:CCl$_3$)$^-$ ion. The process is completed by the protonation of :CCl$_3^-$ (which is a very strong base), to give CHCl$_3$.

In the case of the equilibria (35) the reaction is complete to the right because in the alkaline solution the ionization of the acid CH$_3$CH$_2$COOH is complete to give CH$_3$CH$_2$COO$^-$, to which carbonyl addition does not occur.

The haloform reaction is carried out by adding a solution of sodium hypochlorite or hypobromite to the methyl ketone, or by adding iodine (dissolved in aqueous potassium iodide) to a solution of the ketone in aqueous or methanolic alkali. For diagnostic (as contrasted with preparative) purposes iodine is commonly used because iodoform is a crystalline yellow solid which can be readily isolated and identified by its appearance and melting point; both bromoform and chloroform are liquids at ordinary temperature.

### 12.18 The haloform reaction of compounds that are not methyl ketones

Although the haloform reaction is characteristic of methyl ketones RCOCH$_3$, it is also given by (1) compounds that can be oxidized by the reagent to produce this grouping; (2) compounds that can be hydrolyzed to produce the structural unit —COCH$_3$ or its halogenated forms —COCH$_2$X, —COCHX$_2$; (3) acetaldehyde (and ethanol, which can be oxidized by hypohalites to acetaldehyde).

For example, secondary alcohols containing the structure —CHOHCH$_3$ (e.g., *sec*-butyl alcohol, isopropyl alcohol) can be oxidized by hypohalites to the corresponding ketones.

In addition to these, compounds that produce such alcohols by hydrolysis under the (alkaline) conditions of the reaction will give the haloform reaction. Thus, ethyl, isopropyl, and *sec*-butyl esters, which hydrolyze in alkali to give the respective alcohols, can give positive haloform tests.

It is, of course, easy to distinguish between $-CHOHCH_3$ and $-COCH_3$ in the case of an unknown compound which is found to give a positive haloform test, since the application of a carbonyl reagent such as hydroxylamine or phenyl-hydrazine will show whether or not a carbonyl group is present.

---

**Exercise 5.** Which of the following compounds will give a positive haloform test? $CH_3CH_2CH=CHCOCH_3$; $CH_3CHO$; $CH_3CHOHCH_2CH_2CH_2CH_3$; $CH_3CH_2CHO$; $CH_3CH_2CHOHCH_2CH_3$; $(CH_3)_2C=CHCOCH_2CH_3$; $(CH_3)_2CHCOCH(CH_3)_2$.

---

Finally, it must be noted that a compound such as methyl acetoacetate, $CH_3COCH_2COOCH_3$, which might be expected to give a positive haloform test, does not do so because halogenation occurs on the $-CH_2-$ group rather than on the methyl group. Subsequent alkali hydrolysis (cleavage) yields (when NaOBr is used) dibromoacetic and acetic acids (as their sodium salts):

$$CH_3COCH_2COOCH_3 + NaOBr \longrightarrow CH_3COCBr_2COOCH_3 \xrightarrow{\ H_2O/OH^-\ }$$
$$CH_3COO^- + Br_2CHCOO^- + CH_3OH$$

No haloform is formed from acetic or halogen-substituted acetic acids because their anions are not halogenated by alkaline hypohalites. Consequently, methyl acetoacetate does not give a positive haloform test.

## 12.19 Addition of haloforms to carbonyl compounds

It can be anticipated from the equation (35) that $CHCl_3$ in the presence of a base should be capable of adding to a carbonyl group. This is indeed true:

$$CH_3COCH_3 + CHCl_3 \underset{}{\overset{OH^-}{\rightleftharpoons}} \begin{array}{c} H_3C \\ H_3C \end{array}\!\!>\!\!C\!\!<\!\!\begin{array}{c} OH \\ CCl_3 \end{array} \quad \text{``chloretone''} \tag{36}$$

The addition of bromoform gives the corresponding tribromo compound "brometone." These trihalo-*t*-butyl alcohols have been used in medicine as hypnotics and sedatives; they are similar in structure and action to chloral hydrate, $Cl_3C \cdot CH(OH)_2$, a commonly used hypnotic drug.

This addition of chloroform or bromoform to the carbonyl group is, of course, simply another of the class of addition reactions that proceed by nucleophilic attack upon the carbonyl carbon atom:

$$\begin{array}{c} H_3C \\ H_3C \end{array}\!\!>\!\!C=O \ ^\curvearrowright H:B \longrightarrow \begin{array}{c} H_3C \\ H_3C \end{array}\!\!>\!\!C\!\!<\!\!\begin{array}{c} OH \\ CBr_3 \end{array} \tag{37}$$
$$B:^- \curvearrowright H\!\!-\!\!CBr_3 \qquad\qquad\quad + H:B + :B^-$$

### 12.20 Reduction of carbonyl compounds

The addition of hydrogen to the carbonyl double bond leads to the formation of primary alcohols from aldehydes,

$$RCH{=}O \xrightarrow{[H]} RCH_2OH$$

and secondary alcohols from ketones:

$$\underset{R}{\overset{R}{>}}C{=}O \xrightarrow{[H]} \underset{R}{\overset{R}{>}}CHOH \tag{38}$$

The reduction of carbonyl compounds has been discussed in Chapter 8, in which the use of catalytic hydrogenation and of lithium aluminum hydride was described. Another reagent related to $LiAlH_4$ is sodium borohydride, $NaBH_4$. This reagent is useful for the reduction of carbonyl groups of aldehydes and ketones but not those of acid derivatives (e.g., esters). Lithium aluminum hydride, on the other hand, reduces acids and acid derivatives as well as aldehydes and ketones.

---

**Exercise 6.** Refer to the reduction of an aldehyde with $LiAlH_4$ (see Chapter 8) and write the reaction for reduction by $NaBH_4$.

---

The reduction of carbonyl compounds with the aid of an aluminum alkoxide, described briefly in Section 8.21, can now be discussed in greater detail. Let us take as an example the reduction of crotonaldehyde by isopropyl alcohol, with aluminum isopropoxide as the catalyst. Aluminum isopropoxide is a Lewis acid (Section 4.12) and can form a coordination compound with the aldehyde by accepting into its vacant $3p$ orbital one of the unshared electron pairs on the carbonyl oxygen atom:

$$CH_3CH{=}CHCH{=}\overset{..}{\underset{..}{O}}: + Al(OiPr)_3 \rightleftharpoons CH_3CH{=}CHCH{=}\overset{+}{O}{-}\overset{-}{\underset{..}{Al}}(OiPr)_3 \tag{39}$$

The resulting complex has the following structure, in which the carbonyl carbon atom is strongly electrophilic as a result of the coordination with the aluminum atom (in the steps shown, the —OiPr groups that are not involved will be written —OR);

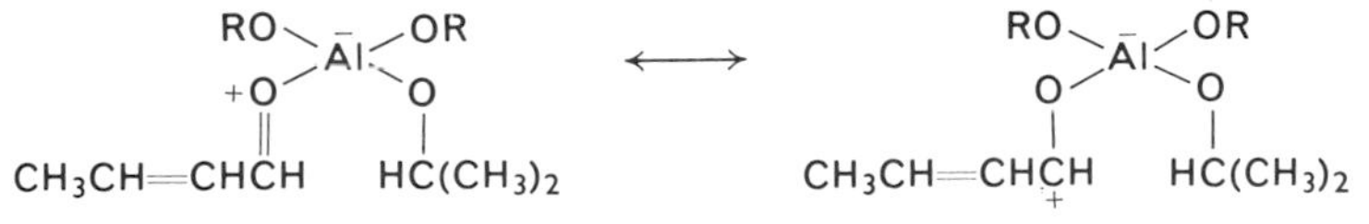

$$
\begin{array}{ccc}
\underset{CH_3CH{=}CHCH}{\overset{RO}{\underset{+O}{>}}}\overset{-}{Al}\underset{HC(CH_3)_2}{\overset{OR}{<}} & \longleftrightarrow & \underset{CH_3CH{=}CH\overset{+}{C}H}{\overset{RO}{\underset{O}{>}}}\overset{-}{Al}\underset{HC(CH_3)_2}{\overset{OR}{<}}
\end{array} \tag{40}
$$

Transfer of a hydrogen atom, with its binding electron pair, by the process indicated by the arrows, leads to the equilibrium

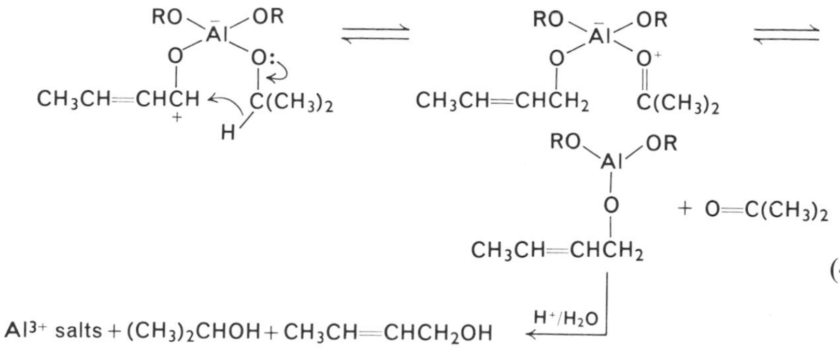

$$\text{(41)}$$

which can be shifted to the right (with the formation of the reduction product, crotyl alcohol) by the continuous removal of the acetone that is formed by distilling it out of the reaction mixture. The crotyl alcohol that is formed is, of course, present in the form of the mixed aluminum alkoxide, and is isolated after destruction of the complex by acid hydrolysis.

A typical application of this reaction, which is known as the *Meerwein-Ponndorf-Verley* reduction, is in the preparation of vitamin $A_1$ by the reduction of the corresponding aldehyde, retinene$_1$:

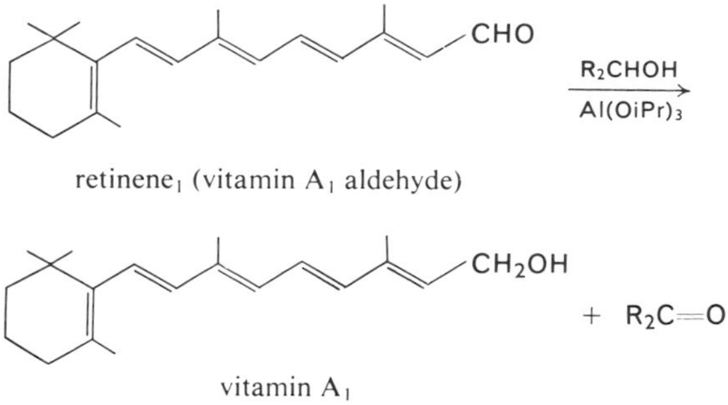

retinene$_1$ (vitamin $A_1$ aldehyde)

$$\text{(42)}$$

vitamin $A_1$

Retinene could not be reduced to vitamin A by catalytic hydrogenation because of the susceptibility of the carbon-carbon double bonds to reduction.

The use of lithium aluminum hydride and sodium borohydride, reagents developed since the discovery of the Meerwein-Ponndorf-Verley reaction, has in many cases supplanted the latter method.

## 12.21   The pinacol reduction

The reaction of acetone with magnesium (which has been activated by amalgamation with mercury by treatment with mercuric chloride) yields a solid magnesium derivative which is decomposed by hydrolysis into a magnesium salt (of the acid used for the hydrolysis) and *pinacol:*

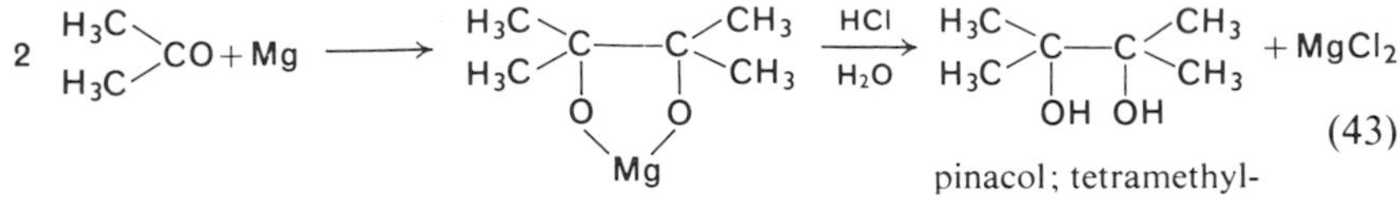

$$(43)$$

pinacol; tetramethyl-
ethylene glycol

The "pinacol reduction" is a one-electron reduction (per molecule of ketone), and proceeds by the acceptance of the two electrons of magnesium by two molecules of acetone, followed by pairing of the lone electrons to form a carbon-carbon bond:

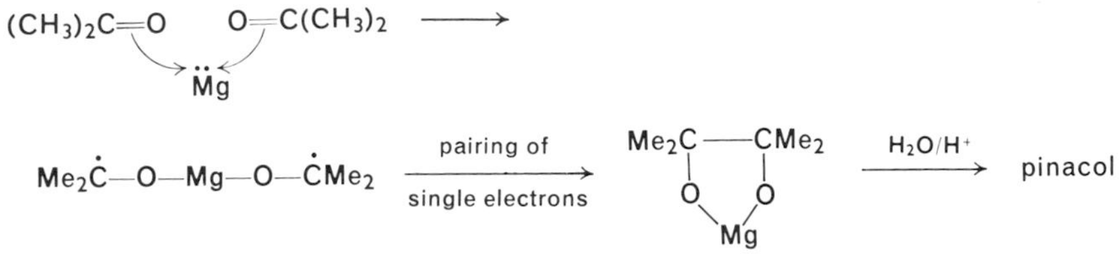

The reduction products formed from the ketones $R_2CO$ are often called the "pinacols" of the corresponding ketones:

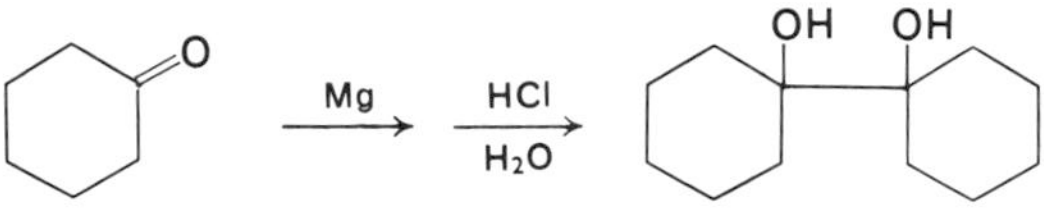

$$(44)$$

cyclohexanone

dicyclohexyl-1,1'-diol
"the pinacol of cyclohexanone"

---

**Exercise 7.** Formulate the pinacol reduction of *(a)* diethyl ketone; *(b)* 2 7-octanedione.

---

The pinacol reduction of compounds which contain two carbonyl groups can lead to ring formation if the ring so formed has five or six members:

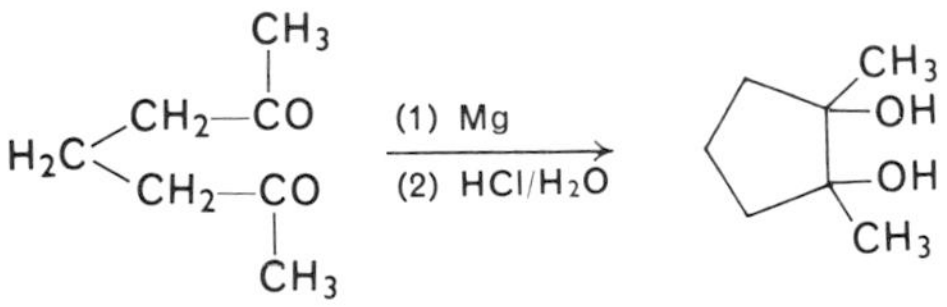

$$(45)$$

2,6-heptanedione            1,2-dimethylcyclopentane-1,2-diol

## 12.22  Dehydration of pinacols: the pinacol rearrangement

When pinacol is treated with hot, dilute sulfuric acid the elements of water are removed, and a ketone, *pinacolone*, is formed:

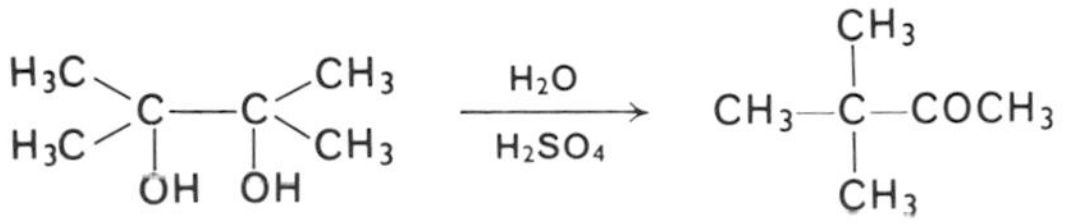

$$(46)$$

It is apparent that in this reaction a rearrangement of the carbon skeleton has occurred.

An analysis of this reaction in terms of what we have already learned of the behavior of alcohols with acidic reagents will clarify the course of the rearrangements. Let us examine the stages by which it proceeds.

The initial stage of the reaction is a simple acid-base reaction in which the basic hydroxyl group accepts a proton from the strong acid $H_3O^+$, to lead to the protonated glycol:

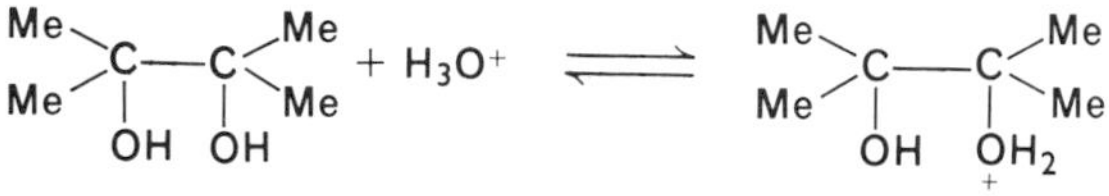

(47)                              (48)

The ion (48), a protonated tertiary alcohol, readily ionizes, losing $H_2O$ and forming the carbonium ion (49):

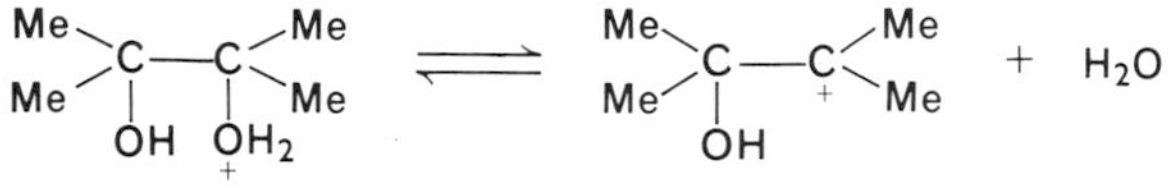

(48)                              (49)

The migration of a methyl group in (49) from $(CH_3)_2C-$ to $-\overset{+}{C}(CH_3)_2$ now

occurs. This shift of the methyl group is seen to be a very favorable process because it results in the generation of a *better* carbonium ion (50) (i.e., one with greater resonance stabilization):

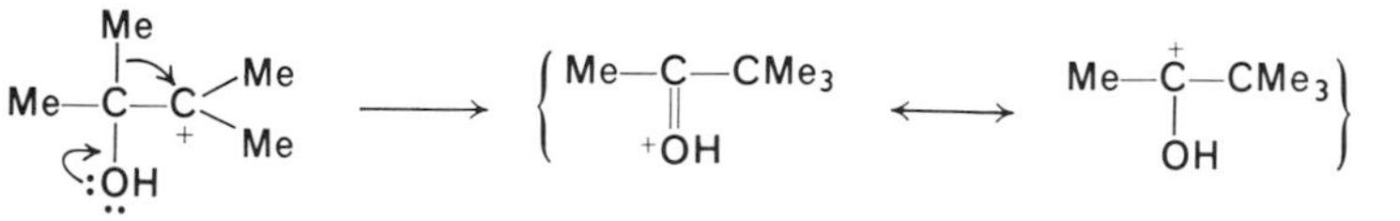

(49)                              (50)

The new ion (50) is seen to be the protonated form of *pinacolone*, and, being a carbonyl compound, possesses the resonance stabilization of the protonated carbonyl group, as described by the way in which (50) is written.

The carbonium ion intermediate shown in the description of the course of the pinacol rearrangement is not isolable; it would be represented by a high-energy minimum in an energy profile of the reaction. Indeed, it appears from stereochemical evidence that the migration of the methyl group occurs as a concomitant of the departure of the water molecule, because the configuration of the pinacol determines that of the ketone that is formed, as shown by the following diagram:

(51)

Were the carbonium ion to become "free" it would be expected that rotation about the C—C single bond would take place and that both enantiomers of the ketone would be formed. The experimental result is that the carbon atom to which the group (R‴ in the equation) migrates is inverted as shown in Equation 51.

### 12.23   The Wittig synthesis of olefins

An important synthetic use of aldehydes and ketones is in the preparation of olefinic compounds by the Wittig synthesis. The overall reaction, illustrated first in general terms, is the following:

*(a)* Preparation of the Wittig reagent:*

$$(C_6H_5)_3P + RCH_2I \longrightarrow (C_6H_5)_3\overset{+}{P}CH_2R\}I^- \tag{52}$$

$$(C_6H_5)_3\overset{+}{P}CH_2R + RLi \longrightarrow \{(C_6H_5)_3\overset{+}{P}\overset{\cdot\cdot}{C}HR \longleftrightarrow (C_6H_5)_3P{=}CHR\} + RH + Li^+ \tag{53}$$

The structure of $R_3P{=}CHR$ is a valid one, in contrast to $R_3N{=}CHR$, which is not, because phosphorus can utilize additional orbitals in the *M* shell and accommodate ten electrons in its valence shell.

*(b)*   Reaction of the Wittig reagent with an aldehyde or ketone:

$$\underset{R}{\overset{R}{>}}C{=}O + (C_6H_5)_3\overset{+}{P}{-}\overset{-}{C}HR' \longrightarrow$$

$$\underset{R}{\overset{R}{>}}\underset{\underset{R'CH{-}\overset{+}{P}(C_6H_5)_3}{|}}{C{-}O^-} \longrightarrow \underset{R}{\overset{R}{>}}C{=}CHR' + (C_6H_5)_3PO \tag{54}$$

The reaction probably proceeds through a four-center transition state that can be pictured as

$$\underset{R}{\overset{R}{\diagdown}}\underset{\underset{R'CH{=\!=}\overset{}{P}(C_6H_5)_3}{|\quad\quad|}}{C{=\!=\!=}O}$$

An interesting extension of the Wittig olefin synthesis involves the use of chloromethyl ether in place of the alkyl halide shown in the above equations. The course of the reaction, including the initial preparation of the Wittig reagent and its addition to a ketone, is shown in the following equations; the final product is an aldehyde:

---

* The group $C_6H_5$— in these reactions is the phenyl group, and the reagent $(C_6H_5)_3P$ is triphenylphosphine. Triphenylphosphine can be prepared by the reaction of phenyl-magnesium bromide, $C_6H_5MgBr$, with phosphorus trichloride, $PCl_3$.

$$CH_3OCH_2Cl \xrightarrow[\text{(2) BuLi}]{\text{(1) } (C_6H_5)_3P} (C_6H_5)_3P\!\!=\!\!CHOCH_3 \qquad (55)$$

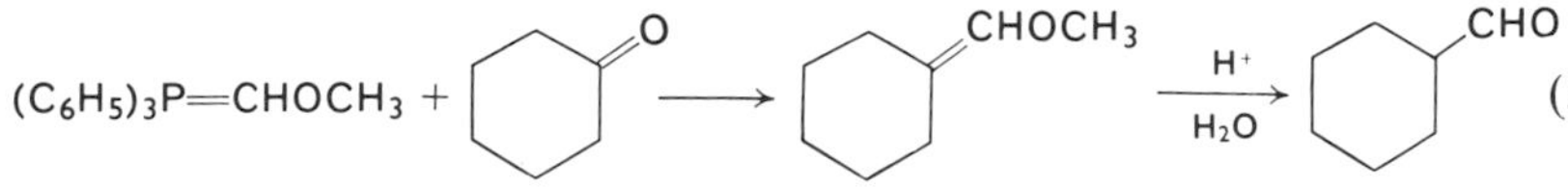

$$(C_6H_5)_3P\!\!=\!\!CHOCH_3 + \overset{O}{\bigcirc} \longrightarrow \overset{CHOCH_3}{\bigcirc} \xrightarrow[\text{H}_2O]{\text{H}^+} \overset{CHO}{\bigcirc} \qquad (56)$$

---

**Exercise 8.** Formulate the steps in the acid-catalyzed hydrolysis of $\overset{\;\;CHOCH_3}{\bigcirc}$ to

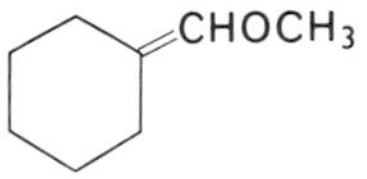

yield $\overset{CHO}{\bigcirc}$ . It may be assumed that protonation of the double bond will be the first

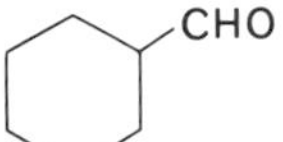

step. Will this lead to the ion $\overset{CHOCH_3}{\bigcirc}$ or $\overset{CH_2OCH_3}{\bigcirc}$ Why?

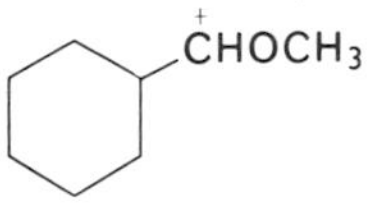

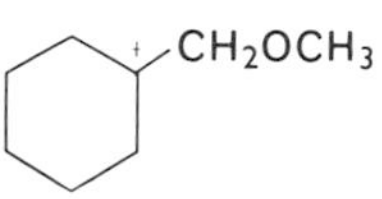

---

Another variant of the general Wittig synthesis is that in which an α-bromo ester is used. The Wittig reagent that results is formed according to the following equation:

$$(C_6H_5)_3P + BrCH_2COOR \longrightarrow (C_6H_5)_3\overset{+}{P}CH_2COOR\}Br^- \xrightarrow{OH^-} (C_6H_5)_3P\!\!=\!\!CHCOOR \quad (57)$$

---

**Exercise 9.** Devise a synthesis of $\overset{CH=CH-COOCH_3}{\bigcirc}$ starting with cyclohexa-none.

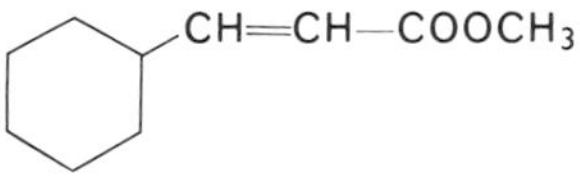

---

## 12.24 Applications to Synthesis

1. Prepare 4-heptanone, starting with compounds containing three carbon atoms or less. As with other syntheses, it is convenient to work this out in the reverse order of steps.

*(a)* 4-Heptanone can be prepared by oxidation of the corresponding secondary alcohol:

$$CH_3CH_2CH_2\underset{\underset{OH}{|}}{C}HCH_2CH_2CH_3 \xrightarrow{CrO_3} CH_3CH_2CH_2COCH_2CH_2CH_3$$

*(b)* The required 4-heptanol can be prepared by the Grignard synthesis:

$$CH_3CH_2CH_2CHO + CH_3CH_2CH_2MgBr \longrightarrow \text{4-heptanol}$$

*(c)* The *n*-butyraldehyde required in *(b)* can be prepared by the oxidation of 1-butanol, which can be prepared from 1-propanol by the following route:

$$n\text{-PrOH} \xrightarrow{HBr} n\text{-PrBr} \xrightarrow[\text{ether}]{Mg} n\text{-PrMgBr} \xrightarrow{H_2CO} n\text{-C}_3H_7CH_2OH$$

2. Prepare 1,6-hexanediol from 1,2-dimethylcyclohexene.

*(a)* The diol can be prepared from the corresponding acid by reduction with lithium aluminum hydride:

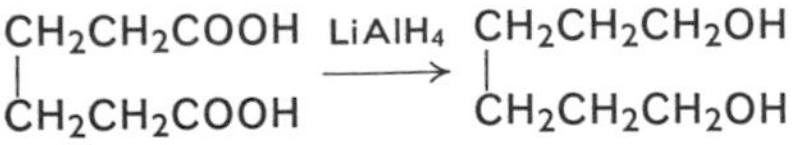

$$\begin{array}{ccc}
CH_2CH_2COOH & \xrightarrow{LiAlH_4} & CH_2CH_2CH_2OH \\
| & & | \\
CH_2CH_2COOH & & CH_2CH_2CH_2OH
\end{array}$$

*(b)* Adipic acid, used in *(a)*, can be prepared by the hypohalite oxidation of 2,7-octanedione:

$$CH_3COCH_2CH_2CH_2CH_2COCH_3 \xrightarrow{NaOBr} HOOC(CH_2)_4COOH + 2CHBr_3$$

*(c)* The diketone can be prepared by ozonization of the starting material, 1,2-dimethylcyclohexene.

3. Prepare 2-methylbutanal, starting with methyl ethyl ketone.

*(a)* 2-Methylbutanal is the product of the pinacol rearrangement of the following glycol:

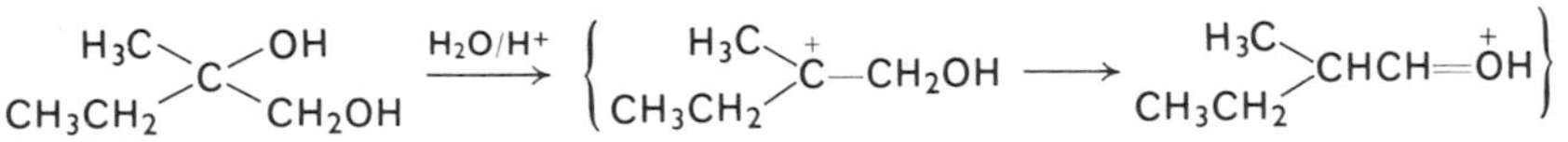

*(b)* The required glycol can be prepared by $LiAlH_4$ reduction of the corresponding hydroxy acid. The acid can be prepared by hydrolysis of the cyanohydrin of the starting ketone. These reactions are straightforward and can be formulated by the student.

4. Prepare trimethylacetic acid, starting with acetone. The series of reactions used for this synthesis can be summarized as follows (the student should write them out in detail):

$$\text{acetone} \xrightarrow{Mg/Hg} \text{pinacol} \xrightarrow{H_2O/H_2SO_4} \text{pinacolone} \xrightarrow{NaOCl} \text{trimethylacetic acid} + CHCl_3$$

5. Prepare cycloheptanone, starting with cyclohexanone. The following series of steps involve reactions that have been discussed in this and previous chapters. They should be worked out in detail by the student:

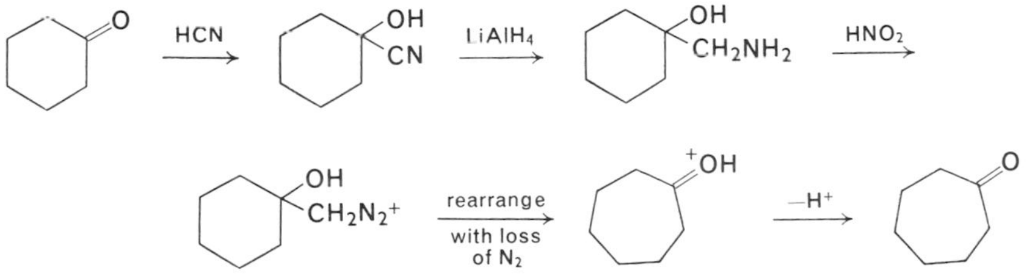

(not isolated)

Note: the ring-enlargement step in the above series bears a close mechanistic relationship to the pinacol rearrangement discussed earlier in this chapter. Show the details of this step.

## Exercises

(Exercises 1–9 will be found within the text.)

**10** Write the structural formulas of the following compounds: *(a)* *n*-butyraldehyde; *(b)* 3-methylpentanal; *(c)* methyl vinyl ketone; *(d)* propanal; *(e)* trimethylacetaldehyde; *(f)* allylacetaldehyde; *(g)* 2,3-dibromobutanal; *(h)* diisopropyl ketone; *(i)* 5-nonanone.

**11** Name each of the compounds of Exercise 10 in another way.

**12** Write the equations showing the reaction of acetone acting as a base, with the following acids. *(a)* HCl; *(b)* $MgCl_2$; *(c)* $AlCl_3$; *(d)* $BF_3$.

**13** Write the structures of the protonated forms (the conjugate acids) of the following compounds: *(a)* dimethyl ether; *(b)* acetaldehyde; *(c)* acetone; *(d)* cyclopentanone.

**14** Complete the following equations using structural formulas.

*(a)* acetone + HCN

*(b)* acetaldehyde + sodium bisulfite

*(c)* cyclohexanone + sodium bisulfite

*(d)* isobutyraldehyde + hydroxylamine

*(e)* diethyl ketone + hydrazine

*(f)* cyclopentanone + Girard's reagent ($Cl^-(CH_3)_3\overset{+}{N}CH_2CONHNH_2$)

*(g)* 2-hexanone + NaOBr

**15** Write the structures of the cyclic hemiacetal (or hemiketal) forms of *(a)* 5-hydroxy-pentanal; *(b)* 6-hydroxy-2-hexanone.

**16** 4-Aminobutanal is unstable, tending to lose the elements of water to form a compound $C_4H_7N$. What is the reaction involved and what is the structure of this product?

**17** 2-Aminocyclohexanone readily changes into a compound $C_{12}H_{18}N_2$, with the loss of two molecules of water. What is the structure of this product?

**18** Which of the following compounds will give iodoform with sodium hypoiodite: *(a)* acetaldehyde; *(b)* 3-methyl-2-butanone; *(c)* 3-pentanone; *(d)* acetic acid; *(e)* methyl-isopropylcarbinol; *(f)* methyl cyclopropyl ketone; *(g)* butanal; *(h)* 1-methylcyclo-pentanol; *(i)* ethanol; *(j)* methanol?

**19** Describe and write the equations for the experiments you would conduct to demonstrate that acetone has the constitution $CH_3COCH_3$.

**20** Write the structures for the lettered compounds in each of the following:

*(a)* A compound A when treated with sodium hypoiodite yields iodoform and succinic acid $HOOCCH_2CH_2COOH$.

*(b)* A neutral compound B ($C_4H_{10}O_2$) does not reduce Fehling's solution or Tollens' reagent but when treated with dilute HCl gives acetaldehyde and methanol.

*(c)* A compound C ($C_4H_6O_2$) gives a positive iodoform reaction and when oxidized with chromic acid (or any of a number of other oxidizing agents) gives 2 moles of acetic acid per mole of (C) and no other organic products.

**21** A compound A ($C_{11}H_{22}$) when ozonized yields B ($C_6H_{12}O$) and C ($C_5H_{10}O$). Mild oxidation of C yields D ($C_5H_{10}$) $O_2$. Compound B reacts with NaOCl to yield chloroform and the sodium salt of D. Reduction of B with $LiAlH_4$ yields E ($C_6H_{14}O$). Ozonolysis of F yields G($C_2H_4O$) and H($C_4H_8O$). Compound H does not react with Tollens' reagent or Fehling's solution but does form an oxime and a semicarbazone. Write these reactions using structural formulas throughout.

*Chapter thirteen* / **Acidity of Hydrogen Linked to Carbon. Aldol Condensations**

## 13.1 Structural effects upon acidity

In Chapter 4 the general principles of acid-base reactions were discussed, along with some of the factors that affect the acidity of hydrogen.* One of the important factors is the electron-attracting ability of the atom or group to which the hydrogen is attached. The last four elements of the first period form the simple hydrides $CH_4$, $NH_3$, $H_2O$, and HF. Of these, HF is an acid in the usual sense of the term; but $H_2O$ and $NH_3$, whose conjugate bases $OH^-$ and $NH_2^-$, are well-known entities, are also acids in the terms of the Brønsted-Lowry concept. Methane, on the other hand, although theoretically capable of undergoing proton transfer to a hypothetical base, B:⁻,

$$CH_4 + B{:}^- \rightleftharpoons {:}CH_3^- + B{:}H \qquad (1)$$

is in fact so weak an acid that this reaction is not experimentally observed. Although methane has no demonstrable acidic properties, metallo-organic derivatives of methane, such as $CH_3Na$, $CH_3Li$, and $CH_3K$, can be prepared in other ways. As would be expected, the $:CH_3^-$ ion is a powerful base capable of extracting a proton from most hydrogen-containing molecules. Other paraffin hydrocarbons are comparable to methane in acidity; the $pK_a$'s of compounds of this type are over 40.

---

* The term "acidity of hydrogen" refers to the degree of acidity of a compound from which a hydrogen atom can be removed, as a proton, by reaction with a base. In principle, any compound H:A can be transformed into the anion, $:A^-$, by removal of the proton by a base; in practice the acidity of hydrogen atoms in H: covers a very large range according to the nature of A.

*Inductive effects upon acidity.* The acidity of hydrogen attached to carbon is strongly influenced by other substituents attached to the same carbon atom, and by the nature of the hybridization of the C—H bond orbital. The acidity of acetylenic hydrogen (RC≡C—H) has been mentioned earlier (Section 9.24). Acetylene, H—C≡CH, with a $pK_a$ of about 26, is a very weak acid but is enormously more acidic than ethane. The additional inductive effect of nitrogen in H—C≡N causes a marked increase in the acidity of the hydrogen; hydrocyanic acid has a $pK_a$ of 9.4.

The ability of acetylene to form the nucleophilic ion, HC≡C:⁻ is shown by the addition of acetylene to ketones in the presence of a strong base:

$$HC{\equiv}CH + (CH_3)_2C{=}O \quad \xrightarrow[\text{NH}_3]{\text{NaNH}_2} \quad \begin{array}{c} H_3C \\ H_3C \end{array}\!\!C\!\!\begin{array}{c} C{\equiv}CH \\ OH \end{array} \tag{2}$$

$$HC{\equiv}CH + \text{(cyclohexanone)} \quad \xrightarrow[\text{tBuOH}]{\text{tBuOK}} \quad \text{(1-ethynylcyclohexanol)} \tag{3}$$

The base-catalyzed addition of acetylene to the carbonyl group is to be compared with the base-catalyzed addition of HCN to the carbonyl group.

Acetylenes are sufficiently acidic to react with Grignard reagents to form acetylenic halomagnesium derivatives, which can then be used as Grignard reagents:

$$CH_3C{\equiv}CH + CH_3MgBr \longrightarrow CH_3C{\equiv}CMgBr + CH_4 \tag{4}$$

$$CH_3C{\equiv}CHMgBr + (CH_3)_2C{=}O \longrightarrow CH_3{-}\!\!\begin{array}{c} OH \\ | \\ C \\ | \\ CH_3 \end{array}\!\!{-}C{\equiv}CCH_3 \tag{5}$$

2-methyl-3-pentyn-2-ol

---

**Exercise 1.** With the use of any starting materials having three carbon atoms or less, show how you could prepare $CH_2{=}CH{-}CH_2{-}C{\equiv}C{-}\underset{\underset{\displaystyle OH}{|}}{CH}{-}CH_3$.

---

Inductive effects upon acidity of the H—C bond in $sp^3$ hybridization are, while not strongly acidifying influences, clearly recognizable. Chloroform, bromoform, and fluoroform ($CHCl_3$, $CHBr_3$, $CHF_3$) are demonstrably acidic. The addition of $CHBr_3$ to the carbonyl group, through the intermediate formation of the ion: $CBr_3^-$, was described in Section 12.19. The acidity of the H—C bond in the haloforms must be due in part to the strong inductive effect of the electronegative halogen atoms; but that this is only part of their effect is shown by the fact that $CHCl_3$ is a stronger acid than $CHF_3$. Since the inductive effect of fluorine is greater than that of chlorine (fluoroacetic acid $CH_2FCOOH$, $pK_a$ 2.6, is a stronger acid than chloroacetic acid, $CH_2ClCOOH$, $pK_a$ 2.9), it might have been anticipated that the H—C bond in fluoroform would be more acidic than that in chloroform. It is probable that the ion $:CCl_3^-$ can be stabilized by resonance involving the $3d$ orbitals of chlorine:

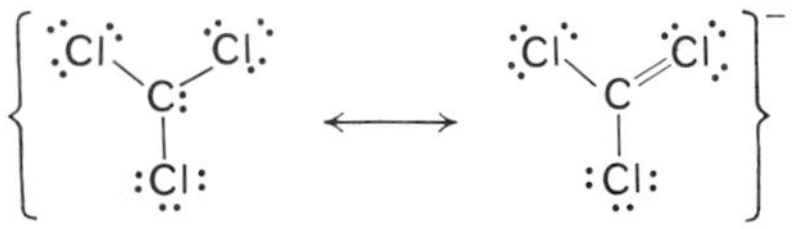

Chlorine, in the second period, has available orbitals in the $M$ shell. For fluorine, in the first period, no such "expansion" of the octet can provide stabilization to the $:CF_3^-$ ion.

Purely inductive effects upon the acidity of C—H bonds are uncommon; the most important influences for stabilization of carbon anions, $-\overset{|}{\underset{|}{C}}:^-$, are those provided by delocalization of the negative charge by resonance, or *resonance stabilization*.

***Resonance effects upon acidity.*** Reduction of electron density upon the carbon atom of the anion $-\overset{|}{\underset{|}{C}}:^-$ by delocalization of the electron pair over several atoms provides for stabilization of the ion and greatly enhances the acidity of the corresponding $-\overset{|}{\underset{|}{C}}:H$ compound.

The acidity of hydrogen *alpha* to carbonyl groups can be accounted for by stabilization of the corresponding anion by charge delocalization over the system $(\overset{..}{C}-C=\overset{..}{O}: \leftrightarrow C=C-\overset{..}{O}:)^-$:

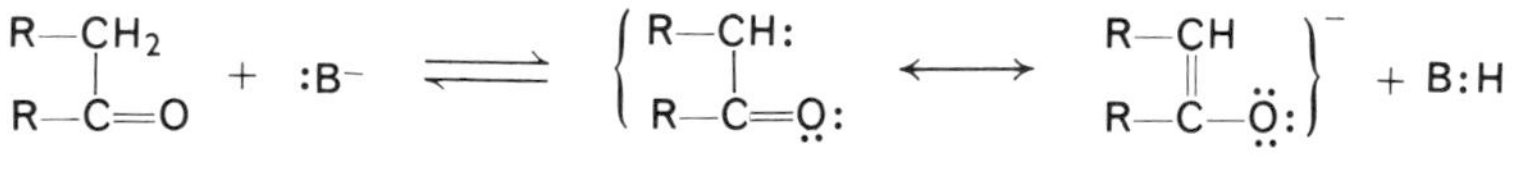

$$\begin{matrix} R-CH_2 \\ | \\ R-C=O \end{matrix} + :B^- \rightleftharpoons \left\{ \begin{matrix} R-CH: \\ | \\ R-C=\overset{..}{O}: \end{matrix} \longleftrightarrow \begin{matrix} R-CH \\ \| \\ R-C-\overset{..}{\underset{..}{O}}: \end{matrix} \right\}^- + B:H \qquad (6)$$

It can be seen that it is only *alpha*-hydrogen that is in a position to assume acidic character for this reason. Hydrogen atoms at the $\beta$- or more distant positions are more nearly comparable to those on a paraffinic hydrocarbon, and have no appreciable acidic character:

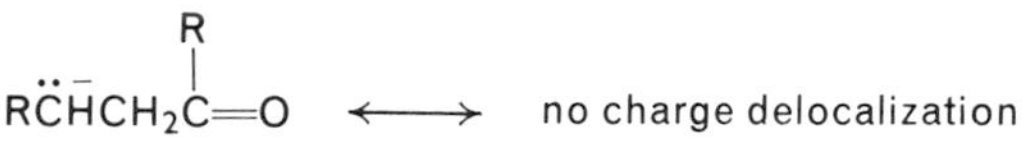

$$R\overset{..}{\underset{}{C}}\overset{-}{H}CH_2\overset{\underset{\displaystyle R}{|}}{C}=O \longleftrightarrow \text{no charge delocalization}$$

An energy diagram illustrating the difference between the two possible anions derived from propionaldehyde (propanal) is shown in Figure 13-1. It is seen that the $\beta$-anion is very much less stable than the $\alpha$-anion; it has very little more resonance stabilization than the anion of a saturated hydrocarbon (i.e., essentially none).

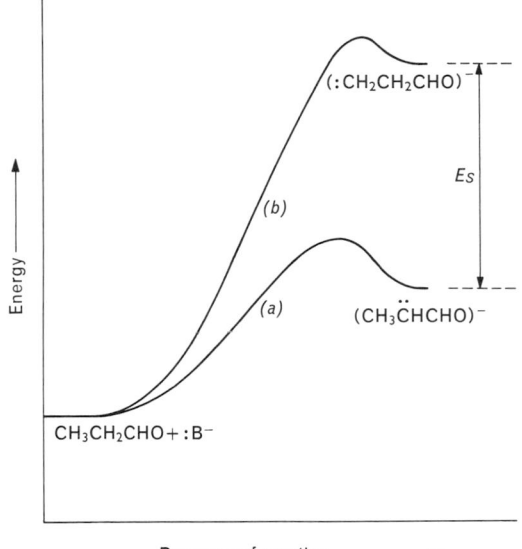

**Figure 13–1.** Energy diagram of ionization of $CH_3CH_2CHO$ by *(a)* removal of $\alpha$-proton, and *(b)* removal of $\beta$-proton. $E_s$ represents resonance stabilization of $\alpha$-anion.

The transition states illustrated in Figure 13-1 might be represented as follows for *(a)*,

$$\left\{ \begin{array}{c} \overset{\displaystyle H}{\underset{\displaystyle}{\mathrm{CH_3-CH\cdots C\cdots O}}} \\ \mathrm{B\!:\cdots H} \end{array} \right\} \tag{7}$$

in which the charge delocalization in the anion is a contributing factor in lowering the energy of the transition state. Although the difference in acidity of the $\alpha$- and $\beta$-protons is expressed in terms of an *equilibrium constant,* and thus in terms of energy differences between ionized and unionized forms, the *rates* of the ionization reactions would also be related to these energy differences because of the manner in which charge delocalization would participate in stabilizing the transition state in *(a).*

Although the energy values shown in Figure 13-1 are arbitrary and not explicit, it is apparent from their relationships that the equilibrium in the expression

$$\mathrm{RCH_2CHO + \!:B^- \; \rightleftharpoons \; \{R\ddot{C}HCHO\}^- + B\!:H} \tag{8}$$

as shown in Figure 13-1 is far to the left. Aldehydes and ketones are very weak acids. Acetone, for example, has $pK_a = 20$, and so the concentration of $CH_3COCH_2\!:^-$ in aqueous solution, with the base $OH^-$, would be very low. Nevertheless, as we shall see in the following sections, the anion *is* present in the solution, and if it reacts further, and so is "removed", the equilibrium permits it to be restored. Consequently, acetone can react to an extent much greater than its exceedingly low acidity might indicate.

## 13.2 Activation of α-hydrogen by nitro and other groups

Although the reactions of aldehydes and ketones are the most commonly encountered, the acidity of α-hydrogen atoms is a factor in the reactivity of other kinds of compounds in which charge delocalization in the α-carbon anion is possible.

The nitro group, $-NO_2$, is a more powerfully activating group* than the carbonyl group and nitromethane ($pK_a$ about 10) is a much stronger acid than acetone. The ionization of the α-hydrogen is aided by the inductive effect of positively charged nitrogen (in the $N^+$—$O^-$ dipole), and the carbon anion is stabilized by resonance:

$$CH_3-\overset{+}{N}\!\!\underset{O^-}{\overset{O}{\diagdown}} + B:^- \;\rightleftharpoons\; \left\{ :\overset{-}{C}H_2-\overset{+}{N}\!\!\underset{O}{\overset{O}{\diagdown}} \;\longleftrightarrow\; CH_2\!=\!\overset{+}{N}\!\!\underset{O^-}{\overset{O^-}{\diagdown}} \right\} + B:H \qquad (9)$$

There are a number of other activating groups that are characterized by their ability to participate in charge delocalization of the same kind as that described in the case of ketones, aldehydes, and nitro compounds; among these are the functional groups of esters, acid anhydrides, nitriles, and sulfones. In the following equations are shown the anions formed from simple representatives of the above classes of compounds by reaction with a strong base, $B:^-$. In practice, the bases used for reactions of this kind may be $OH^-$, $CH_3O^-$, other alkoxides, $NH_2^-$, and, in some reactions, tertiary amines. Specific cases will be dealt with in later sections.

Ethyl acetate:

$$CH_3COOEt + B:^- \;\rightleftharpoons\; \left\{ :CH_2COOEt \;\longleftrightarrow\; CH_2\!=\!\overset{\overset{\displaystyle :\ddot{O}:}{\displaystyle |}}{C}\!-\!OEt \right\}^- + B:H$$

Acetic anhydride:

$$\begin{matrix} CH_3C\!=\!O \\ \diagup \\ \end{matrix}\!\!\!\overset{O}{\underset{\diagdown}{}}\!\!\!\begin{matrix} \\ CH_3C\!=\!O \end{matrix} + B:^- \;\rightleftharpoons\; \left\{ \begin{matrix} :CH_2C\!=\!O \\ \diagup \\ \end{matrix}\!\!\!\overset{O}{\underset{\diagdown}{}}\!\!\!\begin{matrix} \\ CH_3C\!=\!O \end{matrix} \;\longleftrightarrow\; \begin{matrix} CH_2\!=\!C\!-\!\ddot{O}: \\ \diagup \\ \end{matrix}\!\!\!\overset{}{\underset{\diagdown}{}}\!\!\!\begin{matrix} \\ CH_3C\!=\!O \end{matrix} \right\}^- + B:H$$

Acetonitrile:

$$CH_3C\!\!\equiv\!\!N: + B:^- \;\rightleftharpoons\; \left\{ :CH_2C\!\!\equiv\!\!N: \;\longleftrightarrow\; CH_2\!=\!C\!=\!\ddot{N}: \right\}^- + B:H$$

---

\* The term "active hydrogen" is often used in the same sense as "acidic hydrogen", and the hydrogen atoms on carbon α- to carbonyl groups are said to be "activated" by the carbonyl group. In this terminology, a nitro group is a more powerfully activating group than the carbonyl group.

Methyl sulfone:*

$$CH_3-\overset{\overset{\displaystyle O}{\|}}{\underset{\underset{\displaystyle O}{\|}}{S}}-CH_3 + :B^- \;\rightleftharpoons\; \left\{ :CH_2-\overset{\overset{\displaystyle :\ddot{O}}{\|}}{\underset{\underset{\displaystyle O}{\|}}{S}}-CH_3 \;\longleftrightarrow\; CH_2=\overset{\overset{\displaystyle :\ddot{O}:}{\|}}{\underset{\underset{\displaystyle O}{\|}}{S}}-CH_3 \right\}^{-} + B:H$$

---

**Exercise 2.** If the compound acetamide, $CH_3C\overset{\displaystyle O}{\underset{\displaystyle NH_2}{\diagup}}$ , were treated with a strong base such as sodamide with the removal of one proton, would the resulting anion be $^-:CH_2CONH_2$ or $CH_3CONH:^-$? Why?

---

### 13.3  Active hydrogen in hydrocarbons

It might now be asked if the carbon-carbon double bond can activate $\alpha$-hydrogen by stabilization of the anion in the following way:

$$\left\{ -\overset{\displaystyle \ddot{C}}{\underset{\displaystyle |}{|}}-\overset{\displaystyle C}{\underset{\displaystyle |}{}}=\overset{\displaystyle C}{\underset{\displaystyle |}{}}- \;\longleftrightarrow\; -\overset{\displaystyle C}{\underset{\displaystyle |}{}}=\overset{\displaystyle C}{\underset{\displaystyle |}{}}-\overset{\displaystyle \ddot{C}}{\underset{\displaystyle |}{}}- \right\}^{-} \tag{10}$$

The answer is that this effect, although real, is relatively unimportant in activating hydrogen $\alpha$- to $C\!=\!C$. Because of the inherently low degree of acidity of the $H\!-\!C$ bond and the lack of any appreciable inductive effect by the $C\!=\!C$ bond, the total activation is very small. Although it is true that the methyl group of $CH_2\!=\!CHCH_3$ would be very much more acidic than that of $\underline{CH_3}CH_2CH_3$, the absolute value of its acidity ($pK_a$ probably $> 30$) is very small.

When *two* carbon-carbon double bonds flank the carbon atom bearing the ionizable hydrogen, it is then possible for the additional resonance stabilization to bring about a sufficient lowering of the energy of the anion to permit a clearly demonstrable acidity of a $-CH_2-$ group [Equation (11)]:

$$\left\{ -\overset{}{C}=\overset{}{C}-\overset{}{\ddot{C}}-\overset{}{C}=\overset{}{C} \;\longleftrightarrow\; -\overset{}{C}=\overset{}{C}-\overset{}{C}=\overset{}{C}-\overset{}{\ddot{C}}- \;\longleftrightarrow\; -\overset{}{\ddot{C}}-\overset{}{C}=\overset{}{C}-\overset{}{C}=\overset{}{C}- \right\}^{-}$$

$$\tag{11}$$

---

* There is still not general agreement about just what role the $-SO_2R$ group plays in activating $\alpha$-hydrogen. If the two $S\!-\!O$ bonds are semipolar bonds, the effect of the $-SO_2R$ group could be regarded as a powerful inductive effect:

$$-\overset{\overset{\displaystyle H}{|}\;\overset{\displaystyle O}{\uparrow}}{\underset{\underset{\displaystyle O}{\downarrow}}{C}}-\overset{}{S}-R \;\xrightarrow{B:^-}\; -\overset{\overset{\displaystyle O}{\uparrow}}{\underset{\underset{\displaystyle O}{\downarrow}}{\ddot{C}}}-\overset{}{S}-R$$

However, the ability of sulfur, a second row element, to use $M$ shell $d$ orbitals makes it possible to regard the effect as a resonance phenomenon, involving the formal process

$$\left\{ -\overset{}{\ddot{C}}-\overset{}{S}-\ddot{O}: \;\longleftrightarrow\; -\overset{}{C}=\overset{}{S}-\ddot{O}: \right\}$$

Thus, the compound cyclopentadiene ($pK_a$ about 16) has "active" hydrogen:

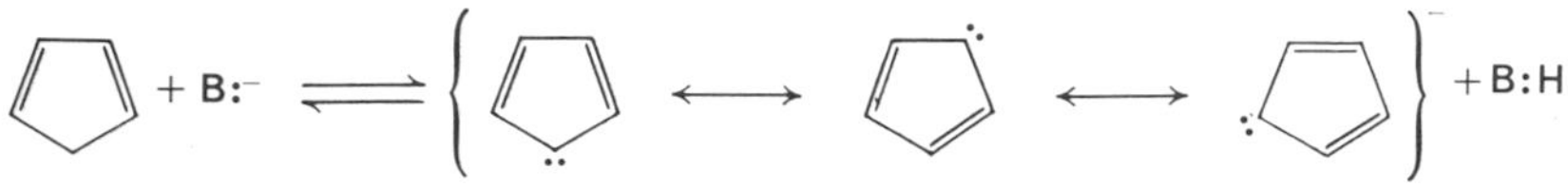

or  (a symmetrical anion).

Cyclopentadiene, like acetylene, reacts with Grignard reagents to give cyclopentadienylmagnesium halides which can be used as Grignard reagents in reactions with, for example, aldehydes and ketones (see Section 13.1 for the comparable behavior of acetylenes).

## 13.4   Multiple activation of carbon-linked hydrogen

The presence of two or three carbonyl, nitro, or other activating groups on the carbon atom holding the ionizable hydrogen atom causes greatly enhanced acidity because of the inductive effect of two or three activating groups and the increased opportunity for charge delocalization, as in Equation (12):

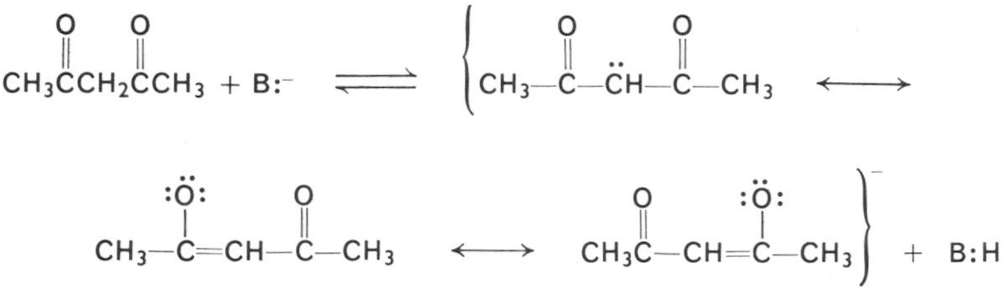

These multiple activating effects cause large increases in acidity of $\alpha$-hydrogen as the $pK_a$ values in Table 13-1 show:

**Table 13-1.**   *Approximate pK$_a$ values for some active hydrogen compounds*

| Compound | pK$_a$ (approximate) |
|---|---|
| $CH_3COCH_3$ | 20 |
| $(CH_3CO)_2CH_2$ | 9 |
| $(CH_3CO)_3CH$ | 6 |
| $CH_3NO_2$ | 10 |
| $CH_2(NO_2)_2$ | 4 |
| $CH(NO_2)_3$ | 1 (a "strong" acid) |

## 13.5 The carbon anion as a nucleophile. The aldol condensation

The carbon anions that are formed from compounds with active hydrogen atoms are the conjugate bases of very weak acids, and so are strong bases and effective nucleophiles. When acetaldehyde is treated with sodium hydroxide in aqueous solution, the anion formed in the equilibrium

$$CH_3CHO + OH^- \rightleftharpoons {}^-{:}CH_2CHO + H_2O$$

can attack the carbonyl group of the undissociated aldehyde in a typical carbonyl addition reaction:

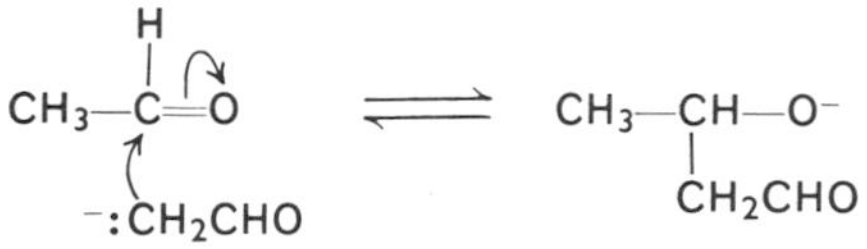

$$CH_3-\overset{\overset{\text{H}}{|}}{\underset{\underset{-{:}CH_2CHO}{\uparrow}}{C}}=O \rightleftharpoons CH_3-CH-O^- \quad\quad (13)$$
$$\underset{CH_2CHO}{|}$$

Protonation of the initial adduct by the acid-base reaction with solvent

$$CH_3\overset{\overset{O^-}{|}}{C}HCH_2CHO + H_2O \rightleftharpoons CH_3\overset{\overset{OH}{|}}{C}HCH_2CHO + OH^- \quad\quad (14)$$

yields the four-carbon atom compound *aldol* ($\beta$-hydroxy-*n*-butyraldehyde or 3-hydroxybutanal). The name aldol is derived from the structural features of the compound: it is an *ald*ehyde and an alcoh*ol*.

This condensation is the simplest example of a large family of closely related reactions which are referred to by the term *aldol condensation*, and of an even larger number, which, although they are not called aldol condensations, are mechanistically allied. Aldol condensations may take place between

*(a)* aldehyde and aldehyde (as in the above example);

*(b)* aldehyde and ketone;

$$CH_3CHO + CH_3COCH_3 \xrightarrow[\text{NaOH}]{\text{aq.}} CH_3\underset{\underset{OH}{|}}{C}HCH_2COCH_3$$

*(c)* ketone and ketone;

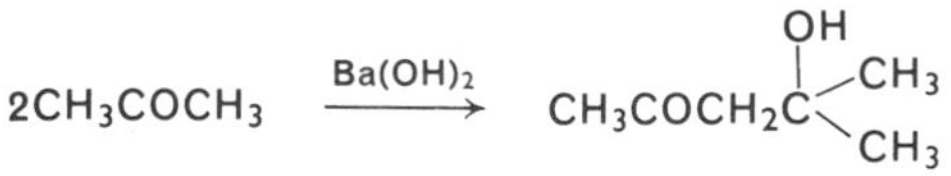

$$2CH_3COCH_3 \xrightarrow{\text{Ba(OH)}_2} CH_3COCH_2\overset{\overset{OH}{|}}{C}\Big\langle{\overset{CH_3}{CH_3}}$$

diacetone alcohol

*(d)* compounds other than aldehydes and ketones that react in analogous ways. These reactions will be described as they are pertinent to discussions in later chapters.

When two *different* carbonyl compounds condense with one another, there should be an appreciable difference between either (1) the activity of the $\alpha$-hydrogen atoms or (2) the reactivity of the carbonyl groups of the two reactants.

If this difference does not exist, mixtures of products will result. For instance, in propionaldehyde and *n*-butyraldehyde, the α-methylene groups are very similar in type and molecular environment; and the carbonyl groups are also not much different. As a consequence, the aldol condensation in which a mixture of these two aldehydes is used will result in all four possible products. Such a mixture would be difficult to separate into pure compounds, and so such a reaction would not be a useful one.

---

**Exercise 3.**   Formulate the four condensations between propionaldehyde and *n*-butyral-dehyde according to the detailed scheme shown for the aldol condensation of acetaldehyde.

---

An extremely reactive carbonyl group is found in formaldehyde; moreover, formaldehyde has no α-hydrogen atoms and thus can act only as the carbonyl component of an aldol condensation:

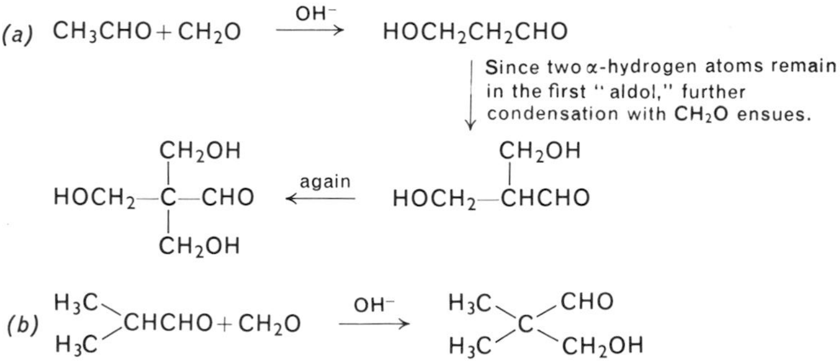

*(a)* $CH_3CHO + CH_2O \xrightarrow{OH^-} HOCH_2CH_2CHO$

Since two α-hydrogen atoms remain in the first " aldol," further condensation with $CH_2O$ ensues.

$$HOCH_2-\underset{\underset{CH_2OH}{|}}{\overset{\overset{CH_2OH}{|}}{C}}-CHO \xleftarrow{again} HOCH_2-\underset{}{\overset{\overset{CH_2OH}{|}}{C}}HCHO \qquad (15)$$

*(b)* $\underset{H_3C}{\overset{H_3C}{>}}CHCHO + CH_2O \xrightarrow{OH^-} \underset{H_3C}{\overset{H_3C}{>}}C\underset{CH_2OH}{\overset{CHO}{<}}$

Acetaldehyde, with a more reactive carbonyl group than acetone, acts as the carbonyl component in the reaction

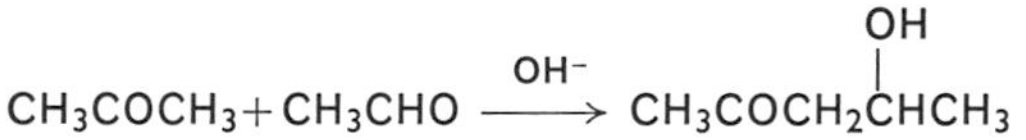

$$CH_3COCH_3 + CH_3CHO \xrightarrow{OH^-} CH_3COCH_2\overset{\overset{OH}{|}}{C}HCH_3 \qquad (16)$$

The carbonyl group of the aldehyde undergoes addition much more readily than that of the ketone, and so the reaction proceeds principally as shown. Some aldol, by self-condensation of the acetaldehyde, would be formed, but no diacetone alcohol, unless the acetone were in excess.

## 13.6   Dehydration of aldols

Although aldol condensations give rise to β-hydroxy carbonyl compounds as described above and can be generalized as

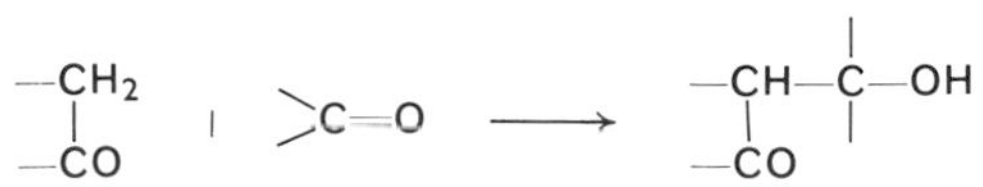

$$\underset{-CO}{\overset{-CH_2}{|}} \quad _| \quad >C=O \longrightarrow \underset{-CO}{\overset{-CH-\overset{|}{C}-OH}{|}} \qquad (17)$$

the usual final result of aldol condensations is the consequence of the subsequent dehydration of the aldol to give an unsaturated carbonyl compound:

$$-CH-\overset{|}{C}-OH \xrightarrow[]{-H_2O} -C{=}C\overset{}{\underset{}{\diagup}} \qquad (18)$$
$$-\overset{|}{CO} \qquad\qquad -CO$$

Indeed, this step frequently occurs with such ease that if often requires special (mild) conditions in the condensation reaction to permit isolation of the aldol; for example:

$$+ \ CH_3COCH_3 \longrightarrow \qquad\qquad (19)$$

citral    pseudoionone

$$CH_3CH_2CHO + CH_3CHO \longrightarrow CH_3CH{=}\overset{\overset{\displaystyle CH_3}{|}}{C}CHO \qquad (20)$$

tiglic aldehyde

$$+ \ CH_3COCH_3 \longrightarrow \qquad\qquad (21)$$

furfuraldehyde    furfuralacetone

---

**Exercise 4.**  Furfuralacetone can condense further with another molecule of furfuraldehyde. Formulate this.

---

When dehydration of the aldol ensues, the equilibrium is usually shifted in favor of the final product, as a result of the increased stability of the system $-C{=}C-C{=}O$ that is formed. The readiness with which aldols are dehydrated is also shown by the ease with which this step can be carried out by acid catalysts. Diacetone alcohol is readily converted into mesityl oxide (an old name that is still used for 4-methyl-3-penten-2-one, despite its inappropriateness for a compound that is in no sense an oxide):

$$\overset{H_3C}{\underset{H_3C}{\diagup}}\overset{}{C}CH_2COCH_3 \xrightarrow[\text{acid}]{\text{trace of}} \overset{H_3C}{\underset{H_3C}{\diagup}}C{=}CHCOCH_3 \qquad (22)$$

diacetone alcohol    mesityl oxide

$$CH_3\overset{\overset{\displaystyle OH}{|}}{C}HCH_2CHO \xrightarrow[]{I_2^*} CH_3CH{=}CHCHO$$

acetaldol,    crotonaldehyde
or "aldol"

---

* The effectiveness of iodine in catalyzing this dehydration indicates that $I_2$ is electrophilic, and thus acts as a Lewis acid. The formation of $I_3^-$ ($I_2 + I^- \rightarrow I_3^-$) is consistent with this.

The ease with which acid-catalyzed dehydration occurs is a result of the ease with which ("active") α-hydrogen is removed to complete the elimination of water from the protonated aldol:

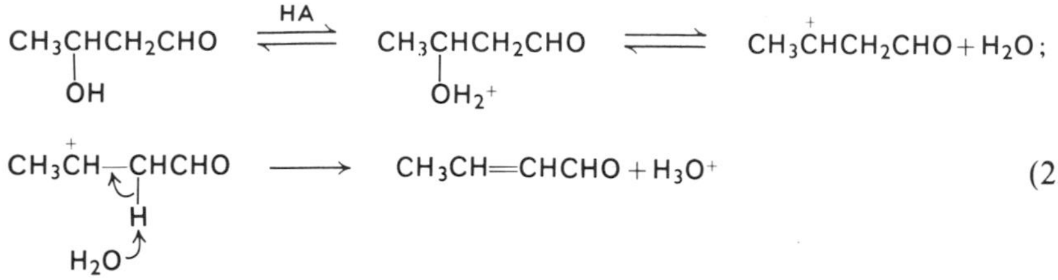

$$\underset{\underset{OH}{|}}{CH_3CHCH_2CHO} \xrightarrow{HA} \underset{\underset{OH_2^+}{|}}{CH_3CHCH_2CHO} \rightleftharpoons CH_3\overset{+}{C}HCH_2CHO + H_2O;$$

$$\underset{\underset{\underset{H_2O \nearrow}{H \downarrow}}{|}}{CH_3\overset{+}{C}H \curvearrowleft CHCHO} \longrightarrow CH_3CH{=}CHCHO + H_3O^+ \qquad (23)$$

This dehydration, it will be seen, is an elimination reaction in which the removal of the proton in the final stage is greatly facilitated by the activity of the α-hydrogen atom.

It should be noticed how the formation of an unsaturated carbonyl compound in the aldol condensation resembles the formation of oximes, semicarbazones, Schiff's bases, and so on. In the aldol condensation, the attack upon the carbonyl group is made by the nucleophile $\left\{ -\overset{\cdot\cdot}{\underset{|}{C}}-\overset{|}{C}O- \right\}^-$, whereas in the formation of one of the above-mentioned nitrogen derivatives, the nucleophile is :NH₂—R:

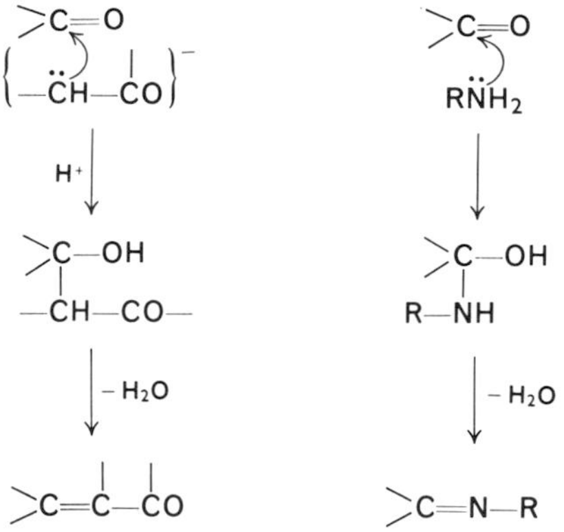

$$\underset{\left\{ -\overset{\cdot\cdot}{C}H-\overset{|}{C}O \right\}^-}{\overset{>C=O}{\Big\downarrow}}$$

$$\underset{RNH_2}{\overset{>C=O}{\phantom{x}}}$$

$$\underset{-CH-CO-}{\overset{>C-OH}{\Big| \, H^+}}$$

$$\underset{R-NH}{\overset{>C-OH}{\Big\downarrow}}$$

$$\underset{>C=C-CO}{\overset{}{\Big\downarrow -H_2O}}$$

$$\underset{>C=N-R}{\overset{}{\Big\downarrow -H_2O}}$$

### 13.7 Reversibility of the aldol condensation

*The aldol condensation is reversible* and proceeds to an equilibrium, as the equations that we have written for it indicate. The reaction reaches an equilibrium that may in some cases be very unfavorable for the formation of a desired product. For example, when acetone aldolizes to form diacetone alcohol (Section 13.5), the amount of diacetone alcohol formed is very small, the equilibrium lying largely on the side of acetone. It is possible to utilize a special technique for increasing the yield of the product in such a case, by removing it from the catalyst and thus preventing the reversal of the condensation. If acetone is refluxed and the vapors are condensed and allowed to flow over the catalyst (barium hydroxide) suspended in a porous container, the equilibrium concentration of diacetone alcohol is reached in contact with the catalyst. The resulting solution is then allowed to return to the boiling flask where acetone, but not the high-boiling diacetone alcohol, is again vaporized and returned to the catalyst. After some time, the concentration of diacetone alcohol that has accumulated in the boiling flask is sufficient to allow its isolation.

If pure diacetone alcohol is treated with aqueous or alcoholic alkali it will rapidly revert to the equilibrium mixture of diacetone alcohol and acetone, containing chiefly the latter. In the presence of the base, the removal of a proton from the alcoholic hydroxyl group

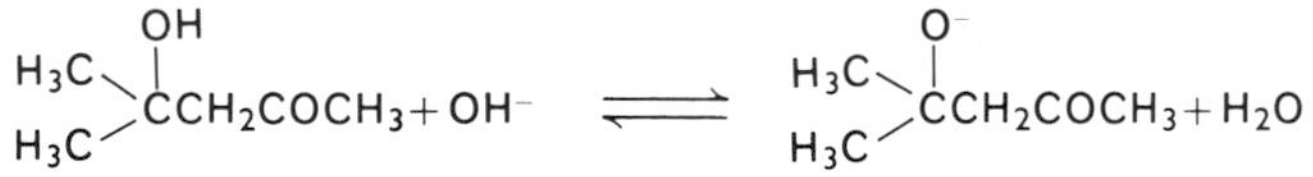

$$\begin{matrix} H_3C \\ \phantom{x} \\ H_3C \end{matrix} \overset{OH}{\underset{}{\underset{}{C}}} CH_2COCH_3 + OH^- \rightleftharpoons \begin{matrix} H_3C \\ \phantom{x} \\ H_3C \end{matrix} \overset{O^-}{\underset{}{\underset{}{C}}} CH_2COCH_3 + H_2O \tag{24}$$

gives rise to an ion, which will be recognized as the addition product that represents the first stage in the self-condensation of acetone, and is a member of the equilibrium,

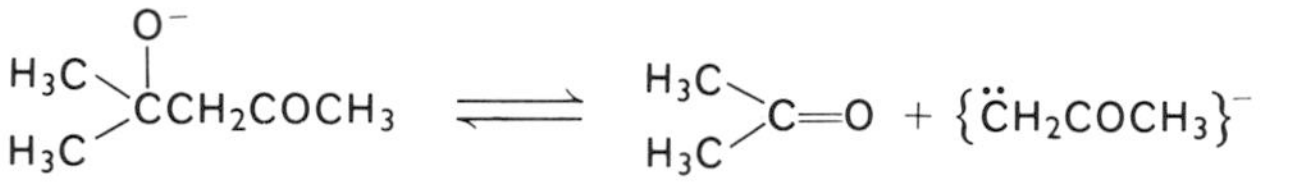

$$\begin{matrix} H_3C \\ \phantom{x} \\ H_3C \end{matrix} \overset{O^-}{\underset{}{\underset{}{C}}} CH_2COCH_3 \rightleftharpoons \begin{matrix} H_3C \\ \phantom{x} \\ H_3C \end{matrix} C{=}O + \{\ddot{C}H_2COCH_3\}^- \tag{25}$$

Thus, the equilibrium can be established from either direction.

Although the $\alpha,\beta$-unsaturated carbonyl compounds that are formed by dehydration of aldols are somewhat more stable than the aldols themselves, they too are subject to cleavage by bases. For example, citral can be cleaved in the following way:

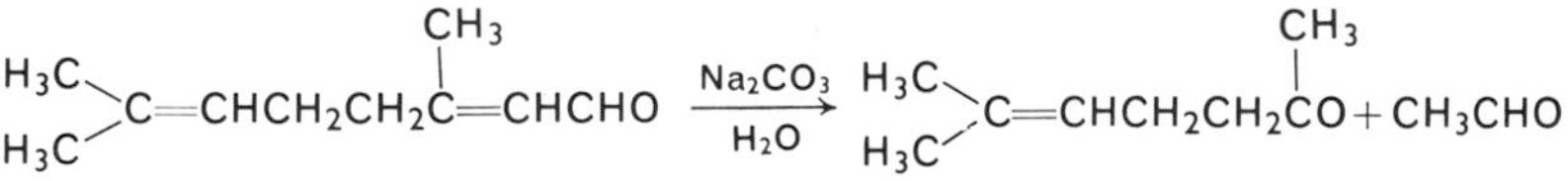

$$\begin{matrix} H_3C \\ \phantom{x} \\ H_3C \end{matrix} C{=}CHCH_2CH_2\overset{CH_3}{\underset{}{\underset{}{C}}}{=}CHCHO \xrightarrow[H_2O]{Na_2CO_3} \begin{matrix} H_3C \\ \phantom{x} \\ H_3C \end{matrix} C{=}CHCH_2CH_2\overset{CH_3}{\underset{}{\underset{}{C}}}O + CH_3CHO$$

citral $\qquad\qquad$ 6-methyl-5-hepten-2-one $\qquad$ (26)

The cleavage of the unsaturated carbonyl compound is the consequence of the re-formation of the aldol by addition of water, through attack of a hydroxyl ion to the double bond [compare the Michael reaction (Section 13.10)]:

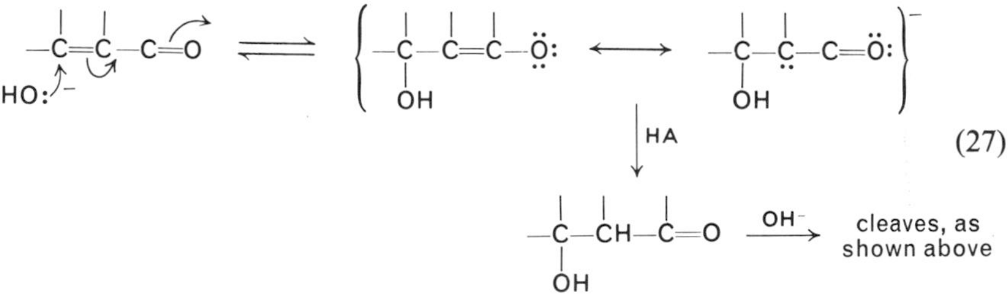

$$\underset{HO:^-}{-\overset{|}{C}{=}\overset{|}{C}{-}C{\overset{\frown}{=}}O} \rightleftharpoons \left\{ -\overset{|}{\underset{OH}{C}}{-}\overset{|}{C}{=}\overset{|}{C}{-}\ddot{O}: \longleftrightarrow -\overset{|}{\underset{OH}{C}}{-}\overset{|}{C}{-}C{=}\ddot{O}: \right\}^- \tag{27}$$

$$\downarrow HA$$

$$-\overset{|}{\underset{OH}{C}}{-}\overset{|}{C}H{-}\overset{|}{C}{=}O \xrightarrow{OH^-} \begin{matrix} \text{cleaves, as} \\ \text{shown above} \end{matrix}$$

## 13.8   Cyclization by aldol condensations

When the two reacting centers—the active methylene group and the carbonyl group—are present in the same molecule, aldol condensation with ring closure will occur if the ring that is formed is five- or six-membered:

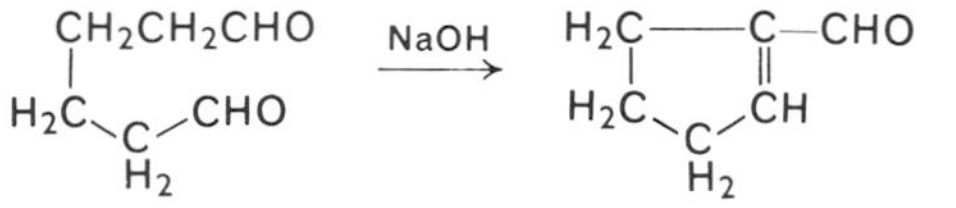

$$
\begin{array}{ccc}
\text{CH}_2\text{CH}_2\text{CHO} & & \text{H}_2\text{C}\text{—}\text{C}\text{—}\text{CHO} \\
| & \xrightarrow{\text{NaOH}} & | \quad\quad \| \\
\text{H}_2\text{C}\underset{\underset{\text{H}_2}{\text{C}}}{\diagdown}\text{CHO} & & \text{H}_2\text{C}\underset{\underset{\text{H}_2}{\text{C}}}{\diagdown}\text{CH}
\end{array}
\tag{28}
$$

$$
\begin{array}{cc}
\text{adipic dialdehyde} & \text{1-cyclopentenaldehyde} \\
& \text{(1-formylcyclopentene)}
\end{array}
$$

These reactions will be more fully discussed in Chapter 17, in which additional examples of ring closures of the aldol type, involving esters of carboxylic acids, will be encountered.

## 13.9   Structure diagnosis with the aid of the aldol condensation

A reaction that is sometimes of value in structure determination is that between an aromatic aldehyde, such as benzaldehyde, and a compound that contains the

grouping $-\text{CH}_2-\overset{|}{\text{C}}=\text{O}$. In general, this reaction is represented by the following partial structures:

$$\text{(29)}$$

benzaldehyde

Suppose, for example, we obtain sufficient experimental evidence to decide that an unknown compound has either of the two following structures, but remain with the task of deciding between them:

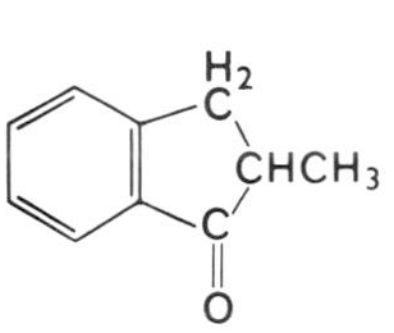

(A)            or            (B)

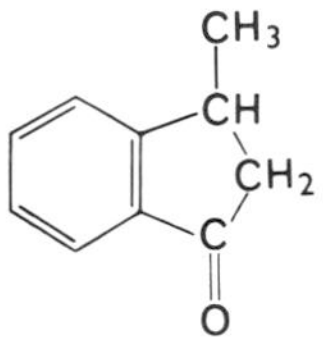

Reaction of the compound with benzaldehyde, with sodium ethoxide as the catalyst, will allow the choice to be made. If the unknown compound has structure (B), the benzal derivative will form:

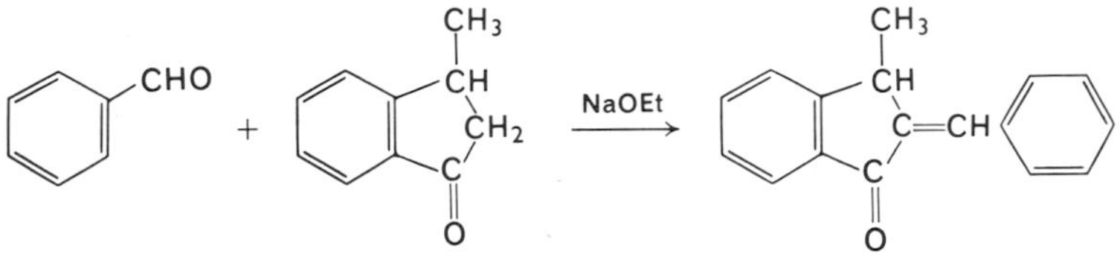

If the unknown compound has structure (A), no benzal derivative will form.

## 13.10 α,β-Unsaturated carbonyl compounds and the Michael condensation

The α,β-unsaturated carbonyl system represented by the partial structure

$$-\overset{|}{C}=\overset{|}{C}-\overset{|}{C}=O$$

possesses, as a whole, certain of the characteristics of the carbonyl group alone. For example, just as the carbonyl group, by virtue of the polarization of the carbon-oxygen bond, possesses an electrophilic carbon atom at which nucleophilic attack can take place, the α,β-unsaturated carbonyl system can accept nucleophilic attack at the end of the four-atom system:

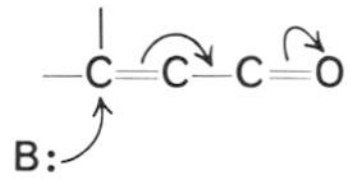

When the attacking nucleophilic reagent (B: in the above expression) is a carbon anion, the reaction is called the "Michael condensation." The Michael condensation is clearly a reaction of the general aldol class; it may be described as a "conjugate aldol condensation."

A simple illustration of the Michael condensation is the reaction between acrolein and diethyl malonate:

$$CH_2=CHCHO + CH_2(COOEt)_2 \xrightarrow{\text{NaOEt}} \underset{\underset{CH(COOEt)_2}{|}}{CH_2CH_2CHO} \tag{30}$$

The Michael condensation is base-catalyzed; the function of the sodium ethoxide in the above reaction is to provide the carbanion of the active methylene group:

$$CH_2(COOEt)_2 + OEt^- \rightleftharpoons HOEt + \{:CH(COOEt)_2\}^-$$

The resulting anion then attacks the α,β-unsaturated aldehyde:

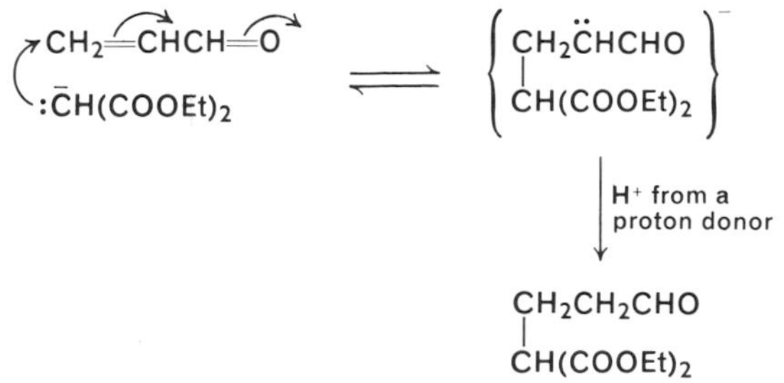

$$\underset{\underset{CH(COOEt)_2}{|}}{CH_2CH_2CHO}$$

The Michael reaction is reversible, just as the normal aldol condensation is. This occasionally leads to complicated reactions in which the product of a Michael condensation can, by reversal in another way, cleave to give products that are different from the original reactants. Let us consider the following Michael condensation:

$$C_6H_5CH{=}\overset{\overset{\displaystyle COOEt}{|}}{C}{-}COOEt + CH_3COCH_2COOEt \rightleftarrows C_6H_5CH{-}CH\underset{COOEt}{\overset{COOEt}{<}} \quad (32)$$

$$\underset{\underset{COCH_3}{|}}{\overset{CH{-}COOEt}{|}}$$

It will be seen that the product of this reaction is the same one that would be obtained by the following condensation:

$$C_6H_5{-}\overset{\overset{\displaystyle \downarrow}{|}}{\underset{\underset{COCH_3}{|}}{\overset{\overset{CH}{\parallel}}{C}{-}COOEt}} + CH_2(COOEt)_2 \rightleftarrows C_6H_5CH{-}CH\underset{COOEt}{\overset{COOEt}{<}} \quad (33)$$

$$\underset{\underset{COCH_3}{|}}{\overset{CH{-}COOEt}{|}}$$

As a consequence, the reaction mixture from either pair of reactants would be an equilibrium mixture containing all four.

An example of a Michael condensation as a step in a synthetic reaction series is described in Section 17.7.

---

**Exercise 5.**   Write the equations for the Michael condensation between the following:
(a) $CH_3CH{=}CHCOOEt$ and $CH_2(COOEt)_2$
(b) $CH_2{=}CHCOOEt$ and $CH_3COCH_2COOEt$
(c) $CH_2{=}C(COOEt)_2$ and $CH_2(COOEt)_2$

---

### 13.11   Cyanoethylation

A special example of the Michael condensation is that in which addition takes place, not to an $\alpha,\beta$-unsaturated carbonyl compound, but to the closely related acrylonitrile. The cyano group is capable of acting in a manner comparable to that of a carbonyl group, as described in the preceding section, and permits the attack of a nucleophilic reagent at the end of the conjugated system:

$$B{:}\overset{\frown}{\phantom{}}CH_2{=}CH{-}CN \longrightarrow (B{-}CH_2{-}CH{-}CN)^- \overset{H^+}{\longrightarrow} B{-}CH_2{-}CH_2{-}CN$$

$$\text{acrylonitrile} \quad (34)$$

Cyanoethylation is difficult to control; when active hydrogen compounds that contain more than one $\alpha$-hydrogen atom are used, the reaction tends to proceed until all of the active hydrogen atoms are replaced by cyanoethyl groups.

Acetone gives a tricyanoethyl derivative with ease, and a tetracyanoethyl derivative can be prepared; cyclohexanone gives a tetracyanoethyl derivative:

$$CH_3COCH_3 + CH_2{=}CHCN \xrightarrow{\text{alcoholic KOH}} CH_3COC(CH_2CH_2CN)_3$$

$$\downarrow CH_2{=}CHCN \qquad (35)$$

$$NCCH_2CH_2CH_2COC(CH_2CH_2CN)_3$$

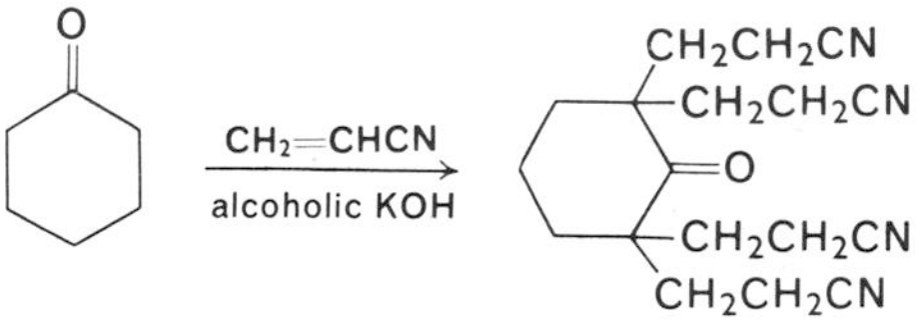

(36)

It is seldom practicable to attempt to prepare a monocyanoethylated product when more than one active hydrogen is present.

A useful reaction of acrylonitrile that is closely allied to the Michael reaction, but not called by that term, is that in which an amine adds to the conjugated system by attack of the nucleophilic nitrogen atom. The *N*-cyanoethyl amine that is so formed can be reduced to a diamino compound or be hydrolyzed to a $\beta$-amino acid:

$$CH_3NH_2 + CH_2{=}CHCN \longrightarrow CH_3NHCH_2CH_2CN \xrightarrow[\text{hydrolysis}]{\text{acid}} CH_3NHCH_2CH_2COOH$$

$$\xrightarrow{\text{reduction}} CH_3NHCH_2CH_2CH_2NH_2 \qquad (37)$$

## 13.12 Halogenation of carbonyl compounds

The $\alpha$-hydrogen atoms of ketones, aldehydes, and acid halides are readily replaced by halogen. Halogenation can be carried out under either acidic or basic conditions. Under proper control of the conditions of the reaction and of the amount of halogen used, it can often be performed in such a way as to permit the preparation of mono- or poly-$\alpha$-halogen compounds of known structure. Acetone can be brominated in aqueous or acetic acid solution to give bromoacetone in good yield. Under these conditions the reaction is acid-catalyzed, because HBr is produced in the course of the bromination. Other ketones can be monobrominated in the same way:

$$CH_3COCH_3 + Br_2 \xrightarrow[\text{solution}]{\text{acetic acid}} CH_3COCH_2Br + HBr \qquad (38)$$

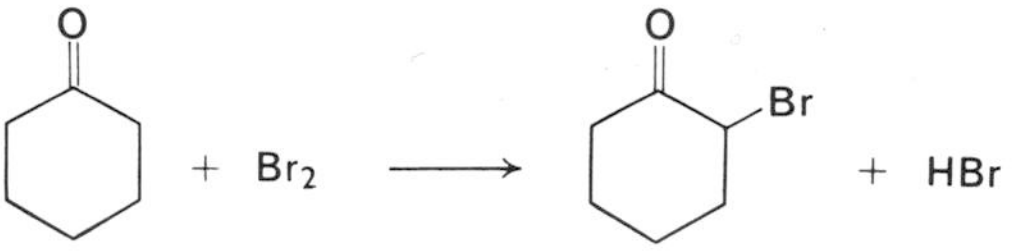

(39)

$$\text{(cyclopentanone)} + \text{Br}_2 \longrightarrow \text{(bromocyclopentanone)} + \text{HBr} \qquad (40)$$

Polyhalogenation can be accomplished if excess halogen is used, but mixtures are often formed. Acetone, upon chlorination, yields chiefly 1,1-dichloro-2-propanone ($\alpha,\alpha$-dichloroacetone), but the product contains some 1,3-dichloro-2-propanone. Cyclohexanone has been reported to give chiefly 2,6-dibromocyclo-hexanone when brominated, but chlorination gives a mixture of the 2,2- and 2,6-dichloro compounds. These results emphasize the uncertainties that attend the halogenation of ketones of these types; when the carbonyl group is flanked by $\alpha$-hydrogen atoms the course of dibromination or dichlorination can lead to mixtures of the kind described.

Halogenation in alkaline solution tends to proceed until all of the $\alpha$-hydrogen atoms have been replaced. Pinacolone, for example, gives $\alpha,\alpha,\alpha$-tribromo-pinacolone:

$$\underset{\underset{CH_3}{|}}{\overset{\overset{CH_3}{|}}{CH_3-C-COCH_3}} \quad \xrightarrow[\text{NaOH}]{Br_2} \quad \underset{\underset{CH_3}{|}}{\overset{\overset{CH_3}{|}}{CH_3-C-COCBr_3}} \qquad (41)$$

The reaction of methyl ketones ($RCOCH_3$) with halogen in alkaline solution ordinarily leads to the formation of the corresponding haloform (chloroform, bromoform, iodoform). This reaction was described in Section 12.17.

The halogenation of ketones under acidic conditions proceeds by way of the enol form of the ketone. In a careful study of the halogenation of acetone, the following observations were made: (1) the bromination of acetone in aqueous solution is a first-order reaction; (2) the rate of reaction is proportional to the concentration of acetone but independent of the bromine concentration; and (3) the rate of iodination under the same conditions is equal to the rate of bromi-nation. These facts indicate that the rate of the overall reaction is the rate at which acetone is converted into some intermediate, which then reacts rapidly with the halogen to give the final product. The intermediate is the enol form of the ketone:

$$CH_3-\overset{\overset{O}{\|}}{C}-CH_3+H:A \;\rightleftharpoons\; CH_3-\overset{\overset{+OH}{\|}}{C}-CH_3+:A^- \;\rightleftharpoons\;$$

$$CH_3-\overset{\overset{OH}{|}}{C}=CH_2+H:A \quad \text{(slow)}$$

$$\qquad (42)$$

$$CH_3-\overset{\overset{OH}{|}}{C}-CH_2 \;\; Br-Br \longrightarrow CH_3-\overset{\overset{+OH}{\|}}{C}-CH_2Br + Br^- \longrightarrow$$

$$CH_3-\overset{\overset{O}{\|}}{C}-CH_2Br+HBr \quad \text{(fast)}$$

Halogenation under alkaline conditions proceeds by way of the anion formed by abstraction of $\alpha$-hydrogen by the base:

$$CH_3-\overset{\overset{O}{\|}}{C}-CH_3+OH^- \;\rightleftharpoons\; CH_3-\overset{\overset{O}{\|}}{C}-CH_2{:}^-+H_2O \tag{43}$$

$$CH_3COCH_2{:}^-+Br_2 \;\longrightarrow\; CH_3COCH_2Br+Br^-$$

Since the $\alpha$-hydrogen atom of $RCOCH_2Br$ is more acidic than that of $RCOCH_3$ (because of the inductive effect of the halogen), further bromination takes place on the same carbon atom. When the starting ketone is $RCOCH_3$, the final product, prior to cleavage to give the haloform, is $RCOCBr_3$.

The halogenation of acid halides is called the *Hell-Volhard-Zelinsky reaction:*

$$CH_3C\overset{\displaystyle\nearrow O}{\underset{\displaystyle\searrow Br}{}} + Br_2 \;\longrightarrow\; BrCH_2COBr+HBr \tag{44}$$

An application of this reaction is described in Section 31.6.

## 13.13 Carbon alkylation of carbonyl compounds

*Alkylation of carbon anions.* One of the most characteristic reactions of nucleophilic reagents is their reaction with alkyl halides in the nucleophilic displacement reaction. Numerous examples of this reaction have been given in earlier chapters. It would be expected that the nucleophilic carbon anion could take part in this kind of reaction, with a resulting alkylation of the $\alpha$-carbon atom:

$$RCOCH_3 + B{:}^- \;\rightleftharpoons\; RCOCH_2{:}^- + B{:}H$$

$$\tag{45}$$

$$RCOCH_2{:}^- \;\longrightarrow\; CH_3\!-\!I \;\longrightarrow\; RCOCH_2CH_3+I^-$$

The alkylation of carbonyl compounds, and especially of ketones, does proceed in the manner illustrated in Equation (45). A strong base, such as sodamide or potassium *t*-butoxide, is required, and because of the ease with which many alkyl halides (especially secondary and tertiary alkyl halides) are dehydrohalogenated by strong bases, the reaction is limited in practice to halides that are not prone to undergo the elimination reaction; examples of these are methyl and allyl halides.

There are, however, a number of practical disadvantages to this reaction: (1) The presence of the strong base can bring about aldol condensation of the ketone which is being alkylated:

$$\tag{46}$$

(2) The alkylation is often not selective. An unsymmetrical ketone, such as $CH_3CH_2COCH_3$, can form two different monoanions, and both can undergo alkylation:

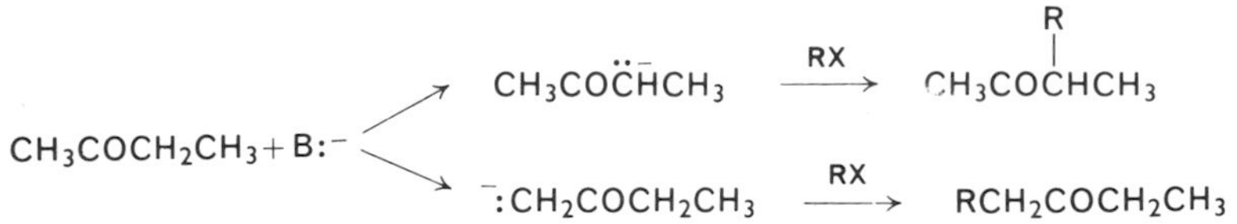

$$(47)$$

(3) Polyalkylation can occur:

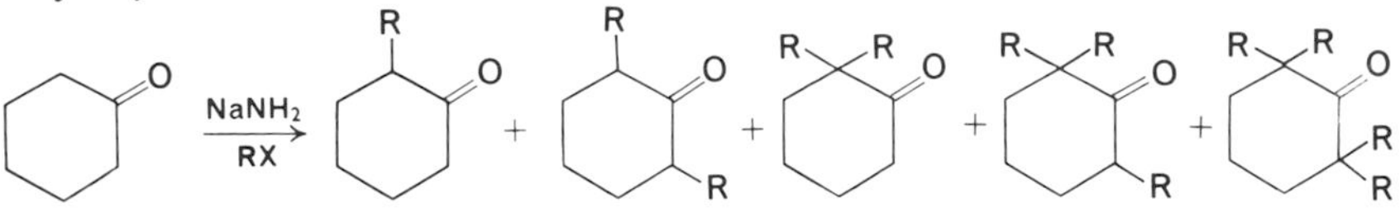

$$(48)$$

When complete alkylation is desired, the third difficulty, polyalkylation, is minor. For example, hexamethylacetone can be prepared by exhaustive methylation of lower homologues:

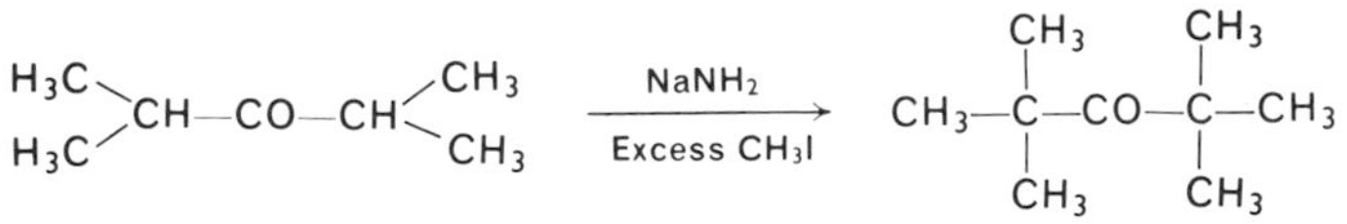

$$(49)$$

When the separation of the mixture of the alkylation products is practicable, the alkylation reaction can be of preparative value despite the fact that it does not lead to a single product. An intermediate in an early synthesis of vitamin A—2,2,6-trimethylcyclohexanone—was prepared by the methylation of 2-methylcyclohexanone:

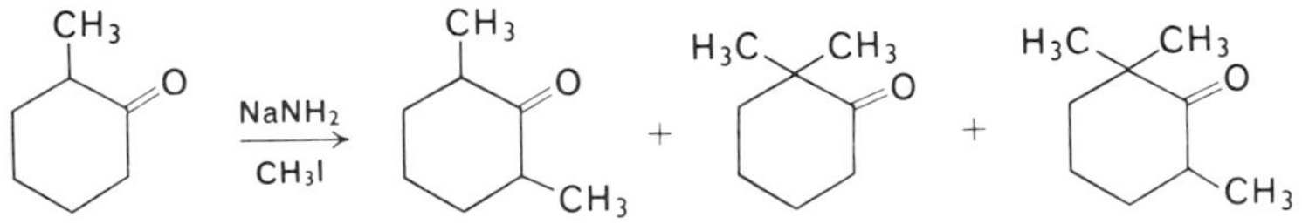

$$(50)$$

(separated by distillation)

There are alternative methods of alkylating carbonyl compounds. Some of these will be described in the next chapter.

*Enamines.* A valuable procedure for controlled alkylation of α-carbon atoms in ketones and aldehydes is an indirect method that involves the intermediate preparation of a derivative called an *enamine*. An enamine is formed by the reaction of a ketone with a secondary amine:

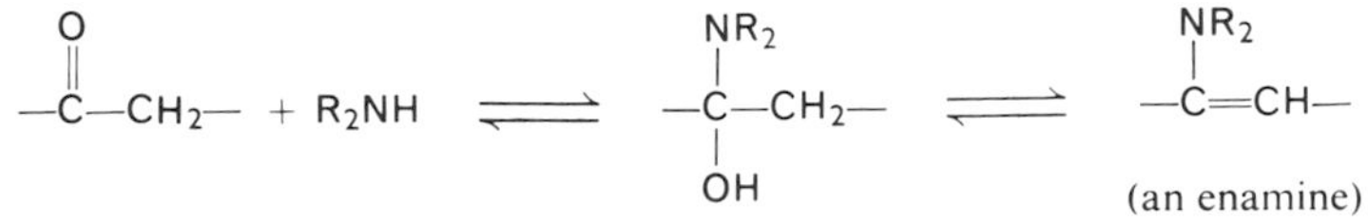

$$(51)$$

(an enamine)

The enamine (*en-* for the double bond + *amine*) is a nucleophile whose nucleophilic character can be expressed at both carbon and nitrogen because the unshared electron pair is delocalized as in the following expression:

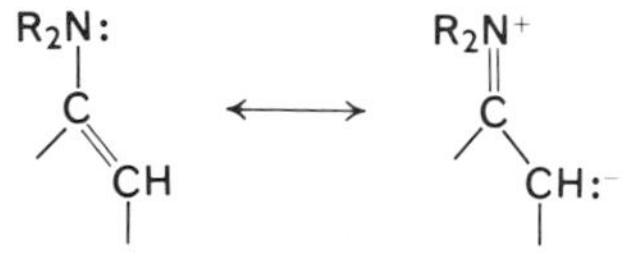

Both alkylation reactions are observed:

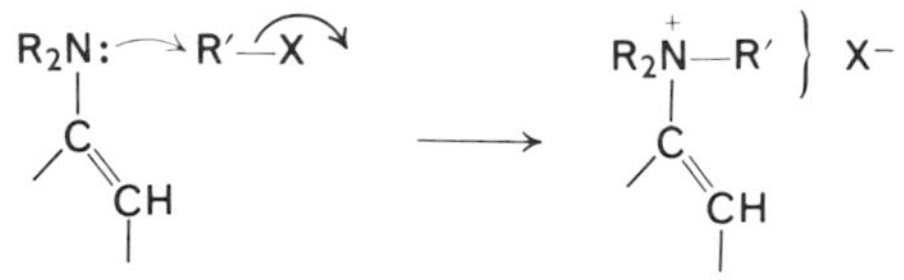

$$(52)$$

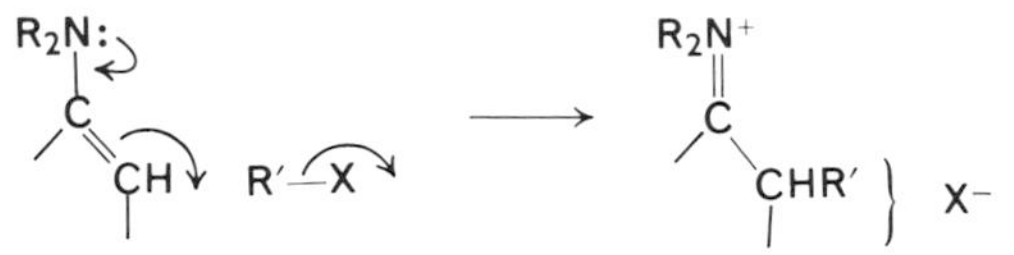

$$(53)$$

but it is found that carbon alkylation (53) occurs as the predominant reaction; the course shown in (52), the alkylation of the amino nitrogen, would be valueless [unless, as may be the case, the reaction (52) were reversible while that shown in Equation (53) were not].

The advantage of alkylation through the enamine, as compared with direct alkylation of the carbon anion, is that polyalkylation does not usually occur, and under most conditions the monoalkylated ketone is obtained in good yield. Some examples that have been described are the following (the amine, $R_2NH$, used in these examples is pyrrolidine; see Section 7.4):

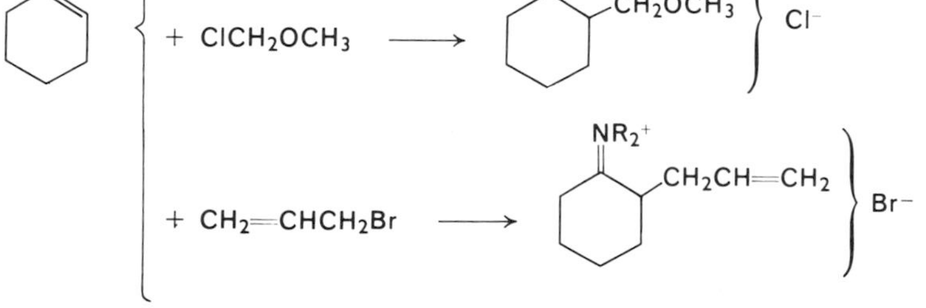

$$(54)$$

$$(55)$$

$$(56)$$

The alkylated enamines shown as the products in the above equations are readily hydrolyzed to the ketones:

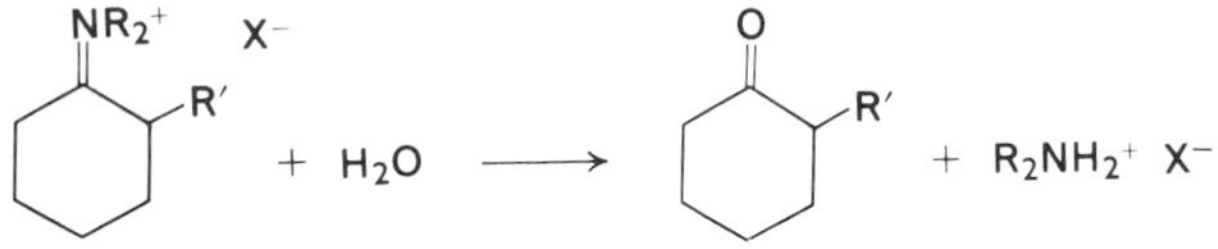

$$+ H_2O \longrightarrow \qquad + R_2NH_2{}^+ \ X^- \tag{57}$$

---

**Exercise 6.**   Formulate the steps in the hydrolysis of the alkylated enamine to the ketone [Eq. (57)]. Remember that the nitrogen of $=\overset{+}{N}R_2$ cannot accept an electron pair (why?) and the resonance $>C=\overset{+}{N}R_2 \longleftrightarrow >\overset{+}{C}-NR_2$ exists.

---

Enamines are versatile reagents that are useful in other ways than those illustrated. Some of their other uses will be shown in later chapters, but a full treatment of their chemistry is beyond the scope of this text.

## 13.14   Synthetic applications of aldol condensations

The synthetic importance of condensations of the aldol type depends upon the fact that a new carbon-carbon bond is formed, permitting the synthesis of complex molecular structures from simpler ones.

The unsaturated ketones and aldehydes that are the usual products of aldol condensations can be reduced to the corresponding saturated compounds or further reduced to alcohols:

$$CH_3CH=CHCHO \longrightarrow CH_3CH_2CH_2CHO \longrightarrow CH_3CH_2CH_2CH_2OH \tag{58}$$

Reduction of aldol itself yields 1,3-butanediol, which can be converted to 1,3-butadiene by dehydration.

The commercial production of a number of important chemicals is based upon aldol condensations. Some examples of industrial processes and of some reactions leading to useful intermediates in synthesis are the following:

$$CH_3CHO + CH_3CH_2CHO \xrightarrow{\ NaOH\ } \underset{\underset{CH_3}{|}}{CH_3\overset{\overset{OH}{|}}{C}HCHCHO} \xrightarrow{\ -H_2O\ }$$

$$\underset{\underset{CH_3}{|}}{CH_3CH=CCHO} \xrightarrow[\text{cat.}]{\ H_2\ } \underset{\underset{CH_3}{|}}{CH_3CH_2CHCH_2OH} \tag{59}$$

$$\text{2-methyl-1-butanol}$$

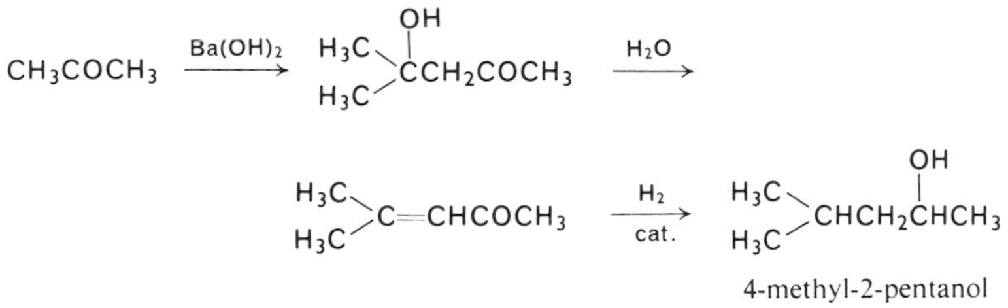

$$CH_3COCH_3 \xrightarrow{Ba(OH)_2} \underset{H_3C}{\overset{H_3C}{>}}\overset{\overset{OH}{|}}{C}CH_2COCH_3 \xrightarrow{H_2O}$$

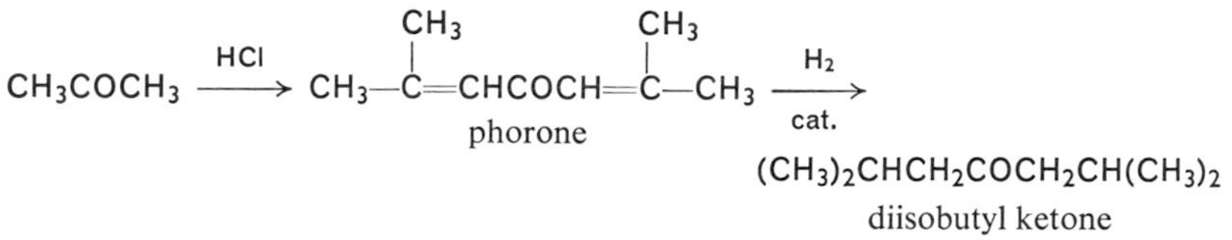

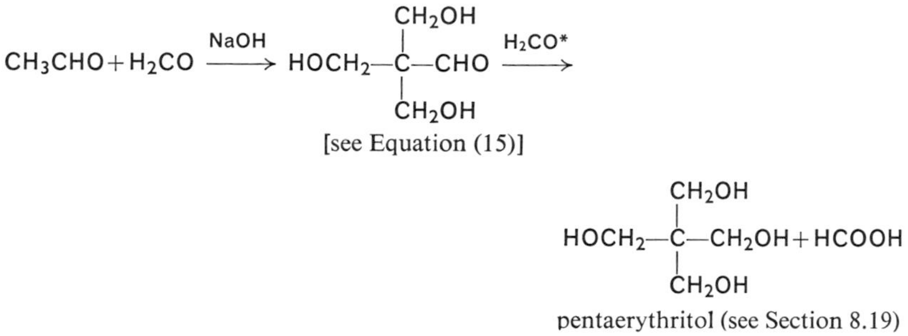

$$\underset{H_3C}{\overset{H_3C}{>}}C{=}CHCOCH_3 \xrightarrow[\text{cat.}]{H_2} \underset{H_3C}{\overset{H_3C}{>}}CHCH_2\overset{\overset{OH}{|}}{C}HCH_3 \quad (60)$$

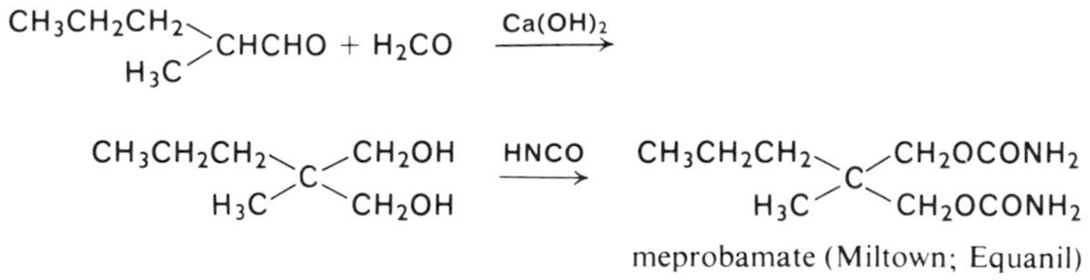

4-methyl-2-pentanol

$$CH_3COCH_3 \xrightarrow{HCl} CH_3{-}\overset{\overset{CH_3}{|}}{C}{=}CHCOCH{=}\overset{\overset{CH_3}{|}}{C}{-}CH_3 \xrightarrow[\text{cat.}]{H_2}$$

phorone

$$(CH_3)_2CHCH_2COCH_2CH(CH_3)_2 \quad (61)$$

diisobutyl ketone

$$CH_3CHO + H_2CO \xrightarrow{NaOH} HOCH_2{-}\overset{\overset{CH_2OH}{|}}{\underset{|}{C}}{-}CHO \xrightarrow{H_2CO^*}$$
$$\underset{CH_2OH}{}$$

[see Equation (15)]

$$HOCH_2{-}\overset{\overset{CH_2OH}{|}}{\underset{|}{C}}{-}CH_2OH + HCOOH \quad (62)$$
$$\underset{CH_2OH}{}$$

pentaerythritol (see Section 8.19)

$$\underset{H_3C}{\overset{CH_3CH_2CH_2}{>}}CHCHO + H_2CO \xrightarrow{Ca(OH)_2}$$

$$\underset{H_3C}{\overset{CH_3CH_2CH_2}{>}}\underset{CH_2OH}{\overset{CH_2OH}{C<}} \xrightarrow{HNCO} \underset{H_3C}{\overset{CH_3CH_2CH_2}{>}}\underset{CH_2OCONH_2}{\overset{CH_2OCONH_2}{C<}} \quad (63)$$

meprobamate (Miltown; Equanil)

$$CH_3COCH_3 \xrightarrow[(2)\ I_2(-H_2O)]{(1)\ Ba(OH)_2} \underset{H_3C}{\overset{H_3C}{>}}C{=}CHCOCH_3 \xrightarrow{NaOCl} \underset{H_3C}{\overset{H_3C}{>}}C{=}CHCOOH$$

$\beta,\beta$-dimethylacrylic acid

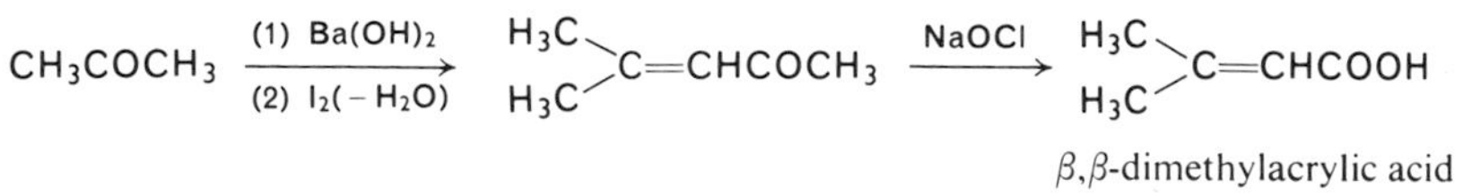

$$(64)$$

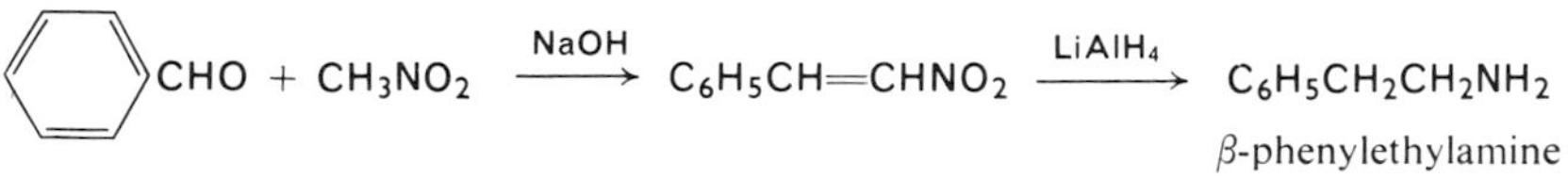

$$\underset{}{\bigcirc}CHO + CH_3NO_2 \xrightarrow{NaOH} C_6H_5CH{=}CHNO_2 \xrightarrow{LiAlH_4} C_6H_5CH_2CH_2NH_2$$

$\beta$-phenylethylamine

$$(65)$$

---

* The formaldehyde acts as the reducing agent to convert $-\overset{|}{\underset{|}{C}}{-}CHO$ into $-\overset{|}{\underset{|}{C}}{-}CH_2OH$, and is oxidized to formic acid.

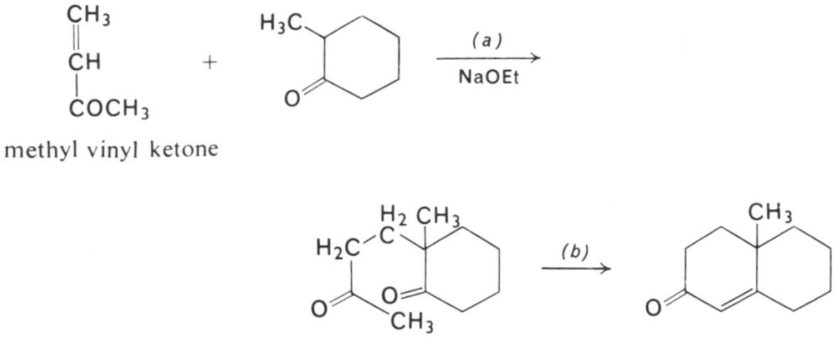

methyl vinyl ketone

(66)

*(a)* Michael condensation
*(b)* aldol condensation of —CO<u>CH</u>₃ with ring ketone

## 13.15   Summary remarks

The reactions of carbonyl compounds described in Chapter 12 and this chapter are some of the most commonly encountered reactions in organic chemistry, and many examples of addition to carbonyl groups and reactions at α-carbon atoms will be met with in the chapters to come. The fundamental character of these reactions has been discussed in terms of relatively simple examples and will be recognized in the more complex examples that are yet to be described.

Reactions of the "aldol family" will be found to be of a number of kinds that have not yet been mentioned. It should always be borne in mind that these related reactions, which are known under such names as the Claisen, Knoevenagel, Perkin, Thorpe, Michael, Dieckmann, Stobbe reactions, proceed by mechanistic courses in which the principles of the simple aldol condensation can be discerned.

The subject of the next chapter is the behavior of another class of substances which, while not ordinarily referred to as carbonyl compounds, are indeed members of the class. These are the carboxylic acids and their derivatives— esters, amides, anhydrides, halides—whose structure is represented by R—C—X,
$$\underset{O}{\overset{||}{}}$$

in which the C—X bond is not C—H (as in aldehydes) or C—C (as in ketones), but is C—O, C—N, or C-halogen.

### Exercises

(Exercises 1–6 will be found within the text.)

**7** Write the acid-base equilibria showing the conversion of each of the following compounds into its monoanion by the strong base sodium ethoxide: *(a)* acetone; *(b)* 3,3-dimethyl-2-butanone; *(c)* 2,2-dimethyl-3-pentanone; *(d)* diethyl ketone; *(e)* 2,4-pentanedione; *(f)* propionaldehyde; *(g)* isobutyraldehyde; *(h)* cyclopentanone; *(i)* 2,6-dimethyl-3,5-heptanedione.

**8** Write all of the equilibria involved in the base-catalyzed aldol condensation of propionaldehyde to form 2-methyl-3-hydroxypentanal.

**9** Show the steps in the base-catalyzed cleavage ("reverse aldol") of *(a)* diacetone alcohol; *(b)* 3-hydroxybutanal.

10 Show the reactants required for the preparation of the following compounds by the aldol condensation: *(a)* 3-hydroxyhexanal; *(b)* 4-hydroxy-2-butanone; *(c)* pentaery-thritol; *(d)* 2-(hydroxymethyl)-cyclohexanone.

11 Explain in detail why propionaldehyde is transformed by the action of a strong base into anion *(a)* and *not* into anion *(b)*.

$\quad$ *(a)* $(CH_3\ddot{C}HCHO)^-$ $\qquad\qquad$ *(b)* $(\ddot{C}H_2CH_2CHO)^-$

12 Write structures for the contributing forms that represent the structures of the following

$\quad$ *(a)* $(:CH_2CHO)^-$; $\quad$ *(b)* $\quad$ $(CH_3CO\ddot{C}HCOCH_3)^-$; $\quad$ *(c)* $\quad$ $(CH_2{=}CH\ddot{C}HCHO)^-$;

$\quad$ *(d)* $(CH_2{=}CH\ddot{C}HCH{=}CH_2)^-$; *(e)* $CH_2{=}CHCH{=}\overset{+}{O}H$.

13 Write the structure of the predominant product that would be formed by the base-catalyzed aldol condensation of each of the following pairs: *(a)* acetone, acetone; *(b)* acetone, acetaldehyde; *(c)* cyclohexanone, acetaldehyde; *(d)* isobutyraldehyde, formaldehyde; *(e)* propionaldehyde 2,4-pentanedione; *(f)* nitromethane, acetaldehyde.

14 The aldol condensation of vinylacetaldehyde, $CH_2{=}CHCH_2CHO$, with acetaldehyde, in the presence of alkali as a catalyst, yields the product

$CH_3CH{=}CH{-}CH{=}CHCHO$

Explain this result and formulate the sequence of steps that leads to it.

15 The reaction of 3,3-dimethyl-2,4-pentanedione with alcoholic sodium ethoxide gives ethyl acetate and methyl isopropyl ketone. Formulate this reaction, recalling the principles of nucleophilic attack upon carbonyl groups, and point out the relationship of the cleavage of the diketone to the "reversal" of the aldol condensation.

16 A useful method for demonstrating the presence of the grouping $-CH_2{-}\overset{|}{C}{=}O$ in an organic compound consists in treating the compound with benzaldehyde, $C_6H_5CHO$, and a trace of alkali (such as sodium ethoxide). The aldol condensation at the active methylene group gives rise to a "benzal" derivative, of the partial structure $-\overset{|}{C}{=}CHC_6H_5$. Show the formation of the benzal derivatives of the following:

$-\overset{|}{C}{=}O$

$\quad$ *(a)* acetone; *(b)* cyclopentanone; *(c)* diethylketone; *(d)* pinacolone.

17 How could you distinguish by the application of a chemical test between *(a)* acetone and diethyl ketone; *(b)* methyl isopropyl ketone and isobutyraldehyde; *(c)* diisopropyl ketone and di-*n*-propyl ketone; *(d)* di-*n*-butyl ether and 2-butanone; *(e)* 2-pentanone and 3-pentanone?

18 Explain why acetylacetone $CH_3COCH_2COCH_3$ is a stronger acid than acetonylacetone, $CH_3COCH_2CH_2COCH_3$. Write the structures of the monoanions formed by the reactions of these two ketones with a base such as sodium ethoxide.

19 What product would you predict would be formed by the intramolecular aldol condensation of 2,8-nonanedione?

20 Using acetone as the starting material, devise practical syntheses for the following. Show steps and reagents.

$\quad$ *(a)* $(CH_3)_2CHCH_2COOH$

$\quad$ *(b)* $CH_3CHOHCH_2CH_2OH$

$\quad$ *(c)* $\begin{array}{c} H_3C \\ H_3C \end{array}\!\!>\!\!\overset{\displaystyle O{-}CH_2}{\underset{\displaystyle O{-}\overset{|}{\underset{CH_3}{C}H}}{C}}\!\!<\!\!\begin{array}{c} \\ CH_2 \\ \end{array}$

**21** A compound (A), $C_6H_8O$, gave cyclohexanone upon catalytic reduction. It had no high-intensity ultraviolet absorption above 200 m$\mu$, but had a weak ($\epsilon$ about 100) maximum at about 290 m$\mu$. When A was treated with alcoholic alkali it was converted into an isomeric ketone (B). Compound B also gave cyclohexanone upon catalytic hydrogenation and had $\lambda_{max} = 255$ m$\mu$ with $\epsilon$ more than 10,000. Write possible structures for A and B, and suggest what further experiments might be performed to establish the structure of A.

**22** A compound (A), $C_7H_{12}O_3$, showed no high intensity ultraviolet absorption above 200 m$\mu$. When A was treated with alcoholic KOH there were formed two products: one was acetic acid; the other (B) was a compound $C_5H_8O$ with an ultraviolet absorption maximum at 225 m$\mu$ ($\epsilon$ about 10,000). Both A and B gave iodoform when treated with iodine in alkaline solution.

**23** Pulegone, 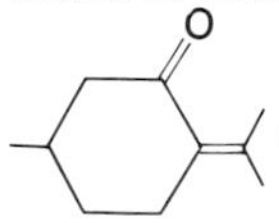 , shows an absorption maximum of 252 m$\mu$ (log $\epsilon$ about 3.7).

When a solution of pulegone in alcoholic KOH is allowed to stand, the intensity of the ultraviolet maximum at 252 m$\mu$ falls off with time and eventually disappears. Why?

## Chapter fourteen

# Carboxylic Acids
# and Acid Derivatives

## 14.1 Carboxylic acids

The most highly oxidized state of carbon in organic combination—i.e., attached to another carbon atom—is found in the *carboxyl group*, the functional group of the largest class of organic acids. *Carboxylic acids* are the end-products of the oxidation series

$$RCH_3 \longrightarrow RCH_2OH \longrightarrow RCHO \longrightarrow RCOOH$$

primary · · · · · · · · aldehyde · · · · · · carboxylic
alcohol · · · · · · · · · · · · · · · · · · · · · · acid

and can be prepared synthetically in this way (Section 14.3). Acids are end-products of many types of oxidative degradations of organic compounds.

The carboxylic acids RCOOH, in which R is an alkyl group, are the *aliphatic*, or *fatty acids*. The simplest member of this group is formic acid, HCOOH. Many aliphatic acids bear trivial names; because of their wide occurrence in nature, many of them bear names which are derived from the animal or plant sources from which they are obtained.

The naming of acids of all kinds by the I.U.C. system follows the general rules already described in Chapter 6. Substitution names, based upon the trivial names, are also used. In Table 14-1 are given a number of representative carboxylic acids with names that are customarily used. It will be noted that the designation of the carbon atoms by Greek letters follows the same rules used in the case of aldehydes and ketones: $\overset{\delta}{C}-\overset{\gamma}{C}-\overset{\beta}{C}-\overset{\alpha}{C}-COOH$. This terminology is useful when *types* of compounds are discussed. For example, one may refer to $\beta$-hydroxy acids, $\delta$-amino acids, $\alpha$-bromo acids, $\beta,\gamma$-unsaturated acids, and so on.

**Table 14-1.** *Some representative carboxylic acids*

| Acid | Common Name | I.U.C. Name | pKa |
|---|---|---|---|
| HCOOH | formic acid | methanoic acid | 3.77 |
| $CH_3COOH$ | acetic | ethanoic | 4.75 |
| $CH_3CH_2COOH$ | propionic | propanoic | 4.88 |
| $CH_3$<br>    CHCOOH<br>$CH_3$ | isobutyric | 2-methylpropanoic | 4.85 |
| $(CH_3)_3CCOOH$ | pivalic | 2,2-dimethylpropanoic | 5.02 |
| $[CH_3(CH_2)_nCOOH]$ | | | |
| $n=3$ | n-valeric | pentanoic | 4.81 |
| 5 | pelargonic | heptanoic | 4.96 |
| 8 | capric | decanoic | — |
| 10 | lauric | dodecanoic | — |
| 12 | myristic | tetradecanoic | — |
| 14 | palmitic | hexadecanoic | — |
| 16 | stearic | octadecanoic | — |
| $CH_3CH{=}CHCOOH$ | crotonic | 2-butenoic | — |
| $CH_3OCH_2COOH$ | — | methoxyacetic* | — |
| $CH_2{=}CHCH_2COOH$ | — | vinylacetic* | — |
| $CH_2{=}CHCH_2CH_2COOH$ | — | allylacetic* | — |
| $CH_3CH{=}CH{-}CH{=}CH{-}COOH$ | — | 2,4-hexadienoic | — |
| $CH_3OCH_2COOH$ | — | methoxyacetic | 3.5 |
| $BrCH_2COOH$ | — | bromoacetic* | 2.87 |
| $ClCH_2CH_2COOH$ | — | β-chloropropionic* | 4.1 |
| $HOCH_2COOH$ | glycolic | hydroxyacetic* | 3.83 |
|     Br<br>    &#124;<br>$CH_3CHCHCHCOOH$<br>   $CH_3$  $CH_3$ | — | 2,4-dimethyl-3-bromopentanoic | — |

\* A substitution name.

## 14.2 The acyl group

The general term for the grouping $R{-}\overset{\displaystyle O}{\overset{\|}{C}}{-}$ is "acyl", and the name of a particular acyl group is derived from that of the carboxylic acid by changing the ending *-ic* to *-yl*. Thus, $CH_3CO$— is the acetyl group; $(CH_3)_2CHCO$— is isobutyryl; $HCO$— is formyl (and it will be noted that this group is the functional group of aldehydes). The compounds listed in Table 14-2 contain acyl groups as substituents.

<table>
<tr><td colspan="2">**Table 14-2.** *Compounds containing acyl groups as substituents*</td></tr>
<tr><td>Compound</td><td>Name</td></tr>
<tr><td>$CH_3COCl$</td><td>acetyl chloride</td></tr>
<tr><td>$CH_3COCH_2COCH_3$</td><td>acetylacetone (diacetylmethane)</td></tr>
<tr><td>$(CH_3CO)_3CH$</td><td>triacetylmethane</td></tr>
<tr><td>$(CH_3)_2CHCOCH_2COOH$</td><td>isobutyrylacetic acid</td></tr>
<tr><td>$CH_3CH_2CHCHCHCH_2COBr$<br>         $|$    $|$<br>      $CH_3$  $CH_3$</td><td>3,5-dimethylheptanoyl bromide</td></tr>
</table>

"Acylation" refers to the substitution of a hydrogen atom by an acyl group. The following examples illustrate this terminology:

*O-acylation* (O-acetylation):

$$CH_3CH_2OH + CH_3COCl \longrightarrow CH_3CH_2OCOCH_3 + HCl \qquad (1)$$

*N-acylation* (N-benzoylation):

$$C_6H_5NH_2 + C_6H_5COCl \longrightarrow C_6H_5NHCOC_6H_5 \qquad (2)$$

  aniline     benzoyl          acetanilide
            chloride    (N-benzoylaniline)*

*C-acylation:*

$$CH_3COCH_2COCH_3 \xrightarrow[CH_3COCl]{NaOEt} CH_3CO\overset{\displaystyle COCH_3}{\underset{\displaystyle |}{C}}HCOCH_3 \quad \text{(Chapter 15).} \qquad (3)$$

In the same terminology, the hydrolysis of esters and amides is often referred to as "deacylation."

## 14.3  Preparation of carboxylic acids

*Oxidation.*  Carboxylic acids can be prepared by the oxidation of alcohols and aldehydes. The oxidation of an aliphatic alcohol to an acid with the same number of carbon atoms is often of limited usefulness because oxidation often proceeds further with degradation of the molecule:

$$RCH_2CH_2CH_2OH \xrightarrow{CrO_3} \{RCH_2CH_2COOH \longrightarrow RCH_2COOH \longrightarrow RCOOH\}$$

---

* In naming compounds in this text the name most commonly used will be the one given first in most cases. When two or more names are given, the one in parentheses is given to illustrate the application of a form of terminology even though the name itself may not be used for the particular compound in the example.

An important application of oxidative degradation of this kind, which is analytical rather than preparative, is described in Section 14.18.

Aldehydes, which are more readily oxidized than alcohols, can be converted into acids under sufficiently moderate conditions to avoid further degradation. A useful reagent for the oxidation of an aldehyde under mild conditions is silver oxide (or the silver-ammonia complex, $Ag(NH_3)_2^+$):

$$RCHO + Ag_2O \longrightarrow RCOOH + Ag \tag{4}$$

Silver oxide is of particular value when the compound to be oxidized contains, besides the formyl group, other oxidizable groups, such as double bonds or hydroxyl groups.

The oxidation of unsaturated compounds, involving a cleavage of the molecule at the carbon-carbon double bond, can be carried out in a number of ways. Direct oxidation which chromic acid, potassium permanganate, or nitric acid (in special cases) is illustrated by the following examples:

$$CH_3CH{=}CHCH_3 \xrightarrow{\text{CrO}_3} 2CH_3COOH \tag{5}$$

$$(CH_3)_2C{=}CHCH_3 \xrightarrow{\text{KMnO}_4} (CH_3)_2CO + CH_3COOH \tag{6}$$

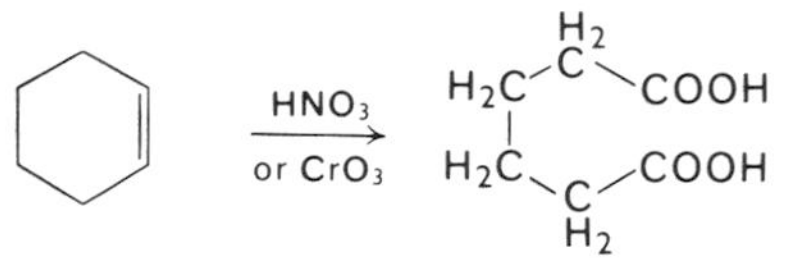

$$\text{cyclohexene} \xrightarrow[\text{or CrO}_3]{\text{HNO}_3} \text{adipic acid} \tag{7}$$

Ozonolysis of olefinic double bonds is often an initial stage in the oxidation of unsaturated compounds (Section 9.20). Olefins that contain the structural element RCH= yield aldehydes, RCHO, which can be oxidized to the carboxylic acids.

***Other methods.*** Nitriles (Section 14.12) can be hydrolyzed to carboxylic acids

$$RCN + H_2O \xrightarrow[\text{or alkali}]{\text{acid}} RCOOH + NH_3 \tag{8}$$

and so synthetic routes to nitriles can be routes leading to the preparation of the corresponding acids. The formation of nitriles by the displacement of halogen from primary alkyl halides has been described earlier:

$$RCH_2Br + CN^- \longrightarrow RCH_2CN + Br^- \tag{9}$$

Aldehyde oximes undergo dehydration (elimination of water) when heated with acetic anhydride:

$$RCHO \xrightarrow{\text{NH}_2\text{OH}} RCH{=}NOH \xrightarrow[\text{anhydride}]{\text{acetic}} RCN \tag{10}$$

This procedure is equivalent in its overall result to the oxidation of the aldehyde to the acid.

---

**Exercise 1.** In the sequence RCHO → oxime → nitrile → acid, the aldehyde is converted to the higher oxidation state of the acid. What is reduced in the process?

---

An important synthetic method for the preparation of carboxylic acids is the reaction of Grignard reagents with carbon dioxide:

$$RMgX + CO_2 \longrightarrow RC\overset{O}{\underset{OMgX}{\diagdown}} \xrightarrow{H_2O\ H^+} RCOOH + Mg^{++}\ \text{salts} \qquad (11)$$

This procedure is widely applicable, and can be used for the preparation of aliphatic and aromatic carboxylic acids. Indeed, the reaction of Grignard reagents with carbon dioxide is so general that it is used as a method for recognizing that an organomagnesium compound is in hand, and in some cases for determining the structure of the Grignard reagent. Some examples of this reaction, called the "carbonation" of the Grignard reagent, are the following:

$$(CH_3)_3CCl \xrightarrow[\text{ether}]{Mg} (CH_3)_3CMgCl \xrightarrow[(2)\ H_2O]{(1)\ CO_2} (CH_3)_3CCOOH \qquad (12)$$
$$\text{pivalic acid}$$

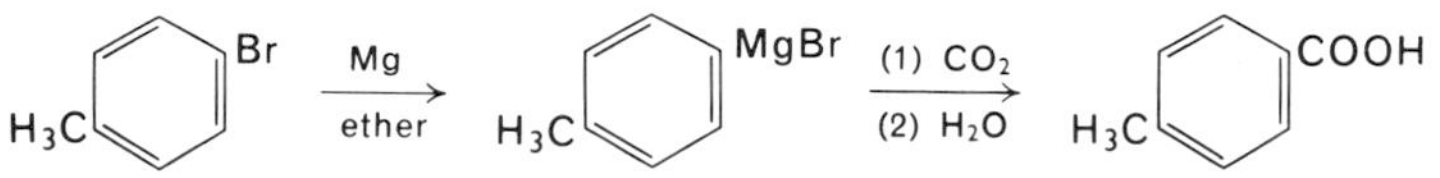

$$\text{p-bromotoluene} \qquad\qquad\qquad \text{p-toluic acid}$$

$$CH_3C{\equiv}CH \xrightarrow{EtMgBr} CH_3C{\equiv}CMgBr \xrightarrow[(2)\ H_2O]{(1)\ CO_2} CH_3C{\equiv}CCOOH \qquad (14)$$
$$\text{propyne} \qquad\qquad\qquad\qquad\qquad \text{tetrolic acid} \atop \text{(2-butynoic acid)}$$

The haloform reaction (Section 12.17) is not only a diagnostic procedure for structure analysis; it is also a useful preparative method for the synthesis of acids:

$$CH_3COCH_3 \xrightarrow[(2)\ I_2]{(1)\ Ba(OH)_2} (CH_3)_2C{=}CHCOCH_3 \xrightarrow{NaOCl} (CH_3)_2C{=}CHCOOH + CHCl_3 \qquad (15)$$

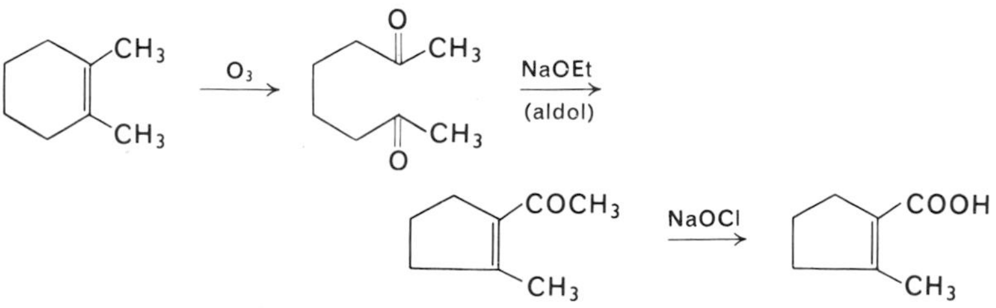

**Exercise 2.** Formulate the steps in the ring-closure of 2,7-octanedione to 1-acetyl-2-methylcyclopentene [Equation (16)].

$$(CH_3)_2\underset{\underset{OH}{|}}{C}\!-\!\underset{\underset{OH}{|}}{C}(CH_3)_2 \xrightarrow[\text{rearrangement}]{\text{pinacol}} (CH_3)_3CCOCH_3 \xrightarrow{\text{NaOCl}} (CH_3)_3CCOOH \qquad (17)$$

Many other methods for the preparation of carboxylic acids are available to the organic chemist. Some of these will be described in later chapters.

## 14.4 Esters

The preparation of esters by the acid-catalyzed reaction between an alcohol and an acid has already been discussed (Sec. 8.16). For this to be a practical synthetic method the equilibrium, which may be unfavorable to the complete conversion of the alcohol or the acid into the ester, must be shifted to the side of the desired product. This is usually accomplished by using one of the reactants—usually the alcohol—in large excess, with the result that the conversion of the acid into the ester is substantially complete. Esterification by this means is ordinarily carried out with the use of the inexpensive methanol or ethanol:

$$RCOOH + CH_3OH \text{ (large excess)} \underset{\phantom{H^+}}{\overset{H^+}{\rightleftharpoons}} \underset{\substack{\text{essentially complete} \\ \text{conversion}}}{RCOOCH_3 + H_2O} \qquad (18)$$

The special reagent diazomethane, $CH_2N_2$, converts carboxylic acids into their methyl esters in excellent yields and under mild conditions. The reagent is used in ether solution, from which the isolation of the ester is easy, and the only other product of the reaction is nitrogen:*

$$RCOOH + CH_2N_2 \longrightarrow RCOOCH_3 + N_2 \qquad (19)$$

Diazomethane is prepared from any of several starting materials; two of these are shown in the following equations:

$$CH_3\!-\!\underset{\underset{N=O}{|}}{N}CONH_2$$

$$\xrightarrow{\text{KOH}}$$

$$\xrightarrow{\text{KOH}}$$

$$H_3C\!-\!\!\!\bigcirc\!\!\!-SO_2\underset{\underset{N=O}{|}}{N}\!-\!CH_3$$

$$CH_2N_2 \left\{ :\bar{C}H_2\!-\!\overset{+}{N}\!\!\equiv\!\!N: \longleftrightarrow CH_2\!=\!\overset{+}{N}\!=\!\ddot{\underset{..}{N}}: \right\} \qquad (20)$$

---

* Diazomethane is a very toxic substance and must be used with precautions against exposure, especially exposure to the eyes. It is used in solution, usually in ether.

The reaction of diazomethane with a carboxylic acid may be formulated as follows. It is evident that the last step represents a typical nucleophilic displacement reaction with expulsion of $N_2$ as the leaving group:

$$RCOOH + CH_2N_2 \rightleftharpoons RCOO + CH_3N_2^+$$

$$RCOO^- \frown CH_3 \frown N_2^+ \longrightarrow RCOOCH_3 + N_2 \tag{21}$$

Other examples of the preparation of esters by the displacement reaction involving attack of a carboxylate ion upon an alkyl halide are exemplified by the following; these are useful preparative procedures:

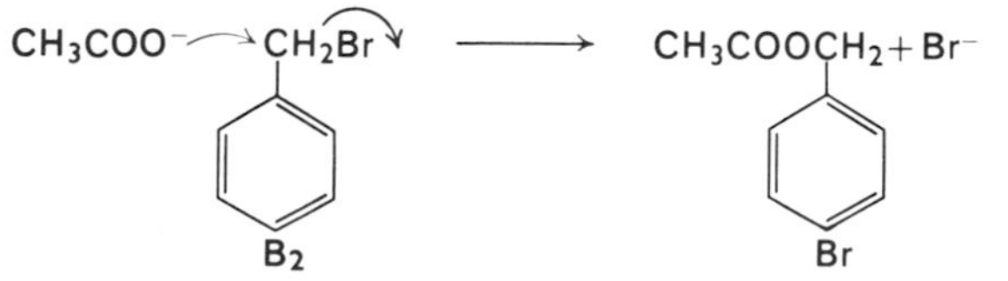

$$CH_3COO^- \frown CH_2Br \longrightarrow CH_3COOCH_2 + Br^- \tag{22}$$

$$B_2 \qquad\qquad Br$$

*p*-bromobenzyl bromide  *p*-bromobenzyl acetate

Other examples of this reaction have already been described (Sections 7.8 and 8.13).

Perhaps the most general method of preparing esters is by means of the reaction of hydroxy compounds (alcohols, phenols) with acid chlorides and anhydrides. These reagents and their use in O-acylation will be discussed in Section 14.9.

## 14.5  Naturally occurring esters: fats and oils

A general reaction of esters is their hydrolysis with aqueous alkali (hydrolysis under acidic conditions has been discussed in connection with acid-catalyzed esterification in Section 8.16). The alkaline hydrolysis of esters is one of the oldest arts of mankind, for when the esters that are used are the naturally occurring fats and oils, the product is a *soap*. The term *saponification*, which relates specifically to this process, is applied broadly, and is commonly used to refer to the general reaction

$$RCOOR' + NaOH \longrightarrow RCOO^-Na^+ + R'OH \tag{23}$$

even when the sodium salt of the acid is not a true soap. Soaps are the sodium salts of long-chain fatty acids.

The natural fats and oils are esters of such acids with glycerol, and are often referred to as "glycerides":

$$
\begin{array}{ll}
CH_2OH & CH_2OCOR \\
CHOH & CHOCOR' \\
CH_2OH & CH_2OCOR'' \\
\text{glycerol} & \text{a fat; the R groups are} \\
& \text{long-chain alkyl groups}
\end{array}
$$

**Table 14-3.** *Some natural fats and oils*

| R in RCOO⁻ | Name of Acid RCOOH | Name of Glycerol Triester | Important Natural Sources |
|---|---|---|---|
| $CH_3(CH_2)_{16}$ | stearic | tristearin | animal tallow, lard |
| $CH_3(CH_2)_{14}$ | palmitic | tripalmitin | tallow, palm oil |
| $CH_3(CH_2)_{12}$ | myristic | trimyristin | nutmeg, coconut |
| $cis$-$CH_3(CH_2)_7CH{=}CH(CH_2)_7$ | oleic | triolein | olive oil, peanut oil |

Natural fats include glycerol esters in which all three of the fatty acid residues are the same (R=R′=R″) and others in which two or three different fatty acids are esterified with a glycerol molecule. Some common natural fats (the term "oils," when used in this context, is used for those fats that are liquid at ordinary temperatures) are given in Table 14-3.

Natural fats and oils are mixtures of glycerides. Tallow contains chiefly myristic, palmitic, stearic, oleic, and linoleic (9,12-octadecadienoic) acids in ester combination with glycerol. In general, the solid fats contain relatively higher preparations of the saturated fatty acids, and oils contain higher proportions of unsaturated acids.

The long-chain fatty acids also occur in nature in ester combination with alcohols other than glycerol; for example, with cholesterol and other sterols. The esters of long-chain fatty acids and long-chain aliphatic alcohols are *waxes*, and are found in plants, often as epidermal coatings on leaves and fruits, where their function is to prevent excessive water loss by transpiration.

The hydrolysis of a fat with alkali yields glycerol and three molecules of the salt of the fatty acid (i.e., a *soap*):

$$RCOOCH_2$$
$$|$$
$$RCOOCH + 3NaOH \longrightarrow 3RCOO^-Na^+ + HOCH_2CHCH_2OH$$
$$|$$
$$RCOOCH_2 \qquad\qquad\qquad\qquad \text{a soap}$$

with $OH$ on the central carbon.

This is an industrial process of great importance, for soaps are valuable substances. They possess detergent (cleansing) properties that depend upon the ability of the soap molecules to form aggregates with fat-soluble materials in which the long, fatty side-chains surround the "dirt" in such a way as to enclose it within a cluster of soap anions, leaving the hydrophilic —$COO^-$ groups as a peripheral water-solubilizing envelope. The resulting dirt-soap complex can then be carried away in water.

The glycerol that is formed in the process of soap-making is an important by-product. It is used in the manufacture of resins and nitroglycerin, in pharmaceutical and cosmetic industries, and as a humectant. Humectants are substances that help to preserve the moisture content of materials. The most common use of glycerol as a humectant is for tobacco. Glycerol, being a trihydroxy compound, is able, by hydrogen bonding, to retain water molecules in loosely bound association. Thus, the presence of glycerol in a material such as smoking tobacco helps to prevent drying and keeps the tobacco in a properly moist condition.

## 14.6 Saponification as a general reaction

The term saponification is used in the generic sense to refer to the general reaction of the alkaline hydrolysis of esters:

$$CH_3C \underset{OCH_2CH_3}{\overset{O}{\diagup}} + NaOH \longrightarrow CH_3COO^- Na^+ + CH_3CH_2OH \tag{24}$$

Saponification of esters is a general reaction that has few limitations. The *rate* of saponification is strongly influenced by structure. As we have seen, and as will be discussed further in the next section, saponification is a reaction that belongs to the general class of carbonyl addition reactions. Therefore, increasing substitution around the $\alpha$-carbon atom of the acid portion of the ester will cause the rate of saponification to decrease. Esters of trimethylacetic acid, for example, saponify slowly, whereas those of an "unhindered" acid, such as formic acid, saponify rapidly. Steric interference to attack upon the ester carbonyl group is also observed when the alkyl group of the —OR portion is highly branched:

$$(CH_3)_3CCOOEt + NaOH/H_2O \longrightarrow (CH_3)_3CCOO^- Na^+ + CH_3OH \text{ (slow)} \tag{25}$$

$$CH_3COOEt + NaOH/H_2O \longrightarrow CH_3COO^- Na^+ + CH_3OH \text{ (fast)} \tag{26}$$

The large differences in the rate of saponification of esters of $\alpha$-substituted and of unsubstituted acids* can be used advantageously in synthetic studies. For example, the following ester can be saponified to yield the ester-acid; the —COOEt group on the fully substituted carbon atom undergoes saponification very much more slowly than the one on the primary carbon atom:

$$\begin{array}{c} CH_3 \\ \diagdown COOEt \\ \diagup \\ CH_2CH_2COOEt \end{array} \xrightarrow{\text{H}_2\text{O/NaOH (1 mole)}} \begin{array}{c} CH_3 \\ \diagdown COOEt \\ \diagup \\ CH_2CH_2COOH \end{array} \tag{27}$$

---

**Exercise 3.** Since the dicarboxylic acid corresponding to the diester shown in [Equation (27)] would undergo esterification very slowly, suggest a practical means of preparing the dimethyl ester from the diacid. The successful method would involve an esterification reaction that did not proceed by way of an attack on the carbonyl carbon atom.

---

* It should not be overlooked that since the esterification reaction between an alcohol and an ester is also a carbonyl addition reaction, the rate of acid-catalyzed *esterification* is also strongly influenced by $\alpha$-substitution. Trimethylacetic (pivalic) acid esterification is very much slower than that of aliphatic acids of the general structure $RCH_2COOH$, in which the $\alpha$-carbon is monosubstituted.

The advantage of a selectivity of this kind is that it opens the way for subsequent manipulations in which the —COOH group is involved and in which the —COOEt group is unaffected.

## 14.7 The mechanism of the saponification reaction

The attack of a nucleophilic reagent such as the hydroxide ion upon the ester group is an example of the general carbonyl addition reaction discussed in Chapter 12:

$$-\overset{|}{\underset{B:}{C}}\!\!\overset{\frown}{=}\ddot{O}: \quad\rightleftharpoons\quad -\overset{|}{\underset{B}{C}}-\ddot{O}:^- \tag{28}$$

Using the ester as the carbonyl component of this process, and $OH^-$ as the nucleophile, we have

$$CH_3-\overset{\overset{\displaystyle OR}{|}}{\underset{\underset{\displaystyle HO:^-}{}}{C}}\!\!\overset{\frown}{=}\ddot{O}: \quad\rightleftharpoons\quad CH_3-\overset{\overset{\displaystyle OR}{|}}{\underset{\underset{\displaystyle OH}{|}}{C}}-\ddot{O}:^- \tag{29}$$

The intermediate anion in (29) is in equilibrium with the initial reagents, $CH_3COOR$ and $OH^-$; but it is apparent that it has a nearly equivalent alternative course open to it: it can expel $OR^-$ as well as $OH^-$

$$CH_3-\overset{\overset{\displaystyle OR}{|}}{\underset{\underset{\displaystyle HO^-}{}}{C}}\!\!=\!O \quad\underset{(b)}{\overset{(b)}{\rightleftharpoons}}\quad \overset{(a)}{\underset{(b)}{}}CH_3-\overset{\overset{\displaystyle OR}{|}}{\underset{\underset{\displaystyle OH}{|}}{C}}\!-\!\ddot{O}:^- \quad\overset{(a)}{\rightleftharpoons}\quad CH_3-\overset{\overset{\displaystyle OR^-}{}}{\underset{\underset{\displaystyle OH}{|}}{C}}\!=\!O \tag{30}$$

Courses *(a)* and *(b)* would differ relatively little since $RO^-$ and $HO^-$ are of comparable nucleophilicity.

The balance of the reaction is now tipped in favor of course *(a)* since the acid that is the initial product of this reaction will be converted into the carboxylate anion in the alkaline solution

$$CH_3-\overset{\overset{\displaystyle}{}}{\underset{\underset{\displaystyle OH}{|}}{C}}\!=\!O + OH^- \longrightarrow CH_3-\overset{\overset{\displaystyle}{}}{\underset{\underset{\displaystyle O^-}{|}}{C}}\!=\!O + H_2O$$

This ionization is essentially complete to the right under the conditions of alkaline saponification, and so the hydrolysis of the ester is complete. Figure 14-1 shows an energy profile that indicates the large overall difference in energy levels between starting reactants and final products, and thus shows why the equilibrium constant of the saponification reaction will be very large.

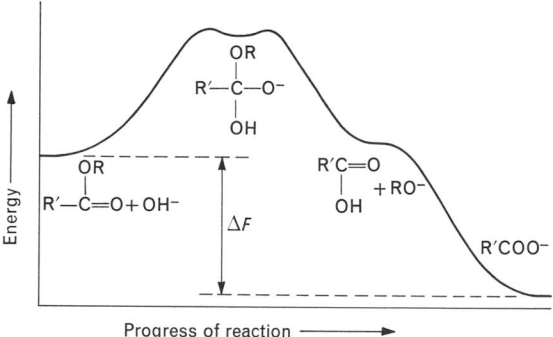

**Figure 14–1.** Energy profile for a saponification of an ester. The equilibrium is far to the side of complete hydrolysis because of the final ionization of RCOOH to RCOO⁻.

## 14.8 Base-catalyzed ester interchange

When an ester is treated with a base other than OH⁻, carbonyl attack takes place in an exactly analogous way; for example, the reaction of an ethyl ester with a solution of sodium methoxide in methanol proceeds as follows:

$$R-\overset{\overset{\displaystyle OEt}{|}}{C}=O + MeO^- \;\rightleftharpoons\; R-\overset{\overset{\displaystyle OEt}{|}}{\underset{\underset{\displaystyle OMe}{|}}{C}}-O^- \;\rightleftharpoons\; R-\overset{\overset{\displaystyle}{}}{\underset{\underset{\displaystyle OMe}{|}}{C}}=O + EtO^- \tag{31}$$

This is called *ester interchange*, and is a very useful method of converting one ester into another in a single step. It is apparent that if the two alkoxide ions involved in the equilibrium are comparable in their steric and electronic characteristics, the equilibrium constant for the overall reaction

$$RCOOR' + R''O^- \;\rightleftharpoons\; RCOOR'' + R'O^- \tag{32}$$

will be close to unity. This is illustrated by Figure 14-2, in which ethoxide ion and methoxide ion are chosen for illustration.

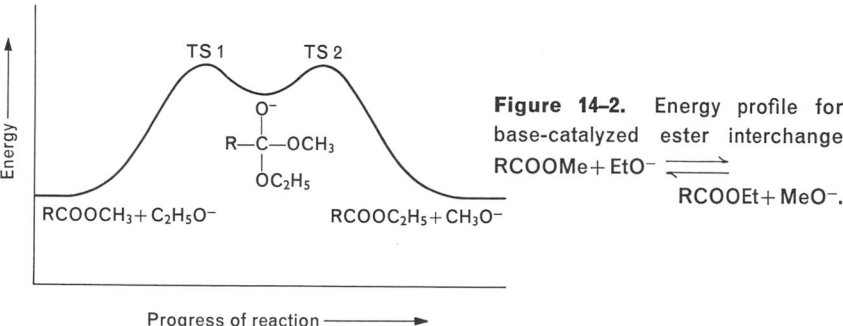

**Figure 14–2.** Energy profile for base-catalyzed ester interchange RCOOMe + EtO⁻ ⇌ RCOOEt + MeO⁻.

In practice, the transesterification reaction can be brought to completion, with essentially quantitative formation of the new ester, by the use of a large excess of the alcohol. If an ethyl ester is dissolved in a large excess of methanol in which a small amount of sodium has been dissolved (to form sodium methoxide), the final mixture at equilibrium will consist almost entirely of the methyl ester. It is not necessary to use an excess of the alkoxide ion, because, as is evident from the above equations, the base is not consumed. For each $MeO^-$ that disappears, an $EtO^-$ appears; and because of the proton-exchange equilibrium

$$EtO^- + MeOH \rightleftharpoons MeO^- + EtOH$$

the final constitution of the ester mixture will be determined by the ratio of methanol to ethanol in the solution, the alkoxide ion acting only as the catalyst.

The transesterification reaction is valuable when it is desired to prepare an ester that is difficult to prepare by other means. Esterification, for example with diazomethane, to give the methyl ester followed by ester interchange with the "new" alcohol, gives the desired ester in good yield. An example is found in the preparation of certain esters of benzyl alcohol with acids that do not lend themselves conveniently to other methods of esterification:

dimethyl
methylmalonate                     dibenzyl methylmalonate

Ester interchange can often be anticipated as an undesired side reaction whenever esters are used in reactions that are carried out in alcoholic solution under basic conditions. It is important to keep this in mind because the simple alcohols are so often used as solvents in carrying out organic reactions. When a compound that is, say, a methyl ester is to be submitted to a reaction (that does not involve the ester group) in alcoholic solution, it would be best to use methanol, rather than ethanol, as the solvent.

## 14.9   Acyl halides and acid anhydrides

Acyl halides have the general structure $RC{\overset{O}{\underset{X}{\diagdown}}}$, where X is a halogen. Acyl chlorides are the most commonly used, and are prepared by the reaction of carboxylic acids with halides of phosphorus ($PCl_5$, $PCl_3$, $POCl_3$), or with thionyl chloride ($SOCl_2$):

$$3CH_3COOH + PCl_3 \longrightarrow 3CH_3COCl + P(OH)_3 \tag{34}$$

$$(CH_3)_2CHCOOH + PCl_5 \longrightarrow (CH_3)_2CHCOCl + POCl_3 + HCl \tag{35}$$

$$BrCH_2COOH + SOCl_2 \longrightarrow BrCH_2COCl + SO_2 + HCl \tag{36}$$

(The degree to which all of the available chlorine in $PCl_5$, $PCl_3$, or $POCl_3$ is utilized in these reactions depends upon the experimental conditions.)

Thionyl chloride is especially advantageous for the preparation of most acyl chlorides because the other products of the reaction are gaseous ($SO_2$, HCl) and excess thionyl chloride (b.p. $79°C$) is readily removed by distillation.

Acyl chlorides are lachrymators that attack mucous membranes. The lower aliphatic acyl chlorides are very reactive; they fume in moist air and react vigorously with water, alcohols, and amines. Acyl chlorides are valuable reagents in organic chemistry; along with acid anhydrides, they are the reagents ordinarily used for O-acylation of alcohols and N-acylation of amines (Section 14.2). Their high degree of reactivity makes them the preferred reagent for the synthesis of acid derivatives of other kinds:

$$CH_3COCl \begin{cases} \xrightarrow[\text{hydrolysis}]{H_2O} CH_3COOH + HCl* \\[2em] \xrightarrow[\text{alcoholysis}]{ROH} CH_3COOR + HCl \\[2em] \xrightarrow[\text{ammonolysis}]{NH_3} CH_3CONH_2 + HCl \\[2em] \xrightarrow{RNH_2} CH_3CONHR + HCl \\[2em] \xrightarrow{\langle\ \rangle OH} CH_3COO\langle\ \rangle + HCl \end{cases} \qquad (37)$$

Acid anhydrides are so called because they are formally derived from two molecules of a carboxylic acid by the loss of a molecule of water. They are not usually prepared in this way, but can be made by the reaction of a sodium salt of a carboxylic acid with an acyl chloride:

$$RCOO^- + RCOCl \longrightarrow \overset{O}{\overset{\|}{RC}} - O - \overset{O}{\overset{\|}{CR}} + Cl^- \qquad (38)$$

Acid anhydrides are somewhat less reactive than acyl halides, but are frequently used in place of acyl chlorides for O- and N-acylations. Acetic anhydride is an inexpensive and readily available industrial chemical and finds extensive use in laboratory work. Many examples of its use, and of the use of other acid anhydrides will be encountered in later chapters.

---

* The HCl that is written as a product of these reactions will, of course be ionized in the usual way; in the presence of an excess of the reagent, it will form $H_3O^+$, $ROH_2^+$, $NH_4^+$, and so on, and chloride ion.

### 14.10 The relative reactivity of acid derivatives

The high degree of reactivity of acyl halides is due to the greatly enhanced electrophilic character of the carbonyl carbon atom as a result of the strong inductive effect of the halogen atom. This effect causes a marked lowering of the activation energy in the initial step of the reaction—the nucleophilic attack upon the carbonyl group—and thus causes a great increase in reaction rate:

$$
\underset{\text{B:}}{R-\overset{\displaystyle O}{\underset{\displaystyle \|}{C}}-Cl} \longrightarrow
\qquad
\begin{array}{l}\text{strong inductive effect of Cl, lowering electron}\\ \text{density on carbon}\\ \text{Result: enhanced attraction by carbon for nucleophile.}\end{array}
\qquad (39)
$$

The relatively much slower reactions of esters in reactions involving nucleophilic attack on the carbonyl carbon atom can be understood in the same terms:

$$
\underset{\text{B:}}{R-\overset{\displaystyle O}{\underset{\displaystyle \|}{C}}-\ddot{O}CH_3}
\qquad
\begin{array}{l}\text{strong ``resonance'' effect of unshared electrons on oxygen,}\\ \text{diminishing electrophilic character of carbon}\\ \text{Result: diminished rate of carbonyl attack, relative to that of the}\\ \text{acyl halide.}\end{array}
$$

Although the —OR group in esters has an inductive effect in the same direction as that of chlorine (the $pK_a$ of methoxyacetic acid is 3.5; see Table 14-1), the ability of the oxygen of —OR to engage in molecular $\pi$-orbital formation embracing the carbonyl carbon atom has the effect of decreasing the electron deficiency on carbon. Similar $\pi$-orbital overlap is much less effective in the case of chlorine, which must use $d$-orbitals.

Acid anhydrides are intermediate in reactivity: they are much more reactive than esters, and less reactive than acyl chlorides, to nucleophilic attack on the carbonyl carbon atom.

---

**Exercise 4.** It will be noted that the relative reactivity of $CH_3COCl$, $CH_3COOCOCH_3$, and $CH_3COOCH_3$ parallels the relative acidities of HCl, $HOCOCH_3$, and $HOCH_3$. Explain why this is consistent with the explanation given in the foregoing discussion.

---

### 14.11 Amides

Equations (29) and (31) show the initial steps in the reaction of an ester with the nucleophiles OH⁻ and RO⁻. If the strongly nucleophilic ammonia molecule (or an amine) is used instead of OH⁻ or RO⁻, the reaction takes the same form:

$$
\underset{\ddot{R}NH_2}{CH_3-\overset{\displaystyle O}{\underset{\displaystyle \|}{C}}-OR'} \overset{(a)}{\rightleftharpoons}
\underset{RNH_2^+}{CH_3-\overset{\displaystyle O^-}{\underset{\displaystyle |}{C}}-OR'} \overset{(b)}{\rightleftharpoons}
\underset{RNH}{CH_3-\overset{\displaystyle OH}{\underset{\displaystyle |}{C}}-OR'} \overset{(c)}{\rightleftharpoons}
$$

$$
\underset{RNH}{CH_3-\overset{\displaystyle O^-}{\underset{\displaystyle |}{C}}-OR'} \rightleftharpoons
CH_3-\overset{\displaystyle O}{\underset{\displaystyle \|}{C}}-NHR + R'O^- \qquad (40)
$$

The result of this series of reactions [in which *(a)* corresponds to the initial step of Equation (29) and *(b)* and *(c)* are simple proton transfer reactions] is the *ammonolysis* of the ester with formation of an *amide*. As these equations suggest, the reverse reaction, the alcoholysis of an amide to yield an ester, can also occur. The actual result of such a reaction will, of course, depend upon the experimental conditions: if the amine (or ammonia) is in large excess, the formation of the amide from the ester can be a practical preparative procedure, and is a method often used for the synthesis of amides.

The much greater reactivity of acyl chlorides and anhydrides makes them the most practical reagents for the preparation of amides. Moreover, the equilibrium in the overall reaction between an amine and an acyl halide is very far to the side of the product, the amide. The reason for this is that the reverse reaction would require nucleophilic attack of the chloride ion upon the amide carbonyl group, and it will be apparent that the very low degree of nucleophilicity of chloride ion would make this reaction negligible. Consequently, the reaction of ammonia and amines with acyl chlorides and acid anhydrides is fast and complete.

Some of the most useful applications of the reactions of ammonia derivatives with acyl halides, acid anhydrides, and esters are found in the preparation of *hydrazides* and *hydroxamic acids*.

The preparation of hydroxamic acids from esters constitutes a valuable diagnostic test for esters:

$$RCOOR' + NH_2OH \longrightarrow RCONHOH + R'OH \tag{41}$$
$$\text{a hydroxamic acid}$$

The *hydroxamic acid test for esters* depends upon the fact that the hydroxamic acids give deeply colored (red to purple) complex salts with ferric ion (ferric chloride is ordinarily used):

$$RCONHOH + Fe^{+++} \longrightarrow \text{colored complex}$$

---

**Exercise 5.** The hydroxamic acid test is not uniquely diagnostic of esters, but will be given by any acid derivative that will react with hydroxylamine with the formation of the hydroxamic acid. Which of the following compounds will give a positive test: *(a)* $CH_3CH_2COCl$; *(b)* $(CH_3CO)_2O$; *(c)* $CH_3CH_2CHO$; *(d)* $CH_3COOCH_2CH_2OCOCH_3$; *(e)* $CH_3CH=CHCOOH$?

---

*Hydrazides*, formed by the reaction of hydrazine ($NH_2NH_2$) with esters, acyl halides, and acid anhydrides, have several important uses. The $-NH_2$ group is capable of reacting with the carbonyl group of aldehydes and ketones, with the formation of *acyl hydrazones*; these are useful derivates for identification of carbonyl compounds:

$$\langle\ \rangle\text{CONHNH}_2 + \begin{matrix} R \\ R \end{matrix}\!\!>\!C\!=\!O \longrightarrow \langle\ \rangle\text{CONHN}\!=\!C\!<\!\begin{matrix} R \\ R \end{matrix} \tag{42}$$

benzhydrazide                    a benzhydrazone

Hydrazides are used as the starting point for the Curtius degradation of acids (Section 28.8).

## 14.12 Nitriles

Nitriles, RC≡N, can be prepared by the methods described in Section 14.3, and also by the dehydration of amides:

$$RC\overset{O}{\underset{NH_2}{\diagdown}} \xrightarrow{P_2O_5} RC{\equiv}N + H_3PO_4 \tag{43}$$

This "dehydration" can also be accomplished by means of sulfonyl chlorides, for example, the readily available reagent, *p*-toluenesulfonyl chloride:

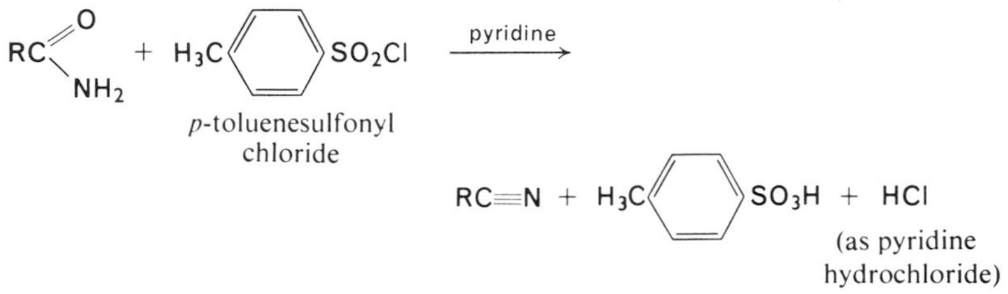

$$RC\overset{O}{\underset{NH_2}{\diagdown}} + H_3C\text{—}\underset{\text{\emph{p}-toluenesulfonyl chloride}}{\bigcirc}\text{—}SO_2Cl \xrightarrow{\text{pyridine}}$$

$$RC{\equiv}N + H_3C\text{—}\bigcirc\text{—}SO_3H + HCl \tag{44}$$
$$\text{(as pyridine hydrochloride)}$$

These dehydrations probably proceed by way of a base-catalyzed elimination reaction; using generalized formulas, the reaction may be written

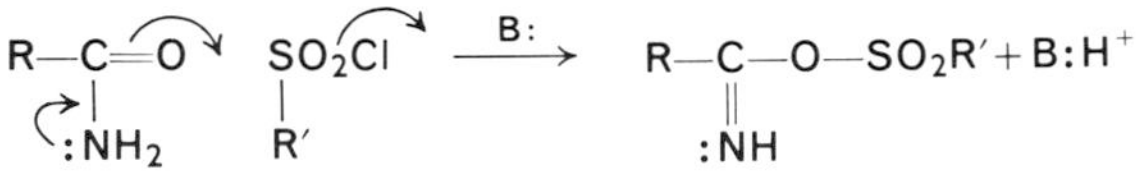

$$R\text{—}\underset{:NH_2}{\overset{}{C}}\text{—}O \quad SO_2Cl \xrightarrow{B:} R\text{—}\underset{:NH}{\overset{\|}{C}}\text{—}O\text{—}SO_2R' + B:H^+ \tag{45}$$

$$R\text{—}\underset{:N\text{—}H \quad :B}{\overset{}{C}}\text{—}OSO_2R' \longrightarrow R\text{—}C{\equiv}N + R'SO_2O^- + B:H^+ \tag{46}$$

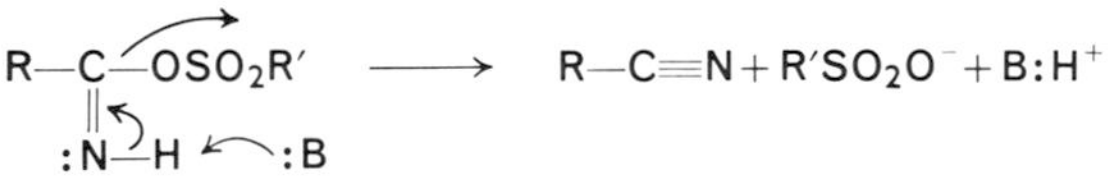

The pyridine (a tertiary amine whose structure has been given in Section 4.12) is the base that extracts the proton in Step (46); and it will be noted that the group expelled in this step is the very effective leaving group, $RSO_2O^-$.

Nitriles are useful for the synthesis of ketones by reaction with Grignard reagents:

$$RC{\equiv}N + R'MgX \longrightarrow R\text{—}\underset{R'}{\overset{}{C}}{=}NMgX \xrightarrow{H_2O}$$

$$R\text{—}\underset{R'}{\overset{}{C}}\text{—}NH \xrightarrow[H^+]{H_2O} R\text{—}\underset{R'}{\overset{}{C}}\text{—}O + NH_4^+ \tag{47}$$

The initial addition product, it will be seen, is the Schiff base of the final ketone with ammonia, and thus is readily hydrolyzed by aqueous acid to yield the ketone and an ammonium salt.

### 14.13 Acyl halides as reagents for characterization and identification of organic compounds

The use of acyl derivatives for the characterization of alcohols and amines finds frequent application. Although the low-molecular-weight acid halides are not useful for the characterization of low-molecular-weight alcohols (because the simple esters are usually liquids), other acid halides are available for the purpose. An acid chloride widely used for the characterization of hydroxyl compounds is 3,5-dinitrobenzoyl chloride. The esters (e.g., the methyl ester) derived from the reaction of this acyl halide with alcohols are usually crystalline solids (Table 14-4):

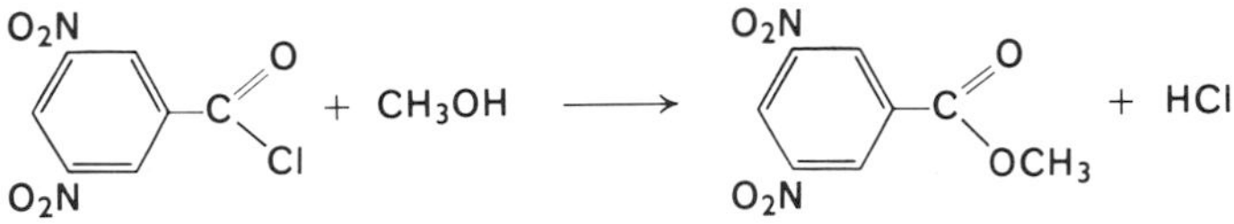

$$\text{3,5-dinitrobenzoyl chloride} + CH_3OH \longrightarrow \text{methyl 3,5-dinitrobenzoate} + HCl \tag{48}$$

3,5-dinitrobenzoyl chloride          methyl 3,5-dinitrobenzoate

**Table 14-4.** *3,5-Dinitrobenzoates of alcohols*

| Alcohol | m.p. of 3,5-Dinitrobenzoate (°C) |
|---|---|
| methyl | 107 |
| ethyl | 93 |
| n-propyl | 74 |
| isopropyl | 122 |
| n-butyl | 64 |
| isobutyl | 86 |

### 14.14 Reduction of acid derivatives to alcohols

The reduction of acid derivatives constitutes a means of retracing the oxidative pathway,

$$RCH_2OH \underset{\text{reduction}}{\overset{\text{oxidation}}{\rightleftharpoons}} RCHO \underset{\text{reduction}}{\overset{\text{oxidation}}{\rightleftharpoons}} RCOOH$$

and thus of preparing compounds at the oxidation levels of aldehydes and alcohols, starting with acids and acid derivatives that may be obtained through indirect methods.

The most general method of reducing acids and acid derivatives is by the use of lithium aluminum hydride, $LiAlH_4$. This reagent, usually used in ether solution, reduces acids, esters, acid halides, and acid anhydrides to the corresponding primary alcohols:

$$4RCOOH + 3LiAlH_4 \longrightarrow (RCH_2O)_4AlLi + 4H_2 + 2LiAlO_2 \tag{49}$$

$$2RCOOR' + LiAlH_4 \longrightarrow [(RCH_2O)_2Al(OR')_2]Li \tag{50}$$

$$2RCOCl + LiAlH_4 \longrightarrow [(RCH_2O)_2AlCl_2]Li \tag{51}$$

$$(RCO)_2O + LiAlH_4 \longrightarrow [(RCH_2O)_2AlO]Li \tag{52}$$

Hydrolysis of the reaction mixture leads to the formation of the alcohol and (if mineral acid is used) the aluminum and lithium salts:

$$(RCH_2O)_4AlLi + H_2O(HCl) \longrightarrow 4RCH_2OH + LiCl + AlCl_3$$

Lithium aluminum hydride is a very widely used reagent for laboratory purposes, but its cost precludes its use in large scale reductions on a commercial scale. Esters can be reduced catalytically

$$RCOOR' + H_2 \text{ (“copper chromite” catalyst)} \longrightarrow RCH_2OH + R'OH \tag{53}$$

or by means of sodium and an alcohol (ethanol, butanol):

$$RCOOR' \xrightarrow[\text{EtOH}]{\text{Na}} RCH_2OH + R'OH \tag{54}$$

The hydrogenation of fats is used for the production of long-chain aliphatic alcohols,

$$\begin{array}{l} CH_2OCO(CH_2)_{10}CH_3 \\ | \\ CHOCO(CH_2)_{10}CH_3 \xrightarrow[H_2]{\text{catalyst}} CH_3(CH_2)_{10}CH_2OH + \text{glycerol} \\ | \\ CH_2OCO(CH_2)_{10}CH_3 \end{array} \tag{55}$$

trilaurin                              lauryl alcohol
(glyceryl trilaurate)            [1-dodecanol (G)]

## 14.15   Reduction of amides and nitriles

An important use of lithium aluminum hydride* is for the reduction of amides to amines:

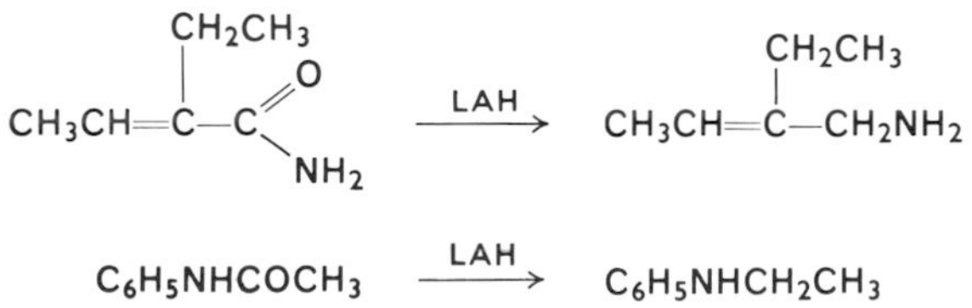

$$\begin{array}{l} \quad\quad CH_2CH_3 \\ \quad\quad\; | \quad\quad O \\ CH_3CH{=}C{-}C{\diagup}^{\diagdown}_{NH_2} \xrightarrow{\text{LAH}} \end{array} \quad \begin{array}{l} CH_2CH_3 \\ | \\ CH_3CH{=}C{-}CH_2NH_2 \end{array}$$

$$\tag{56}$$

$$C_6H_5NHCOCH_3 \xrightarrow{\text{LAH}} C_6H_5NHCH_2CH_3$$

A useful application of this reduction is in the alkylation of amines by reduction of their acyl derivatives.

_____

* We shall often use the abbreviation "LAH" for $LiAlH_4$.

If the acylation is a *formylation*, reduction leads to an N-methyl compound:

$$C_6H_5NH_2 \xrightarrow[\Delta]{HCOOH} C_6H_5NHCHO \xrightarrow{LAH} C_6H_5NHCH_3 \tag{57}$$

$$C_6H_5NHCH_2CH_3 \xrightarrow[\Delta]{HCOOH} C_6H_5N\begin{matrix}CHO\\CH_2CH_3\end{matrix} \xrightarrow{LAH} C_6H_5N\begin{matrix}CH_3\\CH_2CH_3\end{matrix} \tag{58}$$

*Nitriles* can be reduced to primary amines by means of lithium aluminum hydride, catalytic hydrogenation, or sodium-alcohol:

$$RC{\equiv}N \xrightarrow{4[H]} RCH_2NH_2 \tag{59}$$

$$CN(CH_2)_8CN \xrightarrow{LAH} (CH_2)_{10}\begin{matrix}NH_2\\NH_2\end{matrix} \quad [1,10\text{-diaminodecane (G)}] \tag{60}$$

$$C_6H_5CH_2CN \xrightarrow{H_2/Pt} C_6H_5CH_2CH_2NH_2 \tag{61}$$

The behavior of lithium aluminum hydride is that of a donator of $(H{:})^-$ ions to electrophilic centers. Thus, its characteristic reaction with carbonyl compounds is another example of a carbonyl addition reaction very similar in type to the addition of the Grignard reagent:

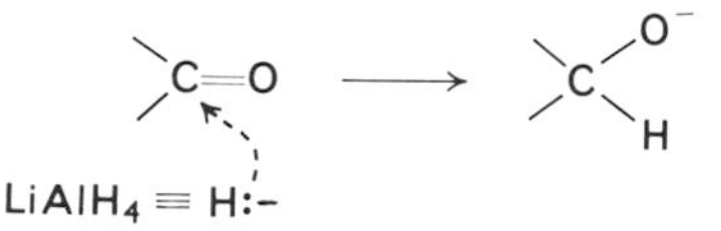

$$\begin{matrix} \diagdown \\ \diagup \end{matrix}C{=}O \longrightarrow \begin{matrix}\diagdown \\ \diagup\end{matrix}C\begin{matrix}O^-\\H\end{matrix}$$

$$LiAlH_4 \equiv H{:}^-$$

The resulting alkoxide ion is present before hydrolysis as the complex aluminum alkoxide shown in foregoing equations. The stoichiometry of the use of $ALiAlH_4$ as a reducing agent shows that it is capable of effecting an eight-electron reduction per mole (thus it acts as a source of the equivalent of four hydride ions):

$$Li^+ (AlH_4)^- \equiv (4H{:}^- + Al^{+++} + Li^+)$$

## 14.16  Reduction of acid derivatives to aldehydes

The reduction of acids and most acid derivatives to aldehydes is not readily accomplished; the only practical methods that have general application involve the reduction of acyl halides:

$$RC\begin{matrix}O\\Cl\end{matrix} \xrightarrow{2[H]} RC\begin{matrix}O\\H\end{matrix} + HCl \tag{62}$$

By the use of a special palladium catalyst, this reaction can be carried out by the direct addition of hydrogen and loss of HCl:

$$RC\begin{matrix}O\\Cl\end{matrix} \xrightarrow{H_2/Pd} RC\begin{matrix}O\\H\end{matrix} + HCl \tag{63}$$

The catalyst is a somewhat "deactivated" palladium black, the activity of which is diminished to the point at which it retains the ability to catalyze the reduction of the acid chloride but is not sufficiently active to bring about the further reduction of the aldehyde to an alcohol. The method is called the *Rosenmund reduction*.

The use of lithium aluminum hydride offers another means for reducing acid chlorides to aldehydes, with the added advantages that other reducible groups (which are subject to catalytic reduction) are not affected. The reagent is prepared by replacing three of the hydrogen atoms of $LiAlH_4$ by *t*-butoxyl groups:

$$LiAlH_4 + 3t\text{-BuOH} \longrightarrow LiAlH(OtBu)_3 + 3H_2$$

The resulting lithium tri-*t*-butoxyalumino hydride is a mild reducing agent, and when used in equivalent molar proportions reduces acyl chlorides to aldehydes:

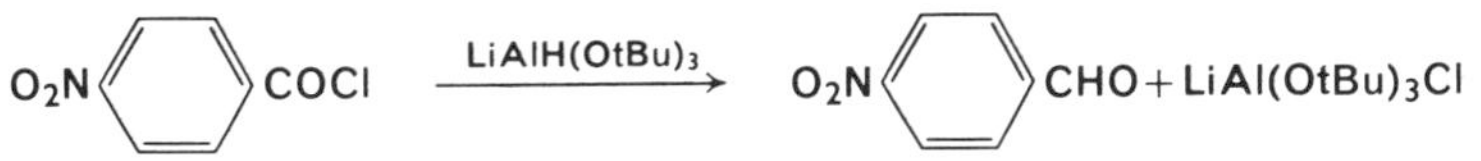

$$O_2N\langle\bigcirc\rangle COCl \xrightarrow{\text{LiAlH(OtBu)}_3} O_2N\langle\bigcirc\rangle CHO + LiAl(OtBu)_3Cl \qquad (64)$$

*p*-nitrobenzoyl chloride

$$(CH_3)_3CCOCl \xrightarrow{\text{LiAlH(OtBu)}_3} (CH_3)_3C \cdot CHO + LiAl(OtBu)_3Cl \qquad (65)$$

pivaloyl chloride

### 14.17 Applications to synthesis

1. Prepare 3,4-dibromobutanoic acid.

   *(a)* The final step in this synthesis is

   $$CH_2{=}CHCH_2COOH + Br_2 \longrightarrow BrCH_2\underset{\underset{Br}{|}}{C}HCH_2COOH$$

   *(b)* Vinylacetic acid can be prepared by the carbonation of an allyl Grignard reagent:

   $$CH_2{=}CHCH_2MgCl + CO_2 \longrightarrow CH_2{=}CHCH_2COOH$$

2. Prepare pentanoic acid, using any reagents of two carbon atoms or less.

   *(a)* The final steps could be:

   $$CH_3CH_2CH_2CH_2Br + KCN \longrightarrow CH_3CH_2CH_2CH_2CN \xrightarrow[\text{H}_2\text{O}]{\text{H}^+}$$
   $$CH_3CH_2CH_2CH_2COOH$$

   *(b)* 1-Bromobutane can be prepared *via* the reaction of ethylmagnesium bromide with ethylene oxide:

   $$CH_3CH_2MgBr + CH_2\overset{\displaystyle O}{\overbrace{\quad}}CH_2 \longrightarrow n\text{BuOH} \xrightarrow{\text{HBr}} n\text{BuBr}$$

3. Prepare 3-methylbutanoic acid (isovaleric acid), starting with acetone:

(*a*) $CH_3COCH_3 \xrightarrow{\text{Ba(OH)}_2}$ diacetone alcohol $\xrightarrow[(-H_2O)]{\text{H}^+} (CH_3)_2C{=}CHCOCH_3$

(*b*) $(CH_3)_2C{=}CHCOCH_3 + NaOCl \longrightarrow (CH_3)_2C{=}CHCOOH + CHCl_3$

(*c*) $(CH_3)_2C{=}CHCOOH \xrightarrow{\text{H}_2/\text{Pt}} (CH_3)_2CHCH_2COOH$

4. Prepare adipic acid, starting with succinic acid.

(*a*) $\begin{array}{l} CH_2COOH \\ | \\ CH_2COOH \end{array} \xrightarrow{\text{LiAlH}_4} \begin{array}{l} CH_2CH_2OH \\ | \\ CH_2CH_2OH \end{array} \xrightarrow{\text{HBr}} \begin{array}{l} CH_2CH_2Br \\ | \\ CH_2CH_2Br \end{array} \xrightarrow{\text{KCN}}$

$\begin{array}{l} CH_2CH_2CN \\ | \\ CH_2CH_2CN \end{array} \xrightarrow{\text{H}_2\text{O/H}^+} \begin{array}{l} CH_2CH_2COOH \\ | \\ CH_2CH_2COOH \end{array}$

5. Prepare [cyclopentane ring with —COOH] starting with cyclohexanone.

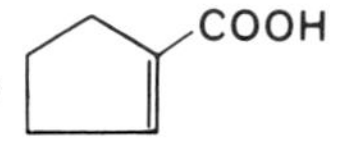

Since the starting material, cyclohexanone, has a six-membered carbocyclic ring, and the desired product has a five-membered ring, it is clear that some kind of ring contraction will have to be effected. One way to do this is to open the cyclohexanone ring and then reclose the resulting open-chain intermediate to form a five-membered ring. Since both the starting material and the final product have a total of six carbon atoms each, the —COOH group of the product will represent one of the original six carbon atoms of the starting material.

It will be seen that an aldol condensation can effect the desired change, in the following way:

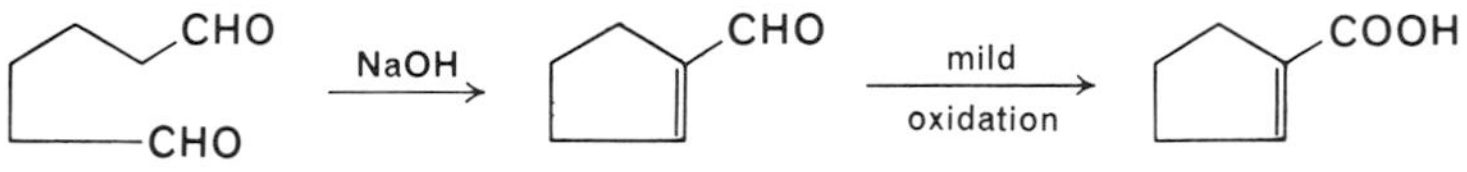

[open-chain dialdehyde with two CHO groups] $\xrightarrow{\text{NaOH}}$ [cyclopentene ring with CHO] $\xrightarrow[\text{oxidation}]{\text{mild}}$ [cyclopentene ring with COOH]

The dialdehyde can be obtained by the ozonolysis of cyclohexene or by the periodic acid cleavage of cyclohexane-1,2-diol; and since this diol is most readily prepared from cyclohexene, the synthetic problem consists of the preparation of the latter compound. This can be accomplished by the series

cyclohexanone $\xrightarrow[\text{or LiAlH}_4]{\text{NaBH}_4}$ cyclohexanol $\xrightarrow[\text{heat}]{\text{H}_3\text{PO}_4}$ cyclohexene

## 14.18  Analytical procedures involving carboxylic acids and derivatives

*Neutralization equivalent.*  The determination of the molecular weight of an organic compound is always of great value in the determination of its composition and structure. Molecular weights can be measured in many ways, some of them simple and reasonably accurate, some of them more accurate but requiring costly and elaborate equipment. Carboxylic acids lend themselves to a rapid and accurate determination of their equivalent weight (the molecular weight per

—COOH) group by virtue of the fact that they can be titrated to sharp end points with the use of standard alkali. The end point may be determined with the use of an indicator, such as phenolphthalein, or with a $pH$ meter. The following examples will illustrate the procedure.

Benzoic acid, $C_6H_5COOH$, has a molecular weight of 122. Its neutralization will require 1 equivalent of NaOH for 122 g or 1 milliequivalent (10 ml of 0.1 $N$ NaOH) for 122 mg.

Thus, a sample of 0.244 g of benzoic acid will require 20 ml of 0.1 $N$ NaOH for neutralization:

$$\text{Neutral equivalent} = \frac{\text{mg of acid used}}{\text{meq of alkali required}} = \frac{244}{20 \times 0.1} = 122$$

Succinic acid is a dibasic acid, and two equivalents of alkali will be required for the neutralization of one mole of the acid. It has a neutral equivalent (59) of one-half of its molecular weight (118). Hence, a sample of 0.356 g of succinic acid will require 30.0 ml of 0.201 $N$ NaOH for neutralization:

$$\text{Neutral equivalent} = \frac{356}{30 \times 0.201} = 59$$

A sample of 0.278 g of an unknown acid required 18.8 ml of 0.156 $N$ NaOH for neutralization:

$$\text{Neutral equivalent} = \frac{278}{18.8 \times 0.156} = 95$$

The unknown acid could be a monobasic acid with a molecular weight of 95, a dibasic acid with a molecular weight of 190, a tribasic acid with a molecular weight of 285, and so on.

***Saponification equivalent.*** The saponification of an ester requires one equivalent of alkali for each —COOR group:

$$RCOOR' + NaOH \longrightarrow RCOO^-Na^+ + R'OH$$

The analytical procedure consists in saponifying a weighed sample of the ester with a measured excess of standard alkali, and titrating the alkali that is not consumed in the reaction.

For example, 0.300 g of an ester is saponified with the use of 5.0 ml of 1.00 $N$ NaOH. After the saponification is complete the mixture is titrated with 0.100 $N$ HCl to a phenolphthalein end point; 30 ml of HCl is required:

Equiv. NaOH initially used: 5 meq
Equiv. HCl required: 3 meq
Equiv. NaOH consumed by the ester: $5 - 3 = 2$ meq

$$\text{Saponification equivalent} = \frac{300 \text{ mg}}{2 \text{ meq}} = 150 \text{ g/equiv.}$$

The compound could be a monoester, with a molecular weight of 150, or a diester with a molecular weight of 300, and so on.

The determination of neutralization equivalent or saponification equivalent cannot, of course, be applied unless the compound contains a titratable or a saponifiable group (usually —COOH or —COOR). It is often possible, however, to convert a compound into a *derivative* that can be titrated or saponified. There are numerous devices that can be resorted to. For example, an alcohol can be treated with succinic anhydride to yield the half-ester of succinic acid:

$$\text{ROH} \; + \; \begin{array}{c} \text{H}_2\text{C}^{-\text{CO}} \\ | \qquad \text{O} \\ \text{H}_2\text{C}_{\diagdown\text{CO}} \end{array} \longrightarrow \begin{array}{c} \text{H}_2\text{C}-\text{COOR} \\ | \\ \text{H}_2\text{C}-\text{COOH} \end{array}$$

This derivative is an acid, and can be titrated. Since the molecular weight of the succinic acid-derived portion of the molecule is known, its neutral equivalent affords the molecular weight of the alcohol.

Another procedure that is sometimes useful is to convert an alcohol into an ester with the use of a known acyl halide, and then to determine the saponification equivalent of the ester. It must be borne in mind that derivatives prepared for such purposes must be carefully purified before being subjected to analysis.

*Carbon-methyl determination.* One of the most useful analytical devices available to the organic chemist is the determination of carbon-linked methyl groups. This method depends upon the fact that when compounds containing C-methyl groups are subjected to oxidation with chromic acid under vigorous conditions, the methyl groups are not oxidized, but are present in the final solution as acetic acid, while the remainder of the molecule is converted largely to carbon dioxide. The acetic acid can be isolated from the solution (by steam-distillation) and titrated. The method thus serves as a means of determining the number of carbon-linked methyl groups in an organic compound, and is commonly referred to as the *C-methyl* or *Kuhn-Roth determination*. The procedure consists of heating a weighed sample of compound to be analyzed with an oxidizing mixture (chromic acid in aqueous sulfuric acid) for a standard period of time (about an hour), distilling the mixture into a receiver, and titrating the contents of the receiver with standard alkali. The results are calculated and expressed as the *C-methyl number*, which is the number of carbon-linked methyl groups per molecule of compound.

Acetic acid is not produced quantitatively from methyl groups in all combinations, but with experience in the behavior of model compounds, the C-methyl number can nearly always be used as a measure of the number of methyl groups that are present. The list in Table 14-5 gives some values for the C-methyl number found for compounds of several types. It will be noted that methyl groups in saturated environments give low values, while those adjacent to a carbonyl group or to a carbon-carbon double bond give values near 1.

**Table 14-5.**   *The C-methyl numbers for some compounds*

| Compound | C-methyl Number |
|---|---|
| acetic acid | 0.97 to 1.0 (recovered) |
| acetone semicarbazone | 0.96 |
| crotonic acid | 0.94 |
| *n*-butyric acid | 0.93 |
| pivalic acid | 0.89 |
| 2-butylhexanoic acid | 1.64 (82% of two) |
| methyl isopropyl ketone | 1.75 (88% of two) |
| methylmalonic acid | 0.89 |
| malonic acid | 0.00 |
| succinic acid | 0.00 |
| methylsuccinic acid | 0.33 (resistant to oxidation) |

## Exercises

(Exercises 1–5 will be found within the text.)

**6** Name each of the following acids in two ways:

(a) $CH_3CH_2COOH$

(b) $BrCH_2COOH$

(c) $CH_3CH_2\overset{\underset{\displaystyle |}{OH}}{C}HCOOH$

(d) $HOCH_2\overset{\underset{\displaystyle |}{CH_3}}{C}HCOOH$

(e) $CH_2{=}CHCH_2CH_2COOH$

(f) $CH_2{=}CHCH_2COOH$

(g) $(CH_3)_2CHCOOH$

(h) $(CH_3)_2CH\overset{\underset{\displaystyle |}{CH_2CH_3}}{C}HCOOH$

(i) $(CH_3CH_2)_3CCOOH$

**7** Write the structural formulas for the following: (a) ethoxyacetic acid; (b) dichloro-ethanoic acid; (c) acetyl chloride; (d) $\alpha,\beta,\gamma$-tribromo-*n*-butyric acid; (e) 2-bromo-3-heptenoic acid; (f) acetyl cyanide; (g) perfluoropropionic acid; (h) 2-hexenoyl chloride.

**8** Write the equations for the reactions of propionic acid with (a) water; (b) ammonia; (c) ethylamine; (d) sodium ethoxide; (e) ethanol (with a trace of sulfuric acid); (f) tri-methylamine.

**9** Describe the saponification of isopropyl acetate with sodium hydroxide by writing the sequence of steps through which the reaction proceeds.

**10** Explain base-catalyzed ester interchange by describing the nature of the reaction between ethyl propionate and methanol in the presence of sodium methoxide.

**11** Write the reaction of isobutyryl chloride with (a) ethanol; (b) isobutyl alcohol; (c) sodium acetate; (d) water; (e) hydrazine; (f) hydroxylamine; (g) methylethylamine.

**12** Write the reactions that would be involved in the following transformations.

(a) *n*-butyl bromide into pentanoic acid

(b) acetyl chloride into acetic anhydride

(c) propionyl chloride into *n*-butylamine

(d) ethylene dibromide into succinic acid (butanedioic acid)

(e) tristearin into a soap

**13** Which of the following compounds will give a positive hydroxamic acid test? Write the relevant reactions: *(a)* ethyl acetate; *(b)* methyl formate; *(c)* sodium acetate; *(d)* acetyl chloride.

**14** Outline a practical experimental procedure that could be used to separate each of the following mixtures into its pure individual components:

*(a)* di-*n*-butyl ether, pentanal, *n*-butyric acid, *n*-octane

*(b)* diethylamine, 1-propanol, acetic acid

*(c)* 2-hexanone, 3-hexanone, propionic acid

**15** Show how trimethylacetic acid can be prepared from acetone in a three-step synthesis.

## *Chapter fifteen* / Ester Condensations. β-Keto Esters and Malonic Esters

### 15.1 Active hydrogen in acid derivatives

The α-hydrogen atoms of carboxylic acid derivatives are, like those of aldehydes and ketones, active hydrogens, susceptible to ionization by the attack of sufficiently strong bases. Although the general reaction may be written in a way that shows the parallelism between the behavior of acid derivatives and aldehydes and ketones

$$RCH_2COX + B{:}^- \rightleftharpoons \{R\ddot{C}HCOX\}^- + B{:}H \tag{1}$$

the reaction is of value only in those cases in which X is —OR (esters) or —OCOR (acid anhydrides). It is apparent that if X is halogen, the base will attack at the highly reactive carbonyl group rather than at the α-hydrogen atom; and if X is $NH_2$ or NHR, the hydrogen atom or nitrogen will be the one removed by the base. Carboxylic acids themselves (X=OH) cannot ionize according to Equation (1) because the —COOH group will undergo ionization far more readily than will the α-C—H linkage.

For these reasons, the most practicable means of carrying out the initial ionization [Equation (1)] is by the reaction of an ester with an alkoxide. In this case, the ester-interchange reaction,

$$RCOOR'' + R'O^- \qquad RCOOR' + R''O^- \tag{2}$$

which would take place, is of no consequence, since the ester is not destroyed.*

* If the alkoxide $R'O^-$ is different from the alkoxyl group —OR″ in the ester, a mixture of esters will result. Except for manipulative difficulties that this may cause, it is of no great consequence.

Treating methyl acetate with a solution of sodium ethoxide would yield

$$CH_3COOCH_3 + CH_3O^- \rightleftharpoons \{:CH_2COOCH_3\}^- + CH_3OH \tag{3}$$

The ester anion [this is a resonance hybrid, but for convenience the structure shown in Equation (3) will be used in the discussion] can attack the carbonyl carbon atom of the undissociated ester that is present in the solution:

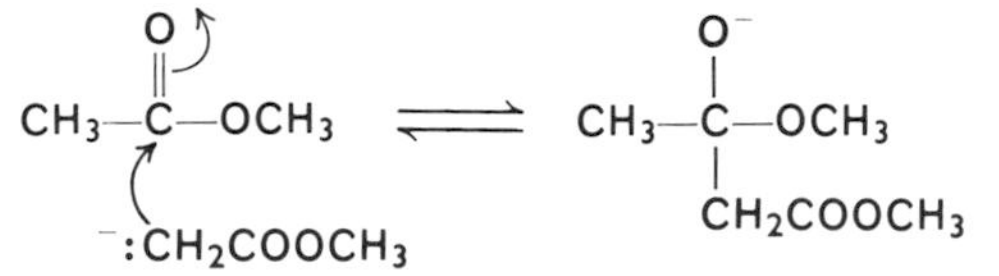

$$
\begin{array}{ccc}
\overset{O}{\underset{|}{CH_3-\overset{}{\underset{}{C}}-OCH_3}} & \rightleftharpoons & CH_3-\overset{O^-}{\underset{|}{C}}-OCH_3 \\
\ \ \ ^-\!:CH_2COOCH_3 & & CH_2COOCH_3
\end{array}
\tag{4}
$$

The intermediate anion produced by the reaction shown in Equation (3) is in equilibrium *(a)* with the starting reactants, and *(b)*, by expulsion of $CH_3O^-$, with the products shown in Equation (5):

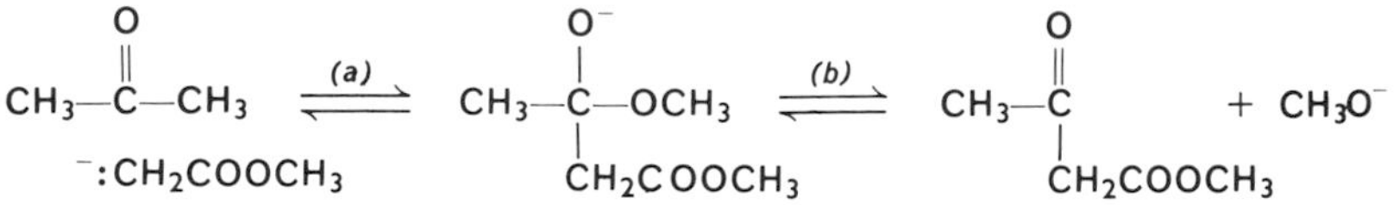

$$
CH_3-\overset{O}{\overset{||}{C}}-CH_3 \overset{(a)}{\rightleftharpoons} CH_3-\overset{O^-}{\underset{|}{C}}-OCH_3 \overset{(b)}{\rightleftharpoons} CH_3-\overset{O}{\overset{||}{C}} \ + \ CH_3O^-
$$

$$
\ \ ^-\!:CH_2COOCH_3 \qquad CH_2COOCH_3 \qquad\qquad CH_2COOCH_3 \tag{5}
$$

Under the influence of the strong base ($CH_3O^-$) used in this condensation, the keto ester produced by path *(b)* of (5) can be converted into an anion by ionization at the highly active $-CH_2-$ group:

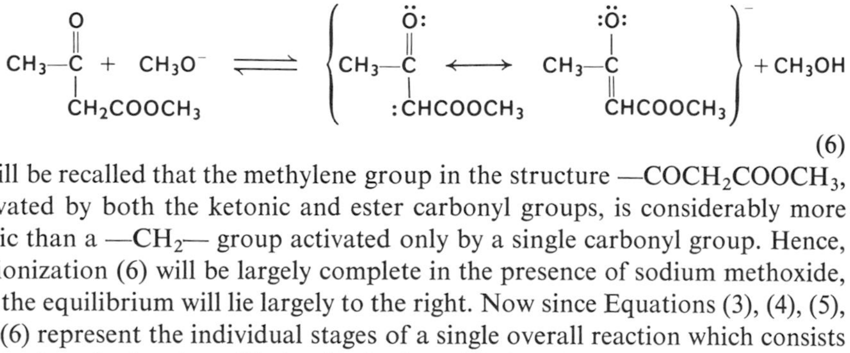

$$
CH_3-\overset{O}{\overset{||}{C}} \ + \ CH_3O^- \rightleftharpoons \left\{ CH_3-\overset{\ddot{O}:}{\underset{|}{C}} \longleftrightarrow CH_3-\overset{:\ddot{O}:}{\underset{||}{C}} \right\}^- + CH_3OH
$$

$$
CH_2COOCH_3 \qquad\qquad :CHCOOCH_3 \qquad CHCOOCH_3 \tag{6}
$$

It will be recalled that the methylene group in the structure $-COCH_2COOCH_3$, activated by both the ketonic and ester carbonyl groups, is considerably more acidic than a $-CH_2-$ group activated only by a single carbonyl group. Hence, the ionization (6) will be largely complete in the presence of sodium methoxide, and the equilibrium will lie largely to the right. Now since Equations (3), (4), (5), and (6) represent the individual stages of a single overall reaction which consists of a series of related equilibria, the final equilibrium, lying as it does to the far side of the ionized keto ester, will regulate the extent of the *overall* reaction

$$2CH_3COOCH_3 \rightleftharpoons CH_3COCH_2COOCH_3 + CH_3OH \tag{7}$$
$$\text{methyl acetoacetate}$$

and the self-condensation of the ester will proceed essentially to completion.*

---

\* In the actual performance of ester condensations such as that shown in the foregoing equations, with a sodium alkoxide as the base, the reaction mixture at the conclusion of the reaction will be a solution or suspension of the sodium salt of the keto ester, which is shown as the final product of Equation (6). The isolation of the product, the keto ester itself, will be accomplished by acidifying the reaction mixture and separating the keto ester by extraction with ether, distillation, or crystallization.

The acylation of other kinds of active methylene groups by esters can occur in the same way. The reaction of acetone with ethyl acetate, catalyzed by sodium ethoxide, proceeds as follows:

$$CH_3COCH_3 + EtO^- \rightleftharpoons \{CH_3COCH_2:\}^- + EtOH \tag{8}$$

$$CH_3\text{—}\overset{\overset{O}{\|}}{C}\text{—}OEt \rightleftharpoons CH_3\overset{\overset{O^-}{|}}{C}\text{—}OEt \rightleftharpoons CH_3\text{—}\overset{\overset{O}{\|}}{C} + EtO^- \rightleftharpoons$$
$$\qquad\quad \overset{|}{:}CH_2COCH_3 \qquad\qquad \overset{|}{CH_2COCH_3} \qquad\qquad \overset{|}{CH_2COCH_3}$$

$$\{CH_3CO\ddot{C}HCOCH_3\}^- + EtOH \tag{9}$$

In the C-acylation of acetone, as in the self-condensation of methyl acetate, the final ionization of the $\beta$-diketone displaces the equilibrium far to the side of the final product and the reaction is essentially complete.

The carbon-acylation of compounds having active methylene groups by reactions such as those illustrated by (7) and (9) is known as the *Claisen ester condensation*. Bases other than sodium alkoxides are often used; a convenient reagent is sodium hydride, NaH, by means of which the ionization of the $\alpha$-hydrogen atom is brought about by the base H:$^-$. The products of this reaction are the desired carbon anion and molecular hydrogen, and complications that may be caused by the presence of alcohols or alkoxides are minimized. Condensation reactions brought about by means of sodium hydride are conducted in aprotic solvents, such as benzene, toluene, or ether.

## 15.2   The use of oxalic and formic esters

When the ester contains no $\alpha$-hydrogen atoms it can act *only* as the carbonyl component in ester condensations. Two types of esters in which this is clearly the case are esters of formic and oxalic acids:

$$H\text{—}C\overset{\displaystyle O}{\underset{\displaystyle OR}{\diagup}} \qquad\qquad RO\overset{\diagup}{\underset{\diagdown}{}}\!\!\!\overset{\displaystyle }{\underset{O}{C}}\text{—}\overset{OR}{\underset{O}{C}}$$

$\quad$ (R = CH$_3$, methyl formate)$\qquad$ (R = C$_2$H$_5$, ethyl oxalate)

Moreover, *formic and oxalic esters have very reactive ester carbonyl groups*, that of the first being unsubstituted, that of the second having the inductive effect of the second —COOR group to increase the reactivity of the $\diagup\!\!C\!\!=\!\!O$ group at which nucleophilic attack takes place. The reactions of formic and oxalic esters with active methylene compounds lead to *formylation* and *oxalylation* of carbon atoms:

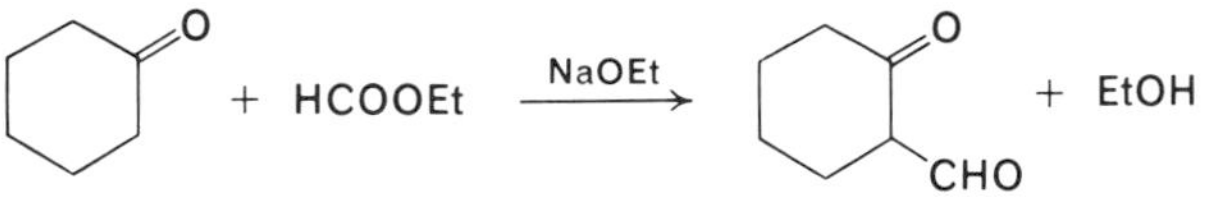

$$+ \text{ HCOOEt } \xrightarrow{\text{NaOEt}} \qquad + \text{ EtOH} \qquad (10)$$

$$\text{CH}_3\text{COCH}_3 + \text{HCOOEt} \xrightarrow{\text{NaOEt}} \text{CH}_3\text{COCH}_2\text{CHO} + \text{EtOH} \qquad (11)$$

$$\text{CH}_3\text{CH}_2\text{CH}_2\text{CH}_2\text{COOEt} + \text{HCOOEt} \xrightarrow{\text{NaOEt}} \text{CH}_3\text{CH}_2\text{CH}_2\text{CHCOOEt} + \text{EtOH}$$
$$\underset{\text{CHO}}{|} \qquad (12)$$

$$+ (\text{COOEt})_2 \xrightarrow{\text{NaOEt}} \qquad + \text{ EtOH} \qquad (13)$$

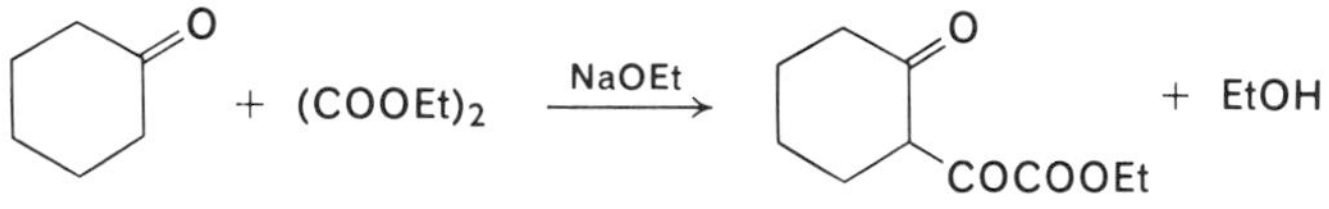

$$+ (\text{COOEt})_2 \xrightarrow{\text{NaOEt}} \qquad + \text{ EtOH} \qquad (14)$$

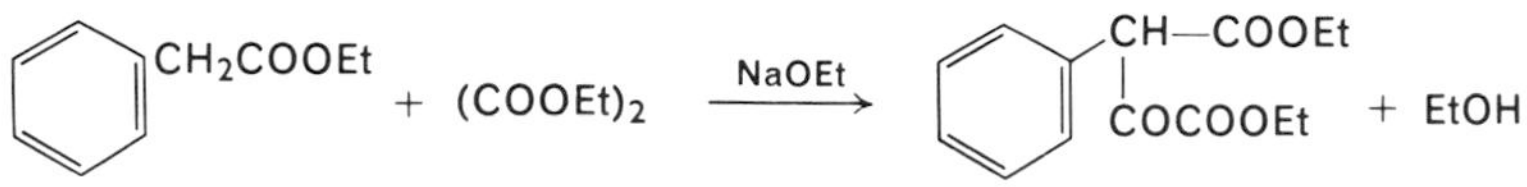

---

**Exercise 1.** Formulate the above condensation reactions, using step-by-step equations such as those used in Section 15.1.

---

## 15.3 Carbethoxylation with ethyl carbonate

Ethyl carbonate is a somewhat less reactive ester than formic and oxalic esters; but, having no $\alpha$-hydrogen atoms, it too serves only to accept anionic attack and thus furnishes the —COOEt group to an active methylene group:

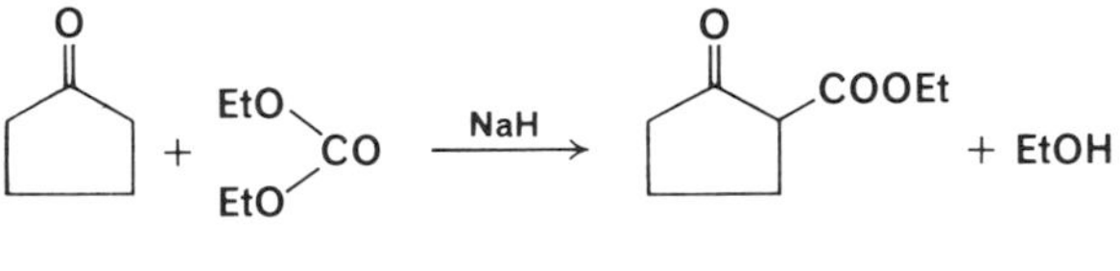

$$+ \begin{array}{c}\text{EtO} \\ \text{EtO}\end{array}\!\!\text{CO} \xrightarrow{\text{NaH}} \qquad + \text{ EtOH} \qquad (15)$$

## 15.4 Carbon acylation of nitriles

Nitriles that contain an $\alpha$-methylene group can also be acylated on carbon by reaction with an ester. For example, phenylacetontrile reacts with ethyl acetate as follows:

$$\text{C}_6\text{H}_5\text{CH}_2\text{CN} \underset{}{\overset{\text{EtO}^-}{\rightleftharpoons}} \{\text{C}_6\text{H}_5\ddot{\text{C}}\text{HCN}\}^- + \text{EtOH} \qquad (16)$$

phenylacetonitrile

$$CH_3C \overset{\displaystyle O}{\underset{\displaystyle \{C_6H_5\overset{..}{C}HCN\}^-}{\diagup}} \overset{\diagdown}{OR} \longrightarrow \overset{\displaystyle CH_3C=O}{\underset{\displaystyle C_6H_5\overset{|}{C}HCN}{}} + RO^- \tag{17}$$

## 15.5  Properties of 1,3-Dicarbonyl compounds. Tautomerism

For a good many years after ethyl acetoacetate was first prepared (about a hundred years ago) there was considerable uncertainty about its structure. It shows many of the reactions that can be accounted for by the structure $CH_3COCH_2COOEt$: it adds hydrogen cyanide and sodium bisulfite; it forms derivatives from carbonyl reagents such as hydroxylamine and phenylhydrazine; and it shows the properties expected of a compound that contains an active —$CH_2$— group.

On the other hand, it also reacts rapidly with bromine, gives an intense color with ferric chloride, and has an acid dissociation constant of about $3 \times 10^{-11}$ (about that of phenol, Section 27.1). Most striking of all, however, is its reaction with diazomethane, the product of which is the methyl ether $CH_3{-}\overset{\displaystyle OCH_3}{\overset{|}{C}}{=}CHCOOEt$

These observations are all consistent with the structure $CH_3{-}\overset{\displaystyle OH}{\overset{|}{C}}{=}CHCOOEt$.

It was not until near the end of the nineteenth century that chemists reached agreement that the interpretation of these experimental observations was simply that ethyl acetoacetate is a mixture of two readily interconvertible isomers, the *keto* and the *enol* forms:

$$\underset{\text{keto form}}{CH_3{-}\overset{\displaystyle O}{\overset{||}{C}}{-}CH_2COOEt} \;\rightleftharpoons\; \underset{\text{enol form}}{CH_3{-}\overset{\displaystyle OH}{\overset{|}{C}}{=}CHCOOEt} \tag{18}$$

Ordinary ethyl acetoacetate (an oily liquid) contains about 7% of the enol form. The percentage of the enol form is different in solutions of the ester in various solvents, and varies with concentration.

The relationship between the keto and enol forms is called *tautomerism*, and the two isomers are referred to as *tautomeric* forms, or *tautomers*. *This is not a case of resonance*; these tautometic forms are isomers: they have different structures and independent existence. In one, a hydrogen atom is attached to oxygen; in the other, to carbon:

$$\overset{\displaystyle O}{\overset{||}{-C}}{-}CH_2{-} \;\rightleftharpoons\; \overset{\displaystyle OH}{\overset{|}{-C}}{=}CH{-}$$

It is possible, by careful manipulation, to separate ethyl acetoacetate into the pure enol and pure keto forms. One way in which this can be done is by distillation in an all-quartz apparatus. The enol form is slightly the more volatile and can be distilled out. It can be preserved for a short time in quartz vessels, but a trace of acid or alkali (even the alkalinity of ordinary glass) causes a rapid conversion of either pure tautomer to the equilibrium mixture.

This equilibration occurs through the intermediate ion produced by the removal of the proton from the —CH₂— group of the keto form or from the —OH group of the enol form. The ion, which is a resonance hybrid,

$$\left\{ \begin{array}{c} \text{CH}_3\text{C}=\text{CHCOOEt} \\ | \\ :\ddot{\text{O}}: \end{array} \longleftrightarrow \begin{array}{c} \text{CH}_3\overset{\cdot\cdot}{\text{C}}\text{CHCOOEt} \\ \| \\ \ddot{\text{O}}: \end{array} \right\}^-$$

is the same whether it is formed from the keto or the enol tautomer, and its reprotonation yields the equilibrium keto-enol mixture.

Each pure isomer shows the reactions that would be expected from the separate structures given above: the enol form reacts instantly with bromine, the keto form reacts more slowly; the enol forms gives an immediate color with ferric chloride, the keto form gives the same color after a short time; the enol form is a stronger acid than the keto form.

Although any compound that contains the grouping —CH₂—CO— may be regarded as capable of existing in both keto and enol forms, most aldehydes and monoketones contain so little of the enol form that we can regard them as having the keto structure only. Acetone, for example, has been estimated to contain only about 0.0001 % of the enol form in the pure liquid.

Keto-enol tautomerism is encountered chiefly in 1,3-diketones, β-keto aldehydes, and β-keto esters. Even 1,1-dicarboxylic acid esters (such as diethyl malonate) are essentially nonenolic. Acetylacetone (2,4-pentanedione) is about 70 % enolic, and many β-ketoaldehydes are essentially pure enols:

$$\underset{\text{acetylacetone}}{\text{CH}_3\text{COCH}_2\text{COCH}_3} \;\rightleftharpoons\; \underset{70\% \text{ enol}}{\overset{\text{OH}}{\underset{|}{\text{CH}_3\text{C}}}=\text{CHCOCH}_3} \tag{19}$$

$$\tag{20}$$

2-formylcyclohexanone      100 % enol

The latter are usually referred to as "hydroxymethylene ketones."

No useful generalizations can be made that will enable us to predict with accuracy the amount of enol in any given compound. We can, however, make a summary of *relative* tendencies toward enolization (Table 15-1).

In simple ketones the keto-enol equilibrium lies nearly completely on the side of the keto form. This explains why synthetic reactions that might be expected to yield the enol give the ketone as the actual product.

**Table 15-1.** *Enolic content of some carbonyl compounds*

| Essentially No Enol | Partly Enolic | Largely Enolic |
|---|---|---|
| Simple ketones | $\beta$-keto esters | $\beta$-diketones |
| acid derivatives | | $\beta$-keto aldehydes |
| $CH_3COCH_3$ | $CH_3COCH_2COOEt$ | $CH_3COCH_2COCH_3$ |
| $CH_3CHO$ | | $C_6H_5COCH=CHOH$ |
| $CH_3COOEt$ | | |

For example, the course of the hydration of an acetylenic compound might be expected to proceed as follows, in analogy with the addition of water to an olefin:

$$RC\equiv CH + H_2O \longrightarrow R-\underset{\underset{OH}{|}}{C}=CH_2 \tag{21}$$

The product is, in fact, $RCOCH_3$; if the enol is indeed an intermediate, the equilibrium

$$RC=CH_2 \underset{\underset{OH}{|}}{\rightleftarrows} RCOCH_3 \tag{22}$$

is completely in favor of the keto form, and this is the actual *product* of the reaction.

### 15.6 Alkylation and acylation of $\beta$-keto esters and malonic esters

Ethyl acetoacetate and diethyl malonate are sufficiently acidic to be converted largely into the corresponding anions by a base such as sodium ethoxide in ethanol solution:

$$CH_3COCH_2COOEt + OEt^- \rightleftharpoons \{CH_3C\ddot{O}CHCOOEt\}^- + EtOH \tag{23}$$

$$H_2C\begin{matrix} \diagup COOEt \\ \diagdown COOEt \end{matrix} + OEt^- \rightleftharpoons \left\{ :CH\begin{matrix} \diagup COOEt \\ \diagdown COOEt \end{matrix} \right\}^- + EtOH \tag{24}$$

*C-alkylation.* These anions can effect nucleophilic displacement by attack upon an alkyl halide,

$$CH_3COCH:\overset{\frown}{\phantom{}}CH_3Br \longrightarrow CH_3COCH-CH_3 + Br^- \tag{25}$$
$$\underset{COOEt}{|} \qquad\qquad\qquad \underset{COOEt}{|}$$

with the useful result that the ester is *alkylated* on carbon. Malonic esters can be alkylated in a similar way:

$$(EtOOC)_2CH:^- + RBr \longrightarrow (EtOOC)_2CH—R + Br^- \tag{26}$$

Why is the acetoacetic ester alkylated on carbon rather than on oxygen? Since the negative charge on the anion is delocalized over the system $\left\{ -\overset{|}{\underset{|}{C}}=C—\overset{..}{\underset{..}{O}}: \longleftrightarrow -\overset{|}{\underset{|}{C}}—C=\overset{..}{\underset{..}{O}}: \right\}^-$, it might be expected that both carbon and oxygen could act as the nucleophilic center. The reason why carbon-alkylation occurs must be that the electron density in the anion is largely upon carbon, or that the negative carbon atom is more nucleophilic than the oxygen atom. A categorical decision on this point cannot be made, for two reasons: (1) the alkylation of some β-dicarbonyl compounds *does* occur on both carbon and oxygen, and (2) the relative proportions of C- and O-alkylation vary with varying conditions and with different alkylating agents.

In general, the more enolic the unionized β-dicarbonyl compound, the greater the tendency for O-alkylation. Thus, the alkylation of 2-hydroxymethylenecyclohexanone with isopropyl iodide gives chiefly the O-isopropyl derivative:

$$\tag{27}$$

A thorough discussion of this rather involved question of C-alkylation versus O-alkylation cannot be undertaken here. For our purposes we shall simply note that for most β-keto esters and diesters of the malonic acid type, alkylation occurs on carbon.

*Carbon acylation.* The reaction of the carbon anions of β-keto esters and malonic esters with acyl halides proceeds in a way that is analogous to their reaction with alkyl halides, and carbon acylation results:

$$\tag{28}$$

Because of the high degree of reactivity of acyl halides with nucleophilic reagents, the carbon acylation reaction is not carried out in alcoholic solution, for the acyl chloride would be consumed by reaction with the alcohol. The desired reaction can be carried out with a suspension of the sodium salt of the β-keto or malonic ester in an anhydrous, aprotic solvent, such as ether or benzene. The preparation of sodium malonic ester, $NaCH(COOEt_2)$, can be accomplished by adding the ester to a suspension of finely divided ("powdered") metallic sodium in ether or benzene. The ester reacts, with the evolution of hydrogen, and a suspension of the insoluble sodium derivative results. Sodium hydride can also be used under the same conditions.

Carbon acylation of $\beta$-keto esters results in the formation of diketo esters:

$$\{CH_3CO\ddot{C}HCOOEt\}^- Na^+ + C_6H_5COCl \longrightarrow \underset{\substack{| \\ COC_6H_5}}{CH_3COCHCOOEt} + NaCl \qquad (29)$$

benzoyl
chloride

ethyl benzoylacetoacetate

and the acylation of $\beta$-diketones yields triketones:

$$\{CH_3CO\ddot{C}HCOCH_3\}^- Na^+ + C_6H_5COCl \longrightarrow \underset{\substack{| \\ COC_6H_5}}{CH_3COCHCOCH_3} + NaCl \qquad (30)$$

diacetylbenzoylmethane
(3-benzoylpentane-2,4-dione)

Further reactions involving the keto diesters, diketo esters, and triketones that are prepared in this way will be described in Section 15.8.

## 15.7 Synthetic uses of β-keto esters and malonic esters

The alkylation of 1,3-dicarbonyl compounds is a general and useful reaction with a wide range of applications to synthesis. Some examples are given below. They should be studied carefully with a view to recognizing how the essential feature of each is represented by the formulations shown in Equations (25) and (26):

$$CH_3COCH_2COOEt \xrightarrow[CH_3CH_2I]{NaOEt} \underset{\substack{| \\ CH_2CH_3}}{CH_3COCHCOOEt} \qquad (31)$$

$$CH_2(COOEt)_2 \xrightarrow[(CH_3)_2CHI]{NaOEt} \underset{H_3C}{\overset{H_3C}{>}}CHCH(COOEt)_2 \qquad (32)$$

$$\underset{\substack{| \\ CH_2COOEt}}{CH_3COCHCOOEt} \xrightarrow[CH_3I]{NaOEt} CH_3CO-\underset{\substack{| \\ CH_2COOEt}}{\overset{\overset{\textstyle CH_3}{|}}{C}}-COOEt \qquad (33)$$

$$CH_3COCH_2COOEt \xrightarrow[CH_3COCH_2Br]{NaOEt} \underset{\substack{| \\ CH_2COCH_3}}{CH_3COCHCOOEt} \qquad (34)$$

$$CH_2(COOEt)_2 \xrightarrow[BrCH_2CH_2Br]{2NaOEt} \underset{COOEt}{\overset{COOEt}{\triangleright}} \qquad (36)$$

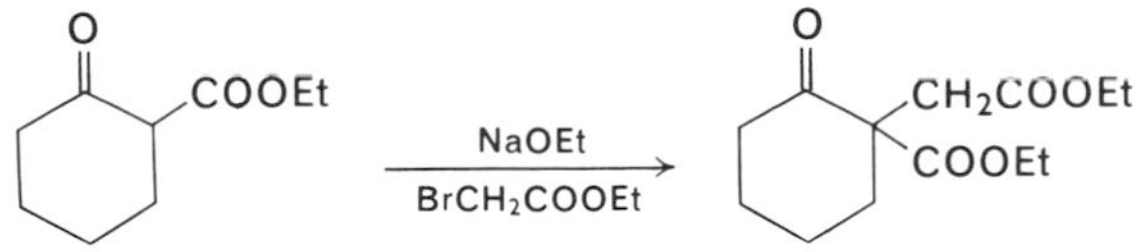

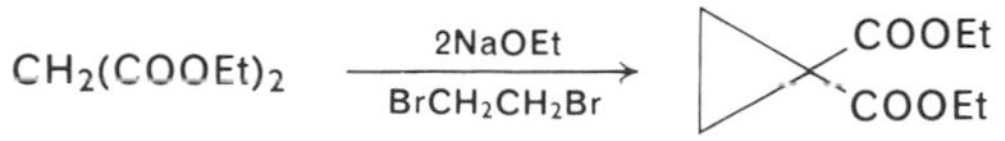

---

**Exercise 2.** Write out each of the foregoing reactions in detail, showing the steps involved and balancing the equations.

---

The alkylation reaction is a general one for β-keto esters of many kinds. In one of the preceding examples, ethyl cyclohexanone-2-carboxylate, a cyclic analog of acetoacetic ester, is used. It will be noted that ethyl cyclohexanone-2-carboxylate is an α-substituted β-keto ester.

In Chapter 17 a number of examples of this alkylation reaction in the formation of rings further illustrate the general method.

## 15.8 Decarboxylation of 1,1-dicarboxylic acids and β-keto acids

The utility of the above alkylation reactions in synthesis is very great indeed. For example, the substituted malonic esters prepared in this way can be saponified to the corresponding malonic acids, and the latter can then be decarboxylated by heating to produce substituted acetic acids. Three examples of this sequence are the following:

$$CH_3CH_2CH_2CH(COOEt)_2 \xrightarrow[\text{(2) H}^+]{\text{(1) NaOH}} CH_3CH_2CH_2CH(COOH)_2$$

$$\xrightarrow{\Delta} CH_3CH_2CH_2CH_2COOH + CO_2 \quad (37)$$

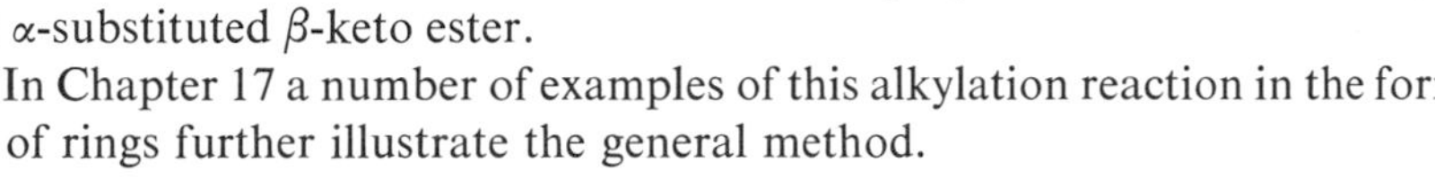

$$(38)$$

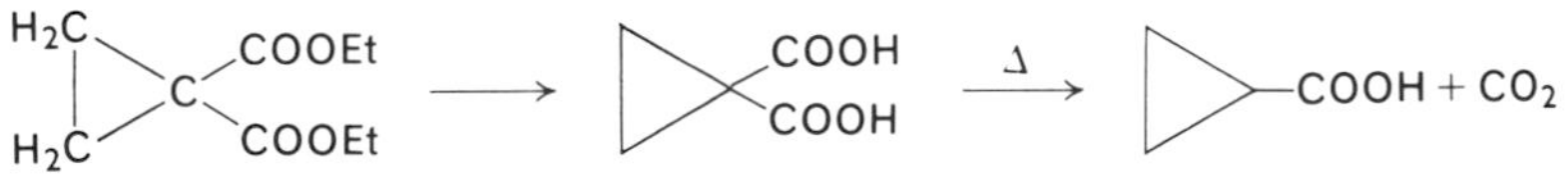

$$(39)$$

The decarboxylation of malonic and substituted malonic acids is a general characteristic of dicarboxylic acids in which the two carboxyl groups are attached to the same carbon atom:

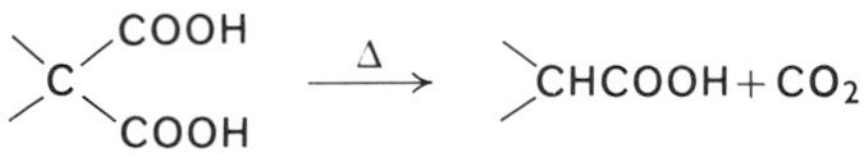

β-Keto acids are similarly unstable and readily lose carbon dioxide on heating to yield ketones:

$$-COCH_2COOH \xrightarrow{\Delta} -COCH_3 + CO_2$$

$$\text{e.g.,} \quad CH_3COCHCOOH \xrightarrow{\Delta} CH_3COCH_2CH_3 + CO_2$$
$$\qquad\qquad\; \overset{|}{CH_3}$$

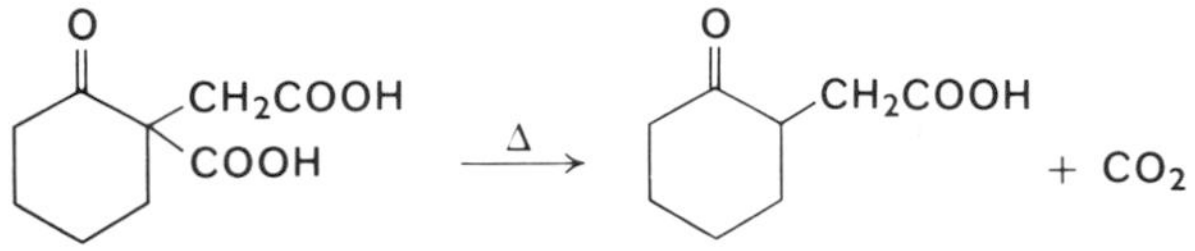

$$+ CO_2 \qquad\qquad (40)$$

---

**Exercise 3.**   In the last example, why does not the second carboxyl group lose $CO_2$ as well, to give 2-methylcyclohexanone as the final product?

---

## 15.9   Cleavage of β-keto esters by alkali

Another reaction of compounds containing two or three carbonyl groups attached to the same carbon atom (β-keto esters, β-diketones, triacylmethanes) that has application to synthetic problems depends upon the fact that, upon treatment with aqueous or alcoholic alkali, compounds of these classes can be cleaved in the manner illustrated by the following examples:*

(a)   $\underset{\underset{CH_2CH_3}{|}}{CH_3COCHCOOEt} \xrightarrow{\text{NaOH}} CH_3COOH + CH_3CH_2CH_2COOH + EtOH$

ethyl α-ethylacetoacetate

(b)

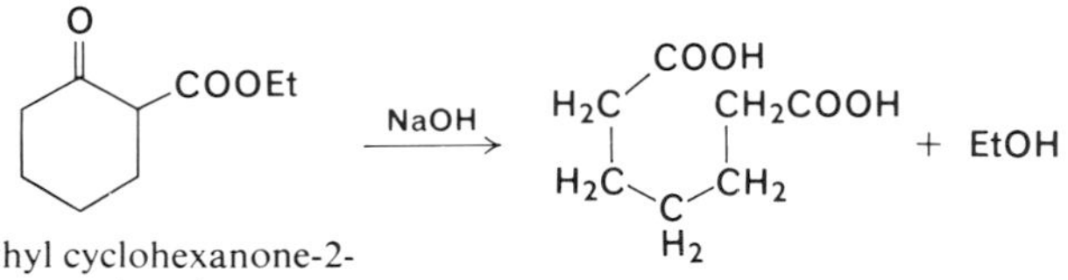

ethyl cyclohexanone-2-
carboxylate

(c)

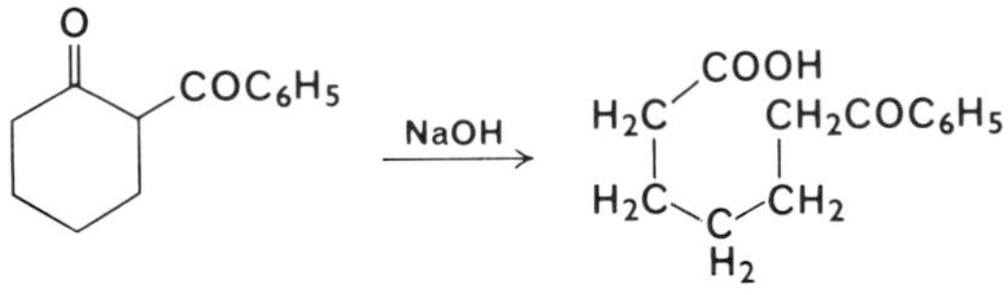

2-benzoylcyclohexanone          6-benzoylhexanoic acid

(d)   $\underset{\underset{CH_2CH_3}{|}}{\overset{\overset{CH_2CH_3}{|}}{CH_3COC}}-COOEt \xrightarrow{\text{NaOH}} \underset{\underset{CH_2CH_3}{|}}{\overset{\overset{CH_2CH_3}{|}}{CH_3COC}}-COOH \xrightarrow[(-CO_2)]{\text{heat}} (CH_3CH_2)_2CHCOCH_3$

ethyl α,α-diethylacetoacetate

It will be noted that two kinds of reactions are shown for substituted aceto-acetic esters: one leading to the formation of two carboxylic acids, the other (d) to a ketone. Although the synthetic utility of these reactions is limited (because

---

   * The final products are the carboxylic acids that would be formed after acidification of the reaction mixture.

the compounds produced can usually be prepared more conveniently and in better yields by other methods of synthesis), they do have occasional applications in which they are superior to alternative synthetic methods. For the purposes of this discussion, they deserve study because they represent examples of carbonyl reactions that further illuminate the nature of the process of nucleophilic attack upon the carbonyl carbon atom.

The cleavage reaction is mechanistically allied to the general carbonyl addition reactions that have been described in this and foregoing chapters. Consequently, the *direction* of cleavage is governed by the relative susceptibility of the two (or, in triketones, the three) carbonyl groups to nucleophilic attack. Consider the reaction of ethyl α-ethylacetoacetate to attack by the hydroxide ion [Equation (41)]. There are two positions at which the reaction can be initiated: *(a)* at the carbonyl carbon atom of the $CH_3CO-$ group; or *(b)* at the carbonyl carbon atom of the $-COOEt$ group:

$$CH_3-\overset{\overset{O}{\|}}{C}-\underset{\underset{CH_2CH_3}{|}}{CH}-\overset{\overset{O}{\|}}{C}-OCH_2CH_3 \qquad (41)$$

$$(a) \qquad (b)$$
$$OH^-$$

Course *(b)* will lead to the saponification of the ester; course *(a)* will lead to the intermediate

$$CH_3\overset{\overset{O}{\|}}{C}-\underset{\underset{CH_2CH_3}{|}}{CH}COOEt \;\rightleftharpoons\; CH_3-\underset{\underset{HO}{|}}{\overset{\overset{O^-}{|}}{C}}-\underset{\underset{CH_2CH_3}{|}}{CH}COOEt \qquad (42)$$
$$HO^-$$

which can cleave by the following course:

$$CH_3-\underset{\underset{HO}{|}}{\overset{\overset{O^-}{|}}{C}}-\underset{\underset{CH_2CH_3}{|}}{CH}-\overset{\overset{O}{\|}}{C}\;OEt \;\longrightarrow\; CH_3COOH+(CH_3CH_2\overset{..}{C}HCOOEt)^- \qquad (43)$$

In an aqueous or alcoholic solution of sodium hydroxide, the carboxylic acid formed in the cleavage step (43) will be ionized to the carboxylate ion; and the carbon-anion of ethyl butyrate (the conjugate base of a very weak acid) will be protonated by the solvent. The final result of the cleavage will be

$$CH_3COCHCOOEt+OH^- \longrightarrow CH_3COO^-+CH_3CH_2CH_2COOEt \qquad (44)$$
$$\underset{CH_2CH_3}{|}$$

This is not reversible, and so the cleavage reaction will proceed to completion. Subsequent saponification of the ethyl butyrate by the alkali used will ensue, and at the completion of the process the solution will contain the sodium salts of acetic and butyric acids.

The direction of cleavage [that is, cleavage according to *(a)* or *(b)* in Equation (41)] depends upon the experimental conditions. Course *(b)*, which is simply saponification of the keto ester, takes place when dilute alkali is used; and course *(a)* occurs when concentrated alkali is used. Cleavage according to *(a)* leads to the α-substituted acetic acid (butyric acid in the example above) and is known as "acid cleavage." Saponification of ethyl α-ethylacetoacetate, [course *(b)*] followed by decarboxylation if the β-keto acid so formed, leads to methyl *n*-propyl ketone, and is known as "ketonic cleavage."

Alkaline cleavage of triketones [such as the one formed according to Equation (30)] occurs by attack of the base at the carbonyl group most susceptible to nucleophilic attack. Thus, cleavage of the following triketone (45) will occur at the $CH_3CO$— group, because this carbonyl group, of the three present, would be the most reactive in the carbonyl addition reaction:

$$CH_3CO\overset{\vdots}{\underset{\vdots}{\overset{\displaystyle CHCOCH(CH_3)_2}{\underset{\displaystyle COCH_2CH_3}{|}}}} \quad \overset{OH^-}{\longrightarrow} \quad CH_3COO^- + CH_3CH_2COCH_2CH(CH_3)_2$$

cleaves here $\hspace{8cm}$ (45)

The reason for the effect of hydroxide ion concentration upon the direction of cleavage of a β-keto ester, as in (41), seems to be the following. The rate of saponification of a simple ester is dependent upon the concentration of the ester and of the hydroxide ion, whereas the rate expression for the cleavage of a β-dicarbonyl compound [course *(a)*, Equation (41)] appears to contain a term that includes a dependence upon the square of the hydroxide ion concentration. Thus, as stronger alkali is used, cleavage according to path *(a)* becomes increasingly important, and at a high concentration of alkali becomes the faster reaction.

The "acid cleavage" [for example, Equations (46), (47), (48)] has disadvantages from the practical point of view, because both kinds of cleavage occur to some extent, and consequently yields may be low. The synthesis of substituted acetic acids by the alkylation, saponification and decarboxylation of malonic acid derivatives [Equations (37), (38), (39)] is more practicable, since only one course can be followed.

The following examples represent some typical synthetic applications of these reactions:

$$CH_3COCH\underset{\displaystyle CH_3}{\overset{\displaystyle COOEt}{|}} \xrightarrow{\text{conc. NaOH}} CH_3COO^- + CH_3CH_2COO^- \hspace{3cm} (46)$$

$$CH_3COCH\underset{\displaystyle CH_2COOEt}{\overset{\displaystyle COOEt}{|}} \xrightarrow{\text{conc. NaOH}} CH_3COO^- + CH_2\underset{\displaystyle COO^-}{\overset{\displaystyle CH_2COO^-}{|}} \hspace{2cm} (47)$$

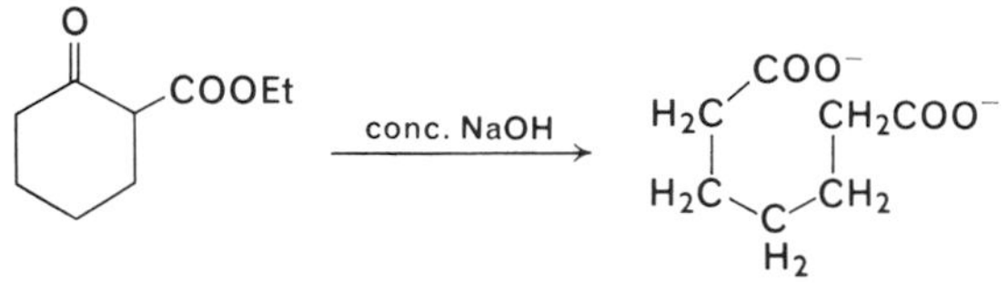

(48)

pimelic acid

(49)

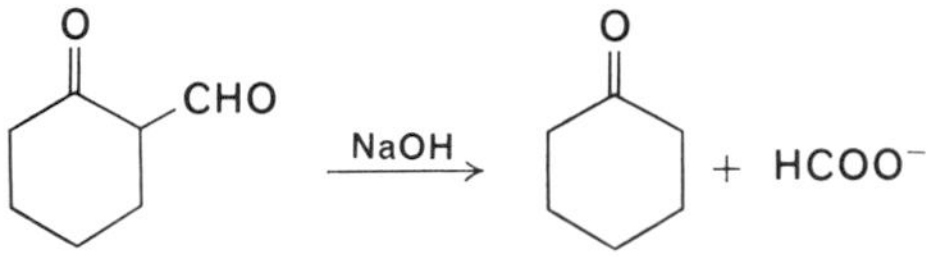

$$CH_3COCHCOOEt \xrightarrow{\text{aq. NH}_3} CH_3CONH_2 + C_6H_5COCH_2COOEt \qquad (50)$$
$$\underset{COC_6H_5}{\vert}$$

ethyl benzoylacetate

(*a*) $Mg(CH(COOEt)_2)_2 + CH_3COCl \longrightarrow CH_3COCH(COOEt)_2 + MgCl_2$ (51)

(*b*) $CH_3COCH(COOEt)_2 \xrightarrow{\text{dil. NaOH}} CH_3COCH(COOH)_2 \xrightarrow{\Delta} CH_3COCH_3 + 2CO_2$

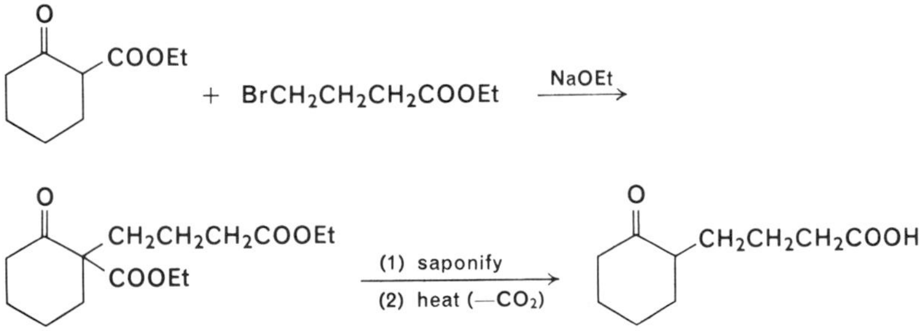

(52)

## 15.10 Dicarboxylic acids

The alkanedioic acids comprise a homologous series starting with oxalic acid and proceeding by increments of —$CH_2$— to longer chain lengths (Table 15-2).

Malonic acid can be prepared by a reaction that is an example of the general method shown in Equation (9), Chapter 14:

$$ClCH_2COOH + KCN \longrightarrow \underset{\underset{CN}{\vert}}{CH_2COOH} \xrightarrow[\text{hydrolysis}]{\text{alkaline}} H_2C \overset{\displaystyle COOH}{\underset{\displaystyle COOH}{}} \qquad (53)$$

If the intermediate cyanoacetic acid is esterified by boiling with ethanol containing sulfuric acid as a catalyst, the cyano group is ethanolyzed, and the diester is produced:

$$\underset{\underset{CN}{\vert}}{CH_2COOH} \xrightarrow[\text{H}_2\text{SO}_4]{\text{EtOH}} H_2C \overset{\displaystyle COOEt}{\underset{\displaystyle COOEt}{}} + (NH_4)_2SO_4 \qquad (54)$$

**Table 15-2.**    *The alkanedioic acids*

| Acid | Common Name | Systematic Name |
|------|-------------|-----------------|
| HOOC—COOH | oxalic acid | ethanedioic acid |
| HOOC(CH$_2$)$_n$COOH | | |
| $n=1$ | malonic | propanedioic |
| 2 | succinic | butanedioic |
| 3 | glutaric | pentanedioic |
| 4 | adipic | hexanedioic |
| 5 | pimelic | heptanedioic |
| 6 | suberic | octanedioic |
| 7 | azelaic | nonanedioic |
| 8 | sebacic | decanedioic |

Succinic acid, and all of the higher members of the series, can be prepared by a similar procedure, starting with the dibromoalkanes (or sulfonic acid esters of the corresponding glycols):

$$\begin{array}{l} \text{CHBr} \\ | \\ \text{CHBr} \end{array} + \text{KCN} \longrightarrow \begin{array}{l} \text{CH}_2\text{CN} \\ | \\ \text{CH}_2\text{CN} \end{array} \xrightarrow{\text{hydrolysis}} \begin{array}{l} \text{CH}_2\text{COOH} \\ | \\ \text{CH}_2\text{COOH} \end{array} \tag{55}$$

$$(\text{CH}_2)_n\begin{array}{l} \diagup\text{CH}_2\text{OSO}_2\text{R} \\ \diagdown\text{CH}_2\text{OSO}_2\text{R} \end{array} + \text{KCN} \longrightarrow (\text{CH}_2)_n\begin{array}{l} \diagup\text{CH}_2\text{CN} \\ \diagdown\text{CH}_2\text{CN} \end{array} \xrightarrow{\text{hydrolysis}}$$

$$(\text{CH}_2)_n\begin{array}{l} \diagup\text{CH}_2\text{COOH} \\ \diagdown\text{CH}_2\text{COOH} \end{array} \tag{56}$$

The series can be ascended with ease, for lithium aluminum hydride reduction of a dicarboxylic acid yields the corresponding glycol, and conversion of the glycol into the dibromide, or a sulfonic acid ester, can be followed by the reaction with KCN as in the above examples:

$$\begin{array}{l} \text{CH}_2\text{COOH} \\ | \\ \text{CH}_2\text{COOH} \end{array} \xrightarrow{\text{LiAlH}_4} \begin{array}{l} \text{CH}_2\text{CH}_2\text{OH} \\ | \\ \text{CH}_2\text{CH}_2\text{OH} \end{array} \xrightarrow{\text{as above}} \begin{array}{l} \text{CH}_2\text{CH}_2\text{COOH} \\ | \\ \text{CH}_2\text{CH}_2\text{COOH} \end{array} \tag{57}$$

Substituted dicarboxylic acids can be prepared by appropriate applications of the malonic ester synthesis; for example:

$$\text{CH}_2(\text{COOEt})_2 \xrightarrow[\substack{(2)\ \text{CH}_3\text{CHCOOEt} \\ | \\ \text{Br}}]{(1)\ \text{NaOEt}} \begin{array}{l} \text{CH}_3\text{CHCH}(\text{COOEt})_2 \\ | \\ \text{COOH} \end{array} \xrightarrow[(2)\ \text{heat}\ (-\text{CO}_2)]{(1)\ \text{saponify}}$$

$$\begin{array}{l} \text{CH}_3\text{CHCH}_2\text{COOH} \\ | \\ \text{COOH} \end{array} \tag{58}$$

methylsuccinic acid

Certain acids of the series are obtainable from natural sources:

$$CH_3(CH_2)_7CH=CH(CH_2)_7COOH \xrightarrow[\text{(cleave at C=C)}]{\text{oxidize}}$$

oleic acid

(e.g., with $HNO_3$)

$$CH_3(CH_2)_7COOH + HOOC(CH_2)_7COOH \quad (59)$$

azelaic acid

$$\overset{\displaystyle \overset{OH}{|}}{CH_3(CH_2)_5CHCH_2CH=CH(CH_2)_7COOH} \xrightarrow[250°]{NaOH/H_2O}$$

ricinoleic acid

$$CH_3(CH_2)_5\overset{\displaystyle \overset{OH}{|}}{CH}CH_3 + (CH_2)_8\!\!\begin{array}{c} \diagup COOH \\ \diagdown COOH \end{array} \quad (60)$$

sebacic acid

The $\alpha,\omega$-dicarboxylic acids* have many properties in common with mono-carboxylic acids. The $pK_a$ values of the long-chain acids (first ionization constant) are between 4 and 5, and thus compare closely with the simple alkanoic acids. In oxalic acid, however, the proximity of the electron-attracting —COOH group increases the acidity of the other carboxyl group, and the acid has a $pK_a$ (first proton) value of 1.3. Malonic acid, in which the carboxyl groups are somewhat further apart, has a $pK_a$ value of 2.8; in succinic acid the inductive effect of the second —COOH group is small ($pK_a=4.2$).

Succinic and glutaric acids readily undergo intramolecular anhydride formation on heating:

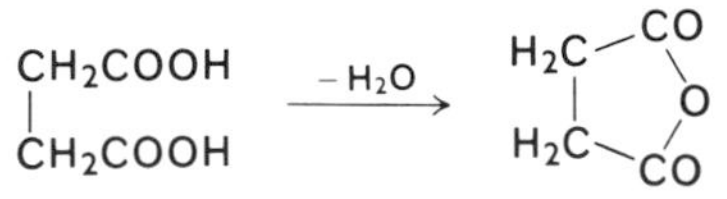

$$\begin{array}{c} CH_2COOH \\ | \\ CH_2COOH \end{array} \xrightarrow{-H_2O} \begin{array}{c} H_2C\!\!\diagup^{CO} \\ | \quad\quad O \\ H_2C\!\!\diagdown_{CO} \end{array} \quad (61)$$

succinic anhydride

$$H_2C\!\!\begin{array}{c} \diagup CH_2COOH \\ \diagdown CH_2COOH \end{array} \xrightarrow{-H_2O} \begin{array}{c} H_2C\!-\!CO \\ \diagup \quad\quad \diagdown \\ H_2C \quad\quad O \\ \diagdown \quad\quad \diagup \\ H_2C\!-\!CO \end{array} \quad (62)$$

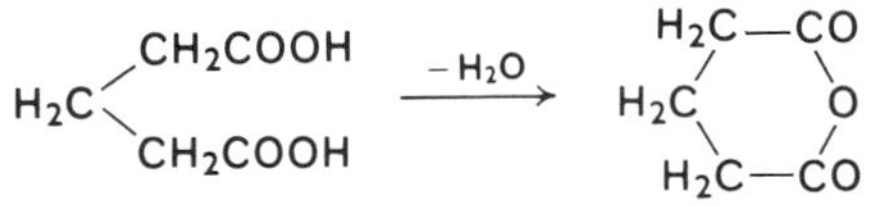

glutaric anhydride

Malonic acid does not form the four-membered cyclic anhydride on heating; instead, it decarboxylates. The higher alkanedioic acids (adipic and above) do

---

* The Greek letter *omega*, $\omega$, is often used to denote a substituent at the extreme end of a carbon chain. Thus, an $\alpha,\omega$-dicarboxylic acid possesses carboxyl groups separated by a chain of —$CH_2$— groups.

not form anhydrides as a result of simple heating. Their anhydrides are known but are prepared in other ways. A discussion of cyclization reactions, including the behavior of the alkanedioic acids, will be found in Chapter 17.

---

**Exercise 4.**  Devise a synthesis of $\beta$-methylglutaric acid, $\overset{\displaystyle CH_3}{HOOCCH_2\overset{|}{C}HCH_2COOH}$, starting with diethyl malonate.

---

## 15.11  Some applications to synthesis

In the following examples, some of the reactions that have been discussed in preceding chapters are used for the synthesis of some complex organic structures. The student is urged to analyze these syntheses in terms of the individual steps that are used, and not to attempt to memorize them as a whole. Their importance here lies only in the way they exemplify what has already been discussed in the foregoing part of this book:

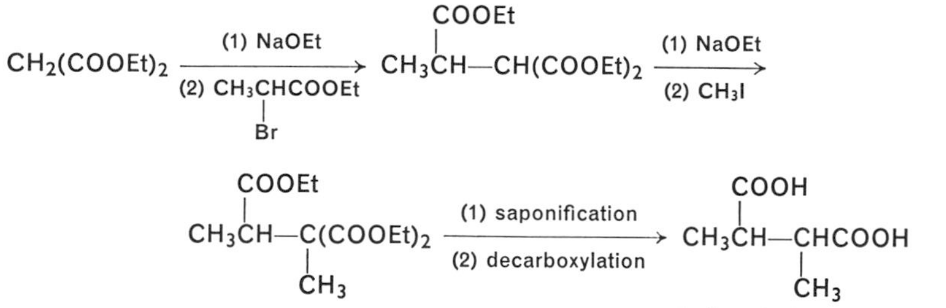

$$CH_2(COOEt)_2 \xrightarrow[\text{(2) CH}_3\text{CHCOOEt}]{\text{(1) NaOEt}} \underset{|}{\overset{\displaystyle COOEt}{CH_3\overset{|}{C}H-CH(COOEt)_2}} \xrightarrow[\text{(2) CH}_3\text{I}]{\text{(1) NaOEt}}$$

$$\underset{\displaystyle CH_3}{\overset{\displaystyle COOEt}{CH_3\overset{|}{C}H-\overset{|}{C}(COOEt)_2}} \xrightarrow[\text{(2) decarboxylation}]{\text{(1) saponification}} \underset{\displaystyle CH_3}{\overset{\displaystyle COOH}{CH_3\overset{|}{C}H-\overset{|}{C}HCOOH}} \quad (63)$$

$$\alpha,\beta\text{-dimethylsuccinic acid}$$

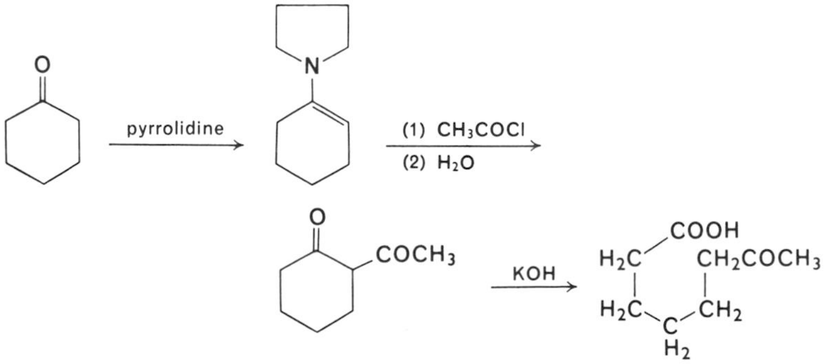

$$\text{(64)}$$

7-ketoöctanoic acid or
6-acetylhexanoic acid

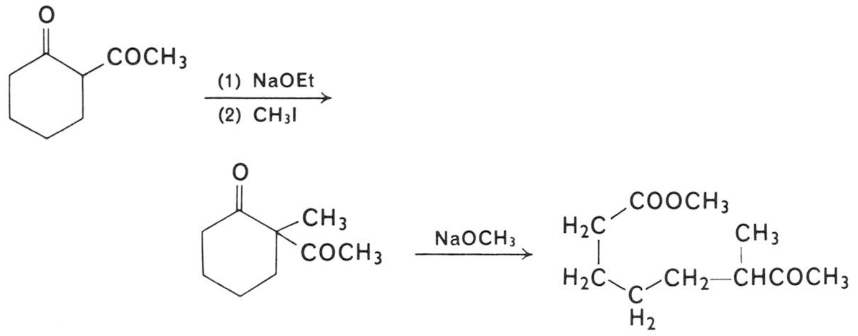

$$CH_3CH_2CH_2COCH_2CH_2CH_3 \xrightarrow{\text{pyrrolidine}}$$

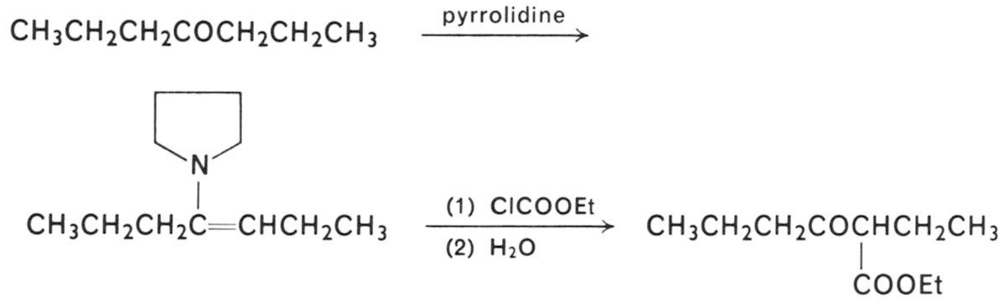

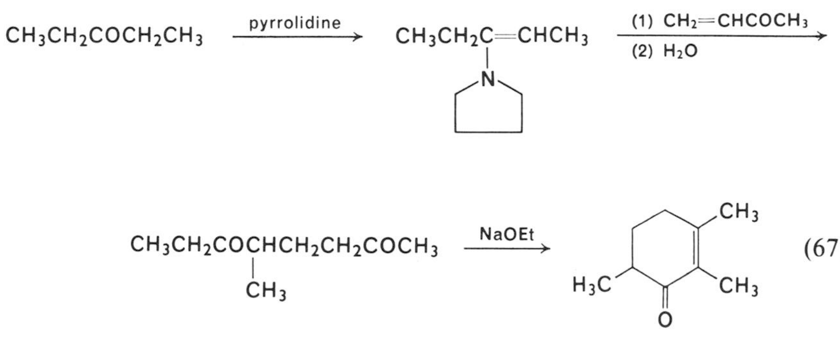

$$CH_3CH_2CH_2C=CHCH_2CH_3 \xrightarrow[\text{(2) H}_2\text{O}]{\text{(1) ClCOOEt}} CH_3CH_2CH_2COCHCH_2CH_3 \quad (66)$$

with N-pyrrolidine substituent and COOEt group

$$CH_3CH_2COCH_2CH_3 \xrightarrow{\text{pyrrolidine}} CH_3CH_2C=CHCH_3 \xrightarrow[\text{(2) H}_2\text{O}]{\text{(1) CH}_2=\text{CHCOCH}_3}$$

$$CH_3CH_2COCHCH_2CH_2COCH_3 \xrightarrow{\text{NaOEt}} \quad (67)$$

with CH_3 substituent; product is a cyclohexenone with CH_3 groups

$$C_6H_5CHO + CH_3COOEt \xrightarrow{\text{NaOEt}} C_6H_5CH=CHCOOEt \quad (68)$$

ethyl cinnamate

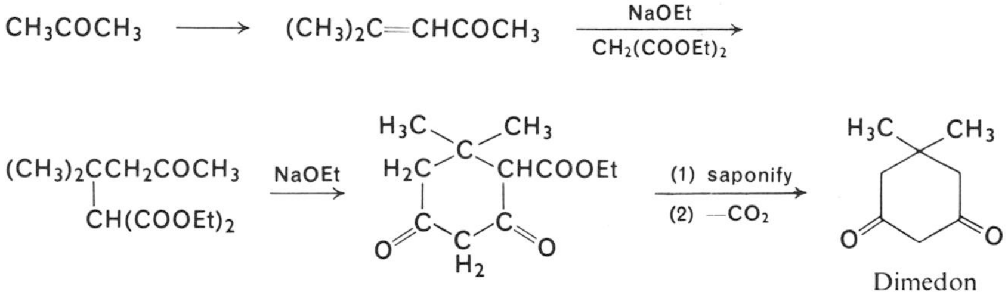

$$CH_3COCH_3 \longrightarrow (CH_3)_2C=CHCOCH_3 \xrightarrow[\text{CH}_2(\text{COOEt})_2]{\text{NaOEt}}$$

$$(CH_3)_2CCH_2COCH_3 \xrightarrow{\text{NaOEt}} \quad \xrightarrow[\text{(2) } -\text{CO}_2]{\text{(1) saponify}}$$

Dimedon

$$(69)$$

$$2CH_2=CHCOOCH_3 + CH_3NH_2 \longrightarrow$$

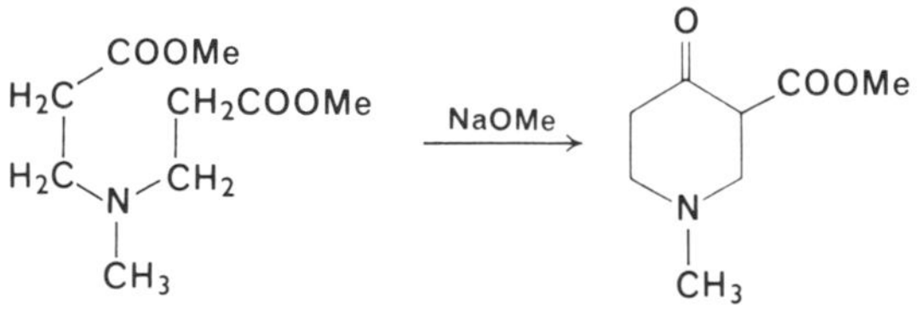

(70)

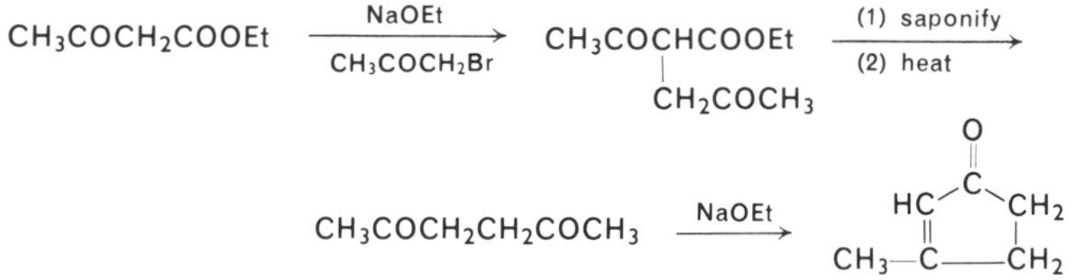

$$CH_3COCH_2COOEt \xrightarrow[CH_3COCH_2Br]{NaOEt} CH_3COCHCOOEt \xrightarrow[\text{(2) heat}]{\text{(1) saponify}}$$

with $CH_2COCH_3$ substituent

(71)

$$CH_3COCH_2CH_2COCH_3 \xrightarrow{NaOEt}$$

Note: In the above examples, the sequence of alkaline saponification and subsequent decarboxylation by heating is often accomplished in another way. If the 1,1-diester or $\beta$-keto ester is heated with aqueous acid (HCl or $H_2SO_4$), acid-catalyzed hydrolysis of the ester and concomitant decarboxylation in the hot reaction mixture accomplishes the desired removal of —COOEt in one operation.

---

**Exercise 5.**   Starting with ethyl acetoacetate, outline syntheses for *(a)* 4-methyl-2-pentanone *(b)* 4-ketopentanoic acid (levulinic acid).

---

### Exercises

(Exercises 1–5 will be found within the text.)

**6** Write the reactions for the following alkylation reactions:
   *(a)* alkylation of ethyl acetoacetate with methyl bromide
   *(b)* alkylation of ethyl malonate with *sec*-butyl bromide
   *(c)* alkylation of ethyl malonate with bromoacetone
   *(d)* alkylation of ethyl cyclohexanone-2-carboxylate with ethyl iodide
   *(e)* alkylation of methyl acetoacetate with ethyl bromoacetate
**7** Devise practical syntheses for the following compounds, starting with any ester, 1,1-diester, or $\beta$-ketoester, and any necessary organic halogen compounds:
   *(a)* *n*-amyl methyl ketone (2-heptanone)
   *(b)* levulinic acid (4-oxopentanoic acid)
   *(c)* ethylsuccinic acid
   *(d)* 2-ethyl-5-methylcyclohexanone
   *(e)* cyclopropylcarbinol
   *(f)* 2-isobutyl-1,4-butanediol [since the desired product is obtainable by the LiAlH$_4$ reduction of isobutylsuccinic acid, the synthetic problem is reduced to the preparation of the latter; see *(c)*]
   *(g)* 2,2-dimethyl-1,3-butanediol

**8** Write the structure of the anion formed by the action of a base on each of the following compounds. In each case remove only the most acidic proton: *(a)* $CH_3CHO$; *(b)* $CH_3COCH_3$; *(c)* $(CH_3)_3CCOCH_3$; *(d)* $CH_3COCH_2COCH_3$; *(e)* $(CH_3CO)_3CH$; *(f)* $CH_2(COOCH_3)_2$; *(g)* $CH_3CH(COOCH_3)_2$; *(h)* $CH(COOEt)_3$; *(i)* dicyclopentyl ketone.

**9** Write the sequence of steps involved in the following ester condensations:
*(a)* ethyl propionate with ethyl propionate (i.e., self-condensation)
*(b)* ethyl formate with acetone
*(c)* ethyl oxalate with cyclohexanone
*(d)* diethyl carbonate with diethyl ketone

**10** When two ester functions are part of the same molecule, condensation between them, with ring formation, can often occur (see Chapter 17). For example, ethyl adipate (the ethyl ester of hexanedioic acid) can be cyclized with the use of sodium ethoxide to yield ethyl cyclopentanone-2-carboxylate. Formulate this ring closure as in the example of the reactions shown in the step-by-step series given in Section 15.1.

**11** What would be the product of the cyclization, as described in Exercise 10, of ethyl $\alpha$-methyladipate?

**12** What would be the product of the reaction between the enamine derived from diethyl ketone and methyl acrylate (methyl propenoate)?

*Chapter sixteen* / **Stereoisomerism. Optical Activity**

## 16.1 Rotation about the carbon-carbon single bond

It is often possible to write two or more projection formulas for a given compound in such a way as to suggest the existence of a number of forms that differ only in the angular conformation of the substituent atoms or groups. For example, 1,2-dibromoethane can be written as

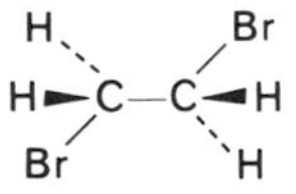

*(a)*

or

*(b)*

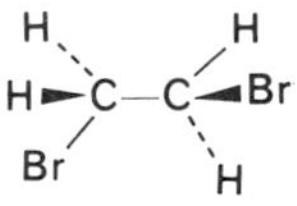

or, indeed, in other ways which show many intermediate positions for the two bromine atoms. The experimental fact is that there is only one 1,2-dibromoethane, and that this possesses a statistically average structure, probably best represented by *(a)*, in which the bulky bromine atoms are separated and occupy positions *trans* to each other. However, so little energy is required to rotate the —$CH_2Br$ groups about the single bond that thermal agitation at ordinary temperatures is sufficient to surmount the energy barriers between the various conformations. We speak, therefore, of *free rotation* about the single bond, using the term to indicate not that the groups joined by the bond spin completely freely with respect to one another, but that the energy barriers to rotation are so low that

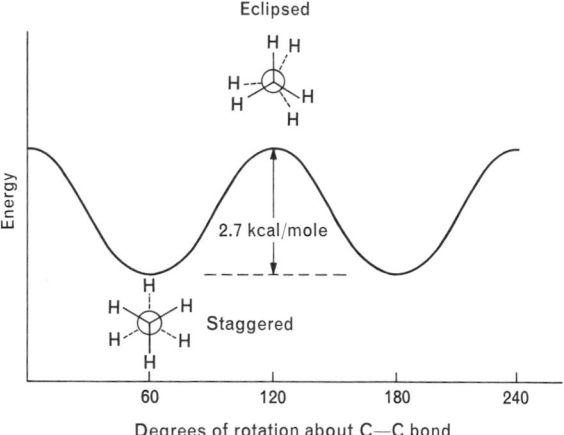

**Figure 16–1.** Conformational energy barriers to rotation about C—C bond in ethane.

we cannot isolate individual rotational isomers. Barriers to rotation exist even in so simple a compound as ethane for which the "staggered" conformation is more stable than the "eclipsed" conformation (Figure 16-1).

Propane and butane consist of populations of molecules, the majority of which, at room temperature, have the conformations *(c)* and *(d)*

*(c)*  propane        *(d)*  butane

In Section 16.19 will be described an example of restricted rotation about a single bond which, for steric reasons, permits the isolation and separate existence of two rotational isomers. Isomerism of this kind is uncommon because, although energy barriers to "free" rotation about single bonds are often large enough to ensure the statistical abundance of the most stable conformation, they are not often large enough to permit the independent existence of rotational conformers.

## 16.2  *Cis-trans* (geometrical) isomerism

The term "geometrical isomerism" is applied to isomerism that depends upon the arrangement of substituents on a double bond or on a cyclic structure. The *cis-trans* isomerism of the carbon-carbon double bond is described in Chapter 9.

Geometrical isomers can occur in other than olefinic compounds. For example, in cyclic compounds such as the following, *cis-trans* isomerism occurs; several kinds of conventional representations are shown:

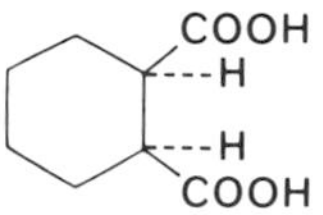

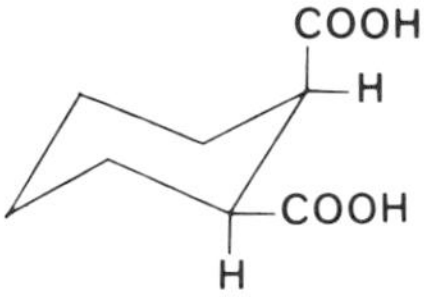

*cis*-cyclohexane-1,2-dicarboxylic acid*

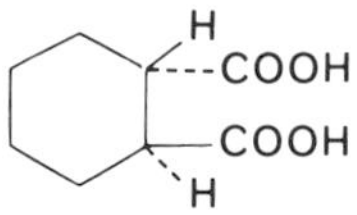

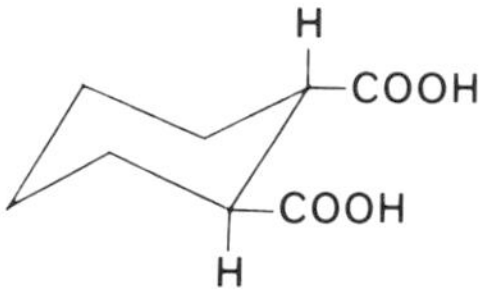

*trans*-cyclohexane-1,2-dicarboxylic acid

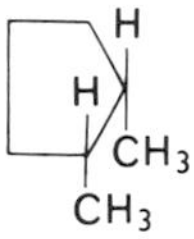

*cis*-1,2-dimethylcyclopentane 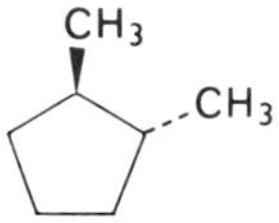         *trans*-1,2-dimethylcyclopentane

---

**Exercise 1.**   Write appropriate projection formulas for all of the possible
(a) 1,2-dibromocyclopropanes
(b) 1,2,4-trichlorocyclohexanes

---

Geometrical isomers have distinct chemical and physical properties; they are different compounds. A number of representative pairs of *cis-trans* isomers are given in Table 16-1; the compounds chosen include a number of carboxylic acids which have been selected to show that the isomeric pairs differ not only in melting point but in their acid dissociation constants as well.

A number of questions can now be asked about geometrical isomerism:

1. How can we distinguish experimentally between the *cis* and *trans* forms of a pair of geometric isomers?
2. How can a synthesis of an olefin be directed to produce the *cis* or *trans* form at will?
3. Which is the more stable form; that is, what are the relative stabilities of *cis* and *trans* forms?
4. How can a *cis* form be converted into a *trans* form, and vice versa?

---

* As in earlier chapters, the conformational representation of the cyclohexane ring shown on the right will ordinarily show only those substituents relevant to the point under discussion. The four —$CH_2$— groups of the ring are implied in the structure shown.

**Table 16-1.**  *Physical properties of geometrical isomers*

| Compound | m.p. | $pK_a$ |
|---|---|---|
| *cis*-1,2-cyclopropanedicarboxylic acid | 139°C | 3.40 |
| *trans*-1,2-cyclopropanedicarboxylic acid | 175° | 3.70 |
| oleic acid *(cis)* | 13° | — |
| elaidic acid *(trans)* | 44° | — |
| cinnamic acid *(trans)** | 133° | 4.44 |
| allocinnamic acid *(cis)* | 68° | 3.96 |
| *cis*-1,2-dichloroethene | (b.p. 48°C) | — |
| *trans*-1,2-dichloroethene | (b.p. 60°C) | — |
| crotonic acid *(trans)* | 72° | 4.70 |
| isocrotonic acid *(cis)* | 15° | 4.44 |
| *cis*-1,4-cyclohexanedicarboxylic acid | 161° | 4.52 |
| *trans*-1,4-cyclohexanedicarboxylic acid | 300° | 5.34 |
| maleic acid *(cis)* | 130° | 1.9 |
| fumaric acid *(trans)* | 287° | 3.0 |

* Cinnamic acid is 3-phenylpropenoic acid ($C_6H_5CH{=}CH{-}COOH$).

## 16.3 The use of chemical behavior in assigning geometrical structure

If a compound can be transformed into another of known configuration by a reaction which does not involve *cis-trans* change, the configuration of the compound in question can be established. For example, maleic and fumaric acids are the *cis-* and *trans*-butenedioic acids. Maleic acid can be transformed with case under mild conditions (heating with acetyl chloride, or distillation) into the cyclic anhydride; fumaric acid cannot be transformed into an anhydride under any but very vigorous treatment, and *then does so only by undergoing isomerization, with the result that maleic anhydride is again formed.* Clearly, maleic anhydride is structurally capable of existence and of easy formation; and an examination of molecular models shows that only the *cis* anhydride can be constructed. The configurations of the two acids must be the following:

$$\begin{array}{c} \text{H} \diagdown\ \diagup \text{COOH} \\ \text{C} \\ \parallel \\ \text{C} \\ \text{H} \diagup\ \diagdown \text{COOH} \end{array} \quad \xrightarrow[\text{on heating}]{\text{easily}} \quad \begin{array}{c} \text{H} \diagdown\ \diagup \text{CO} \\ \text{C} \\ \parallel \quad\quad \text{O} \\ \text{C} \\ \text{H} \diagup\ \diagdown \text{CO} \end{array} \quad \xrightarrow{\text{H}_2\text{O}} \quad \text{hydrolyzed to } \textit{maleic acid} \qquad (1)$$

maleic acid  　　　　　maleic anhydride

$$\begin{array}{c} \text{HOOC} \diagdown\ \diagup \text{H} \\ \text{C} \\ \parallel \\ \text{C} \\ \text{H} \diagup\ \diagdown \text{COOH} \end{array} \quad \longrightarrow \quad \text{no anhydride (except under very vigorous} \qquad (2)$$

　　　　　　　　　　　　　　　treatment, when *maleic anhydride* is formed)

fumaric acid

This conclusion is substantiated by the observation that maleic acid is formed by the oxidation of *p*-benzoquinone, a cyclic compound in which the grouping —CO—CH=CH—CO must have the *cis* configuration:

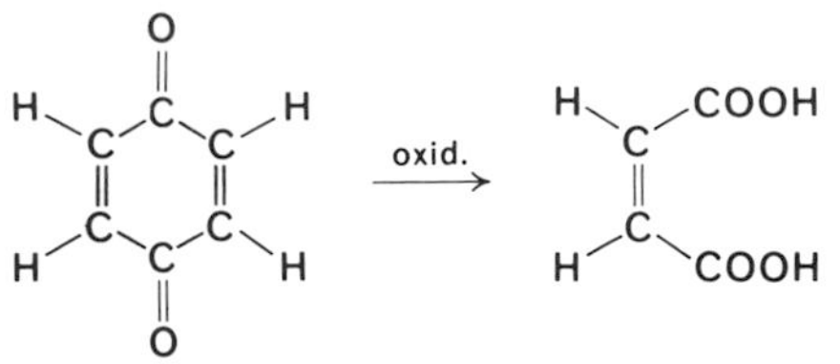

$$(3)$$

p-benzoquinone          maleic acid

In cyclic systems, *cis-trans* relationships can be studied in similar ways: of the two cyclopentane-1,2-dicarboxylic acids one *(cis)* will form the anhydride with ease, the other *(trans)* will not. Furthermore, as we shall learn in the discussion to follow, one *(trans)* can exist in optically isomeric forms *(dextro and levo)*, the other *(cis)* is a *meso* form and cannot be so resolved (Section 16.11):

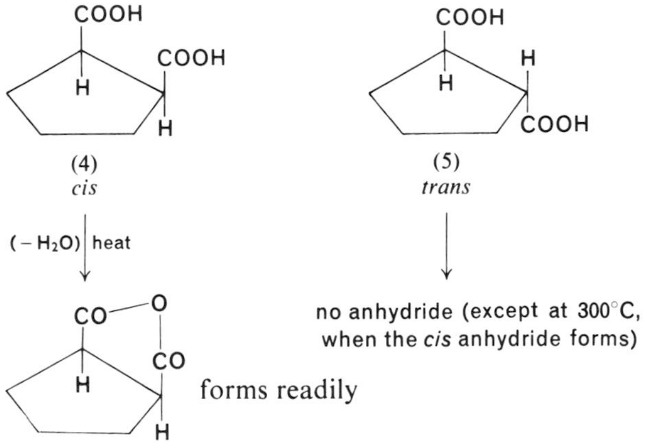

(4)          (5)
*cis*          *trans*

$(-H_2O)$ heat

forms readily

no anhydride (except at 300°C, when the *cis* anhydride forms)

The stereochemical requirements of ring closures of other kinds are also useful in establishing the spatial disposition of groups. For example, the two 2-chloro-1-methylcyclopentanols

can be readily distinguished from one another, since by treatment with alkali the *trans* OH—Cl compound readily forms the oxide by a displacement reaction,

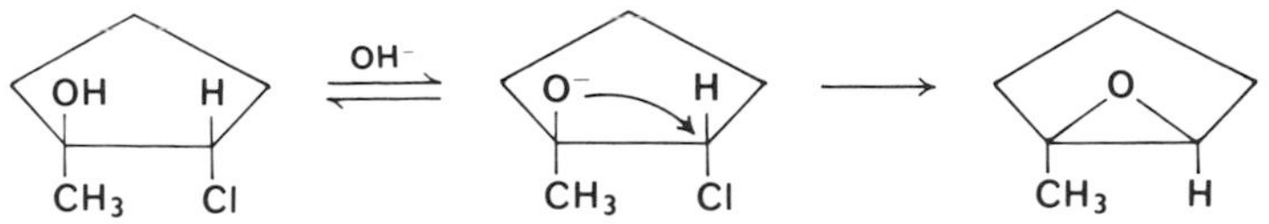

$$(6)$$

1-methylcyclopentene oxide

since only when the OH and Cl are *trans* can the displacement of chloride occur by an attack on the rear side of the carbon atom holding the chlorine. The *cis* compound (OH—Cl) does not form the oxide under the same treatment; instead under somewhat more vigorous treatment, chloride is displaced by the migrating $(CH_3:)$, leading to 2-methylcyclopentanone:

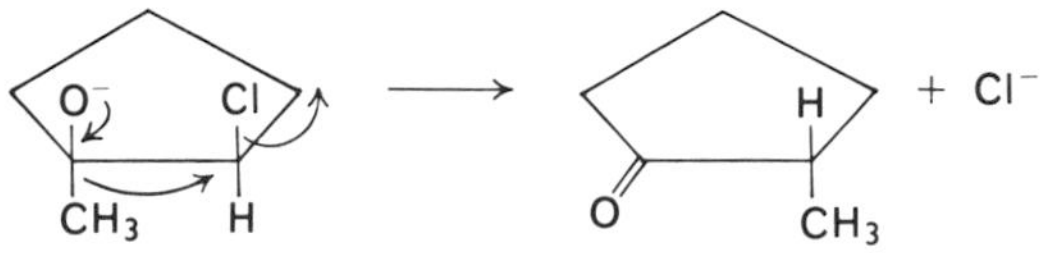

$$(7)$$

## 16.4 The use of physical properties in assigning geometrical structure

Physical constants such as the dipole moment can often be used to determine configuration. The dipole moment of the *cis* compound is usually higher than that of the *trans* compound:

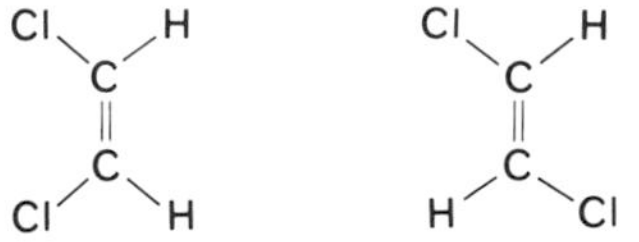

In general the *trans* forms of pairs of *cis-trans* isomers have lower boiling points, higher melting points, and lower solubilities. In most cases an assignment of structure based upon physical properties can be made only if both members of a pair are available for study although dipole moments can often be calculated for the two possible structures with sufficient accuracy to enable a choice on the basis of the measurement of the dipole moment of one of the isomers and comparison of the value found with those calculated.

A widely used physical method of determining configuration is X-ray crystallography. This method depends upon the fact that the diffraction of X-rays by a crystalline substance produces a pattern (on a photographic plate) that can be used to calculate the relative positions of the atoms in the molecule. Modern X-ray crystallographic techniques are highly developed and, with the aid of modern computers in the solution of the intricate mathematical problems involved, are widely used to determine the configurations of very complex compounds.

A simple, but typical, problem of *cis-trans* isomerism that has been solved by the use of X-ray methods is that of the configuration of the compounds angelic acid and tiglic acid. These 2-methyl-2-butenoic acids were shown to have the following configurations:

$$
\begin{array}{cc}
\underset{\text{tiglic acid}}{
\begin{array}{c}
H_3C \quad\; H \\
\diagdown C \diagup \\
\; C \; \\
\diagup \quad \diagdown \\
H_3C \qquad COOH
\end{array}
} &
\underset{\text{angelic acid}}{
\begin{array}{c}
H \quad\;\; CH_3 \\
\diagdown C \diagup \\
\; C \; \\
\diagup \quad \diagdown \\
H_3C \qquad COOH
\end{array}
}
\end{array}
$$

An interesting by-product of this finding was the discovery that the formation and carbonation of the lithium derivative from a bromoalkene proceeds with retention of configuration. Thus, when *trans*-2-bromo-2-butene is converted into the lithium derivative, and this carbonated, the acid that is formed is angelic acid:

$$
\underset{H_3C}{\overset{H_3C}{\diagdown}}C=C\underset{CH_3}{\overset{Br}{\diagup}} \;\xrightarrow{\;Li\;}\; \underset{H}{\overset{H_3C}{\diagdown}}C=C\underset{CH_3}{\overset{Li}{\diagup}} \;\xrightarrow[\text{(2) } H_2O/H^+]{\text{(1) } CO_2}\; \underset{H}{\overset{H_3C}{\diagdown}}C=C\underset{CH_3}{\overset{COOH}{\diagup}}
$$

$$(8)$$

## 16.5 Establishment of configuration by synthetic methods

Methods of synthesis that are stereochemically predictable serve the double purpose of allowing us to synthesize unsaturated compounds of known configuration and to determine configuration. As an example, the synthesis of oleic acid by the catalytic hydrogenation of the corresponding acetylenic acid (Section 9.11) proves that it has the *cis* configuration. The stereoisomer of oleic acid is *elaidic acid*, which has the *trans* configuration. Both oleic and elaidic acid give the same products upon ozonolysis, and both are hydrogenated to stearic acid. Catalytic hydrogenation of butynedioic acid to the butenedioic acid yields maleic acid, another piece of evidence that supports the assignment of the *cis* configuration to maleic acid (Section 16.3):

$$
HOOC-C\equiv C-COOH \;\xrightarrow[\text{(1 mole)}]{Pd/H_2}\; \underset{\underset{\text{maleic acid}}{H \qquad\; H}}{\overset{HOOC \quad COOH}{\diagdown C=C \diagup}}
$$

$$(9)$$

Besides catalytic hydrogenation, other stereospecific addition reactions can be used to prepare compounds of known configuration. The *trans* addition of halogens and other electrophilic reagents to double bonds and the *cis* and *trans* hydroxylation of double bonds (Chapter 9) lead to products of defined stereochemistry.

The stereospecificity of catalytic hydrogenation is not absolute in many cases. Hydrogenation of benzene derivatives, in which the aromatic ring is reduced to a cyclohexane ring, usually yields products that contain more than one geometrical isomer. For example, the catalytic hydrogenation of the ester of phthalic acid leads to the formation of a mixture of the *cis-* and *trans*-cyclohexane-1,2-dicarboxylic acids (after saponification):

$$\text{(10)}$$

Nevertheless, the pure *cis* acid can be prepared by treating the mixture of isomers with hot acetic anhydride, which converts both of the isomers into the *cis* anhydride, from which the *cis* acid can be obtained by hydrolysis:

mixture of *cis-* and *trans-* cyclohexane-1,2-dicarboxylic acids

$$\xrightarrow{\text{acetic anhydride}} \qquad \xrightarrow{H_2O} \text{cis acid}$$

$$\text{(11)}$$

The conversion of a compound of known configuration into another can also be used for configurational determination. For example, if the known *cis*-crotonic acid is converted into 2-butene by the following series of reactions, the butene produced is the *cis* form; conversely, if the butene produced were the reference compound of known configuration, its preparation from the crotonic acid would prove the configuration of the latter:

*cis*-crotonic acid

$$\xrightarrow{\text{LiAlH}_4} \qquad \xrightarrow{\text{RSO}_2\text{Cl}}$$

$$\xrightarrow{\text{LiAlH}_4{}^*} \qquad \text{(12)}$$

In the course of such a series, intermediates can be given configurational assignments. For instance, the 2-buten-1-ol formed from *cis*-crotonic acid in the above series is the *cis* compound.

---

**Exercise 2.** *Cis*-stilbene ($C_6H_5CH{=}CHC_6H_5$) adds bromine to produce a stilbene dibromide ($C_6H_5CHBrCHBrC_6H_5$). When this dibromide is treated with alcoholic KOH, HBr is eliminated and $C_6H_5CH{=}CBr(C_6H_5)$ is formed. Write these reactions using suitable representations that show the stereochemistry of the compounds.

**Exercise 3.** How could you prepare a sample of *trans*-2-butene with fumaric acid as the starting material?

---

* This reaction, the reduction of the group $-CH_2OSO_2R$ to $-CH_3$ by means of lithium aluminum hydride, is a useful method for reducing certain types of alcohols in the manner shown. It may be looked upon as a nucleophilic displacement of the $RSO_2O-$ group by $H{:}^-$ provided by the LiAlH$_4$.

### 16.6   Interconversion of *cis* and *trans* isomers

*Cis-trans* interconversion about double bonds can be brought about in many ways, which depend essentially upon creating some degree of single-bond character in the double bond. This can be accomplished experimentally by the excitation of the molecule by the absorption of light energy, or by transitory addition, for example of a proton, to the double bond. When a solution of maleic acid in aqueous HCl is heated, the acid is isomerized to fumaric acid, probably by way of a reversible protonation. The equilibrium that is established lies far on the side of the thermodynamically more stable *trans* acid:

$$(13)$$

Irradiation of *trans*-stilbene (see Exercise 2) with ultraviolet light causes isomerization to the less stable *cis* form; the energy of the absorbed light is represented by the higher energy state (lesser stability) of the *cis* isomer. The *cis* form, being the less stable of the two isomers, has a higher heat of combustion.

In a mixture of *cis-trans* isomers that is produced by thermal, catalytic, or other means, the final equilibrium will usually lead to a predominance of the *trans* form, the proportion of the two isomers being determined by their relative thermodynamic stability.

### 16.7   Optical isomerism. Polarized light

Ordinary light is an advancing wave front in which the vibrations occur in all directions perpendicular to the direction of propagation. Each vibration may be looked upon as the resultant vector of two vibrations at right angles to one another:

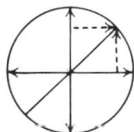

By absorbing or selectively refracting the vibrations in one of the two directions the resulting light ray is vibrating in a single plane only and is said to be "plane polarized":

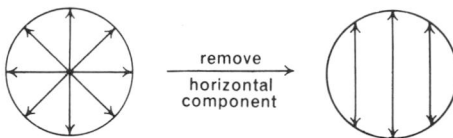

This result can be accomplished in several ways. A suitably oriented crystal of calcite ( a form of $CaCO_3$) displays a different index of refraction for rays with one plane of vibration than for those vibrating in the other plane. This means that the two components travel with different velocities in calcite, and that a properly cut calcite crystal can utilize this difference in separating the two components. The Nicol prism is a device that utilizes this property (called "double refraction") of calcite, and produces plane polarized sight by a combined refraction and reflection of one ray, with the result that the other is allowed to proceed through and emerge from the prism. The Nicol prism consists of two halves of a crystal of calcite, cut with precise angles, and cemented together with a Canada balsam, a resin chosen for the particular value of its refractive index.

Another means of producing polarized light is by the use of a *dichroic* crystal. A dichroic crystal is one that has the property of absorbing the component vibrating in one plane more strongly than that vibrating at right angles to it. A dichroic crystal of proper thickness will effectively remove one ray by completely absorbing it, allowing the other to be transmitted to a useful extent (although it, too, will be absorbed to some, but a lesser, degree). The widely used "Polaroid" material utilizes the dichroism of certain salts of quinine, properly oriented on a transparent support, to produce polarized light.

## 16.8   The measurement of optical rotation. The polarimeter

*An "optically active" substance is one that rotates the plane of polarized light.* Since it is usually necessary to measure the actual magnitude of this rotation, a device called a "polarimeter" is used. The polarimeter is a precision instrument, comprising the parts shown in Figure 16-2.

If the polarizer and analyzer are "crossed" at 90° with respect to one another and the sample removed, no light will reach the eyepiece. If the sample is optically active, and, after insertion, rotates the plane of the polarized light entering it, the analyzer will have to be turned through an angle equal to the extent of the rotation in order again to produce a dark field in the eyepiece. The rotation is measured on a circular scale, calibrated in degrees, to which the analyzer is mounted. If the plane of polarization is rotated to the right, the sample is "dextrorotatory;" if to the left, "levorotatory."

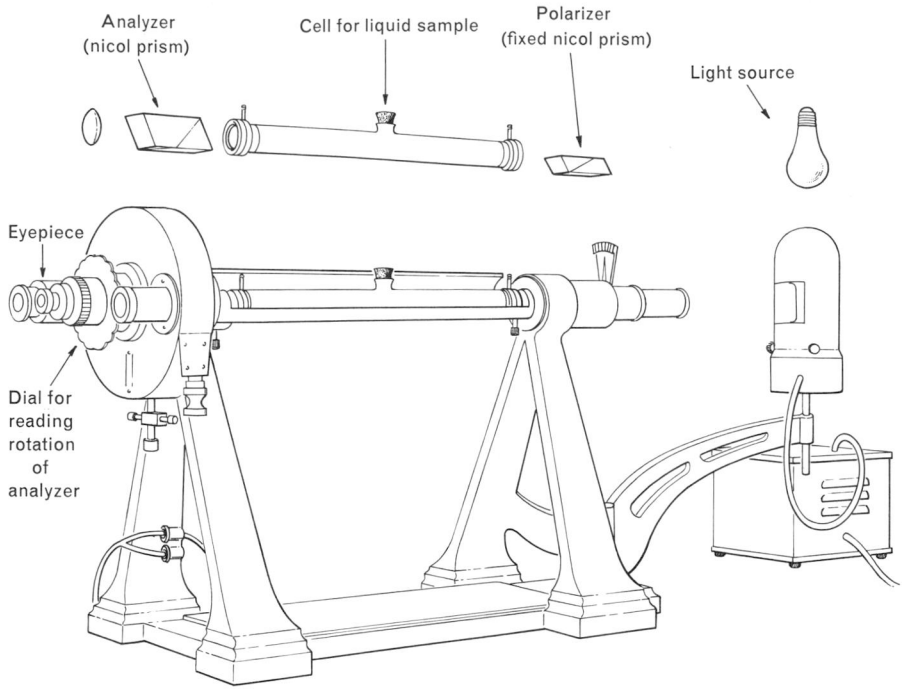

**Figure 16–2.** A polarimeter: schematic diagram, and drawing of an actual instrument.

The degree to which the plane of polarization is rotated by a given substance changes with the wavelength of the light used, and so monochromatic light (e.g., that produced by a sodium lamp or a mercury or cadmium arc) is used in measurements of rotation.

With any given optically active substance, the actual rotation in degrees is proportional to the concentration of the solution and the length of the light path through the solution.

The actual rotation, $\alpha$, is therefore

$$\alpha = kcl$$

If the concentration, $c$, is in grams per cubic centimeter; and if $l$, the length of the tube that contains the sample, is measured in decimeters, the constant, $k$, is called the "specific rotation," for which the symbol $[\alpha]$ is used. Then

$$k = [\alpha] = \frac{\text{observed rotation in degrees} \times \text{volume in cc}}{\text{length in dm} \times \text{weight of substance in g}}$$

Optical rotations are reported as *specific rotations*, the temperature and wavelength of the light used being reported as in the following example; and, in addition, the concentration and solvent used are usually given:

$$[\alpha]_D^{20} = +18.5° \ (c = 0.011 \ \text{g/cc, water})$$

This means that at 20°C, using the D line of sodium, a solution of 1.10 g of a substance in 100 ml of water had a specific rotation of 18.5° to the right (dextrorotatory). The concentration and solvent are given because with many compounds different rotations are observed with changes in concentration, and with different solvents.

## 16.9 Optical activity and molecular structure

Optical activity was first observed in 1811, when it was found that crystals of certain substances—notably quartz—existed in enantiomorphic forms, one of which rotated the plane of polarized light to the right, the other to the left. Enantiomorphic (Greek *enantios*, opposite; *morph*, form) objects are those which are *nonidentical mirror images*; the right- and left-handed members of a pair of gloves are enantiomorphs; the pairs of objects pictured in Figure 16-3 are enantiomorphic pairs.

The discovery in the early years of the nineteenth century of the optical activity of quartz was soon followed by observations of optical activity in solutions of organic compounds (e.g., sugar, tartaric acid) and liquid substances (turpentine). Since not all forms of silica are optically active, the activity of quartz appears to reside in the particular crystal structure of this form of silica; this is borne out by the observation that sections properly cut from two enantiomorphic quartz crystals rotate the plane of polarized light oppositely. Furthermore,

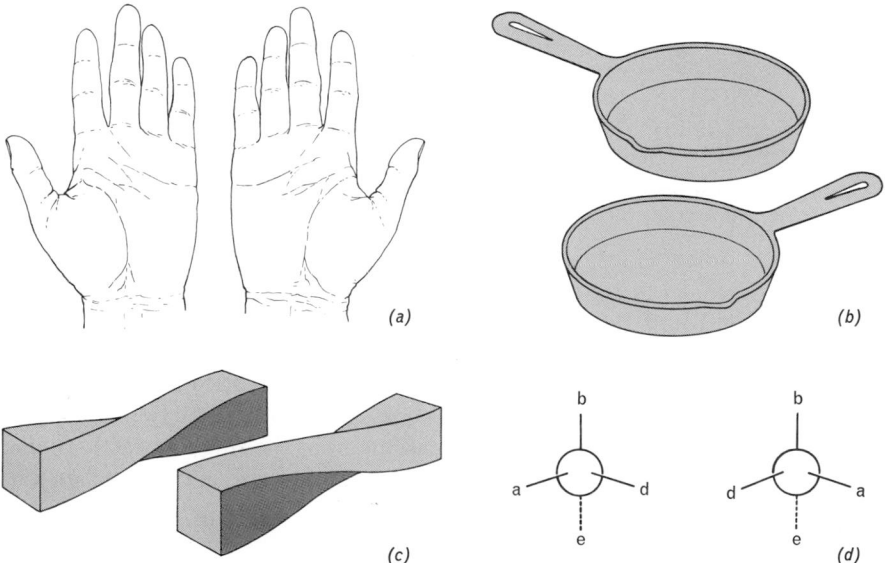

*(a)*

*(b)*

*(c)*

*(d)*

**Figure 16–3.** Enantiomorphic pairs: nonidentical mirror images.

since a homogeneous liquid or solution does not have the discrete structure possessed by a crystal, it must be concluded that the structural requirements for optical activity in these cases are to be found in the *arrangement of the atoms in the individual (solute or pure liquid) molecules.*

The discoveries of many optically active compounds continued throughout the first three-quarters of the nineteenth century, but until van't Hoff and Le Bel proposed their theory of the tetrahedral arrangement of the valence bonds of carbon in 1874, no satisfactory explanation for the phenomenon existed.

Van't Hoff and Le Bel observed that *in every known case of optical isomerism the compound exhibiting this property contained at least one carbon atom to which four different atoms or groups were joined;* e.g.,

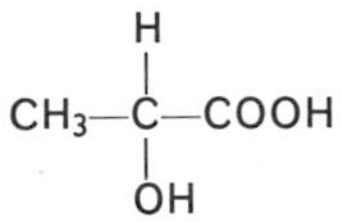

$$\begin{array}{cccc}
\overset{\displaystyle H}{\underset{\displaystyle OH}{CH_3-C-COOH}} &
\overset{\displaystyle COOH}{\underset{\displaystyle CH_2COOH}{HO-C-H}} &
\overset{\displaystyle CH_3}{\underset{\displaystyle COOH}{NH_2-C-H}} &
\overset{\displaystyle H}{\underset{\displaystyle CH_2OH}{CH_3-C-CH_2CH_2}}
\end{array}$$

|          |          |          |                    |
|:--------:|:--------:|:--------:|:------------------:|
| lactic acid | malic acid | alanine | "active" amyl alcohol |

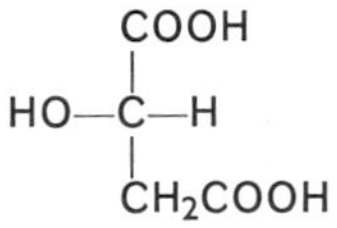

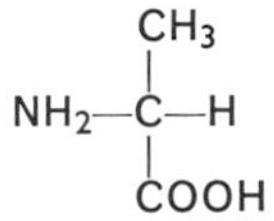

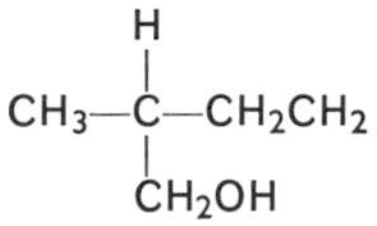

It is significant that at the time of van't Hoff and Le Bel all of the known optically active compounds had been isolated from natural (plant or animal) sources.

The following figures show that for a compound to exist in enantiomorphic forms (I, II), it cannot have two identical groups attached to one carbon atom, as in III and IV:

|          |          |          |          |
|:--------:|:--------:|:--------:|:--------:|
| I        | II       | III      | IV       |

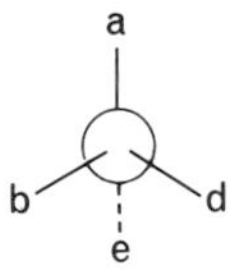

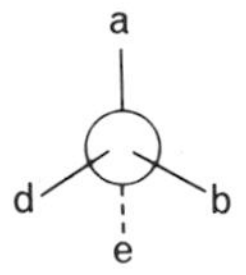

*enantiomorphs*          *not enantiomorphs*

I is the mirror image of II, and the two are *not* superimposable.
*Cabde*

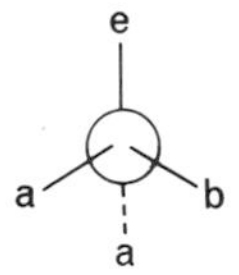

III is the mirror image of IV, but the two *are* superimposable (i.e., identical).
*Caabe*

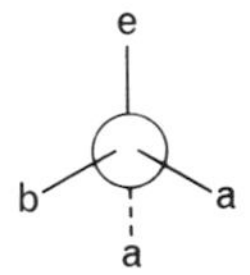

Examination of I and II discloses that these compounds lack any element of symmetry. Compound III ($=$IV) possesses a plane of symmetry that bisects the groups *b* and *e*.\* If groups *b* and *e* are themselves symmetrical, the compound *Caabe* is symmetrical. Groups such as —$CH_3$, —$COOH$, —$NH_2$, —$CH_2OH$, are "symmetrical" because of the free rotation about the single bond that joins them to the rest of the molecule. Thus, if such groups are represented in the

---

\* The only other element of symmetry we shall find it necessary to consider is the *center of symmetry;* an example of this is found in α-truxillic acid (Section 17–18).

following way, the page upon which they are drawn represents the plane of symmetry:

*Models showing symmetry of* —$CH_3$, —$NH_2$, —$COOH$, *and* —$CH_2OH$ *groups.*

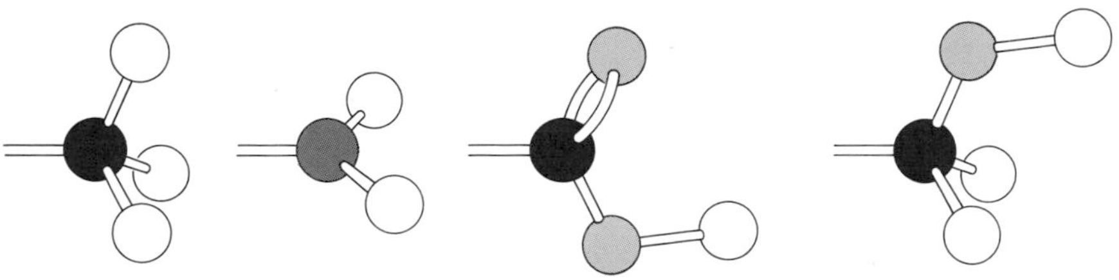

| —CH₃ | —NH₂ | —COOH | —CH₂OH |

Consequently, we can ignore such groups (or regard them as equivalent to symmetrical spheres) in estimating stereochemical possibilities in a structure.

A study of models of the following will show that all compounds of the types $Ca_4$, $Ca_3b$, $Ca_2b_2$, $Ca_2bd$ have a plane of symmetry (providing, of course, that none of the substituents, $a$, $b$, and so on, is itself dissymmetrical):

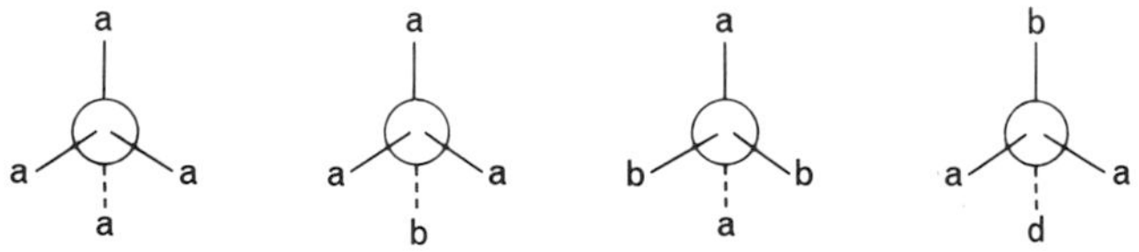

All of these are drawn so that the plane of symmetry is vertical and at right angles to the page, passing through the groups at top and bottom of each model.

The carbon atom in (I) (and II) is called an *asymmetric carbon atom*, and the presence of a single asymmetric carbon atom (in general, the carbon atom in *Cabcd*) is a sufficient—*but not a necessary*—condition for optical activity. *The necessary condition for optical activity is molecular asymmetry: the existence of two isomers as nonsuperimposable forms which bear the relationship of an object to its mirror image.* The *most frequent* condition in which the mirror image and object are enantiomorphic is encountered in the case of compounds possessing one or more *asymmetric carbon atoms*.

---

**Exercise 4.** What kind of isomerism would be expected if in the compound *Caabd* the carbon atom were *(a)* planar, with the four bonds at 90°; and *(b)* pyramidal, with the carbon atom at the apex of a regular pyramid?

---

## 16.10 The Fischer transformation

Fischer showed by experimental means that if any two groups in the compound *Cabde* are interchanged, the enantiomorph results:

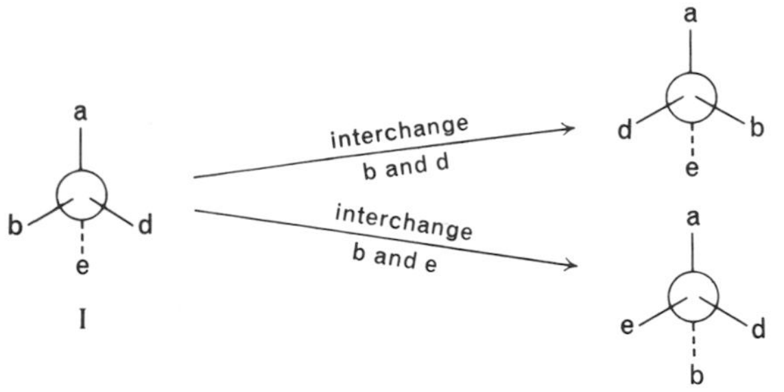

Fischer's actual experiment was as follows:

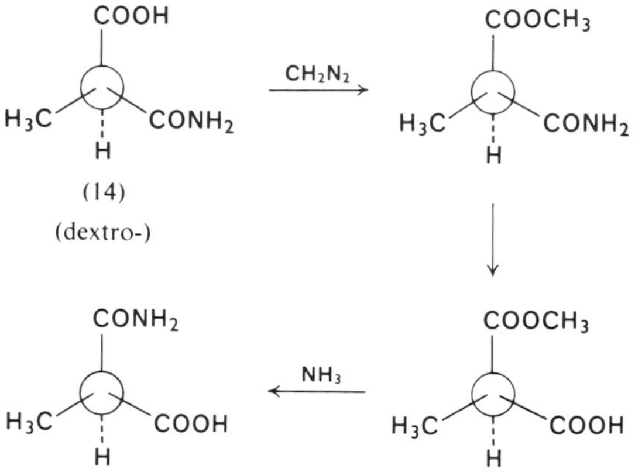

(14)
(dextro-)

(15)
(levo-)

---

**Exercise 5.** Convince yourself that (14) and (15) are enantiomorphs. (Models should be constructed.)

---

The Fischer transformation is a useful device in manipulating formulas; its actual performance in the laboratory is rare. Suppose, for example, we wish to decide whether (16) has the same configuration as (17),

(16)                    (17)

or whether these are enantiomorphic structures. This can be discerned by actually constructing the models (but this is often inconvenient), or by drawing three-dimensional projections. The use of the Fischer transformation is a simple method of arriving at an answer:

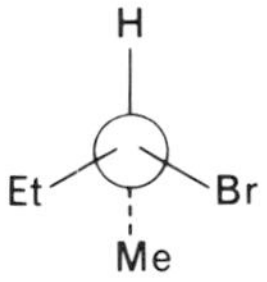

Each change, (16) → (18), (18) → (19), (19) → (17), is made by transposing one pair of the substituents; thus, (16) and (18) are enantiomorphs, but (16) and (19) are identical. Therefore, (16) and (17) are enantiomorphs.

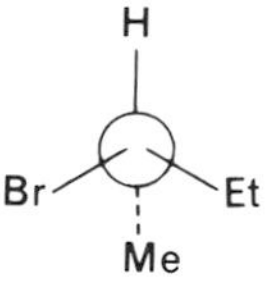

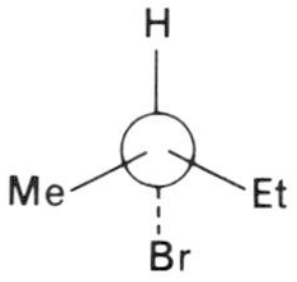

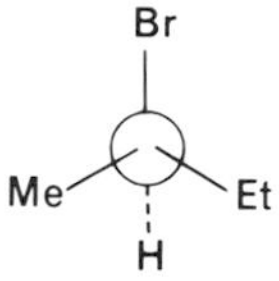

---

**Exercise 6.** Select the identical and the enantiomorphic forms of the following:

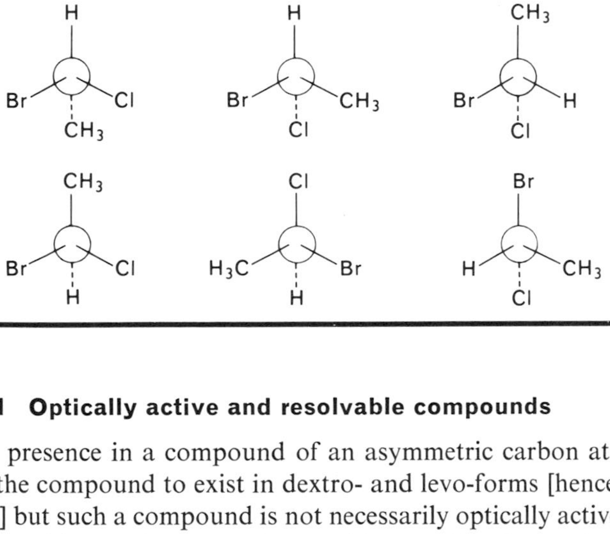

---

## 16.11 Optically active and resolvable compounds

The presence in a compound of an asymmetric carbon atom makes it *possible* for the compound to exist in dextro- and levo-forms [henceforth called (+) and (−)] but such a compound is not necessarily optically active: the actual material prepared by synthesis may be—and in practice almost always is—an equimolar mixture of (+) and (−) forms, and therefore optically inactive. As an example, let us consider the bromination of propionic acid to produce α-bromopropionic acid:*

$$CH_3CH_2COOH \xrightarrow[\text{Br}_2]{\text{red phosphorus}} CH_3\overset{*}{C}HCOOH$$
$$\phantom{CH_3CH_2COOH \xrightarrow[\text{Br}_2]{\text{red phosphorus}} CH_3C}Br$$

---

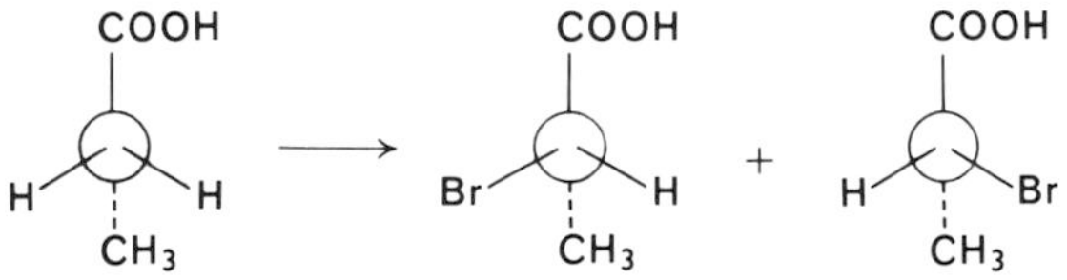

* The symbol * denotes an asymmetric carbon atom.

Since the two hydrogen atoms in the $CH_2$ group of propionic acid are identically situated with respect to the methyl and carboxyl groups (they are, in fact, on opposite sides of the plane of symmetry), there is no distinction between them so far as replacement by bromine is concerned, and consequently the enantiomorphs are both produced in equal amount. *The product of the reaction, α-bromopropionic acid, is therefore optically inactive,* since the dextrorotatory power of one of the enantiomorphs is exactly counteracted by the levorotatory power of the other.

---

**Exercise 7.** With the use of appropriate projection formulas, show how the (±) mixture is formed by *(a)* the addition of ethylmagnesium bromide to acetaldehyde; *(b)* the addition of HBr to 1-pentene.

---

The mixture of (+)- and (−)-α-bromopropionic acids (called a *racemic mixture*; see Section 16.12) can be separated by chemical means into the separate optically active compounds. This process is called "resolution", and is described in Section 16.17. Any compound that contains an asymmetric carbon atom, or other element of molecular dissymmetry, is resolvable, and the term "resolvable" is often used as a way of denoting that a compound does contain such an element of dissymmetry. For example, if 2-butanol is prepared by ordinary means, as by reduction of butanone, the product will consist of equal numbers of (+) and (−) molecules; it is optically inactive, but it is resolvable.

## 16.12 Characteristics of optical isomers

When an asymmetric carbon atom is present in a molecule, the compound may exist as the (+), (−), or (±) form. The (±) form is called the racemic form or racemic modification, and, if crystalline, may be a *racemic mixture*, a *racemic compound*, or a *racemic solid solution*.

The racemic mixture is a mechanical mixture of the separate crystals of the (+) and (−) forms. The crystals of the enantiomorphs may be, but are not necessarily, themselves enantiomorphic. If the crystals are enantiomorphic it is sometimes possible to separate the (+) and (−) forms by picking out the different crystals by hand. This procedure, which is tedious and seldom practicable, accomplishes a *resolution* of the (±) compound. It is rarely used; chemical methods of resolution are nearly always employed.

A racemic solid solution forms one kind of crystal only. Mechanical resolution is therefore impossible. A racemic compound forms crystals which are identical and which contain equal amounts of each enantiomorph. It is always difficult and usually impossible to tell by inspection of the crystals whether a racemic form is a racemic compound, mixture, or solid solution. The question can be answered, however, by a study of the melting-point behavior of the substance. The melting-point diagrams shown in Figure 16-4 are those of the several

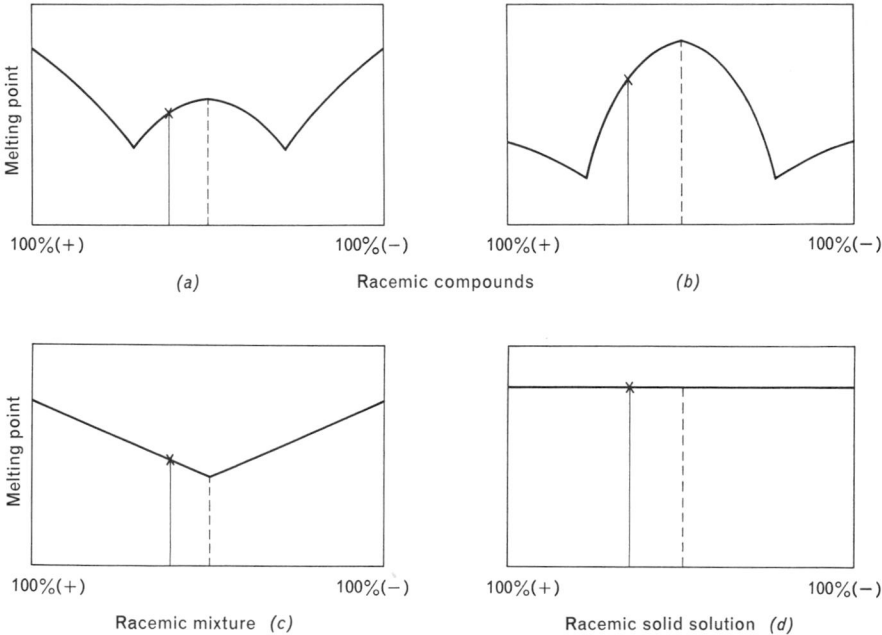

**Figure 16–4.** Melting point-composition diagrams of racemic substances. Point marked by x shows melting points of mixtures of (±) form and some added (+) form.

forms of racemic substances. If the addition of a small amount of the pure (+) or (−) form of the compound lowers the melting point (curves *a* and *b*), the racemate is a racemic compound; if the melting point rises (curve *c*), it is a racemic mixture; and if the melting point remains unchanged (curve *d*), the (±) form is a racemic solid solution.

The answer to the question of the "number of isomers" that are possible for a compound containing an asymmetric carbon atom must be arbitrary, and a matter of terminology. The two enantiomorphs are optical isomers, the (+) and the (−). The (±) modification may or may not have unique physical properties; its melting point may be the same as, or different from, that of the separate enantiomorphs (which have, of course, identical physical properties).

## 16.13 Two asymmetric carbon atoms

A compound such as 3-chloro-2-butanol (the chlorohydrin formed by the addition of HOCl to 2-butene) has two asymmetric carbon atoms:

$$CH_3-\overset{*}{C}H-\overset{*}{C}H-CH_3$$
$$\phantom{CH_3-}\underset{OH}{|}\phantom{-CH}\underset{Cl}{|}$$

Projection formulas for the possible optical isomers of this may be drawn as follows:

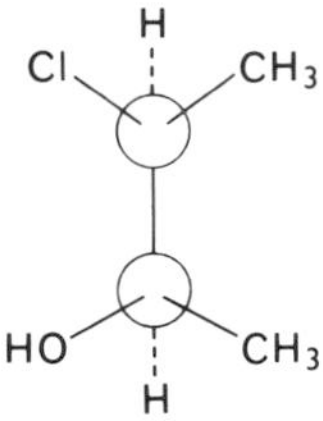

(20) 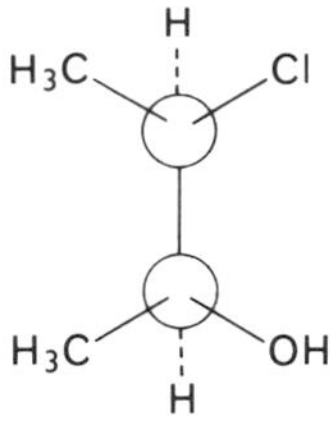    (21) 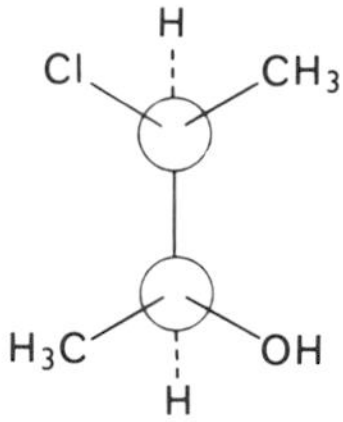    (22) 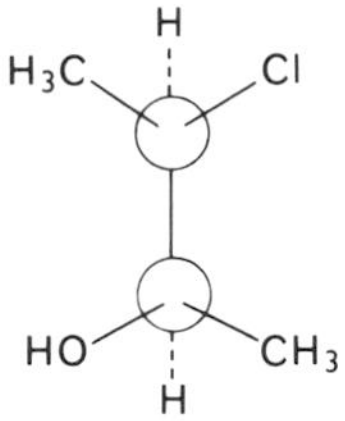    (23)

(±) pair₁    (±) pair₂

A useful convention for the representation of three-dimensional structures of this kind is shown in the following diagrams. It is to be emphasized that *this convention requires that the horizontal lines represent bonds projecting forward toward the viewer*; the vertical lines project away from the viewer. Formulas (20) through (23) are thus represented as follows:

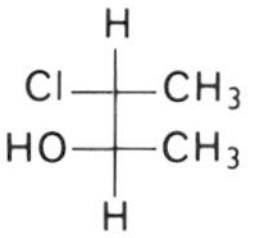

$$
\begin{array}{cccc}
\text{H} & \text{H} & \text{H} & \text{H} \\
\text{Cl}-\!\!-\text{CH}_3 & \text{H}_3\text{C}-\!\!-\text{Cl} & \text{Cl}-\!\!-\text{CH}_3 & \text{H}_3\text{C}-\!\!-\text{Cl} \\
\text{HO}-\!\!-\text{CH}_3 & \text{H}_3\text{C}-\!\!-\text{OH} & \text{H}_3\text{C}-\!\!-\text{OH} & \text{HO}-\!\!-\text{CH}_3 \\
\text{H} & \text{H} & \text{H} & \text{H} \\
(20) & (21) & (22) & (23)
\end{array}
$$

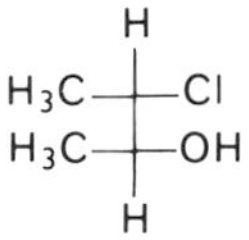

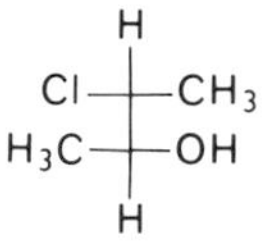

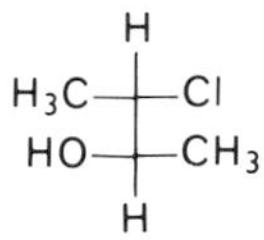

There are thus four optically active forms of 3-chloro-2-butanol. This can be shown in another way. If we call one of the asymmetric carbon atoms $A$ and the other $B$, the following combinations are possible, since each asymmetric carbon atom can exist in a $(+)$ and a $(-)$ configuration:

$$
\begin{array}{ccccc}
 & & \overbrace{\hspace{3cm}}^{(\pm)\,\text{pair}} & & \\
\text{CH}_3\text{CHCl} \quad (A) & +A & +A & -A & -A \\
\text{CH}_3\text{CHOH} \quad (B) & +B & -B & +B & -B \\
 & & \underbrace{\hspace{3cm}}_{(\pm)\,\text{pair}} & &
\end{array}
$$

Although it would take careful study to convince oneself that (24)–(27) are all different ways of writing the same compound, it is apparent from the $\pm A$, $\pm B$ treatment that *four isomers are all that are possible*, and thus that $(24)=(25)=(26)=(27)$ is one of the four, (20)–(23):

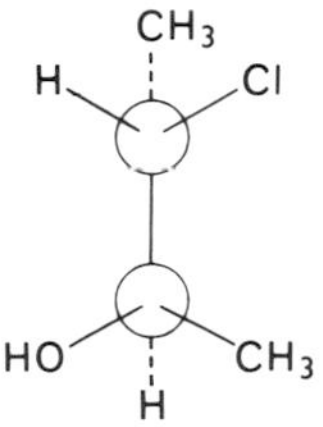

(24) 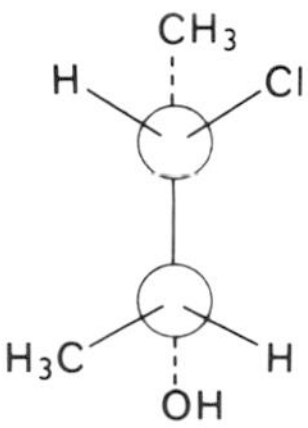    (25) 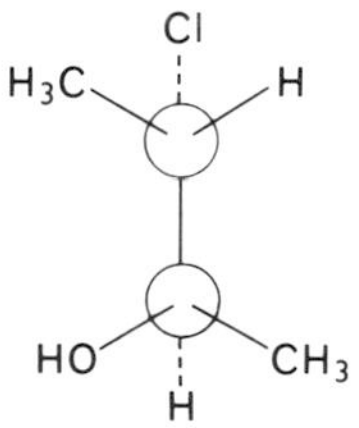    (26) 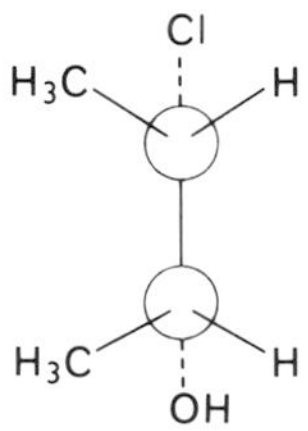    (27)

**Exercise 8.**   Which of the four isomers (20)–(21) is represented by the one drawn in the four representations (24)–(27)?

**Exercise 9.**   Which of the two ($\pm$) pairs (20)–(21) and (22)–(23) is the one that is formed by the addition of HOCl to *cis*-2-butene?

## 16.14   Three asymmetric carbon atoms

When *three* asymmetric carbon atoms are present, as in

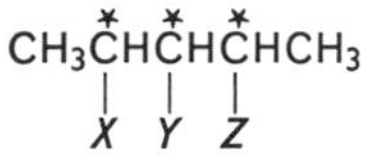

$$CH_3\overset{*}{C}H\overset{*}{C}H\overset{*}{C}HCH_3$$
$$\quad\;\; | \quad\; | \quad\; |$$
$$\quad\;\; X \quad Y \quad Z$$

the following projection formulas can be written

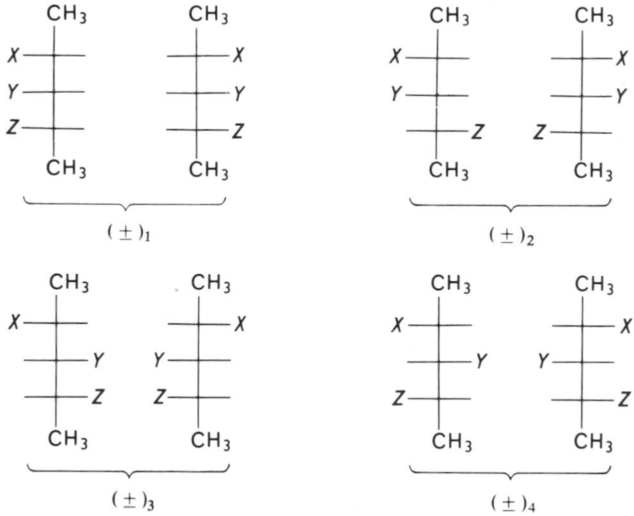

$$(\pm)_1$$

$$(\pm)_2$$

$$(\pm)_3$$

$$(\pm)_4$$

Calling the three asymmetric carbon atoms *A*, *B*, and *D*, we have

| $+A$ | $-A$ | $+A$ | $-A$ | $+A$ | $-A$ | $+A$ | $-A$ |
|------|------|------|------|------|------|------|------|
| $+B$ | $-B$ | $+B$ | $-B$ | $-B$ | $+B$ | $-B$ | $+B$ |
| $+D$ | $-D$ | $-D$ | $+D$ | $-D$ | $+D$ | $+D$ | $-D$ |
| $(\pm)_1$ | | $(\pm)_2$ | | $(\pm)_3$ | | $(\pm)_4$ | |

It is apparent that all the possible combinations are represented in this table.

With four asymmetric carbon atoms, sixteen optically active isomers can exist (eight ($\pm$) pairs), and for $n$ asymmetric carbon atoms there are $2^n$ possible optical isomers.

**Exercise 10.**   How many optically active isomers can exist for each of the following?

(a) 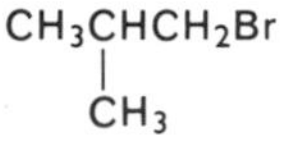 CH₃CHCH₂Br
    |
    CH₃

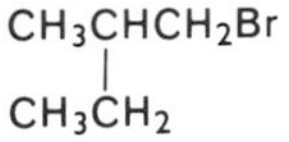

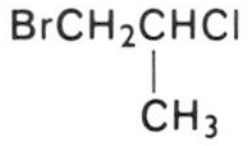

(d) 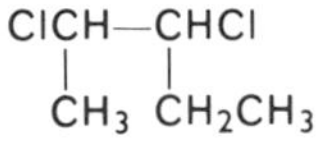 ClCH—CHCl
    |    |
    CH₃  CH₂CH₃

(b) CH₃CHCH₂Br
    |
    CH₃CH₂

(e) 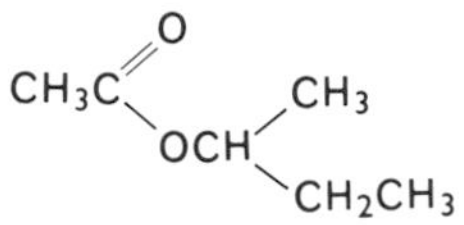 CH₃C(=O)(OCH)—CH₃, CH₂CH₃

(c) BrCH₂CHCl
    |
    CH₃

(f)  cyclopropane-1,2-dicarboxylic acid (*cis* and *trans*).

## 16.15   Diastereomers

The compounds (21) and (23) are stereoisomers, but they are not enantiomorphs. The same is true of the compounds $+A+B+D$, $+A+B-D$, $+A-B-D$, $+A-B+D$ in the preceding section.

This relationship between nonenantiomorphic optical isomers is called "diastereoisomerism," and the isomers are called *diastereoisomers* or more commonly, *diastereomers.*

Diastereomers are usually different in physical properties and often show marked differences in chemical reactivity. *Cis-* and *trans-* 2-chlorocyclohexanol are diastereomers:

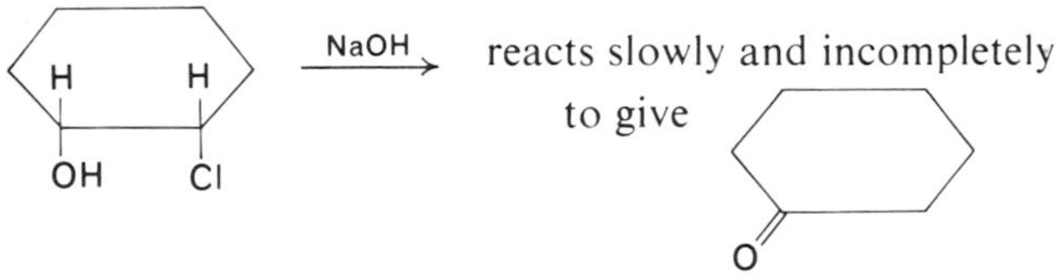

(28)

cyclohexanone

(29)

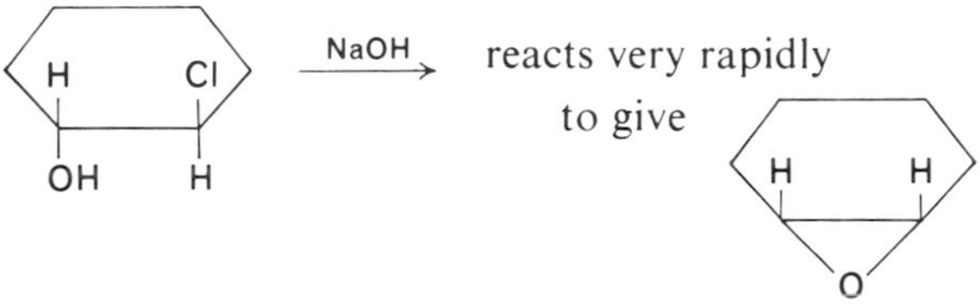

cyclohexene oxide
(1,2-epoxycyclohexane)

Each of the diastereomeric cyclohexene chlorohydrins exists in (+)- and (−)-modifications (of which only one is shown in the above formulas for each isomer). The (+)-*cis*-chlorohydrin, its (−) isomer, and the (±) mixture all react with alkali at identical rates to give the same product, cyclohexanone. The *trans* forms all react at identical rates (but different from that of the *cis* forms) to give the same product, cyclohexene oxide.

The difference in the reactivity of the diastereomeric 2-butene chlorohydrins has been discussed in detail in an earlier chapter (Section 9.18). See also Section 16.3.

In the cyclohexene chlorohydrins, the rigidity of the ring system requires that relative dispositions of the —OH and —Cl groups must remain *cis* or *trans*, and the *cis* compound *cannot* assume the favorable configuration required for rearside attack by oxygen on the carbon atom holding —Cl. The *trans* compound possesses this favorable configuration, and thus yields the oxide very readily. In this case the *cis* compound actually undergoes a different *kind* of reaction from that undergone by its diastereomer.

Diastereomeric compounds can be separated by the usual physical methods such as distillation or fractional crystallization. It should be stressed again that when diastereomeric compounds result from a chemical reaction in which optically inactive reactants are used, the products are optically inactive: each of the diastereomers is a ($\pm$)-mixture, and if optically active compounds are desired the mixtures must be resolved.

---

**Exercise 11.** How many actual products can be isolated from the reaction mixture resulting from each of the following operations? *(a)* addition of bromine to *cis*-2-butene, *(b)* addition of HOCl to *trans*-2-pentene.

**Exercise 12.** How many chlorohydrins would be formed by the addition of HOCl to 1-butene? Would they differ in their rates of reaction with alkali to give 1,2-epoxybutane [2-ethyloxirane *(G)*]?

**Exercise 13.** *(a)* What would be the result of reducing ($\pm$)-3-methyl-2-pentanone to the secondary alcohol? How many products would be formed? *(b)* What would be the result if (+)-3-methyl-2-pentanone were used?

---

## 16.16 "Similar" asymmetric carbon atoms

Tartaric acid contains two asymmetric carbon atoms, but to each of them are attached the same four substituents: H, OH, COOH, and $-CH{<}^{OH}_{COOH}$ . The four apparent ways of arranging the groups in space can be represented as follows:

| COOH | COOH | COOH | COOH |
|:---:|:---:|:---:|:---:|
| H——OH | HO——H | H——OH | HO——H |
| HO——H | H——OH | H——OH | HO——H |
| COOH | COOH | COOH | COOH |
| (30) | (31) | (32) | (33) |

It can be seen that (30) and (31) are indeed enantiomorphs and represent $(+)$ and $(-)$ forms; but (32) and (33) are not enantiomorphs: they are identical, as can be seen by rotating one of them $180°$ in the plane of the page. Compound (32) $(=33)$ is not optically active: it possesses a plane of symmetry. This compound is called *meso*-tartaric acid.

If the top asymmetric carbon atom is called $A$, the lower, which carries the same substituents, must also be designated as $A$. The $(+)$, $(-)$, and *meso* forms can be represented

$$
\begin{array}{cc}
+A \quad -A & +A \quad -A \\
+A \quad -A & -A \quad +A \\
\underbrace{\qquad\qquad} & \underbrace{\qquad\qquad} \\
(\pm) & \textit{meso}
\end{array}
$$

In compounds such as 2,3,4-pentanetriol the central carbon atom is called a "pseudoasymmetric" carbon atom. An examination of the various possible configurations of the hydroxyl groups in this triol will provide a good exercise in determining the number of isomers that can exist. Let us first write down in a systematic way all possible configurations of the three CHOH groups:

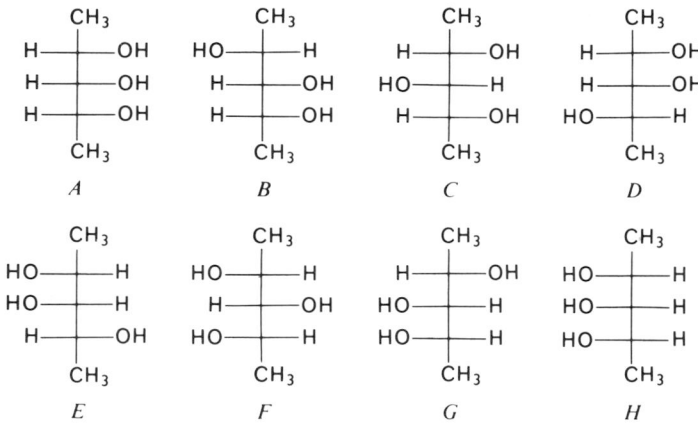

How many of these represent identical compounds? It will be remembered that, according to the convention we use for writing these formulas, we can turn any of them through $180°$ in the plane of the page (but not out of that plane!). Thus, we see that $A$ and $H$ are the same; so are $B$ and $E$, $C$ and $F$, and $D$ and $G$. We are left with only four different configurations from the above set of eight:

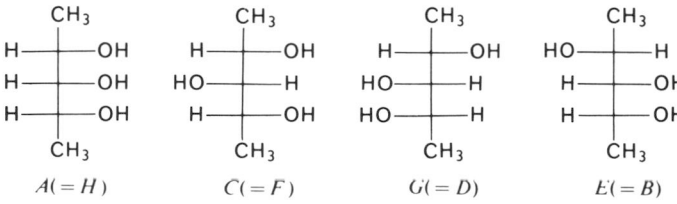

Since the mirror image of *H* is *A*, and these are identical, *H* (=*A*) must be a symmetrical compound. It is a *meso* form. The same is true of *F* (=*C*); this is another *meso* form. In these *meso* compounds it will be noted that a plane of symmetry bisects the molecule through the central —CHOH group.

The pair *G* and *E* is an enantiomorphic pair. Neither has an element of symmetry; they are mirror images, and they are not identical.

Thus, there are four 2,3,4-pentanetriols: two different *meso* compounds, one dextrorotatory compound, and its enantiomorphic levorotatory compound.

It can be seen that if the two groups at the ends of the molecule were not identical, all eight of the configurations corresponding to *A–H* would be different. In this case there would be three asymmetric carbon atoms, and thus $2^n = 8$ possible isomers.

---

**Exericse 14.** Draw all of the possible configurations of 2,3,4-tribromohexane, and find the enantiomorphic pairs. Are there any *meso* forms?

---

With some practice in visualizing formulas as three-dimensional projections the student will find it easy to discern whether elements of symmetry are present. For instance, the following represent nonresolvable *(meso)* compounds:

---

**Exercise 15.** Find the plane of symmetry in each of the above compounds.

---

## 16.17  Resolution of optical isomers*

The separation of a mixture of two enantiomorphs into the separate *d-* and *l-*isomers is called resolution. Since *d-* and *l-*isomers have identical melting points, solubilities, ionization constants (if acids or bases), and since all other physical properties are identical (except the direction in which they rotate the plane of

---

* The symbols *d-* and *l-* are often used in place of (+) and (−) in discussing optical isomerism. Present-day custom tends toward the use of the (+)-(−) symbolism, but it is sometimes more convenient to use the alternative notation. Since the two symbols mean the same thing either may be chosen. The reason for the abandonment of the *d-l* notation is that another kind of configurational notation, in which the letters D- and L- are used, is also employed. While *d-* stands for "dextrorotatory," D- does not. The use of D- and L- will be encountered in Section 16.21.

**Table 16-2.** *Methods of resolution of racemic substances*

| *dl*-Pair | | Resolving Agent | | Diastereomers |
|---|---|---|---|---|
| 1. | *d*-acid<br>*l*-acid | + *d*-base | → | salt: *d*-acid-*d*-base<br>salt: *l*-acid-*d*-base |
| 2. | *d*-amine<br>*l*-amine | + *l*-acid | → | salt: *d*-amine-*l*-acid<br>salt: *l*-amine-*l*-acid |
| 3. | *d*-alcohol<br>*l*-alcohol | + *d*-acid | → | ester: *d*-acid-*d*-alcohol<br>ester: *d*-acid-*l*-alcohol |
| 4. | *d*-alcohol<br><br>*l*-alcohol | + anhydride<br>or chloride<br>of *d*-acid | | as in 3 |
| 5. | *d*-ester acid<br>*l*-ester acid | + *d*-base | → | as in 1 |
| 6. | *d*-ketone<br>*l*-ketone | + *l*-hydrazide | → | *d*-ketone-*l*-hydrazone<br>*l*-ketone-*l*-hydrazone |

polarized light), they cannot be separated by distillation, crystallization, differential partition between solvents, or by any other such means. A rare exception is encountered in the case of those crystalline racemic mixtures which form hemihedral, enantiomorphic crystals that can be separated by hand.

Since diastereomers can be separated by physical means, the conversion of enantiomorphs into *diastereomeric derivatives* allows their separation. Subsequent decomposition of the separated derivatives allows the recovery of each enantiomorph in the optical active state.

The selection of the particular means of converting the *dl*-mixture into a mixture of diastereomers depends upon the chemical nature of the compound to be resolved and the physical properties of the resulting diastereomers. It is usually desirable to prepare derivatives such that one or both diastereomers are crystalline compounds. General examples of resolution procedures are given in Table 16-2.

That *l-A-l-B* and *l-A-d-B* (as in example 2, Table 16-2, where $A=$ acid, $B=$ base) are not enantiomorphic can be appreciated when it is recognized that the mirror image of *l-A-l-B* is *d-A-d-B*. This is also shown in the case of example 3; if we write the projection formulas for the two esters, we have ester $\begin{cases} d\text{-acid} \\ l\text{-alcohol} \end{cases}$      ester $\begin{cases} d\text{-acid} \\ d\text{-alcohol} \end{cases}$

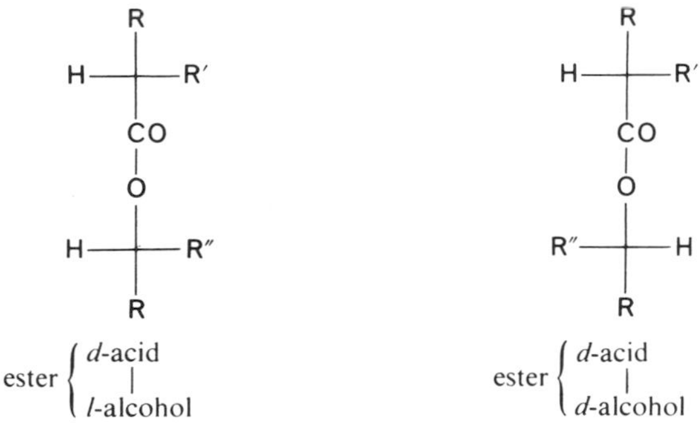

Since salts are readily prepared, are usually crystalline, and are easily reconverted into their component acids and bases, the most widely used methods of resolution are variations of example 1 in Table 16-2. Compounds which are not acids can often be converted into acidic derivatives, these resolved, and the separate diastereomers then reconverted into the original compound to give the separate enantiomers. For example, *sec*-butyl alcohol can be resolved by means of the following scheme:

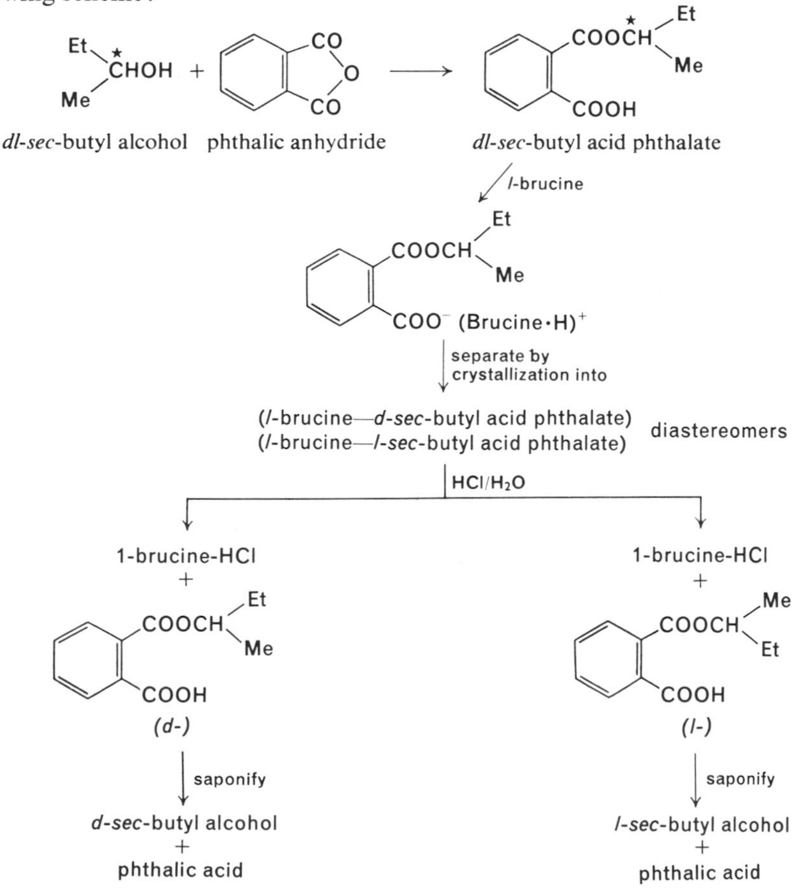

*dl-sec*-butyl alcohol   phthalic anhydride       *dl-sec*-butyl acid phthalate

*l*-brucine

separate by crystallization into

(*l*-brucine—*d-sec*-butyl acid phthalate)  diastereomers
(*l*-brucine—*l-sec*-butyl acid phthalate)

$HCl/H_2O$

1-brucine-HCl
+

(*d-*)

saponify

*d-sec*-butyl alcohol
+
phthalic acid

1-brucine-HCl
+

(*l-*)

saponify

*l-sec*-butyl alcohol
+
phthalic acid

## 16.18 Racemization

Racemization is the process whereby an optically-active (*d-* or *l-*) compound is converted into the mixture of *d-* and *l-*forms, with the result that optical activity disappears and the observed rotation drops to zero. If an optically active compound enters into an equilibrium in which it forms, and is reformed from, a symmetrical compound, racemization takes place.

Compounds in which the asymmetric carbon atom contains a readily ionizable hydrogen atom are usually racemized with ease by the action of alkali. The enolic form that is present in the equilibrium mixture (even if in very small concentration) is a symmetrical molecule, and upon reversion to the ketone will return to both enantiomorphic configurations. For example, optically active 3-methyl-2-pentanone can racemize in the following way:

|  (+) form  |  enol (not asymmetric)  |  (−) form  |

Racemization can occur by other mechanisms than this, but whatever the manner in which it takes place, the existence of a symmetrical intermediate can be found to account for the disappearance of optical activity.

Many compounds are optically very stable. Optically active 3-methylhexane would not be expected to undergo ready racemization, for this would involve the breaking of strong bonds. Such bond breaking would not be expected to occur under conditions short of those that would bring about extensive destruction of the compound. On the other hand, compounds in which the asymmetric carbon atom bears a readily ionizable hydrogen atom, as in the case of the ketone discussed above, usually racemize with ease.

## 16.19 Optical isomerism without an asymmetric carbon atom

Optical activity depends upon an enantiomorphic relationship between the *d-* and *l-*forms. Although by far the greatest number of known optically active compounds owe their asymmetry to the presence of an asymmetric carbon atom, optical activity can exist, in the absence of an asymmetric carbon atom, as a result of *asymmetry of the molecule as a whole*, or as a result of the *asymmetry of an atom other than carbon*.

Molecular asymmetry is found in allenes of the type RCH=C=CHR, the enantiomorphic forms of which are shown in Figure 16-5.

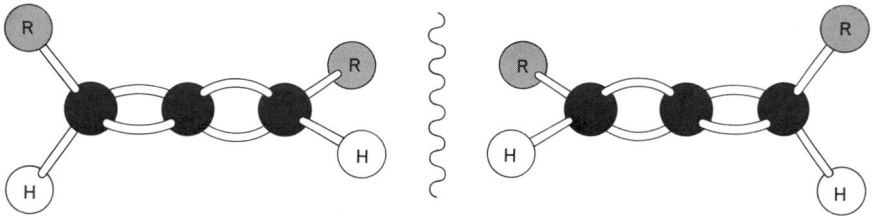

**Figure 16–5.** Enantiomorphic allenes, RCH=C=CHR.

---

**Exercise 16.** Would the allene, $(CH_3)_2C=C=CHCH_2CH_3$ be resolvable?

---

Molecular asymmetry analogous to that possessed by properly constituted allenes is also observed in bicyclic (spirane) systems of the following type:

H, H     H, H<br>
R, R     R, R<br>
enantiomorphic spiranes

In the case of both *cis*- and *trans*-1,4,dimethylcyclohexane the carbon atoms bearing the methyl groups resemble the usual asymmetric carbon atoms. They are not in fact asymmetric, however, since compounds bearing symmetrical substituents in only the 1,4-positions of the cyclohexane ring are not resolvable because they possess a plane of symmetry passing through, and perpendicular to, the ring at the 1 and 4 positions.

*Stereoisomerism due to atoms other than carbon. Oximes.* The oximino grouping $>C=N-OH$ can assume two stereochemical forms:

R, OH     R,<br>
   C=N      C=N,<br>
R′        R′    OH<br>
(34)        (35)

and many oximes exist in two forms, called *syn* and *anti*, corresponding to *cis* and *trans*. In (34), the hydroxyl group is *syn* with respect to R; in (35), it is *anti* with respect to R and *syn* with respect to R′. Assume R and R′ are symmetrical.

---

**Exercise 17.** Would either (34) or (35) be resolvable into enantiomorphic forms? Illustrate with suitable perspective drawings, and explain.

---

*Optical activity in substituted biphenyls.* Optical isomerism which is the result of restricted rotation about a single bond is encountered in certain appropriately substituted biphenyl derivatives. The following generalized structures show the nature of the isomerism that has been been observed in compounds of this type:

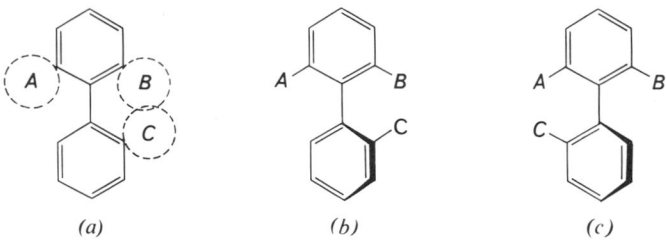

$(a)$ $(b)$ $(c)$

If $A$, $B$, and $C$ are sufficiently bulky, the lower benzene ring cannot rotate freely with respect to the upper ring, and thus forms $(b)$ and $(c)$ can have separate existence, separated by an energy barrier whose height will be a measure of the ease with which substituent $C$ can pass $A$ or $B$. If at any time the two rings were to become coplanar, as in $(a)$, any movement from this position could as easily yield $(b)$ as $(c)$, and equal numbers of these two enantiomorphic forms would be formed, and racemization would result.

In Figure 16-6 is shown a diagram of the energy barriers to the rotation of three hypothetical biphenyl derivatives:

1. Where $A$, $B$, $C$, and $D$ are all small in size (e.g., $H$ or $F$). In this case there is a small energy barrier for the passage of these substituents past each other, and rotation of the rings takes place with comparative ease.

2. Where $A$, $B$, and $C$ are large enough (with $D$ small) to restrict rotation by passage of $C$ past $A$ or $B$, but not to prevent it. Here the enantiomeric forms,

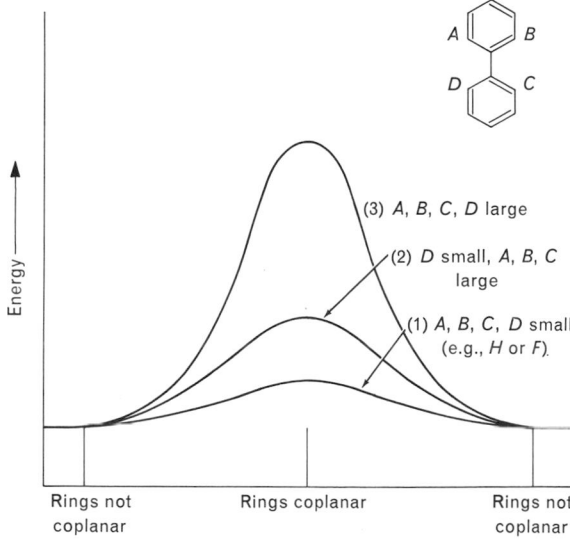

**Figure 16–6.**   Energy diagram tor rotation of a biphenyl about the single bond.

as *(b)* and *(c)* above, can be isolated, but racemization would be expected to occur at a measurable rate.

3. Where *A*, *B*, *C*, and *D* are all large groups whose bulk would prevent their passage by one another. The activation energy for racemization would be high, and the compounds would be optically stable. Racemization would occur very slowly unless sufficient energy were supplied, as by elevation of the temperature.

The restriction of rotation in biphenyl derivatives can also be recognized by the effect of substituents upon the ultraviolet absorption spectrum. Biphenyl itself, in which there is only a small energy barrier to coplanarity of the rings, shows a high-intensity absorption ($\epsilon = 20,000$) in the ultraviolet, whereas benzene absorbs at a rather low intensity ($\epsilon = 250$). The high intensity of absorption of biphenyl is ascribable to the fact that when the rings are coplanar, $\pi$-orbital overlap can encompass the complete system, and $\pi \rightarrow \pi^*$ transition takes place

**Table 16-3.** *Ultraviolet absorption data for some biphenyls*

| Compound | $\lambda_{max}$ | $\epsilon$ | Solvent |
|---|---|---|---|
| 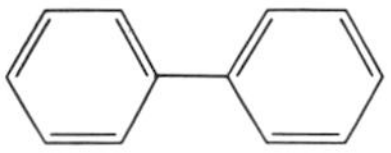 (biphenyl) | 250 | 20,000 | ethanol |
| 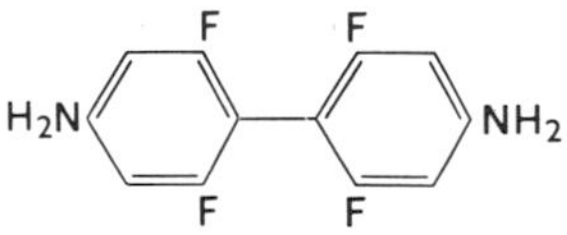 $H_2N$–(difluoro-biphenyl)–$NH_2$ (F, F, F, F) | 261 | 25,000 | ethanol |
| 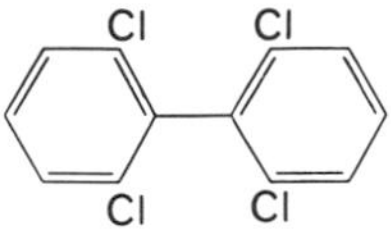 (tetrachloro-biphenyl) Cl, Cl, Cl, Cl | 275 | 630 | ethanol |
| 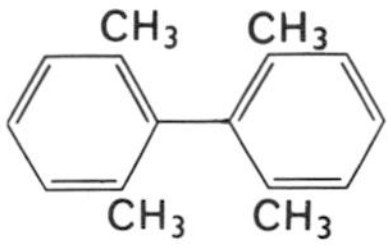 (tetramethyl-biphenyl) CH$_3$, CH$_3$, CH$_3$, CH$_3$ | 264 | 500 | ethanol |

Compare:

| | | | |
|---|---|---|---|
| 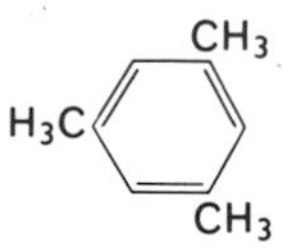 $H_3C$–(trimethylbenzene) CH$_3$, CH$_3$ | 266 | 320 | hexane |

with a relatively high extinction coefficient. When bulky substituents—such as *A*, *B*, *C*, and *D* in Figure 16-6(3)—are present, coplanarity is restricted and each ring absorbs independently, with only a small contribution from the other. The ultraviolet absorption data in Table 16-3 illustrate these effects.

## 16.20  Stereochemistry of amines and ammonium salts

The bond angles in ammonia and amines are found to have values near the tetrahedral value. For example, in dimethylamine the C—N—C bond angle is about $111 \pm 3°$, and triethylamine the C—N—C angle is about $113 \pm 3°$. It might be expected that tertiary amines having three different groups attached to nitrogen would be capable of existing in enantiomorphic forms. We can draw such a pair of enantiomorphs for methylethylamine:

The fact is that no such resolution has been accomplished. The explanation for the nonresolvability of tertiary amines of this kind is that the energy barrier to the interconversion of two such enantiomorphic species, through a planar (and thus symmetrical) intermediate state, is so low that the thermal energy at ordinary temperatures is sufficient to cause the interconversion. If we could carry out the necessary experiments at a sufficiently low temperature, resolution of a tertiary amine could probably be accomplished.

A tertiary amine that has been resolved is the complex compound known as Troeger's base. In this, the cage-like structure holds the molecule with sufficient rigidity such that the enantiomorphs shown in the following formulas can be obtained and preserved as separate, optically active compounds:

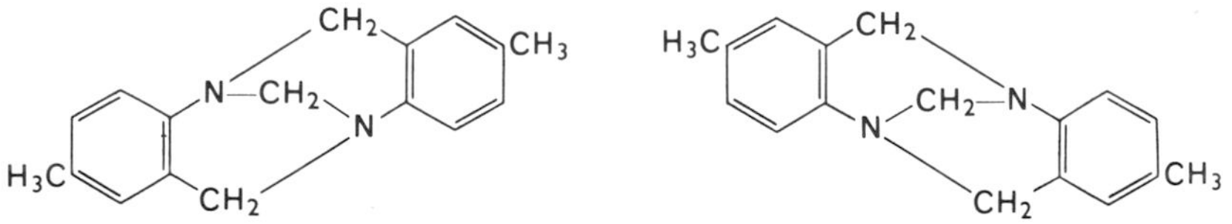

In quaternary ammonium compounds, the four bonds to the nitrogen atom are tetrahedrally disposed. Thus, there exist the same possibilities for stereoisomerism as in the case of the tetrahedral carbon atom. Stable, optically active enantiomorphic forms of ammonium salts of the following kind are known compounds; the four valences are tetrahedrally disposed ($sp^3$ hybridization), and the isomerism is like that due to tetracovalent, asymmetric carbon atoms:

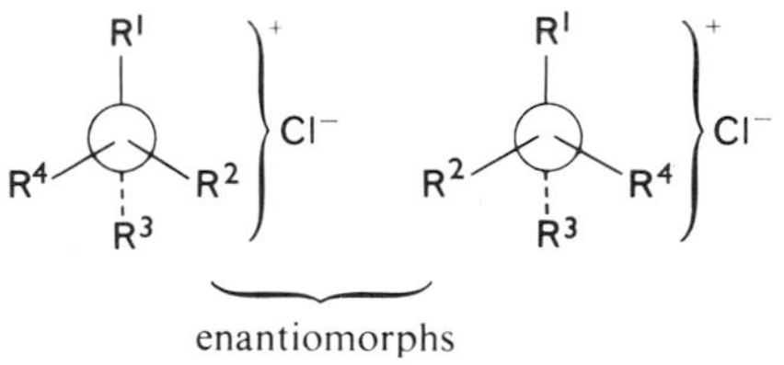

enantiomorphs

## 16.21 Relative and absolute configuration

The terms "dextro" and "levo," applied to optically active compounds, refer to the direction of rotation of the plane of polarized light, as determined experimentally. There is no generally useful way of determining, *a priori*, which of two enantiomorphs will be the dextrorotatory isomer. The configurations of organic compounds are usually related to each other by relating them, by experimental means, to some standard substance, the configuration of which is known.

A widely used reference substance for this purpose is glyceraldehyde. The dextrorotatory isomer of glyceraldehyde is designated as the D-isomer, its enantiomorph the L-isomer:

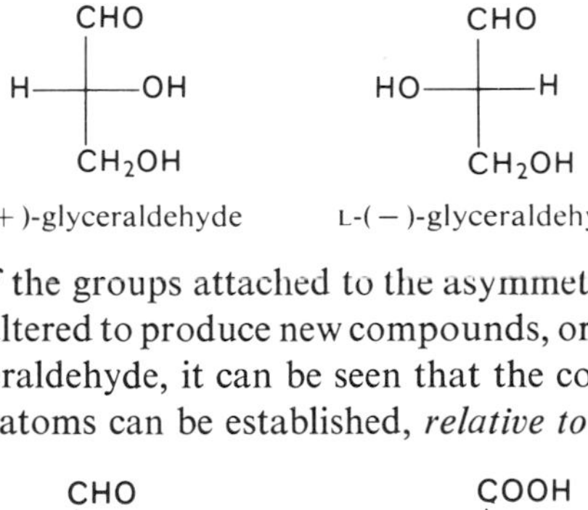

D-( + )-glyceraldehyde    L-( − )-glyceraldehyde

Thus, if the groups attached to the asymmetric carbon atom of D-glyceraldehyde can be altered to produce new compounds, or if other compounds can be degraded to glyceraldehyde, it can be seen that the configurations of relevant asymmetric carbon atoms can be established, *relative to that of D-glyceraldehyde*:

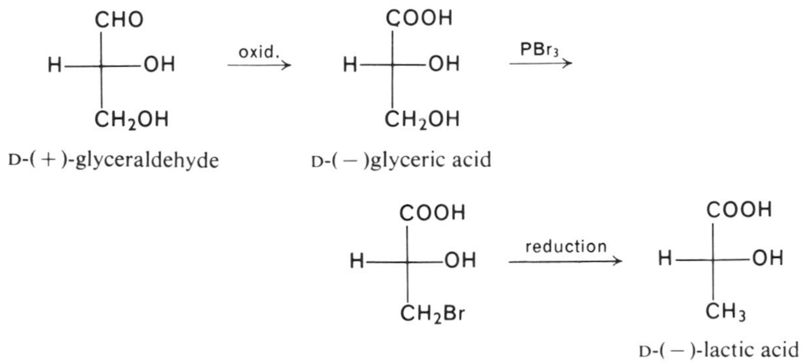

D-( + )-glyceraldehyde    D-( − )glyceric acid

D-( − )-lactic acid

Since nothing in the above transformations has affected the disposition of the four bonds to the central asymmetric carbon atom, all of the compounds in the above series have the same *relative configuration*. It will be noted that D-glyceric acid and D-lactic acid are *levorotatory*, while the D-glyceraldehyde to which they are related chemically is *dextrorotatory*. Thus the sign of rotation is not an indication of configuration. Indeed, it is proper nomenclature to designate both the configuration (D- or L-) and the actual sign of rotation (+ or −) when these are both known.

It is now apparent why the designations *d*- and *l*- are being replaced in current usage by the designations (+) and (−) (see the footnote to Section 16.17). It would be awkward to speak of D-*l*-tartaric acid or of D-*d*-malic acid.

The more complex the compounds that can be related to D-glyceraldehyde, the easier becomes the task of determining relative configurations, since, as the number of reference compounds becomes larger, new relationships can be established rapidly.

There is, of course, an element of arbitrariness in the assignment of relative configurations. For example, D-(+)-malic acid can be written in either of two ways:

$$
\begin{array}{ccc}
\text{COOH} & & \text{CH}_2\text{COOH} \\
\text{H}\!-\!\!\!-\!\text{OH} & \quad\text{or}\quad & \text{HO}\!-\!\!\!-\!\text{H} \\
\text{CH}_2\text{COOH} & & \text{COOH}
\end{array}
$$

D-(+)-malic acid

If D-(+)-glyceraldehyde were to be transformed into malic acid as follows:

$$
\begin{array}{ccccc}
\text{CHO} & & \text{COOH} & & \text{COOH} \\
\text{H}\!-\!\!\!-\!\text{OH} & \longrightarrow & \text{H}\!-\!\!\!-\!\text{OH} & \longrightarrow & \text{H}\!-\!\!\!-\!\text{OH} \longrightarrow \\
\text{CH}_2\text{OH} & & \text{CH}_2\text{OH} & & \text{CH}_2\text{Br}
\end{array}
$$

D-(+)-glyceraldehyde

$$
\begin{array}{ccc}
\text{COOH} & & \text{COOH} \\
\text{H}\!-\!\!\!-\!\text{OH} & \longrightarrow & \text{H}\!-\!\!\!-\!\text{OH} \qquad (37)\\
\text{CH}_2\text{CN} & & \text{CH}_2\text{COOH}
\end{array}
$$

D-(+)-malic acid<br>(natural)

a different malic acid would be formed from that prepared from D-(+)-glyceraldehyde, as shown on the top of the opposite page.

It can be seen that both D-(+)- and L-(−)-malic acid can be derived *chemically* from D-(+)-glyceraldehyde; and therefore it is necessary to add the specification as to how the transformation is brought about. The convention is adopted of designating as D-malic acid that one in which —COOH is derived directly from —CHO of D-glyceraldehyde.

By similar manipulations and chemical interconversions, absolute configurations have been assigned to a great many compounds, since the absolute configuration of (+)-glyceraldehyde is known.

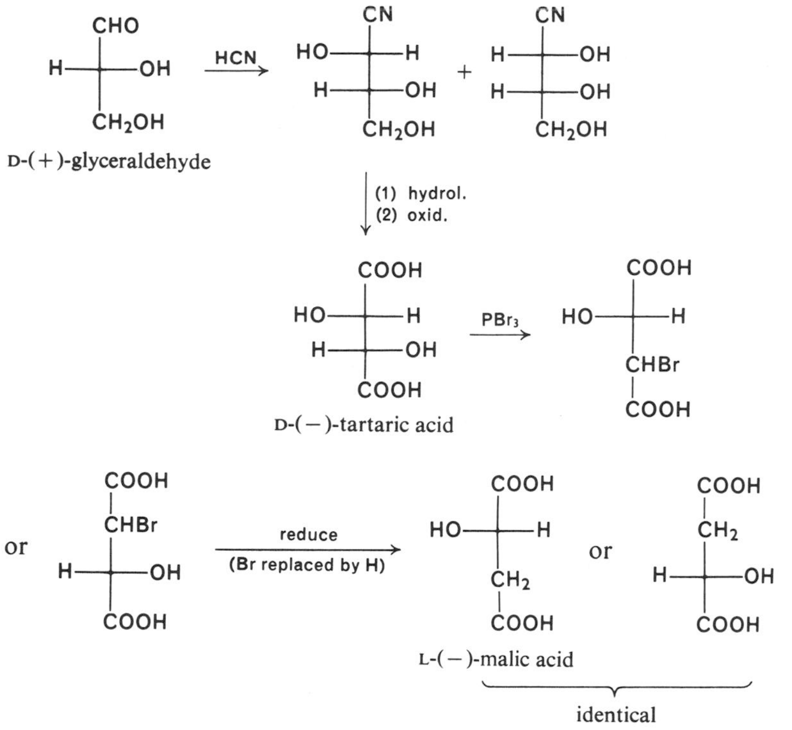

After a system of relative configurations has been established, the determination of the *absolute configuration of any one compound in the series* can establish the absolute configuration of them all. It was a happy turn of events that (+)-glyceraldehyde has been shown by experimental means to have the *absolute configuration*

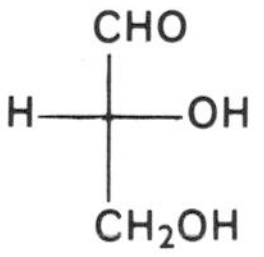

that had been assigned arbitrarily to the dextrorotatory isomer, and hence is indeed D-(+)-glyceraldehyde (Bijvoet, 1953). This result depends upon the demonstration by methods of X-ray analysis that dextrorotatory tartaric acid has the *absolute configuration*

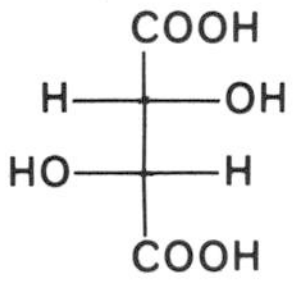

## 16.22   The *R* and *S* system of nomenclature

A general system of configurational nomenclature has been devised by R. S. Cahn (England), C. K. Ingold (England), and V. Prelog (Switzerland) and is now widely used. In the compound *cabde*, the groups can be arranged in two ways:

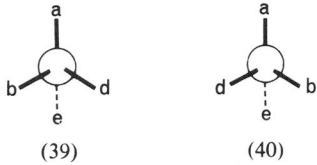

(39)                    (40)

If we now assign a priority order, or sequence rule, that states that *b* takes priority over *a*, *a* over *d*, and *d* over *e*, we can express the difference between (39) and (40) in the following way: (1) draw the three-dimensional figure so that, looking down upon it, it is seen with the lowest priority group pointing away from the observer [as in (41)]:

(39)                    (41)

(2) determine whether the path that is traveled in going from the group of highest priority *(b)* toward the group of next to lowest priority *(d)* is clockwise (rectus, or *R*) or counter-clockwise (sinister, or *S*). In 39 (=41) the order of priority is (by definition) *b > a > d*; thus, (39) has the *R*-configuration, and (40) has the *S*-configuration.

In order for this system to be used in naming actual compounds, a priority order, or a set of sequences rules, must be established. These are the following:

1. Groups attached to the asymmetric carbon atom are arranged in the order of decreasing atomic number of the atom bound to carbon. Thus, in iodobromoacetic acid, IBrCHCOOH, the order is I > Br > C > H, and

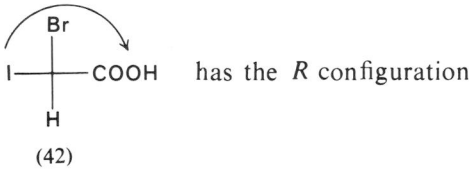

has the *R* configuration

(42)

[It will be noted that the planar projection shown in (42) is just as convenient as the ball-and-stick drawing (41), as indicated by the arrow in (42).]

2. If two or more of the atoms (first set) attached to the asymmetric carbon atom have the same atomic number (this is usually carbon), the priority order is determined by the atoms attached to these (second set). Thus —CH₂OH takes

priority over —$CH_3$; —$CH_2OH$ over —$CH_2NH_2$; —$CH_2Cl$ over —$CH_2OH$; —$CH_2CH_3$ over —$CH_3$ (=—$CH_2$—H).

3. If the second set of atoms are the same in two groups, the number of such atoms determines priority: thus, —$CH(OCH_3)_2$ over—$CH_2OCH_3$; —$CHCl_2$ over —$CH_2Cl$; —$CH(CH_3)_2$ over —$CH_2CH_3$; but —$CH_2Br$ over —$CHCl_2$ or —$CCl_3$ (rule 1).

4. Rule 3 is extended to the third, fourth, and other positions if necessary: thus, —$CH_2CH_2CH_3$ over —$CH_2CH_3$; —$CH_2CH_2CH_2Cl$ over —$CH_2CH_2CH_2CH_2Cl$.

5. When the atom attached to the asymmetric carbon atom is bound to other atoms by a double or triple bond, the atom at the end of the multiple bond is counted twice (double bond) or three times (triple bond): thus $C\equiv CH$ over —$CH=CH_2$; —$CH=O$ over —$CH_2OH$; —$CH=NCH_3$ over —$CHCH_3$

$$\left(\text{i.e., } -CH\begin{smallmatrix} \nearrow N \\ \searrow N \end{smallmatrix} \text{ over } CH\begin{smallmatrix} \nearrow N \\ \searrow C \end{smallmatrix}\right).$$
$$\underset{NHCH_3}{|}$$

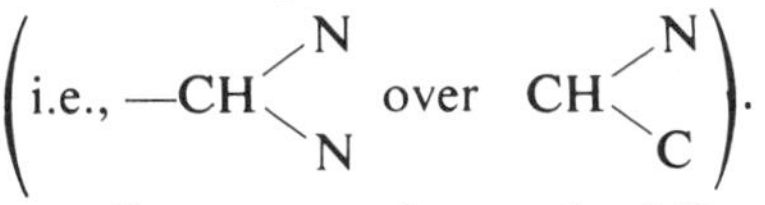

Some examples are the following:

L-(−)-glyceraldehyde: sequence, OH, CHO, $CH_2OH$, H:

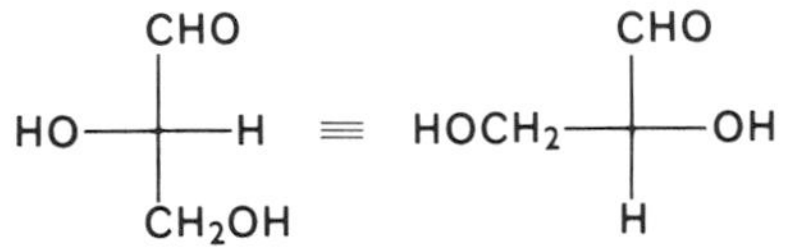

L-(−)-glyceraldehyde = *S*-glyceraldehyde

2-butanol: sequence, OH, $CH_2CH_3$, $CH_3$, H:

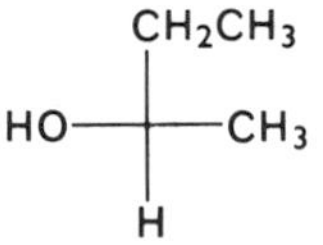

*R*-2-butanol 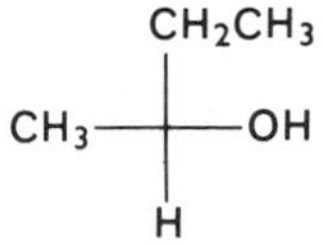          *S*-2-butanol

*trans*-2-chlorocyclohexanol

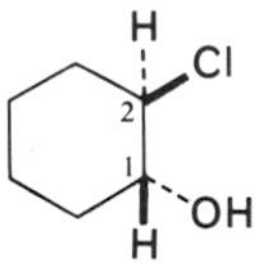

sequence, Cl, CHOH, $CH_2$, H (C-2)

sequence, HO, —CHCl, —$CH_2$—, H(C-1)

(43)

For the configuration shown in (43), asymmetric carbon atom 1 can be arranged,

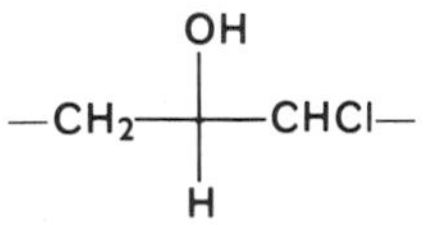

Carbon atom 2 can be arranged,

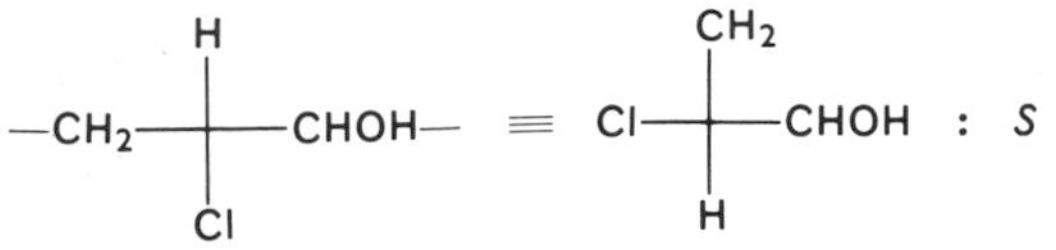

$$-CH_2 \overset{\text{H}}{\underset{\text{Cl}}{|}} -CHOH- \quad \equiv \quad Cl \overset{\text{CH}_2}{\underset{\text{H}}{|}} -CHOH \quad : \quad S$$

Thus, (42) is (1 *R*:2 *S*)-2-chlorocyclohexanol. Its enantiomer is (1 *S*:2 *R*)-2-chlorocyclohexanol

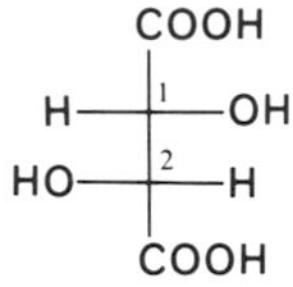

$$\begin{array}{c} \text{COOH} \\ H \overset{1}{\underset{\phantom{2}}{|}} OH \\ HO \overset{2}{\underset{\phantom{2}}{|}} H \\ \text{COOH} \end{array}$$ 

sequence, OH, COOH, CHOHCOOH, H (C-1)

sequence, OH, COOH, CHOHCOOH, H (C-2)

L-(+)-tartaric acid

For carbon 1:

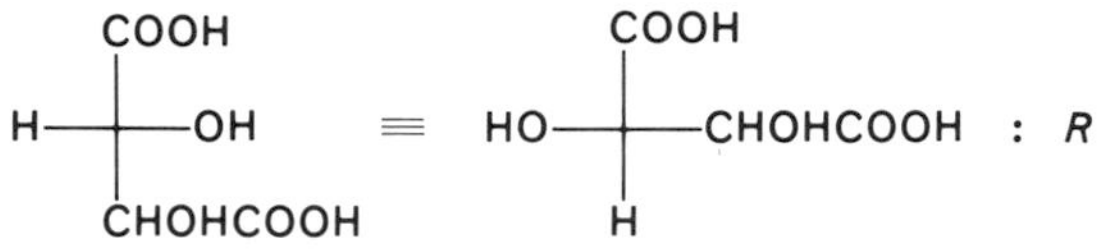

$$\begin{array}{c} \text{COOH} \\ H \overset{}{\underset{}{|}} OH \\ \text{CHOHCOOH} \end{array} \quad \equiv \quad \begin{array}{c} \text{COOH} \\ HO \overset{}{\underset{}{|}} CHOHCOOH \\ H \end{array} \quad : \quad R$$

For carbon 2:

$$\begin{array}{c} \text{CHOHCOOH} \\ HO \overset{}{\underset{}{|}} H \\ \text{COOH} \end{array} \quad \equiv \quad \begin{array}{c} \text{CHOHCOOH} \\ HOOC \overset{}{\underset{}{|}} OH \\ H \end{array} \quad : \quad R$$

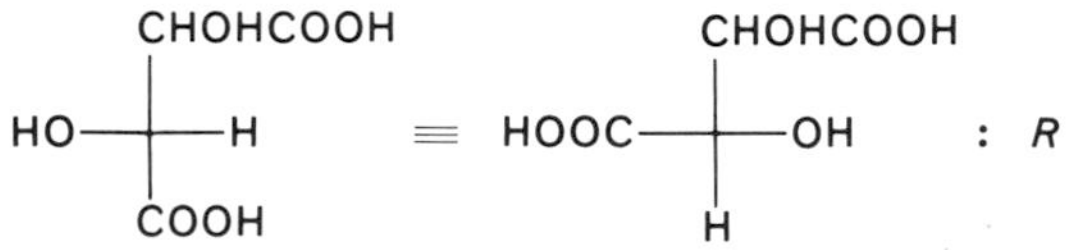

Thus L-(+)-tartaric acid is (2 *R*:3 *R*)-2,3-dihydroxybutanedioic acid.

---

**Exercise 18.** Name the following compounds by the *R/S* system:

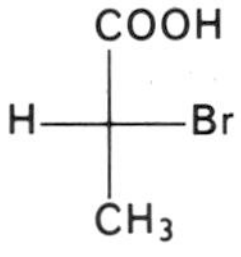

(a) 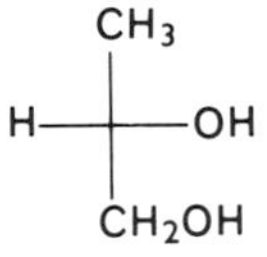 $$\begin{array}{c} \text{CH}_3 \\ H \overset{}{\underset{}{|}} OH \\ \text{CH}_2\text{OH} \end{array}$$

(b) $$\begin{array}{c} \text{COOH} \\ H \overset{}{\underset{}{|}} Br \\ \text{CH}_3 \end{array}$$

(c) [cyclopentane ring with OH and OH substituents]

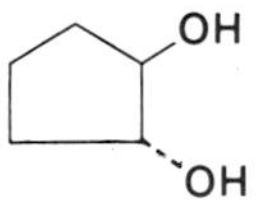

(d) D-(−)-lactic acid

(e) L-(+)-malic acid

---

## 16.23 The significance of molecular asymmetry in biological systems

Living organisms are largely constructed of optically active substances. Except for compounds that are not asymmetric (e.g., water and inorganic salts, which make up the bulk of the mass of living things), most of the compounds that can exist in D- or L-forms are found in nature in one or the other configuration and seldom as the racemic ($\pm$) substances. The most conspicuous example is found in the proteins, which constitute the major part of the organic substance of living organisms. Proteins consist of $\alpha$-amino acids in polymeric amide linkage. The amino acids formed by the hydrolysis of proteins are optically active, and, even more remarkably, belong to the same L-configurational series (glycine, $NH_2CH_2COOH$, which is not an asymmetric molecule, is an exception):

$$\text{proteins} \xrightarrow{\text{hydrol.}} NH_2\text{—}\overset{\displaystyle COOH}{\underset{\displaystyle R}{\rule[0.5ex]{0pt}{2ex}|}}\text{—H}$$

L-amino acids

Why the proteins are constituted from the L-amino acids and not the D-amino acids is not known. Perhaps in the early stages of the evolution of living things the process of protein synthesis began, by chance, with the L-forms, and was then obliged to continue with the utilization of this enantiomorphic series.

As a result of the optical asymmetry of proteins, the metabolic processes of the body are sensitive to configurational factors in foodstuffs that are utilized by the body, and in drugs that act upon it. A general description of how this dependence obtains is given in Figure 16-7. Suppose the tissue upon which a drug, for example, acts, is represented by the asymmetric element $A'$, $B'$, $C'$. Suppose further that an optically active drug requires for its action a combination of three groups $A$, $B$, and $C$ with the corresponding elements of the tissue. Not all drug action is *absolutely* dependent upon configuration, and in some

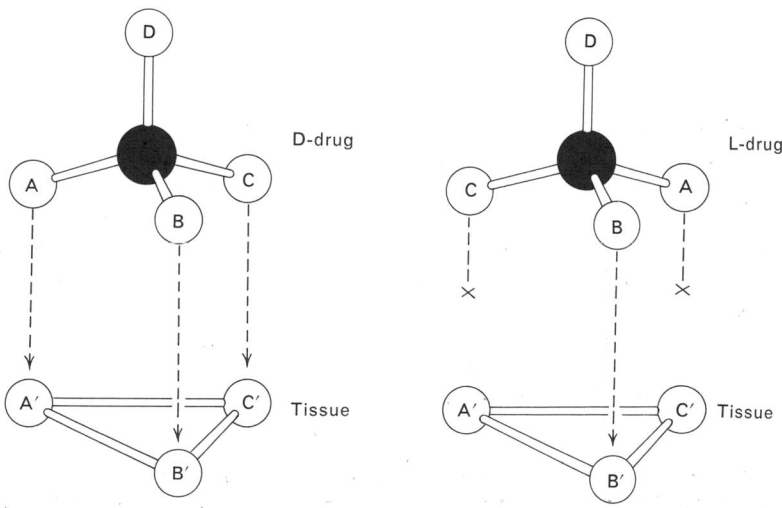

**Figure 16–7.** Fit and nonfit of enantiomorphic forms of an optically active drug to an asymmetric tissue site.

cases D- and L-isomers show little difference in activity. This may mean that a fit of only two sites is needed for activity to be observed.

Optical activity exists in nearly all compounds (whose structures permit it) that are found in nature. This indicates that optically active agents play an intimate role in their synthesis. These agents are the enzymes—the cellular catalysts—whose asymmetry is due to the fact that they are constructed of protein molecules.

Further discussion of the biological processes that have been mentioned but briefly here will be reserved for a later chapter.

---

**Exercise 19.**   It is occasionally observed that a *dl*-drug is *less* than one-half as active as the optically active form. Can you suggest a possible reason for this?

---

### Exercises

(Exercises 1–19 will be found within the text.)

**20** Write projection formulas showing all the stereoisomeric forms of each of the following.

(a) $(CH_3)_2CHCOOH$

(b) $CH_3\overset{\overset{\displaystyle OH}{|}}{C}HCOOH$

(c) $HOCH_2\overset{\overset{\displaystyle OH}{|}}{\underset{\underset{\displaystyle CH_3}{|}}{C}}COOH$

(d) $HOCH_2\overset{\overset{\displaystyle OH}{|}}{C}HCH_2OH$

(e) $CH_3\overset{\overset{\displaystyle Br}{|}}{C}H\overset{\overset{\displaystyle Br}{|}}{C}HCH_3$

(f) $CH_3CH{=}CHCH_2OH$

(g) $CH_3CH{=}\overset{\overset{\displaystyle OH}{|}}{C}HCHCH_3$

(h) $CH_3\overset{\overset{\displaystyle OH}{|}}{C}HCH_2\overset{\overset{\displaystyle OH}{|}}{C}HCH_3$

(i) $CH_3\overset{\overset{\displaystyle Br}{|}}{C}H\overset{\overset{\displaystyle Cl}{|}}{C}HCH_3$

(j) 1,3-dibromocyclobutane

**21** (a) Write the structures for all of the heptanes ($C_7H_{16}$), marking asymmetric carbon atoms with an asterisk and indicating enantiomorphic pairs; (b) do the same for all of the six-carbon alcohols, $C_6H_{13}OH$.

**22** When (+)-3-methylcyclohexanone,

is reduced by the Wolff-Kishner or Clemmensen method, the product is optically inactive. Why?

**23** Draw projection formulas for the products obtained by the addition of bromine to (+)-5-methyl-1-hepten-4-one.

**24** What would be the stereochemical result of reducing (a) D-(+)-glyceraldehyde, (b) 4-hydroxycyclohexanone, with lithium aluminum hydride?

**25** *(a)* Of the possible configurationally isomeric inositols (1,2,3,4,5,6-hexahydroxycyclo-hexanes), only one is capable of existing in enantiomorphic—(+) and (−)—forms. Show its configuration by means of a simple projection formula. *(b)* Draw the structure of the remaining isomers and point out the nature of the symmetry in each.

**26** How would you prepare the following? *(a)* cis-1,2-dihydroxycyclohexane (cyclohexane-1,2-diol), starting from cyclohexanone; *(b)* meso-1,2,3,4-butanetetrol (erythritol), starting from fumaric acid; *(c)* (±)-1,2,3,4-butanetetrol, starting from fumaric acid.

**27** Devise a method for preparing optically active $CH_3CH_2CHCH_2OH$, given as starting
$$\overset{\displaystyle |}{CH_3}$$
materials dl-2-butanol, an optically active amine, and any necessary inorganic reagents.

**28** Given as the only optically active reagent $(+)\text{-}CH_3CH_2\overset{\displaystyle \overset{CH_3}{|}}{C}HCH_2OH$, devise means of resolution of

*(a)* $CH_3\overset{\displaystyle \overset{}{|}}{C}HCOOH$     *(b)* $CH_3CH_2\overset{\displaystyle \overset{}{|}}{C}HCH_2NH_2$     *(c)*
    $\overset{\displaystyle \overset{}{Br}}{}$            $\overset{\displaystyle \overset{}{CH_3}}{}$

(c) structure:
$$\begin{array}{c} CH_2\!-\!CO \\ | \qquad\quad\diagdown \\ \qquad\qquad O \\ | \qquad\quad\diagup \\ CH_3CH\!-\!CO \end{array}$$

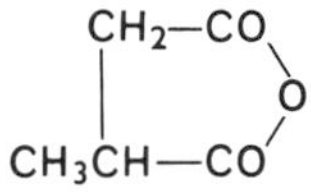

The optically active reagent may be converted into other reagents by suitable manipulation of the —$CH_2OH$ group. Assume diastereomers can be separated.

**29** How many possible stereoisomers of cholesterol (see index) can theoretically exist?

**30** Which of the following allenes would be capable of existing in enantiomorphic forms? Indicate the kind of symmetry that is present in those that are nonresolvable.

*(a)* $CH_3CH\!=\!C\!=\!CH_2$            *(d)* $(CH_3)_2C\!=\!C\!=\!C(CH_3)_2$

*(b)* $CH_3CH\!=\!C\!=\!CHCH_3$        *(e)* $CH_3CHCH\!=\!C\!=\!CH_2$

*(c)* $(CH_3)_2C\!=\!C\!=\!CHCH_3$           $\overset{\displaystyle \overset{}{|}}{Br}$

## Chapter seventeen

# Ring Formation and

# Conformation of Organic Molecules

## 17.1 Ring size and stability

The ability of carbon atoms to form not only long chains but rings as well was not clearly recognized until the 1870's, following Kekulé's proposal of the cyclic formula for benzene. It was not until the 1880's that ring systems of other than six members (i.e., cyclohexane derivatives) were known, but the preparation by Perkin of derivatives of cyclopropane, cyclobutane, and cyclopentane derivatives soon established that these ring systems were capable of existence.

Studies of the chemistry of compounds of these classes showed that whereas the cyclopentane and cyclohexane rings are very stable and are not easily opened, cyclobutane and cyclopropane rings show a tendency to react to give open-chain derivatives. For example, cyclopropane gives propene when heated,

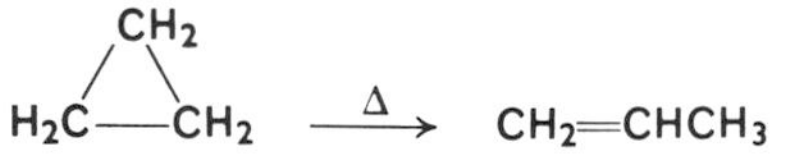

$$\underset{H_2C\text{---}CH_2}{\overset{CH_2}{\diagup\diagdown}} \quad \xrightarrow{\Delta} \quad CH_2\text{=}CHCH_3$$

and reacts with bromine to give 1,3-dibromopropane.

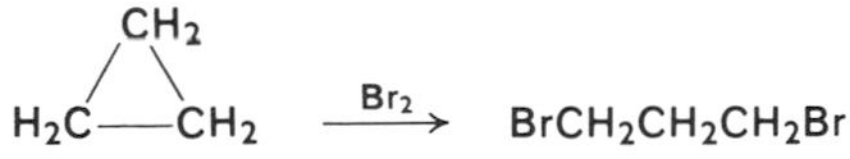

$$\underset{H_2C\text{---}CH_2}{\overset{CH_2}{\diagup\diagdown}} \quad \xrightarrow{Br_2} \quad BrCH_2CH_2CH_2Br$$

Cyclobutane derivatives are somewhat less readily converted to acyclic derivatives; and five-, six-, and higher-membered alicyclic rings are quite stable and are opened no more readily than corresponding open chains are cleaved.

**Table 17-1.** *Deflection from tetrahedral angles for carbon rings*

| C atoms in ring | 3 | 4 | 5 |
|---|---|---|---|
| Deflection in C—C—C bond angle from 109°28′ | $+24.7°$ | $+9.7°$ | $+0.7°$ |

The relative instability of three- and four-membered carbon rings led Baeyer in 1885 to propose his "strain theory." The strain theory was based upon the postulate of a planar ring, and proposed that the distortion of the normal tetrahedral bond angle (for carbon) of 109°28′ to the angles formed by the bonds in the cyclic system set up a "strained" condition in the molecule and contributed to instability. The deflections from the tetrahedral angle can be calculated to be as shown in Table 17-1.

It will be noted that rings of more than five members do not appear in the table (although they were included by Baeyer in his original presentation of the theory). This is because rings of six members and larger are not constrained to assume a planar conformation, but can pucker in such a way as to permit the normal tetrahedral angles to obtain.

The Baeyer strain theory does predict the increasing stability in going from cyclopropane to cyclopentane, and offers a reason for the ease with which five-membered rings are formed. Indeed, since the departure from normal bond angles in five- and six-membered rings containing oxygen and nitrogen is also very small, it is readily seen that compounds such as succinic and glutaric anhydrides, $\gamma$- and $\delta$-lactones, would also be readily formed and stable:

succinic anhydride  $\gamma$-butyrolactone  glutaric anhydride  $\delta$-valerolactone

Since the cyclohexane ring is not planar, it is free from strain that would be caused by distortion of the C—C—C bond angles from the normal tetrahedral value. The puckered cyclohexane ring system can assume two strainless conformations; these are the "chair" and "boat" forms. These two conformations differ slightly in stability for reasons that will be discussed in the following section. The chair form is the more stable and is the conformation normally assumed by cyclohexane and its derivatives. Molecular models of the two conformations of the cyclohexane ring are shown in Figure 17-1.

Very large rings, such as cyclopentadecane ($C_{15}H_{30}$), are very stable, and are not, as was predicted by the earlier strain theory, as highly strained as cyclopropane (see Table 17-1). Again, the explanation is that the ring can assume a

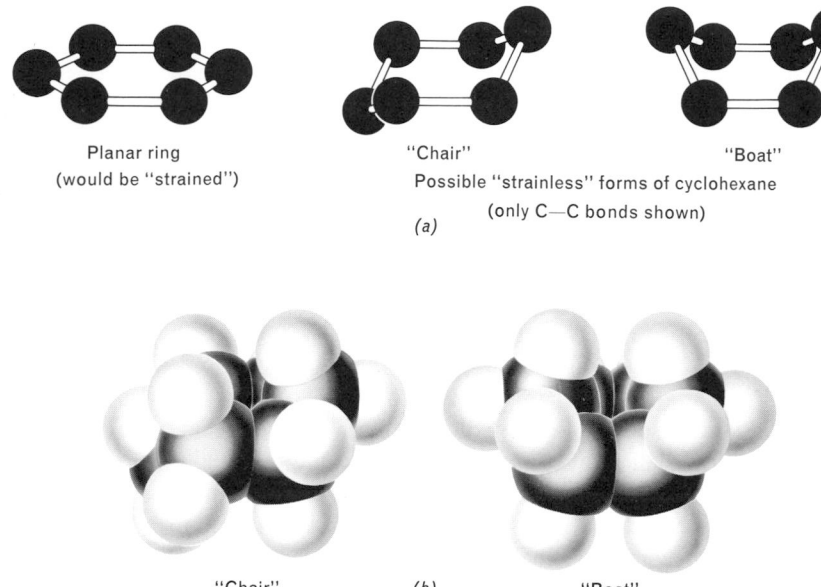

Planar ring
(would be "strained")

"Chair"

"Boat"

Possible "strainless" forms of cyclohexane
(only C—C bonds shown)

*(a)*

"Chair"      *(b)*      "Boat"

**Figure 17–1.** *(a)* Carbon skeleton of cyclohexane, showing several conformations. *(b)* Molecular models of cyclohexane in chair and boat conformations.

nonplanar, staggered conformation in which the bonds are at the normal tetrahedral angles (Section 17.3).

## 17.2    Cyclohexane and its derivatives

The chair conformation of cyclohexane itself is shown by the following figure:

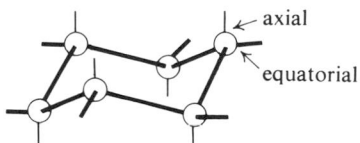

axial

equatorial

The twelve bonds (to hydrogens or other substituents) are disposed in two ways: six of them extend in the direction parallel to an axis drawn perpendicular to the ring, and are called "axial" (a); the remaining six extend parallel to the general plane of the ring, in a kind of "equatorial belt," and are called "equatorial" (e). It can be seen that the chair conformation (1) allows the most effective separation of adjacent substituents,

end-on view
along bond
joining 1-2

(1)

whereas in the boat form (2), adjacent hydrogen atoms would "eclipse" each other:

$$\text{(2)}$$

The interference caused by the eclipsing of adjacent substituents represents a degree of strain that, while not large, is a factor that helps to determine which conformation is the more stable.

When the cyclohexane ring carries a substituent, a new question arises. Taking chlorocyclohexane as an example, and assuming that the ring will be in the more stable chair form, will the chlorine atom be axial or equatorial?

chlorine axial            chlorine equatorial

chlorocyclohexane

It is found experimentally by electron diffraction measurements that the most stable conformation of chlorocyclohexane is the form in which chlorine is equatorial. In general, *in monosubstituted cyclohexanes the substituent is in the equatorial conformation.*

The reason for this is that an axial substituent and the two other axial hydrogen atoms or substituents offer some interference with one another; but when the substituent is equatorial no such interaction is present. In cyclohexane derivatives with more than one substituent, the maximum number of substituents are equatorial. Thus, a *trans*-1,2-disubstituted cyclohexane prefers to assume conformation (3) rather than (4):

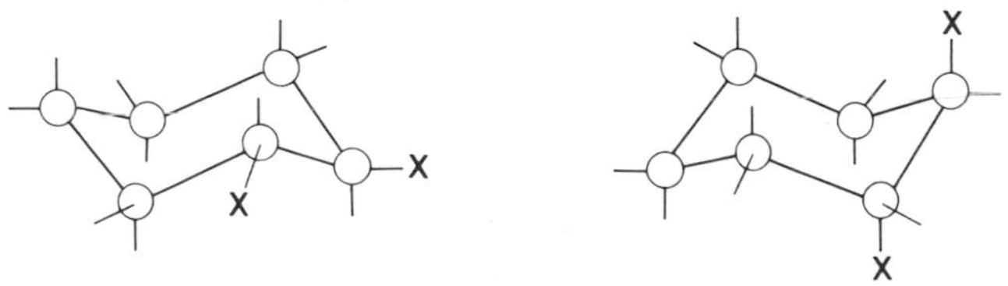

(3) diequatorial                (4) diaxial

In *cis*-1,2-, *trans*-1,3-, and *cis*-1,4-disubstituted cyclohexanes, one substituent is necessarily axial, the other equatorial.

---

**Exercise 1.** Draw conformational structures of *cis*-1,3- and *trans*-1,4-dimethylcyclohexanes, with the methyl groups in equatorial positions.

---

These are *conformations*, not *configurations*. Numbers (3) and (4) have the same *configuration*. There are, of course, an unlimited number of possible *conformations*, since these depend upon the *shape* assumed by the molecule. Usually only one of these exists in fact. The question might be asked why several forms of, for example, *trans*-1,2-dichlorocyclohexane, cannot be isolated. The reason is that one conformation is the most stable (least strained), and the energy differences between the various possible conformations are small; hence, ordinary thermal equilibration is sufficient to allow the most stable isomer [chair; 1(e)2(e)-dichloro] to predominate. However, *if the geometrical requirements for a particular reaction are better satisfied by the less favorable conformation, the latter may be the actual reacting species.*

Rings containing more than six members can also accommodate themselves to a conformation of minimum strain by puckering of the ring. In "medium-sized rings" of eight to ten members there is some crowding by the eclipsing of adjacent hydrogen atoms in even the most favorable conformation, but in "large rings" of more than ten members a nearly strainless conformation can be assumed. However, all rings from cyclohexane derivatives upward are essentially strainless from the point of view of distortion of the tetrahedral bond angles, and, once formed, exhibit none of the instability of the highly strained three- and four-membered rings. Cyclodecane, for example, could assume a staggered conformation like the following:

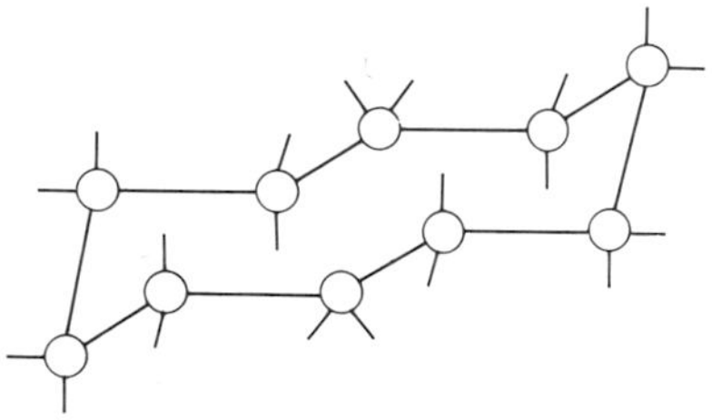

## 17.3  Configuration and conformation in bicyclic and polycyclic systems

When two cyclohexane rings are condensed to form the bicyclic ring system decahydronaphthalene (bicyclo-[4,4,0]-decane) there are several possibilities to consider. First, there are two possible configurational isomers, with *cis* and *trans* fused rings:

*cis*                              *trans*

decahydronaphthalenes (decalins)

Electron diffraction studies have shown that the stable conformations of these isomers are the following (only the relevant hydrogen atoms are shown):

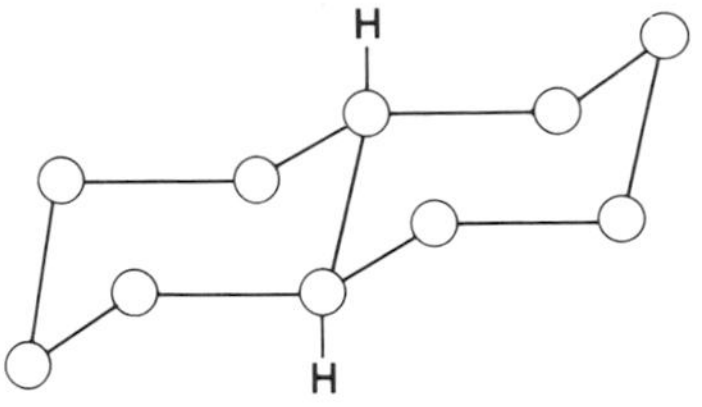

*trans*-decalin 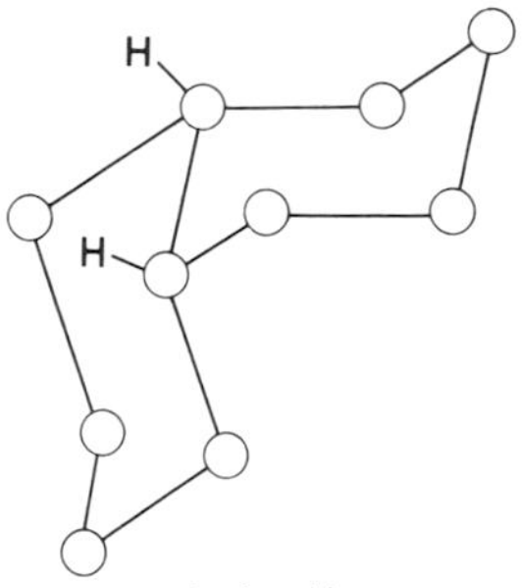         *cis*-decalin

The representation of polycyclic compounds by conformational formulas offers a useful method of visualizing the actual steric relationships. Thus, (6) is a conformational representation of the conventional formula (5).

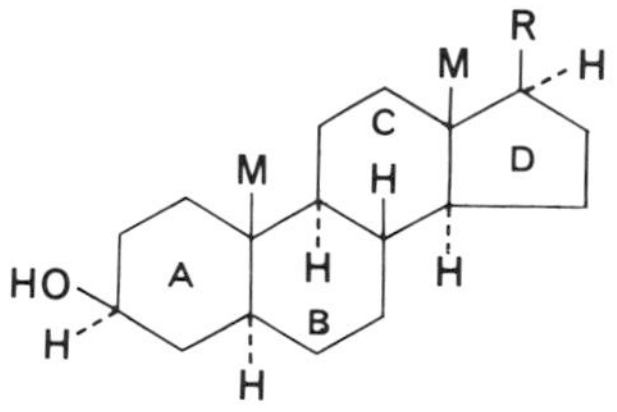

(5)

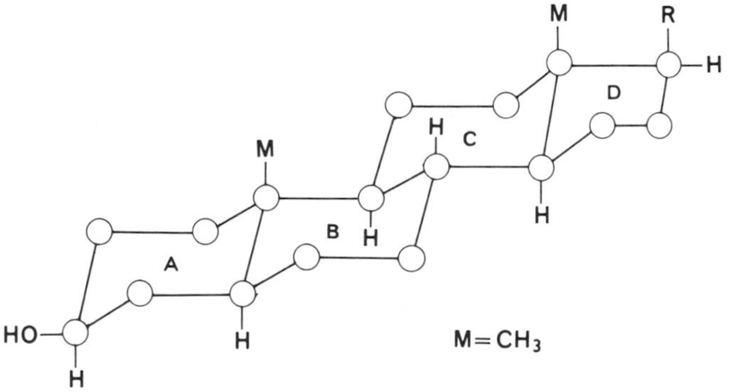

(6)

$M = CH_3$

A/B *trans*; B/C *trans*; C/D *trans*

Fused-ring systems, such as those found in the decalins and in polycyclic compounds such as (6), differ from simple cyclohexane rings in an important respect. Cyclohexane itself, although largely in the chair form because of the greater stability of this conformation, is not rigid, and can assume various nonplanar conformations by changes that proceed through relatively small energy barriers.

In a compound such as (6), however, the ring system is quite rigid, and the hydroxyl group in ring A is constrained to remain in that conformation.

The greater stability of an equatorial hydroxyl group, compared with one in the axial conformation (Section 17.2), coupled with the rigidity of the fused-bicyclic system found in *trans*-decalin, permits the ready equilibration of the configuration of an axial hydroxyl group, as shown in the following:

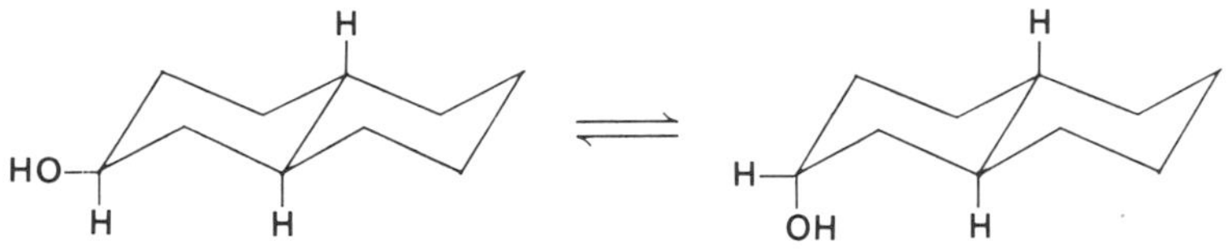

(most)

This interconversion occurs by way of an oxidation-reduction reaction, catalyzed by a trace of the ketone, such as that described in Equation (103), Chapter 8.

## 17.4  Formation of rings

In general, any reaction in which a new bond is formed can be applied to ring formation. There are three factors which influence the success of a ring-closing operation:

1. *The intrinsic stability of the ring that is formed,* as a result of distortions of bond angles that exist in the final ring. It is to be emphasized, however, that "stability" and "instability" are relative terms, and that cyclopropane derivatives, as well as other three-membered ring systems, are not difficult to prepare; but when we attempt to prepare one of them by means of a reaction that can take some alternative course, or when the ring, once formed, is subject to cleavage under the conditions of the reaction, the yield of the desired product may be poor.

2. *The distance, or probability, factor.* It would be expected that small rings would be easier to close than large rings because the ends of the chain that are to react will be closer together in the case of reactions leading to smaller rings. The probability that the ends of long chains will come together to allow ring closure is smaller the longer the chain.

3. *Stability factors associated with effects other than simple bond-angle distortions.* These factors are of greatest importance in medium-sized rings, and account for a minimum in the yields of eight- and nine- and ten-membered rings. They may be most simply described as due to crowding of adjacent hydrogen atoms on the closely packed —$CH_2$— groups of such rings, and can be classed as conformational effects.

## 17.5 Ring closure by dehalogenation of dihalides

One of the simplest methods of synthesizing cycloalkanes is by the reaction of an $\alpha,\omega$-polymethylene dibromide or dichloride with zinc or sodium. The reaction proceeds best with the use of zinc in alcohol solution, and gives the highest yields in the case of cyclopropane:

$$\text{ClCH}_2\text{CH}_2\text{CH}_2\text{Cl} \xrightarrow[\text{EtOH}]{\text{Zn}} \overset{\overset{\displaystyle\text{H}_2}{\underset{\displaystyle |}{\text{C}}}}{\text{CH}_2\!\!-\!\!\text{CH}_2} + \text{ZnCl}_2$$

This reaction may be regarded as an extension of the well-known dehalogenation of 1,2-dihalides to give olefins (Chapter 9). Indeed, ethylene may be regarded as the simplest member of the cycloparaffins, in which the double bond is effectively a two-membered ring. The method has little preparative importance except for simple cyclopropane hydrocarbons. For rings of four, five, six, or more members the yields are low.

## 17.6 Ring closure by aldol condensations and related reactions

The most general and widely employed methods of ring closures are those in which some reaction of the general aldol condensation type is employed. Simple aldol condensation of a compound containing a carbonyl group and an active methylene group as parts of a chain can lead to ring closure:

$$\begin{matrix} \text{R} \\ | \\ \text{-CO} \\ \text{-CH}_2\text{X} \end{matrix} \longrightarrow \begin{matrix} \text{R} \\ | \\ \text{-C} \\ \| \\ \text{-C}\!\!-\!\!\text{X} \end{matrix} \qquad \begin{array}{l} \text{X} = \text{activating group such} \\ \text{as} >\!\!\text{CO, NO}_2\text{, CN} \end{array}$$

An example of this is found in ring closures of diketones in which a five- or six-membered ring can result:

$$\begin{matrix} \overset{\text{CH}_3}{\underset{|}{}} \\ \text{H}_2\text{C}^{\text{CO}} \quad \text{CH}_2\text{COCH}_3 \\ | \qquad\qquad | \\ \text{H}_2\text{C}\text{---}\text{CH}_2 \end{matrix} \xrightarrow{\text{base}} \begin{matrix} \overset{\text{CH}_3}{\underset{|}{}} \\ \text{H}_2\text{C}^{\text{C}}\!\!\approx\!\!\text{C}\!\!-\!\!\text{COCH}_3 \\ | \qquad\qquad | \\ \text{H}_2\text{C}\text{---}\text{CH}_2 \end{matrix} \qquad (7)$$

1-acetyl-2-methylcyclopentene

$$\begin{matrix} \overset{\text{CH}_3}{\underset{|}{}} \\ \text{H}_2\text{C}^{\text{CO}} \quad \text{CH}_3 \\ | \qquad\quad | \\ \text{H}_2\text{C}\text{---}\text{CO} \end{matrix} \xrightarrow{\text{base}} \begin{matrix} \overset{\text{CH}_3}{\underset{|}{}} \\ \text{H}_2\text{C}^{\text{C}}\!\!\approx\!\!\text{CH} \\ | \qquad\quad | \\ \text{H}_2\text{C}\text{---}\text{CO} \end{matrix} \qquad (8)$$

3-methyl-2-cyclopenten-1-one

Rings of three, four, or more than six members are not often successfully prepared by condensations of this kind, since the alternative of *inter*molecular aldolization can dominate the course of the reaction when the only ring that could be formed is one that is strained or sterically unfavorable:

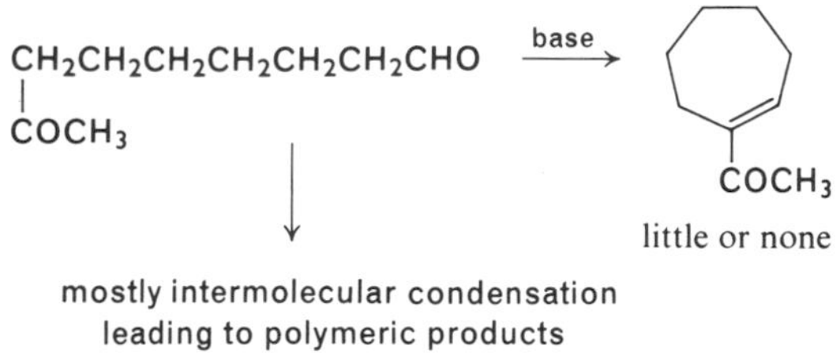

$$CH_2CH_2CH_2CH_2CH_2CH_2CHO \xrightarrow{\text{base}}$$

(9)

$$\underset{|}{COCH_3}$$

COCH₃
little or none

mostly intermolecular condensation
leading to polymeric products

The reversibility of the aldol condensation will insure that, if more than one reaction is possible, the final product will be the one of greatest stability. If one of the possible products is a ring compound that is unstable because of angle strain or strain due to eclipsing or crowding interactions, alternative courses will be taken.

## 17.7 The Dieckmann condensation

One of the most commonly used ring-forming reactions of the aldol family is the *Dieckmann condensation*, which is the name applied to the intramolecular Claisen (or acetoacetic ester type) condensations of diesters and keto esters:

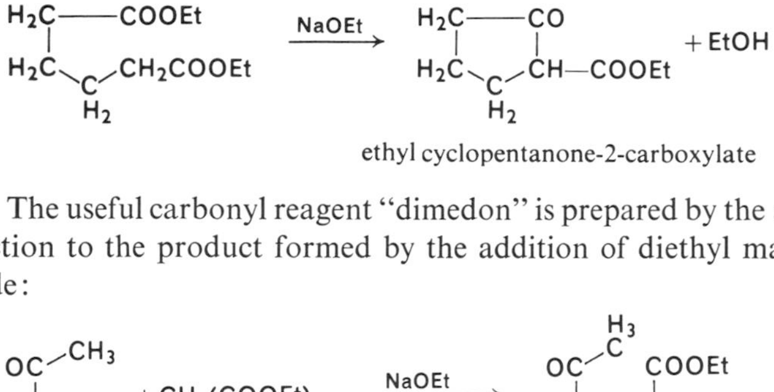

(10)

ethyl cyclopentanone-2-carboxylate

The useful carbonyl reagent "dimedon" is prepared by the application of this reaction to the product formed by the addition of diethyl malonate to mesityl oxide:

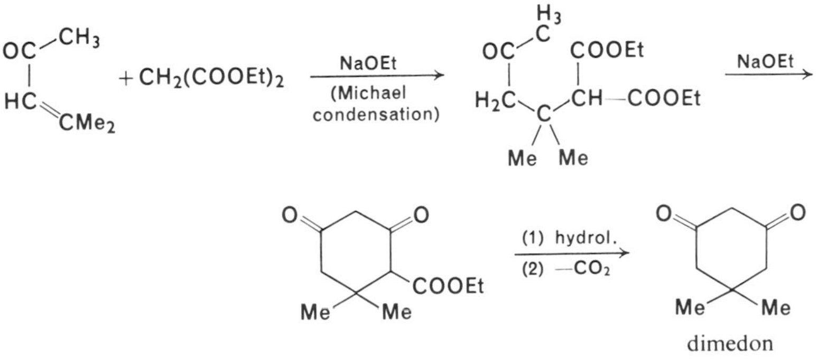

(11)

dimedon

Dimedon* is a valuable reagent; it reacts at the —CH₂— group between the carbonyl groups, forming condensation products with aldehydes:

---

\* Dimedon is also known by the name "methone."

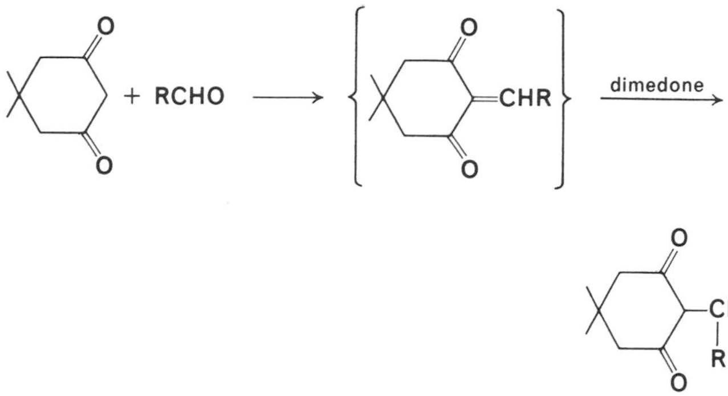

$$(12)$$

These are crystalline compounds that are readily prepared in good yields, and are valuable derivatives for the characterization and identification of aldehydes. The dimedon derivative of formaldehyde (R=H) is formed in quantative yield and is used for the quantitative estimation of formaldehyde.

The variation in the kinds of ring closure that can be effected by ester condensations is almost limitless. Further examples are the following:

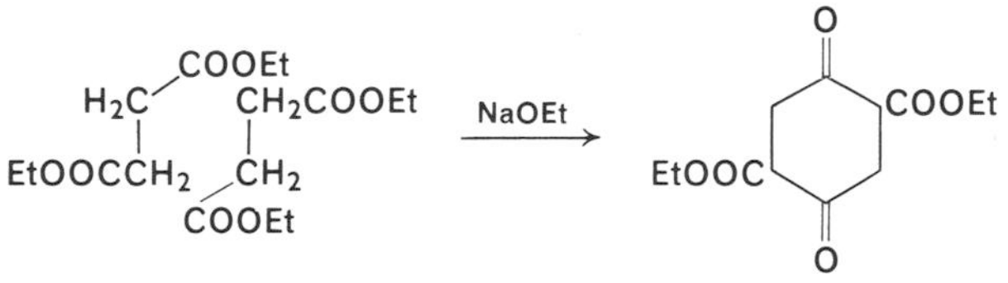

$$(13)$$

It will be noted that two molecules of the ester condense to form the six-membered ring; a Dieckmann-cyclization to give the three-membered ring

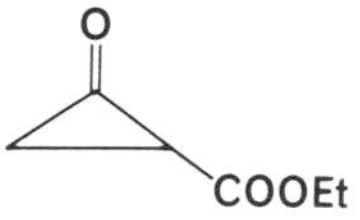

does not occur. The Dieckmann condensation, too, is inherently reversible because of the cleavage of $\beta$-keto esters by the sodium ethoxide used for the condensation, and so the three-membered ring would not survive. A similar comment applies to the following example:

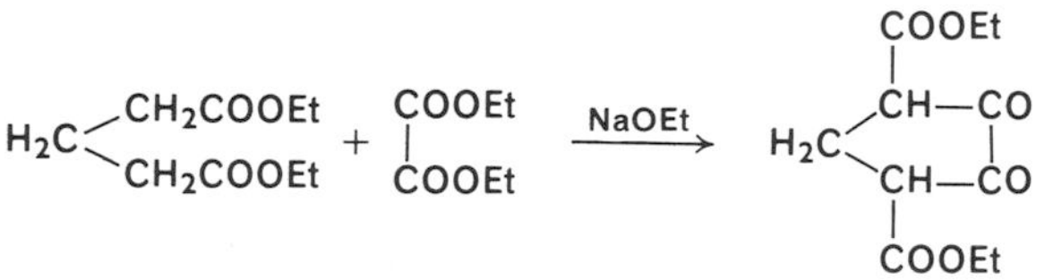

$$(14)$$

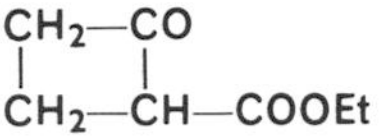

In (14) the ring closure of diethyl glutarate to form the four-membered ring does not occur; the reaction with ethyl oxalate to yield the five-membered ring is the sole result.

The use of a dinitrile in place of a diester is also a method for the formation of cyclic compounds; in a nitrile the $\alpha$-hydrogen atoms are activated by the —CN group, and the —CN group can undergo carbonyl-like addition reactions. In (15) and (16) are shown the nature of the reaction; an example is (17):

$$-CH_2CN \xrightarrow{\text{B:}^-} \{-\ddot{C}HCN\}^- + B{:}H \tag{15}$$

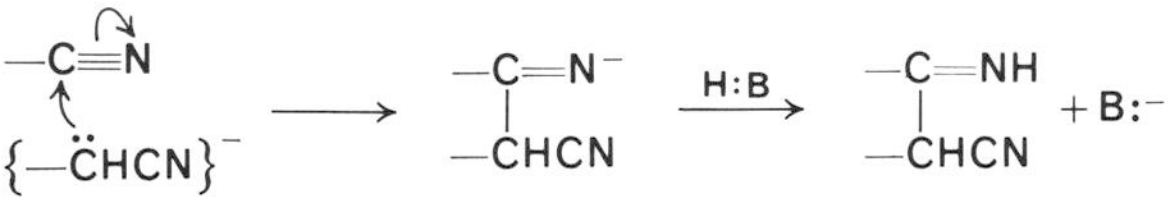

$$\tag{16}$$

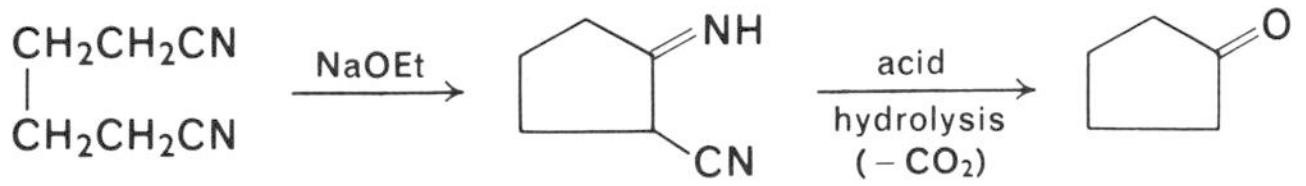

adiponitrile
$$\tag{17}$$

## 17.8  Ring closure by acid-catalyzed additions to carbon-carbon double bonds

We have seen earlier (Section 9.22) that polymerization of certain olefins can be brought about by acid catalysts:

$$>\!\!C\!\!=\!\!CH_2 + HA \rightleftharpoons >\!\!\overset{+}{C}CH_3 \Big) A^-$$

$$>\!\!C\!\!=\!\!CH_2 + >\!\!\overset{+}{C}CH_3 \rightleftharpoons >\!\!\overset{+}{C}\!\!-\!\!CH_2\!\!-\!\!\overset{|}{C}\!\!-\!\!CH_3 \longrightarrow \text{etc.}$$

This is clearly a carbon-carbon bond-forming-process; and, when the bond so formed closes a five- or six-membered ring, it constitutes a useful and widely applied method of forming cyclic compounds:

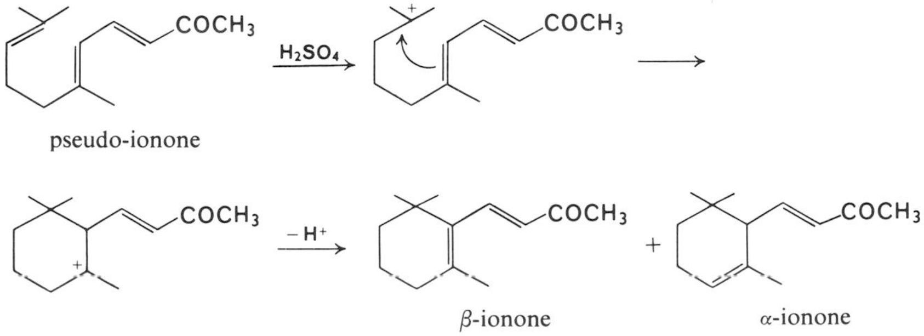

pseudo-ionone

$\beta$-ionone        $\alpha$-ionone
$$\tag{18}$$

$$\text{(19)}$$

## 17.9 The Diels-Alder reaction (diene synthesis)

One of the most prolific and versatile of the reactions at the disposal of the organic chemist is the Diels-Alder reaction. This reaction in its most general form is

diene    dienophile    adduct

The reaction is also known by the name "diene synthesis" because, as the type equation shows, one of the reactants is a 1,3-diene.

In the most general applications of the diene synthesis, the mono-olefinic component (the dienophile) carries one or two substituents of the kind represented by the carbonyl group in acrolein, the nitro group in nitroethylene; maleic anhydride and benzoquinone are also reactive dienophiles:

$$\begin{array}{llll}
\text{CH}_2 & \text{CH}_2 & & \\
\| & \| & & \\
\text{CH}\cdot\text{CHO} & \text{CH}\cdot\text{NO}_2 & & \\
\\
\text{acrolein} & \text{nitroethylene} & \text{maleic} & p\text{-benzoquinone} \\
& & \text{anhydride} &
\end{array}$$

Such olefins as ethylene itself, vinyl acetate, allyl chloride, have also been used as dienophiles, although they undergo the reaction less readily.

Some typical examples of the diene synthesis are the following:

$$\text{(20)}$$

1,3-butadiene          1,2,5,6-tetrahydrobenzaldehyde

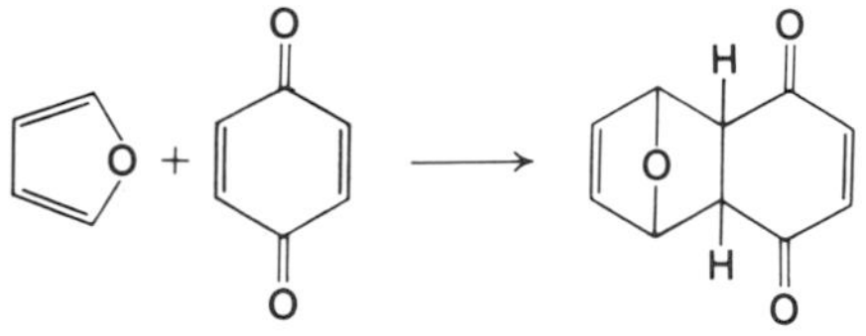

(21)

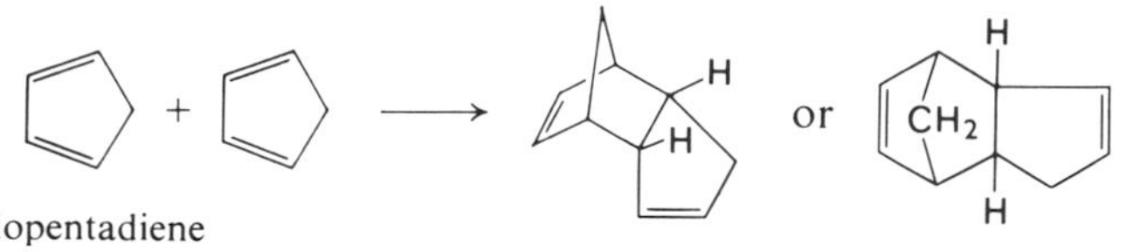

furan

(22)

cyclopentadiene

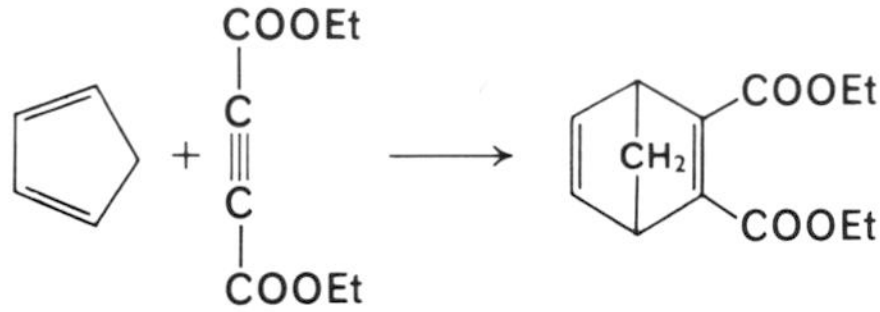

(23)

One of the advantages of the Diels-Alder reaction is that it leads to compounds of predictable stereochemistry. It will be noted that the configuration of the products shown in Equations (21) and (22) is that in which the rings are folded in such a way as to indicate the following course of the addition reaction. For Equation (22):

diene

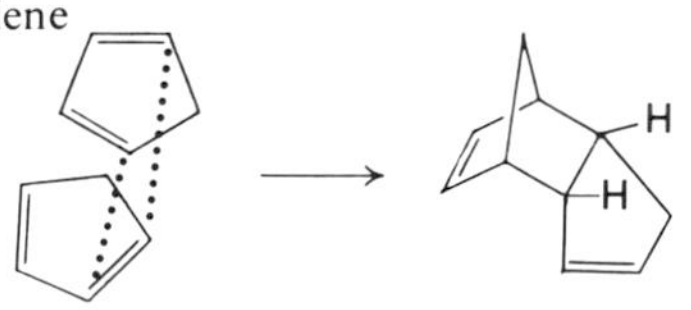

dienophile          *endo*-dicyclopentadiene

Maleic anhydride (dienophile) adds to furan (diene) in a comparable manner, giving the product shown in the following equation:

diene

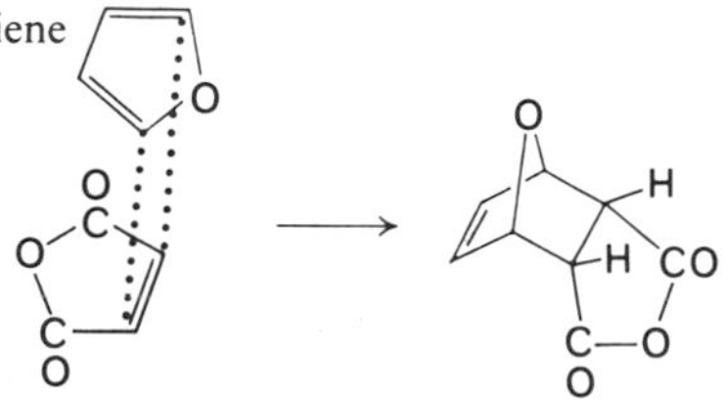

(24)

dienophile          *endo*-adduct

The products shown in Equations (22) and (24) have the *endo* configuration, so called because the five-membered ring is inside, toward the six-membered ring formed by the addition. The adduct with the *endo* configuration is the major product; only minor amounts of the *exo*-compound are formed:

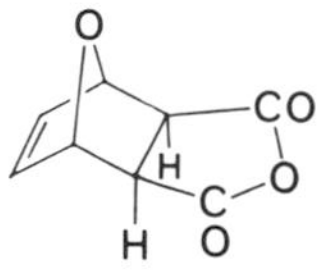

*exo*-adduct

These addition reactions occur in such a way as to provide maximum interaction of the unsaturated systems ($\pi$-orbital interaction) in the transition state. This stereospecificity, coupled with the versatility of the reaction in respect to the kinds of dienes and dienophiles that can be caused to react, makes the Diels-Alder reaction one of great synthetic utility. Among the products of Diels-Alder reactions that have proved to be of considerable practical importance are the insecticidal compounds *aldrin* and *dieldrin*, which are prepared according to the following equations:

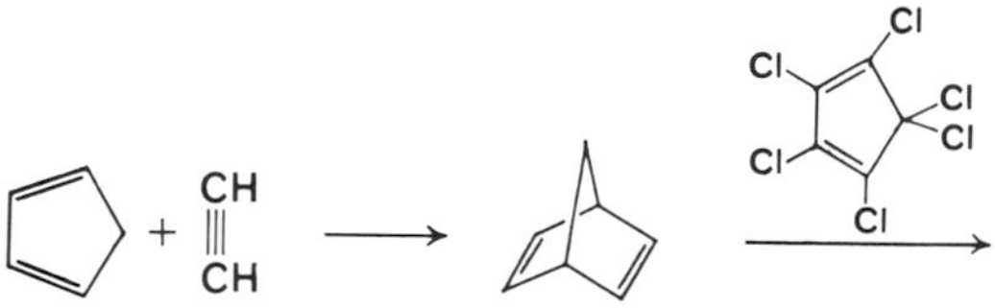

norbornadiene
(bicyclo[2,2,1]-2,5-heptadiene)

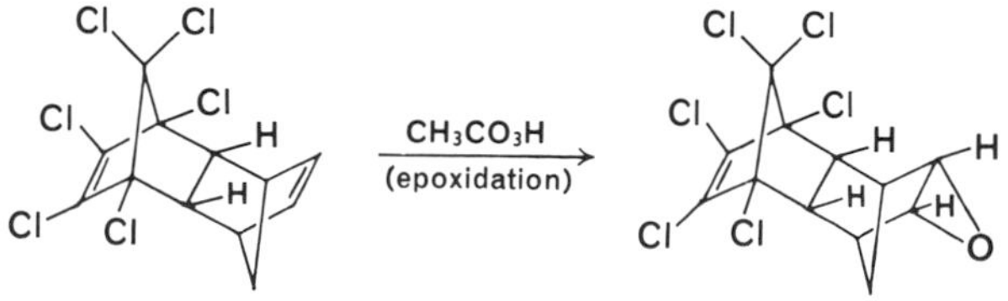

aldrin $\qquad$ dieldrin

## 17.10 Cyclization by carbon alkylation reactions

The alkylation of $\beta$-keto esters and malonic esters leads to carbon-carbon bond formation. When the displaceable halogen atom and the active methylene group are part of the same molecule, cyclization by intramolecular alkylation ensues:

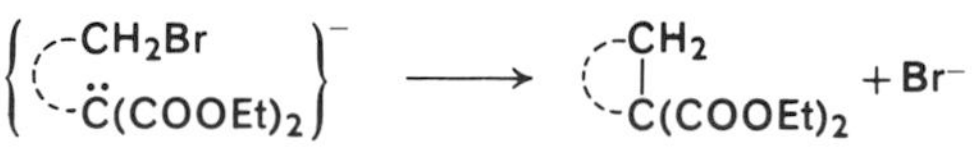

Numerous examples of this reaction are known. It is one of the most effective means of preparing three- and four-membered rings:

(a)  $\begin{array}{l} \text{CH}_2\text{Br} \\ | \\ \text{CH}_2\text{Br} \end{array}$ $+ \overset{+}{\text{Na}}(\text{CH(COOEt)}_2)^- \longrightarrow$ $\begin{array}{l} \text{CH}_2\text{—CH(COOEt)}_2 \\ | \\ \text{CH}_2\text{Br} \end{array}$ $+\text{Na}^+ + \text{Br}^-$

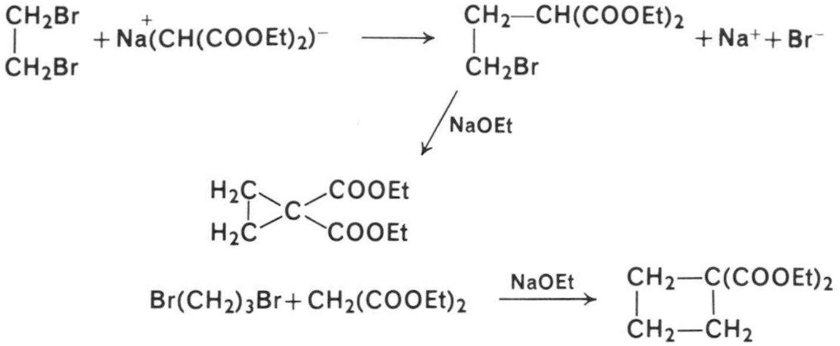

$\Big/ \text{NaOEt}$

(b)  $\begin{array}{c} \text{H}_2\text{C} \diagdown \quad \diagup \text{COOEt} \\ \quad \text{C} \\ \text{H}_2\text{C} \diagup \quad \diagdown \text{COOEt} \end{array}$    (25)

$\text{Br(CH}_2)_3\text{Br} + \text{CH}_2(\text{COOEt})_2 \xrightarrow{\text{NaOEt}} \begin{array}{l} \text{CH}_2\text{—C(COOEt)}_2 \\ | \qquad | \\ \text{CH}_2\text{—CH}_2 \end{array}$

Cyclopentane and cyclohexane derivatives are also readily prepared by this method:

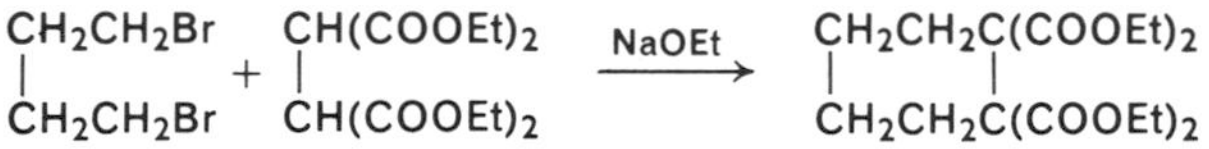

$\begin{array}{l} \text{CH}_2\text{CH}_2\text{Br} \\ | \\ \text{CH}_2\text{CH}_2\text{Br} \end{array} + \begin{array}{l} \text{CH(COOEt)}_2 \\ | \\ \text{CH(COOEt)}_2 \end{array} \xrightarrow{\text{NaOEt}} \begin{array}{l} \text{CH}_2\text{CH}_2\text{C(COOEt)}_2 \\ | \qquad\qquad | \\ \text{CH}_2\text{CH}_2\text{C(COOEt)}_2 \end{array}$    (26)

tetraethyl
ethane-1,1,2,2-
tetracarboxylate

---

**Exercise 2.**   Show how diethyl cyclopentane-1,1-dicarboxylate can be prepared from 1,4-dibromobutane and diethyl malonate.

---

### 17.11  Lactones

The formation of an ester by the reaction of an alcohol and an acid is a reaction that can be applied to ring closure. A hydroxy acid can undergo *intra*molecular ester formation to yield a cyclic ester, or *lactone*:

$\begin{array}{l} \text{C--------COOH} \\ | \\ \text{OH} \end{array} \longrightarrow \begin{array}{l} \text{C--------CO} \\ | \qquad\qquad | \\ \text{O————}\rfloor \end{array}$

This reaction is most successful when the lactone is a five- or six-membered cycle. Attempts to carry out lactone formation with lactic acid lead, not to the three-membered lactone ring, but to the six-membered dilactone formed from two molecules of the acid:

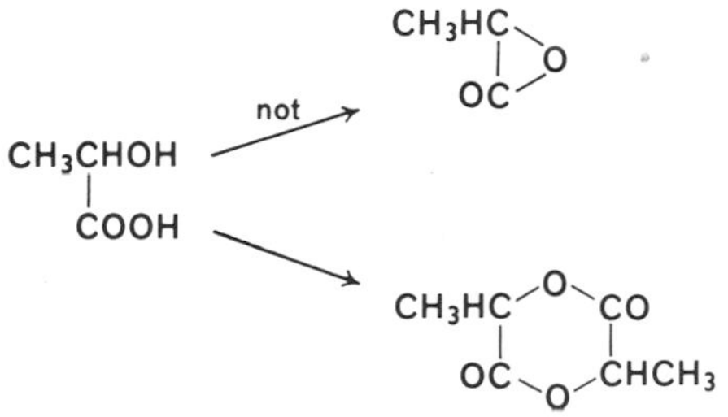

$\begin{array}{l} \text{CH}_3\text{CHOH} \\ | \\ \text{COOH} \end{array}$ $\xrightarrow{\text{not}}$ $\begin{array}{c} \text{CH}_3\text{HC} \diagdown \\ \quad | \quad \diagup \text{O} \\ \text{OC} \end{array}$    (27)

$\xrightarrow{\qquad}$ $\begin{array}{c} \text{CH}_3\text{HC} \diagup^{\text{O}} \diagdown \text{CO} \\ \quad | \qquad\qquad | \\ \text{OC} \diagdown_{\text{O}} \diagup \text{CHCH}_3 \end{array}$

"lactide"

Four-membered lactones are not readily formed, since the $\beta$-hydroxy acids that would be used for their preparation tend to lose the elements of water to give $\alpha,\beta$-unsaturated acids

$$\begin{array}{c} \text{R—CHCH}_2\text{COOH} \\ | \\ \text{OH} \end{array} \xrightarrow[\text{(H}^+)]{-\text{H}_2\text{O}} \text{RCH}=\text{CHCOOH} \tag{28}$$

or, in some cases, to decarboxylate (i.e., lose $CO_2$):

$$\begin{array}{c} \text{RCHCH}_2\text{COOH} \\ | \\ \text{OH} \end{array} \xrightarrow{\text{H}^+} \begin{array}{c} \text{R—CH}\curvearrowleft\text{CH}_2\text{—C}\diagdown\begin{array}{c}\text{O}\\\text{OH}\end{array} \\ | \\ \sqrt{}\ ^+\text{OH}_2 \end{array} \longrightarrow$$

$$\text{RCH}=\text{CH}_2+\text{H}_2\text{O}+\text{CO}_2+\text{H}^+ \tag{29}$$

When four-membered lactones are prepared by other means they prove to be very reactive compounds and undergo ring-opening reactions with great readiness. For example, $\beta$-propiolactone, which is prepared as an industrial chemical by the reaction of formaldehyde with ketene,

$$\begin{array}{c} \text{CH}_2=\text{C}=\text{O} \\ + \\ \text{CH}_2=\text{O} \end{array} \longrightarrow \begin{array}{c} \text{CH}_2\text{—C}=\text{O} \\ | \quad\quad | \\ \text{CH}_2\text{—O} \end{array} \tag{30}$$

$$\beta\text{-propiolactone}$$

is a very reactive substance. It reacts readily with acids to form the polymeric, open-chain ester

$$\begin{array}{c} \text{CH}_2\text{—CO} \\ | \quad\quad | \\ \text{CH}_2\text{—O} \end{array} \xrightarrow{\text{H}^+} \text{HOCH}_2\text{CH}_2\text{CO—O—(CH}_2\text{CH}_2\text{CO—O—)}_x\text{CH}_2\text{CH}_2\text{COOH} \tag{31}$$

and with nucleophilic reagents of various kinds to give $\beta$-substituted propionic acid derivatives.

Four- and 5-hydroxyalkanoic acids form the corresponding lactones with ease:

$$\text{HOCH}_2\text{CH}_2\text{CH}_2\text{COOH} \xrightarrow{-\text{H}_2\text{O}} \begin{array}{c} \text{H}_2\text{C}\text{———}\text{CH}_2 \\ | \quad\quad\quad | \\ \text{O}\diagdown_{\text{C}}\diagup\text{CH}_2 \\ \| \\ \text{O} \end{array} \tag{32}$$

$$\gamma\text{-butyrolactone}$$

quinic acid          quinide

$$\tag{33}$$

The formation of lactones by the intramolecular displacement of halogen (e.g. bromine) by the ionized carboxyl group is analogous to the formation of esters by the intermolecular displacement reaction:

$$RCOO^- + R'Br \longrightarrow RCOOR' + Br^- \quad \text{(intermolecular)}$$

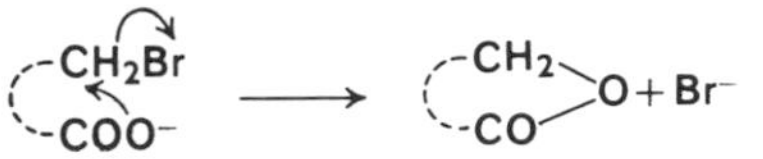


(intramolecular)

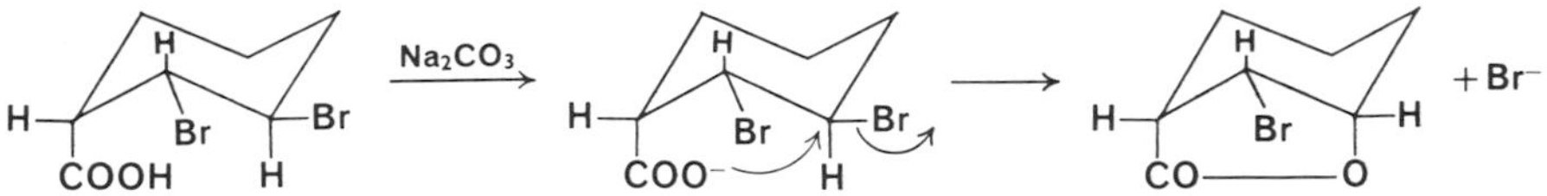


(34)

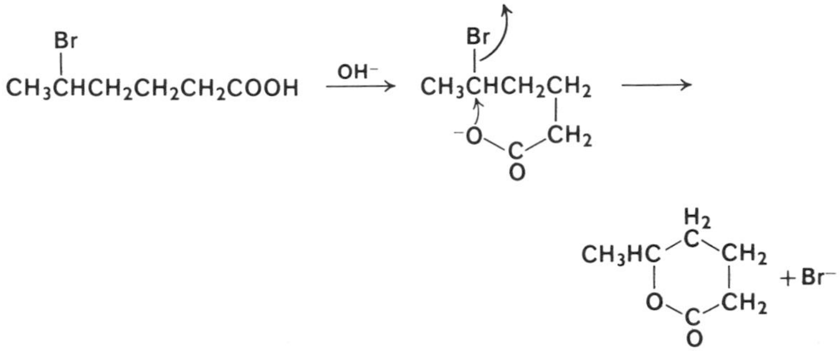


(35)

## 17.12 Large-ring lactones

If an attempt is made to prepare the lactone by heating a hydroxy acid in which the —OH group is farther away from the carboxyl group than the δ-position, e.g., as in 7-hydroxyheptanoic acid, the lactone does not form. Instead, the polymeric ester

$$HOCH_2CH_2CH_2CH_2CH_2CH_2CO{-}(O{-}CH_2CH_2CH_2CH_2CH_2CH_2CO{-})_x{-}O{-}CH_2CH_2CH_2CH_2CH_2CH_2COOH$$

is formed. However, if this ester is heated under reduced pressure at a temperature above the boiling point of the monomeric lactone, the equilibrium

$$HO(CH_2)_6CO-\{-O-(CH_2)_6-CO\}_x O(CH_2)_6COOH \rightleftharpoons (x+2)\underset{O\text{———}CO}{(CH_2)_6} \quad (36)$$

will be established (by ester interchange within the polymeric chain), and the monomeric lactone will slowly distil from the heated polymer.

Large-ring lactones can be prepared from large-ring ketones (Sections 17.15 through 17.17) by an intramolecular oxidative ring expansion. The reagent used is a peroxy acid (for example, peroxybenzoic acid), and the reaction is known as the *Baeyer-Villiger reaction.* The oxidation of acyclic ketones by this method yields esters:

$$R-\underset{\underset{O}{\|}}{C}-R + C_6H_5CO_3H \longrightarrow R-\underset{\underset{O}{\|}}{C}-OR + C_6H_5COOH \quad (37)$$

Applied to cyclic ketones, the reaction follows the same course, except that the cyclic ester, or lactone, results. A very significant feature of the Baeyer-Villiger reaction as applied to large-ring ketones is that large-ring lactones are formed directly. Since the same lactones cannot be formed by direct ring-closure of the corresponding hydroxy acids, this shows that the Baeyer-Villiger reaction does not yield an open-chain (acyclic) intermediate, which then closes to the lactone:

$$(CH_2)_n \overset{CH_2}{\underset{CO}{|}} \xrightarrow{RCO_3H} (CH_2)_n \overset{H_2C}{\underset{C}{<}} O \ ; \ not \ via \ (CH_2)_n \overset{CH_2OH}{<}_{COOH} \qquad (38)$$

A discussion of the details of this reaction will be found in a later chapter.

### 17.13 Cyclic amides (lactams)

The formation of cyclic amides by the heating of amino acids or esters will be successful if the cycle formed contains five or six members:

$$\underset{NH_2}{\overset{|}{CH_2CH_2CH_2COOEt}} \longrightarrow \overset{H_2C\text{———}CH_2}{\underset{HN\diagdown_C\diagup CH_2}{|}} + EtOH \qquad (39)$$
$$\underset{O}{}$$

2-pyrrolidone

As in the cyclization of hydroxy acids, seven- and higher-membered cyclic amides are not formed in this way, and polymeric linear amides result:

$$NH_2(CH_2)_5COOEt \longrightarrow NH_2(CH_2)_5CO\{NH(CH_2)_5{-}CO\}_xNH(CH_2)_5COOEt \qquad (40)$$

The last example shown represents a method for the preparation of a commercially important polymer which has properties similar to those of Nylon. The polymeric amide of 6-aminohexanoic acid is called "Perlon L," or "6-Nylon." It is prepared in practice by a process which is the reverse of that in which a monomeric lactone is formed from a polymeric ester (Section 17.12)—that is, by heating the cyclic seven-membered monomeric lactam:

$$\overset{\text{(ring structure with }=O, N, H)}{} \xrightarrow{\Delta} \text{-------}\{NHCO(CH_2)_5\}_{\bar{x}}\text{------} \qquad (41)$$

Large-ring cyclic lactams cannot be prepared by the direct cyclization of long-chain amino acids, but can be synthesized by an intramolecular ring-expansion reaction known as the Beckmann rearrangement. An example of this reaction is the following; a thorough discussion of its nature will be found in a later chapter:

$$\overset{\text{(cyclohexanone)}}{} \xrightarrow{H_2NOH} \overset{\text{(cyclohexanone oxime)}}{} \xrightarrow{H_2SO_4} \overset{\text{(caprolactam)}}{} \qquad (42)$$

### 17.14 Cyclic acid anhydrides

The principles that have been illustrated above for ring formation, showing that five- and six-membered rings form readily by intramolecular analogues of known intermolecular bond-forming reactions, apply to many other kinds of ring formations.

Acids that contain two carboxyl groups disposed such that five- and six-membered anhydrides can form are readily dehydrated to form the anhydride:

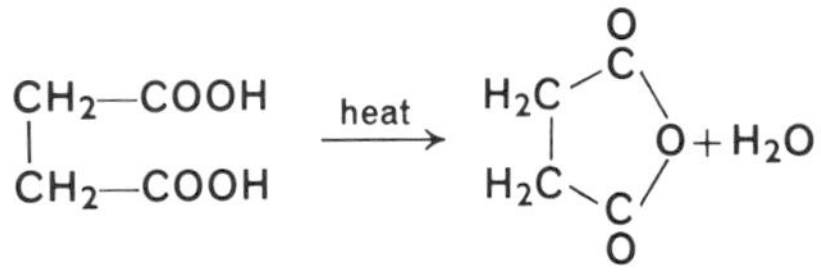

$$\text{(43)}$$

succinic acid          succinic anhydride

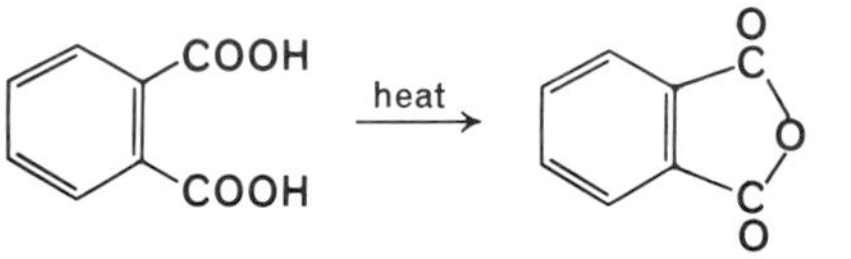

$$\text{(44)}$$

phthalic acid          phthalic anhydride

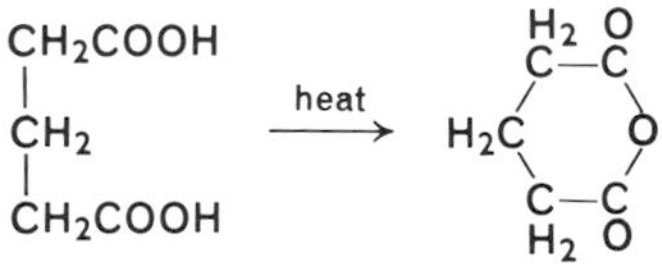

$$\text{(45)}$$

glutaric acid          glutaric anhydride

Treatment of succinic acid with acetyl chloride or acetic anhydride is also a means of transforming it into the anhydride. Oxalic and malonic acids, which might be expected to form three- and four-membered cyclic anhydrides, do not do so. Malonic acid can be converted to an "anhydride" of quite another kind; when it is heated with phosphorus pentoxide two molecules of water are lost and suboxide, $C_3O_2$, is formed:

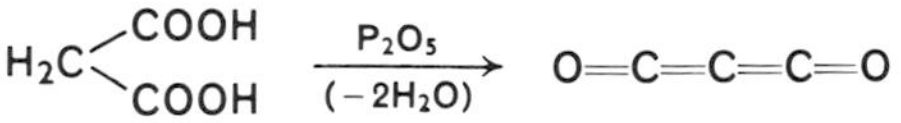

$$\text{(46)}$$

**Exercise 3.** What would you predict the shape of the carbon suboxide molecule to be?

Simple heating of malonic acid, *and of 1,1-dicarboxylic acids in general*, leads to decarboxylation.

The cyclic anhydride of adipic acid,

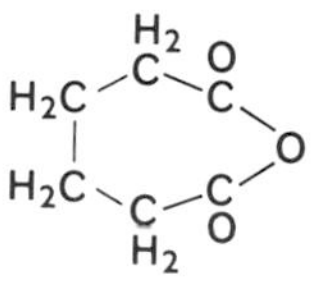

is a seven-membered cyclic compound, and is not formed from adipic acid by the simple methods used in preparing succinic or glutaric anhydride. The anhydride formed from adipic acid by treatment with refluxing acetic anhydride is the polymeric linear anhydride:

$$(CH_2)_4 \begin{array}{c} COOH \\ COOH \end{array} \xrightarrow{Ac_2O} HOOC(CH_2)_4CO-\{-O-CO(CH_2)_4CO-\}_x-O-CO(CH_2)_4COOH$$

$$(47)$$

The polymer can be converted into the monomeric cyclic anhydride by slow distillation under vacuum.

Cyclic anhydrides are useful derivatives for certain synthetic applications. Treatment of succinic anhydride, for instance, with an alcohol, leads to a half-ester of succinic acid:*

$$\begin{array}{c} H_2C^{-C \searrow O}_{\phantom{C}\nearrow} \\ H_2C_{\searrow C}^{\phantom{C}} \\ O \end{array} + RCH_2OH \longrightarrow \begin{array}{c} CH_2COOCH_2R \\ | \\ CH_2COOH \end{array} \qquad (48)$$

Half-esters of this kind are often useful derivatives of alcohols: being acids, as well as esters, they can be titrated with standard alkali and their equivalent weights (and thus the molecular weight of the alcohol) determined. Manipulation of the —COOH group leads to synthetic results not always easy to arrive at by other means. For example,

$$\begin{array}{c} H_2\ O \\ C-C \\ H_2C \phantom{xx} O \\ C-C \\ H_2\ O \end{array} + EtOH \longrightarrow H_2C \begin{array}{c} CH_2COOEt \\ CH_2COOH \end{array} \xrightarrow{SOCl_2}$$

$$H_2C \begin{array}{c} CH_2COOEt \\ CH_2COCl \end{array} \xrightarrow{(CH_3)_2Cd} CH_3COCH_2CH_2CH_2COOEt \qquad (49)$$

ethyl 5-oxohexanoate

The last step of this series of reactions involves the reaction of an acyl chloride with the organometallic compound, dimethylcadmium. Dialkylcadmium compounds are prepared by the reaction of a Grignard reagent with cadmium chloride:

$$2\ RMgCl + CdCl_2 \longrightarrow R_2Cd + 2\ MgCl_2 \qquad (50)$$

Although Grignard reagents react with acyl halides, the reaction ordinarily proceeds past the stage of the intermediate ketone and leads to the tertiary alcohol (Chapter 8). Dialkylcadmium compounds, however, are sufficiently less reactive than Grignard reagents to permit their use in the synthesis of ketones by the above procedure.

---

* The term "half-ester" is a convenient descriptive designation that has no proper place in formal nomenclature. The half-ester formed by the reaction of succinic anhydride with ethanol is properly called *ethyl hydrogen succinate*.

**17.15  Formation of cyclic ketones from dibasic acids**

Under somewhat more vigorous conditions (by distilling a mixture of the acid and acetic anhydride at atmospheric pressure) adipic and pimelic acids are converted into ketones, with concomitant loss of $CO_2$:

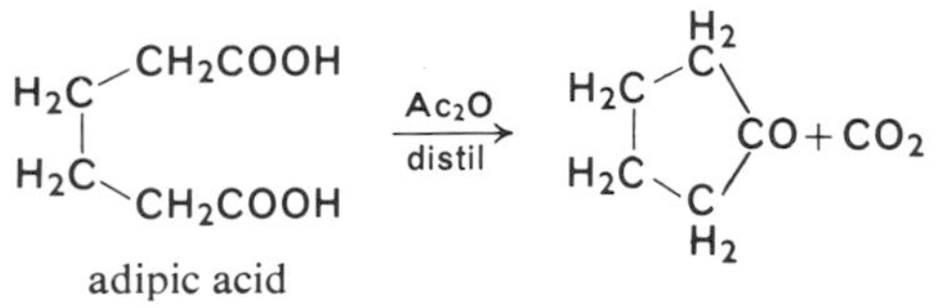

(51)

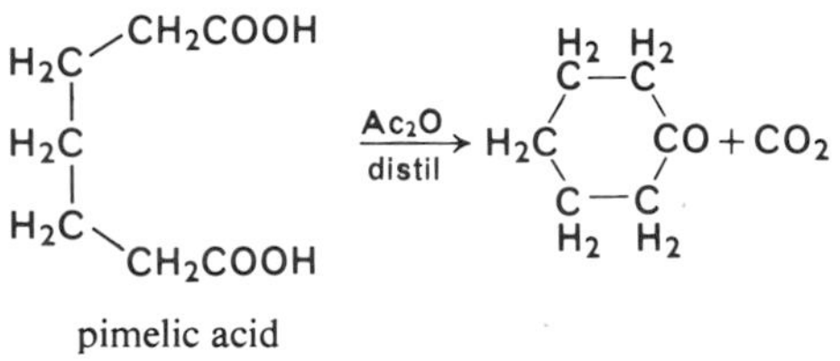

(52)

The differences between the behavior of succinic, glutaric, adipic, and pimelic acids on cyclization have been expressed in a generalization known as Blanc's rule:

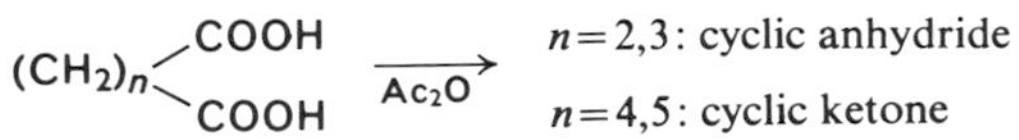

$$(CH_2)_n \underset{COOH}{\overset{COOH}{\Big\langle}} \xrightarrow{Ac_2O} \begin{array}{l} n=2,3: \text{cyclic anhydride} \\ n=4,5: \text{cyclic ketone} \end{array}$$

This has been of use in structure determination, as shown in the following example:

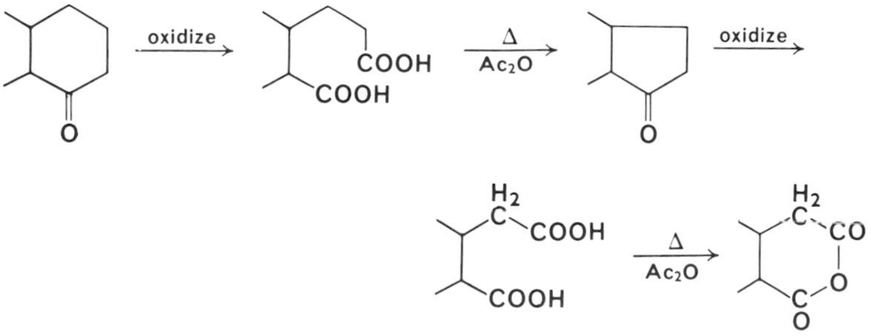

(53)

Since the acid produced in the first oxidation gave the cyclic ketone, and that produced in the second oxidation gave the anhydride, the sizes of the rings are thus known to be those shown in the above sequence. The reader should examine the alternative courses for an initial seven-membered ring ketone and for an initial five-membered ring ketone to assure himself of the validity of the structural conclusions reached by these manipulations. There are some exceptions to the rule, and thus it cannot be relied on absolutely, but exceptions are rare.

## 17.16 Large rings (macrocycles)

Compounds containing many-membered rings (that is, of from 8 to 20–25 members) are interesting for several reasons. They are of theoretical importance because their stability and ease of formation (by special methods that will be described) are evidence that the Baeyer theory of ring strain, based upon the premise of planarity of the rings, is incorrect for rings larger than five members. A curious and interesting aspect of the large-ring compounds is that the musk-like odorous constituents of valuable perfumes are large-ring compounds:

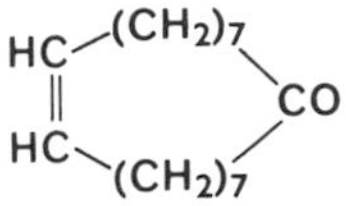

civetone 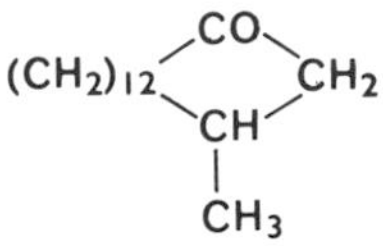          muscone

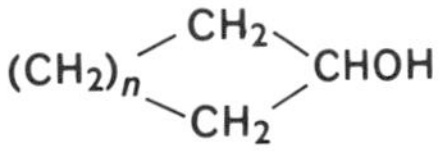

$n = 10, 12, 14, 16$
found in musk of muskrat

It is interesting to note the similarity in structure of these compounds and commonly occurring long-chain fatty acids

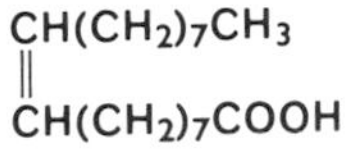

oleic acid
(compare with civetone)

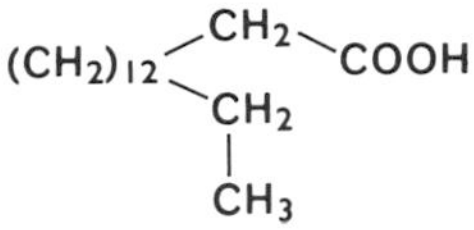

palmitic acid
(compare with muscone)

These relationships suggest a close biogenetical relationship of these naturally occurring macrocyclic ketones and the fatty acids.

## 17.17 The synthesis of large-ring compounds

The earliest systematic study of macrocyclic compounds was stimulated by an interest in the compounds civetone and muscone, which are found in the scent glands of certain animals and which are important ingredients in perfumes. Ruzicka (1887–), after establishing the structures of these compounds by degradation to known substances, embarked upon a study of the means of preparing macrocyclic ketones, and first prepared the cycloalkanones by the simple device of heating heavy-metal salts (lead, barium, thorium) of dibasic fatty acids:

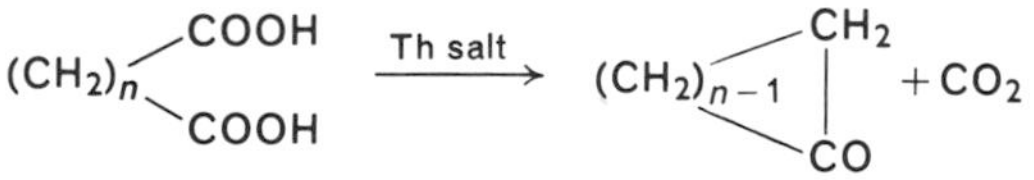

$$(CH_2)_n \overset{\displaystyle COOH}{\underset{\displaystyle COOH}{\Big\langle}} \xrightarrow{\text{Th salt}} (CH_2)_{n-1} \overset{\displaystyle CH_2}{\underset{\displaystyle CO}{\Big\langle}} \Big| + CO_2 \tag{54}$$

In recent years numerous other methods of preparing large rings have been discovered, one of the most successful of these being the *acyloin* synthesis. In noncyclic systems, acyloins are formed by the reaction of esters with metallic sodium:

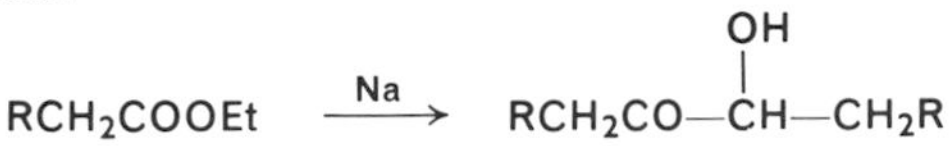

$$RCH_2COOEt \xrightarrow{\text{Na}} RCH_2CO-\overset{\displaystyle OH}{\overset{\displaystyle |}{CH}}-CH_2R \tag{55}$$

an acyloin

The use of the ester of a dibasic acid leads to the cyclic acyloin:

$$(CH_2)_n \begin{matrix} COOEt \\ COOEt \end{matrix} \xrightarrow[\text{xylene}]{Na} (CH_2)_n \begin{matrix} CO \\ | \\ CHOH \end{matrix} \qquad (56)$$

The yields in this reaction are surprisingly high: in the range $n = 7–11$, where Ruzicka's method gave exceedingly small yields, the acyloin synthesis gives yields of about 40%. The $C_{21}$ acyloin ($n = 19$) is formed in 86% yield! The total synthesis of natural *cis*-civetone has been carried out by the use of the acyloin procedure as follows:

$$O = \begin{matrix} (CH_2)_7COOMe \\ (CH_2)_7COOMe \end{matrix} \xrightarrow[\begin{matrix} CH_2OH \\ | \\ CH_2OH \end{matrix}]{} \begin{matrix} O \\ \diagdown \diagup \\ O \end{matrix} \begin{matrix} (CH_2)_7COOEt \\ (CH_2)_7COOEt \end{matrix} \xrightarrow{Na}$$

$$\begin{matrix} O \\ \diagdown \diagup \\ O \end{matrix} \begin{matrix} (CH_2)_7{-}CO \\ | \\ (CH_2)_7{-}CHOH \end{matrix} \xrightarrow[\begin{matrix} (1) \text{ reduce} \\ (2) \text{ HBr} \\ (3) \text{ Zn/EtOH} \end{matrix}]{} \begin{matrix} O \\ \diagdown \diagup \\ O \end{matrix} \begin{matrix} (CH_2)_7{-}CH \\ || \\ (CH_2)_7{-}CH \end{matrix} \xrightarrow{H_2O/H^+} \qquad (57)$$

$$\begin{matrix} CH_2OH \\ | \\ CH_2OH \end{matrix} + O = \begin{matrix} (CH_2)_7{-}CH \\ || \\ (CH_2)_7{-}CH \end{matrix}$$

civetone
(mixture of *cis* and *trans* isomers)

The *cis* form proved to be identical with natural civetone. It is interesting to note that in a ring having seventeen members, the *trans* configuration of the double bond is possible. There is sufficient flexibility in this large ring to permit this configuration. In cyclohexene, a *trans* form is stereochemically impossible. This can best be seen by an examination of suitable molecular models.

The musk character typical of muscone and civetone appears to be associated with the presence of a large ring, and some synthetic macrocyclic lactones, ketones, and anhydrides with fourteen to eighteen members in the ring have odorous characteristics similar to those of the natural musks. It is an interesting observation that the androstenols and androstenone have musk-like or urine-like odors:

$R' = OH, R = H; R' = H, R = OH$ (androstenols)

$R', R = {>}{=}O$ (androstenone)

civetone

## 17.18 Cyclic substances of natural occurrence

Naturally occurring substances with alicyclic (alicyclic: from *ali*phatic, *cyclic*) rings abound in nature; those with single and fused five- and six-membered rings constitute a large class of compounds among which are found the terpenes, polyterpenes, and steroids; these will be discussed separately in Chapter 32.

The occurrence of three- and four-membered rings in nature is limited, except for those compounds which contain a cyclopropane or cyclobutane ring fused with another ring, as in thujone, sabinol, pinene, verbenol, and other terpenoid compounds:*

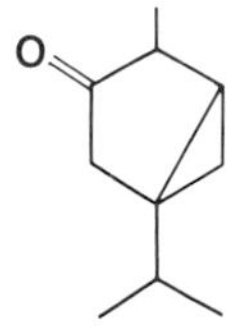

thujone 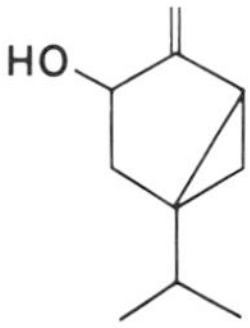          sabinol 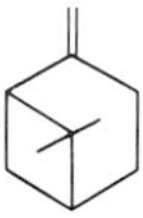          β-pinene 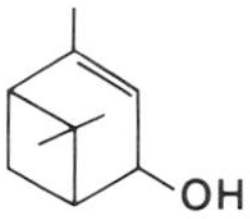          verbenol

A group of four important cyclopropane derivatives found in the flowers of *Chrysanthemum cinerarifolium* are the valuable insecticides known as "pyrethrins." These are structurally related compounds, called pyrethrin I, pyrethrin II, cinerin I, and cinerin II. Pyrethrin I is an ester of the cyclopropane derivative *chrysanthemic acid*, which occurs in the (+)-*trans* form:

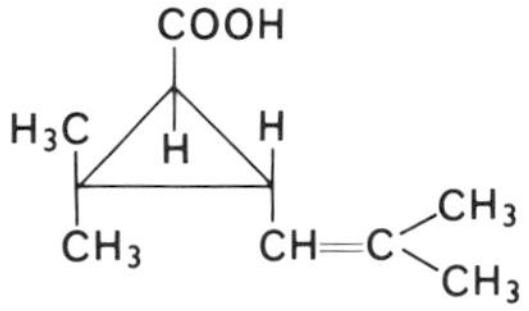

chrysanthemic acid

In pyrethrin II the side chain is

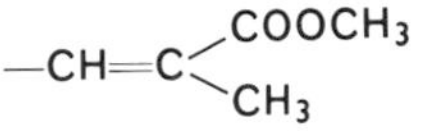

The complete structure of pyrethrin I is

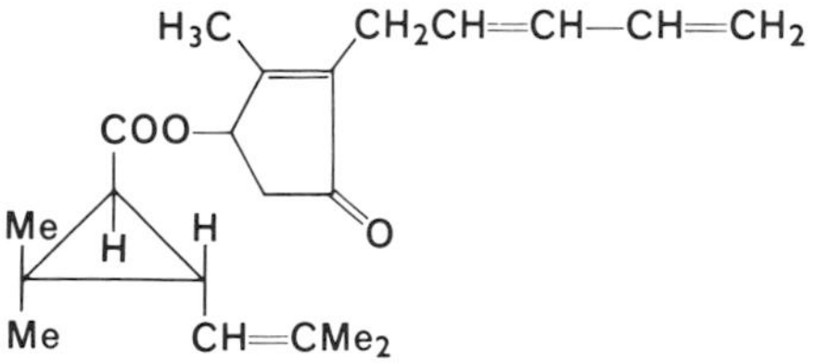

pyrethrin I

---

* In writing structures of terpenes, the abbreviated notation shown is common. In these formulas the lines, as in 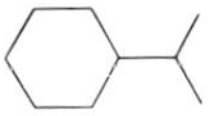 and , are abbreviations for 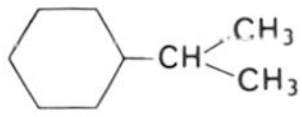 and 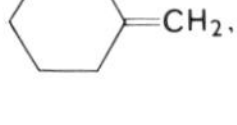 respectively.

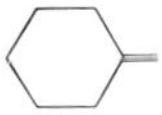

Two cyclopropane derivatives, lactobacillic acid (from *Lactobacillus arabinosus*) and sterculic acid (from the seed oil of *Sterculia foetida*), have the following structures:

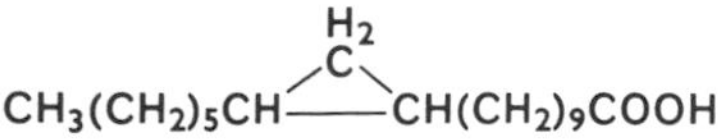

$$CH_3(CH_2)_5CH\overset{\overset{\displaystyle H_2}{C}}{\underset{}{\longrightarrow}}CH(CH_2)_9COOH \qquad CH_3(CH_2)_7C\overset{\overset{\displaystyle H_2}{C}}{\underset{}{=\!\!=}}C(CH_2)_7COOH$$

lactobacillic acid 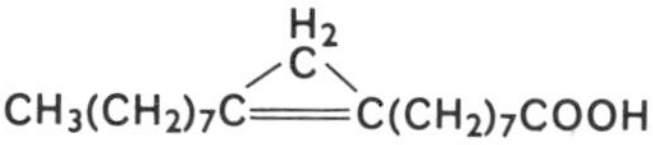　　　　　　　　　　　　　　sterculic acid

Cyclobutane derivatives occur in natural sources, and some of them appear to be formed by the dimerization of unsaturated compounds. The truxinic and truxillic acids found in the leaves of *Erythroxylon coca* (from which cocaine is isolated) are dimers of cinnamic acid:

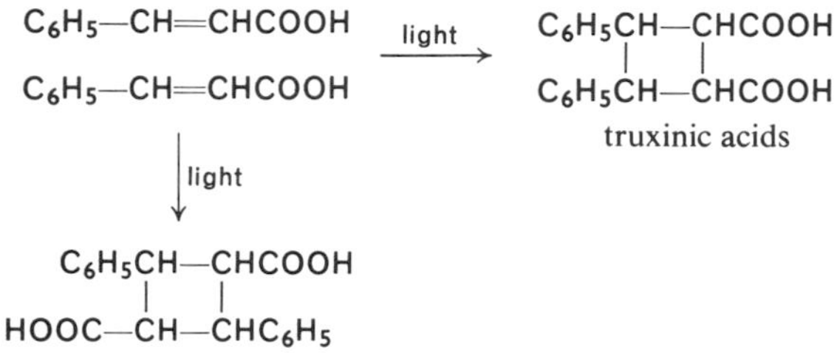

$$\begin{array}{c} C_6H_5-CH\!=\!CHCOOH \\ C_6H_5-CH\!=\!CHCOOH \end{array} \xrightarrow{\text{light}} \begin{array}{c} C_6H_5CH-CHCOOH \\ |\quad\ | \\ C_6H_5CH-CHCOOH \end{array}$$

truxinic acids

$$\Big\downarrow \text{light}$$

$$\begin{array}{c} C_6H_5CH-CHCOOH \\ |\quad\ | \\ HOOC-CH-CHC_6H_5 \end{array} \tag{58}$$

truxillic acids

Both the truxillic and the truxinic acids can exist in stereoisomeric forms which differ in the relative arrangement of the —C₆H₅ and the —COOH groups above and below the plane of the ring. For example, the following configurations are possible for the truxillic acids; all of these are known:

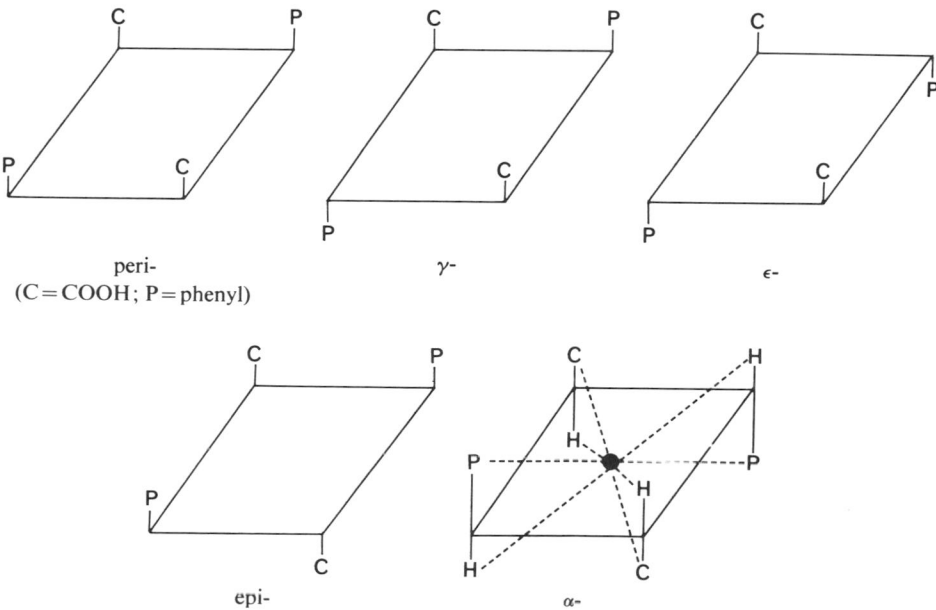

peri-　　　　　　　　　γ-　　　　　　　　　ε-
(C=COOH; P=phenyl)

epi-　　　　　　　　　α-

These formulas are drawn in simplified fashion to make the stereochemical features more obvious. It will be noted that all but the α-acid have planes of symmetry. The α-acid is of special interest: it has no plane of symmetry, yet it is not capable of existing in enantiomeric forms; the mirror images are super-imposable. This molecule has a *center of symmetry*, indicated by the dotted lines in the formula.

---

**Exercise 4.**   How many stereoisomers of truxinic acid are possible, and how many of these are resolvable?

---

The interesting cyclobutane derivative *anemonin* is formed in the steam distillation of the oil from certain plants of the genera *Anemone* and *Ranunculus*. Anemonin is formed during the isolation process by the dimerization of the simple lactone protoanemonin:

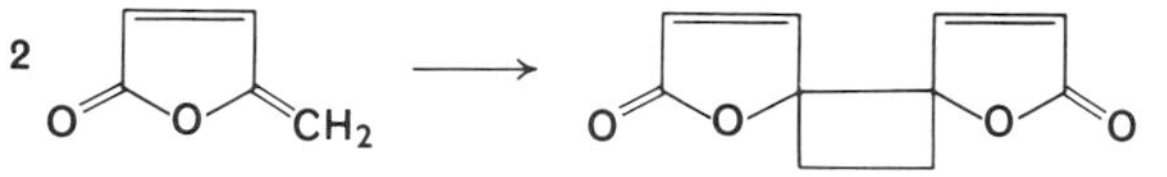

protoanemonin          anemonin

## 17.19   The tropolones

A group of naturally occurring compounds, some of them the products of the metabolism of certain fungi, others occurring as constituents of the wood of certain coniferous trees, contain seven-membered rings, and are known as *tropolones*. Typical of these are the *thujaplicins*:

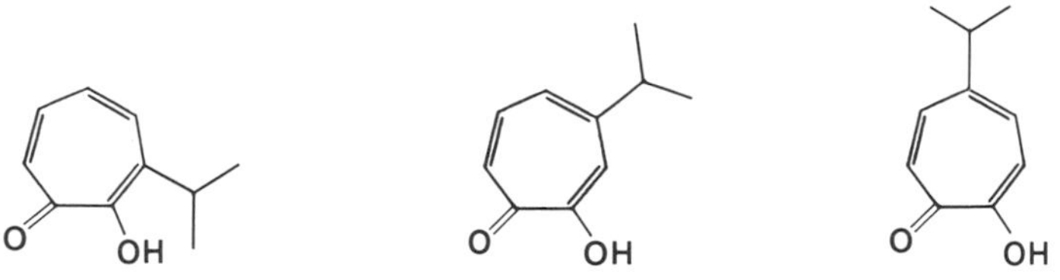

α-                    β-                    γ-

thujaplicins (isopropyltropolones)

The tropolone ring is unique, since it possesses a conjugated system of double bonds that confers a high degree of stability to the structure. The hydroxyl group has a degree of acidity approaching that of the carboxylic acids and it can be seen that the ion formed by removal of the proton from tropolone can be stabilized by resonance among equivalent contributing structures:

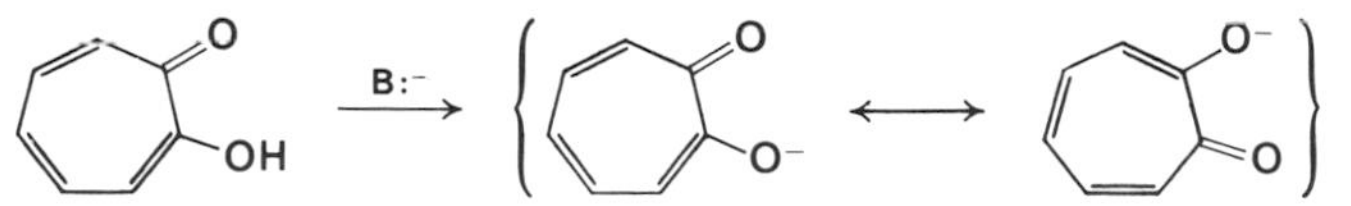

$$\tag{59}$$

Other seven-membered ring compounds are found among the terpenes (Chapter 32); and the tropolone ring system as well as another seven-membered ring, is present in colchicine, a compound present in the crocus, *Colchicum autumnale:*

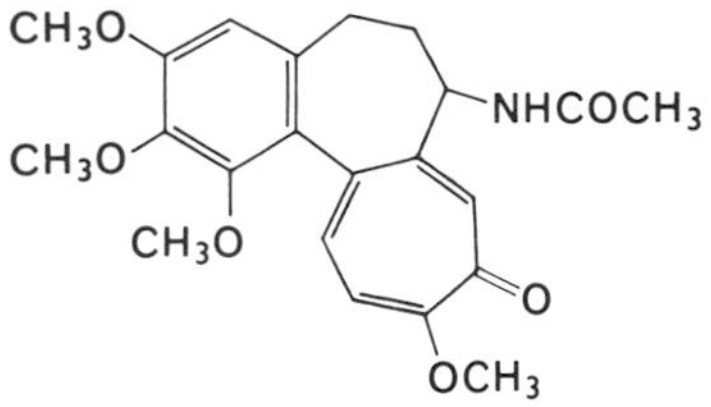

colchicine

Colchicine is a biologically important substance. Besides its use in medicine as an effective remedy for gout, it has the remarkable property of arresting cell division and can be used to produce experimental chromosome doubling in plants and animals.

Rings of "medium" size (more than seven, less than seventeen members) are rare in nature. The terpene caryophyllene is unusual in that it contains a nine-membered ring fused to a cyclobutane ring:

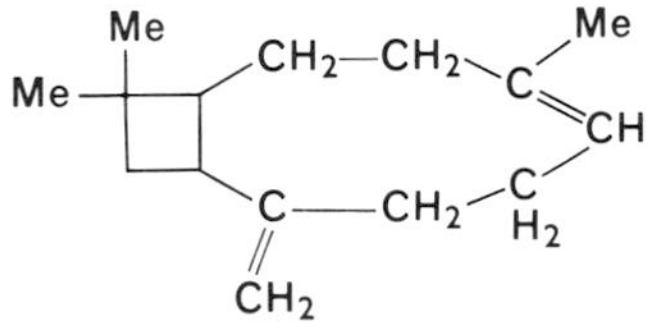

$\beta$-caryophyllene

## 17.20 Applications to synthesis

1. Prepare 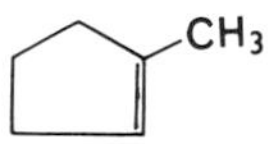 starting with cyclohexanone.

*(a)* The desired product can be prepared from cyclopentanone by the addition of methylmagnesium bromide, followed by the dehydration of the resulting carbinol:

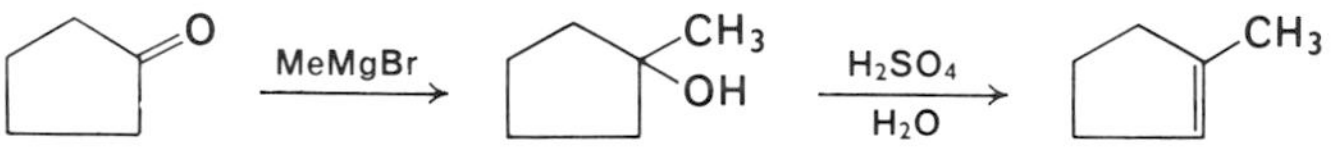

*(b)* Cyclopentanone can be prepared from cyclohexanone by converting the latter into adipic acid (by direct oxidation, or by way of cyclohexanol → cyclohexene), followed by the conversion of the adipic acid into cyclopentanone:

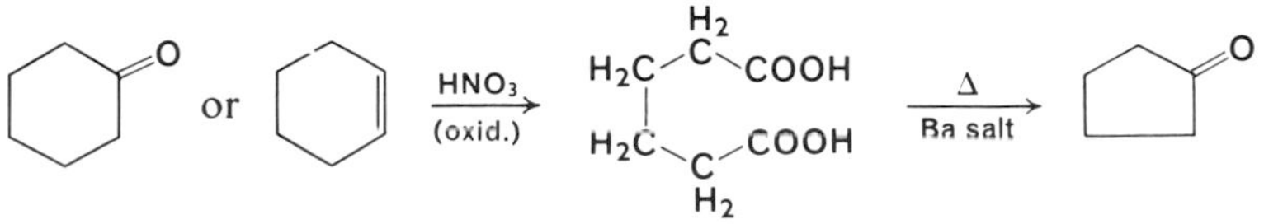

2. Prepare 

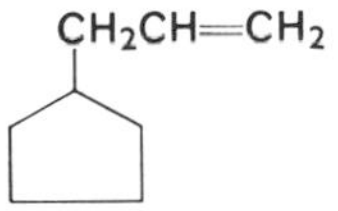

starting with cyclohexanone.

*(a)* Adipic acid, prepared from cyclohexanone, can be cyclized (as its diethyl ester) by the Dieckmann condensation to give ethyl cyclopentanone-2-carboxylate. This is a β-keto ester, and can be alkylated with allyl bromide:

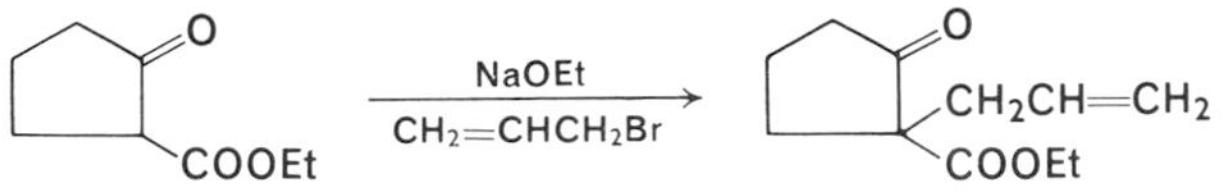

*(b)* Hydrolysis of the allylated ester gives the β-keto acid, which is readily decarboxylated by heat to give the ketone:

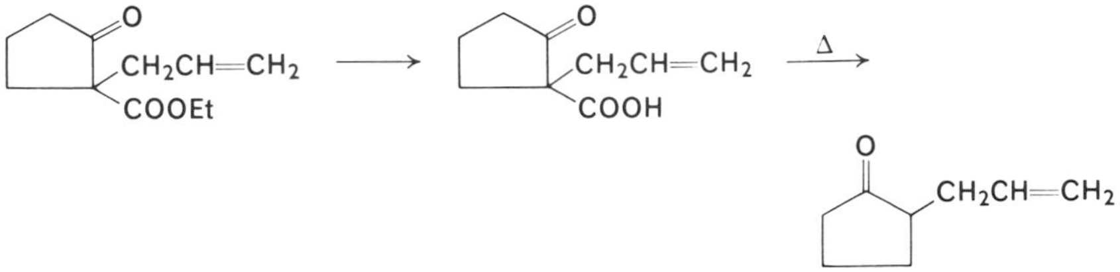

*(c)* The final step is the reduction of the carbonyl group to a methylene group $(CO \rightarrow CH_2)$. Since the allyl group contains a double bond that must be preserved, certain limitations are imposed upon the possible methods of reduction that can be used. One useful method for the specific reduction of a carbonyl group to a methylene group is the desulfurization, by means of Raney nickel, of the thioketal of the carbonyl compound:

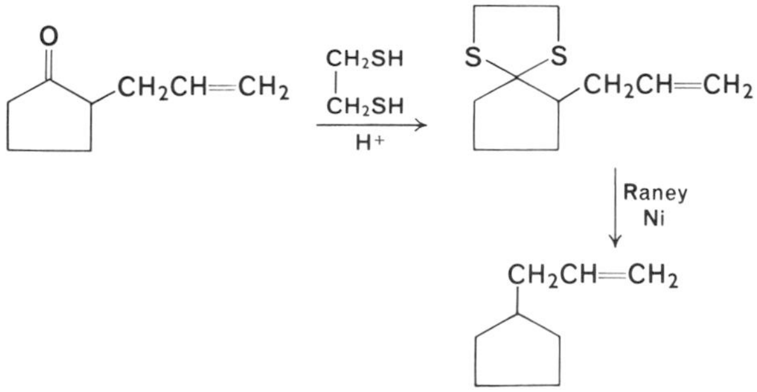

An alternative method of reduction of the allylcyclopentanone would be by the Wolff-Kishner method.

3. Prepare 

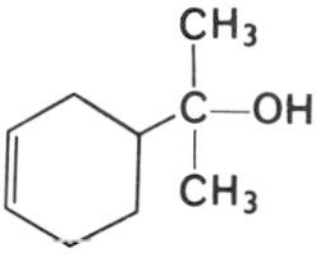

*(a)* The desired product can be prepared by the action of methylmagnesium bromide on either the corresponding methyl ketone or the ester:

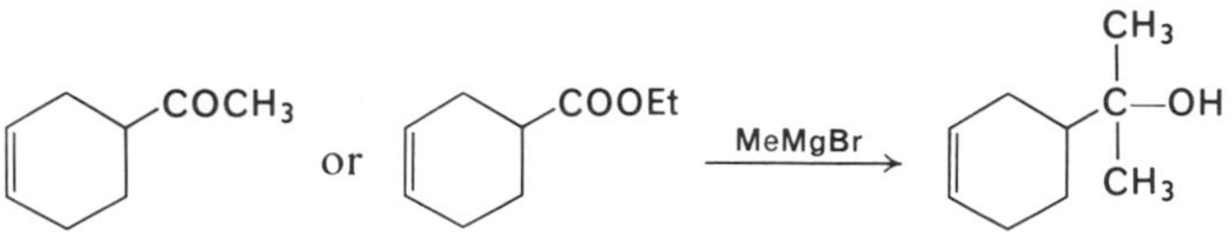

*(b)* The ketone (or the ester) can be prepared by a Diels-Alder reaction; for the ketone, this is the following:

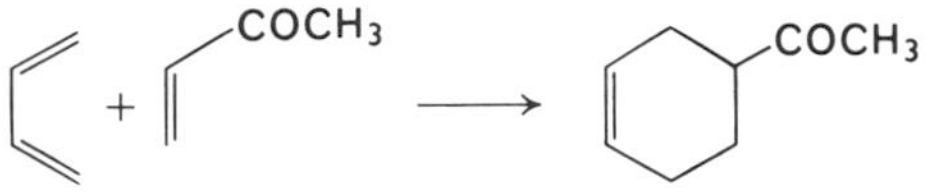

4. Prepare 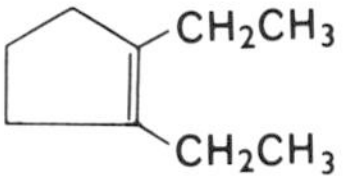 , starting with tetrahydrofuran,

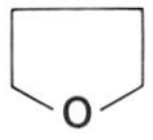

*(a)* The final step can be the dehydration of the following carbinol:

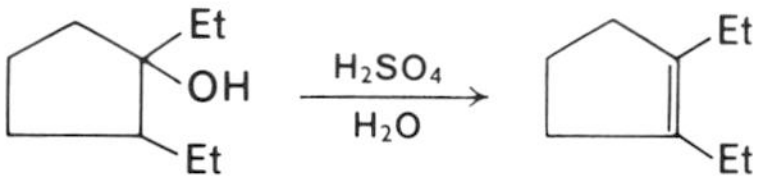

*(b)* The carbinol can be prepared by the addition of ethylmagnesium bromide to 2-ethylcyclopentanone:

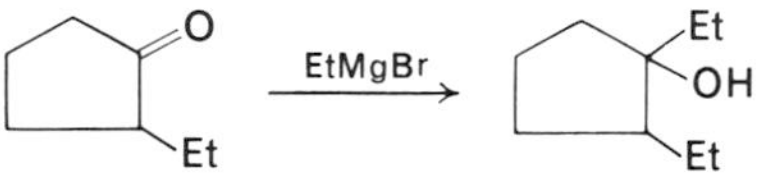

*(c)* 2-Ethylcyclopentanone can be prepared by the C-ethylation of ethyl-cyclopentanone-2-carboxylate, followed by saponification and decarboxylation:

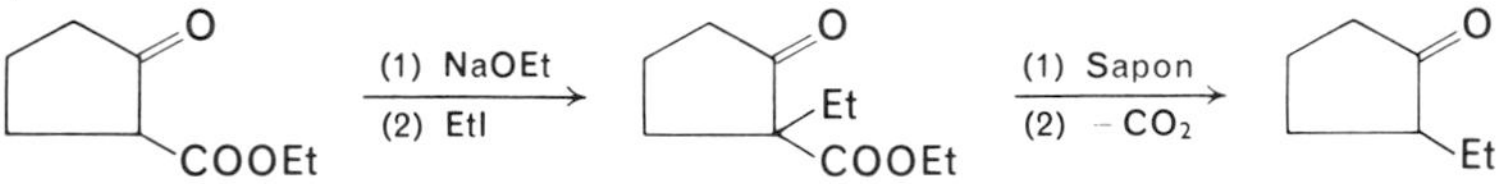

*(d)* The keto ester can be prepared by Dieckmann cyclization of diethyl adipate (as in example 2); and diethyl adipate can be prepared by ethanolysis of 1,4-dicyanobutane:

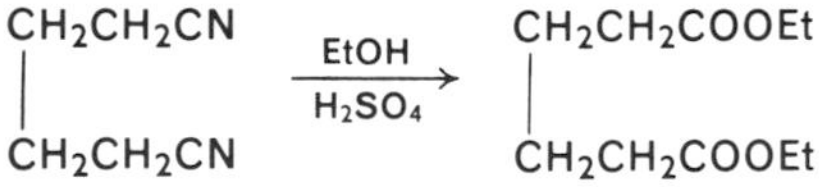

$$CH_2CH_2CN \quad \xrightarrow[H_2SO_4]{EtOH} \quad CH_2CH_2COOEt$$
$$| \qquad\qquad\qquad\qquad | $$
$$CH_2CH_2CN \qquad\qquad\quad CH_2CH_2COOEt$$

*(e)* 1,4-Dicyanobutane can be prepared by the action of potassium cyanide on 1,4-dibromobutane; and the dibromo compound can be prepared from tetrahydrofuran by vigorous treatment of this cyclic ether with concentrated HBr:

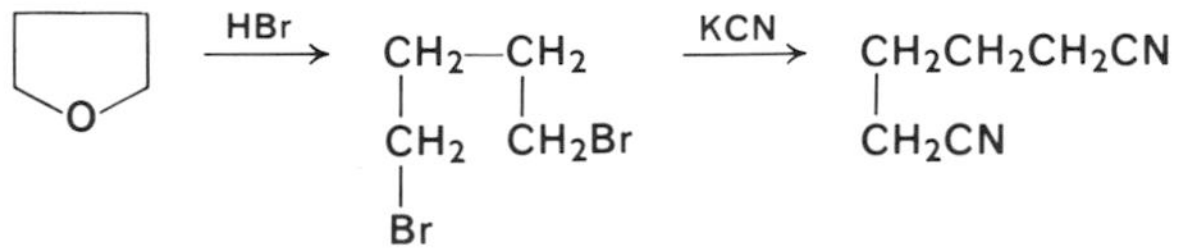

5.  Prepare heptane-1,7-diol starting from cyclohexanone.

*(a)* The final step can be the lithium aluminum hydride reduction of the seven-carbon-atom cyclic lactone (lactones can be reduced by LiAlH$_4$ in a manner comparable to the reduction of acyclic esters):

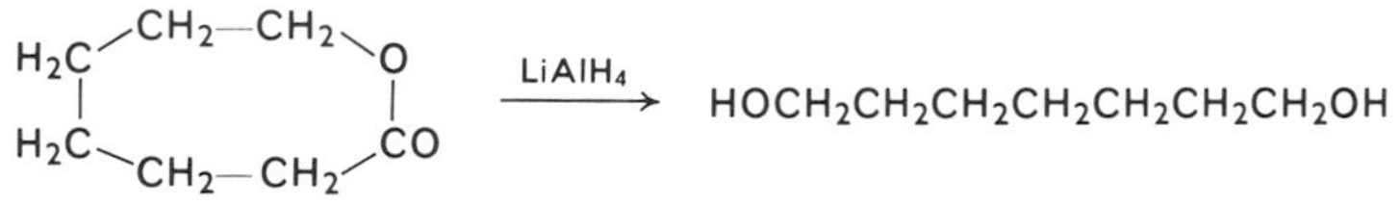

$$\text{(lactone structure)} \xrightarrow{\text{LiAlH}_4} \text{HOCH}_2\text{CH}_2\text{CH}_2\text{CH}_2\text{CH}_2\text{CH}_2\text{CH}_2\text{OH}$$

*(b)* The lactone can be prepared by the Baeyer-Villiger oxidation (a peroxyacid oxidation, Section 17.12) of cycloheptanone.

*(c)* Cycloheptanone can be prepared from cyclohexanone by the following series of reactions, all of which have been described in earlier chapters:

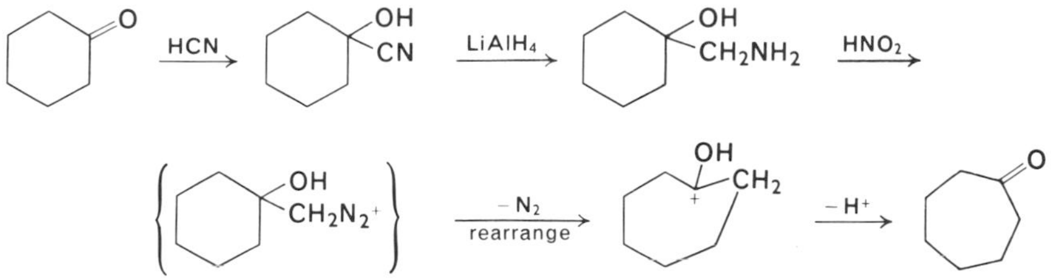

## Exercises

(Exercises 1–4 will be found within the text.)

**5** Write the structures of *(a)* 1,3-cyclohexadiene; *(b)* trans-1,3-dimethylcyclobutane; *(c)* cyclononane; *(d)* 1,3,5,7-cyclooctatetraene; *(e)* dioxane; *(f)* dioxadiene; *(g)* bicyclohexyl; *(h)* 3-bromocyclohexene; *(i)* cis-cyclopropanedicarboxylic acid.

**6** Show three ways of preparing cyclohexane from open-chain starting materials.

**7** Using examples not described in the text, show the formation of a cyclic compound by the use of each of the following: *(a)* an aldol condensation; *(b)* the debromination of a 1,2-dibromo compound with zinc; *(c)* an ester (Claisen) condensation; *(d)* a malonic ester synthesis; *(e)* an acetoacetic ester synthesis; *(f)* a Diels-Alder reaction; *(g)* lactonization of a bromo acid; *(h)* a pinacol reduction.

**8** Draw conformational structures for the most stable forms of *(a)* chlorocyclohexane; *(b)* trans-1,4-dimethylcyclohexane; *(c)* trans-decalin; *(d)* cis-1,3-dibromocyclohexane.

**9** Formulate the Beckmann rearrangement of cyclopentanone oxime to the six-membered cyclic amide (lactam).

**10** What product or products would you expect as the result of the Dieckmann ring closure of the following?

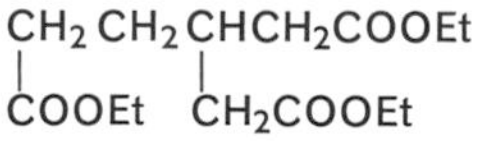

CH$_2$ CH$_2$ CHCH$_2$COOEt
|        |
COOEt   CH$_2$COOEt

**11** Treatment of CH$_2$=CHCH$_2$CH$_2$COOH with aqueous mineral acid causes it to isomerize to a neutral compound. Formulate the reaction, showing the intermediate stages.

**12** When *optically active* cyclopentane-1,2-dicarboxylic acid is transformed by strong heating into its anhydride, the latter is *optically inactive*. Formulate the reaction and account for the stereochemical result.

**13** Write the structures of the dimedone derivatives of *(a)* acetaldehyde; *(b)* isobutyraldehyde; *(c)* methanal.

**14** Formulate the Diels-Alder reaction between benzoquinone and *(a)* 1,3-butadiene; *(b)* cyclopentadiene; *(c)* anthracene; *(d)* 2,4-hexadiene.

# Chapter eighteen / Carbohydrates. The Chemistry of Sugars and Allied Compounds

## 18.1 Introduction

The *carbohydrates* include the sugars and related substances, both simple and complex, natural and synthetic, and constitute what is the most widespread class of compounds occurring in nature. The name "carbohydrate" is an old one, and originated in the observation that many compounds of the class, including the three of commonest occurrence—glucose, $C_6H_{12}O_6$; sucrose, $C_{12}H_{22}O_{11}$; and cellulose, $(C_6H_{10}O_5)_x$—have the empirical composition $C_n(H_2O)_m$, and were at one time referred to as "hydrates of carbon." Although many natural sugars and related compounds do have compositions that correspond to these proportions, a large number of natural and derived sugars differ from it but are nevertheless members of the same class of compounds. Table 18-1 lists some important carbohydrates including several whose empirical compositions are not $C_n(H_2O)_m$.

## 18.2 The occurrence of carbohydrates

Carbohydrates are ultimately derived from the reduction of carbon dioxide by living plants, with the aid of the green pigment chlorophyll. The energy required for the overall process

$$x CO_2 + x H_2O + energy \longrightarrow (CH_2O)_x + x O_2$$

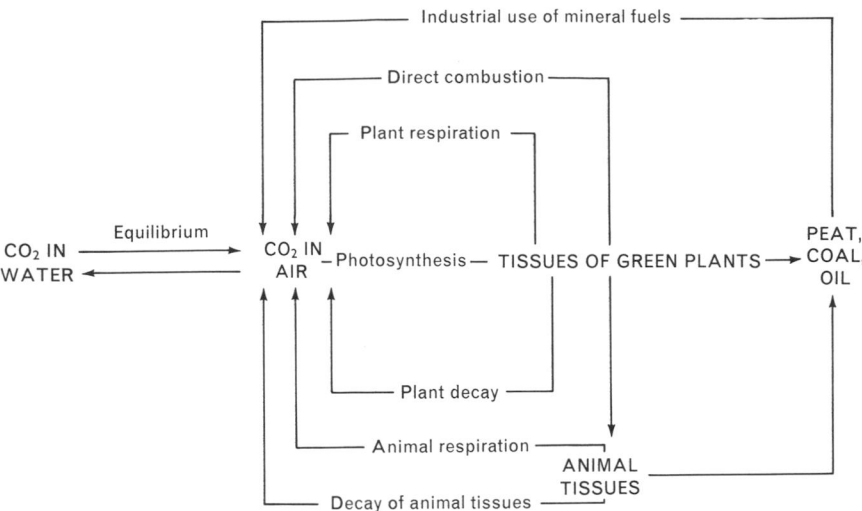

**Figure 18–1.** The biochemical carbon cycle.

is the energy of sunlight, which falls both upon the land, where it is used by the forests and vegetable crops, and upon the sea, where simpler plants (the phytoplankton) in inestimably huge numbers utilize it for photosynthesis. The energy bound up in the carbohydrate molecules is available to man and other animals as the energy of food and fuels. The cycle of fixation of carbon dioxide by photosynthesis and its return to the atmosphere can be represented by the various processes shown in Figure 18-1.

The cycle of photosynthesis and oxidation (combustion of fuels, metabolism of animals and plants, decay) maintains the carbon dioxide concentration in the atmosphere at a fairly constant value of about $0.03\%$ (300 parts per million), a figure which corresponds to a total of about $6 \times 10^{11}$ tons of carbon dioxide. In addition to this, the oceans contain an even greater amount in the form of carbonates and dissolved carbon dioxide. It has been estimated that land plants convert about $10\%$ of the atmospheric carbon dioxide—some $6 \times 10^{10}$ tons— into carbohydrates each year. Most of this is converted into two substances, cellulose and starch, both of which are polymers of glucose. The formation of other sugars, among which are those mentioned in Table 18-1, is the result of secondary metabolic alterations in particular plants.

Most of the natural sugars are very much alike in their main structural features. They are polyhydroxy compounds, the simpler ones being polyhydroxy aldehydes and ketones. They are usually crystalline, water-soluble substances (except for the high-molecular-weight compounds), and many are sweet to the taste.

The most familiar natural sugar is sucrose, which is produced commercially from sugar beets and sugar cane in amounts of more than 50 million tons per year. Sucrose is produced in a high degree of purity, and may be regarded as the

**Table 18-1.**  *Some important carbohydrates*

| Naturally Occurring Carbohydrate | Empirical Formula | Naturally Occurring Carbohydrate | Empirical Formula |
|---|---|---|---|
| D-glucose | $C_6H_{12}O_6$ | L-rhamnose | $C_6H_{12}O_5$ |
| D-fructose | $C_6H_{12}O_6$ | D-glucuronic acid | $C_6H_{10}O_7$ |
| D-mannose | $C_6H_{12}O_6$ | D-mannitol | $C_6H_{14}O_6$ |
| D-galactose | $C_6H_{12}O_6$ | L-ascorbic acid (vitamin C) | $C_6H_8O_6$ |
| D-ribose | $C_5H_{10}O_5$ | D-sedoheptulose | $C_7H_{14}O_7$ |

most abundant pure organic compound. Glucose occurs in the free state, as well as combined in its polymeric forms; it is found in many fruit juices, in body fluids, and in honey. Fructose, also found in honey, occurs as a polymer in the starch-like polyfructoside, inulin, a reserve carbohydrate found in the roots of certain plants.

Two sugars whose importance has been emphasized by the dramatic developments in biology during recent years are the five-carbon-atom compounds D-ribose and D-deoxyribose. These are found in all living organisms, in combination with phosphoric acid and certain nitrogen-containing compounds (Chapter 31), in the nucleic acids RNA (ribonucleic acid) and DNA (deoxyribonucleic acid). These compounds are central to the mechanisms of genetic control of the development and growth of living cells. Some aspects of their chemistry will be dealt with in Chapter 31.

## 18.3  Kinds of carbohydrates

Carbohydrates, or sugars, are conveniently discussed with reference to four main classes of substances:

1. *The simple sugars*, or monosaccharides, most of which are five- and six-carbon-atom compounds. The six-carbon sugars are called "hexoses," the five-carbon sugars "pentoses." Similarly, there are the terms tetrose, triose, heptose, and so forth. The simple sugars are the products of hydrolysis of the three classes 2, 3, and 4, and are themselves not hydrolyzed to simple sugars. Most simple sugars are polyhydroxy aldehydes and ketones.

2. *The oligosaccharides* (Greek, *oligos*, few), which comprise two or more simple sugars joined by acetal formation between the aldehyde or ketone grouping of one and a hydroxyl group of another. When two sugars are so joined, the resulting compound is a *disaccharide;* when three, a *trisaccharide*, and so on. The linkage joining the sugars together is called a *glycosidic* linkage.

3. *The polysaccharides*, in which many molecules of one or more than one kind of simple sugar are joined in glycosidic linkages. Oligosaccharides are simply polysaccharides containing a few sugar units, and there is no sharp line separating oligosaccharides and polysaccharides.

4. *The glycosides*, which are distinguished from the oligo- and poly-saccharides by the fact that they contain a nonsugar molecule joined to a sugar by a glycosidic linkage.

## 18.4  The structure of glucose

Glucose, $C_6H_{12}O_6$, is the most widely occurring sugar. The naturally occurring form is dextrorotatory, and, as will be considered in detail below, is configurationally related to D-(+)-glyceraldehyde and so is called D(+)-glucose. It is the unit from which the universally distributed polysaccharides cellulose and starch are derived, and is the sugar most frequently found in the natural glycosides (Section 18.19). Glucose is intimately involved in the metabolic activities of most living organisms, and, formed by the hydrolysis of sucrose and starch, constitutes the chief energy source in the diet of most human beings.

The six carbon atoms of glucose form an unbranched chain: it is reduced by vigorous reduction with hydriodic acid to *n*-hexane, and can be hydrogenated, with the addition of one molecule of hydrogen, to sorbitol, a 1,2,3,4,5,6-hexane-hexol. The presence of an aldehyde group is shown by the addition of HCN to form a cyanohydrin, and by the formation of an oxime. The aldehyde group is readily oxidized, with the formation of gluconic acid, $C_6H_{12}O_7$, and gluconic acid can be further oxidized to the dibasic acid, glucaric acid, $C_6H_{10}O_8$, demonstrating that the grouping —$CH_2OH$ is also present. Glucose forms a penta-acetate, and sorbitol a hexaacetate. If we ignore for the moment the stereo-chemical questions that are apparent from an examination of the structure, we can formulate these transformations of glucose as shown in Chart 18-1.

The presence of the —CHO (formyl) group in glucose is less clearly demonstrated by certain other of its properties, such as its relative inertness to Schiff's reagent; and the fact that, although it reacts with Fehling's solution and Tollens' reagent, fructose, a *ketone*, does likewise. However, there was little question even in the earliest years of the study of glucose that it is indeed a pentahydroxy *n*-hexyl aldehyde, the only question being the manner in which the properties of the aldehyde group were influenced by the presence of the hydroxyl groups in the molecule.

An early suggestion by Tollens (and others) was that one of the hydroxyl groups was involved in hemiacetal formation with the formyl group, and that glucose was in reality a cyclic compound. That this has indeed proved to be the case will be shown as we proceed; but for the present we shall consider glucose as if it were the open-chain compound, and return to the question of the ring structure at a later time.

**Chart 18-1.** *Transformations of glucose*

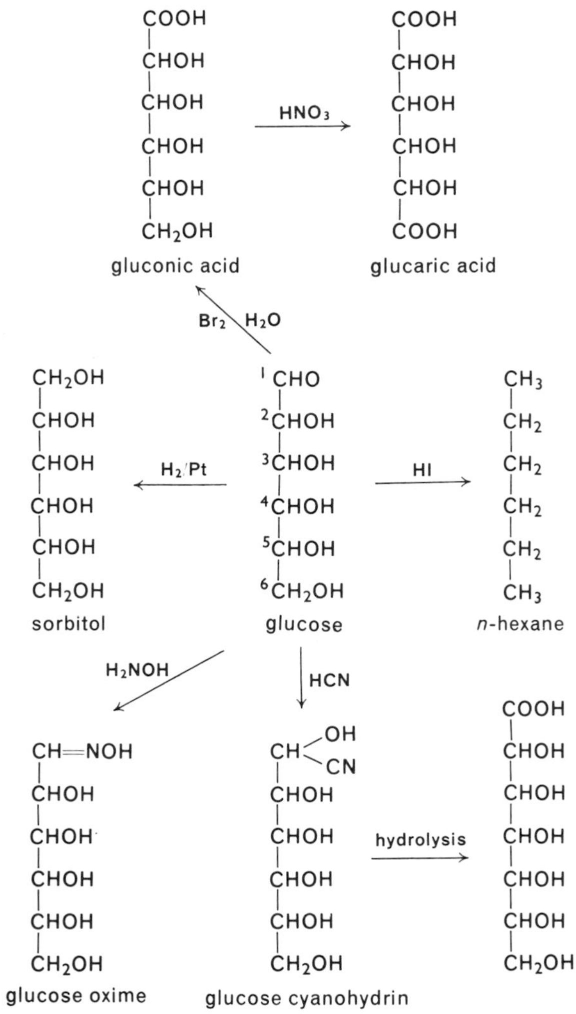

The remaining question is that of the stereochemistry of the sugar molecule. In the molecule of glucose, represented as

$$HOCH_2-\overset{*}{C}HOH-\overset{*}{C}HOH-\overset{*}{C}HOH-\overset{*}{C}HOH-CHO$$
$$\phantom{HOC}6\phantom{H_2-C}5\phantom{HOH-C}4\phantom{HOH-C}3\phantom{HOH-C}2\phantom{HOH-C}1$$

There are four asymmetric carbon atoms. Therefore, D-glucose is one of sixteen possible compounds, which differ in stereochemical configuration. These sixteen compounds are the *aldohexoses*, so called because they are the six-carbon *(hex)* sugars *(ose)*, all of which contain the terminal formyl group *(aldo)*. The determination of the configurations of the simple sugars was the work of Emil Fischer (1852–1919), and was a monumental achievement that laid the foundation stone for the advances made in carbohydrate chemistry up to the present time.

## 18.5   Stereochemical representation of sugars

In order to discuss the stereochemistry of the sugars it is necessary to adopt a method for writing their structures that will show the configurations of the asymmetric carbon atoms.

The simplest aldose that can exist in $(+)$ and $(-)$ forms is glyceraldehyde. If we write the structure of D-$(+)$-glyceraldehyde (Chapter 16) as follows

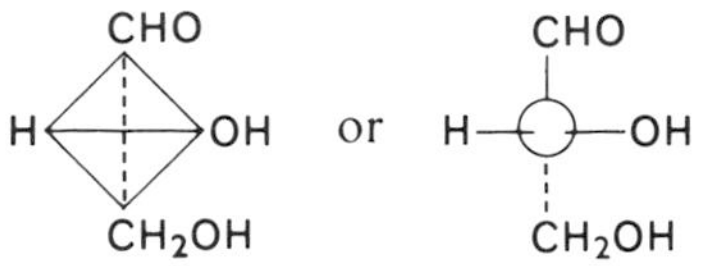

and adopt the convention that the two-dimensional projection

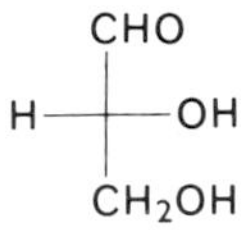

represents the same configuration, this means that in the latter formulation the H and OH groups are to be regarded as projecting forward from the plane of the paper. In the discussions to follow, this convention will be used. This manner of representation is often abbreviated: the H may be omitted, or the configuration of the OH group alone may be shown by a short line. Thus, the sugar D-arabinose may be represented by the following equivalent formulas:

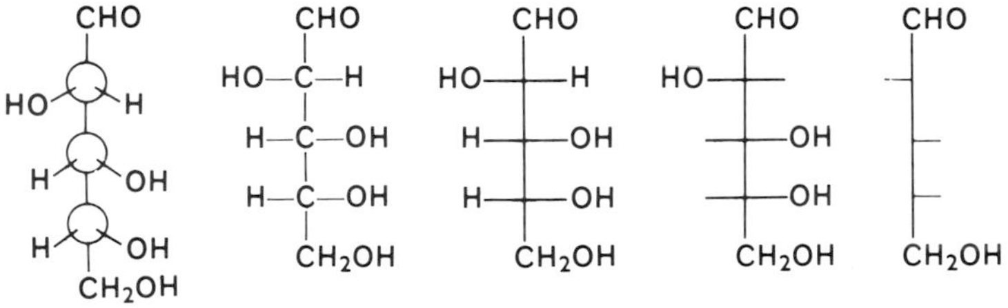

The last two of these are convenient for many purposes. Since these formulas represent projections of three-dimensional structures, drawn according to the convention described above, certain restrictions apply to their manipulation on paper. They can be rotated in the plane of the page, but not out of the plane; that is,

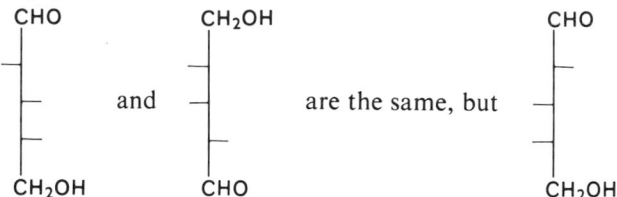

is the enantiomorph of this.

## 18.6   Establishment of the stereochemistry of the sugars

In order to write a complete structure for glucose, it is necessary to specify the configurations of the four asymmetric carbon atoms. This can be accomplished with the aid of four experimental procedures:

1. the formation of osazones;
2. the lengthening of the carbon chain by addition to the terminal —CHO group;
3. the shortening of the carbon chain by removal of the terminal —CHO group;
4. the conversion of the end groups (—CHO and —CH$_2$OH) into identical groups (e.g., by reduction of —CHO to —CH$_2$OH, or by the oxidation of both to —COOH).

## 18.7   Sugar osazones

The study of sugar chemistry was greatly aided by the discovery and application of phenylhydrazine as a reagent (Fischer, 1884). Phenylhydrazine is a general reagent for carbonyl compounds and it will be recalled that it reacts with most aldehydes and ketones to yield derivatives known as *phenylhydrazones* (Chapter 12). Applied to glucose, which, as we have seen, contains the formyl group, phenylhydrazine reacted to form a beautifully crystalline compound, and Fischer discovered that this product contained not one but two phenylhydrazine residues. The derivative of glucose was called *glucosazone*, and the related compounds formed by the action of phenylhydrazine on other sugars are known by the term *osazones* (*ose*, sugar; hydr*azone*).

Glucosazone was found to be hydrolyzable with acid, and the product of the hydrolysis was a compound C$_6$H$_{10}$O$_6$ called *glucosone*. Glucosone gave glucosazone again when treated with phenylhydrazine. The comparison of the formulas of glucose (C$_6$H$_{12}$O$_6$) and glucosone (C$_6$H$_{10}$O$_6$) indicates that in the formation of glucosazone an oxidation has occurred:

$$
\begin{array}{c}
\text{CHO} \\
\text{CHOH} \\
\text{(CHOH)}_3 \\
\text{CH}_2\text{OH}
\end{array}
\xrightarrow{\text{C}_6\text{H}_5\text{NHNH}_2}
\begin{array}{c}
\text{CH}=\text{NNHC}_6\text{H}_5 \\
\text{C}=\text{NNHC}_6\text{H}_5 \\
\text{(CHOH)}_3 \\
\text{CH}_2\text{OH}
\end{array}
\xrightleftharpoons{\text{HCl}}
\begin{array}{c}
\text{CHO} \\
\text{CO} \\
\text{(CHOH)}_3 + 2\text{C}_6\text{H}_5\text{NHNH}_2 \\
\text{CH}_2\text{OH}
\end{array}
$$

glucose                    glucosazone                    glucosone

The reduction of glucosone yielded a new hexose, *fructose*, isomeric with glucose but containing a ketone carbonyl group and not a formyl group. The reduction thus affected the —CHO group of glucosone, reducing it to the —CH$_2$OH grouping:

$$
\begin{array}{c}
\text{CHO} \\
\text{CO} \\
\text{(CHOH)}_3 \\
\text{CH}_2\text{OH}
\end{array}
\xrightarrow[\text{acetic acid}]{\text{Zn}}
\begin{array}{c}
\text{CH}_2\text{OH} \\
\text{CO} \\
\text{(CHOH)}_3 \\
\text{CH}_2\text{OH}
\end{array}
\xrightarrow{\text{C}_6\text{H}_5\text{NHNH}_2}
\text{glucosazone}
$$

glucosone                    fructose

These results lead to an important conclusion: Glucose and fructose have the same configuration at carbon atoms 3, 4, and 5. Moreover, a third sugar, *mannose*, also yields glucosazone with phenylhydrazine. Now since the formation of glucosazone from glucose involves the destruction of asymmetry on carbon 2, mannose and glucose must have opposite configurations at this carbon atom:

$$
\begin{array}{c}
\text{CHO} \\
\text{H—C—OH} \\
\text{(CHOH)}_3 \\
\text{CH}_2\text{OH}
\end{array}
\qquad
\begin{array}{c}
\text{CHO} \\
\text{HO—C—H} \\
\text{(CHOH)}_3 \\
\text{CH}_2\text{OH}
\end{array}
\qquad
\begin{array}{c}
\text{CH}_2\text{OH} \\
\text{CO} \\
\text{(CHOH)}_3 \\
\text{CH}_2\text{OH}
\end{array}
$$

glucose                    mannose                    fructose

glucosazone

thcrefore glucose, mannose, and fructose have the same configuration at C3, 4, and 5.

### 18.8  Lengthening of the chain in sugars. The cyanohydrin synthesis

The formation of cyanohydrins from aldoses, and their hydrolysis to glyconic acids, is known as the Kiliani reaction. Fischer later succeeded in reducing the glyconic acids (as their lactones) to the corresponding aldehydes, thus completing the overall transformation:

$$
\text{R—CHO} + \text{HCN} \longrightarrow
\begin{array}{c}
\text{R—CH—CN} \\
\text{OH}
\end{array}
\longrightarrow
\left\{
\begin{array}{c}
\text{R—CH—COOH} \\
\text{OH}
\end{array}
\right\}
\longrightarrow
\begin{array}{c}
\text{R—CH—CHO} \\
\text{OH}
\end{array}
$$

The Kiliani-Fischer reaction is of great usefulness in sugar chemistry. It affords a means of building up higher sugars from lower ones, since it introduces a new —CHOH grouping and a —CN group that can be reduced to —CHO. The triose, D-(+)-glyceraldehyde, for example, reacts as follows:

$$
\begin{array}{ccc}
 & \overset{\displaystyle CN}{\underset{\displaystyle |}{}} & \overset{\displaystyle CN}{\underset{\displaystyle |}{}} \\
CHO & H\!-\!C\!-\!OH & HO\!-\!C\!-\!H \\
H\!-\!C\!-\!OH \ + \ HCN \longrightarrow & H\!-\!C\!-\!OH \ + & H\!-\!C\!-\!OH \\
CH_2OH & CH_2OH & CH_2OH \\
\text{D-(+)-glyceraldehyde} & (1) & (2)
\end{array}
$$

Since the original asymmetric carbon atom is not involved in the reaction, its configuration remains unchanged. The new —CHOH— grouping, a new asymmetric carbon atom, is formed in *both* configurations (1) and (2). However, because of the presence of the asymmetric center in the starting material, the two products are not formed in exactly equal amounts. Thus, the mixture of (1) and (2) will contain a preponderance of one of these.

Hydrolysis (—CN to —COOH) and oxidation (—CH$_2$OH to —COOH) of the products from glyceraldehyde give *meso-* and D-tartaric acid:

$$
\begin{array}{cccc}
CN & COOH & COOH & CN \\
H\!-\!C\!-\!OH & H\!-\!C\!-\!OH & HO\!-\!C\!-\!H & HO\!-\!C\!-\!H \\
H\!-\!C\!-\!OH \longrightarrow & H\!-\!C\!-\!OH & H\!-\!C\!-\!OH & \longleftarrow \ H\!-\!C\!-\!OH \\
CH_2OH & COOH & COOH & CH_2OH \\
(1) & meso\text{-tartaric} & \text{D-(−)-tartaric} & (2) \\
 & \text{acid} & \text{acid} &
\end{array}
$$

The conversion of the cyanohydrins formed in the Kiliani reaction into aldoses is accomplished by hydrolysis of —CN to —COOH, formation of a lactone between —COOH and one of the hydroxyl groups, and reduction of the lactone to the aldehyde. In the case of D-glyceraldehyde, this series of transformations is as follows:

$$
\begin{array}{cc}
 & \overset{\displaystyle CN}{\underset{\displaystyle |}{}} \qquad \overset{\displaystyle CN}{\underset{\displaystyle |}{}} \\
CHO & H\!-\!C\!-\!OH \qquad HO\!-\!C\!-\!H \\
H\!-\!C\!-\!OH \ \xrightarrow{\ HCN\ } & H\!-\!C\!-\!OH \ + \ H\!-\!C\!-\!OH \\
CH_2OH & CH_2OH \qquad CH_2OH
\end{array}
$$

$$
\begin{array}{cccc}
 & \text{hydrol.} & & \text{same series} \\
COOH & CO\!-\! & CHO & CHO \\
H\!-\!C\!-\!OH & H\!-\!C\!-\!OH \ \Big| \ \xrightarrow{Na/Hg} \ H\!-\!C\!-\!OH & HO\!-\!C\!-\!H \\
H\!-\!C\!-\!OH \longrightarrow & H\!-\!C\!-\!OH & H\!-\!C\!-\!OH & H\!-\!C\!-\!OH \\
CH_2OH & CH_2\!-\!O & CH_2OH & CH_2OH \\
 & & \text{D-erythrose} & \text{D-threose}
\end{array}
$$

The configurations of erythrose and threose are now established, since the cyanohydrin that gives erythrose can also be transformed into *meso*-tartaric acid, showing that the —OH groups are on the same side (as the projection formula is written).

### 18.9   Degradation of sugars

The *removal* of one carbon atom from an aldose by a reversal of the cyanohydrin synthesis is known as the Wohl degradation. In the case of D-arabinose, the procedure, which depends upon the reversibility of the addition of HCN to an aldehyde (Section 12.7), is as follows:

$$
\begin{array}{l}
\text{CHO} \\
|\\
\text{HO—C—H} \\
|\\
\text{H—C—OH} \xrightarrow{\ \text{NH}_2\text{OH}\ } \\
|\\
\text{H—C—OH} \\
|\\
\text{CH}_2\text{OH} \\
\text{D-arabinose}
\end{array}
\qquad
\begin{array}{l}
\text{CH=NOH} \\
|\\
\text{HO—C—H} \\
|\\
\text{H—C—OH} \xrightarrow{\ \text{Ac}_2\text{O}\ } \\
|\\
\text{H—C—OH} \\
|\\
\text{CH}_2\text{OH} \\
\text{D-arabinose} \\
\text{oxime}
\end{array}
\qquad
\begin{array}{l}
\text{CN} \\
|\\
\text{HO—C—H} \\
|\\
\text{H—C—OH} \longrightarrow \\
|\\
\text{H—C—OH} \\
|\\
\text{CH}_2\text{OH}
\end{array}
\qquad
\begin{array}{l}
\text{CHO} \\
|\\
\text{H—C—OH} + \text{HCN} \\
|\\
\text{H—C—OH} \\
|\\
\text{CH}_2\text{OH} \\
\text{D-erythrose}
\end{array}
$$

The Wohl degradation of D-glucose (and of D-mannose) gives D-arabinose.

### 18.10   The configuration of glucose

If we now take stock of the information we have accumulated, it will be apparent that most of the data necessary to establish the configuration of glucose have been described in the foregoing paragraphs.

Since glucose is degraded to arabinose, and arabinose to erythrose, the configurations of C-4 and C-5 are the following:

$$
\begin{array}{l}
\text{CHO} \\
|\\
\text{(CHOH)} \\
|\\
\text{(CHOH)} \\
|\\
\text{H—C—OH} \\
|\\
\text{H—C—OH} \\
|\\
\text{CH}_2\text{OH} \\
\text{D-glucose}
\end{array}
\longrightarrow
\begin{array}{l}
\text{CHO} \\
|\\
\text{(CHOH)} \\
|\\
\text{H—C—OH} \\
|\\
\text{H—C—OH} \\
|\\
\text{CH}_2\text{OH} \\
\text{D-arabinose}
\end{array}
\longrightarrow
\begin{array}{l}
\text{CHO} \\
|\\
\text{H—C—OH} \\
|\\
\text{H—C—OH} \\
|\\
\text{CH}_2\text{OH} \\
\text{D-erythrose}
\end{array}
\xrightarrow{\text{oxid.}}
\begin{array}{l}
\text{COOH} \\
|\\
\text{H—C—OH} \\
|\\
\text{H—C—OH} \\
|\\
\text{COOH} \\
\textit{meso}\text{-tartaric} \\
\text{acid}
\end{array}
$$

[The configurations in parentheses (CHOH) are not explicit in this scheme.]

The configuration of C-2 of D-arabinose can be established by the following evidence. The aldopentose D-ribose gives the same osazone as D-arabinose; this shows that D-ribose and D-arabinose differ only in the configuration at C-2:

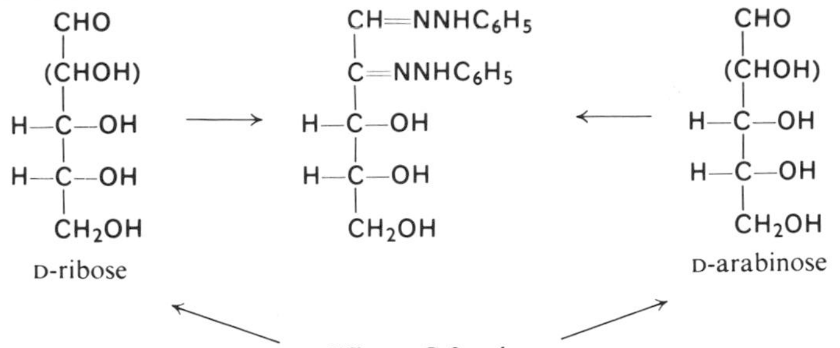

$$
\begin{array}{ccc}
\text{CHO} & \text{CH}=\text{NNHC}_6\text{H}_5 & \text{CHO} \\
| & | & | \\
\text{(CHOH)} & \text{C}=\text{NNHC}_6\text{H}_5 & \text{(CHOH)} \\
| & | & | \\
\text{H—C—OH} & \text{H—C—OH} & \text{H—C—OH} \\
| & | & | \\
\text{H—C—OH} & \text{H—C—OH} & \text{H—C—OH} \\
| & | & | \\
\text{CH}_2\text{OH} & \text{CH}_2\text{OH} & \text{CH}_2\text{OH}
\end{array}
$$

D-ribose  →  ←  D-arabinose

differ at C-2 only

The oxidation of D-ribose gives an *optically inactive* dicarboxylic acid, while D-arabinose is converted by oxidation into an *optically active* dibasic acid:

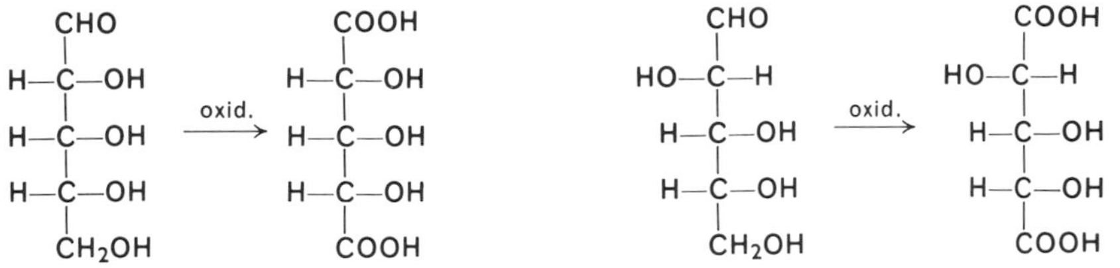

$$
\begin{array}{cccc}
\text{CHO} & \text{COOH} & \text{CHO} & \text{COOH} \\
| & | & | & | \\
\text{H—C—OH} & \text{H—C—OH} & \text{HO—C—H} & \text{HO—C—H} \\
| & | & | & | \\
\text{H—C—OH} & \text{H—C—OH} & \text{H—C—OH} & \text{H—C—OH} \\
| & | & | & | \\
\text{H—C—OH} & \text{H—C—OH} & \text{H—C—OH} & \text{H—C—OH} \\
| & | & | & | \\
\text{CH}_2\text{OH} & \text{COOH} & \text{CH}_2\text{OH} & \text{COOH}
\end{array}
$$

D-ribose — oxid. → D-*ribo*trihydroxy-glutaric acid (*meso*: inactive)     D-arabinose — oxid. → D-*arabo*trihydroxy-glutaric acid (optically active)

Since the configuration of D-arabinose is thus fixed, D-glucose and D-mannose are

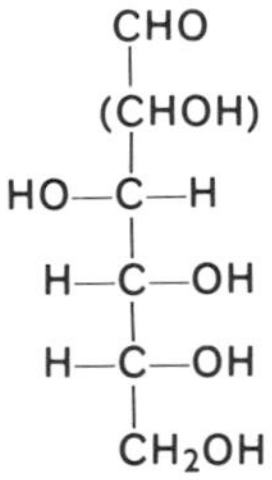

$$
\begin{array}{c}
\text{CHO} \\
| \\
\text{(CHOH)} \\
| \\
\text{HO—C—H} \\
| \\
\text{H—C—OH} \\
| \\
\text{H—C—OH} \\
| \\
\text{CH}_2\text{OH}
\end{array}
$$

D-glucose  
D-mannose  } epimeric at C-2

---

**Exercise 1.** Recapitulate the arguments that lead to the conclusion that D-glucose and D-mannose have the configuration shown, differing only at C-2. From its use in this context, what does the word "epimeric" mean?

---

The final fact that establishes the complete configuration of glucose is that oxidation of D-glucose gives the same saccharic acid [the dicarboxylic acid (3)] as does another aldohexose, L-gulose. This means that in D-glucose and L-gulose the —CHO and —CH₂OH groups are interchanged:

$$
\begin{array}{ccccc}
\text{CHO} & & \text{COOH} & & \text{CH}_2\text{OH} \\
\text{H—C—OH} & & \text{H—C—OH} & & \text{H—C—OH} \\
\text{HO—C—H} & \xrightarrow{\text{oxid.}} & \text{HO—C—H} & \xleftarrow{\text{oxid.}} & \text{HO—C—H} \\
\text{H—C—OH} & & \text{H—C—OH} & & \text{H—C—OH} \\
\text{H—C—OH} & & \text{H—C—OH} & & \text{H—C—OH} \\
\text{CH}_2\text{OH} & & \text{COOH} & & \text{CHO}
\end{array}
$$

which
is
$$
\left\{
\begin{array}{c}
\text{CHO} \\
\text{HO—C—H} \\
\text{HO—C—H} \\
\text{H—C—OH} \\
\text{HO—C—H} \\
\text{CH}_2\text{OH}
\end{array}
\right\}
$$

D-glucose     D-glucaric acid     L-gulose     L-gulose
(D-*gluco*saccharic
acid = L-*gulo*sac-
charic acid)
(3)

There is no other hexose that can give the same saccharic acid that is obtained from D-mannose, since if we interchange —CHO and —CH₂OH on mannose, the result is still D-mannose.

### 18.11   The D-series of sugars

It is now known the D-glyceraldehyde has the absolute configuration

$$
\begin{array}{c}
\text{CHO} \\
\text{H—C—OH} \\
\text{CH}_2\text{OH}
\end{array}
$$

D-( + )-glyceraldehyde

Until recently, however, it was an arbitrary assumption, based upon the optical dextrorotation, that D-glyceraldehyde has this configuration. A suggestion by Rosanoff in 1906 led to the adoption of the convention that the dextrorotatory form of glyceraldehyde be assigned the configuration that is drawn with the —OH group to the right in the tetrahedral projection shown in the above formula, and that other compounds related to (+)-glyceraldehyde by having the same configuration at this carbon atom be designated as members of the *L-series*. Their enantiomorphs are members of the *L-series*.

We can now write the members of the D-family of aldoses by regarding them as being derived by successive cyanohydrin syntheses from D-glyceraldehyde (Chart 18-2).

The designation of the actual rotation of a compound is inserted in parentheses following the letters L or D. The symbols L- and D- are *configurational designations only*, whereas the rotation of a compound is an experimentally determined value, and may be positive, negative, or even zero.

**Chart 18-2.**  *The aldoses of the D-series*

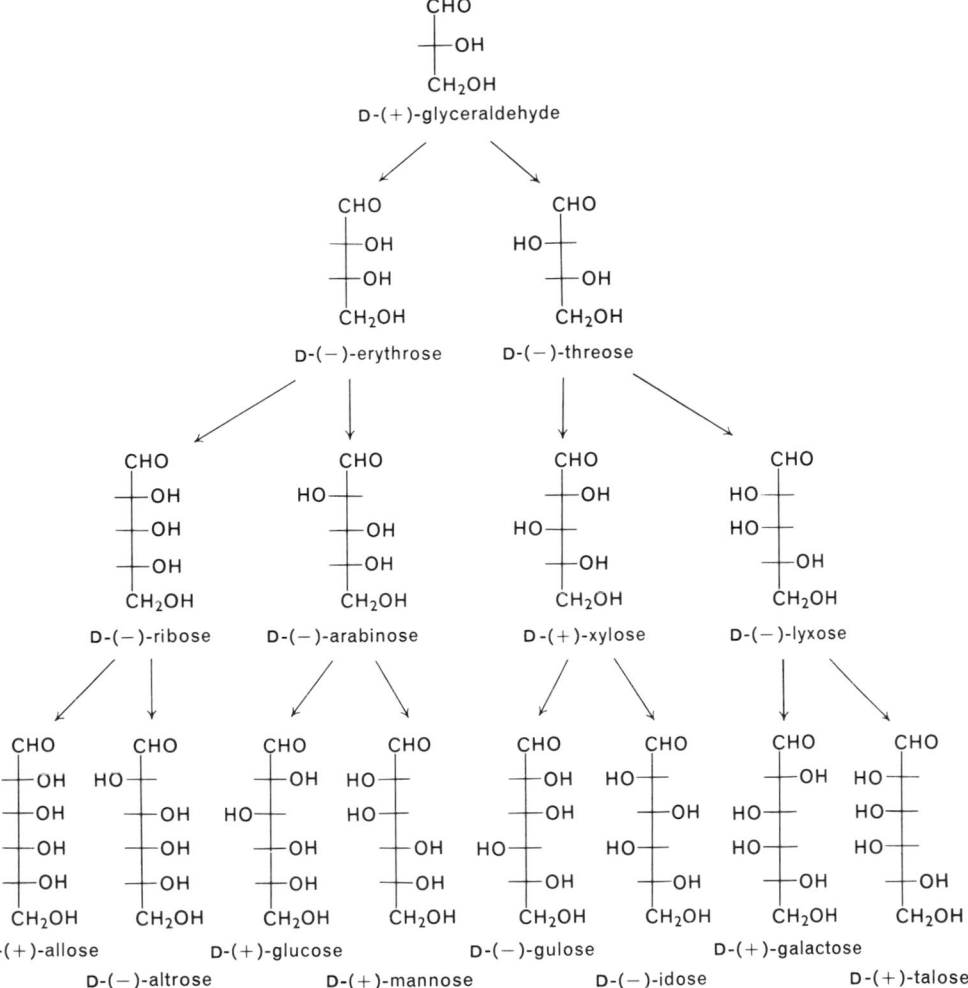

## 18.12   The cyclic structures of the sugars

When D-glucose is heated with methanol in the presence of hydrogen chloride the product is not the dimethyl acetal, as was first expected; two compounds, each containing but one methoxyl group, are formed. These are isomeric, and both can be readily hydrolyzed to give glucose. They are called methyl α- and β-D-glucoside, and are derivatives of a cyclic form of glucose:

$$\text{D-glucose} \xrightarrow[\text{HCl}]{\text{CH}_3\text{OH}} \text{methyl } \alpha\text{-D-glucoside} + \text{methyl } \beta\text{-D-glucoside}$$

$$[\alpha]_D + 157° \qquad\qquad [\alpha]_D - 33°$$

Since the methyl glucosides no longer reduce Tollens' reagent or Fehling's solution (and are thus called *nonreducing*, in contrast to glucose, a *reducing* sugar), the aldehyde group must be involved in the reaction and is bound up in a form that destroys its reducing properties. The formulation of the two glucosides is as follows:

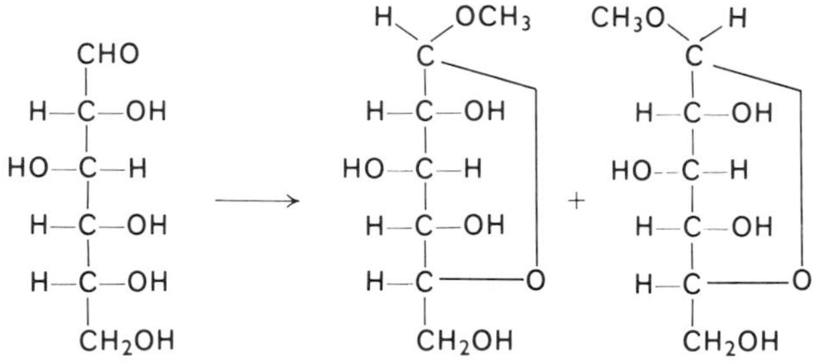

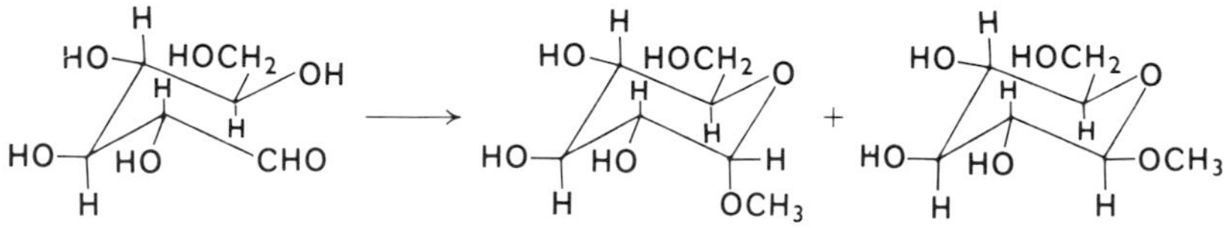

D-glucose          Methyl α-D-glucoside          Methyl β-D-glucoside

Tollens had earlier suggested that glucose itself existed as a cyclic (hemiacetal) form; and it was soon found that two crystalline forms of glucose could be prepared by suitable manipulation of the experimental conditions. Crystallized from aqueous-alcoholic solutions at low temperatures, glucose with a specific rotation of $+113°$ was obtained; but when solutions of glucose in water were allowed to evaporate at about 100°, the glucose that crystallized had a specific rotation of $+19°$. The rotations of both of these forms of glucose changed rapidly when they were dissolved in water, and both reached an equilibrium value of $+52°$. The form with specific rotation $+113°$ was designated as α-glucose, that with specific rotation $+19°$ was called β-glucose. Furthermore, when a solution of glucose in water was brought to crystallization by the addition of alcohol, the crystalline glucose that precipitated had specific rotation of $+52°$, and is the equilibrium mixture of the α- and β-forms:

$$\text{α-D-glucose} \rightleftharpoons \text{D-glucose} \rightleftharpoons \text{β-D-glucose}$$
$$[\alpha]_D + 113° \qquad [\alpha]_D + 52° \qquad [\alpha]_D + 19°$$

This phenomenon is called *mutarotation*, and is caused by the change in configuration of *one* optical center in a molecule while others remain unaltered. In the case of glucose, the change is in the configuration of the new asymmetric carbon atom created by the ring closure shown below:

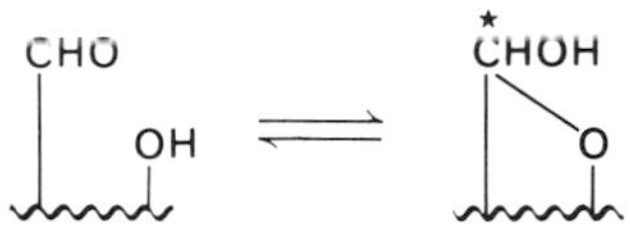

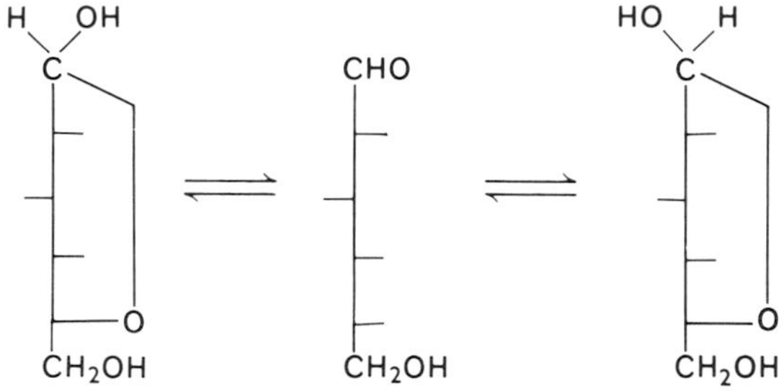

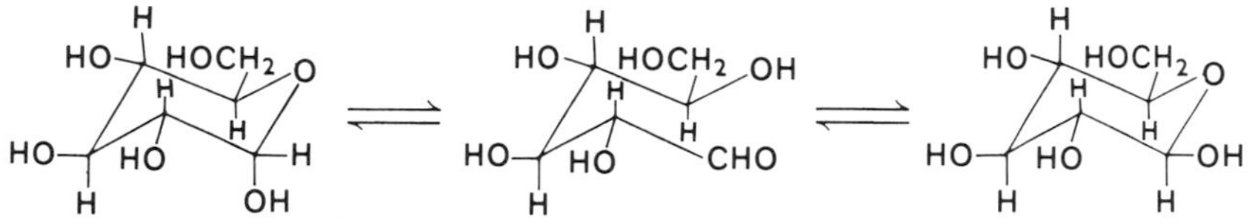

α-D-glucose                                                    β-D-glucose

## 18.13   Haworth projection formulas for sugars

In dealing with the cyclic structures, in which form the sugars commonly exist, it is convenient to use a new kind of formula. The one usually adopted was devised by the British chemist Haworth, and is called the "Haworth formula" of a sugar.

D-Glucose is represented in the following way:

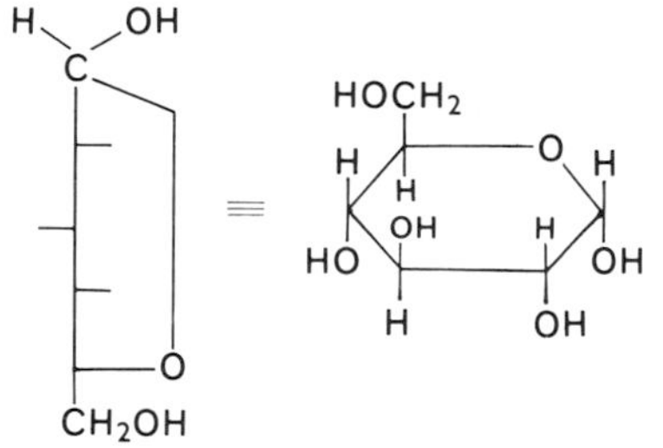

(α-D-glucose)                                    (β-D-glucose)
α-D-glucopyranose                            β-D-glucopyranose

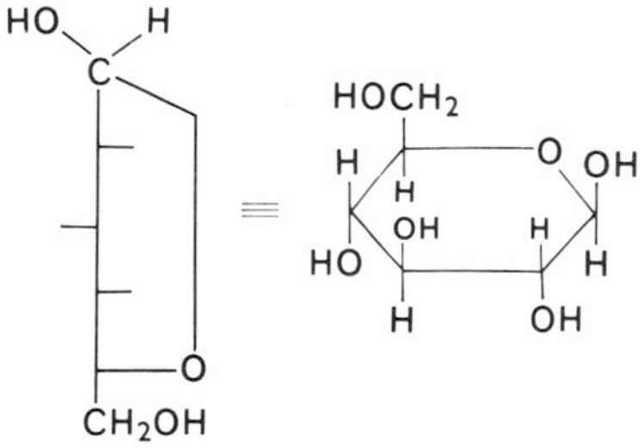

The Haworth formulas are conventionalized perspective views of the ring, which is, of course, not planar but is buckled in the manner shown in the conformational representations used in the preceding section. The added designations *-pyranose*

and -*furanose* are derived from the names for the heterocyclic compounds pyran and furan:

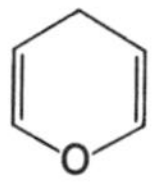

pyran     furan

The following formulas show several ways of representing the structures of sugars:

D-arabinose

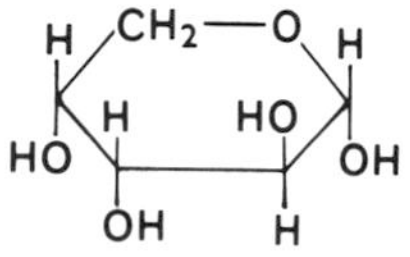

α-D-arabopyranose

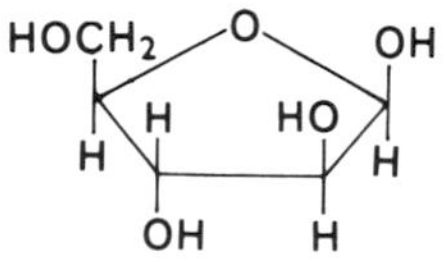

β-D-arabofuranose

α-D-glucopyranose

ethyl β-D-glucopyranoside

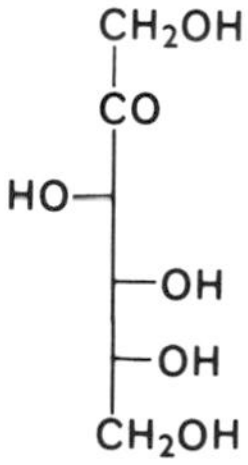

D-fructose

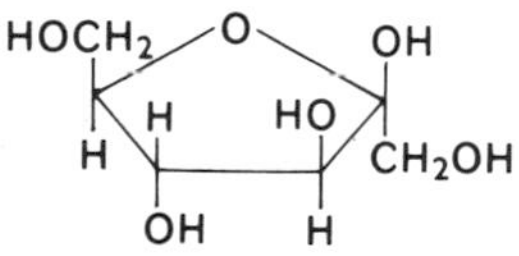

β-D-fructofuranose

β-D-fructopyranose

The Haworth formulas for the pyranose forms of three other of the aldohexoses of the D-series are as follows (the α-forms are shown; in the *anomeric* β-forms the hydroxyl groups on carbon atom one are in the opposite configuration):

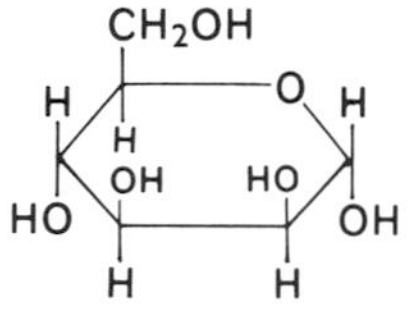

D-mannose 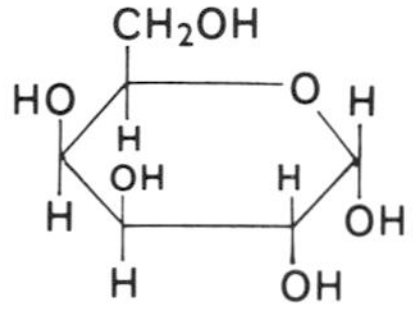                   D-galactose 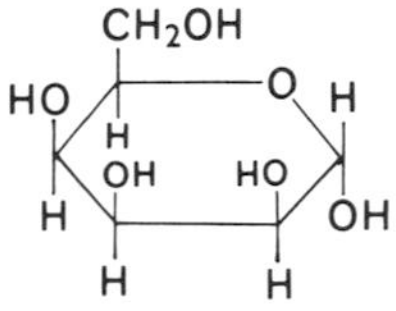                   D-talose

Finally, it is often convenient to simplify the structural respresentations by writing the Haworth structural formulas in the following ways:

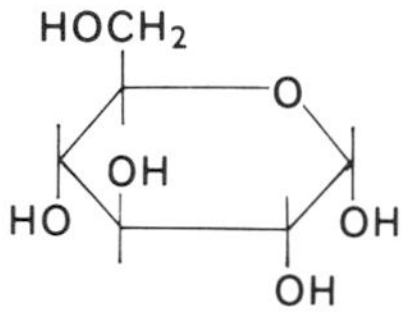

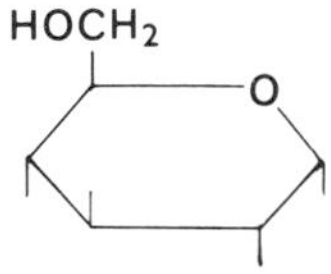

α-D-glucopyranose

---

**Exercise 2.** Complete the series by drawing the Haworth formulas for the α-pyranose forms of D-allose, D-gulose, D-altrose, and D-idose (see Chart 18-2).

---

## 18.14 Determination of the ring size of glycosides

Although the ring structures for the two methyl glucosides account for their formation and behavior, no evidence has yet been given to show where the ring closure has taken place. Of the two most likely structures (writing only β-forms) both are *a priori* possibilities:

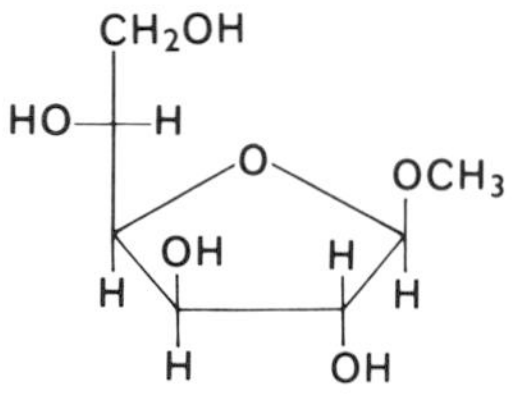

methyl β-D-glucofuranoside 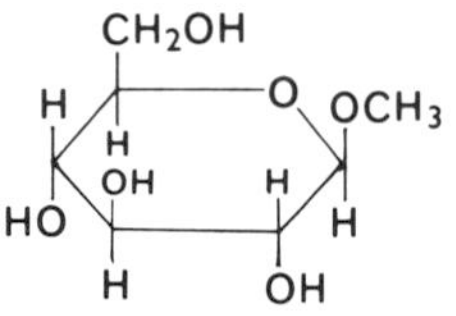          methyl β-D-glucopyranoside

The problem of the ring size of the glucosides was first solved by the degradation of the *completely* methylated glucoside, in which form the ring was fixed in one possible size. The methylation of the alcoholic hydroxyl groups of a sugar can be accomplished by methods which are in effect variants of the Williamson synthesis of ethers. The use of methyl iodide and silver oxide, first used by the Scottish chemists Purdie and Irvine, and now known by their names, or of methyl sulfate and sodium hydroxide, converts methyl glycosides of sugars into their fully O-methylated derivatives. In the following equations the configuration ($\alpha$- or $\beta$-) of the methyl glucoside is not specified, for it will be seen that this methyl group, which is present in the acetal linkage, is later removed by acid hydrolysis; the remaining methyl groups, which are in ether linkages, are unaffected by mild acid hydrolysis:

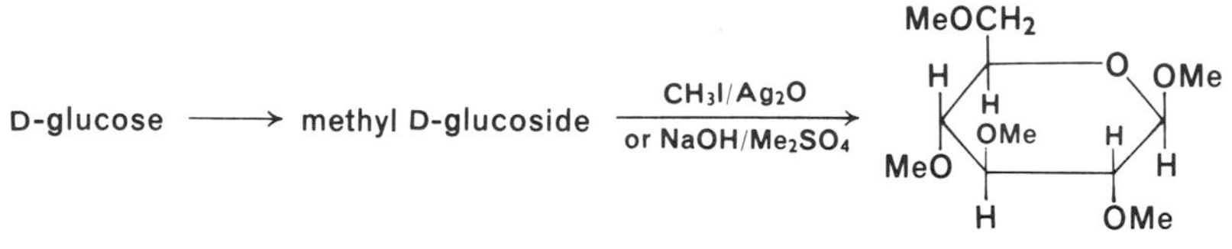

$$\text{D-glucose} \longrightarrow \text{methyl D-glucoside} \xrightarrow[\text{or NaOH/Me}_2\text{SO}_4]{\text{CH}_3\text{I/Ag}_2\text{O}}$$

Methyl tetra-O-methyl-D-glucoside

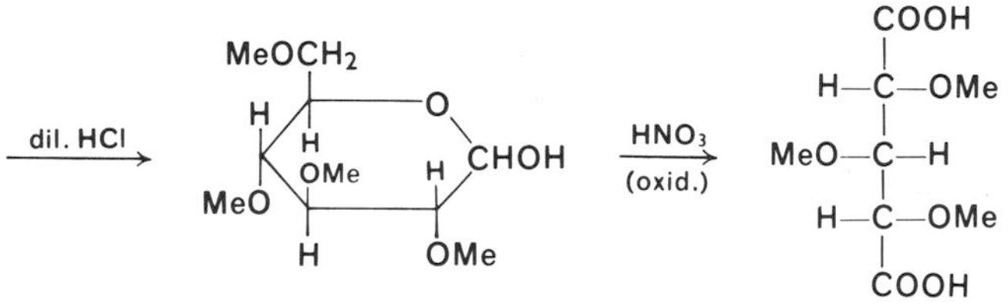

tetra-O-methyl glucose          *xylo*trimethoxyglutaric acid

The formation of the trimethoxyglutaric acid shows that the ring was pyranose, since if the glucoside were in the furanose form,

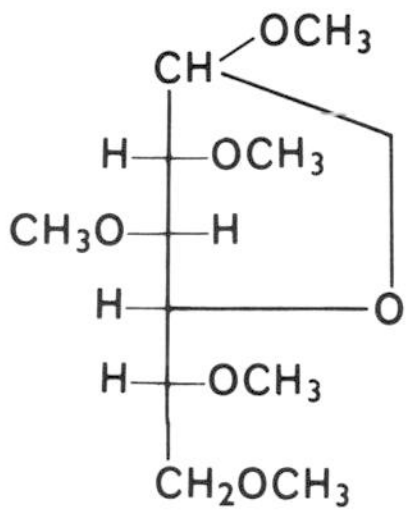

no trimethoxyglutaric acid could be obtained inasmuch as there are no three adjacent methoxyl groups.

***Periodic acid as a reagent for the determination of the ring size of glycosides.***
Periodic acid, $HIO_4$, is an elegant reagent that oxidizes 1,2-glycols, $\alpha$-hydroxy-

**Table 18-2.** *Products of periodate oxidation of five possible ring-isomeric glucosides*

|  | Structure | | | | |
|---|---|---|---|---|---|
|  | A | B | C | D | E |
| moles $HIO_4$ consumed | 3 | 2 | 2 | 2 | 3 |
| moles HCOOH | 2 | 1 | 0 | 1 | 2 |
| moles $CH_2O$ | 1 | 1 | 1 | 0 | 0 |

aldehydes, $\alpha$-hydroxyketones, 1,2-diketones, and $\alpha$-hydroxycarboxylic acids with cleavage of the carbon-carbon bond. It will be noted that when three —CHOH— groups are contiguous, the center one is oxidized to formic acid. Consequently, the determination of *(a)* the number of moles of periodic acid used, and *(b)* the number of moles of formic acid produced gives the information needed for the establishment of the size of the ring present in an unknown glycoside.* The application of periodic acid oxidation to the determination of the ring size of glycosides is illustrated by Table 18-2, which shows the results that would be obtained upon periodic acid oxidation of the five possible ring structures of a hexoside.

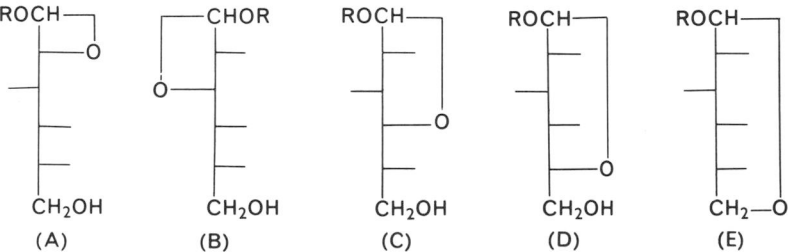

When methyl $\alpha$- or $\beta$-D-glucoside is oxidized with periodic acid, 2 moles of $HIO_4$ are used, and 1 mole of formic acid (and no formaldehyde) is formed. These results are in agreement with the glucopyranoside structure (D).

### 18.15  The configuration of the anomeric carbon atom in $\alpha$- and $\beta$-glucopyranose

The free sugars, $\alpha$- and $\beta$-D-glucose are in equilibrium in solution, as shown by their mutarotation to a mixture with a rotation intermediate between the two. In the crystalline state, each possesses a definite configuration, represented by the following structures:

---

* The term "glycoside" is a general one; when specific sugars are referred to, the terms glucoside, mannoside, arabinoside, and so on, are used.

HOCH$_2$

HO OH OH

α-D-glucopyranose

HOCH$_2$

HO OH OH

β-D-glucopyranose

"anomers"

X-ray measurements show that the crystalline α- and β-glucose are in the pyranose form. The first detectable oxidation products (with bromine water) are δ-lactones, indicating again that the pyranose ring predominates even in solution.

The configuration of the C-1 hydroxyl group was shown by Boeseken to be as indicated in the above formulas. This method is based upon the observation that the conductivity of boric acid (H$_3$BO$_3$) solutions is increased by *cis*-1,2-glycols

$$-\underset{|}{\overset{|}{C}}-OH \quad -\underset{|}{\overset{|}{C}}-O \qquad O-\underset{|}{\overset{|}{C}}- $$
$$\left. \begin{array}{c} -\overset{|}{\underset{|}{C}}-OH \\ -\overset{|}{\underset{|}{C}}-OH \end{array} + H_3BO_3 \rightleftharpoons \begin{array}{c} -\overset{|}{\underset{|}{C}}-O \\ -\overset{|}{\underset{|}{C}}-O \end{array} B \begin{array}{c} O-\overset{|}{\underset{|}{C}}- \\ O-\overset{|}{\underset{|}{C}}- \end{array} \right\}^{-} H^+ + 2H_2O$$

more than by their *trans* isomers. Boeseken found that the conductivity of a solution containing boric acid and α-glucose decreases with time, and that of a solution of boric acid and β-glucose increases with time. This is interpreted to mean that in α-glucose the *cis* relationship of the 1,2-hydroxyl groups (high conductivity) is altered by mutarotation to the *trans* relationship of the β-form. The reverse is true of β-glucose. A complete X-ray analysis of crystalline α-D-glucose has confirmed these conclusions.

### 18.16 Oligosaccharides

The commonest of the oligosaccharides are those in which two sugars are combined by glycoside linkages between the aldehyde group of one and a hydroxyl group of another. These are the *disaccharides*. If a "free" aldehyde group is present (as in lactose) the compound is a *reducing disaccharide*; if the sugars are joined through the two carbonyl groups (as in sucrose) the compound is a *nonreducing disaccharide*. The "reduction" referred to in these names is of Fehling's solution or Tollens' reagent, under the alkaline conditions of which the glycosidic union of the nonreducing disaccharides (compare with methyl glucoside) is stable.

*Lactose ( reducing ):* *

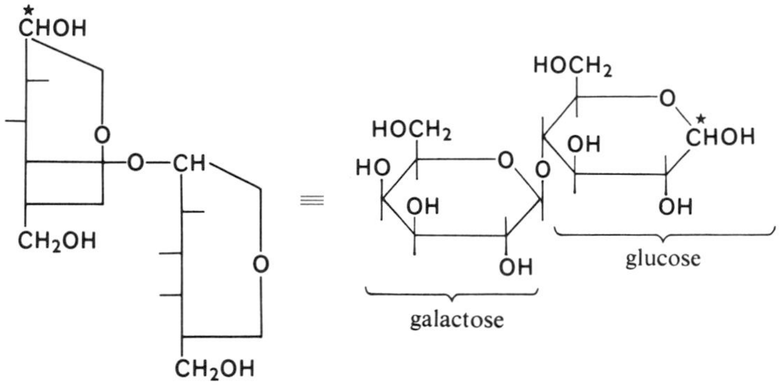

4-O-[β-D-galactopyranosyl]-α or β-D-glucopyranose

*Maltose (reducing):*

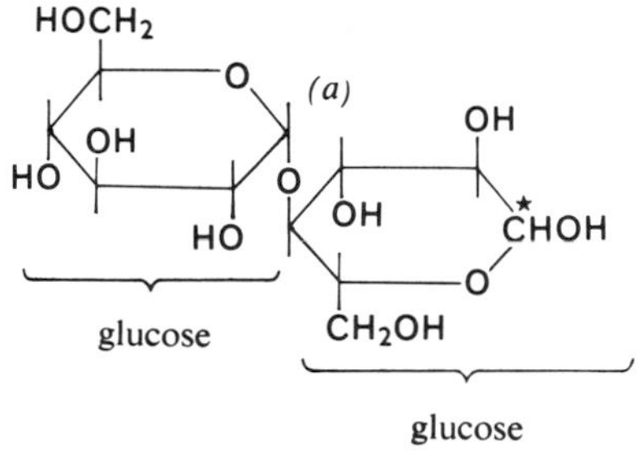

4-O-[α-D-glucopyranosyl]-α
*or*  β-D-glucopyranose

*Sucrose (nonreducing):*

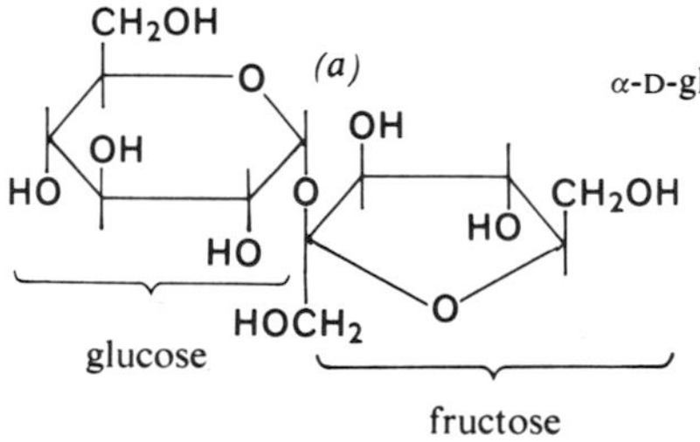

α-D-glucopyranosido-β-D-fructofuranoside
*or* β-D-fructofuranosyl
α-D-glucopyranoside

It is to be noted that there is no free (or potential) reducing group in sucrose: the aldehyde group of glucose and the keto group of fructose, which are responsible for the reducing action of these sugars, are joined together in an ether (acetal) linkage:

---

* The starred carbon atom is the potentially aldehydic carbon atom that is responsible for the reducing power of the disaccharide. It is shown in these formulas in the noncommittal stereochemistry; lactose (in solution) is, like glucose, an equilibrium mixture of the α- and β-anomers. Because this equilibrium involves a transformation in which the aldehyde form appears as an intermediate, the reducing sugars are said to possess "free" aldehyde groups, even though the concentration of this form in solution is very low.

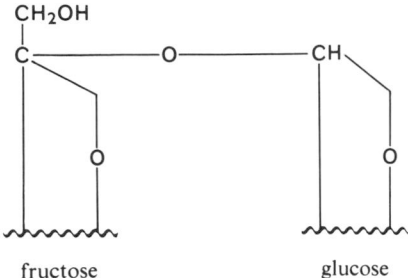

fructose          glucose

Thus, reducing disaccharides are glycosyl *aldoses*, and nonreducing di-saccharides are glycosyl *aldosides* (or ketosides). Thus, lactose is a galactosyl *glucose*; sucrose is a glucosyl *fructoside*.

---

**Exercise 3.** Write the (Haworth) structure for 3-O-[β-D-glucopyranosyl]-α-D-arabino-pyranose.

---

### 18.17  Determination of the structure of oligosaccharides

The structures of the disaccharides possess three features that require demon-stration: (1) the size of the rings (*pyrano*- or *-furano*) of each of the two sugars involved; (2) the position of attachment of the glucoside linkage to each sugar; and (3) the nature (α- or β-) of the glycoside linkage at the aldehydic or keto carbon atom.

Lactose may be examined as an example of structure proof:

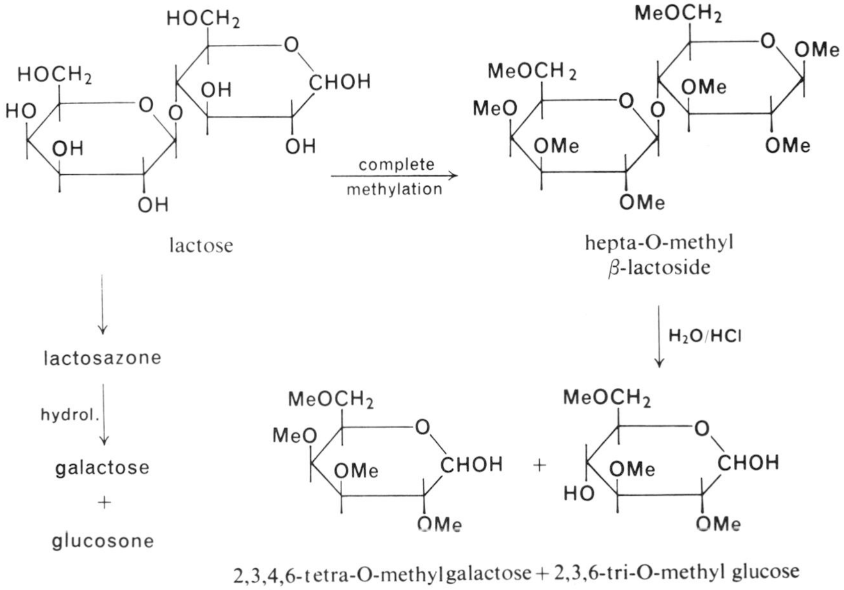

lactose

hepta-O-methyl β-lactoside

lactosazone

hydrol.

galactose
+
glucosone

2,3,4,6-tetra-O-methylgalactose + 2,3,6-tri-O-methyl glucose

The above scheme shows the main structural features: the positions of the methyl groups in the two fragments (tetramethyl galactose and trimethyl glucose) show that the galactose unit is linked to C-4 of the glucose, and the formation of glucosone by hydrolysis of lactosazone confirms the presence of the reducing carbon atom in the glucose residue and the absence of a free aldehyde group in the galactose portion.

Periodic acid oxidation of methyl lactoside gives results that agree with this structure:

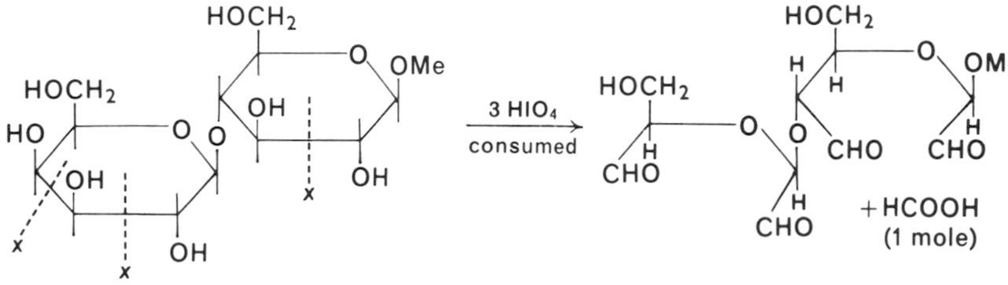

β-methyl lactoside

(*x* -------- : cleavage by HIO₄)

By comparable methods, the structures of other oligosaccharides can be determined.

## 18.18 Determination of configuration by enzymatic action

The remaining structural question that is not answered in the above example is the configuration of the glycosidic carbon atom (carbon atom 1 in the galactose portion of lactose). The same results would be obtained on degradation and HIO₄ oxidation whether lactose were a β- or an α-galactoside.

That lactose has the β-galactosido structure is shown by the fact that it is hydrolyzed by an enzyme that is specific for β-galactosides but inactive toward α-galactosides. Enzymes capable of hydrolyzing β-galactosides are called *β-galactosidases*, and their specificity can be established by testing their ability to bring about the hydrolysis of simple (e.g., methyl) galactosides of known configuration.

The use of enzymes to determine the configuration of the glycosidic carbon atom of glycosides is a useful and widely practiced technique. The *β-glucosidase* obtained from fresh almond kernels (called "emulsin" when in crude form) or that from the intestinal juices of the snail (*Helix* species) is capable of bringing about the hydrolysis of methyl-β-D-glucoside, cellobiose (a disaccharide from cellulose), and of many naturally occurring glucosides, but is incapable of hydrolyzing methyl-α-D-glucoside or maltose (a disaccharide from starch). On the other hand, maltase, an *α-glucosidase* from barley sprouts, is capable of bringing about the hydrolysis only of α-glucosides.

## 18.19   Natural glycosides

An abundance of compounds containing sugars in glycosidic combination with nonsugar molecules is found in nature and particularly in plants. Natural glycosides have been known for a very long time; the benzaldehyde studied by Liebig and Wöhler (Section 1.5) was called "oil of bitter almonds" and was obtained by hydrolysis of a glycoside, *amygdalin*, which is present in the kernels of almonds and related fruits.

The majority of the naturally occurring glycosides are β-glucopyranosides of phenolic compounds; but there are also known glycosides of other hexoses, of some pentoses, and of the methyl-pentoses L-rhamnose and D-fucose. Glycosides of compounds containing nonphenolic hydroxyl groups are also known. A few representative natural glycosides are shown in the following:

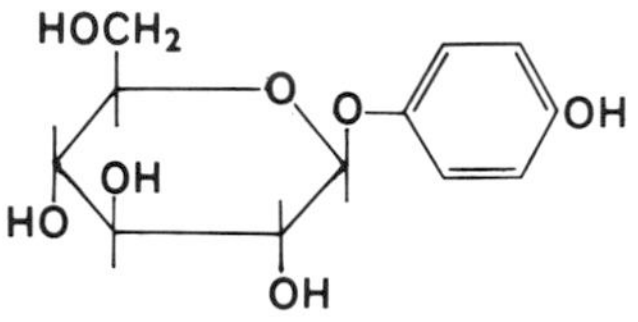

*arbutin*

hydroquinone β-D-glucoside

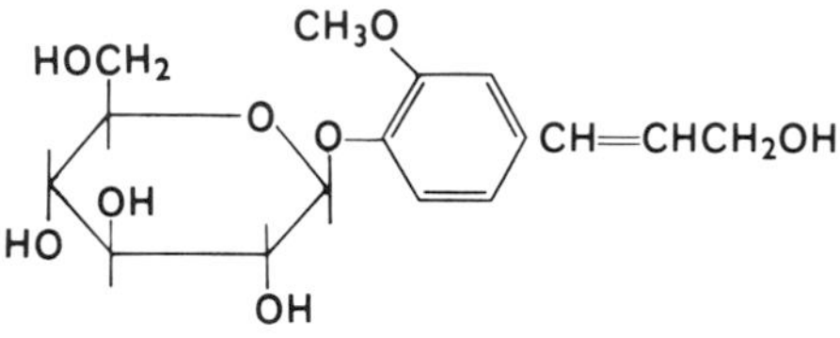

*coniferin*

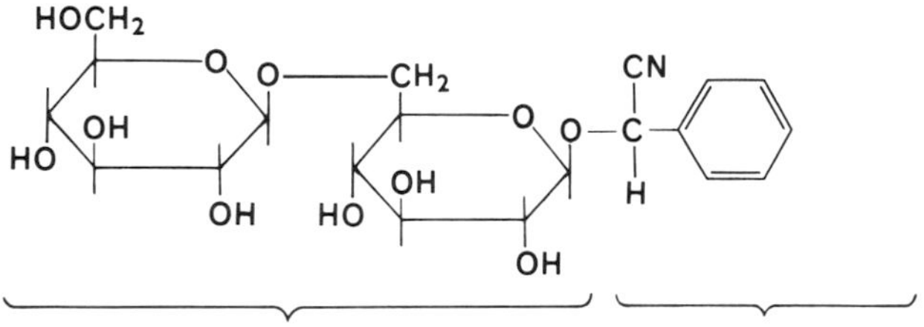

gentibiose                benzaldehyde cyanohydrin
                          (mandelonitrile)

*amygdalin*

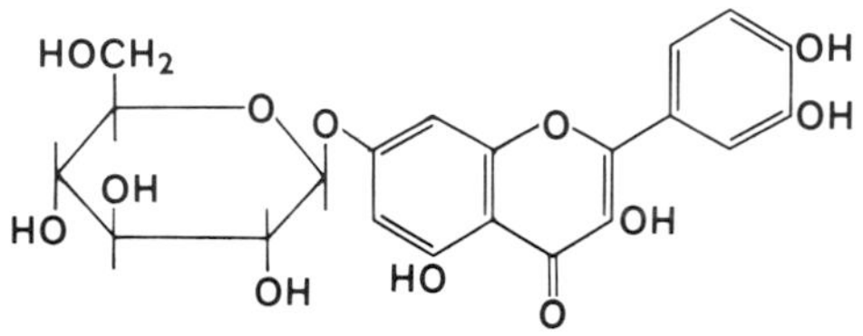

*quercimeritrin*

---

**Exercise 4.**   Write equations for the hydrolysis (aqueous HCl) of the above glycosides.

---

The noncarbohydrate portion of the glycoside is called the *aglycon* (when a glucoside, it is called the *aglucon*); thus, hydroquinone is the aglucon of arbutin, benzaldehyde cyanohydrin the aglycon of amygdalin, and so on.

Other glycosides of benzaldehyde cyanohydrin, and glycosides of the cyanohydrins of other carbonyl compounds, are well known in nature. They are known as "cyanogenetic" glycosides because they liberate hydrocyanic acid upon hydrolysis. Plants containing cyanogenetic glycosides are a cause of poisoning of range animals and have been known to be responsible for serious losses of stock. The release of the hydrocyanic acid is brought about by the hydrolytic action of enzymes that are contained in the plant and are released when the plant is crushed.

## 18.20 Determination of structure of glycosides

The determination of structure of the natural glycosides consists in the following:

1. Hydrolysis to the aglycon and sugar (or sugars).
2. Identification of the aglycon and the sugar (or sugars).
3. If the glycoside contains two or more sugar residues, enzymatic hydrolysis may sometimes be used to remove a disaccharide as a unit; or controlled hydrolysis may remove the sugars one at a time:

$$\text{aglycon} - \text{sugar}_1\text{-sugar}_2 \longrightarrow \text{aglycon} + \underbrace{\text{sugar}_1\text{-sugar}_2}_{\text{a disaccharide}}$$

$$\downarrow$$

$$\text{aglycon-sugar}_1 + \text{sugar}_2$$

$$\downarrow$$

$$\text{aglycon} + \text{sugar}_1$$

The final proof of structure of a natural glycoside is often its total synthesis, although it should be noted that it is frequently possible to prove a structure beyond reasonable doubt by other methods, so that a final synthesis which completes the work is often largely confirmatory. The attachment of a sugar residue to the aglycon is usually performed with aid of the 1-bromo derivative of the acetylated sugar. For example, acetobromoglucose (tetra-O-acetyl-α-glucosyl-1-bromide) is prepared by the action of hydrogen bromide on glucose pentacetate in acetic acid solution:

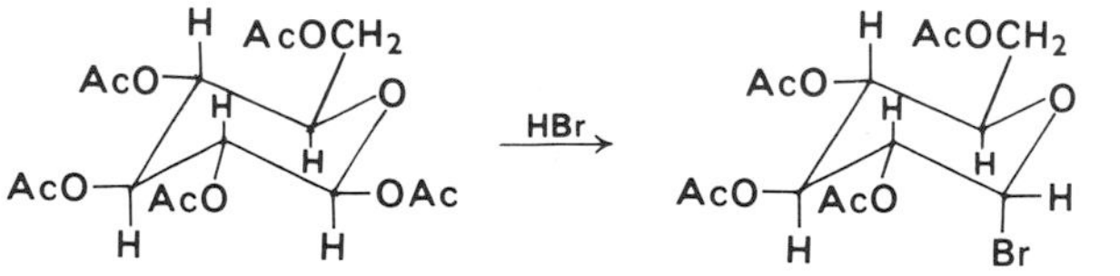

$\beta$-D-glucose pentacetate              acetobromoglucose

Acetobromoglucose is an α-bromo ether, and consequently the bromine atom is highly reactive to displacement reactions. The compound is readily hydrolyzed in the presence of silver carbonate to yield 2,3,4,6-tetra-O-acetylglucose; and the bromine is readily replaced by nucleophilic attack of alcohols and phenols with the formation of glucosides. The replacement takes place with inversion, and the glucosides formed are β-glucosides. The usual procedure is to treat the alcohol or phenol with acetobromoglucose in the presence of alkali or silver oxide:

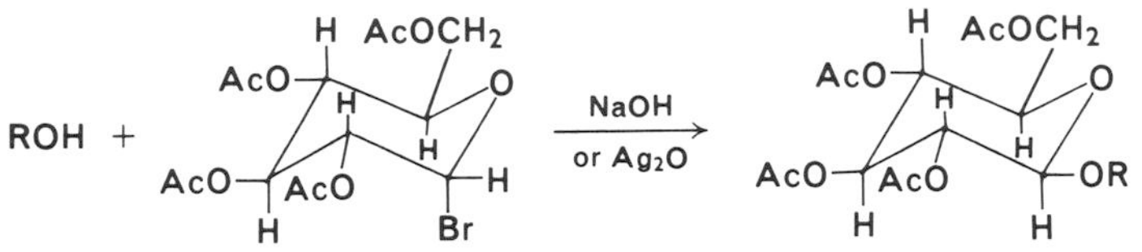

Since the glycosidic grouping (an acetal) is stable to alkali, the acetyl groups can be removed by saponification (or base-catalyzed ester interchange with methanol) to give the β-glucoside.

### 18.21   Structure proof and synthesis of salicin

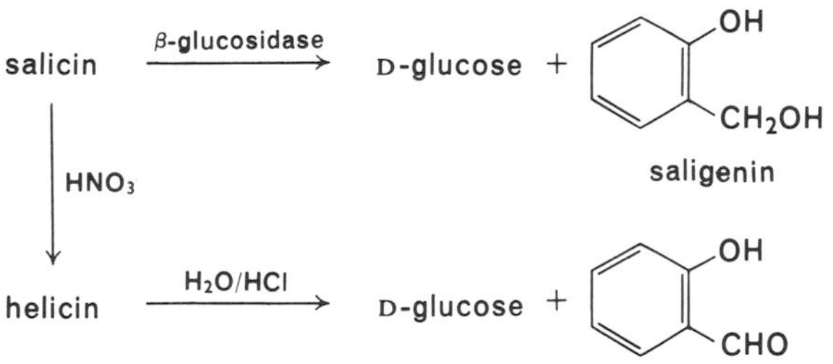

The oxidation of salicin to helicin shows that the glucose is linked to the phenolic hydroxyl group of saligenin, and not to the —CH$_2$OH group. Hence, salicin is

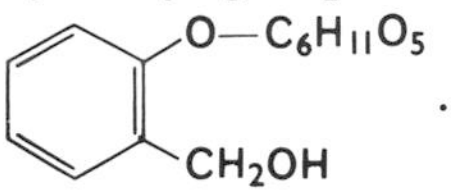

. It can be synthesized by the reaction between acetobromo-

glucose and saligenin, followed by deacetylation of the resulting tetraacetyl salicin. The condensation of acetobromoglucose with saligenin is carried out in the presence of alkali. Hence, the acetobromoglucose reacts with the phenolic hydroxyl group, rather than with the hydroxyl group of the —CH$_2$OH, because the former, being the more acidic of the two, exists in the alkaline reaction mixture in the form of the highly nucleophilic anion.

## 18.22 Other glycosides

Natural glycosides are known that contain sugar residues linked through nitrogen or sulfur to the aglycon. The latter is represented by *sinigrin*, a glucoside found in mustard and horseradish:

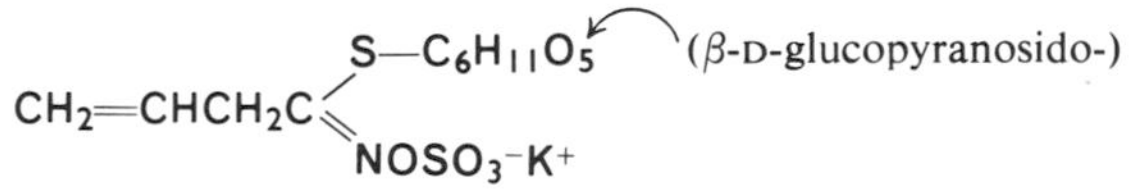

$$CH_2=CHCH_2C \underset{NOSO_3^-K^+}{\overset{S-C_6H_{11}O_5}{\diagdown}} \quad (\beta\text{-D-glucopyranosido-})$$

*sinigrin*

Sulfur-containing glycosides of related structures, differing in the nature of the group attached to carbon (allyl in sinigrin), are found in other plants of the mustard family (Cruciferae). An interesting feature of sinigrin is that when it is hydrolyzed enzymatically a rearrangement occurs, and allyl isothiocyanate, $CH_2=CHCH_2N=C=S$, is formed.

*Nitrogenous glycosides*, in which the sugar is linked to a nitrogen atom, are of great importance, for it is in this class that many compounds of the most profound biological significance are found. These glycosides are known as *nucleosides*, and are found as components of enzymes and of nucleoproteins (in which *nucleic acids* are combined with proteins). An important nucleoside is *adenosine*:

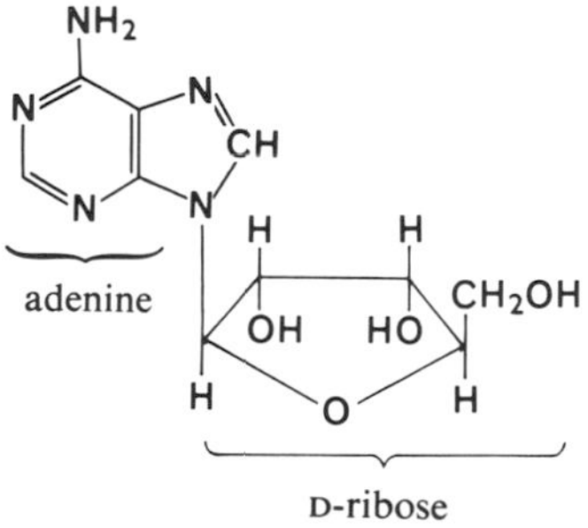

adenine

D-ribose

adenosine

An important sugar found in the nucleic acids derived from cell nuclei is D-2-deoxyribose:

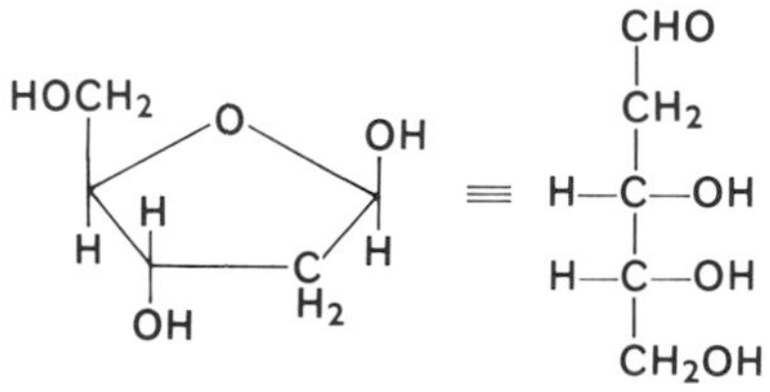

$\beta$-D-2-deoxyribofuranose     D-2-deoxyribose

**Exercise 5.** Write the structure of D-2-deoxyglucopyranose.

Further discussion of the nucleic acids and nucleosides will be deferred until Chapter 31.

## 18.23 Acetone derivatives of sugars

The formation of cyclic acetals and ketals by the reaction of aldehydes and ketones with 1,2- and 1,3-diols has been described in Chapter 8. Sugars having suitably situated hydroxyl groups can form such compounds, and the resulting derivatives are very useful intermediates in carbohydrate chemistry because the acetal linkages are stable to reaction conditions in which alkaline media are used but can later be removed by acid hydrolysis.

The *cis* relationship of adjacent hydroxyl groups is favored for the reaction with a ketone—for example, with acetone. D-Galactose, with acetone and a trace of a mineral acid as catalyst, forms diacetone galactose (1,2-3,4-O-diisopropylidene galactose):

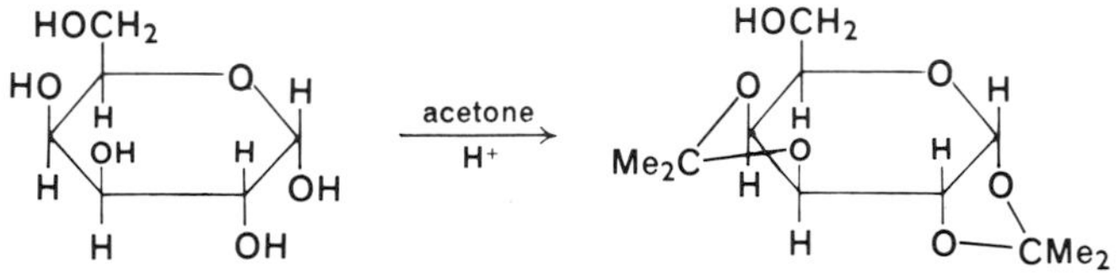

α-D-galactopyranose          diacetone galactose

D-Glucose, however, in which the 3,4-hydroxyl groups are *trans*, reacts in the *furanose form* to give a diacetone glucose:

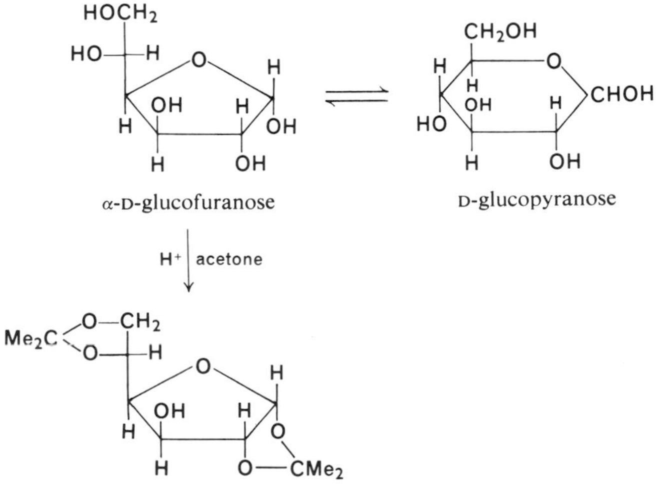

α-D-glucofuranose          D-glucopyranose

diacetone glucose

The use of acetonation to protect some of the hydroxyl groups of a carbohydrate while a reaction is performed on another is illustrated in the synthesis of ascorbic acid (vitamin C):

D-glucose          D-sorbitol          L-sorbose

2,3,4,6-diacetone sorbose          2-keto-L-gulonic acid

L-ascorbic acid
(vitamin C)

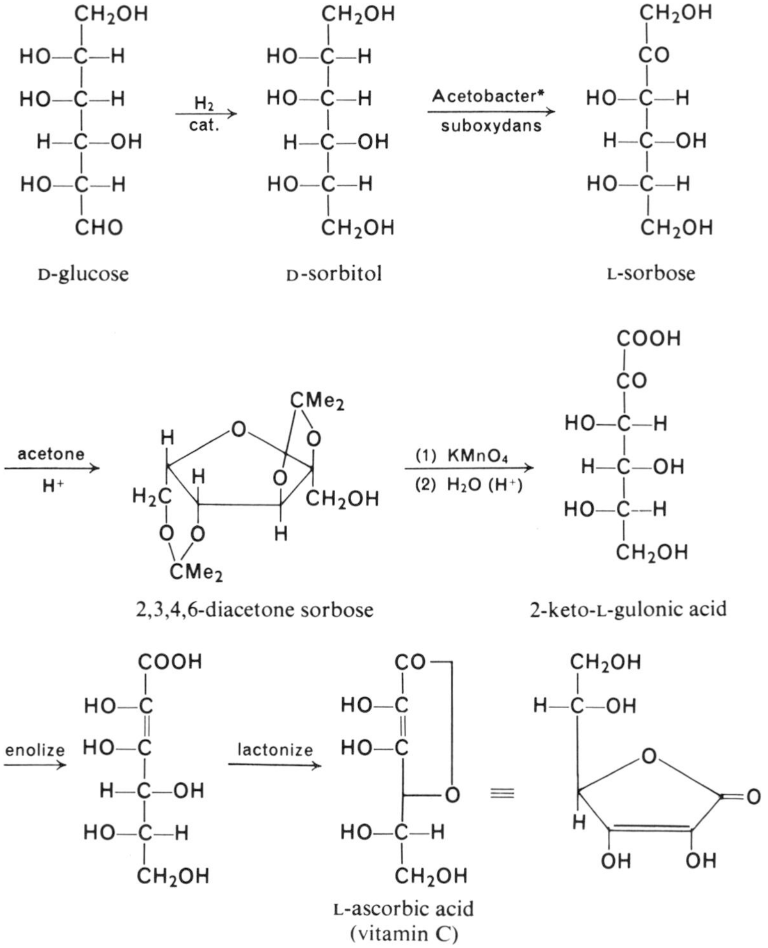

The two ketal linkages of the protecting groups are stable under the alkaline conditions under which the permanganate oxidation is carried out. The two acetone residues are, however, readily removed by acid hydrolysis.

---

**Exercise 6.** Vitamin C is a moderately strong organic acid, with a first ionization constant ($pK_1$) of 4.2. Which of the two enolic hydroxyl groups is ionized when the monoanion is formed?

---

* This is a bacterial oxidation; the organism, *Acetobacter suboxydans*, selectively oxidizes only the —CHOH— group shown.

### 18.24  Identification of sugars

The establishment of the identity of a sugar is often an important part of the determination of the structure of a naturally occurring glycoside. When the glycoside is hydrolyzed, usually by dilute aqueous acid, the sugar is liberated and can be isolated from the hydrolysis mixture. There are a number of ways in which its identity can be established. It is not always necessary to isolate the sugar as a pure, crystalline compound, but when this can be done identification can be made by the determination of its physical properties (melting point, optical rotation) and by direct comparison with an authentic specimen of known identity.

Sugars that are obtained only in dilute solution can be identified either by comparison with known sugars on paper chromatograms or by the preparation of crystalline derivatives. The osazones, prepared with the use of phenylhydrazine, can be used for the latter purpose. Another class of sugar derivatives that are valuable in identification are the *benzimidazoles;* these are formed from the aldonic acids, which are formed by mild oxidation of aldoses, upon reaction with *o*-phenylenediamine:

aldose        aldonic acid        benzimidazole derivative

The benzimidazoles are valuable derivatives. One reason is that the total asymmetry of the sugar molecule is preserved. This is in contrast to the osazones; for example, mannose and glucose give the same osazone, but different benzimidazoles.

Many color tests have been devised to distinguish between different classes of sugars (aldopentoses from aldohexoses, ketoses from aldoses, and so forth), but few of these are applicable to the identification of individual sugars.

### 18.25  Polysaccharides

The vegetation of the earth contains enormous amounts of chemically combined carbon, most of which occurs in the form of two polyglucoses: starch, and cellulose. These polysaccharides are exceedingly complex, and consist of from hundreds to thousands of glucose units linked together in long and complex chains. An exact structure cannot be written for any of the polysaccharides, but many years of investigation of these substances has brought our knowledge of them to the point at which we can describe the structures of many of them with a reasonable degree of completeness.

## 18.26 Cellulose

Cellulose is the most widely distributed plant polysaccharide. It forms the main constituent of the cell walls of plants and, in association with another (non-carbohydrate) polymer, lignin, is the structural material of woody plants. Cellulose occurs in almost pure form in cotton and certain other plant fibers (flax, ramie fiber). Wood cellulose (wood pulp) is prepared by treating wood to remove lignin and hemicelluloses (lower molecular weight, cellulose-like polysaccharides), leaving the cellulose as a pure or nearly pure residue. The economic importance of cellulose can be readily recognized when it is realized how widely paper products and cotton textiles are used.

The composition of cellulose corresponds to $(C_6H_{10}O_5)_x$, where $x$ is a very large number. The substance is a condensation polymer of glucose, since complete hydrolysis gives a nearly quantitative yield of this sugar. Under milder conditions of hydrolysis the oligosaccharides cellobiose, cellotriose, and cellotetrose have been isolated. When to these observations are added the results of methylation studies, which have shown that 2,3,6-trimethyl-D-glucose is nearly the exclusive product, the structure of cellulose can be described: it is a long chain of 1,4-linked $\beta$-D-glucopyranose units: refer to p. 500.

Attempts to measure the molecular size of cellulose have involved many difficulties. By a variety of physical measurements of viscosity, osmotic pressure, ultracentrifugal sedimentation, values have been obtained that indicate that untreated native cellulose has a molecular weight of around 600,000, and thus consists of a chain of about 3,000–4,000 glucose units. The strength of cellulose fibers is due partly to the covalent bonds in these long chains and partly to hydrogen bonding between the hydroxyl groups in adjacent chains.

A better representation of the cellulose chain is that in which the rings are drawn in such a way as to show the conformations of the rings in the chair form:

cellulose

## 18.27 Enzymatic degradation of cellulose

Cellulose is not digested by man, but certain animals (ruminants) are able to utilize it as food because they maintain intestinal colonies of microorganisms that produce enzymes known as "cellulases." The cellulolytic bacteria thus provide simultaneously for their own needs and those of the host animal by converting cellulose into metabolically useful glucose.

Cellulases are widely distributed in nature, being found in the digestive juices of the snail *(Helix pomatia)*, many bacteria and fungi, in the seedlings of some

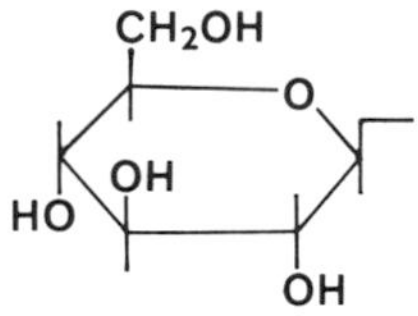

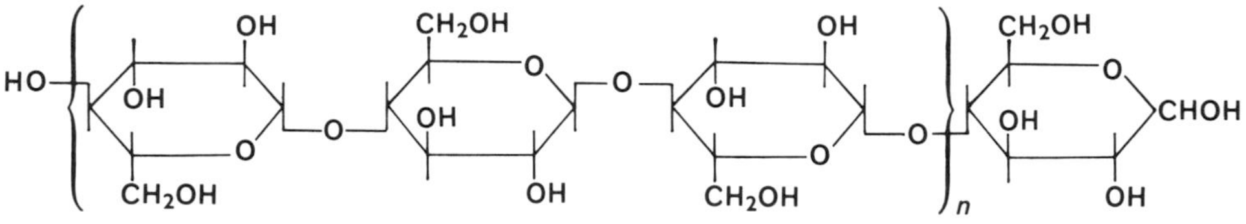

cellulose

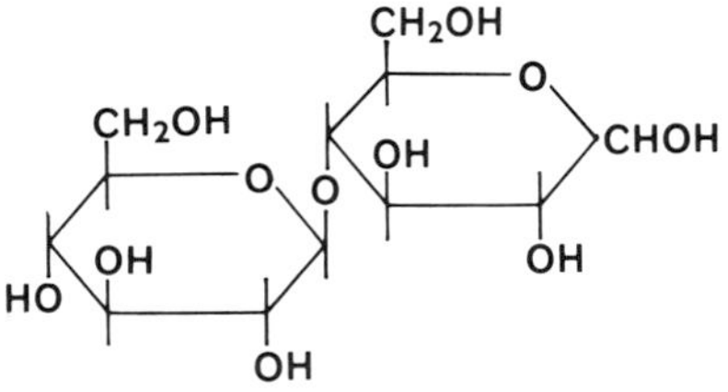

cellobiose

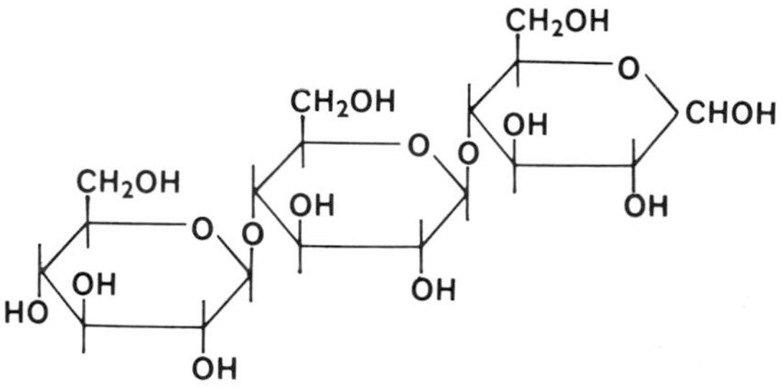

cellotriose

plants, and in some nematodes. The function of cellulases in bacteria and fungi is to provide glucose as an energy source, and plant pathogens may use them as an aid in penetrating the cellulosic membranes of plant cells to enable them to gain entry into their hosts.

The economic importance of the cellulases is very great indeed. The deterioration of cellulosic textiles by bacterial and fungal attack is a serious and costly problem, particularly in tropical regions. Much study has been devoted to methods of reducing the attack of microorganisms upon textiles and paper products. Some of the measures that have been used are impregnation of the textile with fungicides and bactericides, coating of textile materials with substances (e.g., paraffin, plastics) impervious to fungal and bacterial attack, and dyeing with mineral dyes (containing heavy metals). Some attempts have been made to alter the chemical structure of cellulose itself by introducing substituents into the molecule (at the hydroxyl groups). The problem is still one of major importance, and it received particular attention during the last war, when the deterioration of stores and equipment in tropical regions represented a grave logistic problem.

## 18.28 Technical applications of cellulose

Besides the use of cotton, wood pulp, linen, ramie in textile and paper products, chemical alteration of cellulose has provided many interesting and useful materials.

"Nitration" of cellulose is the conversion of cellulose by treatment with nitric and sulfuric acids onto a polynitrate ester. The product, known as "guncotton," is a useful explosive. Nitrocellulose, as the material is called, is a cellulose nitrate, and though the extent to which the hydroxyl groups are esterified varies with the conditions of the process used, fully nitrated cellulose would have the following (partial) structure:

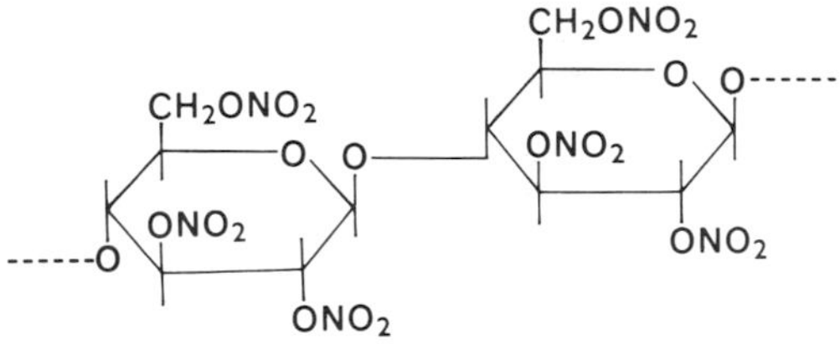

Guncotton approaches this degree of conversion. Less completely nitrated cellulose is known as pyroxylin, a mixture of which with camphor (which reduces the brittleness of the pyroxylin) is the common moldable plastic known commercially as Celluloid.

The most widely used ester of cellulose is the acetate. Cellulose acetate is produced in very large amounts for use as a fiber (rayon) and as a molding plastic. The incorporation of other acyl residues to produce mixed acetate-propionate, acetate-butyrate and other esters produces materials with a wide range of properties and industrial applications. Cellulose *ethers*, in which the —OH groups of cellulose are converted into $OCH_3$— or —$OC_2H_5$ groups, are made by treatment of cellulose with alkali and the appropriate alkyl chloride. The use of chloroacetic acid gives carboxymethyl cellulose, in which —$OCH_2COOH$ groups replace some of the —OH groups. This material (CMC) is a widely used industrial material.

The use of cellulose itself in artificial fibers and films is accomplished by precipitating cellulose from solution in the form of threads or sheets. Cellulose is not soluble in any solvent, but may be brought into solution by chemical means. In the cuprammonium process, cellulose is dissolved in an ammoniacal solution of copper hydroxide (Schweitzer's reagent). The dissolution of cellulose in this reagent involves the formation of a cellulose-copper ion complex. The solution is forced through a tiny orifice into a solution of sulfuric acid; the copper-ammonia cellulose complex is decomposed and the cellulose is precipitated as a thread or filament.

In the "viscose" process, the cellulose is converted into a sulfur-containing ester called a xanthate.* The process may be illustrated in simplified form by using one of the —OH groups of cellulose in the equation

$$(C_6H_{10}O_5)_x \xrightarrow[CS_2]{NaOH} \left\{ C_6H_9O_4\!-\!O\!-\!C\!\!\underset{S^-Na^+}{\overset{S}{<}} \right\}_x$$

cellulose xanthate

The solution of cellulose xanthate can be extruded as a filament (to produce threads and yarns of "viscose rayon") or a sheet (to produce Cellophane). The viscose solution is extruded into dilute sulfuric acid, which causes decomposition of the xanthate and liberation of cellulose:

$$\left( C_6H_9O_4\!-\!OC\!\!\overset{S}{\underset{}{\diagup}}\!\!-\!SNa \right)_x \xrightarrow{H_2SO_4} (C_6H_{10}O_5)_x + CS_2 + Na_2SO_4$$

(Numerous technical details are omitted from these brief descriptions, which provide only the main features of the processes used.)

### 18.29 Starch

Starch, like cellulose, is a polyglucoside; but starch, unlike cellulose, is composed of glucose residues linked by $\alpha$-glucosidic unions. Starch is a reserve polysaccharide: it is the form in which glucose is stored, chiefly in roots and tubers (e.g., potatoes), and from which the glucose may be mobilized for metabolic uses

---

* The formation of xanthates is a general reaction of alcohols:

$$ROH + CS_2 \xrightarrow{NaOH} R\!-\!O\!-\!C\!\!\underset{SNa}{\overset{S}{<}}$$

by enzymatic breakdown. In animals, the reserve polysaccharide is a starch-like substance called glycogen.

Starch is not a single substance. Starches from different sources differ somewhat in their makeup, although there is an essential similarity between them. Starches consist of two polysaccharides, called *amylose* and *amylopectin*. Most starches consist of about 20% of amylose and the remainder amylopectin. These two substances may be separated by swelling the starch granules in warm water, when the amylose diffuses out into solution leaving the amylopectin as an insoluble residue. Both amylose and amylopectin give only D-glucose upon complete hydrolysis, and hence differ only in the manner in which the individual glucose units are combined in the polymers.

*Amylose* is essentially a linear polyglucoside in which 1,4-linked glucose units are joined by α-glucosidic bonds:

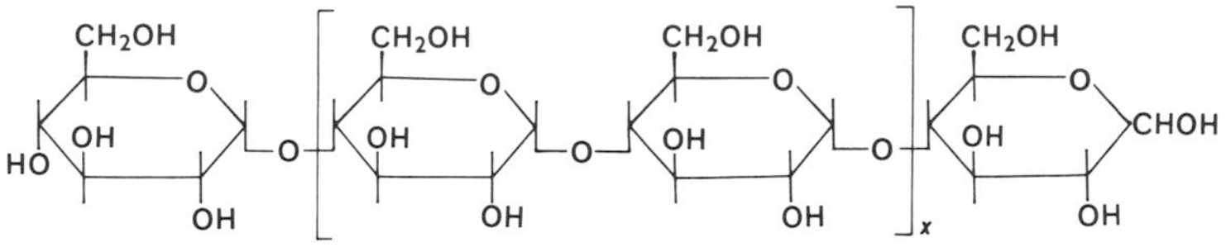

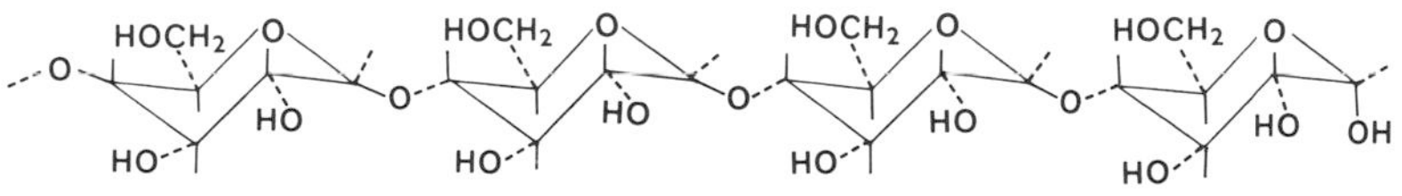

Partial hydrolysis of starch yields maltose:

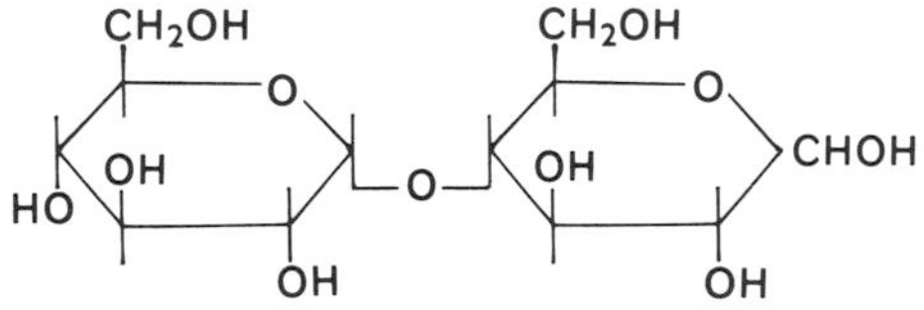

maltose

Maltotriose and maltotetrose, the corresponding tri- and tetrasaccharides (compare with cellotriose) are also obtained under suitable conditions of partial hydrolysis.

Amylose gives a blue color with iodine, whereas amylopectin gives a red to violet color. It has been suggested that the amylose chain is coiled into a helical configuration and that the iodine molecules are able to fit inside this coil in the form of a linear chain of iodine atoms separated by a single iodine-iodine distance.

## 18.30 Amylopectin

Amylopectin differs from amylose in that it contains 1,6-linkages in addition to 1,4-linkages. When the Purdie-Irvine methylation procedure (Section 18.14) is applied to amylopectin there results in addition to 2,3,6-tri-O-methyl glucose (the product characteristic of amylose and cellulose), some 2,3-di-O-methyl glucose, and a remarkably high proportion of 2,3,4,6-tetra-O-methyl glucose. The tetra-methyl glucose is derived from a terminal unit. The conclusion has been reached that amylopectin consists of *branched chains*, each branch consisting of 20–25 glucose units bound by 1,6-linkages to a "backbone" of a $\alpha$-1,4-linked units:

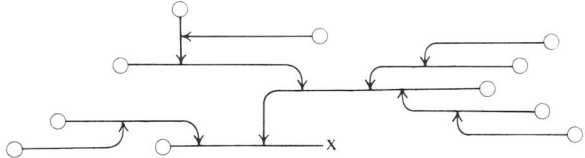

or in schematic form:

probable structure of amylopectin

$x$ = reducing end-group
$\longrightarrow$ = $\alpha$-1,6-linkages, origin of 2,3-di-O-methyl glucose
$\bigcirc$ = end group giving 2,3,4,6-tetra-O-methyl glucose
— = chains of $\alpha$-A,4-linked glucopyranose units

**Exercise 7.**  Show how 2,3,4,6-tetra-O-methylglucose is derived from fully methylated amylopectin.

## 18.31  Enzymatic hydrolysis of starch

Enzymes capable of hydrolyzing starches abound in nature. The mobilization of starch (and glycogen, an amylopectin-like reserve carbohydrate of animals) requires its eventual conversion into glucose (see Chapter 31). In man, the important enzymes for accomplishing this are the salivary amylases and the amylases of the digestive tract (principally from the pancreas). These enzymes convert starches into maltose and the polysaccharide fragments that remain aft er the side chains of amylopectin have been degraded down to the 1,6-branches. Enzymes are known that can degrade amylose chains by removing one terminal nonreducing residue at a time, and others act only on the $\alpha$-1,6-linkages of the amylopectin molecule.

Dextrins are the polysaccharide fragments that result from the enzymatic hydrolysis (by amylases) of the side chains of starch. The term "dextrin" is also applied to the substances that are formed by heating (200–250°C) dry starch or by the partial hydrolysis of starch with dilute acids. The material so formed is soluble in water, yielding a solution known as mucilage, which is used as an adhesive for paper and as a sizing material for textiles.

## 18.32   Other polysaccharides

Although starch and cellulose are the most ubiquitous of the plant polysaccharides many others are known.

The polyuronides are polysaccharides that yield *uronic acids* on hydrolysis. The most important of these are the *pectins*. Pectins are found in most plants, especially in fruits, and are extracted in commercial quantities, chiefly from apples and citrus fruits, for use in making jellies and in altering the consistencies of foodstuffs by virtue of the power of pectins to form gels. Citrus pectin is chiefly a methylated polygalacturonic acid, in which $\alpha$-1,4-linked galacturonic acid units are present. Pectins are commonly found associated with other polysaccharides, notably *arabans* (polyarabinose) and *galactans* (polygalactose).

Alginic acid is a pectin-like polysaccharide (poly-$\beta$-D-mannuronic acid) found in seaweed.

Polymeric pentoses, known as *pentosans*, abound in nature; these may be composed of essentially one sugar, as is xylan, a 1,4-linked polyxylose; or they may consist of two or more monosaccharides in polyglycosidic union. Many plant gums (such as the resinous exudates found on the bark of nonconiferous trees) and mucilages (the viscous juices of certain fleshy leaved plants) are poly-saccharides of this kind. Pentosans occur widely in plants, and very commonly in straws and seed hulls of plants of the grass family (oats, wheat, barley, and others). Oat hulls, straw, and corn cobs are abundant commercial sources of pentosans and are the raw materials for the production of the important industrial chemical, furfural. Furfural is formed by the treatment of pentoses (and pentosans, which yield pentoses by hydrolysis) with hot, dilute sulfuric acid:

$$\begin{array}{cc} \mathrm{HO-CH-CHOH} \\ \underset{\underset{\displaystyle \mathrm{OH}}{|}}{\mathrm{HOCH_2}} \overset{|}{\mathrm{CH-CHO}} \xrightarrow[(-H_2O)]{H_2SO_4} \quad \mathrm{HC} \underset{O}{\overset{\displaystyle\Vert\quad\quad\Vert}{\diagdown}} \overset{\mathrm{CH}}{\underset{\diagup}{\mathrm{C-CHO}}} \end{array}$$

$$\text{a pentose} \qquad\qquad\qquad \text{furfural}$$

## 18.33   Polysaccharides containing nitrogen and sulfur

Polysaccharides that contain nitrogen and sulfur play very important roles in the physiology of animals. Chitin, the structural material of insect and crustacean shells, is a 1,4-linked polysaccharide composed of N-acetyl glucosamine units

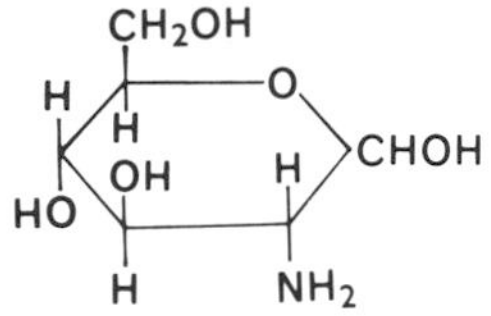

D-glucosamine

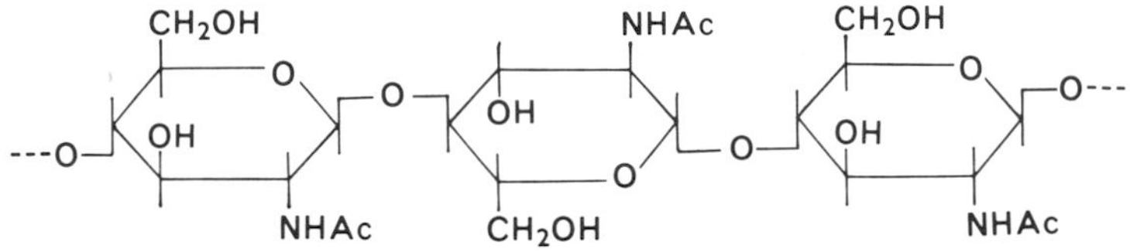

chitin

Hydrolysis of chitin by *chitinase*, an enzyme found in the intestinal tract of the snail, gives N-acetyl-D-glucosamine in high yields, and degradation of chitin with sulfuric acid and acetic anhydride (acetolysis) gives chitobiose hexaacetate, analogous to the cellobiose octaacetate similarly derived from cellulose.

---

**Exercise 8.**   Write the structure of chitobiose hexaacetate.

---

*Heparin* is a complex nitrogen- and sulfur-containing polysaccharide that plays a vital role in the regulation of the clotting power of blood. It is prepared commercially, largely from animal lung tissue, and is widely used clinically as an *anticoagulant*, especially in the treatment of thromboses and cardiac circulatory disease. The structure of heparin is not known with certainty in all details, but a structure that has been proposed is the following:

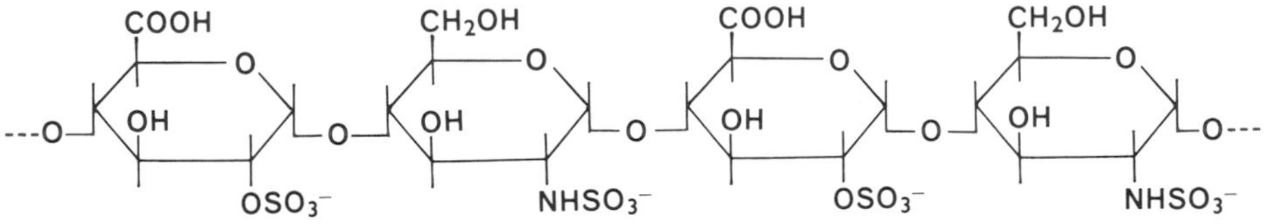

It is probable that an additional one or two of the remaining four hydroxyl groups shown in this tetrasaccharide unit is also present in sulfated form (as —$OSO_3^-$ instead of —OH), and that native heparin has five or six sulfate residues per tetrasaccharide unit. This point is still uncertain, however, and remains to be established by further studies. The molecular size of heparin is not known with certainty; it has been estimated that the molecular weight is about 20,000, but this figure is far from certain.

Seaweed (algal) polysaccharides often contain sulfur in the form of sulfate groupings, like those in heparin, and occur in the plants as salts of magnesium, calcium, sodium, and potassium.

Attempts have been made to introduce sulfuric ester groupings into polysaccharides such as starch and dextrins, xylans, and so forth, with the aim of producing synthetic heparin-like substances. Many such esters do indeed, have anticoagulant properties, and some have been tested in clinical practice, but none has yet been found to be equal or superior to natural heparin, and certain of them have been found to produce toxic side-reactions when administered to humans.

## Exercises

(Exercises 1–8 will be found within the text.)

**9** Write the structures of the two D-aldotetroses, D-threose, and D-erythrose. Write equations showing their *(a)* reaction with phenylhydrazine; *(b)* conversion, through the Kiliani reaction, into the D-aldopentoses; *(c)* reduction to butanetetrols; *(d)* oxidation to the *-onic* acids; *(e)* degradation, through the Wohl reaction, to glyceraldehyde. Show the stereochemistry involved.

**10** How many moles of periodic acid would be consumed by each of the following compounds, and what products would be formed?

*(a)* $HOCH_2CH_2OH$                   *(e)* methyl D-glucofuranoside
*(b)* $CH_3OCH_2CHOHCH_2OH$      *(f)* ethyl tartrate
*(c)* $HOCH_2CHOHCH_2OH$          *(g)* cyclopentane-1,2,3-triol
*(d)* D-glucose                             *(h)* 1,4-butanediol

**11** Show how D-ribose is converted, by way of the cyanohydrin (Kiliani) reaction, into a pair of aldohexoses. Write the configurational formulas for these two aldohexoses and answer the following questions about them. *(a)* Would they give the same or different osazones? *(b)* Would the $C_6$-dicarboxylic acids derived from them by oxidation be optically active? *(c)* Would they reduce Fehling's solution? *(d)* What would be the product or products if they were degraded by the Wohl degradation to aldopentoses?

**12** D-Erythrose, an aldotetrose, can be oxidized to *meso*-tartaric acid, and by the Kiliani synthesis it gives D-ribose and D-arabinose. By the Wohl degradation both D-altrose and D-allose give D-ribose. D-Altrose and D-talose give the same hexitol (hexahydroxyhexane) when reduced. D-Talose and D-galactose (an aldose) give the same osazone. Using this information write the linear projection formulas for D-talose, D-allose, D-altrose, and D-galactose, and show the transformations that have been described.

**13** Methylreductinic acid has the structure Write the structure of the monoanion that would be formed by ionization of the *more acidic* of the two hydroxyl groups. Explain why the OH group you chose is the more acidic of the two.

**14** L-Rhamnose ($C_6H_{12}O_5$) is a naturally occurring methyl aldopentose, also called L-mannomethylose. It forms an osazone, a tetraacetate, and upon vigorous oxidation yields 1 mole of acetic acid. Among the products of a milder oxidation is L-trihydroxyglutaric acid (optically active). Methyl rhamnopyranoside ($\alpha$ or $\beta$) reacts with 2 moles of $HIO_4$ to yield 1 mole of formic acid, no formaldehyde, and a dialdehyde $C_6H_{10}O_4$, which can be oxidized to a dicarboxylic acid, $C_6H_{10}O_6$, with bromine water. When this diacid is hydrolyzed with dilute HCl there is formed methanol, glyoxylic acid ($HOOC \cdot CHO$), and L-(+)-lactic acid. Using the above information write the structure of L-rhamnose (use the linear projection formula), and formulate the reactions described.

**15** A glucoside (A), $C_7H_{14}O_6$, yielded methanol and D-glucose when hydrolyzed with acid. It reacted with periodic acid, consuming 2 moles of the reagent and yielding 1 mole of formaldehyde and no formic acid. Complete methylation of the glucoside (A) yielded a tetra-O-methyl glucose (C). Oxidation of C gave L-1,2-dimethoxysuccinic acid, some methoxyacetic acid, but no glutaric acid derivatives. Write the Haworth projection formula for A and show the reactions described.

# Chapter nineteen / Aromatic Compounds and the Structure of Benzene

## 19.1 The aromatic structure

During the first fifty years of the nineteenth century repeated attempts were made by chemists to devise structural theories into which could be organized the rapidly accumulating body of empirical knowledge of organic compounds. It was not, however, until about mid-century that order began to appear in the chaos of contending theories, culminating in 1858 in Kekulé's express formulation of the tetravalency of carbon and his and Couper's proposal of the use of formulas in which valence bonds were used to join the constituent atoms.

Throughout this period there were known to chemists a large number of compounds, mostly of vegetable origin, which, because of their characteristic odors, were known collectively as "aromatic" compounds. These compounds, which included such substances as oil of wintergreen, gum benzoin, coumarin, various fragrant oils and balsams, and certain derived compounds, such as cinnamic acid, benzoic acid, cymene, and aniline, received relatively less attention than those of the large group of homologous derivatives of aliphatic compounds The relationship of the aromatic compounds to the aliphatic substances was not understood, although it was recognized that the two classes had certain characteristics in common. For example, it was known that the aromatic compounds contained definite groups of atoms that passed through series of chemical transformations intact, and were "radicals" in the same way that ethyl and methyl were "radicals." It will be recalled that Liebig and Wöhler developed early compounds theory of organic compounds from their researches on the derivatives of benzoic acid, and had found that the benzoyl radical (i.e., group) was a common constituent of benzoic acid, benzoyl chloride, methyl benzoate, benzamide, and benzaldehyde.

Benzoic acid was first discovered by Scheele in 1775, and derives its name from gum benzoin, a naturally occurring resin, from which it is formed on heating. When benzene was prepared by Mitscherlich in 1834 by the decarboxylation of benzoic acid, and found to be identical with the "dicarburetted hydrogen" of Faraday, the name benzol (now "benzene") was applied to it by Liebig (*benz-* from the original source, gum benzoin; *-ol* for oil). Faraday, who in 1825 isolated the hydrocarbon from the liquid residues deposited in the gas mains of London, found that its empirical composition could be represented by (CH). The formula $C_6H_6$ was subsequently established by vapor-density measurements.

An essential unity within the compounds of the aromatic group was soon recognized. Numerous direct interrelationships were known to exist between them, for as early as 1834 Mitscherlich had prepared nitrobenzene from benzene, and the reduction of nitrobenzene to aniline had been accomplished by 1842. Aniline was already a known substance, having been isolated in 1826 as a product of the destructive distillation of indigo. In 1845 benzene was isolated from coal tar, an event that may be regarded as the beginning of the now enormous synthetic chemical industry based upon the products derived from coal tar.

It was clear to the chemists of Kekulé's time that benzene possessed a structure that could not be expressed in the terms applicable to the compounds of the aliphatic series. Although benzene has the same proportions of carbon and hydrogen as acetylene, its behavior is very different from that of this highly unsaturated and reactive substance. Benzene is resistant to the action of potassium permanganate and many other oxidizing agents that attack acetylenic and olefinic compounds with ease; it does not react with bromine or chlorine at ordinary temperatures unless special catalysts are present; and it is unaffected by aqueous acids that attack olefins and acetylenes. These facts were known to Kekulé, who pointed out further that all of the decomposition products and derivatives of the aromatic compounds show a group resemblance, having at least six carbon atoms (as in aniline, $C_6H_5NH_2$; phenol, $C_6H_5OH$; nitrobenzene, $C_6H_5NO_2$), as did benzene ($C_6H_6$) itself, a fact which suggested that benzene occupied a key position in the group and represented the essential nucleus from which the others were derived. Moreover, of the monosubstitution products of benzene, there appeared to be one and only one of each kind: there was only one compound, aniline, having an amino group attached to the benzene nucleus; only one, phenol, having a hydroxyl group attached to the benzene nucleus; only one bromobenzene, one nitrobenzene, and so forth. Although Kekulé possessed no proof of the equivalence of the six carbon atoms of benzene, he accepted this as a postulate and proposed that the six carbon atoms of benzene form a closed chain, or ring, and that each carbon atom is attached to one and only one hydrogen atom:

Since Kekulé himself had concluded that carbon is tetravalent, and that its valence is constant in all of its compounds, he found it necessary to add the provision that a third bond must be added to each of the carbon atoms of the ring, and wrote the structure with three double and three single bonds in alternate positions around the ring. Although there were at the time of its proposal many valid criticisms of Kekulé's formula—criticisms with which we shall deal in detail—its essential soundness was immediately recognized. Within a few years Körner proposed a structure for pyridine, and Erlenmeyer, one for naphthalene; both were modeled on the Kekulé structure for benzene:

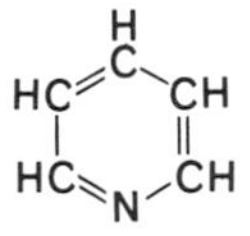

pyridine 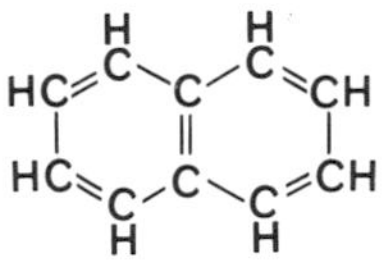          naphthalene

After the publication of Kekulé's structure for benzene the development of the chemistry of aromatic compounds was rapid and fruitful. The term "aromatic" had by that time lost its original meaning and was no longer confined to substances characterized by special odors or origins. Aromatic compounds are theoretically derived from benzene, and contain as a nucleus the benzene ring; and that area of organic chemistry which deals primarily with aromatic compounds is distinguished from the area of aliphatic chemistry by the special properties of the nucleus of benzene and its derivatives.

An objection to the formulation of benzene as 1,3,5-cyclohexatriene was soon raised by Kekulé's contemporaries. Although this structure adequately accounts for the symmetry of the ring evidenced by the existence of only a single substitution product of the formula $C_6H_5X$, it leads to the prediction that two disubstituted products, corresponding to the following formulas, should exist:

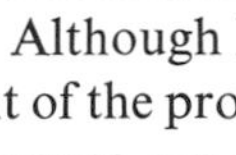

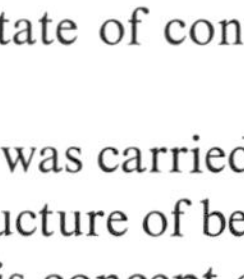

A revision of the original formula, designed to overcome this difficulty, was proposed in 1872 by Kekulé, who suggested a "dynamic formula" in which the double bonds of the ring were in a state of constant oscillation and thus possessed no individual identity.

Although Kekulé's work was carried out about fifty years before the restatement of the problem of the structure of benzene in terms of the modern concept of resonance, we shall see that his concept of the "oscillation" of the double bonds bears a resemblance to, but is not identical with our present-day concepts of the symmetry of the benzene ring.

Early attempts to modify the structural representation of benzene to account for its properties and the isomerism observed in its derivatives resulted in the proposals of Claus (1867), Dewar (1867), Ladenburg (1869), and Baeyer and Armstrong (1887):

| Claus | Dewar | Ladenburg | Baeyer-Armstrong |

Although the Dewar structure was not actually proposed as an alternative to the Kekulé formula for benzene, the term "Dewar benzene" or "Dewar structure" is still used in referring to compounds with the carbon skeleton represented by this formula. In recent years derivatives of the Ladenburg structure (now called *prismane*) and the Dewar structure have been synthesized. They do not have the properties of the corresponding benzene derivatives; however, the compounds possessing the Dewar structure readily isomerize to the benzene derivatives.

That the Ladenburg formula represented the structure of benzene was soon disproved by the demonstration that 1,2-, 1,3-, and 1,4-disubstituted benzene derivatives could be reduced to the corresponding 1,2-, 1,3-, and 1,4-disubstituted cyclohexane derivatives. The other structures advanced suffered from the disadvantages that they required the invention of new kinds of bonds having no counterparts in other compounds, and thus were simply expressions of the impossibility of writing a structure that accommodated the known facts of the symmetry of benzene, the nature of the isomerism shown by its substitution products, and its failure to display the usual properties of unsaturated compounds. We shall see that the differences between aromatic unsaturation and the unsaturation of aliphatic olefinic compounds, when considered in terms of the electronic structures of the molecules, are differences, not of *kind*, but of *degree* only.

The foregoing discussion leads us to ask the question: Is benzene indeed *not* 1,3,5-cyclohexatriene; or is it indeed this compound, whose properties are different from those that would be anticipated for it? The answer to this question can be found in the synthesis of cyclohexatriene by conventional methods, for example as in the following:*

---

\* Note concerning the conventions used in writing the structures of cyclic compounds: The aromatic ring will always be represented by the six-membered ring with three double bonds. When no double bonds are shown, the ring (whatever its size) may be taken to be saturated: that is, if it is six-membered, it is a cyclohexane derivative. In unsaturated but nonaromatic rings, double bonds will be explicitly shown. Thus, bromobenzene, bromocyclohexane, and 3-bromocyclohexene are represented, respectively, as follows:

The product of this synthesis should indeed be cyclohexatriene; *and it is benzene.* Thus the unique properties of benzene are the consequence of the presence of the three conjugated double bonds within a six-membered ring, and must be accounted for not in terms of artificial devices for the representation of the structure but in a consideration of the physical and chemical properties of the compound.

Let us recapitulate the features of aromatic compounds that must be accommodated in an acceptable structural representation:

1. *Benzene is completely symmetrical and the six positions of the ring are equivalent.*
2. *All of the carbon-carbon bond distances in benzene are equal.*
2. *The benzene ring is planar, the carbon-carbon bond angles being equal and exactly 120°.*

These chemical and physical facts can be accounted for as follows. The hybridization of three of the carbon orbitals in a planar disposition and at angles of 120° suggests that the bonds are $sp^2$ in type, in which case there remains a fourth orbital disposed at right angles to the plane of the three $sp^2$ bonds. The six carbon atoms of the benzene ring would thus be represented, as a first approximation, as

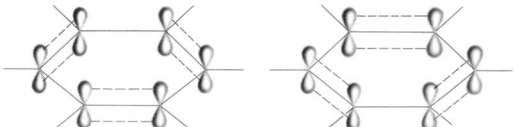

The six carbon atoms and six hydrogen atoms of the benzene ring possess a total of thirty valence electrons. Twenty-four of these are found in the twelve bonds shown in the above (incomplete) formula, and six remain to be accounted for. One of these can be assigned to each of the remaining $p$-orbitals of the six carbon atoms, and the three double bonds of the Kekulé formula can be formed by overlapping of the six adjacent $p$-orbitals in such a way as to form three $\pi$-orbitals that contain a pair of electrons in each. It is apparent, however, that if the six adjacent $p$-orbitals are completely equivalent and if the ring is symmetrical, there is no more justification for completing this pairing by overlap between 1,2/3,4/5,6 than by overlap between 6,1/2,3/4,5 (Figure 19-1). In short, all of the $p$-orbitals can overlap with one another in such a way that no discrete *pairs* of orbitals will overlap, and so three distinct carbon-carbon double bonds will not be formed. Rather, the six $p$-electrons of the six carbon atoms will form *a continuous $\pi$-electron cloud that is symmetrical and coplanar with the plane of the six carbon atoms.* Although

**Figure 19-1.** Equivalent ways in which $p$-orbital overlap may occur in benzene.

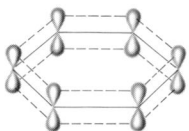 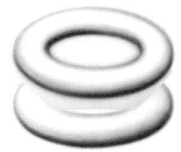

**Figure 19–2.** A representation of the overlap of the six *p*-electrons in benzene to give symmetrical molecular orbitals.

Figure 19-2 shows the disposition of the atoms of the benzene molecule and the cloud of electrons in which are found the six electrons in the $\pi$-orbitals, it must be recalled that since a single orbital can contain no more than two electrons, the $\pi$-orbital shown in the figure is not to be regarded as representing a single orbital containing all six of the electrons. There must be three $\pi$-orbitals, which cannot be conveniently represented by pictures of this sort, but which can be regarded as present in the "cloud of $\pi$-electrons" depicted in Figure 19-2.

It is now apparent that the "Kekulé structures" are those that result from the arbitrary pairing of *p*-orbitals in the two possible ways. But since no one of these can be regarded as correctly representing the structure of benzene, *it is impossible to write a formula for benzene using conventional valence bond notation that accurately depicts its structure.* A reasonable approximation is found in writing the various structures formed by pairing of *p*-electrons in the several most probable ways and joining these several formulas by the double-headed arrow. This symbol indicates that the several

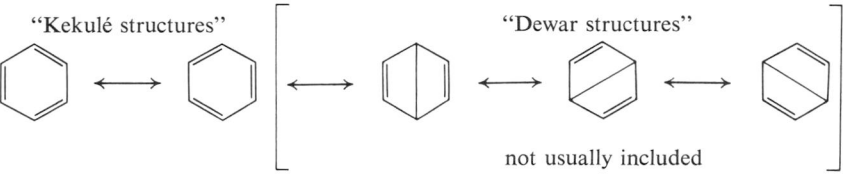

formulas represent forms that contribute to the structure, but that the *structure is properly represented by no one of them, and is a resonance hybrid of all of them.*

In benzene, additional pairings such as 1,4/2,3/5,6, representing one of three possible Dewar structures, are ordinarily omitted because of the relatively smaller contribution made by the overlap of orbitals separated by so great a distance as are the para positions, and thus the two Kekulé structures remain as the two major contributing forms.

## 19.2    Conventional representation of benzene rings

How, then, shall we write the structure of the aromatic ring? We must conclude that there is no one conventional valence bond formula that is a proper representation of the structure, and that any attempt to approach a proper representation will require the invention of a new kind of symbol. The problem is solved by arbitrarily writing the formula as one of the Kekulé forms. It will be recalled that two other ways of representing the benzene ring were mentioned in an earlier chapter (Section 10.4). Although these are occasionally used, the Kekulé formula is generally preferred in present-day chemical literature, and will be used in this book. Benzene will thus be written as

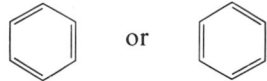

Since this is a conventional representation, the "double bonds" may be written in either arrangement.

We shall find, however, that this solution is a most practical one, because in writing equations that describe the behavior of the aromatic nucleus it will be convenient to depict stages in a reaction in which the electrons become localized. When, for example, an electron-seeking reagent ($E^+$) utilizes two of the six $\pi$-electrons of the benzene ring, the remaining four are no longer symmetrically delocalized; this can be represented with the aid of the structures,

## 19.3 Resonance energy and the stability of the aromatic structure

We may take it as axiomatic that if a molecule may assume more than one possible structure, it will, in the absence of exciting influences, assume that structure which is the most stable (see Chapter 10). In the foregoing discussion we have concluded, from considerations based upon the chemical and physical properties of benzene, that it cannot have a discrete system of three olefinic double bonds, but that the six electrons that correspond to such a system of double bonds are uniformly distributed among all six of the carbon atoms. It is a reasonable inference from this conclusion that benzene is really a more stable substance than it would be if it were *actually* 1,3,5-cyclohexatriene in which the three double bonds of that structure were localized. This inference can be examined by recourse to data that deal, not with bond lengths and angles, but with experimental measurements of the energy of the system.

## 19.4 Evidence from thermochemical measurements

When simple olefins are reduced to the corresponding alkanes, energy is liberated. Careful measurements of the heats of hydrogenation of unsaturated compounds have been carried out by performing the hydrogenations in a calorimeter; some data are shown in Table 19-1.

**Table 19-1.**  *The heats of hydrogenation of some unsaturated olefin compounds*

| Olefin | Heat of Hydrogenation (kcal/mole) | Difference between Calculated and Experimental |
|---|---|---|
| 1-butene | 30.3 | — |
| cis-2-butene | 28.6 | — |
| 2-pentene (mixture of cis and trans) | 28.0 | — |
| cyclopentene | 26.9 | — |
| cyclohexene | 28.6 | — |
| 1,4-pentadiene | 60.8 | 0 |
| 1,5-hexadiene | 60.5 | 0 |
| 1,3-butadiene | 57.1 (calculated 60.6) | 3.5 |
| 1,3-pentadiene | 54.1 (calculated 58.3) | 4.2 |
| 1,3-cyclohexadiene | 55.4 (calculated 57.2) | 1.8 |
| 1,3,5-cycloheptatriene | 72.8 (calculated 79.5) | 6.7 |
| benzene | 49.8 (calculated 85.8) | 36.0 |

The data in Table 19-1 are derived from calorimetric measurements of the reaction, for example, $cis$—$CH_3CH{=}CHCH_3 + H_2 \rightarrow CH_3CH_2CH_2CH_3 + 28.6$ kcal. The examples of the two dienes in which the double bonds are isolated from each other (1,4-pentadiene and 1,5-hexadiene) show that the heats of hydrogenation of such isolated double bonds are strictly additive: both of these dienes have heats of hydrogenation very nearly exactly double that of 1-butene (30.3). A different result is obtained in the case of conjugated dienes, 1,3-Butadiene, 1,3-pentadiene, and 1,3-cyclohexadiene have heats of hydrogenation that are less by about 2 to 4 kcal than the sum of the heats of hydrogenation of the individual double bonds. This means that these dienes are closer to the energy level of the final state than they would be expected to be; that is, that they are more stable by 2 to 4 kcal than they would be if their double bonds were independent, as are those in the nonconjugated dienes. The suspicion is aroused that their representation as dienes is not quite accurate, and that the system of two double bonds is actually altered in character by the conjugation.

The heat of hydrogenation of benzene is very different from that calculated on the assumption that three cyclohexene-like double bonds are present. The difference between the observed (49.8) and calculated (85.8) values is 36 kcal/mole. This value represents the degree to which benzene is more stable than it would be if it possessed three double bonds that acted independently of one another. It is also evident that there is a factor that depends upon the fact that benzene contains the three "double bonds" in a six-membered ring, for 1,3,5-cycloheptatriene shows a difference of only 6.7 kcal/mole between the heats of hydrogenation observed and calculated for three double bonds of the *cis*-2-butene type. We see that no two structures for the cycloheptatriene can be written which correspond to the two Kekulé structures for benzene: if $\pi$-orbitals are constructed to encompass atoms 2,3 and 4,5, those on atoms 1 and 6 are not adjacent and cannot form an effective third $\pi$-orbital because of the considerable distance between them:

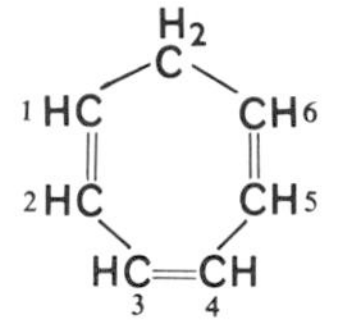

In the case of benzene, however, the two Kekulé structures are completely equivalent. We can summarize this by saying that cycloheptatriene is *most nearly correctly* represented by structure A, although structure B may contribute to a small degree; whereas in the case of benzene neither Kekulé structure can make more contribution to the actual structure than the other. This can be expressed in another way by saying that in the cycloheptatriene the delocalization of the electrons from the 1,3,5 arrangement occurs to some degree, whereas in benzene the delocalization is complete.

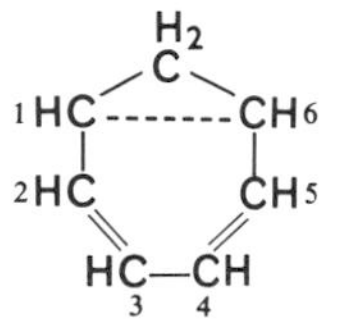

The value of 36 kcal/mole is the *resonance energy* of benzene, as calculated from heat of hydrogenation data. Other calculations, based upon thermochemical data of other kinds, give a value for the resonance energy of benzene that is slightly different from but close to this value. If the heat of formation of benzene is calculated from heats of combustion data, it is found that the difference between the experimentally determined value and that calculated for one of the Kekulé structures (that is, for a discrete 1,3,5-cyclohexatriene) is also about 36 kcal/mole (Section 10.4). We shall not be concerned with the ways in which such calculations can be refined; it is clear that benzene is considerably more stable than it would be were the cyclohexatriene valence bond structure the correct one, and that the extent to which its actual structure is more stable than a Kekulé structure is measurable, and is about 36 kcal/mole.

## 19.5 Formation of the aromatic structure

Most aromatic compounds are prepared from starting materials that are already aromatic, that is, from materials that contain the benzene ring. Relatively few aromatic compounds are prepared from nonaromatic starting materials, for, as we shall see in the chapters ahead, the introduction of substituents into the aromatic nucleus provides a means of preparing an almost unlimited array of aromatic compounds.

An important exception to the generalization that few benzene derivatives are prepared from nonaromatic precursors is found in the industrial production of toluene by the catalytic cyclization-dehydrogenation of paraffin hydrocarbons (e.g., *n*-heptane) in the petroleum industry (Section 23.1). This is an important commercial source of aromatic hydrocarbons, but it is not a reaction that is adaptable to the laboratory preparation of benzene derivatives. It does, however, provide in vast quantities the raw materials from which other aromatic compounds can be synthesized.

Many other examples are known in which aliphatic or alicyclic compounds are readily converted into benzene derivatives. These reactions, few of which constitute general methods of synthesis of aromatic compounds, do provide convincing evidence that the formation of the aromatic ring takes place with great readiness, an indication of the great gain in stability that derives from the resonance energy of the aromatic ring system. When a cyclization reaction leads to a compound that can be aromatized by the loss of two hydrogen atoms, a rearrangement of a double bond, a dehydrohalogenation, or some other simple change, the final step that leads to aromatization is usually easy to accomplish. For example, the Diels-Alder condensation of cyclohexadiene with dimethyl acetylenedicarboxylate gives the bicyclic compound shown in the following equation:

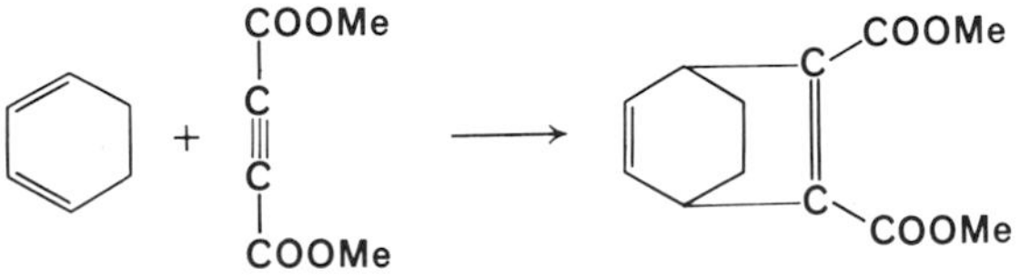

This compound loses ethylene upon heating, with the formation of the benzene derivative, dimethyl phthalate:

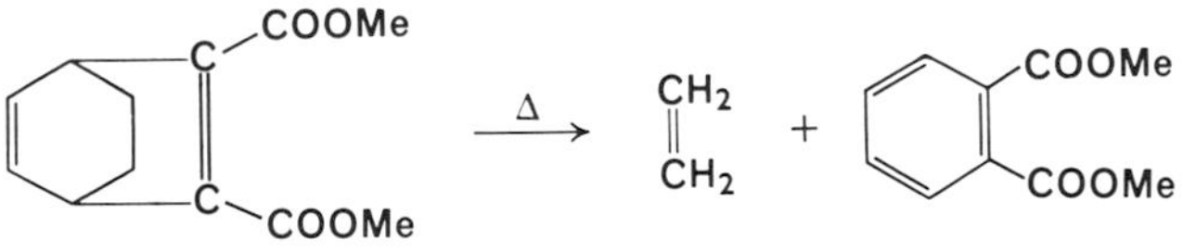

The Dieckmann condensation of diethyl succinate yields the diketo diester in the following way:

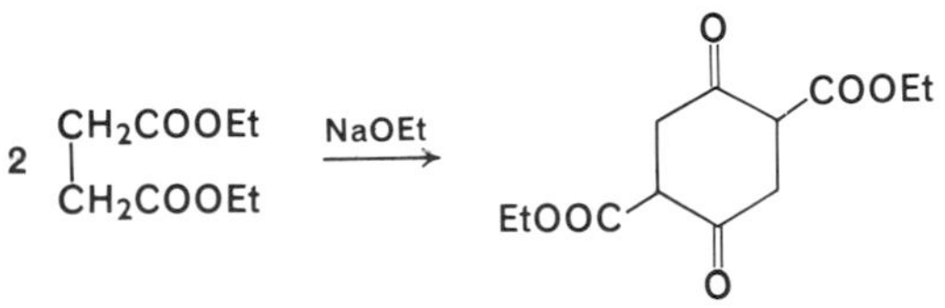

This compound readily loses two hydrogen atoms when treated with bromine, to give the aromatic compound:

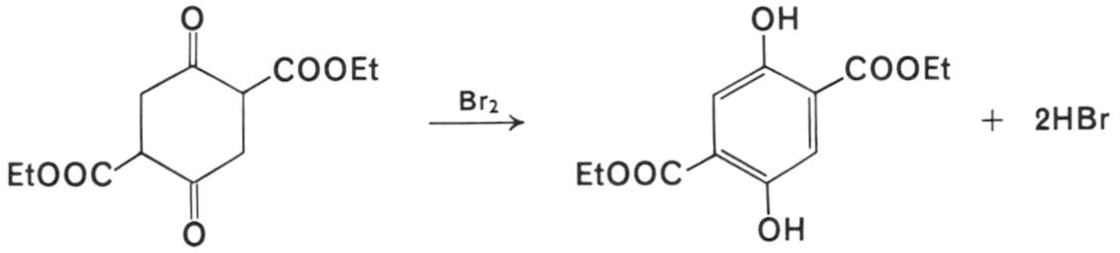

The cyclic ketone, carvone, can be isomerized by treatment with strong acid (HCl) to give the aromatic compound, carvacrol:

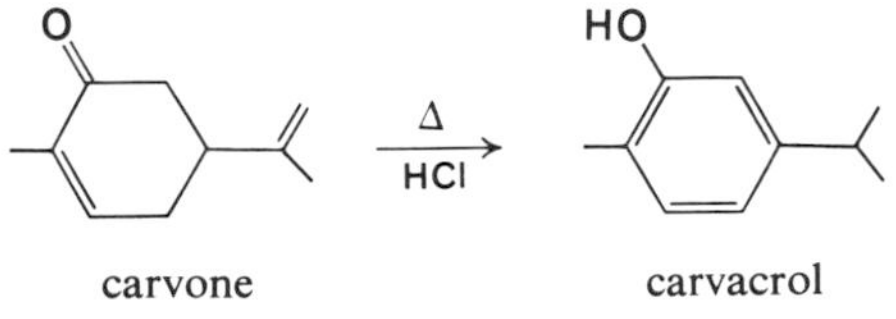

carvone                    carvacrol

The conversion of dihydrobenzene and tetrahydrobenzene derivatives (i.e., cyclohexene and cyclohexadiene derivatives) into the aromatic compounds by the removal of hydrogen can be accomplished in several ways. In the example given above, this was done by the action of bromine. Another useful method is dehydrogenation by heating with sulfur or selenium, or with palladium (usually as a charcoal-supported palladium black). With selenium or sulfur, the hydrogen is removed as hydrogen sulfide or hydrogen selenide; with palladium, the hydrogen is removed as such and is evolved as gaseous hydrogen or, in some cases, transferred to some hydrogen acceptor:

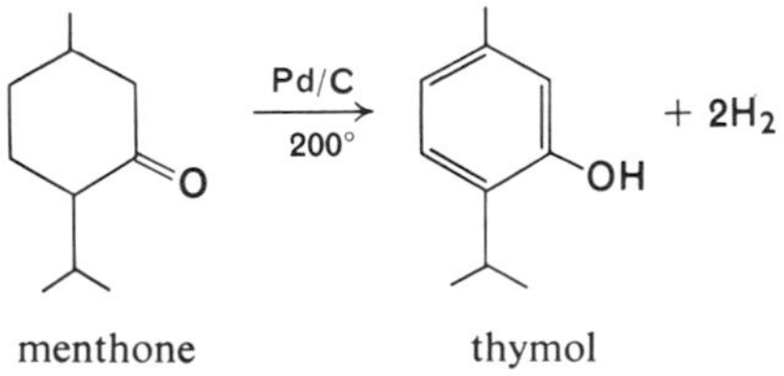

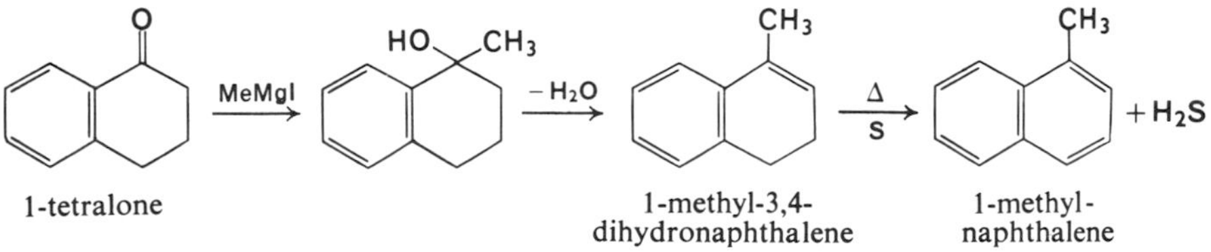

menthone                    thymol

1-tetralone                    1-methyl-3,4-          1-methyl-
                               dihydronaphthalene     naphthalene

These examples show that the aromatic and nonaromatic structures are related by simple changes. We shall find, however, that with the appearance of the aromatic structure a number of unique aspects of behavior are encountered. These special modes of behavior are not, however, peculiar to aromatic compounds; but they represent *characteristic* aspects of the chemistry of aromatic compounds that justify our devoting special treatment to the chemistry of benzene derivatives. As we become more familiar with aromatic compounds it will become apparent that the division between aromatic and nonaromatic compounds is somewhat arbitrary, and that there are many situations in which we can discern no clear line separating the characteristics of one class from those of the other. The chemistry of aromatic compounds differs from that of nonaromatic compounds chiefly in degree, and not in kind.

/ **Electrophilic Substitution into the Aromatic Ring**

## 20.1 Naming of benzene derivatives

Many of the simple, monosubstituted benzenes have been known for so long a time that the trivial names by which they have been known are still used: for example,

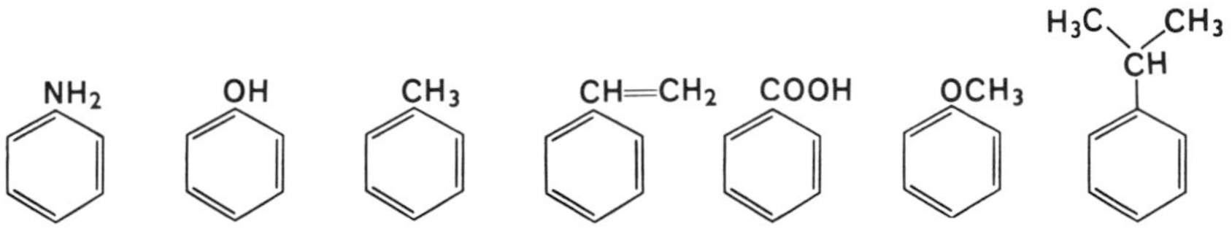

aniline    phenol    toluene    styrene    benzoic acid    anisole    cumene

Many monosubstituted benzenes are named by simple substitution names:

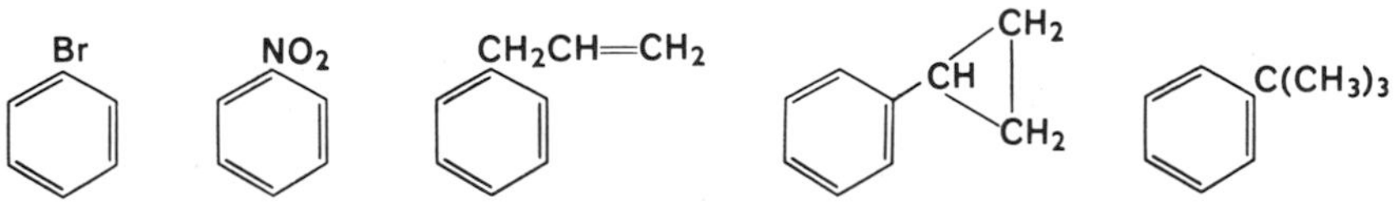

bromo-    nitrobenzene    allylbenzene    cyclopropylbenzene    *t*-butylbenzene
benzene

Disubstituted benzenes are usually named with the use of the prefixes *ortho* (*o*-) for 1,2-disubstitution; *meta* (*m*-) for 1,3-substitution; and para (*p*-) for 1,4-substitution.

Both substituents may be named:

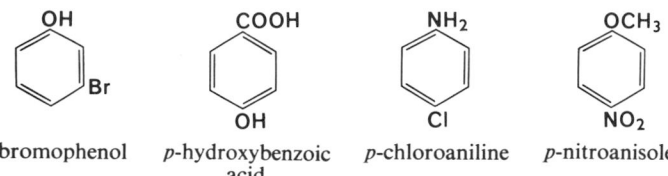

o-dibromobenzene    m-dinitrobenzene    p-dichlorobenzene

or the name may be derived from the trivial name of the parent monosubstituted compound:

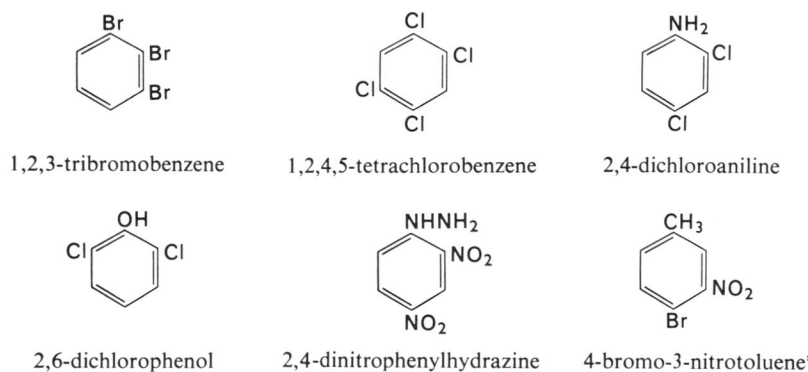

m-bromophenol    p-hydroxybenzoic    p-chloroaniline    p-nitroanisole
                       acid

Polysubstituted benzene derivatives are usually named by numbering the substituents systematically around the ring:

1,2,3-tribromobenzene    1,2,4,5-tetrachlorobenzene    2,4-dichloroaniline

2,6-dichlorophenol    2,4-dinitrophenylhydrazine    4-bromo-3-nitrotoluene*

It is to be noted that when the name is derived from a trivial name (aniline, toluene, phenol) the number *one* is given to the substituent that characterizes the parent compound.

While numerous di-, tri- and tetra- substituted benzenes are known by trivial names

mesitylene    phloroglucinol    p-cymene    o-xylene    durene

---

\* The name 3-nitro-4-bromotoluene would also be acceptable, but since the system of indexing used in *Chemical Abstracts* is an alphabetical one, the name 4-bromo-2-nitrotoluene would ordinarily be chosen.

and most of these will become familiar to the student, it is not improper (although it is sometimes awkward) to use systematic names, with numbering of the substituents, for such common compounds.

## 20.2 Orientation in the benzene ring

The word "orientation" refers to the relative positions of the substituents on the aromatic ring. When a second substituent is introduced into a benzene ring that already bears a substituent, the directive influence of the first substituent in determining the position that will be occupied by the incoming group is known as its *orienting effect*, or *orienting influence*, and we speak of the *orientation* of the substituents on the ring.

The earliest method of establishing the orientation of the substituents on a disubstituted benzene was devised by Körner (1839–1925), and is an *absolute* method. As an example of this method, consider the three possible dibromo-benzenes (*o*-, *m*-, and *p*-). Two of these are liquids (m.p. 6.7°C and $-7$°C) and one is a crystalline solid (m.p. 87°C). Which of these is the *para*-, which the *meta*-, and which the *ortho*- isomer? Suppose we now introduce a third bromine atom into each of these. As the following equations show, one of the dibromobenzenes gives one, and only one, tribromobenzene; this is the isomer with m.p. 87°, and it must be *p*-dibromobenzene, since it is only in this compound that all four of the vacant positions are identical:

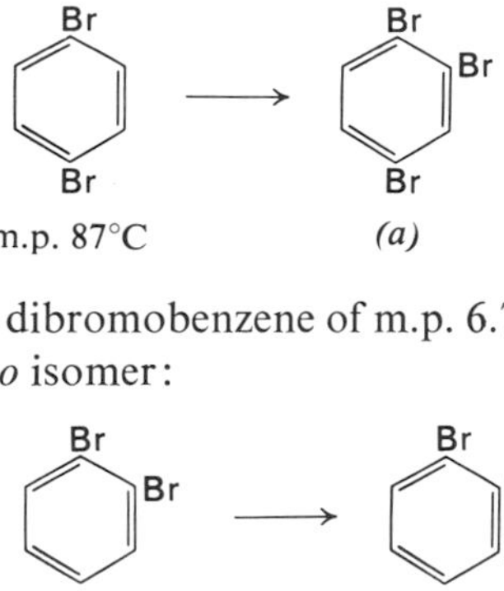

m.p. 87°C  (a)  (1)

The dibromobenzene of m.p. 6.7°C gives *two* tribromobenzenes; this must be the *ortho* isomer:

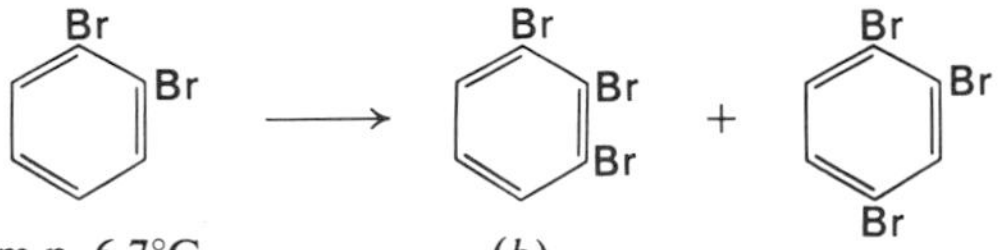

m.p. 6.7°C  (b)  (2)

The third dibromobenzene gives *three* dibromobenzenes:

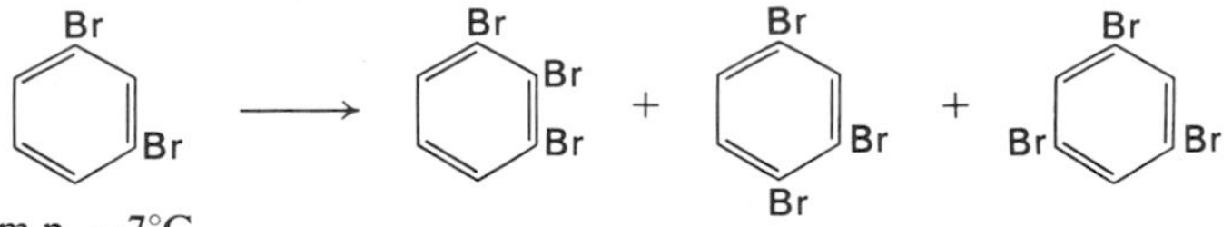

m.p. $-7$°C  (c)  (3)

These experiments prove that the compound with m.p. 87°C is *p*-dibromobenzene, the one with m.p. 6.7°C is *o*-dibromobenzene, and the one with m.p. −7°C is *m*-dibromobenzene. They also prove that the tribromobenzene *(a)* is 1,2,4-tribromobenzene. This same compound is produced from *o*-dibromobenzene, and so *(b)* must be 1,2,3-tribromobenzene. It is easy to see that the third possible tribromobenzene must be *(c)*, which, in addition to *(a)* and *(b)*, is formed from *m*-dibromobenzene.

---

**Exercise 1.** In Körner's original experiment, nitration was used to introduce a nitro group ($-NO_2$) as the third substituent. Work out the scheme showing how the three xylenes (1,2-, 1,3- and 1,4-dimethylbenzenes) can be identified by mononitration to give all of the possible mononitroxylenes. Can the orientation of the resulting nitroxylenes be established in the same way that was used to identify the three tribromobenzenes? How many nitroxylenes are formed? Are any of the nitroxylenes formed from more than one of the xylenes?

---

### 20.3 Orientation by reference to compounds of known structure

After a number of di- and trisubstituted benzenes have been oriented in the way described above they can be used as reference substances to establish the orientation of other compounds. For example, if the known *m*-dibromobenzene is subjected to the *Wurtz-Fittig reaction,*

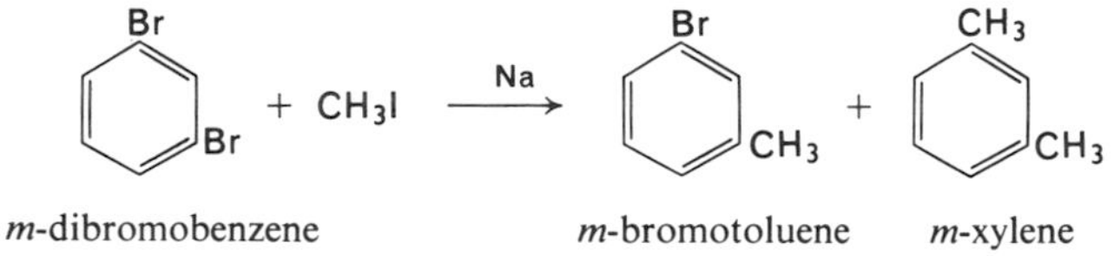

$$\text{m-dibromobenzene} \qquad \text{m-bromotoluene} \quad \text{m-xylene} \tag{4}$$

the products (*m*-bromotoluene and *m*-xylene, in proportions that will depend upon the conditions used in carrying out the reaction) will have the orientations shown in the equation.

The oxidation of the *m*-xylene with potassium permanganate to the dicarboxylic acid will lead to isophthalic acid:

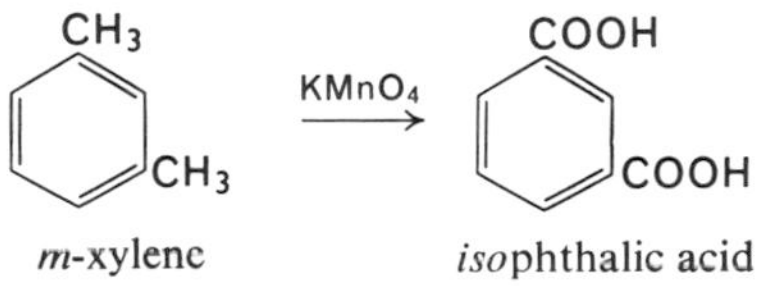

$$\text{m-xylene} \qquad\qquad \textit{iso}\text{phthalic acid} \tag{5}$$

This result can be checked. Ladenburg, in his early studies on the question of the equivalence of the positions on the benzene ring, showed that mesitylene is 1,3,5-trimethylbenzene. This is also in agreement with what would be expected from the manner of its synthesis by the aldol condensation from acetone:

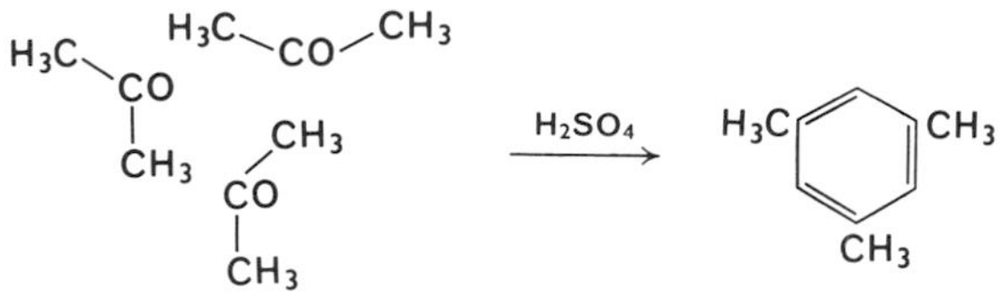

$$(6)$$

If mesitylene is oxidized in such a way that the oxidation is incomplete, among the products will be found some 3,5-dimethylbenzoic acid:

mesitylene    3,5-dimethylbenzoic acid

(plus further oxidation products)    $(7)$

Now if this acid is decarboxylated, the product must be *m*-xylene; and oxidation of the *m*-xylene prepared in this way to the dicarboxylic acid will yield isophthalic acid, identical with the acid obtained from *m*-dibromobenzene as described above:

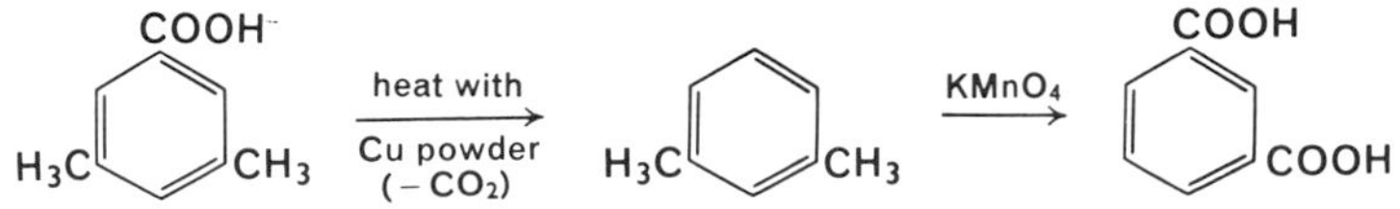

*m*-xylene    *iso*phthalic acid

$$(8)$$

Of the remaining two benzenedicarboxylic acids, one can be converted easily into the monomeric anhydride by simple heating:

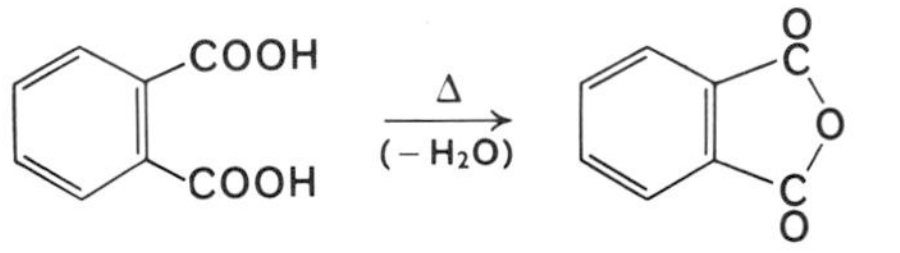

phthalic acid    phthalic anhydride

$$(9)$$

Since only the *ortho* dicarboxylic acid can undergo this reaction, the remaining dicarboxylic acid (of the three possible) must be the *para* compound, terephthalic acid.

The distinction between an *ortho*-substituted compound and its *meta* and *para* isomers can often be made by observing the differences between the three compounds in reactions that involve ring formation. Phthalic acid, as has been shown above, forms the cyclic anhydride, a reaction that cannot occur in the case of isophthalic or terephthalic acid. Similarly, the reactions of *o*-phenylenediamine with 1,2-dicarbonyl compounds

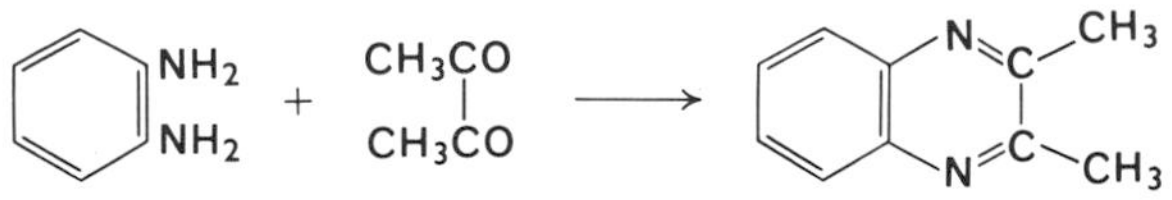

*o*-phenylenediamine    2,3-dimethylquinoxaline

$$(10)$$

and with carboxylic acids, e.g.,

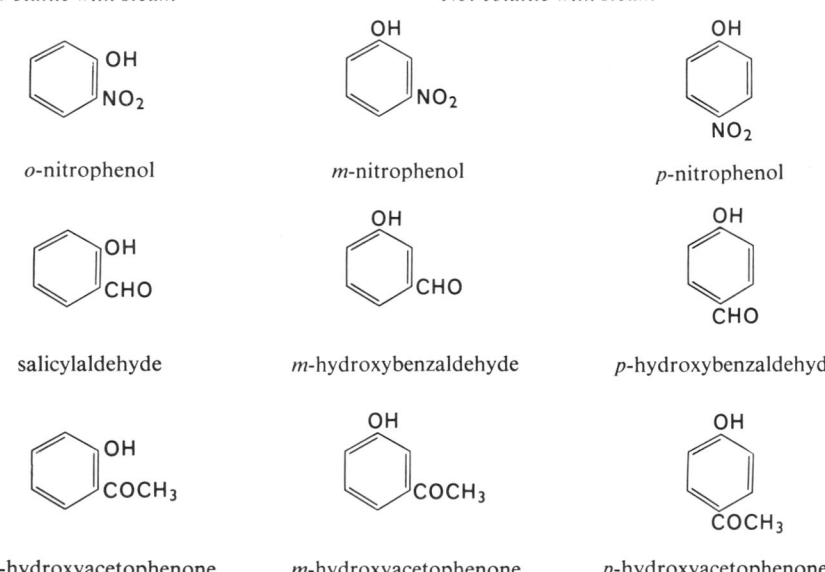

$$\text{2-methylbenzimidazole} \tag{11}$$

have no counterparts in the case of *m*- and *p*-phenylenediamine.

The steam-volatility of *o*-nitrophenol, salicylaldehyde, *o*-hydroxyaceto-phenone is the consequence of the intramolecular hydrogen bond formation in these compounds. The corresponding *m*- and *p*-substituted compounds are not steam-volatile.

hydrogen bonding in salicylaldehyde

*Volatile with steam*                    *Not volatile with steam*

*o*-nitrophenol          *m*-nitrophenol          *p*-nitrophenol

salicylaldehyde    *m*-hydroxybenzaldehyde    *p*-hydroxybenzaldehyde

*o*-hydroxyacetophenone    *m*-hydroxyacetophenone    *p*-hydroxyacetophenone

The determination of the orientation of a benzene derivative in the ways that have just been described, in which the ultimate reference standard is a compound of known orientation, is a "relative" method of orientation. The method first used by Körner to distinguish between the three dibromobenzenes is an "absolute" method since it involved no prior information concerning the orientation of any reference compound. With the wealth of information available today absolute methods of orientation are no longer necessary.

## 20.4 Substitution reactions

Although the term "substitution" could be used to describe any reaction in which one group replaces another, with the new group taking the position formerly occupied by the group replaced, it is usual to limit its meaning to the replacement of hydrogen. Typical substitution reactions are halogenation, nitration, alkylation, and acylation, in which a hydrogen is replaced by a halogen, a nitro group, an alkyl group, or an acyl group.

*Aliphatic hydrocarbons.* Substitution reactions, while known, are neither common nor characteristic of aliphatic hydrocarbons, but are among the most characteristic and useful of the reactions of aromatic hydrocarbons. The halogenation of a paraffin hydrocarbon proceeds by a free radical process; in the case of chlorination, the activation energy supplied by heat or light causes an initial dissociation of the chlorine molecule:

$$Cl_2 \xrightarrow[\text{(heat or light)}]{\text{energy}} 2Cl\cdot \tag{12}$$

The highly reactive atomic chlorine attacks the hydrocarbon

$$Cl\cdot + HR \longrightarrow HCl + R\cdot \tag{13}$$

$$R\cdot + Cl_2 \longrightarrow RCl + Cl\cdot \tag{14}$$

to yield the chloroalkane and a new chlorine atom, which can continue the process. This is called a "chain reaction"; and unless the chlorine atoms are destroyed by some alternative process, a relatively few initial dissociations can lead to a great many substitution reactions. But since the reactivity of chlorine atoms is so great, and the hydrogen atoms on a paraffin hydrocarbon are so nearly alike in character, the reaction leads to a mixture of the various possible substitution products. Moreover, the initial monohalogenation products are also susceptible to attack and polyhalogenated products are formed. The chlorination of so simple a hydrocarbon as *n*-hexane, for example, would lead to a complex mixture that would have little usefulness. The halogenation of paraffin hydrocarbons is a reaction of limited utility, and is employed only when the mixture of halogenation products can be used as such, or, in the case only of methane, when the individual pure compounds can be easily separated. Methyl chloride, dichloromethane, chloroform, and carbon tetrachloride are prepared by the chlorination of methane either photochemically or thermally.

The nitration of paraffin hydrocarbons is also a reaction that takes place only at high temperatures, and leads to mixtures of nitroparaffins, some of which arise by fission of carbon-carbon bonds:

$$CH_3CH_2CH_3 \xrightarrow[\text{400°C}]{HNO_3} CH_3CH_2CH_2NO_2 + (CH_3)_2CHNO_2 + CH_3CH_2NO_2 + CH_3NO_2 \tag{15}$$

(vapor phase)

***Aromatic compounds.*** In contrast to aliphatic hydrocarbons, benzene, a typical aromatic hydrocarbon, undergoes substitution reactions under mild and controllable conditions to yield either monosubstitution products, or polysubstitution products of predictable orientation. The following characteristic aromatic substitution reactions will be considered in detail in subsequent sections in this chapter:

*(a)* Halogenation (e.g., bromination)

$$\text{C}_6\text{H}_6 + Br_2 \xrightarrow{FeBr_3} \text{C}_6\text{H}_5Br + HBr \tag{16}$$

*(b)* Nitration

$$\text{C}_6\text{H}_6 + HNO_3 \xrightarrow{H_2SO_4} \text{C}_6\text{H}_5NO_2 + H_2O \tag{17}$$

*(c)* Alkylation

$$\text{C}_6\text{H}_6 + RCl \xrightarrow{AlCl_3} \text{C}_6\text{H}_5R + HCl \tag{18}$$

*(d)* Acylation

$$\text{C}_6\text{H}_6 + RCOCl \xrightarrow{AlCl_3} \text{C}_6\text{H}_5COR + HCl \tag{19}$$

Other substitution reactions, many of them variants and analogues of these, will also be described in chapters to come; but they will be considered in less detail than the above four since the principles that govern them will be found to be the same as for these.

It will be seen that in each of the above four types of reaction there is present, besides the reagent that supplies the substituent ($Br_2$, $HNO_3$, RCl, RCOCl), an additional reagent ($FeBr_3$, $H_2SO_4$, $AlCl_3$). This additional reagent, which is either a strong acid ($H_2SO_4$) or a Lewis acid, plays an important role in the reaction, and an understanding of its catalytic function is a part of the understanding of the nature of the total substitution process.

Substitution into benzene itself is not complicated by questions regarding the location of the entering group, since all six positions are equivalent. Hence, our discussion of the substitution reactions of benzene will deal chiefly, with the mechanism of the reaction.

Nitration, halogenation, alkylation, and acylation of benzene are *electrophilic substitution reactions:* all of them take place by attack upon the benzene ring of an electron-deficient species or a positively charged ionic species. It is also important to recognize the reciprocal nature of this relationship: the electrophilic substitution reaction also occurs by a nucleophilic attack of the aromatic ring upon the electron-seeking reagent.

When a second group is introduced into an aromatic ring that already bears a substituent, a new question arises: will the entering group take a position *ortho*, *meta*, or *para* to the first substituent? Several other questions arise as well, and all of them can be summarized as follows:

When a new substituent is introduced into a monosubstituted benzene, in a reaction that may be generalized as

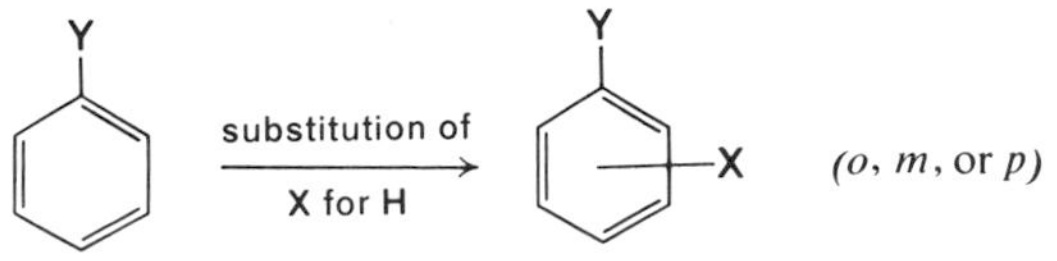

$$(o, m, \text{or } p) \qquad\qquad (20)$$

The following questions may be asked

*(a)* Does X take a position *ortho*, *meta*, or *para* to Y?

*(b)* Does the nature of Y (for example, if Y is Br, $NO_2$, $OCH_3$, $CH_3$) influence the position taken by X?

*(c)* Does the substitution of X proceed more or less easily than in benzene itself? And, as a corollary to this question,

*(d)* Does the nature of Y influence the ease with which the substitution of X proceeds? For example, is nitrobenzene brominated more or less readily than toluene, and how do these compare with benzene?

*(e)* What is the mechanism of the reaction in terms of the nature of the attack upon the ring, the characteristics of the attacking reagent, and the role of substituents in aiding or impeding the substitution?

## 20.5 Nitration of aromatic compounds

Before we consider the nitration of a substituted benzene, let us examine the nitration of benzene itself:

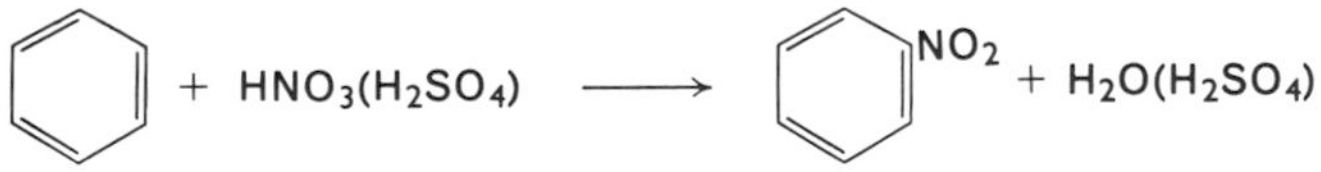

$$+ \; HNO_3(H_2SO_4) \; \longrightarrow \; \overset{NO_2}{\phantom{.}} \; + \; H_2O(H_2SO_4) \qquad\qquad (21)$$

This reaction proceeds readily when a mixture of concentrated nitric and sulfuric acids is used, but very slowly when nitric acid alone is used. Sulfuric acid alone, under the conditions of this reaction, has no effect upon benzene; consequently, its role in the nitration reaction is that of a catalyst—to convert nitric acid into a form that is capable of reacting with benzene.

Sulfuric acid is a strong acid, and reacts with nitric acid in the acid-base equilibrium (see Chapter 4):

$$H_2SO_4 + HONO_2 \; \rightleftharpoons \; H_2\overset{+}{O}NO_2 + HSO_4^-$$

Freezing-point measurements show that solutions of nitric acid in sulfuric acid have an "$i$-factor" of 4, which indicates that this expression is incomplete and must be expanded as follows:

$$H_2SO_4 + HONO_2 \rightleftharpoons H_2\overset{+}{O}NO_2 + \underline{HSO_4^-}$$

$$\underline{H_3\overset{+}{O}} + \underline{HSO_4^-} \underset{}{\overset{H_2SO_4}{\rightleftharpoons}} H_2O + \underline{NO_2^+}$$

The four underlined products of the reaction account for the observed "$i$-factor" of four.

The presence of the nitronium ion in sulfuric-nitric acid ("nitration") mixtures has been established by spectroscopic studies (Raman spectra). The existence of the nitronium ion is independently confirmed in other ways; for example, crystalline nitronium salts (nitronium perchlorate, $NO_2^+ClO_4^-$; nitronium nitrate, $NO_2^+NO_3^-$) have been prepared as pure substances.

The $NO_2^+$ (nitronium) ion formed in this reaction represents a powerful electrophile. It is, of course, a Lewis acid, and can accept an electron pair from the benzene ring in a step that is conveniently formulated as

$$\text{(22)}$$

The loss of a proton to some proton-acceptor in the solution (e.g., $HSO_4^-$, $HNO_3$) leads to the reconstitution of the aromatic nucleus and thus to the retention of the stabilization inherent in the aromatic structure:

$$\text{(23)}$$

We shall see in the subsequent discussion that the presence of certain substituents can increase the reactivity of the ring in the nitration reaction, and in such cases much milder conditions are sufficient for the substitution. The range of effective nitrating agents runs from aqueous nitric acid to solutions of fuming nitric acid in fuming sulfuric acid. Under these various conditions the attacking reagent may be either a nitronium ion, or a nitronium-ion carrier, which we can formulate as $NO_2$—X. The more readily X can be displaced from $NO_2$—X by the nucleophilic attack of the ring, the better the nitrating ability of the reagent. Thus, for nitronium-ion carriers, we can write the more general nitration reaction as

$$\text{(24)}$$

The series $NO_2$—OH, $NO_2OCOR$, $NO_2$—$ONO_2$, $NO_2OH_2^+$, $NO_2^+$ (i.e., $NO_2X$ where $X = OH^-$, $RCOO^-$, $NO_3^-$, $H_2O$) represents a number of nitrating agents capable of supplying $NO_2^+$ to a nucleophilic molecule. That the nitric acid molecule *itself* is an effective $NO_2^+$-donor is doubtful, since its action would require the displacement of $:OH^-$ from $NO_2$—$OH$; it is unlikely that a weakly nucleophilic aromatic nucleus could effect this displacement.

The attack of $NO_2^+$ upon the benzene ring resembles, in its initial stage, certain reactions of nitric-sulfuric acid mixtures upon olefins. Ethylene, for example, reacts as follows:

$$CH_2{=}CH_2 \xrightarrow[\text{H}_2\text{SO}_4]{\text{HNO}_3} \begin{array}{l} CH_2NO_2 \\ | \\ CH_2ONO_2 \end{array} \tag{25}$$

The initial complex, $(CH_2CH_2NO_2)^+$, reacts with nitric acid to yield the addition product, rather than, as in the case of benzene, losing a proton. In the nitration of benzene, the reconstitution of the aromatic nucleus by loss of a proton represents an energetic advantage for the substitution reaction.

The rate-determining step in the nitration of benzene is the formation of an intermediate complex in which $NO_2^+$ has formed a bond with a carbon atom of the benzene ring, with the use of two of the six $\pi$-electrons of the ring. The subsequent loss of a proton from the intermediate leads to nitrobenzene. There is evidence to suggest that the reaction starts by the formation of a complex (called a $\pi$-complex) in which there is a weak association between the electron-deficient reagent and the $\pi$-electron system of the ring. The formation of the species shown as the resonance hybrid in Equations (22), (23), and (24) follows this initial complex formation. An energy diagram for a general case in which an electrophilic reagent, $X^+$, attacks benzene, with the formation of $C_6H_5X$, is shown in Figure 20-1.

Among the experimental observations that support the view that the breaking of the C—H bond is *not* the rate-determining step in nitration is the fact that when monotritiated benzene $(C_6H_5T)$ is nitrated, the ratio of the products $C_6H_4TNO_2$ to $C_6H_5NO_2$ is 5:1. It is known from other studies that the rate of breaking a C—T bond is much slower than that of a C—H bond. Consequently, were the C—T or

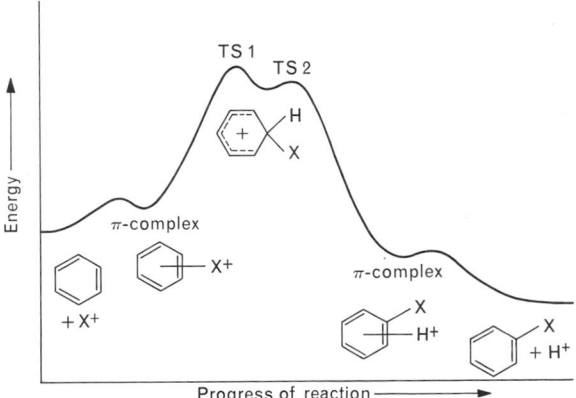

**Figure 20–1.** Energy profile for aromatic substitution: $C_6H_6 + X^+ \rightarrow C_6H_5X + H^+$.

C—H bond-breaking step rate-determining, the relative amount of replacement of T by $NO_2$ should be much less than the statistical (and observed) one-fifth of the total.

In Figure 20-1, the energy of activation (or, approximately, the energy of the intermediate $C_6H_6X^+$) determines the rate of the overall reaction. Any structural influence—for example, the presence of a substituent on the ring—that stabilizes this intermediate (lowers its energy) will increase the rate of substitution relative to that of benzene. This will be considered in detail in a later section.

## 20.6   Other electrophilic substitution reactions

*Halogenation.*   We may preface our consideration of the reaction between bromine and benzene by recalling the reaction between bromine and an olefin (Chapter 9). The addition of bromine to ethylene, for example, is a stepwise process of which the first step can be described in the equivalent statements:

> an electrophilic attack of bromine upon ethylene, or
> a nucleophilic attack of ethylene upon bromine.

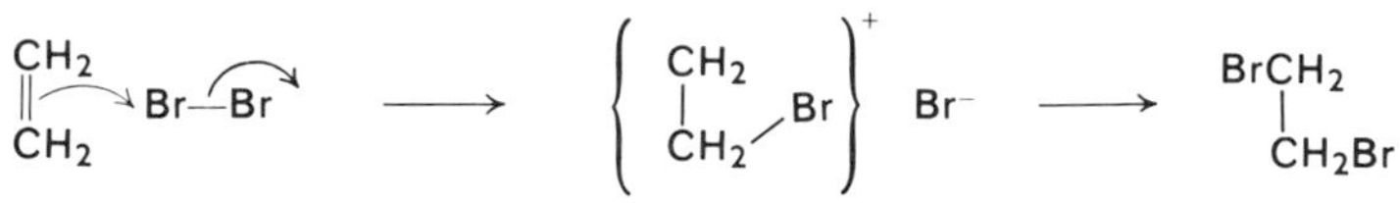

The first stage in the bromination of benzene proceeds in a manner that parallels the first step of this reaction:

Now benzene itself (as distinguished from certain substituted benzenes) reacts very slowly with bromine, but in the presence of a suitable catalyst the reaction proceeds readily. Catalysts that are effective in promoting the reaction are Lewis acids such as $FeBr_3$, and $AlBr_3$. The purpose of the Lewis acid is to develop positive character in one of the atoms of the halogen molecule

$$Br—Br + FeBr_3 \rightleftharpoons \{Br\}^+ \{FeBr_4^-\}$$

and thus to increase the *electrophilic* character of the halogen. The symbol $Br^+$ is shown in braces, since it is not possible to generalize concerning the degree to which *free* $Br^+$ is present; but it is apparent that between the states of the completely covalent bromine molecule and the completely ionized molecule there must be states, which can be formulated as

$$\overset{\delta+}{Br}\cdots\overset{\delta-}{Br}\cdots FeBr_3$$

in which varying degrees of electrophilic character are found in one of the bromine atoms.

Thus, we may generalize the equation showing the catalytic effect of $FeBr_3$ upon aromatic bromination

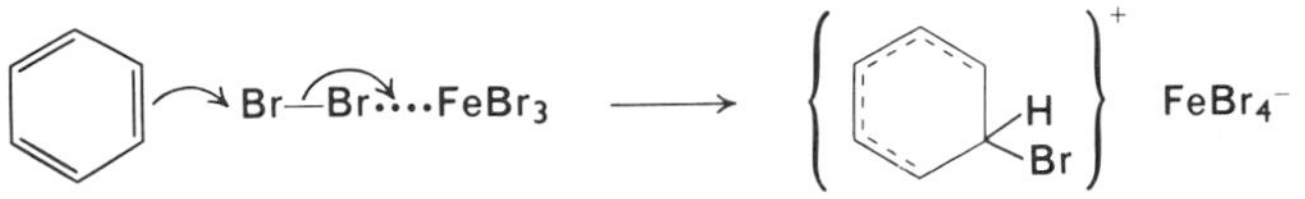

without assuming actual $Br^+$ character in the reagent prior to the attack upon the ring.

The remaining step in the overall reaction is, as it was in the nitration reaction, the loss of the proton and the recovery of the aromatic structure:

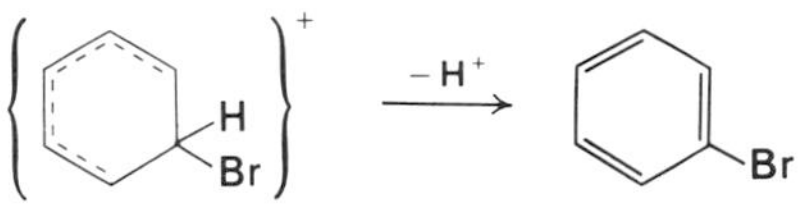

(Again, it should be borne in mind that this proton is lost by coordination with some basic species, e.g., $Br^-$, and is not simply ejected as $H^+$.)

The difference between halogen addition to an olefin and halogen substitution into the aromatic ring is thus seen to reside chiefly in the subsequent fate of the positively charged fragment formed in the first stage of the reaction. The difference lies chiefly in the tendency for the aromatic system to reconstitute itself. The olefinic double bond, having no comparable large resonance stabilization, disappears in the process with the formation of an addition, rather than a substitution, product.

If other sources of $Br^+$ can be found, other brominating agents will be available. It has been observed that hypobromous acid, HOBr, is not itself an effective brominating agent; but in an acidic medium, the protonation of HOBr

$$HOBr + HA \rightleftharpoons H_2\overset{+}{O}Br + A^-$$

leads to the very active brominating agent $H_2\overset{+}{O}Br$. This may be regarded as a "bromonium ion carrier," just as $H_2\overset{+}{O}NO_2$ is a nitronium ion carrier, and it reacts as follows:

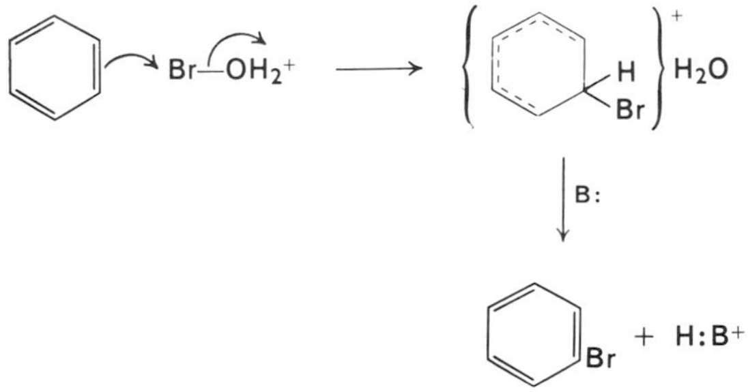

Other effective brominating reagents consist of mixtures of bromine with salts of silver or mercury; in these reagents the formation of silver or mercury halides provides the active (electrophilic) reactant, $Br^+$, for the aromatic substitution step.

It is to be emphasized again that the bromination reaction is a *nucleophilic attack of the ring upon the bromine-carrying reagent*, as well as the *electrophilic attack of the latter upon the ring*. Thus it can be seen that if the ring contains substituents that increase its nucleophilic properties (i.e., increase its electron-supplying ability), the less powerfully electrophilic will the reagent required for the reaction need to be.

*Alkylation.*     The reaction of an alkyl halide with benzene in the presence of a Lewis acid (typically, aluminum chloride), is known as the Friedel-Crafts reaction. It is a typical electrophilic substitution reaction which can be formulated in its simplest form as

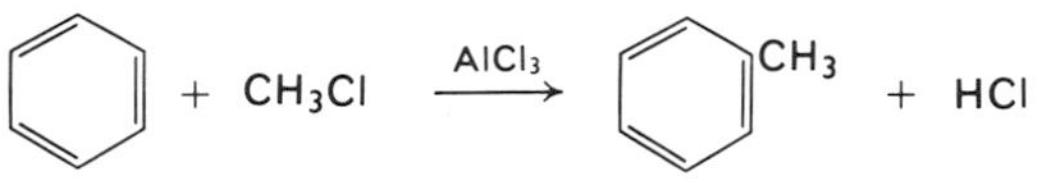

The function of the aluminum chloride is to supply the electron-deficient attacking species—in this case, the fragment $CH_3^+$:

$$CH_3Cl + AlCl_3 \rightleftharpoons \{CH_3^+\}AlCl_4^-$$

Again, the braces indicate that the exact degree to which a free methyl-carbonium ion exists in the solution cannot be specified. The substitution reaction can now be rewritten in greater detail:

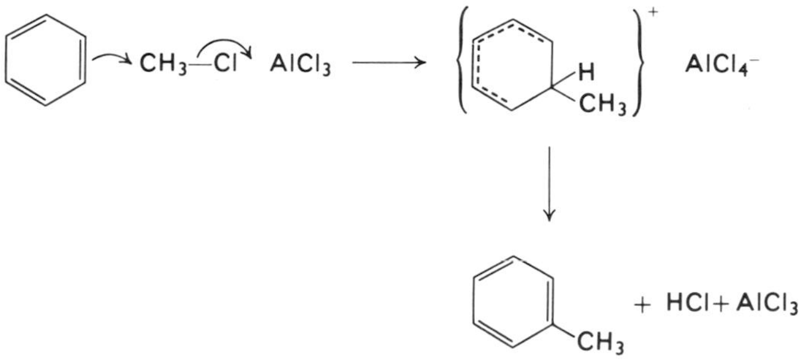

In the Friedel-Crafts alkylation reaction a considerable variation in the nature of the alkyl group is possible. However, the general equation that can be written

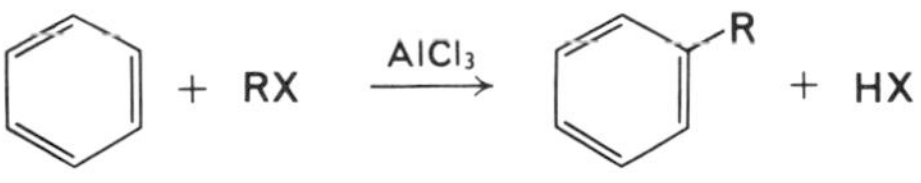

is not always correct in detail, for *rearrangement of the alkyl group is sometimes observed:*

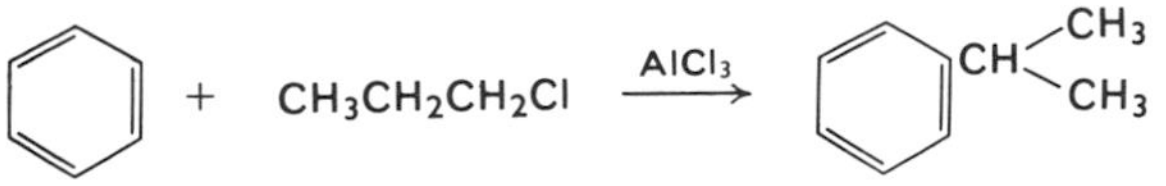

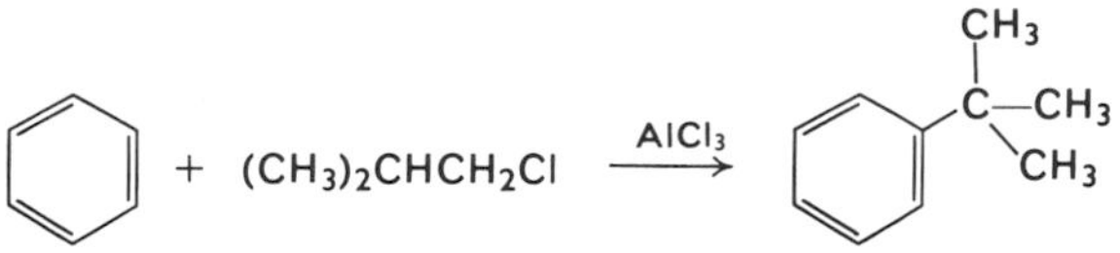

*This rearrangement is in complete accord with our picture of the nature of the reaction.* If the alkyl halide is converted into a complex in which the alkyl group is dissociated to form a carbonium ion, for example,

$$(CH_3)_2CHCH_2Cl + AlCl_3 \ \rightleftharpoons \ \{(CH_3)_2CHCH_2\}^+ \ AlCl_4^-$$

$$\downarrow$$

$$\{(CH_3)_3C\}^+$$

we can anticipate that rearrangement of primary to secondary or tertiary, and secondary to tertiary, carbonium ions will occur, after which the rearranged alkyl group will engage in the substitution step. Indeed, it is found that *n*-propyl bromide can be isomerized to *iso*propyl bromide by the action of aluminum bromide when benzene is not present. Thus, the preparation of *n*-alkylbenzenes by the Friedel-Crafts reaction is not practicable (except for ethylbenzene): mixtures of isomers are formed, in which rearrangement products, derived from the *n*-alkyl halide, are present.

Since the necessary condition for the alkylation is not the alkyl halide as such but the alkyl carbonium ion to which it gives rise, it would be anticipated that alkylation of benzene could be brought about by other reagents that could generate alkyl carbonium ions. Since olefins, alcohols, ethers, and esters can, in the presence of acids, give rise to carbonium ions (or charged complexes that serve as carbonium-ion donors), these reagents should serve as alkylating agents. This is indeed the case; for example,

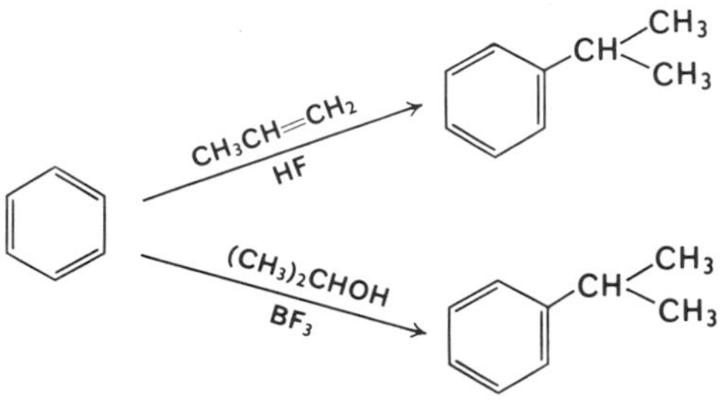

The Friedel-Crafts reaction is not generally practicable for the preparation of monoalkyl benzene derivatives. Beside the rearrangements that occur when complex alkyl halides are used, the presence of alkyl substituents on the aromatic ring

increases its nucleophilicity, and polyalkylation is a frequent result. The preparation of monoalkylbenzenes and benzene derivatives is carried out in other ways; these will be described later.

*Acylation.* The use of an acyl halide in place of an alkyl halide in the Friedel-Crafts reaction leads to the formation of an acyl benzene:

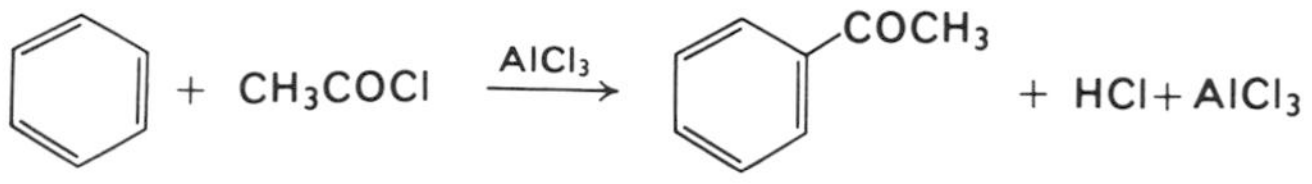

acetophenone

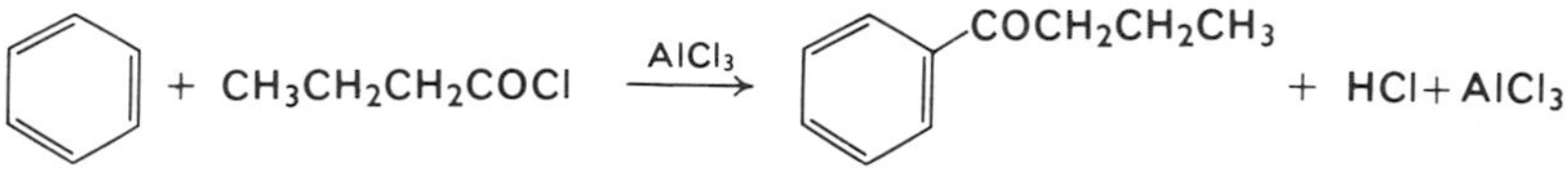

*n*-butyrophenone

Acid anhydrides can be used in place of acyl halides, the —OCOR group undergoing displacement:

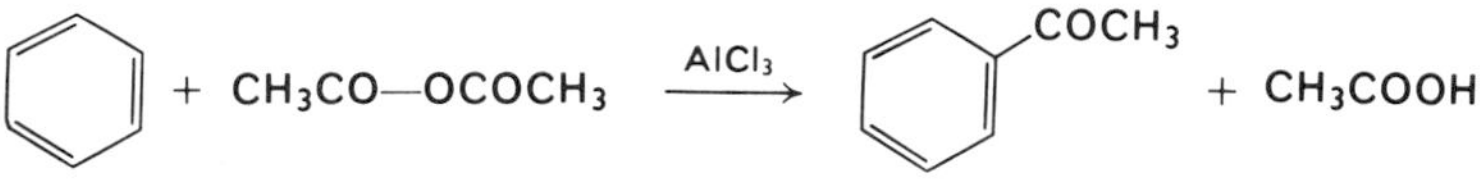

The course of these reactions is again that of an electrophilic attack by an electron-deficient fragment, in this case the "acylium" ion formed by a mechanism similar to that which leads to a carbonium ion in the analogous alkylation reaction:

$$CH_3COCl + AlCl_3 \rightleftharpoons \{CH_3CO\}^+ \ AlCl_4^-$$

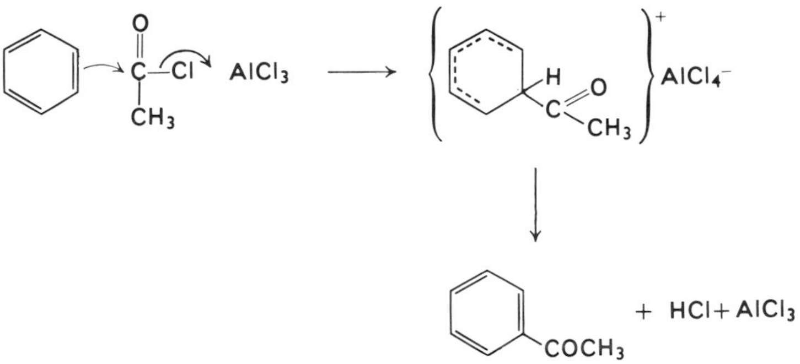

In confirmation of the conclusion that the attacking species is an acylium ion, it is found that acetyl perchlorate, $(CH_3CO)^+(ClO_4)^-$, is a very effective acetylating reagent:

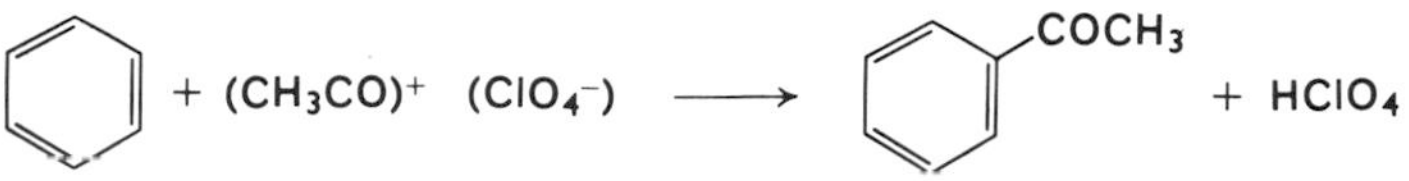

Again, the true reactant, $(CH_3CO^+)$, can be generated in more than one way.

The acylation reaction, in contrast to alkylation, is of wide and general usefulness in synthesis.

---

**Exercise 2.** Write the reactions for the acylation of benzene with *(a)* propionyl chloride; *(b)* isobutyryl chloride, *(c)* bromoacetyl bromide.

**Exercise 3.** Can you suggest what product would be formed by the action of aluminum chloride on β-phenylpropionyl chloride?

---

A revealing insight into the nature of the Friedel-Crafts acylation reaction is found in the reaction of pivaloyl (trimethylacetyl) chloride with benzene, with toluene, and with anisole. The nucleophilicity of anisole is greater than that of toluene, and that of toluene greater than that of benzene (Chapter 21). Hence, the reactivity of these three compounds in electrophilic substitution reactions is anisole ≫ toluene > benzene.

The reaction of benzene with pivaloyl chloride and aluminum chloride yields not trimethylacetophenone, but *t*-butylbenzene. This can be accounted for by the great stability of the *t*-butyl carbonium ion, which can be formed from the intermediate acylium ion by loss of carbon monoxide:

$$(CH_3)_3CCOCl + AlCl_3 \longrightarrow ((CH_3)_3C\overset{+}{C}O)AlCl_4^- \longrightarrow (CH_3)_3C^+ + CO$$

The *t*-butyl carbonium ion then attacks the benzene ring to give *t*-butylbenzene.

Toluene gives some *p*-*t*-butyltoluene and some of the expected ketone; but anisole gives only *p*-methoxypivalophenone.

The explanation for these results is that the rates of the two reactions, *(a)* and *(b)* in the following equation:

$$ArH + ((CH_3)_3C\overset{+}{C}O)AlCl_4^- \xrightarrow{(a)} ArCOC(CH_3)_3$$

$$\downarrow (b)$$

$$CO + ((CH_3)_3C^+)AlCl_4^- \longrightarrow Ar-C(CH_3)_3$$

are such that with a highly nucleophilic ArH, reaction *(a)* takes place sufficiently rapidly so that decomposition according to path *(b)* does not occur. In the case of toluene, which is of intermediate reactivity, reactions *(a)* and *(b)* proceed at comparable rates and both products are formed; and benzene, which is the least nucleophilic of the three, reacts slowly according to path *(a)*, permitting path *(b)* to dominate the course of the overall reaction.

*Sulfonation.*    Sulfur trioxide is a powerful Lewis acid, and reacts with aromatic hydrocarbons to form sulfonic acids. The structure of $SO_3$ can be most simply represented as follows:*

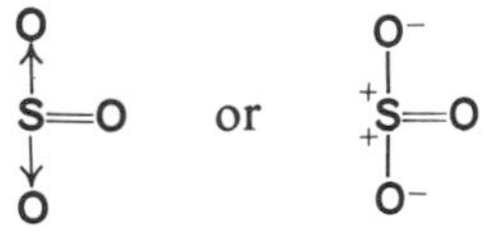

The sulfur atom, its electron density greatly diminished by the two electronegative oxygen atoms attached by semipolar bonds, is strongly electron-demanding, and attacks the aromatic ring in the manner already discussed:

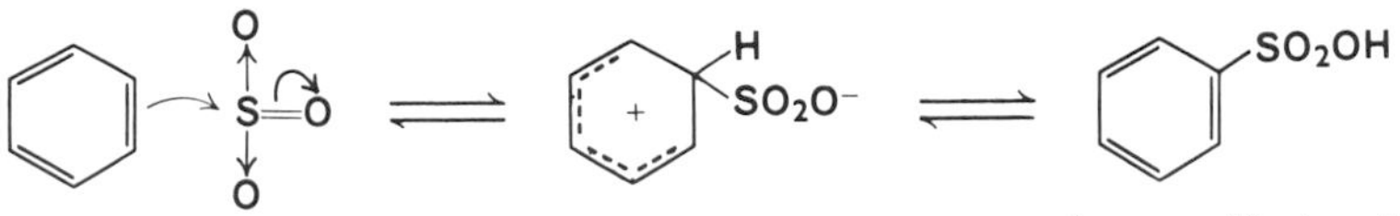

benzensulfonic acid

The reaction is reversible, and aromatic sulfonic acids can by hydrolyzed by the action of hot, aqueous mineral acids, with removal of the —$SO_3H$ group.

Aromatic sulfonic acids are useful in several ways. The sulfonyl chlorides, for example *p*-toluenesulfonyl chloride, prepared by the reaction of sulfonic acids with phosphorus halides, are valuable reagents; they react with alcohols and amines to form O- and N-sulfonyl derivatives. The use of O-sulfonyl derivatives in displacement reactions, in which the —$OSO_2R$ group is the displaced group, has already been alluded to a number of times in earlier chapters.†

Another important use of sulfonated aromatic compounds is in the synthesis of phenols. This will be discussed in a later chapter.

*Miscellaneous.*    As we shall see in later chapters, there are a great many electrophilic substitution reactions, mechanistically allied to those just discussed, that are limited to special types of aromatic nuclei. This is because electrophilic character (i.e., the degree of electrophilicity) can vary over a wide range, and not all electrophilic reagents can attack all aromatic rings. The reciprocal relationship between the nucleophilic character of the aromatic compound and the electrophilicity of the reagent suggests that strongly nucleophilic aromatic nuclei will attack weak electrophiles, and weakly nucleophilic aromatic nuclei will be attacked only by powerful electrophiles.

---

* But see the comment in the footnote, Section 13.2.

† *p*-Toluenesulfonyl chloride is so generally used that it is usually referred to by the abbreviated name "tosyl chloride," and the symbol TosCl used to represent it. Its reaction with an alcohol is commonly written,

ROH + TosCl $\longrightarrow$ R—OTos + HCl

"Tosylates," such as R—OTos, are often used instead of the corresponding halides (RBr, RCl) in reactions involving displacement (of Br⁻, Cl⁻ or TosO⁻).

The degree of nucleophilicity of the aromatic ring is governed by the substituents that it contains. The discussion so far has been based upon the reactions of benzene itself; but benzene derivatives, in which one hydrogen has been replaced by some other group, vary widely in the ease with which they undergo electrophilic attack. The next chapter will deal with the effects of substituents upon substitution into the benzene ring.

### Exercises

(Exercises 1–3 will be found within the text.)

**4** Show all the possible mononitration products of each of the following.

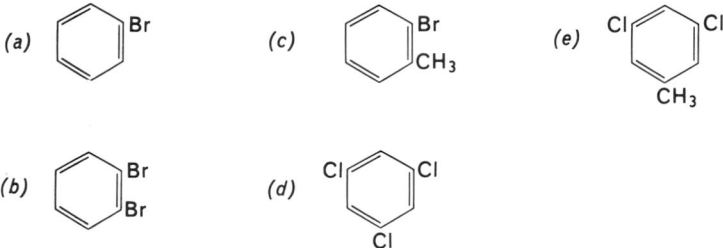

**5** Write the structures of all of the possible tetramethylbenzenes. How many bromotetramethylbenzenes can exist?

**6** How many dibromocyclooctatetraenes can exist? Write their structural formulas.

**7** Suppose the following reaction could be carried out on each of the three known trimethylbenzenes.

$$C_6H_3(CH_3)_3 \xrightarrow{\text{oxidize}} C_6H_3(CH_3)_2COOH \xrightarrow{\text{decarboxylate}} C_6H_4(CH_3)_2$$

Describe how this sequence of reactions could be used as a basis for determining the orientation of the three methyl groups in the three trimethylbenzenes (an "absolute" method).

**8** One of the three bromo-*m*-xylenes was treated with methyl iodide and metallic sodium with the formation of a trimethylbenzene. The same trimethylbenzene was formed when *p*-xylene was brominated, and the resulting bromo-*p*-xylene subjected to the Wurtz-Fittig reaction with methyl iodide and sodium. What was the structure of the original bromo-*m*-xylene?

**9** How could you distinguish experimentally between *p*-xylene (1,4-dimethylbenzene) and ethylbenzene, using as the only experimental procedure the introduction of a nitro group by direct nitration?

**10** Write the equations for the nitration, bromination, Friedel-Crafts acetylation, and sulfonation of (*a*) 1,4-dimethylbenzene (*p*-xylene); (*b*) 1-methyl-4-ethylbenzene (*p*-ethyltoluene).

**11** What alkylbenzene (consider only monoalkylation) would you expect to be formed by the Friedel-Crafts alkylation of benzene with aluminum chloride and (*a*) ethyl chloride; (*b*) *n*-butyl chloride; (*c*) *t*-butyl chloride; (*d*) allyl chloride; (*e*) 2-methylpropene; (*f*) 2-butene.

# Chapter twenty-one / The Effect of Substituents upon the Electrophilic Substitution Reaction

## 21.1 The nitration of anisole and nitrobenzene

Substitution reactions of benzene itself are, of course, limited in number; but the introduction of substituents into aromatic rings that already bear one or more substituents can give rise to an unlimited number of aromatic compounds. It is now our primary concern to inquire into the effect of a substituent already present in the ring upon *(a)* the ease of substitution (the rate of the reaction), relative to benzene; and *(b)* the position that is taken by the entering substituent. This will be done by an examination of the *nitration* of two typical monosubstituted benzenes, *anisole* and *nitrobenzene.*

The experimental observations can be summarized in the two following statements:

*(a)* Anisole reacts readily with nitric acid, under much milder conditions than are required for the nitration of benzene itself. The product is largely *p*-nitroanisole along with smaller amounts of *o*-nitroanisole. No appreciable amount of *m*-nitroanisole is formed:

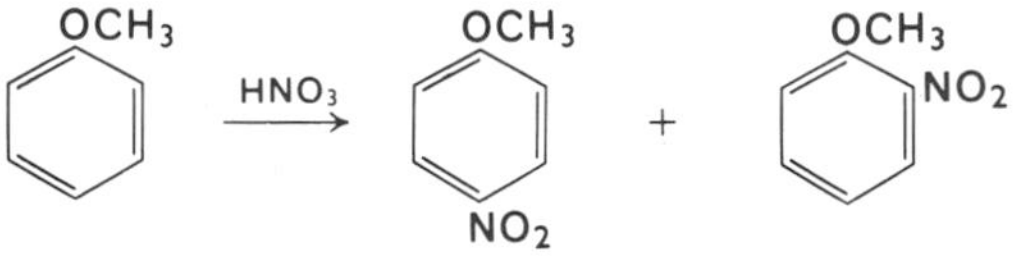

anisole  *p*-nitroanisole  *o*-nitroanisole

The methoxyl group is thus an *ortho-para-directing* group, since it directs the entering nitro group to the *ortho* and *para* positions.

*(b)* Nitrobenzene is nitrated with difficulty: fuming nitric-sulfuric acid mixtures and an elevated temperature are required, the reaction taking place much less readily than the nitration of benzene itself. The product is almost exclusively *m*-dinitrobenzene:

The nitro group is thus a *meta-directing group*.

The distinction between these two reactions is striking; the ease with which anisole and nitrobenzene undergo nitration is widely different, and the position taken by the entering nitro group is not the same in the two compounds. The fact that more vigorous conditions are required for the nitration of nitrobenzene than for the nitration of anisole is a reflection of a difference between the two reactions that can be stated in another way, namely, that the nitration of anisole is *faster* than that of nitrobenzene. Thus, the distinction between the two reactions is first in their rates and secondly in the positions in which substitution occurs. It will be seen that both of these factors are the consequences of the effect of the substituent (the methoxyl group or the nitro group) upon the course of the reaction.

Since anisole is nitrated in the *para* position rather than in the *meta* position, it is clear that the transition state for *para* attack must be more stable than that for *meta* attack.* The reason for the lower activation energy for the *para* substitution will be seen in the following consideration of the nature of the intermediates for these two alternative reaction routes.

We shall seek the answers to these questions by examining the reaction of anisole with nitric acid as it proceeds from the reactants, through a transition state, to the final products.

## 21.2 The nitration of anisole: reactivity

In order to simplify the discussion, the attacking species will be taken to be the nitronium ion, $NO_2^+$. The first stage of the reaction has been described as†

---

\* As will become apparent in the course of the discussion to follow, the arguments adduced for *para* attack are similar to those for *ortho* attack, but both of these are sharply distinguished from those for *meta* attack. Descriptions of the effects of *ortho-para*-directing substituents are usually couched in terms of the *para*-substitution reaction. The ratio of *para* to *ortho* substitution varies with the nature of the *ortho-para*-directing group and with the kind of reaction, and is subject to alteration by factors which are not directly relevant to the particular questions being discussed at this point.

† The resonance hybrid I in the braces does not represent the true transition state, but it will be used as a substitute for the transition state in the discussion. This is justified because the resonance stabilization shown by this intermediate is also to be found in the transition state, and thus that the energy-lowering factors in the activated complex of the reactants can be examined in this nearly equivalent stage.

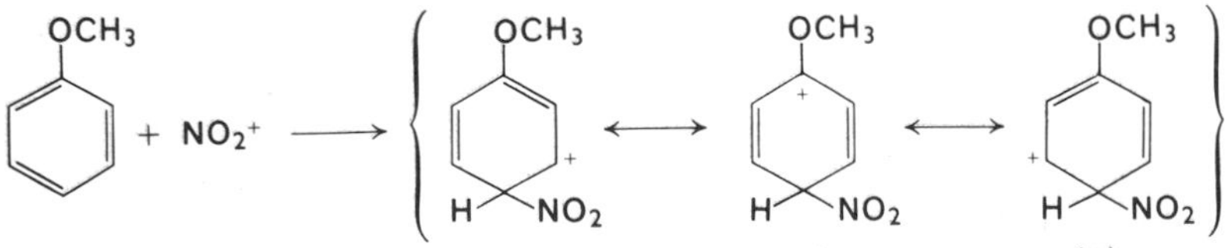

(a)    (b)    (c)

I

An examination of I now discloses that another important contribution to the state of this system should be added: that supplied by the oxygen atom with its unshared electron pairs. Examining I(b) more closely, we see that the positive charge on the carbon atom holding the methoxyl group can be transferred to oxygen by the formal change

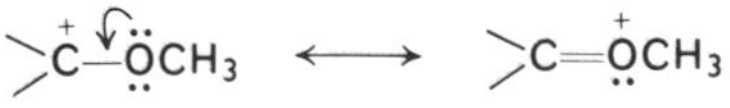

Thus, the resonance hybrid I is incomplete as it is written above, and is actually II, in which a *fourth contributing structure*, II(d), is added:

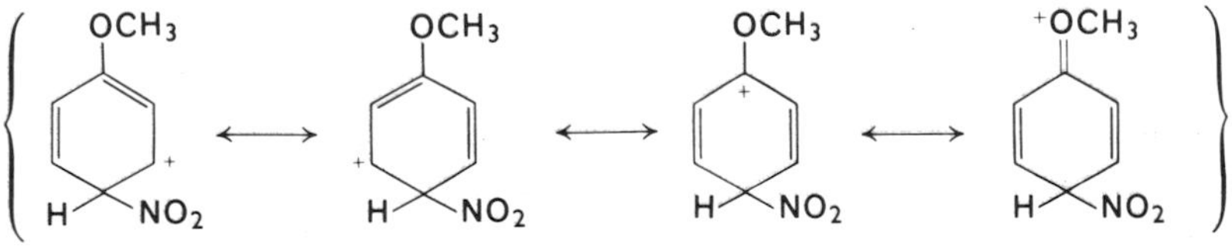

(a)    (b)    (c)    (d)

II

Let us now recall the course of the nitration of benzene, and examine the system at the same stage as is represented by I; in this case the same three contributing structures make up the hybrid

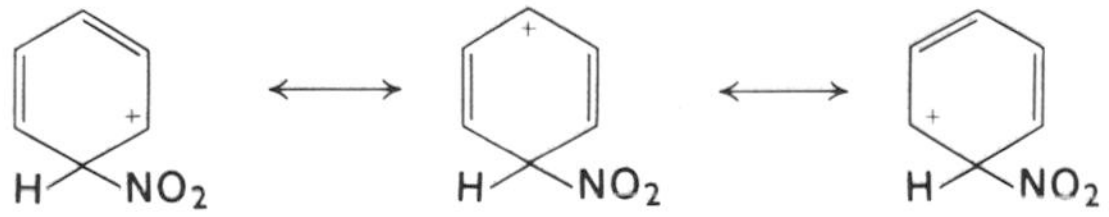

but *here there can be no fourth form corresponding to* II(d) *involved in the resonance.*

If we make the simplifying assumption that the energy differences between reactants and products do not differ by a large amount in these two reactions, the one factor that is quite different in the two cases is the resonance stabilization of the transition state. Since that through which anisole and $NO_2^+$ pass is more highly stabilized (i.e., has a greater number of important contributing structures) than that passed through by benzene and $NO_2^+$, the nitration of anisole is much faster than the nitration of benzene. Figure 20-1 should be referred to again.

## 21.3 The nitration of anisole: position of substitution

The *position* of nitration can now be discussed. We shall examine this question by comparing the two reactions: attack of anisole by $NO_2^+$ at the *meta* position, and attack by $NO_2^+$ at the *para* position. If it can be shown that one of these reactions passes through a transition state that has a greater degree of stabilization, the reaction will proceed the faster by that path; and if the difference in the two rates is considerable, that reaction will predominate.

*Meta* attack by $NO_2^+$ will proceed as follows:

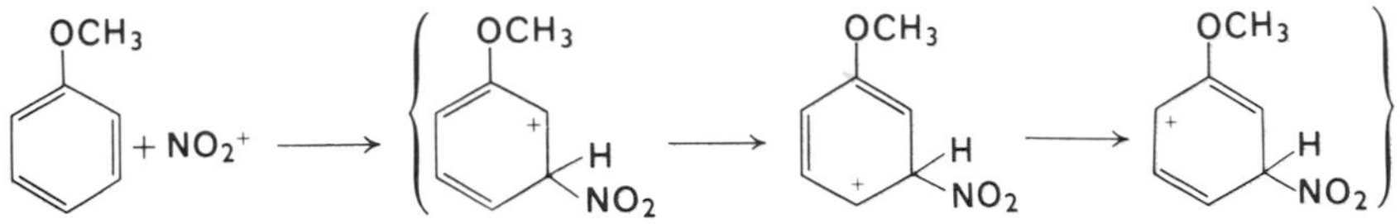

It can be seen that no form can be written in which the methoxyl group participates by furnishing one of the oxygen electron pairs; that is, there is no important contributing structure in which the structure $\overset{|}{\underset{|}{C}}=\overset{+}{O}CH_3$ is involved.* Thus, *meta* attack upon anisole is not supported by participation of the methoxyl group; but *para* attack *is* supported by important contributions of the methoxyl group [as shown in the structure II(a)–(d)].

It should be noted that *ortho* attack is comparable to *para* attack, since participation of the methoxyl group is again possible in this case.

---

**Exercise 1.** Show the participation of the methoxyl group in the resonance hybrid formed by *ortho* attack of $NO_2^+$ upon anisole, formulating the reaction as above.

---

There is another way of describing the rule of the methoxyl group in aiding *para* substitution. It will be recalled that we have described electrophilic substitution into the aromatic ring as either an *electrophilic attack upon the ring* (as *by* $NO_2^+$), or *a nucleophilic attack by the ring* (as *upon* $NO_2^+$). It is reasonable to infer from this that if a substituent can supply electrons to the positions on the ring at

---

* This statement is correct, although forms such as 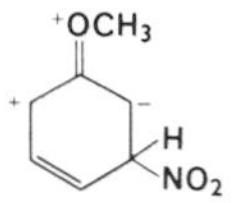 can be written. But since such structures involve a charge separation, and, what is more important, imply contributions such as 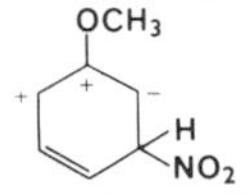 in which adjacent $+$ charges are written, they are of high energy and thus make negligible contributions that can be ignored.

which the electrophilic demand is being made, substitution at those positions will be aided. For anisole this can be pictured as

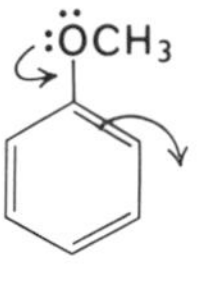

electron availability at
*ortho* position

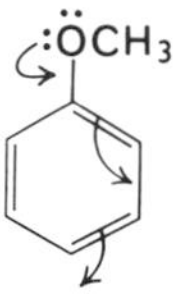

electron availability at
*para* position

In other words, the methoxyl group increases the nucleophilic character of the *ortho* and *para* positions. This should be interpreted to mean that *the demand made by the electrophilic reagent can best be supplied at the ortho and para* positions. It is important to keep in mind the reciprocal nature of the relationship between the electron-supplying reactant—the ring—and the electron-demanding reactant.

### 21.4   The nitration of nitrobenzene

The initial attack by $NO_2^+$ upon nitrobenzene can be examined in the same way that the nitration of anisole was studied. Let us inspect the consequences of *meta* attack on the one hand, and *para* attack on the other:

*meta* attack

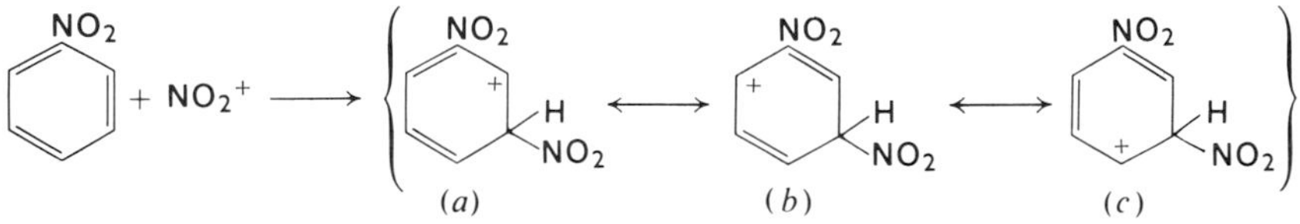

*para* attack

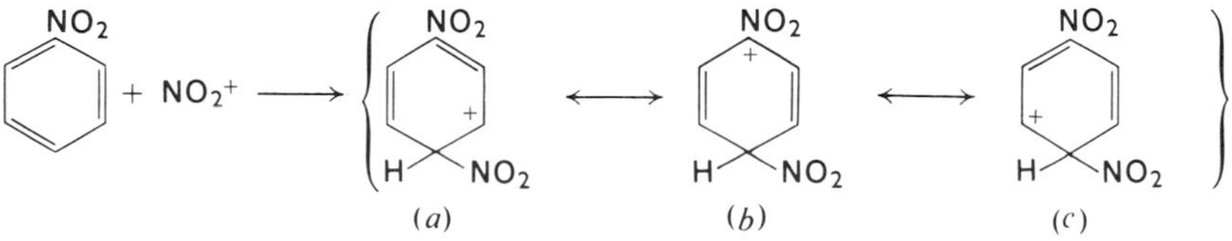

Now the nitro group cannot supply electrons *to* the ring, as the methoxyl group can, because it possesses no unshared pair of electrons on the nitrogen atom:

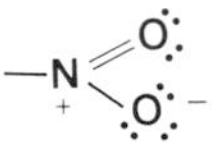

Indeed, the formal charges on nitrogen and oxygen constitute a dipole with the negative end at oxygen and the positive end at nitrogen. Thus, the nitrogen atom has a large degree of positive character, and consequently the form IV*(b)* will be a less important form than III*(a)–(c)* or IV*(a)* and IV*(c)* because it possesses

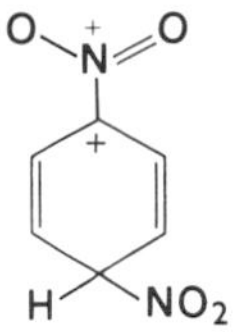

IV*(b)*

two adjacent positive charges, an energetically less likely condition than that in structures in which the positive charges are separated by one or more atoms.

The consequence of this argument is that for *meta* attack (III) there are three major contributing structures; for *para* attack (IV), only two. Therefore, resonance stabilization of the *meta*-transition state will exceed that of the *para*-transition state, and so the reaction will proceed largely over the lower activation peak and lead to *meta* substitution.

## 21.5  Deactivation of the ring by the nitro group

The slowness of the nitration reaction is a consequence of the reluctance of the ring to supply the demand made by the attacking electrophilic $NO_2^+$ group. Not only can the nitro group not supply electrons to the ring, but because of the resonance in the nitro group, resulting in a positively charged nitrogen atom, electrons are withdrawn from the ring:

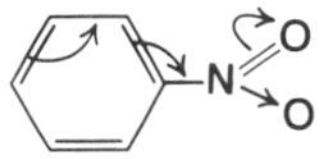

It is apparent that this withdrawal results in a greater electron deficiency at the *ortho* and *para* positions

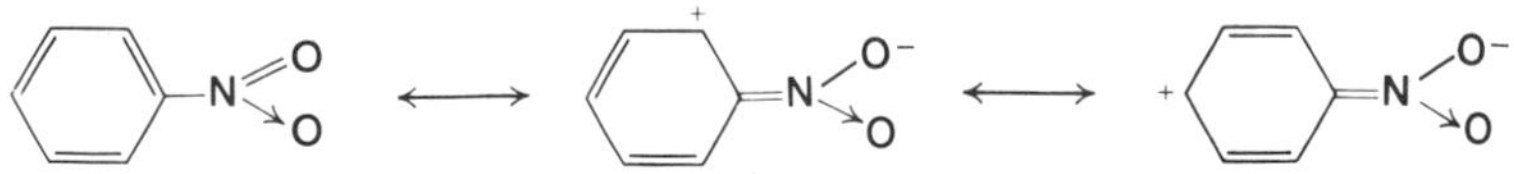

than at the *meta* position. Therefore, no reinforcement of electron supply can be made at any position of the ring, and electron deficiency is greater at *ortho-para* than at *meta*. Consequently, the electron-demanding, attacking $NO_2^+$ group is *nowhere aided in its attack*, but finds less difficulty in securing electron supply at the *meta* positions. The result of these factors is that nitrobenzene is nitrated with difficulty (i.e., slowly) and undergoes nitration at the *meta* position. The nitro group is deactivating and *meta*-directing.

## 21.6  Inductive and resonance effects in substitution

The nitration of chlorobenzene results in *ortho-para* substitution; yet chlorobenzene is less reactive than benzene. Thus, chlorine is *ortho-para*-orienting but *deactivating*, in contrast to the methoxyl group, which is *ortho-para*-orienting but strongly activating. The effect of the chlorine is to reduce electron availability to the attacking electrophile; but when the creation of a position of electron availability is called into play by the reagent, this occurs at the *ortho-para* positions:

$$:\ddot{C}l - \bigcirc - NO_2{}^+ \longrightarrow \quad p\text{-nitrochlorobenzene}$$

It will be recalled that the addition of a halogen acid to an olefin is initiated by electrophilic attack upon the carbon-carbon double bond:

$$\begin{matrix} \diagup \\ C \\ \| \quad \curvearrowright H{-}Br \\ C \\ \diagdown \end{matrix} \rightleftharpoons \left. \begin{matrix} \diagup \\ C \\ | \\ CH \\ \diagdown \end{matrix} \right\} Br^-$$

Markovnikov's rule, which describes the direction of such an addition reaction in the case of an unsymmetrical olefin, is an expression of an orientation effect. This is more clearly demonstrated in the addition of HBr to vinyl bromide, the product of which is 1,1-dibromoethane

$$CH_2{=}CHBr + HBr \longrightarrow CH_3CHBr_2$$

Vinyl bromide reacts a great deal more slowly in this reaction than does propylene, but both vinyl bromide and propylene add HBr in accordance with Markovnikov's rule. The inductive effect of bromine reduces electron availability to the attacking electrophile; but the ability of bromine to participate in charge delocalization of the intermediate ion

$$\left\{ CH_3\overset{+}{C}HBr \quad \longleftrightarrow \quad CH_3CH{=}\overset{+}{Br} \right\}$$

allows bromine to have the same kind of orienting effect as the methyl group.

We might predict that since propylene undergoes HBr addition at a faster rate than ethylene, toluene should undergo electrophilic substitution at a faster rate than benzene. This is indeed correct. Moreover, the methyl group in toluene is *ortho-para*-orienting.

The ability of the methyl group to participate in stabilization of the transition state for *para* substitution, and the ability of the methyl group to stabilize an adjacent carbonium carbon atom can be seen to be expressions of the same property:

$$\left\{ {>}\overset{+}{C}{-}CH_3 \quad \longleftrightarrow \quad {>}C{=}CH_2\overset{+}{H} \right\}$$

$$\left\{ CH_3{-}\overset{+}{\bigcirc}{<}^H_{NO_2} \quad \longleftrightarrow \quad \overset{+}{H}CH_2{=}\bigcirc{<}^H_{NO_2} \right\}$$

The inductive effect of the halogen atom withdraws electrons from the ring; the inductive effect of the methyl group is in the opposite direction.

All of the substituents that we shall discuss affect both electron supply and the positions of electron availability in the aromatic ring to which they are attached. A discussion of the role they play in electrophilic substitution reactions is facilitated by the use of a terminology that permits a concise and convenient description of the way in which substituents affect electron availability to an electrophilic reagent. Two terms that are useful are the *resonance* (R) effect and the *inductive* (I) effect.

## 21.7   The resonance (R) effect

The resonance effect* of a substituent may be such as to provide electrons to the ring ($+$R effect), or to withdraw electrons from the ring ($-$R effect).

The $+$R effect is that shown by atoms or groups whose unshared pairs of electrons can participate in contributions of the following kind:

Thus, a $+$R substituent *supplies* electrons to the ring to the *ortho* and *para* positions. Three $+$R substituents have already been discussed: —OCH$_3$, —CH$_3$, —Cl. Others of this group are: —NR$_2$, —NHCOR, —OH, —O$^-$, Br, I, F, —SR, —OR, —OCOR. All of these groups or atoms possess one or more unshared electron pairs on the atom adjacent to the ring (with the exception of the alkyl group, which has been discussed). They are *ortho-para*-directing and, except for halogen, activating.

The —R effect is shown by atoms or groups for which forms such as the following effect a *withdrawal* of electrons from the *ortho* and *para* positions:

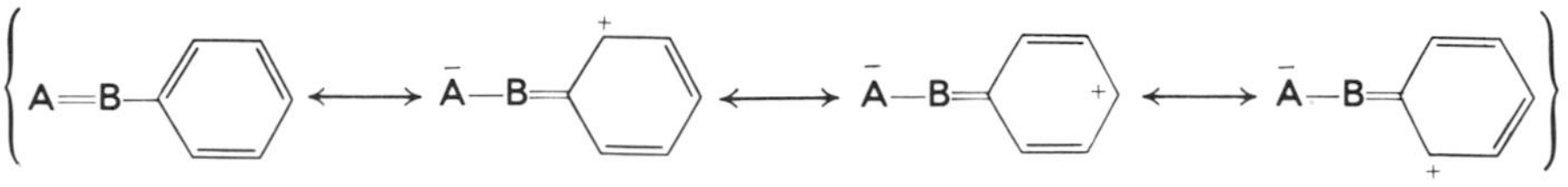

---

* The term resonance effect is now preferred for what used to be called the *tautomeric* (T) effect. The latter term was used before the concept of resonance and the distinction between resonance and tautomerism were well established. Since the R effect describes the way in which a substituent participates in the stabilization of the transition state for substitution, and because we can describe this state by writing the structures that contribute to this resonance hybrid, the term resonance effect is an appropriate one. Some organic chemists, particularly in Great Britain, use still another designation, and speak of *mesomeric* (M) effects. A discussion of the subtle differences that may exist in the interpretation of the terms T, R, and M effect is beyond the scope of our treatment, and we can regard these terms as equivalent.

The nitro group is a typical —R group, others being —COOH, —COOR, —COR, CHO, CN. In these examples, the generalized group A=B— in the above expression is represented by O=N— in the nitro group, O=C— in the carbonyl substituents, and N≡C— in the cyano group.

---

**Exercise 2.**   Rewrite the general expressions shown above in specific terms of each of the substituent +R and −R groups and atoms mentioned.

---

It is to be noted that none of the −R groups shown has an unshared pair of electrons on the atom adjacent to the ring. Thus, *they cannot furnish electrons to the ring* when called upon to do so by the attack of an electrophilic reagent. These groups are typically deactivating, *meta*-orienting substituents.


## 21.8   The inductive (I) effect

The inductive effect of a substituent is the result of its attraction for the electrons in the bond that joins it to the ring. Inductive effects, and the relationship of these effects to the electronegativity of atoms, have been discussed briefly in Section 3.11.

Inductive effects are designated as $-I$ or $+I$, depending upon whether the result is the reduction ($-I$) or reinforcement ($+I$) of electron density in the ring. The recognition of inductive effects comes from two experimental criteria: the strengths of acids, and the direction and magnitudes of dipole moments. The influence of structure upon acid strength has been discussed briefly in Chapter 4. Acid strength in carboxylic acids, measured by the equilibrium constant for the transfer of the proton to water, is a measure of the ease with which the proton can dissociate from the electron pair that binds it to oxygen.

In a series of acids RCOOH, differing only in the nature of R, the resonance effects associated with this ionization are a constant factor, and differences in $K_a$ are primarily due to differences in inductive effects in the group R. Consider the examples of acetic acid and chloroacetic acid, which differ markedly in their $K_a$ values

$$H—CH_2COOH \qquad \overset{\longleftarrow}{ClCH_2COOH}$$
$$pK_a = 4.75 \qquad\quad pK_a = 2.80$$

The fact that the chloroacetic acid is much the stronger of the pair means that the electron pair binding hydrogen to oxygen in the —OH group is drawn more closely to the oxygen atom. The $-I$ effect of the chlorine atom, by attracting the electron pair in the Cl—C bond away from the carboxyl group, creates a decrease in electron density that, in turn, withdraws the electrons in the C—O bond closer to C, and those in the O—H bond closer to O. The result is that the proton is

**Table 21-1.** *Inductive effects of the substituent groups of some substituted acetic acids*

| Acid | p$K_a$ | I Effect of Substituent |
|---|---|---|
| $HCH_2COOH$ | 4.75 | H = reference point |
| $CH_3CH_2COOH$ | 4.90 | +I |
| $CH_3OCH_2COOH$ | 4.30 | −I |
| $ICH_2COOH$ | 3.0 | −I |
| $BrCH_2COOH$ | 2.9 | −I |
| $ClCH_2COOH$ | 2.8 | −I |
| $HOOCCH_2COOH$ | 2.8 (p$K_1$) | −I |
| $(CH_3)_3\overset{+}{N}CH_2COOH$ | 1.8 | −I |

bound less strongly and thus is more readily removed than is the proton of acetic acid.

Propionic acid is weaker than acetic acid. This indicates that the methyl group exerts a +I effect

$$CH_3CH_2COOH \qquad H\text{---}CH_2COOH$$
$$\overrightarrow{\phantom{CH_3CH_2COOH}}$$
$$pK_a = 4.90 \qquad\qquad pK_a = 4.75$$

increasing electron density in the direction of the —OH group, thus binding the proton more firmly and thus decreasing the ease with which it can be removed by a base.

Measurements of the strengths ($K_a$ values) of substituted acetic acids give a measure of inductive effects of the substituent groups (Table 21-1).

It will be noted, in the example of the three chlorobutyric acids given in Table 4-3, that the inductive effect diminishes rapidly as the substituent is moved farther from the carboxyl group.

The inductive effect does not involve contributions from additional valence bond structures in which electron pairs are written in available orbitals of different atoms. This is best seen in the example of $(CH_3)_3\overset{+}{N}CH_2COOH$, in which the nitrogen atom *cannot* accommodate additional electrons.

Inductive effects can be recognized in another property of organic compounds, *the dipole moment*. The dipole moment is a manifestation of electrical asymmetry, resulting in a separation of charge within the molecule and a resulting tendency for the molecule to line up along the direction of an electrostatic field. The value of the dipole moment is the product of the magnitude of the charge and the distance of charge separation, and is experimentally measurable to a high degree of accuracy.

From the discussion of resonance it can be seen that in compounds in which structures that involve charge separation make significant contributions to a

**Table 21-2.** *Dipole moments of substituted methanes*

| Compound | Dipole Moment (D) |
|---|---|
| $CH_3$—H | 0.00 |
| $CH_3NH_2$ | 1.32 |
| $CH_3OH$ | 1.69 |
| $CH_3F$ | 1.81 |
| $CH_3Cl$ | 1.83 |
| $CH_3Br$ | 1.79 |

resonance hybrid, electrical dissymmetry will result; and such compounds will show a dipole moment. For instance, ketones and aldehydes have dipole moments; propionaldehyde and butyraldehyde have dipole moments of 2.73 D and 2.72 D (Debye units), respectively, and formaldehyde has a dipole moment of 2.27 D. These moments indicate a charge separation that can be accounted for by the resonance $\{RCH{=}O \leftrightarrow RCH{-}O\}$.

Trimethylamine oxide has the large dipole moment of 5.0 D, the consequence of the formal charge in the N—O semipolar bond. This moment need not be looked upon as the result of contributions of other structures, for trimethylamine oxide possesses a structure best represented by the single Lewis formula. Dipole moments for other substances in which no formal charges exist and for which single valence bond formulas best represent their structures (e.g., $CH_3Cl$) indicate an electron imbalance that must represent an unequal distribution of the electron pair shared by two atoms in a covalent bond.

If the molecule is symmetrical, as is ethane or benzene, there is no dipole moment. But the replacement of one of the hydrogen atoms of methane by hydroxyl, amino, or halogen results in the creation of sufficient dissymmetry in the C—O, C—N, or C—Hal bond to give rise to a dipole moment. Measurements of the dipole moments of saturated compounds show that progressively greater electron attraction is exerted as we proceed along the elements in a row toward the halogen. Some dipole moments for substituted methanes are given in Table 21-2.

## 21.9 Dipole moments of substituted benzenes

The *direction* of the inductive effect of an atom or group, which has been designated by the symbols $+I$ and $-I$, can be determined by observations of the dipole moments of molecules in which more than one such group is present in the *para* positions of a benzene ring. A group of known direction of inductive effect must be used as a reference. The nitro group, in which the formal positive charge on nitrogen establishes the direction of electron imbalance away from nitrogen and toward oxygen, gives rise to a dipole moment of 3.1 D in nitromethane and 3.95 D in nitrobenzene. Toluene possesses a dipole moment of 0.40 D; and *p*-nitrotoluene shows 4.40 D. This shows that the methyl and nitro groups induce dipole moments opposite in direction with respect to the ring, since the moments reinforce each other:

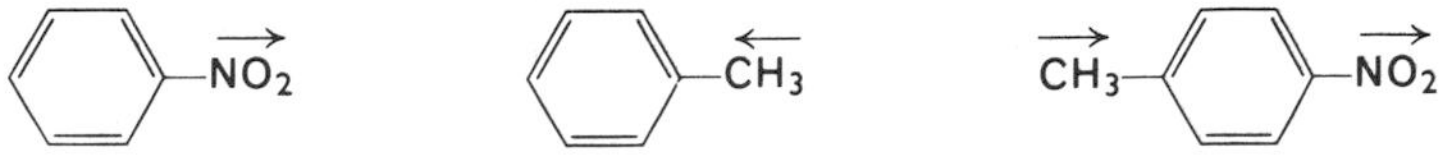

| 3.95 D | 0.40 D | 4.40 D |

Some dipole moments of benzene derivatives are the following:

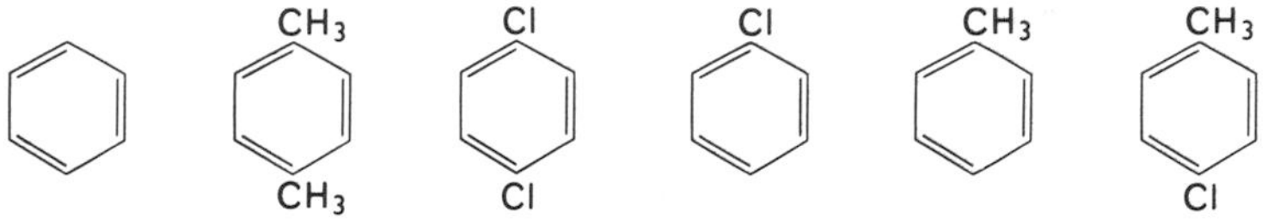

| 0.00 D | 0.00 D | 0.00 D | 1.55 D | 0.40 D | 1.90 D |

| 3.95 D | 2.50 D | 1.53 D | 6.10 D | 1.16 D | 1.52 D |

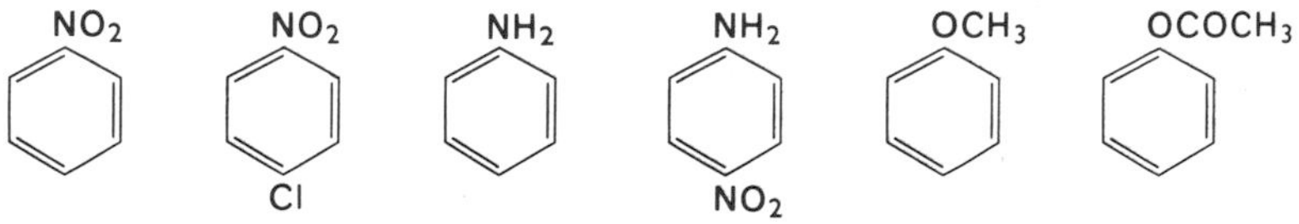

Two of the examples shown deserve special mention. It will be noted that the moment of chlorobenzene is 1.55 D, and thus is less than that of methyl chloride (1.83 D). This indicates that the chlorobenzene has a structure to which forms of the kind shown in Section 21.2 (under the resonance effect) must contribute. This resonance provides electrons to the ring, an effect opposite to the inductive effect of the halogen which withdraws electrons from the ring to the halogen atom; thus, the opposing effects result in a diminution of the moment from that of methyl chloride in which no contribution involving the unshared electrons of the chlorine is possible.

In *p*-nitroaniline the moment (6.10 D) is greater than the sum (5.48 D) of the moments of aniline (1.53 D) and nitrobenzene (3.95 D). An important contribution of the structure

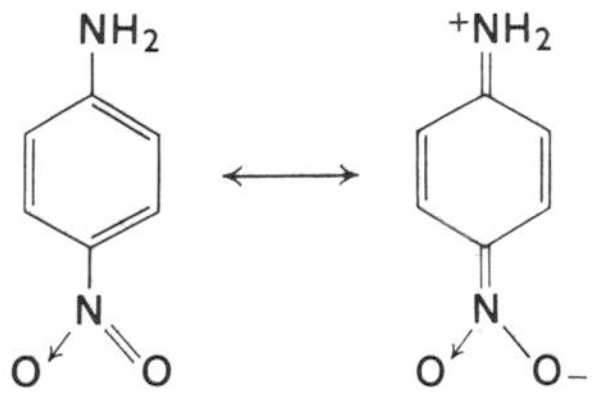

is therefore indicated.

The conclusion from these data is that dipole moments of benzene derivatives are the result of the combination of the inductive withdrawal or supply of electrons by electrostatically induced dissymmetry, and the increase or decrease of electron density caused by resonance.

### 21.10    Inductive and resonance effects in orientation

In most monosubstituted benzenes the orienting effect of the substituent is largely governed by the *preponderance* of the resonance or inductive effect in the overall influence of the substituent on the ring. For example, dipole moment data and measurement of the strength of methoxyacetic acid indicate that the methoxyl group has a negative ($-I$) inductive effect. Nevertheless, the methoxyl group (in anisole) is not only *ortho-para*-orienting but is strongly activating. This indicates that the resonance effect of the methoxyl group makes electrons so readily available to the ring that the restraint that might be expected to be imposed by the $-I$ effect is overwhelmed.

In the case of halogen, the inductive and resonance effects are also in the opposite direction, but in this case the inductive effect causes a deactivation of the ring to electrophilic attack, whereas the resonance effect determines the *position* of the attack, with the result that substitution occurs *ortho-para*.

The foregoing discussions can be summarized as follows.

1. The inductive effect controls chiefly the electron availability in the ring. It does not involve changes in the covalence of the atoms, since it is principally a result of the unequal sharing of the electrons in covalent bonds between unlike atoms.

2. The resonance effect permits electron release or withdrawal by changes in the covalence of the participating atoms in the transition state. It is a resonance phenomenon, and as such is transmitted over conjugated unsaturation, with the result that its consequences are observed not only in the availability (supply or withdrawal) of electrons but also in the *positions* of this availability or deficiency.

3. The combined result of the inductive and resonance effects of a substituent is an overall electron attraction or electron repulsion. In most cases, electron-attracting groups are *meta*-directing and deactivating, and electron-repelling groups are *ortho-para*-directing and activating. The case of chlorobenzene is thus not anomalous, for the *ability* of the halogen atom to aid the supply of electrons to the *ortho-para* positions by its $+R$ effect is associated with restraint upon the electrons by the $-I$ effect.

4. The predominance of the resonance effect in substituents such as hydroxyl, alkoxyl, and amino results in *ortho-para* substitution and *activation*.

### 21.11    The electrostatic nature of the inductive effect

The presence of the powerful inductive effect of a positively charged atom attached to the benzene ring results in practically exclusive *meta* orientation and strong deactivation:

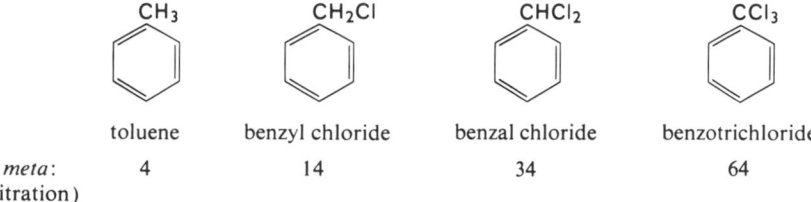

trimethylanilinium iodide                    (exclusive *meta* orientation)

In terms of the transition state picture of the reaction, it can be observed that the strong electron withdrawal in the N—C bond impedes the attack of the electrophilic substituent; and the improbability of effective contributions such as

(which would be involved in *para* attack) lead to the overall result of *meta* substitution and deactivation of the ring.

The electrostatic picture of the inductive effect is illuminated further by the following observations. If the $-\overset{+}{\text{N}}\text{Me}_3$ group is removed from the ring, as in the following series of compounds, the relative amount of *meta* substitution in nitration falls off with increasing distance between the ring and the positive center:

| $\overset{+}{\text{N}}\text{Me}_3$ | $CH_2\overset{+}{\text{N}}\text{Me}_3$ | $CH_2CH_2\overset{+}{\text{N}}\text{Me}_3$ | $CH_2CH_2CH_2\overset{+}{\text{N}}\text{Me}_3$ |
|---|---|---|---|

% *meta:* 100      88      19      5
(nitration)

This effect of the removal of the $\overset{+}{\text{N}}\text{Me}_3$ from the ring is reminiscent of the effect of the halogen atom upon the strengths of $\alpha$-, $\beta$-, and $\gamma$-chlorobutyric acids. The effect falls off with increasing distance, sharply at first, then somewhat more slowly.

A related example shows the effect of increases in the inductive effect by increasing substitution of a $-I$ atom on the carbon atom attached to the ring

| $CH_3$ | $CH_2Cl$ | $CHCl_2$ | $CCl_3$ |
|---|---|---|---|

     toluene      benzyl chloride      benzal chloride      benzotrichloride

% *meta:*      4      14      34      64
(nitration)

The following examples should be studied by the student and the data accounted for in terms of what has just been discussed:

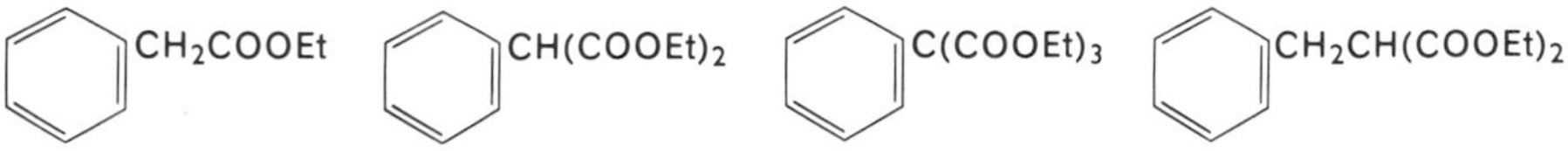

% *meta*: 10    23    57    8
(nitration)

## 21.12  Steric inhibition of resonance

Structural contributions to a resonance-stabilized transition state may have steric consequences that will affect the role of a substituent in a way that is not immediately obvious from inspection of the ordinary formulas. For instance, we have seen that the electron-attracting effect of a nitro group involves the resonance,

(A)    (B)

In the contributing structure (B), the nitro group is coplanar with the ring: the oxygen atoms, the nitrogen atom, and the carbon atoms of the ring are all in one plane. In the case of a 2,6-disubstituted nitrobenzene, steric interference with this planarity results in a decreased contribution of structure (B) and so would diminish the electron-attracting power of the nitro group. The result of this would be to reduce the deactivating effect of the nitro group.

This result is demonstrated in the nitration of durene (1,2,4,5-tetramethylbenzene). Durene cannot be mononitrated, dinitrodurene being the product formed upon nitration. This result shows that the introduction of the first $-NO_2$ group does not deactivate the ring to further nitration as it does in the case of benzene, the dinitration of which, it will be recalled, requires vigorous treatment with fuming nitric and sulfuric acids:

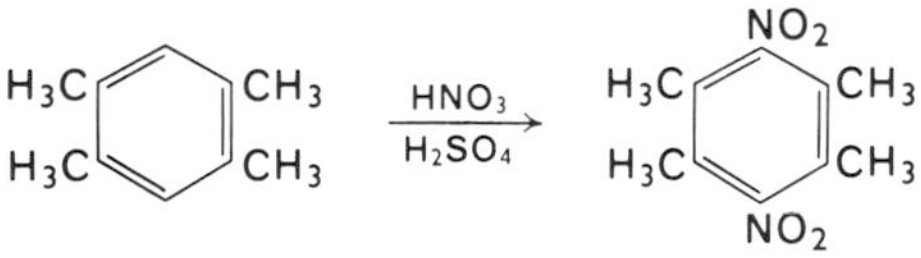

durene    dinitrodurene

Thus, in the mononitro derivative the structure

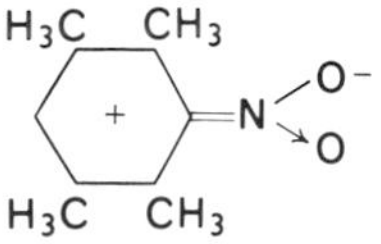

cannot contribute effectively because of the steric hindrance of the methyl groups to bringing the oxygen atoms into the plane of the ring (Figure 21-1).

Consequently, deactivation of the ring by electron-withdrawal by the nitro group is greatly diminished, and the activation to substitution that is provided by

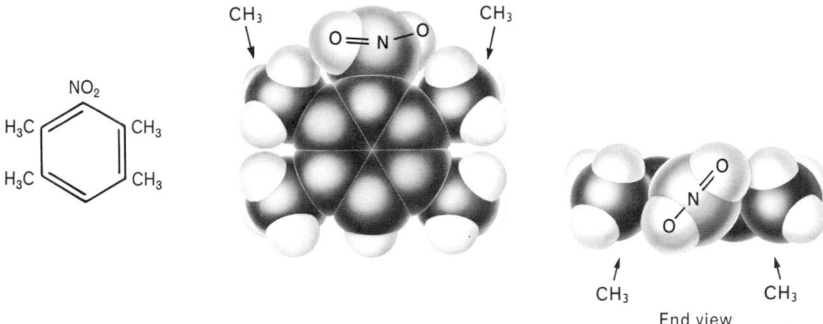

**Figure 21–1.** Molecular model of nitrodurene. End view shows steric interference of *ortho*-methyl groups to effective participation of nitro group in resonance with ring.

the four methyl groups is sufficient to overcome the deactivating inductive effect of the nitro group.

Resonance involving the dimethylamino group is similarly affected by 2,6-substitution:

$$\text{(C)} \qquad \longleftrightarrow \qquad \text{(D)}$$

If the groups X are bulky groups such as methyl, the form (D) cannot make effective contribution. As would be expected, 2,6,N,N-tetramethylaniline is much less reactive toward electrophilic reagents than is N,N-dimethylaniline.

These steric effects can be recognized in dipole moment data:

| 3.95 D | 3.39 D | 6.87 D | 4.11 D |
|---|---|---|---|
| | $(CH_3NO_2 = 3.10\ D)$ | | |

That the reduction in the dipole moments of the tetramethyl derivatives is not due to the purely electrostatic effects of the methyl groups is shown by the following values:

| 1.52 D | 1.55 D |
|---|---|

**Table 21-3.** *Orientation by substituents in nitration\**

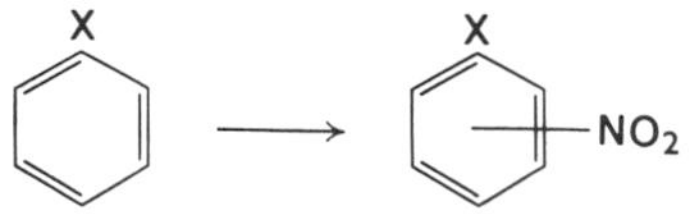

| X | % meta | % ortho-para† |
|---|---|---|
| $\overset{+}{N}Me_3$ | 100 | — |
| $\overset{+}{A}sMe_3$ | 98 | — |
| $NO_2$ | 93 | 7/0 |
| COOH | 82 | 17/1 |
| CN | 80 | 20 |
| CHO | 79 | 21 |
| $SO_3H$ | 72 | 21/7 |
| $CCl_3$ | 64 | 7/29 |
| $COCH_3$ | 55 | 45/0 |
| $CH_2Cl$ | 12 | 41/47 |
| OH | — | 55/45 |
| $NHCOCH_3$ | — | 19/79 |
| $NH_2$ | — | 100 |
| Cl | — | 30/70 |
| Br | — | 38/62 |
| I | — | 41/59 |
| F | — | 12/88 |
| $CH_3$ | 4 | 59/37 |
| $OCH_3$ | — | 100 |
| $C_6H_5$ | — | 100 |
| $CH_2COOH$ | — | 100 |
| $CH_2CN$ | — | 100 |
| $CH{=}CHCOOH$ | — | 100 |
| $C{\equiv}CCOOR$ | 8 | 92 |

\* These values are not to be regarded as "absolutes" since they are gathered from various sources and are subject to some variations of conditions, isolation, and separation procedures. They do, however, offer a basis for comparison by placing the various substituents in their proper categories.

† Where the *ortho-para* ratio is known it is given as a fraction (e.g., for —$CCl_3$ it is 7/29). The total (*ortho+para*) for this example is 36%. Single figures give total *ortho +para*.

In this case forms such as $\overset{+}{Br}$ =⟨ Me Me − Me Me ⟩ are seen not to be subject to these steric effects, since there are no significant steric requirements for double-bond character in the C—Br bond that are affected by *ortho* substituents.

### 21.13 Summary of orienting effects in monosubstituted benzenes

In Table 21-3 are gathered data on the orienting effect of a number of commonly encountered substituents. A close examination of these data, in the light of the foregoing discussion of resonance and inductive effects of substituents, will afford further evidence of the operation of these orienting influences.

### 21.14 The *ortho-para* ratio

It will be noted in the data of Table 21-2 that the *ortho-para* ratio in nitration varies over a wide range. A factor that is not shown in the table but which also influences this ratio is the nature of the substituting (attacking) reagent. Other factors are the sizes of the orienting substituent and the attacking reagent, and the inductive effect of the orienting substituent.

A bulky *ortho-para*-orienting substituent will offer steric interference to the introduction of a group in a position *ortho* to it. In the same way, the introduction of a bulky substituent will be less favored in the *ortho* position because of steric interaction in the transition state. These observations are best amplified by an examination of some typical data (Table 21-4).

It must be emphasized that a given *ortho-para* ratio results from the combined operation of *all* of the electrical and steric effects in any one reaction, and so it is only the more conspicuous consequences that are clearly seen. For further discussion of this subject, advanced textbooks and monographs should be consulted.

**Table 21-4.** *Effect of nature of substituent and entering group upon ortho-para ratio*

| Substance | Orienting Substituent | *Ortho-para* Ratio by Group Introduced | | | |
|---|---|---|---|---|---|
| | | Cl | NO$_2$ | Br | SO$_3$H |
| toluene | CH$_3$ | — | 56/41 | 40/60 | 32/62 |
| chlorobenzene | Cl | 39/55 | 30/70 | 11/87 | 0/100 |
| bromobenzene | Br | 45/53 | 38/62 | 13/85 | 0/100 |
| phenol | OH | 50/50 | 40/60 | 10/90 | — |

### 21.15 Alteration of orienting and activating effect of a substituent: the "protection" of hydroxyl and amino groups

The anionic oxygen atom, as in the phenoxide ion, is a powerfully activating *ortho-para*-directing substituent. This is because the negatively charged oxygen with its six unshared electrons can very effectively supply electrons to the ring in the attack of an electrophilic reagent. In derivatives of phenols, in which the oxygen is attached to an alkyl, aryl, or acyl group,* the mobility of the electrons of the oxygen atom is reduced, and compounds of this kind are, although still highly reactive and *ortho-para*-orienting, less susceptible to electrophilic attack than the phenoxide anion:

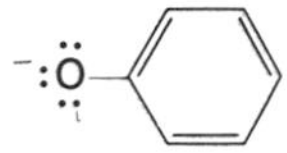

highly reactive in electrophilic substitution

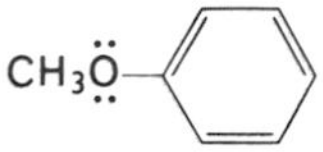

very reactive, but less so than phenoxide ion

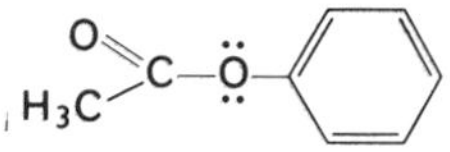

least reactive of these three

In both the methoxyl and acetyl groups, the oxygen has but two unshared electron pairs, in contrast to the three of the phenoxide oxygen atom. Moreover, in the acetyl group these electrons are influenced by an *opposing* resonance

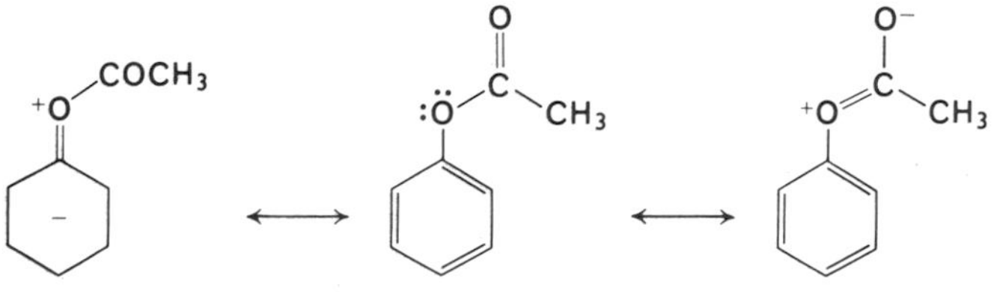

that markedly diminishes the ability of the oxygen atom to supply electrons to the ring. Thus, though the acetoxyl group is an activating, *ortho-para*-directing substituent, its activating effect is less than that of methoxyl and much less than that of the anionic oxygen of the phenoxide ion.

A similar deactivation of the amino group is effected by acetylation. Acetanilide is much less reactive to nitration, halogenation, and so on, than is aniline. Again, the "opposing" resonance, symbolized by

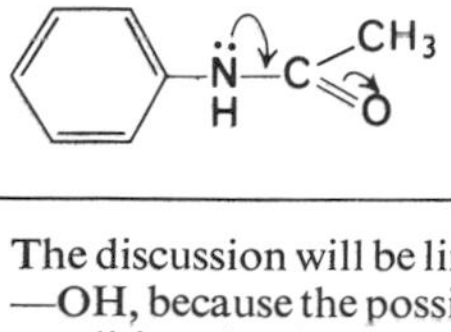

---

* The discussion will be limited to the examples of —$OCH_3$ and —$OCOCH_3$ and will exclude —OH, because the possibility of the ionization of the —OH group under particular reaction conditions is a factor not always accountable for in sufficiently precise terms.

diminishes the availability to the ring of the electrons of the amide nitrogen atom.

The infrared and ultraviolet absorption of some amines and phenols and their acetyl derivatives, and their ionized forms have been given in Section 11.6. It will be seen from these data that the diminution, by acylation, of the electron-donating ability of a hydroxyl group or an amino group has a pronounced effect upon the light absorption properties of the compounds described there. For example, the carbonyl absorption of $p$-aminoacetophenone is observed at 1677 cm$^{-1}$, while that of $p$-acetaminoacetophenone is at 1686 cm$^{-1}$. Carbonyl absorption at 1677 cm$^{-1}$ is indicative of a greater degree of single-bond character than absorption at 1686 cm$^{-1}$, indicating that the contribution of the form

(C=O stretch, 1677 cm$^{-1}$).

in the free amine is of greater importance in the structure of the compound than the contribution of the form

(C=O stretch, 1686 cm$^{-1}$)

to the structure of the acetylated amino compound.

The strong shift in the ultraviolet absorption maximum of $p$-hydroxyaceto-phenone (276 m$\mu$) to 325 m$\mu$ upon ionization to the phenoxide ion is also indicative of a much larger contribution of the electrons on the anionic oxygen atom to delocalization involving the carbonyl group.

The deactivating effects of the acylation of hydroxyl and amino groups are very useful in synthetic work. Phenol and aniline are so highly reactive toward electrophilic attack that the use of powerful electrophiles, such as mixtures of nitric and sulfuric acids, halogens, and Friedel-Crafts reagents, results in poly-substitution or in the formation of intractable mixtures as a result of the heat generated by uncontrollable reactions. Acetanilide, however, in contrast to aniline or N-alkylated anilines, can be smoothly and controllably nitrated and bromi-nated to $p$-nitro- and $p$-bromoacetanilide. The preparation of $p$-nitroaniline is carried out in the following way

| aniline | acetanilide | $p$-nitroacetanilide | $p$-nitroaniline |

rather than by the direct nitration of aniline. $p$-Bromoaniline can be made by a similar sequence in which bromination of acetanilide is the first step.

## 21.16 Competitive orienting effects of two substituents

The nitration of *p*-nitrotoluene gives 2,4-dinitrotoluene. Since the second nitro group enters *meta* to the first and *ortho* to the *ortho-para*-directing methyl group, it is not possible to decide from this result alone whether the orienting influence at work is that of the methyl or the nitro group. If *m*-nitrotoluene is nitrated, the products are a mixture of 3,4-, 2,5-, and 2,3-dinitrotoluenes. None of the products is 3,5-dinitrotoluene, indicating that the controlling directive influence is that of the methyl group:

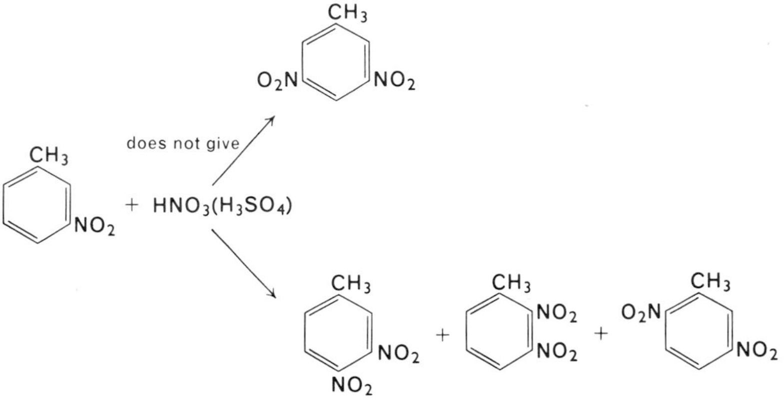


When a *meta*-directing substituent and an *ortho-para*-directing substituent both are present, the controlling directive influence will be that of the *ortho-para*-directing group. Since the *ortho-para*-directing group can *actively aid the reaction by supplying electrons upon the demand* of the attacking electrophile, whereas the *meta*-directing group deactivates by decreasing electron availability, the predominant effect of the former can be understood.

When two *ortho-para*-directing groups are present in a ring, the controlling directing power will be exerted by the group which is the more capable of responding to the electron demand made by the substituting reagent:

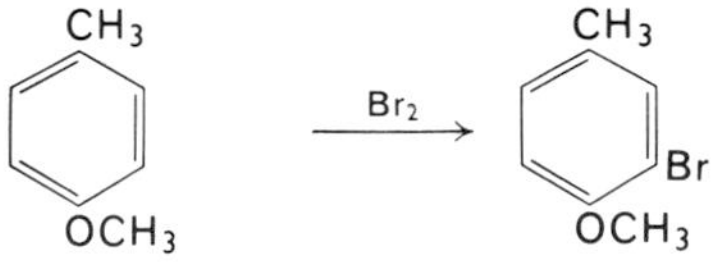


predominant product

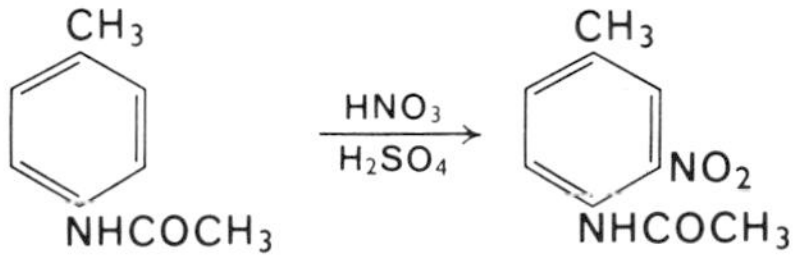


aceto-*p*-toluidide

It is noted that in aceto-*p*-toluidide the substitution occurs *ortho* to the acetamino group; but when the activating power of the nitrogen is further diminished by diacylation, the methyl group dominates:

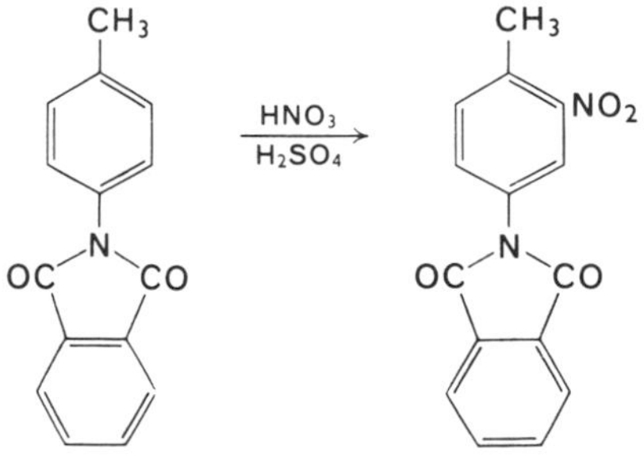

N-*p*-tolylphthalimide

But that the diacylated amino group is still an *ortho-para*-directing group and not a *meta*-directing group is shown by

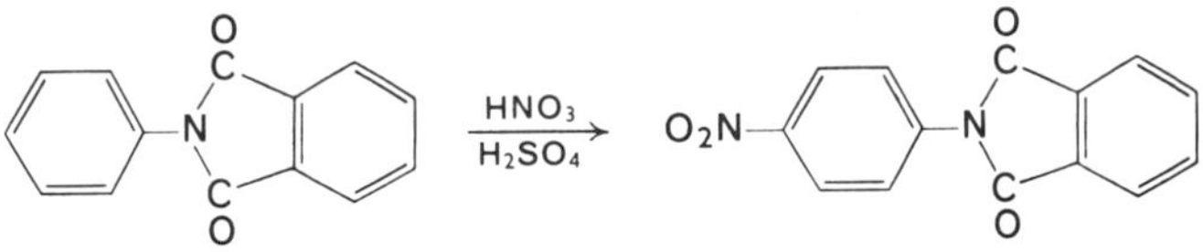

N-*p*-nitrophenylphthalimide

Other cases, which the student should examine in the light of the foregoing discussion, are the following:

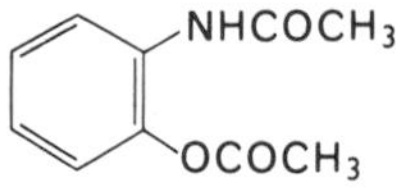

| | Percentage Yield in Nitration to Mononitro Derivative* | |
| --- | --- | --- |
| | *o-p* to N- | *o-p* to O- |
| (*a*) | 76 | 13 |
| (*b*) | 69 | 4 |
| (*c*) | 71 | 0 |

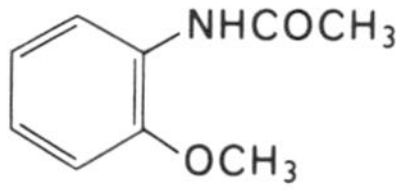

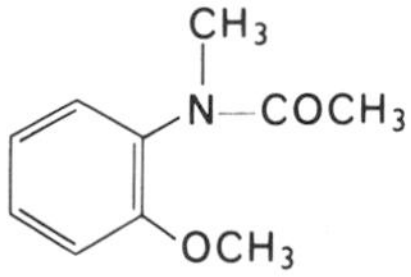

* These figures represent *actual yields* of products isolated.

## 21.17  Other orienting influences

It has been pointed out above that the *ortho-para* ratio in substitution reactions varies with the substituting reagent. Temperature is also a factor in determining the *ortho-para* ratio. The sulfonation of toluene gives mostly the *para* toluene-sulfonic acid, but varying amounts of the *ortho* and *meta* isomers are formed, depending upon the temperature at which the sulfonation is carried out. The pronounced effect of temperature upon the course of this reaction is shown by the following results, obtained in the sulfonation of toluene at 0°C and 100°C:

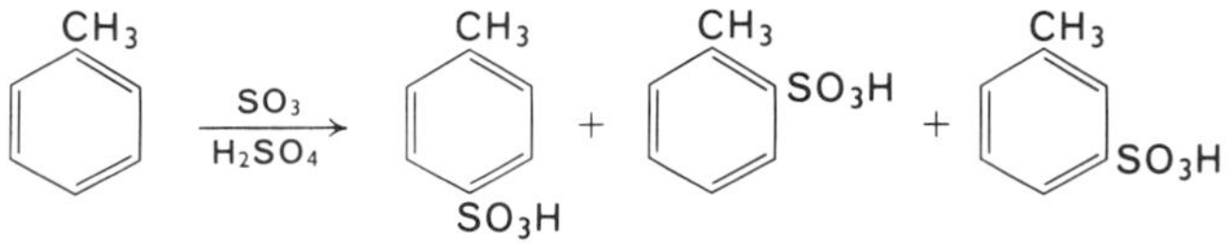

| Temp. | para | ortho | meta |
|-------|------|-------|------|
| 0°C | 52.5% | 45% | 2.5% |
| 100°C | 73% | 17% | 10% |

The effect of temperature upon the *ortho-para* ratio in sulfonation can be explained by the reversibility of the sulfonation reaction, and the fact that the *para* isomer is the most stable. Since the methyl group is *ortho-para*-directing, the initial substitution takes place predominantly at the two *ortho* positions and the *para* position. At the higher temperature, the establishment of the equilibrium leads to the formation of a predominance of the more stable *para* isomer, with the observed shift in the *ortho-para* ratio.

## 21.18 Applications to synthesis

1. Suppose we wish to prepare 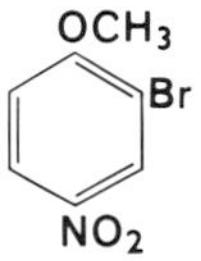 starting with anisole. There are two alternatives: (1), bromination, followed by nitration; and (2), nitration, followed by bromination. The second of these is preferable, for the following reasons. Bromination of anisole would be expected to give largely *p*-bromoanisole, and relatively little *o*-bromoanisole. Suppose, nevertheless, that one were to separate the *o*-bromoanisole and subject it to nitration. Now, further complications arise. It will be seen that with two *ortho-para*-directing groups (OCH$_3$ and Br) in the molecule, the nitro group can enter in any of several positions. Thus, even if 2-bromo-4-nitroanisole (the desired compound) were the predominant product, as it would be, problems of separation and purification would have to be dealt with.

The other alternative is preferable because the predominant product of nitration of anisole is *p*-nitroanisole. Bromination of this will give *only* the desired 2-bromo-4-nitroanisole.

2. Suppose it is desired to prepare 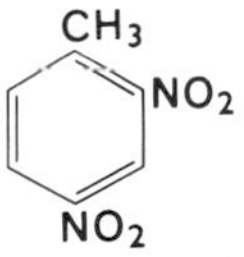 starting with benzene. Here the choice is clear, because if the first step were the dinitration of benzene it would be impossible to introduce the methyl group into this compound in the desired

position. Indeed, because of the great deactivation caused by two nitro groups, the Friedel-Crafts alkylation reaction will not proceed at all. *Even nitrobenzene, with only a single nitro group to deactivate the ring, is not reactive in the Friedel-Crafts alkylation reaction.*

Thus, the desired compound could be prepared from benzene only by the route: benzene → toluene → 2,4-dinitrotoluene.

It should be added that the preparation of toluene from benzene by the Friedel-Crafts alkylation reaction is not satisfactory from a practical standpoint because polyalkylation, with the formation of polymethylbenzenes, would occur.

3. To prepare 3-bromo-4-aminotoluene it would be impracticable to use aniline as the starting compound because the introduction of a methyl group into aniline or an acylated aniline by the Friedel-Crafts reaction would not be successful; nor would bromination of aniline or acetanilide give a practical yield of the *ortho*-bromo derivative. The use of toluene as the starting material, however, would be practicable, for the mononitration of toluene gives the readily separable *o*- and *p*-nitrotoluenes. Reduction of the nitro group, acetylation of the amine, bromination, and deacetylation leads to the desired compound:

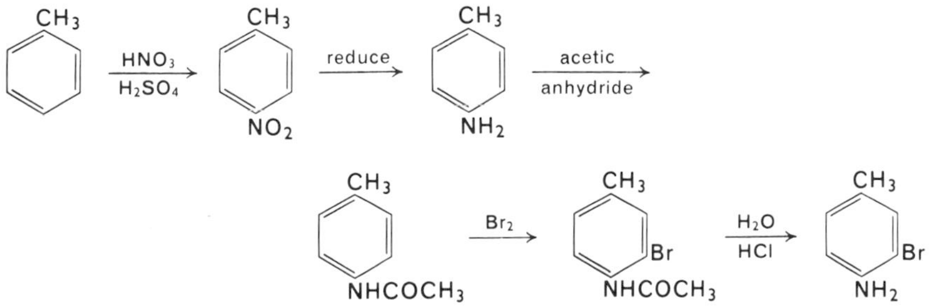

The organic chemist has at his disposal many monosubstituted and disubstituted benzene derivatives, available from commercial sources. *p*-Toluidine is an important industrial chemical, and in actual practice it would be used as the primary starting material in a laboratory synthesis of the bromotoluidine.

### Exercises

(Exercises 1 and 2 will be found within the text.)

3 Show by the use of appropriate symbols the manner in which the resonance effect of the following substituents (on a benzene ring) would operate: *(a)* —$OCH_3$, *(b)* —$NMe_2$; *(c)* —Br; *(d)* —CHO; *(e)* —$NHCOCH_3$; *(f)* —COOEt; *(g)* —NO.

4 Explain why the dipole moment of nitrodurene is only about 0.56 D lower than that of nitrobenzene, whereas *p*-nitrodimethyaniline and 1-nitro-4-dimethylaminodurene differ by 2.76 D.

5 With reference to your discussion in Exercise 4, why are the dipole moments of bromobenzene and bromodurene nearly the same?

**6** Each of the following compounds will give largely only a single monosubstitution product upon nitration. Write the structure of the *mono*-nitro derivative of each.

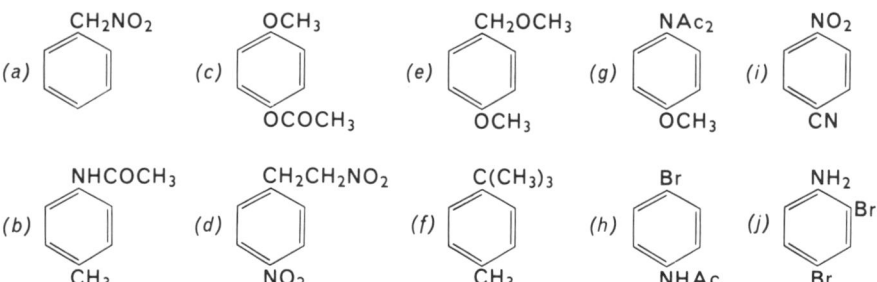

**7** Which of the following compounds would undergo predominantly *meta* electrophilic substitution?

**8** Formulate a practical synthesis for each of the following compounds:
*(a)* p-bromonitrobenzene, from benzene
*(b)* 2-bromo-4-nitrotoluene, from toluene
*(c)* m-bromonitrobenzene, from benzene
*(d)* p-bromoacetophenone, from benzene
*(e)* 3-bromo-4-aminotoluene, from p-toluidine

**9** For electrophilic substitution into cinnamic acid we can represent a − R effect by the following symbols:

Why, then, does cinnamic acid undergo *ortho-para* substitution?

## 22.1 Preparation of aromatic halogen compounds

The methods that are useful for the preparation of alkyl halides are not generally applicable to the preparation of aryl halides. Although alcohols will react with hydrogen halides, phosphorus halides, and thionyl chloride to yield the corresponding halides by replacement of the hydroxyl group, most phenols cannot be converted into aryl halides by these reagents. Di- and trinitrophenols will react with $PCl_5$ and $PBr_5$ to yield the corresponding di- and trinitro halogenated benzenes, but these reactions are not successful with phenols of other kinds.

Nuclear-halogenated aromatic compounds are usually prepared by direct halogenation, with due regard to the orientation problems that may be involved. By methods that remain to be discussed in chapters to follow, substituents can be modified after halogen has been introduced, or further substitution can be carried out on the halogenated compound.

The direct iodination of the aromatic nucleus proceeds satisfactorily when the hydrogen iodide that is formed in the reaction is removed, thus shifting the unfavorable equilibrium that obtains in the presence of the strongly reducing hydrogen iodide:

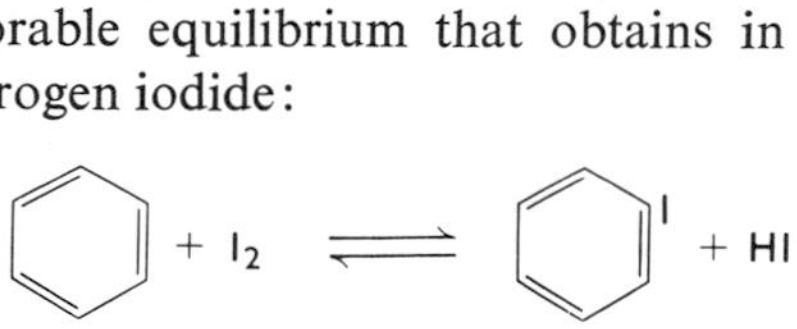

$$\bigcirc + I_2 \rightleftharpoons \bigcirc^I + HI$$

$$2HI + (O) \longrightarrow H_2O + I_2$$

Oxidizing agents that are used are nitric acid, iodic acid, and mercuric oxide. Iodobenzene can be prepared in good yield by direct iodination under these conditions. However, aromatic iodo compounds are more commonly prepared through the Sandmeyer reaction. This reaction will form the subject of a later discussion; it is also applicable to the preparation of chloro- and bromo- derivatives, and a related reaction is used for the introduction of fluorine into the aromatic ring.

## 22.2 Reactivity of aryl halides

Aromatic compounds that contain halogen bound to the aromatic nucleus can be placed in two general classes:

1. Those, like chlorobenzene, the bromotoluenes, and others in which there is present either halogen alone or halogen and $+R$ substituents, which are characterized by inertness of the halogen to displacement by nucleophilic reagents:

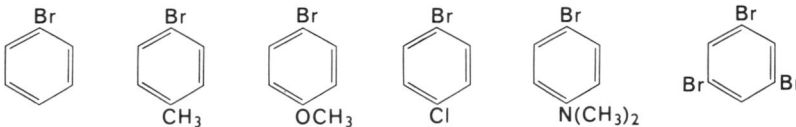

not reactive toward nucleophilic reagents; halogen not displaced by $OH^-$, $NH_3$, $CN^-$, $:CH(COOEt)_2{}^-$, and comparable nucleophiles under usual conditions of ionic displacement reactions.

2. Those, like $p$-bromonitrobenzene, 2,4-dinitrochlorobenzene, and others in which the halogen atom is *ortho* or *para* to powerfully $-R$ substituents, which undergo displacement of the halogen by nucleophilic reagents:

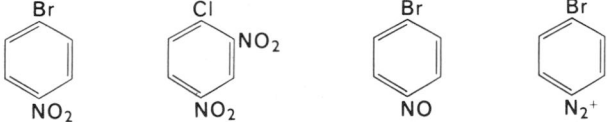

reactive toward nucleophilic reagents; halogen displaced by nucleophilic attack of $OH^-$ $OR^-$, $NH_3$ and comparable nucleophiles under usual conditions of ionic displacemen reactions.

## 22.3 The hydrolysis of chlorobenzene

The hydrolysis of chlorobenzene to phenol does not take place by the treatment of chlorobenzene with boiling sodium hydroxide solution. Phenol is, nevertheless, produced by the hydrolysis of chlorobenzene in an industrial process, but the reaction takes place under drastic conditions (high temperature and pressure) that are in marked contrast to those under which alkyl halides are hydrolyzed. It is to be recalled, however, that certain alkyl halides, and vinyl halides in particular, are resistant to the displacement by hydroxyl, hence the unreactive character of chlorobenzene in this respect is not to be regarded as a reflection of properties peculiar to an aromatic, as contrasted with an aliphatic, compound.

We have seen that the displacement of halogen by hydroxyl in a primary alkyl halide proceeds by way of a rear-side attack of the nucleophilic hydroxyl ion upon the carbon atom to which the halogen is attached. Moreover, the nature of the carbon-chlorine bond in the alkyl halide is such that the electron pair is displaced toward the electronegative chlorine atom, with a consequent diminution of electron density on carbon that favors nucleophilic attack.

But in chlorobenzene, a rear-side attack on the carbon atom holding the halogen is sterically impossible; and the carbon-chlorine bond is a shorter, stronger bond than that in an alkyl halide by virtue of the degree of double-bond character induced by contributions of the following kind to the structure of the compound (see Section 21.7):

Another kind of attack which the hydroxide ion might make upon chlorobenzene is pictured in the following way:

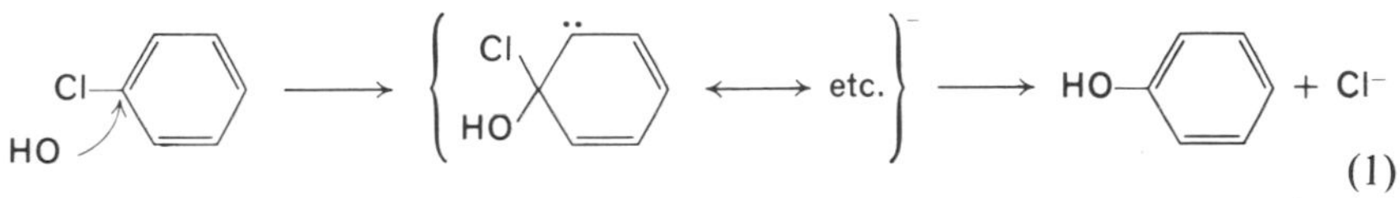

$$(1)$$

A reaction of this kind appears to be one of very high activation energy. The reaction of chlorobenzene with alkali does not proceed under any but the most vigorous conditions. When replacement of the halogen atom by hydroxyl does occur, as in the industrial preparation of phenol, the high temperature under which the process is conducted may permit the reaction to proceed by this route. However, as will be described in the following section, there is a third possible course for the displacement reaction, and it may be this third route that is followed when chlorobenzene reacts with alkali at elevated temperatures.

## 22.4 The "benzyne" intermediate in replacement reactions of aromatic halides

Certain aromatic halides react with strong bases such as alkali metal amides in liquid ammonia to yield substituted anilines. The most remarkable feature of these reactions is that the entering $-NH_2$ group does not always occupy the position at which the halogen was present in the original compound. Indeed, mixtures are often obtained:

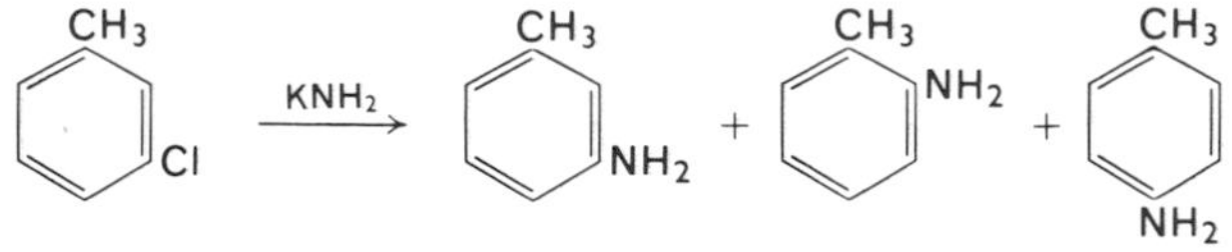

$$(2)$$

Careful studies of this reaction have led to the conclusion that it proceeds by way of an elimination of the elements of hydrogen halide, with the formation of a highly reactive intermediate that is formulated as containing a triple bond. For this reason, the intermediate is called "benzyne." The reaction of chlorobenzene with potassium amide to form aniline would be formulated in the following way:

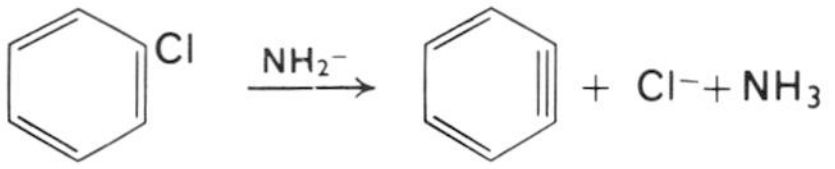

$$\text{(Cl)} \xrightarrow{\text{NH}_2{}^-} \text{(benzyne)} + \text{Cl}^- + \text{NH}_3 \tag{3}$$

The intermediate "benzyne" or "dehydrobenzene" is evidently a very unstable and highly reactive species, for it has not been isolated. The electron pair represented by the conventional symbol for the triple bond would actually occupy a $\pi$-like orbital formed by overlap of the two $sp^2$ hybrid orbitals present in the original halobenzene in the C—H and C—Cl bonds. An alternative symbol for benzyne is the following; it is probably a more realistic representation, for it is unlikely, on steric grounds, that a triple bond of the usual kind is present in the molecule:

The tendency for the formation of the stable $sp^2/\pi$-orbital structure of the benzenoid nucleus is such that the benzyne intermediate reacts at once with ammonia to yield aniline:

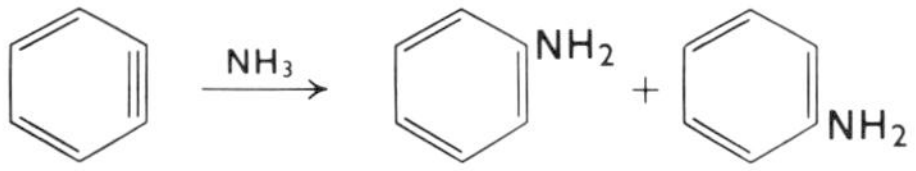

$$\text{(benzyne)} \xrightarrow{\text{NH}_3} \text{(NH}_2) + \text{(NH}_2) \tag{4}$$

That aniline is produced by the addition of ammonia in both possible ways can be demonstrated by using radioactive ($^{14}$C) chlorobenzene, labeled at the carbon atom that carries the chlorine:

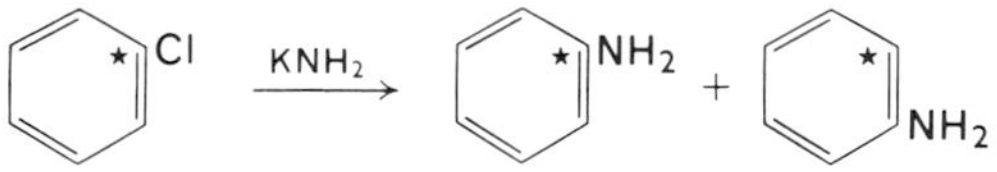

$$\text{(Cl)} \xrightarrow{\text{KNH}_2} \text{(NH}_2) + \text{(NH}_2) \tag{5}$$

By degradation of the aniline formed in this reaction it is possible to show that one-half of the radioactivity is in the carbon holding the —NH$_2$ group, the other half in the position *ortho* to this. The formation of all three toluidines from the reaction of *m*-chlorotoluene with potassium amide is, of course, in complete accord with this interpretation of the reaction.

The reaction of *o*-chlorotoluene with sodium hydroxide at a high temperature yields both *o*-cresol and *m*-cresol. This is evidence that a benzyne intermediate is formed in this reaction as well:

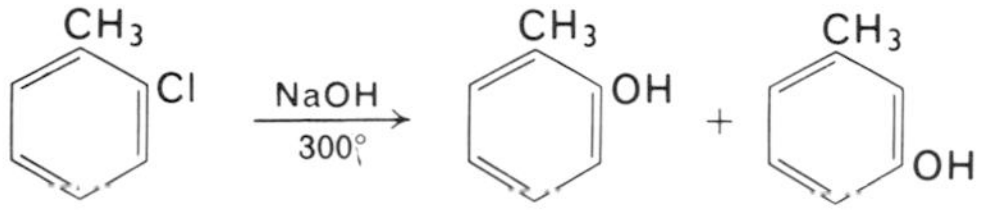

$$\text{(CH}_3)(\text{Cl}) \xrightarrow[300°]{\text{NaOH}} \text{(CH}_3)(\text{OH}) + \text{(CH}_3)(\text{OH}) \tag{6}$$

## 22.5 Nucleophilic displacement of halogen from *p*-nitrobromobenzene

When the halogen atom on an aromatic ring is in the *ortho* or *para* position to a nitro group, replacement of the halogen by nucleophilic attack proceeds with ease, and without the "rearrangement" observed in replacements that proceed by way of the benzyne intermediate. In the case of *p*-nitrobromobenzene the following reactions occur:

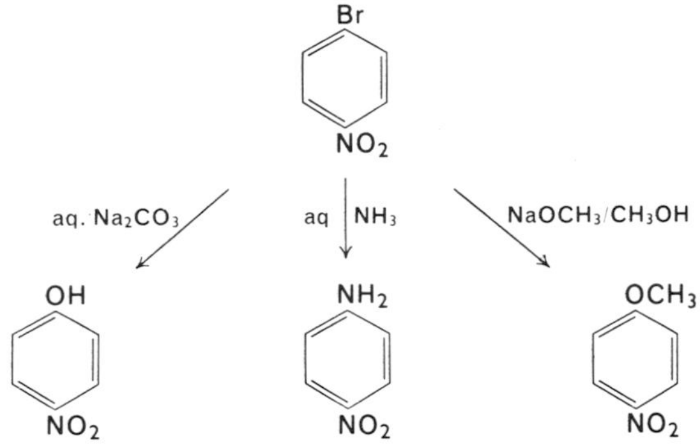

(7)

The presence of two or three nitro groups in the positions *ortho* and *para* to the halogen atom confers a high degree of reactivity, and such compounds as 2,4-dinitrochlorobenzene, 2,4-dinitrobromobenzene, 2,4,6-trinitrobromobenzene, and 2,4,6-trinitrochlorobenzene undergo nucleophilic displacement reactions with ease:

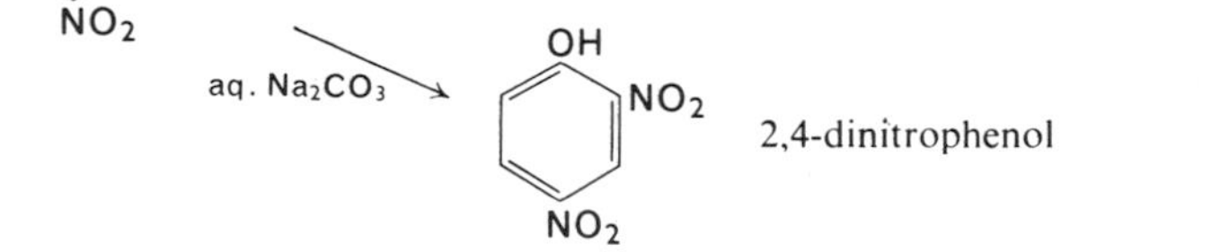

(8)

In 2,4,6-trinitrochlorobenzene the reactivity of the halogen atom is so great that the compound bears the trivial name "picryl chloride," derived from picric acid, the name commonly used for 2,4,6-trinitrophenol. The name picryl chloride suggests an order of reactivity comparable to that of an acid chloride.

The reason for the exceptional reactivity toward nucleophilic displacement of a halogen atom activated by an *ortho* or *para* nitro group is that the attack of the nucleophilic reagent (e.g., $CH_3O^-$) is facilitated by the ability of the nitro group to accommodate the negative charge provided by the attacking reagent:

$$CH_3O \underset{CH_3\overset{-}{O}}{\overset{Cl}{\diagdown}} \diagup N \diagdown O \rightleftharpoons \left\{ \underset{CH_3O}{\overset{Cl}{\diagdown}} N \diagup O \right\}^{-} \longrightarrow$$

$$CH_3O \diagup \diagdown NO_2 + Cl^- \qquad (9)$$

*p*-nitroanisole

The resonance contribution of the nitro group in permitting delocalization of the negative charge provides stabilization of the transition state and lowering of the activation energy. When two or three nitro groups can participate in such charge delocalization the transition state is correspondingly more stable and the reaction correspondingly much faster. The completion of the reaction, with expulsion of halide ion and retention of the methoxyl group, reflects the relatively weakly nucleophilic character of the halide ion as compared with the alkoxide ion. Thus, the attack of bromide ion upon *p*-nitroanisole would not be expected to proceed with replacement of the methoxyl group by bromine.

It will now be clear why chlorobenzene itself reacts so slowly with such bases as hydroxide ion, methoxide ion, and amines; no comparable stabilization of the negative charge in the ring is provided, and so the reaction is slow. With sufficiently strong bases (such as amide ion, $NH_2^-$) the reaction takes the alternative course through the benzyne intermediate.

Nucleophilic attack with replacement of groups or atoms other than halogen is possible when nitro groups are present in *ortho* and *para* positions. *p*-Nitroaniline can be prepared by the treatment of *p*-nitroanisole with ammonia at 200°C,

$$\underset{NO_2}{\overset{OCH_3}{\bigcirc}} \xrightarrow[\Delta]{aq.\ NH_3} \underset{NO_2}{\overset{NH_2}{\bigcirc}} + CH_3OH \qquad (10)$$

and the replacement of the nitro group (which is eliminated as nitrite ion) is accomplished in a comparable way:

$$\underset{NO_2}{\overset{NO_2}{\bigcirc}} \xrightarrow{aq.\ NH_3} \underset{NO_2}{\overset{NH_2}{\bigcirc}} + NH_4NO_2 \qquad (11)$$

One of the most interesting recent applications of nucleophilic displacements of this type is found in the use of 2,4-dinitrofluorobenzene in the structure determination of polypeptides (Chapter 31). The replacement of the fluorine atom by a free amino group occurs under mild conditions, leading to the formation of a dinitrophenyl derivative of the polypeptide:

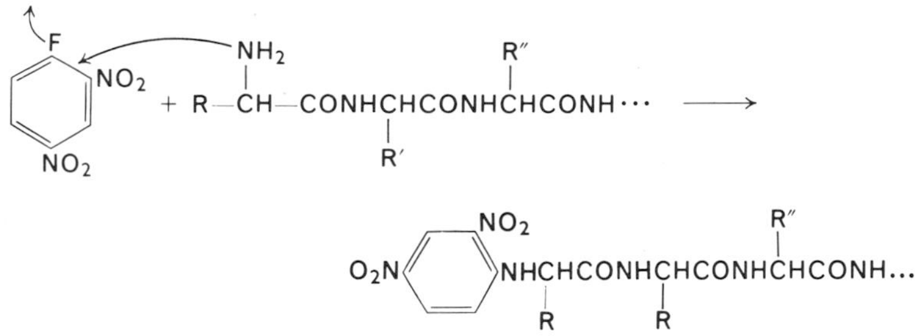

$$(12)$$

Hydrolysis of the N-2,4-dinitrophenyl derivative yields fragments containing the 2,4-dinitrophenyl group:

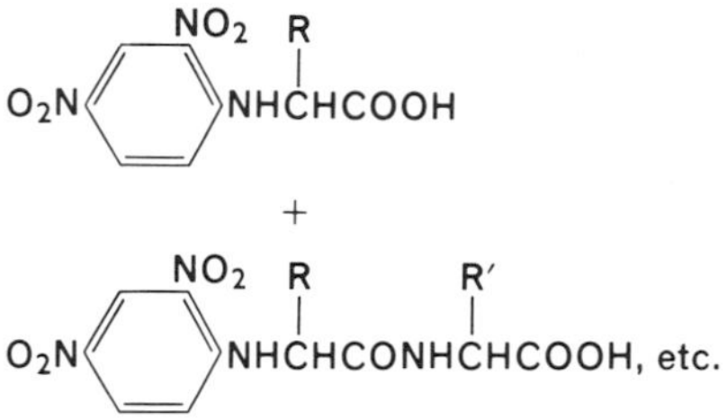

These can be separated by chromatographic methods and either identified or further hydrolyzed to identifiable fragments.


## 22.6 Steric inhibition in nucleophilic displacement

The nucleophilic replacement of the nitro group in the case of 1,3-dimethyl-2,5-dinitrobenzene, provides another example of the steric inhibition of resonance:

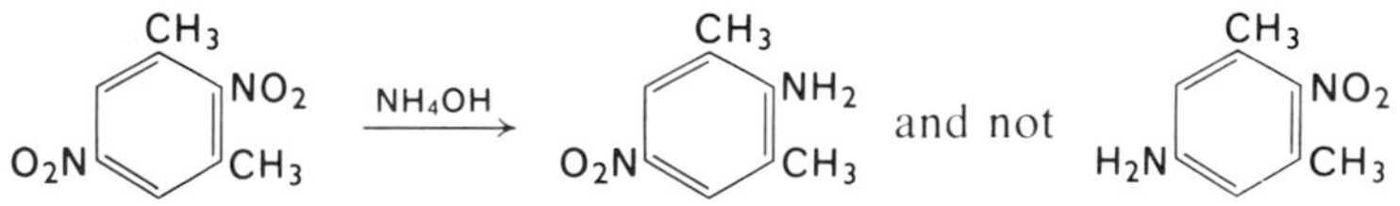

$$(13)$$

The ability of the 2-nitro group to engage in resonance stabilization of the intermediate

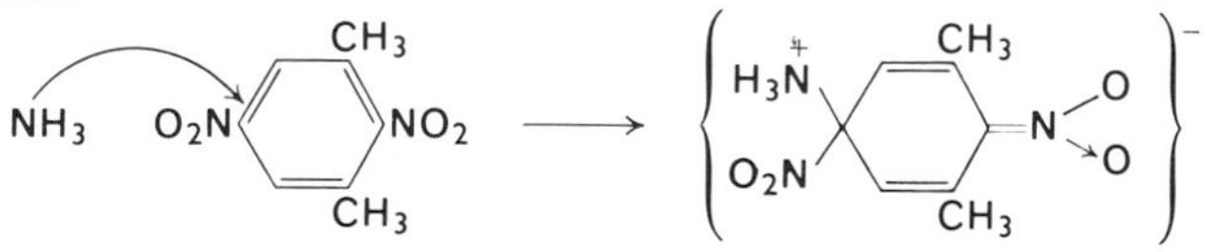

$$(14)$$

is inhibited by the *ortho*-methyl groups (Figure 22-1); but attack in the other position, leading to

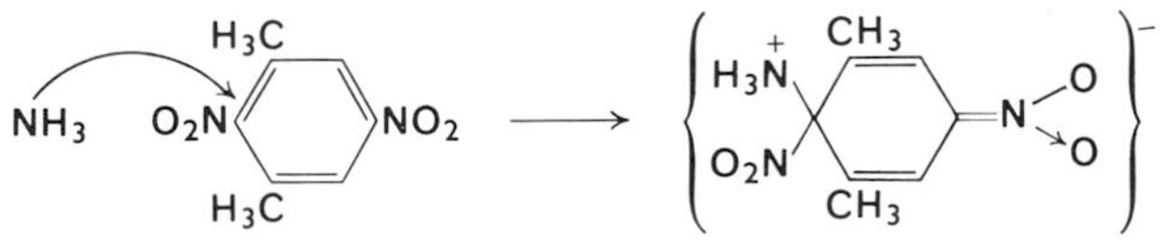

$$(15)$$

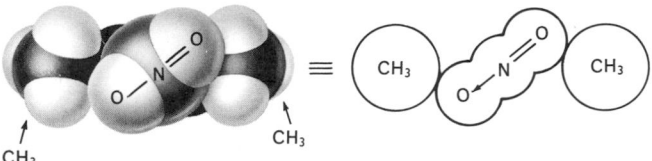

**Figure 22–1.** Model showing how *ortho*-methyl groups interfere with ability of nitro group to become coplanar with aromatic ring.

is not inhibited in this way (here the $\overset{+}{N}H_3$ and $NO_2$ groups in the 2-position are not coplanar with the ring, since in this intermediate stage the 2-carbon atom is tetrahedral).

## 22.7   Nucleophilic displacement of hydrogen

The ability of *m*-dinitrobenzene itself to accept a nucleophilic reagent forms the basis for a widely used biological analytical procedure, the Zimmermann reaction. This, a diagnostic and analytical method for the general structure —CO—CH₂—, is used for the detection and estimation of 17-keto-steroids, and is carried out by treating the steroid compound with solutions of *m*-dinitrobenzene and alcoholic alkali. The anion of the ketone

$$\text{—CH}_2\text{—CO} + \text{B:}^- \;\rightleftharpoons\; \{\text{—}\overset{..}{\text{C}}\text{H—CO—}\}^- + \text{B:H}$$

can attack *m*-dinitrobenzene in the following way

$$\text{(16)}$$

to give an addition compound which, by subsequent changes (possibly involving displacement of H:⁻), leads to the formation of deeply colored products that can be estimated colorimetrically.

The formation of 2,4-dinitrophenol by the reaction of *m*-dinitrobenzene with powdered potassium hydroxide further illustrates the nucleophilic replacement of hydrogen, but the reaction is not of general preparative usefulness.

## 22.8   Structural factors affecting nucleophilic aromatic substitution

The following tabulation summarizes the common groups (1) activating the ring toward nucleophilic displacement, and (2) capable of being displaced. For attack by —OCII₃⁻ (as a typical nucleophilic displacement reaction),

$$CH_3O^- + Y\!\!-\!\!\langle\bigcirc\rangle\!\!-\!\!X \longrightarrow CH_3O\!\!-\!\!\langle\bigcirc\rangle\!\!-\!\!X + :Y^- \qquad (17)$$

the tabulation shows the order of the replaceability of Y and of activating effect of groups X.

| | Y | X | |
|---|---|---|---|
| most readily displaceable | F | $-N_{2^+}$ | most activating |
| | $NO_2$ | $-NO$ | |
| | Cl, Br, I | $-NO_2$ | |
| | $-OAlkyl$ | $-SO_2CH_3$ | |
| | $-SO_2Alkyl$ | | |
| | | $\overset{+}{N}Me_3$ | |
| | $-NR_2$ | | |

It will be noted that $-\overset{+}{N}Me_3$, a powerfully deactivating and *m*-orienting group in electrophilic substitution, is not correspondingly activating in nucleophilic substitution. The $-\overset{+}{N}Me_3$ group exhibits a powerful inductive attraction of electrons but cannot contribute to the stabilization of the intermediate state by means of structures analogous to those involved in the case of activation by the nitro group.

The validity of the mechanism for nucleophilic substitution that is discussed above is supported by the observation that in some cases the intermediate addition compound can be isolated: for example, the same compound is formed by the reaction of sodium methoxide with 2,4,6-trinitrophenetole and by the reaction of sodium ethoxide with 2,4,6-trinitroanisole:

$$ (18) $$

We can now summarize, in a representative energy diagram, the nucleophilic aromatic substitution reaction for the four typical substitution patterns that have been discussed: anisole, *p*-nitroanisole, 2,4-dinitroanisole, and 2,4,6-trinitroanisole. Figure 22-2 shows energy profiles for the reaction of these four compounds with the ethoxide ion. Anisole, in which there is no nitro group for stabilization of the intermediate anionic complex, reacts so slowly as to appear not to react at all; the activation energy for formation of the unstabilized intermediate is high. With increasing substitution of nitro groups in the *ortho* and *para* positions there is increasing opportunity for charge delocalization in the transition state, with lowering of the activation energy and increasing stability of the intermediate; and in the trinitro compound, the intermediate is a stable compound that can be isolated. In these diagrams the simplifying assumption is made that the energies of the initial

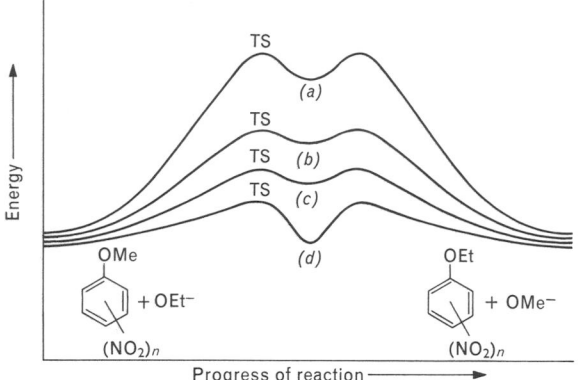

**Figure 22–2.** Energy profiles for the nucleophilic displacement reaction in ethanol: $MeO—Ar(NO_2)n + EtO^- \rightarrow EtO—Ar(NO_2)n + MeO^-$. *(a)* $n=0$ (anisole); *(b)* $n=1$ (*p*-nitroanisole); *(c)* $n=2$ (2,4-dinitroanisole); *(d)* $n=3$ (2,4,6-trinitroanisole).

and final systems are equal; since the two substituents involved in the replacement are the ethoxyl and methoxyl groups, this is probably nearly correct.

## 22.9   The nitroso group as a substituent

The nitroso (—N̈=O) group orients *ortho-para* in electrophilic substitution, and activates *ortho* and *para* halogen to nucleophilic substitution. For this reason, it offers a most interesting example of the operation of resonance effects in these two types of aromatic substitution.

The *ortho-para* directing power of the nitroso group depends upon its ability to *furnish* the unshared electron pair on nitrogen to the ring upon *demand* of the electrophilic reagent:

O=N̈⏜⬡⬡NO₂⁺ ⟶ O=N̈=⬡ ⟵⟶ etc.

---

**Exercise 1.**   Complete this formulation to the final substitution product.

---

Although the resonance effect of the nitroso group can be conceived of as operating in either direction,

O=N̈—⬡          O=N̈—⬡
← −R            → +R

it is only the latter of these (+ R) that is called upon by electrophilic attack.

However, when the attack is nucleophilic, the ability of the nitroso group to *accommodate* the electron pair ($-R$) is called into play:

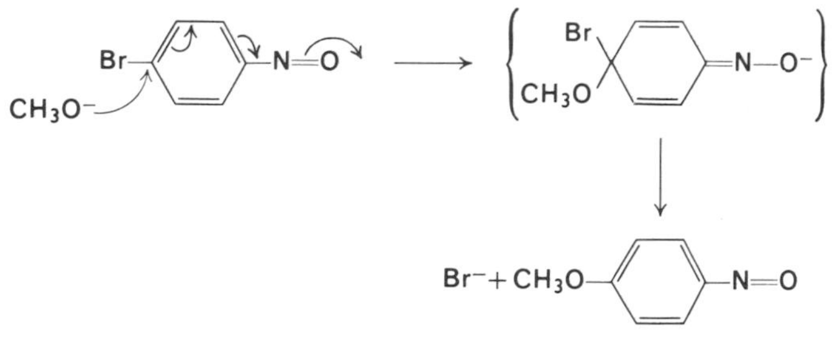

$$(19)$$

It can be seen that an appreciation of the mechanisms of the reactions involved allows an understanding of the behavior of the nitroso group, which acts in two ways that may appear superficially to be mutually exclusive. The nitroso group is thus acting quite normally in its ability to activate the *ortho-para* positions to *both* electrophilic and nucleophilic attack.

## 22.10  Other properties of aromatic halogen derivatives

The formation of Grignard reagents and lithium derivatives of aryl halides such as bromobenzene, bromonaphthalene, *p*-iodotoluene, proceeds with ease. Chlorobenzene reacts only slowly with magnesium unless more vigorous conditions are used (higher temperatures), but aryl bromides and iodides form Grignard reagents readily and in excellent yields. This property of aryl halides lends them to a wide variety of synthetic uses. The behavior of Grignard reagents discussed earlier may be regarded as representative of that of the aryl Grignard reagents, of which phenylmagnesium bromide is typical.

Compounds containing halogen as a substituent in an alkyl side chain behave in general as do alkyl halides. The usual displacement reactions of alkyl-bound halogen proceed in the expected way. Benzyl halides are exceptional, chiefly in their greater degree of reactivity. Benzyl chloride, bromide, and iodide are readily hydrolyzed with aqueous bases, ammonolyze with amines to benzylamines, and react with alkali cyanides to give phenylacetonitrile. Benzylmagnesium halides are readily formed, although the high degree of reactivity of the iodide and bromide lead to yield-reducing side reactions. Benzylmagnesium chloride can be prepared in good yield. Alkylation of benzylmagnesium chloride with sulfuric and sulfonic acid esters and allyl halides proceeds satisfactorily and constitutes a valuable means of preparing higher alkylated benzenes:

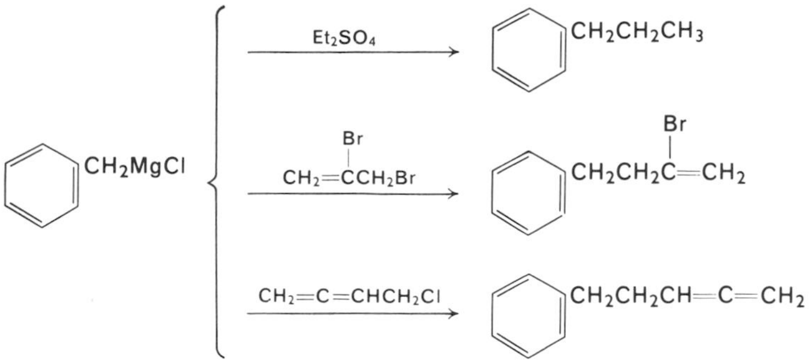

$$(20)$$

## 22.11 The Wurtz-Fittig reaction

The reaction of alkyl halides with metallic sodium to give hydrocarbons formed by coupling of the two alkyl groups is known as the *Wurtz reaction*:

$$2CH_3CH_2CH_2CH_2Br + 2Na \longrightarrow CH_3CH_2CH_2CH_2CH_2CH_2CH_2CH_3 + 2NaBr \quad (21)$$

Aryl halides do not undergo this coupling reaction; but when an aryl halide and an alkyl halide are allowed to react with metallic sodium, the alkylbenzene is formed:

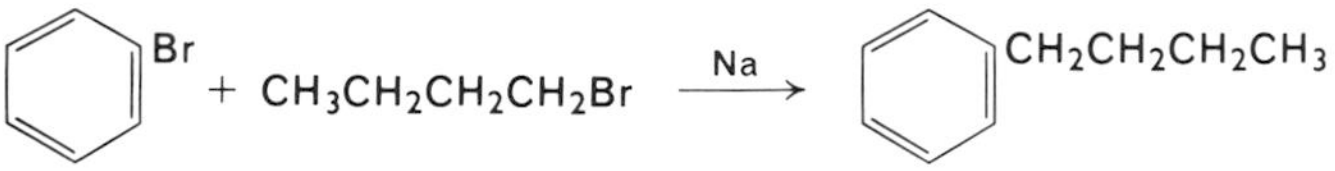

$$(22)$$

This reaction is known as the *Wurtz-Fittig reaction*, and is a useful method for the preparation of some alkylbenzenes. The course of both reactions is by way of intermediate sodium derivatives, the alkylsodium and the arylsodium. The Wurtz reaction involves nucleophilic displacement of halogen from the alkyl halide by attack of the alkyl anion provided by $R:^-Na^+$. Since the aryl halide is not susceptible to the nucleophilic displacement of the halogen, the aryl anion cannot react with it to form a biaryl; but the aryl anion can effect a displacement by attack upon the alkyl halide, and the alkylbenzene is produced in satisfactory yield. The reaction, which is practicable only with primary alkyl halides, yields products in which the alkyl group has not undergone rearrangement, for displacement by the $S_N2$ mechanism results in attachment of the aryl group at the carbon atom from which the halogen is displaced.

## 22.12 The Ullmann reaction

The coupling of aromatic nuclei by the reaction of iodo- or bromobenzenes with copper (or sometimes silver) powder is known as the *Ullmann* reaction. *o*-Iodotoluene yields 2,2'-dimethylbiphenyl,

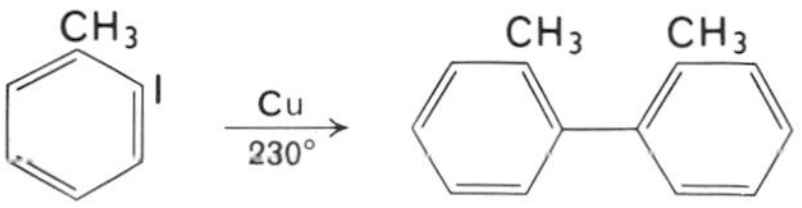

$$(23)$$

and 1-bromonaphthalene yields 1,1′-binaphthyl:

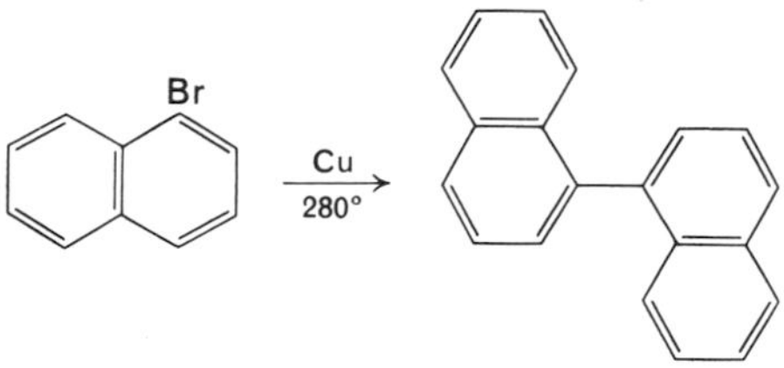

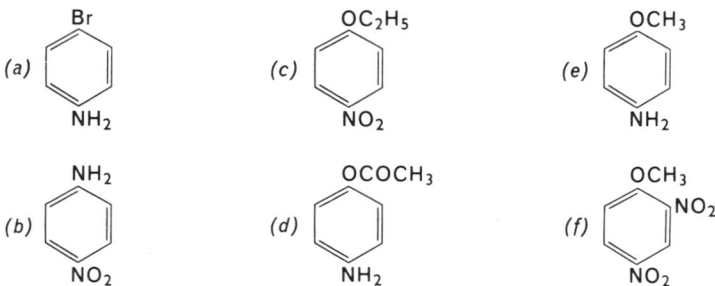

(24)

Complex substituted biphenyl derivatives can be prepared by this method; in many cases the yield is not high but there is no more satisfactory alternative method:

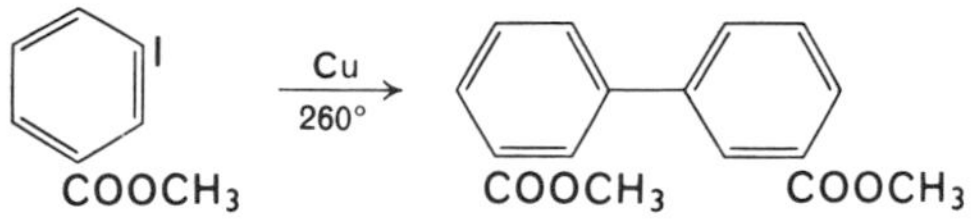

(25)

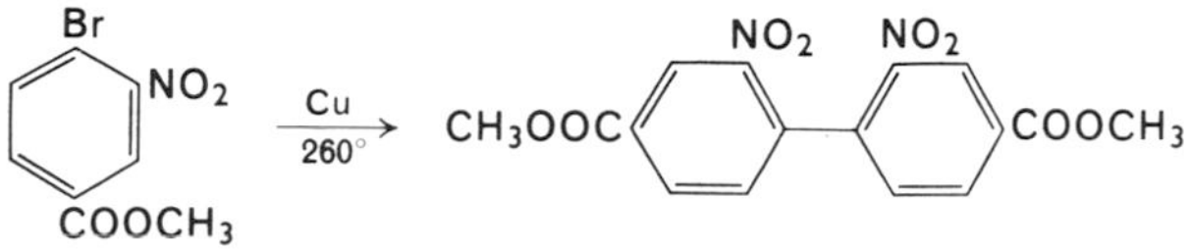

(26)

The preparation of biphenyl derivatives capable of existence in optically active forms is accomplished by the use of the Ullmann reaction. The product of the coupling reaction is, of course, the racemic biphenyl, which is then resolved by usual methods into the enantiomeric forms (Section 16.19).

### Exercises

(Exercise 1 will be found within the text.)

**2** With a monosubstituted benzene as the starting material, show how the following compounds can be prepared.

(a)   (c)   (e)

(b)   (d)   (f)

**3** Write the reaction between 2,4-dinitrofluorobenzene and *(a)* α-aminopropionic acid; *(b)* aminoacetamide; *(c)* benzylamine; *(d)* aqueous alkali.

**4** Complete the following equations, using structural formulas.

(*a*) 1,3-dimethyl-2,5-dinitrobenzene + methylamine $\longrightarrow$

(*b*) 2,3-dinitrotoluene + sodium methoxide $\longrightarrow$

(*c*) *p*-dinitrobenzene + hydrazine $\longrightarrow$

(*d*) *p*-nitrosonitrobenzene + ammonia $\longrightarrow$

(*e*) 3,4-dibromonitrobenzene + aniline $\longrightarrow$

**5** Why is benzyl chloride more reactive in displacement reactions (with nucleophilic reagents) than *n*-heptyl chloride? Would *p*-methoxybenzyl chloride show even greater, or less, reactivity? Why?

**6** Ethyl benzenesulfonate can be used in place of diethyl sulfate in the alkylation of a Grignard reagent. Suggest a method for the preparation of

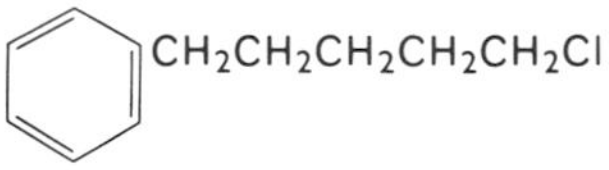

$CH_2CH_2CH_2CH_2CH_2Cl$

having 4-chloro-1-butanol and benzenesulfonyl chloride available.

**7** Devise practical ways of carrying out the following transformations.

(*a*) benzene $\longrightarrow$ benzylamine

(*b*) toluene $\longrightarrow$ phenylacetic acid

(*c*) toluene $\longrightarrow$ benzene-1,4-dicarboxylic acid

(*d*) benzene $\longrightarrow$ triphenylcarbinol

(*e*) bromobenzene $\longrightarrow$ *p*-xylene

**8** Show how the following compounds could be prepared from toluene:

(*a*)

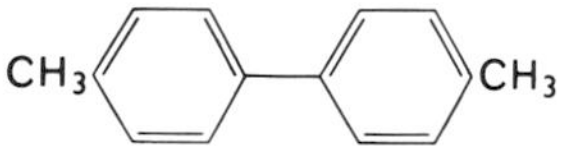

(*b*)

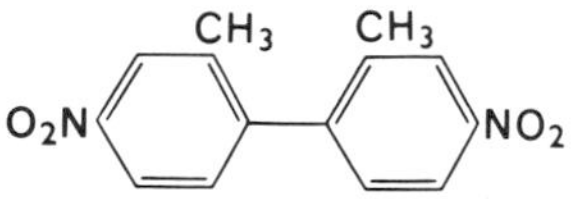

**9** Starting with benzene or toluene, and any necessary aliphatic and inorganic reagents, and with the use of reactions that have been described up to this point in the text, devise practical syntheses for the following compounds:

(*a*)

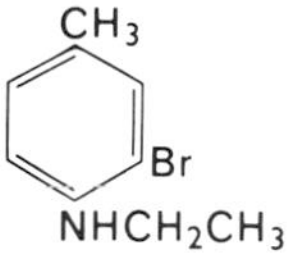

(*b*)

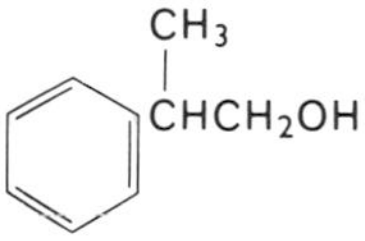

(*c*)

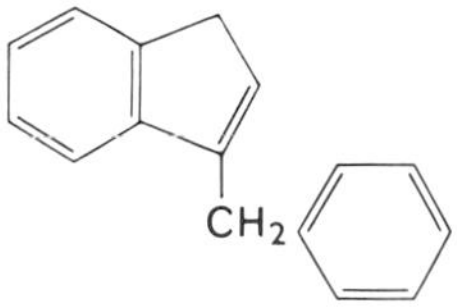

(*d*)

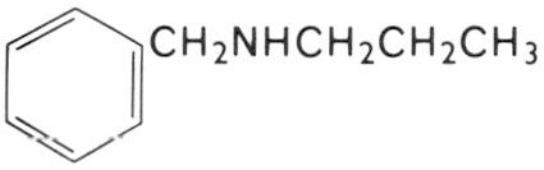

(*e*)

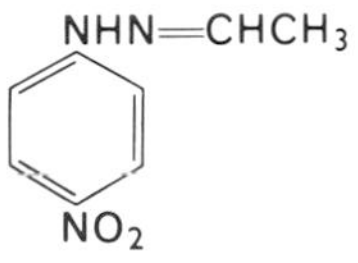

(*f*)

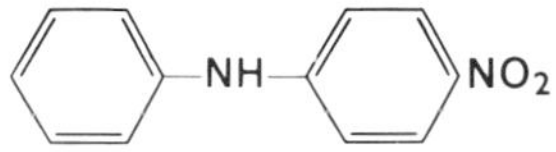

# Chapter twenty-three / Aromatic Hydrocarbons and Carbonyl Compounds

## 23.1 General methods of preparation of alkylated benzenes

The simple aromatic hydrocarbons are important articles of commerce and are produced in large quantities by the chemical industry. Benzene, toluene, and the xylenes are obtained from coal tar and are found in small amounts in some petroleums but at the present time they are produced largely by chemical transformation of petroleum fractions. Treatment of petroleum fractions in the boiling range of gasoline and light oils with catalysts (chromium and aluminum oxides; platinum) at high temperatures causes cyclization and dehydrogenation, with the formation of "reformed" petroleum fractions containing appreciable proportions of aromatic hydrocarbons. The cyclo-dehydrogenation of $n$-heptane can be formulated as follows:

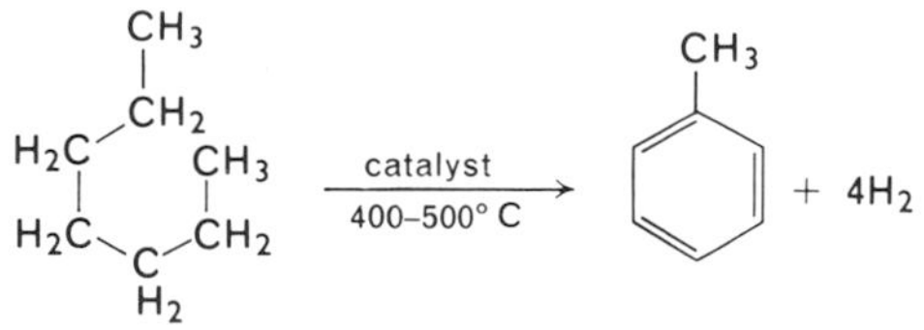

$$\begin{array}{c} \text{CH}_3 \\ | \\ \text{CH}_2 \\ \text{H}_2\text{C} \quad \text{CH}_3 \\ | \qquad | \\ \text{H}_2\text{C} \quad \text{CH}_2 \\ \text{C} \\ \text{H}_2 \end{array} \quad \xrightarrow[\text{400–500° C}]{\text{catalyst}} \quad \text{C}_6\text{H}_4\text{CH}_3 \; + \; 4\text{H}_2 \qquad (1)$$

Other alkanes give benzene and other alkylated benzenes, and the mixture that is produced by the reforming operation is separated into the pure aromatic compounds by combinations of fractional distillation, preferential extraction and, in some cases, crystallization. The combined production of benzene, toluene, and the xylenes by these processes exceeds 1.5 billion gallons per year and has nearly superseded their production from coal tar.

Other aromatic hydrocarbons are produced industrially by synthetic processes, starting with benzene. Ethylbenzene is prepared by the alkylation of benzene with ethylene with the use of an aluminum chloride catalyst

$$\text{(benzene)} + CH_2\!\!=\!\!CH_2 \xrightarrow{AlCl_3} \text{(ethylbenzene)} CH_2CH_3 \tag{2}$$

Catalytic dehydrogenation of ethylbenzene yields styrene, an important raw material for the preparation of polymers, including "synthetic rubber":

$$\text{(ethylbenzene)} CH_2CH_3 \xrightarrow[650°]{ZnO/Cr_2O_3} \text{(styrene)} CH\!\!=\!\!CH_2 + H_2 \tag{3}$$

styrene

Laboratory syntheses of aromatic hydrocarbons can be accomplished by a wide variety of methods. Polymethylbenzenes are prepared in two ways.

1. Friedel-Crafts alkylation. The alkylation of benzene, toluene, or xylene with methyl chloride and toluene yields a complex mixture of polymethylbenzenes from which individual compounds can be isolated by fractional distillation and crystallization:

$$\text{(xylene)}\begin{Bmatrix} CH_3 \\ CH_3 \end{Bmatrix} \xrightarrow[AlCl_3]{CH_3Cl} \begin{array}{l}\text{mixture of tri-, tetra-, penta,-}\\ \text{and hexamethylbenzenes}\end{array} \tag{4}$$

mixture of xylenes

1,2,4,5-Tetramethylbenzene (m.p. 80°) (durene) can be separated from the tetramethylated fraction by crystallization.

2. The preparation of polymethylbenzenes of unambiguous orientation can be accomplished with the use of the reaction between an aryl Grignard reagent and methyl sulfate:

$$H_3C\text{—}\text{(ring)}\text{—}MgBr + (CH_3)_2SO_4 \longrightarrow H_3C\text{—}\text{(ring)}\text{—}CH_3 \tag{5}$$

$$H_3C\text{—}\text{(ring)}\begin{smallmatrix}CH_3\\ \\MgBr\\CH_3\end{smallmatrix} + Me_2SO_4 \longrightarrow H_3C\text{—}\text{(ring)}\begin{smallmatrix}CH_3\\ \\CH_3\\CH_3\end{smallmatrix} \tag{6}$$

isodurene

The introduction of other side chains by the reaction of benzylmagnesium chloride with alkylation agents has been described in Section 22.10.

Alkylbenzenes with side chains more complex than ethyl are prepared by a variety of methods, many of them conventional applications of reactions that we have considered in earlier chapters. Some examples of these, that require no comment, are the following:

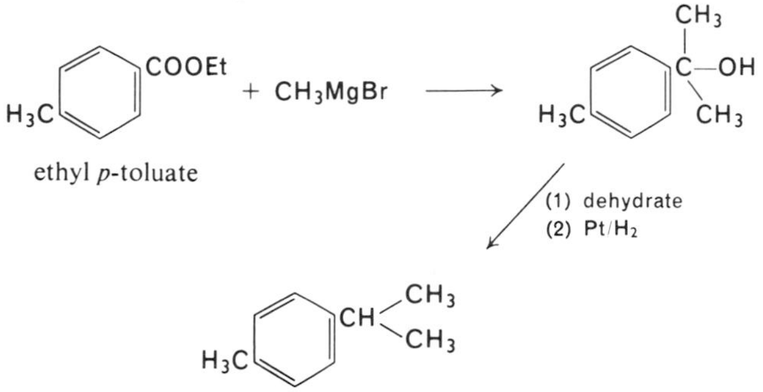

ethyl *p*-toluate

(1) dehydrate
(2) Pt/H₂

(7)

*p*-cymene

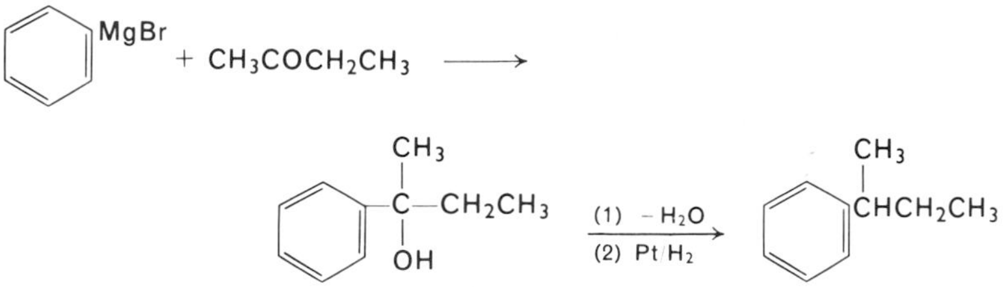

(8)

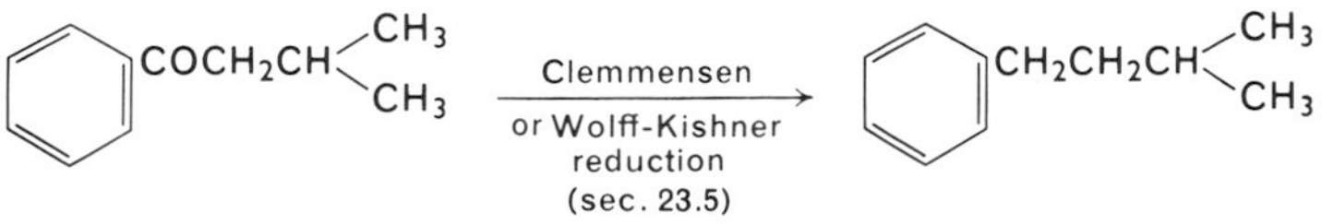

*iso*valerophenone                                    *iso*pentylbenzene

Clemmensen
or Wolff-Kishner
reduction
(sec. 23.5)

(9)

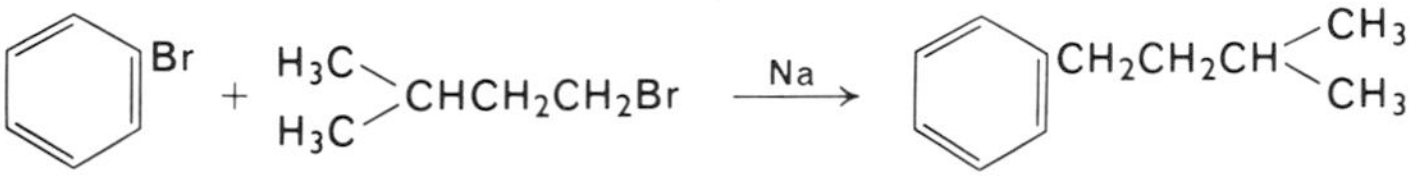

Na

(10)

AlCl₃

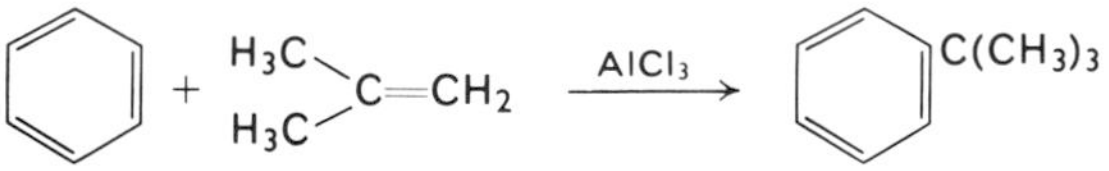

(11)

The introduction of higher alkyl groups by the Friedel-Crafts alkylation of benzene is limited by the occurrence of rearrangements in the alkyl group; this has been discussed in Chapter 20. Methylation, ethylation, *t*-butylation, and isopropylation are reactions that are not subject to this disadvantage; but when polyalkylation is attempted with alkyl halides and olefins certain unusual orientation problems arise. These are described in the next section.

### 23.2   *Meta* alkylation

At high temperatures (i.e., about 100°C) or in the presence of excess catalyst the result of the Friedel-Crafts alkylation of benzene often leads to *meta* orientation of the alkyl groups:

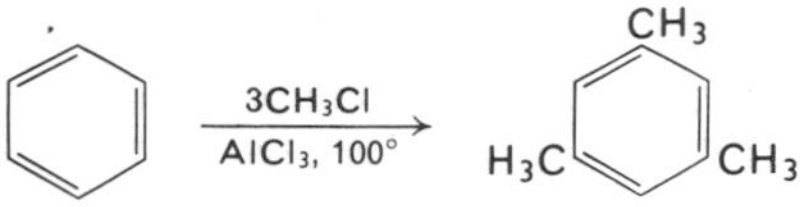

$$\text{(12)}$$

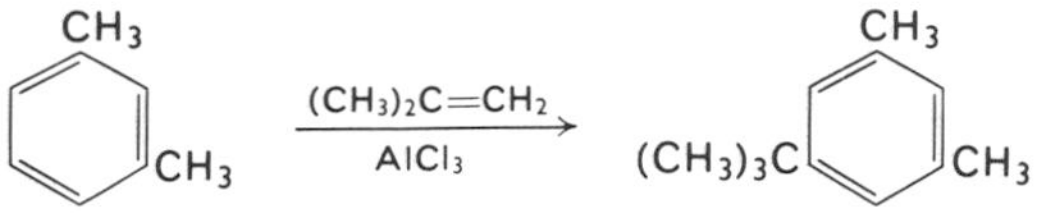

$$\text{(13)}$$

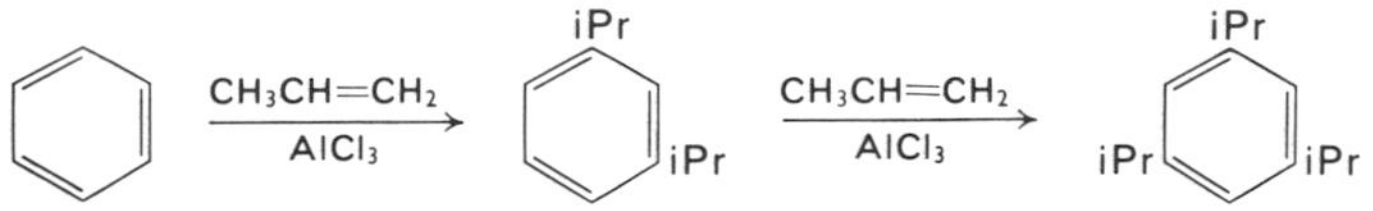

$$\text{(14)}$$

This orientation is contrary to what is expected, since the alkyl group first intro-duced—or, in the case of *m*-xylene, those already present—should orient the in-coming group to the *ortho-para* positions. The apparent anomaly has been ex-plained in three ways:

1. The known reversibility of the Friedel-Crafts reaction could lead to the removal of an alkyl group (that is, its replacement by hydrogen). Thus, the *meta* substitution that always accompanies the *ortho-para* substitution builds up the concentration of the *meta* isomer, as in the following equation:

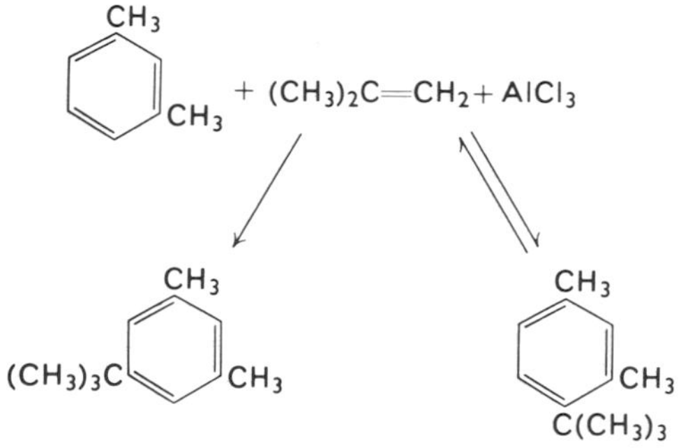

$$\text{(15)}$$

The dealkylation of an alkylated benzene by reversal of the Friedel-Crafts alkylation reaction is, of course, subject to the same orienting influences as other electrophilic substitution reactions. Dealkylation may be looked upon as electro-philic replacement of an alkyl group by a proton; in general terms:

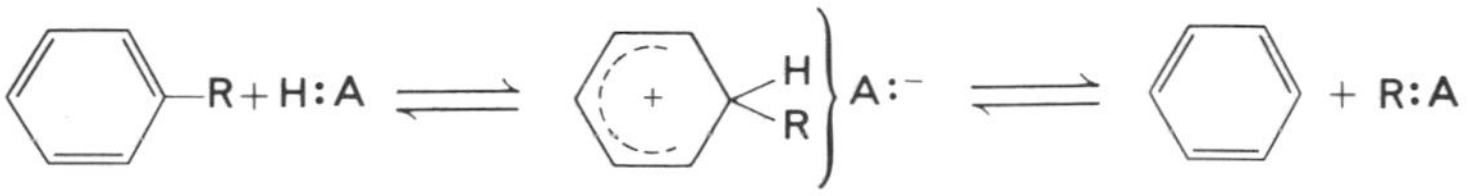

$$\text{(16)}$$

The electrophilic attack of a proton upon 1,3-dimethyl-4-*t*-butylbenzene would take place at the position activated by the *ortho-* and *para*-situated alkyl groups; namely, at the 4-position:

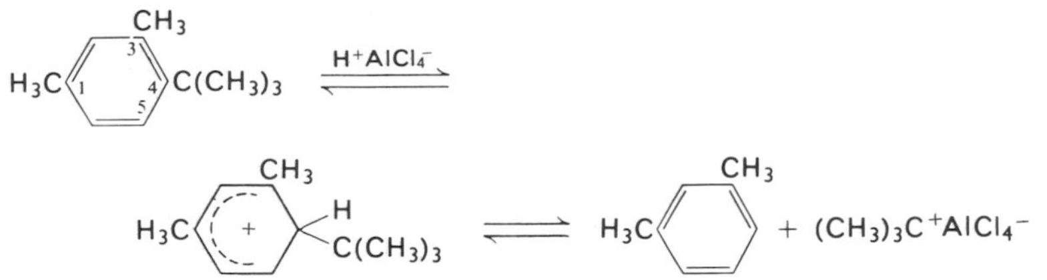

The result of this reaction is the removal of the *t*-butyl group. Thus, a substituent may enter and leave the 4-position over a period of time, but a substituent that enters the 5-position is not as rapidly equilibrated with hydrogen, for in the 1,3,5-substituted derivative it is still the 4-position (now occupied by hydrogen) which is subject to electrophilic attack.

2. Successive *ortho-para* introductions followed by removal of one group can lead to eventual *meta*-substitution:

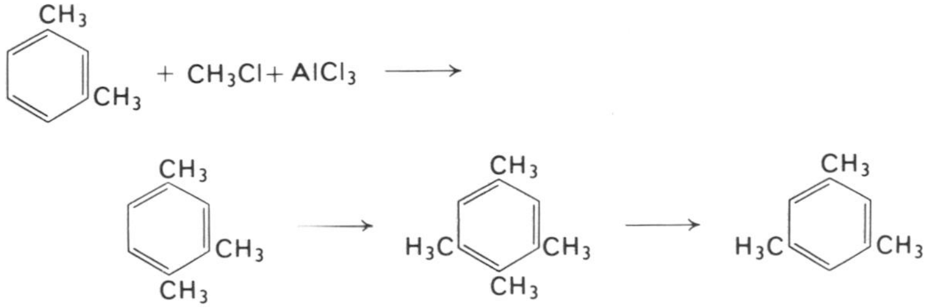

---

**Exercise 1.** Formulate the removal of the 2-methyl group from 1,2,3,5-tetramethylbenzene interpreting the reaction as an electrophilic displacement of —$CH_3$ by the attack of $H^+$.

---

3. A rearrangement of an alkyl group can take place in the transition-state complex:

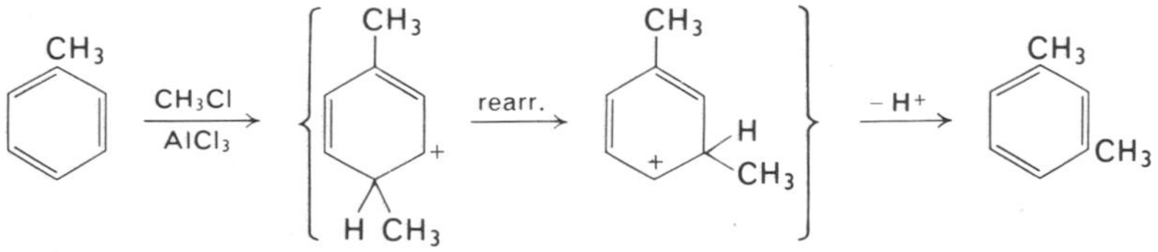

## 23.3 Reactions of aromatic hydrocarbons

*Side-chain oxidation.* The oxidation of compounds of the general constitution

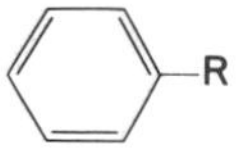

under vigorous conditions, usually with potassium permanganate, leads to the conversion of the side chain to a carboxyl group. Toluene and monoalkylated benzenes are converted into benzoic acid; xylenes are oxidized to the corresponding phthalic acids; mesitylene to benzene-1,3,5-tricarboxylic acid; pentamethylbenzene to benzenepentacarboxylic acid. Isopropylbenzene and $t$-butylbenzene are rather resistant to this kind of oxidative degradation, the former being oxidized to phenyldimethylcarbinol, which is further oxidized slowly. However, under sufficiently drastic conditions or prolonged treatment, complete oxidation to benzoic acid can be accomplished. The nature of R can vary over a wide range of structural types, so long as there is a carbon atom attached to the ring. There are few exceptions to the reaction as formulated in the most general way:

$$\text{(20)}$$

When the ring bears, in addition to the side chain, a nitro or sulfonic acid group or a halogen atom, the oxidation is still a practicable route to the corresponding substituted benzoic acid. However, the presence of nuclear hydroxyl or amino groups renders the ring itself susceptible to oxidation and deep-seated oxidation (leading to the destruction of the benzene ring) results:

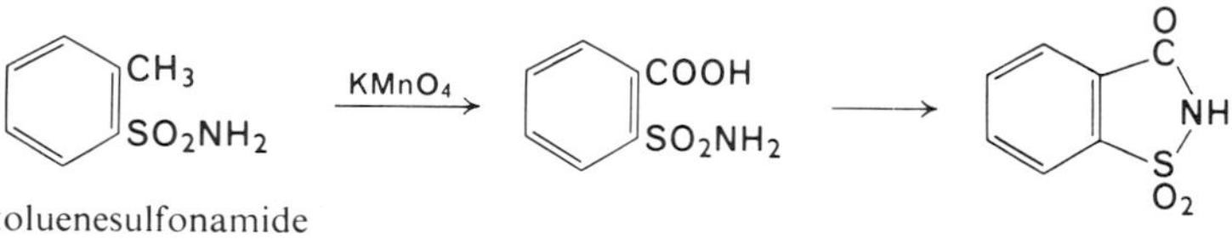

$$\text{(21)}$$

$o$-toluenesulfonamide

saccharin

sodium and ammonium salts are "soluble saccharins", and are intensely sweet substances.

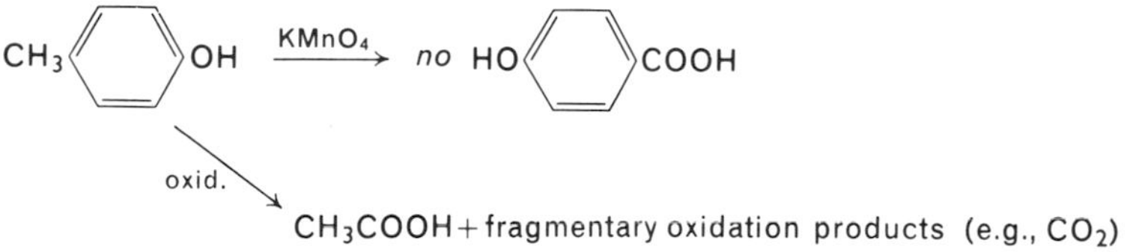

$$\text{(22)}$$

CH$_3$COOH + fragmentary oxidation products (e.g., CO$_2$)

The oxidation of 2,4,6-trinitrotoluene, followed by decarboxylation of the trinitrobenzoic acid, is a method for the preparation of 1,3,5-trinitrobenzene, which is not easily obtainable by nitration of benzene (or of $m$-dinitrobenzene):

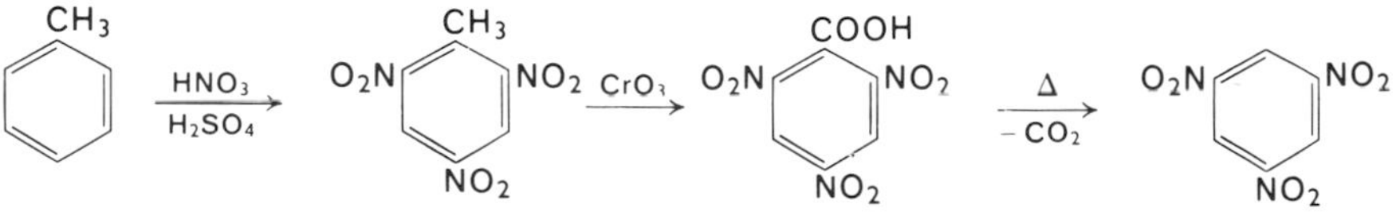

2,4,6-trinitrotoluene
(TNT)

1,3,5-trinitrobenzene
(TNB)

$$\text{(23)}$$

Oxidation to carboxylic acids has considerable value in structure determination. For example, a compound $C_8H_9Br$ could be either a bromoxylene or an ethylbromobenzene. If oxidation were to give a bromobenzoic acid the compound would thus be shown to be a bromo derivative of ethylbenzene, and the identity of the bromobenzoic acid (*ortho*, *meta*, or *para*) would prove its structure. A bromoxylene would yield a bromophthalic acid on oxidation.

The deep-seated and extensive degradation that is possible in the oxidation of aromatic compounds to carboxylic acids is most strikingly shown in the oxidation of graphite with fuming nitric acid, with the formation of benzenehexacarboxylic acid (mellitic acid):

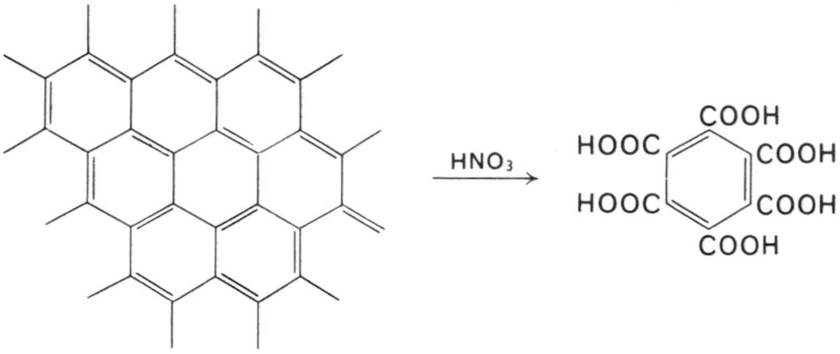

$$\xrightarrow{\text{HNO}_3} \tag{24}$$

A portion of the graphite "molecule"    mellitic acid

***Side-chain halogenation.***   We have already discussed the nuclear bromination or chlorination of aromatic hydrocarbons. The halogenation of the side chain occurs under other conditions, namely, in the absence of nuclear halogenation catalysts such as ferric salts, and with the aid of higher temperatures and sunlight or ultraviolet radiation:

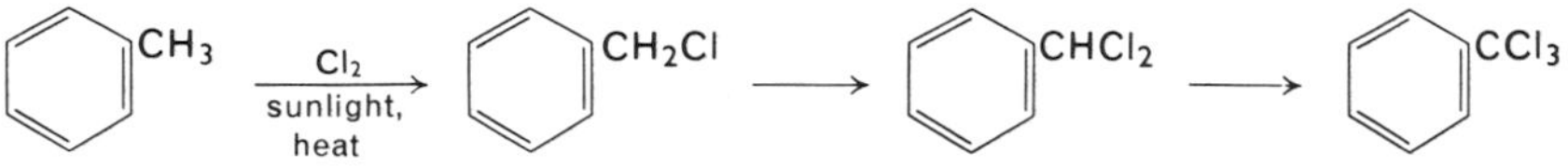

$$\xrightarrow[\text{heat}]{\underset{\text{sunlight,}}{\text{Cl}_2}} \tag{25}$$

benzyl chloride          benzal chloride          benzotrichloride

Side-chain halogenation can also be accomplished with sulfuryl chloride in the presence of a peroxide as a catalyst:

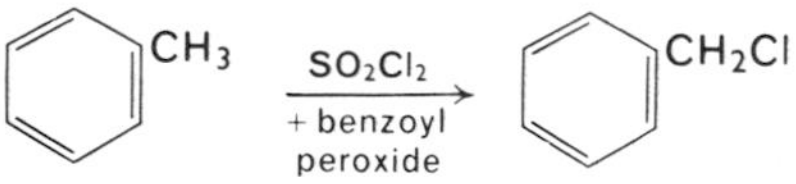

$$\xrightarrow[\substack{+\text{ benzoyl} \\ \text{peroxide}}]{\text{SO}_2\text{Cl}_2} \tag{26}$$

The conditions under which side-chain chlorination occurs indicate that a quite different mechanism is operating from that in nuclear halogenation. The probable course of the reaction is one in which chlorine *atoms* are involved:

$$\text{Cl}_2 \rightleftharpoons 2\text{Cl}\cdot \quad \text{(induced by light or free radicals from the peroxide)} \tag{27}$$
$$\text{Ar}-\text{CH}_3 + \text{Cl}\cdot \longrightarrow \text{ArCH}_2\cdot + \text{HCl}$$
$$\text{ArCH}_2\cdot + \text{Cl}_2 \longrightarrow \text{ArCH}_2\text{Cl} + \text{Cl}\cdot; \quad \text{etc.}$$

That the effective reacting species is not the chlorine molecule is indicated by the need for radiant energy or a source of free radicals, without which the reaction does not proceed; and that $Cl^+$ ions or their equivalent are not involved in side-chain halogenation is shown by the fact that when conditions are such that these can be formed, *nuclear* halogenation is the exclusive course of the reaction of toluene with chlorine.

It is the $\alpha$-hydrogen atoms that are replaced by halogens in this specific and controllable way. For example:

$$\text{(benzene ring)}CH_2CH_2CH_3 \xrightarrow[\Delta,\ light]{Br_2} \text{(benzene ring)}\overset{\overset{Br}{|}}{C}HCH_2CH_3 \tag{28}$$

The hydrogens attached to carbon atoms farther from the ring than the $\alpha$-carbon atom behave like those in alkanes, and under vigorous conditions general and random halogenation occurs.

The $\alpha$-bromination of *n*-propylbenzene is clearly to be ascribed to the lower activation energy of the reaction path leading through the $\alpha$-radical, compared with those leading through the $\beta$- or $\gamma$-radicals. The greater stability of the intermediate that bears the unpaired single electron on the carbon atom adjacent to the ring must be the result of electron delocalization in which the aromatic ring participates. It is clear that no such delocalization can exist in the $\beta$- or $\gamma$-radical:

$$\left\{ \text{(ring)}-\dot{C}HCH_2R \longleftrightarrow \text{(ring)}=CHCH_2R \longleftrightarrow \right.$$

$$\left. \text{(ring)}=CHCH_2R \longleftrightarrow \text{(ring)}=CHCH_2R \right\}$$

The resonance stabilization of the $\alpha$-radical will be reflected in the energy requirement for the transition state leading to its formation, and thus in a lower activation energy and a faster rate of reaction, relative to the rates of $\beta$- and $\gamma$-bromination.

*Addition of chlorine to benzene.*    Under ordinary conditions and in the absence of nuclear halogenation catalysts, benzene is quite inert to halogens. In the presence of ultraviolet light, chlorine adds to benzene to yield a mixture of stereoisomeric hexachlorocyclohexanes:

$$\text{(benzene)} + 3Cl_2 \xrightarrow[light]{u.v.} \text{(hexachlorocyclohexane structure)} \tag{29}$$

1,2,3,4,5,6-hexachlorocyclohexane

There are eight possible configurations of a 1,2,3,4,5,6-hexachlorocyclohexane, of which one can occur in two enantiomorphic forms (making a total of seven *meso*, one *d-*, and one *l-* forms). Not all of these have been obtained as pure compounds, but five have been studied with considerable care. The interest in these compounds is due to the fact that one of them (the $\gamma$-form) is the very effective insecticide known as Lindane or Gammexane.

On the basis of the more stable chair form of cyclohexane, the disposition of the six chlorine atoms can be shown for the nine isomers. The axial bonds will be designated *(a)*, the equatorial bonds *(e)*. The $\gamma$-1,2,3,4,5,6-hexachlorocyclohexane has been found to have the configuration represented by *(aaaeee)*, or

$$(30)$$

Lindane

This isomer has a plane of symmetry and thus is a nonresolvable *meso* form. Of the other four known isomers, only $\alpha$ *(aaeeee)* has no element of symmetry and is resolvable into *d-* and *l-*forms. The configurations of $\beta$, $\delta$, and $\epsilon$ are: $\beta$ *(eeeeee)*; $\delta$ *(aeeeee)*; $\epsilon$ *(aeeaee)*. The isomers are separated by fractional crystallization.

---

**Exercise 2.** Draw the structures of the $\beta$, $\delta$ and $\epsilon$ isomers in the chair form. Locate the plane of symmetry in each. Draw the structures of the enantiomorphic [(+) and (−)] forms of the $\alpha$-isomer.

---

*Ozonolysis of the benzene ring.* Benzene reacts with ozone to form a triozonide. The reaction is slower than that with alkenes since the activation energy for the attack of the first ozone molecule must be increased by an amount that corresponds to the loss of the delocalization energy of the aromatic ring. After the addition of one ozone molecule, the resulting mono-ozonide is a cyclohexadiene, and the addition of the next two molecules of ozone proceeds rapidly. Decomposition of the triozonide in the usual way (Chapter 9) leads to the formation of glyoxal, $O{=}CH{-}CH{=}O$. Early studies of the structure of the benzene ring, prior to the development of our present concepts of the character of the six $\pi$-electrons, aimed at answering the question of the "location" of the double bonds, included a study of the products formed in the ozonolysis of *o*-xylene. The products included all of the products that would be expected to result from both Kekulé structures, and were originally interpreted as lending support to the view that *o*-xylene contained molecules with both structures, in rapid oscillation. The formation of ozonolysis products from both Kekulé forms can, however, be explained simply and without the necessity for the assumption that two different kinds of *o*-xylene exist. The attack of the first molecule of ozone need not be confined exclusively to one course, and if the initial addition takes place in two ways, to yield

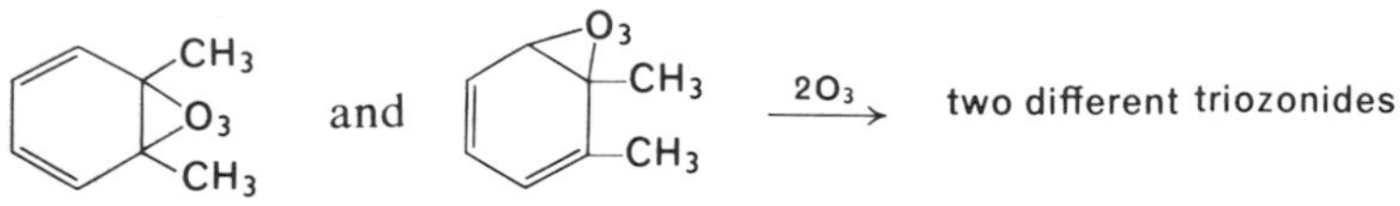

$$\text{(structures)} \quad \text{and} \quad \text{(structures)} \xrightarrow{2O_3} \quad \text{two different triozonides} \qquad (31)$$

the observed experimental findings are readily explained.

Ozonolysis of aromatic rings is often a valuable experimental technique for "destroying" the aromatic nucleus in a molecule, leaving the nonolefinic portions of the compound intact. It has been used in stereochemical studies; an example is the following:

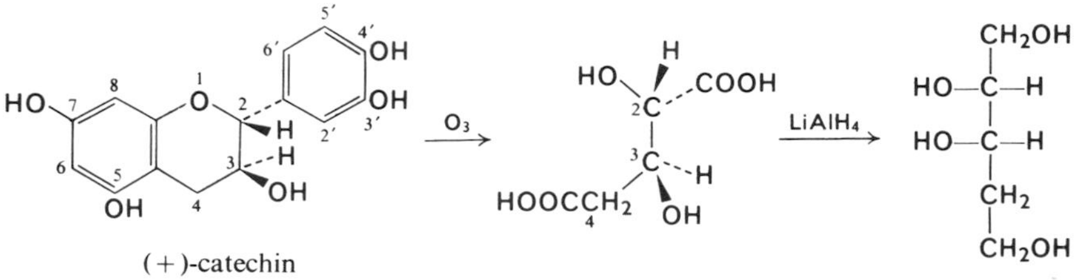

(+)-catechin

$$(32)$$

The stereochemistry of the tetrahydroxypentane, and thus of (+)-catechin, was established by its synthesis from 2-deoxy-D-ribose:

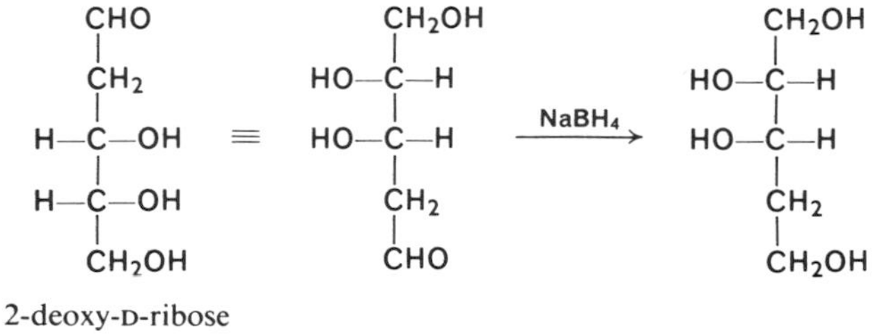

2-deoxy-D-ribose

$$(33)$$

---

**Exercise 3.**   Write the complete reaction for the ozonolysis of (+)-catechin (making any reasonable assumptions as about the structures of intermediate products that contain fragments of the benzene rings).

**Exercise 4.**   Write the complete systematic name for (+)-catechin, and for the dihydroxyglutaric acid derived from it by ozonolysis, using the *R*: *S* system of stereochemical nomenclature. The name for catechin may be based upon the basic three-ring structure, lacking all hydroxyl groups, called flavan, and numbered as shown above; that is, the name would be $2R$ (or *S*):$3R$ (or *S*)-(+)-3,3′,4′,5,7-pentahydroxy flavan.

## 23.4 Aryl alkyl ketones

The most general way of preparing aryl alkyl ketones of the type Ar—CO—R is by the Friedel-Crafts reaction of a benzene derivative and an acid halide (usually the chloride) or anhydride, with aluminum chloride as the catalyst. The method is represented by the following typical examples:

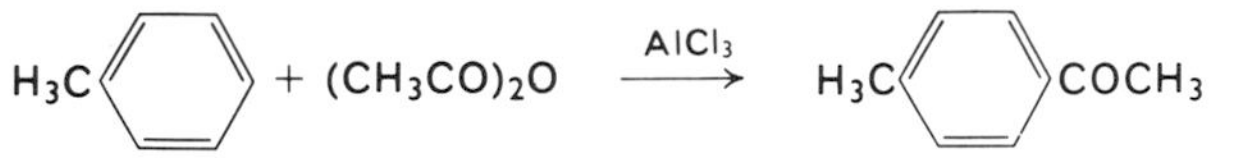

$$H_3C-\!\!\bigcirc\!\!- + (CH_3CO)_2O \xrightarrow{AlCl_3} H_3C-\!\!\bigcirc\!\!-COCH_3 \tag{34}$$

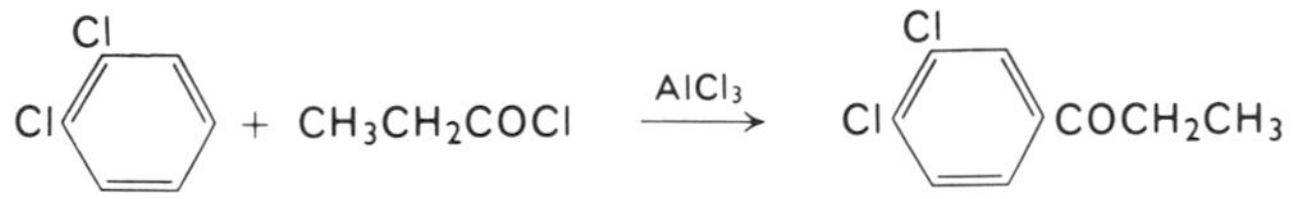

$$Cl_2-\!\!\bigcirc\!\!- + CH_3CH_2COCl \xrightarrow{AlCl_3} Cl_2-\!\!\bigcirc\!\!-COCH_2CH_3 \tag{35}$$

$$H_3C-\!\!\bigcirc\!\!(CH_3) + (CH_3)_2CHCH_2COCl \xrightarrow{AlCl_3} H_3C-\!\!\bigcirc\!\!(CH_3)COCH_2CH(CH_3)_2 \tag{36}$$

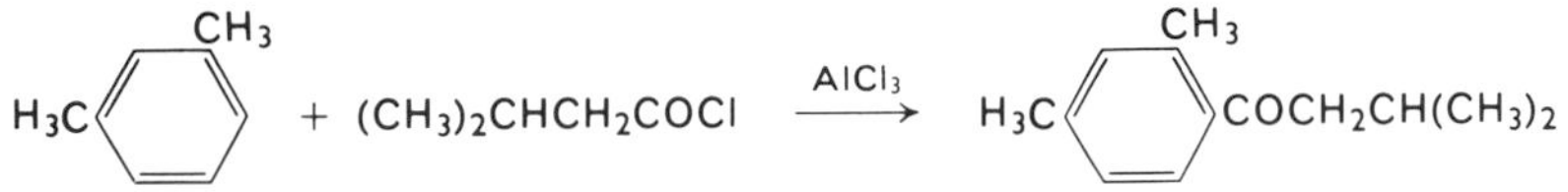

**Exercise 5.** Why does acylation, in contrast to alkylation, stop after the introduction of the first acyl group?

The Friedel-Crafts acylation can also be applied to ring closure:

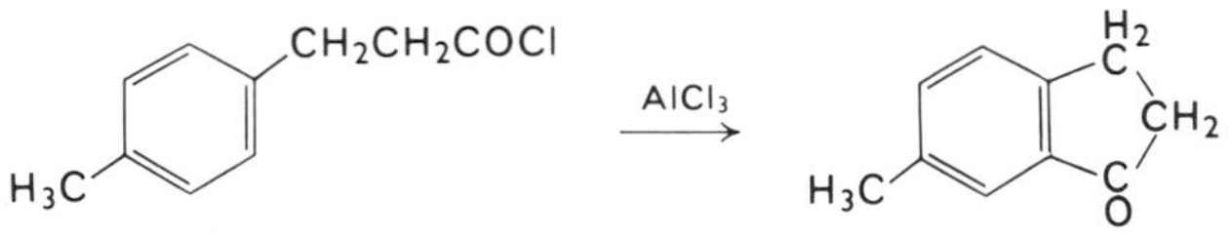

$$\xrightarrow{AlCl_3} \tag{37}$$

6-methyl-1-hydrindone

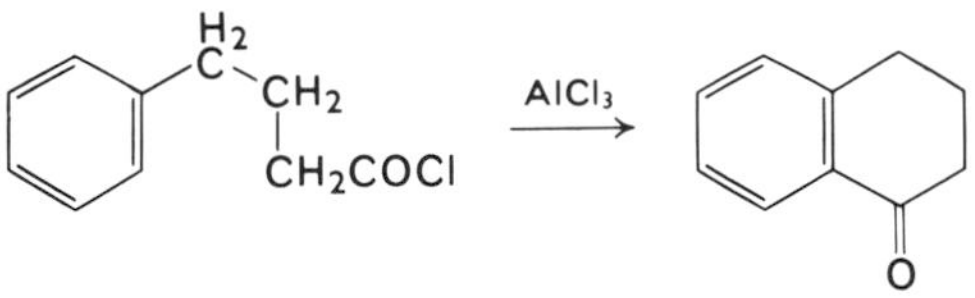

$$\xrightarrow{AlCl_3} \tag{38}$$

1-tetralone

Deactivating (i.e., *meta*-orienting) substituents on the aromatic ring usually render the Friedel-Crafts acylation reaction impracticable. Indeed, nitrobenzene is an excellent solvent for the reaction.

Grignard reagents prepared from aryl halides react with alkyl cyanides (nitriles) to yield addition products which, when hydrolyzed, afford aryl alkyl ketones:

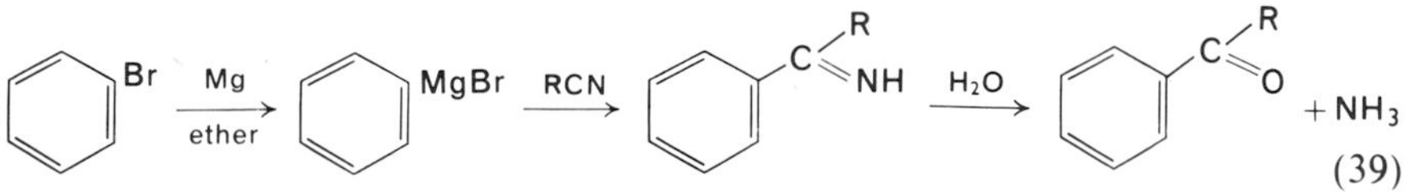

$$\bigcirc\!\!-Br \xrightarrow[ether]{Mg} \bigcirc\!\!-MgBr \xrightarrow{RCN} \bigcirc\!\!-C\!\!\underset{NH}{\overset{R}{<}} \xrightarrow{H_2O} \bigcirc\!\!-C\!\!\underset{O}{\overset{R}{<}} + NH_3 \tag{39}$$

This reaction, like all reactions in which Grignard reagents are required, is limited by the fact that Grignard reagents cannot be prepared from all aromatic halogen compounds. It is most useful for the preparation of ketones containing alkyl or alkoxyl groups as substituents on the ring.

---

**Exercise 6.**   What is an alternative route to aryl alkyl ketones by the use of the Grignard reagent?

---

## 23.5   Reduction of acyl groups to alkyl groups

We have seen earlier that the Friedel-Crafts *alkylation* of the aromatic ring is not generally useful for the introduction of side chains of more than two carbon atoms because isomerization (*n*-propyl to isopropyl, isobutyl to *t*-butyl, and so on) occurs. Alkylbenzenes containing higher alkyl groups of known structure can be prepared by the introduction of an acyl group and subsequent reduction of the $C{=}O$ group to $CH_2$.

---

**Exercise 7.**   In the acylation of the aromatic nucleus with RCOCl, why does not isomerization of R take place?

---

The reduction of the acyl group can be carried out in a number of ways, two of which are widely used. The *Clemmensen* reduction consists in heating the ketone with a mixture of amalgamated zinc and hydrochloric acid. The *Wolff-Kishner* method consists in heating the ketone with hydrazine and a solution of sodium hydroxide, usually in a high-boiling solvent such as diethylene glycol.

Numerous modifications of the experimental conditions for the Wolff-Kishner reduction have been devised, the most useful of which appears to be the use of potassium *t*-butoxide in the dipolar, aprotic solvent dimethyl sulfoxide. The Clemmensen reduction has the disadvantage that the strong hydrochloric acid that is used as the reaction medium can affect acid-sensitive groups that may be present in the molecule; for example, dealkylation of aryl alkyl ethers may occur at the same time that the desired reduction is being brought about (Section 8.14).

$$\text{(40)}$$

The Clemmensen and Wolff-Kishner reductions are applicable to ketones of other classes as well:

$$CH_3CH_2CH_2CH_2CH_2COCH_2CH{\bigg\langle}{\begin{array}{l} CH_3 \\ CH_3 \end{array}} \longrightarrow CH_3CH_2CH_2CH_2CH_2CH_2CH_2CH{\bigg\langle}{\begin{array}{l} CH_3 \\ CH_3 \end{array}}$$

## 23.6 Aromatic aldehydes

The simplest aromatic aldehyde is benzaldehyde, which, it will be recalled, is found in nature in the form of the glycoside of its cyanohydrin. Benzaldehyde is an important industrial chemical, and is prepared commercially by the hydrolysis of benzal chloride (sometimes called benzylidene chloride), which is one of the products of the side-chain chlorination of toluene:

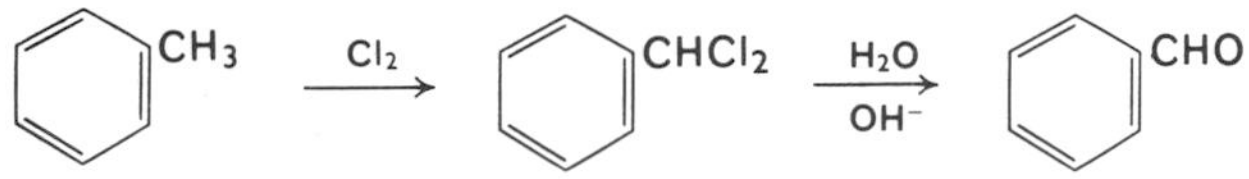

$$\text{C}_6\text{H}_5\text{CH}_3 \xrightarrow{\text{Cl}_2} \text{C}_6\text{H}_5\text{CHCl}_2 \xrightarrow[\text{OH}^-]{\text{H}_2\text{O}} \text{C}_6\text{H}_5\text{CHO} \qquad (41)$$

Among other methods of preparation of aromatic aldehydes are the Rosenmund reduction of aromatic acyl chlorides (Section 14.16), the controlled oxidation of benzyl alcohols (Section 8.24), and the reaction of aromatic Grignard reagents with orthoformic esters. Orthoformic esters are 1,1,1-trialkoxymethanes, an example of which is methyl orthoformate, $\text{CH(OCH}_3)_3$. The reaction of this with a Grignard reagent such as phenylmagnesium bromide takes the following course:

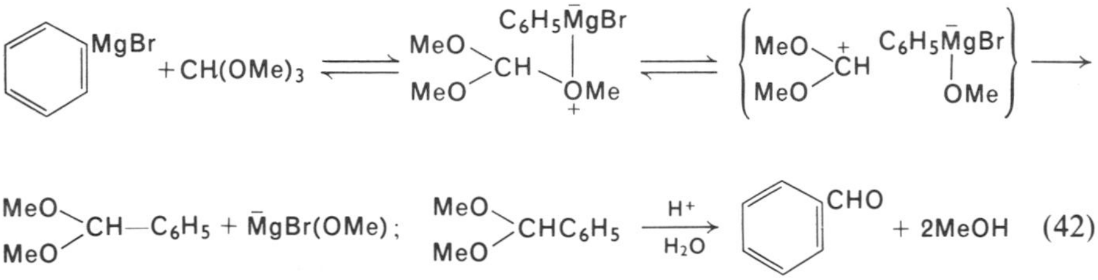

The initial phase of this reaction, the formulation of which shows the probable sequence of stages through which it proceeds, involves the coordination of the Grignard reagent (which is a Lewis acid) with one of the methoxyl groups. Dissociation of the C—O bond is strongly aided by resonance stabilization of the dimethoxycarbonium ion, both oxygen atoms of which can contribute to the delocalization of the positive charge. The shift of the phenyl group, as $\text{C}_6\text{H}_5:^-$, to the electron-deficient carbon atom completes the reaction. The final step is the hydrolysis of the acetal that is formed, a reaction that needs no further comment here.

A method for the direct introduction of the formyl group into the aromatic nucleus is the *Gattermann-Koch* synthesis, in which a mixture of carbon monoxide and hydrogen chloride is used, with aluminum chloride as the catalyst:

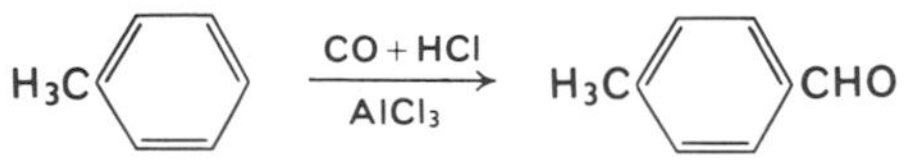

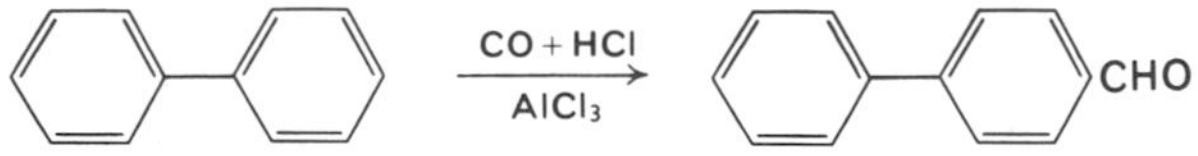

This is an electrophilic substitution reaction, closely allied to the Friedel-Crafts acylation reaction. The protonation of carbon monoxide gives what is clearly the acylium ion that would be formed by abstraction of a chloride ion from the hypothetical formyl chloride. Formyl chloride is a known substance, but is extremely unstable and cannot be isolated and preserved at ordinary temperatures:

$$:C\!=\!\ddot{O}:+HCl(AlCl_3) \longrightarrow H:\overset{+}{C}\!=\!\ddot{O}:\} \ AlCl_4^- \tag{45}$$

$$\left\{H-C\begin{matrix}\nearrow O \\ \searrow Cl\end{matrix} + AlCl_3\right\}$$

## 23.7  Reactions of aromatic aldehydes and ketones. Aldol condensations

Aromatic aldehydes and ketones show most of the characteristic properties of carbonyl compounds, but are generally less reactive than their aliphatic counterparts. They undergo the usual addition reactions with Grignard reagents; they form oximes, semicarbazones, arylhydrazones, and other carbonyl derivatives; they are reducible to the corresponding alcohols; and aromatic aldehydes are readily oxidized to the corresponding carboxylic acids.

Aldol condensations of aromatic carbonyl compounds* are limited, in comparison with aliphatic aldehydes and ketones, by the fact that aryl alkyl ketones undergo carbonyl addition reactions more slowly than their aliphatic counterparts; and aromatic aldehydes, lacking $\alpha$-hydrogen atoms, can react only as the recipient of anionic attack. They cannot undergo aldol self-condensations.

Benzaldehyde, for example, undergoes aldol condensation with acetone to yield benzalacetone or, depending upon the conditions and proportions of the reagents used, dibenzalacetone:

$$\bigcirc\!\!\!-CHO + CH_3COCH_3 \xrightarrow{NaOH} \bigcirc\!\!\!-CH\!=\!CHCOCH_3 \tag{46}$$

benzalacetone

In a corresponding reaction, benzaldehyde condenses with acetophenone to give benzalacetophenone,

$$\bigcirc\!\!\!-CHO + CH_3CO\!-\!\bigcirc \xrightarrow{NaOH} \bigcirc\!\!\!-CH\!=\!CHCO\!-\!\bigcirc \tag{47}$$

benzalacetophenone

---

* The use of the term "aromatic carbonyl compounds" implies that the carbonyl group is attached directly to the aromatic ring, as in acetophenone or benzaldehyde. If the carbonyl group is removed from the ring the properties of the compound are those typical of aliphatic carbonyl compounds and need no special discussion.

and condenses with nitromethane to give $\beta$-nitrostyrene:

$$\langle\text{C}_6\text{H}_5\rangle\text{CHO} + \text{CH}_3\text{NO}_2 \xrightarrow{\text{NaOH}} \langle\text{C}_6\text{H}_5\rangle\text{CH}{=}\text{CHNO}_2 \qquad (48)$$

$\beta$-nitrostyrene

The condensation of benzaldehyde (and substituted benzaldehydes) with malonic acid is a convenient synthetic route to cinnamic acids:

$$\text{Br}\langle\text{C}_6\text{H}_4\rangle\text{CHO} + \text{CH}_2(\text{COOH})_2 \xrightarrow{\text{pyridine}}$$

$$\text{Br}\langle\text{C}_6\text{H}_4\rangle\text{CH}{=}\text{C(COOH)}_2 \xrightarrow{\Delta} \text{Br}\langle\text{C}_6\text{H}_4\rangle\text{CH}{=}\text{CHCOOH} + \text{CO}_2 \qquad (49)$$

## 23.8 The Cannizzaro reaction

Although benzaldehyde does not undergo the aldol condensation with itself, it does react with strong alkali in another way in the *Cannizzaro reaction*. This is a *disproportionation* reaction, in which one molecule of the aldehyde is oxidized, with concomitant reduction of another:

$$2\,\langle\text{C}_6\text{H}_5\rangle\text{CHO} \xrightarrow{\text{NaOH}} \langle\text{C}_6\text{H}_5\rangle\text{COO}^-\text{Na}^+ + \langle\text{C}_6\text{H}_5\rangle\text{CH}_2\text{OH} \qquad (50)$$

Aliphatic aldehydes that lack $\alpha$-hydrogen atoms will also undergo the Cannizzaro reaction. Trimethylacetaldehyde undergoes the reaction, and so does formaldehyde:

$$\text{HCHO} \xrightarrow{\text{NaOH}} \text{HCOO}^-\text{Na}^+ + \text{CH}_3\text{OH} \qquad (51)$$

When formaldehyde and benzaldehyde are mixed and treated with alkali under the conditions that bring about the Cannizzaro reaction, the benzaldehyde undergoes reduction, and the formaldehyde is oxidized. This is called the "crossed" Cannizzaro reaction, and has found use as a means of reducing aromatic aldehydes:

$$\langle\text{C}_6\text{H}_5\rangle\text{CHO} + \text{HCHO} \xrightarrow{\text{NaOH}} \langle\text{C}_6\text{H}_5\rangle\text{CH}_2\text{OH} + \text{HCOONa} \qquad (52)$$

It will be recalled that the preparation of pentaerythritol (Section 13.5) involves a crossed Cannizzaro reaction as the final step. The condensation of acetaldehyde with formaldehyde produces first tris-(hydroxymethyl)-acetaldehyde; this lacks $\alpha$-hydrogen atoms, and cannot undergo further aldol condensation. Instead, it is reduced by the formaldehyde to the tetrahydroxy compound.

Alkali-induced oxidation-reduction reactions that are closely related to the Cannizzaro reaction and involve an intramolecular disproportionation are typified by the reaction of glyoxal with alkali:

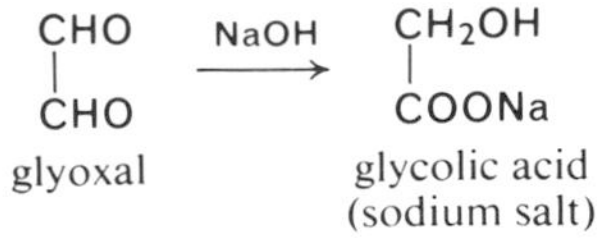

$$\begin{array}{ccc} \text{CHO} & \xrightarrow{\text{NaOH}} & \text{CH}_2\text{OH} \\ | & & | \\ \text{CHO} & & \text{COONa} \\ \text{glyoxal} & & \text{glycolic acid} \\ & & \text{(sodium salt)} \end{array} \tag{53}$$

One interpretation of the Cannizzaro reaction is that it proceeds by way of the transfer of a hydride ion from the anion formed by initial addition of $OH^-$ to the aldehyde:

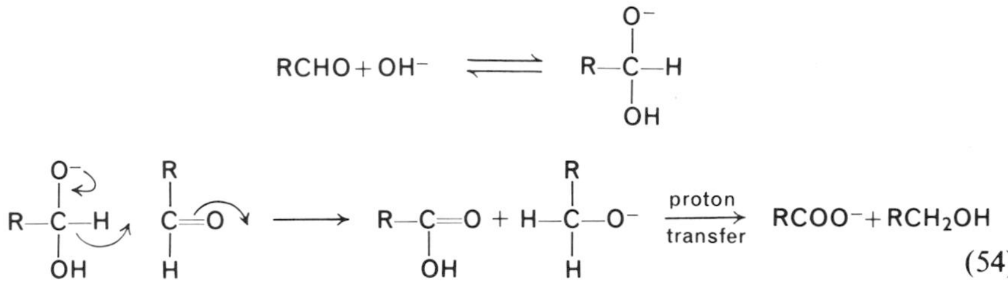

$$\begin{array}{ccc} & & O^- \\ & & | \\ \text{RCHO} + OH^- & \rightleftharpoons & R-C-H \\ & & | \\ & & OH \end{array}$$

$$\begin{array}{c} O^- \quad R \\ | \quad\quad | \\ R-C-H \quad C=O \longrightarrow R-C=O + H-C-O^- \xrightarrow[\text{transfer}]{\text{proton}} RCOO^- + RCH_2OH \\ | \quad\quad | \quad\quad\quad\quad | \quad\quad | \\ OH \quad H \quad\quad\quad OH \quad H \end{array} \tag{54}$$

### 23.9   The benzoin condensation

Two molecules of benzaldehyde condense in the presence of potassium cyanide, with the formation of the hydroxy ketone benzoin:

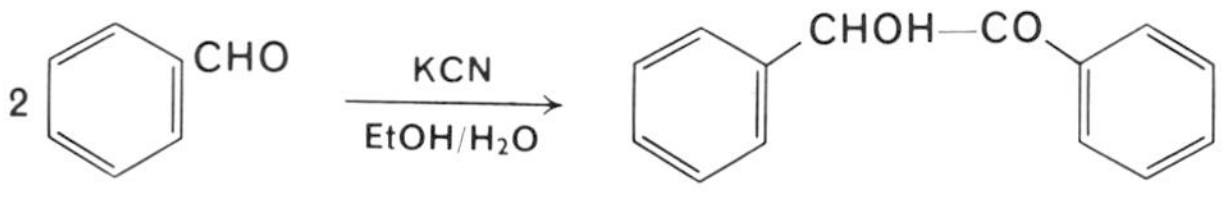

$$2 \quad \text{C}_6\text{H}_5\text{CHO} \xrightarrow[\text{EtOH/H}_2\text{O}]{\text{KCN}} \text{C}_6\text{H}_5\text{—CHOH—CO—C}_6\text{H}_5 \tag{55}$$

benzoin

It will be noticed that the KCN does not appear in the overall reaction; it plays the role of a catalyst for the condensation, and as such, its function is unique and specific. Other bases do not promote the benzoin condensation, and it is therefore evident that cyanide ion does not act simply by virtue of its basic character.

The special role of the cyanide ion can be understood if we follow the course of the reaction step by step.

1. The usual addition of $CN^-$ to benzaldehyde is the first step:

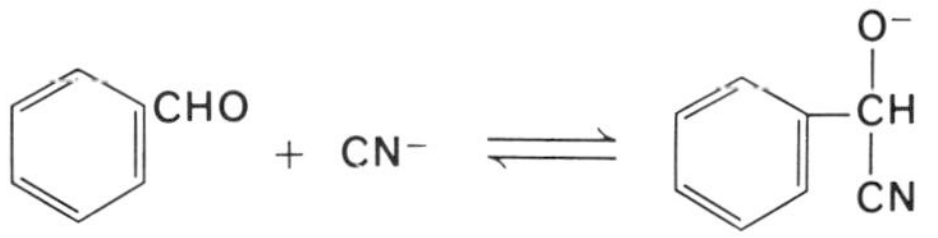

$$\text{C}_6\text{H}_5\text{CHO} + CN^- \rightleftharpoons \begin{array}{c} O^- \\ | \\ \text{C}_6\text{H}_5\text{—CH} \\ | \\ \text{CN} \end{array}$$

2. It will be seen that the addition product formed in this step possesses an $\alpha$-hydrogen atom activated by both the phenyl group and the cyano group. Thus,

this hydrogen atom is sufficiently acidic to permit the following proton exchange to occur:

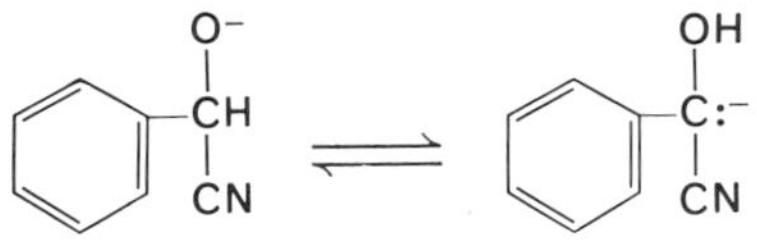

3. The carbon anion so formed can now add to another molecule of benzaldehyde in an aldol-like reaction:

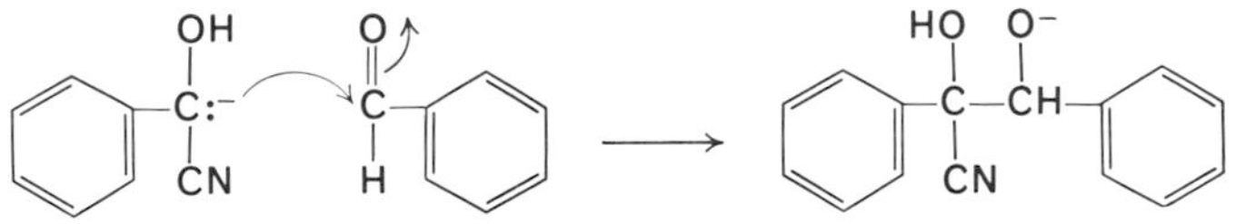

4. This last product is clearly (the anion of) the product that would be obtained by the addition of HCN to benzoin, and, since the addition of HCN to a carbonyl group is a reversible reaction, it will revert to benzoin and $CN^-$:

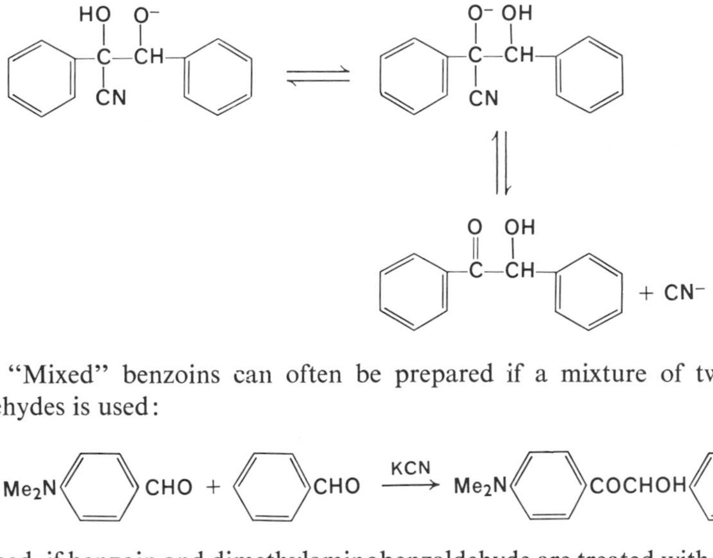

"Mixed" benzoins can often be prepared if a mixture of two aromatic aldehydes is used:

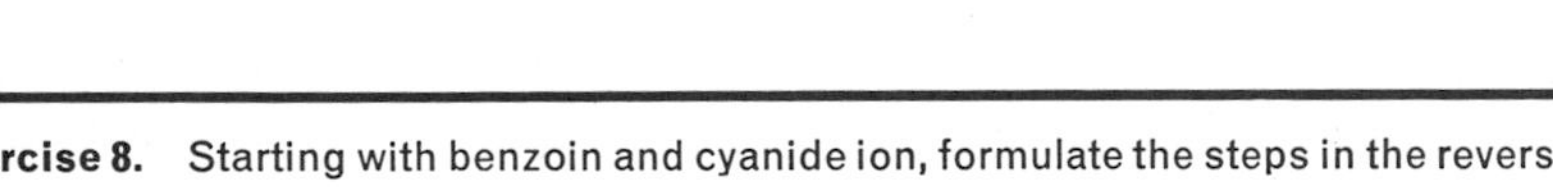

$$\text{(56)}$$

Indeed, if benzoin and dimethylaminobenzaldehyde are treated with an alcoholic solution of KCN, the same mixed benzoin is formed. This indicates that the benzoin condensation is reversible, as the formulation given above would lead us to expect.

---

**Exercise 8.** Starting with benzoin and cyanide ion, formulate the steps in the reversal of the benzoin condensation.

---

## 23.10    The Perkin reaction

The base-catalyzed aldol-type condensation of aromatic aldehydes and acetic anhydride and other acid anhydrides containing $\alpha$-$CH_2$—groups is known as the *Perkin reaction*. It is a useful method for the preparation of $\alpha,\beta$-unsaturated acids, of which cinnamic acid is a typical example:

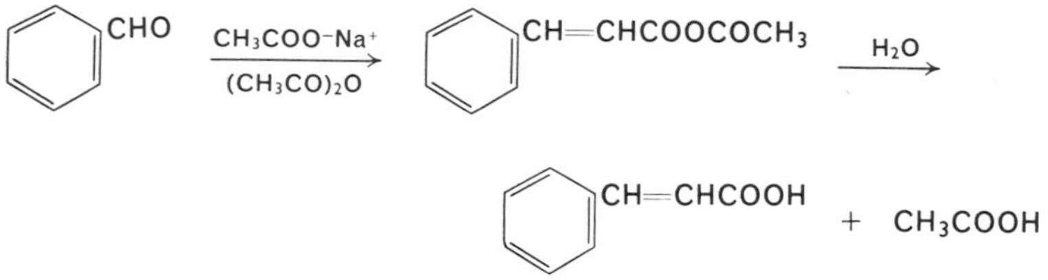

The immediate product, present in the initial reaction mixture, is a complex anhydride, which is hydrolyzed during the manipulations involved in isolating the final product.

The usual basic catalyst for the Perkin condensation is the anhydrous sodium salt of the acid corresponding to the acid anhydride used, but other basic catalysts are also effective. Anhydrous sodium carbonate and trialkylamines have been used with equally satisfactory results.

Interesting variations of the Perkin condensation involve the use of carbonyl compounds other than benzaldehyde. An example is the condensation of phthalic anhydride (which in this instance acts as the carbonyl component in the condensation reaction), acetic anhydride, and phenylacetic acid, with the use of sodium acetate as the basic catalyst:

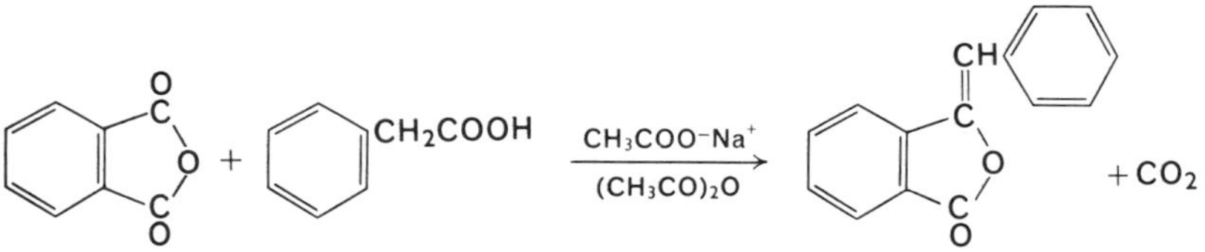

benzalphthalide        (58)

Decarboxylation, concomitant with the condensation, yields benzalphthalide as the product.

## 23.11    Applications to synthesis

Some examples of synthetic sequences in which some of the above-described methods are employed are shown on the opposite page. It is suggested that a sequence like that in example (60) but starting with *p*-xylene, be worked out by the reader. The product will be 1,4,5-trimethylnaphthalene.

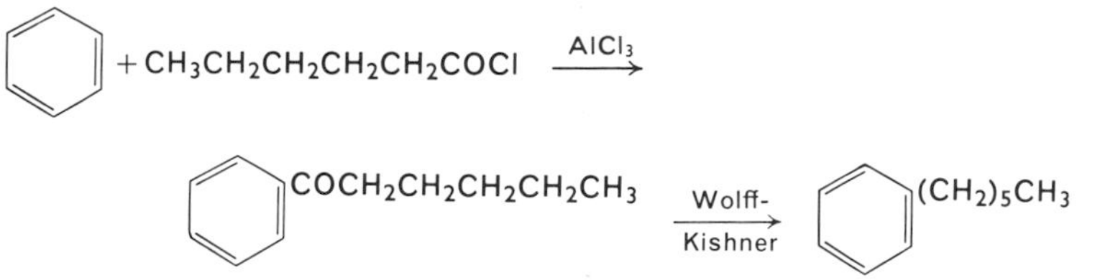

phenyl *n*-pentyl ketone

*n*-hexylbenzene
(1-phenylhexane (G))

(59)

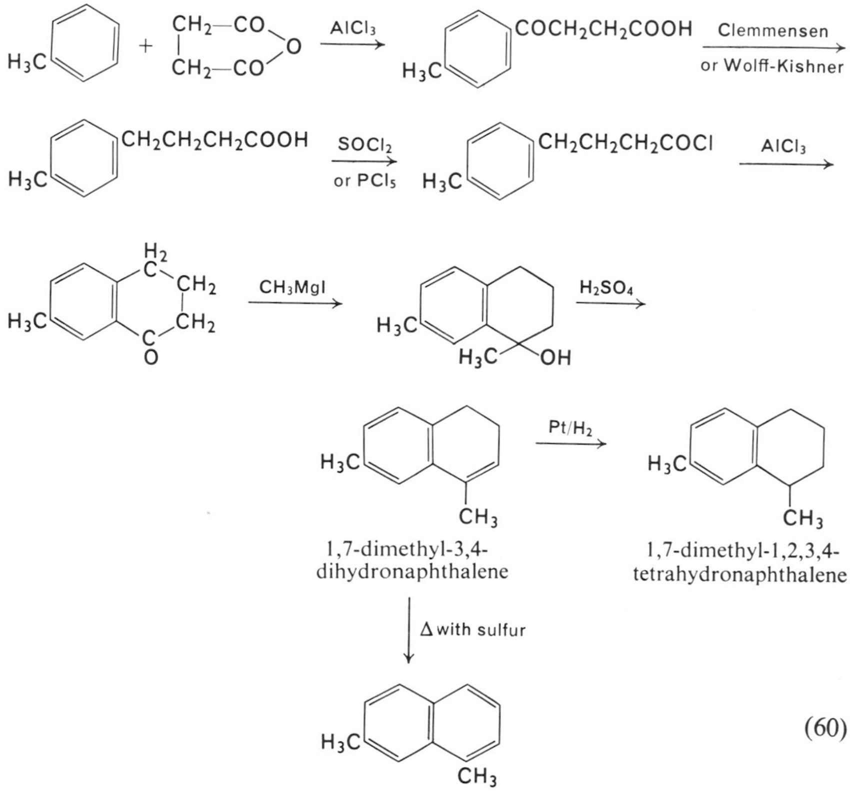

1,7-dimethyl-3,4-dihydronaphthalene

1,7-dimethyl-1,2,3,4-tetrahydronaphthalene

(60)

1,7-dimethylnaphthalene

The introduction of the acetyl group by the Friedel-Crafts acylation reaction, followed by oxidation of the acetyl group (e.g., by the haloform reaction), is a useful method for the preparation of carboxylic acids:

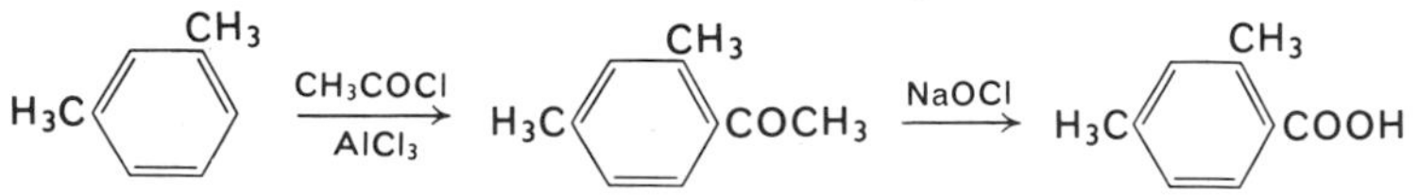

(61)

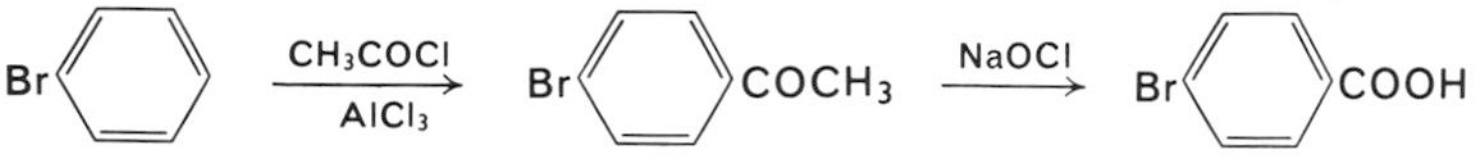

(62)

Another useful synthetic procedure is *chloromethylation.* Reactive aromatic hydrocarbons, such as mesitylene, when treated with formaldehyde and hydrochloric acid, yield chloromethyl derivatives, according to the following equation:

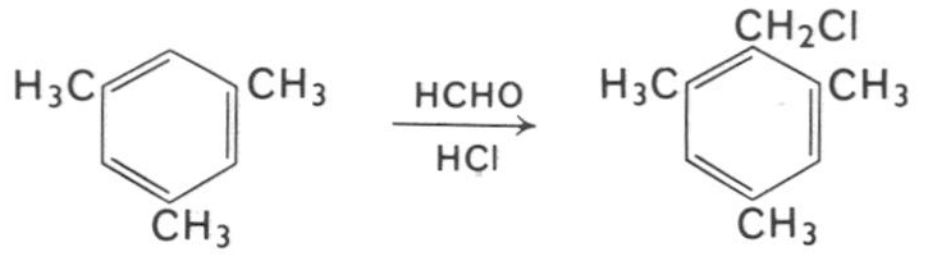

The reader should show that this is a typical electrophilic substitution reaction, by formulating its steps in detail. Hint: What is the structure of protonated formaldehyde?

### Exercises

(Exercises 1–8 will be found within the text.)

**9** Show how the following transformations can be effected: *(a)* p-xylene to 2,5-dimethylacetophenone; *(b)* toluene to p-xylene; *(c)* acetophenone to 1,4-diethylbenzene; *(d)* benzoic acid to 1-hydrindone (1-indanone); *(e)* phenylacetone to 2-benzylhexane (1-phenyl-2-methylhexane).

**10** An optically active compound (A), $C_{10}H_{12}$, could be oxidized by hot aqueous $KMnO_4$ to phthalic acid (benzene-1,2-dicarboxylic acid). What is a possible structure for A?

**11** A compound (A), $C_{10}H_{10}O$, did not react with bromine or $KMnO_4$ in the cold, and formed an oxime and a semicarbazone but not a bisulfite addition compound. It reacted with methylmagnesium iodide to yield B, $C_{11}H_{14}O$. Treatment of B with hot 15% sulfuric acid yielded C, $C_{11}H_{12}$. Compound C was unsaturated and when ozonized yielded D, $C_{11}H_{12}O_2$. Compound D readily lost the elements of water when treated with sodium ethoxide in ethanol, and formed E, $C_{11}H_{10}O$. Vigorous oxidation of A, B, C, D, or E yielded phthalic acid. Write structures for A–E that account for the above results.

**12** An aromatic hydrocarbon $C_7H_8$ reacted with chlorine at 100°C in the presence of light to yield a monochloro derivative, $C_7H_7Cl$. The chloro compound reacted readily with potassium cyanide to give $C_8H_7N$, and hydrolysis of the latter with sulfuric acid-water gave an acid, $C_8H_8O_2$. Vigorous oxidation of the acid gave a new acid, $C_7H_6O_2$. Write these reactions.

**13** How could you distinguish between the compounds in each of the following pairs? (Use chemical tests, not differences in physical properties such as boiling point.) *(a)* 1-methylcyclohexene and toluene; *(b)* p-bromotoluene and benzyl bromide; *(c)* phenylacetone and ethyl phenyl ketone; *(d)* bromodurene and p-bromo-n-butylbenzene; *(e)* p-methylacetophenone and p-ethyltoluene.

# Chapter twenty-four / Aromatic Nitro Compounds

## 24.1 Preparation of aromatic nitro compounds

Aromatic amines and nitro compounds form a large class of compounds of inestimable importance in organic chemistry. Nitro compounds have many direct uses of their own, but their chief importance is in their ready formation by direct nitration of benzene derivatives and the ease with which they can be reduced to substituted anilines and other aromatic amines. The coal-tar dye industry is based principally upon the aromatic amines, and is so called because the hydrocarbons found in coal tar are converted into the many final products through nitration and reduction to aromatic amines. The great technical importance of these compounds is due largely to the abundance of low-cost starting materials (among them benzene, toluene, xylenes, phenols, naphthalene) and to the ease with which these raw materials can be converted into an enormous array of dyes, pharmaceuticals, vitamins, polymers, and other products indispensable to the convenience, pleasure, and well-being of mankind.

Aromatic nitro compounds are usually prepared by direct nitration. This reaction has been discussed in detail in an earlier chapter (Chapter 20). Because of the *meta*-directing effect of the nitro group, di- and trinitration usually lead to compounds in which the nitro groups are 1,3- and 1,3,5- with respect to one another, as shown in the formula on the next page.

The 2,3,4- and 2,4,5-isomers can be removed easily, since the labile (3- in one, 5- in the other) nitro groups are readily replaced by —$SO_3Na$ by treatment with sodium sulfite to form the water-soluble sodium sulfonates.

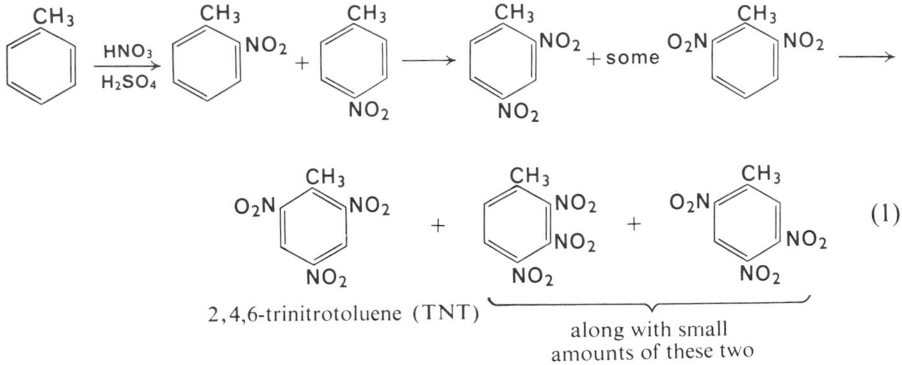

$$2,4,6\text{-trinitrotoluene (TNT)} \qquad \text{along with small amounts of these two} \tag{1}$$

---

**Exercise 1.**  Formulate these replacement reactions and explain their nature.

---

1,3,5-Trinitrobenzene (TNB) cannot be prepared practically by direct nitration of benzene. It can be prepared by decarboxylation of the acid that is formed from 2,4,6-trinitrotoluene by oxidation (Section 23.3).

Phenols, which are very reactive in electrophilic substitution, are nitrated with ease, and dilute (aqueous) nitric acid converts phenol into a mixture of o- and p-nitrophenols. These compounds are readily separated by steam distillation, the success of this operation depending upon the fact that o-nitrophenol is volatile with steam, whereas p-nitrophenol is not (Section 20.3).

Further nitration of partially nitrated phenol leads to the formation of *picric* acid (2,4,6-trinitrophenol). Another important trinitrophenol is *styphnic* acid, 2,4,6-trinitroresorcinol:

picric acid                     styphnic acid

Nitro compounds can also be prepared by methods which do not involve direct nitration. A useful procedure is the oxidation of an aromatic amino group to a nitro group. For example, the preparation of 2,5-dinitrobenzoic acid is carried out by the following reactions:

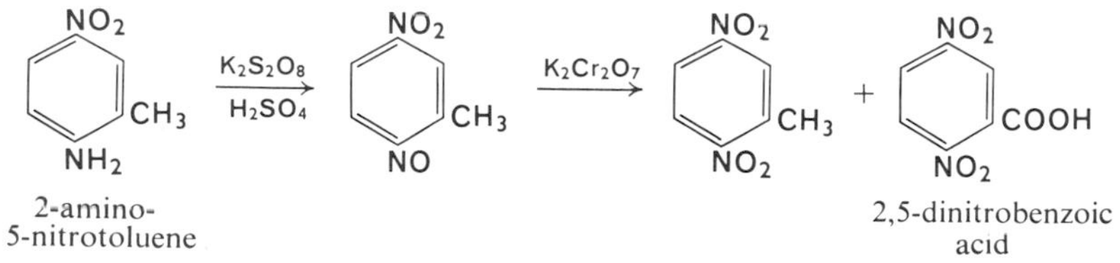

The direct oxidation of —NH$_2$ to —NO$_2$ can be accomplished by the use of peroxytrifluoroacetic acid (F$_3$C·CO$_3$H):

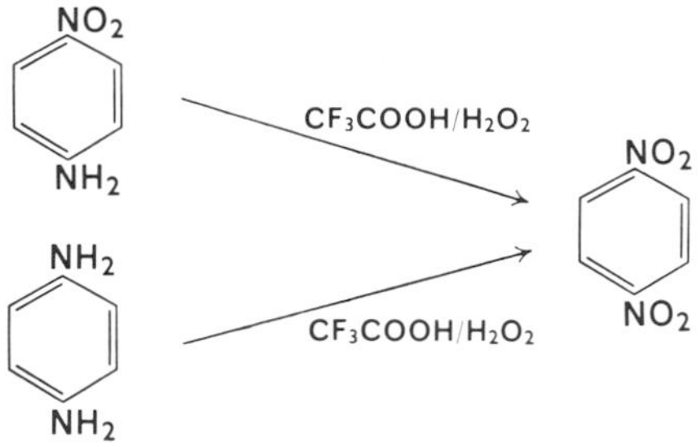

(3)

(86-89%) (4)

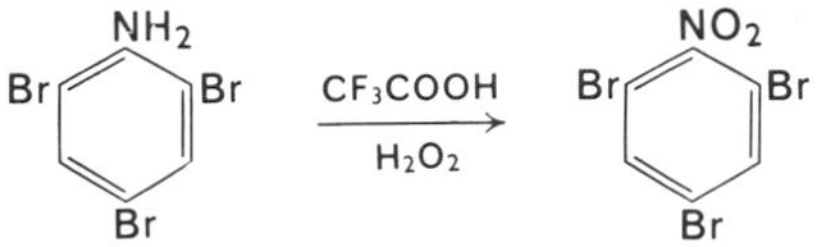

(100%) (5)

---

**Exercise 2.** Devise a synthesis for 3-bromo-4-nitrotoluene from *p*-toluidine.

---

The replacement of an amino group by a nitro group can also be accomplished by diazotization (Chapter 26) and treatment of the diazonium fluoborate with sodium nitrite and copper powder. *p*-Dinitrobenzene can be prepared from *p*-nitroaniline by this method. It can be seen from the above examples that the conversion of an amino group into a nitro group permits the synthesis of nitro compounds that cannot be prepared by direct nitration.

## 24.2 Uses of aromatic nitro compounds

The tendency for compounds containing one or more nitro groups to form well-defined crystalline solids makes many nitro compounds useful as reagents for the preparation of derivatives used for the characterization and identification of organic substances:

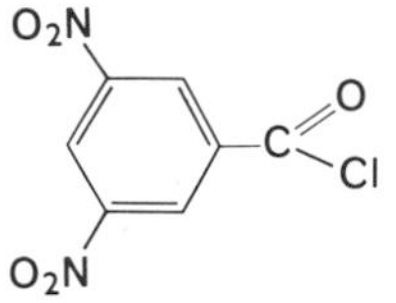

+ alcohols ⟶ alkyl 3,5-dinitrobenzoates (6)

(methyl 3,5-dinitrobenzoate, m.p. 107°C)

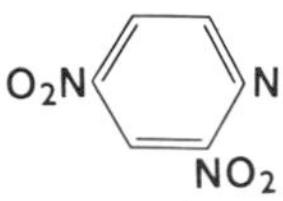

3,5-dinitrobenzoyl chloride

O$_2$N⟨ ⟩NHNH$_2$ + aldehydes, ketones ⟶ 2,4-dinitrophenylhydrazones (7)

(acetone 2,4-dinitrophenylhydrazone, m.p. 128°C)

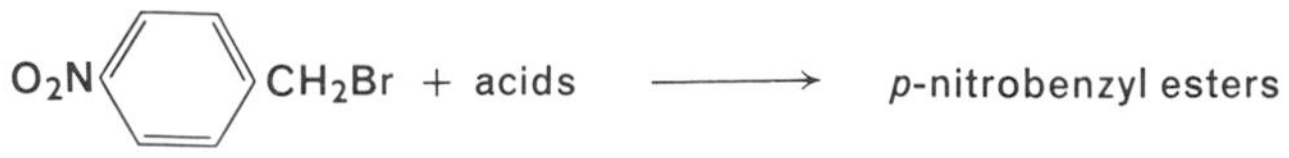

$$O_2N-\langle\ \rangle-CH_2Br\ +\ acids\ \longrightarrow\ \textit{p}\text{-nitrobenzyl esters} \qquad (8)$$

(as sodium salts)     (p-nitrobenzyl acetate, m.p. 78°C)

The nitration of aromatic hydrocarbons is useful for identification purposes, for di- and poly-nitro derivatives have relatively high melting points: *m*-dinitrobenzene melts at 90°C, 2,4-dinitrotoluene at 70°C, 2,4,6-trinitro-*m*-xylene at 182°C.

The use of 2,4-dinitrofluorobenzene for the structure proof of proteins and polypeptides has been mentioned earlier (Section 22.5).

An application of nitro compounds that is of special interest is found in the use of 3-methoxy-2,6-dinitro-4-*t*-butyltoluene, 2,4,6-trinitro-1,3-dimethyl-5-*t*-butylbenzene, and several other related compounds in perfumery as "synthetic musks":

---

**Exercise 3.** Suggest a synthesis of 2,4,6-trinitro-1,3-dimethyl-5-*t*-butylbenzene, starting from *m*-xylene.

---

The *n*-propyl ether of 2-amino-4-nitrophenol has the remarkable property of being 4,000 times sweeter than cane sugar.

Nitro compounds are among the most widely used explosives. TNT, 1,3,5-trinitrobenzene, picric acid, trinitroxylene, N,2,4,6-tetranitro-N-methylaniline are high explosives of great technical and military importance. Picric acid and its salts were formerly used as industrial and military explosives, but have been largely superseded by newer and more satisfactory explosives.

### 24.3 Picrates and styphnates: complex formation with polynitro compounds

An important use of picric and styphnic acids in laboratory practice is in the preparation of the *picrates* and *styphnates* of amines. These compounds are salts; their value lies in the fact that they are usually nicely crystalline compounds and are very useful for the isolation, purification, and identification of amines:

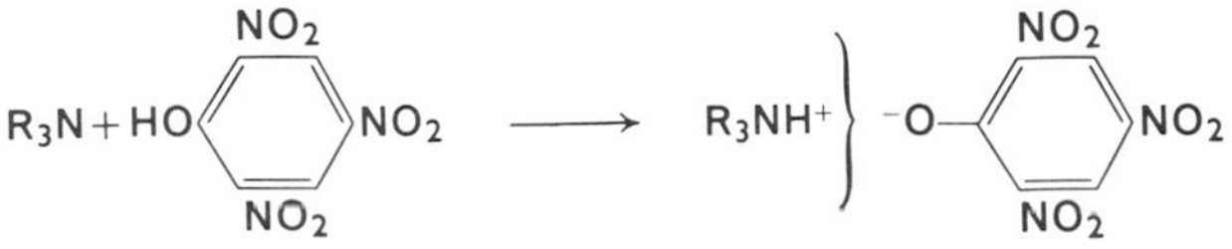

Crystalline "picrates" are also formed from the reaction of picric acid with polynuclear aromatic hydrocarbons and their derivatives. These are molecular complexes, not salts of the kind that are formed with amines. They possess definite compositions (usually one mole of picric acid per mole of hydrocarbon), and find use in the characterization of such compounds as naphthalene and phenanthrene derivatives and other polycyclic aromatic compounds:

phenanthrene picrate

(10)

(11)

Picrates such as those of phenanthrene and naphthalene derivatives have a special usefulness in addition to their value as crystalline derivatives of definite and characteristic melting points: the picric acid that they contain can be titrated with standard alkali, with the important result that an equivalent weight can be obtained for the complex. Since the molecular weight of picric acid is known, the equivalent (or molecular) weight of the hydrocarbon can be obtained.

The ability of picric acid to form complexes of this kind is not dependent upon the presence of the phenolic hydroxyl group, for trinitrobenzene forms similar complexes.

The exact nature of the complexes formed between hydrocarbons and polynitro compounds has been the subject of much discussion and speculation. The following experimental facts must be taken into consideration: the bonding energy between the two molecules in the complex is small, and the complexes are easily dissociable; most of the complexes are deeply colored, many of them brilliantly so; and the formation of the complexes is facilitated by increasing number of nitro groups in the polynitro compound, and by increasing basicity in the hydrocarbon. Hexamethylbenzene forms a crystalline complex with 1,3,5-trinitrobenzene, but benzene and toluene do not. It appears that there is some charge transfer between the two members of the complexing pair, not to an extent that constitutes covalent bonding, but to a greater extent than can be ascribed to simple electrostatic interaction between polarized molecules. Complexes of this kind, and others of related types that are formed between strongly electrophilic and strongly nucleophilic aromatic compounds, are usually referred to as "charge transfer complexes" or "$\pi$-complexes."

## 24.4   Activation of the methyl group by the nitro group

The methyl group of *o*- and *p*-nitrotoluenes is an "active" methyl group with properties comparable to those of methyl groups activated by carbonyl and cyano groups in aliphatic compounds. This property of the methyl group is greatly enhanced in 2,4-dinitro- and 2,4,6-trinitrotoluene.

   *o*-Nitrotoluene can be acylated with oxalic ester under the influence of alkali metal alkoxides:

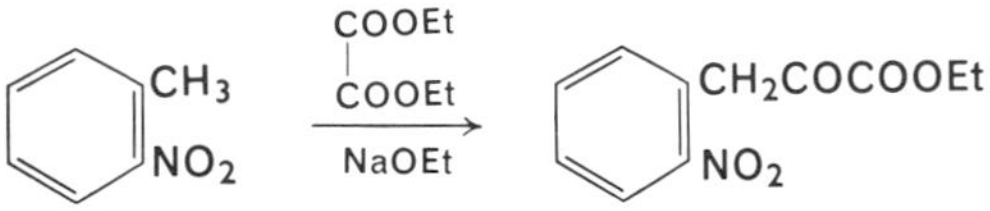

$$\text{(12)}$$

ethyl o-nitrophenylpyruvate

*p*-Nitrotoluene undergoes a similar reaction.

   With the additional activation furnished by two or three nitro groups, the methyl group is capable of undergoing a base-catalyzed condensation of the aldol type with aldehydes and aromatic nitroso compounds:

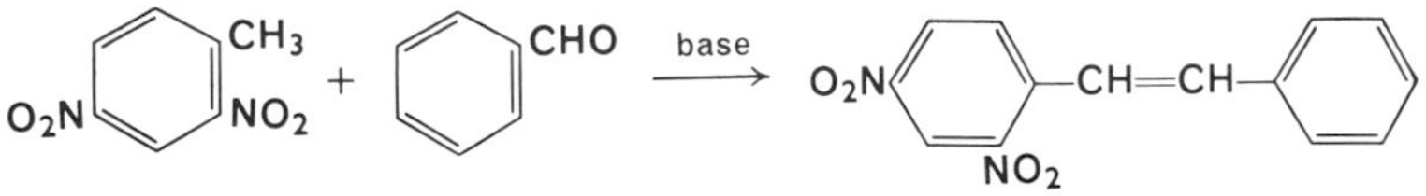

$$\text{(13)}$$

2,4-dinitrostilbene

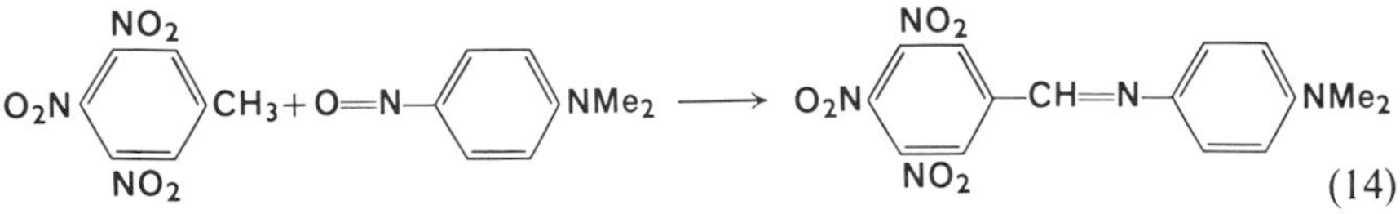

$$\text{(14)}$$

   The enhancement of the acidity of the methyl group in the nitrotoluenes is related to the resonance stabilization of the anion that is formed by the removal of a proton:

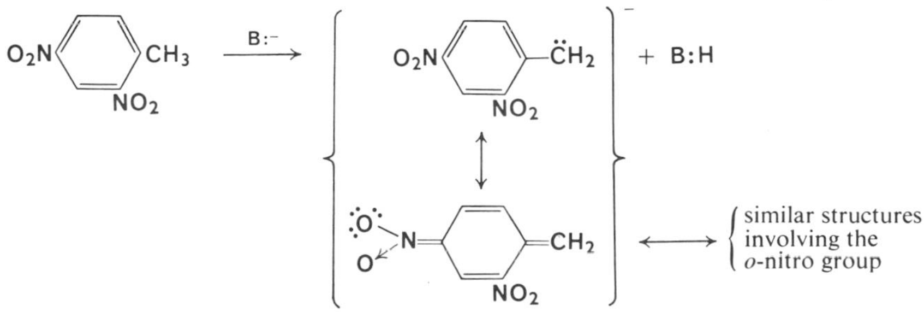

similar structures
involving the
*o*-nitro group

---

**Exercise 4.**   Write the structures involving the *o*-nitro group in the above anion.

---

In the condensations with benzaldehyde, *p*-nitrosodimethylaniline, and ethyl oxalate, the addition reaction involves the carbonyl, or nitroso, group and the anion shown above. For the condensation of 2,4-dinitrotoluene with a nitroso compound,

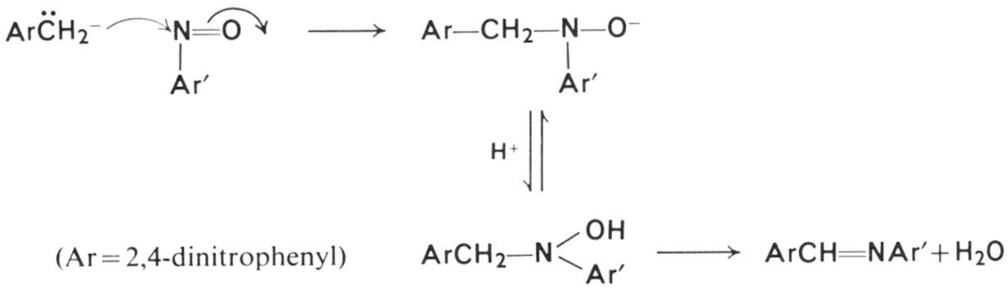

$$(Ar = 2,4\text{-dinitrophenyl}) \quad ArCH_2{-}N{\overset{OH}{\underset{Ar'}{\diagdown}}} \longrightarrow ArCH{=}NAr' + H_2O \quad (15)$$

These condensations are clearly analogues of the related condensations of the aldol type.

The condensation product of 2,4,6-trinitrotoluene and *p*-nitrosodimethylaniline, shown earlier in this section, is a Schiff's base. It is evident that it is the Schiff's base that would be formed by the condensation of 2,4,6-trinitrobenzaldehyde and *p*-dimethylaminoaniline, and thus that it can be hydrolyzed to these compounds:

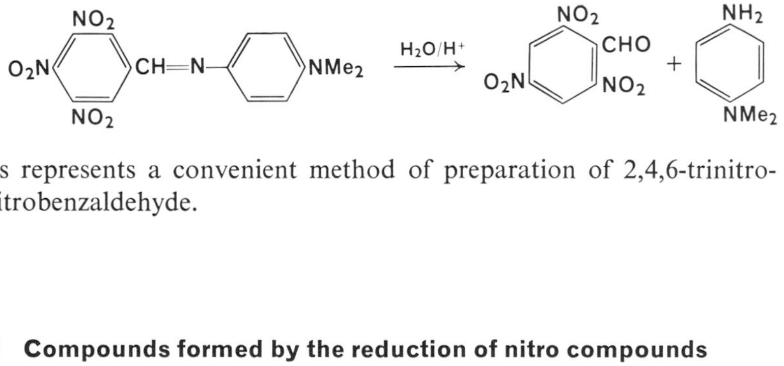

$$(16)$$

This represents a convenient method of preparation of 2,4,6-trinitro- or 2,4-dinitrobenzaldehyde.

## 24.5  Compounds formed by the reduction of nitro compounds

The nitro group represents the highest, the amino group the lowest, oxidation state of nitrogen attached to carbon. Between these extremes two others are known, as shown in the following examples of the benzene derivatives:

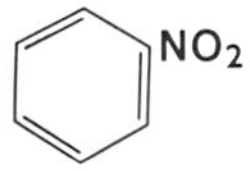

nitrobenzene 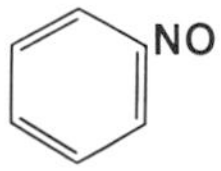          nitrosobenzene 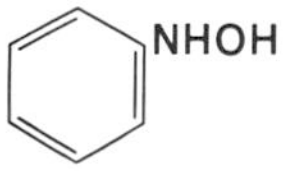          phenylhydroxylamine

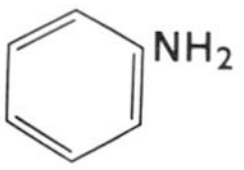

aniline

## 24.6 Aromatic nitroso compounds

The reduction of nitrobenzene to nitrosobenzene cannot be carried out by generally practical means: under certain conditions nitroso compounds are formed by the action of reducing agents on the corresponding nitro compounds, but the yields are poor because of the ease with which further reduction occurs, and the reaction is of very limited value. Nitrosobenzene is best prepared by the oxidation of phenylhydroxylamine, or by the oxidation of aniline; other halogen- and nitro-substituted anilines can be oxidized to the corresponding nitroso compounds.

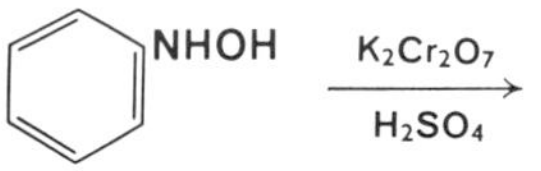

$$\text{(17)}$$

Nitrosobenzene has the interesting property of being a colorless, crystalline solid but of melting to a green liquid and forming green solutions. The colorless solid is a dimeric form; the green liquid is a monomeric form. This is a property of the nitroso group, since the tertiary nitroso compound, $Me_3CNO$, behaves similarly; it is colorless in the solid state, blue in the liquid state (melted).

Aromatic nitroso compounds undergo condensation reactions with primary amines and with active methylene compounds. The latter reaction was described in Section 24.4, where it is represented by the condensation of di- and tri-nitrotoluenes with $p$-nitrosodimethylaniline. The reaction of nitrosobenzene with aniline gives azobenzene:

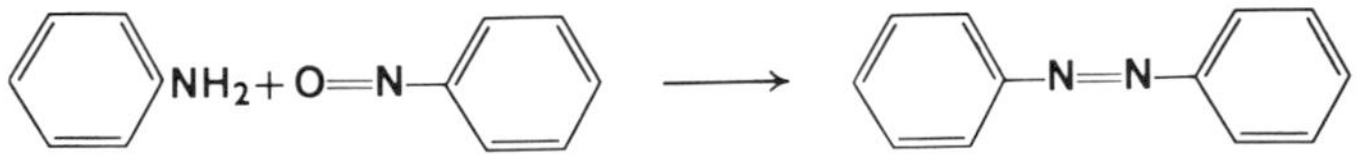

$$\text{(18)}$$

azobenzene

The behavior of the nitroso group in reactions of this kind resembles that of the carbonyl group in the formation of Schiff's bases; e.g.,

$$\text{(19)}$$

The capacity of the nitroso group to undergo addition reactions with amines can be represented by the partial equation

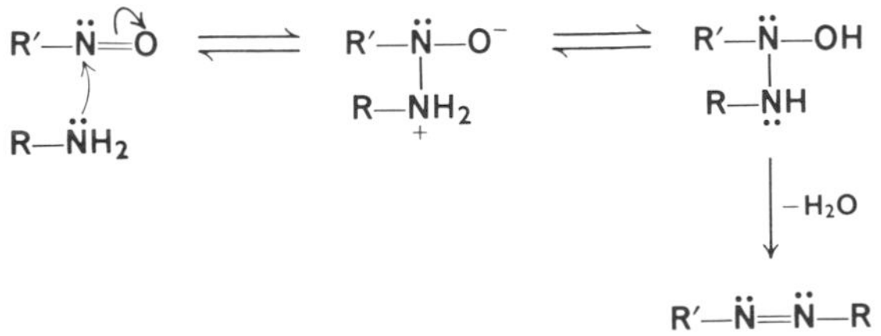

$$\text{(20)}$$

and is a manifestation of the same property that is shown in the reaction of nitroso compounds with active methylene compounds, exemplified by (15) and expressed generally by the following partial equation:

$$\text{\textgreater}CH_2 + -N{=}O \longrightarrow \text{\textgreater}C{=}N{-} \xrightarrow{H_2O} \text{\textgreater}C{=}O + -NH_2$$

active
methylene

An inspection of these equations discloses that the overall result is an oxidation of the active methylene group to a carbonyl group, with concomitant reduction of $-NO$ to $-NH_2$. The condensation of 2,4,6-trinitrotoluene with *p*-nitrosodimethylaniline, and the hydrolysis of the condensation product to 2,4,6-trinitrobenzaldehyde, described in Section 24.4, are examples of this oxidation process.

---

**Exercise 5.** Account for the fact that hydrolysis of the $\text{\textgreater}C{=}N{-}$ linkage gives $\text{\textgreater}CO$ and $-NH_2$ rather than $\text{\textgreater}CH_2$ and $-NO$.

---

## 24.7 Direct nitrosation of phenols and amines

The most commonly encountered nitroso compounds are those formed by the direct nitrosation of the aromatic nucleus. This reaction, an electrophilic substitution, requires a sufficiently activated aromatic ring and is useful chiefly for phenols and amines having free *o*- or *p*-positions. The reaction is carried out by acidifying a solution containing the phenol and sodium nitrite, or by adding sodium nitrite to a solution of the amine hydrochloride:

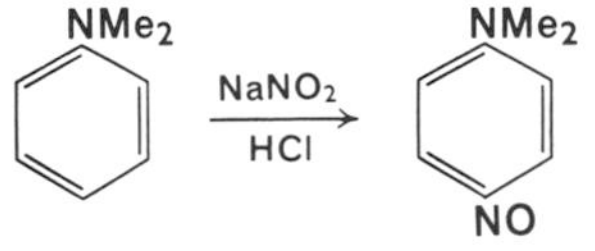

$$\tag{21}$$

*p*-nitrosodimethylaniline

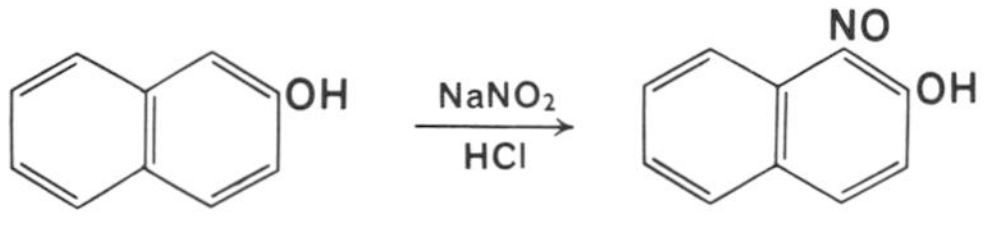

$$\tag{22}$$

β-naphthol · · · · · · 1-nitroso-2-naphthol

Nitrosophenols are of special interest because they exist as mixtures of the tautomeric forms shown in the following example of *p*-nitrosophenol:

*p*-nitrosophenol · · · · *p*-benzoquinone monoxime

Indeed, the same compound is obtained by the nitrosation of phenol and by the reaction of hydroxylamine with *p*-benzoquinone:

$$O=\overset{}{\underset{}{\bigcirc}}=O \quad \xrightarrow{NH_2OH} \quad \left\{ \begin{array}{l} O=\overset{}{\underset{}{\bigcirc}}=NOH \qquad\qquad (23) \\ \\ \Updownarrow \\ \\ HO\text{-}\overset{}{\underset{}{\bigcirc}}\text{-}N=O \qquad\qquad (24) \end{array} \right.$$

$$HO\text{-}\overset{}{\underset{}{\bigcirc}} \quad \xrightarrow{HNO_2} \quad$$

The interconversion of the nitrosophenol and the quinone monoxime can be understood when it is recognized that the anion that would be formed by ionization of the proton from either of these is the hybrid ion:

$$HO\text{-}\overset{}{\underset{}{\bigcirc}}\text{-}N=O \quad \xrightarrow{-H^+} \quad \left\{ ^-O\text{-}\overset{}{\underset{}{\bigcirc}}\text{-}N=O \quad \longleftrightarrow \quad O=\overset{}{\underset{}{\bigcirc}}=N\text{-}O^- \right\}$$

$$\Big\uparrow {\scriptstyle -H^+}$$

$$O=\overset{}{\underset{}{\bigcirc}}=N\text{-}OH \qquad\qquad (25)$$

Recapture of a proton by this ion will give the equilibrium mixture of the two tautomers.

## 24.8   Arylhydroxylamines

Phenylhydroxylamine, the next lower reduction state from nitrosobenzene, can be prepared by the reduction of nitrobenzene with zinc in a buffered medium (ammonium chloride solution). It is probable that nitrosobenzene is the first reduction product formed from nitrobenzene and is so easily reduced that the reduction proceeds further under nearly all conditions.

Since phenylhydroxylamine can be oxidized by an ammoniacal solution of silver nitrate (Tollens' reagent), with the formation of a silver mirror (or black metallic silver), the reduction of nitro compounds to arylhydroxylamines, followed by testing with Tollens' reagent, serves as a useful *diagnostic test for the nitro group*.

Di- and trinitrophenylhydroxylamines can be prepared by treatment of the appropriate chloro or methoxy compounds with hydroxylamine; for example:

$$\underset{NO_2}{\overset{OCH_3}{\underset{}{\bigcirc}}}\overset{NO_2}{} \quad \text{or} \quad \underset{NO_2}{\overset{Cl}{\underset{}{\bigcirc}}}\overset{NO_2}{} \quad \xrightarrow{H_2NOH} \quad \underset{NO_2}{\overset{NHOH}{\underset{}{\bigcirc}}}\overset{NO_2}{} \qquad\qquad (26)$$

Arylhydroxylamines have little general usefulness.

## 24.9 Amines

The most important reduction products of aromatic nitro compounds are the amines. These are the products of the action of most reducing agents on nitro compounds. Because of the great importance of the aromatic amines, many methods of carrying out the reductions of which they are the products have been discovered. The following equations describe a number of important technical and laboratory methods:*

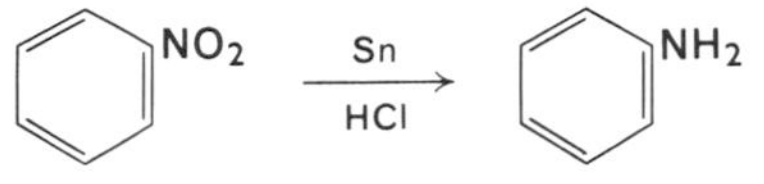

$$\text{(27)}$$

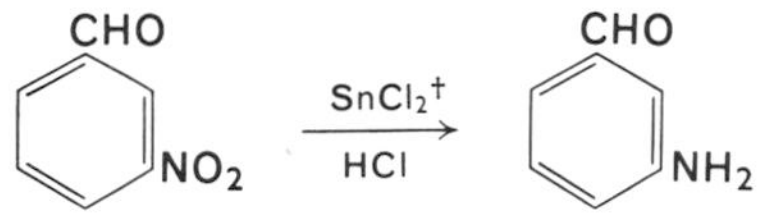

$$\text{(28)}$$

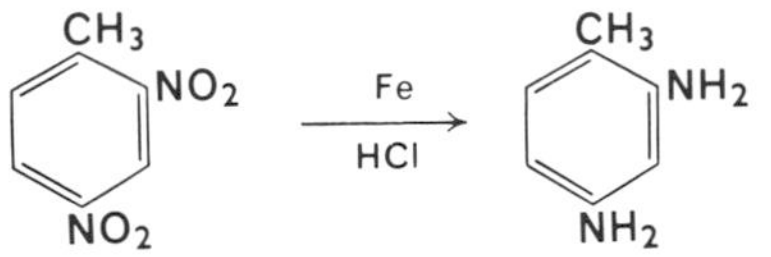

$$\text{(29)}$$

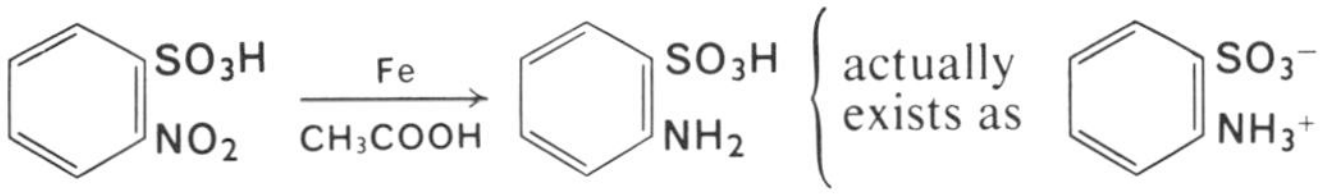

$$\text{(30)}$$

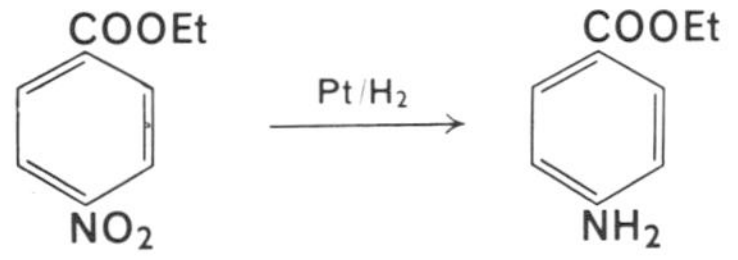

$$\text{(31)}$$

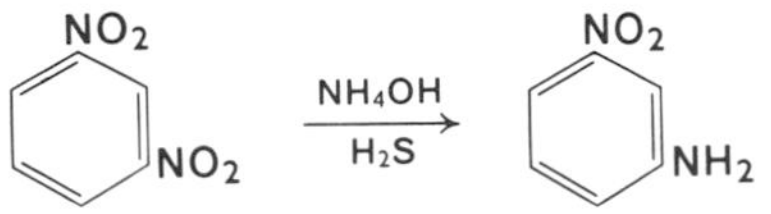

$$\text{(32)}$$

The technical reduction of nitrobenzene is carried out on a large scale, and, as a matter of economy, an inexpensive reducing agent must be used. The reduction is carried out technically with iron (scrap and waste iron) in the presence of a small amount of HCl. The effective reagents are thus iron and water, the products aniline and oxides of iron.

---

* The procedures described in the equations are not mutually exclusive, with the exception of the last, which is a special method of reducing one nitro group of a pair. The examples given are those of specific preparations, in which the choice of the reducing agent was governed by the particular considerations of convenience, yield, and availability of the starting material.

† The use of stannous chloride is often advantageous, since the theoretically required amount of reducing agent can be somewhat more easily used than when metallic tin is employed. When metallic tin is used in excess, the tin is oxidized chiefly to the stannous state, with the result that the extent to which the possible reducing power of the metal ($Sn \rightarrow 4e + Sn^{4+}$) is utilized may be uncertain and not subject to precise control.

## 24.10 Reduction of nitro compounds under alkaline conditions

The reduction of nitrobenzene in an alkaline medium leads to different results. The intermediate reduction products (nitrosobenzene and phenylhydroxylamine), in the presence of alkali, engage in condensation reactions with one another, with nitrobenzene, and with aniline to form "bimolecular" reduction products containing two aromatic rings and two nitrogen atoms.

The course of these reductions is summarized in the following chart:

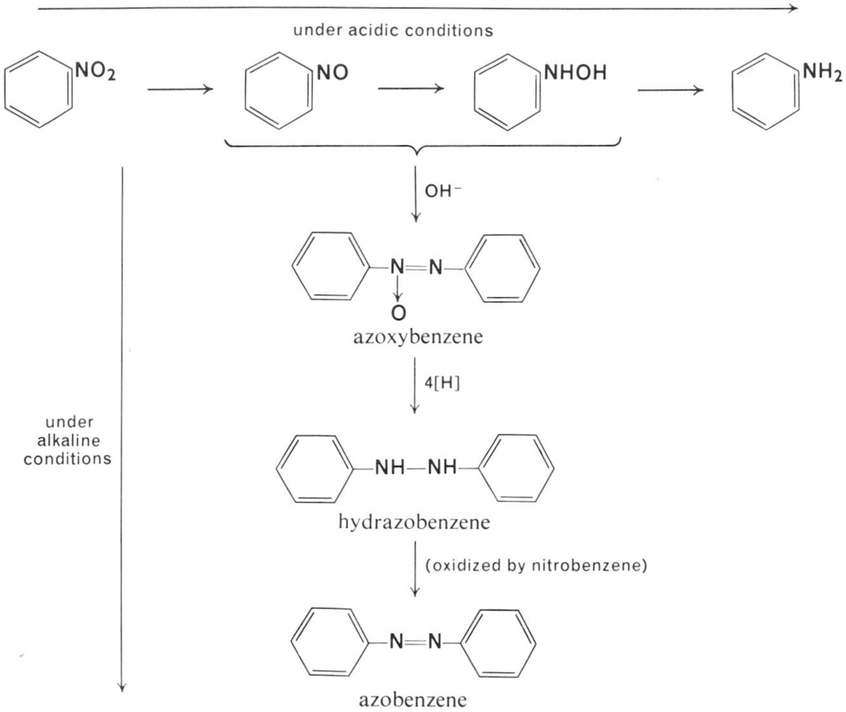

under acidic conditions

$$\text{C}_6\text{H}_5\text{NO}_2 \longrightarrow \text{C}_6\text{H}_5\text{NO} \longrightarrow \text{C}_6\text{H}_5\text{NHOH} \longrightarrow \text{C}_6\text{H}_5\text{NH}_2$$

OH⁻

azoxybenzene

4[H]

under alkaline conditions

hydrazobenzene

(oxidized by nitrobenzene)

azobenzene

1. Azoxybenzene is the chief product of the reduction of nitrobenzene with an alkaline solution of glucose.

2. Hydrazobenzene is prepared by the reduction of nitrobenzene with zinc and alkali.

3. Azobenzene is prepared by reduction of nitrobenzene with stannous chloride and alkali, but it is most conveniently prepared by the oxidation of hydrazobenzene with air. It can also be prepared by the reduction of azoxybenzene with iron in aqueous solution.

### Exercises

(Exercises 1–5 will be found within the text.)

6 Write the equations for the mononitration of each of the following compounds. Show the predominant product only: *(a)* bromobenzene; *(b)* p-xylene; *(c)* p-nitrotoluene;

*(d)* anisole; *(e)* acetanilide; *(f)* o-bromoacetanilide; *(g)* m-xylene; *(h)* p-*tert*-butyltoluene; *(i)* benzoic acid; *(j)* p-toluic acid; *(k)* 1,2,3-trimethoxybenzene.

7 Show by equations how the following compounds can be prepared. Any monosubstituted benzene may be used as starting material. *(a)* p-nitroaniline; *(b)* 3,4-dinitrobromobenzene; *(c)* 2,5-dinitrotoluene; *(d)* 2,4-dinitroaniline.

8 Show the mechanistic details of the reaction between sodium propionate and p-bromobenzyl bromide.

9 Another reagent for the preparation of crystalline derivatives of carboxylic acids is p-bromophenacyl bromide. The derivatives are esters. Show by means of an equation how this reagent is used to prepare the derivative of acetic acid.

10 Write the equations that describe the preparation of 2,4-dinitrobenzaldehyde, with toluene and N,N-dimethylaniline as the organic starting materials.

11 Suggest a method by which mesoxalic acid (as the diethyl ester) could be prepared from diethyl malonate:

$$CH_2(COOEt)_2 \longrightarrow CO(COOEt)_2$$

12 Write the reaction showing the equilibrium between the tautomeric forms of 1-nitroso-2-naphthol (compare with p-nitrosophenol).

13 Read the entry "Explosives" in a good encyclopedia, and from it make a list of aromatic nitro compounds that are used as explosives.

14 How could p-methoxyazobenzene $\left( CH_3O-\langle\text{⬡}\rangle-N{=}N-\langle\text{⬡}\rangle \right)$ be prepared, starting with benzene.

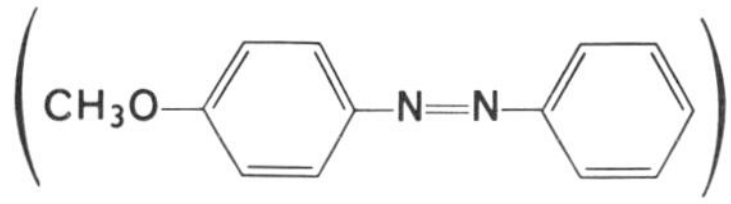

Chapter twenty-five / **Amines**

## 25.1 Properties of amines

Aromatic amines are typified by aniline and its substitution products. They are nearly always prepared by the reduction of nitro compounds, which in turn are the products of the direct nitration of aromatic compounds. Amines, such as aniline, the toluidines, and xylidines, are of great technical importance, since they are prepared from the abundant hydrocarbons of coal tar, and form the basis for a large and diversified chemical industry:

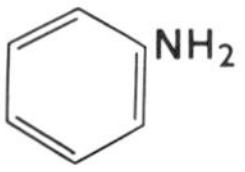


aniline

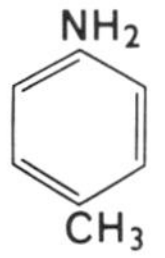


*p*-toluidine

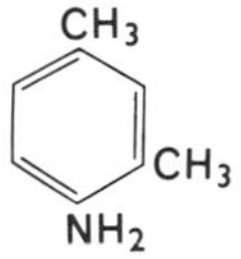


4-amino-*m*-xylene

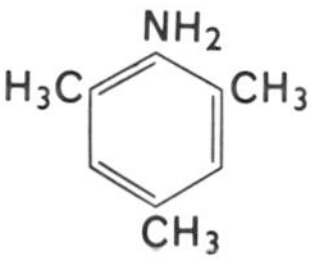


mesidine

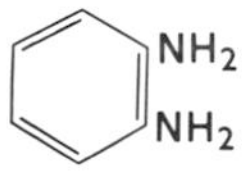


*o*-phenylenediamine

Aryl amines are basic, and form salts with acids. Alkylation of the amino group leads to the formation of mono- and dialkylamines, such as N-methylaniline and N,N-dimethylaniline

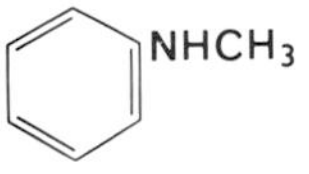

N-methylaniline

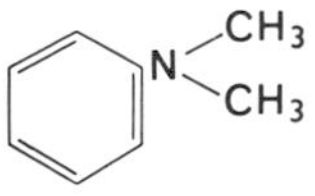

N,N-dimethylaniline

and acylation affords N-arylamides:

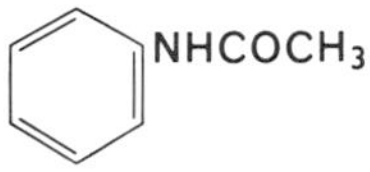

acetanilide

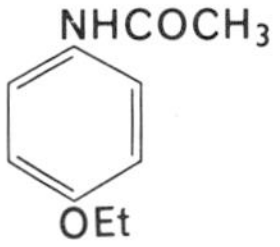

phenacetin

In these respects the aromatic amines resemble their aliphatic analogues. The two classes of amines differ in their degree of basic strength, the aromatic amines being considerably weaker as bases than the aliphatic amines. In Table 25-1 are given the $pK_a$ values for a number of typical representatives of each class.

It is apparent that aniline hydrochloride (anilinium chloride) is an acid of about the same strength as acetic acid ($pK_a = 4.75$); thus, it can be titrated with alkali with the use of an indicator such as phenolphthalein. Titration of the salt of an aromatic amine is a valuable analytical method for determining the equivalent weight of the compound.

The considerably lower strength of the aromatic amines as bases, as compared with the aliphatic amines, is the result of electron withdrawal from the nitrogen atom by the participation of such structures as

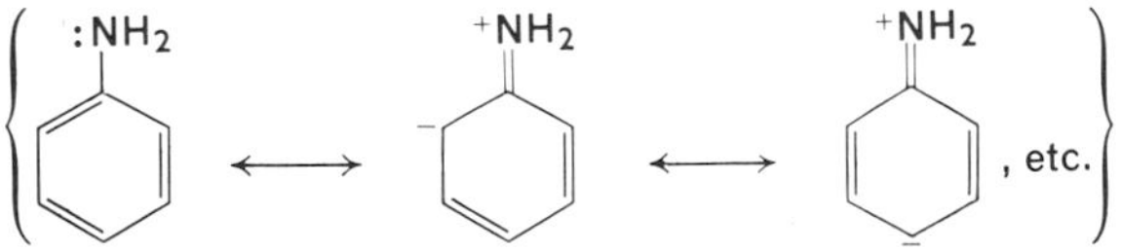

**Table 25-1.** *Ionization constants of some typical amines*

| Amine | $pK_a$* | $K_b$ |
|---|---|---|
| aniline | 4.62 | $4.2 \times 10^{-10}$ |
| o-toluidine | 4.40 | $2.5 \times 10^{-10}$ |
| p-toluidine | 4.10 | $1.2 \times 10^{-10}$ |
| N-methylaniline | 4.85 | $7.1 \times 10^{-10}$ |
| N,N-dimethylaniline | 5.04 | $1.1 \times 10^{-9}$ |
| methylamine | 10.64 | $4.4 \times 10^{-4}$ |
| ethylamine | 10.67 | $4.7 \times 10^{-4}$ |
| dimethylamine | 10.71 | $5.1 \times 10^{-4}$ |

\* $pK_a$ is the acid dissociation constant of the ammonium form of the base $RNH_3^+ \rightleftharpoons RNH_2 + H^+$.

in the resonance hybrid. Another way of expressing this is to say that the additional resonance energy of the aromatic amine that is due to participation of the electrons of the nitrogen atom is lost when a proton is added to the amino group. Protonation of methylamine, on the other hand, does not involve such a loss of resonance energy, and consequently the aliphatic amine is more readily protonated; that is, it is a stronger base:

$$CH_3\ddot{N}H_2 \xrightarrow{HA} CH_3\overset{+}{N}H_3 \} A^- \quad \text{(involves no loss of resonance stabilization)}$$

A further marked decrease in basic strength results from the substitution of one or both of the amino hydrogen atoms by aryl groups. Diphenylamine is a very weak base, and triphenylamine is, by the usual criteria, a neutral compound. Diphenylamine ($pK_a = 0.8$) is soluble in strong HCl, but upon dilution with water the free amine precipitates from solution.

---

**Exercise 1.** Explain why diphenylamine hydrochloride is converted to free diphenylamine by the action of water.

---

Triphenylamine is soluble only in very strong acids, such as concentrated sulfuric acid, and thus shows basic properties no more pronounced than such oxygen bases as ethers and alcohols.

The presence of electron-attracting substituents, such as nitro groups, in the ring reduces the basicity of the aromatic amines. $p$-Nitroaniline is a very weak base ($pK_a = 2.0$); 2,4-dinitroaniline is insoluble in aqueous acids and thus acts as a neutral substance; and 2,4,6-trinitroaniline is called *picramide* because of the ease with which it is formed from picryl chloride (2,4,6-trinitrochlorobenzene), and can be hydrolyzed to picric acid. The electron-withdrawing character of the nitro group reduces the electron availability of the amino nitrogen atom by the combined inductive and resonance effects. That the inductive effect alone is an important factor is shown by the low value of the $pK_a$ for $m$-nitroaniline in which direct resonance interaction between —$NH_2$ and —$NO_2$ is not important:

| $pK_a$ | 4.62 | 2.0 | 0.20 | 2.6 |

The result of the combination of the inductive effect with the resonance effect is shown in the case of $o$-nitroaniline, in which the proximity of the —$NO_2$ and

—$NH_2$ groups allows the inductive effect of the nitro group to exert its maximum action.

Other electron-withdrawing substituents weaken the base strengths of substituted anilines. The three chloroanilines have $pK_a$ values of 1.95 (*ortho*), 3.60 (*meta*), and 2.00 (*para*).

The above discussion of the base strength of amines has been couched in terms of electron "availability" at the nitrogen atom of the —$NH_2$ group. It might be objected that—for example, in *p*-nitroaniline—the electrons, although delocalized by the participation of the nitro group, are still "available," and thus that their delocalization does not account for the very weak basicity of this amine. It will be recalled, however, that resonance stabilization involves the increased stability that derives from such delocalization, and hence that to protonate the amine, and by so doing to localize the electron pair in forming the H—N bond of the —$NH_3{}^+$ group, means that the resonance stabilization is sacrificed.

The base strength of the amine is expressed by the equilibrium constant of its reaction with a proton donor (water, when $K_b$ values are used); and the equilibrium constant is a measure of the *difference* in energy of the system in its initial (unprotonated) and final (protonated) states. A loss of delocalization energy means an increase in energy upon protonation ($\Delta F$ positive); and since large values of $K_b$ correspond to large decreases in free energy (high values of $-\Delta F$), a positive value of $\Delta F$ corresponds to very small values of $K_b$. The greater the loss of resonance stabilization, then, the smaller the value of $K_b$, and the weaker the base.

Base-weakening inductive effects can be accounted for in the same terms. The electron withdrawal (for example, by *o*-chloro-) creates electron deficiency, or a degree of positive character, at the carbon atom to which the —$NH_2$ group is attached. As a result, the creation of a positive charge on nitrogen by the protonation of the amino group will involve the energy requirement associated with the creation of adjacent centers of like charge, thus raising the energy of the protonated form (compared with that of a compound otherwise the same but lacking the $-I$ substituent), and thus lowering the value of $K_b$.

## 25.2    Alkylation of aniline derivatives

The amino group of aniline is nucleophilic and can be alkylated by alkyl halides, sulfates, and sulfonates. In view of the general parallelism between basic strength and nucleophilic character, it is to be expected that aniline would react more slowly with, say, methyl iodide than will methylamine; and that negatively substituted anilines, such as *p*-nitroaniline, would be alkylated with difficulty or not at all.

The reaction between aniline and primary alkyl halides proceeds readily and, depending upon the proportion of alkyl halide that is used, may lead to mono-, di-, or trialkylation. With 1:1 proportion of amine to halide, monoalkylation can be obtained as the main result:

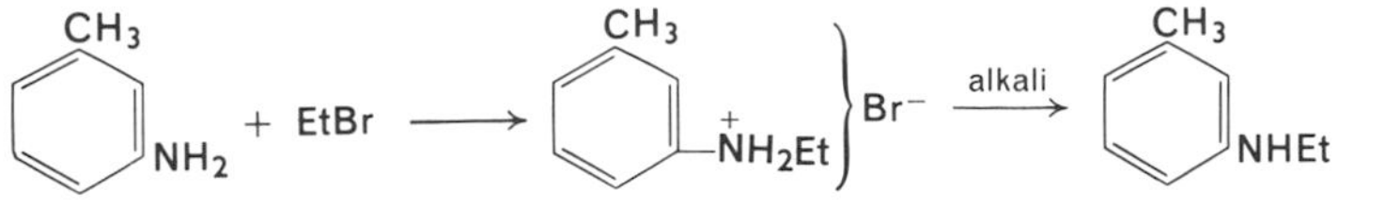

(1)

N-ethyl-*m*-toluidine

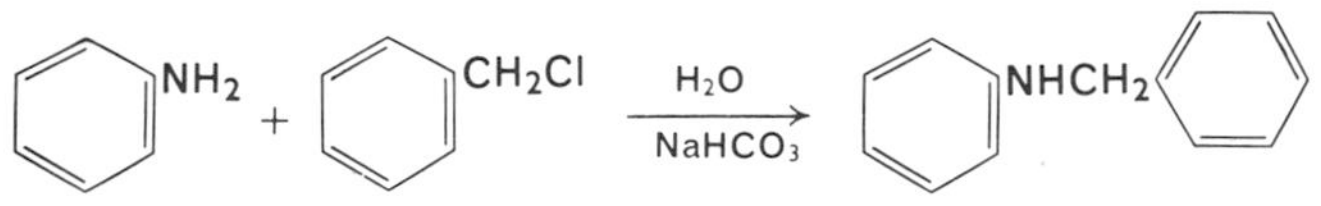

(2)

N-benzylaniline

Mono- and dimethylaniline are compounds of great technical importance and are ordinarily made by the action of methyl alcohol and hydrochloric acid upon aniline under pressure in a sealed vessel (an autoclave).

In the preparation of monoalkylated anilines the formation of dialkylated products usually cannot be avoided entirely, and separation of the products is necessary. This can be carried out by treatment of the crude product, containing mono- and dialkyl compounds, with nitrous acid. The N-nitroso compound derived from the monoalkyl aniline is a nonbasic compound and can be extracted from the acidic reaction mixture with ether or benzene. The dialkylaniline forms an acid-soluble nuclear nitrosation product, and any nonalkylated amine left in the reaction mixture is destroyed. (The reaction between amines and nitrous acid will be described in Chapter 26.) In the case of N-ethyl-*m*-toluidine, described above, the N-nitroso-N-ethyl-*m*-toluidine so formed is transformed to the desired N-ethyl-*m*-toluidine by reduction:

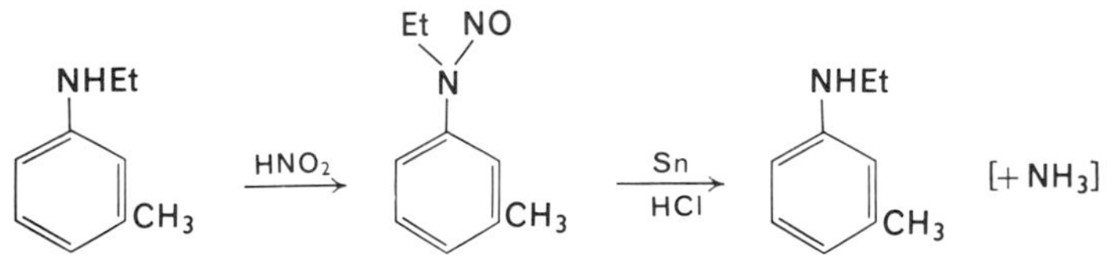

(3)

An alternative method of separation is one in which the secondary amine is acetylated, and the acetyl derivative separated from the tertiary amine by distillation. This procedure has the disadvantage that any primary amine present in the mixture cannot easily be separated, since its acetyl derivative and the acetylated secondary amine would be likely to have boiling points rather close to each other.

A useful and general method of dimethylating a primary amine and monomethylating a secondary amine is with the use of formaldehyde and formic acid (which acts as the reducing agent):

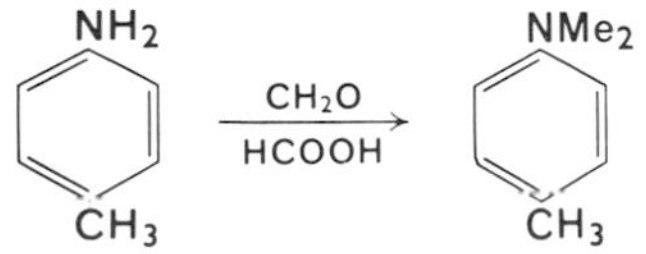

(4)

N,N-dimethyl-*p*-toluidine

Secondary amines are also prepared in good yields by the catalytic hydrogenation of the Schiff's bases formed by condensation of aromatic amines and aldehydes:

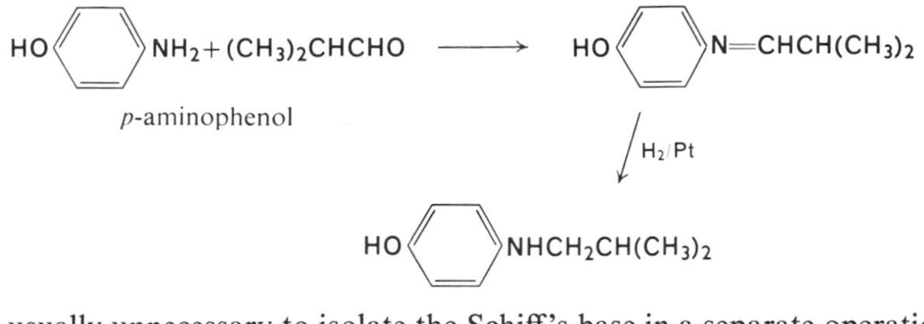

$p$-aminophenol

$$HO\text{—}C_6H_4\text{—}NHCH_2CH(CH_3)_2 \tag{5}$$

It is usually unnecessary to isolate the Schiff's base in a separate operation. The alkylated amine is formed when the amine, the aldehyde or ketone, and a platinum catalyst are shaken together under a hydrogen atmosphere, usually in alcoholic solution.

The carbon-nitrogen bond of N-benzyl amines is readily cleaved by hydrogen in the presence of a palladium catalyst, with the regeneration of the amine (and the formation of toluene from the benzyl group). This property provides a useful means for the protection of an amino group during the performance of reactions upon another part of the molecule; after the desired transformations have been carried out, the benzyl group can be removed:

$$R_2NCH_2C_6H_5 \xrightarrow{H_2/catalyst} R_2NH + C_6H_5CH_3 \tag{6}$$

---

**Exercise 2.** Could $p$-N,N-dimethylaminophenol be prepared from $p$-aminophenol by means of dimethyl sulfate and alkali?

---

## 25.3 Acylation of aromatic amines

The reaction of aniline and substituted anilines with acyl halides and anhydrides leads to the formation of N-acylanilines. This reaction is a general one for aliphatic and aromatic primary and secondary amines and requires no additional discussion:

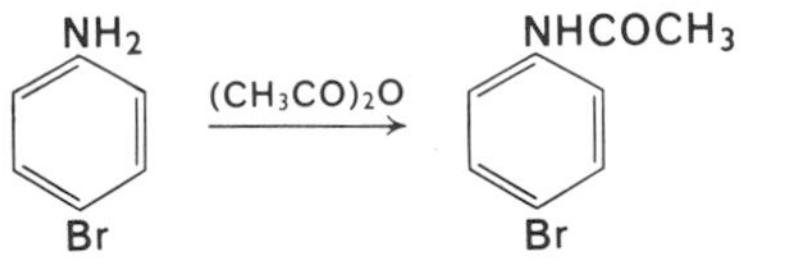

$$\tag{7}$$

$p$-bromoacetanilide

$$\tag{8}$$

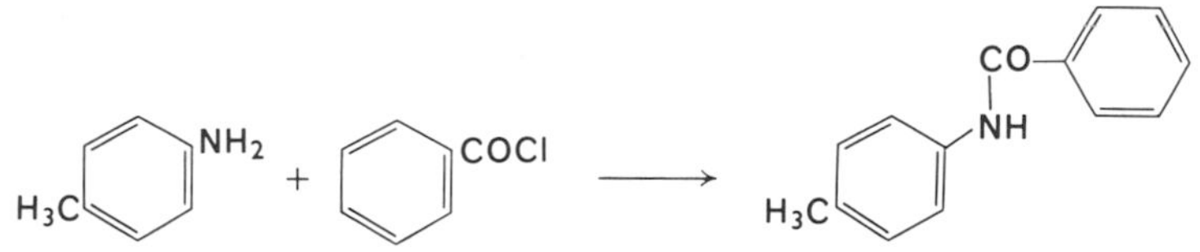

N-$p$-tolylbenzamide
(or benz-$p$-toluidide)

N-Formylation can be accomplished by heating the amine with 90–100% formic acid, or with ethyl formate:

$$\underset{R'}{\overset{R}{\diagdown}}NH + HCOOEt \longrightarrow \underset{R'}{\overset{R}{\diagdown}}N-CHO + EtOH$$

This is a useful method for the methylation of an amine, for the N-formyl derivative can be reduced to N-methyl by means of lithium aluminum hydride:

$$\underset{R'}{\overset{R}{\diagdown}}N-CHO + LiAlH_4 \longrightarrow \underset{R'}{\overset{R}{\diagdown}}N-CH_3$$

Other N-acylated amines can be reduced to the corresponding N-alkyl amines in the same way; for example, $\diagdown N-COCH_3$ gives $\diagdown N-CH_2CH_3$.

Certain acylated aromatic amines are of importance as drugs. *Acetanilide* and *phenacetin* are mild analgesics and antipyretics, and are found in proprietary remedies used for pain and mild fever. Acetanilide, on continued or excessive use, can cause a serious blood disorder called "methemoglobinemia," a transformation of hemoglobin into methemoglobin. Phenacetin (*p*-ethoxyacetanilide) is subject to the same disadvantages as acetanilide, but to a somewhat lesser degree; also it is somewhat more effective as an antipyretic and analgesic agent.

The preparation of N-acetyl derivatives of aromatic amines can be accomplished by the Beckmann rearrangement of the oximes of the corresponding aryl methyl ketones. For example:

| acetophenone | acetophenone oxime | acetanilide |

(9)

The Beckmann rearrangement is discussed again in Chapter 28.

The chief use of N-acylation of the aromatic amines is to reduce the activating power of the amino group so that controllable substitution reactions can be carried out (Chapter 21); and for the characterization and identification of amines. Acetyl and benzoyl derivatives of primary and secondary amines are easy to prepare in good yield, easy to purify by recrystallization, and usually possess sharp and characteristic melting points. They are equally useful for the characterization of aliphatic and aromatic amines.

The reaction of primary aromatic amines with phosgene, ClCOCl, leads to the formation of derivatives containing the group —NHCOCl. Aniline gives phenylcarbamyl chloride, $C_6H_5NHCOCl$, the chief use of which is in its conversion, by loss of hydrogen chloride on heating, into phenyl isocyanate, $C_6H_5N=C=O$. Phenyl isocyanate is a valuable reagent for the characterization and identification of several classes of compounds. It reacts with amines to yield *substituted ureas*, with alcohols to yield *urethanes*, and with Grignard reagents to yield amides:

$$C_6H_5N{=}C{=}O + ROH \longrightarrow C_6H_5NHCOOR \tag{10}$$

a urethane; an
N-phenylcarbamate

$$C_6H_5N{=}C{=}O + RNH_2 \longrightarrow C_6H_5NHCONHR \tag{11}$$

an N,N′-disubstituted
urea

$$C_6H_5N{=}C{=}O + RMgX \longrightarrow \left( \underset{\overset{|}{OMgX}}{C_6H_5N{=}C{-}R} \right) \xrightarrow{H_2O} \underset{\overset{\|}{O}}{C_6H_5NHC{-}R} \tag{12}$$

The chief utility of the last reaction lies in the fact that it serves as a means of identification of alkyl halides: these are converted into the Grignard reagents, allowed to react with phenyl isocyanate, and the amide that is formed, usually a crystalline compound, identified by melting point or other physical properties.

## 25.4 N-Arylsulfonyl derivatives of amines. The Hinsberg method for the separation and characterization of amines

The use of a sulfonyl chloride for acylation leads to the formation of an amide of the sulfonic acid. Benzenesulfonyl, *p*-bromobenzenesulfonyl, and *p*-toluenesulfonyl chlorides are most commonly used, the resulting sulfonyl derivatives being of importance for characterization and identification, and for the separation of amines:

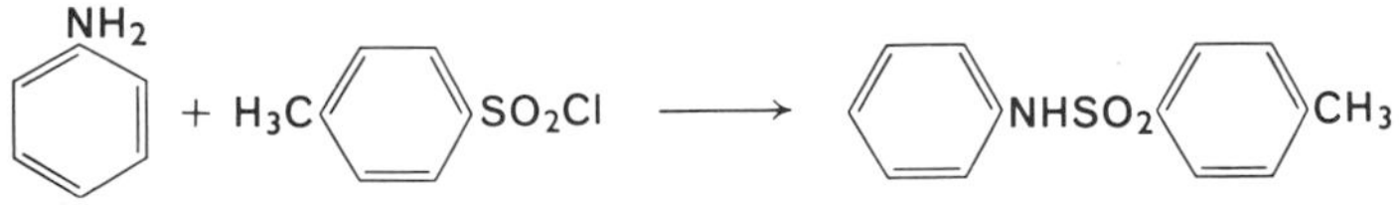

$$\text{(benzene-NH}_2) + H_3C\text{—(benzene)—}SO_2Cl \longrightarrow \text{(benzene)—}NHSO_2\text{—(benzene)—}CH_3 \tag{13}$$

*p*-toluenesulfonyl
chloride

*p*-toluenesulfonanilide
(N-phenyl-*p*-toluenesulfonamide)

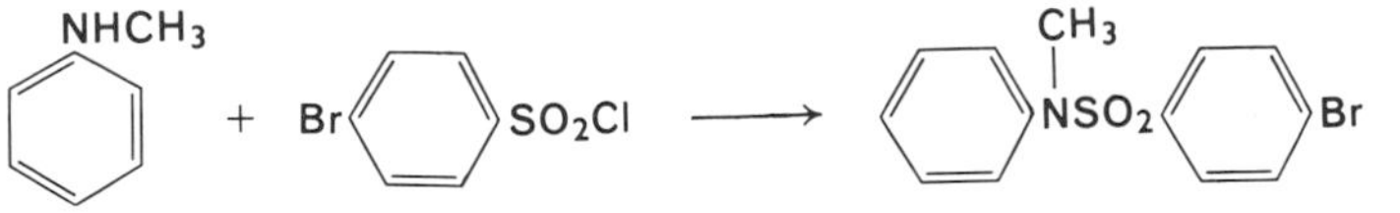

$$\text{(benzene-NHCH}_3) + Br\text{—(benzene)—}SO_2Cl \longrightarrow \text{(benzene)—}\underset{CH_3}{N}SO_2\text{—(benzene)—}Br \tag{14}$$

*p*-bromobenzenesulfonyl
chloride

N-methyl-
*p*-bromobenzenesulfonanilide

Tertiary amines, which have no replaceable hydrogen atom, do not form N-acyl derivatives.

The separation of amines by means of their arylsulfonyl derivatives depends upon the fact that *the derivative of the primary amine possesses a hydrogen atom sufficiently acidic to form a salt with aqueous alkali:*

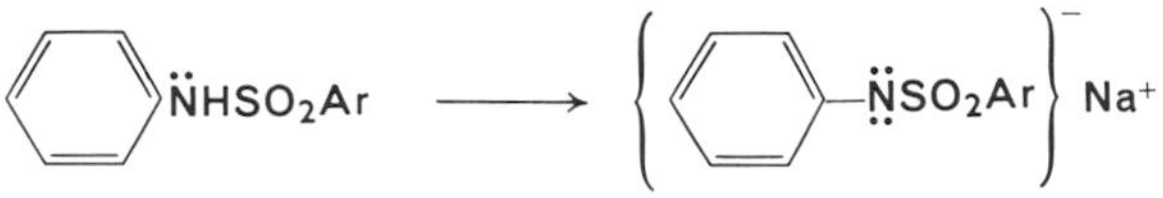

$$\text{(benzene)—}\ddot{N}HSO_2Ar \longrightarrow \left\{ \text{(benzene)—}\ddot{N}SO_2Ar \right\}^{-} Na^{+} \tag{15}$$

If a mixture of aniline, methylaniline, and dimethylaniline is shaken with benzenesulfonyl chloride in aqueous (5%) sodium hydroxide, the following will result.

1. The tertiary amine will be unaffected.
2. The secondary amine will form a derivative of the structure

$$ArN\overset{CH_3}{\underset{SO_2Ar}{\diagdown}}$$

3. The primary amine will form $ArNHSO_2Ar$, which will be soluble in the solution as the sodium salt $\{ArNSO_2Ar\}^- Na^+$.\*

Extraction of the alkaline mixture with ether will remove the ether-soluble tertiary amine and the alkali-insoluble benzenesulfonamide of the secondary amine. Subsequent extraction of the ether extract with dilute HCl will remove the tertiary amine as the water-soluble hydrochloride, leaving the benzenesulfonamide of the secondary amine in the ether. Acidification of the aqueous alkaline solution will precipitate the (usually crystalline) benzenesulfonamide of the primary amine.

This method of separation of a mixture of amines is known as the *Hinsberg* method; it is not confined to aromatic amines, and could be used to separate, for example, a mixture of *n*-butylamine, diethylamine, and tri-*n*-propylamine.

---

**Exercise 3.** Formulate the reactions in the separation of the mixture just mentioned by the Hinsberg method.

---

Recovery of the amines that are isolated in the form of their benzene-sulfonamides can be accomplished by hydrolysis with sulfuric acid. The hydrolysis is slow and is sometimes attended with partial destruction of the amines by the action of the hot, strong (usually 60–80%) sulfuric acid.

### 25.5 Uses of sulfonamides: "sulfa" drugs

The acidity of the sulfonyl derivatives of primary amines is probably the result of the strong electron-attracting properties of the $—SO_2Ar$ grouping, exercised principally through an inductive effect of the positively charged sulfur atom:

$$R—\overset{H}{\underset{\cdot\cdot}{N}}—S\overset{\nearrow O}{\underset{\searrow O}{—}}Ar$$

This acidic character of the $—SO_2NH—$ grouping causes it to bear a qualitative resemblance to the $—COOH$ group (although the latter is the more acidic of

---

\* The symbol "Ar" is often used to denote an aryl group, just as "R" is used for such alkyl groups as methyl, ethyl, and propyl.

the two), and it has been suggested that the antibacterial activity of sulfanilamide and other "sulfa" drugs is the result of their antagonism to the metabolic function of *p*-aminobenzoic acid:

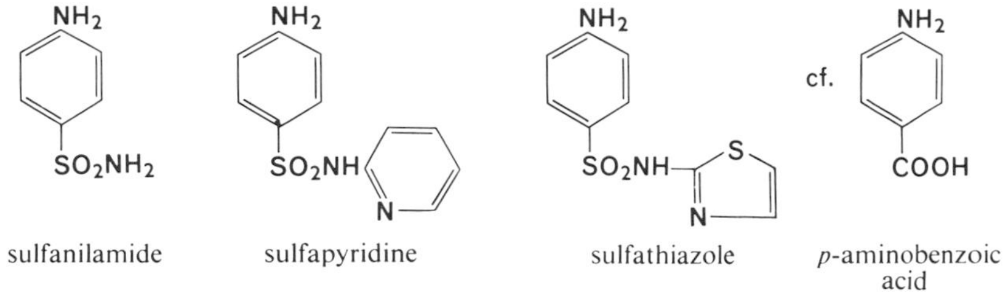

| sulfanilamide | sulfapyridine | sulfathiazole | *p*-aminobenzoic acid |

*p*-Aminobenzoic acid will inhibit the activity of sulfanilamide *in vitro;* and organisms that do not require folic acids, of which *p*-aminobenzoic acid is a precursor, are not sensitive to sulfa drugs:

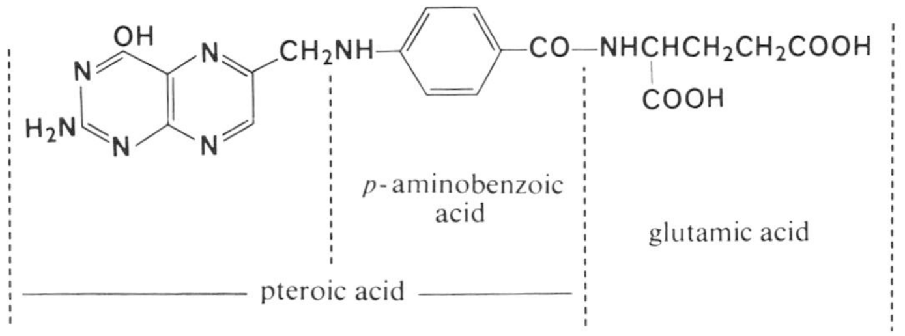

folic acid (pteroylglutamic acid)

The introduction of sulfonamide derivatives into another quite different area of medical therapy is found in the discovery that certain compounds of this class are orally active antidiabetic agents of clinical usefulness. Two of these are:

## 25.6  Infrared and ultraviolet spectral properties of amines and amides

The amino and acylated amino groupings have no conspicuous or useful absorption in the ultraviolet region, and ultraviolet absorption spectra of amines and amides are useful only insofar as these groupings affect the absorption of the molecule as a whole. In Chapter 11 it was pointed out that the amino group is auxochromic; the participation of its unshared electrons in the delocalization associated with a conjugated system causes profound effects upon the ultraviolet and

**Table 25-2.** *Some compara-*
*tive C=O stretching frequencies*
*for two groups of related amides,*
*esters, and acyl chlorides*

| Compound | C=O Stretching Frequency, $cm^{-1}$ |
|---|---|
| $CH_3COCl$ | 1822 |
| $CH_3COOEt$ | 1743 |
| $CH_3CONHC_6H_5$ | 1685 |
| $C_6H_5COCl$ | 1773 |
| $C_6H_5COOEt$ | 1723 |
| $C_6H_5CONH_2$ | 1675 |

visible absorption due to such systems. These effects are illustrated by several examples given in Chapter 11.

The stretching frequencies of the N—H bonds in amines and amides are found in the same region as those of the O—H bands and, in the absence of other information, are often not uniquely diagnostic. Amides, on the other hand, show characteristic carbonyl stretching absorptions which are quite different in position from the carbonyl absorptions of related acid derivatives. The figures in Table 25-2 show some comparative C=O stretching frequencies for two groups of related amides, esters, and acyl chlorides. It is apparent that amides show carbonyl absorptions at much lower wave numbers than do esters and acyl halides. It is apparent from these data, and from what was said in Chapter 11 on the relationship between the carbonyl stretching frequency and the character of the C=O bond, that the carbonyl group of amides has a much greater degree of single-bond character than that of esters, and esters than that of acyl chlorides. The carbonyl group of amides derives its single-bond character from the contribution of the unshared electron pair in the resonance expressed by the formula

$$R-C{\overset{\textstyle O}{\underset{\textstyle NH_2}{\big<}}} \quad \longleftrightarrow \quad R-C{\overset{\textstyle O^-}{\underset{\textstyle \overset{NH_2}{+}}{\big<}}}$$

It is clear from the infrared absorption characteristics of amides, esters, and acyl chlorides that the contribution of the dipolar form to the structure of the resonance hybrid is greater in the case of amides than in the other two compounds. This is, of course, consistent with the greater basicity of nitrogen, which is a reflection of its electron-donating capacity.

## Exercises

(Exercises 1–3 will be found within the text.)

**4** Write the structures of the following compounds: *(a)* p-chloroaniline; *(b)* 2,4,6-tribromo-aniline; *(c)* m-aminobenzoic acid; *(d)* 3-amino-o-xylene; *(e)* N,N-di-n-propyl-p-toluidine; *(f)* propionanilide; *(g)* N-methylacet-o-toluidide.

**5** Complete the following reactions (use structural formulas).
   *(a)* aniline + HBr $\longrightarrow$
   *(b)* N-methylaniline + acetic acid $\longrightarrow$
   *(c)* anilinium chloride + sodium methoxide $\longrightarrow$
   *(d)* acetanilide + boiling aqueous NaOH $\longrightarrow$
   *(e)* p-chloroaniline + acetyl chloride $\longrightarrow$

**6** Why is N-nitroso-N-methylaniline so weak a base as to be classed as a "neutral" compound?

**7** How could you use aniline and propionaldehyde as the starting materials for the preparation of di-n-propylamine?

**8** If the theory of the antibacterial activity of sulfa drugs is as it is outlined in this chapter, which of the following would you expect to be *inactive* drugs:

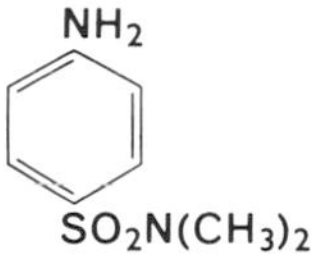

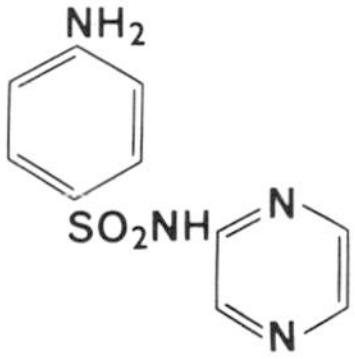

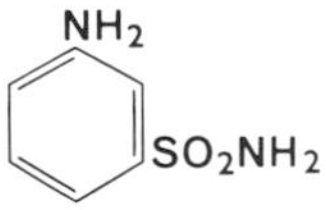

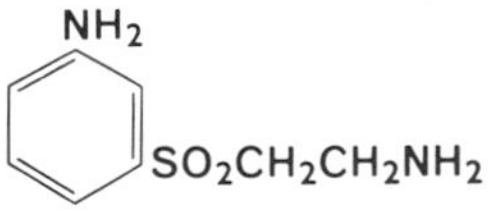

**9** How could you convert 3,5-dimethylbenzoic acid into 3,5-dimethylaniline (5-amino-m-xylene)? [Hint: see Equation (9).]

**10** How could you make use of the nucleophilic nature of the anions (i.e., the sodium salts) of N-p-toluenesulfonyl derivatives of anilines as an aid in the preparation of N-alkylated anilines? Write the series of steps that would be involved in the preparation of N-ethyl-p-toluidine by this means.

**11** Two isomeric compounds ($C_{10}H_{13}NO$) gave benzoic acid upon oxidation with potassium permanganate. Compound A was soluble in dilute aqueous HCl and showed an infrared stretching frequency in the carbonyl region at 1710 cm$^{-1}$. Compound B was insoluble in dilute aqueous acid, and showed a carbonyl stretching frequency at 1687 cm$^{-1}$. Write structures for A and B that account for these observations. Can more than one structure be written for each? What further experiments can you suggest to establish the correct structures of A and B?

# *Chapter twenty-six* / Reactions of Amines with Nitrous Acid. Diazo Compounds

## 26.1 Reactions of amines with nitrous acid: general

One of the most interesting and useful reactions of amines, both aromatic and aliphatic, is that with nitrous acid. Reactions of this kind lead to various consequences, depending upon whether the amine is primary, secondary, or tertiary, aliphatic or aromatic. We shall examine each of these, and it will be seen that in every case *the initial stage of the reaction involves the nucleophilic attack of the amine upon the nitrous acid;* the subsequent events are governed by the structure of the amine. Certainly the most important of these reactions is that between primary aromatic amines and nitrous acid, leading to the aromatic diazonium compounds. This reaction, called "diazotization," applied to synthetic processes both in the laboratory and in the chemical industry, is so various and capable of such a degree of versatility that we shall devote special attention to it. Certain aspects of the behavior of amines of other kinds are discussed in the following sections.

Nitrous acid can be described by the structure $H-\underset{\cdot\cdot}{\ddot{O}}-\ddot{N}=\ddot{O}:$. It is, of course, a hydrogen acid, and by removal of the proton yields the nitrite ion, $NO_2^-$:

$$\left\{ :\underset{\cdot\cdot}{\ddot{O}}-\ddot{N}=\ddot{O}: \longleftrightarrow :\ddot{O}=\ddot{N}-\underset{\cdot\cdot}{\ddot{O}}: \right\}^-$$

Whereas the nitrite ion is a nucleophile, nitrous acid is typically electrophilic in character, attack by a nucleophilic reagent upon the nitrogen atom (2) resembling the first step in carbonyl addition reactions (1):

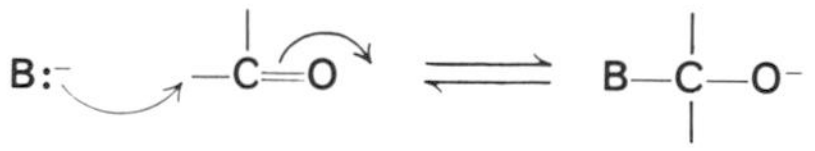

$$\text{B:}^- \quad -\overset{|}{\underset{}{C}}\!=\!O \quad \rightleftharpoons \quad \text{B}-\overset{|}{\underset{}{C}}-O^- \tag{1}$$

$$\text{B:}^- \quad -\overset{..}{N}\!=\!O \quad \rightleftharpoons \quad \text{B}-\overset{..}{N}-O^- \tag{2}$$

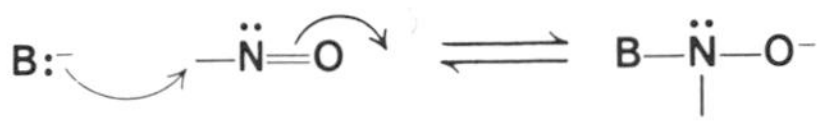

The role of the step (2) in the reactions of amines with nitrous acid will be amplified by the examples to be found in the discussion to follow.

## 26.2 Tertiary amines

The reaction of a tertiary aliphatic amine with nitrous acid,

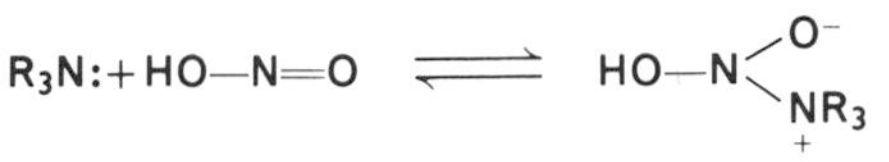

$$R_3N: + HO-N\!=\!O \quad \rightleftharpoons \quad HO-N\!\!\begin{array}{c} O^- \\ \diagdown \\ NR_3 \\ + \end{array} \tag{3}$$

is simply the reaction of the type (2) of the base $R-\overset{R}{\underset{R}{\overset{|}{N}}}:$ with the electrophilic

grouping ($-\overset{..}{N}\!=\!\overset{..}{O}: \longleftrightarrow -\overset{..}{N}-\overset{..}{\underset{..}{O}}:$), and is reversible. It leads to no consequences in the form of a stable product, and upon the addition of a base the amine is recovered unchanged.

A tertiary aromatic amine reacts with nitrous acid to undergo nuclear nitrosation (unless the *ortho* and *para* positions are occupied by substituents). This is a typical electrophilic substitution reaction, and can be formulated in accordance with our general view of this reaction as it has been developed in our study of other electrophilic substitution reactions. In the simplest terms, the nitrosating agent can be regarded as a nitrosonium ion carrier such as $H_2\overset{+}{O}-NO$, the protonated form of nitrous acid:

$$HONO + HA \quad \rightleftharpoons \quad H_2ONO^+ + A^-$$

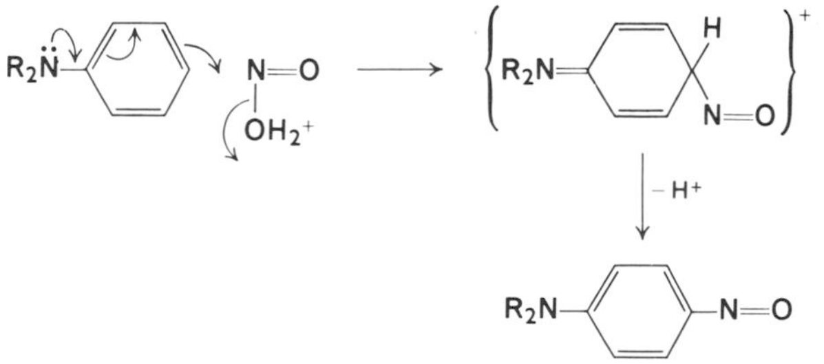

$$\tag{4}$$

The nitrosonium ion is much less electrophilic than the nitronium ion, $NO_2^+$, and attacks only strongly nucleophilic aromatic nuclei. Phenols are readily nitrosated (Chapter 25), as are tertiary aromatic amines; but aromatic rings bearing less strongly activating substituents than hydroxyl or amino groups, or deactivating substituents, are not attacked by nitrosating agents.

Attack of the nitrosating agent upon the nitrogen atom of a tertiary aromatic amine, according to Equation (3), will, of course, also occur; but since this reaction is reversible and leads to no stable product the nuclear attack of Equation (4) occurs to give the final product of the reaction. It is not possible to describe all nitrosating reagents in terms of a single nitrosonium ion-carrying species, for because of the equilibria that can exist in a strongly acid solution of nitrous acid, the nitrosonium ion carrier may be the protonated nitrous acid, $N_2O_3$ ($O_2N$—$NO$), BrNO (in HBr), or even the nitrosonium ion itself, $NO^+$. Here, as in cases of halogenation, nitration, and acylation, it is not always possible to be categorical about the actual identity of the attacking electrophile. Thus, the summary equation (4) given above must be regarded as representative only.

$p$-Nitrosodialkylanilines are easily prepared by the addition of sodium nitrite to a solution of the dialkylaniline in hydrochloric acid.

$p$-Nitrosodimethylaniline is a brilliant green compound; its hydrochloride is yellow. When it is treated with hot sodium hydroxide solution, dimethylamine is liberated, and $p$-nitrosophenol (as the sodium salt) is formed:

$$O{=}N{-}\langle\bigcirc\rangle{-}NMe_2 + OH^- \longrightarrow O{=}N{-}\langle\bigcirc\rangle{-}OH + (CH_3)_2NH \qquad (5)$$

This reaction constitutes a useful method for the preparation of some secondary amines by the following sequence: dialkylation of aniline, nitrosation, and alkaline cleavage in the above manner. The displacement of the dialkylamine will be recognized as a typical nucleophilic displacement reaction (see Chapter 22), carried to completion in this case by the formation of the anion of the $p$-nitrosophenol. Treatment of $p$-nitrosodimethylaniline with sodium methoxide results in the liberation of dimethylamine and the formation of $p$-nitrosoanisole.

---

**Exercise 1.**  Write the reaction of $p$-nitrosodimethylaniline and sodium methoxide, and compare it with alcoholysis of an amide with sodium methoxide.

**Exercise 2.**  Starting with N-benzylaniline, devise a synthesis for ethylbenzylamine, $CH_3CH_2NHCH_2C_6H_5$.

---

## 26.3  Secondary amines

The reaction of a secondary aliphatic or aromatic amine (e.g., methylaniline) with nitrous acid results in N-nitrosation:

$$(CH_3)_2NH + HONO \longrightarrow (CH_3)_2N{-}NO \qquad (6)$$

N-nitrosodimethylamine

$$\langle\bigcirc\rangle{-}NHCH_3 + HONO \longrightarrow \langle\bigcirc\rangle{-}N\big\langle{}^{NO}_{CH_3} \qquad (7)$$

N-nitroso-N-methylaniline

---

**Exercise 3.**   Why does not nuclear nitrosation occur?

---

The N-nitroso amine is no longer basic, in contrast to the amine from which it is formed; it is not an amine, but an amide of nitrous acid, and, like amides in general, is a very weakly basic substance.

It will be recalled that in the preparation of N-ethyl-*m*-toluidine (Chapter 25) the secondary amine was purified by converting it into the neutral, ether-soluble N-nitroso derivative, which could be separated from the basic unethylated and diethylated toluidines, and from which the amine was regenerated by reduction. Reduction of N-nitroso secondary amines, with cleavage of the N—N bond, is accomplished by the use of reducing agents such as stannous chloride or metal-mineral acid combinations:

$$
\underset{}{\overset{}{\text{C}_6\text{H}_5\text{N(NO)CH}_3}} \xrightarrow[\text{HCl}]{\text{Sn}} \underset{}{\overset{}{\text{C}_6\text{H}_5\text{N(H)CH}_3}} + \text{NH}_3 \tag{8}
$$

Mild reducing agents such as sodium amalgam or zinc and dilute acetic acid convert N-nitroso amines into the corresponding 1,1-disubstituted hydrazines:

$$
\underset{}{\overset{}{\text{C}_6\text{H}_5\text{N(NO)CH}_3}} \xrightarrow[\text{or Zn/AcOH}]{\text{Na/Hg}} \underset{}{\overset{}{\text{C}_6\text{H}_5\text{N(NH}_2)\text{CH}_3}} \tag{9}
$$

N-methyl-N-phenylhydrazine

N-Methyl-N-phenylhydrazine is a useful reagent for the characterization of sugars, forming derivatives called methylphenylhydrazones.

N-Nitroso derivatives of secondary aromatic amines rearrange under the influence of HCl or HBr to give C-nitroso compounds [Equation (10)].

This "rearrangement" is really an intermolecular reaction between the amine and NO⁺ (which, in HCl, is in equilibrium with the species NOCl), and is a nuclear nitrosation reaction comparable in all respects to that between a nitrosating agent and a tertiary amine. N-Nitrosation and nuclear nitrosation are thus the separate consequences of the interrelated equilibria expressed in the following equations, and the formation of the ring-nitrosated product is only the consequence of the greater thermodynamic stability of that compound. Under mildly acidic conditions, the greater nucleophilicity of the nitrogen atom leads, in a rapid reaction, to the N-nitroso derivative. When the acidity of the medium is increased, the increased availability of the more electrophilic nitrosating agent, NO⁺ or NOCl, permits the establishment of the final, more favorable, equilibrium, and the nuclear nitrosation is the end result. Concomitant protonation of the methylamino group stabilizes the *p*-nitroso compound.

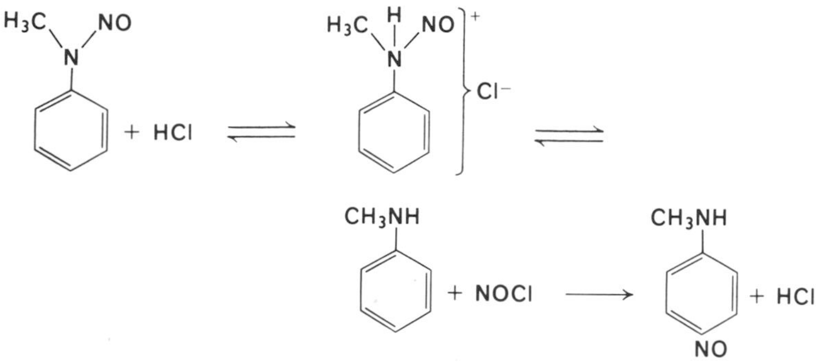

## 26.4  The Liebermann nitroso reaction: a diagnostic test for secondary amines

The denitrosation of N-nitroso compounds forms a basis for a test for these substances, called the *Liebermann nitroso reaction*. Treatment of an N-nitroso amine with phenol in concentrated sulfuric acid gives a solution which turns red on dilution and blue on addition of alkali. The course of this reaction involves the following: the nitrosation of the phenol by nitrous acid formed by denitrosation of the nitrosoamine; and the condensation of the nitrosophenol with a second molecule of phenol to yield the indophenol, which is red in the unionized form and gives a blue anion with alkali:

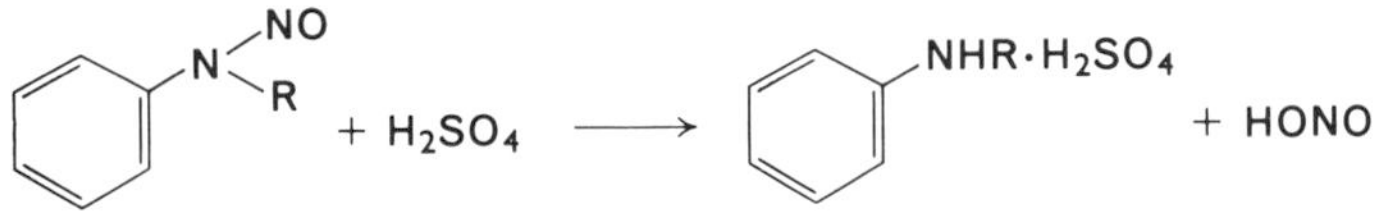

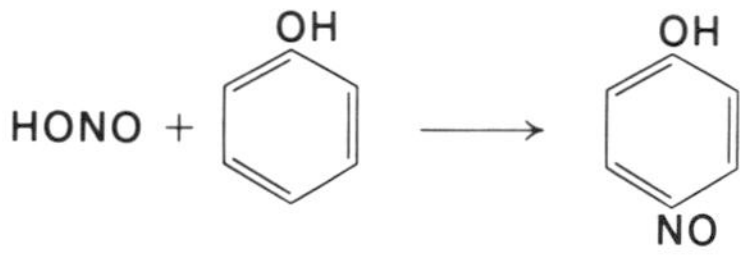

When phenol is used as the test reagent, the Liebermann reaction is also given by *p*-nitrosophenol and by nitrous acid or nitrites; it is also used as a test for phenols, an N-nitrosoamine being used as the test reagent in this case.

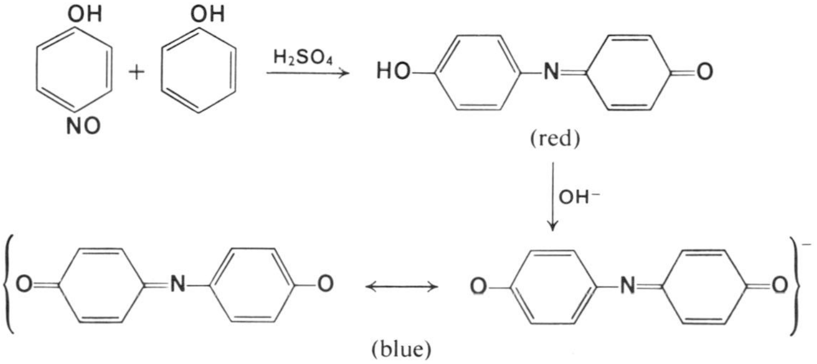

**Exercise 4.** Step *(c)* [Equation (13)] is a typical electrophilic substitution reaction. Formulate the course of this reaction through the sequence of stages from electrophilic attack upon phenol to the formation of the final product, the indophenol. Note that the essential feature of the reaction is the formation of a bond between the nitrogen atom of the nitrosophenol and the carbon atom *para* to the OH group of the phenol.

## 26.5 The reaction of primary amines with nitrous acid

The reaction between a primary amine and nitrous acid leads, probably through a transitory stage involving an N-nitrosoamine, to a *diazonium salt*. The overall reaction, written first without regard to the nature of the process, is

$$R\text{—}NH_2 + HONO \xrightarrow{\text{HA}} R\text{—}N_2{}^+\}A^- + 2H_2O$$

The reaction is often carried out by adding sodium nitrite to a solution of the amine in HCl, with the result that the final solution contains the diazonium chloride ($A^- = Cl^-$).

When R in the above equation is an aliphatic group, the diazonium salt is very unstable and decomposes at once by the loss of nitrogen and combination of R with a nucleophilic reagent (e.g., water). Depending upon the character of the alkyl group (primary, secondary, tertiary) the decomposition of $RN_2{}^+$ can occur by an $S_N1$ or an $S_N2$ process. If water were the nucleophilic reagent, the reaction would proceed as follows:

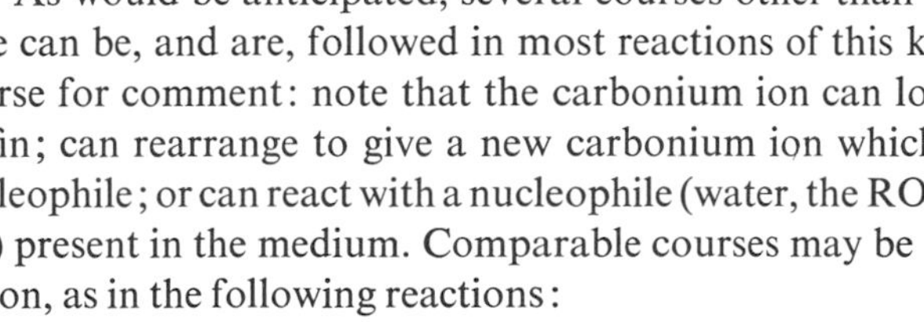

$$RN_2{}^+ \quad \overset{(S_N1)}{\nearrow} \quad \{R^+\} + N_2 \xrightarrow{\ H_2O\ } ROH + H^+ \qquad (14)$$
$$\underset{(S_N2)}{\searrow} \quad \overset{H_2O}{\phantom{x}} \quad ROH + H^+ + N_2$$

As would be anticipated, several courses other than the simple ones written here can be, and are, followed in most reactions of this kind. Let us use the $S_N1$ course for comment: note that the carbonium ion can lose a proton to yield an olefin; can rearrange to give a new carbonium ion which can then react with a nucleophile; or can react with a nucleophile (water, the ROH that is formed, halide ion) present in the medium. Comparable courses may be followed by the $S_N2$ reaction, as in the following reactions:

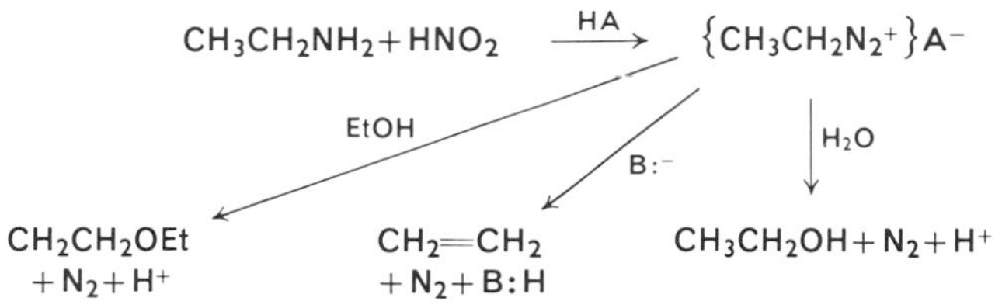

$$CH_3CH_2NH_2 + HNO_2 \xrightarrow{\text{HA}} \{CH_3CH_2N_2{}^+\}A^-$$

$$\overset{\text{EtOH}}{\swarrow} \qquad \overset{B:^-}{\swarrow} \qquad \overset{H_2O}{\downarrow} \qquad (15)$$

$$CH_2CH_2OEt \qquad CH_2{=}CH_2 \qquad CH_3CH_2OH + N_2 + H^+$$
$$+ N_2 + H^+ \qquad + N_2 + B:H$$

The extreme instability of $RN_2^+$, as compared with an alkyl halide, $RX$, is the consequence of the ease with which the very stable nitrogen molecule is lost.

It can be seen that the diazotization of aliphatic amines is simply another way to produce carbonium ions, with the consequences that are to be expected as a result.

---

**Exercise 5.** Predict the result of treating each of the following amines with nitrous acid in aqueous solution. *(a)* $(CH_3)_3CNH_2$; *(b)* $(CH_3)_3CCH_2NH_2$; *(c)* $CH_3NH_2$; *(d)* $(CH_3)_2C{-}C(CH_3)_2$;

with $HO$ and $NH_2$ on the last two carbons.

*(e)* $CH_3CH{=}CHCH_2NH_2$.

---

## 26.6 The nature of the diazotization reaction and the structure of the diazonium ion

The structure of the diazonium ion is best discussed for the case of the aromatic diazonium salts, since these are relatively stable compounds which can be prepared and preserved in solution and can often be isolated as pure substances.

The reaction of nitrous acid with a primary aromatic amine appears, from studies of the mechanism of the reaction, to involve the free amine and the species $N_2O_3$ ($O{=}N{-}O{-}N{=}O$, the anhydride of nitrous acid). The initial step can be regarded as a nucleophilic attack of $-NH_2$ on $N_2O_3$ with displacement of $NO_2^-$:

$$Ar{-}\overset{\underset{\displaystyle H}{|}}{\underset{\underset{\displaystyle H}{|}}{N}}{:}\,{\to}\,N{-}ONO \longrightarrow Ar{-}\overset{+}{N}H_2{-}NO + NO_2^- \tag{16}$$

Subsequent proton exchange in the product of the initial attack (16) and loss of a molecule of water leads to the formation of the diazonium ion:

$$Ar\overset{+}{N}H_2N{=}O \rightleftarrows Ar\overset{+}{N}H{-}\overset{\cdot}{N}{-}OH \longleftrightarrow Ar\overset{+}{N}H{=}N{-}OH \rightleftarrows$$

$$Ar N{=}N{-}\overset{+}{O}H_2 \rightleftarrows Ar N{=}\overset{+}{N} + H_2O \tag{17}$$

---

**Exercise 6.** Show the electronic details of these reactions by rewriting the above equilibria with complete electronic structures.

---

The relative stability of the aromatic diazonium ion is the result of the stabilization associated with the resonance among the contributing structures of the hybrid:*

$$\tag{18}$$

---

* In writing the structure of the ion (18) the charges are placed on discrete atoms for the sake of clarity in presentation.

Contributions pictured in (18) distinguish the aromatic diazonium salts from the aliphatic, since the latter are incapable of enlisting effective cooperation of the R group in contributions of this kind.

## 26.7 The diazotization process and the nature of diazonium salts

The diazotization reaction is ordinarily carried out by the addition of sodium nitrite to an ice-cooled solution of the aromatic amine in aqueous hydrochloric or sulfuric acid (usually about 3 equivalents of acid per mole of amine). The amine hydrochloride or sulfate may be present initially as a suspension, but it dissolves as the diazotization proceeds. The final solution must be kept cold (0–5°C) to avoid decomposition of the diazonium salt, and under these conditions the salt may be preserved until it is used in subsequent manipulations.

Solid diazonium salts are seldom isolated but upon occasion may be prepared in pure form by carrying out the diazotization in alcohol (using an alkyl nitrite as the source of nitrous acid) and precipitating the diazonium salt by the addition of ether, in which it is insoluble. The dry diazonium salts are very unstable, and often decompose with explosive violence. Certain complex double salts of diazonium compounds are quite stable. The fluoborates are often stable enough to be dried and stored:

$$ArN_2{}^+Cl^- + HBF_4 \longrightarrow ArN_2{}^+BF_4{}^- + HCl$$

(a diazonium fluoborate;<br>insoluble, stable)

The diazotization of aromatic amines occupies a prominent place in organic chemistry, both from the standpoint of the industrial applications of the process and the remarkable versatility of the diazonium salts in many kinds of synthetic processes. A summary of the important transformations that the diazonium salts undergo, illustrated with the simplest example, benzenediazonium chloride, is given in the following chart:

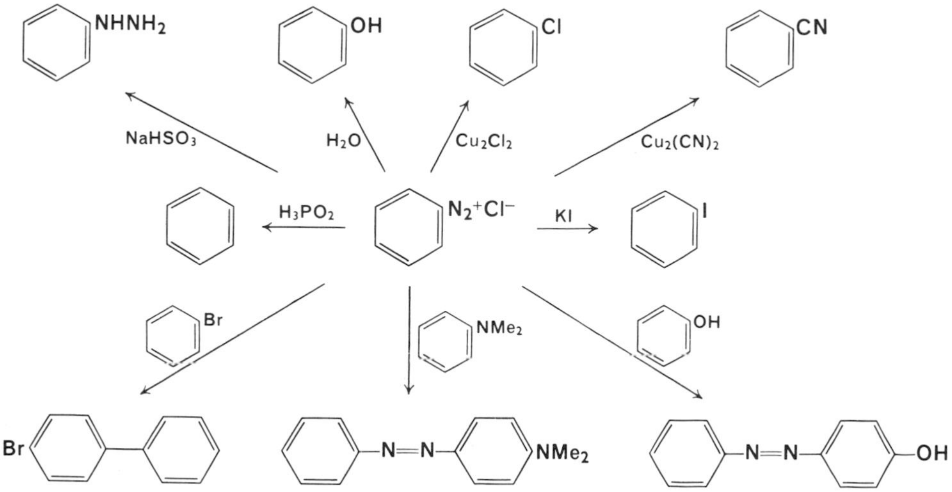

These reactions will be discussed in the sections to follow.

### 26.8 The diazonium ion as an electrophilic reagent

The aromatic diazonium ion is clearly *electrophilic*; it is deficient in electrons, a fact that can most easily be seen by examining the contributing structure Ar—$\overset{..}{N}$=$\overset{+}{N}$:, in which the terminal nitrogen atom can accommodate another pair of electrons and thus can coordinate with a nucleophile.

Aromatic diazonium salts are capable of attacking a sufficiently active aromatic nucleus, a typical electrophilic substitution being the result:

$$ArN_2^+\}Cl^- + \langle\!\!\!\bigcirc\!\!\!\rangle OH \longrightarrow Ar—N=N—\langle\!\!\!\bigcirc\!\!\!\rangle OH + HCl \tag{19}$$

This reaction is *not different in kind* from the many electrophilic substitution reactions we have already examined. It differs from nitration, acylation, and halogenation *in degree*, in that the diazonium ion is not a powerful electrophile and thus can attack only reactive aromatic nuclei. Hence, the *coupling reaction* is practically confined to the reaction of dizaonium salts with phenols and amines; some phenolic ethers and, in a few cases, reactive aromatic hydrocarbons, such as mesitylene, can also undergo coupling, but this is not general.

The activating influence of, for example, an amino substituent, is clearly the

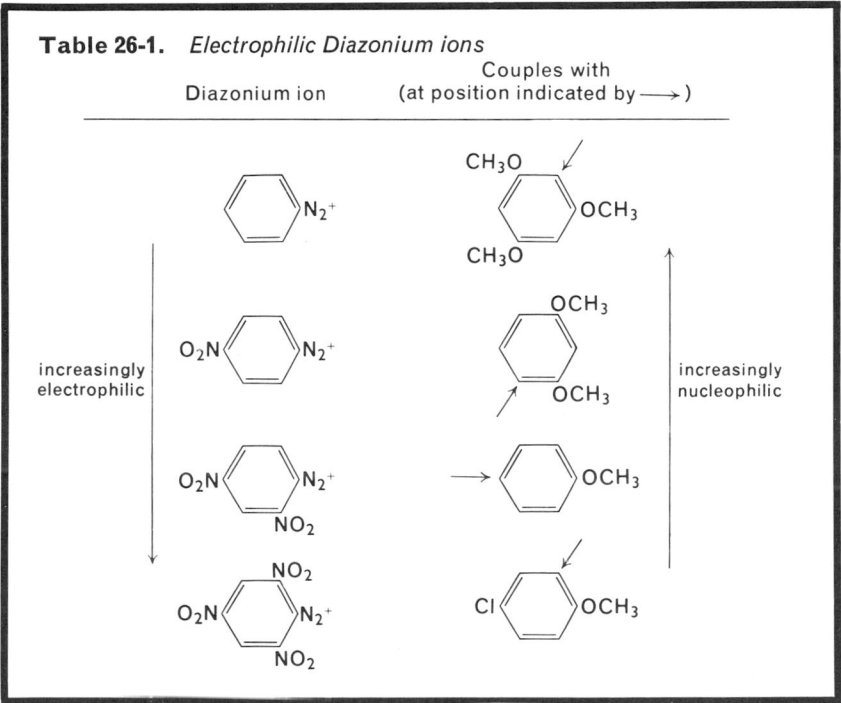

**Table 26-1.** *Electrophilic Diazonium ions*

same as the activating effect of an amino substituent in electrophilic substitutions in general:

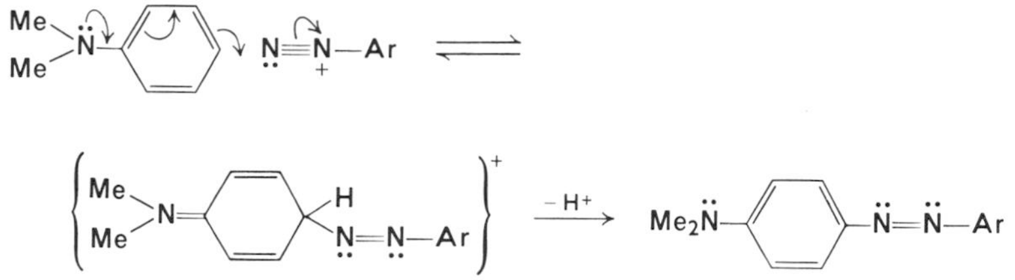

Since the electrophilic character of the diazonium ion will be increased by electron withdrawal from the nitrogen atoms, electron-attracting substituents in the aromatic ring of $ArN_2^+$ increase the coupling ability of the ion. *p*-Nitro-, 2,4-dinitro- and 2,4,6-trinitrobenzenediazonium salts are increasingly more reactive in coupling, and require correspondingly less active nuclei with which to react. In Table 26-1 is shown a series of increasingly more electrophilic diazonium ions. It can be seen that the benzenediazonium ion will couple only with the most reactive nucleus (phloroglucinol trimethyl ether); but with an increasing number of nitro groups in the diazonium ion, less reactive (i.e., nucleophilic) nuclei are required for successful coupling. Indeed, the trinitro compound couples with the ring bearing the deactivating halogen atom. The effect of the nitro group, for example, can be shown by writing the electrophilic attack in the following summary fashion:

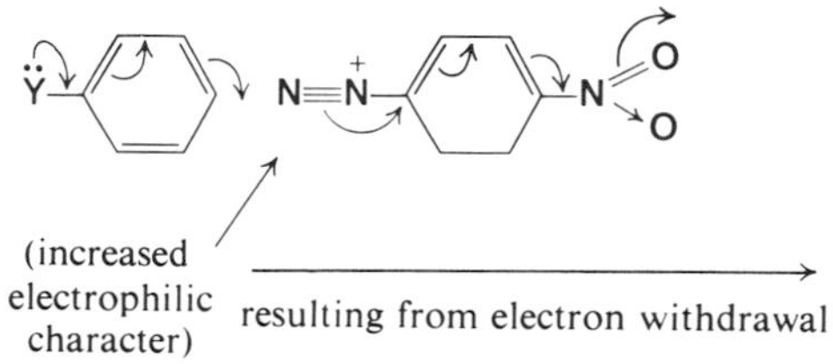

## 26.9 Effect of *p*H on the coupling reaction

Coupling is best carried out in solutions that are neither too strongly acidic nor alkaline. The reasons for this can be seen if we examine a typical case, that of the coupling of benzenediazonium chloride with dimethylaniline. *At high pH*, the diazonium ion will be present in very low concentration, since most of it will have been converted to $ArN{=}N{-}OH$ and $ArN{=}N{-}O^-$. Neither $ArN{=}NOH$ (the diazohydroxide) nor $ArN{=}N{-}O^-$ (the diazotate ion) is an electrophilic species and thus will not couple with the amine. *At low pH*, the dimethylaniline will be largely protonated to $Ar\overset{+}{N}Me_2H$, and thus the activating effect of the $-NMe_2$

group will be destroyed, since the nitrogen atom will no longer possess an unshared pair of electrons:

*high* pH:   ArN=N—OH   +   [Me—N(..)—Me aryl structure]   no coupling
or ArN=N—O⁻

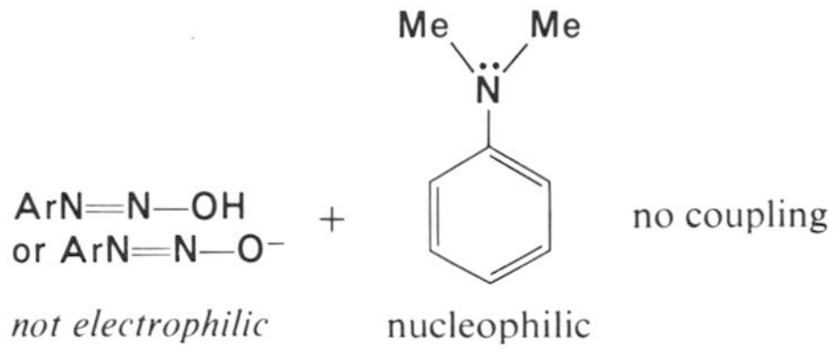

*not electrophilic*   nucleophilic

*low* pH:   Ar—N⁺≡N   +   [Me—N⁺(H)—Me aryl structure]   no coupling

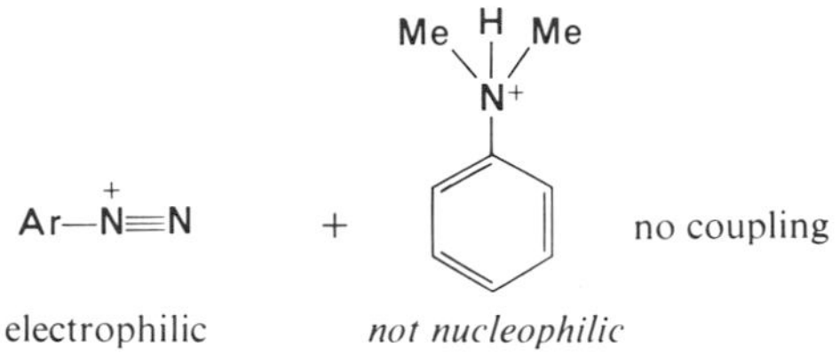

electrophilic   *not nucleophilic*

*intermediate* pH:   ArN⁺≡N   +   [Me—N(..)—Me aryl structure]   coupling

electrophilic   nucleophilic

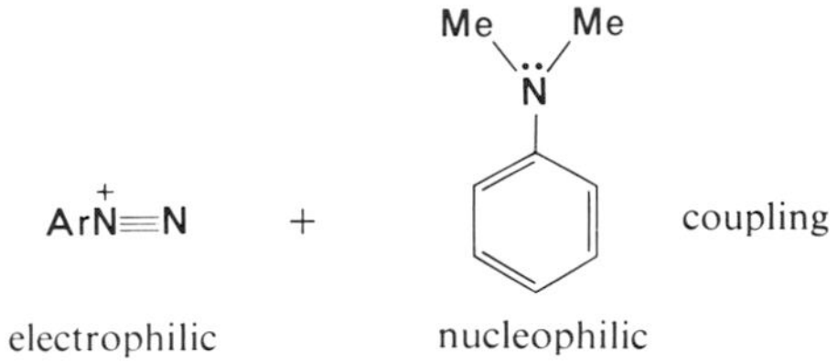

The effect of pH upon the coupling reaction is well illustrated by the coupling of the naphthalene derivative called "H-acid," with diazonium salts. At slightly acidic pH values, at which a large proportion of the amino groups are free (unprotonated), but at which ionization of the OH group is suppressed, the amino group exerts the stronger directing influence. At high pH, the hydroxyl group is ionized, and the strongly activating —O⁻ substituent is the stronger directing influence:

couples in weakly acidic solution   couples in basic solution

H₂N   OH

HO₃S   SO₃H

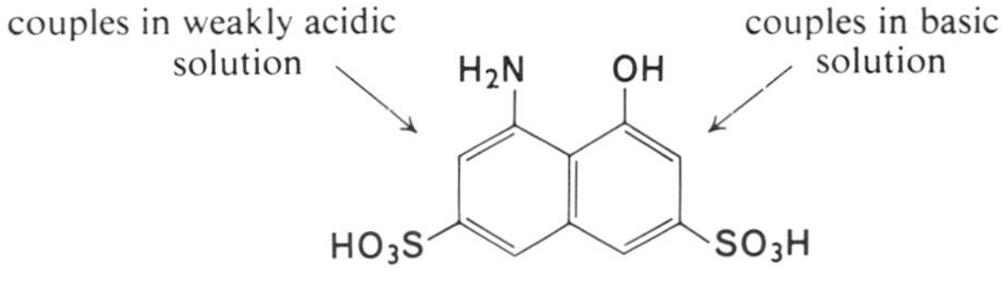

H-Acid

## 26.10   Azo compounds

The products of the diazonium coupling reactions contain the grouping —N=N—, and are called *azo compounds*:

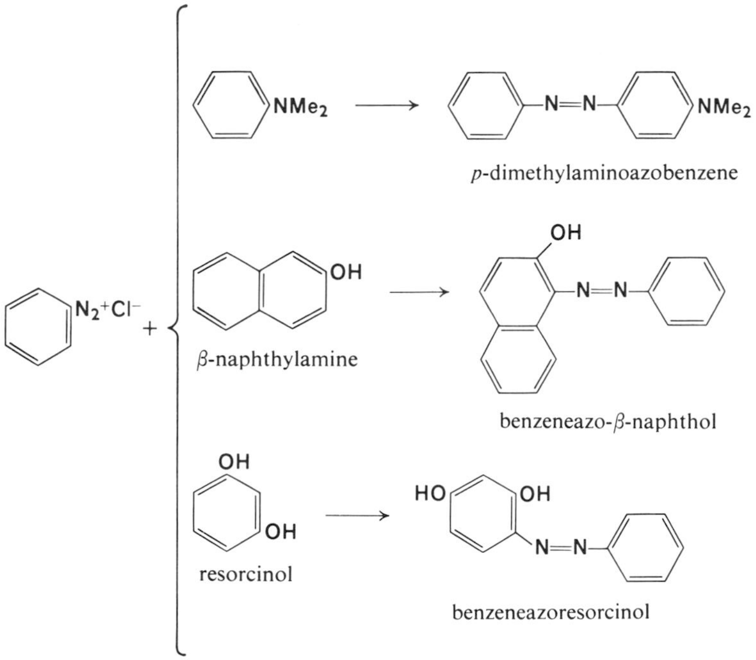
NMe2
N=N
NMe2
p-dimethylaminoazobenzene
N2+Cl−
+
OH
β-naphthylamine
OH
N=N
benzeneazo-β-naphthol
OH
OH
resorcinol
HO
OH
N=N
benzeneazoresorcinol

(21)

(22)

(23)

Azo compounds are colored; the simple examples are usually yellow, orange, or red, but with increasing complexity in the two partners involved in the coupling reaction an almost unlimited range of brilliantly colored substances can be formed. The important *azo dyes* are prepared by the coupling reaction, and are known in a large variety of colors and shades. Some examples of commercial azo dyes are shown in the following:

*Amine diazotized* *Coupled with (at ↓)* *Product*

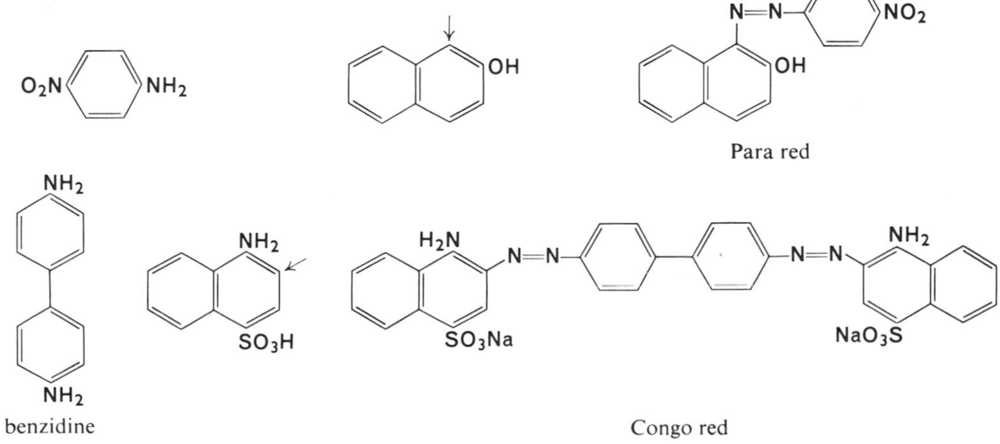
O2N
NH2
OH
N=N
NO2
OH
Para red
NH2
NH2
benzidine
NH2
SO3H
H2N
N=N
N=N
NH2
SO3Na
NaO3S
Congo red

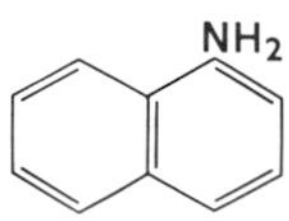

α-naphthylamine 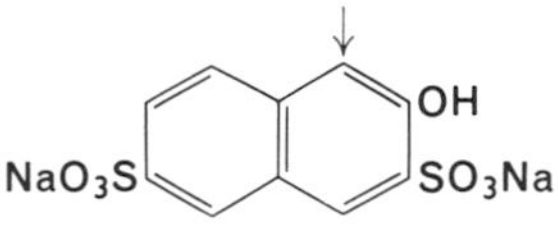 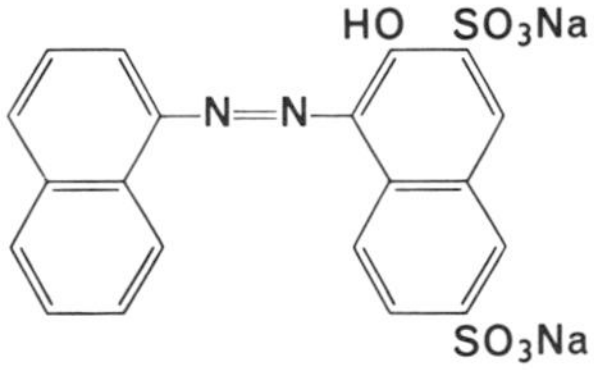                                                        Fast red B

Azo compounds containing free amino or hydroxyl groups are often useful as indicators, changing color with alteration of *pH*. The well known indicator, methyl orange, is formed by the coupling of diazotized sulfanilic acid with dimethylaniline:

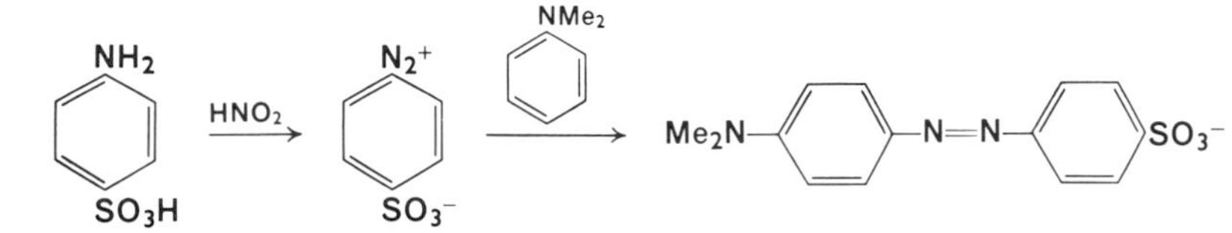

sulfanilic acid                                              methyl orange        (24)

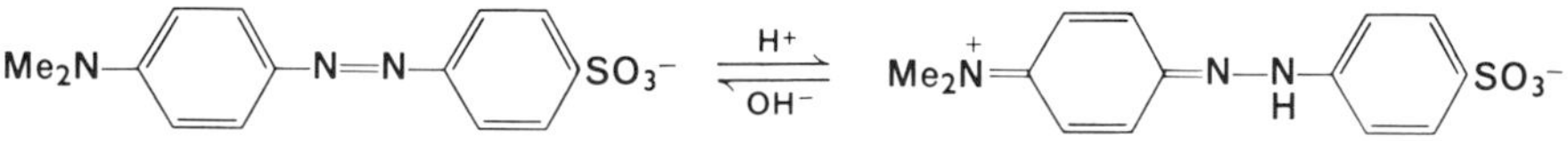

yellow-orange                                                        red
*p*H 4.4                                                            *p*H 3.1        (25)

## 26.11   Reactions related to the azo coupling reaction

Diazonium salts react with nucleophilic compounds other than amino- or hydroxyl-activated aromatic compounds. The *Japp-Klingemann reaction* is closely allied to the azo coupling reaction, and illustrates both an attack of the electrophilic diazonium ion upon a β-keto ester and a subsequent cleavage of a β,β-disubstituted β-keto ester. The reaction is carried out at intermediate *pH*, at which both OH⁻ and H⁺ may be utilized to play the roles shown in the following equations:

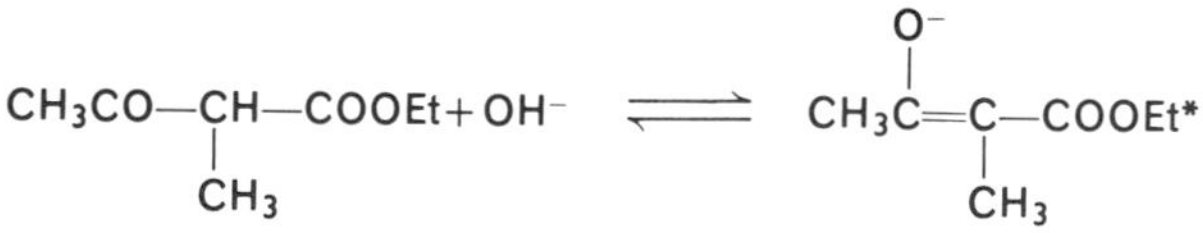

$$CH_3CO-\underset{\underset{CH_3}{|}}{CH}-COOEt+OH^- \;\rightleftharpoons\; CH_3\overset{\overset{O^-}{|}}{C}=\underset{\underset{CH_3}{|}}{C}-COOEt^* \qquad (26)$$

$$CH_3\overset{\overset{O^-}{||}}{C}-\underset{\underset{\overset{+}{N}=NAr}{|}}{\overset{\overset{CH_3}{|}}{C}}-COOEt \;\longrightarrow\; CH_3\overset{\overset{O}{||}}{C}-\underset{\underset{N=NAr}{|}}{\overset{\overset{CH_3}{|}}{C}}-COOEt \qquad (27)$$

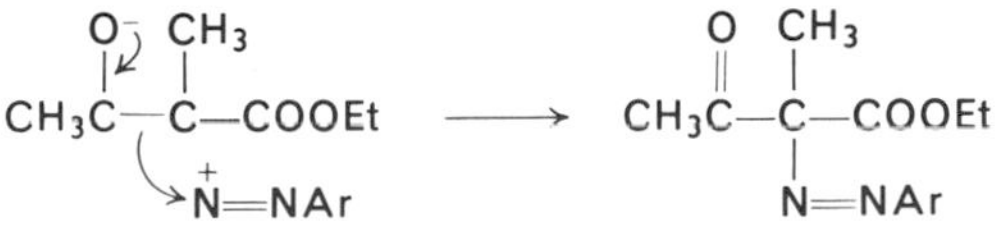

---

\* In writing equations in which an ion such as the anion derived from a β-keto ester is involved, an arbitrary selection of one of the forms that contribute to the resonance hybrid is usually made for simplicity and clarity in the formulation.

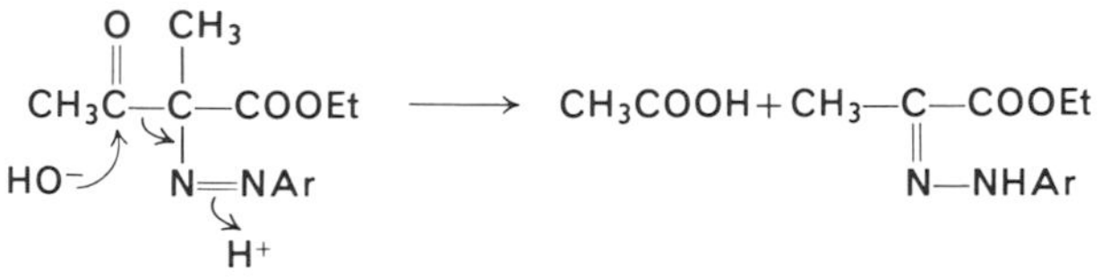

$$CH_3\overset{\overset{O}{\|}}{C}\underset{\underset{\underset{H^+}{N=NAr}}{|}}{\overset{\overset{CH_3}{|}}{\underset{HO^-}{C}}}-COOEt \longrightarrow CH_3COOH + CH_3\overset{\overset{CH_3}{|}}{\underset{\underset{N-NHAr}{\|}}{C}}-COOEt \tag{28}$$

Another example of a Japp-Klingemann reaction, the details of which should be worked out by the reader, is the following:

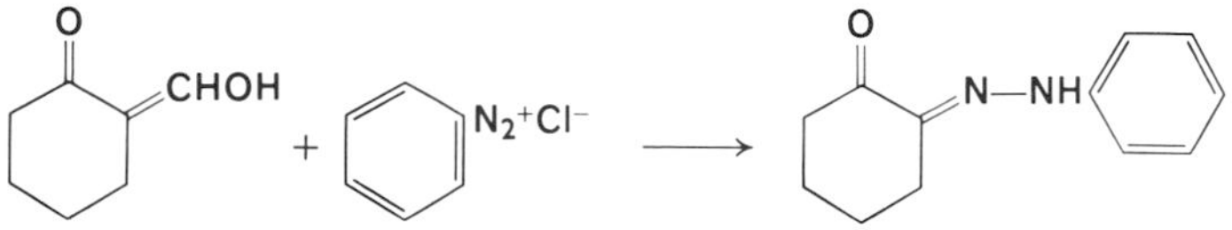

(29)

An examination of the products prepared in these two examples of the Japp-Klingemann reaction discloses that they are phenylhydrazones of an α-keto ester and of an α-diketone.

---

**Exercise 7.** The reaction of p-methoxyphenylacetylene with benzenediazonium chloride at intermediate pH yields as the ultimate product the following compound:

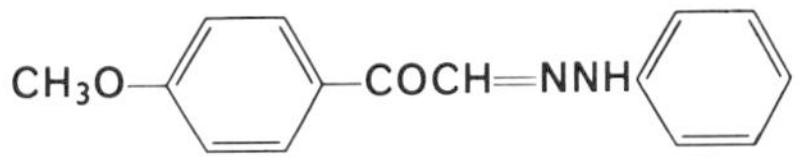

Show the course of this reaction by formulating the steps through which it proceeds.

---

## 26.12 Prontosil and sulfa drugs

One of the most fascinating stories of the role of organic chemistry in medicine involves a group of antibacterial azo dyes, of which an important member is 4′-sulfonamido-2,4-diaminoazobenzene, Prontosil:

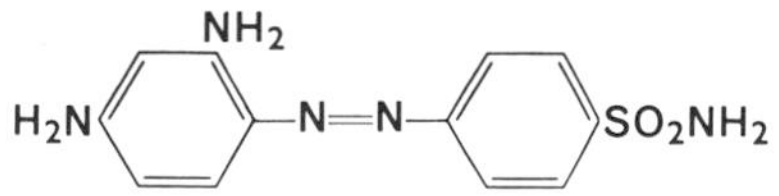

Prontosil

The discovery of the antibacterial action of Prontosil (and of a number of other azo compounds containing the 4′-sulfonamido group) was the result of the systematic testing of hundreds of such compounds, as part of a program of study which was based upon an early idea of Paul Ehrlich that it might be possible to find a dye that would selectively stain, or dye, a bacterial cell and thus destroy it. Prontosil was one such dye tested; and though it was found to be active against streptococcal infections *in vivo*, it was quite ineffective outside the body.

It was soon recognized that the active drugs of this type contained the grouping  , and investigators in France (Tréfouël, Bovet,

Fourneau) reasoned that perhaps the activity of the drug had nothing to do with its properties as a dye, but may have been the result of its conversion, by reduction, into *p*-aminobenzenesulfonamide:

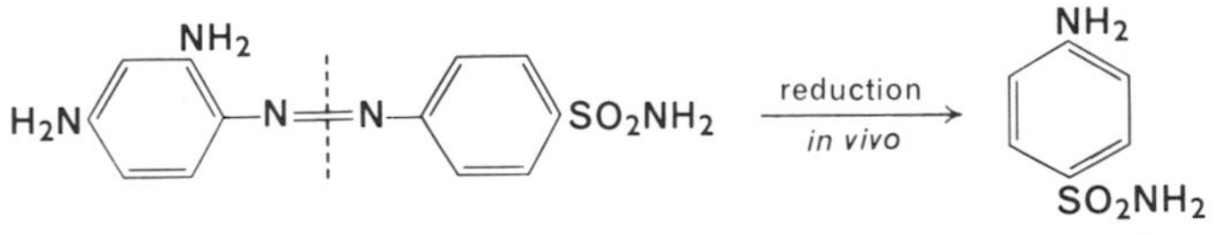

$$(30)$$

sulfanilamide

This reasoning led to the testing of sulfanilamide, and it was a brilliant verification of the hypothesis that this was indeed the active drug. Since that time, attention has turned from azo dyes and is now devoted to the study of substituted sulfanilamides, and today an extensive list of such drugs is in clinical use as effective anti-bacterial agents against a wide variety of bacterial diseases (see Section 33.2).

Thus, though azo dyes of the Prontosil type have been superseded by the sulfa drugs, they have an important place in the historical development of this still indispensable class of therapeutic agents.

## 26.13   Reduction of azo compounds

The *in vivo* reduction of the azo linkage in Prontosil is an example of a well known and valuable reaction of azo compounds. Reduction by various means—catalytic, metal-acid combination, sodium hydrosulfite ($Na_2S_2O_4$)—leads to rupture of the —N=N— linkage, with the formation of two —$NH_2$ groups.

$$Ar-N=N-Ar' \xrightarrow{\text{reduction}} ArNH_2 + Ar'NH_2$$

By a combination of coupling with a diazonium salt and reduction of the azo compound, the introduction of amino groups can be readily accomplished:

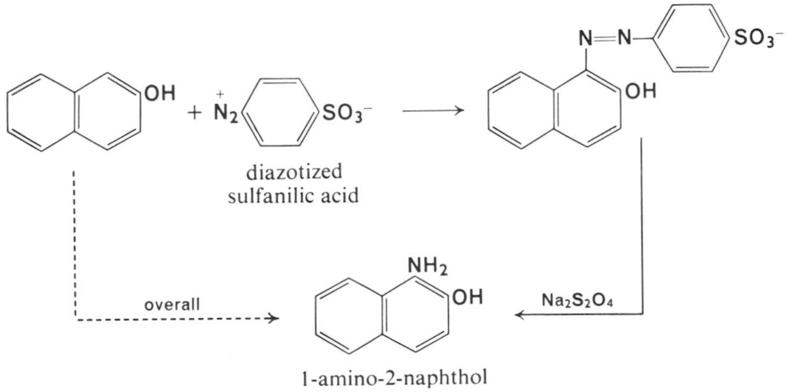

diazotized
sulfanilic acid

$$(31)$$

1-amino-2-naphthol

An application of the reduction of azo compounds to amines to the technical preparation of a drug is found in the conversion of phenol into *p*-phenetidine and thence phenacetin:

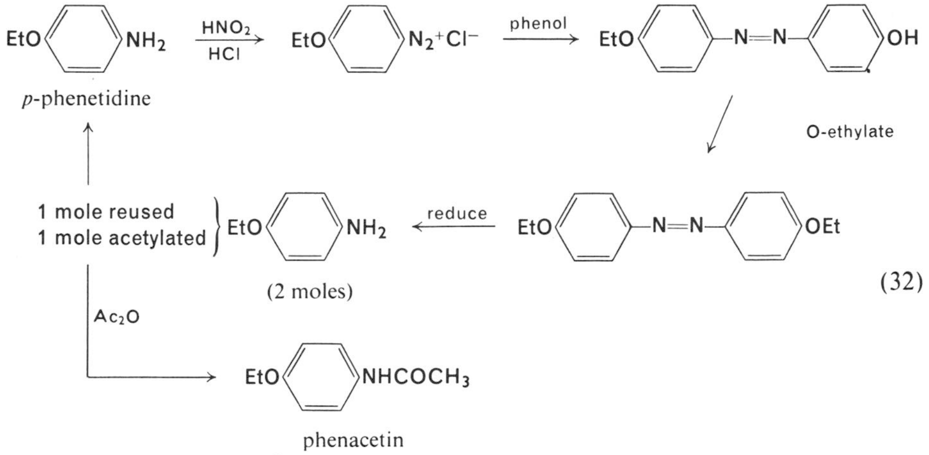

(32)

## 26.14 The replacement of the diazonium group with loss of nitrogen

The great utility of the diazotization reactions in synthetic operations depends upon the ability of diazonium salts to undergo replacement of the $-N_2^+$ grouping by halogen, cyano, hydroxyl, or hydrogen. This kind of reaction is called the "replacement" reaction of diazonium salts, and is typified by the following.

*Replacement by halogen (Sandmeyer reaction).* Treatment of a diazonium salt with one equivalent of potassium iodide in aqueous solution results in the evolution of nitrogen and the formation of the corresponding aryl iodide:

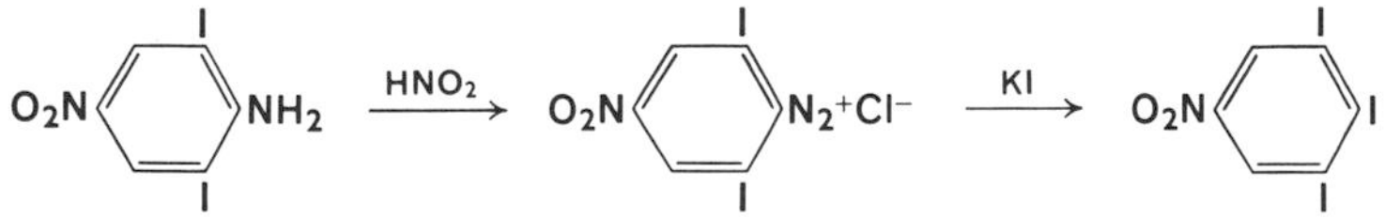

(33)

---

**Exercise 8.** What product would be formed if 3,4,5-triiodonitrobenzene were treated with potassium phenoxide?

---

The replacement of the diazonium group by chlorine and bromine does not take place under similar treatment of the diazotized amine with potassium chloride or bromide. However, if cuprous chloride or cuprous bromide are used, the replacement occurs in good yield:

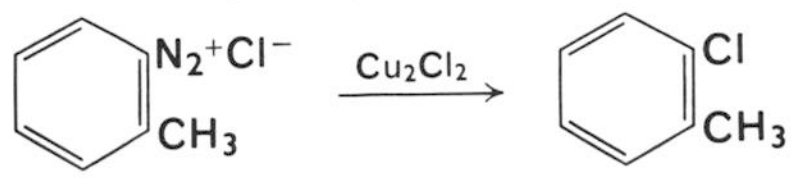

(34)

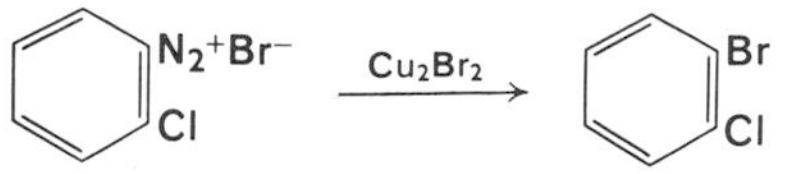

$$\text{(35)}$$

Replacement of —$N_2^+$ by fluorine is best carried out by a special method which involves heating of the dry diazonium fluoborate:

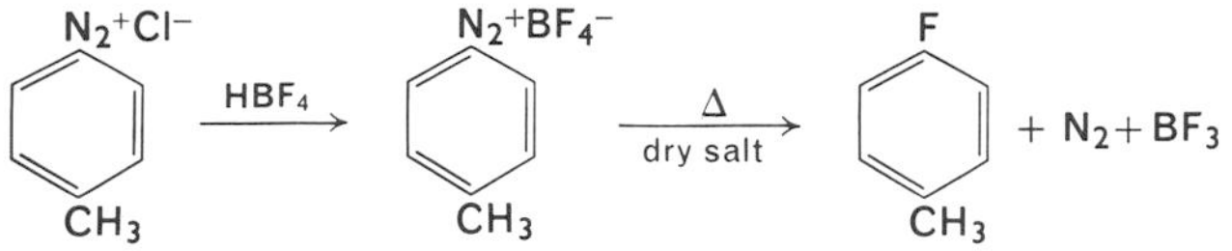

$$\text{(36)}$$

***Replacement by the cyano group.*** Treatment of the diazonium salt with cuprous cyanide yields the corresponding aryl cyanide:

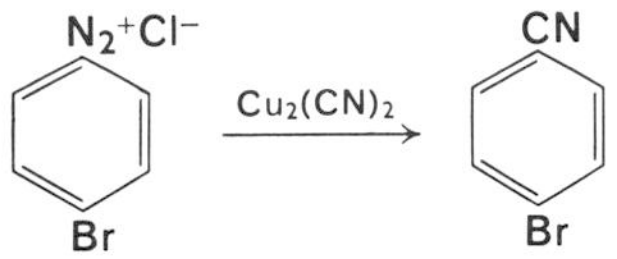

$$\text{(37)}$$

The usefulness of this reaction can be recognized when it is recalled that the cyano group can be hydrolyzed to the carboxyl group, and can be converted into —$CH_2NH_2$ by reduction or into —COR by treatment with the Grignard reagent, RMgX.

***Replacement by the hydroxyl group.*** Addition of the diazonium salt solution to hot, dilute acid results in the formation of a phenol:

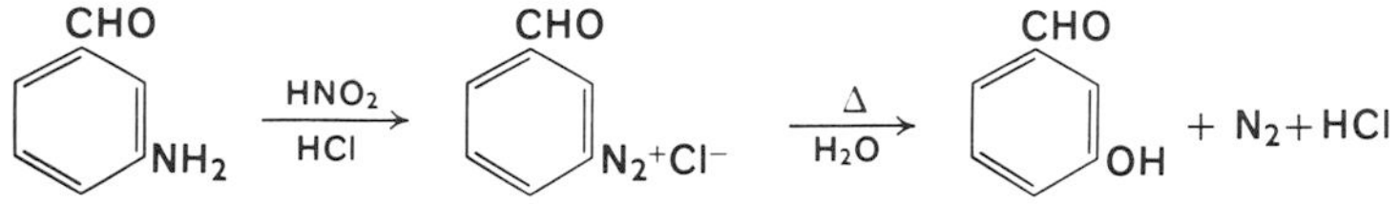

$$\text{(38)}$$

Since phenols couple with diazonium salts it is usually advantageous to add the diazonium salt solution slowly to the aqueous acid under conditions that permit the phenol to be removed by steam distillation as it is formed. The yields of phenols in this reaction vary with the degree and nature of substitution in the ring and with the nature of the salt (the anion) used. In general, the reaction is not as satisfactory in terms of yields as the other replacement reactions that have been described.

***Replacement by hydrogen.*** A very useful reaction of diazonium compounds is their "reduction" with either ethanol or, better, hypophosphorous acid. The result of this procedure is the removal of —$N_2^+$ and its replacement by hydrogen:

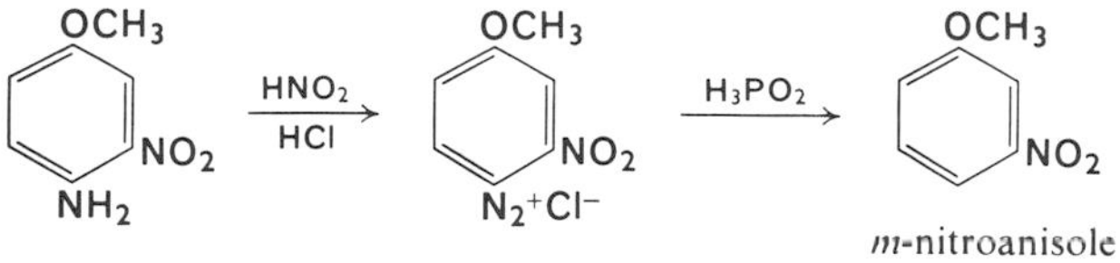

$$\text{(39)}$$

*m*-nitroanisole

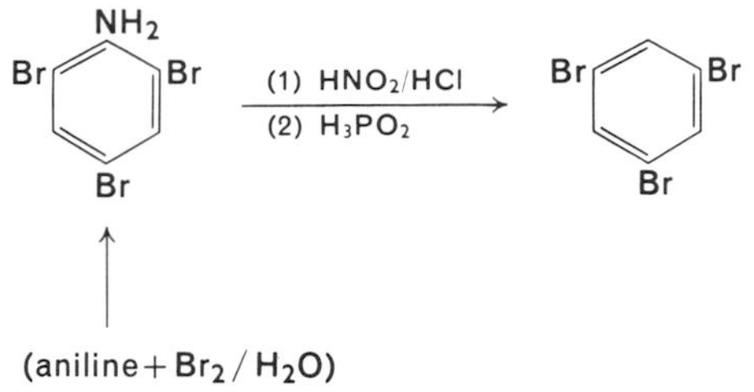

(40)

(aniline + Br$_2$ / H$_2$O)

---

**Exercise 9.** Devise a synthesis of *m*-nitroanisole starting with *m*-dinitrobenzene.

---

## 26.15  Replacement reactions in the synthesis of aromatic compounds

The utility of the various replacement reactions so far described in synthetic manipulations is very great indeed. Since nitration is so widely applicable a process, and the reduction of the nitro group to the amino group is so readily accomplished, the replacement of —N$_2^+$ by halogen, cyano, hydrogen, and so on, offers a ready means of preparing a wide variety of substituted benzenes. Moreover, the replacement by hydrogen allows us to introduce a nitro or amino group for purposes of directing orientation and then to remove it at a later time.

Suppose we wish to prepare *m*-nitrotoluene. Since nitration of toluene does not give the *m*-nitro compound, it cannot be made directly; moreover, the methyl group cannot be introduced into nitrobenzene by a Friedel-Crafts reaction, since nitrobenzene is unreactive in this reaction. It can be made by the following method from commercially available *p*-nitrotoluene:

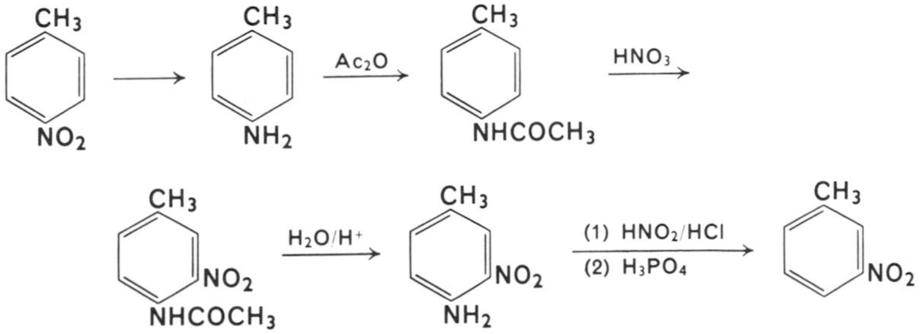

(41)

The preparation of 1,3,5-tribromobenzene from 2,4,6-tribromoaniline is shown above, and depends upon the ready preparation of the starting material by the direct bromination of aniline.

Another example is found in the preparation of *o*-toluic acid from the commercially available *o*-nitrotoluene:

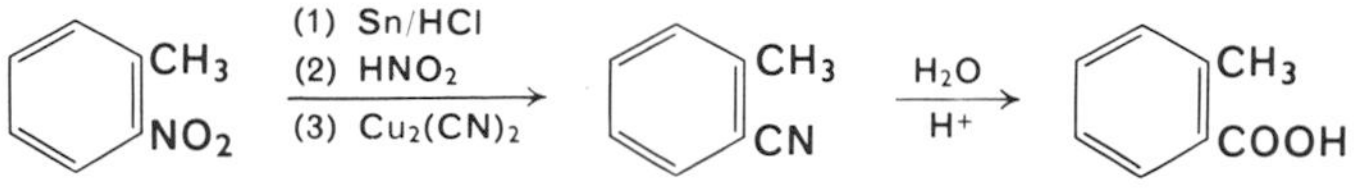

(42)

*o*-toluic acid

**Exercise 10.**  Show the preparation of the following, starting from *p*-nitrotoluene.

CH$_3$ ... Br   |   CH$_3$ ... Br, Br   |   CH$_3$ ... COCH$_2$CH$_3$   |   COOH ... Cl

## 26.16  The Meerwein reaction. Arylation of α,β-unsaturated acid derivatives

In the presence of cuprous chloride, aromatic diazonium salts decompose into free aryl radicals by a process that probably involves a one-electron change in the oxidation state of the cuprous ion:

$$ArN_2^+ + CuCl_2^- \longrightarrow Ar\cdot + N_2 + CuCl_2 \tag{43}$$

In the presence of an α,β-unsaturated nitrile or ester, the aryl radical adds to the double bond, with the formation of a radical which then reacts with the cupric chloride to yield the final product:

$$Ar\cdot + CH_2{=}CH{-}CN \longrightarrow ArCH_2{-}\overset{\cdot}{C}H{-}CN \xrightarrow{CuCl_2} ArCH_2\overset{Cl}{\underset{|}{C}}H{-}CN + CuCl \tag{44}$$

acrylonitrile

$$Ar\cdot + C_6H_5CH{=}CHCOOEt \longrightarrow C_6H_5\overset{\cdot}{C}H{-}\underset{\underset{Ar}{|}}{CH}COOEt \xrightarrow{CuCl_2}$$

ethyl cinnamate

$$C_6H_5\overset{Cl}{\underset{|}{C}}H{-}\underset{\underset{Ar}{|}}{CH}COOEt + CuCl \tag{45}$$

The noteworthy feature of this reaction is that the aryl group adds to the β-carbon atom of acrylonitrile, but to the α-carbon atom of ethyl cinnamate.

The formation of the free radical at the α-position of the nitrile permits resonance stabilization by participation of the cyano group. The free radical that would result by addition of the aryl radical to the α-carbon atom of acrylonitrile

$$\cdot CH_2{-}\underset{\underset{Ar}{|}}{CH}{-}CN$$

lacks opportunity for resonance stabilization.

In the case of ethyl cinnamate, however, the formation of the free radical with the free electron at the β-position permits resonance stabilization by participation of the phenyl group. It appears that the free radical α- to the phenyl group is more effectively stabilized than that α- to the —COOEt group.

## 26.17 The coupling of aromatic nuclei (Gomberg reaction)

The reaction of aromatic diazohydroxides with aromatic nuclei is a reaction different in type from either the coupling reaction or the displacement reaction. It is carried out by adding alkali to a well-stirred mixture of the diazonium solution and an aromatic compound, such as toluene:

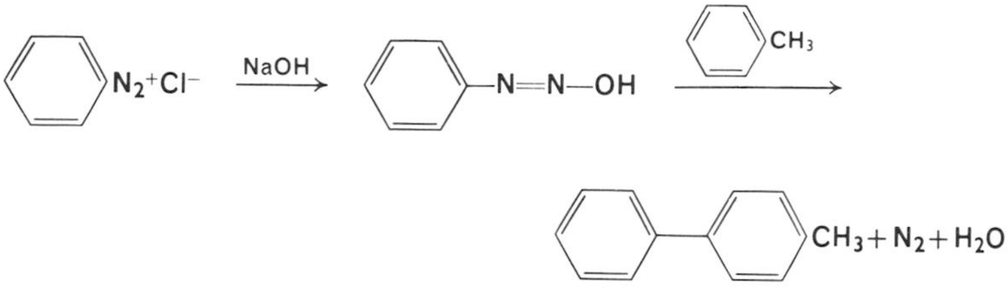

$$CH_3 + N_2 + H_2O \quad (46)$$

Under the alkaline conditions of the reaction, the diazonium compound is converted into the diazohydroxide. This, being a nonionic compound, is extracted into the toluene layer where it undergoes decomposition, with loss of nitrogen, and coupling ensues. It is probable that the diazohydroxide decomposes to yield an *aryl radical*, and that ions are not the reacting species:

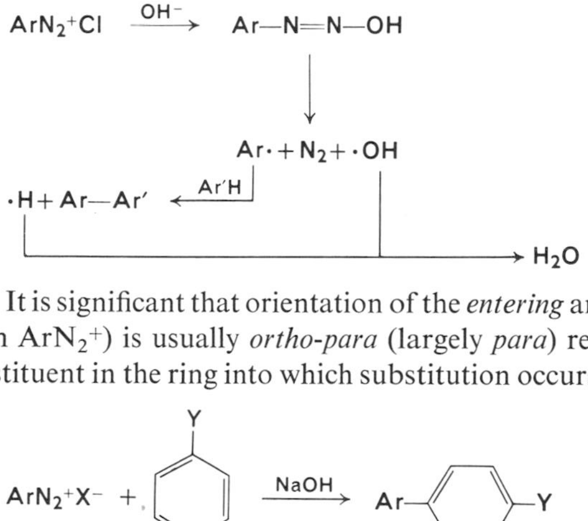

It is significant that orientation of the *entering* aryl group (i.e., the one derived from $ArN_2^+$) is usually *ortho-para* (largely *para*) regardless of the nature of the substituent in the ring into which substitution occurs:

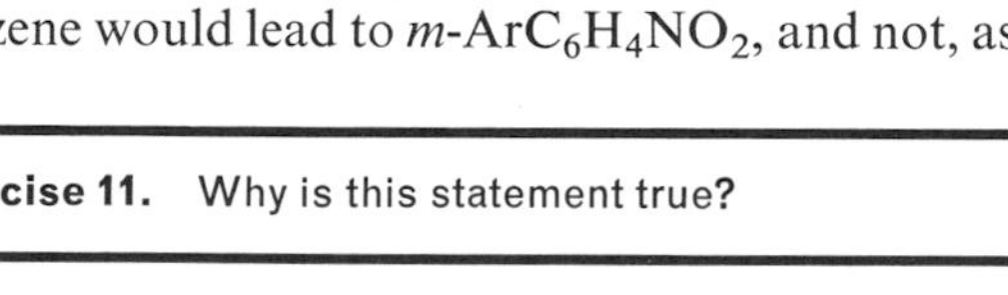

$$(Y = Br, CH_3, NO_2)$$

If the attacking species were $Ar^+$ we would expect that reaction with nitrobenzene would lead to $m\text{-}ArC_6H_4NO_2$, and not, as is observed, $p\text{-}ArC_6H_4NO_2$.

---

**Exercise 11.** Why is this statement true?

---

The yields of biphenyls prepared in this way are often poor, but the reaction is so readily performed that it has considerable utility in the synthesis of derivatives of biphenyl.

### 26.18 Diagnostic uses of diazonium reactions

A useful application of diazotization and coupling to yield an azo compound is found in a test for the presence of the nitro group or of the primary amino group in an aromatic ring. If the unknown substance is reduced, diazotized, and treated with an alkaline solution of $\beta$-naphthol, the formation of a red azo compound is evidence that the original compound contains a nitro group. Similarly, an aromatic amine can be classified as a primary amine if it can be diazotized and coupled with $\beta$-naphthol to form a red dye.

When this process is adapted for colorimetric measurement of the intensity of the red color, it can be employed as an analytical method for measurement of aromatic nitro or primary amino compounds present in micro amounts.

**Exercise 12.** Devise a quantitative method for determining the amount of nitrobenzene in the atmosphere of an industrial plant in which nitrobenzene is being manufactured.

### 26.19 Syntheses involving orientation problems

The various means of introducing, removing, and altering substituents in aromatic rings serve as the means for preparing a nearly unlimited array of compounds. Many of the synthetic operations that are possible are illustrated in the examples given throughout the text, and in the following paragraphs are described the solutions to a few specific problems in synthesis:

1. Synthesize

This cannot be prepared by bromination of benzoic acid, and no satisfactory method is available for the direct introduction of —COOH into bromobenzene. The desired compound can be prepared by oxidation of *o*-bromotoluene. The preparation of *o*-bromotoluene by Friedel-Crafts alkylation of bromobenzene is impractical because the orientation of the first methyl group would be largely *para*, and polyalkylation (and probably partial debromination) would ensue. The nitration or bromination of toluene to a mixture of *o*- and *p*-nitro- or -bromotoluene would be practicable, and would be an acceptable method, since the separation of the *o*- and *p*-nitro- or -bromotoluenes could be accomplished. A method that involves no isomer separation would be that shown on the opposite page.

**Exercise 13.** How could *o*-bromobenzoic acid be prepared starting with *o*-nitrotoluene?

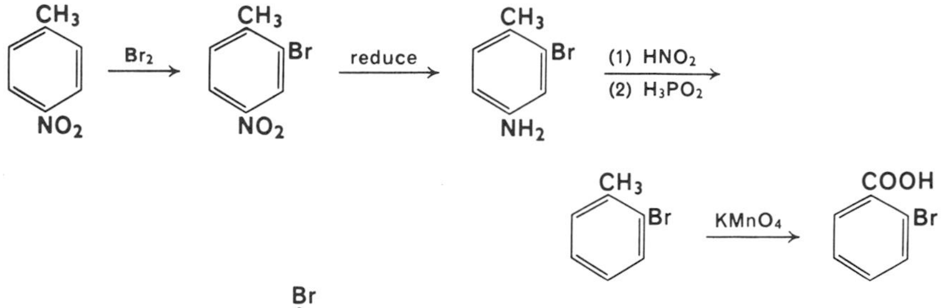

2. Synthesize

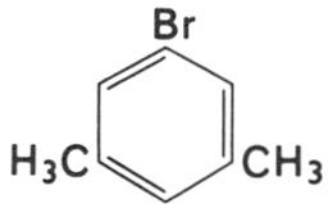

(a)

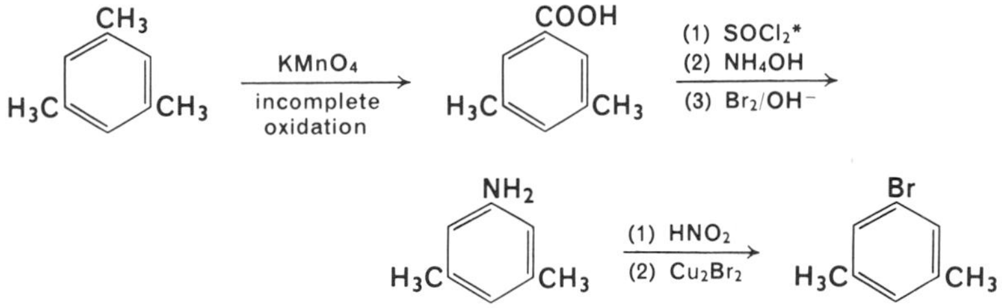

(b)

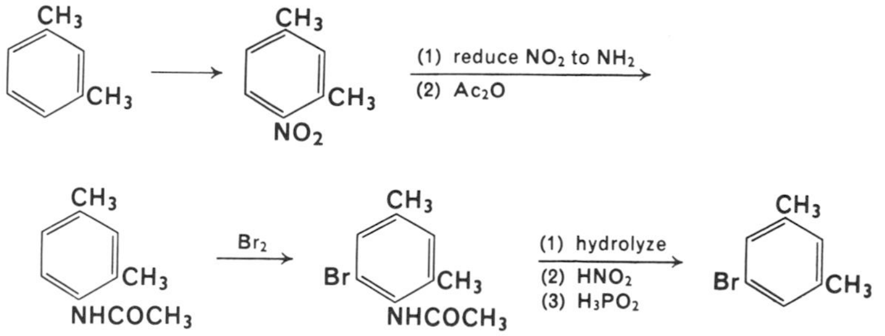

3. Synthesize

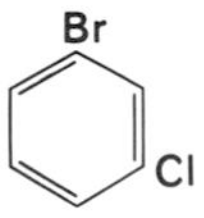

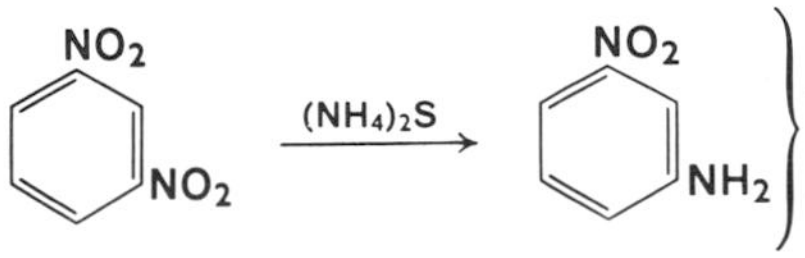

replace —$NH_2$ by —Br as in 2(a),
then reduce —$NO_2$ to —$NH_2$
and replace by —Cl in the same
way (using $Cu_2Cl_2$)

---

* This transformation of a carboxylic acid into an amine is called the Hofmann reaction. It is described in Section 28.8.

4. Synthesize

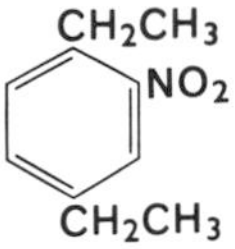

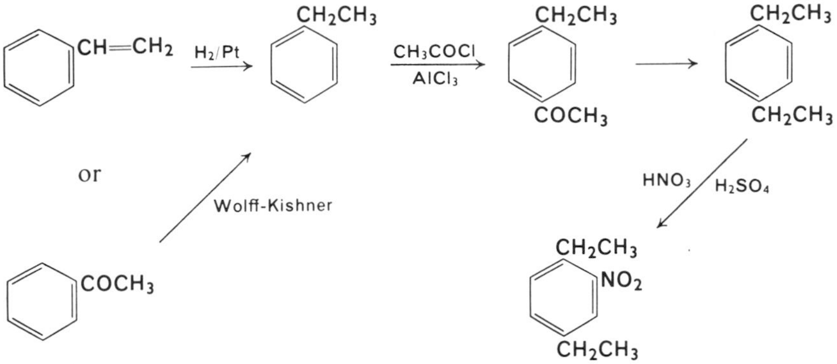

The nitro group would be introduced as the last step, since it would not survive the reduction steps, and it would not be practicable to attempt to acetylate *o*-ethylnitrobenzene, even if this were available.

5. Synthesize

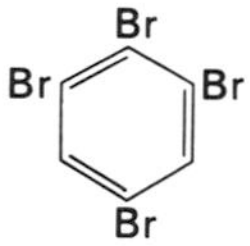

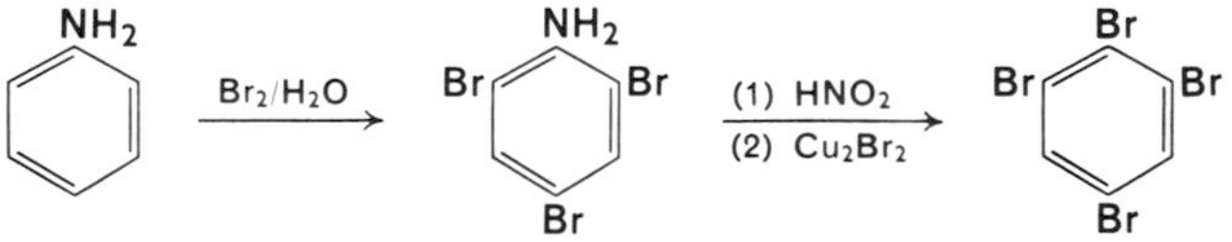

6. Synthesize

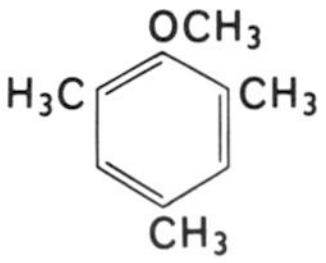

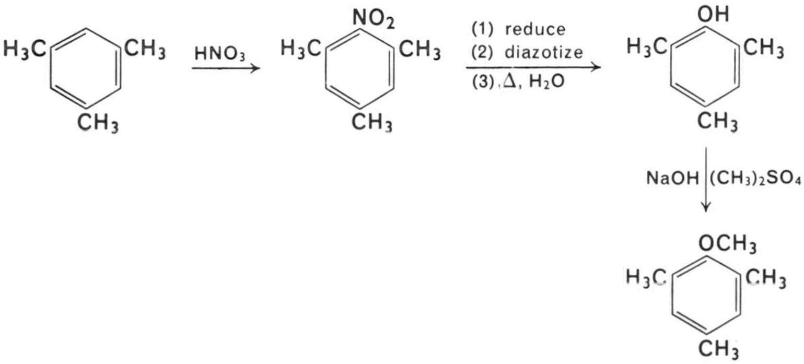

## Exercises

(Exercises 1–13 will be found within the text.)

**14** Write the equation for the reaction of nitrous acid (i.e., sodium nitrite and aqueous mineral acid) with each of the following: *(a)* N-methylaniline; *(b)* di-*n*-butylamine; *(c)* triethylamine; *(d)* p-chloroaniline; *(e)* m-toluidine; *(f)* N,N-diethylaniline.

**15** Starting with any monosubstituted benzene, devise syntheses for

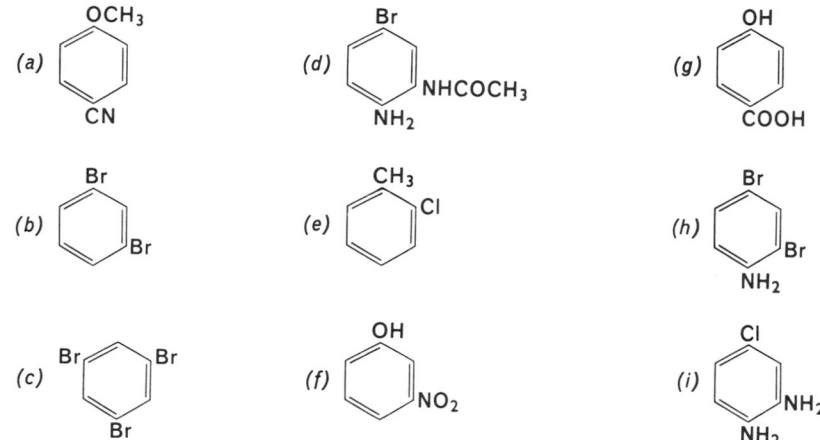

**16** Of the following compounds, select those that will couple with benzenediazonium chloride to yield an azo compound. Write the reactions: *(a)* benzene; *(b)* toluene; *(c)* phenol; *(d)* N,N-diethylaniline; *(e)* anisole; *(f)* resorcinol; *(g)* β-naphthol; *(h)* o-nitrotoluene; *(i)* benzoic acid; *(j)* o-cresol.

**17** Show how you could prepare the following compounds, starting with benzene and using the diazo reaction at some stage in the synthesis: *(a)* benzoic acid; *(b)* anisole; *(c)* p-hydroxyazobenzene; *(d)* p-dibromobenzene; *(e)* p-aminophenol.

**18** Look up in *Organic Syntheses* the description of the diazotization of an aromatic amine. Using this description as a model, write out complete and detailed experimental directions for the diazotization of 17.0 g of o-bromoaniline.

**19** Show how you could distinguish between the compounds in each of the following pairs. Use chemical methods, and describe the experimental result of each test (color reaction, solubility, and so on): *(a)* N-methyl-o-toluidine and N,N-dimethylaniline; *(b)* m-toluidine and cyclohexylamine; *(c)* acetanilide and p-aminoacetophenone; *(d)* aniline hydrobromide and p-bromoaniline; *(e)* phenol and anisole; *(f)* m-aminophenol and m-toluidine; *(g)* p-bromoaniline and p-aminobenzyl bromide.

**20** How could you separate a mixture of aniline, acetanilide, and m-aminophenol into the pure components?

**21** Describe the method you would use to prove the structure of m-toluidine. (Degradative, analytical, and synthetic means may be employed.)

**22** Write equations showing the chemical changes that would take place when an acidic solution of benzenediazonium chloride is made increasingly more alkaline, until it is finally strongly alkaline.

**23** What would you expect to obtain as the product if a solution of p-aminobenzylamine were diazotized and the resulting solution allowed to warm up until nitrogen evolution was at an end?

**24** Although N,N-dimethylaniline couples readily with benzenediazonium chloride, 2,6,N,N-tetramethylaniline does not. Explain.

## Chapter twenty-seven

**Phenols and**

# Hydroxycarbonyl Compounds

## 27.1 Characteristic properties of phenols

Phenols are aromatic hydroxy compounds in which the hydroxyl group is attached directly to the aromatic ring:

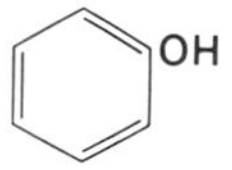

phenol 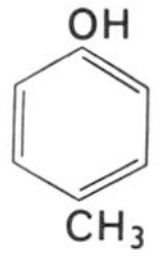       *p*-cresol 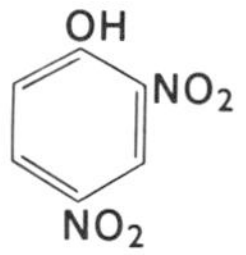       2,4-dinitrophenol 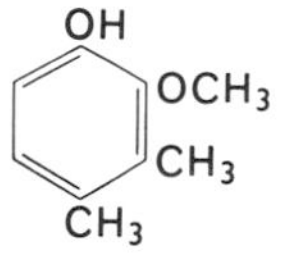       2-methoxy-3,4-dimethylphenol

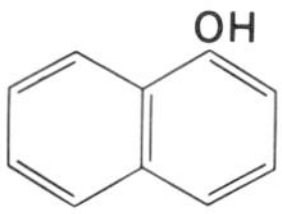

α-naphthol
1-naphthol

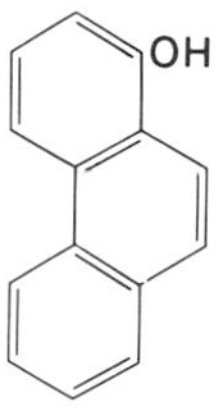

1-phenanthrol

     Phenols bear a formal resemblance to, and indeed undergo some reactions in common with, alcohols; but they deserve special consideration because of the profound effect of the ring upon the properties of the hydroxyl group and the influence of the hydroxyl group upon the properties of the ring. It is not proper simply to regard phenols as "alcohols" of the aromatic series.

The differences between, for example, phenol and 1-hexanol are very great. Phenol is appreciably acidic ($K_a = 10^{-10}$); consequently, the basic character (nucleophilic character) of the hydroxyl group is diminished. In neutral or acid solution, in which the unionized hydroxyl group reacts, 1-hexanol is more reactive than phenol toward reagents such as acid halides and anhydrides and other substances that react by accepting the unshared electron pair on the oxygen atom of —OH.

Under basic conditions, however, phenol is an excellent nucleophile, the nucleophilic species being the *phenoxide ion*:

$$\text{C}_6\text{H}_5\text{—OH} + \text{NaOH} \rightleftharpoons \text{C}_6\text{H}_5\text{—}\ddot{\text{O}}\text{:}^- \ \text{Na}^+ + \text{H}_2\text{O} \tag{1}$$

Phenols can be distinguished from alcohols, if both are water-insoluble, by the solubility of the phenols in aqueous alkali; but since simple phenols are very weak acids, they do not dissolve in aqueous sodium bicarbonate solution and so can be distinguished from carboxylic acids, which are bicarbonate-soluble. These same properties make possible a convenient means of separating phenols and carboxylic acids: if an alkaline solution of a phenol and a carboxylic acid is saturated with carbon dioxide, the phenol separates, while the acid remains in solution as the sodium salt. After removal of the phenol by filtration or extraction, acidification of the remaining aqueous solution with a mineral acid liberates the carboxylic acid, and permits its isolation.

Phenol is a very toxic substance and is corrosive to living tissue by virtue of its destructive action on proteins. Its common name, "carbolic acid," is indicative of this property rather than of its acid strength. Phenol was first used, in the nineteenth century, for antisepsis and was the first substance to be used generally in surgical operations and in the disinfection of hospital premises and equipment. Although phenol has been superseded in medicine by other, more effective antiseptics, commercial preparations containing mixtures of simple phenols (e.g., cresols) are still marketed for household disinfectant uses. The term "phenol coefficient" refers to a number that compares the effectiveness of bactericidal substances against a standard solution of phenol.

Other therapeutic uses of certain substituted phenols will be described later in this chapter.

## 27.2  Synthesis of phenols

Several methods for introducing a hydroxyl group into the aromatic ring have been discussed in other chapters, and are only to be recalled here:

1. The synthesis of phenol itself on a technical scale by the high pressure and temperature hydrolysis of chlorobenzene (Section 22.3).

2. The transformation of the aromatic amino group into a hydroxyl group by diazotization and hydrolysis of the diazonium salt (Section 26.14).

3. Replacement of aromatic halogen, when it is in *ortho* and *para* positions to an activating group such as nitro or nitroso, by hydrolysis (Section 22.5). In this category is also found the hydrolysis of *p*-nitrosodimethylaniline to *p*-nitrosophenol.

A general method of preparing phenols is the fusion of aromatic sulfonic acids with alkali. This is a valuable industrial method, and is used for the manufacture of a number of phenols of technical importance. Several examples of the reaction are the following:*

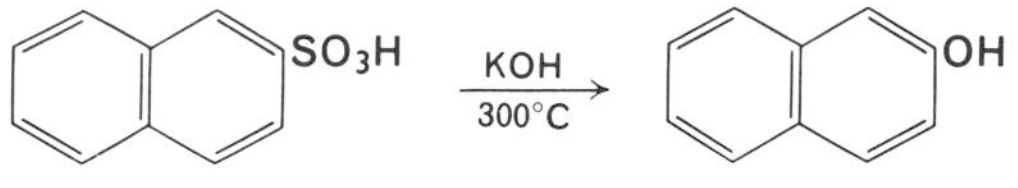

(2)

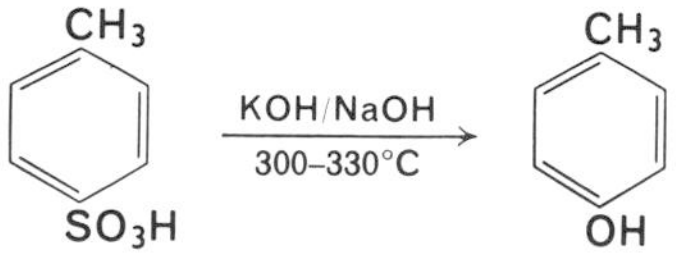

(3)

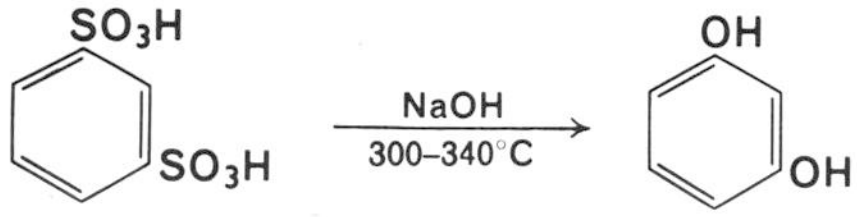

(4)

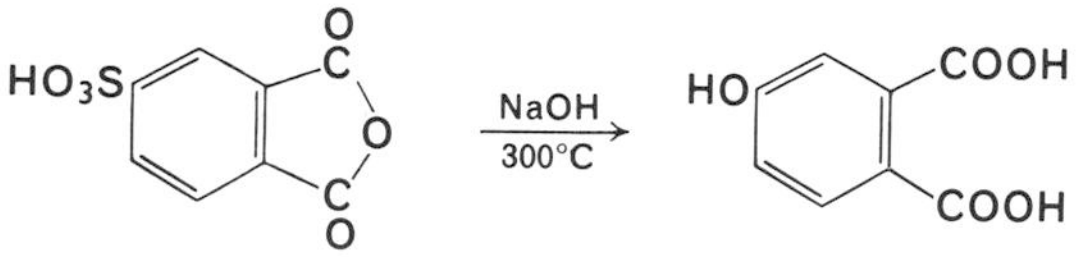

(5)

This method has important industrial applications in the preparation of phenols, particularly for use in coupling with diazotized amines in the production of azo dyes. It is limited by the availability of the necessary sulfonic acid, and by the possible sensitivity of other substituents to the severe conditions used. A presently important technical method of preparing phenol involves the interesting rearrangement of cumene hydroperoxide that is described in Chapter 28. The cumene hydroperoxide used in the process is prepared by the oxidation of isopropylbenzene.

Other methods of preparing phenols fall into numerous diverse classes of reactions. There are few useful methods of introducing a hydroxyl group directly into the aromatic nucleus (i.e., replacement of hydrogen); syntheses of phenols nearly always involve the replacement of some substituent by hydroxyl.

The direct oxidation of phenols with potassium persulfate does allow the introduction of an additional hydroxyl group into the ring. This is of most usefulness in the synthesis of polyhydroxy compounds of rather complex structures,

---

* In these reactions it is, of course, the sodium or potassium salt of the sulfonic acid that is present in the alkaline reaction mixture. Further, the immediate product of the reaction is the alkali metal salt of the phenol. A final acidification step would be used to liberate the free phenol.

chiefly because it is often more convenient to introduce a single hydroxyl group into a molecule of moderately complex structure than to build up the final structure in other ways:

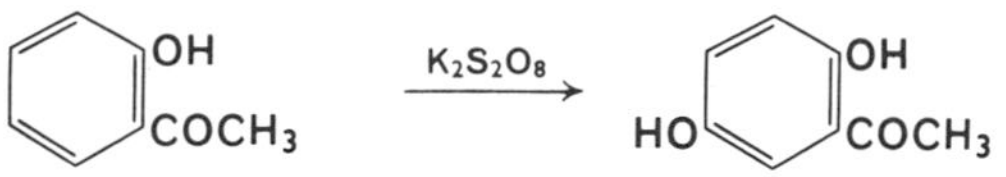

$$\text{(6)}$$

*o*-hydroxyacetophenone

In the above equation, only the final product is shown. An intermediate sulfuric acid ester of the final phenol is formed; this is seldom isolated, but is hydrolyzed with aqueous HCl to the products, as shown in Equation (7):

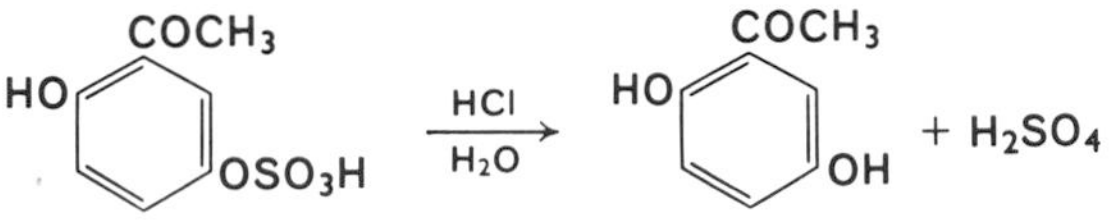

$$\text{(7)}$$

---

**Exercise 1.** The exact mechanism of this oxidation reaction has not been established with certainty. Both free radical and ionic mechanisms have been proposed. Using the simpler expression for persulfuric acid, $HO-OSO_2OH$, devise a rational course for the reaction to give the sulfate ester of Equation (7) by an ionic mechanism.

---

## 27.3 Effect of substituents upon the acidity of phenols

We have seen that the substitution of electron-attracting groups into the nucleus of aniline decreases the basic strength of the amine. It would be anticipated that the same effect would result in the case of similar substitution of phenol, and this is what is found. However, since phenol is ordinarily characterized not by the basicity of its —OH group but by the acidity of the hydrogen atom of this group, and in view of the reciprocal relationship between acidity and basicity of a function, the result is that *base-weakening substituents in aniline are acid-strengthening substituents in phenol:*

| Phenol | $K_a$ |
|---|---|
| unsubstituted | $1.3 \times 10^{-10}$ |
| 3-nitro | $1.0 \times 10^{-8}$ |
| 4-nitro | $6.5 \times 10^{-8}$ |
| 2,4-dinitro | $8.3 \times 10^{-5}$ |
| 2,4,6-trinitro | $4.2 \times 10^{-1}$ |

---

**Exercise 2.** Express the above dissociation constants as $pK_a$ values.

---

2,4-Dinitrophenol is a somewhat stronger acid than acetic acid, and 2,4,6-trinitrophenol is nearly in the class of "strong" acids. Indeed, the latter is commonly called picric acid. It is to be noted that *m*-nitrophenol is a weaker acid than *p*-nitrophenol. This shows that the effect of the nitro group upon the acidity is due to more than the inductive effect, since this should be more effective in the *m*-position (closer to —OH) than in the *p*-position. The resonance effect of the nitro group in stabilizing the anion

is important in contributing to the ease with which the proton can be removed.

---

**Exercise 3.** Inspect the structure of *m*-nitrophenol and show that no comparable structures, in which the negative charge resides on the nitro group, can be written for the anion.

---

The discussion of the basicity of aromatic amines is applicable to the acidity of phenols, and should be reviewed here (Section 25.1).

## 27.4   Oxygen alkylation and oxygen acylation of phenols

Alkylation of phenols can be accomplished readily by the use of alkyl halides, alkyl sulfates, or alkyl sulfonates in the presence of aqueous alkali:

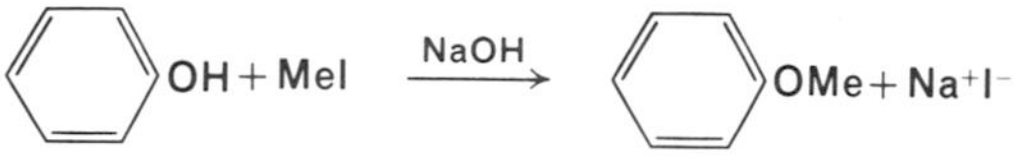

(8)

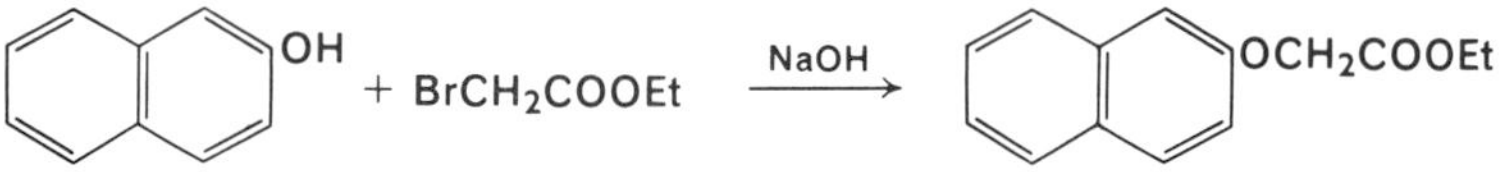

(9)

An alternative method of carrying out the alkylation of many phenols is by the use of an alkyl sulfate in the presence of anhydrous potassium carbonate in acetone as a solvent:

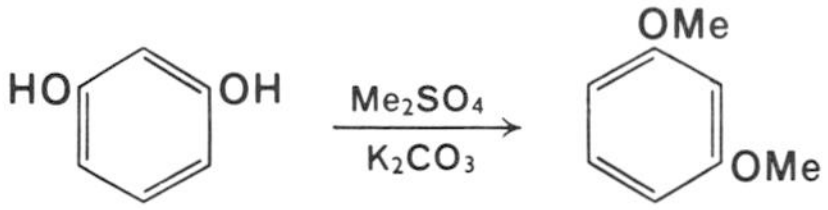

(10)

The similarity of these alkylations, which are simple nucleophilic displacement reactions, to the Williamson ether synthesis (Chapter 8) is apparent. The difference is one of degree, and lies in the experimental conditions used. The Williamson synthesis of dialkyl ethers requires the use of an alkoxide, usually prepared by the addition of sodium or potassium to an excess of the alcohol. Because of the very low degree of acidity of the alcoholic —OH group, the use of an alcohol and aqueous alkali is not effective for the preparation of aliphatic ethers, since the concentration of $RO^-$ would be low and the alkylating agent would be largely consumed by reaction with the alkali.

The acylation of phenols can be accomplished by the use of acid chlorides or anhydrides; the presence of a trace of sodium acetate (when acetic anhydride is the acylating agent) or pyridine; or a strong mineral acid (sulfuric, perchloric) catalyzes the reactions, the base by providing phenoxide ion, the acid by increasing the electrophilic character of the acid anhydride:

$$ArOH + OAc^- \rightleftharpoons ArO^- + HOAc$$
$$\downarrow AcCl$$
$$\longrightarrow ArOAc + Cl^- \qquad \text{base-catalyzed acylation} \qquad (11)$$

$$Ac_2O + H_2SO_4 \rightleftharpoons Ac_2\overset{+}{O}H + HSO_4^-$$
$$ArOH + Ac_2OH^+ \longrightarrow ArOAc + HOAc + H^+ \qquad \text{acid-catalyzed acylation} \qquad (12)$$

In the O-acylation of phenols with the (less reactive) aromatic acid chlorides (Schotten-Baumann reaction), aqueous alkali is used; again, the phenoxide ion formed is the actual nucleophile that reacts with the acid derivative.

The use of pyridine as the basic catalyst in the O-acylation of alcohols with aromatic acid chlorides probably involves the formation of an intermediate N-acyl pyridinium salt, whose reaction with the alcohol in a second step leads to the ester:

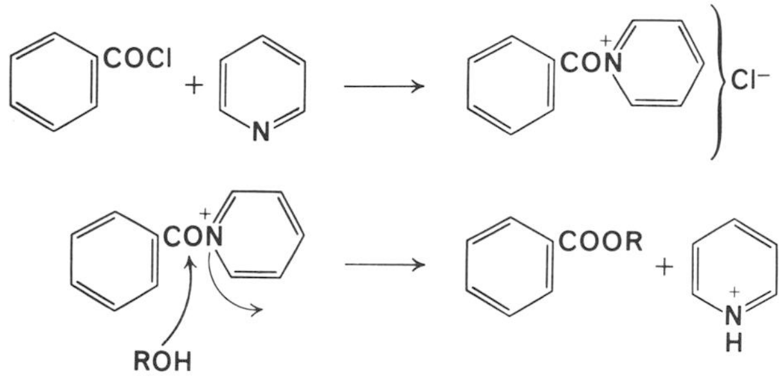

$$(13)$$

## 27.5 Cleavage of phenyl alkyl ethers

Phenyl alkyl ethers can be cleaved (hydrolyzed) by concentrated, aqueous hydriodic and hydrobromic acids. While this procedure is not an important preparative method for phenols (since phenyl alkyl ethers are not commonly available raw materials), it is of importance in laboratory operations involved in the proof of the structures of organic compounds. The cleavage probably proceeds by way of a nucleophilic attack of the halide ion on the protonated phenyl ether; the reaction is illustrated for the case of the cleavage of anisole by HBr:

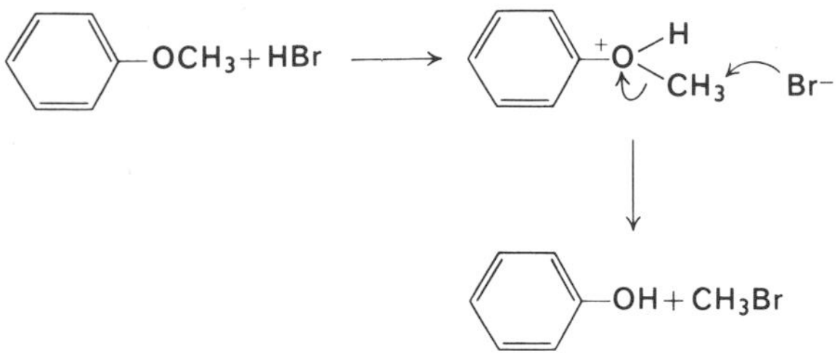

(14)

The cleavage of methyl and ethyl ethers of phenols by hydriodic acid is the basis for an important analytical method, called the *Zeisel alkoxyl determination*. A weighed sample of the ether is heated with concentrated HI, and the methyl or ethyl iodide is passed into an alcoholic solution of silver nitrate. The precipitated silver iodide is weighed, and the percentage of alkoxyl thus determined. The Zeisel method of alkoxyl determination is capable of a degree of accuracy that is comparable with that obtained in such other organic analytical methods as carbon-hydrogen, nitrogen, and halogen determinations.

The O-alkylation of phenolic hydroxyl groups is often carried out as a prelude to the performance upon another part of the molecule of reactions which would not be possible in the presence of free hydroxyl groups. In most cases, O-acylation would not be an acceptable alternative, for the hydrolysis of the O-acyl linkage under many reaction conditions—particularly in alkaline media—would defeat the purpose of the protective maneuver. On the other hand, although an O-methyl ether would be stable under many experimental conditions, the later removal of the protecting methyl group often requires undesirably drastic conditions. The ideal "protecting" group that would be stable under alkaline reaction conditions, or when Grignard reagents are used, would be an ether that would be cleaved under mildly acidic conditions, or that could be removed by other means. Two such protecting groups that are commonly used are O-benzyl groups, and those formed by reaction of the phenolic hydroxyl groups with dihydropyran.

Dihydropyran, a commercially available reagent, is a vinyl ether. In the presence of a proton donor (a trace of mineral acid may be used) it is readily protonated in the equilibrium (15):

(15)

The electrophilic tetrahydropyranyl ion is capable of reacting with a hydroxyl group in the equilibrium (16):

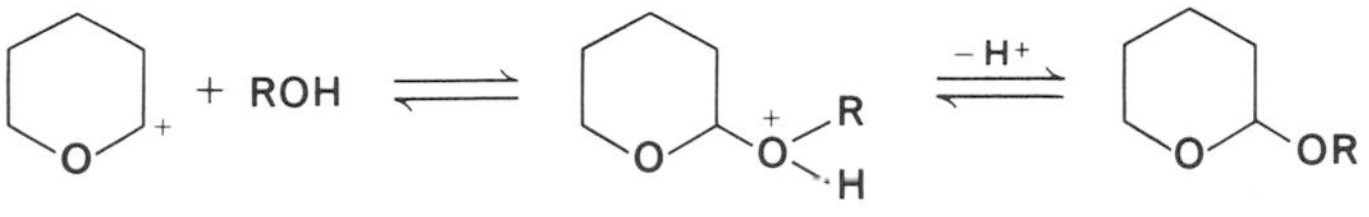

(16)

When dihydropyran is used in excess the tetrahydropyranyl ether is formed in substantially quantitative yield and, after neutralization of the trace of acid catalyst, can be isolated and used in subsequent reactions. It will be noted that the tetrahydropyranyl ether is an acetal, and thus is stable to alkaline reaction conditions; but it is very readily hydrolyzed by aqueous acid, with regeneration of the original hydroxyl group. The reaction is applicable to both phenolic and alcoholic hydroxyl groups, and can also be used to prepare tetrahydropyranyl esters of carboxylic acids.

An example of the use of tetrahydropyranylation is found in a convenient synthesis of certain phenolic carboxylic acids. When resorcinol dimethyl ether is allowed to react with *n*-butyllithium, a nuclear hydrogen atom is replaced with the formation of the aryllithium compound:

$$CH_3O \underset{\phantom{x}}{\bigcirc} OCH_3 \quad \xrightarrow{n\text{-BuLi}} \quad CH_3O \underset{\phantom{x}}{\overset{Li}{\bigcirc}} OCH_3 \quad + \; n\text{-butane} \tag{17}$$

Carbonation, followed by hydrolysis, yields 2,6-dimethoxybenzoic acid. Demethylation of this acid, with the formation of 2,6-dihydroxybenzoic acid, can be accomplished with the use of hydrogen iodide or hydrogen bromide, but the yields are often unsatisfactory. An alternative course, which is easily carried out and gives good yields, involves the use of the *bis*tetrahydropyranyl ether, which reacts with butyllithium in the same way as does the dimethyl ether. After the lithium derivative is carbonated, acid hydrolysis of the reaction mixture causes removal of the tetrahydropyranyl groups, and 2,6-dihydroxybenzoic acid is obtained without the need for a separate dealkylation step.

Phenolic O-benzyl ethers are much more readily cleaved by acids than are methyl ethers, but have the special advantage that the removal of the benzyl groups is easily accomplished by catalytic hydrogenolysis:

$$ArOCH_2C_6H_5 \quad \xrightarrow[\text{Pd}]{H_2} \quad ArOH + C_6H_5CH_3 \tag{18}$$

## 27.6 Nuclear reactivity of phenols

The nucleophilic properties of a saturated alcohol are of necessity confined to the oxygen atom of the hydroxyl group, with its unshared electrons. In phenols, however, the hydroxyl group can reinforce the nucleophilic properties of the nucleus, and thus is a powerful influence in aiding electrophilic attack on the ring:

$$HO \overset{..}{\underset{..}{\bigcirc}} \downarrow \; X^+$$

nucleophilic   electrophilic
response          demand

The orienting and activating effect of the hydroxyl group, and of ethers and esters derived from it, have been discussed in detail in Chapter 21, and in the discussion of the coupling reaction with diazonium salts (Chapter 26).

Di- and trihydroxybenzenes, such as catechol, resorcinol, hydroquinone, pyrogallol, hydroxyhydroquinone, and phloroglucinol,

catechol          resorcinol          hydroquinone          pyrogallol

hydroxyhydroquinone          phloroglucinol

are much more reactive to nucleophilic substitution than is phenol. Indeed, substitution into the ring of resorcinol and phloroglucinol is extraordinarily facile.

Resorcinol, with two *meta*-disposed hydroxyl groups, is subject to electrophilic attack at the 4-position by reagents that are too weakly electrophilic to attack benzene or monophenols. It is apparent that the electron-donating powers of the two hydroxyl groups are *both* directed to the 4-position, since it is *ortho* to one and *para* to the other.

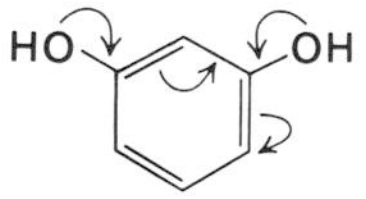

and hence, in the transition state for electrophilic substitution, both —OH groups can participate in stabilization of the positively charged complex:

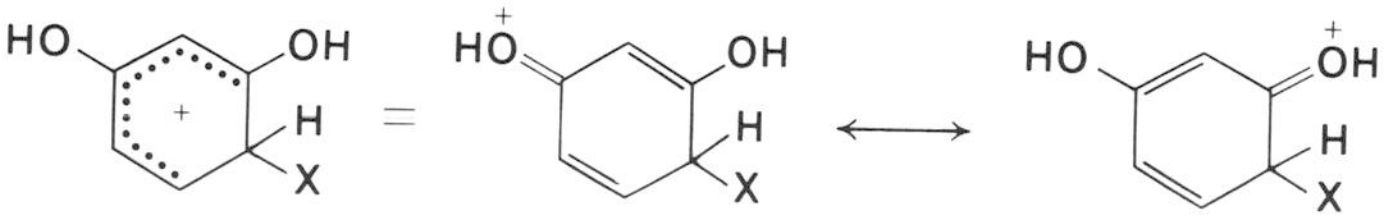

Resorcinol can undergo nuclear acylation by acetic acid in the presence of sulfuric acid, zinc chloride, or boron trifluoride:

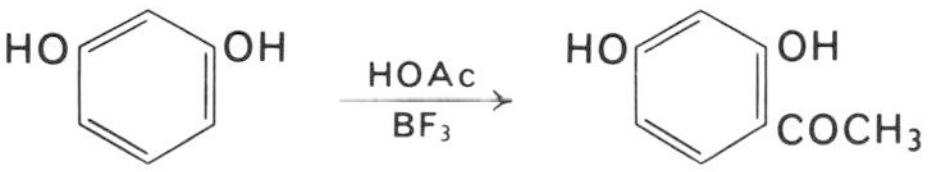

$$\text{(19)}$$

resacetophenone

Less highly activated benzene derivatives (e.g., benzene, phenol) will not undergo nuclear acetylation under these conditions.

Nuclear acylation of resorcinol can also be accomplished by the use of an alkyl cyanide, zinc chloride, and hydrogen chloride. The reaction is carried out in dry ether solution at ice-bath temperature (*Hoesch reaction*):

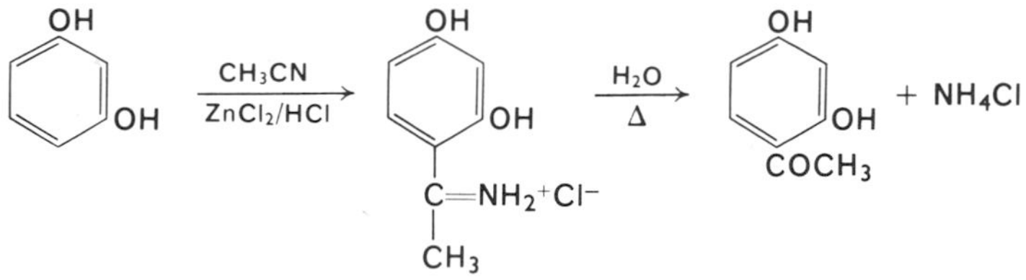

The close mechanistic relationship of the Hoesch synthesis to the Friedel-Crafts acylation can be seen in a consideration of the details of the reaction.

1. In the presence of the strong acids HCl and $ZnCl_2$ (conveniently regarded together as $H_2ZnCl_4$) the nitrile undergoes protonation:

$$CH_3CN + HA \rightleftharpoons CH_3C\overset{+}{\equiv}NH + A^-$$

2. Nucleophilic attack by the resorcinol nucleus upon the electrophilic carbon atom of $CH_3C\overset{+}{\equiv}NH$ leads to an intermediate of the usual kind:

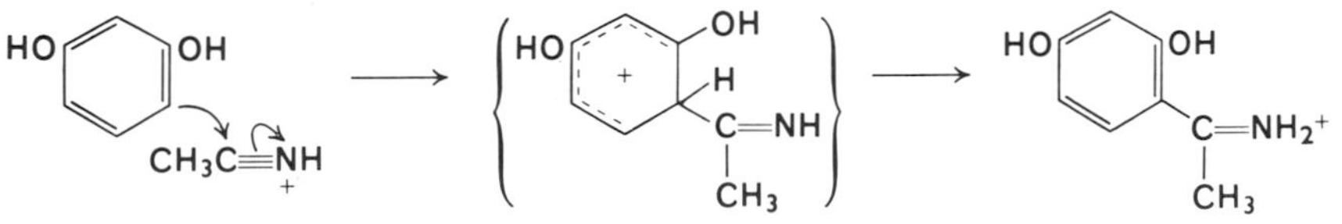

The product of the reaction is the protonated imine (Schiff's base) of the $-COCH_3$ group with ammonia, and is readily hydrolyzed to the free ketone.

The Hoesch reaction is applicable to the preparation of $\omega$-substituted acetophenones by the use of the appropriately substituted acetonitrile (the final ketone is shown in each case):

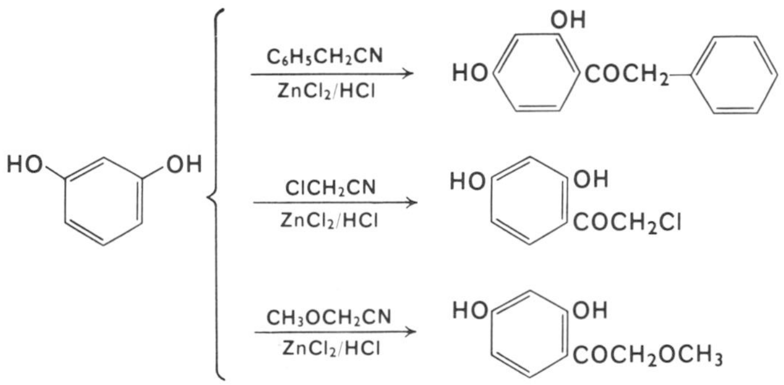

Nitrosation and diazo coupling occur readily with resorcinol.

Phloroglucinol, in which three hydroxyl groups are capable of aiding directly the nucleophilic character of the ring, is more reactive than resorcinol, and undergoes the Hoesch reaction under the same or milder (e.g., without the use of $ZnCl_2$) conditions.

## 27.7 The Reimer-Tiemann reaction: carbenes

Phenols undergo the Reimer-Tiemann reaction with chloroform in the presence of sodium hydroxide to yield hydroxybenzaldehydes. Phenol itself gives salicylaldehyde:

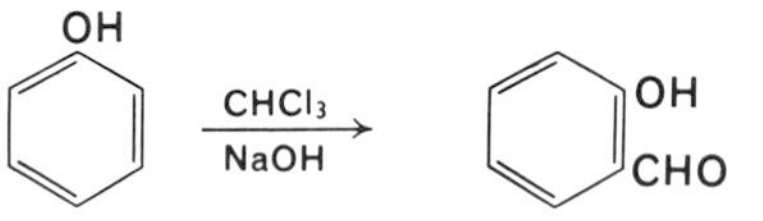

$$(22)$$

salicylaldehyde

That an intermediate containing the —$CHCl_2$ group, which is ultimately hydrolyzed to —CHO, is formed is indicated by the observation that *p*-cresol yields in addition to 5-methylsalicylaldehyde, 4-methyl-4-dichloromethyl-2,5-cyclohexadienone:

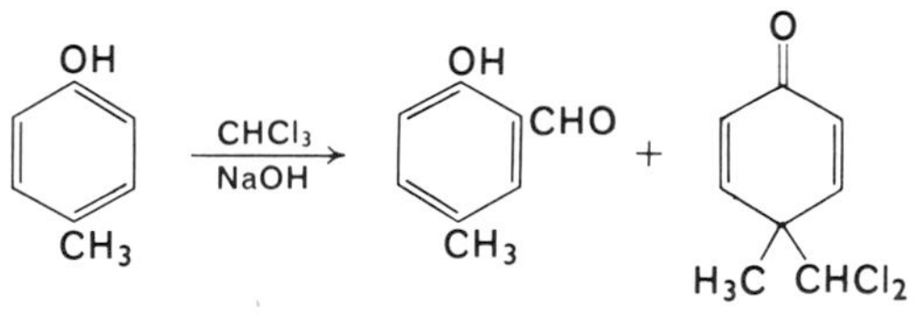

$$(23)$$

2,4-Xylenol reacts as follows:

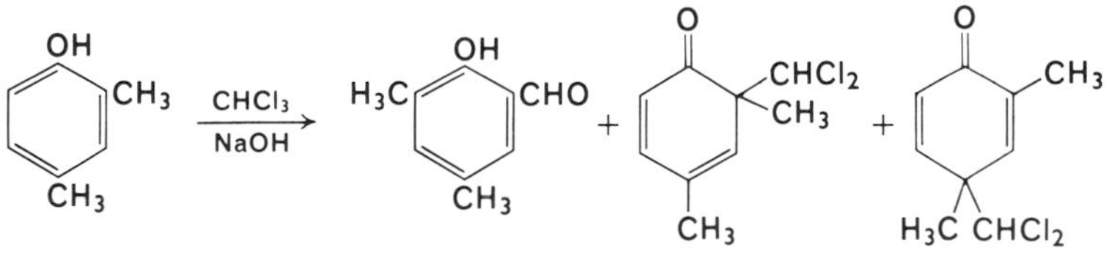

$$(24)$$

Although the Reimer-Tiemann reaction was for a long time formulated as a nucleophilic attack of the phenoxide ion upon chloroform,

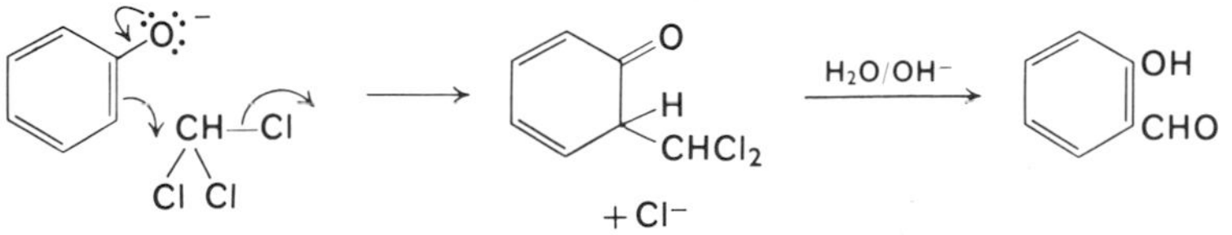

it has been established recently that the reaction involves the divalent carbon compound, $:CCl_2$, dichlorocarbene:

$$CHCl_3 + OH^- \rightleftharpoons \ :CCl_3^- + H_2O$$

$$:CCl_3^- \longrightarrow Cl^- + :CCl_2$$

dichloro-<br>carbene

Dichlorocarbene, in which the carbon atom lacks two electrons of having an octet, would be expected to be exceedingly reactive as an electrophile, and would react with the phenoxide ion as follows:

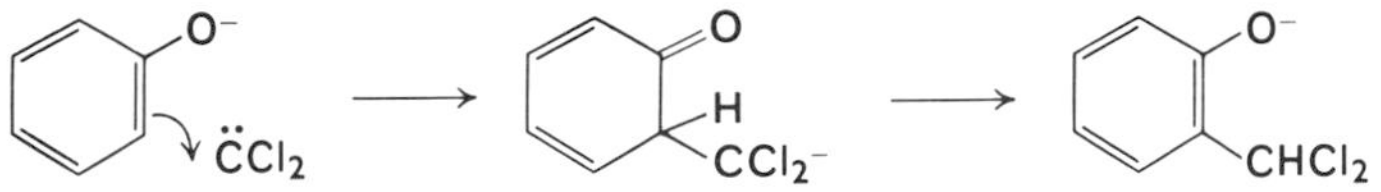

It can be seen that the difference between these two mechanisms lies chiefly in the details of at which stage and under what circumstances the chloride ion is lost. This is an important detail from the point of view of the theoretical aspects of the reaction. In particular, the existence (however transitory) of the divalent carbon compound is of great theoretical interest. Other evidence, not directly related to the reaction we are discussing here, supports the view that intermediate "carbenes" are indeed capable of existence.

Carbenes, like other compounds in which carbon exists in an "abnormal" valence state (carbonium ions, carbanions, free radicals), are very reactive. Although the most thoroughly studied carbene is probably the dichlorocarbene formed from chloroform by the action of strong bases, carbenes of other kinds can be formed in a variety of ways. Carbene itself, or methylene, has been observed as a highly reactive intermediate when diazomethane, $CH_2N_2$, is subjected to irradiation with ultraviolet light or is decomposed by pyrolysis. Depending upon the method used for the preparation of the compound, which may be represented simply by the formula $:CH_2$, the two electrons may be an electron pair, with opposite spins; or they may be single electrons, with like spins, in which case the compound is a diradical. The first of these is designated as a "singlet" state, the second, the diradical, as a "triplet." Singlet methylene is a highly reactive compound. It undergoes "insertion" reactions, in which the $CH_2$ group is introduced into a bond, as in the reaction

$$CH_2{=}C{-}CH_3 + :CH_2 \longrightarrow CH_3CH{=}C{-}CH_3 + CH_2{=}C{-}CHCH_2CH_3$$
$$\quad\ \ \overset{|}{CH_3} \qquad\qquad\qquad\qquad \overset{|}{CH_3} \qquad\qquad \overset{|}{CH_3}$$

and adds to carbon-carbon double bonds, to form cyclopropanes:

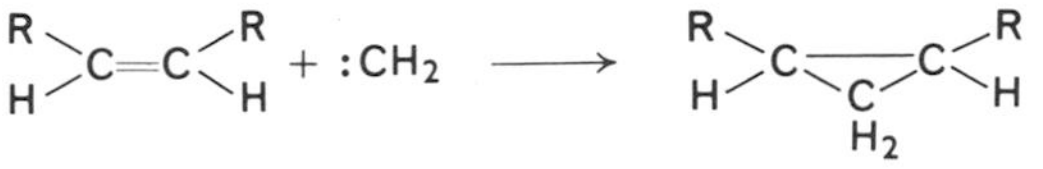

Carbene produced by the photochemical decomposition of diazomethane appears to be in the singlet state, for the addition is, as shown in the above equation, stereospecific. The degree of stereospecificity can, however, be altered by the experimental conditions, indicating that the change singlet → triplet can occur.

Addition of triplet methylene to a double bond would lack stereospecificity because the two-step character of diradical addition would lead through a monoradical that could undergo isomerization by rotation about the single bond:

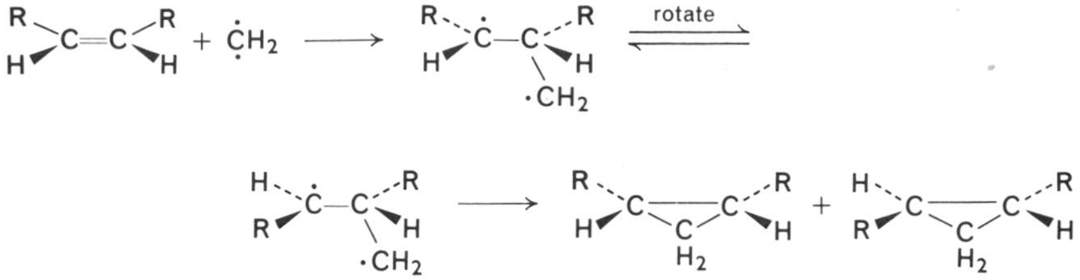

The addition of carbenes to carbon-carbon double bonds is not limited to olefinic double bonds; aromatic rings are attacked, with the formation of products that can be formulated as arising from the addition of the carbene to a double bond of the Kekulé form of the benzene ring.

---

**Exercise 4.** The reaction of dichlorocarbene (from chloroform and potassium *t*-butoxide) with indole proceeds in the following way:

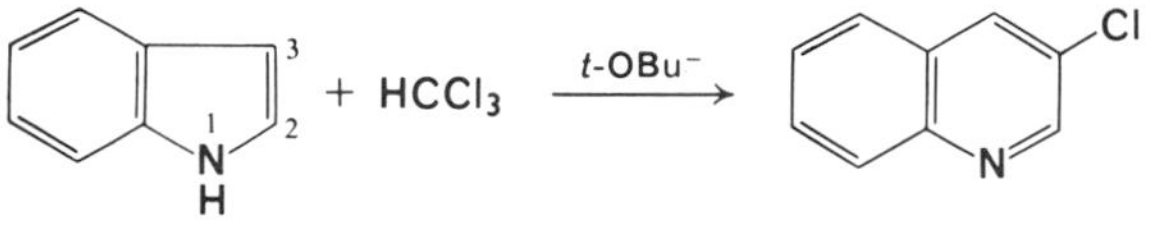

indole                    3-chloroquinoline

Devise a reasonable course for this transformation, starting with the addition of the dichloro-carbene to the 2,3-double bond of indole.

---

## 27.8 The carboxylation of phenols

Phenol, as the sodium salt, reacts with carbon dioxide at 120–140°C (under pressure) to yield the sodium salt of salicylic acid (the Kolbe reaction):

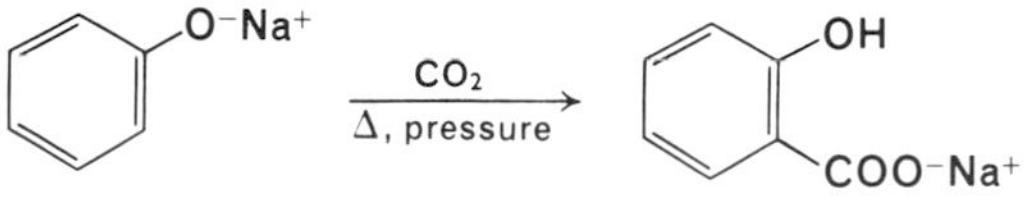

$$(25)$$

salicylic acid
(as sodium salt)

At a higher temperature (220°C) potassium phenoxide reacts with carbon dioxide to yield largely *p*-hydroxybenzoic acid:

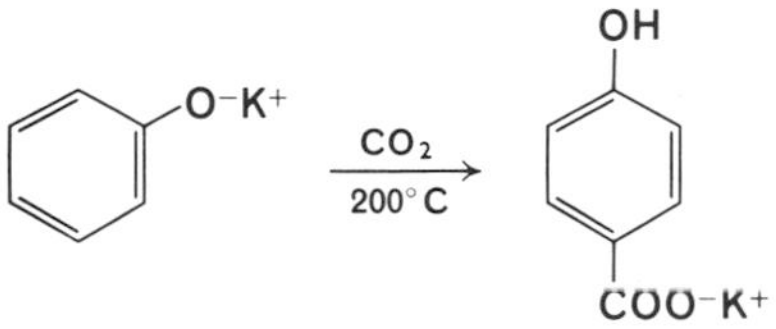

$$(26)$$

Although the carbon atom of $CO_2$ is electrophilic (recall the addition of RMgX, and the fact that $OH^-$ adds to give bicarbonate ion), its ability to effect an electrophilic attack upon an aromatic nucleus is not demonstrated with any but a highly nucleophilic ring. The carboxylation of phenoxide ions is thus an excellent demonstration of the reciprocal relationship between the electrophilic "attacking" reagent and the nucleophilic ring. There is, then, no reason to raise the question: Is the carboxylation of phenoxide ion a nucleophilic attack *by* the ring upon $CO_2$, or an electrophilic attack *upon* the ring by $CO_2$? It is, of course, either (and both) of these, just as the nitration of benzene is equally an attack by $NO_2^+$ *upon* the ring, or upon $NO_2^+$ *by* the ring.

The carboxylation of more nucleophilic phenols takes place with correspondingly greater ease. Phloroglucinolcarboxylic acid is prepared by heating. phloroglucinol with aqueous potassium bicarbonate; pyrogallol behaves similarly. The *decarboxylation* of polyhydroxybenzoic acids takes place readily, a behavior that is easily accounted for, since the decarboxylation reaction in these cases is an electrophilic displacement of $CO_2$ by hydrogen ion:

$$(27)$$

It will be apparent that the carboxylation and decarboxylation reactions follow the same course.

## 27.9 Therapeutic uses of salicylic acid derivatives

Salicylic acid is a very important compound. Its acetyl derivative is *aspirin*, one of the most widely used and useful drugs known. Aspirin is a mild analgetic agent, a febrifuge, and an effective drug in the treatment of rheumatoid arthritis (salicylic acid and its salts are also used for the latter disease). About 30 million pounds (40 billion five-grain tablets) of aspirin are consumed yearly in the United States alone.

In recent years the drug 4-aminosalicylic acid (*p*-aminosalicylic acid, PAS) has come into extensive clinical use in the treatment of tuberculosis. It has a tuberculostatic action, and is commonly used in conjunction with streptomycin and isonicotinic acid hydrazide (isoniazid). PAS can be prepared by the Kolbe carboxylation of *m*-aminophenol:

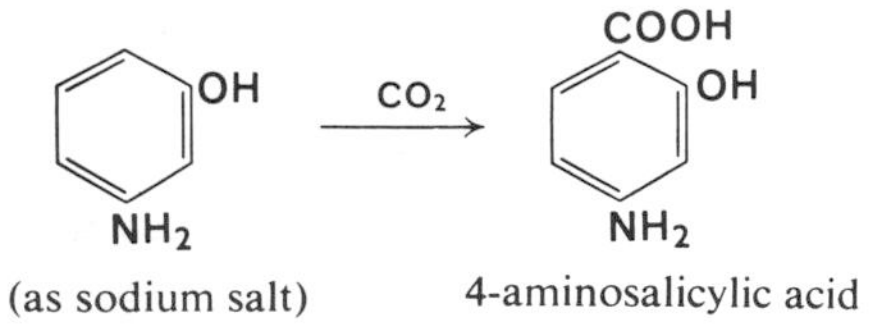

$$ (28) $$

(as sodium salt)    4-aminosalicylic acid

Other salicylic acid derivatives of physiological significance are phenyl salicylate ("salol") and methyl salicylate (oil of wintergreen). The latter occurs in nature in certain plants:

salol                    oil of wintergreen

Gentisic acid and γ-resorcyclic acid have also been used for some of the purposes for which salicyclic acid has been employed (rheumatic diseases):

gentisic acid          γ-resorcyclic acid

## 27.10   Naturally occurring phenols

Phenols, and in particular phenolic carbonyl compounds, occur widely in nature. They are particularly prevalent in plants, where they are found as the sap-soluble pigments of flowers and fruits, as well as a large and varied group of glycosides (Chapter 18), in which one or more of the phenolic hydroxyl groups is combined with a sugar in glycosidic linkage.

A well-known phenolic glucoside is *arbutin*, the mono-β-glucoside of hydroquinone; the monomethyl ether often accompanies it in the plant (e.g., *Arbutus uva-ursi*) in which it is found.

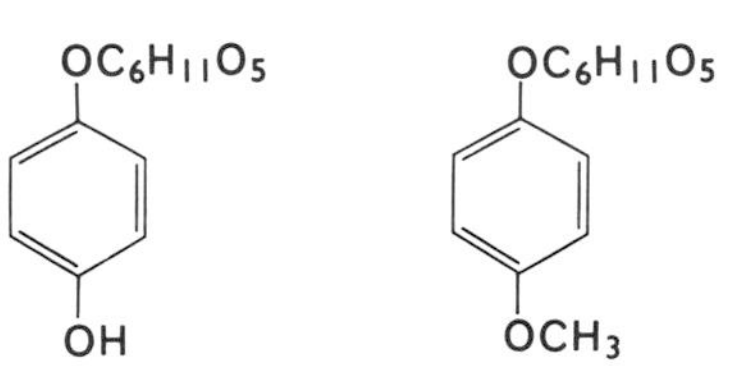

arbutin                methylarbutin

**Exercise 5.**   Write the structure of arbutin using the Haworth formula for the sugar. Arbutin is a β-D-glucopyranoside.

By far the greater number of natural phenols are substituted phenols or their glycosides. Salicin has been described earlier (Chapter 18). *Vanillin*, an important flavoring material, is one of man's most extensively used condiments. It occurs in nature as the odorous constituent of the vanilla bean (an orchid), and as the glucoside in some other plants.

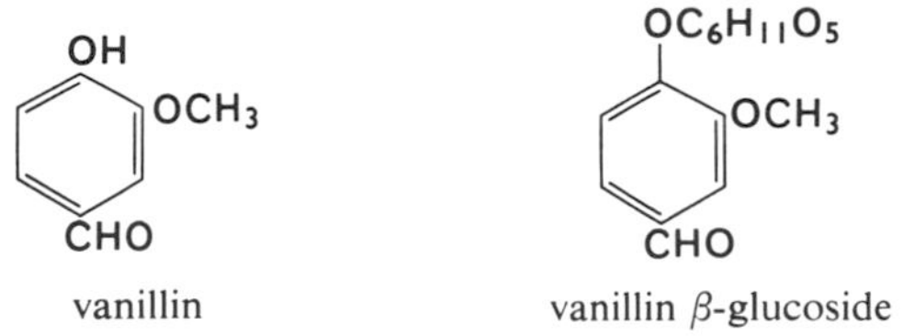

vanillin            vanillin β-glucoside

Vanillin is made industrially in large amounts by the Reimer-Tiemann reaction, starting with guaiacol (catechol monomethyl ether), from which it is formed along with *o*-vanillin:

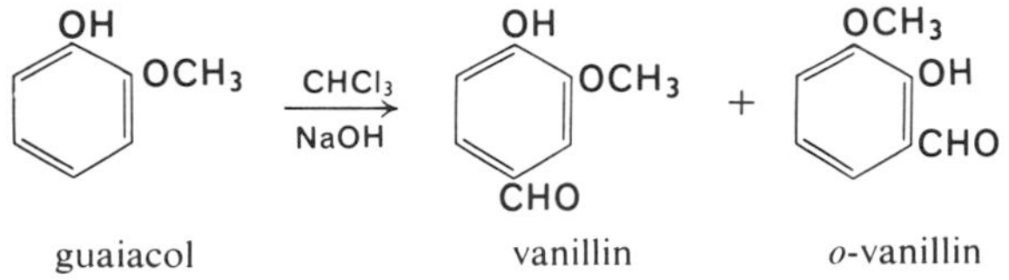

$$\tag{29}$$

guaiacol       vanillin       *o*-vanillin

Another source of vanillin is lignin, a byproduct of paper pulp manufacture. Lignin is a complex, polymeric structural material of woody plants, and contains as a structural unit the fragment

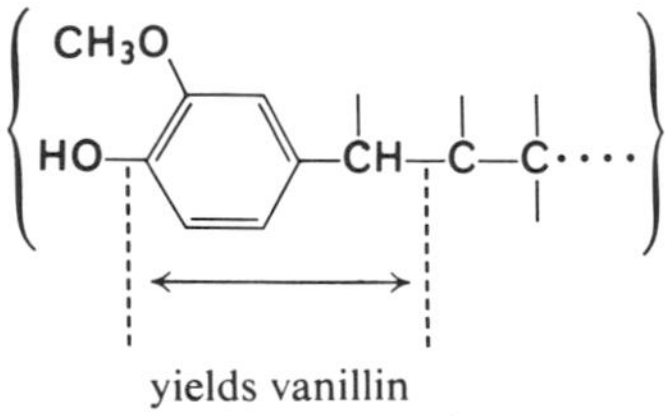

yields vanillin

(combined in a manner too complex to permit its detailed consideration here). Upon oxidation of lignin, vanillin is formed in sufficiently high yield to make the process economically feasible. Vanillin is also prepared by oxidation (chromic acid, ozone, or nitrobenzene) of *isoeugenol*. Isoeugenol is prepared by isomerization of *eugenol*, found in clove oil:

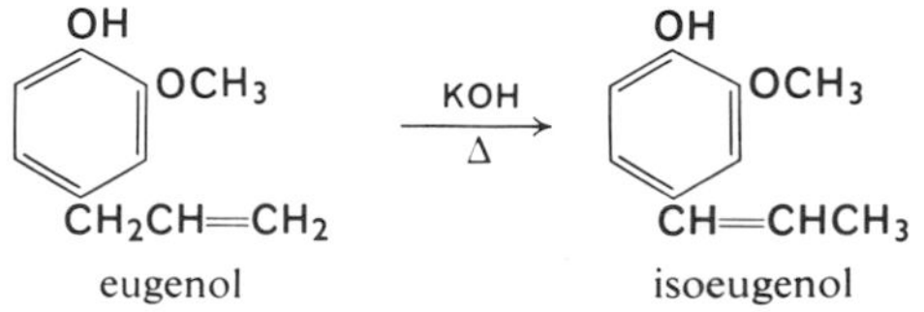

$$\tag{30}$$

eugenol            isoeugenol

Phenolic compounds are not synthesized from simple precursors by animals as they are by plants (in which the ultimate source of all carbon compounds is

$CO_2$). Plants require no accessory supplies of carbon compounds, but animals require a variety of organic substances at an advanced level of synthetic elaboration. Among these are compounds containing aromatic rings; and from such starting materials the animal organism is able to perform further structural alterations to produce, among other kinds of compounds, phenols (see Chapter 32).

## 27.11 Phenolic plant pigments

The phenolic compounds of plants embrace a wide and varied range of structural types, from the simple phenols so far described (arbutin, salicin, vanillin) to very complex compounds. Among the latter are many of the naturally occurring pigments of plants, typified by the red *anthocyanidin* pigment, cyanidin (chloride), and the yellow *flavone* pigment, quercetin:

cyanidin chloride          quercetin

Both of these compounds occur as glycosides, the sugar residues usually being attached at one of the starred hydroxyl groups. Numerous related compounds, including anthocyanins (the term used for the anthocyanidin glycosides) and flavones with fewer or with more hydroxyl groups than the two examples given above, occur in plants. Anthocyanin pigments are responsible for most of the red, violet, and blue colors of flowers and fruits. The monoglucoside of cyanidin is *chrysanthemin*, named from its occurrence in scarlet chrysanthemums, and the diglucoside is *cyanin*, which, curiously, is the pigment of the blue cornflower and the red rose. The difference in color of a single pigment in different plant tissues is probably due to the presence in plant tissues of metallic ions (iron, aluminum), whose complexes with the anthocyanin pigments display a range of color.

## 27.12 Recognition and characterization of phenols

Simple phenols are characterized by their solubility in alkali and insolubility in sodium bicarbonate, although it must be recalled that certain substituted phenols (e.g., *p*-nitrophenol) are much stronger acids than phenol itself (Section 27.3). Most—but not all—phenols give strikingly colored complexes with ferric salts. The "ferric chloride color" is a valuable diagnostic test for phenols, but is to be applied with the recognition that other types of compounds also give colors with this reagent. The ferric chloride color of hydroxamic acids has been mentioned earlier; and enols of many kinds give ferric chloride colors (e.g., acetoacetic esters). Occasionally the kind of color given by a phenol with ferric chloride is a clue to its structure. Catechol and many other *o*-dihydroxy benzene derivatives give green colors. Salicylic acid gives a deep purple color; and many *o*-hydroxy aldehydes and ketones give more intensely red to purple colors than do simple phenols. These subordinate criteria must be used with caution, but when the experimentalist who is using the test is aware of the limits of the context in which his work lies, they are often valuable. For example, if an unknown compound were thought to be either

the appearance of a purple ferric chloride color would be a strong indication that the first of these structures is the correct one, since the characteristic grouping of salicylic acid is present in this, but not the other, structure.

Another useful diagnostic test for phenols having unsubstituted *ortho* or *para* positions is in the use of a diazotized amine. The formation of an azo dye is readily recognized by the intense color that usually characterizes these compounds. The converse of this—the use of $\beta$-naphthol in the test for a primary aromatic amine—has been dealt with in Chapter 26.

Phenoxyacetic acids, prepared by the reaction of phenols with chloroacetic acid in alkaline solution, are valuable derivatives for the characterization and identification of phenols:

(31)

The reaction by which these compounds are prepared is closely analogous to the alkylation with a bromoacetic ester, described in Section 27.4. The phenoxyacetic acids are usually crystalline compounds with characteristic melting points. They possess the additional and valuable property of being carboxylic acids that can be titrated with standard alkali. Thus, they provide a means of determining the equivalent weight (the molecular weight in the case of monophenols) of the phenol. The advantage of knowing the molecular weight of an unknown phenol as an aid in the establishment of its structure is obvious.

---

**Exercise 6.**  An unknown phenol, which contained bromine but no other elements except C, H, and O, formed a phenoxyacetic acid when treated with chloroacetic acid in alkali. Titration of 0.245 g of the derivative with 0.100 *N* NaOH required 10.0 ml of the standard alkali. What is a possible structure for the phenol?

---

### 27.13   Phenolphthalein and fluorescein

When phenols are heated with phthalic anhydride in the presence of a strong acid catalyst (e.g., zinc chloride), condensation to form "phthalein" dyes results. In the case of phenol, the reaction proceeds as follows:

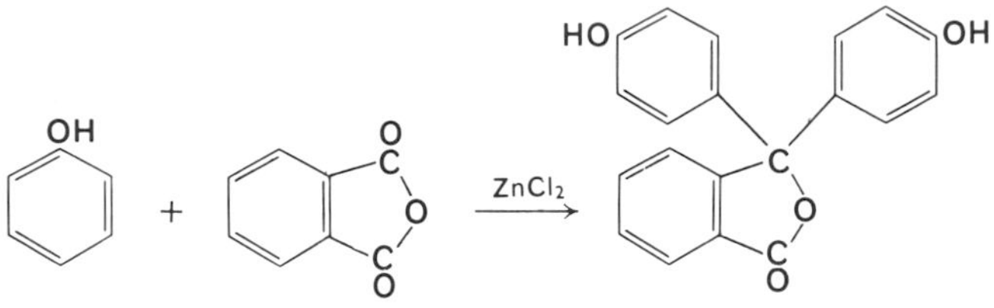

$$(32)$$

phenolphthalein

When resorcinol is used in place of phenol, the product is fluorescein:

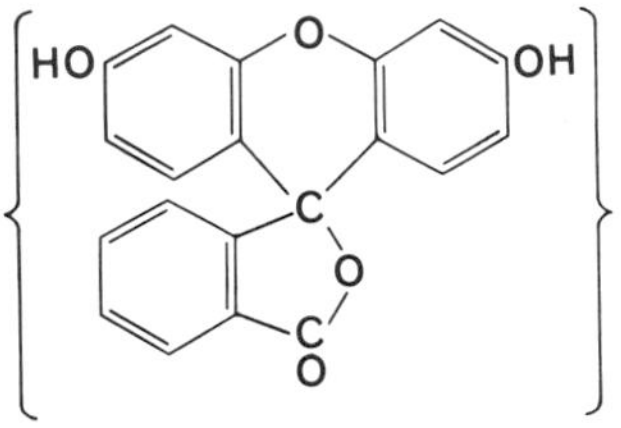

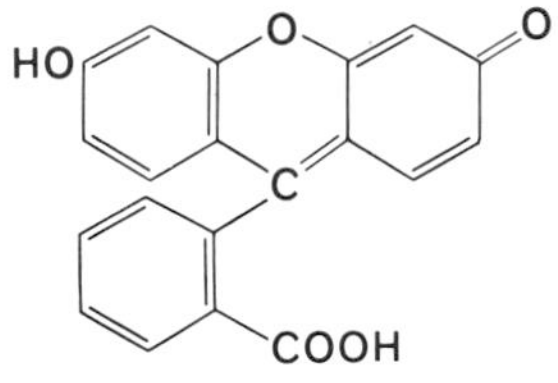

fluorescein

(33)                    (34)

Since phenolphthalein is colorless, the structure written for it above is satisfactory; but fluorescein is a colored compound, and thus probably has the isomeric quinone-like structure (34) and not the structure (33).

The use of phenolphthalein as an acid-base indicator depends upon its color change in alkaline solution from colorless to red. The structural change accompanying the ionization is as follows:

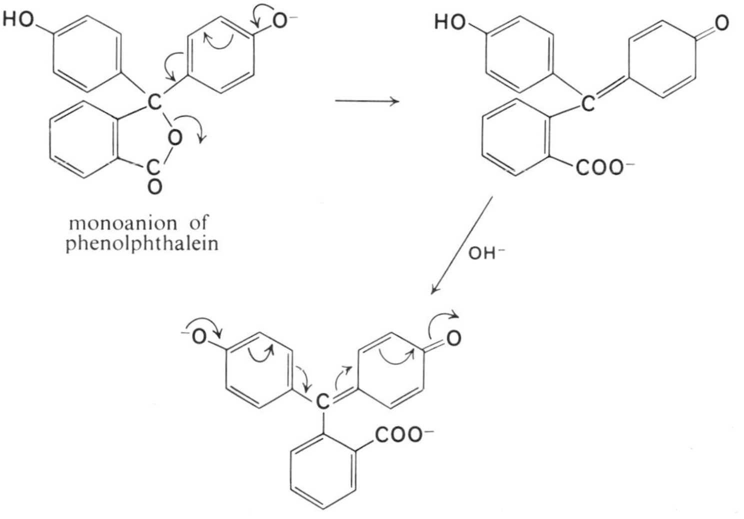

monoanion of
phenolphthalein

$OH^-$

$$(35)$$

resonance hybrid: *red*

With an increase in the concentration of hydroxide ion (high *p*H) the red color disappears and the solution again becomes colorless. This change results from the attack of the hydroxide ion at the carbon atom shown in the following:

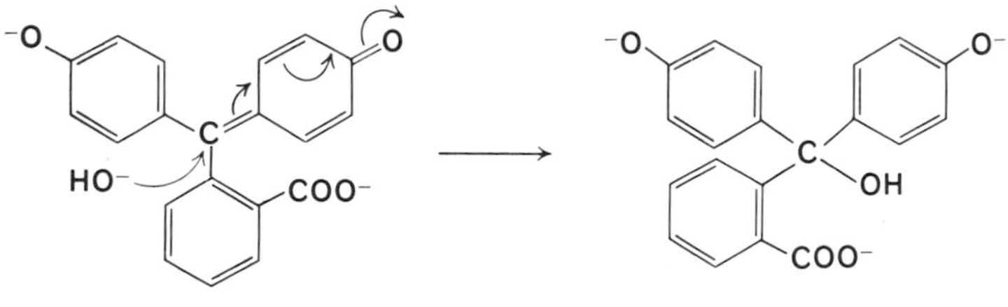

red                                    colorless

It can be understood why a high *p*H is needed to bring this about, since a trinegative ion must be produced (that is, the negative OH ion must attack the doubly negatively charged phenolphthalein ion).

## 27.14   Quinones and hydroquinones

Aromatic 1,2- and 1,4-dihydroxy compounds are readily oxidized to the corresponding *quinones*:

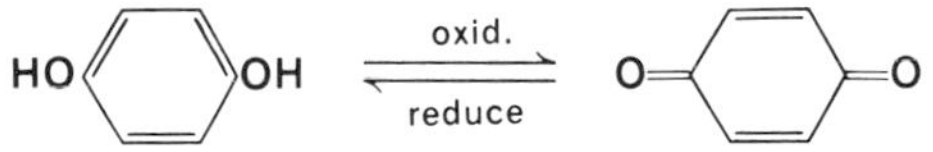

hydroquinone          *p*-benzoquinone
                          (quinone)

$$(36)$$

Conversely, quinones are readily reduced to hydroquinones (quinols). Quinones are also obtainable in other ways: by oxidation of *o*- and *p*-aminophenols, and in many cases by direct oxidation of monohydric phenols and aromatic amines. Indeed, the most practicable method of preparing *p*-benzoquinone itself is by the oxidation of aniline.

The ready oxidation of *p*-aminophenols to quinones permits one to prepare quinones, and thus hydroquinones, from simpler phenols by (1) diazonium coupling to give an azo compound, (2) reduction of the azo compound, and (3) oxidation of the amine so formed; for example:

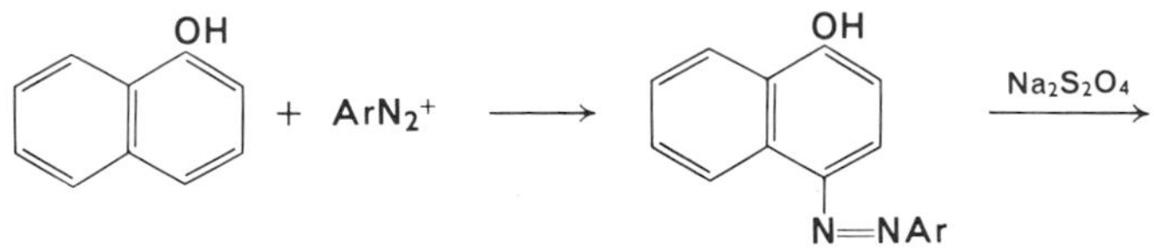

$$(37)$$

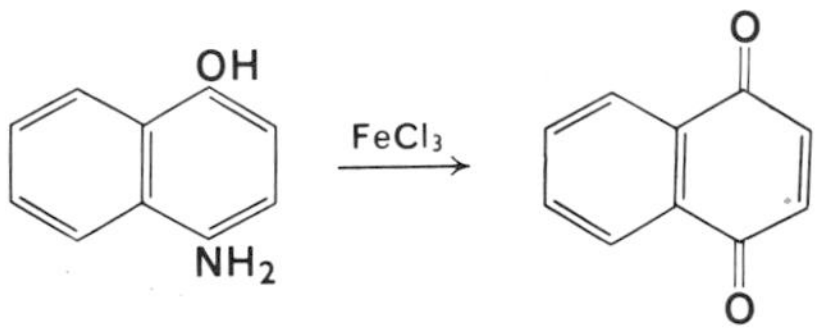

1,4-naphthoquinone

$$(38)$$

*Ortho*-quinones, although less stable than *p*-quinones, can be prepared by analogous methods, as by the oxidation of *o*-dihydroxy and *o*-hydroxyamino compounds. The oxidation of catechol by silver oxide gives *o*-benzoquinone, a red crystalline compound.

Quinones are not aromatic compounds; rather, they are $\alpha,\beta$-unsaturated carbonyl compounds, and their reactions are best interpreted in this way. The addition of amines to quinones is a reaction that will be recognized as a nucleophilic addition to the conjugated system. The reaction between aniline and 1,4-naphthoquinone proceeds as follows:

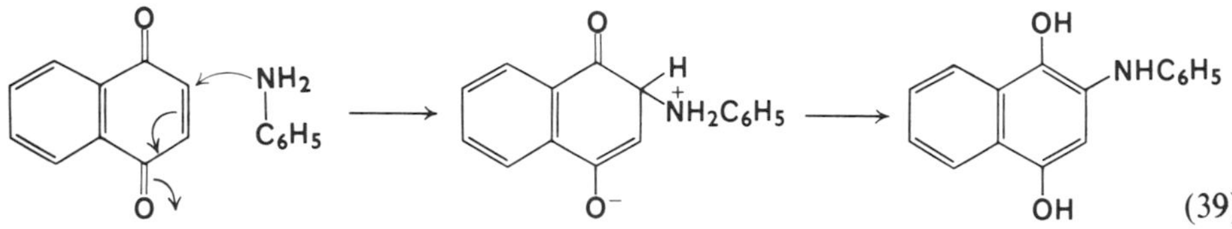

(39)

The hydroquinone derivative that is the product of the addition reaction can be oxidized by the original quinone, and the position of the following equilibrium depends upon the relative oxidation-reduction potentials of the two quinone-hydroquinone systems:

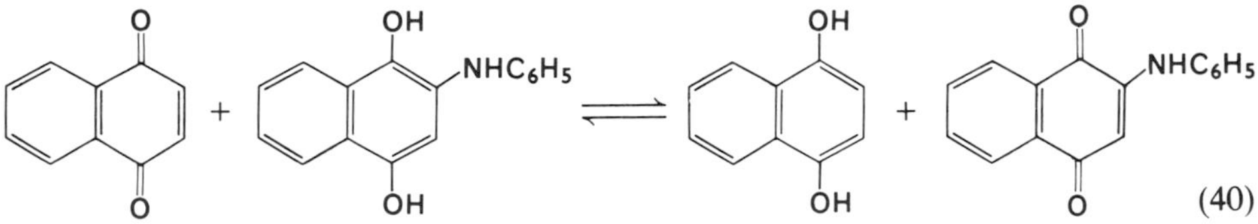

(40)

With *p*-benzoquinone, two molecules of the amine add in successive stages to yield as the final product the disubstituted quinone:

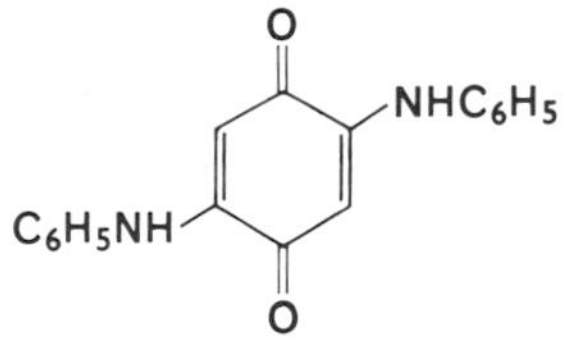

The compound Mytolon, a clinically useful muscle relaxant (curare-like) compound, is prepared in this way:

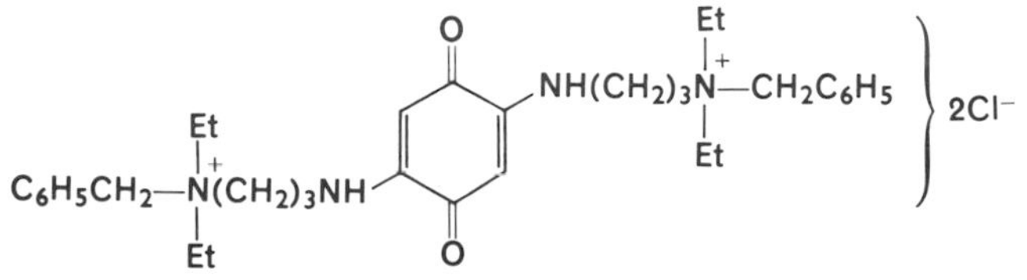

Mytolon

Compounds more weakly nucleophilic than amines add to quinones under the catalytic effect of strong acids. The addition of methanol to *p*-benzoquinone proceeds in the following way:

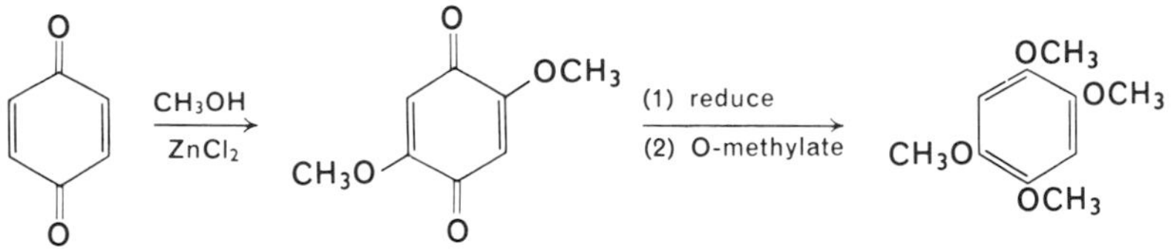

(41)

As shown in the second stage of this reaction, the method is useful for the preparation of tetraalkoxybenzenes.

Acid-catalyzed addition of acetic anhydride to *p*-benzoquinone (the Thiele reaction) leads to the formation of 1,2,4-triacetoxybenzene. In this case, the addition of one molecule of acetic anhydride completes the reaction, which does not proceed past this stage because the intermediate is not a hydroquinone, but its acetate, and thus is not reoxidized to a quinone:

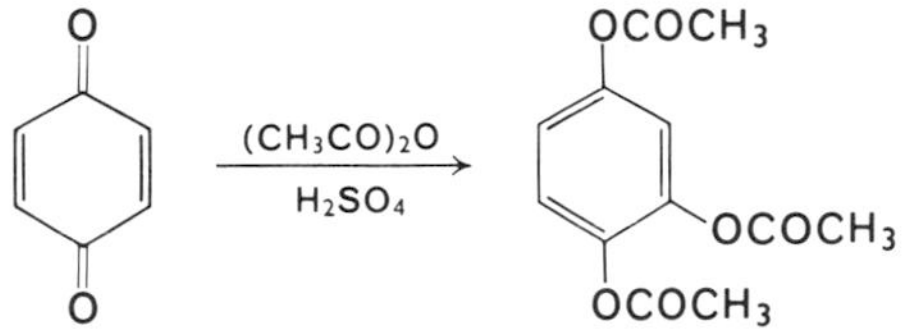

(42)

The course of the reaction can be described by the following steps, the final step (not shown) of which is the acetylation of the diacetoxy phenol:

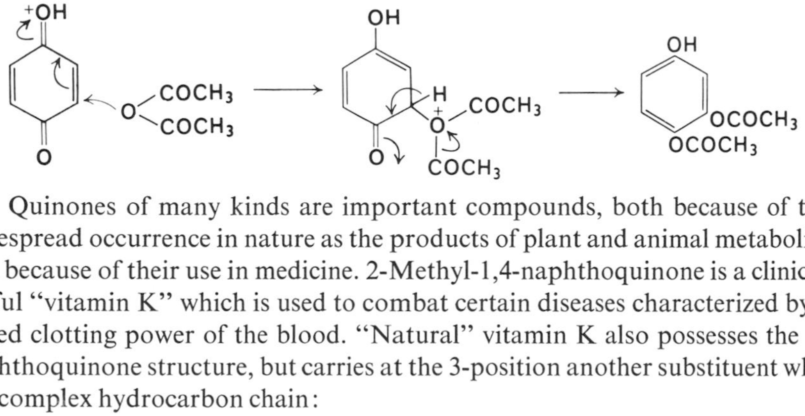

Quinones of many kinds are important compounds, both because of their widespread occurrence in nature as the products of plant and animal metabolism, and because of their use in medicine. 2-Methyl-1,4-naphthoquinone is a clinically useful "vitamin K" which is used to combat certain diseases characterized by reduced clotting power of the blood. "Natural" vitamin K also possesses the 1,4-naphthoquinone structure, but carries at the 3-position another substituent which is a complex hydrocarbon chain:

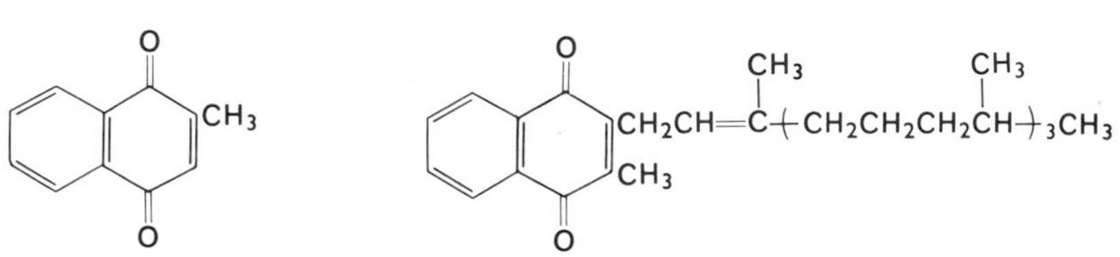

(synthetic) vitamin K

(natural) vitamin $K_1$
2-methyl-3-phytyl-1,4-naphthoquinone

Natural vitamin $K_1$ is found in green plants and is an important dietary constituent that plays a role in the maintenance of the coagulant properties of blood.

A group of quinones, typified by the one shown below and called Coenzyme
$Q_{10}$, but differing in the length of the chain of five-carbon-atom units, are widely
distributed in living cells, in which they play important roles in metabolism, probably by electron transfer involving reversible quinone-hydroquinone interconversion:

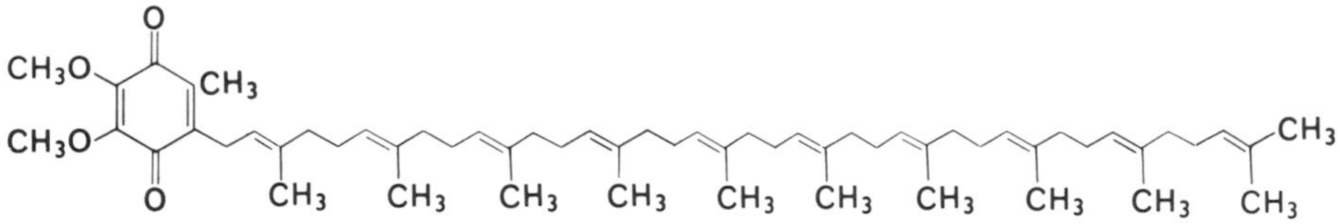

Coenzyme $Q_{10}$

Other quinones occur in plants, lichens, fungi, and marine organisms, in many of
which they are responsible for yellow, orange, and red colors.

### Exercises

(Exercises 1–6 will be found within the text.)

**7** Arrange the following compounds in order of decreasing strength as acids. *(a)* phenol;
*(b)* o-nitrophenol; *(c)* m-bromophenol; *(d)* o-bromophenol; *(e)* 2,4,6-trimethylphenol;
*(f)* p-hydroxybenzoic acid; *(g)* 2,4,6-tribromophenol; *(h)* 2,4-dinitrophenol; *(i)* benzyl
alcohol.

**8** Devise practical syntheses for the following, using as the aromatic starting material any
monosubstituted benzene derivative. *(a)* p-bromophenol; *(b)* resorcinol (m-dihydroxy-
benzene); *(c)* aspirin; *(d)* o-methoxybenzoic acid; *(e)* p-aminoanisole; *(f)* p-n-butyl-
phenol; *(g)* 2-hydroxy-5-methylbenzaldehyde; *(h)* 3,4-dimethoxybenzoic acid.

**9** Devise procedures for separating the following mixtures into their pure components.
Use chemical means of separation. *(a)* phenol, benzoic acid, benzyl alcohol; *(b)* anisole,
salicylaldehyde, benzaldehyde; *(c)* aniline, picric acid, p-cresol; *(d)* β-naphthol, β-
naphthylmethyl ether, β-naphthoic acid; *(e)* salicylic acid, salicylaldehyde, phenol,
methyl salicylate.

**10** A phenolic compound (A), $C_7H_8O_2$, is completely destroyed when oxidized with $KMnO_4$.
When A is treated with dimethyl sulfate in alkali it is transformed into B, $C_8H_{10}O_2$.
Oxidation of B with $KMnO_4$ gives in excellent yield an acid C, $C_8H_8O_3$. Write a possible
structure for A. Are any other structures possible for A? If so how could you distinguish
them?

**11** Contrast the chemical behavior of phenol and benzyl alcohol, using at least four different
types of reactions to illustrate your discussion.

**12** A glucoside (A), $C_{12}H_{16}O_7$, is soluble in dilute aqueous alkali, from which it is recovered
unchanged upon acidification. It can be hydrolyzed with hot, dilute HCl, or by emulsin,
to yield D-glucose and a phenolic aglucone (B). Exhaustive methylation of A yields an
alkali-insoluble product, C. The methylated glucoside (C) can be hydrolyzed to yield
2,3,4,6-tetra-O-methylglucose and a phenol, D, $C_7H_8O_2$. Both B and D can be methylated
with dimethyl sulfate and alkali to a neutral compound, E, $C_8H_{10}O_2$. Bromination of E
gives *only one* monobromo derivative. Write the structures of the lettered compounds.

**13** Picein, $C_{14}H_{18}O_7$, is a naturally occurring glucoside. It is nonreducing, but forms an oxime and a phenylhydrazone. After hydrolysis with a $\beta$-glucosidase or with dilute acid, it yields D-glucose and a compound (A), $C_8H_8O_2$. Compound A is soluble in alkali but not in sodium bicarbonate solution, and when treated with dimethyl sulfate and alkali is transformed into a neutral compound (B), $C_9H_{10}O_2$. Compound B gives a positive iodoform reaction, and from this reaction is isolated, along with iodoform, an acid (C), $C_8H_8O_3$. The acid C can be synthesized by permanganate oxidation of $p$-cresyl methyl ether. Picein reacts with $HIO_4$, consuming 2 moles of the reagent and forming 1 mole of formic acid and no formaldehyde. Draw the Haworth projection formula for picein and write the reactions described.

## Chapter twenty-eight

**Molecular Rearrangements**

# in the Reactions of Organic Compounds

### 28.1 Electron-deficient atoms

One of the problems that confronts the organic chemist, and one to which he must always be alert, arises from the fact that the relative arrangement of atoms in an organic compound may alter in the course of a reaction, and a *rearrangement* may occur. Rearrangements may be simple, as in the shift in a double bond from one position to another,

$$CH_3CH_2CH{=}CH_2 \underset{\xleftarrow{\hspace{1cm}}}{\overset{H_2SO_4}{\rightleftharpoons}} CH_3CH{=}CHCH_3$$

or somewhat more complex, as in the pinacol rearrangement,

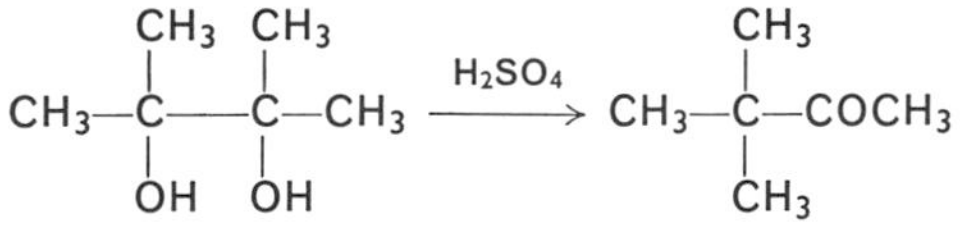

$$
\begin{array}{c}
\overset{\displaystyle CH_3}{\underset{\displaystyle OH}{\overset{|}{\underset{|}{C}}}}\!\!-\!\!\overset{\displaystyle CH_3}{\underset{\displaystyle OH}{\overset{|}{\underset{|}{C}}}}\!\!-\!\!CH_3 \xrightarrow{H_2SO_4} CH_3\!-\!\overset{\displaystyle CH_3}{\underset{\displaystyle CH_3}{\overset{|}{\underset{|}{C}}}}\!\!-\!\!COCH_3
\end{array}
$$

or may involve extensive changes in structure in which more than one atom or group changes its point of attachment (Section 32.14).

The fact that rearrangements do occur is of great value to the chemist. Many rearrangements have considerable practical value in permitting ready synthetic routes to desired products that would otherwise require involved and lengthy methods of preparation. But another usefulness is found in the occurrence of rearrangements: by the study of their nature and course a theoretical insight can be

gained into the fundamental aspects of organic reactions. Thus, far from compli-
cating the study of organic chemistry, the occurrence of reactions in which rearrangements occur serves as a basis upon which to correlate and elucidate the
principles of organic reactions, and helps to disclose common fundamental modes
of behavior of organic compounds.

We have seen in our study so far that many organic reactions owe their incep-
tion to the deficiency of electrons existing on an atomic nucleus, and the consequent
satisfaction of this deficiency by a donor, or nucleophilic, reagent, which is capable
of providing electrons to the deficient atom and thus of forming a new bond.
Examples that may be recalled briefly are the electrophilic character of carbon in
the carbon-halogen bond, the electron-poor carbon atom of the carbonyl group,
the boron atom of the neutral tricovalent boron compounds, and the very electrophilic carbonium carbon atom.

It will also be recalled that in some cases the generation of an electron-
deficient carbon atom, such as by ionization of a C—Cl or C—$\overset{+}{O}H_2$ bond, is
followed by a molecular rearrangement which results in the transfer of the electron
deficiency to a carbon atom better able to sustain it—i.e., by a rearrangement to a
more stable carbonium ion.

Most of the rearrangements with which we shall be concerned will consist in
the breaking of a carbon-carbon bond, and the "migration" of the group so
liberated to form a new bond with another carbon atom or with an oxygen or
nitrogen atom:

$$\begin{array}{cc} \overset{\displaystyle C}{\underset{\displaystyle \vert}{\vphantom{|}}} & \\ -\overset{\vert}{\underset{\vert}{C}}-\overset{\vert}{\underset{\vert}{C}}- & \longrightarrow \quad -\overset{\vert}{\underset{\vert}{\overset{+}{C}}}-\overset{\vert}{\underset{\vert}{C}}- \end{array} \qquad (1)$$

$$-\overset{\vert}{\underset{\vert}{C}}-\overset{..}{\underset{..}{O}}:^{+} \quad \longrightarrow \quad -\overset{+}{\underset{\vert}{C}}-\overset{..}{\underset{..}{O}}: \qquad (2)$$

$$-\overset{\vert}{\underset{\vert}{C}}-\overset{..}{N}: \quad \longrightarrow \quad \left\{ -\overset{+}{\underset{\vert}{C}}-\overset{..}{\underset{..}{N}}:^{-} \right\} \quad \longrightarrow \quad -C=\overset{..}{N}-C \qquad (3)$$

The common feature of the three kinds of rearrangements is that the atom B in
equation (4) is electron deficient, and the migration of A with its binding pair of
electrons satisfies this deficiency:

$$\overset{\displaystyle A}{\underset{\displaystyle \vert}{X}}-\overset{..}{\underset{..}{B}}: \quad \longrightarrow \quad X-\overset{\vert}{\underset{..}{\overset{A}{B}}}: \qquad \text{[general expression for equations (1), (2), (3)]} \qquad (4)$$

It would appear to be a corollary of this that if the compound or ion A—X—B,
in which B is electron deficient, rearranges to X—B—A, leaving X electron

deficient, X is thus better able to tolerate the electron deficiency. The gain in stability of X—B—A over A—X—B is thus the driving force for the rearrangement.

## 28.2 Carbonium-ion rearrangements

Rearrangements following the generation of a carbonium ion have been alluded to in previous chapters. Their review and summary in this section will reemphasize their common basis and amplify, by additional examples, some of the more important and commonly encountered of these reactions.

The generation of an electron-deficient carbon atom (a carbonium ion) can be accomplished in a number of ways; for example:

$$-\overset{|}{\underset{|}{C}}-OH + HA \;\rightleftharpoons\; \left. -\overset{|}{\underset{|}{C}}-OH_2^+ \right\} A^- \;\rightleftharpoons\; -\overset{|}{\underset{|}{C}}{}^+ + H_2O + A^- \tag{5}$$

$$>\!C\!=\!C\!<\; +\; HA \;\rightleftharpoons\; \left. >\!\overset{+}{C}\!-\!CH\!< \right\} A^- \tag{6}$$

$$-\overset{|}{\underset{|}{C}}-NH_2 + HNO_2 \;\overset{HX}{\longrightarrow}\; \left. -\overset{|}{\underset{|}{C}}-N_2^+ \right\} X^- \;\longrightarrow\; -\overset{|}{\underset{|}{C}}{}^+ + N_2 + X^- \tag{7}$$

$$-\overset{|}{\underset{|}{C}}-Cl + Ag^+ \;\longrightarrow\; -\overset{|}{\underset{|}{C}}{}^+ + AgCl \tag{8}$$

$$>\!C\!=\!O + HA \;\rightleftharpoons\; \left\{ >\!C\!=\!\overset{+}{O}H \;\longleftrightarrow\; >\!\overset{+}{C}\!-\!OH \right\} A^- \tag{9}$$

It will be recalled from the discussion of the behavior of organic hydroxy compounds (see Chapter 8) that while the protonation of the alcoholic hydroxyl group is a consequence of its basic nature, and always occurs to a degree that depends upon the strength of the protonating acid (HA), the formation of a "free" carbonium ion according to (5) does not invariably follow, but occurs only when the resulting carbonium ion possesses stabilizing features such as the resonance contributions of the attached methyl groups in $(CH_3)_3C^+$, or of the alkoxyl group

and aromatic ring in $CH_3O\!-\!\!\!\bigcirc\!\!\!-CH_2^+$. We can now recognize another influence

that leads to the loss of $H_2O$ from $-\overset{|}{\underset{|}{C}}-\overset{+}{O}H_2$; namely, a rearrangement to a new

system of greater stability. For instance, although the acid-catalyzed reactions of saturated primary alcohols do not ordinarily proceed by way of discrete carbonium ions, dehydration of the primary alcohol neopentyl alcohol by the action of acid occurs in a reaction (10) that is probably initiated by a dissociation of the protonated alcohol:

$$(CH_3)_3CCH_2OH + H_2SO_4 \rightleftharpoons (CH_3)_3CCH_2\overset{+}{O}H_2\} \ HSO_4^-$$

$$CH_3-\overset{+}{\underset{\underset{\textstyle CH_3}{|}}{C}}-CH_2CH_3 \quad \xleftarrow{\text{rearrange}} \quad (CH_3)_3CCH_2^+ + H_2O$$

$$\xrightarrow[\ -H^+\ ]{} \quad \overset{H_3C}{\underset{H_2C}{\diagdown}}CCH_2CH_3 \ + \ \overset{H_3C}{\underset{H_3C}{\diagdown}}C{=}CHCH_3 \tag{10}$$

In accordance with the reactions indicated in (5)–(9), the treatment of neopentylamine with nitrous acid (11), or of neopentyl chloride with a silver salt (12), will produce a comparable result:

$$(CH_3)_3CCH_2NH_2 \xrightarrow[\ (-N_2)\ ]{HNO_2} \tag{11}$$
$$\xrightarrow{Ag^+} (CH_3)_3CCH_2^+ \longrightarrow \begin{cases} \text{rearranged} \\ \text{products} \\ \text{as in (10)} \end{cases}$$
$$(CH_3)_3CCH_2Cl \tag{12}$$

This kind of rearrangement, which involves either the grouping $C-\overset{\overset{\textstyle C}{|}}{\underset{\underset{\textstyle C}{|}}{C}}-C-OH$

or some other system that can lead under the influence of acid catalysts (ZnCl$_2$,

H$_2$SO$_4$, and others) to $C-\overset{\overset{\textstyle C}{|}}{\underset{\underset{\textstyle C}{|}}{C}}-\overset{+}{C}$, and which proceeds by way of

$$C-\overset{\overset{\textstyle C}{|}}{\underset{\underset{\textstyle C}{|}}{\overset{+}{C}}}-C \longrightarrow C-\overset{\overset{\textstyle C}{|}}{\underset{\underset{\textstyle C}{|}}{\overset{+}{C}}}-C,$$

is often called the *Wagner-Meerwein* rearrangement. It is most often encountered in connection with the chemistry of terpenes, which Wagner and Meerwein studied extensively.

That rearrangements such as that of the neopentyl group occur, even though the generation of a primary carbonium ion is energetically very unfavorable, indicates that the migration of the methyl group occurs as the leaving group departs, and that the neighboring group participates in the process of ionization. We may formulate the general case in the following terms:

$$-\overset{\overset{\textstyle R}{|}}{\underset{\underset{\textstyle X}{|}}{C}}-\overset{|}{\underset{|}{C}}- \xrightarrow{-X^-} -\overset{\overset{\textstyle R}{|}}{\underset{|}{C}}-\overset{|}{\underset{|}{\overset{+}{C}}}- \longrightarrow -\overset{\overset{\textstyle \cdot R\cdot}{\overset{+}{\diagup\diagdown}}}{C}{\cdots}\overset{|}{\underset{|}{C}}- \longrightarrow -\overset{+}{\underset{|}{C}}-\overset{\overset{\textstyle R}{|}}{\underset{|}{C}}- \xrightarrow{Y^-} -\overset{|}{\underset{|}{C}}-\overset{\overset{\textstyle R}{|}}{\underset{\underset{\textstyle Y}{|}}{C}}-$$

Further refinement of this concept would require much more extended comment than is possible here. It can be said that, depending upon the nature of X, R, and Y, and the degree of substitution on the two carbon atoms shown in this generalized expression, the stages shown may be (1) discrete intermediates, (2) stages represented by energy minima in an energy diagram for the progress of the reaction, or (3) stages that represent only a formal dissection of a reaction course that leads through a single intermediate. Thus, Equations (10), (11), and (12) should be regarded as didactic rather than realistic expressions for the overall course of the reaction.

Other examples of rearrangements that proceed by way of carbonium ions are:

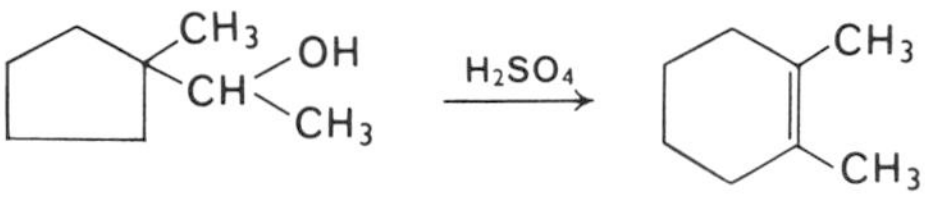

(13)

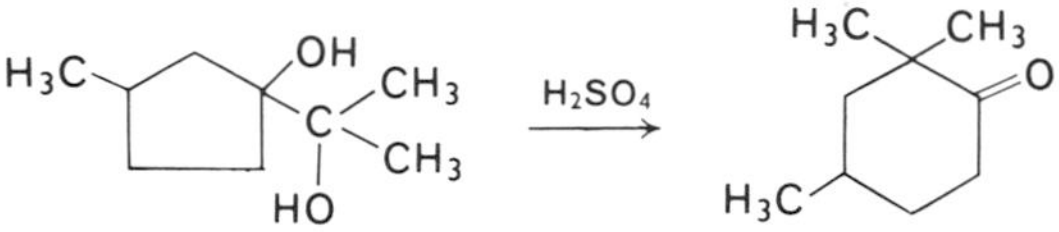

(14)

Reaction (14) is an example of the pinacol rearrangement, which is mechanistically closely related to the Wagner-Meerwein rearrangement.

---

**Exercise 1.**   Formulate the stages in the rearrangements in (13) and (14).

---

In the rearrangement that takes place in (10) the driving force is the transformation of the primary carbonium ion into the more stable tertiary carbonium ion by the migration of the methyl group. In the pinacol rearrangement [e.g., (14)] the protonation (15) and loss of a water molecule to give (16) is followed by a rearrangement that leads to the generation of the resonance-stabilized conjugate acid of a carbonyl group

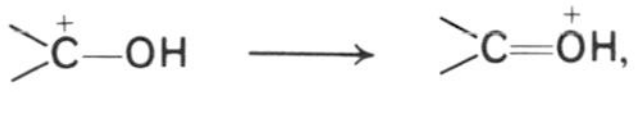

$$>\overset{+}{C}-OH \longrightarrow >C=\overset{+}{O}H,$$

in (17):

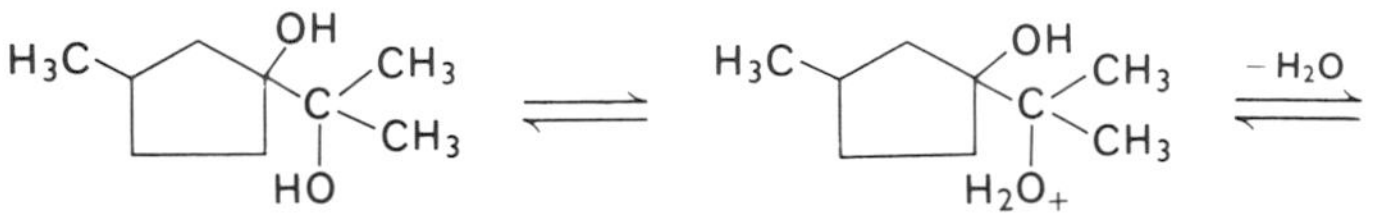

(15)

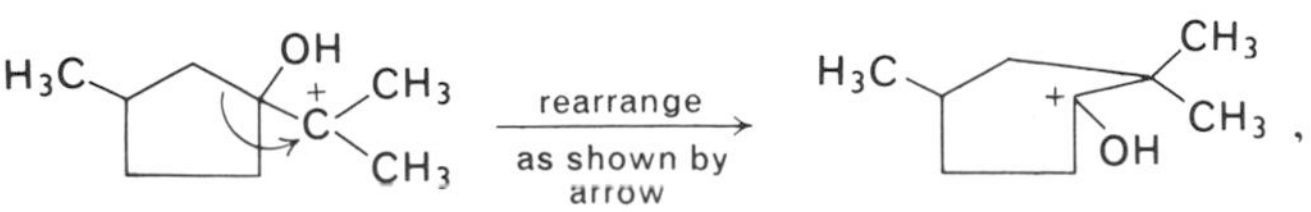

(16)    which is (17):

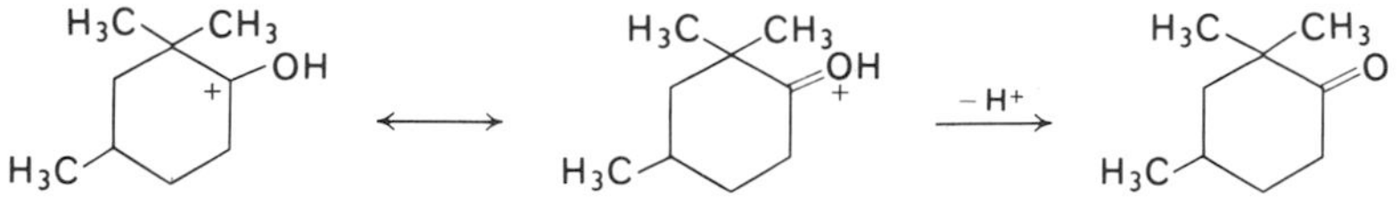

(17)

It is evident from what has been said that in any reaction in which a carbonium ion is generated, the possibility of the occurrence of a rearrangement of the Wagner-Meerwein or pinacol type must be kept in mind. Suppose, for example, it were desired to carry out the reaction

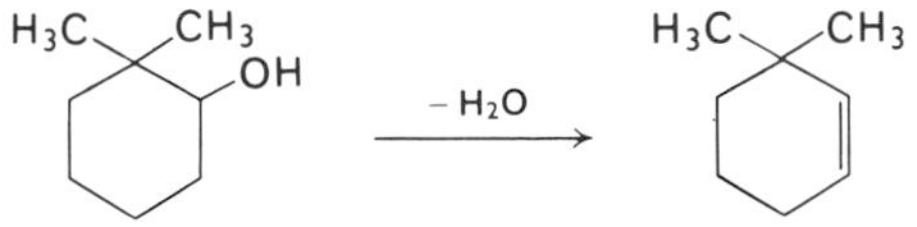

(18)

2,2-dimethylcyclohexanol          3,3-dimethylcyclohexene

The usual methods of dehydration of alcohols by treatment with strong acid catalysts proceed by way of protonated species that can lead to carbonium ion intermediates. Indeed, an attempt to bring about the dehydration of 2,2-dimethylcyclohexanol by treating the alcohol with hot, dilute sulfuric acid leads to the rearranged olefin:

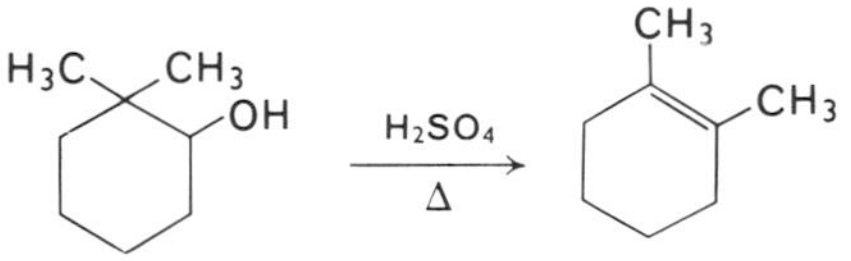

(19)

## 28.3 Dehydration without rearrangement

The dehydration of 2,2-dimethylcyclohexanol to 3,3-dimethylcyclohexene can be accomplished by the use of a reaction in which a carbonium ion is not an intermediate. A procedure for carrying this out, known as the xanthogenate method, involves the pyrolytic decomposition of a thioester, prepared as follows:

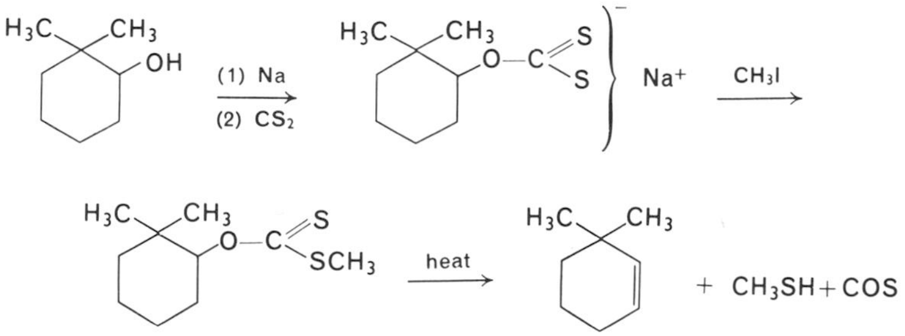

+ $CH_3SH + COS$          (20)

The formation of a cyclic transition state

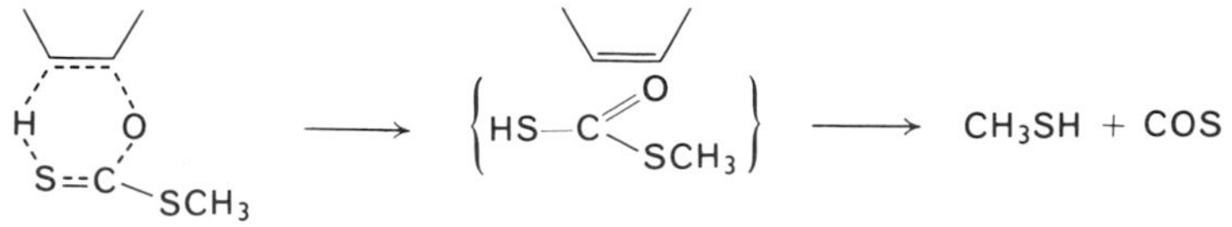

$CH_3SH + COS$          (21)

provides for a concerted rupture of the C—O bond, simultaneous with the deprotonation. The pyrolysis of esters of other kinds (e.g., benzoates and acetates) is also employed.

## 28.4   Aliphatic diazonium compounds

The action of nitrous acid upon aliphatic primary amines (Section 26.5) is a reaction that is very prone to lead to rearrangements, since the tendency for the aliphatic diazonium compounds to decompose to yield nitrogen and carbonium ions is very high indeed. When the amino group is adjacent to a carbinol carbon atom, as in 1-aminomethylcyclohexanol (the synthesis of which is shown), pinacol-like rearrangement occurs:

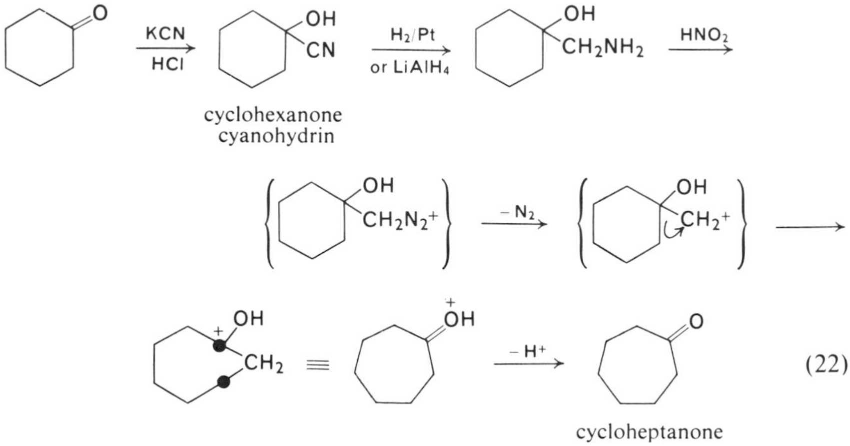

cyclohexanone
cyanohydrin

(22)

cycloheptanone

Ring contractions as well as expansions are observed in cyclic systems:

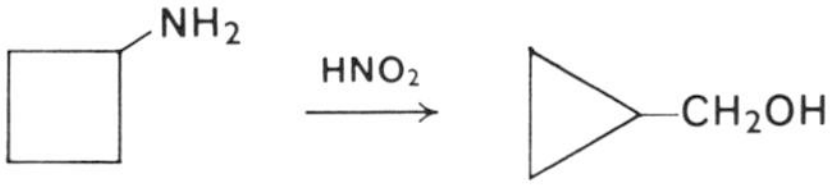

(23)

## 28.5   Base-induced rearrangements

The addition of a strong base (e.g., $OH^-$) to a carbonyl group, as in the following partial expression,

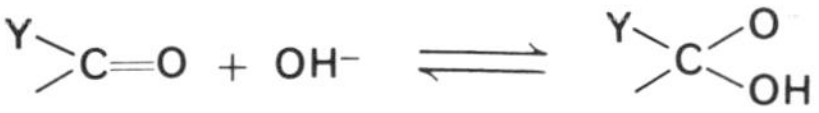

may lead to a number of consequences other than the simple reversal of the addition with expulsion of $OH^-$ (24). That is, the anionic $—O:^-$ can regain its carbonyl character in two ways:

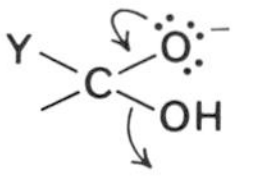

(24) 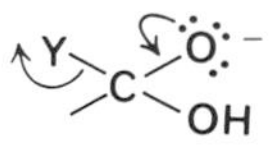                    (25)

Examples of the course, pictured in the same general way as (25) that we have already encountered, and which are not rearrangements as we are now using the term, include many of the reactions of acid derivatives (e.g., hydrolysis of esters, where $Y = OR$) and the cleavage of $\beta$-dicarbonyl compounds; for example:

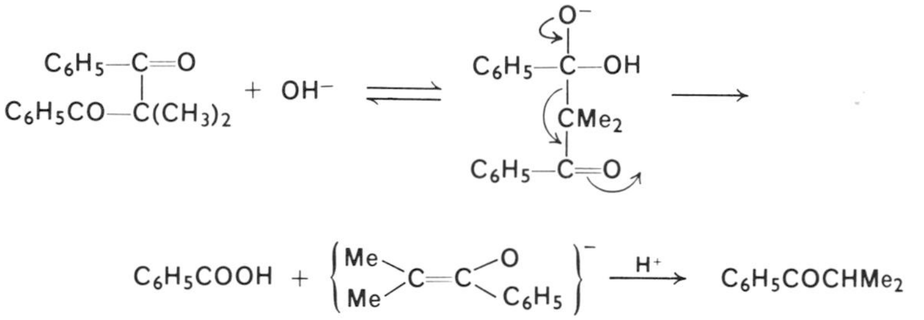

$$C_6H_5COOH + \left[ \begin{matrix} Me \\ Me \end{matrix} C=C \begin{matrix} O \\ C_6H_5 \end{matrix} \right]^- \xrightarrow{\ H^+\ } C_6H_5COCHMe_2 \qquad (26)$$

(25)

A third course to which (25) applies consists in a *migration* of Y to an adjacent electron-deficient center. A typical example of this is the *benzilic acid rearrangement*, so called because its prototype is the rearrangement of benzil into benzilic acid:

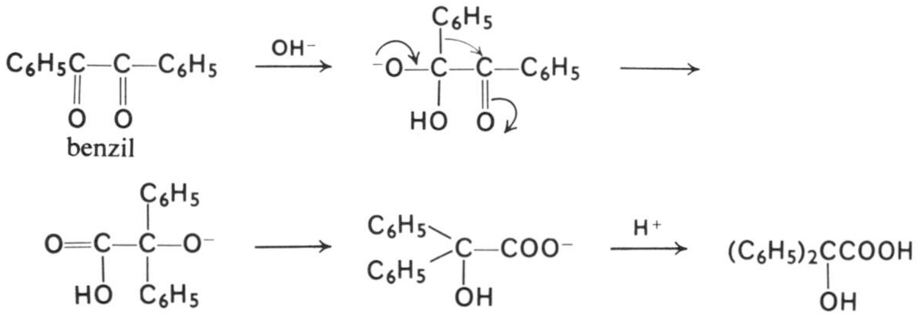

$$(C_6H_5)_2CCOOH \qquad (27)$$

benzilic acid

In this case the driving force for the rearrangement is the combination of the influences of the electrophilic carbonyl carbon atom *to which* the phenyl group migrates, and the anionic oxygen which furnishes the electron pair to the carbon *from which* the phenyl group has migrated.

In the discussion of the Cannizzaro reaction (Section 23.8) it was remarked that glyoxal undergoes a base-catalyzed disproportionation to give glycolic acid. This was described as an intramolecular Cannizzaro reaction. It is clearly a simple example of the benzilic acid rearrangement as well, in which the migrating "group" is a hydride ion. From this we can conclude (1) that the benzilic acid rearrangement and the Cannizzaro reaction are mechanistically closely allied, and (2) that the glyoxal disproportionation might be classed as either one. An examination of the rearrangement of benzil to benzilic acid reveals that this is indeed an oxidation-reduction reaction: benzil, with two carbonyl groups, is transformed into benzilic acid, with one hydroxyl group and one carboxyl group.

Ring contractions can occur when the migrating "group" is a member of a cyclic system. For example, in the benzilic acid rearrangement of phenanthrene-9,10-quinone (phenanthraquinone), the product is a fluorene derivative:

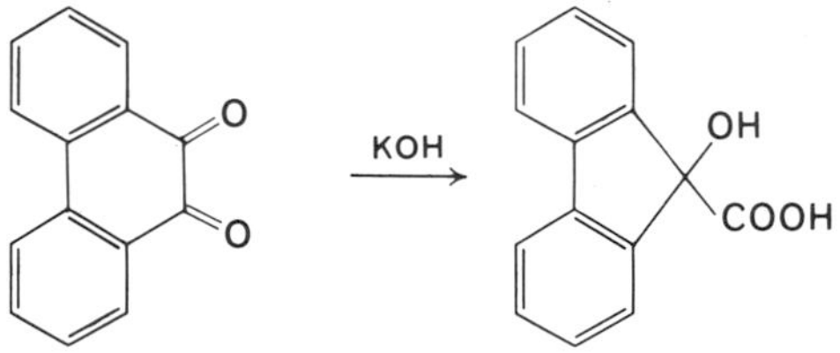

(28)

phenanthrene-9,10-quinone   fluoren-9-ol-9-carboxylic acid

The *Wolff rearrangement* of diazoketones (also called diazomethyl ketones) to carboxylic acid derivatives may be included in this general category. Diazoketones are prepared by the action of diazomethane upon acid chlorides:

$$R{-}C(=O)Cl + CH_2N_2 \longrightarrow RCOCHN_2 + HCl \qquad (29)$$

The structure of this diazoketone is that of a resonance hybrid of the important contributing forms

$$\left\{ RCOCH{=}\overset{+}{N}{=}\ddot{N}{:}^{-} \longleftrightarrow RCO\overset{-}{\ddot{C}}H{-}\overset{+}{N}{\equiv}N{:} \right\}$$

Diazoketones of this kind react with water, alcohols, and ammonia, usually in the presence of finely divided silver, to give, respectively, acids, esters, and amides that contain one more —$CH_2$— group than the acid (RCOOH) from which the acid chloride was derived. The overall method is commonly known as the *Arndt-Eistert* synthesis, and is a valuable way of extending a carbon chain. By successive Arndt-Eistert reactions a synthetic sequence such as the following can be accomplished:

$$RCOOH \longrightarrow RCH_2COOH \longrightarrow RCH_2CH_2COOH$$

The rearrangement step occurs in the following way. The loss of nitrogen from the diazoketone will leave an electron-deficient carbon atom, to which the group R can migrate, as in equation (1), Section 28.1:

$$RCOCHN_2 \xrightarrow{-N_2} \left\{ R{-}C(=O){-}\ddot{C}H \right\} \xrightarrow{\text{shift of R:}} \underset{\text{a ketene}}{RCH{=}C{=}O}$$

The resulting *ketene* (which can be isolated in some cases) the reacts with water, alcohol, or ammonia, depending upon the manner in which the reaction is carried out, to give the final product:

$$RCH=C=O \begin{cases} +\,EtOH & \longrightarrow RCH_2COOEt \\ +\,H_2O & \longrightarrow RCH_2COOH \\ +\,RNH_2 & \longrightarrow RCH_2CONHR \end{cases}$$

An illuminating example of a compound that undergoes rearrangement under both acidic and basic conditions is the following:

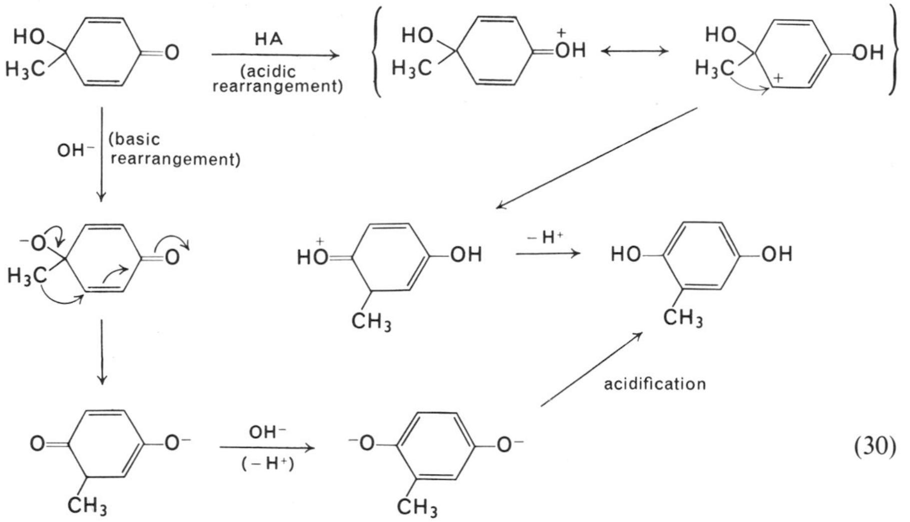

(30)

## 28.6 Rearrangements involving electron-deficient oxygen

The participation of electron-deficient oxygen atoms in organic reactions is a feature of many reactions in which oxidation is the overall process. In general terms, the change has been pictured in (2). The subsequent fate of the rearranged product, which is still "onium" in nature, will depend upon structural factors that can best be described by examining the following examples.

The treatment of cumene hydroperoxide with acid leads to the formation of acetone and phenol and is an industrial process for the large-scale production of these important chemicals:

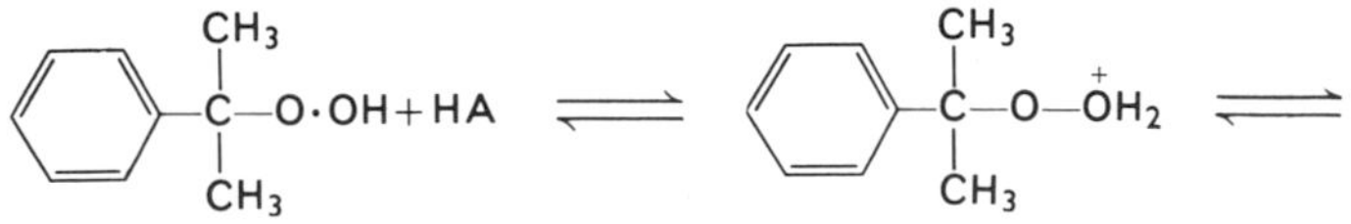

(31)

(from cumene by air
oxidation in the
presence of a catalyst)

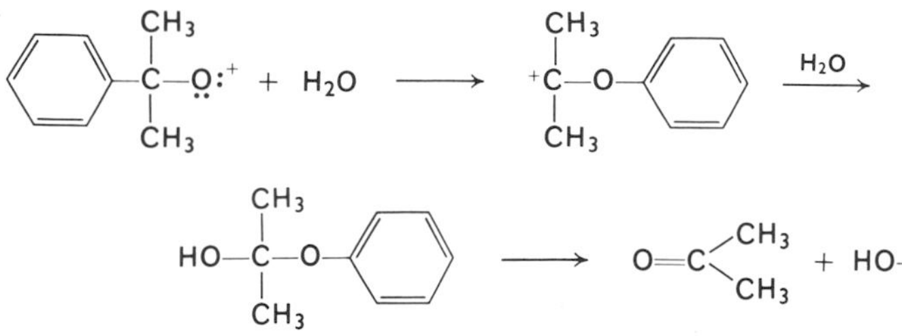

hemiketal of phenol
and acetone

$$\qquad\qquad\qquad\qquad\qquad\qquad\qquad\qquad\qquad\qquad (32)$$

It is emphasized here, as has been done in earlier discussions, that a formulation in which the step of dissociation of $H_2O$ from the protonated intermediate is shown as a discrete process is a convenience, since it is not always possible to distinguish that process from the nearly equivalent one in which the shift of the migrating group is *a part of the step* in which $H_2O$ departs. That is, the equation might better be written

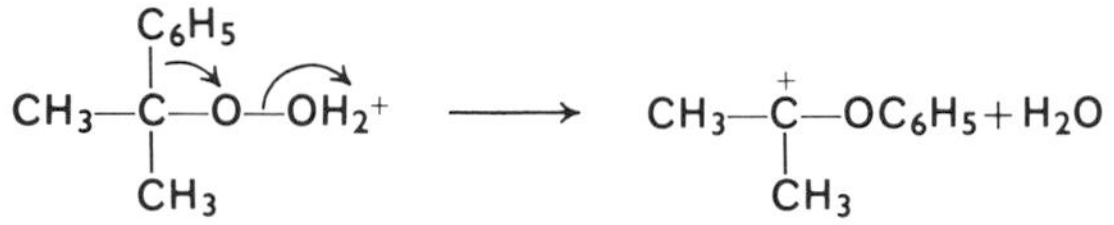

However, since the incipient dissociation, or tendency for the dissociation of $H_2O$ from $—O—\overset{+}{O}H_2$, lends to the residual oxygen atom positive character, the writing of the equation for the reaction in discrete steps serves to emphasize the mechanistic basis for the change.

### 28.7    Oxidation of ketones to esters by means of peroxy acids

A reaction that appears to have a mechanistic basis similar to that of the above example is the oxidation of ketones to esters (or lactones, in the case of cyclic ketones) by peroxy acids. The reaction, in two examples, is as follows:

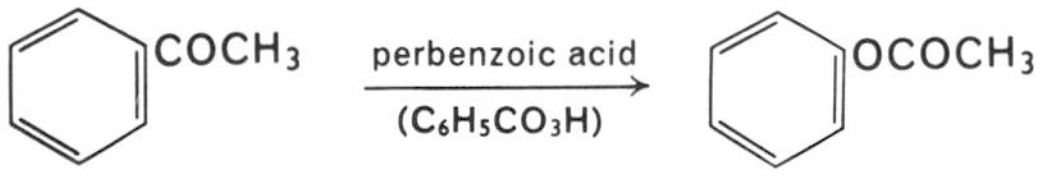

acetophenone                                    phenyl acetate

$$\qquad\qquad\qquad\qquad\qquad\qquad\qquad\qquad\qquad\qquad (33)$$

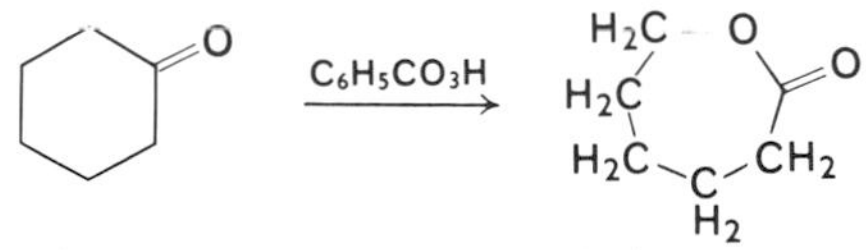

cyclohexanone                    6-hydroxyhexanoic
                                 acid lactone

$$\qquad\qquad\qquad\qquad\qquad\qquad\qquad\qquad\qquad\qquad (34)$$

The addition of the peroxy acid [perbenzoic acid will be used as the example, but other peroxy acids, such as peroxyacetic, peroxytrifluoroacetic, persulfuric (Caro's acid) are also used for reactions of the kind we are considering] to the ketonic carbonyl group may be written in the following way:

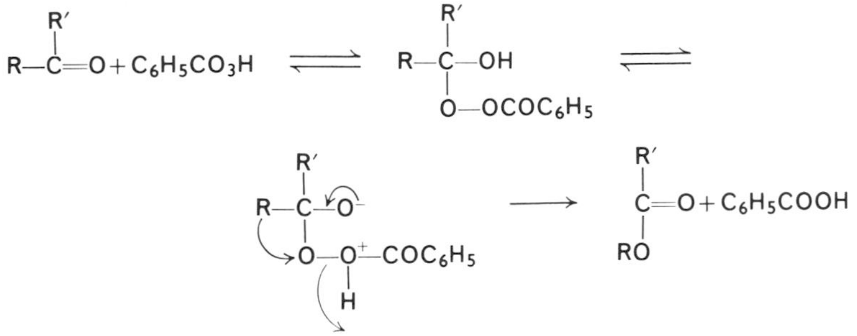

$$R-\underset{\underset{}{\overset{\overset{R'}{|}}{C}}}{=}O + C_6H_5CO_3H \rightleftharpoons R-\underset{\underset{O-OCOC_6H_5}{|}}{\overset{\overset{R'}{|}}{C}}-OH \rightleftharpoons$$

$$R-\underset{\underset{O-O^+-COC_6H_5}{\overset{|}{\underset{H}{|}}}}{\overset{\overset{R'}{|}}{C}}O^- \longrightarrow \underset{RO}{\overset{\overset{R'}{|}}{C}}=O + C_6H_5COOH \quad (35)$$

The oxidation of ketones to esters in this way is a very useful reaction. It serves as a means of displacing an acyl group from the aromatic ring, with the formation of a phenol (as the ester); for example,

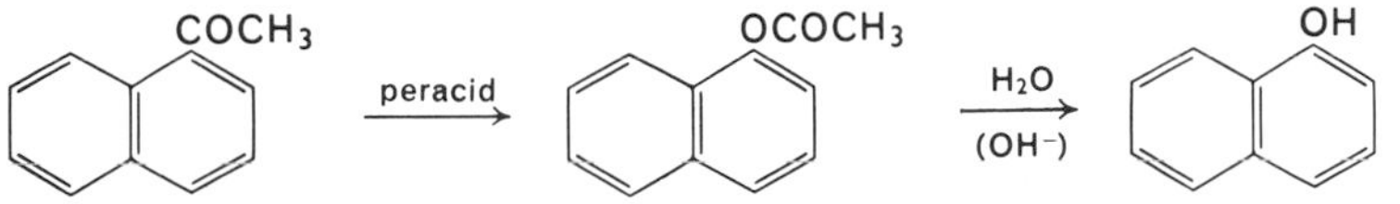

$$(36)$$

It has also been used to produce large-ring lactones from large-ring ketones:

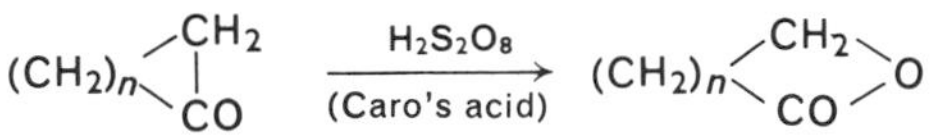

$$(37)$$

The formation of a multi-membered lactone from the corresponding ketone is evidence that the reaction does not proceed by oxidation of the ketone to a hydroxy acid, and subsequent lactonization. In such a case a cyclic ketone of $n=4$ or more in (37) would be expected to yield a polymeric linear lactone.

## 28.8 Rearrangements involving electron-deficient nitrogen

The *Hofmann degradation* of amides results in the overall transformation $RCONH_2 \rightarrow RNH_2$, and clearly involves the formation of a carbon-nitrogen bond between the carbon atom of the R group and the amide nitrogen atom. The reaction is carried out by treatment of the amide with bromine in alkaline solution (to yield an intermediate N-bromoamide which can be, but seldom is, isolated), the final stage being brought about by alkaline hydrolysis under the alkaline conditions employed.

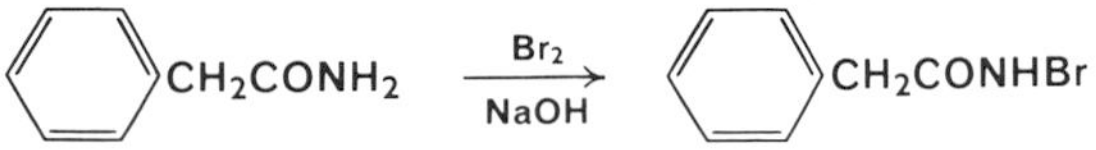

$$(38)$$

The final, or rearrangement step, can be described by the simplified equation

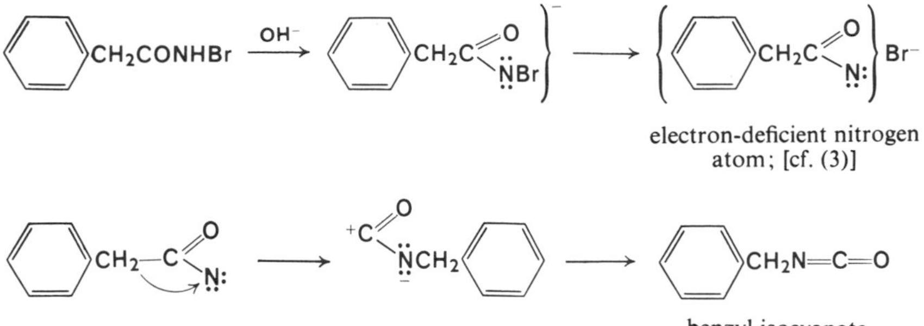

electron-deficient nitrogen
atom; [cf. (3)]

benzyl isocyanate

$$(39)$$

The isocyanate (which can be isolated, but which is more easily prepared in the Curtius reaction to be described) is hydrolyzed by alkali to yield the amine and carbon dioxide (as carbonate):

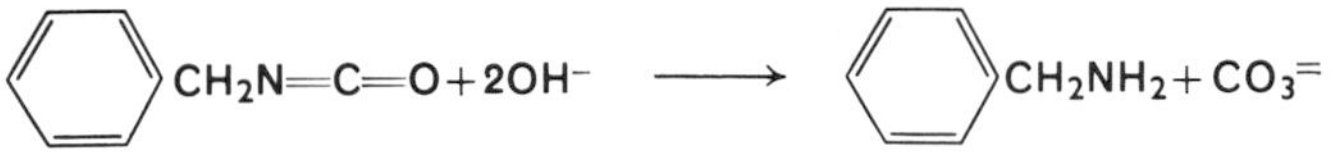

$$(40)$$

The *Curtius reaction* is quite different from the Hofmann reaction in the manner of its performance, but it is *mechanistically very closely allied to it*; it consists in the treatment of an acyl hydrazine (a hydrazide) with nitrous acid, with the formation of an acyl azide. The azide rearranges to the isocyanate on heating in an inert solvent (e.g., benzene); or, if the reaction is carried out in alcoholic solution, the isocyanate reacts with the solvent to form a *urethane*:

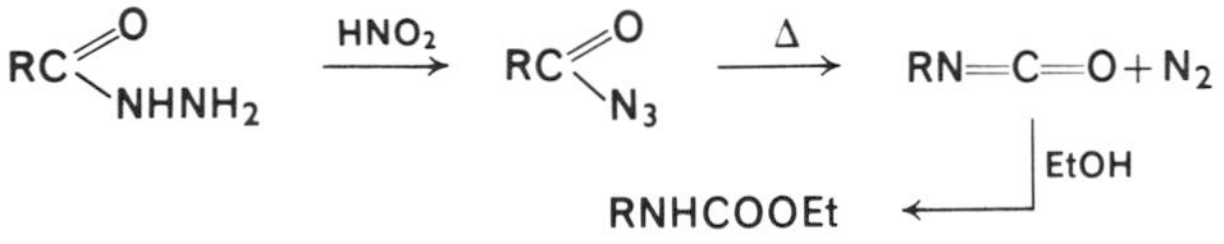

$$(41)$$

A third reaction of the same kind is the *Lossen rearrangement* of hydroxamic acids. Hydroxamic acids, or their O-acyl derivatives, rearrange upon heating, with the loss of —OH or O-acyl and migration of the R group to nitrogen:

$$R-CO-NH-O-COR' \longrightarrow R-N=C=O+R'COOH \qquad (42)$$

All of these rearrangements can be seen to proceed by a common course, which can be summarized in the expression

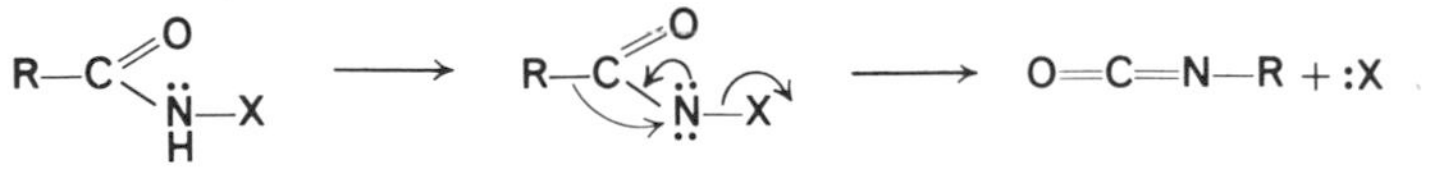

$$(43)$$

in which :X is $Br^-$ in the Hofmann reaction, $N_2$ in the Curtius reaction, and $OCOR^-$ in the Lossen reaction.

An example of the Lossen rearrangement that occurs in nature is the formation of isothiocyanates from mustard oil glycosides, described in Section 16.22. In this case the compound that undergoes the rearrangement is a derivative of the hydroxamic acid derived from a thio acid, RCOSH.

A noteworthy feature of the Hofmann and Curtius reactions is that if the R group in RCOOH is asymmetric its configuration is retained:

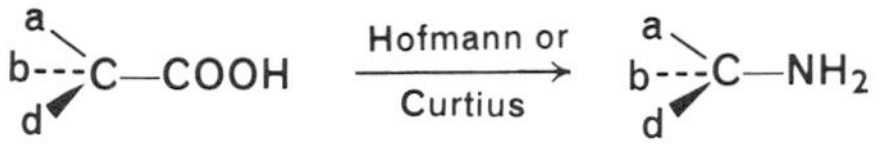

$$\underset{d}{\overset{a}{\underset{b}{\big|}}}\text{C—COOH} \xrightarrow[\text{Curtius}]{\text{Hofmann or}} \underset{d}{\overset{a}{\underset{b}{\big|}}}\text{C—NH}_2$$

This has been demonstrated experimentally by the Hofmann rearrangement of the following amide, in which the optical activity is due to the restricted rotation about the bond joining the two aryl groups:

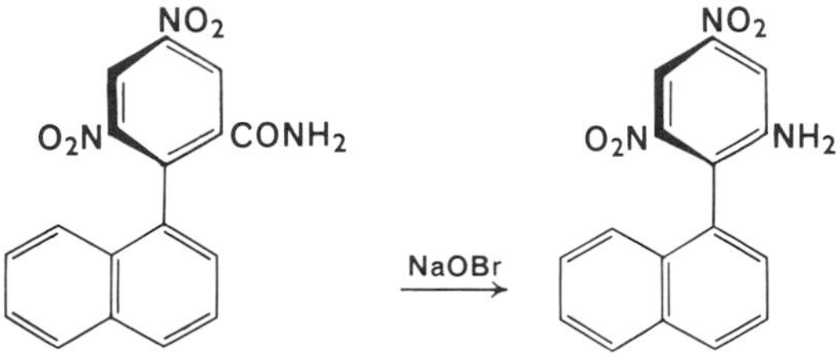

(44)

Had the biaryl system become free at any time during the rearrangement, rotation and racemization would have occurred.

---

**Exercise 2.**   Show how acetophenone can be converted into *(a)* phenol, and *(b)* aniline, by means of reactions described in this chapter. Write the equations.

---

## 28.9   The Beckmann and Schmidt rearrangements

The Beckmann rearrangement is a useful reaction which consists in the conversion of an oxime into an amide by the migration of a group from carbon to nitrogen:

The Beckmann rearrangement can be brought about by a variety of reagents, all of them characterized by a common feature: they all act to permit the removal of the —OH group from nitrogen in such a way as to create an electron deficiency on nitrogen. In the following equations, illustrating this step, a discrete, positively charged nitrogen atom is written; but as will be discussed in greater detail further on, this is a simplification which will require additional elaboration.

Strong acids, such as sulfuric acid, can aid the departure of the OH group by protonation:

$$\underset{R}{\overset{R}{>}}C=NOH \overset{HA}{\rightleftharpoons} \underset{R}{\overset{R}{>}}C=N-OH_2^+ \longrightarrow \underset{R}{\overset{R}{>}}C=N^+ + H_2O \qquad (45)$$

phosphorus halides can aid the departure of the OH group by replacement of OH by Cl, or by conversion of the oxime into a phosphoric ester:

$$\underset{R}{\overset{R}{>}}C=NOH \overset{PCl_5}{\longrightarrow} \underset{R}{\overset{R}{>}}C=N-OPCl_4 \text{(or Cl)} \longrightarrow$$

$$\underset{R}{\overset{R}{>}}C=N^+ + PCl_4O^- \text{(or Cl}^-) \qquad (46)$$

and in some cases tosylation (Section 20.6) brings about the rearrangement by converting OH into the excellent leaving group, —OTos:

$$\underset{R}{\overset{R}{>}}C=NOH \overset{TosCl}{\underset{pyridine}{\longrightarrow}} \underset{R}{\overset{R}{>}}C=N-OTos \longrightarrow \underset{R}{\overset{R}{>}}C=N^+ + TosO^- \qquad (47)$$

The result of any of these various procedures is to alter the hydroxyl group of =N—OH to a group X which is prone to ionize as :X⁻. The complete equation for the Beckmann rearrangement, including a final hydrolysis, is represented by the following:

$$\underset{\text{NOH}}{\overset{R}{\underset{C}{\overset{R'}{>}}}} \longrightarrow \underset{\text{N}-X}{\overset{R}{\underset{C}{\overset{R'}{>}}}} \longrightarrow \underset{\text{N}^+}{\overset{R}{\underset{C}{\overset{R'}{>}}}} + X^- \longrightarrow \underset{R-N}{\overset{+C}{\overset{R'}{>}}} \overset{H_2O}{\longrightarrow} \underset{RNH}{\overset{O}{\overset{C}{<}}}\overset{R'}{} $$

$$(48)$$

The Schmidt reaction involves the reaction of a ketone with hydrazoic acid, with the result that an amide is formed with the simultaneous evolution of nitrogen. The Schmidt reaction can be described by the following sequence of steps:

$$\underset{R}{\overset{R'}{>}}C=O+HA \rightleftharpoons \underset{R}{\overset{R'}{>}}\overset{+}{C}-OH \overset{HN_3}{\longrightarrow} \underset{\underset{+}{HN}-N\equiv N}{\overset{\overset{OH}{|}}{R'-C-R}} \overset{\text{loss of}}{\underset{N_2}{\longrightarrow}}$$

$$\underset{\underset{+}{HN^+}}{\overset{\overset{OH}{|}}{R'-C-R}} \longrightarrow \underset{R'NH}{\overset{HO}{\overset{+}{C}}}\overset{R}{} \overset{-H^+}{\longrightarrow} RC\overset{O}{\underset{NHR'}{<}} \qquad (49)$$

As before, it can be seen that the loss of nitrogen from the —N₃ fragment leads to an electron deficient nitrogen atom:

$$H:\overset{|}{\underset{N}{N}}-\overset{+}{N}\equiv N: \longrightarrow H:\overset{|}{\underset{N}{N}}^+ + N_2$$

Again, it should be borne in mind that the details of the reactions that have been formulated above describe the *nature and course* of these reactions, and are not in all cases the *mechanisms* of the reaction. For example, the course of the Schmidt reaction can be recognized as involving a shift of the R group to an electron-deficient nitrogen, whether the loss of nitrogen precedes the migration of R as a discrete phase, or whether the nitrogen leaves as the group migrates:

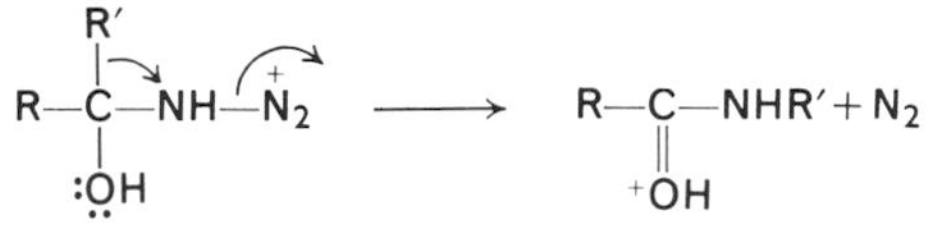

$$R-\underset{\underset{:\ddot{O}H}{|}}{\overset{\overset{R'}{|}}{C}}-NH-\overset{+}{N_2} \longrightarrow R-\underset{\overset{\|}{+OH}}{C}-NHR'+N_2$$

The difference between these details represents a difference in *mechanism;* the reason such details have not been made explicit in the discussion throughout this chapter is partly because a given reaction may follow one course or another, depending upon the particular structure of the substance involved, and partly because in some cases such details are not known with certainty. In the case of the Beckmann rearrangement, the simplified "mechanism" involving an intermediate $>C=N^+$ would lead to erroneous predictions of the sterochemical course of the rearrangement; this is discussed in Section 28.10.

The degradation of carboxylic acids to amines by the use of hydrazoic acid is also called the Schmidt reaction. An alternative but equivalent procedure is to treat the acid chloride with sodium azide. In either case, the acid is converted into the acyl azide, just as in the Curtius reaction, and the subsequent course of the reaction is the same:

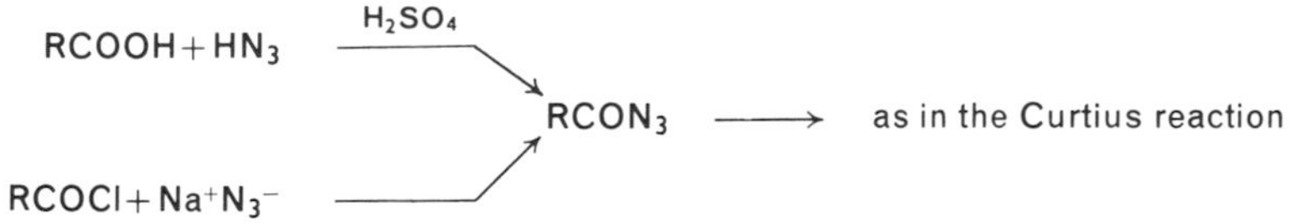

$$\begin{array}{c} RCOOH+HN_3 \quad \overset{H_2SO_4}{\diagdown} \\[2mm] \qquad\qquad\qquad RCON_3 \longrightarrow \text{ as in the Curtius reaction} \quad (50) \\[2mm] RCOCl+Na^+N_3^- \quad \diagup \end{array}$$

## 28.10 The intramolecular nature of the Beckmann and Schmidt rearrangements

The intramolecular nature of the Beckmann and Schmidt rearrangements is shown by the fact that cyclohexanone (via Schmidt) or its oxime (via Beckmann) is converted into the seven-membered cyclic amide (lactam):

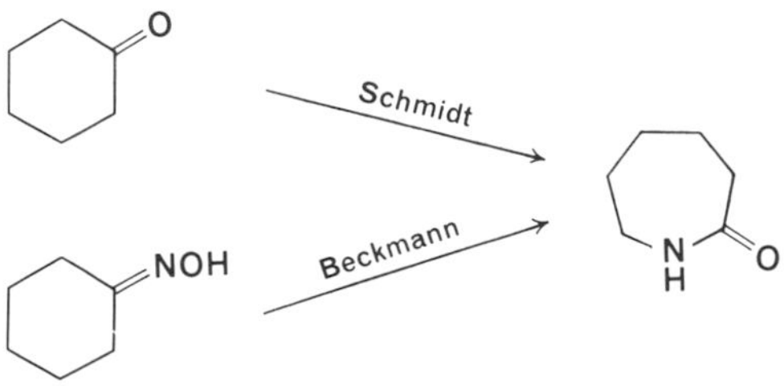

$$(51)$$

$$(52)$$

This result demonstrates that the amide is not formed from an intermediate amino acid.

---

**Exercise 3.**   Devise syntheses for *(a)* methylethylamine; *(b)* benzanilide; and *(c)* 5-amino-pentanoic acid, using the Beckmann or Schmidt reaction in the process.

---

Cycloheptanone oxime undergoes the Beckmann rearrangement to yield the eight-membered lactam, and the macrocyclic ketones of 8–20 members can be converted into large-ring lactams.

A synthetic sequence involving a Beckmann rearrangement and leading finally to an amine can be carried out with aid of lithium aluminum hydride (LAH), which can reduce the amide linkage —CO—NH— to —CH$_2$NH—. For example, piperidine can be prepared from cyclopentanone,

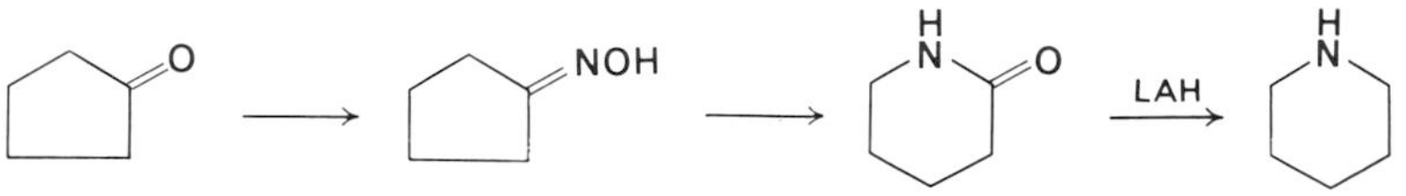

$$(53)$$

piperidine

and propiophenone can be converted to N-*n*-propylaniline:

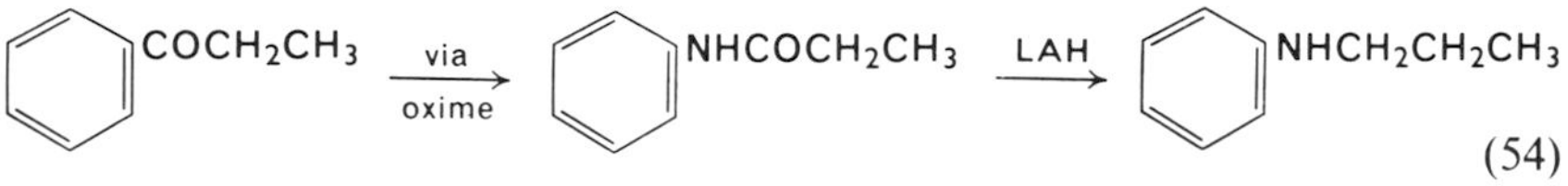

$$(54)$$

## 28.11   The stereochemistry of the Beckmann rearrangement

In the unsymmetrical ketoxime R—C—R′ there would appear to be two possible

$$\underset{\text{NOH}}{\overset{\|}{}}$$

results of a Beckmann rearrangement: RCONHR′ and R′CONHR. With a given oxime, however, but one product is obtained, and it can be concluded that the course of the rearrangement is directed by the configuration of the oximino group. It will be recalled that an unsymmetrical ketone (R—CO—R′) can give rise to two oximes:

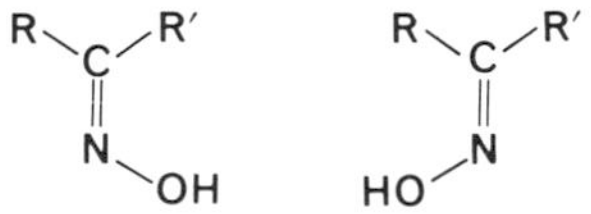

It has been shown experimentally that the Beckmann rearrangement proceeds with a *trans* migration, the migrating group entering from the rear as the —OH (or derived) group leaves the nitrogen atom:

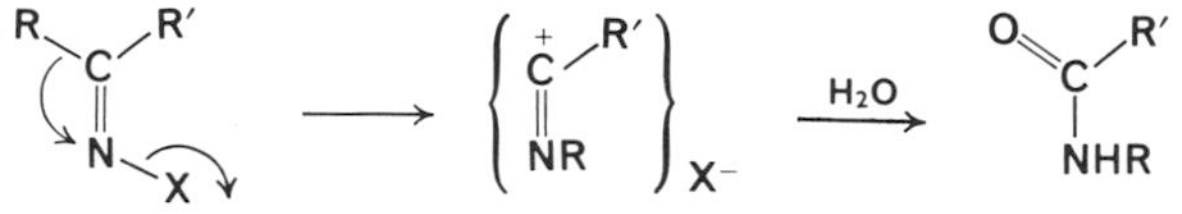

With reference to the remarks concerning the course and mechanism of the reaction (Section 28.9) it can now be seen that a discrete stage

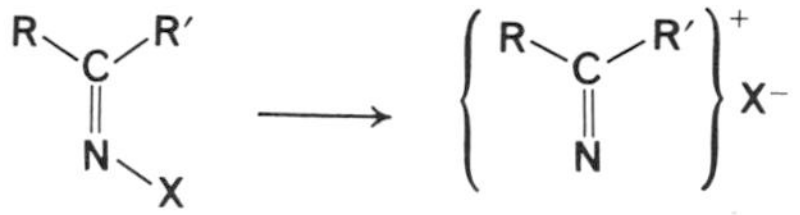

of the reaction is unlikely, since it would be anticipated that a species $R_2C\!\!=\!\!N^+$ would not retain a fixed configuration, and that both R and R′ migration would occur. It thus appears that the *shift of a group from C to N occurs as X departs*, thus preserving the stereochemical specificity of the reaction.

One of several experimental demonstrations of the stereochemical course of the Beckmann rearrangement is the following. Ethyl 1-acetyl-2-naphthol-3-carboxylate forms two oximes, which differ in the stereochemistry at the C=N bond. One of the oximes can be resolved into enantiomorphic, optically active forms; the other is not resolvable. It will be seen from the structures of these two oximes that the compound in which the naphthyl group and the N-hydroxyl group are *syn* disposed will suffer restriction to the rotation of the oximated acetyl group (the blocking substituents in the naphthalene nucleus are the 2-hydroxyl group and the hydrogen atom in the 8-position), while in the isomeric oxime the N-hydroxyl group and the 2-hydroxyl group are not in a position to interfere with free rotation. The resolvable oxime, in which the methyl group is *trans* to the N-hydroxyl group undergoes Beckmann rearrangement with migration of the methyl group (55). The nonresolvable oxime undergoes Beckmann rearrangement with migration of the aromatic group (56):

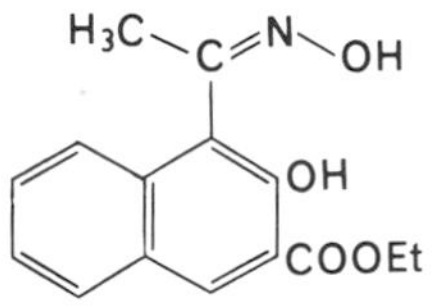

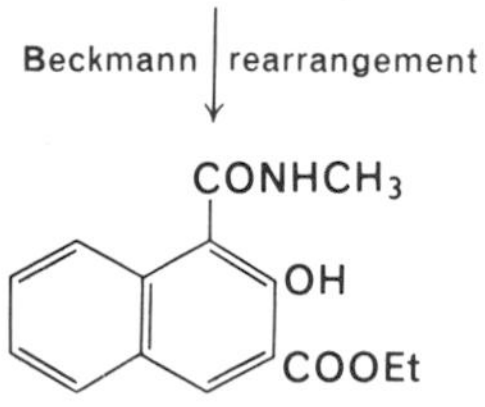

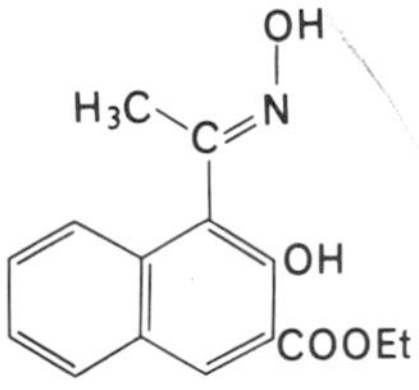

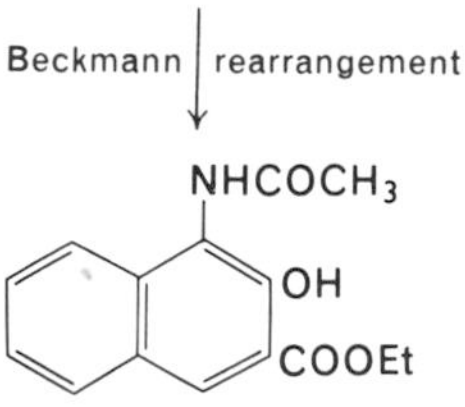

The fact that certain ketones yield oximes that give only one Beckmann rearrangement product is accounted for by the fact that one of the two possible oximes is formed preferentially, probably for steric reasons. For example, aryl methyl ketones always yield arylamine derivatives. This indicates that in

$$\underset{H_3C}{\overset{Ar}{\diagdown}}C{=}NOH,\quad \text{the hydroxyl group is oriented } \textit{away from the bulky aryl group,}$$

and thus the aryl group migrates to nitrogen in the rearrangement.

### 28.12   Rearrangements in reactions of diazomethane

Many of the characteristic reactions of diazomethane depend upon the process, at some stage, of expulsion of molecular nitrogen. The preparation of methyl esters involves this step (Section 14.4). The nucleophilic character of diazomethane, the structure of which may be written $^-{:}CH_2N_2{}^+$, enables it to attack carbonyl groups in a manner illustrated by its addition to an aldehyde:

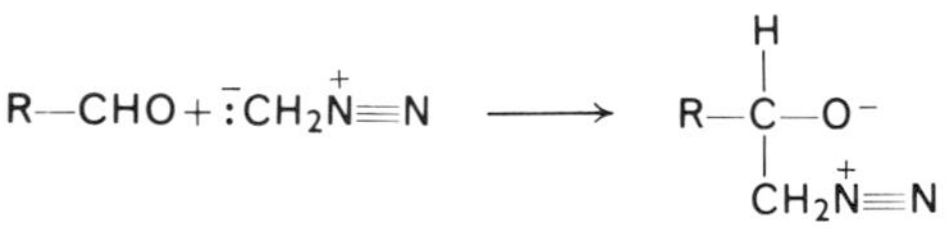

$$R{-}CHO+{:}\overset{+}{CH_2}\overset{}{N}{\equiv}N \longrightarrow R{-}\overset{\overset{\displaystyle H}{|}}{\underset{\underset{\displaystyle \overset{+}{CH_2}{\equiv}N}{|}}{C}}{-}O^- \tag{57}$$

The intermediate formed in this addition can react further in three ways; and, depending upon the structure of the carbonyl compound that is used, any one or all three of these courses may be observed experimentally. In general terms:

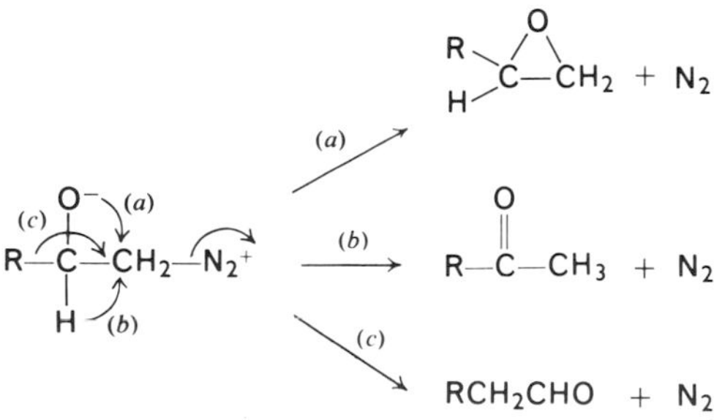

$$
\begin{array}{l}
\underset{(c)}{\overset{O^-}{\diagup}}\\[-2pt]
R{-}\overset{}{\underset{\underset{H}{|}}{C}}{-}\overset{}{CH_2}{-}N_2{}^+
\end{array}
\xrightarrow{(b)}
\begin{cases}
\underset{H}{\overset{R}{\diagdown}}\overset{O}{\overset{\diagup\diagdown}{C{-}CH_2}} + N_2 & (a)\\[8pt]
R{-}\overset{\overset{\displaystyle O}{\|}}{C}{-}CH_3 + N_2 & (b)\\[8pt]
RCH_2CHO \;+\; N_2 & (c)
\end{cases}
\tag{58}
$$

The reaction of 3,4-methylenedioxybenzaldehyde (piperonal) with diazomethane yields the ketone (path $b$) and the ethylene oxide (path $a$) formed by a second addition to the aldehyde formed by path $c$:

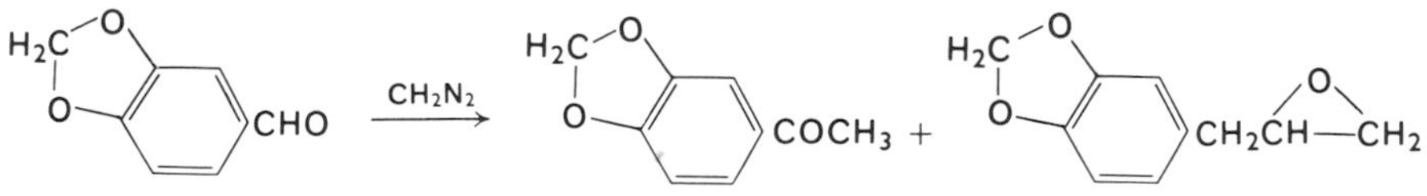

$$\text{piperonal} \xrightarrow{\;CH_2N_2\;} \text{(products)} \tag{59}$$

An application of this reaction in the sugar series is the conversion of *aldehydo*-D-arabinose tetraacetate into the tetraacetate of 1-deoxy-*keto*-D-fructose:

$$
\begin{array}{c}
\text{CHO} \\
|\\
\text{(CHOAc)}_3 \\
|\\
\text{CH}_2\text{OAc}
\end{array}
\quad \xrightarrow{\text{CH}_2\text{N}_2} \quad
\begin{array}{c}
\text{CH}_3 \\
|\\
\text{CO} \\
|\\
\text{(CHOAc)}_3 \\
|\\
\text{CH}_2\text{OAc}
\end{array}
\qquad (60)
$$

aldehydo-D-arabinose     1-deoxy-keto-D-fructose<br>
tetraacetate        tetraacetate

One of the most useful applications of the reaction of diazomethane with ketones is a ring-enlargement reaction that permits a convenient and practicable method of preparing cycloalkanones. The reaction with cyclohexanone proceeds as follows:

$$(61)$$

63% yield

The cyclooctanone that is formed results from a secondary addition of diazomethane to the cycloheptanone that is the initial product.

---

**Exercise 4.** Diazomethane also adds to $\alpha,\beta$-unsaturated ketones in a reaction which, in its initial phase, resembles the Michael reaction. Nitrogen is not lost, however, and the final product contains a five-membered ring containing two nitrogen atoms. Formulate the addition of diazomethane to benzalacetophenone.

---

## 28.13  Rearrangements of carbonyl compounds induced by bases

The generation of an $\alpha$-carbanion by the attack of a strong base upon a carbonyl compound can in certain cases lead to a sequence of events that results in a molecular rearrangement. Halogenated ketones react with bases in various ways: $\beta$-halogenated ketones undergo a ready elimination reaction:

$$
\underset{\text{X}}{\overset{}{\text{RCOCH}_2\overset{|}{\text{C}}\text{HR}}}
\quad \underset{\longleftarrow}{\overset{\text{B:}^-}{\rightleftharpoons}} \quad
\left\{ \text{RCOCH} - \overset{\text{X}}{\overset{|}{\text{CH}}} - \text{R} \right\}^-
\quad \longrightarrow \quad \text{RCOCH}=\text{CHR} + :\text{X}^-
$$

$$(62)$$

$\gamma$-Halogenated ketones often react with bases with the formation of cyclopropane derivatives. These may be stable compounds that can be isolated, as in the case of carone, formed by the action of alcoholic potassium hydroxide on dihydrocarvone hydrobromide:

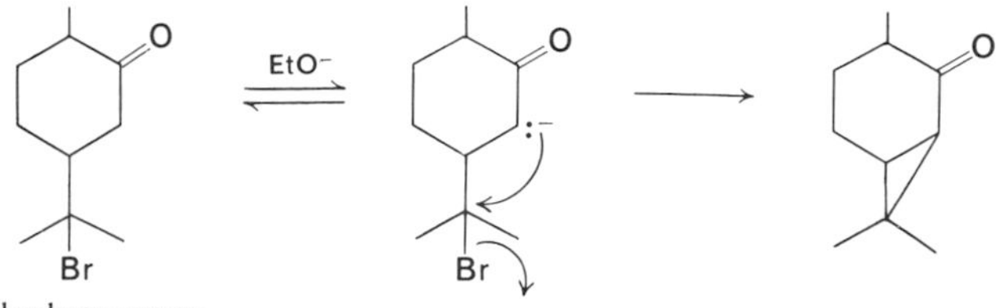

dihydrocarvone          carone
hydrobromide

$$(63)$$

Carvone hydrobromide, however, undergoes a second reaction by the action of the alcoholic alkali upon the cyclopropane formed in a similar manner. The methylene group present in the system $-CH_2-C=C-C=O$ is an active methylene group, for the anion $\{-\ddot{C}H-C=C-C=O\}^-$ is stabilized by resonance. As a consequence, the reaction course shown in the following series of steps leads to the eventual formation of the seven-membered ring compound, eucarvone:

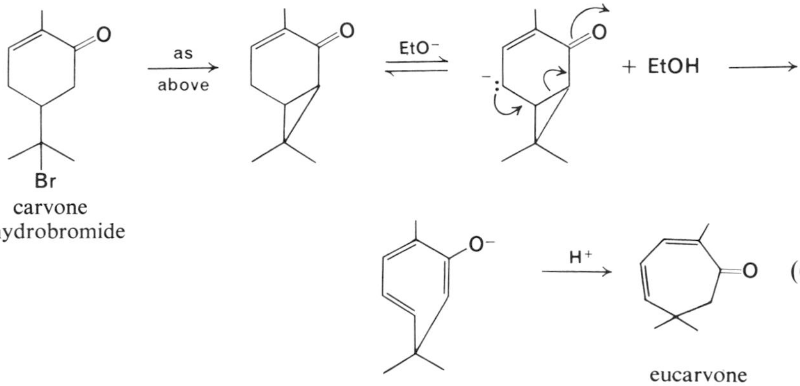

carvone
hydrobromide

eucarvone

$$(64)$$

The *Favorskii rearrangement* bears a close resemblance to the reactions just discussed. A representative example of this reaction is found in the action of alkali upon 2-chlorocyclohexanone. In this case (but not in every case of so-called Favorskii rearrangements) the formation of an intermediate cyclopropanone derivative has been shown to occur, the subsequent cleavage of which by alkali leads to the formation of a cyclopentanecarboxylic acid derivative:

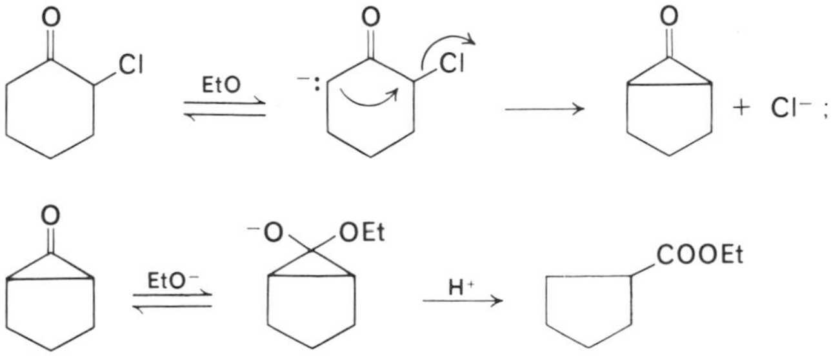

$$(65)$$

An alternative formulation for the course of the Favorskii rearrangement of 2-chlorocyclohexanone appears reasonable:

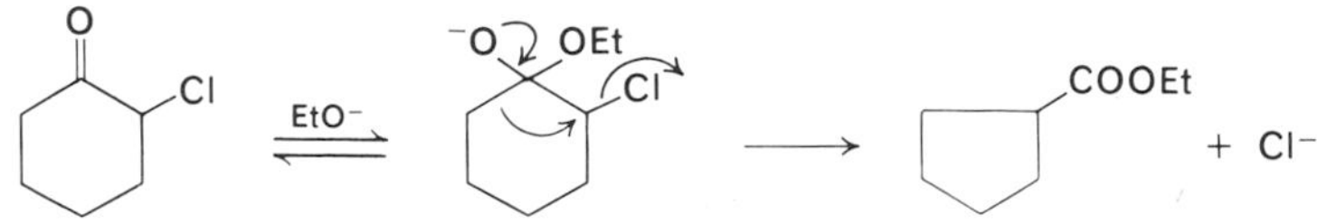

$$\text{(66)}$$

Experiment has shown, however, that when the 2-chlorocyclohexanone is labeled with radioactive carbon ($^{14}$C) at the 2-position, the resulting cyclopentane-carboxylic acid contains half of the radioactivity on the $\alpha$-carbon atom and half on the $\beta$-carbon atom. This result shows that a symmetrical intermediate, as in (65), is formed in the reaction.

Reactions closely allied to the Favorskii rearrangement, but in which a cyclopropanone intermediate cannot be formed, are known; these probably proceed by another mechanism, perhaps one like that shown in Equation (66) above.

### 28.14   Ylides. Sommelet and Stevens rearrangements

*Ylides* are compounds containing an anionic carbon atom in a position adjacent to a quaternary nitrogen, phosphorus, or sulfur atom:

$$R_2\overset{..}{C}\!\!-\!\overset{+}{N}R_3 \qquad\qquad R_2\overset{..}{C}\!\!-\!\overset{+}{P}R_3 \qquad\qquad R_2\overset{..}{C}\!\!-\!\overset{+}{S}R_2$$

Phosphorus ylides, used in the synthesis of olefins by the Wittig reaction, have been described in Section 12.23.

Nitrogen ylides cannot assume the alternative pentacovalent structure that is one of the contributing forms to the structures of phosphorus and sulfur ylides

$$R_2\overset{..}{C}\!\!-\!\overset{+}{P}R_3 \quad\longleftrightarrow\quad R_2C\!\!=\!\!PR_3$$

for the use of a fifth orbital of nitrogen is energetically proscribed. We should therefore expect to find that nitrogen ylides are exceptionally powerful nucleophiles by reason of the unshared electron pair on the anionic carbon atom.

Nitrogen ylides can be formed by the attack of a strong base (sodamide or an organolithium compound, in which the effective bases are $:NH_2^-$ and $:R^-$) upon a proton adjacent to a quaternary nitrogen atom. The ylide formed from benzyltrimethylammonium salts (only the cation is shown in the following equations) arises and reacts in the following way:

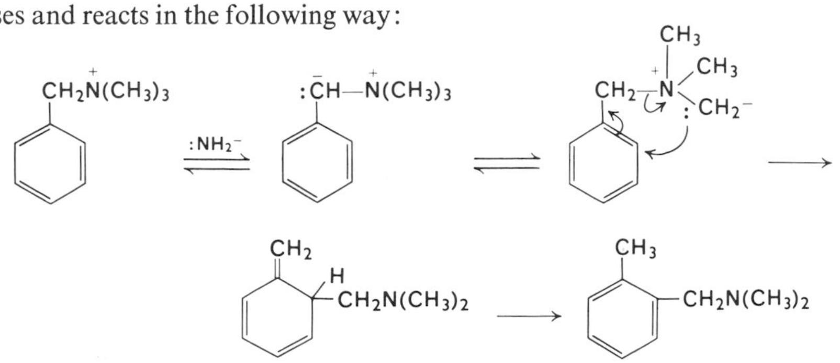

$$\text{(67)}$$

This reaction is a nucleophilic substitution into the benzene ring. Repetition of this process by quaternization of the *o*-methylbenzyldimethylamine and treatment of the quaternary salt with sodamide introduces a second methyl group. The process may be repeated:

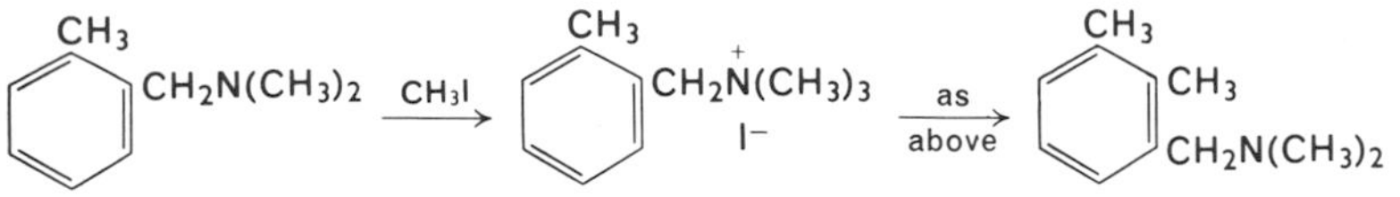

$$\text{(68)}$$

The *Stevens rearrangement* bears a resemblance to the reactions of nitrogen ylides but differs in that the initial ylide is stabilized by additional contributions that are due to the presence of a carbonyl group in the position adjacent to the anionic carbon atom. This additional activation afforded to the *alpha* hydrogen atoms makes it possible to carry out the Stevens rearrangement with the aid of aqueous alkali:

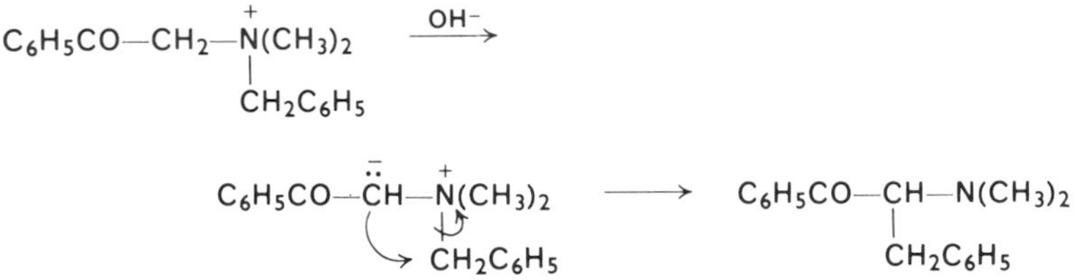

$$C_6H_5CO-CH_2-\overset{+}{N}(CH_3)_2 \quad \xrightarrow{OH^-}$$
$$\underset{CH_2C_6H_5}{\mid}$$

$$C_6H_5CO-\overset{..}{C}H-\overset{+}{N}(CH_3)_2 \longrightarrow C_6H_5CO-CH-N(CH_3)_2 \quad \text{(69)}$$
$$\underset{CH_2C_6H_5}{\mid} \qquad\qquad \underset{CH_2C_6H_5}{\mid}$$

When the migrating group is asymmetric (for example, α-methylbenzyl instead of benzyl) and the compound optically active, the rearranged product is found to retain its optical activity, and the asymmetric carbon retains its original configuration:

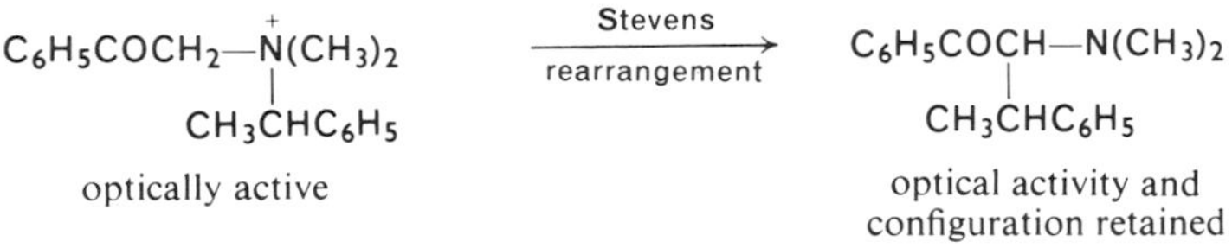

$$C_6H_5COCH_2-\overset{+}{N}(CH_3)_2 \xrightarrow[\text{rearrangement}]{\text{Stevens}} C_6H_5COCH-N(CH_3)_2 \quad \text{(70)}$$
$$\underset{CH_3CHC_6H_5}{\mid} \qquad\qquad\qquad \underset{CH_3CHC_6H_5}{\mid}$$

$$\text{optically active} \qquad\qquad \text{optical activity and}$$
$$\text{configuration retained}$$

### Exercises

(Exercises 1–4 will be found within the text.)

5 Predict the result of the sulfuric acid-catalyzed dehydration of the following alcohols (show all expected *olefinic* products, but it will not be necessary to attempt to assess their relative proportions).

(a) $$\begin{array}{c} CH_3 \\ \mid \\ H_3C-C-CHOHCH_2CH_3 \\ \mid \\ CH_3 \end{array}$$

(b) $$\begin{array}{c} CH_3 \\ \mid \\ CH_3CH_2CHCH_2OH \end{array}$$

(c)

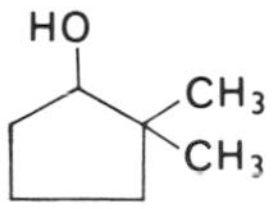

(d)

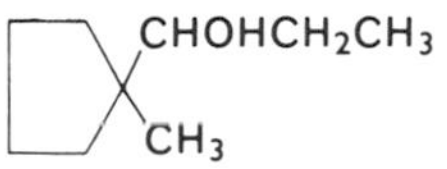

(e)

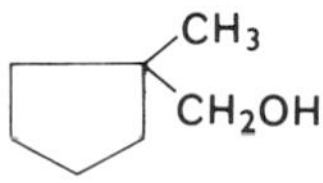

**6** What products would you expect from the diazotization of methylamine in aqueous solution? In ethanol solution?

**7** Formulate the benzilic acid rearrangement of the following $\alpha$-diketones.

*(a)* $CH_3COCOCH_3$      *(b)*      *(c)*

*(d)* $CH_3$ —〈 〉— COCO —〈 〉— Br

**8** What product would result from the treatment of      with sulfuric acid?

**9** How could you prepare 2,5-dimethylphenol starting with 2,5-dimethylpropiophenone?

**10** Starting with succinic acid, devise a synthesis of 5-hydroxypentanoic acid.

**11** *(a)* Of what use is the Hofmann (or Curtius) degradation of amides in the problem of the introduction of substituents into the aromatic ring? *(b)* Show how you could convert acetophenone into benzenediazonium chloride.

**12** Formulate the following transformations.

*(a)* acetophenone $\longrightarrow$ acetanilide

*(b)* phenyl isopropyl ketone $\longrightarrow$ N-isobutylaniline

*(c)* adipic acid $\longrightarrow$ piperidine

*(d)* benzoic acid $\longrightarrow$ hydrocinnamic acid ($\beta$-phenyl propionic acid).

**13** When the Schmidt reaction is performed upon cyclohexanone, one product is "Metrazole," a convulsant drug, with the structure

Suggest the manner of its formation. (Hint: How would $HN_3$ react with the carbonyl group of an amide?)

**14** The lactone 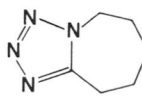      reacts with sodium methoxide in methanol to give the

sodium salt of      $CHC_6H_5$. Formulate the course of this reaction.

## Chapter twenty-nine

# Heterocyclic Compounds. Alkaloids

## 29.1 Classification of heterocyclic compounds

The classical reference work of organic chemistry is Beilstein's *Handbuch der Organischen Chemie* (Encyclopedia of Organic Chemistry). In "Beilstein" organic compounds are organized into three large divisions: acyclic, homocyclic, and heterocyclic.

*Acyclic* compounds are those in which the carbon atoms are arranged only in straight or branched chains; no rings are present. Acyclic compounds are typified by the aliphatic hydrocarbons, alcohols, ketones, amines, and so forth.

*Homocyclic* compounds are those in which a ring, either alicyclic or aromatic, is present. For example, the cycloalkanes and their derivatives, benzene and naphthalene and their derivatives are classified in this category. Although the term "homocyclic" can refer to a ring containing but one kind of atom of any kind, in practice, only rings of carbon atoms need be considered.

*Heterocyclic* compounds are cyclic compounds in which a ring containing more than one kind of atom is present. Heterocyclic rings may contain, in addition to carbon, one or more atoms of nitrogen, oxygen, or sulfur.*

Because of the great number of combinations of carbon, nitrogen, oxygen, and sulfur atoms that can be present in rings of various sizes, the heterocyclic compounds include by far the greatest number of different *kinds* of compounds of any

---

\* Hetero atoms other than nitrogen, oxygen, and sulfur are known; for example, phosphorus and selenium; but these are of relatively less common occurrence.

of the three main divisions. To illustrate, the following examples are given to show a few typical heterocyclic compounds. Others have been encountered in earlier chapters. It should be noted that when a compound contains an alicyclic chain, a homocyclic ring, and a heterocyclic ring in its structure, it is classed, and listed in *Beilstein*, as a heterocyclic compound:

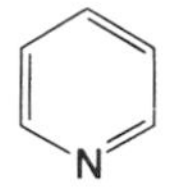

| pyridine | 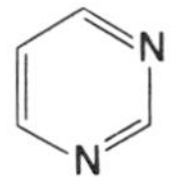 pyrimidine |  furan | 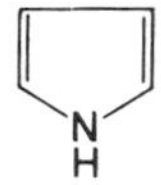 pyrrole | 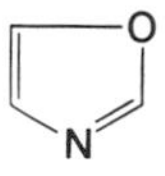 oxazole |

| oxazolidine | 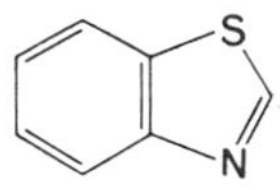 benzothiazole | 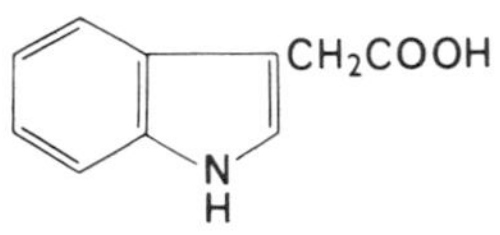 3-indoleacetic acid |

| purine | 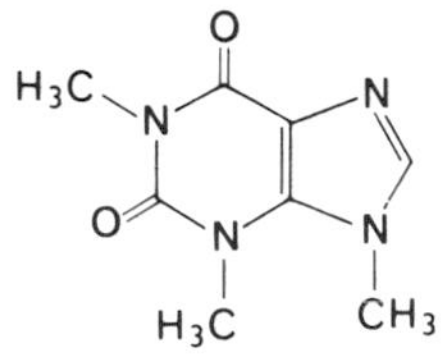 caffeine |

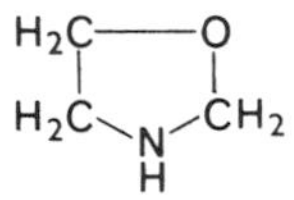

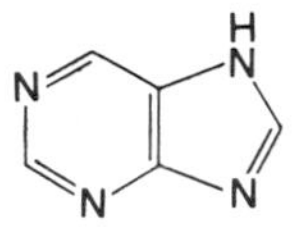

There are a number of kinds of compounds that are, strictly speaking, hetero-cyclic compounds, but which ordinarily are not treated as such. These include such compounds as lactones and acid anhydrides, which are related, respectively, to the corresponding open-chain hydroxyacids and dicarboxylic acids by simple hydrolysis. Cyclic acetals and hemiacetals are similarly related to open-chain aldehydes and ketones, and indeed in this class of substances are found the very large group of the sugars and their derivatives, which are generally treated separately as a discrete group. It is clear that classification rests upon somewhat arbitrary grounds; for our purposes we shall deal chiefly with those heterocyclic compounds whose ring systems are characterized by an integrity and stability resembling that found in the aromatic series, and which can persist unchanged through hydrolytic and oxidative manipulations in which substituents on the ring are altered, removed, or introduced.

### 29.2 Oxygen-containing heterocyclic compounds: methods of formation

Saturated oxygen heterocycles are cyclic ethers, and can be formed by the same kinds of reactions that lead to acyclic ethers. A number of examples of cyclic ether formation have already been encountered in earlier chapters.

The formation of ethylene oxides by the intramolecular displacement of halide ion in the cyclic counterpart of the Williamson ether synthesis

$$\begin{array}{c}\overset{\text{Cl}}{\underset{\text{OH}}{>}}\text{C}-\text{C}< \quad\xrightarrow{\text{OH}^-}\quad >\text{C}-\text{C}< \\ \text{O}\end{array} \qquad (1)$$

has been discussed in Section 9.18. The commercially important ethylene oxide* is manufactured by this method from ethylene chlorohydrin, which in turn is prepared from ethylene.

Another important group of oxygen heterocycles includes furan and its derivatives: furfuraldehyde, furfuryl alcohol, and tetrahydrofuran. The central compound of this group, and the raw material from which the others are prepared, is furfuraldehyde, made by the combined acid hydrolysis and dehydration of the pentose-containing polysaccharides present in certain vegetable materials, such as oat hulls and corn cobs (Chapter 18). Furan, prepared by the catalytic removal of the formyl group from furfuraldehyde, can be hydrogenated to tetrahydrofuran:

$$\text{(furfuraldehyde)}\quad\xrightarrow[\substack{-\text{CHO} \\ \text{group}}]{\text{loss of}}\quad\text{(furan)}\quad\xrightarrow[\text{cat.}]{\text{H}_2}\quad\text{(tetrahydrofuran)} \qquad (2)$$

Tetrahydrofuran is also manufactured by the acid-catalyzed ($H_3PO_4$) dehydration of 1,4-butanediol. This reaction is the cyclic counterpart of the preparation of diethyl ether from ethanol:

$$\begin{array}{cc}\text{CH}_2-\text{CH}_2 \\ \mid \qquad \mid \\ \text{CH}_2\text{OH}\ \ \text{CH}_2\text{OH}\end{array}\quad\xrightarrow{\text{H}^+}\quad\text{(tetrahydrofuran)} + \text{H}_2\text{O} \qquad (3)$$

In general, reactions that lead to the formation of carbon-oxygen bonds in noncyclic compounds can, when applied to bifunctional molecules, lead to ring formation. Three-, five-, and six-membered rings are easiest to prepare, and constitute the most important groups of the oxygen heterocycles.

## 29.3   Chemical behavior of oxygen heterocycles

Ethylene oxide is the raw material from which many important compounds are prepared. The strained three-membered ring is readily opened by nucleophilic attack, and in consequence of this high degree of reactivity, ethylene oxide undergoes many reactions that lead to ethylene glycol derivatives. Hydrolysis and alcoholysis yield ethylene glycol and the various ethylene glycol ethers that have become important industrial solvents:

---

* The preparation of ethylene oxide by the direct oxidation of ethylene is the most important industrial method for producing the compound.

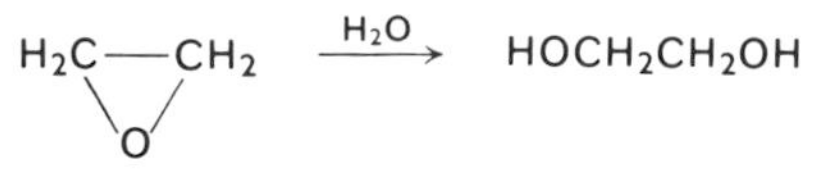

$$H_2C\!-\!CH_2 \xrightarrow{\ H_2O\ } HOCH_2CH_2OH \qquad\qquad (4)$$

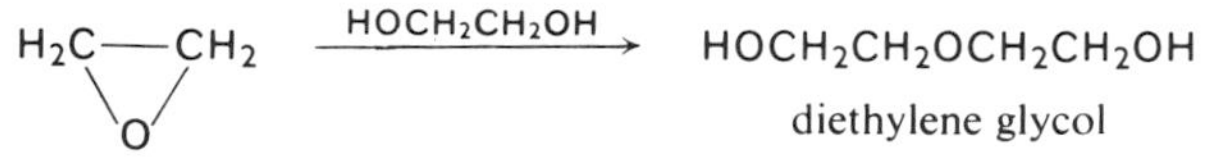

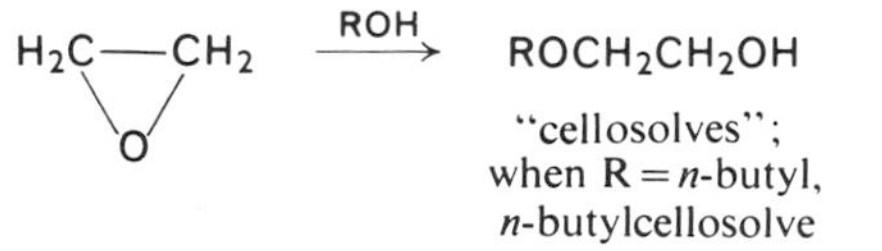

$$H_2C\!-\!CH_2 \xrightarrow{\ HOCH_2CH_2OH\ } HOCH_2CH_2OCH_2CH_2OH \qquad\qquad (5)$$
$$\text{diethylene glycol}$$

$$H_2C\!-\!CH_2 \xrightarrow{\ ROH\ } ROCH_2CH_2OH \qquad\qquad (6)$$
"cellosolves";
when R = *n*-butyl,
*n*-butylcellosolve

Substituted ethylene oxides often rearrange under acidic conditions:

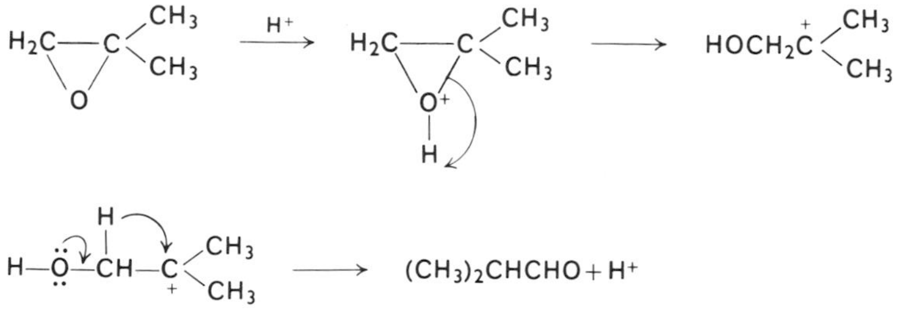

$$(7)$$

The reaction of ethylene oxides with sodium thiocyanate (NaCNS) or seleno-cyanate (NaCNSe) provides an informative example of a sequence of interrelated equilibrium reactions in which the position of the final equilibrium depends essentially upon the relative nucleophilicity of sulfur and oxygen. The overall reaction

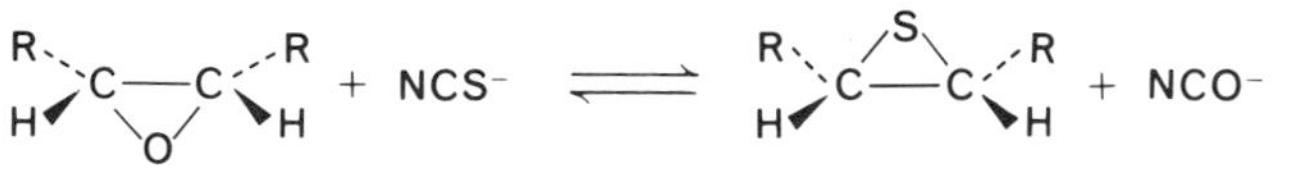

$$(8)$$

can be described in detail by the following equilibria:*

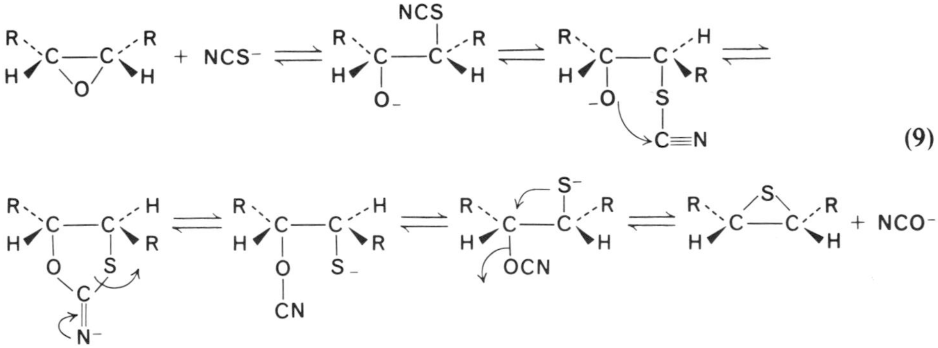

$$(9)$$

The position of the final equilibrium is such that the product of the reaction is the ethylene sulfide. It is to be noted that the position of the sulfur-containing ring is opposite to that of the original oxide ring.

---

\* The student should examine Equation 9 with care, noting the stereochemical disposition of the groups at each step.

When sodium selenocyanate is used, the final product is the cyclic selenide. This compound spontaneously loses selenium and the olefin is formed:

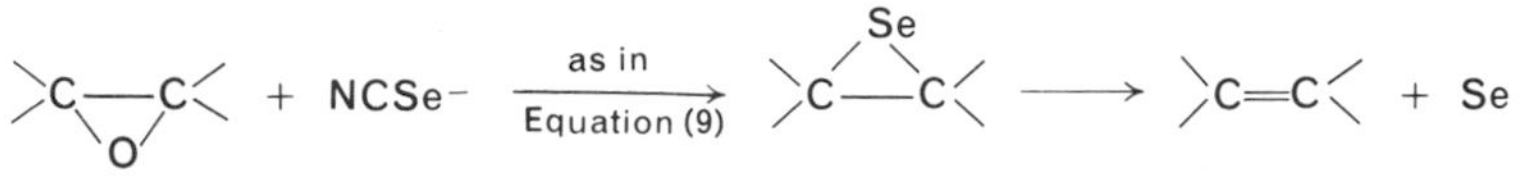

$$\text{>C—C< + NCSe}^- \xrightarrow[\text{Equation (9)}]{\text{as in}} \text{>C—C< } \longrightarrow \text{ >C=C< + Se} \qquad (10)$$

The latter reaction can be used as a preparative means of regenerating an olefin from its oxide; or, since the precipitated selenium forms a dark red-brown deposit, it can be used as a test for the presence of the ethylene oxide grouping.

Since the ease with which ethylene oxide rings are opened is in part the result of the strain associated with the three-membered ring, it is to be expected, and is in fact observed, that four-membered oxide rings would also be subject to ready ring-opening, and that the five- and six-membered cyclic oxides would be stable. This is the case. Tetrahydrofuran and tetrahydropyran are stable substances, having much the same chemical characteristics as the analogous acyclic ethers such as diethyl ether. Tetrahydrofuran is a widely used solvent in modern laboratory practice. It has the chemical inertness of diethyl ether but possesses somewhat different solvent characteristics that lend it to certain special uses. It is a useful solvent in reactions involving the Grignard reagent, in lithium aluminum hydride reductions, and in the preparation of organolithium compounds.

Although tetrahydrofuran is a stable, ether-like compound, its ring can be opened, just as acyclic ethers can be cleaved, by the use of strong, hot hydrobromic acid:

$$CH_3CH_2OCH_2CH_3 \xrightarrow[\Delta]{HBr} 2CH_3CH_2Br \qquad (11)$$

$$\text{(tetrahydrofuran)} \xrightarrow[\Delta]{HBr} BrCH_2CH_2CH_2CH_2Br \qquad (12)$$

Tetrahydrofuran can be dehydrated catalytically (by passing its vapors over an acid catalyst at an elevated temperature), and serves as a source of butadiene:

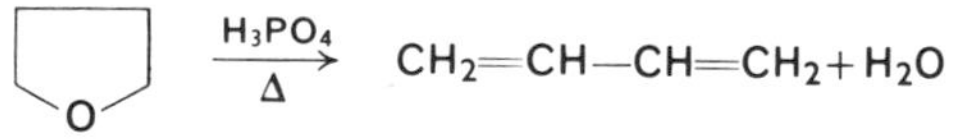

$$\text{(tetrahydrofuran)} \xrightarrow[\Delta]{H_3PO_4} CH_2\text{=}CH\text{—}CH\text{=}CH_2 + H_2O \qquad (13)$$

Another useful, commercially available reagent is dihydropyran:

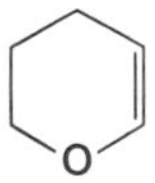

The use of dihydropyran in the protection of phenolic hydroxyl groups has been described in Section 27.5. Another example, in which an aliphatic hydroxyl compound is used, is the following:

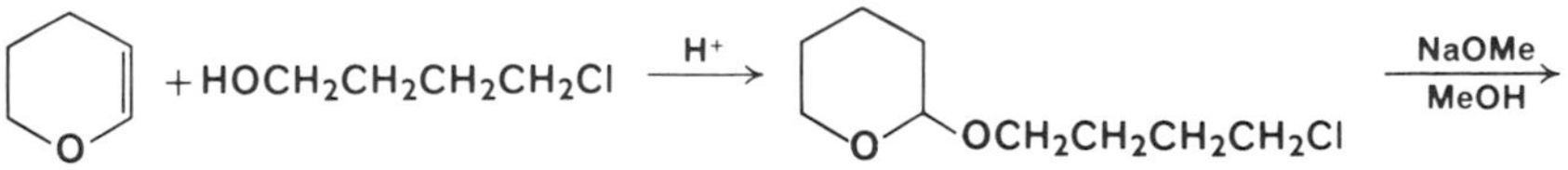

$$\text{(dihydropyran)} + HOCH_2CH_2CH_2CH_2Cl \xrightarrow{H^+} \text{(tetrahydropyranyl ether)}OCH_2CH_2CH_2CH_2Cl \xrightarrow[MeOH]{NaOMe}$$

$$(14)$$

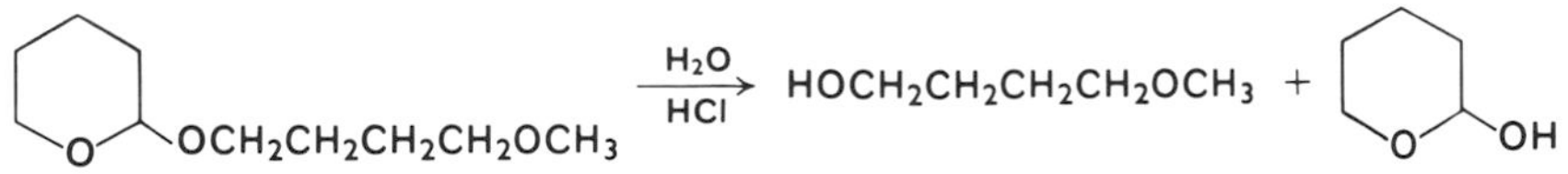

$$\text{(structure)} \xrightarrow[\text{HCI}]{\text{H}_2\text{O}} \text{HOCH}_2\text{CH}_2\text{CH}_2\text{CH}_2\text{OCH}_3 + \text{(structure)}$$

$$(15)$$

If an attempt were made to replace the chlorine in the chloro-alcohol by the methoxyl group by direct reaction with sodium methoxide, ring closure would occur. By "protecting" the hydroxyl group the desired replacement can be effected. Tetrahydropyranyl ethers of chloro- and bromo-alcohols can also be converted into Grignard reagents* (this cannot be accomplished with the free alcohols):

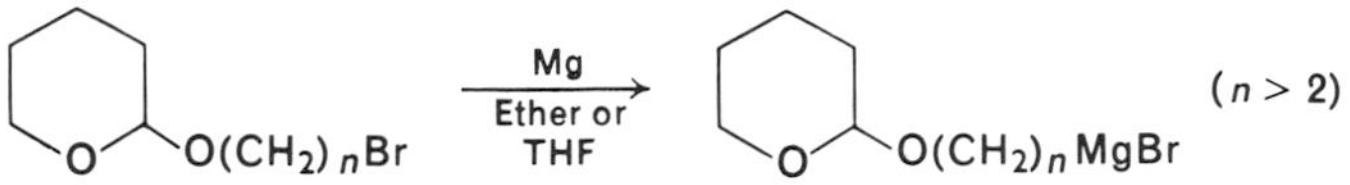

$$\text{(structure)} \xrightarrow[\substack{\text{Ether or} \\ \text{THF}}]{\text{Mg}} \text{(structure)} \qquad (n > 2) \qquad (16)$$

## 29.4 Cyclic oxonium compounds

The ability of oxygen to sustain a positive charge makes it possible to prepare cyclic oxonium compounds of extraordinary stability. The stability of these compounds depends upon their possession, in addition to the oxonium atom, of a six-membered ring containing two carbon-carbon double bonds. The positive charge can be delocalized by a resonance stabilization that resembles the stabilization of the aromatic nucleus. The *pyrilium* salts (the name is derived from pyrone, the six-membered oxygen heterocycle) are usually isolated as chlorides, picrates, perchlorates, or ferrichlorides. An example is 2,4,6-trimethylpyrilium perchlorate:

$$\gamma\text{-pyrone} \qquad\qquad \alpha\text{-pyrone}$$

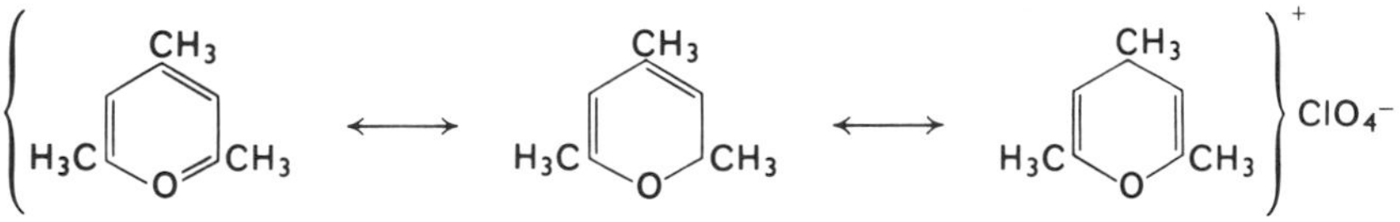

2,4,6-trimethylpyrilium perchlorate

The greatest interest that attaches to pyrilium salts lies in the fact that the red to blue coloring matters of plants—the anthocyanin pigments—are compounds

---

* Except in the case of 2-haloethanol derivatives. Compounds that contain halogen and alkoxyl groups in the 1,2-positions relative to each other react with magnesium to eliminate the elements of RO—MgBr and give the corresponding olefin.

of this class. Cyanin (chloride*) is the commonest of these: it is the pigment of many red flowers and fruits, as well as of the blue cornflower. The difference in flower color associated with a single pigment is the result of the influence of accessory factors in the flower petal—the *p*H of the cell sap, the presence of metal ions—and of co-occurring substances that alter the colloidal state of the cell constituents:

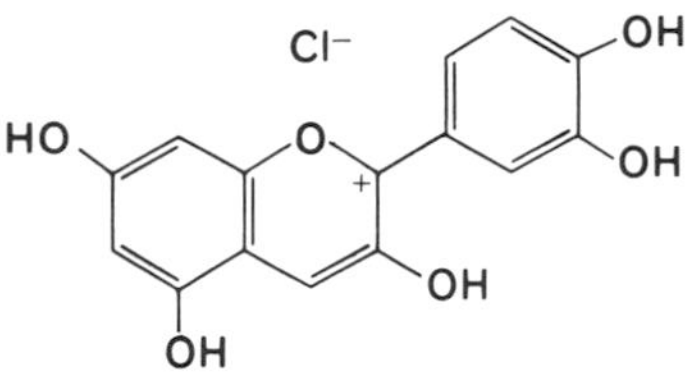

cyanidin chloride
(*cyanin* is the 3,5-diglucoside)

## 29.5  Synthesis of the pyrilium ring system

Pyrilium salts can be formed from, and in neutral solution are in equilibrium with, the corresponding "pseudobases".

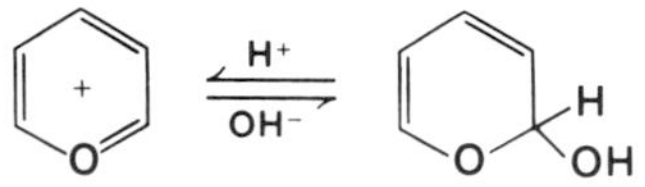

pyrilium cation          pseudobase

Thus, synthetic methods for pyrilium salts consist of methods that lead to the hydroxy compounds of the pseudobase type. One such method, which has been applied to the preparation of a natural anthocyanidin (Section 29.4), consists of the addition of a Grignard reagent to a coumarin:

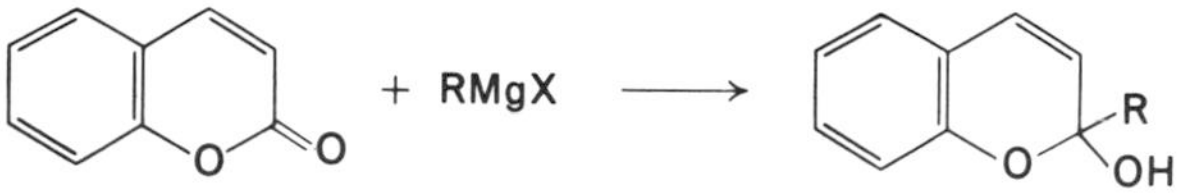

(17)

The initial addition product is a pseudobase, which when treated with acid is converted into the pyrilium salt. Compounds of the type represented by the example (2-arylbenzopyrilium salts) at the top of page 703 are called "flavylium" salts.

Another general method for preparing the flavylium ring system is by the aldol condensation of an *o*-hydroxybenzaldehyde with an acetophenone. The first-formed benzalacetophenone derivative exists in the cyclic hemiketal form; this is

---

* The anion with which the pigment is associated in the cell cannot be explicitly designated. The chloride is the salt usually isolated.

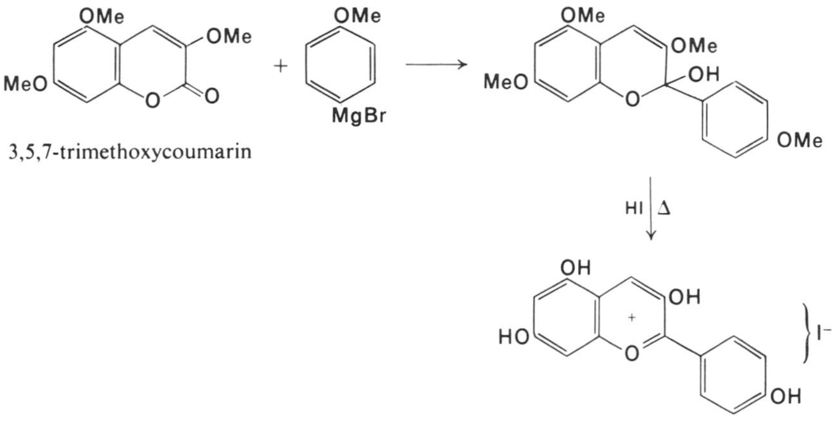

3,5,7-trimethoxycoumarin

pelargonidin iodide, a flavylium salt

(18)

the pseudobase of the corresponding flavylium compound, and can be converted to the flavylium salt by treatment with acid:

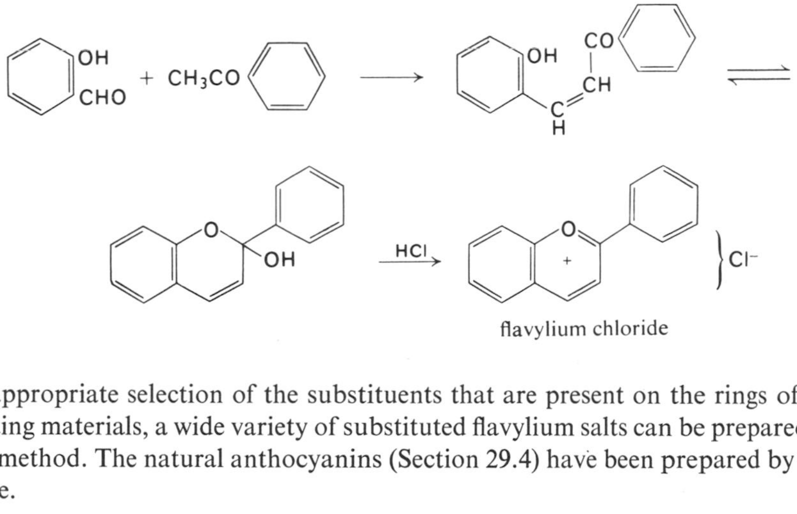

flavylium chloride (19)

By appropriate selection of the substituents that are present on the rings of the starting materials, a wide variety of substituted flavylium salts can be prepared by this method. The natural anthocyanins (Section 29.4) have been prepared by this route.

## 29.6  Other six-ring oxygen heterocycles

The six-membered oxygen heterocycles are a large and important group of naturally occurring compounds. Included in this class are many of the natural pigments of the plant world, many of them derived from the parent structure, flavone. Some representatives of the class are illustrated on the top of the next page.

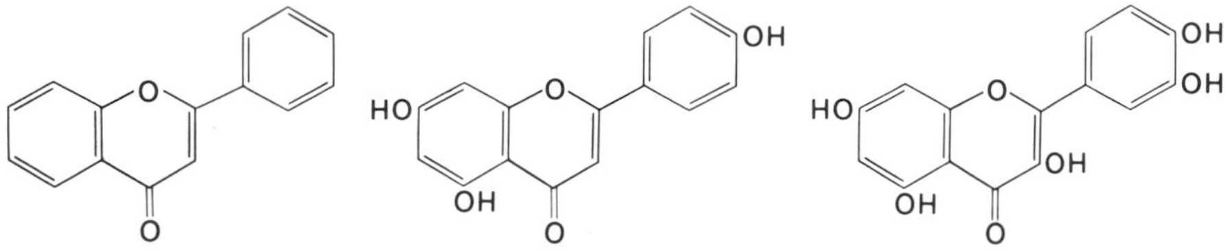

flavone

apigenin
(in parsley, celery)

quercetin
(widely distributed in plants)

The naturally occurring flavones are hydroxyflavones; most of them occur in nature in the form of glycosides.

The synthesis of a typical natural flavone, luteolin, is shown in the following equations. A study of these reactions will disclose the fact that nearly all of them are examples of reactions that have been encountered in earlier parts of this book in other, simpler forms. One additional comment will be appropriate here: it will be noticed that phloroglucinol trimethyl ether is prepared by the simultaneous deacetylation-methylation of phloroglucinol triacetate. This is necessary in the case of phloroglucinol, since direct methylation of this compound with methyl sulfate and alkali leads to C-methylation.

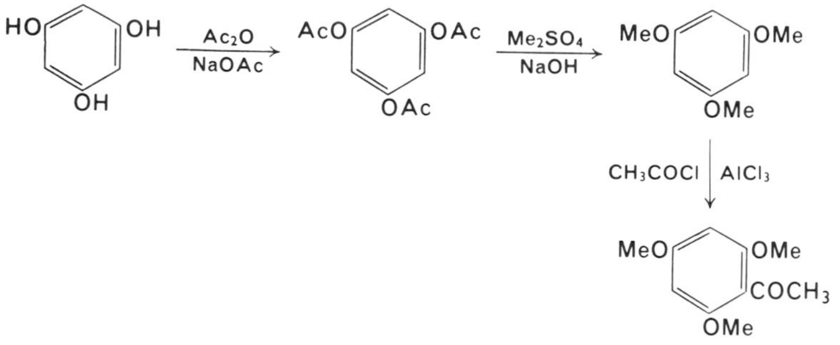

(20)

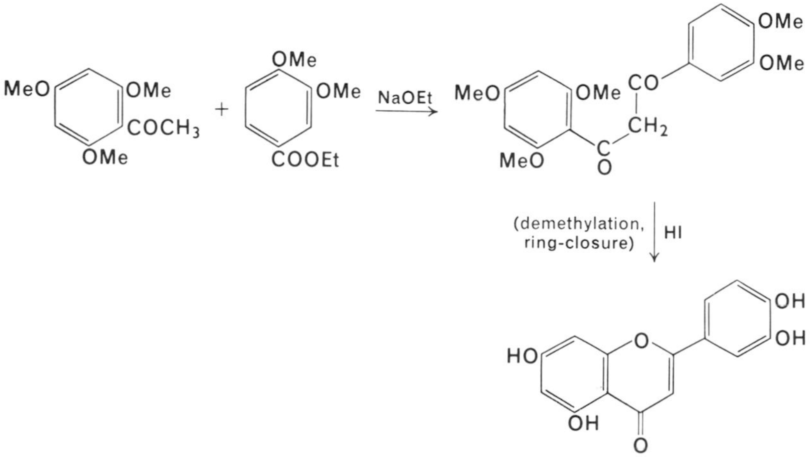

(21)

luteolin

Each step of these synthetic sequences should be examined with care until its nature is understood.

## 29.7 Nitrogen-containing heterocycles

The alkylation of a primary amine with an alkyl halide leads to the formation of a tertiary amine; and if, instead of an alkyl halide, an $\alpha,\omega$-dihalide is used, ring formation occurs:

$$CH_2 \begin{array}{c} CH_2-CH_2Br \\ \\ CH_2-CH_2Br \end{array} + 2RNH_2 \longrightarrow \begin{array}{c} H_2C-CH_2 \\ CH_2 \\ H_2C-CH_2 \end{array} {}^+N-R + RNH_3{}^+ + 2Br^- \qquad (22)$$

Entirely analogous to this reaction is the ring closure of halogen-substituted amines, or of sulfonyl derivatives of amino alcohols:

$$\begin{array}{c} CH_2CH_2X \\ | \\ CH_2CH_2NHR \end{array} \longrightarrow \left. \begin{array}{c} CH_2CH_2 \\ | \\ CH_2-CH_2 \end{array}{\raise2pt{\hbox{$\diagup$}}}\kern-4pt{}^+N{\raise2pt{\hbox{$<$}}}\begin{array}{c} H \\ R \end{array} \right\} X^- \quad (X = Cl,\ Br,\ OSO_2R)$$

$$(23)$$

$$R_2NHC_2CH_2Br \longrightarrow \left. \begin{array}{c} CH_2 \\ | \\ CH_2 \end{array}{\raise2pt{\hbox{$>$}}}{}^+N{\raise2pt{\hbox{$<$}}}\begin{array}{c} R \\ R \end{array} \right\} Br^- + \left. \begin{array}{c} R \quad R \\ \diagup N \diagdown \\ H_2C \quad\quad CH_2 \\ | \quad\quad\quad\quad | \\ H_2C \diagdown\raise-2pt{\hbox{$N$}}\diagup CH_2 \\ R \quad R \end{array} \right\}{}^{++} 2Br^-$$

$\beta$-Haloethylamines containing a tertiary amino group undergo ring closure to cyclic quaternary ammonium salts. These three-membered imonium ions are extraordinarily susceptible to ring opening by nucleophilic attack, resembling in this respect ethylene oxides. The closure and reopening of three-membered cyclic ammonium ions is of great importance in biology and medicine. The well-known wartime "mustard gas" has as its counterparts a series of substances known as "nitrogen mustards," the typical reactions of which are described along with the comparable reactions of mustard gas, in the following discussion. "Mustard gas" reacts by way of a unimolecular nucleophilic displacement to give rise to an intermediate three-membered cyclic sulfonium ion, which then reacts with an external nucleophile with ring-opening (in the following equations the symbol B:$^-$ is used for a nucleophilic reagent):

$$ClCH_2CH_2S{\raise2pt{\hbox{$\diagup$}}}\begin{array}{c} CH_2-Cl \\ | \\ CH_2 \end{array} \rightleftharpoons ClCH_2CH_2{}^+S{\raise2pt{\hbox{$<$}}}\left. \begin{array}{c} CH_2 \\ | \\ CH_2 \end{array} \right\} Cl^-$$

mustard gas:
$\beta,\beta'$-dichlorodiethyl sulfide

$$\Big\downarrow B:^-$$

$$ClCH_2CH_2SCH_2CH_2B + Cl^- \qquad (24)$$

Repetition of the above reactions permits replacement of the second chlorine atom in the same way. The susceptibility of the three-membered sulfonium ring to opening by nucleophilic attack

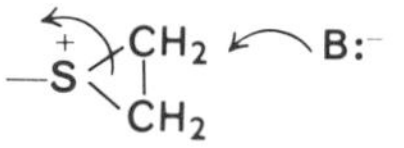

$$-S \overset{+}{\underset{CH_2}{\overset{CH_2}{<}}} \quad B:^-$$

finds parallels in several reactions that we have already encountered in earlier chapters: the opening of protonated ethylene oxide rings; the three-membered "bromonium" ion intermediate in the addition of bromine to the carbon-carbon double bond; and the ready ring-opening of cyclopropane derivatives.

The course of the reaction of $\beta$-haloethylamines can be seen to be the counterpart of the reactions of the chloroethyl sulfides:

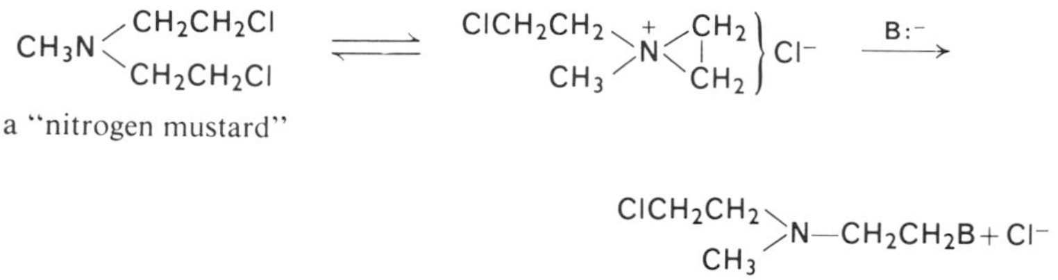

$$CH_3N \overset{CH_2CH_2Cl}{\underset{CH_2CH_2Cl}{<}} \quad \rightleftharpoons \quad \left\{ ClCH_2CH_2 \overset{+}{\underset{CH_3}{N}} \overset{CH_2}{\underset{CH_2}{<}} \right\} Cl^- \quad \xrightarrow{B:^-}$$

a "nitrogen mustard"

$$ClCH_2CH_2 \overset{}{\underset{CH_3}{>}} N - CH_2CH_2B + Cl^- \quad (25)$$

It is believed that the profound physiological effects of "mustard gas" and of many $\beta$-chloroethylamines are the result of interactions of the above kind between the "drug" and an enzyme or other protein component of the tissue. If $B:^-$ in the above equations represents a nucleophilic component of the tissues of an organism (that is, of some cellular component), the result of the series of transformations shown will be *the alkylation of the cellular material*. It is obvious that such a result will bring about profound alterations in the functioning of the system (e.g., an enzyme) of which such a nucleophilic structure is a part; and such an alteration will be expected to be permanent (or to persist until the structure in question is destroyed by metabolic processes and replaced by a newly synthesized structure).

The drug known as *Dibenamine*, N,N-dibenzyl-$\beta$-chloroethylamine, has the property of producing a blockade of the effects of epinephrine at physiological structures that are affected by the latter substance. The effects of Dibenamine, once established, are remarkably persistent, a fact that leads to the supposition that the drug alkylates the structural element of the tissue which normally interacts with epinephrine to produce the effects of this hormone.

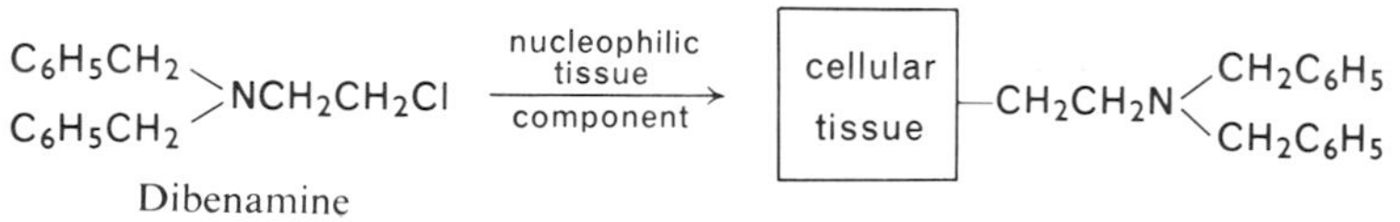

$$C_6H_5CH_2 \overset{}{\underset{C_6H_5CH_2}{>}} NCH_2CH_2Cl \quad \xrightarrow[\text{component}]{\text{nucleophilic tissue}} \quad \boxed{\begin{array}{c}\text{cellular} \\ \text{tissue}\end{array}} - CH_2CH_2N \overset{CH_2C_6H_5}{\underset{CH_2C_6H_5}{<}} \quad (26)$$

Dibenamine

It is of special interest to note that a large number of physiologically active substances possess the ability to act as alkylating agents. It has been suggested that such toxic compounds as diazomethane, dimethyl sulfate, and methyl bromide react with some cellular structure (the exact nature of which is not always known) to form a methylated derivative containing a covalent tissue—$CH_3$ bond. It is of further interest to note that certain substances which have the property of being *bifunctional* alkylating agents have been found to be active as anti-cancer drugs. The nitrogen mustards, useful in the treatment of certain leukemias, are representative of this kind of compound.

Certain synthetic applications of the bifunctional alkylating ability of compounds of the nitrogen mustard class are of special significance here because they demonstrate another means of forming heterocyclic rings. For instance, the reaction of the sodium derivative of benzyl cyanide (phenylacetonitrile) with N-methyl-$\beta$,$\beta'$-dichlorodiethylamine leads to the formation of a substituted piperidine:

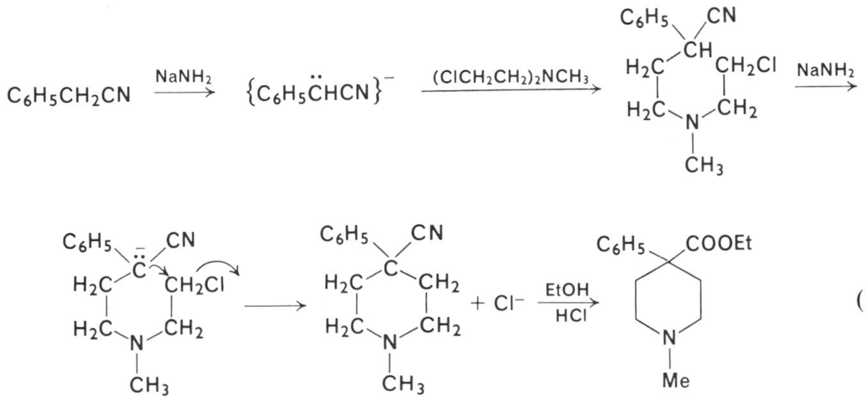

C6H5CH2CN
NaNH2
{C6H5CHCN}−
(ClCH2CH2)2NCH3
C6H5 CN
CH
H2C CH2Cl
H2C N CH2
CH3
NaNH2
+ Cl−
EtOH
HCl
C6H5 COOEt
N
Me

(27)

Demerol, Pethidine

Ethanolysis of the cyanopiperidine, as shown in this equation, leads to the formation of the corresponding ester, which, under the proprietary names "Pethidine" and "Demerol," is an important analgetic with analgesic properties comparable to those of morphine.

## 29.8 The formation of piperidine derivatives by ring closure

Piperidine and its derivatives can be prepared by methods similar to those thus far described. It will be noted that in the foregoing examples the ring has been closed by two quite different methods, one involving the formation of one or two C—N bonds, the other involving the ring closure by C—C bond formation between the 3- and 4-carbon atoms:

**Exercise 1.** Point out examples in the text that correspond to these two type reactions.

Closure of a ring between the 2,3- and 3,4-carbon atoms can also be used to form piperidine rings. Some typical syntheses are the following:

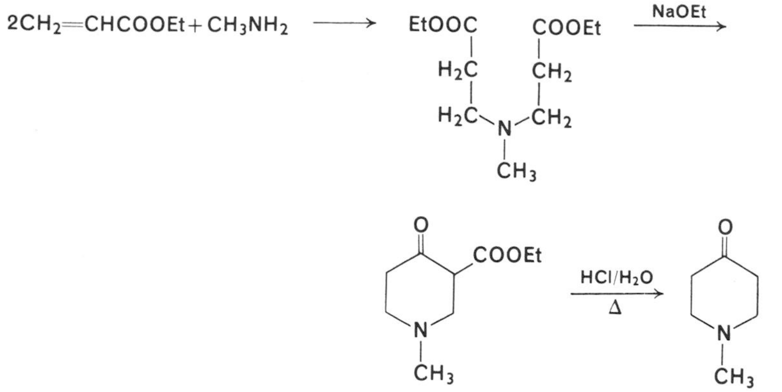

$$2CH_2{=}CHCOOEt + CH_3NH_2 \longrightarrow$$

(28)

1-methyl-4-piperidone

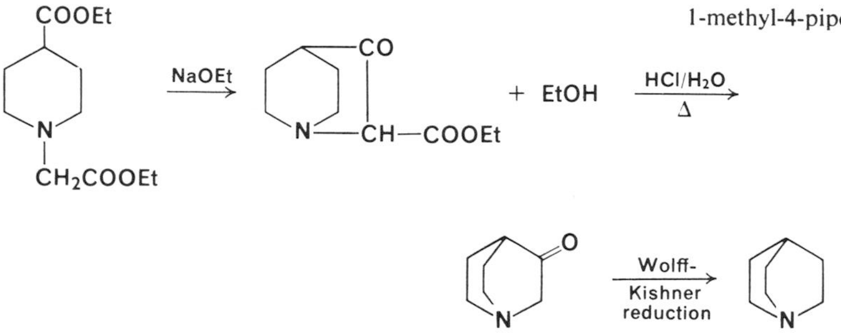

(29)

3-quinuclidone          quinuclidine

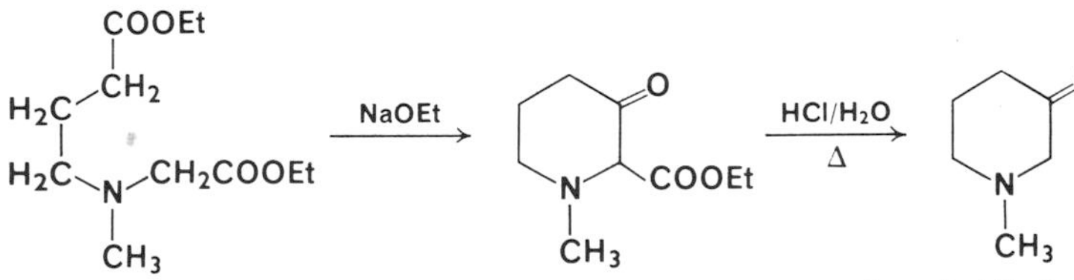

(30)

1-methyl-3-piperidone

Each of the above reaction series involves a Dieckmann condensation as the first step. Hydrolysis and decarboxylation of the $\beta$-keto ester yields in each case the piperidone derivative. Further transformations of these cyclic ketones can be carried out by means of reactions that resemble those of cyclohexanone derivatives, and require no special comment.

## 29.9   The Mannich reaction

The reaction of a carbonyl compound, an amine, and formaldehyde, known as the Mannich reaction, led to an early synthesis of a six-membered nitrogen-containing heterocyclic compound. The Mannich reaction is of extraordinarily wide and varied application, and appears in various modifications in numerous syntheses of nitrogen-containing compounds, both cyclic and acyclic. Although the Mannich reaction is not peculiar to heterocyclic syntheses, its application in several areas of heterocyclic chemistry makes it most appropriate for discussion in this chapter.

A simple example of a Mannich reaction that illustrates the mechanistic course by which it proceeds is that between acetophenone, formaldehyde, and dimethylamine:

$$C_6H_5COCH_3 + CH_2O + (CH_3)_2NH \longrightarrow C_6H_5COCH_2CH_2N(CH_3)_2 \tag{31}$$

The reaction is acid-base catalyzed; that is, it proceeds under conditions of $pH$ that are neither excessively acidic or basic. In the typical case, the amine is used in the form of its hydrochloride, under which conditions both the amine (a base) and its protonated form (an acid) are present in the solution.

Numerous mechanisms for the Mannich reaction have been proposed, but nearly all of these differ only in details concerning the sequence in which the three reactants participate in a few fundamental equilibria. The following course is a summary of these various proposals that appears to constitute a satisfactory picture of the mechanism:

(a) $CH_2O + (CH_3)_2NH \rightleftharpoons HOCH_2N(CH_3)_2 \overset{H^+}{\rightleftharpoons} CH_2\overset{+}{=}N(CH_3)_2 + H_2O$

(b) $C_6H_5COCH_3 + :B \rightleftharpoons C_6H_5COCH_2:^- + B:H^+$

(c) $C_6H_5COCH_2:^- + CH_2\overset{+}{=}N(CH_3)_2 \rightleftharpoons C_6H_5COCH_2CH_2N(CH_3)_2$

A synthesis of the pyrrolidine alkaloid, hygrine, can be accomplished by an application of the Mannich reaction to heterocyclic ring closure:

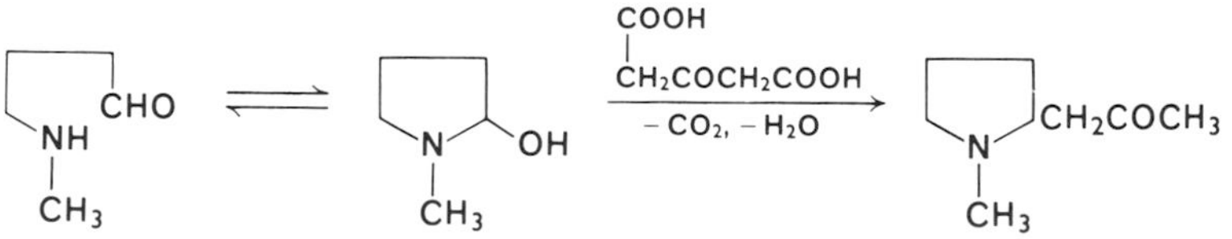

$$\tag{32}$$

hygrine

Arecoline, the alkaloid of the betel palm, *Areca catechu*, has been synthesized by a Mannich reaction, in this case with the use of a primary amine:

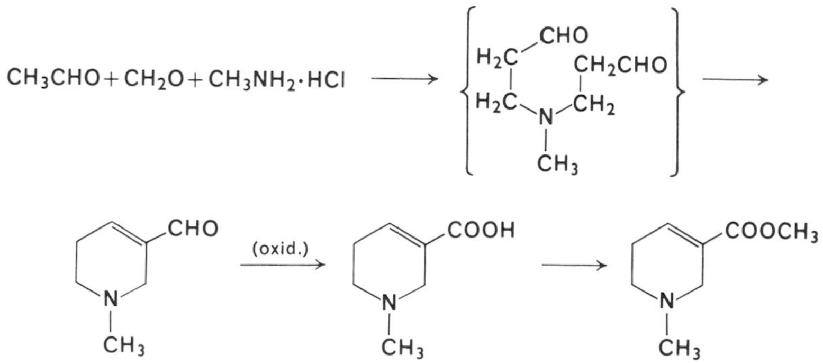

$$\tag{33}$$

arecoline

It will be noted that the immediate product of the Mannich reaction in Equation (33) is the dialdehyde, which cyclizes to the piperidine derivative by an aldol condensation.

Phenols undergo the Mannich reaction, playing the role of the nucleophile in step *(c)* of the mechanism described above. In condensed form, the reaction may be written

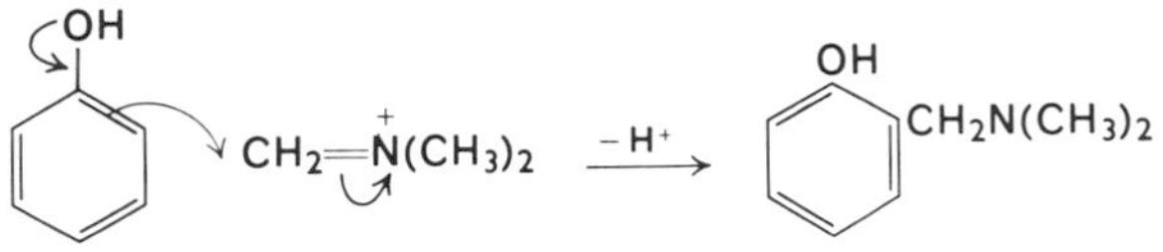

$$\tag{34}$$

This is clearly another example of an electrophilic substitution into a highly active aromatic nucleus.

### 29.10  Pyridine

The electronic structure of pyridine resembles that of benzene, and thus it is not surprising to find that pyridine has aromatic properties that resemble those of the aromatic hydrocarbons:

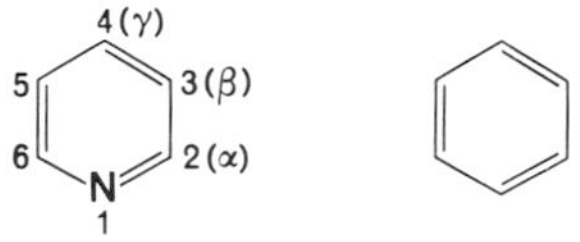

pyridine          benzene

The resemblance between pyridine and benzene lies in their ability to undergo electrophilic substitution reactions; but whereas the nitration, e.g., of benzene takes place readily at any of the six equivalent carbon atoms, that of pyridine is more difficult and, of course, the five available positions of substitution in pyridine are not equivalent. In fact, nitration occurs largely at the $\beta$ (3-) position:

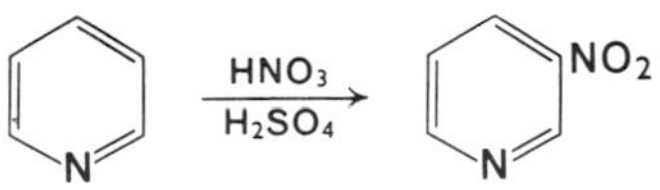

In considering this reaction, an additional point must be recognized: pyridine, with an unshared pair of electrons on the nitrogen atom, is a base (about as strong a base as aniline) and thus is largely protonated under the strongly acid conditions of nitration. The protonation of pyridine imparts to the $\alpha$- and $\gamma$-carbon atoms a high degree of positive character:

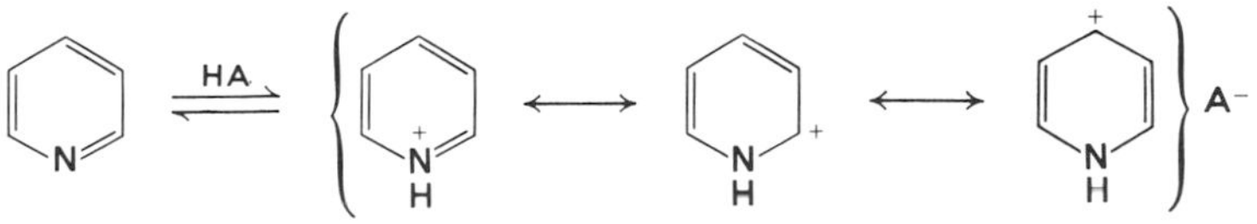

The attack of the electrophilic reagent (e.g., $NO_2^+$) would thus be less likely to occur at an $\alpha$- or $\gamma$-carbon atom, since these are electron deficient, than at the $\beta$-carbon atom. But attack of an electrophilic reagent at any point in the positively charged ring would be difficult, and thus the substitution reaction is slow.

It will be noted that the $\alpha$- and $\beta$-positions of pyridine bear a formal resemblance to the *ortho* and *meta* positions, respectively, of nitrobenzene:

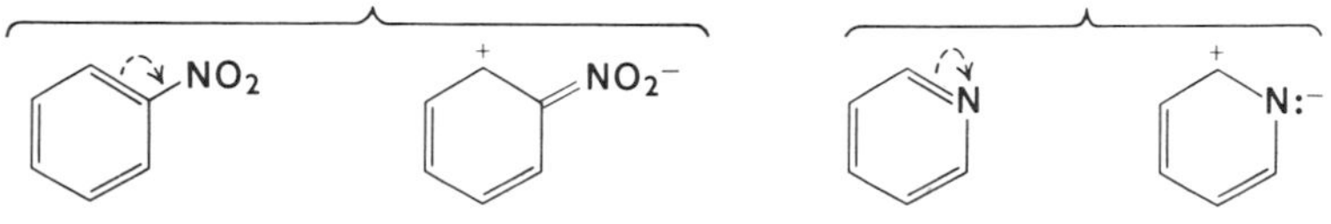

---

**Exercise 2.** What kind of reactivity would you expect to find in 2-bromopyridine?

---

Sulfonation and halogenation of pyridine also occur slowly under the usual conditions employed with benzene derivatives. High-temperature halogenation proceeds readily to give substitution at all three possible positions of the ring ($\alpha$, $\beta$, and $\gamma$), and probably proceeds by way of free halogen atoms. Thus, the orientation would not be expected to correspond to that observed in electrophilic substitution, as indeed it does not.

The preparation of substituted pyridine derivatives is often achieved by indirect methods. Substituents in the 3-position are often introduced by the alteration of the carboxyl group of nicotinic acid, a substance readily obtained from nicotine, a by-product of the tobacco industry:

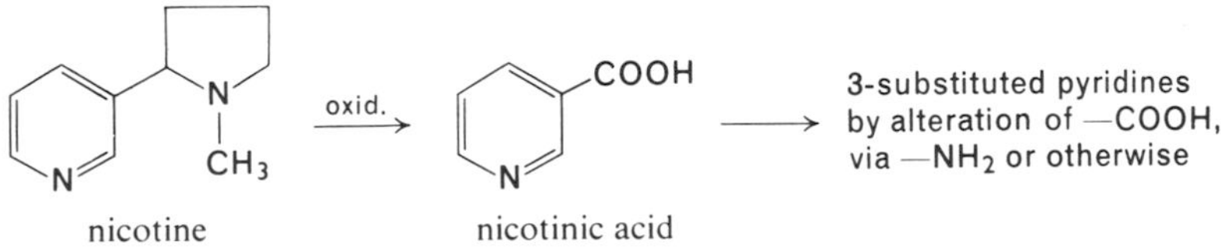

| nicotine | nicotinic acid | (35) |

## 29.11 Nucleophilic substitution in pyridine

The electron-deficient nature of the 2- and 4-positions of pyridine makes these positions susceptible to ready *nucleophilic attack*. One of the most useful of the reactions of this kind is that between pyridine and sodamide, in which nucleophilic attack of the amide anion results in the formation of 2-aminopyridine:

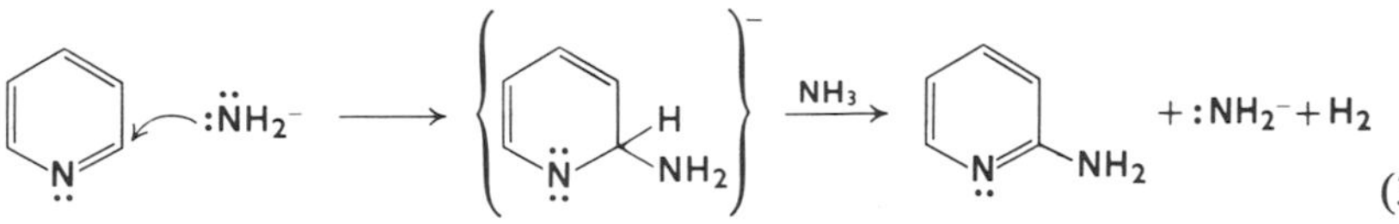

$$(36)$$

It will be noted that the 2-hydrogen atom is expelled in the oxidation state of the hydride ion (H:⁻).

When the electrophilic character of the α-position is enhanced by quaternization of the nitrogen atom, nucleophilic attack is very ready. Indeed, the coordination of hydroxide ion in aqueous alkaline solution does occur, in the equilibrium reaction,

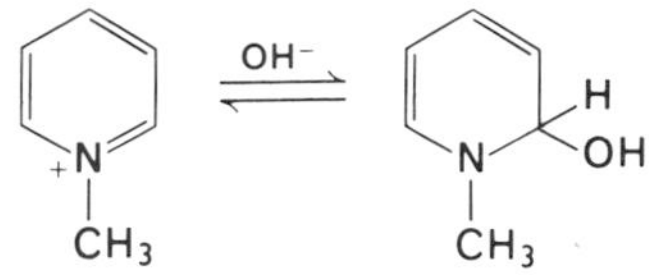

The resulting "carbinolamine" is not isolable since the equilibrium lies largely on the side of the "aromatic" pyridinium ion. But when an oxidizing agent, such as potassium ferricyanide, is present the addition compound is oxidized to the 2-keto compound, N-methyl-2-pyridone:

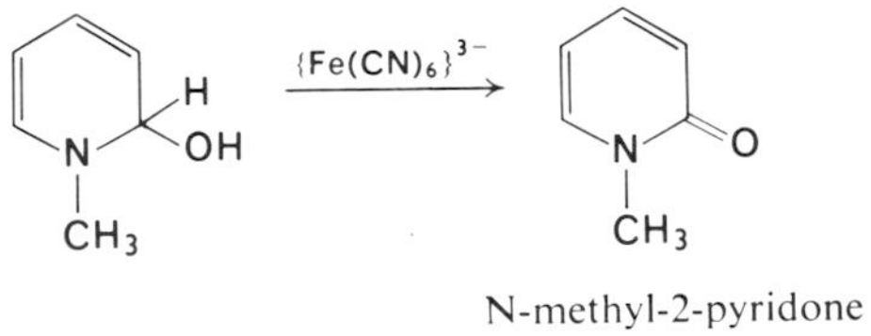

(37)

N-methyl-2-pyridone

Preparation of 2-pyridone (2-hydroxypyridine) is accomplished by hydrolysis of 2-chloro (or -bromo) pyridine. This reaction, too, is a nucleophilic substitution reaction, and probably involves a nucleophilic attack upon the ring that resembles (in regard to mechanism) that which leads to dispacement of nitro-activated halogen from an aromatic nucleus:

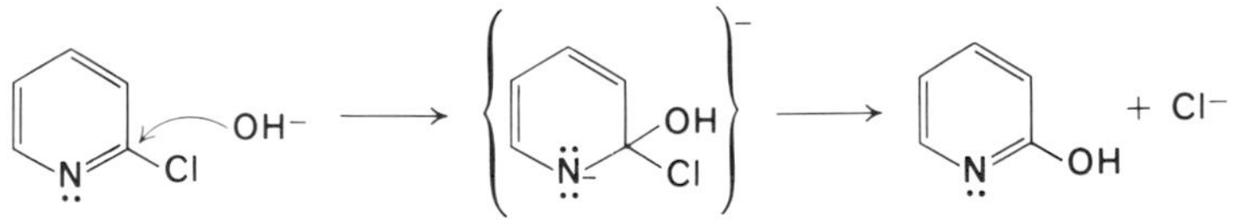

(38)

This reaction again discloses the resemblance of the α-position of the pyridine ring to the position *ortho* or *para* to a nitro group in an aromatic nitro compound.

## 29.12 "Total" synthesis of pyridine derivatives

The synthesis of substituted pyridine derivatives from acyclic precursors can be accomplished in several ways. Among these are found methods in which base-catalyzed aldol or Schiff's base condensations are involved; an example is the following:

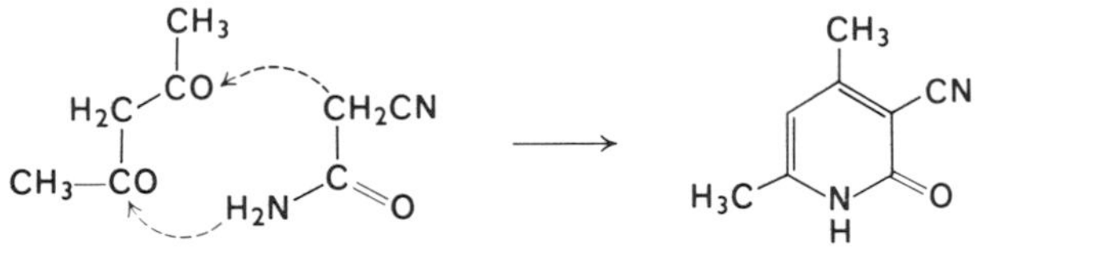

$$\text{acetylacetone} + \text{cyanoacetamide} \longrightarrow \text{3-cyano-4,6-dimethyl-2-pyridone} \qquad (39)$$

---

**Exercise 3.** Formulate the steps in this condensation, showing a reasonable series of intermediate stages.

---

Another general method of synthesis that is adaptable to the preparation of substituted pyridines involves the condensation of ammonia, an aldehyde, and either a $\beta$-keto ester or a $\beta$-diketone. The following is a typical example:

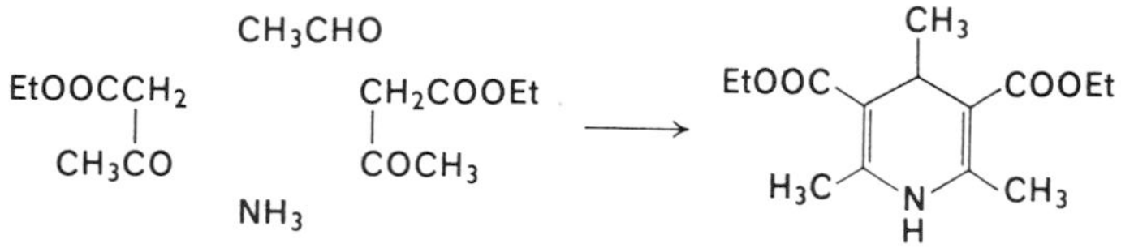

$$(40)$$

The immediate product of this condensation is a dihydropyridine derivative. The dehydrogenation (by oxidation) of this to the corresponding pyridine is accomplished with ease because of the great gain in stability that accompanies generation of the "aromatic" structure of the pyridine ring system.

## 29.13 Occurrence of pyridine derivatives in nature

The pyridine ring, and its reduced (piperidine, dihydropyridine, and tetrahydropyridine) forms are of widespread natural occurrence, and are synthesized by plants, possibly by way of amino acids as precursors. Although the heterocyclic rings of pyridine derivatives are not synthesized by higher animals, several of them are of the greatest importance in the cellular economy of animal life.* Perhaps the most important pyridine derivatives are those which form the prosthetic groups (see Chapter 31) of many enzymes whose function is the catalysis of oxidation reactions in the cells of both plants and animals. These are called NAD (nicotinamide adenine dinucleotide) and NADP (nicotinamide adenine dinucleotide phosphate). NADP is a phosphate ester of NAD, and bears the additional phosphoric acid group on the adenine-linked ribose unit:

---

* And thus must be ingested by animals in their diet, in the form of "vitamins."

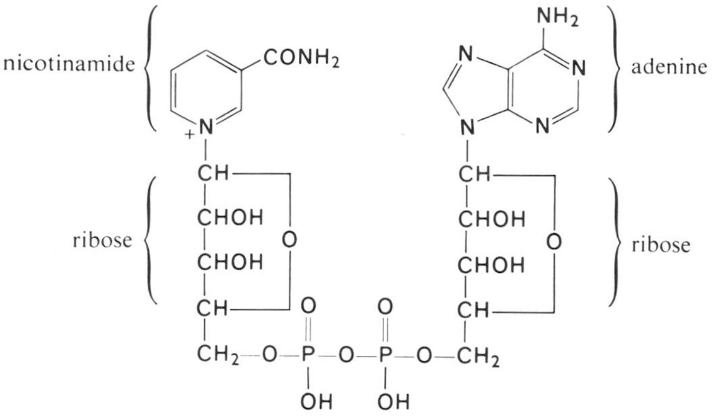

nicotinamide adenine dinucleotide (NAD)

In combination with specific proteins, NAD and NADP form enzymes that participate in oxidation-reduction reactions in living cells. The NAD portion of the enzyme alcohol dehydrogenase, for example, accepts hydrogen (or electrons and protons) from ethanol to form acetaldehyde and the reduced enzyme, which contains NADH:

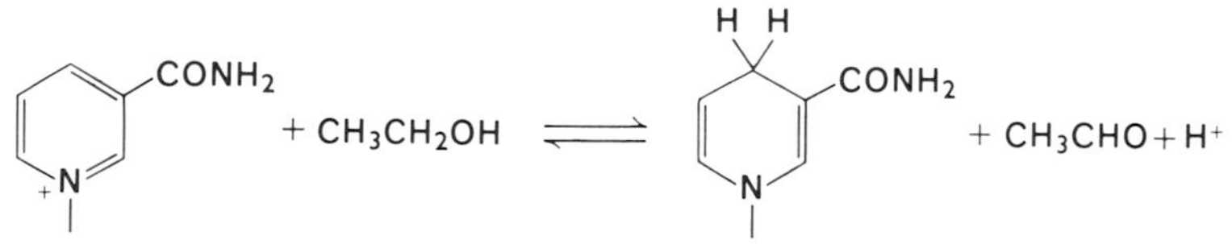

(as part of —NAD— protein)     (as part of —NADH— protein)

$$(41)$$

In coupled cellular reactions, NADH is reoxidized by transfer of its hydrogen to cooperating enzymes, and is transformed into NAD, which can oxidize another molecule of substrate.

## 29.14  Quinoline and isoquinoline

Quinoline and isoquinoline are the two possible *benzopyridines:*

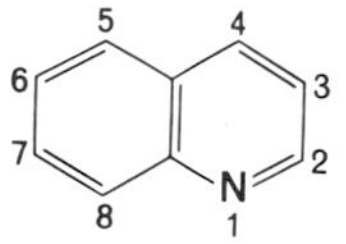

quinoline 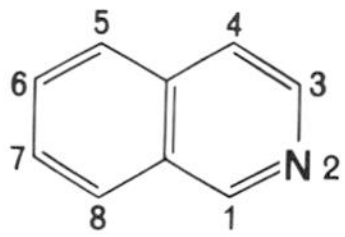          isoquinoline

The properties of the heterocyclic ring of quinoline resemble those of pyridine: the 2-position is susceptible to nucleophilic attack, and 2-halogen substituents are readily displaced by nucleophilic reagents such as amines, alkoxides, or hydroxide ion. Electrophilic substitution can occur in either the hetero ring or the carbocyclic ring; and the ease and position of substitution is subject to considerable variation with changes in the kind of substitution reaction and the conditions of the reaction. For example, the nitration of quinoline in concentrated $H_2SO_4$ leads to substitution in the 5- and 8-positions. In acetic acid solution nitration occurs in the 3-position:

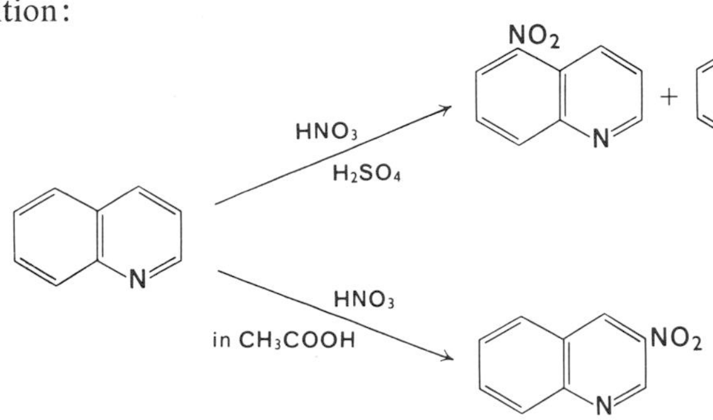

$$(42)$$

A notable feature of the pyridine ring system is its stability to oxidation. Alkylpyridines are readily oxidized to pyridinecarboxylic acids as in the oxidation of nicotine to nicotinic acid, which has been described. Quinoline is oxidized to pyridine-2,3-dicarboxylic acid, the benzene ring being destroyed:

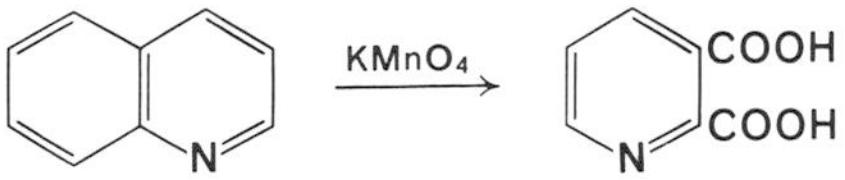

$$(43)$$

quinolinic acid
(pyridine-2,3-dicarboxylic acid)

## 29.15 Synthesis of the quinoline ring system

In most syntheses of quinoline derivatives, the heterocyclic ring is formed by a ring closure, starting with a compound that already contains the carbocyclic (benzene) ring. One of the most versatile of the quinoline syntheses is the *Skraup* synthesis, in which an aromatic primary amine with a free *ortho*-position is heated with glycerol, sulfuric acid, and an oxidizing agent, which may be nitrobenzene. For the synthesis of quinoline itself, the reaction proceeds as follows:

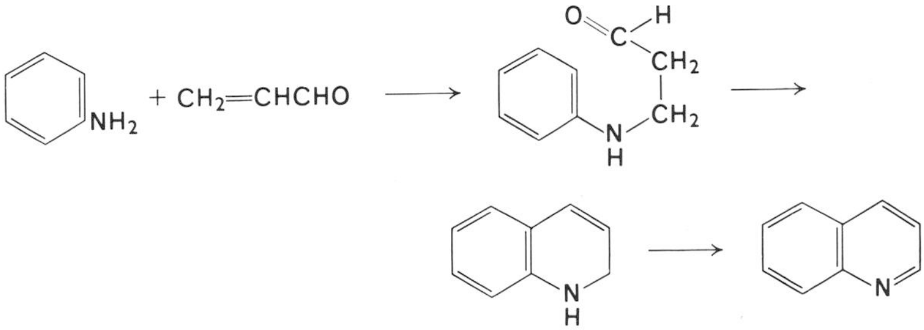

$$(44)$$

The acrolein that appears as the reactant in this formulation is derived from glycerol by dehydration in the course of the reaction. The final oxidation step is accomplished by the nitrobenzene that is used as one of the reagents. When *m*-substituted anilines are used, ring closure can take place to give 5- and 7-substituted quinolines. Often both are obtained, but when the *m*-substituent is a strongly *o-p*-directing group, such as the methoxyl group, the 7-substituted quinoline is the chief, and often the exclusive, product.

A synthesis that resembles the Skraup synthesis is that in which an aromatic amine is condensed with an aliphatic aldehyde in the presence of a mineral acid. This is called the *Döbner-von Miller* synthesis, and is illustrated by the preparation of quinaldine (2-methylquinoline):

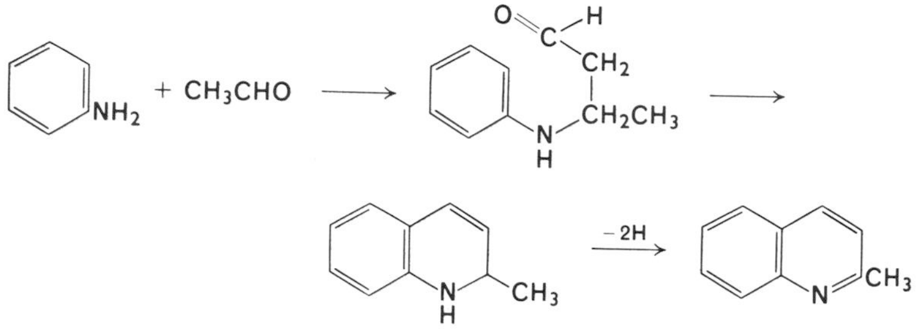

$$\text{(45)}$$

The intermediate anilino compound may be regarded as having been formed by way of crotonaldehyde, which is formed by the aldol condensation of the acetaldehyde. In this process the final dehydrogenation occurs by the transfer of hydrogen to a hydrogen acceptor—perhaps the Schiff's base formed from the starting amine and aldehyde.

### 29.16  N-Oxides

The formation of amine oxides (Section 3.20) from tertiary amines has its counterpart in the oxidation of pyridine and quinoline to the N-oxides. The reagent usually employed for this oxidation is hydrogen peroxide in acetic acid solution:

$$(CH_3)_3N + H_2O_2 \longrightarrow (CH_3)_3N \rightarrow O \tag{46}$$

trimethylamine oxide

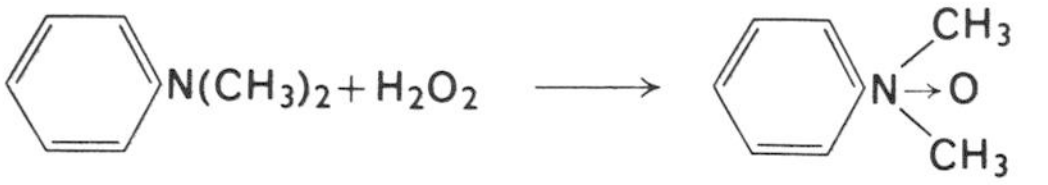

$$\text{(47)}$$

dimethylaniline oxide

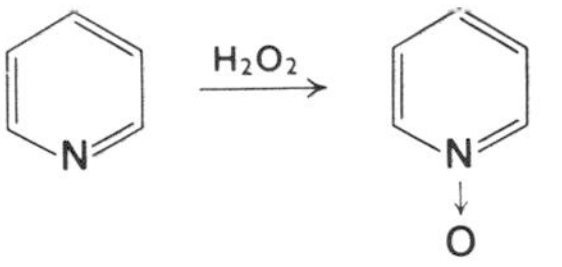

$$\text{(48)}$$

pyridine N-oxide

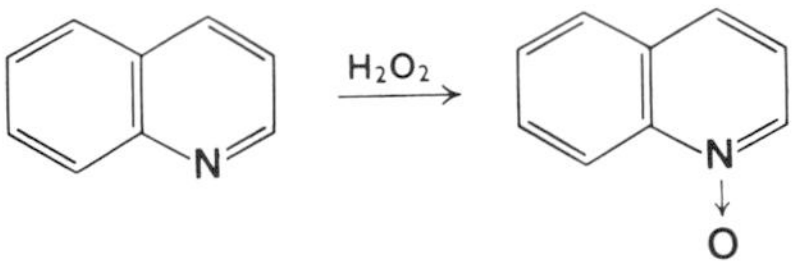

$$\text{quinoline N-oxide} \qquad (49)$$

The N-oxidation of the heterocyclic nitrogen atom has an interesting effect upon the behavior of the ring toward substitution reagents. Pyridine N-oxide is more reactive than pyridine toward *both* nucleophilic and electrophilic substitution. Further, while pyridine undergoes substitution at the 3-position with electrophilic reagents, pyridine N-oxide reacts at the 4-position. The explanation for this is that the oxygen atom of the N—O grouping can supply electrons to the 4-position to accommodate the electrophilic demand:

Pyridine N-oxide is readily nitrated to give 4-nitropyridine N-oxide in good yield.

The increased reactivity of the 2- and 4-positions of pyridine N-oxide toward nucleophilic attack is also readily understood. In this case, the positively charged nitrogen atom of the N—O grouping can accommodate the electrons supplied by the attacking nucleophile:

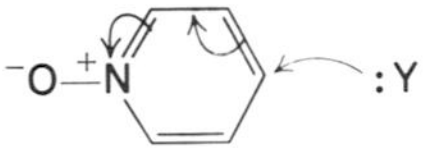

Hence, the ready replacement of the halogen atom from 4-bromopyridine N-oxide by a nucleophilic reagent takes place with greater ease than in the case of 4-bromopyridine itself:

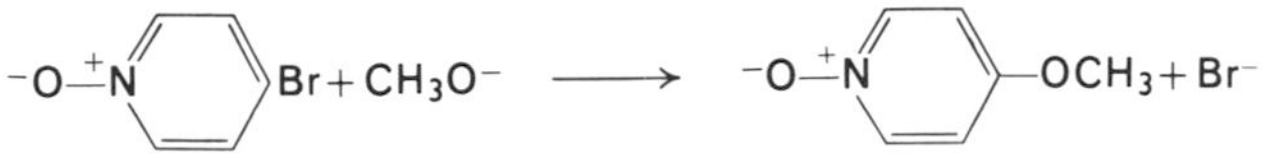

$$\qquad (50)$$

Since the removal of the oxygen atom from pyridine (and other heterocyclic) N-oxides can be readily effected by reduction with iron and acetic acid, N-oxidation can serve as a useful device for the synthesis of 4-substituted pyridine derivatives.

## 29.17 Alkaloids

Many plants contain physiologically active, basic substances, known as "alkaloids," most of which are nitrogen heterocycles and many of which are important drugs. Such well-known drugs as morphine, atropine, strychnine, cocaine, papaverine, curare, and physostigmine are naturally occurring heterocyclic compounds. The field of alkaloid chemistry is a vast and complex one, and we can examine no more than a few representative compounds of the chief groups.

## 29.18 Alkaloids containing a pyridine or reduced pyridine ring

Alkaloids of the pyridine group represent some of the simplest natural heterocyclic compounds. One of these has already been mentioned; it is the well-known substance (−)-*nicotine*, the principal alkaloid of the tobacco plant. Nicotine is extremely toxic, and in sublethal doses has a powerful effect, principally upon the nervous ganglia, stimulating in minute doses, paralyzing in larger doses. It has no useful pharmacological (i.e., clinical) application, but finds wide use as an insecticide, for which purpose it is produced in large quantites as a by-product of the tobacco industry. The structure of nicotine has been established by degradation, and by the following synthesis:

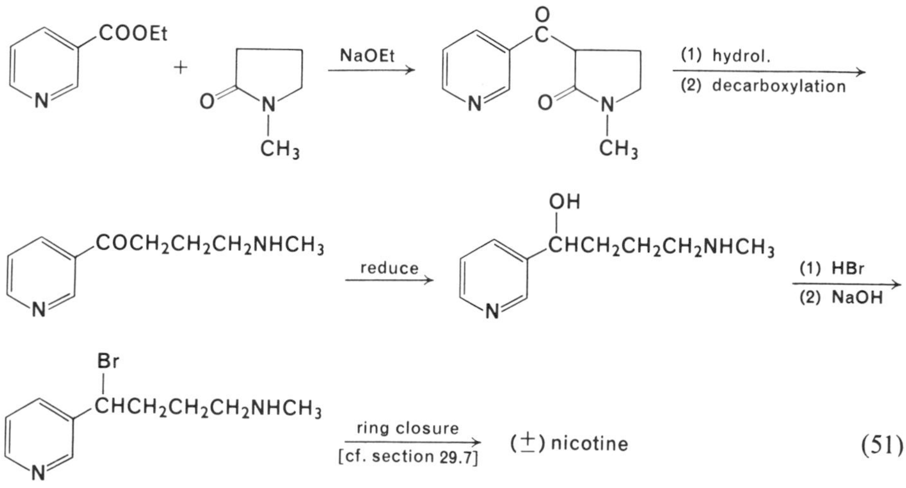

It has been shown (by conversion of nicotine to the N-methylpyridone, followed by the oxidative destruction of the pyridone ring) that the asymmetric carbon atom has the configuration of that in L-proline.

Other simple pyridine-derived alkaloids are coniine, the active (and toxic) principle of the water hemlock, *Conium maculatum*; atropine (Chapter 33), from *Atropa* and other genera of the Solanaceae; and cocaine, from *Erythroxylon coca:*

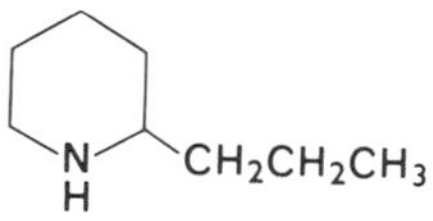

coniine
(−)-2-*n*-propylpiperidine

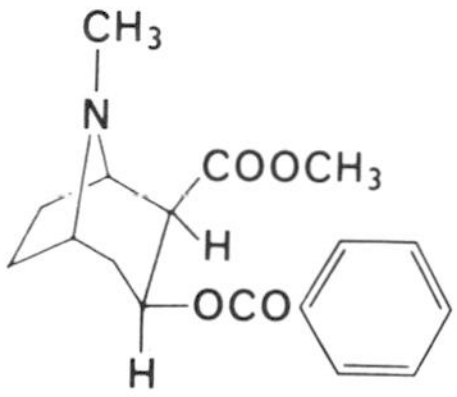

(−)-cocaine

## 29.19   Isoquinoline alkaloids

A very large and diverse group of alkaloids is one in which the compounds have structures based upon the ring system of isoquinoline. The most important and widespread of these are based upon 1-benzylisoquinoline:

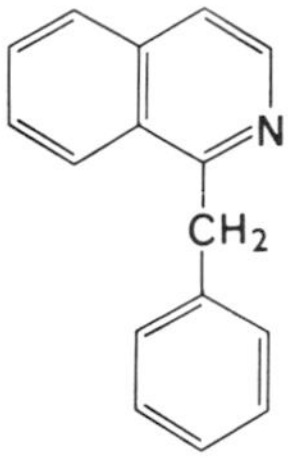

1-benzylisoquinoline 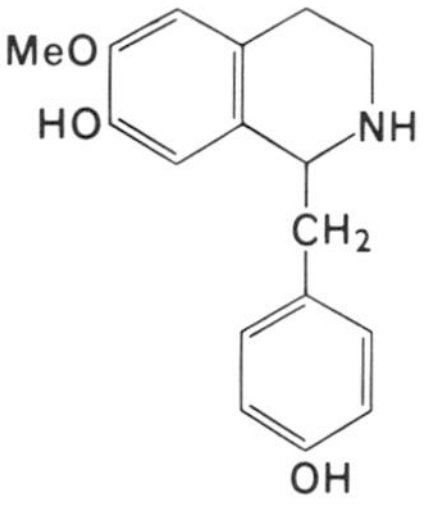                  coclaurine

A number of alkaloids related to one another in this way are found in the latex of the seed capsule of the opium poppy, *Papaver somniferum;* among them are morphine and papaverine:

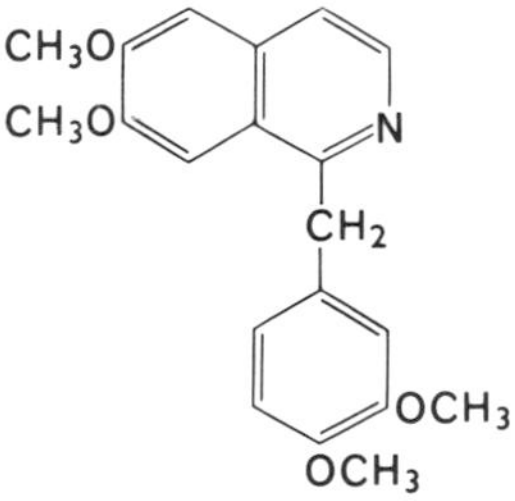

papaverine 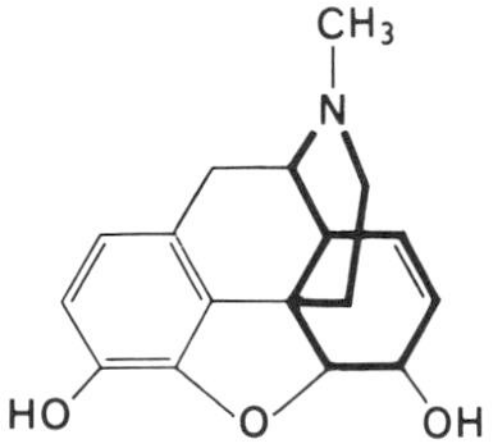                        morphine
                          (isoquinoline ring emphasized)

Apomorphine is a degradation product of morphine; it is a member of the "aporphine" group, many derivatives of which occur in nature:

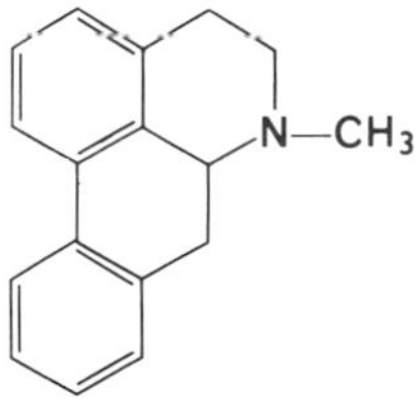

aporphine 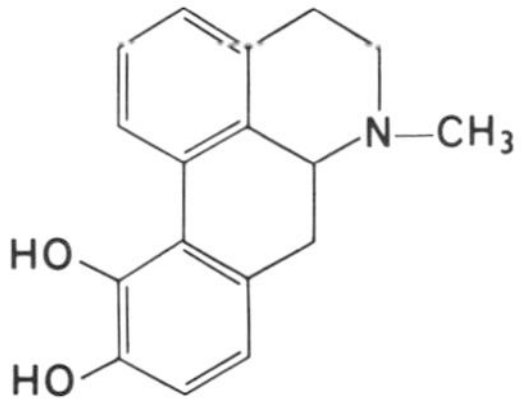                        apomorphine

Synthetic methods for the preparation of alkaloids of the benzylisoquinoline group are various because of the many forms in which the basic isoquinoline structure appears in the numerous alkaloids of this class. One of the routes to the isoquinoline ring system depends upon a ring closure of the following kind:

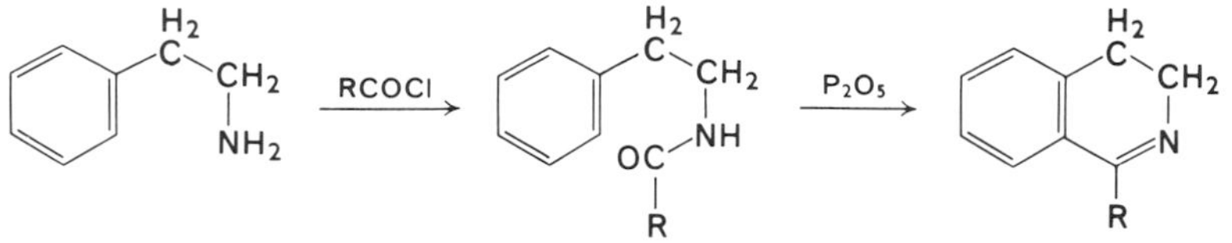

β-phenethylamine                    1-R-3,4-dihydroisoquinoline

$$\text{(52)}$$

The ring-closure step is an acid-catalyzed, electrophilic substitution into the aromatic ring and, as would be expected, is facilitated by electron-releasing (e.g., $CH_3O-$) substituents in the aromatic nucleus.

---

**Exercise 4.** If the isoquinoline synthesis of the kind shown here is conducted with the use of 3,4-dimethoxyphenethylamine and R=3,4-dimethoxybenzyl, the product is 3,4-dihydro-papaverine, convertible to papaverine by dehydrogenation (heating with palladium or platinum black). Formulate these reactions.

---

### 29.20  Quinoline alkaloids

The quinoline nucleus does not occur widely in nature, but is the basic structural entity of the important alkaloid quinine:

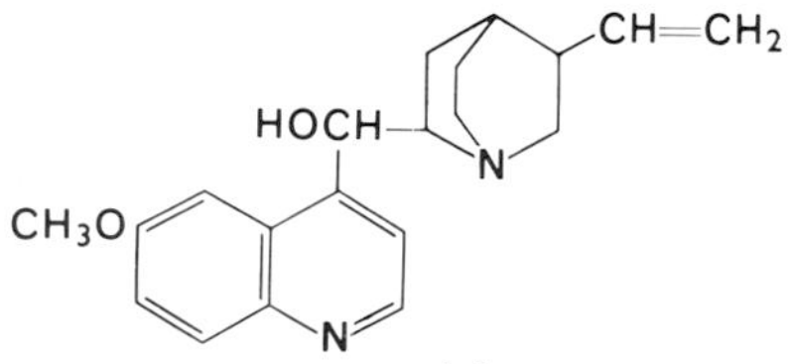

quinine

It is to be noted that quinine also possesses a quinuclidine ring (Section 29.8) as a part of its structure.

Quinine is of great practical as well as historical importance. For centuries it has been the drug used for the control of the symptoms of malaria, one of the most widespread diseases of man. Quinine is being largely replaced by synthetic drugs.

### 29.21  Indole

Indole, a benzopyrrole, is the structural basis for a wide variety of biologically important substances. Indole itself, and its 3-methyl derivative, skatole, are found in feces:

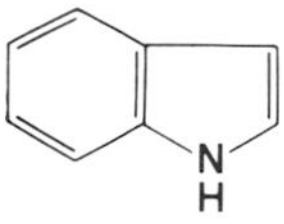

indole 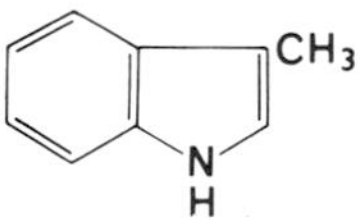                    skatole

Indole is a very reactive nucleophile, and undergoes electrophilic substitution reactions with ease. Both the 2- and the 3-positions are reactive, but unsubstituted indoles undergo substitution in the 3-position. Indole reacts with dimethylamine and formaldehyde in a Mannich reaction to give 3-dimethylaminomethylindole, *gramine*. Gramine undergoes ready nucleophilic displacement of the dimethyl-amino group; when it is treated with potassium cyanide the product is 3-indolyl-acetonitrile. Hydrolysis of the nitrile gives 3-indolylacetic acid (IAA), a naturally occurring plant growth regulator (auxin):

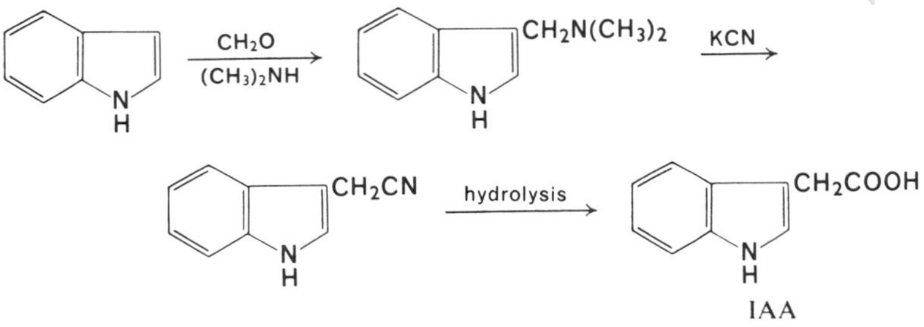

(53)

An important naturally occurring indole derivative is indican, the $\beta$-gluco-side of 3-hydroxyindole, found in plants of the genus *Indigofera*. Acid or enzymatic hydrolysis of the glucoside yields glucose and 3-hydroxyindole, or indoxyl, the air oxidation of which yields the blue dye, indigo:

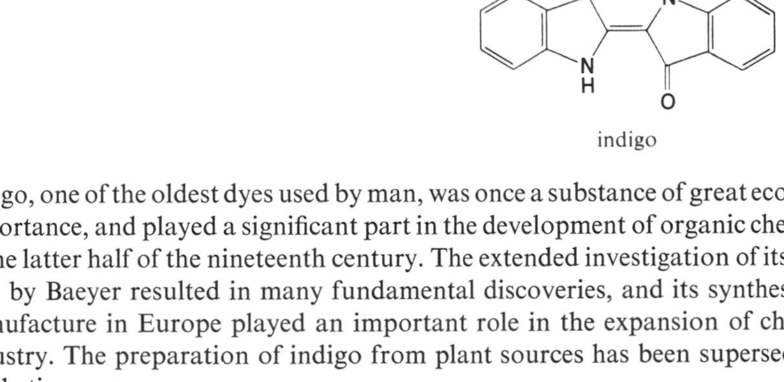

indican          indoxyl

(54)

indigo

Indigo, one of the oldest dyes used by man, was once a substance of great economic importance, and played a significant part in the development of organic chemistry in the latter half of the nineteenth century. The extended investigation of its structure by Baeyer resulted in many fundamental discoveries, and its synthesis and manufacture in Europe played an important role in the expansion of chemical industry. The preparation of indigo from plant sources has been superseded by synthetic processes.

A violet dyestuff used in ancient times, 6,6′-dibromoindigo, is obtained from certain species of mollusc of the genus *Murex*. The dye, called Tyrian Purple, is not present in the animal, but is formed by a combination of oxidation and photo-chemical transformation of a colorless precursor that is obtained from a specialized gland in the mollusc. The yield of Tyrian Purple is very small—about one milligram per gland—and so the cost of producing it made it accessible only to the wealthy and titled; thus the phrase, "born to the purple."

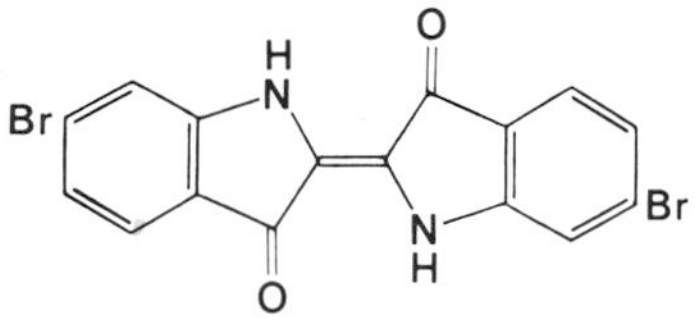

Tyrian Purple

## 29.22   Indole alkaloids

Alkaloids based upon the *indole nucleus* range in complexity from the simple compound gramine, to more complex alkaloids such as yohimbine, reserpine, and strychnine:

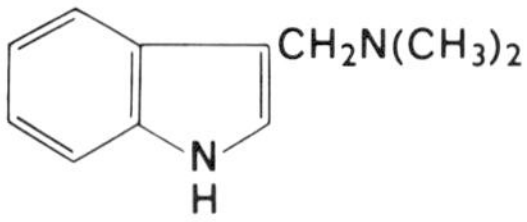

gramine

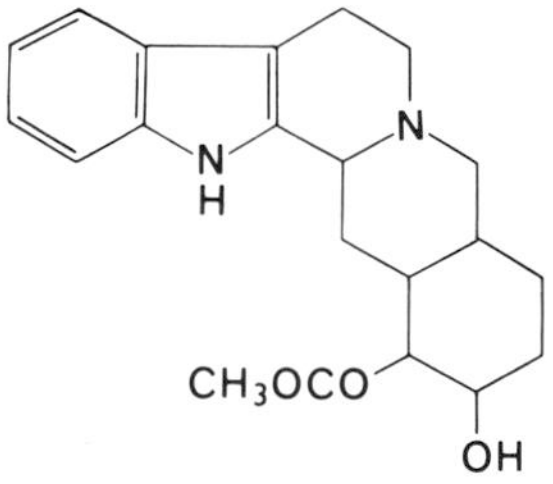

yohimbine

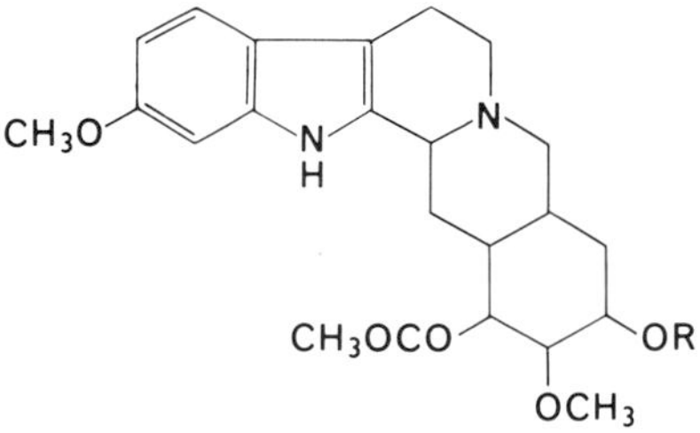

reserpine
(R = 3,4,5-trimethoxybenzoyl)

The indole alkaloids probably arise in the plant cell by biochemical transformations starting from the amino acid tryptophane:

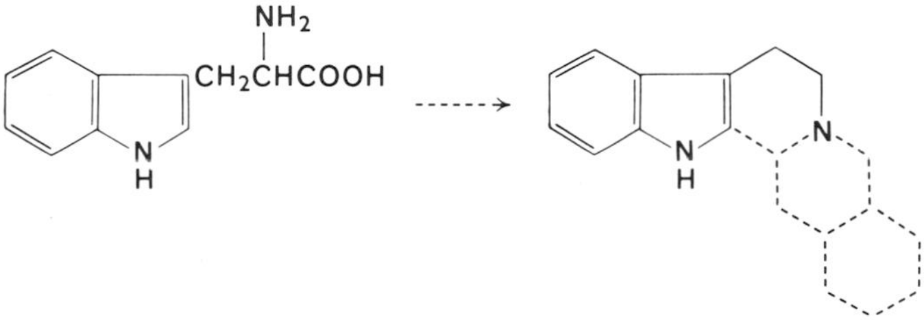

tryptophane                    alkaloids of the yohimbine class

Here, the dotted lines represent additional fragments that go to make up the final alkaloid, but the actual cellular reactions that occur between the starting material —tryptophane in this example—and the final alkaloid are still unknown in all details.

A group of indole derivatives of great importance in medicine are derivatives of lysergic acid, and are known under the collective designation "ergot alkaloids" because of their occurrence in ergot, a fungus of cereal grains of the grass family (e.g., rye). Lysergic acid has the structure,

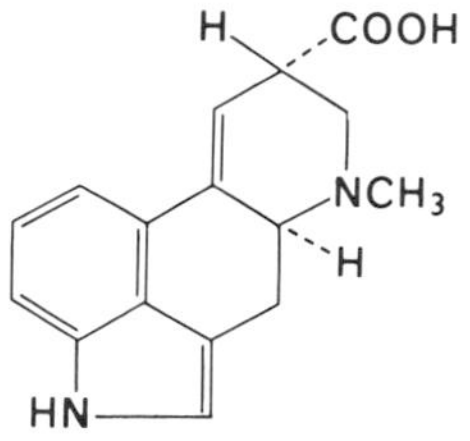

lysergic acid

and occurs in nature in the form of various amides, formed between the acid and the amino group of either an amine or an amino acid. Ergometrine (ergonovine) is the amide of lysergic acid with L-(+)-2-amino-1-propanol. Ergonovine is used in medicine in the treatment of migraine and in the induction of uterine contractions in childbirth.

The diethyl amide of lysergic acid, known as LSD, has the remarkable property of inducing psychotic responses that resemble, but are not identical with, schizophrenia. It is the object of much investigation, not only because of the possibility that it may prove useful in psychiatric treatment, but because of abuses that have made it a serious socio-legal problem.

### Exercises

(Exercises 1–4 will be found within the text.)

5 When 2,5-hexanedione (acetonylacetone) is treated with excess methylmagnesium bromide, and the reaction mixture decomposed with aqueous HCl, the product isolated is not the expected diol, but a compound $C_8H_{16}O$. Formulate this reaction.

6 Write the reaction showing the result of heating tetramethylethylene oxide with hot, dilute $H_2SO_4$. The product has the composition $C_6H_{12}O$ (and thus is isomeric with the starting oxide).

7 What product would result from the reaction between N,N-dimethyl-β-chloroethylamine [$(CH_3)_2NCH_2CH_2Cl$] and dimethylamine?

8 When the hydrochloride of 2-chloromethyl-N-methylpyrrolidine

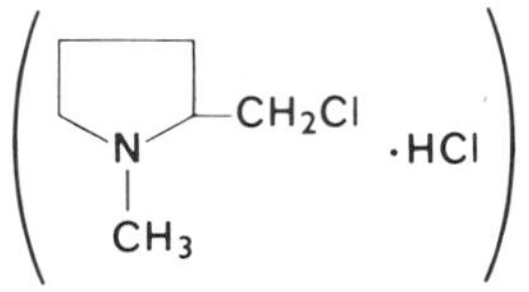

is treated with alkali, 3-chloro-N-methylpiperidine is formed.

Formulate this change. (A cyclic imonium ion is an intermediate.)

**9** Formulate the following reactions:

   *(a)* 2-bromopyridine + hot, alcoholic NaOEt $\longrightarrow$

   *(b)* 2-chloropyridine + $NH_2NH_2$ $\longrightarrow$

   *(c)* 2,3-dibromopyridine + methanolic NaOMe $\longrightarrow$

   *(d)* N-ethylpyridinium chloride + alkaline potassium ferricyanide $\longrightarrow$

   *(e)* quinoline + $NaNH_2$ $\longrightarrow$

**10** Write the reactions involved in the synthesis of papaverine, starting with 3,4-dimethoxy-phenylacetonitrile as the only organic compound.

**11** A reagent that is used for the synthesis of $\alpha$-amino acids is ethyl acetamidomalonate, $CH_3CONHCH(COOEt)_2$. For example, it can be alkylated with benzyl chloride to give a product which can be hydrolyzed and decarboxylated to give $\beta$-phenylalanine. Write the reactions that would be used in a synthesis of tryptophane, starting with indole and using ethyl acetamidomalonate in one of the steps.

Chapter thirty / # Nuclear Magnetic Resonance and Mass Spectrometry

## 30.1 Physical methods of study of organic compounds

In recent years the progress of organic chemistry has been greatly aided by the development of a number of techniques for the study of certain physical properties of organic compounds, the interpretation of which is central to the problem of determining their structure. The development of convenient and accessible instruments for the routine measurement of ultraviolet and infrared absorption spectra has had a profound influence upon the development of the science by making it possible to establish, by a single measurement requiring only small quantities of material, the presence of structural features that would otherwise be recognized only as the result of elaborate chemical procedures.

New techniques are continually being added to those already available to the organic chemist, and with the development of reliable commercial instruments new methods are coming into increasingly wide use in the laboratory. Among these may be mentioned the study of molecular asymmetry by means of *optical rotatory dispersion*, of free radicals by *electron paramagnetic resonance*, of total molecular structure by computer-aided methods of *X-ray crystallography*, and of structural elements of molecules by *nuclear magnetic resonance spectroscopy* and *mass spectrometry*. It is not possible to discuss all of these in useful detail here, and attention will be directed to those that have so far found the most general application to the solution of the problems of molecular structure: nuclear magnetic resonance and mass spectrometry.

## 30.2 Nuclear magnetic resonance

A bare nucleus, such as a proton, is a spinning, charged body which, because of its charge, generates a magnetic field. It may be likened to a tiny bar magnet, the axis of which corresponds to the axis of rotation. If an external magnetic field is applied, the proton will tend to line up with the field in one of two ways: in a stable orientation, in which the north pole of the proton's field points away from the north pole of the static external field (parallel); or in an unstable orientation of higher energy, in which the proton is lined up against the field (antiparallel). The difference in energy of these two orientations is small, and at ordinary temperatures there is only a small excess of protons in the more stable, parallel orientation.

A proton that is not lined up exactly parallel to the applied field will describe a motion known as *precession*, in which the rotational axis will describe a circle at right angles to the applied field. The wobbling of a spinning top or of a gyroscope whose axis of rotation is not perpendicular is a precessional motion. Quantum-mechanical calculations show that the frequency of the precessional rotation is equal to the frequency of electromagnetic radiation that is necessary to cause the nucleus to "flip" from parallel to antiparallel orientation. This transition is brought about by the application of a rotating magnetic field at right angles to the static field, and in phase with the precessing nucleus. When transition occurs, the frequency of precession and the frequency of the rotating magnetic field are equal, and they are in *resonance*. The resulting absorption of energy is detected by a radio frequency receiver, and is recorded.

The energy required for the transition of the nucleus from one energy level to the other is expressed by the equation

$$\text{Energy} = h\nu = 2\,\mu H$$

where $\nu$ is the frequency of the electromagnetic radiation at which transition occurs under a constant static field strength, $H$, and $\mu$ is the nuclear magnetic moment, a characteristic for the nucleus under consideration ($H^1$, $C^{13}$, $F^{19}$, and so on). With a static applied field of about 14,000 gauss, the electromagnetic radiation is in the region of radio frequency of about 60 megacycles.

Figure 30-1 is a schematic drawing of the system used for the measurement of nuclear magnetic resonance. It is possible to bring about resonance between the nuclei and the rotating magnetic field in either of two ways: by keeping the applied radio frequency constant, and varying the static field; or by "sweeping" with a varying electromagnetic field at a constant static field. The former method is the one commonly used in the most widely used commercial instruments.

If the resonance frequency of a nucleus were simply a function of the applied field and of the magnetic properties of the nucleus, all of the protons* in an organic

---

* For reasons outside the scope of this discussion, nuclei of carbon-12, carbon-14, oxygen-16, sulfur-32, and nitrogen-14 either have no magnetic moment or cannot be detected with the instruments in general use. Nuclear magnetic resonance spectroscopy is largely confined to the hydrogen atoms of organic molecules, although fluorine-19 and carbon-13 are also studied.

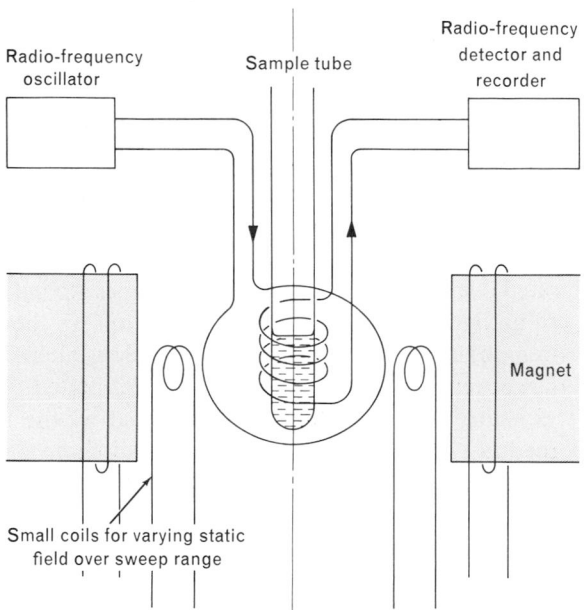

**Figure 30–1.** Essential components of nuclear magnetic resonance spectrometer.

molecule would absorb at the same frequency, and the phenomenon would be of little practical importance. Fortunately, however, not all of the hydrogen atoms in an organic compound are in the same molecular environment, and the electrons in the bonds between the hydrogen atoms and other atoms in the molecule (usually the carbon-hydrogen, oxygen-hydrogen, and nitrogen-hydrogen bonds) screen the nucleus in such a way as to alter the effective magnetic field at the nucleus so that it is not exactly the same as the applied field. A consequence of this is that hydrogen atoms in different environments are *shielded* to different degrees, and thus the various hydrogen atoms in an average organic molecule show resonance over a spectrum of frequency. The measurement of the nuclear magnetic resonance characteristics of a molecule is known as a *nuclear magnetic resonance spectrum*.

The spread of frequency over which protons of different kinds absorb is for most organic molecules about 700 cycles per second when the static field strength is 14,000 gauss. Since the measurement of the absolute values of frequencies which differ by only a few cycles per second from the principal frequency of about $60 \times 10^6$ cycles per second is difficult to carry out with the accuracy that would be required, frequencies are measured *relative* to an arbitrarily chosen value. The frequency that is chosen for reference is that of the protons of some reference compound, all of which are identical in their environment and absorb at one frequency. Any of a number of organic compounds could be used as the standard, but the most satisfactory reference compound for general use is tetramethylsilane, $(CH_3)_4Si$ (TMS), which gives a single, sharp absorption for the twelve equivalent hydrogen atoms at a field that is higher than that for most other protons of organic molecules.

The most convenient way of using the reference compound is to add it directly to the solution of the substance under study. TMS is quite inert and seldom reacts with other compounds.

The procedure for carrying out the measurement of a nuclear magnetic resonance (n.m.r.) spectrum is the following: the compound to be studied, dissolved in a solvent which, ideally, contains no protons, is placed between the poles of the magnet. The TMS is added directly to the solution in an amount sufficient to give a prominent signal. The commonly used n.m.r. instruments are designed to use a chart paper calibrated over a range of about 700 cycles per second (cps), and the TMS signal is ordinarily adjusted to fall at the position on the chart marked 0 cps. The field strength of the main static magnetic field is altered continuously in a "sweep" that covers the 700 cps range of the calibrated chart.* When resonance occurs, a sharp change in the signal detected by the radio-frequency receiver is recorded as a peak on the chart paper. The difference in frequency (in cps) between the TMS peak and that of a proton absorption peak is known as the *chemical shift* of that proton.

Chemical shifts may be recorded as cps; or, more commonly, in the units $\delta$ or $\tau$. These are defined by the relationships

$$\delta = \frac{H - H_r}{H_r} = \frac{(\nu_{proton} - \nu_{TMS}) \times 10^6}{\text{oscillator frequency, cps}} \text{ (parts per million)}$$

where $H_r$ is the external field strength and $H$ the field strength at which the proton signal appears. The TMS signal is at $\delta = 0$, and a proton signal that appears at 600 cps, with an instrument operating at $60 \times 10^6$ cps, has $\delta = 10$. *Tau* values are simply equal to $10 - \delta$. Values of $\delta$ are in parts per million (ppm); they are dimensionless.

The most commonly used solvents are deuterochloroform, $CDCl_3$, and other deuterated compounds such as hexadeuteroacetone and hexadeuterodimethyl sulfoxide. Trifluoroacetic acid, the proton of which resonates outside of the range in which most other kinds of protons are found, and which is an excellent solvent for many compounds, is also useful.

Chemical shifts are dependent upon the molecular environment of the proton in question, and for most organic compounds the protons are found at $\delta$ values of about 0.5 to 12, the majority falling in the range $\delta =$ about 0.5 to 8. In Table 30-1 are given some representative values for the n.m.r. absorption of a number of different kinds of organically bound protons.

The intensity of the absorption at any frequency is proportional to the number of protons, and is measured by the area of the relevant absorption peak. The actual number of protons in an organic molecule can be measured with considerable accuracy by measuring the areas under the absorption peaks (for example, by cutting each peak out of the chart paper and weighing the cut-out

---

* Field strength may be expressed either in terms of magnetic units (gauss) or electromagnetic frequency units (cps).

**Table 30-1.** *Representative values of chemical shifts for protons in organic compounds (tetramethylsilane, $\delta = 0.00$)*

| Compound | Chemical shift, $\delta$, ppm. | Compound | Chemical shift, $\delta$, ppm. |
|---|---|---|---|
| **Methyl Groups (CH₃—C)** | | **Methyl Groups (CH₃—X)** | |
| Si(CH₃)₄ | 0.00 | CH₃OCH₃ | 3.27 |
| CH₃CH₂CH₂CH₃ | 0.85 | —OCH₃ | 3.73 |
| CH₃CH₂OH | 1.22 | | |
| CH₃CH₂Cl | 1.48 | CH₃COOCH₃ | 3.67 |
| | | N(CH₃)₃ | 2.12 |
| CH₃ | 1.60 | CH₃Cl | 3.00 |
| | | CH₃Br | 2.62 |
| | | CH₃I | 2.16 |
| | | CH₃OH | 3.47 |
| CH₃COCH₃ | 2.17 | **Vinyl and Aromatic Protons** | |
| CH₃COOH | 2.10 | | |
| CH₃CN | 2.00 | | 7.37 |
| CH₃ | 2.32 | (CH₃)₂C=CH₂ | 4.60 |
| | | (CH₃)₂C=CHCH₃ | 5.20 |
| **Methylene Groups (C—CH₂—C)** | | | 5.30 |
| | 1.43 | | 5.57 |
| CH₂=CHCH₂CH₂CH=CH₂ | 2.12 | (a) (b) | (a) 6.24 (b) 4.54 |
| CH₂ | 2.15 | **Miscellaneous** | |
| CH₂ | 2.00 | Aldehydes, | 9.5–10 |
| | | Acids, 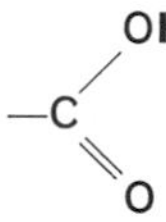 | 11–11.5 |

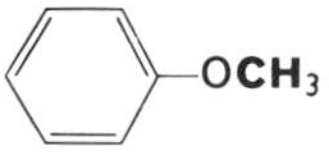

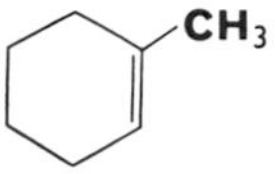

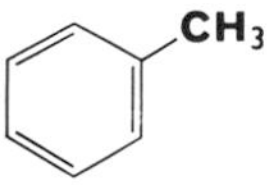

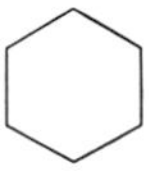

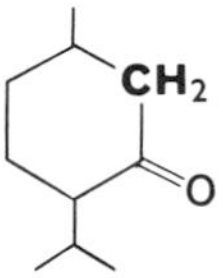

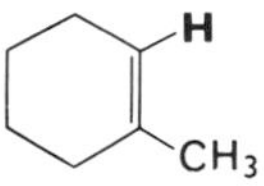

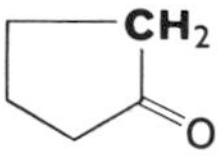

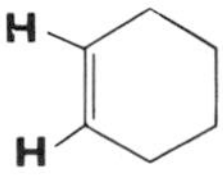

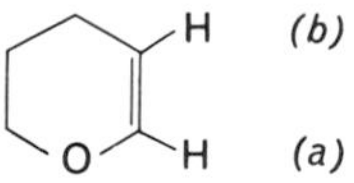

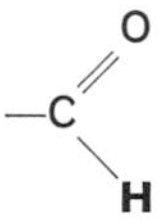

sections). If any one peak is known to, or can be assumed to represent a known number of protons, the number of protons in each of the other peaks can be determined. Modern instruments perform this measurement automatically by recording directly on the chart paper a tracing that rises in steps as each proton absorption frequency is passed. The total rise of this line represents the total number of protons in the compound under study, and the number of protons in each "step" can be estimated if any one is known. Such integration records are shown in some of the figures to follow.

### 30.3  Interpretation of nuclear magnetic resonance spectra

The interpretation of nuclear magnetic resonance spectra is largely empirical, and is based primarily upon a knowledge of the characteristic chemical shifts for protons in different kinds of structural combination. Values such as those given in Table 30-1 are derived from extensive studies of compounds of known structure; they serve only as guides, for the detailed interpretation of n.m.r. spectral features rests upon considerations of numerous secondary structural features. For example, the protons of the methyl group of acetone show absorption at $\delta = 2.1$ ppm, while those of the methyl group of acetophenone are found at $\delta = 2.6$ ppm. In both cases, the methyl group is present in the structural unit —$COCH_3$, but the effects of the remainder of the structure can be seen to influence the position of the absorption band.

Information gained from ultraviolet and infrared absorption spectra often provides valuable assistance in the interpretation of n.m.r. spectra. For example, the presence of a sharp, three-proton absorption peak in the region of $\delta = 3.7$ ppm usually indicates the presence of a methoxyl group, —$OCH_3$. But the methyl group of both methyl esters and methyl ethers is found in this region, and so the signal cannot ordinarily be interpreted in the absence of additional information. If the characteristic carbonyl absorption of the ester grouping in the infrared is absent, such an n.m.r. signal can be attributed to the presence of a methyl ether grouping.

*Spin-spin interactions.*   The absorption bands observed for single protons or for groups of protons in identical environments very often appear, not as single, sharp peaks in the recorded n.m.r. spectrum, but as groups of separate peaks; i.e., they exhibit "fine structure." In instruments incapable of high resolution, such groups of peaks may appear as broad, unresolved absorption bands, but present-day instruments are capable of sufficiently high resolution to resolve such bands into their separate components. Figure 30-2 is the n.m.r. spectrum of acetaldehyde. It might be anticipated that acetaldehyde would show two proton absorption peaks: one, with an area corresponding to three protons, for the methyl group; and one, with an area corresponding to one proton, for the hydrogen atom of the formyl group. It is apparent that while the two "peaks" of areas one and three are present,

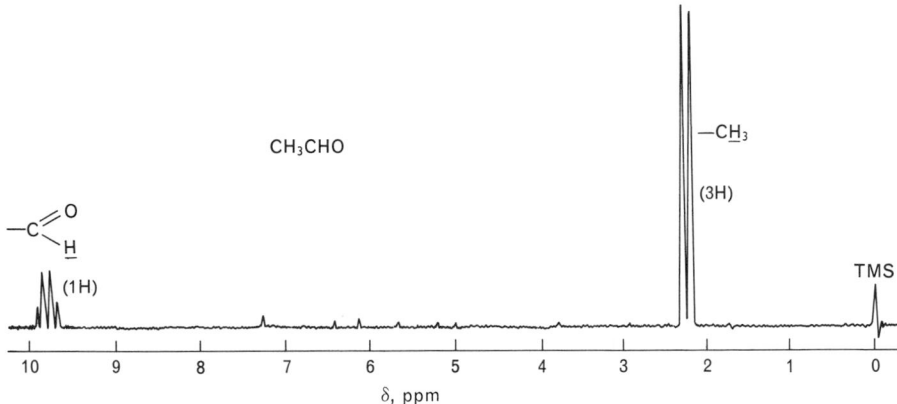

**Figure 30–2.** N.m.r. spectrum of acetaldehyde. The three-proton signal for the methyl group is a doublet, as the result of spin-spin coupling with the proton on the adjacent formyl group. The signal for the formyl hydrogen atom is a quadruplet as the result of spin-spin coupling with the three protons of the adjacent methyl group. Note the low-field position ($\delta = 9.8$) of the formyl hydrogen atom. (The n.m.r. spectra used for illustration in this chapter were taken from the Spectra Catalog, Vol. 1, 1962, courtesy Varian Associates, Palo Alto, California.)

both of these are multiplets: that of area three is a doublet; that of area one is a quadruplet.

The n.m.r. spectrum of ethyl chloride, shown in Figure 30-3, shows two multiplets: one, of total area three, is a triplet; the other, of area two, is a quadruplet.

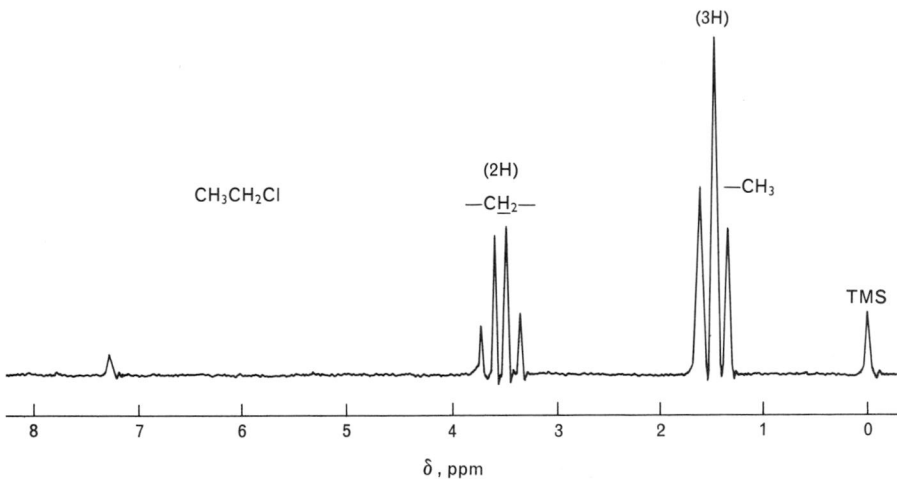

**Figure 30–3.** N.m.r. spectrum of ethyl chloride. The quadruplet signal (two protons) for the CH₂ group, adjacent to Cl, is found at much lower field than the triplet signal (three protons) for the methyl group.

Moreover, a detailed analysis of the triplet and of the quadruplet signals in the ethyl chloride spectrum discloses that the relative areas of the individual peaks are $1:2:1$ in the triplet, and $1:3:3:1$ in the quadruplet.

The reason for the splitting of the signals for the $—CH_2—$ and $—CH_3$ groups in the ethyl chloride spectrum can be found in an examination of the structure of the compound as it affects the environments of the protons of these groupings. Let us first consider the methyl group, the protons of which give rise to the triplet, of total area 3, at $\delta = 1.48$ (the "center of gravity" of the triplet signal). Adjacent to the $—CH_3$ group are the two protons of the $—CH_2—$ group, the spins of which may have three different arrangements, which can be represented by the symbols

each of which represents a particular orientation of a proton magnet with respect to the applied field. Of these three orientations of the methylene group protons, one will reinforce, one will oppose, and two will neither oppose nor reinforce the field by which the $—CH_3$ protons are affected. Thus, there is one chance that the $—CH_3$ resonance will appear earlier, one that it will appear later, and two that it will be unaffected. Thus, three peaks, in the observed ratio $1:2:1$, appear, and the signal for the methyl group is made up of the three-line signal shown in Figure 30-3.

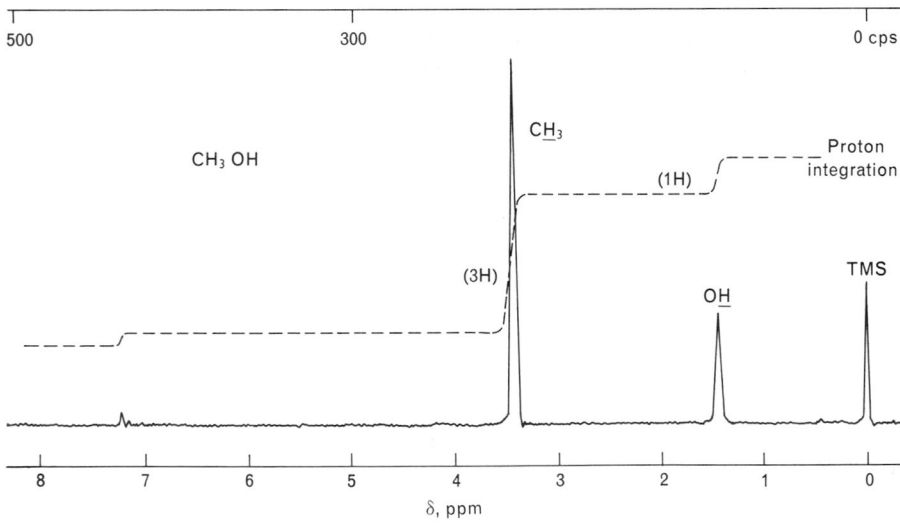

**Figure 30–4.** N.m.r. spectrum of methanol, TMS as internal standard. The spectrum shows two peaks: a three-hydrogen singlet for the $—CH_3$ group, a one-hydrogen singlet for the proton of the $—OH$ group. The dotted line is the automatically recorded proton-integration curve. The small signal at about 7.3 ppm is that of the proton in $CHCl_3$, usually present as an mpurity in the $CDCl_3$ used as the solvent.

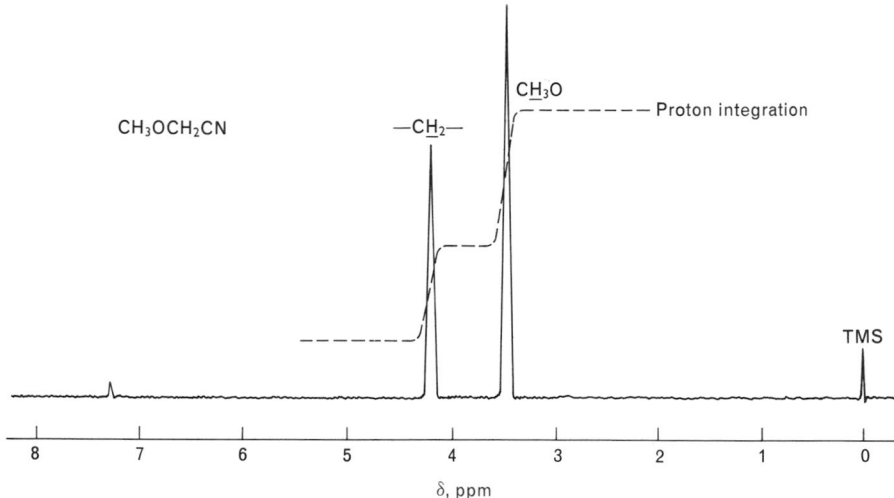

**Figure 30–5.** N.m.r. spectrum of methoxyacetonitrile. Two signals, both singlets, with areas two (for CH₂) and three (for CH₃) are seen. It will be noted that the signal for the methylene group is at a lower field than that for the methyl group; this is the result of the additional deshielding of the methylene group by CN.

The —CH₂— absorption will be affected by the neighboring methyl group in a similar way. The three —CH₃ protons can possess the spin orientations

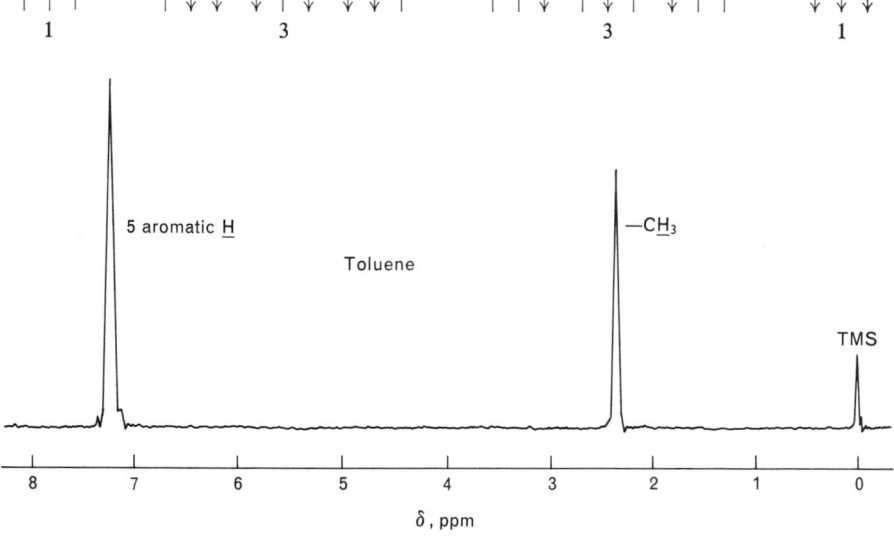

**Figure 30–6.** N.m.r. spectrum of toluene. The five aromatic protons have essentially the same chemical shift and appear as a five-proton singlet. The only other signal is that of the methyl group.

and will affect the absorptions of the —$CH_2$— protons in such a way as to give the four-line signal with the four peaks in the ratio $1:3:3:1$.

The signals for the methyl group and the formyl hydrogen atom in the n.m.r. spectrum of acetaldehyde (Figure 30-2) can be interpreted similarly. The methyl group gives rise to a doublet because of the influence of the adjacent proton of the —CHO group; and the formyl group proton, adjacent to the three protons of the methyl group, appears as the quadruplet.

The phenomenon represented in n.m.r. spectra by the splitting of the signals for equivalent protons into multiplet signals is known as *spin-spin coupling*. The appearance of such multiplet signals, caused by spin-spin coupling, and the separation, in cycles per second (designated by the symbol J, and called the coupling constant), of the individual peaks in such signals, provides valuable information in organic structure determination. The signals for the aromatic protons in Figures 30-7 and 30-8 are typical of splitting of signals for *ortho*-disposed aromatic protons and provide excellent evidence that the methyl and methoxyl groups (Figure 30-7) and the chloro and acetyl groups (Figure 30-8) are in the 1,4- rather than in the 1,2- or 1,3-positions. It is beyond the scope or intent of this presentation to discuss spin-spin coupling in great detail; examples of it may be discerned in some of the figures. The figures to follow have been chosen to provide examples of n.m.r. spectra of a variety of organic compounds containing protons in a variety of

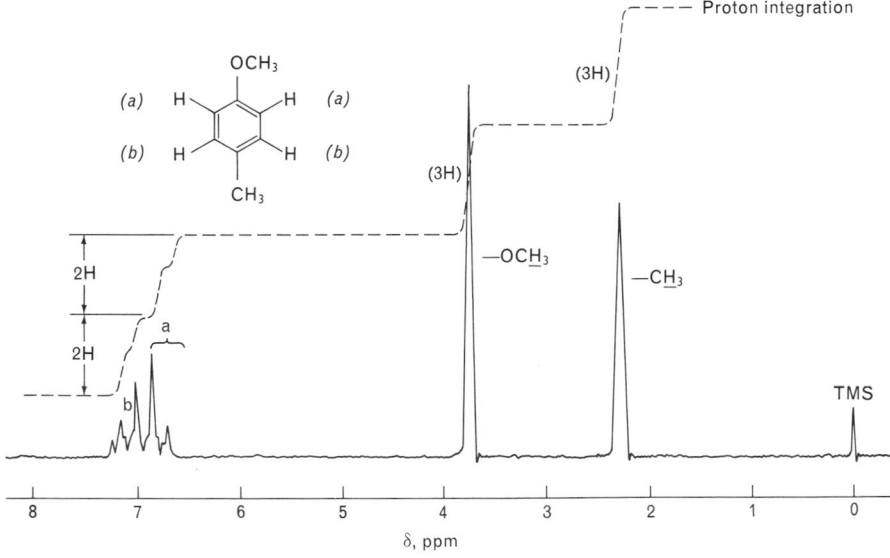

**Figure 30–7.** N.m.r. spectrum of *p*-methoxytoluene. The nuclear methyl group and the methyl group of the methoxyl group appear as three-proton singlets. The ring protons form a coupling pattern characteristic of *ortho*-hydrogen atoms with different chemical shifts. Protons *ortho* to alkoxyl groups are found at higher field (lower values of δ) than those *ortho* to alkyl groups; this permits the quadruplet pattern found at δ= about 7 to be interpreted as shown. The proton integration tracing shows the two protons marked *(a)*, the two marked *(b)*, and the two three-proton signals for the two methyl groups.

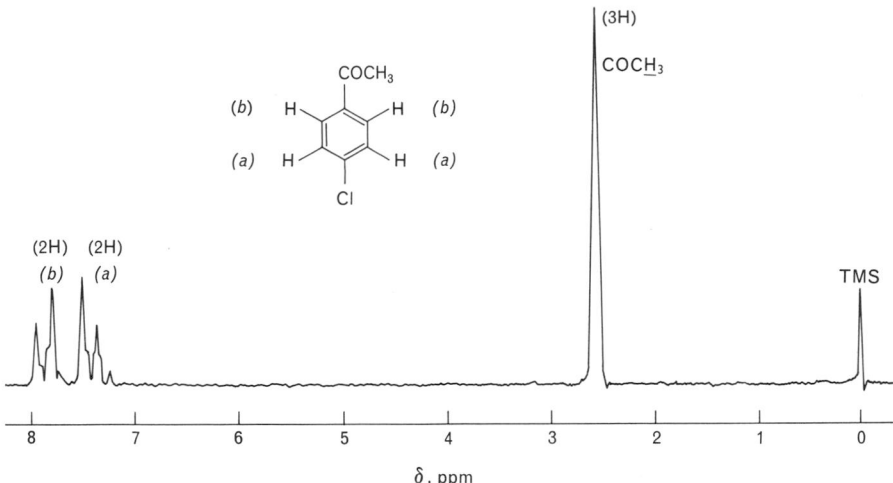

**Figure 30–8.** N.m.r. spectrum of *p*-chloroacetophenone. The methyl group of —COCH₃ is responsible for the three-proton singlet at $\delta = 2.58$. Note the downfield shift of the nuclear protons ($\delta$ about 7 to 8) as compared with those of *p*-methoxytoluene (Figure 30-7). The deshielding effects (downfield shift) of the *ortho* carbonyl and halogen groups are clearly evident by this comparison. The coupling pattern of the *ortho*-disposed protons, resulting in the two doublets, is similar in form (and about the same in the separation of the peaks) to that shown for the protons of *p*-methoxytoluene.

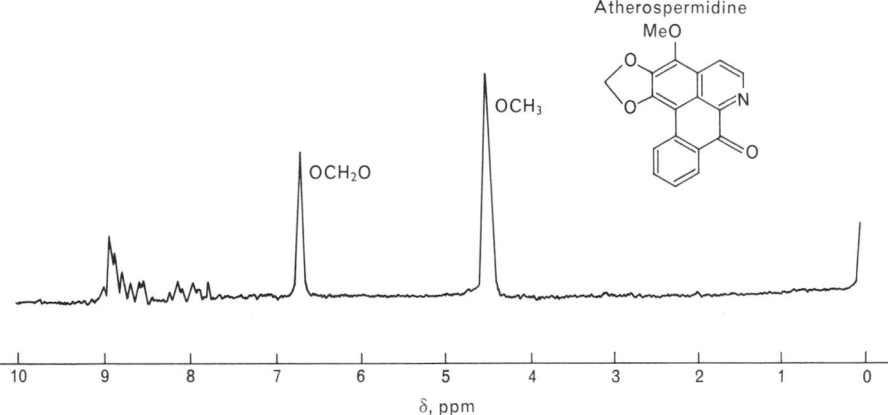

**Figure 30–9.** N.m.r. spectrum of a naturally occurring alkaloid, atherospermidine. Note the simplicity of this spectrum, which is that of a rather complex molecule. The methoxyl group and methylenedioxy (—CH₂—) form sharp three- and two-proton singlets, respectively. The six aromatic protons are coupled in a complex manner and are difficult to interpret in detail. It is apparent, however, that the six protons in the multiplets between $\delta = 8$ and 9 ppm represent aromatic protons. Compare this spectrum with that of liriodenine (Figure 30-10), which is identical with atherospermidine except for the presence of the methoxyl group in the latter. The proton in this position in liriodenine is not coupled with any other proton(s), and appears as the prominent singlet at $\delta =$ about 7.7 ppm. The remaining six protons of liriodenine and atherospermidine appear in substantially the same region and show nearly the same coupling characteristics.

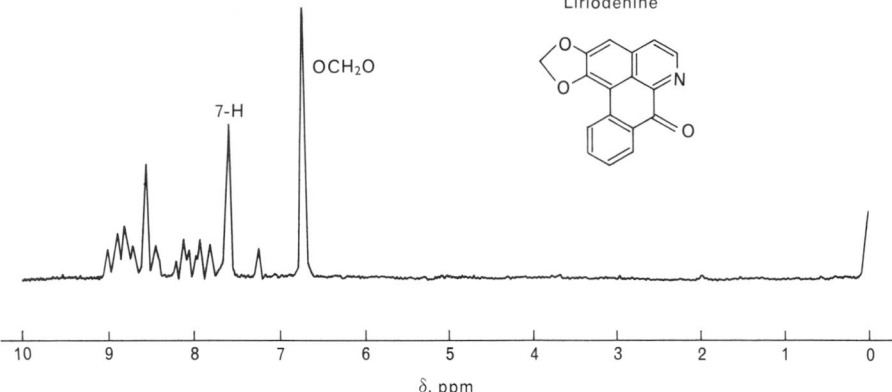

**Figure 30-10.** N.m.r. spectrum of liriodenine. The only nonaromatic protons are those of the methylenedioxy group, which appear as a sharp two-proton singlet. The aromatic protons are found at values of 7–9 ppm, the sharp one-proton singlet being due to the single proton adjacent to the methylenedioxy grouping. See comments under Figure 30-9.

combinations. The legend associated with each spectrum describes briefly the chief characteristics of the spectrum and assigns to the several peaks the protons that they represent.

## 30.4 Mass spectrometry

The *mass spectrometer* is an instrument designed *(a)* to bring about the ionization and fragmentation of an organic molecule by bombarding it with high-energy particles, such as electrons, and *(b)* to analyze the resulting mixture of ionic fragments by determining their mass/charge ratio and measuring their relative abundance.

A mass spectrometer provides a means of producing an intense beam of electrons, a means of introducing a sample of the compound to be studied into the electron beam, and an analyzing and recording system in which the intensity of each species of ion is measured and recorded on a suitable chart. The *mass spectrum* is a chart of the mass/charge ratio (most species of ions produced will have unit charge) against the intensity of each species.

The bombardment of an organic molecule with high energy particles can result in the removal of an electron from the molecule:

$$M + e \longrightarrow M^+ + 2e \tag{1}$$

The resulting positively charged ion is called the *molecular ion*, and since its mass (*m/e*, where $e = 1$) can be measured accurately in a mass spectrometer, *mass spectrometry provides a valuable way of measuring molecular weights.* If the ion $M^+$ were stable, the mass spectrometer would give little information besides the molecular weight. But since the electron beam used has much greater energy than is needed to cause the initial ionization (1), the ion produced by impact has sufficient excess energy to decompose further into smaller fragments. These fragments may be positive ions (which are detected and recorded), uncharged free radicals, and neutral molecules. The latter two are not charged and are not detected by the measuring system of the instrument.

The fragmentation of the ions produced by the electron impact proceeds in a manner that is in accord with the usual principles of the relative stability of intermediate ions and radicals: tertiary carbonium ions, allylic carbonium ions, protonated carbonyl groups, and other stabilized-onium ions are usually produced by fragmentations in preference to alternative, less stable ions.

As an example of the fragmentation that occurs, let us examine the behavior of some simple molecules in the mass spectrometer. Methanol, under electron bombardment, undergoes loss of one of the nonbonding electrons on oxygen:

$$CH_3\ddot{\underset{\cdot\cdot}{O}}-H+e \longrightarrow CH_3-\overset{\cdot+}{\underset{\cdot\cdot}{O}}-H+2e \tag{2}$$
$$m/e=32=M^+$$

Further decomposition of $M^+$ proceeds by loss of a hydrogen atom, with formation of the oxonium ion:

$$\underset{\overset{\mid}{CH_2}-\overset{\cdot+}{\underset{\cdot\cdot}{O}}-H}{\overset{H}{}} \longrightarrow (CH_2=\underset{\cdot\cdot}{O}H)^+ + H\cdot \tag{3}$$
$$m/e=31$$

Further breakdown of the initial ion can occur in the following way:

$$(CH_3-\overset{\cdot}{\underset{\cdot\cdot}{O}}H)^+ \xrightarrow{-2H\cdot} H_2C=\overset{\cdot+}{\underset{\cdot\cdot}{O}}$$
$$m/e=30$$
$$\xrightarrow{-3H\cdot} H-C\equiv\overset{\cdot\cdot}{O}:^+$$
$$m/e=29$$

More complex compounds give more complex fragmentation patterns. Ethanol and higher primary alcohols can decompose in the following way:

$$RCH_2OH \longrightarrow RCH_2\overset{\cdot}{\underset{\cdot\cdot}{O}}H^+ \xrightarrow{-H\cdot} RCH=OH^+$$
$$\downarrow -R\cdot$$
$$CH_2=OH^+$$
$$m/e=31$$

Secondary and tertiary alcohols are very prone to undergo fission because the carbonium ($\equiv$oxonium) ions produced after the initial ionization are increasingly stable with increasing substitution and several courses of fission are available. In the decomposition of the molecular ion of 2-butanol, the loss of $C_2H_5\cdot$, $CH_3\cdot$ or $H\cdot$ can occur, with the production of $M$-1, $M$-15 and $M$-29 ion peaks:

$$CH_3CH_2\underset{\underset{CH_3}{|}}{CHOH} \longrightarrow CH_3CH_2\underset{\underset{CH_3}{|}}{\overset{\cdot+}{C}HOH}$$

$$m/e = 74 = M^+$$

$$-C_2H_5\cdot \qquad \Big|-CH_3\cdot \qquad -H\cdot$$

| $(CH_3CH{=}OH)^+$ | $(CH_3CH_2CH{=}OH)^+$ | $(CH_3CH_2\underset{\underset{CH_3}{|}}{C}{=}OH)^+$ |
|---|---|---|
| $m/e = 45$ | $m/e = 59$ | $m/e = 73$ |
| $(M\text{-}29)$ | $(M\text{-}15)$ | $(M\text{-}1)$ |

In tertiary alcohols the molecular ion is so much less stable than the product resulting from the loss of an alkyl radical that often no molecular ion peak ($M^+$) is observed; its subsequent fission occurs in the following way:

$$CH_3{-}\underset{\underset{CH_3}{|}}{\overset{\overset{CH_3}{|}}{C}}{-}OH \longrightarrow (CH_3{-}\underset{\underset{CH_3}{|}}{\overset{\overset{CH_3}{|}}{C}}{-}\overset{\cdot\cdot}{O}H)^+ \xrightarrow{-CH_3{}^\bullet} (CH_3{-}\underset{\underset{CH_3}{|}}{C}{=}OH)^+$$

$$m/e = 59$$
$$(M - 15)$$

The $m/e = 59$ ion peak is common to 2-methyl-2-alkanols, for the following decomposition can occur:

$$(CH_3{-}\underset{\underset{CH_3}{|}}{\overset{\overset{R}{|}}{C}}{-}OH)^+ \xrightarrow{-R^\bullet} (CH_3{-}\underset{\underset{CH_3}{|}}{C}{=}OH)^+$$

$$m/e = 59$$

As the length of the chains in the alkyl groups of primary, secondary, and tertiary alcohols increase, increasingly complex fission patterns (mass spectra) are observed. Figures 30-11 and 30-12 show the mass spectra of ethanol and 1-octanol. It is evident that many more prominent ion species are produced in the case of the longer chain alcohol.

The sensitivity of the mass spectrometer to single unit differences in the mass/charge number is responsible for the appearance of a number of small, but easily detectable, peaks corresponding to ions containing the isotopes $^2H$, $^{13}C$, $^{17}O$, $^{18}O$, $^{15}N$, and so on. Although most of these isotopes occur naturally in very low abundances ($^2H$, 0.015%; $^{13}C$, 1.11%; $^{17}O$, 0.04%; $^{18}O$, 0.20%), some of the common atoms contain high proportions of isotopes. For example, chlorine consists of $^{35}Cl$ and $^{37}Cl$ in the ratio of about three to one. The mass spectrum of ethyl chloride thus shows two distinct molecular ions, differing by two mass units.

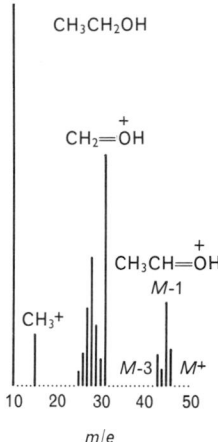

CH₃CH₂OH

$\overset{+}{CH_2}{=}OH$

**Figure 30–11.** Mass spectrum of ethanol.

$\overset{+}{CH_3CH}{=}OH$
*M*-1

CH₃⁺

*M*-3    *M*⁺

Among the prominent peaks in the ethyl chloride spectrum are $(C_2H_5{}^{35}Cl)^+ = 64$; $(C_2H_5{}^{37}Cl)^+ = 66$; $(CH_2{=}{}^{35}Cl)^+ = 49$; $(CH_2{=}{}^{37}Cl)^+ = 51$; $({}^{35}Cl)^+$; $({}^{37}Cl)^+$; and $(C_2H_5{}^+) = 29$.

The introduction (by chemical synthesis) of deuterium ($^2H$) into an organic compound, and comparison of the mass spectra of the original and deuterated compounds, can give valuable clues about the nature of a fragmentation. For example, the mass spectrum of methyl formate, $HCOOCH_3$, shows the most abundant ion to have $m/e = 31$. This might be interpreted to mean that fragmentation occurred to give $(CH_3O)^+$, for which $m/e = 31$. However, when methyl formate-*d* ($DCOOCH_3$) was studied, it was observed that the prominent peak at

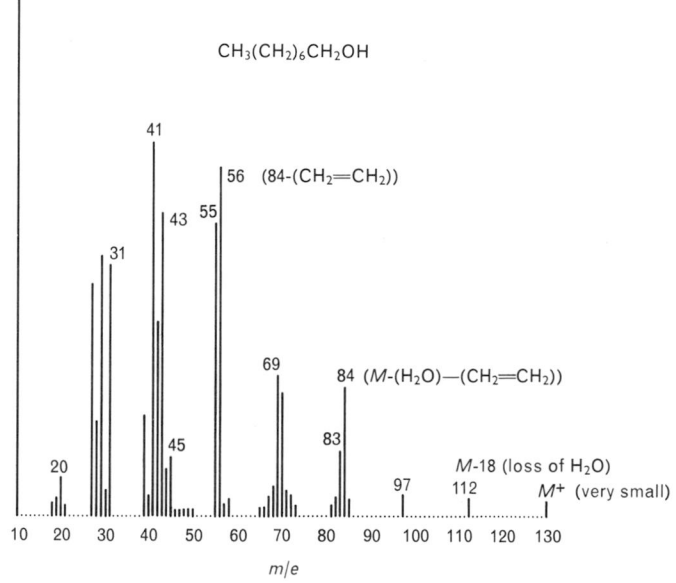

**Figure 30–12.** Mass spectrum of 1-octanol.

$m/e=31$ was replaced by one at $m/e=32$. This showed that the peak at 31 contained the hydrogen of the formyl group, and arises as follows:

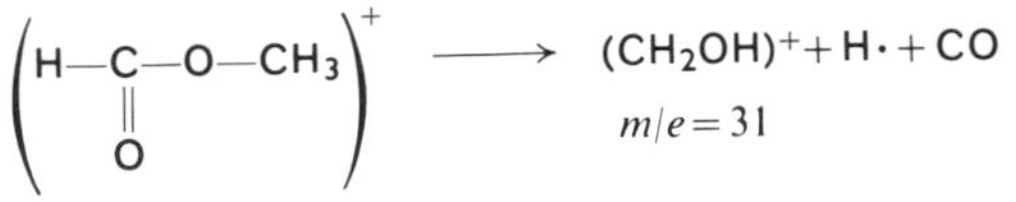

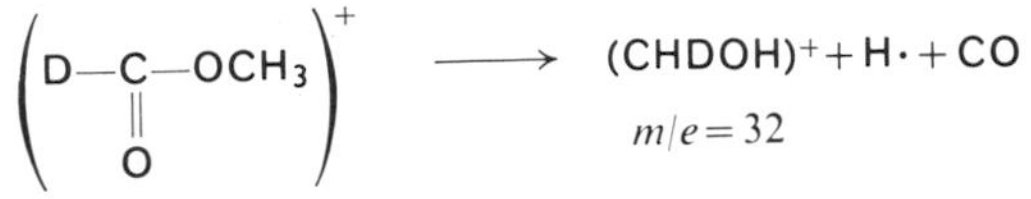

$$\left(H-\underset{\underset{O}{\|}}{C}-O-CH_3\right)^+ \longrightarrow (CH_2OH)^+ + H\cdot + CO$$
$$m/e=31$$

$$\left(D-\underset{\underset{O}{\|}}{C}-OCH_3\right)^+ \longrightarrow (CHDOH)^+ + H\cdot + CO$$
$$m/e=32$$

The mass spectrum of ethyl formate again shows the peak at $m/e=31$, formed by the loss of $CH_3\cdot$ instead of $H\cdot$,

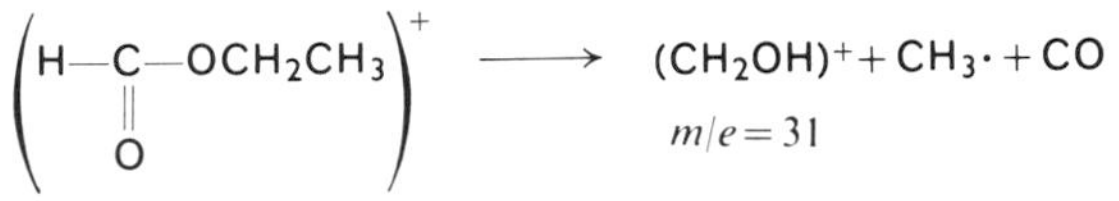

$$\left(H-\underset{\underset{O}{\|}}{C}-OCH_2CH_3\right)^+ \longrightarrow (CH_2OH)^+ + CH_3\cdot + CO$$
$$m/e=31$$

and isopropyl formate yields the corresponding fragment of $m/e=45$:

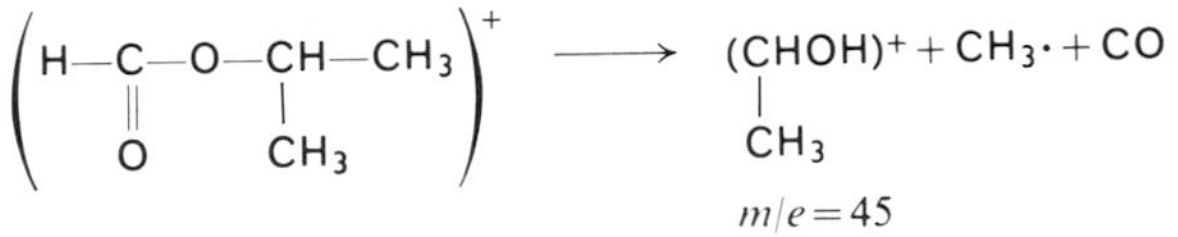

$$\left(H-\underset{\underset{O}{\|}}{C}-O-\underset{\underset{CH_3}{|}}{CH}-CH_3\right)^+ \longrightarrow (\underset{\underset{CH_3}{|}}{CHOH})^+ + CH_3\cdot + CO$$
$$m/e=45$$

An example of the interpretation of the mass spectrum of a complex organic compound is found in the mass spectrum of N,N-dimethyltryptophane methyl ester. Prominent ion species are seen at $m/e=246$ ($M^+$), 187, 144, 130 and 116. These are accounted for the following fissions:

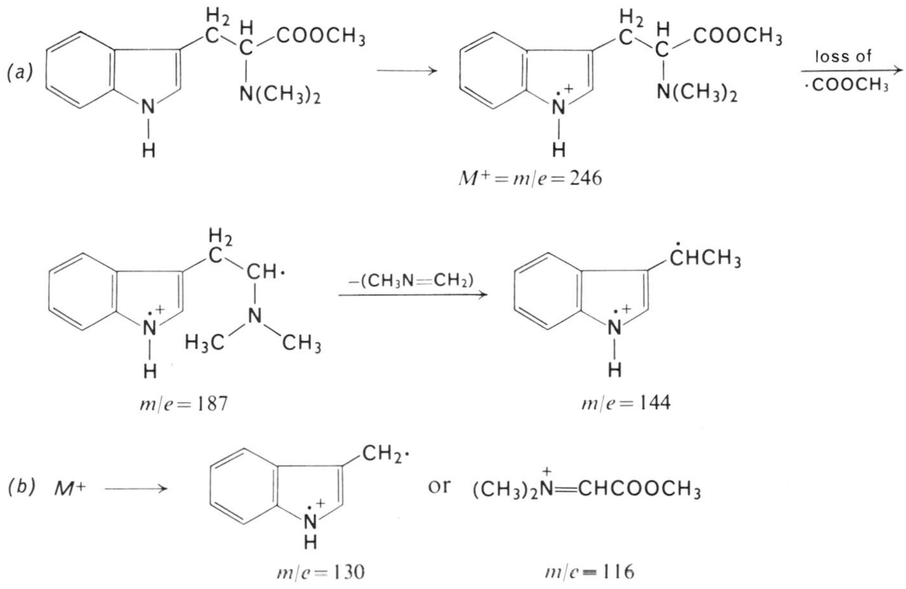

(a)

$$M^+ = m/e = 246$$

loss of
$\cdot COOCH_3$

$$m/e=187$$

$-(CH_3N=CH_2)$

$$m/e=144$$

(b) $M^+ \longrightarrow$

$$m/e=130$$

or     $(CH_3)_2\overset{+}{N}=CHCOOCH_3$

$$m/e=116$$

The application of mass spectrometry to the elucidation of organic structures is rapidly assuming an importance comparable to that of nuclear magnetic resonance spectrometry. An extensive literature, including a number of excellent monographs, on the interpretation of mass spectra is bringing the technique within the purview of most organic chemists.

# Chapter thirty-one / Organic Compounds of Biological Importance. Primary Metabolic Processes

## 31.1 The living cell

Although a great deal of the subject matter of organic chemistry is concerned with the chemical activities of living organisms, much of this deals with organic compounds that are the products of cellular activity. The pigments, sugars, oils, alkaloids, terpenes, and sterols produced by plants and animals are synthetic end-products of metabolic processes. Of even more fundamental interest and importance are the structures of the living matter itself that constitutes the cellular material of plant and animal organisms, and a relatively limited number of chemical reactions that are fundamental to the metabolic and synthetic activities of the organism.

Most of the chemical reactions that take place in living organisms take place within the cells of which the plant or animal is largely composed. Although cells are differentiated into those characteristic of the various tissues and organs, certain of the chemical activities of cells are characterized by universal features that are common to the cells of all living organisms. As we shall see in this and the next chapter, the thousands of organic compounds found in plants, animals, microorganisms, and fungi are formed from a relatively few compounds, of rather simple structure. By genetically controlled, enzymatically catalyzed reactions, many of which are oxidation, reduction, and condensation reactions of extraordinary simplicity, this limited pool of starting materials is converted into a wide range of compounds of complex structures.

The living cell, the site of these synthetic activities, is composed of a cell membrane that surrounds a viscid, aqueous material called the *cytoplasm*, in which is found the *nucleus* and a number of smaller bodies which play the most important roles in cellular chemical reactions. The nucleus contains the hereditary material of the organism, the chromosomes, and is composed largely of *nucleoprotein*. The cytoplasmic material is largely *protein*, and contains a large number of specialized protein molecules called *enzymes*, the catalysts of the chemical reactions of cellular metabolism.

In this chapter we shall describe the chemical nature of proteins, of the special kinds of proteins called enzymes, and of the *nucleic acids*, components of nucleoproteins.

## 31.2  Proteins

Proteins are long-chain polyamides, derived from $\alpha$-amino acids, combined in peptide linkages at the amine and carboxyl groupings, and giving rise to the constituent amino acids on hydrolysis:

$$\cdots CO-\underset{\substack{| \\ \text{R}}}{CH}-NH-CO-\underset{\substack{| \\ \text{R}'}}{CH}-NH-CO-\underset{\substack{| \\ \text{R}''}}{CH}-NH-CO-\underset{\substack{| \\ \text{R}'''}}{CH}-NH\cdots$$

$\underbrace{\phantom{\cdots CO-CH-NH}}_{\alpha\text{-amino acid}}$ $\underbrace{\phantom{-CO-CH-NH-}}_{\text{peptide linkage}}$

Proteins are organic compounds of very large molecular size. Although no molecular weight can be said to be characteristic, proteins are known that range from about 12,000 to several million in molecular weight; that is, contain from about 100 to 50,000 individual amino acids.

The details of the structure of a protein can be ascertained only by analyzing the protein for its constituent amino acids, which are obtained by complete hydrolysis. Proteins may be hydrolyzed by acid which causes complete rupture of the peptide linkages:

$$\cdots NH\underset{\substack{| \\ \text{R}}}{CH}CONH\underset{\substack{| \\ \text{R}'}}{CH}CONH\underset{\substack{| \\ \text{R}''}}{CH}CO\cdots \xrightarrow[\text{H}_2\text{O}]{\text{HCl}} \underset{\substack{| \\ \text{NH}_2}}{R}CHCOOH + \underset{\substack{| \\ \text{NH}_2}}{R'}CHCOOH + \cdots$$

The amino acid mixture that is formed can be separated into its components in various ways: by the use of specific precipitants, by fractional crystallization (of very limited usefulness), or by chromatographic methods. Analysis of many proteins, from both plant and animal sources, has shown that all are made up of a surprisingly small number of amino acids: a total of about twenty-five to thirty* are known at the present time. Not every protein contains all of the known amino

---

\* New amino acids are discovered from time to time, and the list of the natural amino acids is being added to slowly. In addition, more than fifty amino acids, not combined in proteins, have been found in plants.

acids; and some proteins contain very high proportions of certain ones. For example, silk fibroin contains 43.6% of glycine and 2.2% of glutamic acid; egg albumin contains 3.1% of glycine and 16.5% of glutamic acid; and gelatin contains 25.9% of glycine and 11.2% of glutamic acid.

## 31.3 Amino acids

The common amino acids derived from proteins by hydrolysis are listed in Table 31-1.

The natural amino acids derived by the hydrolysis of proteins are optically active* (except for glycine) and possess configurations that are related to L-glyceraldehyde; thus, they are members of the L-series:

$$
\begin{array}{cc}
\text{CHO} & \text{COOH} \\
\text{HO}-\!\!\!\!-\text{H} & \text{H}_2\text{N}-\!\!\!\!-\text{H} \\
\text{CH}_2\text{OH} & \text{CH}_2\text{SH}
\end{array}
$$

L-( − )-glyceraldehyde    L-( + )-cysteine

The configurational interrelationships of the natural amino acids have been established by numerous series of interconversions, e.g.,

$$
\begin{array}{ccc}
\text{COOH} & \text{COOH} & \text{COOH} \\
\text{H}_2\text{N}-\!\!\!\!-\text{H} \xrightarrow{\text{PCl}_5} & \text{H}_2\text{N}-\!\!\!\!-\text{H} \xrightarrow{\text{H}_2\,\text{Pd}} & \text{H}_2\text{N}-\!\!\!\!-\text{H} \\
\text{CH}_2\text{OH} & \text{CH}_2\text{Cl} & \text{CH}_3
\end{array}
$$

L-( − )-serine        L-( + )-alanine    (1)

NaSH ↙        NH₃ ↓

$$
\begin{array}{ccc}
\text{COOH} & \text{COOH} & \text{COOH} \\
\text{H}_2\text{N}-\!\!\!\!-\text{H} & \text{H}_2\text{N}-\!\!\!\!-\text{H} \xleftarrow[\substack{\text{(Hofmann}\\\text{degr.)}}]{\text{NaOBr/OH}^-} & \text{H}_2\text{N}-\!\!\!\!-\text{H} \\
\text{CH}_2\text{SH} & \text{CH}_2\text{NH}_2 & \text{CH}_2\text{CONH}_2
\end{array}
$$

L-( + )-cysteine        L-( − )-asparagine

D-Amino acids are also found in nature, but not as constituents of normal proteins. Some antibiotics contain D-amino acids as part of their structures: e.g., penicillamine ($\beta,\beta$-dimethylcysteine) has the D-configuration.

## 31.4 Isoelectric point

The structures of the amino acids given in Table 31-1 are shown with —NH₂ and —COOH groups. This is a formal representation, for in fact amino acids exist in various states of ionization, depending upon the $p$H.

---

* Under some conditions of hydrolysis (±)-amino acids may be formed as a result of racemization by the acid or alkali used.

**Table 31-1.** *The common amino acids derived from proteins*

| R in R—CH—COOH<br>       &#124;<br>       $NH_2$ | Name of Amino Acid |
|---|---|
| H | glycine |
| $CH_3$ | alanine |
| $(CH_3)_2CH$ | valine |
| $(CH_3)_2CHCH_2$ | leucine |
| $CH_3CH_2CH$<br>     &#124;<br>     $CH_3$ | isoleucine |
| $HOCH_2$ | serine |
| $CH_3CH$<br> &#124;<br> $OH$ | threonine |
| $HSCH_2$ | cysteine |
| $HOOCCHCH_2S—SCH_2$<br>    &#124;<br>    $NH_2$ | cystine |
| $CH_3SCH_2CH_2$ | methionine |
| $HOOCCH_2$ | aspartic acid |
| $H_2NCOCH_2$ | asparagine (the amide of aspartic acid) |
| $HOOCCH_2CH_2$ | glutamic acid |
| $H_2NCOCH_2CH_2$ | glutamine (the amide of glutamic acid) |
| $H_2NCH_2CH_2CH_2CH_2$ | lysine |
| $H_2NCH_2CHCH_2CH_2$<br>      &#124;<br>      $OH$ | hydroxylysine |
| $H_2N—C—NHCH_2CH_2CH_2$<br>     &#124;&#124;<br>     $NH$ | arginine |

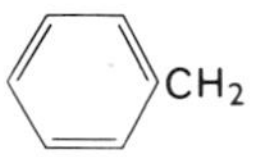

phenylalanine

tyrosine

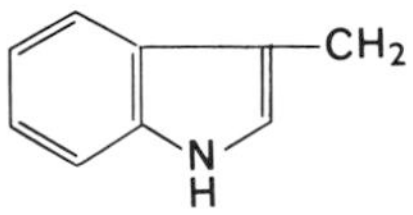

tryptophane

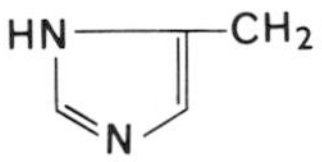

histidine

In addition to these:

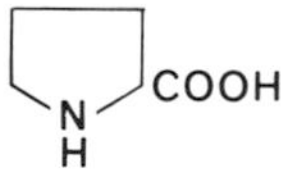

proline

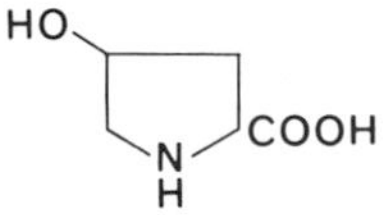

hydroxyproline

A monoamino monocarboxylic acid, such as glycine, will exist as the ion A at low $pH$, as the ion C at high $pH$, and as the dipolar ion B, called a *zwitterion*, at intermediate $pH$:

$$^+H_3NCH_2COOH \underset{H^+}{\overset{OH^-}{\rightleftarrows}} {}^+H_3NCH_2COO^- \underset{H^+}{\overset{OH^-}{\rightleftarrows}} H_2NCH_2COO^- \tag{2}$$

low $pH$ (ca. 1)          $pH$ 6          high $pH$ (ca. 11)

    A                 B                  C

The $pH$ at which the amino acid exists essentially entirely as the zwitterion is called the *isoelectric point*. Basic amino acids, such as lysine, have isoelectric points at high $pH$ values (for lysine, 9.74); acidic amino acids, such as glutamic acid, have isoelectric points at low $pH$ values (for glutamic acid, 3.22). The monoamino monocarboxylic acids, such as glycine, alanine, and leucine, have isoelectric points in the range 5.5 to 6.5 (for alanine, 6.00).

---

**Exercise 1.**    Why is the isoelectric point of lysine at a higher $pH$ value than that of alanine?

---

The dipolar character of amino acids in the solid state is indicated by the high melting points of the crystalline compounds. Their properties resemble those of inorganic salts (although their melting points are not as high as those of typical inorganic salts), and it may be inferred that crystalline amino acids possess solid structures in which the doubly charged ions are strongly oriented and form stable crystal lattices.

## 31.5   Chemical behavior of amino acids

Many of the reactions of amino acids are those to be anticipated from their structures, and need not be formulated here in detail. The amino group can be acylated and alkylated to yield the corresponding N-substituted amino acids. The reaction of the amino group with nitrous acid is of special importance because it forms the basis for an analytical method used for the quantitative estimation of amino acids (the Van Slyke method). The reaction, which proceeds as in the following equation, results in the evolution of one mole of nitrogen per mole of primary amino groups

$$R-CH-COOH + HNO_2 \longrightarrow R-CH-COOH + N_2 + H_2O \tag{3}$$
$$\underset{NH_2}{|} \qquad\qquad\qquad \underset{OH}{|}$$

Measurement of the nitrogen in an accurately calibrated burette permits the calculation of the "amino nitrogen" in a sample of an amino acid or a polypeptide.

A characteristic and useful property of amino acids is their reaction with ninhydrin (triketohydrindene hydrate), which results in the formation of carbon dioxide, ammonia, and an aldehyde that contains one less carbon atom than the amino acid. The reaction is also accompanied by the formation of an intense blue or purple color; this is useful for the colorimetric estimation of amino acids and for their qualitative detection in solution or on paper chromatograms. The reaction of ninhydrin with $\alpha$-amino acids proceeds as follows:

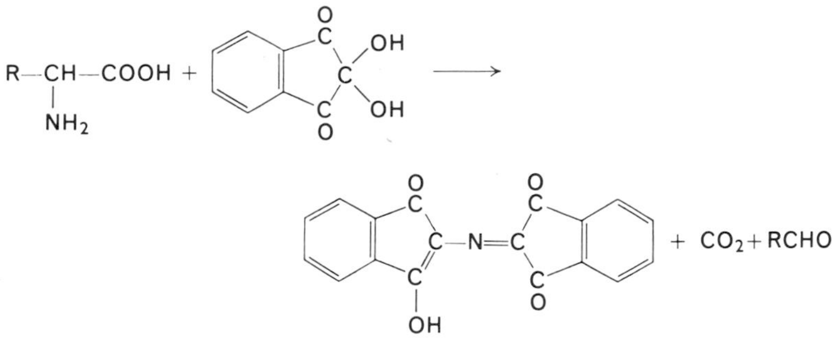

$$\text{R—CH—COOH} + \text{phthalic anhydride} \longrightarrow$$

$$\text{(diketopiperazine-type structure)} + CO_2 + RCHO \quad (4)$$

The carboxyl group of the amino acids can be esterified or, after protection of the amino group by acylation, converted into the acyl halide. The use of amino acid halides in peptide synthesis will be discussed in a later section.

## 31.6 Synthesis of amino acids

Because of the importance of amino acids in biochemistry and medicine, a great deal of research has been devoted to the development of methods for their synthesis. Most commercially produced amino acids are prepared synthetically, although a few [such as $(-)$-glutamic acid from wheat gluten or soy beans] are still isolated from natural sources.

There are several methods of synthesis that are of general application.

*The amination of α-halogen acids.* The preparation of the required halogenated acids can be accomplished by the Hell-Volhard-Zelinsky method (Section 13.12), in which the acid is brominated in the presence of a small amount of phosphorus tribromide. The bromination actually takes place on the acyl bromide:

$$RCH_2COOH + PBr_3 \longrightarrow RCH_2COBr \quad (5)$$

$$RCH_2COBr + Br_2 \longrightarrow \underset{\underset{Br}{|}}{R}CHCOBr \quad (6)$$

Subsequent acid-acyl bromide interchange permits the bromination to continue until all of the original acid is converted into the α-bromo derivative:

$$\underset{\underset{Br}{|}}{R}CHCOBr + RCH_2COOH \rightleftharpoons \underset{\underset{Br}{|}}{R}CHCOOH + RCH_2COBr \quad (7)$$

The reaction of the α-bromo acid (α-chloro acids may also be used) with ammonia yields the α-amino acid:

$$\underset{\underset{Br}{|}}{R}CHCOOH \xrightarrow{NH_3} \underset{\underset{NH_2}{|}}{R}CHCOOH + NH_4Br \quad (8)$$

In an alternative method, in which the α-halogen acid is also used (as the sodium or potassium salt), potassium phthalimide is used in place of ammonia to give the N-phthalimido derivative. This is hydrolyzed by aqueous acid to yield the amino acid and phthalic acid. This procedure is of general applicability to the synthesis of amines, and is known as the *Gabriel synthesis* of amines:

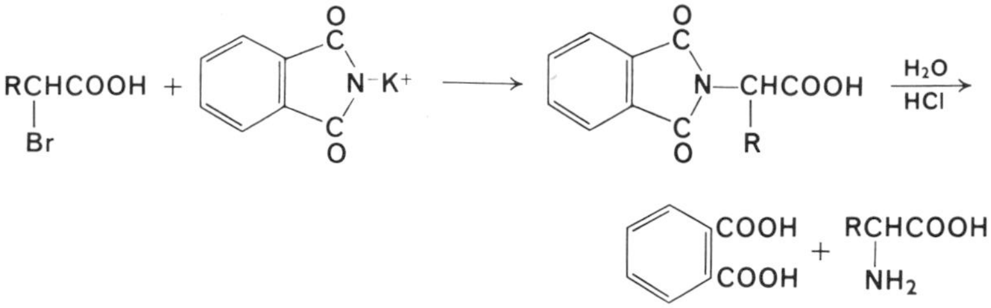

$$\begin{array}{c} \text{COOH} \\ \text{COOH} \end{array} + \begin{array}{c} \text{RCHCOOH} \\ | \\ \text{NH}_2 \end{array} \qquad (9)$$

***The Strecker synthesis.*** The addition of HCN and ammonia to an aldehyde yields the $\alpha$-amino nitrile. This can be hydrolyzed to the amino acid. A synthesis of methionine is shown as an illustration of the use of this procedure:

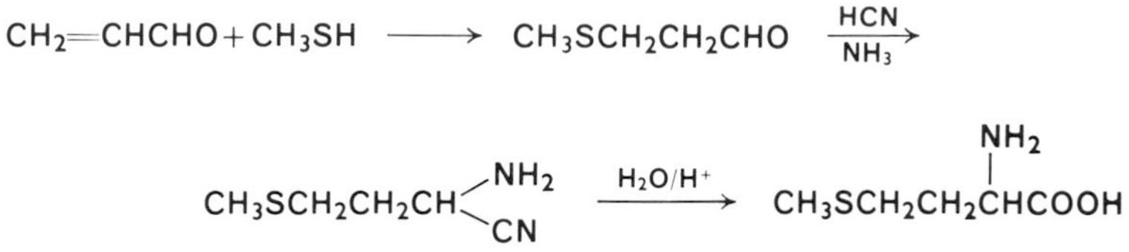

$$CH_2{=}CHCHO + CH_3SH \longrightarrow CH_3SCH_2CH_2CHO \xrightarrow[NH_3]{HCN}$$

$$CH_3SCH_2CH_2CH\begin{array}{c} \nearrow NH_2 \\ \searrow CN \end{array} \xrightarrow{H_2O/H^+} CH_3SCH_2CH_2\overset{\overset{\displaystyle NH_2}{|}}{C}HCOOH \qquad (10)$$

Methionine is prepared commercially in large quantities for use as a poultry feed additive.

***The azlactone synthesis.*** Hippuric acid (N-benzoylglycine) is converted by heating with acetic anhydride into the *azlactone*. This has an active methylene group, and can condense with an aromatic aldehyde to yield a benzal derivative. Reduction and subsequent hydrolysis of the benzal derivative yields the amino acid. The following preparation of tyrosine is an example of the use of this synthetic method:

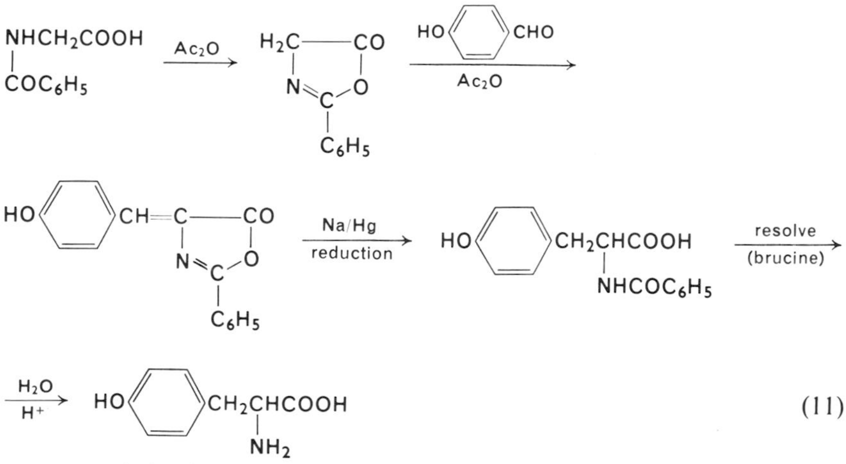

$$\begin{array}{c} \text{HO} \end{array}\!\!\!\!\!\text{CH}_2\text{CHCOOH} \qquad (11)$$
$$\begin{array}{c} | \\ \text{NH}_2 \end{array}$$
$$(+) \text{ and } (-) \text{ tyrosine}$$

***The α-acetamidomalonic ester synthesis.*** The N-acetyl derivative of α-amino-malonic ester has an active α-hydrogen atom, and can undergo alkylation, just as can malonic ester itself. This alkylation reaction provides a very versatile means of preparing α-amino acids, for the resulting malonic ester derivative can be hydro-lyzed and decarboxylated smoothly to the amino acid. The synthesis of aspartic acid is shown as an example of the method:

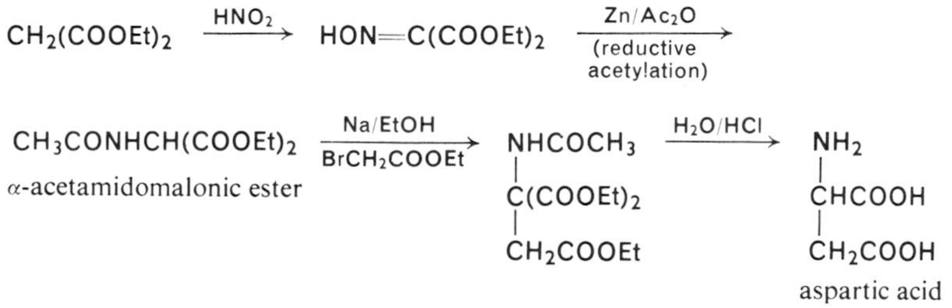

$$CH_2(COOEt)_2 \xrightarrow{HNO_2} HON=C(COOEt)_2 \xrightarrow[\substack{(reductive \\ acetylation)}]{Zn/Ac_2O}$$

$$CH_3CONHCH(COOEt)_2 \xrightarrow[BrCH_2COOEt]{Na/EtOH} \begin{array}{c} NHCOCH_3 \\ | \\ C(COOEt)_2 \\ | \\ CH_2COOEt \end{array} \xrightarrow{H_2O/HCl} \begin{array}{c} NH_2 \\ | \\ CHCOOH \\ | \\ CH_2COOH \end{array} \quad (12)$$

α-acetamidomalonic ester

aspartic acid

***Special methods of synthesis.*** Various special syntheses have been devised for many of the natural amino acids. As one example, the synthesis of proline has been carried out as follows:

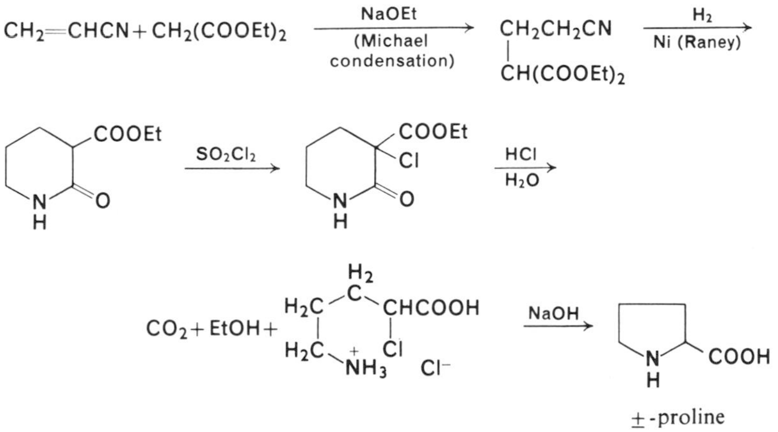

$$CH_2=CHCN + CH_2(COOEt)_2 \xrightarrow[\substack{(Michael \\ condensation)}]{NaOEt} \begin{array}{c} CH_2CH_2CN \\ | \\ CH(COOEt)_2 \end{array} \xrightarrow{H_2 \atop Ni\ (Raney)}$$

$$CO_2 + EtOH + \begin{array}{c} H_2 \\ H_2C^{-C}\diagdown CHCOOH \\ | \quad\quad | \\ H_2C_{\diagdown \underset{+}{N}H_3} \ Cl \\ \quad\quad Cl^- \end{array} \xrightarrow{NaOH} \begin{array}{c} \\ \\ \end{array} \quad (13)$$

±-proline

---

**Exercise 2.** In some of the examples that have been given above, certain steps have not been described in detail. Study these syntheses with care until the individual steps are all clearly understood. All of them find their counterparts in earlier parts of the book.

---

It is to be emphasized that these synthetic methods provide the racemic (±)-amino acids. The preparation of the natural (L-) amino acids requires a reso-lution at some stage in the synthesis. One such resolution step is indicated above in the preparation of tyrosine.

### 31.7 Amino acid sequence in proteins and other polypeptides*

After the composition of a protein has been established by hydrolysis and the subsequent separation and estimation of the constituent amino acids, a most important question remains: In what order are the amino acids arranged to form the polypeptide chain? An answer to this question is clearly essential to an understanding of protein structure, because of the enormous number of ways in which a given number of amino acids can be arranged in a protein molecule. The difficulties in solving this problem are very great indeed, but advances have been made toward a solution, and the complete sequences of amino acids in a number of polypeptides and proteins have been worked out. One successful approach makes use of 2,4-dinitrofluorobenzene (DNFB), which reacts with free amino groups of a protein or polypeptide to yield N-2,4-dinitrophenyl (DNP) derivatives:

$$
\begin{array}{c}
\overset{\displaystyle R}{\underset{\displaystyle |}{}} \\
\text{------ protein chain ------COCH—NH}_2 \\
\text{(terminal —NH}_2\text{ residue)} \\
\downarrow \text{DNFB} \\
\overset{\displaystyle R}{\underset{\displaystyle |}{}} \\
\text{------ protein chain ------COCH—NH}\!\!\left\langle\bigcirc\right\rangle\!\!\text{NO}_2 \\
\text{O}_2\text{N}
\end{array}
\qquad (14)
$$

DNP-derivative

Incomplete hydrolysis of the protein and separation of the fragments containing the 2,4-dinitrophenyl group leads to the isolation of N-2,4-dinitrophenyl di-, tri-, ... peptides, the identification of which defines a portion of the protein chain. Suppose, for example, we consider the simple model pentapeptide glycyl-alanyl-valyl-leucyl-glycine:

$$
\begin{array}{cccc}
& \overset{\text{H}_3\text{C}\diagdown \diagup \text{CH}_3}{} & & \\
\text{CH}_3 & \text{CH} & \text{CH}_2\text{CH(CH}_3)_2 & \\
| & | & | & \\
\text{NH}_2\text{CH}_2\text{CONHCHCONHCHCONHCHCONHCH}_2\text{COOH}
\end{array}
$$

$$
\text{Gl} \text{——}\!\!+\!\!\text{——} \text{Al} \text{——}\!\!+\!\!\text{——} \text{Val} \text{——}\!\!+\!\!\text{——} \text{Leu} \text{——}\!\!+\!\!\text{——} \text{Gl}
$$

Treatment of this pentapeptide with DNFB would yield DNP-Gl-Al-Val-Leu-Gl, partial hydrolysis of which would give DNP-Gl, DNP-Gl-Al, DNP-Gl-Al-Val, as well as Al-Val, Val-Leu, Al-Val-Leu, Val-Leu-Gl, Leu-Gl, and so on. DNP-Gl-Al-Val can be identified by hydrolysis, to give DNP-Gl and a dipeptide that must be

---

* No sharp distinction can be made between proteins and high-molecular-weight polypeptides, but polypeptides of molecular weights of the order of up to several thousand are not ordinarily called proteins.

Al-Val or Val-Al; and by treatment of this dipeptide with DNFB, followed by hydrolysis, which will yield DNP-Al and Val, not DNP-Val and Al. Experiments of this kind give information from which the complete structures of polypeptides can be worked out.

As the size of the popypeptide increases, a great deal more effort is required to work out the complete sequence of amino acids. Insulin, the antidiabetic hormone, a "small" protein or "large" polypeptide of molecular weight 6000, consists of two separate peptide chains, joined by —S—S— linkages. Application of the DNFB method has made it possible to work out the complete structure. One of the chains has the composition. Glycyl-isoleucyl-valyl-glutamyl-glutaminyl-cysteyl-cysteyl-alanyl-seryl-valyl-cysteyl-seryl-leucyl-tyrosyl-glutaminyl-leucyl-glutamyl-asparaginyl-tyrosyl-cysteyl-asparagine.

## 31.8 Synthesis of peptides

The goal of the many studies that have been directed toward the synthesis of chains of peptide-linked amino acids is the development of methods of preparation of polypeptides that are versatile and adaptable enough to permit the synthesis of these compounds with any number and sequence of amino acids. Ultimately, such synthetic methods will lead to the preparation of the natural proteins; they have already permitted the synthesis of important low-molecular-weight polypeptides of great biological importance.

In the sixty years that have elapsed since Emil Fischer first began work on this problem, a great many methods have been devised for the linking of amino acids together by peptide bonds. Some of these are elaborations of the basic methods first discovered by Fischer; others make use of new and original procedures. In the following discussion only a few of the general types of synthesis will be described.

The requirements for a completely satisfactory synthetic procedure are stringent: the amino acid residues must retain their optical asymmetry, for the natural polypeptides are optically active; it must be possible to add amino acid units in a predetermined order; and the synthesis should permit the extension of the polypeptide chain to yield high-molecular-weight compounds. No completely acceptable methods to achieve all of these aims are known at the present time, although a considerable degree of success has been achieved.

The central problem in peptide synthesis is the protection of the amino group of an amino acid or of a peptide in order that its carboxyl group can condense with the amino group of another amino acid, and not with its own. Coupled with this is the problem of removing the protecting group without hydrolyzing the peptide linkage.

Early attempts to use the N-benzoyl and N-ethoxycarbonyl (—NHCOOEt) groups did not prove satisfactory, for attempts to remove these acyl residues usually resulted in hydrolysis of the peptide. A notable advance was made by M. Bergmann, who introduced the use of the reagent benzyl chlorocarbonate.

This compound is a half-ester, half-acid chloride of carbonic acid, and is prepared by the reaction between benzyl alcohol and phosgene:

$$\text{C}_6\text{H}_5\text{CH}_2\text{OH} + \text{ClCOCl} \longrightarrow \text{C}_6\text{H}_5\text{CH}_2\text{OCOCl} + \text{HCl} \tag{15}$$

Benzyl chlorocarbonate, an acid chloride, can acylate an amino group in the usual way; for example, with glycine:

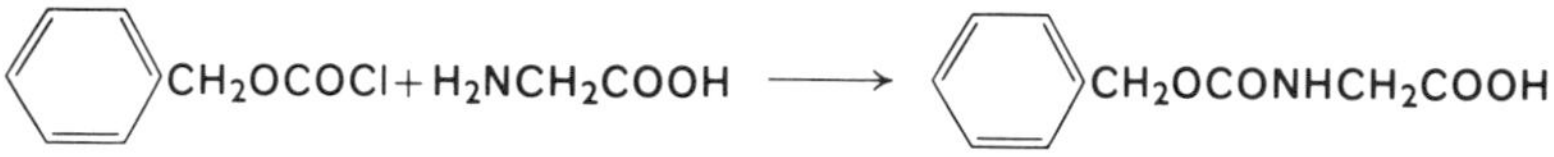

$$\text{C}_6\text{H}_5\text{CH}_2\text{OCOCl} + \text{H}_2\text{NCH}_2\text{COOH} \longrightarrow \text{C}_6\text{H}_5\text{CH}_2\text{OCONHCH}_2\text{COOH} \tag{16}$$

The "protected" glycine can now be converted into the acid chloride by reaction with thionyl chloride:

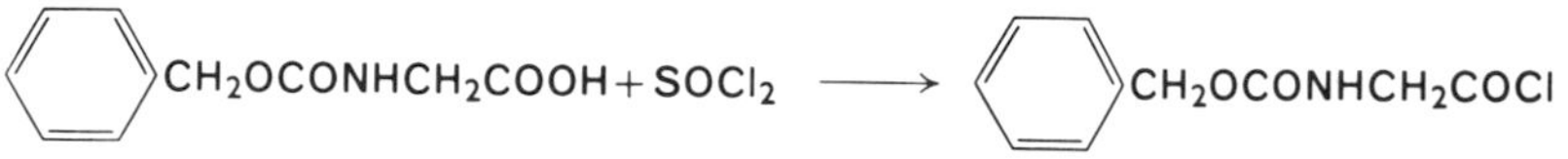

$$\text{C}_6\text{H}_5\text{CH}_2\text{OCONHCH}_2\text{COOH} + \text{SOCl}_2 \longrightarrow \text{C}_6\text{H}_5\text{CH}_2\text{OCONHCH}_2\text{COCl} \tag{17}$$

and the acylamino acid chloride can react with another amino acid (for example, phenylalanine) to yield a dipeptide:

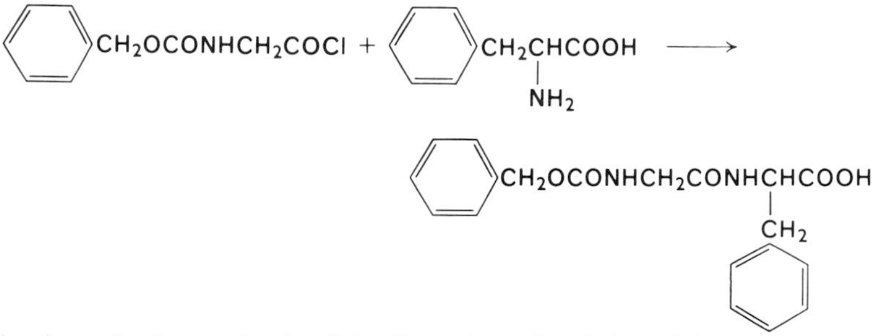

$$\text{C}_6\text{H}_5\text{CH}_2\text{OCONHCH}_2\text{COCl} + \underset{\underset{\text{NH}_2}{|}}{\text{C}_6\text{H}_5\text{CH}_2\text{CHCOOH}} \longrightarrow$$

$$\underset{\underset{\underset{\text{C}_6\text{H}_5}{|}}{\text{CH}_2}}{\text{C}_6\text{H}_5\text{CH}_2\text{OCONHCH}_2\text{CONHCHCOOH}} \tag{18}$$

The final step in the synthesis of the dipeptide, glycylphenylalanine, is the removal of the "protecting" benzyloxycarbonyl group. This is easily accomplished by catalytic hydrogenolysis in the presence of a palladium catalyst:*

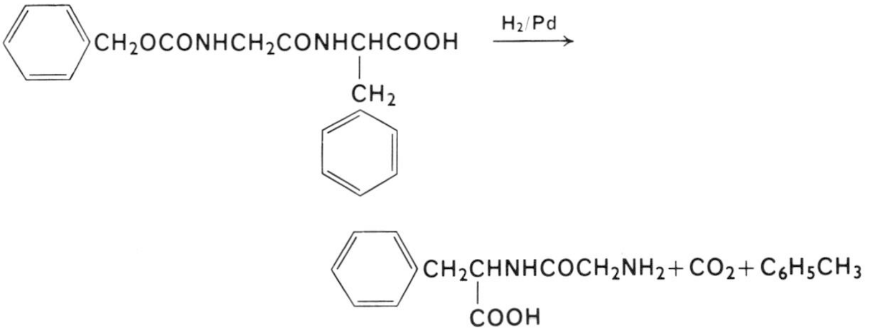

$$\underset{\underset{\underset{\text{C}_6\text{H}_5}{|}}{\text{CH}_2}}{\text{C}_6\text{H}_5\text{CH}_2\text{OCONHCH}_2\text{CONHCHCOOH}} \xrightarrow{\text{H}_2/\text{Pd}}$$

$$\underset{\underset{\text{COOH}}{|}}{\text{C}_6\text{H}_5\text{CH}_2\text{CHNHCOCH}_2\text{NH}_2} + \text{CO}_2 + \text{C}_6\text{H}_5\text{CH}_3 \tag{19}$$

glycylphenylalanine

---

* A group that is now widely used as a blocking group instead of the benzyloxycarbonyl group is the *t*-butyloxycarbonyl group. This is very sensitive to acid decomposition and can be removed under very mild conditions.

Another protecting group that has been found useful is the phthaloyl group, introduced by means of the reaction between an amino acid and phthalic anhydride:

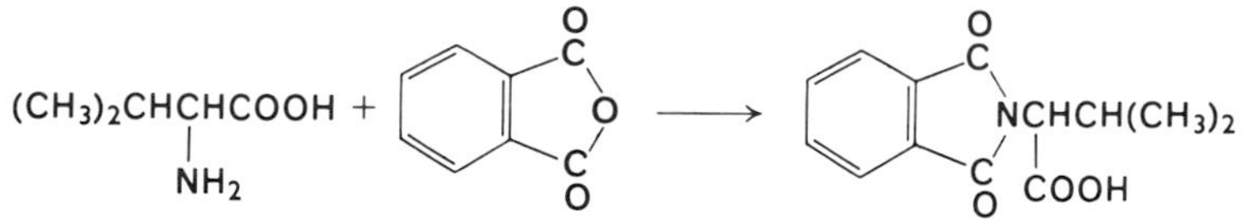

$$(CH_3)_2CHCHCOOH + \underset{\text{valine}}{\underset{NH_2}{|}} \quad \longrightarrow \quad \text{N-phthaloylvaline} \qquad (20)$$

valine

N-phthaloylvaline

The final removal of the phthaloyl group is accomplished by the use of hydrazine, which forms 1,4-dihydroxyphthalazine, with liberation of the amino peptide:

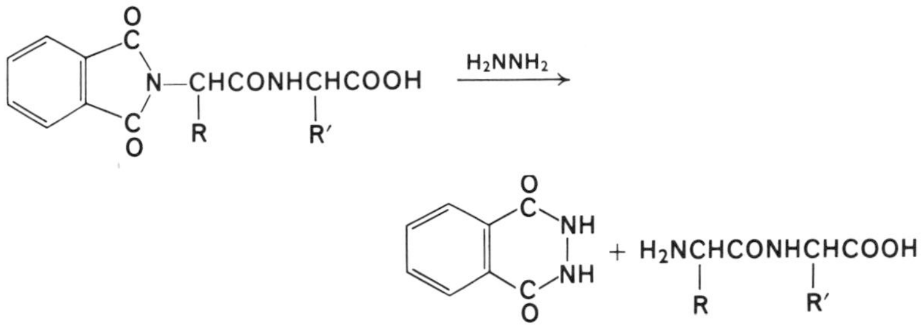

$$N-CHCONHCHCOOH \xrightarrow{\ H_2NNH_2\ }$$

$$+ \ H_2NCHCONHCHCOOH \qquad (21)$$

Polypeptides are named in the manner illustrated by the following examples:

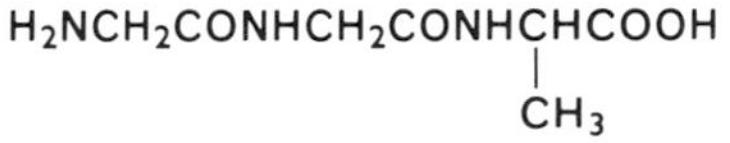

$$H_2NCH_2CONHCH_2CONHCHCOOH$$
$$\underset{CH_3}{|}$$

glycylglycylalanine

$$CH(CH_3)_2$$
$$|$$
$$CH_2$$
$$|$$
$$CH_3CHCONHCH_2CONHCHCONHCHCOOH$$
$$\underset{NH_2}{|} \qquad\qquad\qquad \underset{CH_2COOH}{|}$$

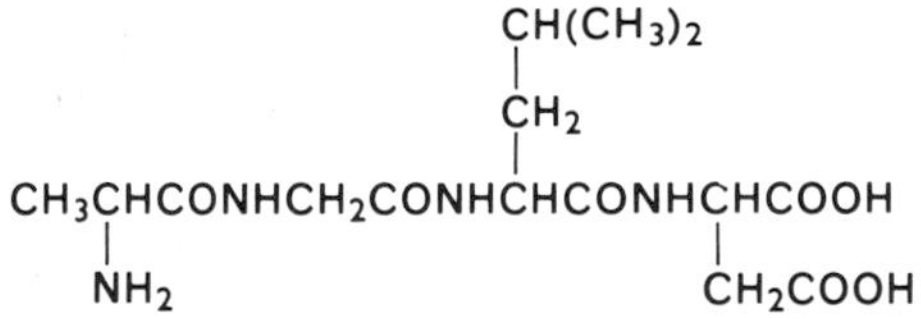

alanylglycylleucylaspartic acid

---

**Exercise 3.** Show all the steps that would be used to synthesize the tetrapeptide, glycylalanyltyrosylvaline.

---

## 31.9 Configuration of protein chains

The description of a protein as a "linear" polypeptide is correct only in the fact that the amino acids are joined in a linear sequence. The actual molecules of most proteins (except for certain fibrous types) form less extended structures, being folded and coiled into compact globular or rounded bundles. The structural features responsible for this folding are chiefly the hydrogen bonds that exist between two sites on a protein chain—for example, between a —NH group and a —C=O group of two different peptide linkages—and the coplanar nature of the individual peptide bonds:

$$\left\{ \underset{H}{\overset{}{>}}\ddot{N}-C\underset{}{\overset{O}{<}} \longleftrightarrow \underset{H}{\overset{}{>}}\overset{+}{N}=C\underset{}{\overset{O^-}{<}} \right\}$$

A consequence of the helical form of a protein chain is that the side chains, i.e., the R groups of $RCH\underset{COOH}{\overset{NH_2}{<}}$ of the amino acid residues, project from the outside of the spiral, thus forming an external "surface" consisting of the aliphatic, aromatic, and heterocyclic residues of the amino acids.

### 31.10  Enzymes

The biological catalysts that control the chemical steps in cellular metabolism are the special class of proteins known as *enzymes*. The study of intermediary metabolism is largely a study of the enzyme systems that are concerned in oxidations, reductions, hydrolyses, and condensation reactions (many of which are of the aldol type). The manner in which foodstuffs are degraded with the production both of energy and of the building blocks of cellular synthesis is now fairly well understood, and many enzymes are known as individual substances, some of them in crystalline form. It must be remembered, however, that the *structures* of proteins are still largely unknown; and since enzymes are proteins, their descriptions are perforce couched chiefly in terms of their chemical behavior and of such gross properties as their molecular weight, amino acid composition (but not sequence), shape, and behavior in an electrostatic field.

Many enzymes possess, in addition to the protein portion, a small fragment of nonprotein nature, often referred to as the *prosthetic group* or *coenzyme*. The best known enzymes of this class are concerned with the breakdown and utilization of glucose, one of the important foodstuffs for living matter of all kinds, from bacteria to man. Since the complex of reactions that are involved in carbohydrate metabolism and its related processes include over a hundred known and describable steps, all of which are enzyme controlled, we shall confine our attention here to a group of coenzymes that we recognize as members of what is called the *vitamin-B complex*. In Chapter 29 was described the biochemical role of one of these, nicotinic acid amide (nicotinamide). Other members of the B vitamins are thiamine ($B_1$), riboflavin ($B_2$), pyridoxine ($B_6$), cyanocobalamine ($B_{12}$), in addition to a number of other substances. Although the exact manner in which the biological role of some of the vitamins is performed is not known with certainty, a great deal is known about the reactions in which glucose is broken down to carbon dioxide and water, with the concomitant liberation of energy and the synthesis of amino acids, fats, sterols, and so forth. In this process several of the B vitamins, acting as the prosthetic groups of specific enzymes, play well-defined roles.

## 31.11  Hydrogen transfer

The role of the pyridine nucleotides in substrate dehydrogenation can be restated briefly with the aid of several examples:

$$CH_3CH_2OH \xrightleftharpoons{\substack{\text{alcohol} \\ \text{dehydrogenase}}} CH_3CHO \qquad (22)$$

$$\text{D-glucose-6-phosphate} \xrightleftharpoons[\substack{\text{dehydrogenase}}]{\substack{\text{Gl-6-Ph-}}} \text{6-phospho-D-gluconic acid} \qquad (23)$$

$$
\begin{array}{l}
\text{COOH} \\
|\\
\text{CHOH} \\
|\\
\text{CHCOOH} \\
|\\
\text{CH}_2\text{COOH}
\end{array}
\quad
\substack{\text{isocitric}\\ \text{dehydrogenase}}
\xrightleftharpoons{\phantom{xxx}}
\begin{array}{l}
\text{COOH} \\
|\\
\text{CO} \\
|\\
\text{CHCOOH} \\
|\\
\text{CH}_2\text{COOH}
\end{array}
\qquad (24)
$$

isocitric acid        oxalosuccinic acid

$$CH_3CHOHCH_2COOH \xrightleftharpoons{\substack{\beta\text{-HBA}\\ \text{dehydrogenase}}} CH_3COCH_2COOH \qquad (25)$$

β-hydroxybutyric acid
(HBA)

In each of these examples the substrate (on the left) is oxidized to yield the product shown. The other product of the reaction is in each case the reduced enzyme, hydrogen transfer taking place to the pyridine ring of the nicotinamide portion of the prosthetic group, or *coenzyme*. Alcohol dehydrogenase and β-HBA dehydrogenase contain NAD; the other two enzymes shown above contain NADP as coenzymes.

## 31.12  Riboflavin

In the reoxidation of NADH or NADPH, the hydrogen (or electrons) are transferred to a prosthetic group found in a number of enzymes called *flavoproteins*, containing in combined form the compound *riboflavin* (vitamin $B_2$):

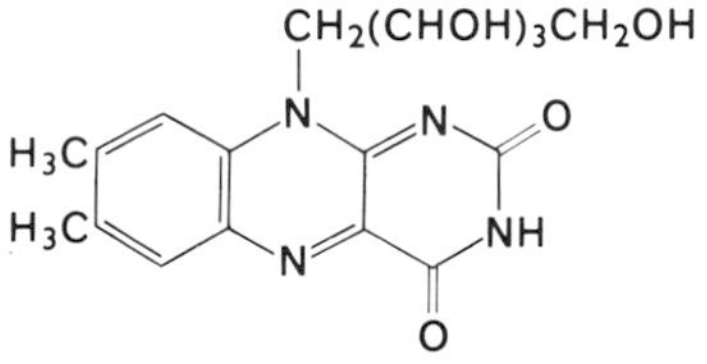

riboflavin

Riboflavin was first isolated from milk, and its structure established, in 1936. In its role as a coenzyme, it exists in a phosphorylated form, often in combination with adenosine as flavin-adenine-diphosphate, also called flavin-adenine-dinucleotide (FAD):

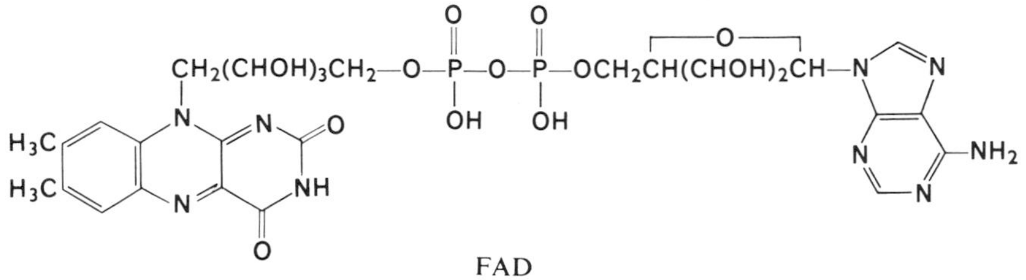

FAD

Riboflavin, the form in which the compound is administered as a vitamin supplement, is synthesized commercially in large quantities. One method of synthesis is the following:

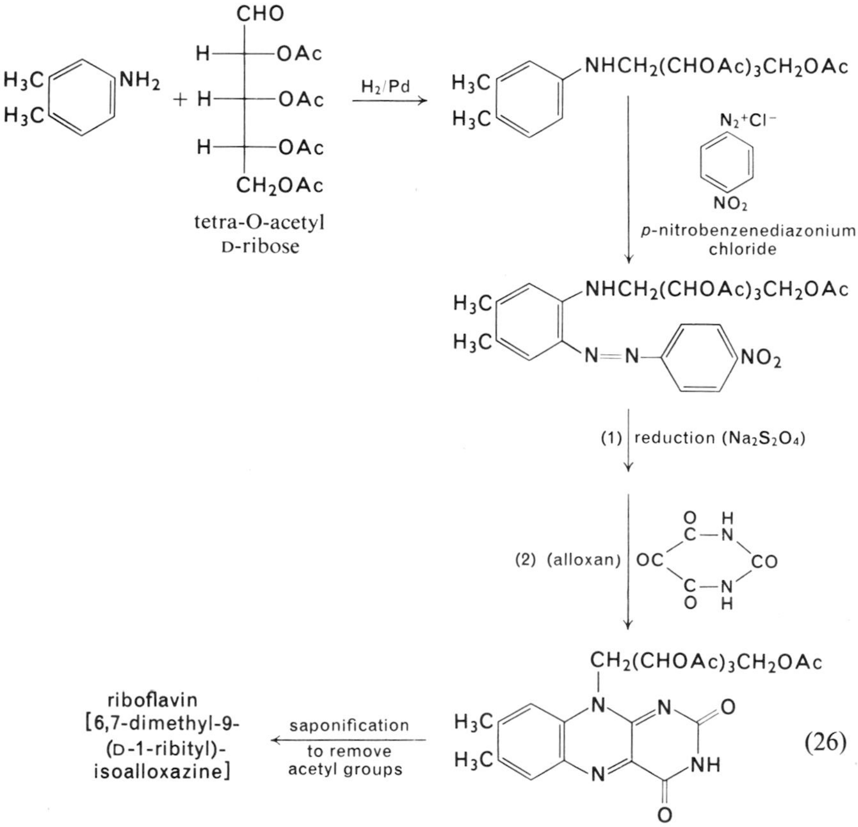

$$(26)$$

The synthetic analog containing two chlorine atoms in place of the two methyl groups [6,7-dichloro-9-(D-1-ribityl)-isoalloxazine] is antagonistic to the action of riboflavin, representing a case of inhibition by a structural analog (see Chapter 33). The 6-methyl, 7-methyl, and 6-ethyl-7-methyl analogs are riboflavin-like in action but are only about half as active.

## 31.13  Thiamin

The vitamin known as $B_1$ (thiamin) is found in enzymes involved in the decarboxylation of $\alpha$-keto acids; thus, it plays an essential role in the oxidation of carbohydrate by way of glycolysis and the citric acid cycle (see Section 31.15). Thiamin deficiency is characterized by a disease known as beriberi and by other disturbances of nervous system activity. The vitamin has the structure

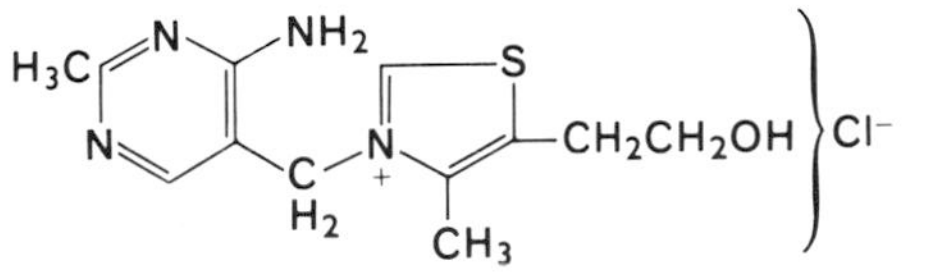

Thiamin plays its biological role in phosphorylated form as thiamine pyrophosphate (cocarboxylase):

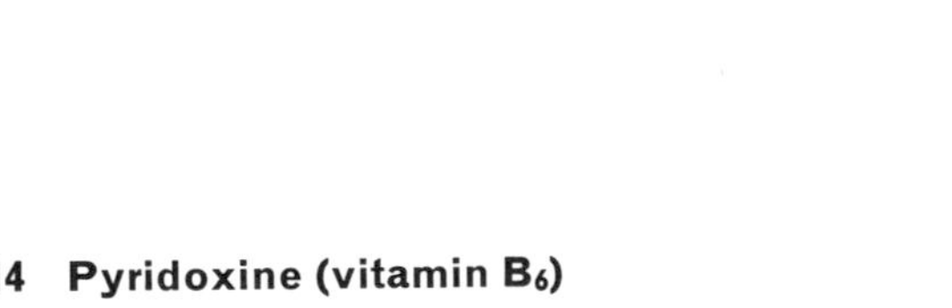

cocarboxylase

## 31.14  Pyridoxine (vitamin B₆)

The coenzymes responsible for the transfer of $-NH_2$ from an $\alpha$-amino acid to an $\alpha$-keto acid are pyridoxine, pyridoxal, and pyridoxamine,

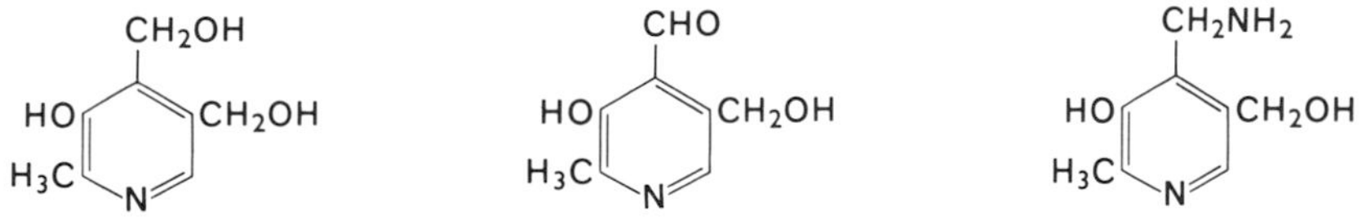

pyridoxine

pyridoxal

pyridoxamine

known collectively as *vitamin B₆*. One of the important functions of the enzymes known as transaminases, which contain pyridoxal-5-phosphate as the coenzyme, is the reaction typified by the general expression

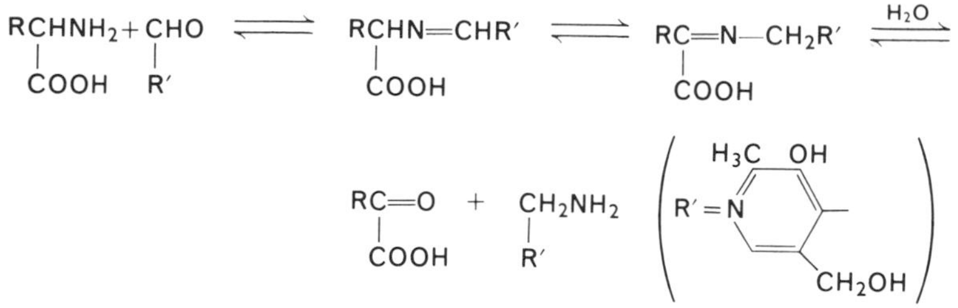

$$\underset{\substack{| \\ \text{COOH}}}{\text{RCHNH}_2} + \underset{\substack{| \\ \text{R}'}}{\text{CHO}} \;\rightleftharpoons\; \underset{\substack{| \\ \text{COOH}}}{\text{RCHN}}{=}\text{CHR}' \;\rightleftharpoons\; \underset{\substack{| \\ \text{COOH}}}{\text{RC}}{=}\text{N}{-}\text{CH}_2\text{R}' \;\xrightarrow{\text{H}_2\text{O}}$$

$$\underset{\substack{| \\ \text{COOH}}}{\text{RC}}{=}\text{O} \;+\; \underset{\substack{| \\ \text{R}'}}{\text{CH}_2\text{NH}_2} \qquad \left( \text{R}'{=}\text{N} \begin{array}{c} \text{H}_3\text{C}\quad\text{OH} \\ \\ \text{CH}_2\text{OH} \end{array} \right) \qquad (27)$$

## 31.15   Glycolysis and the citric acid cycle

The utilization of glucose for the energetic and synthetic requirements of living organisms starts with the formation of the 6-phosphate ester

$$\text{D-glucose} + \text{ATP} \;\xrightarrow{\text{hexokinase}}\; \text{D-glucose-6-phosphate} + \text{ADP} \qquad (28)$$

D-glucose                          D-glucose-6-phosphate

by the phosphorphylating agent, adenosine triphosphate (ATP):

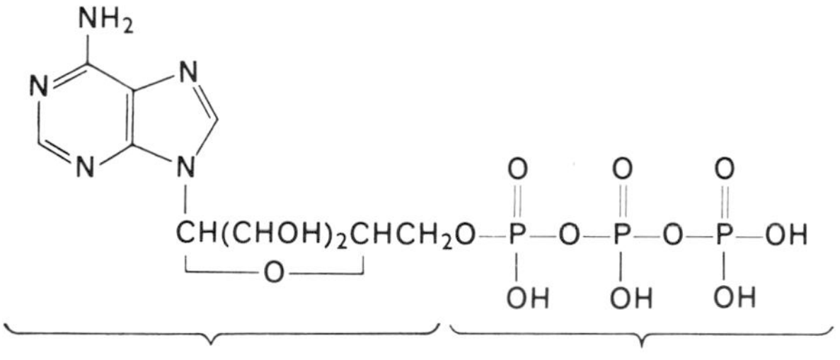

$$\underset{\substack{\underbrace{\qquad\qquad}\\ \text{adenosine}}}{}\;\underset{\substack{\underbrace{\qquad\qquad}\\ \text{triphosphate}}}{} $$

CH(CHOH)₂CHCH₂O—P—O—P—O—P—OH

adenosine            triphosphate

ATP

Subsequent reactions are as follows:*

---

$$\overset{\text{O}}{\underset{\|}{}}$$

* —Ph will be used to represent the —P(OH)₂ grouping. Moreover, although at physiological *p*H values the acidic hydroxyl groups will be ionized, the group will ordinarily be written for convenience as the undissociated structure.

glucose $\xrightarrow{\text{ATP}}$ glucose-6-Ph $\longrightarrow$ fructose-6-Ph

*or* glycogen $\xrightarrow{\text{(via glucose-1-Ph)}}$      $\downarrow$ ATP

fructose-1,6-diPh

$$
\begin{array}{lll}
\text{CH}_2\text{OPh} & & \text{CH}_2\text{OPh} \\
|\ & \longleftrightarrow & | \\
\text{CHOH} & & \text{CO} \\
|\ & & | \\
\text{CHO} & & \text{CH}_2\text{OH}
\end{array}
$$

3-phosphoglyceraldehyde        dihydroxyacetone phosphate

$\xrightarrow[-\text{H}_2]{+\text{H}_2\text{O}} (\text{NAD} \rightarrow \text{NADH})$

$$
\begin{array}{lll}
\text{CH}_2\text{OPh} & & \text{CH}_2\text{OPh} & & \text{CH}_2\text{OPh} \\
| & \xrightarrow{\text{ADP}} & \text{ATP}+ | & \longrightarrow & | & \xrightarrow{\text{H}_2\text{O}} \\
\text{CHOH} & & \text{CHOH} & & \text{CHOPh} \\
| & & \text{COOH} & & \text{COOH} \\
\text{C}{=}\text{O} & & & & \\
\ \ \ \backslash \text{OPh}
\end{array}
$$

$$
\begin{array}{lll}
\text{CH}_2\text{OH} & & \text{CH}_2 & & \text{CH}_3 \\
| & \xrightarrow{-\text{H}_2\text{O}} & \| & \xrightarrow{\text{ADP}} & \text{ATP}+ | \\
\text{CHOPh} & & \text{C}-\text{OPh} & & \text{CO} \\
| & & | & & | \\
\text{COOH} & & \text{COOH} & & \text{COOH}
\end{array} \tag{29}
$$

        phosphoenol       pyruvic acid
      pyruvic acid (PEP)

*Pyruvic acid occupies a central role in glucose breakdown.* In some organisms, e.g., yeast, under anaerobic conditions, the pyruvic acid is decarboxylated to acetaldehyde and the latter reduced to ethanol by an enzyme whose cofactor is reduced NAD:

$$
\text{CH}_3\text{COCOOH} \xrightarrow{-\text{CO}_2} \text{CH}_3\text{CHO} \diagdown \diagup \text{NADH} \tag{30}
$$
$$
\text{CH}_3\text{CH}_2\text{OH} \diagup \diagdown \text{NAD}
$$

alcohol dehydrogenase

In the glycolysis that occurs in muscle under anaerobic conditions, pyruvic acid is reduced to lactic acid.

Under aerobic conditions the degradation proceeds in another way, leading to the oxidation of pyruvic acid and its transformation into fats, amino acids, and other end products of cellular synthetic reactions. The manner in which the transformation of pyruvic acid takes place has been studied in great detail by numerous workers, who have contributed many details to what was first proposed by Krebs as the *citric acid cycle*. Krebs suggested a scheme whereby pyruvic acid is continuously fed into a cyclic process, the end products of which are carbon dioxide and water, and the by-products of which are synthetic intermediates for the elaboration of complex compounds.

The citric acid cycle is represented as follows:

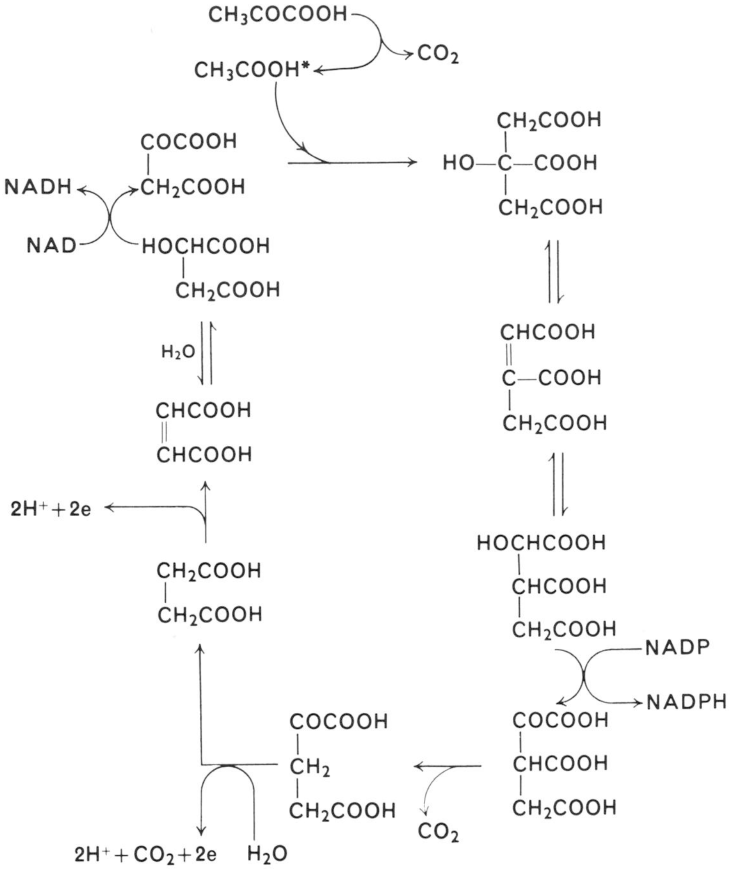

$$(31)$$

Various of the substances found in the citric acid cycle are of importance in the intermediary metabolism of amino acids, fats, steroids, and, in plants, of terpenes, alkaloids, and other secondary plant substances.

Further discussion of the synthetic activities of living organisms in which these low-molecular-weight compounds are utilized will be found in the following chapter.

---

* As acetyl coenzyme A; see next section.

## 31.16 Acetyl coenzyme A

The oxidative decarboxylation of pyruvic acid takes place with the participation of an enzyme which includes a factor known as coenzyme A. Coenzyme A is a derivative of pantothenic acid, which had been recognized as a vitamin before its role in coenzyme A was discovered:

$$\underset{\substack{\phantom{X}\\ \text{pantothenic acid (pantoyl-}\beta\text{-alanine)}}}{HOCH_2\underset{\underset{CH_3}{|}}{\overset{\overset{CH_3}{|}}{C}}{\overset{\overset{OH}{|}}{---CH}}-CONHCH_2CH_2COOH}$$

Coenzyme A (abbreviated CoA—SH) consists of pantothenic acid in combination with a phosphorylated adenosine diphosphate (on the pantoyl portion) and with $\beta$-mercaptoethanolamine (on the $\beta$-alanine portion), and thus is

$$Ph-ADP-OCH_2\underset{\underset{CH_3}{|}}{\overset{\overset{CH_3}{|}}{C}}-CHOHCONHCH_2CH_2CONHCH_2CH_2\;SH$$

Coenzyme A (CoA—SH)

Acetyl coenzyme A is formed from pyruvic acid as follows:

$$NAD + CH_3COCOOH + CoA-SH \longrightarrow CH_3CO-SCoA + CO_2 + NADH$$

The chemical properties of the acetyl group in thioester combination in the acetylated coenzyme A are such that $CH_3CO$—$SCoA$ has a methyl group of enhanced activity for aldol-like condensations, and a carbonyl group of enhanced electrophilic character. Thus, acetyl coenzyme A can undergo condensation reactions in which the two-carbon-atom unit of acetic acid is incorporated into more complex organic compounds.

One step in the utilization of acetyl coenzyme A in some cellular syntheses is its conversion, by reaction with carbon dioxide, into a malonic acid derivative. This reaction, mediated by an enzyme that requires the cofactor biotin, greatly enhances the reactivity of the $\alpha$-hydrogen atoms by providing the additional activation of the second carboxyl group:

$$CH_3CO-SCoA + CO_2 \underset{\text{(biotin)}}{\overset{\text{enzyme}}{\rightleftharpoons}} HOOC-CH_2CO-SCoA \qquad (32)$$

Condensation of the malonic acid derivative (as the nucleophilic component) with the carbonyl group of the acyl coenzyme A (as the electrophilic component) is clearly the biochemical equivalent of a Claisen (acetoacetic ester) condensation:

$$CH_3CO-SCoA + \underset{\underset{\substack{COOH\\ \text{malonyl CoA}}}{|}}{CH_2CO-SCoA} \longrightarrow CH_3CO-\underset{\underset{COOH}{|}}{CHCO}-SCoA + HSCoA \qquad (33)$$

acetyl CoA

Continuation of this process, with concomitant loss of carbon dioxide from the β-keto acid, or, more probably, concomitant with reaction (33), leads to the successive combination of two-carbon units into a chain; and, as will be shown in the discussion of this process in the next chapter, the initial acyl coenzyme A need not be acetyl coenzyme A but may be a derivative of some other acid:

$$\text{RCO—SCoA} + n\ \underset{\overset{|}{\text{COOH}}}{\text{CH}_2\text{CO—SCoA}} \xrightarrow[\text{—HSCoA}]{\text{—CO}_2} \text{RCOCH}_2\text{COCH}_2\text{CO}\cdots\text{CH}_2\text{CO—SCoA} \quad (34)$$

The synthesis of fatty acids proceeds in the manner shown by these equations, with the added step of reduction of the carbonyl group in each stage of the condensation. A condensed representation of the overall process is the following:

$$\text{CH}_3\text{COCH}_2\text{CO—SCoA} \underset{}{\overset{\text{H}_2}{\rightleftharpoons}} \overset{\overset{\text{OH}}{|}}{\text{CH}_3\text{CHCH}_2\text{CO—SCoA}} \underset{}{\overset{\text{—H}_2\text{O}}{\rightleftharpoons}}$$

$$\text{CH}_3\text{CH}\!=\!\text{CHCO—SCoA} \overset{\text{H}_2}{\rightleftharpoons} \text{CH}_3\text{CH}_2\text{CH}_2\text{CO—SCoA}$$

$$\Big\Downarrow \text{malonyl CoA}$$

$$\text{CH}_3\text{CH}_2\text{CH}_2\text{COCH}_2\text{CO—SCoA}$$

$$\Big\Updownarrow \text{(as above)}$$

$$\text{CH}_3\text{CH}_2\text{CH}_2\text{CH}_2\text{CH}_2\text{CO—SCoA}$$

$$\Big\Updownarrow \qquad\qquad\qquad (35)$$

repetitions of these steps to give
long-chain fatty acids

It will be noted that there are two hydrogen transfer steps in the series

$$\text{RCOCH}_2\text{CO—SCoA} \longrightarrow \text{RCH}_2\text{CH}_2\text{CO—SCoA}$$

These are (1) the reduction of —CO— to —CHOH—, and (2) the reduction of —CH=CH— to —CH$_2$CH$_2$—. The enzymes involved in these reactions require NAD in one case and a flavin in the other. Thus, the whole reaction can be summarized as

$$\text{RCO—SCoA} + \text{NADH} + \text{reduced flavoprotein} + \text{CH}_3\text{CO—SCoA.} \rightleftharpoons$$
$$\text{RCH}_2\text{CH}_2\text{CO—SCoA} + \text{NAD} + \text{flavoprotein} + \text{CoA—SH} \quad (36)$$

In summary of the foregoing discussion of glucose degradation and the subsequent utilization of the products of this breakdown, we can write a condensed scheme:

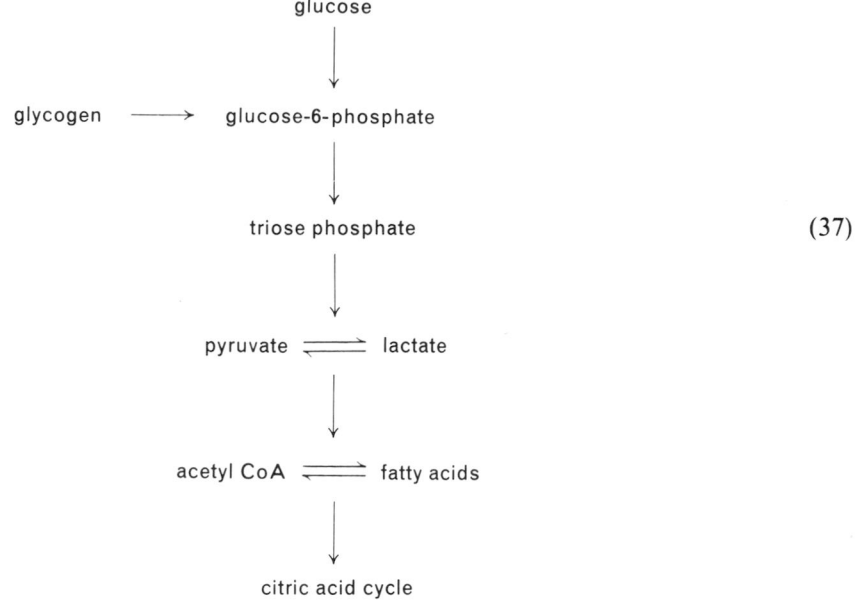

$$\text{glucose} \downarrow$$

$$\text{glycogen} \longrightarrow \text{glucose-6-phosphate}$$

$$\downarrow$$

$$\text{triose phosphate} \tag{37}$$

$$\downarrow$$

$$\text{pyruvate} \rightleftharpoons \text{lactate}$$

$$\downarrow$$

$$\text{acetyl CoA} \rightleftharpoons \text{fatty acids}$$

$$\downarrow$$

$$\text{citric acid cycle}$$

Further elaboration of this scheme, and a discussion of the ways in which these components of the primary metabolic systems of living organisms are used in the synthesis of the many products of "secondary metabolism" will be the subject of the next chapter.

## 31.17 Nucleic acids

The nucleus of the cell is differentiated from the nonnuclear material, chemically as well as structurally. The principal constituent of the nucleus is the chromosomal material that comprises the genetic units of heredity. This genetic material is found to be nucleoprotein in nature. Nucleoproteins are combinations of protein with *nucleic acids*, the latter being complex substances comparable to proteins in molecular weight and consisting of three chief constituents: sugar, purine or pyrimidine bases, and phosphoric acid units. The chromosomal nucleic acids contain deoxyribose as the sugar, in combination with four heterocyclic units: adenine, guanine, cytosine, and thymine:*

adenine     guanine     cytosine     thymine

---

\* Small amounts of other bases are found in some deoxynucleic acid (DNA) samples, but these are variable and of unknown significance.

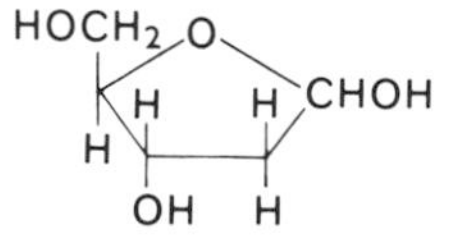

HOCH₂ O

2-deoxy-D-ribose

Hydrolysis of a nucleic acid proceeds in steps:

nucleic acid ——→ nucleotides

nucleotide ——→ nucleoside + phosphoric acid

nucleoside ——→ purine or pyrimidine + sugar

## 31.18   Structure of nucleic acids

The structures of nucleic acids are known with less certainty than are those of proteins (about which exact information is only beginning to be obtained). A schematic description of a nucleic acid structure may be given as follows:

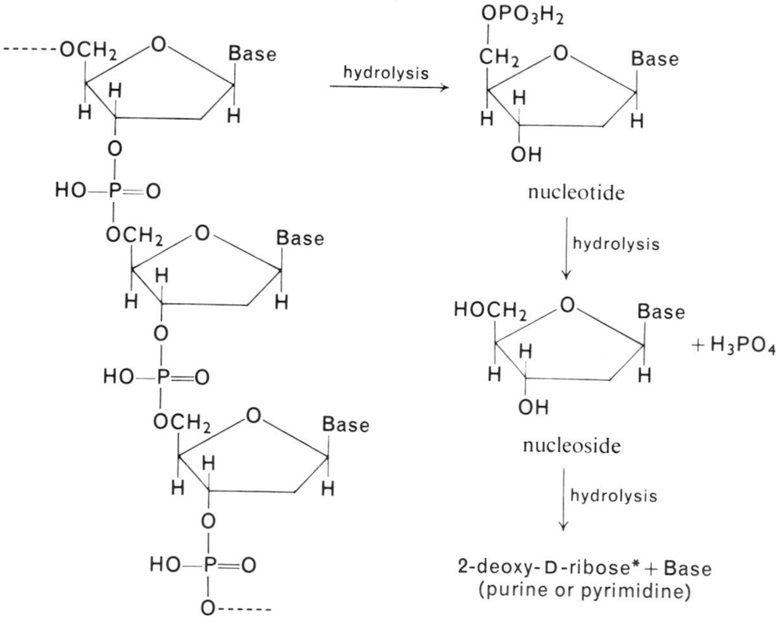

hydrolysis

nucleotide

hydrolysis

nucleoside

hydrolysis

2-deoxy-D-ribose* + Base
(purine or pyrimidine)

$+ H_3PO_4$

deoxyribonucleic acid (DNA)

The exact sequence of the purine and pyrimidine bases in nucleic acids is not yet known. It is clear that an enormous number of different nucleic acids can be

---

* Deoxyribose is characteristic of *nuclear deoxyribo*nucleic acids; *ribo*nucleic acids (RNA), in which the sugar is ribose, are also known.

constructed by altering the sequence of the four bases in a polynucleotide chain of, say, 100 units. Indeed, it is believed that in this possibility for variation in structure lies the ultimate explanation for the fact that the genetic material can contain so great a number of individual hereditary determinants (genes), all of them consisting of combinations of a few common building units.

### 31.19 The structure of deoxyribonucleic acids and the replication of genetic material

The replication of living cells is accomplished with the reproduction of the nuclear material; that is to say that the genetic information contained in the chromosomes is passed on from cell to cell as the organism grows by cell divisions and cellular differentiation. It must be concluded that there exists a mechanism by which the synthesis of nucleic acid is controlled in such a way that the sequence of purine and pyrimidine bases—which characterizes a given molecule of DNA—is identical in the two daughter cells that are formed by cell division.

A model of the DNA molecule, proposed by Watson and Crick in 1953, that provides a satisfactory answer to the problem of how such replication is carried out, is the following. The essential idea of the Watson-Crick model of the DNA molecule is that the molecule consists of two separate chains of polynucleotide material intertwined with each other in the form of a double helix. The separate chains are held together by hydrogen bonds between the purine and pyrimidine groups, and these are always paired in such a way that *adenine is bonded to thymine, and guanine to cytosine*:

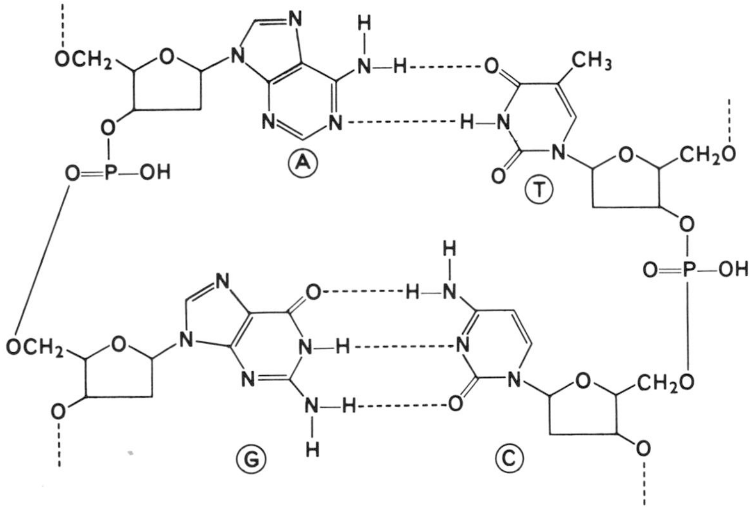

$(A)$ = adenine $\quad$ $(G)$ = guanine

$(T)$ = thymine $\quad$ $(C)$ = cytosine

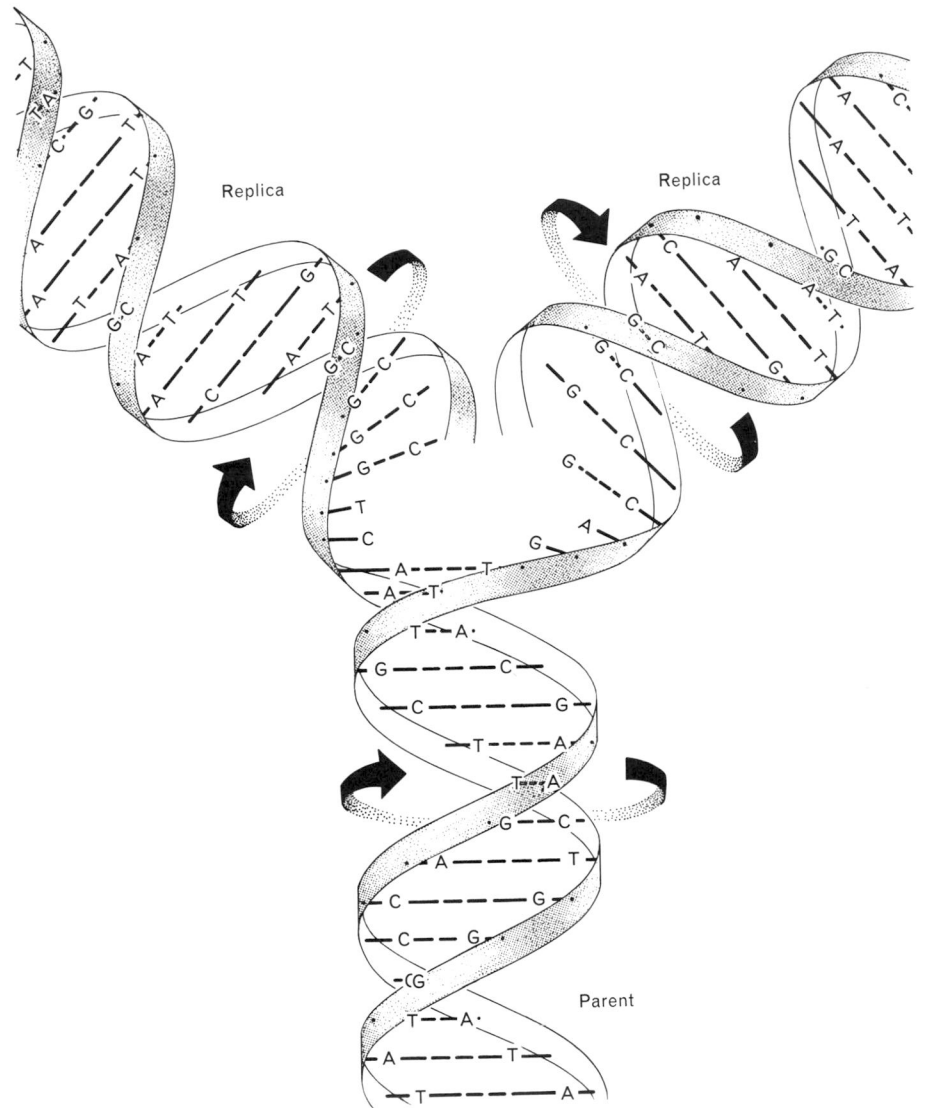

**Figure 31–1.** The replication of DNA, according to Watson and Crick. (From Gunther S. Stent, *Molecular Biology of Bacterial Viruses*, W. H. Freeman and Company, San Francisco and London, 1963.)

A DNA molecule can be represented schematically in the following way:

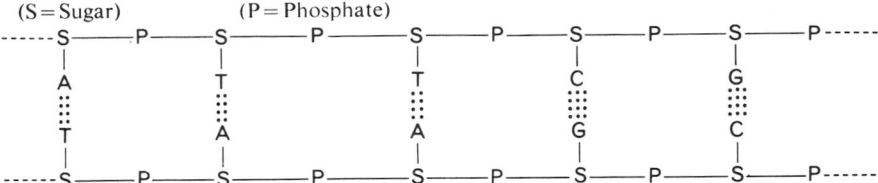

The two polynucleotide chains are intertwined in the double helix shown in Figure 31-1.

Figure 31-1 shows two things: the structure of the DNA molecule (the lower part of the structure); and the process by which the DNA molecule is duplicated. The formation of the new DNA molecule consists of the synthesis of two new polynucleotide chains in such a way that the sequence of purine and pyrimidine residues is directed by the necessity for each base to pair with its complementary base; thus, each strand is "coded" by one strand of the original chain, and the individual nucleotides are "selected" so that A pairs with T, T with A, C with G, and G with C. Consequently, each original double helix gives rise to two new double helices, each identical, and each containing one-half of the original DNA molecule.

The expression of the genetic information contained in the chromosomal DNA is seen in the products of cellular chemical activity. This metabolic activity is controlled by specific enzymes, of which living cells, depending upon the complexity of the organism in which they are found, may contain many thousands. These enzymes are proteins; they are synthesized in the cell under the control of the nuclear DNA by a process which parallels closely that by which the duplication of DNA takes place.

Enzyme synthesis takes place under the direct influence of ribonucleic acids (RNA), the structures of which are in accordance with the "code" contained in the DNA molecule. RNA, which contains ribose instead of deoxyribose, and uracil in place of thymine,

uracil

is a polynucleotide, the synthesis of which occurs at the DNA molecule in such a way that the sequence of the four bases, adenine, uracil, cytosine, and guanine, bears a direct correspondence with the base sequence in that part of the DNA molecule which serves as the template for its formation. Thus, the RNA molecules contain "copies" of the genetic information carried by the DNA, and the enzymes whose syntheses they direct are specifically related to the genetic characteristics of the organism.*

---

* The details of the manner in which protein synthesis is controlled by RNA are beyond the scope of this text. Excellent descriptions of these important processes are to be found in the articles in *The Living Cell: Readings from "Scientific American,"* W. H. Freeman and Company, San Francisco, 1965.

# Chapter thirty-two / Synthesis of Organic Compounds in Living Organisms. Secondary Metabolic Processes

## 32.1 Naturally occurring compounds

The world of nature abounds in organic compounds of nearly every conceivable structural class, the study of which constitutes one of the most fascinating and fruitful areas of study open to the organic chemist. The cells of living organisms—plants, fungi, bacteria, lichens, insects, other animals—are the sites of intricate and complex synthetic activities that result in the formation of many varieties of compounds, many of them of great practical importance to mankind.

It is within the plant kingdom, and particularly in the higher, green plants that we find the broadest spectrum of synthetic capabilities. Synthesis in plants is the more striking because the starting materials are the simple substances—water, carbon dioxide, nitrogen (both elemental and as inorganic salts), and phosphorus compounds. From these are elaborated organic compounds ranging in complexity from simple carboxylic acids, low-molecular-weight aliphatic compounds, even simple hydrocarbons, to polynuclear aromatic compounds, sterols, and alkaloids. Only a few of these can be considered in detail within the limits of a general textbook, and only the broad outlines of the synthetic processes that lead to their formation can be presented. It is the purpose of this chapter to describe some of the main features of synthesis in living organisms (biosynthesis), and to show how a few universally distributed starting materials and reaction types lead to so diverse an array of products.

The primary synthetic process of nature is photosynthesis, by which green plants utilize the energy of sunlight for the incorporation of the carbon of carbon dioxide into carbohydrates. The further metabolic alteration of sugars in the manner described in Chapter 31 leads to the formation of a pool of simple organic compounds, many of them low-molecular-weight carboxylic acids, amino acids, and aromatic compounds, which, by specific, genetically controlled, enzymatically catalyzed organic reactions lead to the formation of the complex compounds in which the plant kingdom abounds. These reactions, which utilize a pool of available starting materials that are quite universal in their distribution, but which lead to compounds that are often unique to one plant or one plant family, are referred to as the *secondary metabolism* of plants.

Three major groups of organic compounds constitute the bulk of the compounds produced by the secondary metabolism of plants.

1. *Terpenes, steroids and carotenoids.* These are the products of reactions in which the primary reactant is acetic acid (as acetyl coenzyme A), which is converted to a primary five-carbon-atom building unit from which the more elaborate final products are formed.

2. *Compounds containing aromatic—usually phenolic—groups in simple benzenoid, naphthalene-derived or anthracene-derived structures.* These, too, are the result of the combination of two-carbon-atom units: again, acetyl coenzyme A.

Another group of aromatic compounds whose origin is not acetic acid, but is through the combination of a four-carbon-atom sugar with a derivative of pyruvic acid, is widely distributed in nature. It might be classified separately from the acetate-derived aromatic compounds, but since compounds of this group often contain structural features that are derived in part from acetyl coenzyme A and in part by the carbohydrate pathway, it will be convenient to discuss all of the aromatic compounds under the single rubric.

3. *Alkaloids.* Although this is a class of compounds of widely various structural types, a very large proportion of them are formed in nature by transformations that start with simple amino acids. A brief introduction to the biosynthesis of alkaloids is to be found in Chapter 29.

## 32.2 Terpenes

Terpenes and the related (*terpenoid*) compounds to be discussed in this and the following sections are among the most widely distributed of naturally occurring compounds. Terpenoid compounds are often called "isoprenoid" compounds because of the occurrence in their structures of a five-carbon-atom unit having the following arrangement of the carbon atoms:

$$C-C-C-C \qquad\qquad CH_2{=}C-CH{=}CH_2$$
$$\underset{\textstyle C}{\mid} \qquad\qquad\qquad \underset{\textstyle CH_3}{\mid}$$

isoprene

The recurring five-carbon-atom unit of isoprenoid compounds has the structure of the carbon skeleton of the simple hydrocarbon, isoprene. Indeed, it was early supposed that the compounds of this class were actually formed in nature by the dimerization or polymerization of isoprene itself, a supposition that may have been due in part to the fact that rubber, an isoprenoid compound, yields isoprene as one product of its thermal (pyrolytic) decomposition. We shall see that this simple hypothesis is not tenable, and that the formation of isoprenoid structures can be accounted for in other, more satisfactory, ways.

The terpenes are the simplest of the naturally occurring isoprenoid compounds. They occur widely in plants, and certain complex members of the group are found in animals. The simple terpenes are ten-carbon-atom compounds, and, being of relatively low molecular weight and hence volatile, are responsible for the characteristic odors of certain plants. The fragrance of mint, of eucalyptus leaves, and of pine needles is due to terpenes; such familiar substances as menthol and camphor belong to this class of compounds.

Terpenes may be open-chain, monocyclic, or bicyclic compounds:*

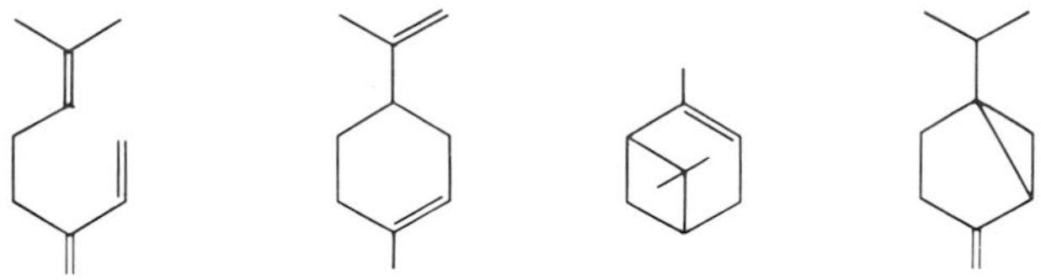

| myrcene | limonene | α-pinene | sabinene |

The term "terpene" refers strictly to a hydrocarbon of the composition $C_{10}H_{16}$. In practice the term applies to all of the ten-carbon-atom isoprenoid compounds, including those containing oxygen, and thus embraces such alcohols and ketones as the following:

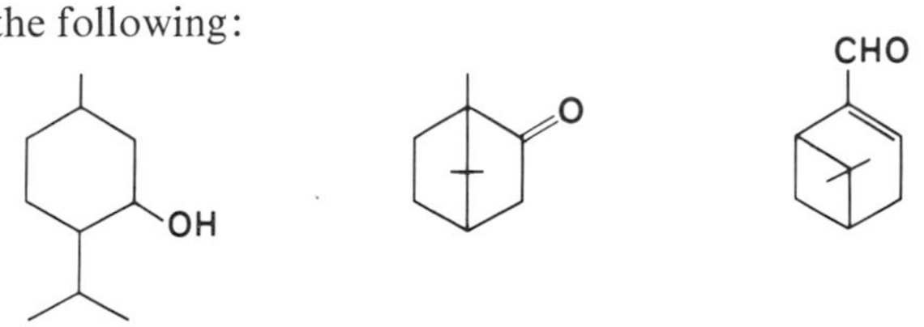

| menthol | camphor | myrtenal |

The structural unity found in the terpenes, whether they be simple open-chain olefins such as myrcene, or complex bicyclic compounds such as pinene, camphor, or sabinene, is found in their carbon skeletons, which are constructed of two isoprene units, joined in the following way:

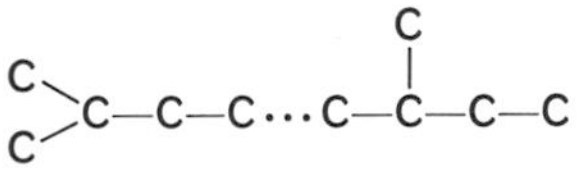

---

* The reader's attention is again called to the conventional representations used in formulas of this type. Single terminal lines, and terminal double lines refer, respectively, to methyl and methylene groups. The formula shown for myrcene, for example, represents the structure $CH_3C\!\!=\!\!CHCH_2CH_2C\!\!-\!\!CH\!\!=\!\!CH_2$. Functional groups (carbonyl, hydroxyl, alde-

$\quad\quad\quad\quad\quad\;\;CH_3\quad\quad\quad\;\;CH_2$

hyde, carboxyl, amino, and so on) are written in the usual ways. (See Section 17.18.)

This way of joining two "isoprene" units is referred to as the head-to-tail union, or sometimes as the "regular" union of the $C_5$ units. The reason for speaking of this as the "regular" manner of joining the units will become clear when the course of biosynthesis of isoprenoid compounds is discussed (Section 32.14):

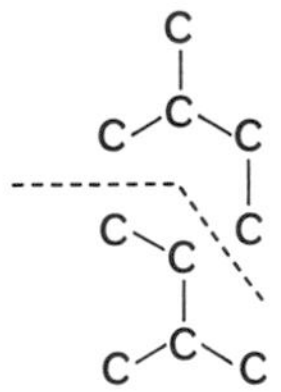

limonene, menthol 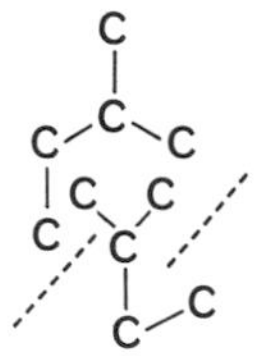                     camphor, pinene

This structural correlation is known as the isoprene rule. It has been found very useful in the study of terpenes and terpenoid compounds, chiefly as a guiding principle to aid in directing structural investigation. The isoprene rule is not, of course, a rule in the sense that it is a natural law, but is simply a summary of experience. There are departures from it, and apparent violations of it. Such exceptions are often valuable clues to the course of biosynthesis of the compounds in question, as we shall see in the following discussion.

The terpenes are but the simplest members of a large class that can be described as *terpenoid* compounds, all of which contain multiples of the $C_5$ unit; that is:

Terpenes: two "isoprene" units: $C_{10}$
Sesquiterpenes: three isoprene units: $C_{15}$
Diterpenes: four isoprene units: $C_{20}$
Triterpenes: six isoprene units: $C_{30}$
Polyterpenes: many isoprene units: $(C_5)_n$ ($n$ is very large)

## 32.3  Sesquiterpenes

The acyclic sesquiterpenes follow the "regular" isoprene rule, as in farnesol:

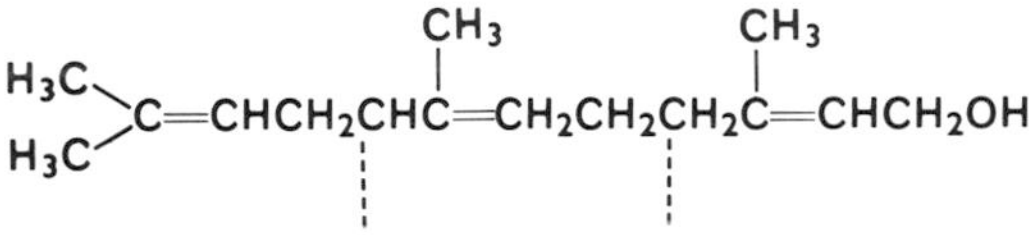

farnesol (isoprene units shown by dotted lines)

Acyclic terpenoid compounds are of far less common occurrence than those containing one or more rings. In the sesquiterpenes the most commonly encountered types are those constructed upon a reduced naphthalene nucleus:

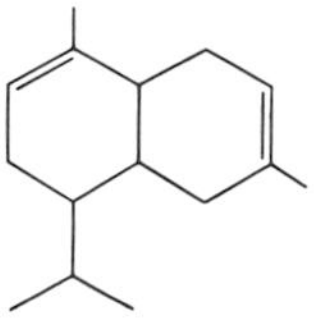

cadinene              $\beta$-selinene

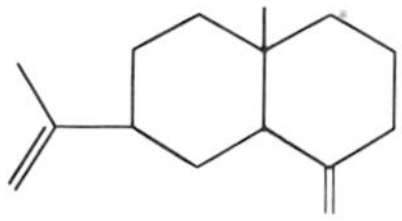

Although the isoprene rule is not immediately apparent in the structures of cadinene and selinene, a comparison with farnesol, rewritten in the following way, makes the relationship clear:

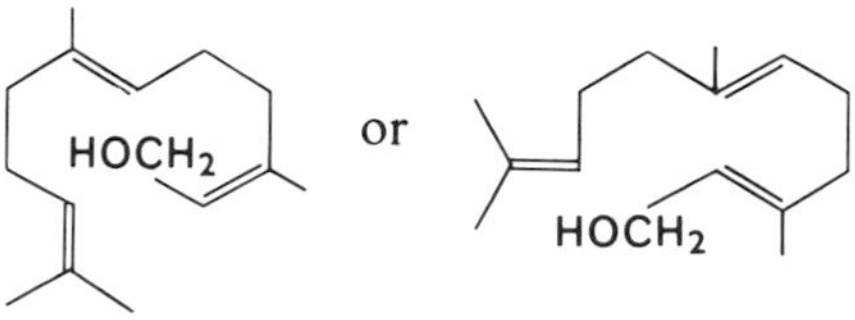

farnesol (*trans*- and *cis*-)

The structure of cadinene, and of numerous other sesquiterpenes built upon the same basic skeleton, was first disclosed by their conversion into 1,6-dimethyl-4-isopropylnaphthalene by dehydrogenation with sulfur or selenium:

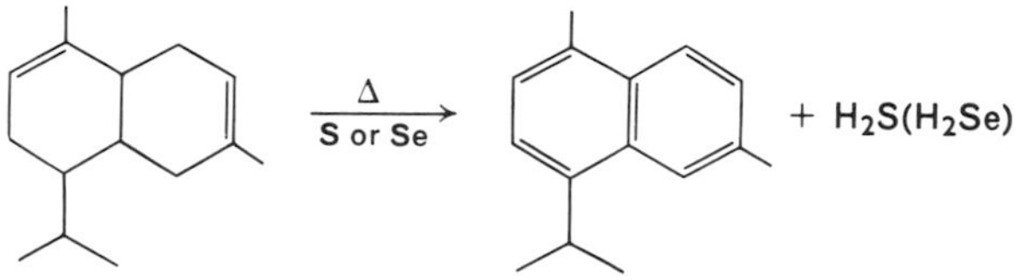

$$+ H_2S(H_2Se) \tag{1}$$

cadinene           cadalene
(1,6-dimethyl-4-isopropylnaphthalene)

Sesquiterpenes of the selinene type give *eudalene* when dehydrogenated with sulfur or selenium:

$$\tag{2}$$

$\beta$-selinene           eudalene
(1-methyl-7-isopropylnaphthalene)

It will be noted that a methyl group is lost in the formation of eudalene; since cadinene undergoes aromatization without loss of carbon, it may be concluded that the carbon atom that is lost in the dehydrogenation of selinene is one that must be removed in order to make possible the formation of the completely aromatic structure. If one attempts to reconstruct the carbon skeleton of the original sesquiterpene, assuming that the methyl and isopropyl groups in eudalene are unaltered in their position, by adding a carbon atom to the eudalene skeleton, there are several *a priori* possibilities:

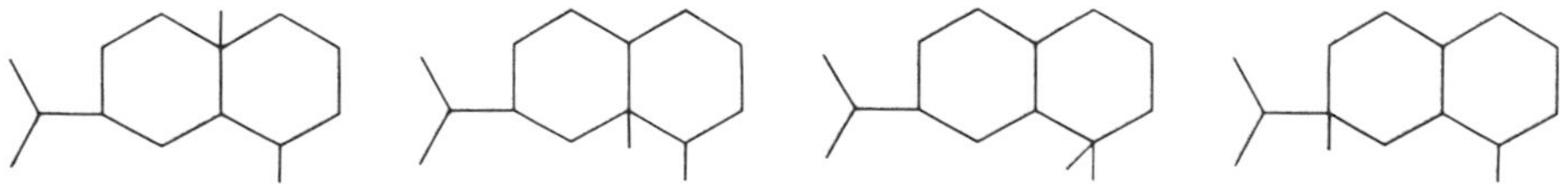

(I)        (II)        (III)        (IV)

Structures (I), (II), and (III) would all be expected to give eudalene on dehydrogenation; (IV) would be an unlikely possibility since 1,7-dimethylnaphthalene ought to be formed as well, and is not found. Of the three possible skeletons, (I), (II), and (III), only (I) follows the "regular" isoprene rule,

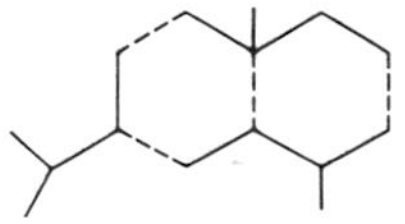

(I)

and it was very early selected as representing the carbon skeleton of selinene and other eudalene-producing sesquiterpenes. Not until some years later was this structure rigorously established to be indeed correct. No natural sesquiterpene based upon the skeleton (III) or (IV) is known.

---

**Exercise 1.** How would the presence of two $\mathord{>}C{=}CH_2$ groupings in selinene be demonstrated experimentally?

---

An interesting departure from the isoprene rule is found in the sesquiterpene ketone, eremophilone:

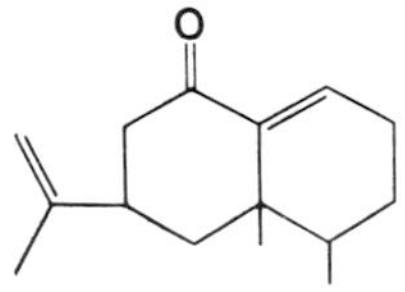

eremophilone

Eremophilone possesses skeleton (II), which *cannot* be constructed of isoprene units. It has been suggested that the eremophilone skeleton arose as a result of a rearrangement in the course of its synthesis in the plant, the "natural" arrangement being converted to the "unnatural" arrangement at some early (but not defined) stage of its formation. Rearrangements in the course of the synthesis of terpenoid compounds in nature, with the migration of methyl groups, will be described in Section 32.14.

## 32.4 Diterpenes

The structures of nearly all of the diterpenes follow the isoprene rule [e.g., (V)]:

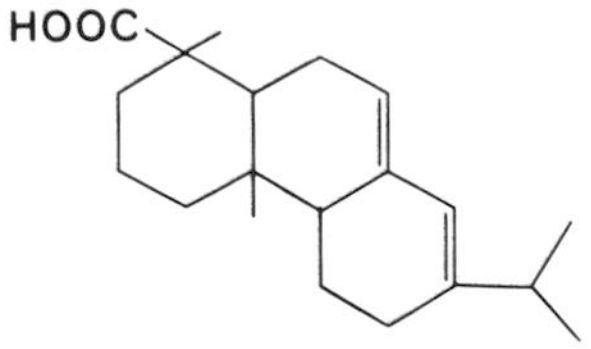

abietic acid

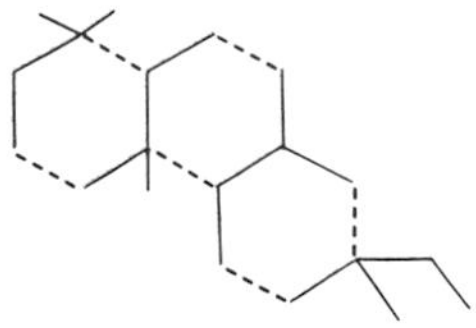

(V)

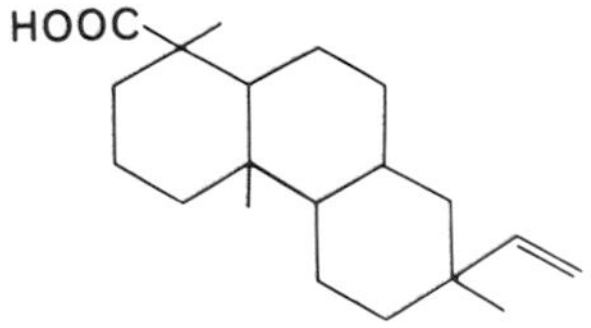

dextropimaric acid

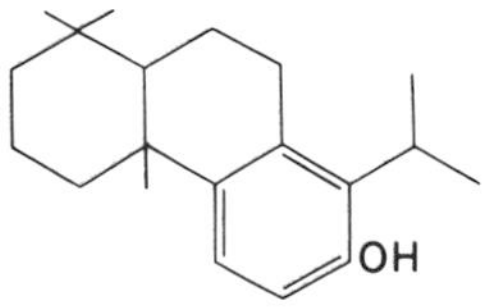

totarol

Structures such as that of abietic acid and totarol, which are "aberrant" in terms of the regular isoprene rule, probably arise by a transformation of a normal isoprenoid precursor by a rearrangement at some stage in the biosynthesis. The present state of our knowledge concerning terpenoid biosynthesis permits this presumption, and makes it unnecessary to postulate a completely different path of synthesis for compounds whose structures show small variations from the "normal."

---

**Exercise 2.** Write the skeleton of dextropimaric acid to show the isoprene units.

---

The establishment of the structures of the diterpenoid compounds was a long and arduous task, requiring years of effort by many chemists. Again, as in the case of the sesquiterpenes, dehydrogenation proved to be a valuable tool in the work, since recognizable products containing most of the carbon atoms can be obtained from many of the diterpenes:

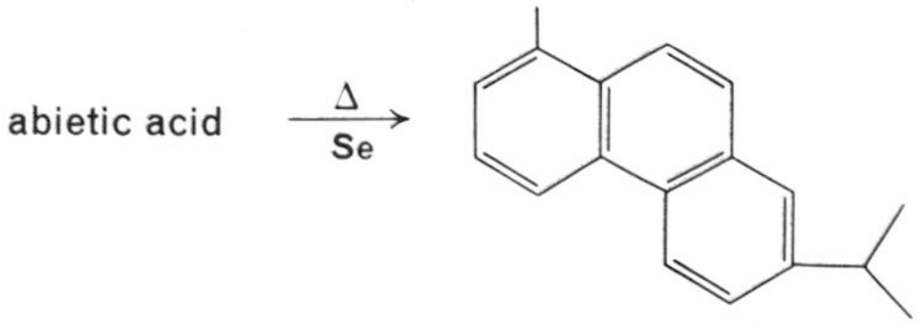

abietic acid $\xrightarrow[\text{Se}]{\Delta}$ 

(3)

retene
(1-methyl-7-isopropylphenanthrene)

A diterpenoid compound found in all green plants is *phytol*, an acyclic alcohol present in ester linkage in the universal plant pigment chlorophyll. Phytol contains four isoprene units linked in regular (head-to-tail) fashion:

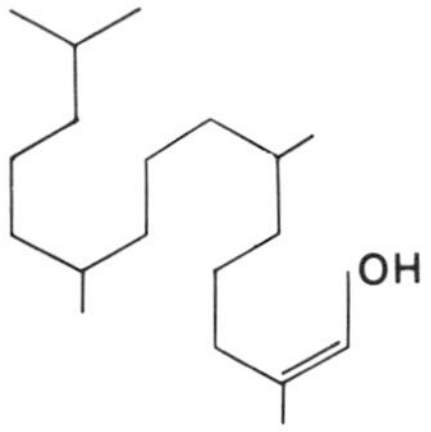

phytol ($C_{20}H_{40}O$)

## 32.5  Triterpenes

Triterpenes are $C_{30}$ compounds, and occur commonly in plants in a wide variety of forms. Of the polycyclic triterpenes, those of the $\alpha$-amyrin group are typical of the pentacyclic (containing five rings) triterpenes:

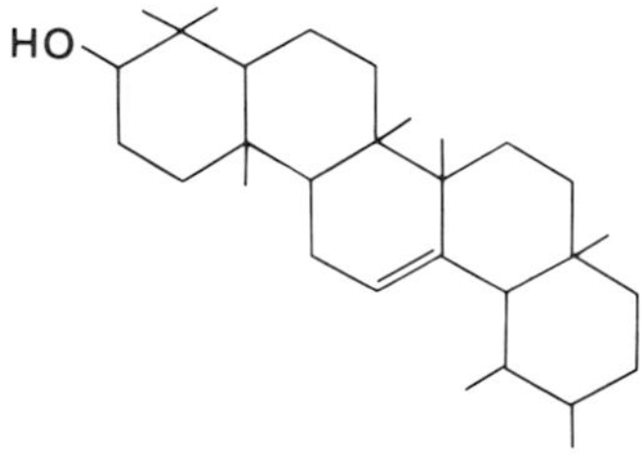

α-amyrin

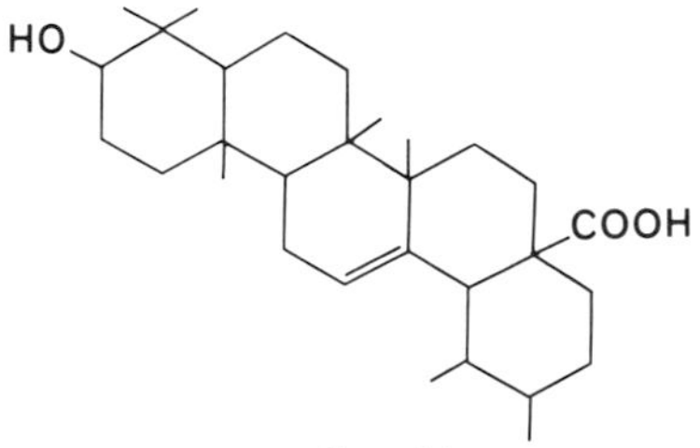

ursolic acid

Of particular interest are the tetracyclic (containing four rings) triterpenes, of which we shall select as the compound for discussion *lanosterol* (VI), first isolated from wool fat:

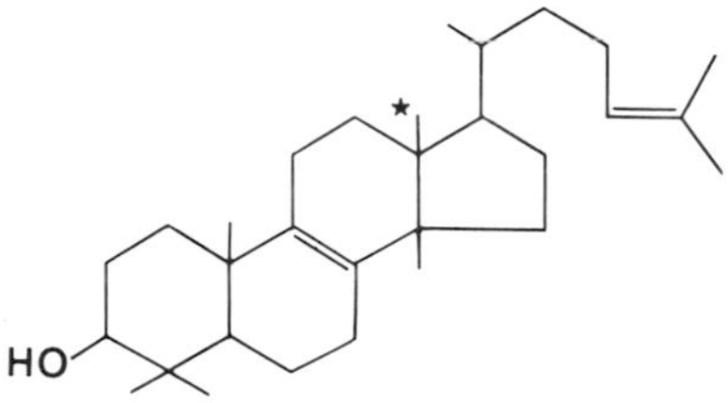

lanosterol (VI)

First of all, for reasons which will become clear, it is necessary to consider the acyclic (no rings) triterpene, *squalene*, found in high concentration in shark liver oil, which has the structure (VII):

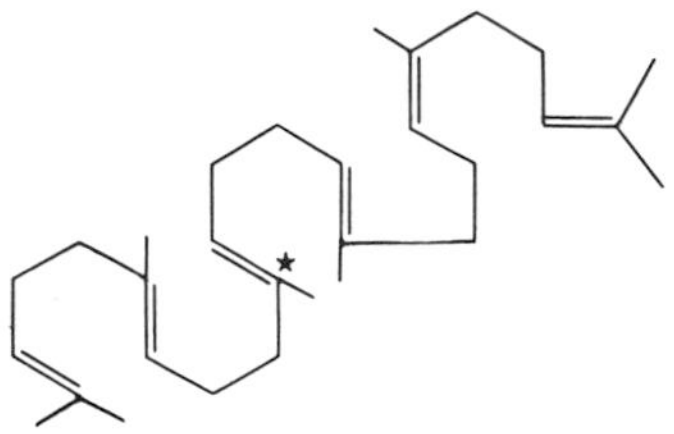

squalene   (VII)

There are two points of interest to emphasize in the structure of squalene. One of them is that whereas its structure is isoprenoid it is composed not of a regular succession of head-to-tail linked isoprene units but of two *regular* $C_{15}$ units linked tail-to-tail in the center of the chain:

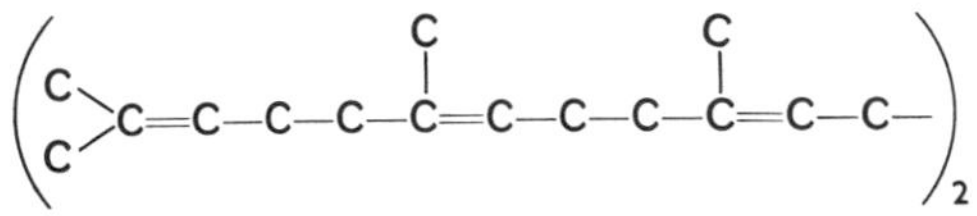

Of even greater interest is the fact that: *squalene has been shown to be the* in vivo *(in animals) precursor of lanosterol, and lanosterol of cholesterol* (see below). The structure of squalene, as it is written in (VII), coincides, except for ring closure, with that of lanosterol in every respect except one: the methyl group marked by a star in lanosterol (VI) and that similarly marked in squalene are not in the same position. In the *in vivo* conversion of squalene to lanosterol, a rearrangement takes place (Section 32.14).

## 32.6  Sterols

Among the most abundant of naturally occurring compounds are the sterols and related compounds (called by the general name "steroids"). The most widely occurring of these is cholesterol, a constituent of the cells of all higher animals:

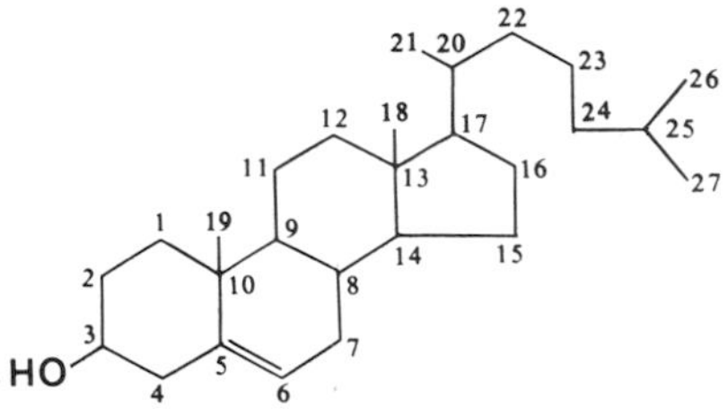

cholesterol 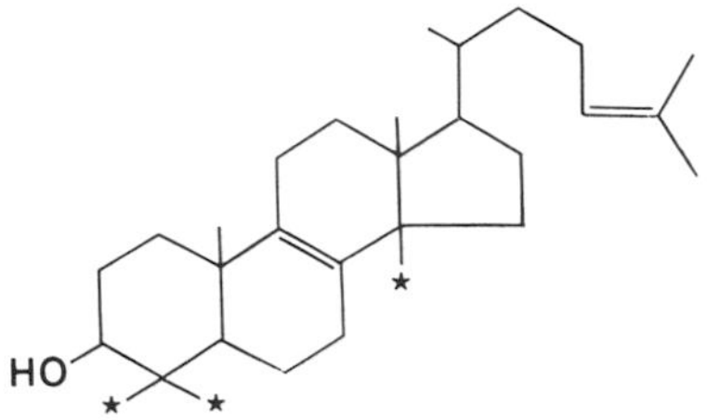                                    lanosterol

The structure of lanosterol is written again, for comparison with that of cholesterol. The similarity is evident; and all that is needed to convert lanosterol into cholesterol (so far as the carbon skeleton is concerned) is for lanosterol to lose the three methyl groups marked by stars. This is what happens in the cellular synthesis of cholesterol, the removal of the methyl groups probably being brought about by their oxidation to —COOH, followed by decarboxylation. Thus, we have the process of synthesis:

$$\text{squalene} \xrightarrow{\text{cyclization}} \text{lanosterol} \xrightarrow{\text{demethylation}} \text{cholesterol} \longrightarrow \text{derived steroids}$$
(see below)

## 32.7 The origin of squalene

The ultimate origin of squalene, and thus of cholesterol, is acetic acid, which, as acetyl coenzyme A, is built up into isoprenoid building blocks by condensation reactions to be discussed in detail in a later section. The course of the synthesis of the five-carbon unit, expressed here only in terms of the carbon skeleton, may be represented as follows:

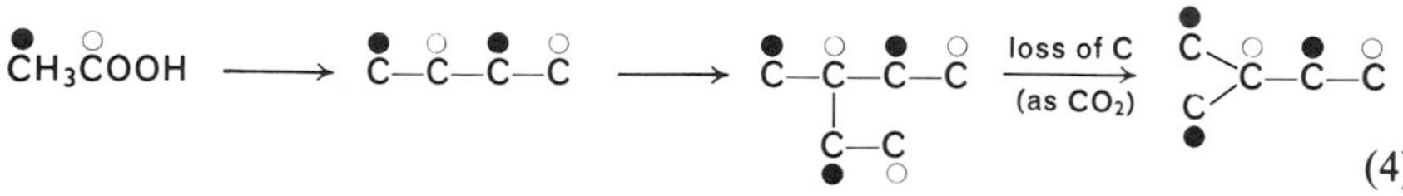

(4)

A poly-isoprenoid chain built up in this way would then have the carbon atoms marked ● derived from the methyl group, and those marked ○ derived from the carboxyl group, of acetic acid. Thus, cholesterol should be constructed as follows:

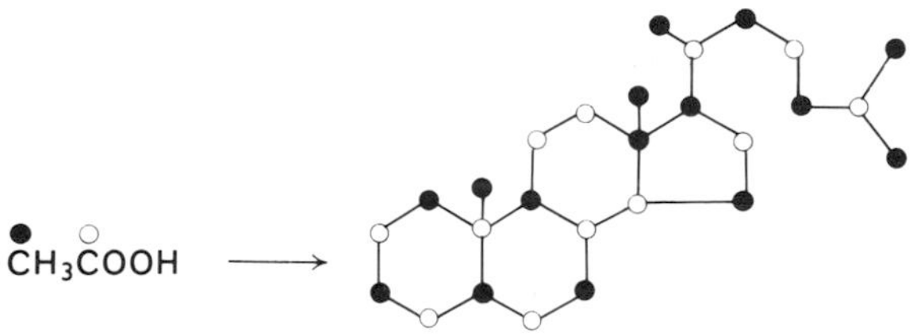

(5)

(VIII)

When radioactive acetic acid, labeled with carbon-14 on the methyl group, is fed to rats, the cholesterol synthesized in and isolated from their livers is found to have the labeled atoms exactly in the positions shown in (VIII); and when carboxyl-labeled acetic acid is fed, the pattern of labeling again follows the predicted pattern.

## 32.8 Steroids of animal organisms

Cholesterol appears to be the sterol from which the numerous steroids that take part in the metabolic processes of higher animals are derived. These steroids belong to three general classes:

1. *The bile acids.* These are a group of hydroxylated steroid carboxylic acids represented by *cholic acid*. The bile acids, as conjugates with glycine or taurine

($NH_2CH_2SO_3H$), are of importance in the process of emulsification and hence the absorption of fatty materials in the intestinal tract:

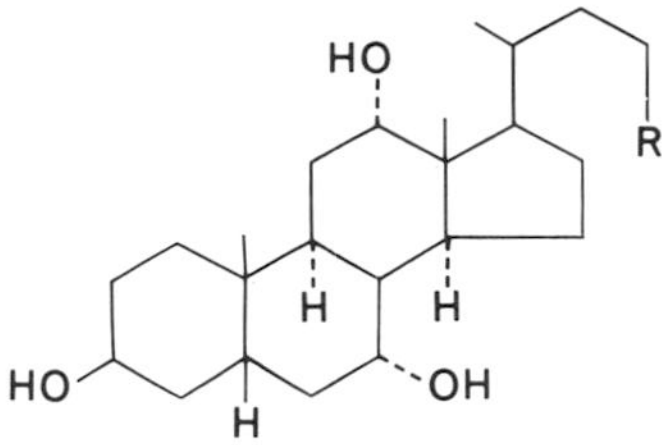

cholic acid (R = COOH)
R = CONHCH₂COOH
glycine conjugate

Numerous bile acids are known, various species of animals often producing highly characteristic structures, most of which differ from one another in the number and position of hydroxyl groups on the fundamental *cholanic acid* skeleton:

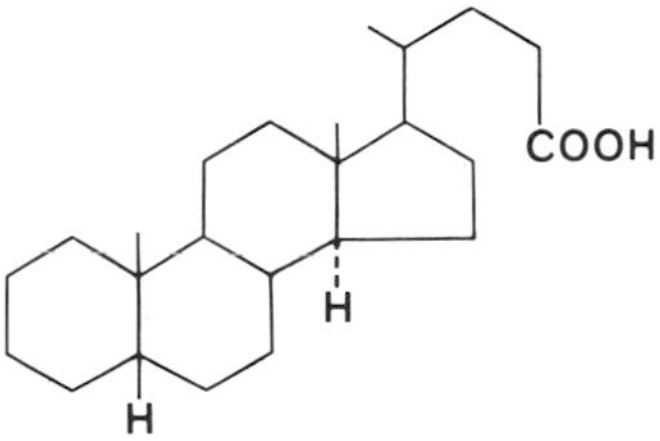

cholanic acid

2. *The sex hormones.* These are steroidal compounds elaborated in the sexual glands of higher animals and transported to sites of action (for example, the ovaries, the uterus, the seminal vesicles) at which their effects are produced. The sex hormones fall into three main classes: (1) the estrogenic hormones, represented by estradiol-17β, which are concerned with the development of secondary female sex characters and with the control of the uterine cycle; (2) the gestogenic hormones, represented by progesterone, which bring about proliferation of the lining of the uterus to prepare it for implantation of the fertilized ovum; and (3) the androgens, represented by testosterone, which induce the development of the secondary male sexual characteristics:

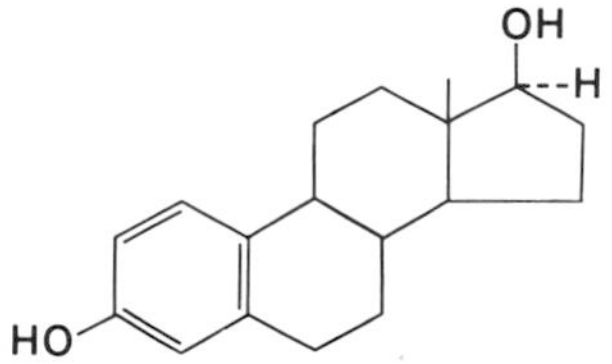

estradiol-17β                    progesterone                    testosterone

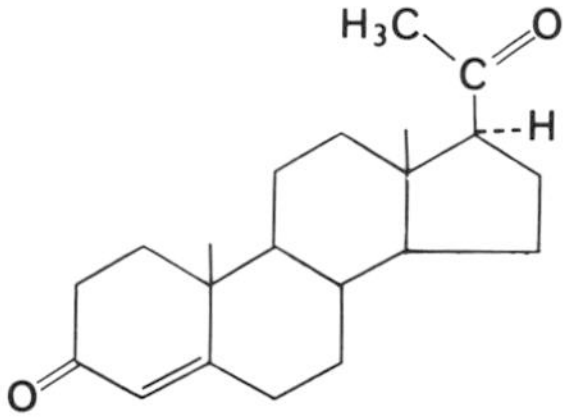

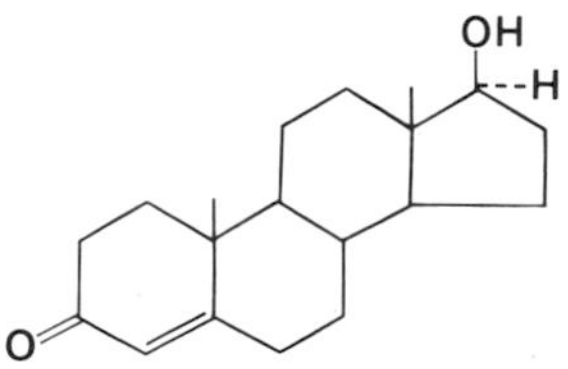

Progesterone is derived directly from cholesterol in the body. In view of the fundamental place occupied by cholesterol in the synthetic pathway, acetate → squalene → lanosterol → cholesterol, it is probable that cholesterol is the parent substance from which other steroids are derived. However, this direct interconversion has not been demonstrated in all cases, and until this has been accomplished, conclusions concerning biogenesis, that are based upon structural similarities alone, should not be regarded as established.

Certain synthetic compounds bearing close structural resemblances to progesterone have been found to suppress ovulation, and have come into widespread use as components of oral contraceptive drugs. Two of these are known by the proprietary names Norlutin and Norethynodrel:

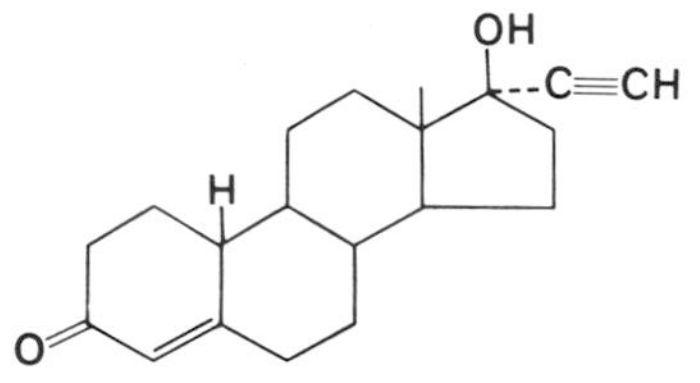

Norlutin 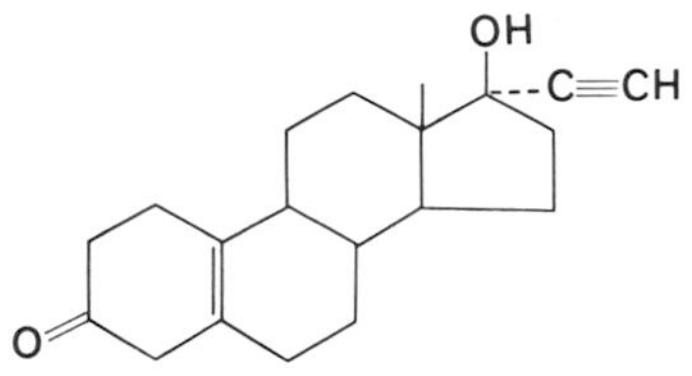                    Norethynodrel

3. *The steroidal hormones of the adrenal cortex.* The cortex (the outer layers of tissue) of the adrenal glands of mammals produces a large number of steroidal compounds, seven of which are of great importance in regulating physiological processes. Animals deprived of the adrenal glands (by disease, or by excision for experimental purposes) suffer from a multiplicity of symptoms of a disease that, in man, is known as Addison's disease. The lives of such animals can be maintained by the administration of extracts of adrenal glands or of the pure compounds known as the *adrenocortical hormones.*

The metabolically most important of these steroids are characterized by two common structural features: an α-hydroxyacetyl grouping at the 17-position, and a 3-one-4-ene grouping in the A ring:

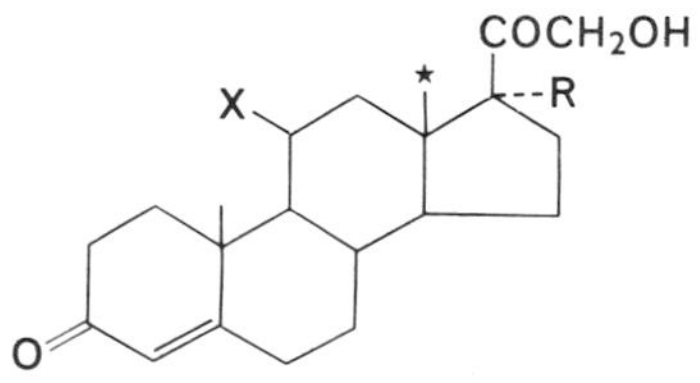

| | |
|---|---|
| X=H; R=H | deoxycorticosterone |
| X=OH; R=H | corticosterone |
| X= =O; R=H | 11-dehydrocorticosterone |
| X=H; R=OH | 17-hydroxydeoxycorticosterone |
| X=OH; R=OH | 17-hydroxycorticosterone |
| X= =O; R=OH | cortisone |
| X=OH; R=H, and CH₃* replaced by CHO: | aldosterone |

Synthetic analogues of the cortical hormones have found an important place in medicine. Three of these are prednisone, prednisolone, and dexamethasone, which are valuable drugs in the treatment of rheumatic and other inflammatory diseases:

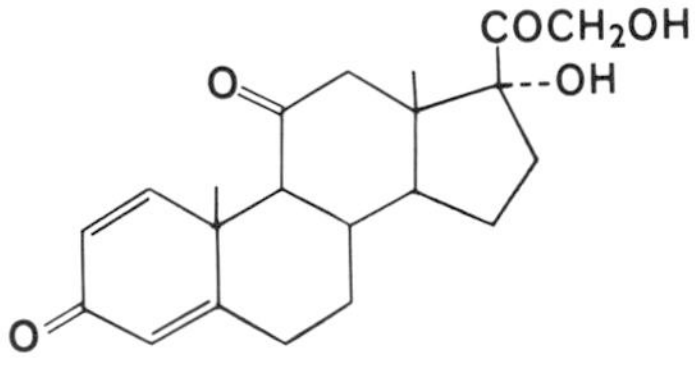

prednisone

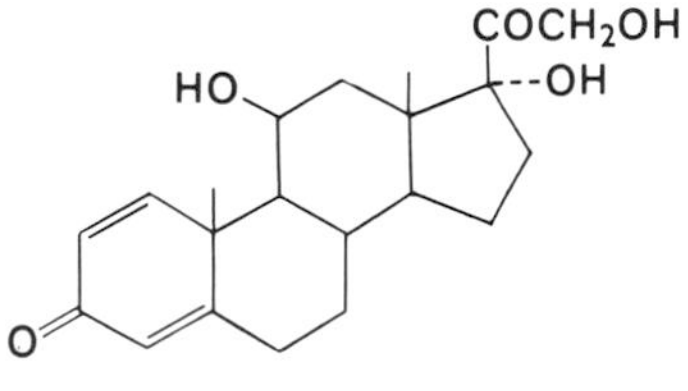

prednisolone

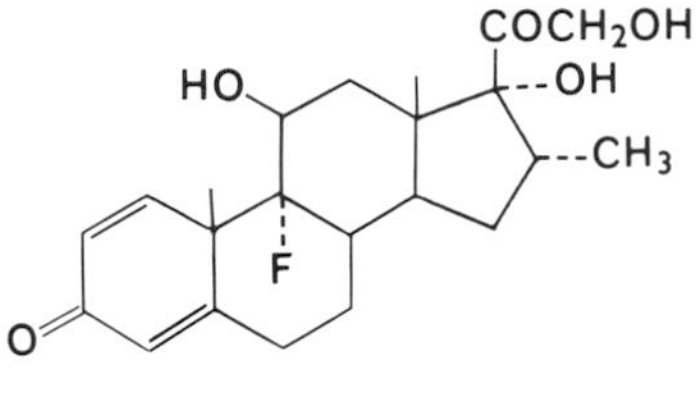

dexamethasone

## 32.9 Plant steroids

Cholesterol is a sterol of the animal kingdom; although it has been reported to occur in plant materials, its occurrence in the plant kingdom is rare. The typical sterols of plants differ from cholesterol in the presence of an additional alkyl group in the $C_{17}$-side chain. Ergosterol, from yeast, and stigmasterol, a widely distributed plant sterol, have the following structures:

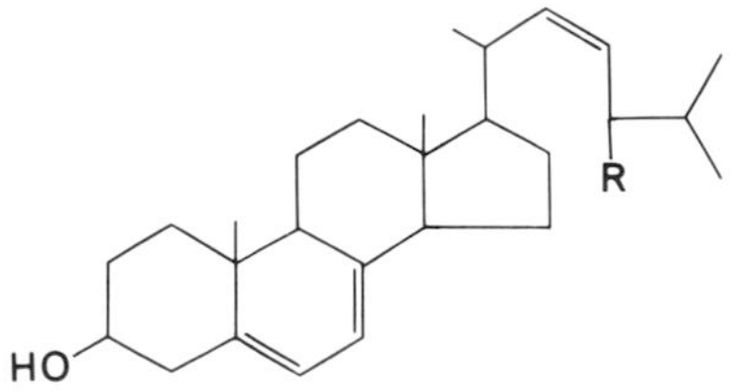

ergosterol
$R = CH_3$

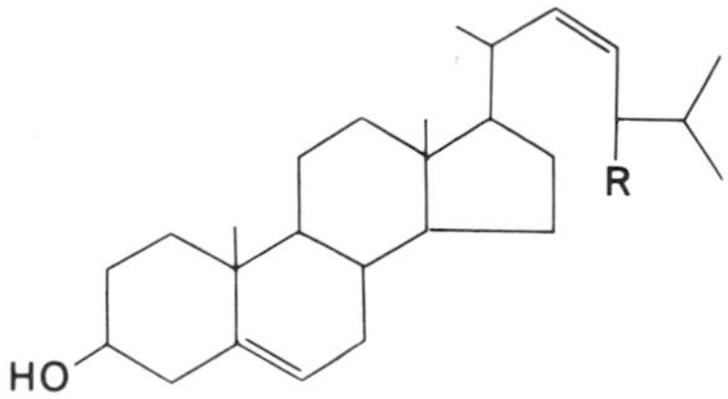

stigmasterol
$R = C_2H_5$

These plant sterols are not utilized by animals, and cannot substitute for cholesterol as a source of dietary sterol. Although the biosynthesis of plant sterols has not been investigated in as great detail as that of cholesterol, it is probable that the main features of their synthesis in plants follows the course described for the biosynthesis of cholesterol in mammalian organisms, with the additional step of the introduction of the extra alkyl group of the side chain.

## 32.10   Other plant steroidal compounds

Plants produce steroidal compounds in a wide variety of structural modifications. The *steroidal saponins*,* so called because of the ability of plant juices in which they occur to produce soapy solutions in water, and the *cardiac glycosides*, or plant heart poisons, are represented by the following examples:

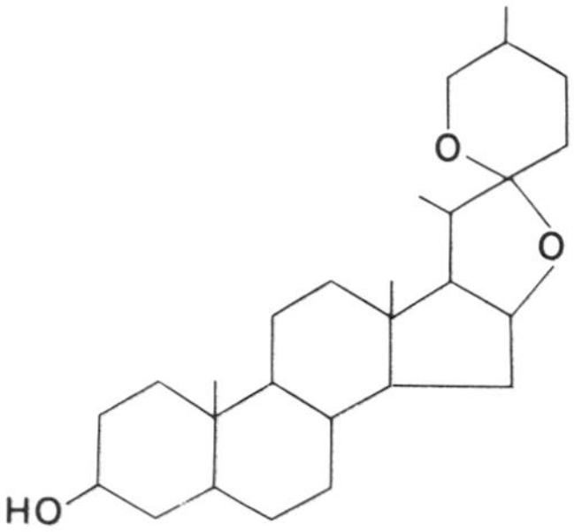

sarsasapogenin (a *sapogenin;*
the *saponin* is a 3-glycoside
of the sapogenin)

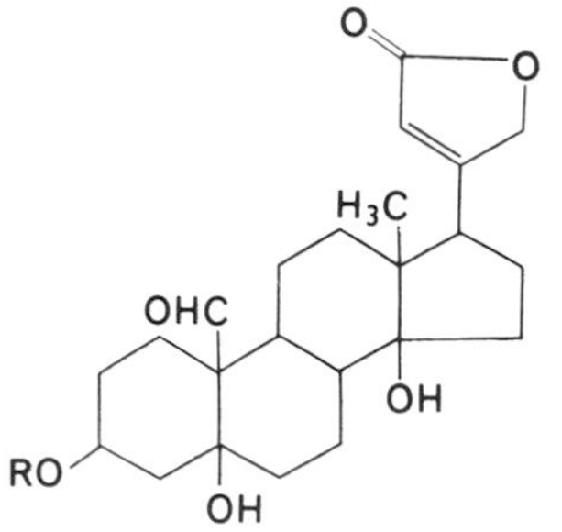

strophanthin (a *cardiac glycoside;*
R = [glucose-cymarose]). In
strophanthidin, R = H

## 32.11   Vitamin D

The deficiency disease known as "rickets," a condition characterized by improper bone formation caused by the inadequate deposition of calcium phosphate during the process of bone growth, is curable by the administration of the products obtained by irradiating ergosterol with ultraviolet light. A number of products may be formed from ergosterol, depending upon the conditions of the irradiation, but the active principle, called vitamin $D_2$, has the following structure:

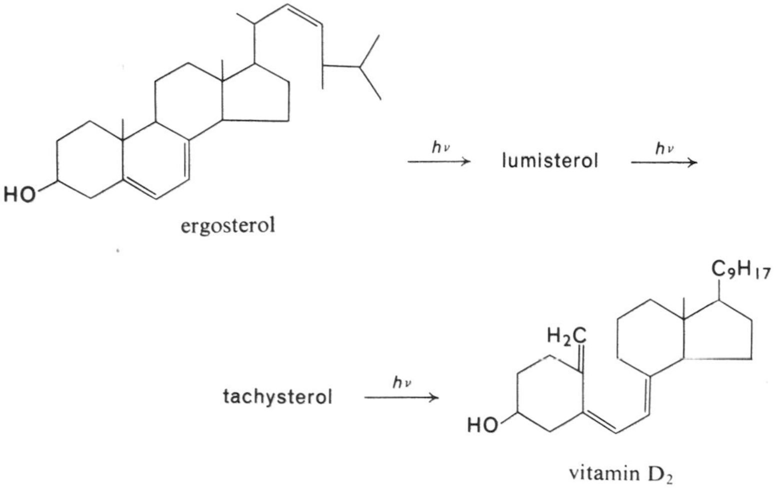

$$\xrightarrow{h\nu} \text{lumisterol} \xrightarrow{h\nu}$$

ergosterol

tachysterol $\xrightarrow{h\nu}$

(6)

vitamin $D_2$

---

* Saponins of the triterpene series of compounds are also known.

Ergosterol is called a "provitamin D." It is not the only sterol that can undergo transformation into a "vitamin D" by irradiation. 22-Dihydroergosterol (ergosterol with the side-chain double bond saturated) and 7-dehydrocholesterol are also provitamins-D. The vitamin derived from ergosterol is called ergocalciferol; that from dehydrocholesterol (vitamin D₃) is called cholecalciferol. By irradiating foodstuffs, the provitamins contained in them can be transformed into useful amounts of the vitamin. Similarly, irradiation of persons suffering from the vitamin deficiency can produce, from the provitamins in their bodies, therapeutic amounts of the vitamin.

## 32.12 Carotenoids

The fat-soluble yellow pigments of plants, found chiefly in the leaf plastids, in flowers, and in fruits, are compounds containing long chains of conjugated double bonds with an isoprenoid structure. The following fragmentary structure represents the essential nature of the polyene chain, with the isoprene units indicated:

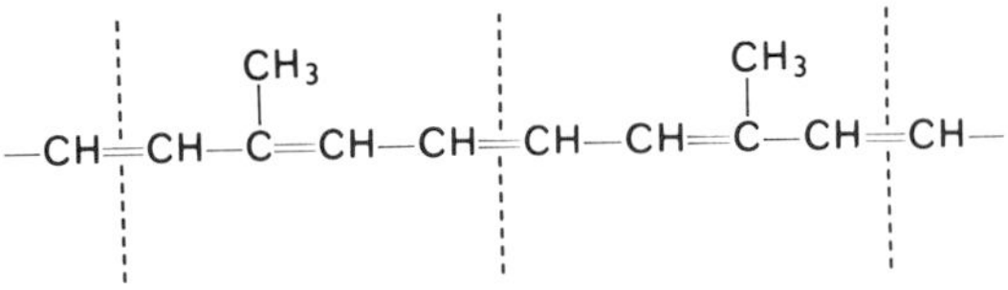

The most widely distributed of the carotenoids is β-*carotene*, the structure of which is

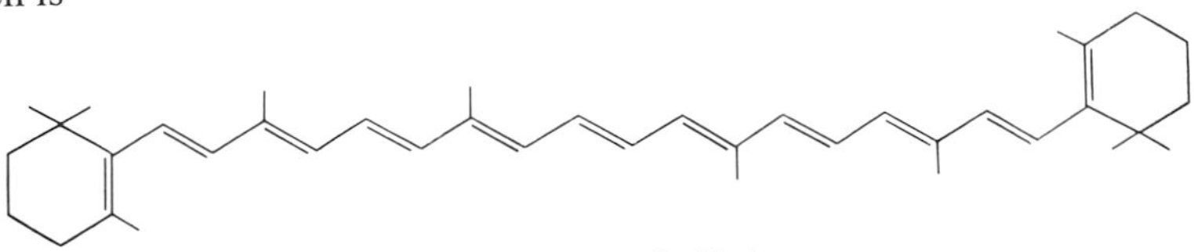

β-carotene ($C_{40}H_{56}$)

*Lycopene*, the pigment responsible for the red color of tomatoes, is the completely acyclic isomer of β-carotene:

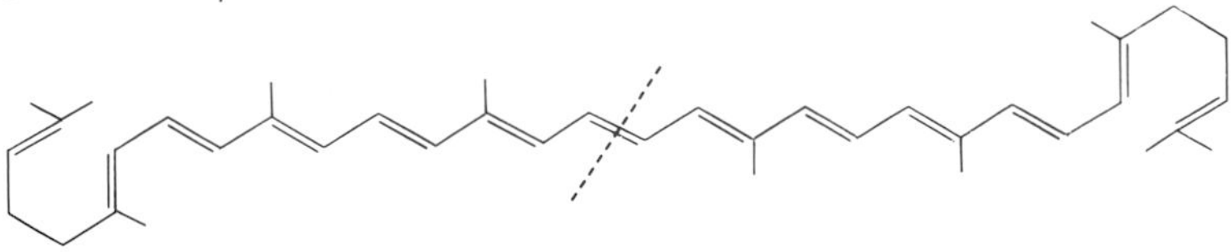

lycopene

It will be noted that the $C_{40}$ carbon skeleton of the carotenoids is constructed of two diterpenoid fragments linked in a tail-to-tail manner in such a way that the molecule is symmetrical about the center. The dotted line drawn in the formula of lycopene is placed at the center of the chain. Squalene, it will be recalled, is similarly constructed of two "regular" isoprenoid $C_{15}$ fragments.

**Table 32-1.** *Electronic absorption spectra of some carotenoid compounds. (Values for solutions in carbon disulfide)*

|  | Absorption Maxima, mμ | | |
|---|---|---|---|
| α-carotene* | 509 | 477 | |
| β-carotene | 520 | 485 | 450 |
| γ-carotene* | 533 | 496 | 463 |
| lycopene | 548 | 507 | 477 |

\* For the structures of α- and γ-carotene, see Section 32-13.

The carotenoids provide striking examples of the effect of long chains of conjugated carbon-carbon double bonds upon the light-absorbing properties of organic compounds, and were the first compounds in which this property was clearly recognized and thoroughly investigated. Table 32-1 gives the values of the prominent absorption maxima for a number of well-known carotenoid compounds, and shows that the electronic transitions, which are of the $\pi \to \pi^*$ type, occur at wavelengths which extend well into the visible region.

### 32.13   Vitamin A

One of the most important aspects of the chemistry of the carotenoids is their relationship to vitamin A, the vitamin responsible for the integrity of epithelial tissue and playing a central role in the process of vision. Vitamin A is a diterpenoid compound, having the carbon skeleton of one-half of the β-carotene molecule. The terminal carbon atom is present as a carbinol group:

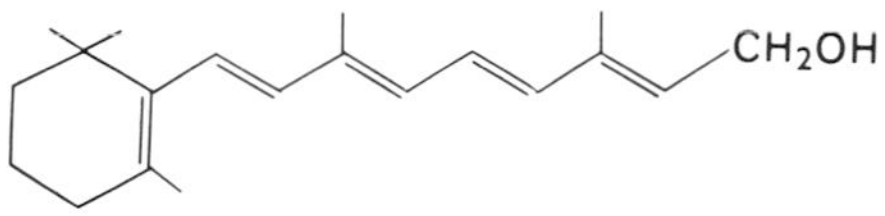

vitamin A ($C_{20}H_{30}O$)

Vitamin A is found in liver, and in high concentrations in the livers of certain fishes (e.g., shark), where it is deposited after its formation by the cleavage of the $C_{40}$ molecules of certain ingested carotenoids.

Not all carotenoids are precursors of vitamin A; those that are (known as provitamins A) have as the essential structure a β-ionone ring at one end of the carotenoid chain:

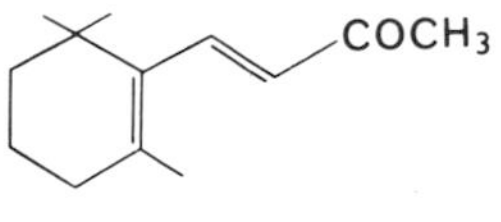

β-ionone 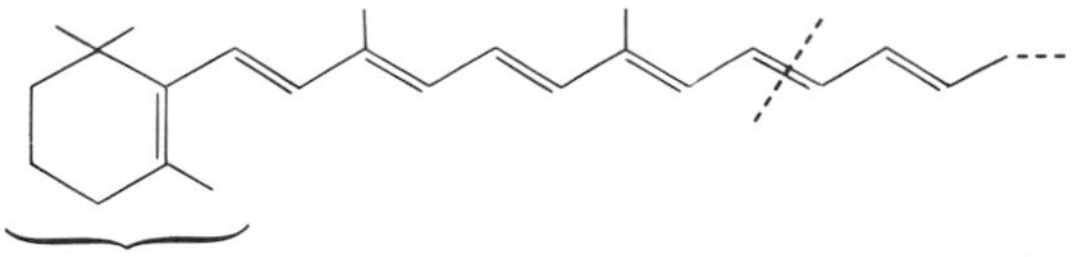          β-ionone ring

β-Carotene is the most efficient provitamin A, since the "β-ionone ring" is present at both ends of the molecule. A number of other naturally occurring carotenoid compounds also have vitamin-A activity; examples are α-carotene, γ-carotene, and cryptoxanthin (the central part of the following structures, which has been omitted for brevity, is like that in β-carotene):

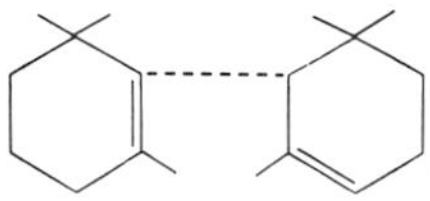

α-carotene 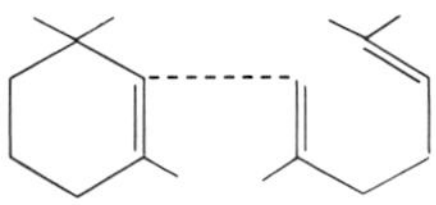                γ-carotene 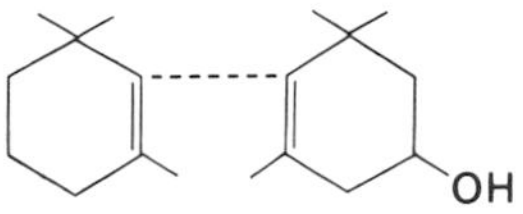                cryptoxanthin

The role of vitamin A in the visual process is of particular interest. It is well known that one of the symptoms of vitamin-A deficiency is the condition known as "night blindness," the impairment of visual acuity in dim light. The reason for the night blindness syndrome is that the retina is deficient in the photosensitive pigment *rhodopsin*, which is a complex of a protein and vitamin A aldehyde (retinene). The action of light upon rhodopsin causes "bleaching" of the pigment from a red to an orange, with a concomitant production of a nervous impulse that results in the sensation of vision. In the subsequent "dark adaptation" phase rhodopsin is regenerated (by a process of which the details are not completely known) and is readied for renewed action. The role of vitamin A in vision is thus to provide the means for the formation of rhodopsin; a summary of the process is given in the following scheme:

vitamin A $\longrightarrow$ (7)

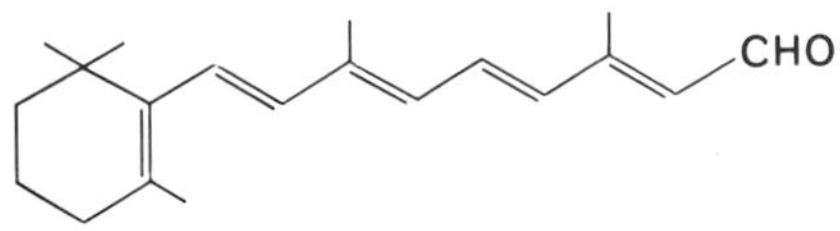

$\dashrightarrow$ retinene

$\downarrow$

rhodopsin   (retinene attached to a protein)

$\downarrow$ light

{ nervous impulse + "bleached rhodopsin"}

$\downarrow$

dark ------------ retinene + protein ("opsin")

## 32.14   Biogenetic relationships in terpenoid compounds

The relationship between the terpenes, the di-, tri-, and polyterpenes, the steroids, and the carotenoids can now be recognized. It is now generally accepted that they are all related by being the products of synthetic pathways that, at least in the early stages, follow a common course. This course involves the transformation of a two-carbon, or acetic acid, unit into a five-carbon "building block" from which the complex terpenoid, steroid, and carotenoid compounds are constructed:

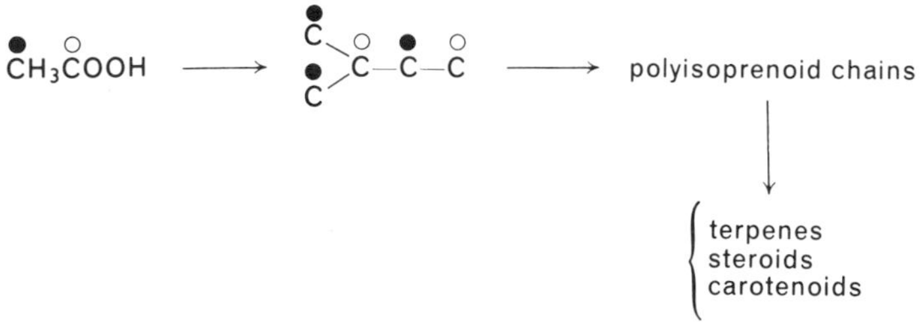

$$CH_3COOH \longrightarrow \quad C-C-C \longrightarrow \text{polyisoprenoid chains} \tag{8}$$

$$\downarrow$$

$$\left\{ \begin{array}{l} \text{terpenes} \\ \text{steroids} \\ \text{carotenoids} \end{array} \right.$$

The key compound in this synthetic sequence is mevalonic acid, which is formed in the following way:

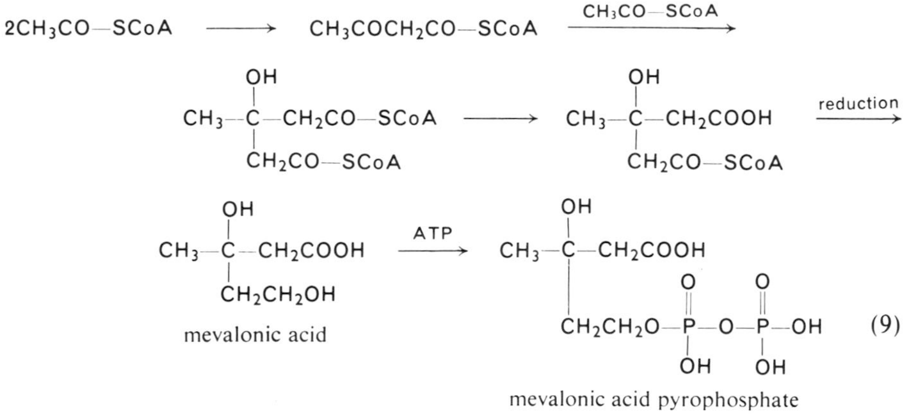

$$2CH_3CO-SCoA \longrightarrow CH_3COCH_2CO-SCoA \xrightarrow{CH_3CO-SCoA}$$

$$CH_3-\overset{OH}{\underset{CH_2CO-SCoA}{\underset{|}{C}}}-CH_2CO-SCoA \longrightarrow CH_3-\overset{OH}{\underset{CH_2CO-SCoA}{\underset{|}{C}}}-CH_2COOH \xrightarrow{\text{reduction}}$$

$$CH_3-\overset{OH}{\underset{CH_2CH_2OH}{\underset{|}{C}}}-CH_2COOH \xrightarrow{ATP} CH_3-\overset{OH}{\underset{CH_2CH_2O-\overset{O}{\overset{\|}{P}}-O-\overset{O}{\overset{\|}{P}}-OH}{\underset{|}{C}}}-CH_2COOH \tag{9}$$

mevalonic acid

mevalonic acid pyrophosphate

Phosphorylated mevalonic acid then undergoes a simultaneous dehydration-decarboxylation to yield "isopentenyl pyrophosphate":*

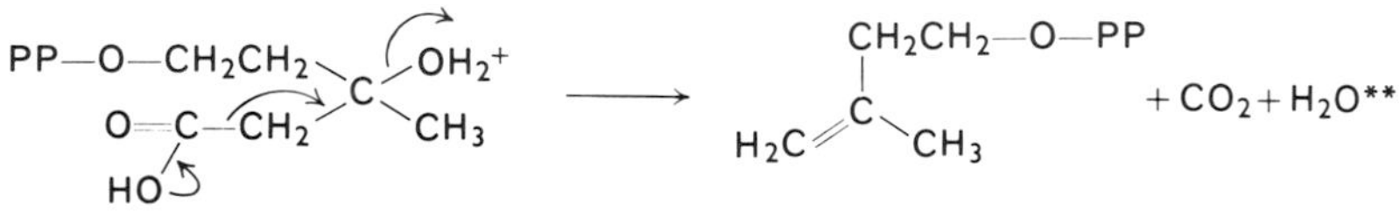

$$PP-O-CH_2CH_2 \overset{+}{\underset{HO}{\overset{OH_2^+}{\underset{O=C-CH_2}{C\diagdown CH_3}}}} \longrightarrow H_2C\overset{CH_2CH_2-O-PP}{\underset{CH_3}{C}} + CO_2 + H_2O^{**} \tag{10}$$

---

* The symbol PP will be used for the pyrophosphate grouping in the subsequent discussion.
** This transformation is formulated as proceeding by a course that involves an initial protonation of the hydroxyl group. While this is mechanistically satisfactory, it is probable that the actual process occurring in the cell involves the conversion of the hydroxyl group to a phosphate or pyrophosphate group to enhance its "leaving group" character.

This compound can undergo enzymatically induced isomerization to $\gamma$-$\gamma$,dimethylallyl pyrophosphate:

$$\underset{H_2C}{\overset{H_3C}{>}}C-CH_2CH_2-O-PP \quad \longrightarrow \quad \underset{H_3C}{\overset{H_3C}{>}}C=CHCH_2-O-PP \qquad (11)$$

These five-carbon-atom compounds are now structurally capable of undergoing polymerization into $C_{10}$, $C_{15}$, $C_{20}$, $C_{30}$, and $C_{40}$ compounds by a process that can be formulated in the following manner. The ready displacement of the pyrophosphate fragment, with the development of carbonium ion character in the dimethylallyl fragment, permits the alkylation shown in the following equations to occur:

geranyl pyrophosphate

The reaction of geranyl pyrophosphate with another molecule of the isopentenyl ester yields the fifteen-carbon-atom compound, farnesyl pyrophosphate, and another reaction of the same kind yields the $C_{20}$ compound. It is now apparent that continuation of the process can lead to a series of pyrophosphate esters in increments of five-carbon atoms, as shown on the next page.

An examination of the structures of the terpenoid compounds that are the final products shows that numerous additional oxidation and reduction steps are involved in the elaboration of the final structures.

The transformation of these acyclic precursors into cyclic compounds occurs by an acid-catalyzed reaction that can be initiated by a proton donor or by a species which, while it cannot be described with certainty, can be regarded as the biological equivalent of protonated oxygen, or $OH^+$. For example, the conversion

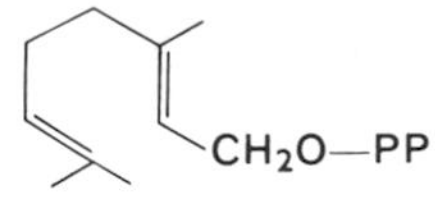

geranyl pyrophosphate

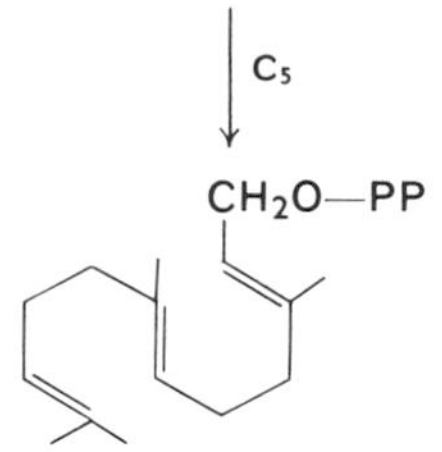

terpenes ($C_{10}$ compounds)  (13)

farnesyl pyrophosphate

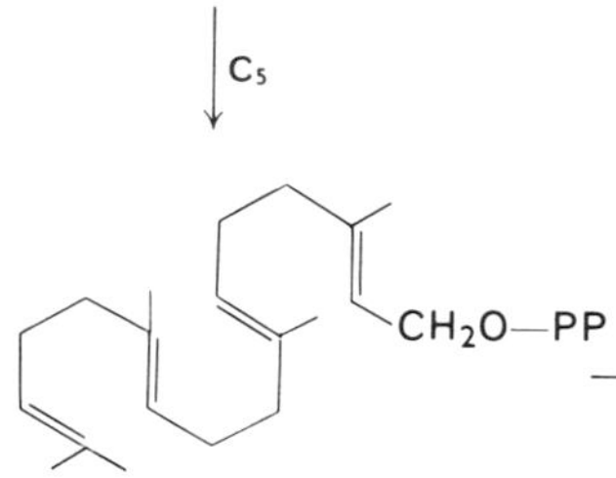

sesquiterpenes ($C_{15}$ compounds)  (14)

diterpenes ($C_{20}$ compounds)  (15)

of the $C_{20}$-pyrophosphate ester into bi- and tricyclic diterpenes can be described by the following reactions:

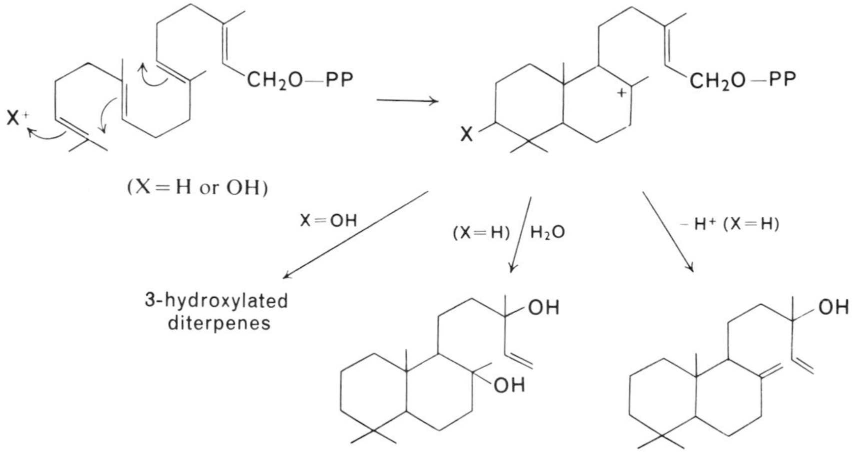

($X = H$ or $OH$)

$X = OH$

($X = H$) / $H_2O$

$-H^+$ ($X = H$)

3-hydroxylated diterpenes

Sclareol

Manool  (16)

**Exercise 3.** Formulate a reasonable course by which the bicyclic carbonium ion intermediate in the above equations can undergo further transformation to give a tricyclic and a tetracyclic diterpene.

Squalene is formed by the (reductive) coupling of two molecules of farnesol (as the pyrophosphate, but in a manner that is not given here in detail):

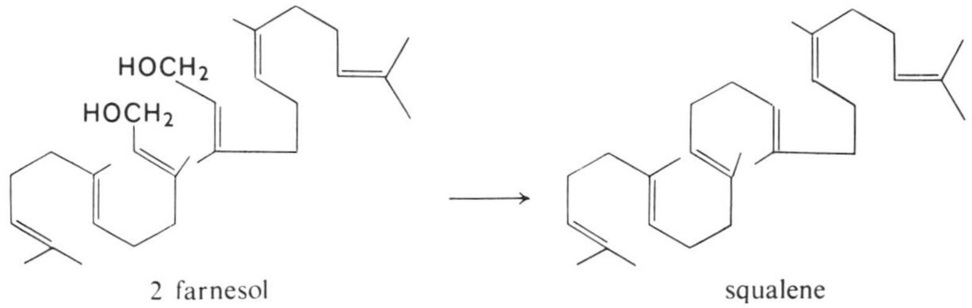

2 farnesol        squalene        (17)

The cyclization of squalene to the polycyclic precursor of the triterpenes and steroids can be brought about by the attack of the hypothetical Lewis acid $OH^+$, as formulated above for the case of the diterpenes:

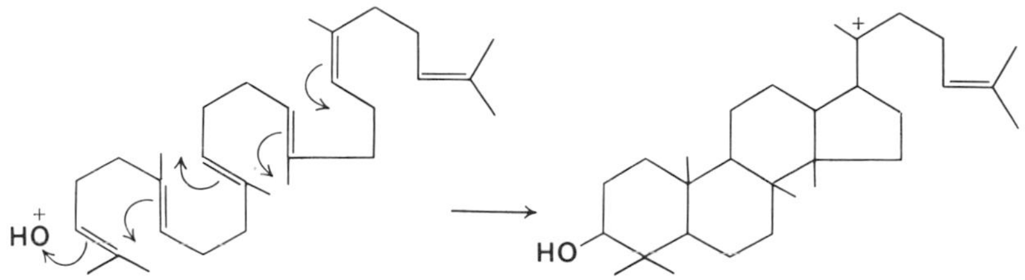

(18)

The ionic intermediate that is shown as the immediate result of the ring-closure step is converted into lanosterol by the abstraction of a proton, with accompanying shifts of methyl groups and hydrogen atoms; these are best shown by rewriting the structure in the conformational representation ($\bullet = CH_3$):

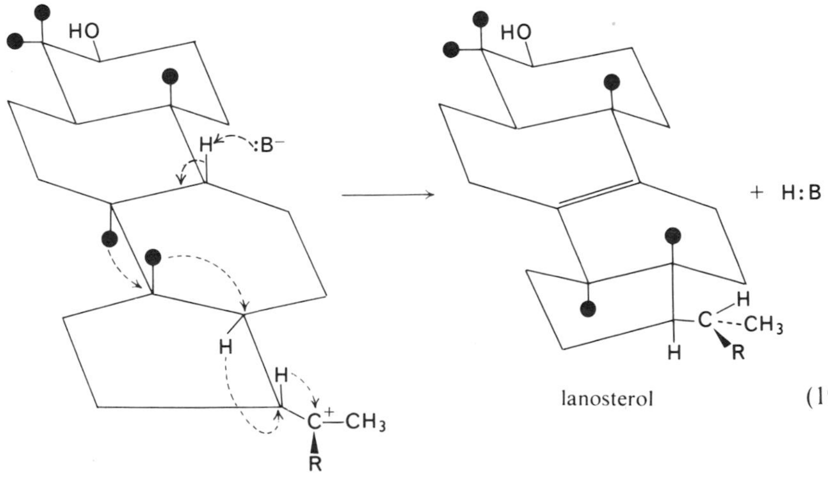

lanosterol        (19)

Subsequent alterations of lanosterol into cholesterol have been described in Section 32.6.

The biosynthesis of carotenoids is believed to involve the coupling of two $C_{20}$ isoprenoid fragments in a manner similar to that in which squalene is formed from farnesol, followed by dehydrogenation to the final conjugated polyene.

### 32.15   Rubber

Natural rubber is a high-molecular-weight, linear polyisoprenoid hydrocarbon of the composition $(C_5H_8)_n$ and the partial structure

$$-----CH_2C\!\!=\!\!CHCH_2\left[\!\!\begin{array}{c} CH_2C\!\!=\!\!CHCH_2 \\ | \\ CH_3 \end{array}\!\!\right]_x \!\!\begin{array}{c} CH_2C\!\!=\!\!CHCH_2----- \\ | \\ CH_3 \end{array}$$

It is seen that the rubber molecule is a long chain of "isoprene units" linked in the head-to-tail fashion. The molecular weights of natural rubbers are very large, with values of $n$ of the order of 5000 (molecular weight 300,000). Rubber is characterized by its elasticity. When it is stretched, rubber assumes an ordered structure. The X-ray diffraction pattern of stretched rubber can be interpreted in terms of an all-*cis* arrangement of the double bonds. Rubber differs in this respect from two other naturally occurring, high-molecular-weight polyisoprenoid compounds, gutta-percha and balata, which have their double bonds in the *trans* configuration, and are horny, nonelastic substances.

Rubber occurs in nature as a colloidal disperson, containing about 40 % of the hydrocarbon as particles suspended in an aqueous medium; this is the latex, or milky sap, of plants from which rubber can be obtained. The most important commercial source of rubber is the rubber tree, *Hevea brasiliensis*, which is cultivated in large plantations, mostly in Malaya and New Guinea. The latex is collected by scoring the bark of the tree and allowing the latex to drip into a cup. The collected latex is placed in large vats containing partitions that divide the vat into compartments about an inch in width. The latex is coagulated by the addition of dilute acetic acid, and the sheets of soft, white, coagulated rubber are lifted from the vat, pressed to remove excess water, and dried in smoke ovens. The final crude rubber is a pale-brown, semitransparent, elastic "crepe" rubber, in which form it is furnished to rubber processing mills.

The double bonds in crude rubber are capable of undergoing many of the typical reactions of olefinic double bonds. Rubber is readily attacked by ozone; it can be hydrogenated; it will add bromine and chlorine; and it is subject to oxidative deterioration when exposed to oxygen or air for prolonged periods of time, particularly when at elevated temperatures. Raw rubber is thus subject to gradual deterioration, becoming hard and brittle and losing its elasticity. It has been found empirically that the useful properties of rubber can be enhanced and preserved if the rubber is subjected to the process known as *vulcanization*. The vulcanization of rubber is a complex process that forms the subject of a vast technology. In its essentials, it consists in the addition of about 3% of sulfur to the "masticated" (mechanically disintegrated) crude rubber, followed by heating to about 140° for short periods of time. Various additives (zinc oxide, carbon black) and accelerators (tetramethylthiuram disulfide, $Me_2NCSS—SCSNMe_2$) are added to the vulcanization mixture to improve the efficiency of the process and to provide rubber of superior or specialized properties. The use of larger amounts of sulfur (20–35%) results in the formation of "hard rubber."

The exact nature of the chemical reactions that take place between sulfur and the rubber hydrocarbon during the vulcanization process is not known. It is likely that sulfur forms both intra- and intermolecular carbon-sulfur bonds, with the formation of a certain amount of cyclic structures within chains of the rubber molecules, and the production of "cross links" between separate chains. The improvement in the properties of rubber upon vulcanization is due in part to the disappearance of some of the carbon-carbon double bonds, with a consequent increase in stability toward oxidative deterioration, and in part to the fact that the rubber molecules are tied together to some extent by cross-linking of the chains. Elongation and alignment of the chains is still possible, but an overall structural integrity is present in vulcanized rubber that is not found in the crude material, with the result that extensive disruption of the molecular organization of vulcanized rubber does not so readily occur in the course of its use.

## 32.16 The biosynthesis of compounds containing aromatic rings

Compounds containing one or more aromatic (benzenoid) rings are the most widely occurring products of secondary plant metabolism.* The largest group of aromatic compounds in nature comprises those that contain phenolic hydroxyl groups or ethers of these. Aromatic compounds are almost exclusively the products of plant metabolism (including fungi and many microorganisms); those that are produced in animal organisms are relatively few in number and are, for the most part, formed by modifications of aromatic compounds ingested as food, or by aromatization of cyclic, nonaromatic precursors.

Most naturally occurring compounds that contain aromatic rings are formed by way of two biosynthetic pathways: (1) the "acetate" pathway, which utilizes acetyl coenzyme A (in a manner different from that involved in terpenoid biosynthesis); (2) the "shikimic acid" pathway, the starting compounds for which are carbohydrate-derived.

---

* This statement refers to the number and structural diversity of the compounds. Chlorophyll (which is present in all green plants), sugars, starch and cellulose (which are produced in large amounts by photosynthesis and are present throughout the plant kingdom), and amino acids and proteins (which are the fundamental constituents of living cells) are universally distributed; but these substances are not characterized by so wide a diversity of chemical structure as the compounds to be dealt with here.

## 32.17   The acetate pathway of biosynthesis

A biosynthetic route to the formation in plants of a wide variety of aromatic compounds, including derivatives of benzene, naphthalene, and anthracene, consists in the formation and cyclization of a chain of two-carbon units which, at some stage in the synthetic sequence, may consist of a poly-$\beta$-ketomethylene ("polyketide") chain which cyclizes by aldol-like condensation reactions:

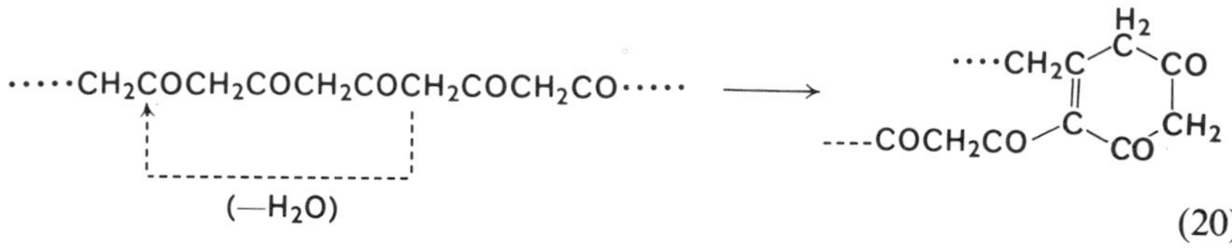

$$\cdots\cdots CH_2COCH_2COCH_2COCH_2CO\cdots\cdots \longrightarrow$$

$$(-H_2O) \tag{20}$$

Some specific examples of compounds formed by the "acetate pathway," in which the presumptive precursors are shown, are the following:

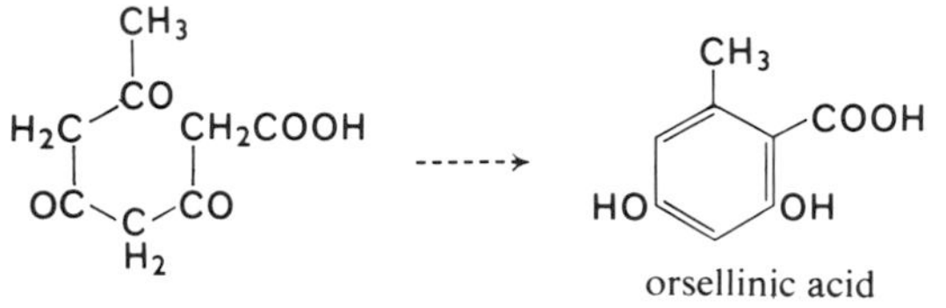

$$\tag{21}$$

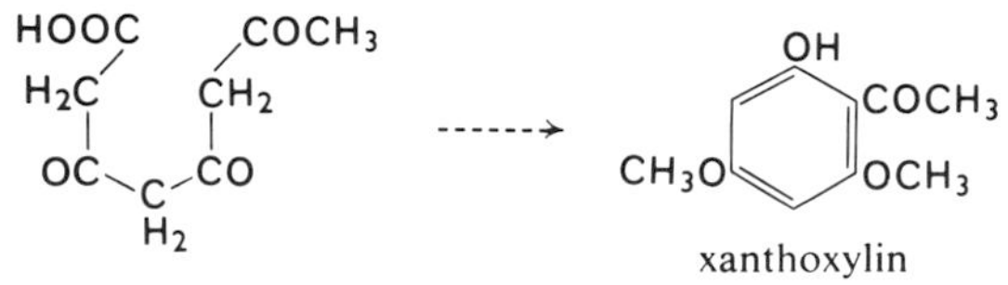

orsellinic acid

$$\tag{22}$$

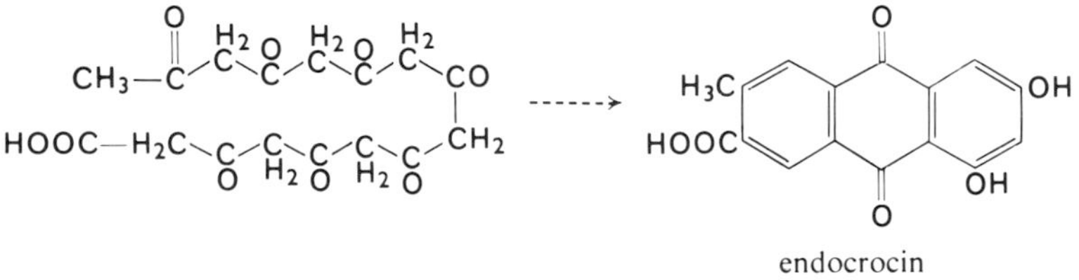

xanthoxylin

$$\tag{23}$$

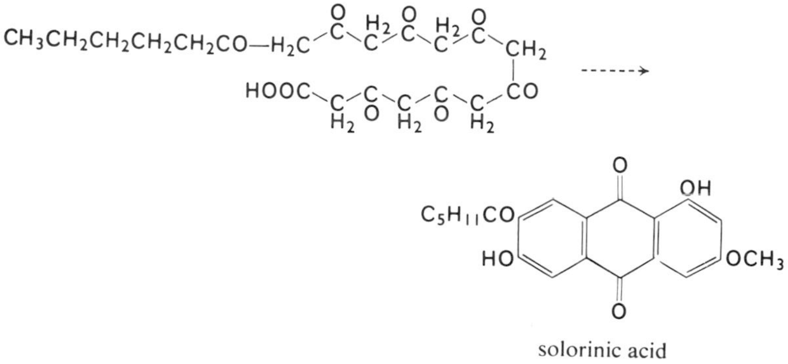

endocrocin

$$\tag{24}$$

solorinic acid

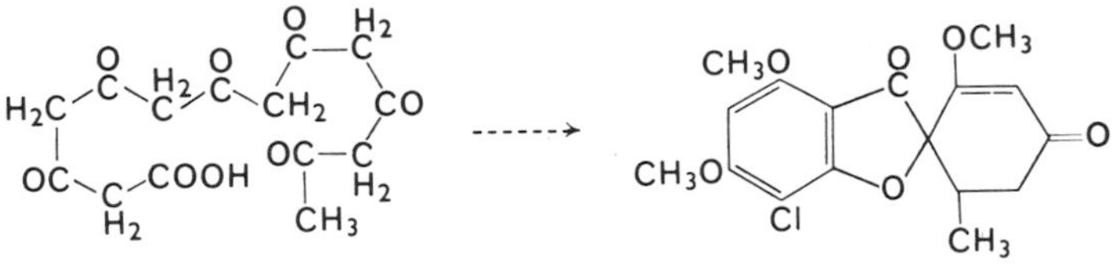

$$(25)$$

griseofulvin

It will be noticed that in these examples a number of processes are involved in the formation of the final products: reduction, oxidation, methylation, and, in one case, the introduction of a nuclear chlorine atom. The dotted arrows in the above formulations indicate that the relationship between the precursor and the final compound is that between the carbon skeletons and the positions of oxygenation. The correlation between the positions of the phenolic hydroxyl groups and the carbonyl groups of the polyketomethylene chain is most striking, and is one of the strongest arguments in support of this synthetic hypothesis.

The formation of the polyketomethylene chain, $CH_3(COCH_2)_nCOOH$, proceeds by self-condensation of "activated" acetyl units. The activation has been shown to proceed by the transformation of acetyl coenzyme A into a malonic acid derivative by reaction with carbon dioxide under the influence of an enzyme whose prosthetic group is biotin:

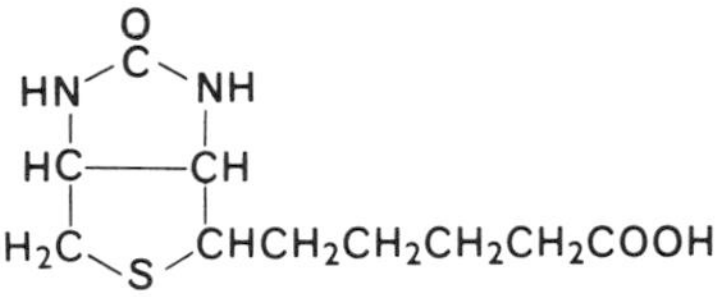

biotin

$$CH_3CO\!-\!SCoA \quad \xrightarrow[\text{biotin/enzyme}]{CO_2} \quad \underset{\underset{COOH}{|}}{CH_2CO\!-\!SCoA} \tag{26}$$

malonyl coenzyme A

It will be apparent that the carboxylation of the methyl group has converted it into a highly active methylene group, particularly susceptible to Claisen-type (acetoacetic ester-type) condensation. The formation of the polyketomethylene chain can now occur as follows:

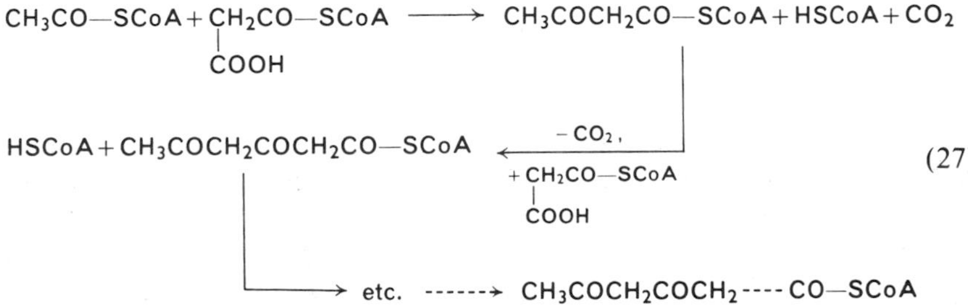

An important alternative to the extension of $CH_3CO$—SCoA by condensation with malonyl coenzyme A involves the same kind of reaction *with some acyl coenzyme A other than acetyl*. Chain formation of this kind can lead to polyketomethylene chains which start with a group other than methyl:

$$RCO-SCoA + \text{malonyl coenzyme A} \xrightarrow[\text{Equation (27)}]{\text{as in}} RCO(CH_2CO)_nCH_2CO-SCoA \quad (28)$$

Thus, the general synthesis described in these expressions becomes a very versatile one, leading to the formation of many different compound types. It is probable that solorinic acid, Equation (24), shown above, has its origin in a process in which the precursor is the CoA derivative of hexanoic acid; this, by chain extension with malonyl coenzyme A, leads to the polyketomethylene precursor that gives rise to the final hexanoyl anthraquinone derivative.

A large and important group of naturally occurring compounds that are formed by this general pathway are fifteen-carbon-atom compounds called "flavonoid" compounds. These are formed by the pathway shown in the following expression, in which the acyl precursor is derived from cinnamic acid (the biosynthetic origin of which is described in Section 32-18):

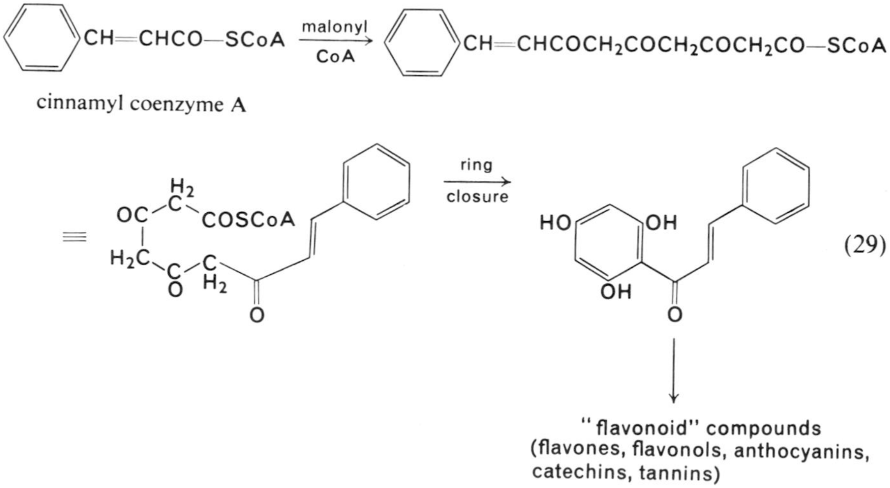

## 32.18  The shikimic acid pathway

The formation of the aromatic ring in many naturally occurring compounds takes place by the cyclization of carbohydrate-derived precursors through the intermediate formation of a series of cyclohexanecarboxylic acid derivatives, the six-membered ring of which ultimately becomes aromatic. The fundamental structural unit of the compounds synthesized by this route is a nine-carbon-atom unit consisting of an aromatic ring attached to a three-carbon-atom side chain. This unit is often referred to as a $C_6$-$C_3$ unit, and compounds derived from it are often called "phenylpropanoid" compounds. Alterations of the $C_6$-$C_3$ unit by extension or degradation of the side chain give rise to a great variety of compounds; some typical naturally occurring compounds formed by this route are the following:

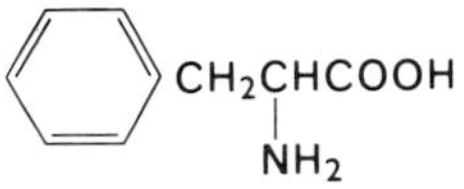

phenylalanine

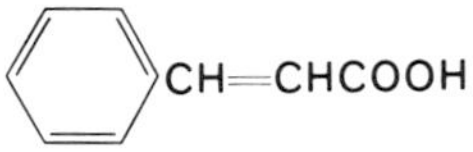

cinnamic acid

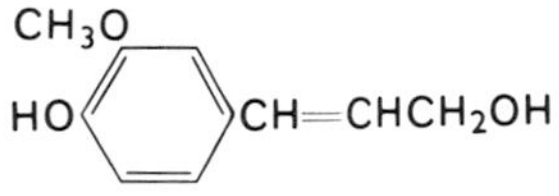

coniferyl alcohol

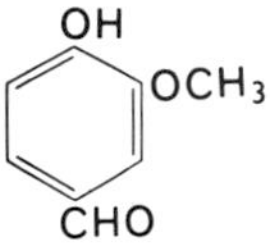

vanillin

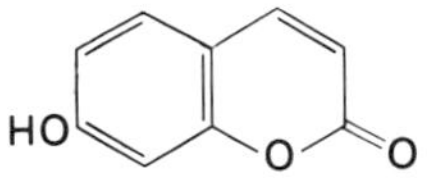

umbelliferone

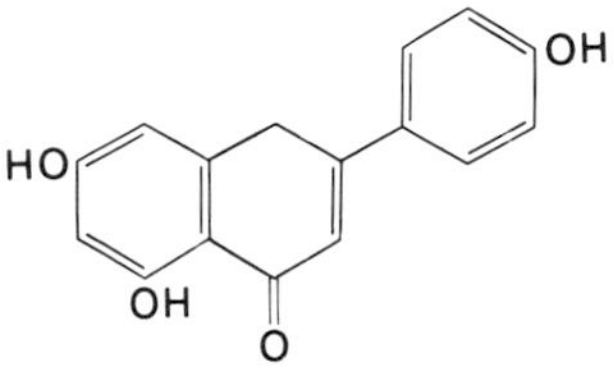

apigenin

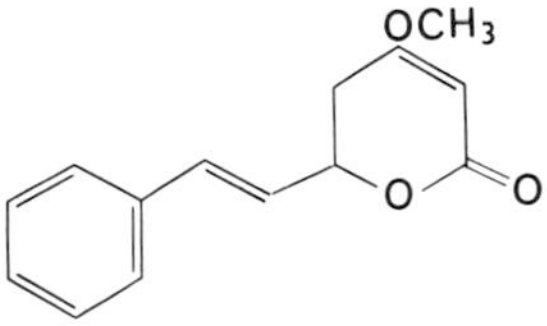

kawain

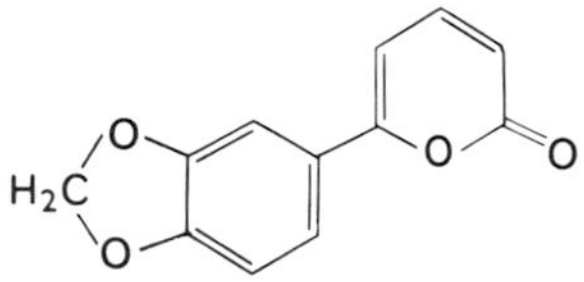

paracotoin

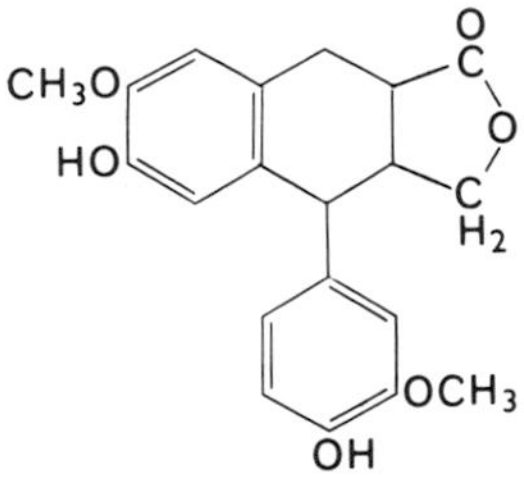

conidendrin

A key compound in the synthesis of the $C_6$-$C_3$ unit is *shikimic acid*; this leads, by way of *chorismic acid*, to *prephenic acid*, which, it will be seen, contains the $C_6$-$C_3$ unit and is the immediate precursor of the final phenylpropanoid compounds. The following equations show the series of reactions that constitute the shikimic acid route of biosynthesis:

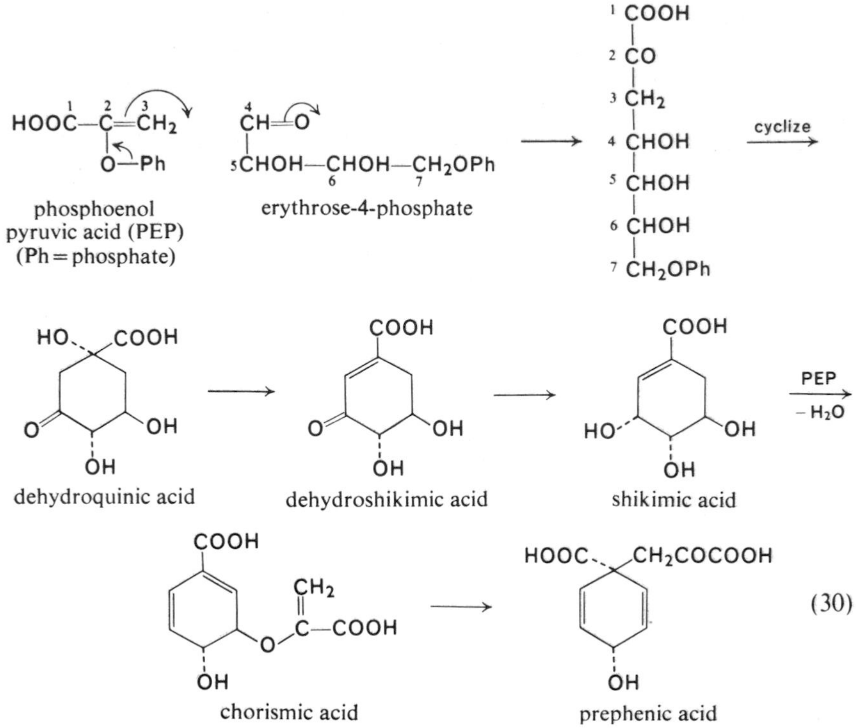

$$(30)$$

Prephenic acid is so called because it readily undergoes dehydration and decarboxylation to yield phenylpyruvic acid; and phenylpyruvic acid, by reaction with ammonia and reduction (in the presence of appropriate enzymatic catalysis), yields phenylalanine:

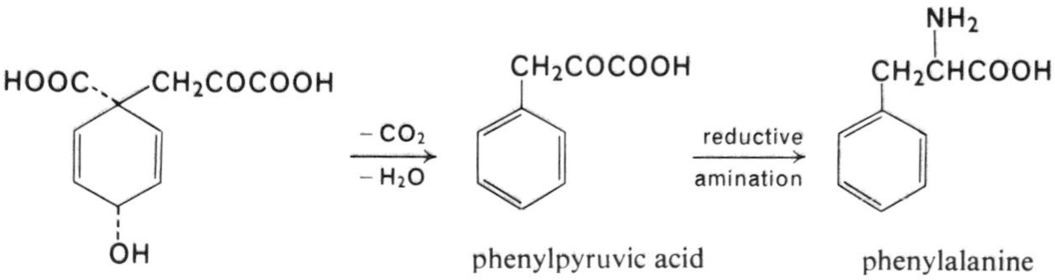

$$(31)$$

Although a great deal is known at the present time about the gross features of biosynthetic pathways—the probable precursors, and the origin of the skeletal structures of the final products of biosynthesis—little is known about the details of the many reactions that intervene between the starting compound (for example, phenylalanine) and the elaborate end product of the synthesis (for example, apigenin). Such intermediate stages as are involved in the building up of chains by the acetate pathway, the formation of isoprenoid intermediates by the mevalonic acid pathway, the nature of the reactions that lead to ring closures of the several kinds that have been discussed, and such incidental reactions as the manner in which methyl groups, sugars, and certain other substituents are attached are understood; but many of the structural alterations that lead to the introduction of oxygen, the creation and removal of double bonds, the loss of fragments of the molecule, and many others are understood neither in respect to the time in the synthetic sequence at which they occur nor the precise mechanism by which they are brought about.

### Exercises

(Exercises 1–3 will be found within the text.)

**4** What product would be formed in each of the following reactions: *(a)* ozonolysis of sabinene; *(b)* reaction of myrtenal with lithium aluminum hydride; *(c)* reaction of menthol with benzoyl chloride and alkali; *(d)* hydrogenation (catalytic) of farnesol; *(e)* reaction of dextropimaric acid with bromine (in $CCl_4$ solution); *(f)* reaction of totarol with benzenediazonium chloride; *(g)* oxidation of phytol with manganese dioxide; *(h)* oxidation of cholesterol by the Oppenauer method?

**5** How could the result of ozonolysis be used to distinguish between $\beta$- and $\gamma$-carotene?

**6** How many geometric isomers of vitamin A are theoretically possible? Draw their structures.

**7** Suggest the pathway by which pinene, upon treatment with HCl, is converted into bornyl chloride,

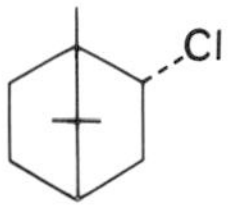

**8** Farnesol (A) can be cyclized by treatment with acid to bisabolene (B). Formulate the course of this cyclization.

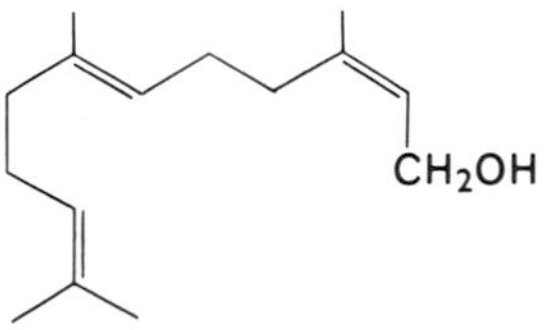

A 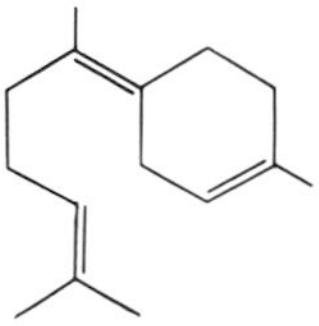                    B

**9** If limonene were isolated from a plant that had been fed radioactive acetic acid (labeled with carbon-14 in the methyl group), where would you expect to find the radioactive carbon atoms in the terpene? Can you suggest a scheme of degradation of the limonene which would allow the identification of each of the ten-carbon atoms as radioactive or nonradioactive?

**10** The following compound has been isolated from a seed oil:

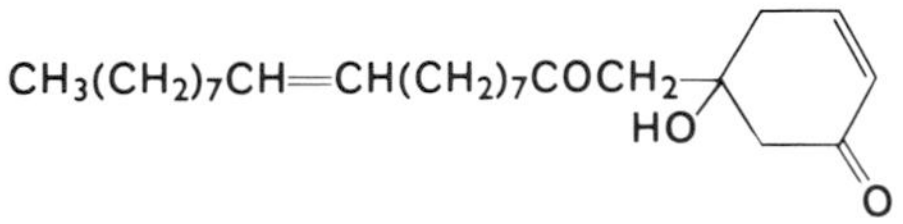

$$CH_3(CH_2)_7CH{=}CH(CH_2)_7COCH_2{-}$$

Comment upon the significance of this structure to the acetate theory of biogenesis, and suggest the route by which it is synthesized in the plant. Assume a loss of $CO_2$ at a late stage of the synthetic pathway.

## Chapter thirty-three

/ **Physiologically Active Compounds. Organic Compounds as Medicinals**

## 33.1 Drugs and chemotherapy

The service of chemistry to medicine represents one of the most fascinating facets of the story of the application of scientific knowledge to the welfare of mankind. Stemming from the use of medicinal plants by herbalists as far back in history as the time of ancient Greece, the utilization of plant materials and of the pure compounds isolated from plants has developed in recent years into a systematic search of the plant kingdom for new drugs. It is a remarkable fact that, up to the present time, nearly all of the important drugs derived from plants have been known since long before the dawn of modern chemistry; and it is equally remarkable that among these drugs are found many that still remain indispensable in modern medical practice. Morphine, cocaine, atropine, and other belladonna alkaloids, strychnine, digitalis, caffeine, ephedrine, adrenaline, and reserpine are drugs widely used in modern clinical practice. All of them are derived from natural sources; and though nearly all of them have been synthesized in the laboratory, most are still isolated from natural sources.

The use of drugs in medicine may be discussed under two chief categories: the first of these includes drugs that are used in the treatment and cure of specific diseases, and is described by the term "chemotherapy." Thus, one speaks of the chemotherapy of tuberculosis, of malaria, of syphilis, of infections caused by certain bacteria, and so on. Chemotherapy involves the use of chemical compounds of definite structure, and is distinct from the treatment of disease that involves the use of serums and vaccines. The second category of drugs includes those that have characteristic effects upon the animal organism but which are not specific remedies for particular diseases. Among drugs of this kind are morphine (for analgesia), cocaine (for anesthesia), barbiturates (for sedation and hypnotic action), atropine (as an antispasmodic), antihistamines (for allergic responses), and others. Although it is true that the use of one of these drugs can assist in recovery from a specific bacterial or viral infection, the action of the drug is not directed at the pathogenic organism; in chemotherapy, however, the invading bacteria are the direct targets of the drug's action.

## 33.2  Chemotherapy

The problem of chemotherapy is to attack and destroy an invading organism without at the same time injuring or destroying the cells of the infected host. The first conspicuous success in the field of chemotherapy was that achieved by Ehrlich with his discovery in 1910 of the specific action of certain organo-arsenic compounds against the spirochete of syphilis. Many years passed before new advances were made in chemotherapeutic drugs, the first of these advances being the sulfa drugs, which were discovered as the result of studies that were directed at the outset by the principles first put forward by Ehrlich. In Chapter 26 there is given a brief account of the discovery of the sulfa drugs, a discovery for which the foundation was laid by Ehrlich's hypothesis that it should be possible to stain (and thus destroy) selectively a bacterial cell by the use of an organic dyestuff. The spectacular success of the sulfa drugs in combating many bacterial diseases showed clearly that chemotherapy offered a new means of attack upon disease.

## 33.3  Antibiotics

The discovery that certain fungi produce powerfully antibacterial substances led to a new avenue of approach to chemotherapy. Antibiotics are compounds produced by one organism that are toxic or lethal to another organism; and in ordinary usage, the targets of antibiotic therapy are bacterial organisms.

The first successful antibiotics to be introduced into clinical practice were the *penicillins*, compounds produced by the mold *Penicillium*:

penicillin G
(benzyl penicillin)*

---

* Other penicillins have groups other than benzyl attached to the —CONH—; for example, 2-pentenyl penicillin (Penicillin F).

The total synthesis of penicillin has been achieved. Penicillin is still produced commercially from the natural source, but the synthetic method may prove useful in permitting the preparation of penicillins of modified structure. The value of having a range of structurally different penicillins lies in the fact that some bacteria have been found to become penicillin-insensitive by a process of adaptation through genetic selection in the presence of the drug. Thus, an organism "fast" to one penicillin may be susceptible to another.

A recent success in the preparation of a new "penicillin" illustrates one way in which progress may be made in the constant search for antibiotics having new and desirable properties. It has been found that a *Penicillium*, when grown under properly controlled conditions, is able to synthesize, not penicillin, but the non-acylated portion of the penicillin molecule, 6-aminopenicillanic acid:

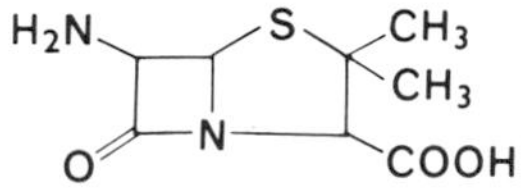

By a study of the culture conditions under which the mold is grown it was found possible to produce 6-aminopenicillanic acid in commercial quantities and to make it available for intensive study. It is evident that this compound can be converted by acylation of the amino group into an almost unlimited number of "synthetic" penicillins. Many hundreds of these have been prepared and studied for their antibiotic activity. One of them, the N-2,6-dimethoxybenzoyl derivative, prepared in 1960, has been found to have the highly desirable property of being effective against penicillin-resistant staphylococci:

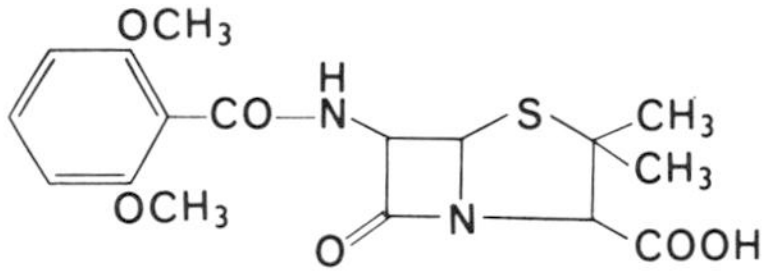

It is believed that the "resistance" of the bacteria is due to their production of penicillinase, an enzyme that destroys penicillin by hydrolyzing the $\beta$-lactam linkage. The new penicillin appears to be resistant to the destructive action of penicillinase, but retains its antibiotic power.

It is to be expected that further study will result in the discovery of new N-acyl-6-aminopenicillanic acids with valuable antibiotic properties.

The search for new antibiotics has led to a large-scale study of fungal and related organisms (actinomycetes, streptomycetes) of many kinds, and to the discovery of a large number of effective chemotherapeutic agents. These compounds, selected by exhaustive tests against experimental infections in animals, include such important therapeutic agents as Chloromycetin, Streptomycin, Terramycin, Aureomycin, Erythromycin, as well as the penicillins mentioned above.

Chloromycetin is a compound of simple chemical structure, and contains two structural features rare in nature: a nitro group on a benzene ring, and covalently bound chlorine:

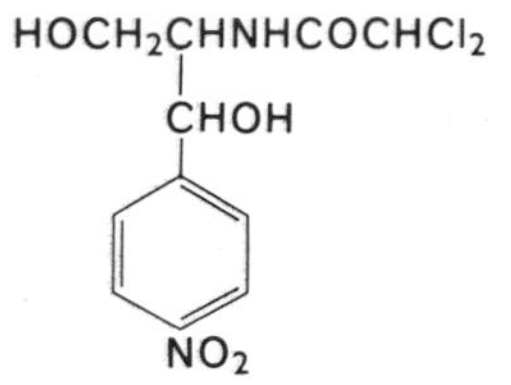
HOCH2CHNHCOCHCl2
CHOH
NO2

Chloromycetin (chloramphenicol)

The structure of Chloromycetin was determined in the following way.

1. On acid hydrolysis, Chloromycetin ($C_{11}H_{12}O_5N_2Cl_2$) gives dichloroacetic acid and a basic compound, $C_9H_{12}O_4N_2$.

2. Although Chloromycetin does not react with periodic acid, the base $C_9H_{12}O_4N_2$ consumes two moles of $HIO_4$ to give formaldehyde, ammonia, formic acid, and *p*-nitrobenzaldehyde.

---

**Exercise 1.** Formulate the reactions that are described in this degradation of Chloromycetin.

---

Chloromycetin has been synthesized in several ways, one of which is the following:

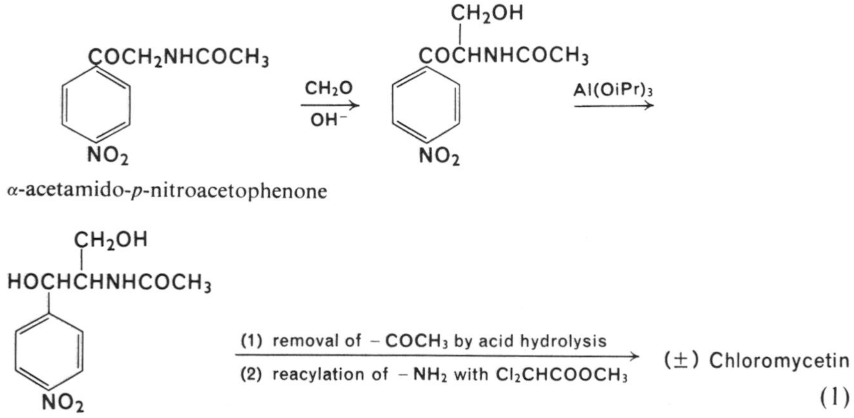
COCH2NHCOCH3
NO2
α-acetamido-p-nitroacetophenone
CH2O
OH−
CH2OH
COCHNHCOCH3
NO2
Al(OiPr)3
CH2OH
HOCHCHNHCOCH3
NO2
(1) removal of −COCH3 by acid hydrolysis
(2) reacylation of −NH2 with Cl2CHCOOCH3
(±) Chloromycetin

(1)

Resolution of the synthetic (±) mixture can be accomplished at the stage prior to the introduction of the dichloroacetyl group.

Terramycin and Aureomycin are the trade names for two clinically effective, closely related drugs derived from "tetracycline":

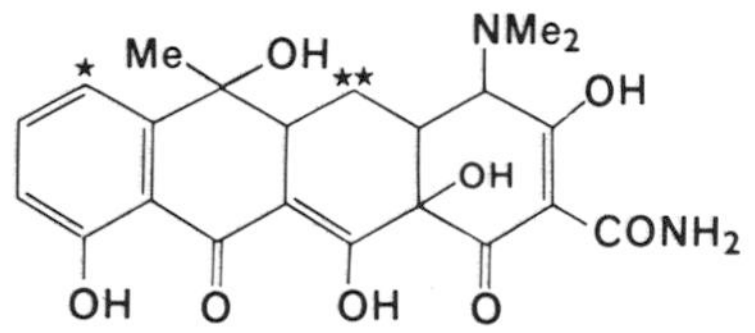
Me
OH
NMe2
OH
OH
CONH2
OH
O
OH
O

tetracycline

Terramycin has a hydroxyl group at (**), and is *oxytetracycline*, whereas Aureomycin bears a chlorine atom at (*); it is *chlorotetracycline*.

Streptomycin has a carbohydrate-like structure; along with dihydrostreptomycin (—$CH_2OH$ in place of —CHO) it is an important drug in the chemotherapy of tuberculosis:

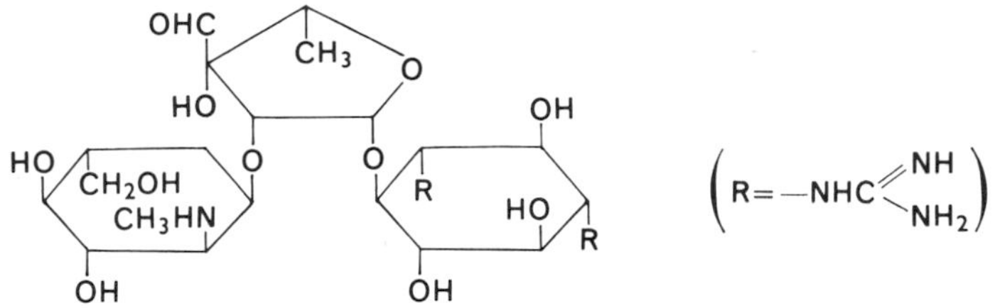

Streptomycin

The clinical chemotherapy of tuberculosis includes the use of two simple synthetic compounds in combination with Streptomycin. These are isonicotinic acid hydrazide (isoniazid) and *p*-aminosalicylic acid (PAS):

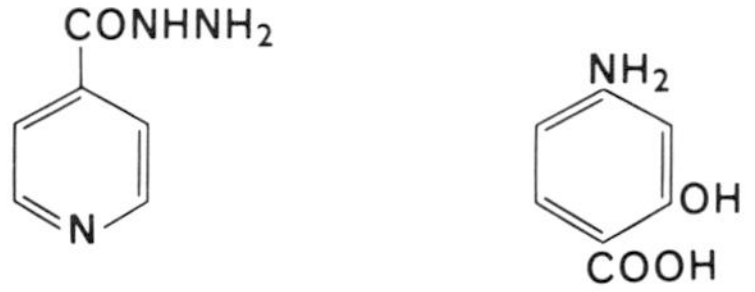

isoniazid        PAS

## 33.4 Macrolide antibiotics

Of the many classes of new antibiotic compounds that have been discovered in the past decade is a group of compounds, produced by various strains of Streptomycetes, that are unique in their possession of a many-membered lactone ring as their essential structural feature. These compounds are called "macrolides." One of the most important, from the standpoint of clinical effectiveness, is *erythromycin*, $C_{37}H_{67}NO_{13}$. Its structure has been shown to be the following:

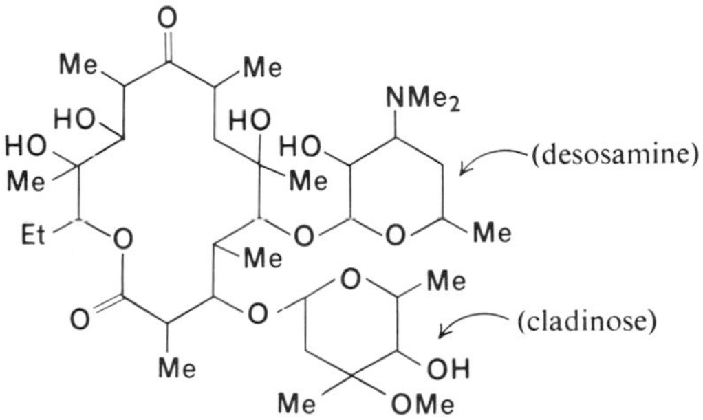

It will be noted that erythromycin contains, in glycosidic linkage, two unusual sugar residues: desosamine (a nitrogen-containing sugar) and cladinose. Desosamine is found in several other antibiotics of the macrolide group.

It is of special interest to note that the lactone portion of erythromycin is synthesized by the organism from propionic acid units. The structure is redrawn here in a way that emphasizes the repeating three-carbon unit:

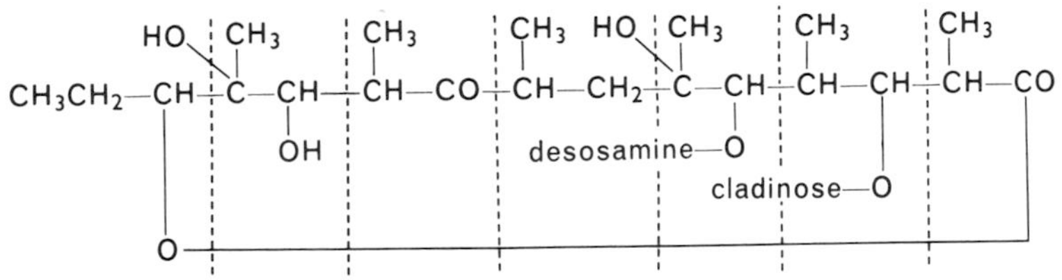

The question of the biosynthesis of erythromycin is clearly one that would at first appear to have two possible answers: (1) by the "acetate" pathway, through a polyketomethylene intermediate, with carbon-methylation of the chain:

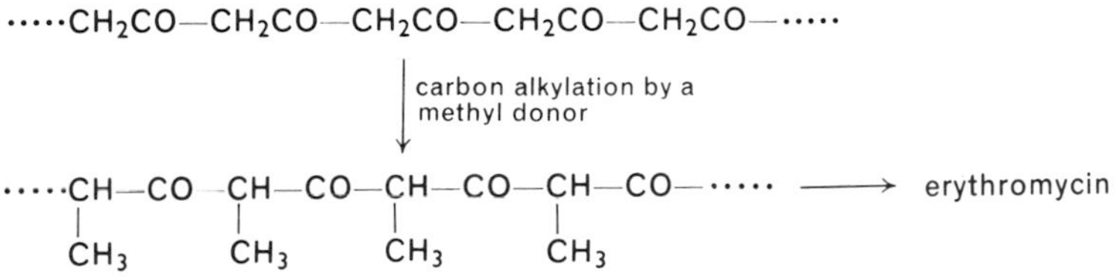

$$\cdots\cdots CH_2CO-CH_2CO-CH_2CO-CH_2CO-CH_2CO-\cdots\cdots$$

or (2) by condensation, not of acetyl units (with intermediate formation of malonyl coenzyme A), but of propionyl units, presumably with intermediate formation of methylmalonyl coenzyme A. This question has been resolved by experimental study, and it has been found that the biosynthesis of erythromycin proceeds by the condensation of propionic acid units, and not by carbon-methylation of an acetate-derived chain.

## 33.5 Structural antagonism

One of the important consequences of the discovery of the sulfa drugs was the recognition of a principle of drug action that arose from a study of the way in which sulfanilamide exerted its antibacterial action. This is the principle of *structural antagonism*. It has been observed that the action of a normal cellular constituent or metabolite is in some instances antagonized or inhibited by a chemical compound of closely similar, yet different, structure. The sulfa drugs, for example, interfere with the normal metabolic utilization of *p*-aminobenzoic acid:

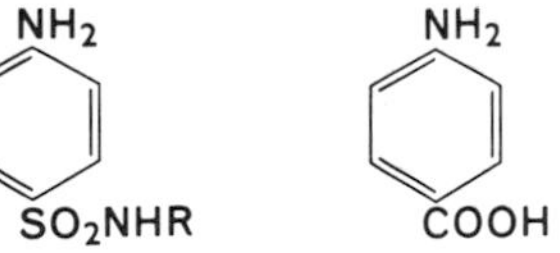

The synthetic compound *pyrithiamine*, in which a pyridine ring replaces the thiazole ring of the natural vitamin, acts as an antivitamin $B_1$:

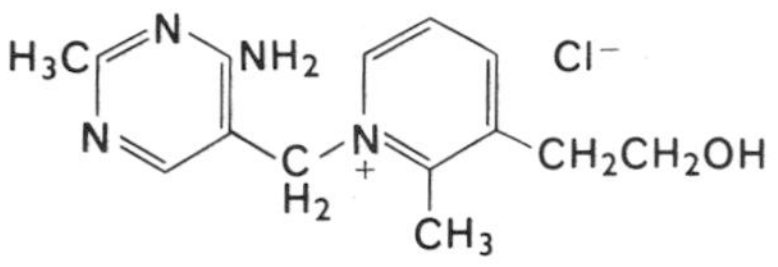

pyrithiamine chloride

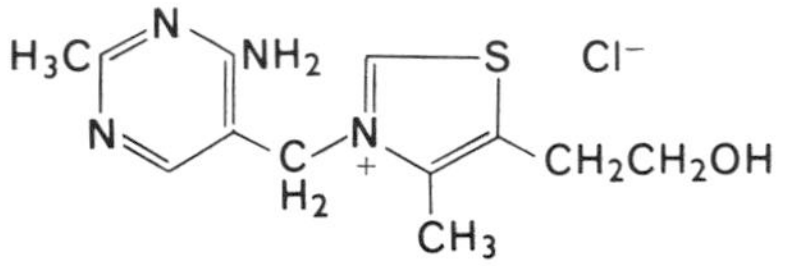

thiamin (vitamin $B_1$) chloride

The search for structurally analogous inhibitors has guided much of the work aimed at the discovery of new drugs and chemotherapeutic agents. Although not all effective drugs act as direct inhibitors of some structurally similar natural metabolite, and not all structural analogues are effective drugs, the principle has been a useful guide in past research and will continue to be applied in future studies. It must be borne in mind, however, that the preparation of structural analogues is not a touchstone for the discovery of new and effective drugs; indeed, few useful drugs can be said to have been discovered by a calculated application of this principle. Nevertheless, it represents one of the few rational approaches to an understanding of drug action and will no doubt continue to provoke fruitful research.

## 33.6   Antimalarial drugs

One of the serious military problems of the last war was the protection of fighting forces from the disabling consequences of malaria, a problem that was heightened by the fact that extensive operations were carried out in those areas of the world where the disease is prevalent. At the same time, supplies of quinine, the drug used for centuries against malaria, were cut off by enemy control of quinine-producing areas, and so it became necessary to search for a substitute as effective as the natural drug. To this end an extensive program of research was undertaken by the allied nations, and thousands of synthetic compounds were prepared and tested against the malarial parasites (*Plasmodium* species).

One approach to the preparation of active antimalarial drugs was to synthesize quinoline derivatives having structural features resembling those of quinine but being less complex in nature. Early studies on synthetic antimalarial drugs had led to the discovery in 1928 and 1933 of two clinically effective substances, Plasmoquin and Quinacrine (atebrin):

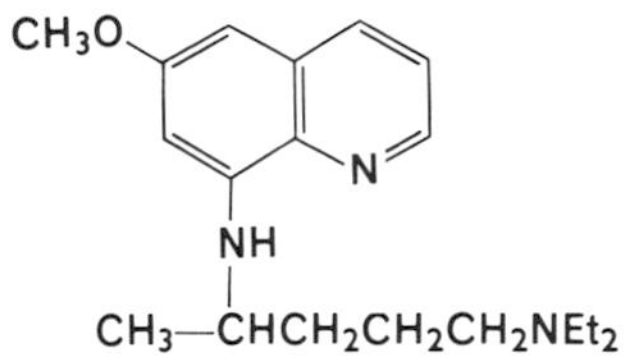

Plasmoquin 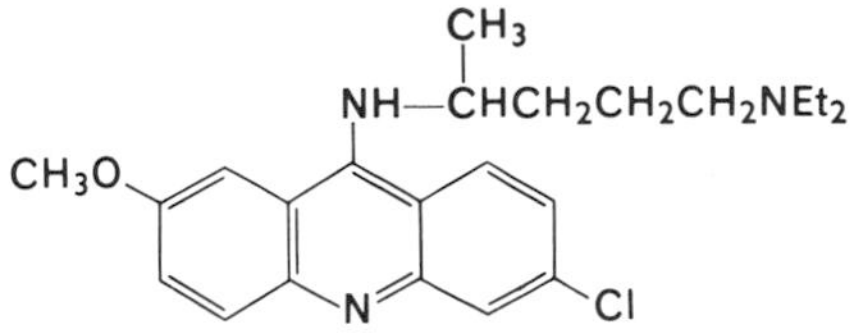                                    Quinacrine (atebrine)

In 1943 Chloroquin was discovered and put into clinical use:

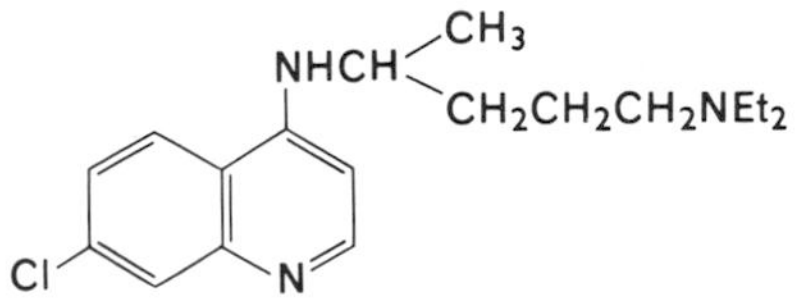

Chloroquin

It will be noted how the same or very similar structural features are found in the most effective antimalarial drugs: the quinoline nucleus, a methoxyl group in the 6-position (quinoline numbering), a chlorine atom in the 7-position, and a particular side chain. Quite a different drug, very effective and with a different mode of attack upon the malaria parasite, is Paludrin:

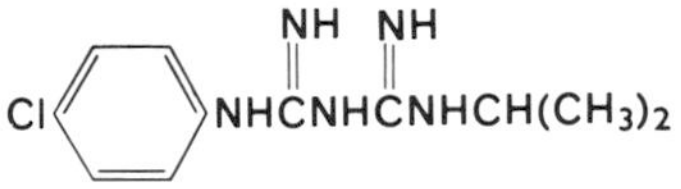

Paludrin

The development of *drug-resistant* microorganisms by adaptation, mutation, and natural selection presents a serious problem in modern medicine. Continued use of many chemotherapeutic agents, particularly when they are used in less than optimal dosage, has in many instances induced the development of a strain of the microorganism that is resistant to the drug. Indeed, there is evidence that resistance developed by one organism can be transferred to another. One example of drug resistance has been observed in the case of antimalarial drugs. In several parts of the world there are now extant strains of *Plasmodium falciparum* which are quite resistant to chloroquin, the most generally used of the antimalarial drugs. These parasites are also resistant to other antimalarial drugs, and human malaria caused by these resistant strains presents a very difficult clinical problem. The organic chemist has an important role to play in such circumstances, and has the responsibility to search for new drugs to which the parasite is susceptible. At the present time this task is being pursued by a number of organic chemists, who are engaged in the synthesis of novel chemical structures that possess antimalarial activity. That resistance to a new drug will develop in time is now to be anticipated; and so the continuing synthesis and testing of new drugs is a never-ending task.

## 33.7 Drugs affecting the nervous system

The human nervous system comprises central and peripheral portions, which are closely interdependent and under chemical control. Three compounds, simple in structure, are principally responsible for regulating the functioning of the nervous system; they are *acetylcholine*, *epinephrine* (also called adrenaline), and the closely related *norepinephrine*:

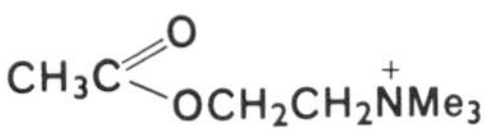

$$CH_3C\!\!\diagdown\!\!\begin{smallmatrix}O\\\phantom{x}\end{smallmatrix}\!\!\diagup OCH_2CH_2\overset{+}{N}Me_3$$

HO—⟨ ⟩—CHCH₂NHR with HO and OH substituents

acetylcholine (cation)

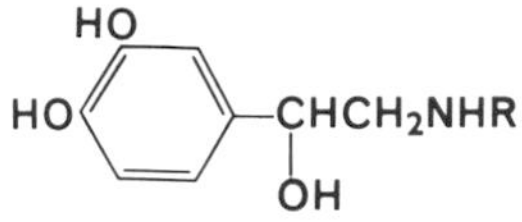

R = CH₃   epinephrine (adrenaline)
R = H      norepinephrine (noradrenaline)
(natural forms are levorotatory)

*Epinephrine.*   Epinephrine and norepinephrine are involved in the functioning of the *sympathetic nervous system*. Their physiological effects are many and various, and the pharmacology of epinephrine, norepinephrine, and epinephrine-like synthetic compounds is chiefly concerned with reproducing or antagonizing the effects of the natural compounds.

Epinephrine is secreted by the adrenal medulla and released at sympathetic nerve endings. Among its physiological effects (when administered by injection) are a rise in blood pressure, a rise in blood sugar (due to mobilization of liver glycogen), and a relaxation of bronchial muscles. Its vasoconstrictor action makes it useful in the control of bleeding, and its effect on the bronchioles renders it of use in the relief of asthma.

A synthesis of epinephrine is as follows:

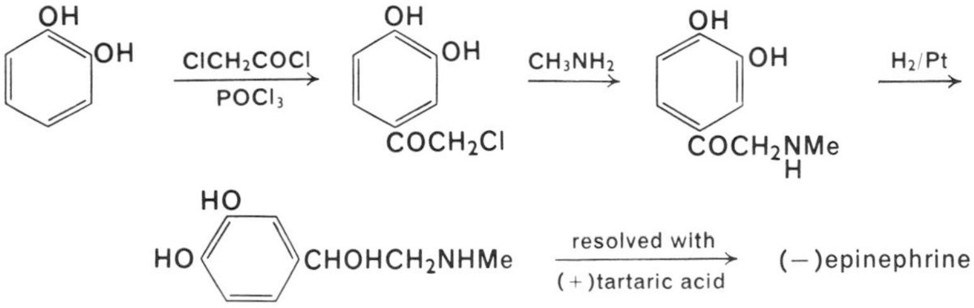

(±) epinephrine → resolved with (+)tartaric acid → (−)epinephrine     (3)

The antagonism of epinephrine, either that released at a nerve ending with the consequent stimulation of an effector organ, or that released into the bloodstream from the adrenal glands, has clinical applications of several kinds, and an effective epinephrine-antagonizing drug is potentially therapeutically useful.

An effective blockage of epinephrine effects is caused by certain N,N-disubstituted β-chloroethylamines. One of these (Dibenamine) is described in Chapter 29.

Numerous epinephrine-like compounds are used in medicine. These are more closely analogous to *ephedrine* rather than to epinephrine itself, and display pharmacological effects similar to those of ephedrine:

OH
/\    |
|  |—CHCHNHCH₃
\/    |
      CH₃
ephedrine

Ephedrine occurs in plants of the genus *Ephedra,* and was used (in the form of the plant extract) in Chinese medicine for centuries before its discovery and study by chemists. Since its discovery hundreds of phenethylamine derivatives have been prepared and tested pharmacologically, and a number of these are in widespread clinical use. Among these are the well-known agents for relief of nasal congestion and bronchial constriction. The effect of ephedrine and its synthetic cogeners resemble some of the actions of epinephrine, and for this reason the group of substances is known by the term "sympathomimetic" drugs: i.e., drugs that mimic the effect of stimulation of the sympathetic nervous system. Among the therapeutically important sympathomimetic agents are

OH                                    HO    OH
HO—⟨ ⟩—CHCH₂NHCH₃              ⟨ ⟩—CHCH₂NHCH₃

    Synephrine                       Neosynephrine

              NH₂
              |
    HO—⟨ ⟩—CH₂CHCH₃

              Paredrine

***Acetylcholine and acetylcholinesterase.***   Acetylcholine plays a part in the transmission of nervous impulses, both across synaptic junctions of nerves and from the endings of *parasympathetic* nerves to the organs they control. As in the case of epinephrine, the pharmacology of parasympathetic nerve action is concerned with mimicking the action of acetylcholine, and blocking or inhibiting the effects of endogenous acetylcholine.

Since the voluntary (skeletal) muscles are innervated by parasympathetic nerves, a deficiency in the supply of acetylcholine to the nerve-muscle junction results in dysfunction, or weakness, of the muscles. A clinical condition known as "myasthenia gravis" (extreme muscular weakness) can be treated by therapy aimed at increasing the available supply of acetylcholine at the neuromuscular junction. This can be accomplished in two ways: (1) by administering a substance having acetylcholine-like action at the nerve ending, or (2) by causing the natural supply of acetylcholine to be maintained at a higher level. It might be supposed that the first of these objectives could be achieved by the administration of acetylcholine itself; but this is ineffective because there is present in the body (and in high concentration in nerve tissue) an enzyme that rapidly destroys acetylcholine by hydrolyzing the ester linkage:

$$Me_3\overset{+}{N}CH_2CH_2OCOCH_3 \xrightarrow[\text{esterase}]{\text{acetylcholine-}} Me_3\overset{+}{N}CH_2CH_2OH + CH_3COOH \qquad (4)$$

$$\text{acetylcholine} \qquad\qquad\qquad \text{choline} \qquad \text{acetic acid}$$

Indeed, the clinical effects of acetylcholine deficiency may be due to overactivity or abnormally high concentrations of the enzyme at the relevant sites. The desired therapeutic object can be obtained, however, if a drug is used which mimics the action of acetylcholine but is resistant to the action of the hydrolyzing enzyme. Several such drugs are known; for example,

$$\underset{\substack{\text{acetyl-}\beta\text{-methylcholine}\\ \text{Mecholyl}}}{(CH_3)_3\overset{+}{N}CH_2\overset{\overset{\displaystyle CH_3}{|}}{C}HOCOCH_3} \qquad \underset{\text{Carbachol}}{(CH_3)_3\overset{+}{N}CH_2CH_2OCONH_2}$$

The second approach to the therapeutic problem of acetylcholine deficiency is in many cases the more effective of the two, and consists in administering a drug that *inhibits the activity of acetylcholinesterase*, thus preserving acetylcholine from destruction and permitting it to exert a more prolonged effect. A widely used drug of this kind is physostigmine [( − )-eserine], a natural alkaloid found in the calabar bean (*Physostigma* species):

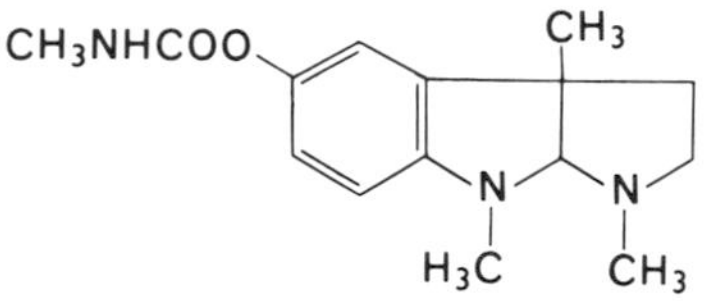

physostigmine; ( − ) eserine

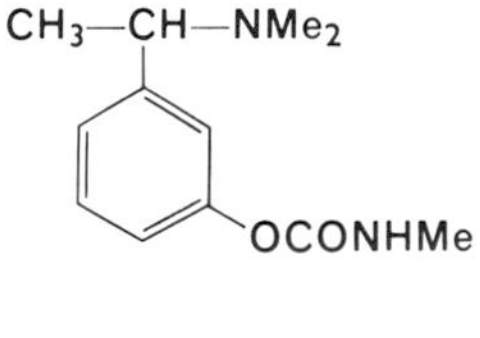

Miotine

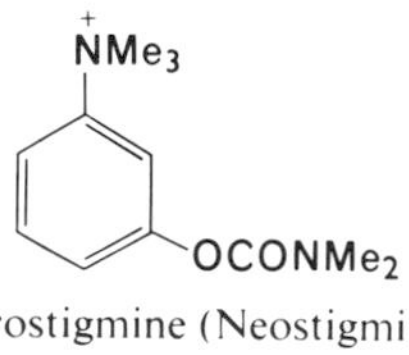

Prostigmine (Neostigmine)

Physostigmine is a potent acetylcholinesterase inhibitor, and thus, when administered, produces the physiological effects of acetylcholine.

Numerous synthetic drugs, modeled upon the physostigmine structure, have been prepared and some of these have important clinical uses. Two examples, shown above, are *Miotine* and *Prostigmine*. The name Miotine is derived from the ability of acetylcholine (and of the synthetic drug) to produce miosis—a contraction of the pupil of the eye. Observation of the miotic effect has been used as a method of assaying the parasympathomimetic activity of drugs of this kind.

One of the most important features of the pharmacology of anticholine-sterases has been the discovery that certain compounds of this class, notably certain phosphorus-containing compounds, are exceedingly toxic substances, and are of potential usefulness as weapons of war. Among these are diisopropyl phosphofluoridate, DFP:

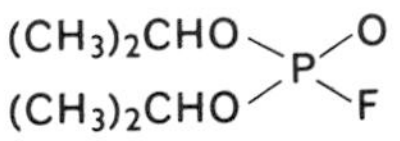

$$(CH_3)_2CHO \diagdown \underset{P}{\diagup} \diagup O$$
$$(CH_3)_2CHO \diagup \overset{P}{\diagdown} F$$

DFP

This and certain other phosphorus compounds of allied structure are also of great economic importance as insecticides. Their lethal action upon insects has been shown to be due to their ability to inactivate acetylcholinesterase.

## 33.8 Parasympatholytics: antispasmodics

The term "parasympatholytic" means relieving or preventing the effects of para-sympathetic nerve stimulation; or, in effect, counteracting the physiological consequences of acetylcholine release. Since one of the most prominent results of parasympathetic hyperactivity is a contraction, or spasm, of "smooth" muscle, such as that of the intestine, parasympatholytic compounds are known by the general term *antispasmodics*. It should be emphasized, however, that the term antispasmodic is a general one, and can properly be applied also to substances that relieve spasm induced in other ways than by parasympathetic hyperactivity.

The oldest known useful antispasmodic drugs are the naturally occurring alkaloids of certain plants of the family Solanaceae. The term "belladonna" alkaloids is used for these compounds,* the most important of which are hyoscy-amine, atropine, and scopolamine (hyoscine). The following structures show the stereochemical details of these compounds:

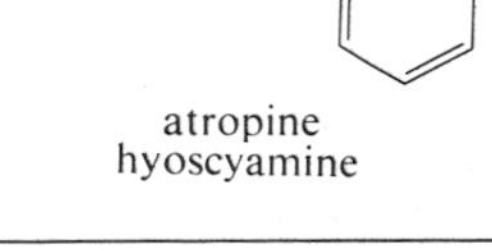

atropine
hyoscyamine

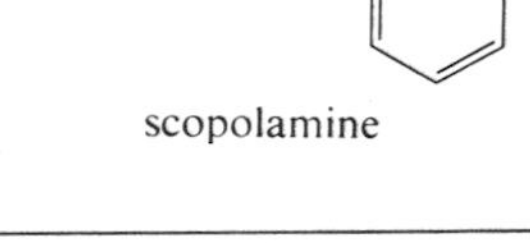

scopolamine

---

* One of the results of parasympathetic action, and of parasympathomimetic drugs, is contraction of the pupils of the eyes. Parasympatholytic drugs induce a widening of the pupil; thus the term belladonna, from the enlargement of the pupil and beautification of the woman whose eyes are so affected.

Atropine is racemic [it is ($\pm$)-hyoscyamine, the asymmetry of the latter residing in the $\alpha$-carbon atom (shown by *) of the acyl radical (tropic acid)]. It is to be noted that tropine, the amino alcohol portion of the alkaloid,

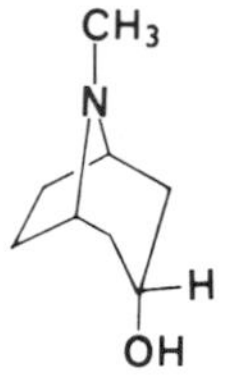

tropine              $\psi$-tropine

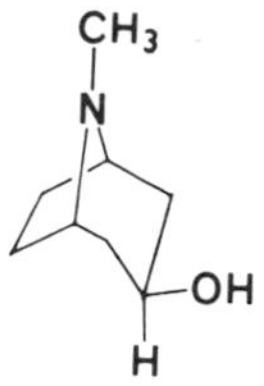

possesses a plane of symmetry, and cannot exist in enantiomorphic forms. A stereoisomer of tropine exists; it is epimeric at the $>$CHOH group: this is $\psi$-tropine (pseudotropine). It, too, possesses a plane of symmetry and is not resolvable.

---

**Exercise 2.**   Can tropine benzoate be resolved? In how many stereoisomeric forms can scopolamine exist?

---

### 33.9   Synthetic antispasmodic drugs

Synthetic antispasmodic compounds are often compared with atropine as the standard of activity. They are usually esters of amino alcohols, thus resembling the natural "model," as is shown by the following three types of synthetic antispasmodic drugs:

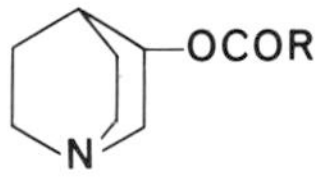

4-quinuclidinol esters 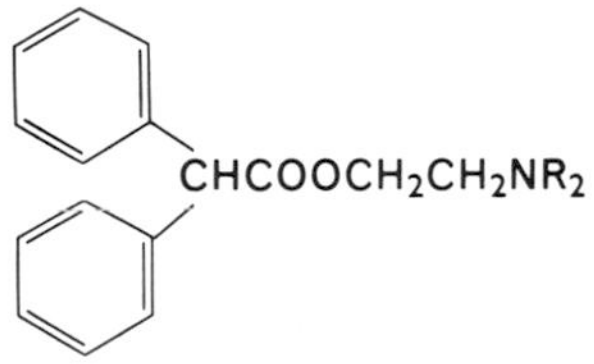              Trasentin (R = Et)*

---

\* Most synthetic drugs of the kinds shown in this chapter bear more than one name. The drugs are usually marketed under different proprietary names in the various foreign countries in which they are used and, in addition, bear nonproprietary, trivial names that are utilized when they are the subjects of scientific communications. For the drugs mentioned in this chapter I have arbitrarily selected well-known proprietary names, and have omitted the designation ® that is ordinarily used to indicate that the name is a registered trademark.

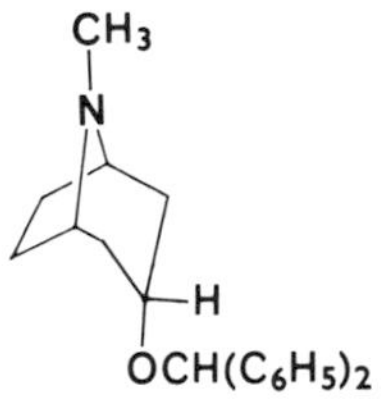

3-tropylbenzhydryl ether

Some synthetic antispasmodics have to an enhanced degree the ability to reduce the secretion of acid in the stomach, and thus are of importance in the therapy of gastric ulcer. This property is coupled with effectiveness in reducing gastric hypermotility. Two such drugs are Banthine and Tricyclamol:

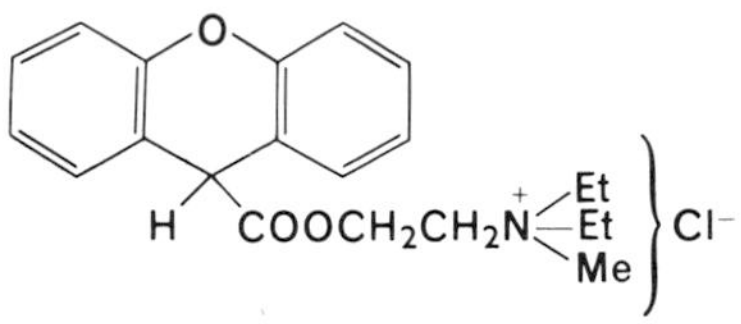

Banthine (chloride)

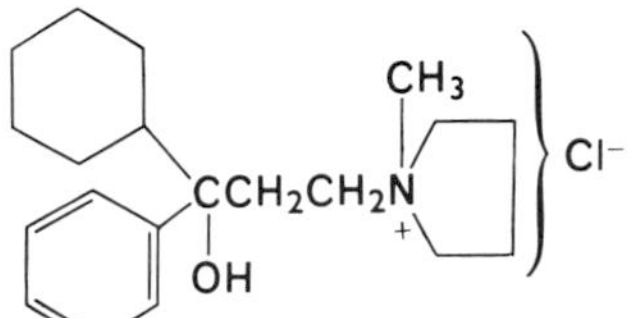

Tricyclamol (chloride)

## 33.10 Antihistamines

Antihistaminic drugs are special kinds of antispasmodics, in which antagonism to the effects of a specific spasmogen, histamine, is much more pronounced than antagonism to acetylcholine-induced effects. Antihistamines have an overall molecular "architecture" that is similar to atropine-like drugs, possessing, as do the latter, a tertiary amino grouping separated by three to four atoms from a bulky moiety such as one or more aromatic nuclei:

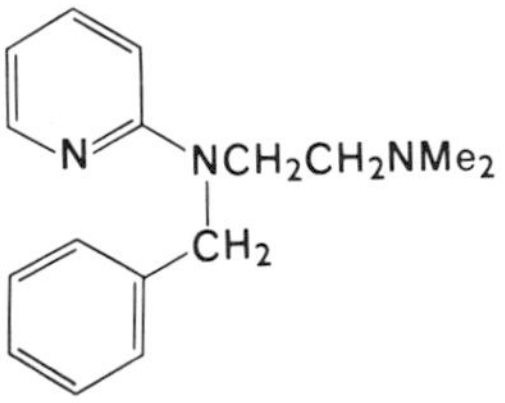

Pyribenzamine

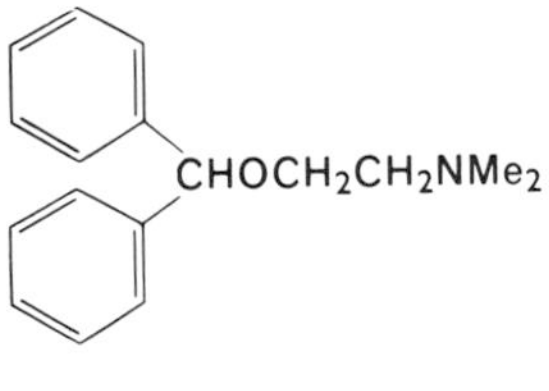

Benadryl

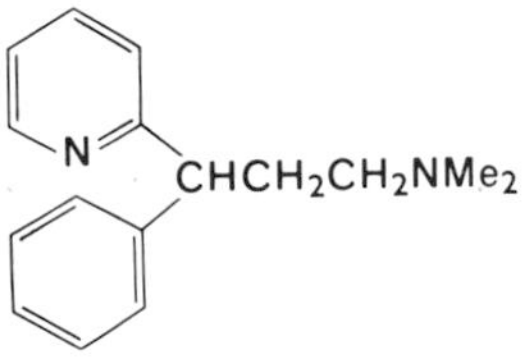

Trimeton
Antihistaminic Drugs

## 33.11   Structural similarities

It will be noted that many of the drugs considered so far have a number of superficial similarities, and may be represented by the type

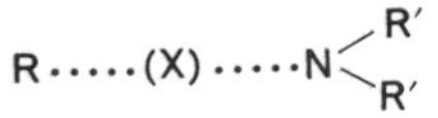

where the specific character of the R, R′, and X (i.e., the type and length of the chain) determine whether the drug is antiacetylcholine or acetylcholine-like, antihistaminic, local anesthetic, and so on. The detailed mechanism of action of many drugs is either unknown or known only with respect to the gross aspects of their behavior. Consequently, the reasons why small structural changes in a drug molecule destroy its physiological usefulness or alter the character of its action are usually not understood. For example, the compound promethazine is a very effective antihistaminic compound, while chlorpromazine has relatively less antihistaminic activity but is a powerful and clinically valuable tranquilizer. It will be noted that the gross structures of these two drugs are very similar, yet their physiological effects are quite different:

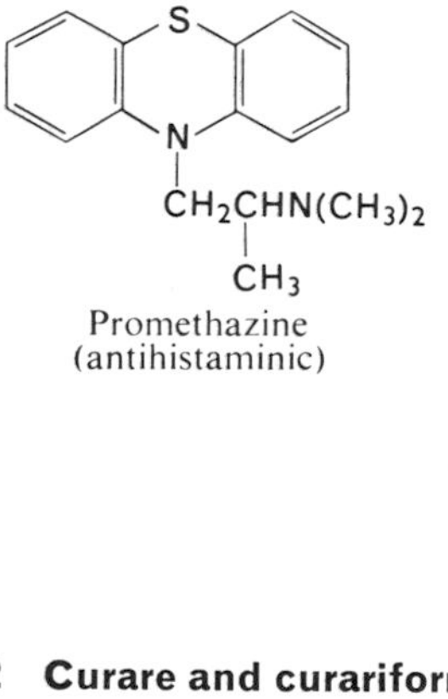

Promethazine
(antihistaminic)

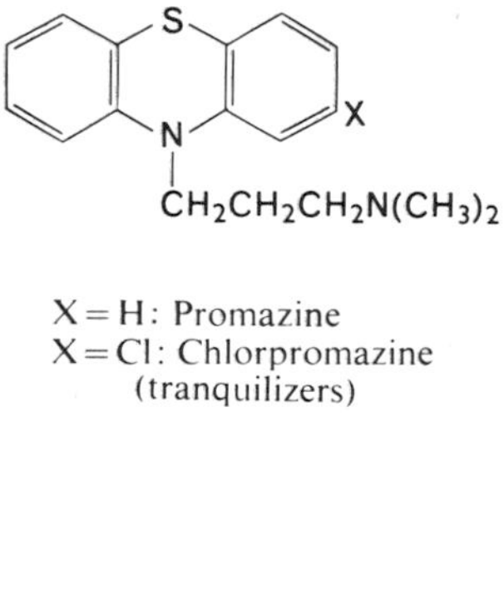

$X = H$: Promazine
$X = Cl$: Chlorpromazine
(tranquilizers)

## 33.12   Curare and curariform drugs

The use of certain plant extracts as paralyzing arrow and dart poisons, chiefly by South American Indian tribes, has been recognized for a long time, but it has been only in recent years that systematic chemical examination of these substances has been carried out. One of the active chemical compounds responsible for the action of these poisons has been characterized, and is called "tubocurarine," isolated from "tube curare". It is illustrated at the top of the facing page.

(+)-Tubocurarine (the natural form is dextrorotatory) is a muscular paralyzant, and in sufficient dosage causes death by paralysis of the respiratory muscles. It is a valuable clinical drug, inducing a muscular relaxation that permits surgery to be performed more effectively and at a lower degree of general anesthesia. It can be seen from the structural formula that (+)-tubocurarine has two quaternary nitrogen atoms separated by a fairly rigid structural assembly. Acting upon the

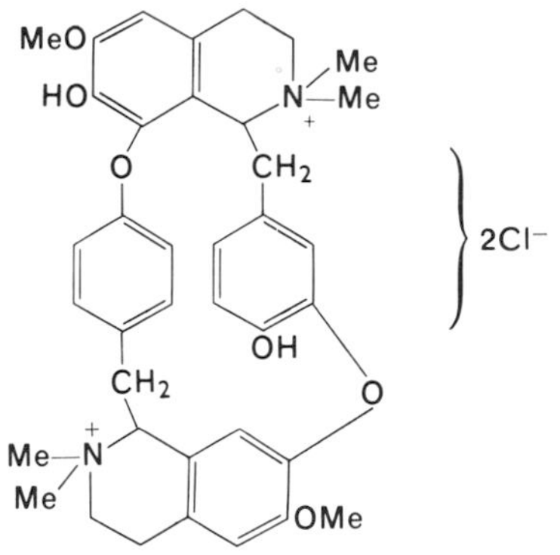

(+) tubocurarine chloride

assumption* that the two quaternary ammonium centers were the essential structural features responsible for the activity, chemists studied simpler bis-ammonium compounds, and found that the very simple compound, deca-methylene-bis-trimethylammonium chloride is a very effective curarelike- drug:

$$(CH_3)_3\overset{+}{N}(CH_2)_x\overset{+}{N}(CH_3)_3\}2Cl^- \qquad x=10, \text{ Decamethonium}$$

Decamethonium, as this drug is called, has enjoyed wide clinical application. Other compounds of similar architecture have also proved to be active curariform drugs. The compound

$$(CH_3)_3\overset{+}{N}CH_2CH_2OCOCH_2CH_2COOCH_2CH_2\overset{+}{N}(CH_3)_3$$

is a diester of choline with succinic acid. It can be seen that this compound contains a total of ten —CH₂—, —CO, and —O— groups between the nitrogen atoms, and thus is about the same "size" as Decamethonium. Hundreds of bis-ammonium compounds have been synthesized in the search for improved curare substitutes, and it is found that some degree of curariform activity is possessed by nearly all compounds that have two quaternary nitrogen atoms which are approximately the same distance apart as those in tubocurarine. See also page 668.

Since the effect of curariform substances is upon the ability of acetylcholine to transmit the nerve impulse from nerve to muscle, it would be anticipated that drugs that mimic the action of acetylcholine would antagonize (relieve) the curare-induced paralysis. This is indeed so in the case of tubocurarine itself. However, for reasons that are outside the scope of this discussion, the paralysis induced by Decamethonium cannot readily be reversed by parasympathomimetic drugs.

---

* This "assumption" is supported by a wealth of existing corollary information on the effects of quaternary ammonium salts on nerve action.

### 33.13  Insecticides

One of the important classes of insecticidal compounds are the chlorinated hydro-carbons. In this group are found such substances as lindane (Section 23.10), dieldrin and aldrin (Section 17.9), and DDT:

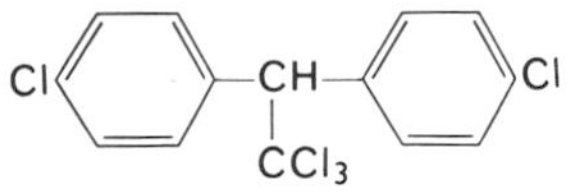

DDT

It was mentioned (Section 33.7) that one of the important classes of insecti-cides includes compounds whose pharmacological classification is that of anti-cholinesterases. There are, however, many effective insecticidal compounds, of many diverse structural types, the mechanism of whose action is not so clearly understood.

---

**Exercise 3.**   The compound DDT is synthesized by the condensation of chlorobenzene and chloral hydrate with the use of concentrated sulfuric acid. Formulate this reaction, and explain how it proceeds.

---

## Chapter thirty-four

/ # Natural and Synthetic Polymers

## 34.1 Characteristics of polymers

A *polymer* is a high-molecular-weight compound that is formed by the combination of a large number of molecules of one or more low-molecular-weight compounds. The small units of which the polymer is composed are described by the term *monomer*. The term polymer is ordinarily reserved for those substances whose structures are made up of a sufficient number of repeating monomeric units to produce molecular weights of the order of several thousand or higher. When the degree of polymerization is small and is known with certainty, the terms dimer, trimer, tetramer, and so on, are used.

In earlier chapters of this book we have encountered a number of compounds that are classed as polymers. Cellulose and starch are polymeric substances that contain D-glucose as the repeating monomeric unit. Rubber is a linear polymer of isoprene, even though it is not formed in nature from isoprene itself (Section 32.15). Proteins are high-molecular-weight substances that are formed by the linear condensation of α-amino acids by repeated amide linkages (Section 31.2). The acid-catalyzed polymerization of isobutylene (Section 9.22) leads to polyisobutylene. Self-esterification of hydroxy acids, in which lactone formation is unlikely for steric reasons, leads to linear polymeric esters in which the monomeric units are joined by repeating —COOC— linkages (Section 17.12). The self-condensation of 6-aminohexanoic acid yields, not the seven-membered lactam, but the polyamide known as Nylon-6 (Section 17.13).

If we examine these and other polymers to be described in the following pages, we can discern two ways in which monomers can be united to form large polymeric molecules: by a condensation reaction, with the elimination of a small molecule such as water; by direct addition to yield a polymer that has the same empirical composition as that of the monomer.

## 34.2   Condensation polymers

In condensation polymers, the polycondensation product is usually formed by the loss of water between the monomeric units, and so the polymer differs in empirical composition from the monomer. An example of this is seen in the case of cellulose, the relationship of which to glucose is shown by the expression

$$n\text{C}_6\text{H}_{12}\text{O}_6 \longrightarrow \text{HO}(\text{C}_6\text{H}_{10}\text{O}_5)_n\text{H} + (n-1)\text{H}_2\text{O}$$

Most condensation polymers are formed by the loss of the elements of water between *bifunctional* (or in some cases, trifunctional) molecules. Let us examine a representative example of polycondensation by comparing the polymerization reaction, which involves bifunctional reactants, with its simple monofunctional counterpart. The esterification reaction can serve as a model for study.

A monobasic acid and a monohydric alcohol combine, with the loss of water, to form an ester:

$$\text{RCOOH} + \text{R}'\text{OH} \longrightarrow \text{RCOOR}' + \text{H}_2\text{O}$$

If either the acid or the alcohol is bifunctional, a diester can be formed.

$$2\text{RCOOH} + \text{HOCH}_2\text{CH}_2\text{OH} \longrightarrow \text{RCOOCH}_2\text{CH}_2\text{OCOR} + 2\text{H}_2\text{O}$$

$$\text{HOOCCH}_2\text{CH}_2\text{COOH} + 2\text{ROH} \longrightarrow \text{ROCOCH}_2\text{CH}_2\text{COOR} + 2\text{H}_2\text{O}$$

In none of these cases can the reaction continue, for all of the available functions have been utilized in ester formation. Suppose now that *both* the acid and the alcohol are bifunctional. Let us use succinic acid and ethylene glycol as the examples. The formation of an ester can now be visualized as taking two courses:

$$
\begin{array}{lll}
\begin{array}{l}\text{CH}_2\text{COOH}\\ |\\ \text{CH}_2\text{COOH}\end{array}
+
\begin{array}{l}\text{CH}_2\text{OH}\\ |\\ \text{CH}_2\text{OH}\end{array}
\xrightarrow{\ (a)\ }
&
\begin{array}{l}\text{CH}_2\text{—CO—O—CH}_2\\ |\qquad\qquad\quad |\\ \text{CH}_2\text{—CO—O—CH}_2\end{array}
\end{array}
$$

$$(1)$$

$$\xrightarrow{\ (b)\ }$$

$$\text{HOOCCH}_2\text{CH}_2\text{COOCH}_2\text{CH}_2\text{OCOCH}_2\text{CH}_2\text{COOCH}_2\text{CH}_2\text{OH}$$

$$(2)$$

Esterification by route *(a)* would occur only if the eight-membered ring (1) were formed; and we have seen (Section 17.4) that ring-closure reactions that lead to rings of more than six members do not ordinarily take place, unless specially designed conditions are employed. The alternative course of linear condensation is favored; that is, route *(b)*. The linear polyester, of which a partial representation is (2), is clearly capable of further reaction with ethylene glycol at the HOOC— end of the chain, and with succinic acid at the —CH$_2$OH end; hence these are referred to as "growing" ends of the chain. Continued interactions of the same kind can lead eventually to long chains and high molecular weights:

---OCOCH$_2$CH$_2$COOCH$_2$CH$_2$OCOCH$_2$CH$_2$COOCH$_2$CH$_2$OCOCH$_2$CH$_2$-
$\qquad\qquad\qquad\qquad\qquad$COOCH$_2$CH$_2$OCOCH$_2$CH$_2$COO---

The degree to which polymerization of this kind will continue, and thus the ultimate molecular weight of the polyester that is formed, will depend upon conditions that can be manipulated at will: the proportions of acid and alcohol that are used; the solvent and concentration of reactants; the temperature at which the reaction is conducted; and the time it is allowed to proceed before interruption.

The example of succinic acid and ethylene glycol has been given as a simple model that illustrates the principle of condensation polymerization. Some actual examples of commercially important condensation polymers of the polyester, and other, types are described in the following sections.

### 34.3 Polyesters: Dacron

The most important of the polyesters is Dacron (called Terylene in England). Dacron is the condensation product of terephthalic acid and ethylene glycol, and is properly described by the name *polyethylene terephthalate*. In the commercial process, dimethyl terephthalate is heated with ethylene glycol to form, by ester interchange, di(hydroxyethyl) terephthalate. This compound is then polymerized by heating it until the desired molecular weight is obtained:

CH$_3$OCO—⟨ ⟩—COOCH$_3$ + 2HOCH$_2$CH$_2$OH $\longrightarrow$ 2MeOH +

HOCH$_2$CH$_2$OCO⟨ ⟩COOCH$_2$CH$_2$OH $\xrightarrow{\Delta}$

HOCH$_2$$\left[$CH$_2$OCO⟨ ⟩COOCH$_2$$\right]_n$ CH$_2$OH + HOCH$_2$CH$_2$OH

Dacron (av. mol. wt. about 15,000)

The resulting polymer, a horny solid, is formed into fibers by melting it and forcing it through tiny orifices called spinnerets. The filaments that emerge from the spinnerets harden on cooling, and are picked up and wound, with stretching, onto bobbins. By adjusting the orifices in the spinnerets, fibers of various diameters can be formed. Another important polyester will be described in Section 34.7.

## 34.4   Polyamides: the nylons

The oldest important synthetic fiber, and still the most widely used, is nylon. Nylon is a polyamide, formed by the polycondensation of hexamethylenediamine and adipic acid. This nylon is called nylon-66 because it is formed from two six-carbon-atom units:

$$HOOC(CH_2)_4COOH + H_2N(CH_2)_6NH_2 \longrightarrow$$
$$HOOC(CH_2)_4CO\{NH(CH_2)_6NHCO(CH_2)_4CO\}_nNH(CH_2)_6NH_2$$
Nylon-66 (av. mol. wt. 12,000–15,000)

Nylon is formed into fibers by a process similar to that described in the preceding section for Dacron.

Nylon-6 is formed by the self-condensation of 6-aminohexanoic acid (Section 17.13). In practice, the starting monomer is not the amino acid itself, but the seven-membered lactam that is produced by the Beckmann rearrangement of cyclohexanone oxime (Section 28.9). When the lactam is heated in the presence of a small amount of water, which acts as the catalyst, polyamide formation ensues:

$$\xrightarrow{H_2O} H_2N(CH_2)_5COOH \xrightarrow{-H_2O}$$

$$H_2N(CH_2)_5\{CONH(CH_2)_5\}_nCOOH$$

Nylon-6 resembles nylon-66, and is used for many of the same purposes.

It should be apparent that many different polyesters and polyamides could be prepared by methods similar to those described above if the appropriate monomeric starting materials were available. Whether or not any process one might devise would be commercially feasible would depend upon the cost and availability of the monomers and on the properties of the polymers that would be formed. No generalities can be offered on this matter. The commercial success of Dacron and Nylon is due not only to the very desirable properties possessed by the polymers and the fibers that are made from them, but also to the abundance and low cost of the raw materials from which they are made. Terephthalic acid is produced by the oxidation of *p*-xylene, a product of the petroleum refining industry. Ethylene glycol is produced in large amounts from ethylene. The nylons have as their basic raw material phenol, which is hydrogenated to cyclohexanol and this in turn dehydrogenated to cyclohexanone. Oxidation of cyclohexanone yields adipic acid, and reduction of adiponitrile provides hexamethylenediamine.

## 34.5   Phenol-formaldehyde and urea-formaldehyde resins

Another important class of condensation polymers is derived from formaldehyde, which, when written as the hydrate, $HOCH_2OH$, is a bifunctional reagent which can be used to form a —$CH_2$— linkage between two molecules by the loss of the elements of water. In the formation of a phenol-formaldehyde polymer, the units that are joined by the —CH$_2$— units are phenol molecules, which can react with formaldehyde according to the following partial (and approximate) expression:

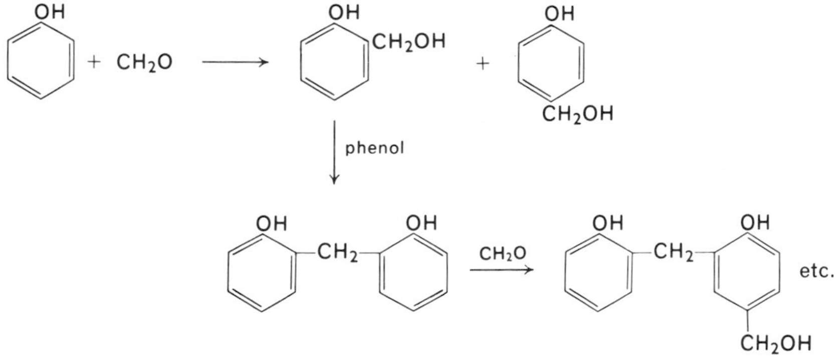

Phenol, which has highly reactive *ortho* and *para* positions, can undergo condensations of this kind that result in the linkage of chains of phenol molecules together to form large and complex polymers (see also Section 34.6).

The condensation of urea and formaldehyde can occur in a comparable way. The first step may be formulated in the following way:

$$H_2NCONH_2 + CH_2O \longrightarrow HOCH_2NHCONHCH_2OH$$

Further reaction in the same way leads at length to a polymer of the following kind. It will be noticed that methylol (—CH$_2$OH) groups can be formed by additional condensations between formaldehyde and the NH groups of the polymeric chain:

As the discussion in the following section will show, the above formulations for the phenol-formaldehyde and the urea-formaldehyde polymers are incomplete. They represent only a stage on the way to the final polymer. Subsequent stages in the formation of the final polymer are described in the next section.

### 34.6 Cross-linking: the thermosetting of resins

The methylol groups that protrude from the chain of the urea-formaldehyde polymer, shown in the preceding section, can react with NH groups on another such chain, with the loss of water, thus uniting two chains by a "cross-linking" —CH$_2$— group:

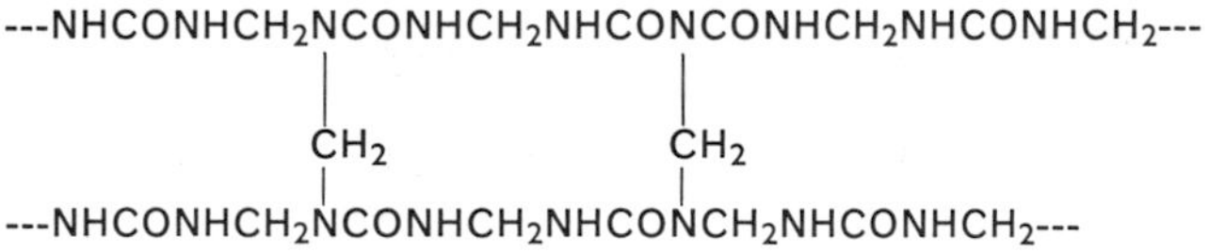

The resulting polymer now has a large, three-dimensional structure, in which the covalent bonds between the individual molecular backbones impart a rigid character to the final polymeric molecule. The cross-linking reaction is ordinarily carried out as the final stage in the manufacture of a product from the resin: for example, in making a molding. The partially polymerized, methylol-containing material, usually in the form of a "molding powder," is placed in the mold (or between sheets that are to be bonded together) and is subjected to heat and pressure. The material liquefies and fills the details of the mold, then undergoes the cross-linking reaction and hardens to an insoluble, infusible resin. This process is described by the term "thermosetting."

Phenol-formaldehyde resins can be cross-linked in a similar way. Indeed, it is possible to prepare as the primary material a raw resin of a structure such as the following:

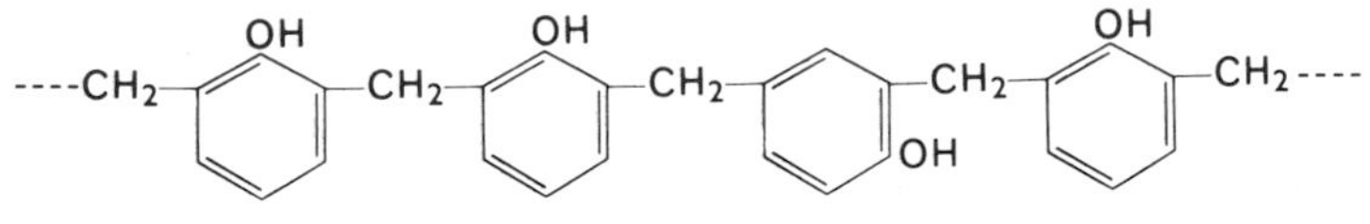

This, it will be seen, cannot cross-link because no free —$CH_2OH$ groups are present. If such a material is now heated in the presence of formaldehyde, cross-linking —$CH_2$— groups are supplied, and the resin sets to the final, infusible form. The common material known as Bakelite is a cross-linked phenol-formaldehyde resin. A portion of the characteristic cross-linked structure would look something like the following:

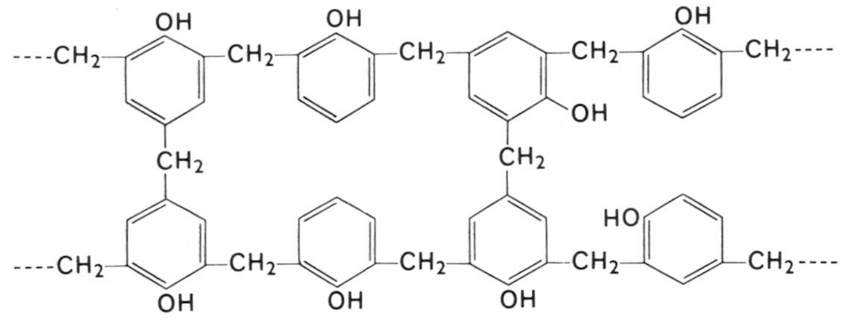

## 34.7 Cross-linked polyesters: alkyd resins

If, instead of ethylene glycol and a dibasic acid, a trifunctional alcohol such as glycerol is used, cross-links may be produced by ester formation between polymer chains. The widely used and valuable thermosetting resins known as *alkyds* are produced by a reaction of this kind. The usual monomers are phthalic anhydride (although other dibasic acids, such as maleic and succinic, are also used) and glycerol. Partial polyester formation may be represented as follows:

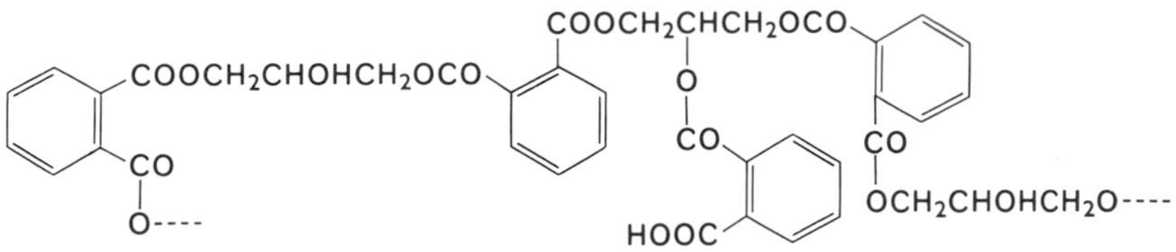

When a polymer of this kind is heated, further reaction occurs between the un-used carboxyl and hydroxyl groups, and thermosetting occurs. The partially polymerized alkyd is a viscous liquid which can be applied as a coating. When the coated article is baked, the resin sets to a hard, durable film:

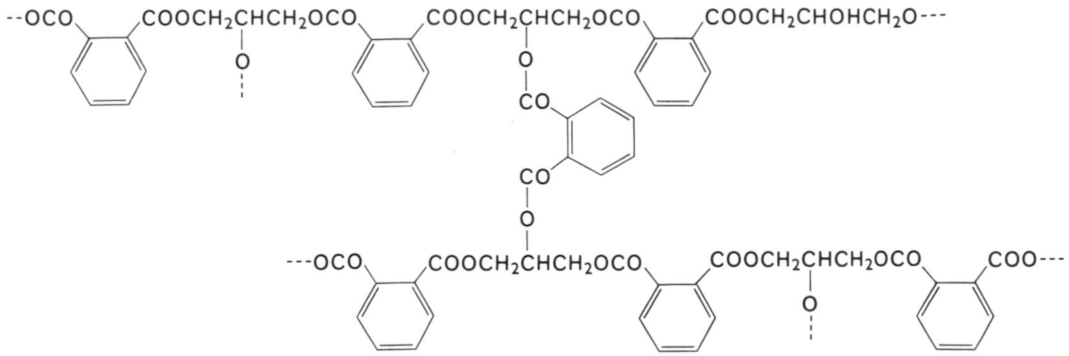

### 34.8 Cross-linking of rubber during vulcanization

The vulcanization of rubber was described briefly in Section 32.15. It will be re-called that vulcanization in the presence of small amounts (1–5%) of sulfur pro-duces a rubber that has gained improved elasticity and less stickiness, but which is still soft and pliable. Under these conditions only a small amount of cross-linking, by the joining of adjacent chains by carbon-sulfur bonds, occurs. When larger amounts of sulfur (20–35%) are used, extensive cross-linking leads to the forma-tion of the hard, brittle substance known as "hard rubber." Vulcanization is a general process, applicable to polymers of other kinds, but which are also polyenes. Some synthetic polymers of this kind, formed by a process of polymerization different from that described so far, will be described in following sections.

### 34.9 Natural condensation polymers

A reexamination of the structures of the natural polysaccharides (Chapter 18) and proteins (Chapter 31) will disclose that these have the structures of typical con-densation polymers. The proteins differ in one respect from the kinds of condensa-tion polymers we have been discussing; the monomeric units of the proteins differ from one another, and the polymer is the result of the condensation of a number of different α-amino acids. Nevertheless, the manner in which they are joined is the same as in the case of polymers that are formed by the repeated linkage of a single amino acid. The peptide linkages of a protein are amide linkages formed between bifunctional molecules.

Other polymers that are found in natural sources are lignin (Section 27.10), a polymeric condensation product formed (probably by an oxidation process) from a monomer that appears to be coniferyl alcohol; and the materials, called *cutins*, which form the cuticular membranes of plants. The structures of the cutins are still incompletely known. They appear to be cross-linked polymers formed from bi- and trifunctional hydroxy fatty acids. Their characteristic properties, which permit them to provide plants with glassy, insoluble external coatings, are typical of cross-linked polymers. Cutins from a number of plants have been studied chemically. Alkaline hydrolysis yields mixtures of hydroxylated fatty acids, some of which have been identified.

## 34.10  Addition polymers

The second important general manner in which monomeric compounds can combine to give polymeric substances is by a process of addition. In this kind of polymerization no smaller fragment is eliminated, and the polymer has the same empirical composition as the monomer.

When isobutylene is polymerized by the catalytic influence of an acid (boron trifluoride is used commercially), the process may be represented by the summary expression:

$$n CH_2{=}C{\Large\langle}^{CH_3}_{CH_3} \longrightarrow \left( -CH_2-\underset{\underset{CH_3}{|}}{\overset{\overset{CH_3}{|}}{C}}- \right)_n$$

The details of this reaction have been described in Section 9.22, where it was shown that the reaction involves the addition of successive monomer molecules to a "growing" carbonium ion. This kind of polymerization is of limited practical importance, although polyisobutylene has been found to be a useful polymer for certain applications.

The most widely used method of polymerization of olefinic compounds utilizes the catalytic action of a compound that can decompose to give rise to free radicals. The addition of a radical to a double bond generates a new radical, and this in turn can add to another double bond to generate still another radical. The process is a chain reaction, for a limited supply of free radicals can initiate and sustain the polymerization of a great many molecules of the monomer. The chains that are formed in this way can be terminated by the combination of two radicals; by the loss of a radical from one chain to another (disproportionation); or in other ways. Hence a supply of radicals from the catalyst maintains the polymerization and permits the control of the process to give products having the desired properties.

A typical polymerization catalyst is benzoyl peroxide, which can decompose to give free radicals of the following kinds:

$$C_6H_5CO-O-O-COC_6H_5 \longrightarrow C_6H_5CO-O\cdot \longrightarrow CO_2 + C_6H_5\cdot$$

The action of the catalyst in initiating the polymerization can be represented by the expression

$$CH_2{=}CHX + R\cdot \longrightarrow RCH_2\overset{\cdot}{C}HX$$

in which the free radical is formulated in the general form $R\cdot$ ; the olefin is in the general form $CH_2{=}CHX$. Propagation of the chain now ensues as follows:

$$CH_2{=}CHX + RCH_2\overset{\cdot}{C}HX \longrightarrow RCH_2\underset{X}{\overset{|}{C}}HCH_2\overset{\cdot}{C}HX \xrightarrow{\ CH_2{=}CHX\ }$$

$$RCH_2\underset{X}{\overset{|}{C}}HCH_2\underset{X}{\overset{|}{C}}HCH_2\overset{\cdot}{C}HX, \text{ etc.}$$

Termination of the chain in one of the ways mentioned above can be controlled in such a way as to regulate the degree of polymerization and thus the properties of the final product.

## 34.11  Polymerization of dienes

An early clue to the manner in which synthetic polymers might be prepared was the observation that natural rubber could be decomposed by heating, with the formation of isoprene, and that isoprene could be polymerized to give products that had a superficial resemblance to rubber. Although the polymerization of isoprene into a material that is probably nearly identical to natural rubber has been achieved only in the last few years, rubber-like polymers have been produced commercially for many years. Neoprene, one of the first successful "synthetic rubbers," is prepared by the polymerization of 2-chloro-1,3-butadiene (chloroprene):

$$CH_2{=}\underset{Cl}{\overset{|}{C}}{-}CH{=}CH_2 \longrightarrow {-}{-}{-}CH_2C{=}\underset{Cl}{\overset{|}{C}}HCH_2CH_2C{=}\underset{Cl}{\overset{|}{C}}HCH_2{-}{-}{-}$$

Crude Neoprene is comparable to crude rubber, and can be vulcanized to produce a rubber-like material that has found many important uses. Neoprene is a tough, stable, elastic material that shows excellent resistance to the action of organic liquids. It cannot compete successfully with natural rubber for general use, but its relatively high cost is no bar to its application in special circumstances.

During the last war, when rubber supplies were cut off by enemy control of the rubber-producing areas of the world, intensive efforts were undertaken to find a practical substitute for rubber. The material known as GR-S (Government Rubber Styrene) was found to be suitable for the manufacture of automobile tires, and was produced in large quantities. GR-S is a *copolymer* of butadiene and styrene, of the partial structure

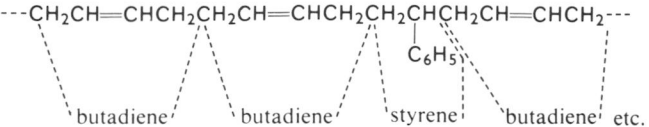

GR-S (also called Buna-S) can be vulcanized with sulfur.

It should be noted that the formulation of GR-S shown above is an arbitrary one; some 1,2-addition can occur, along with the 1,4-polymerization shown above; and the location and spacing of the styrene residues cannot be stated with certainty.

Copolymers of butadiene with other vinyl monomers are known, and some of them are useful rubber-like materials. The copolymers of butadiene and acrylonitrile (vinyl cyanide) yield the rubber substitutes called by such names as Perbunan and Chemipol. The copolymers of isoprene and isobutylene form useful rubbers that are characterized by a low degree of permeability to gases. They find application in the manufacture of automobile tire inner tubes, and are known as "butyl rubbers."

### 34.12   Vinyl polymers

A great many relatively simple monomers of the general constitution $CH_2\!\!=\!\!CHX$ find use in the production of a wide range of useful "vinyl polymers." Simple polymers and copolymers of compounds of this general structure, where X may be Cl, $OCOCH_3$, CN, phenyl, and COOMe, constitute a large and important class of substances whose properties range from those of clear, glass-like plastics (Lucite, from methyl $\alpha$-methylacrylate), films and sheets (from other acrylic acid esters), plastic foams (Polystyrene, from styrene), the highly resistant plastics (Teflon, from tetrafluoroethylene), to those of the useful fiber, Orlon, from acrylonitrile. Although the special technology of plastics manufacture is a large and specialized area of chemical technology, the general principles are summarized in the discussion found in the foregoing sections. The details of the ways in which polymerization is carried out, the kinds of initiators that are used, the control of the processes, and the relationships between properties of the polymers and the chemical nature of the molecules, is far too large a subject to be dealt with further here.

### 34.13   Polyethylene, polypropylene, and polyisoprene

Simple olefins such as ethylene and propylene can be polymerized by free-radical processes such as those described in Section 34.10, but the polymers prepared in this way often possess physical properties that are less than ideal. The most successful of the polymers of this kind has been polyethylene, produced by free-radical polymerization with the aid of high pressure and temperature. Polyethylene molecules consist of long chains of —$CH_2CH_2$— units and have average molecular weights of about 30,000. When prepared by free-radical polymerization processes, polyethylene contains occasional points of chain branching, which destroys the symmetry and linearity of the polymer chain; such branching can occur by the intervention of the following processes during the growth of the polymer chain:

(*a*)  $\cdots CH_2CH_2CH_2CH_2CH_2CH_2 \cdot \xrightarrow{\text{shift of H·}} \cdots CH_2CH_2CH_2CH_2\overset{\cdot}{C}H-CH_3 \xrightarrow{CH_2=CH_2}$

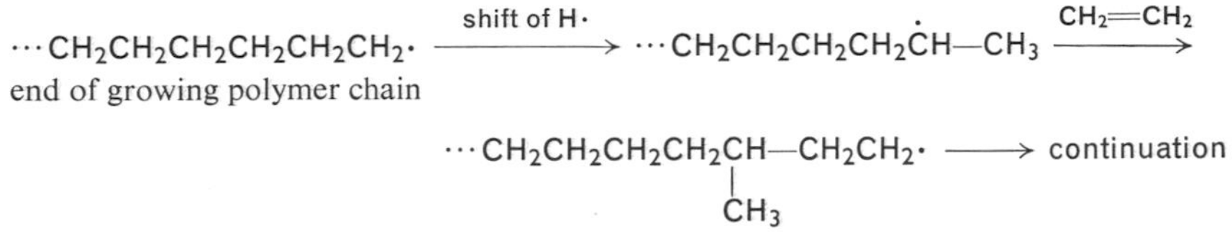

end of growing polymer chain

$$\cdots CH_2CH_2CH_2CH_2CH-CH_2CH_2 \cdot \longrightarrow \text{continuation}$$
$$\underset{CH_3}{|}$$

(*b*)  $\cdots CH_2CH_2CH_2CH_2CH_2CH_2CH_2 \cdots \xrightarrow[\text{radical, R·}]{\text{attack by}} \cdots CH_2CH_2CH_2CH-CH_2CH_2CH_2 \cdots$

$$\xrightarrow{CH_2=CH_2} \cdots CH_2CH_2CH_2CH-CH_2CH_2CH_2 \cdots \longrightarrow \text{continuation}$$
$$\underset{CH_2CH_2 \cdot}{|}$$

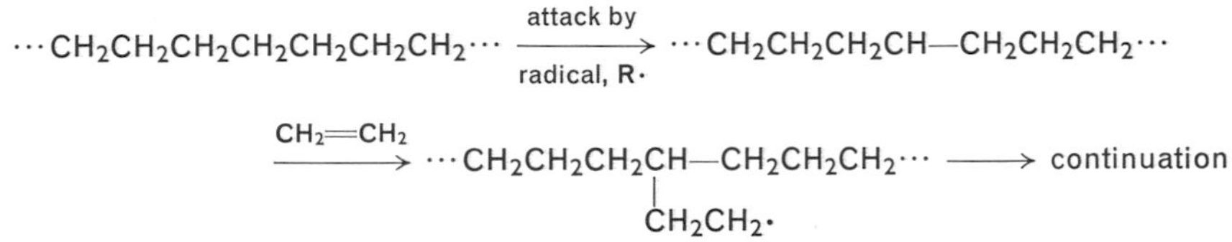

Such departures from strict linearity result in a decrease in density, hardness, and tensile strength, because the more nearly the polymer molecule approaches an un-branched linear structure, the greater the degree of orientation of the polymer chains in the solid polymer. The straight chains, without protruding branches, can pack into a compact, orderly arrangement in which the weak, but numerous, inter-molecular (van der Waals) forces can operate most effectively. The occurrence of branches in the chain destroys to some degree the attractive forces between the polymer chains and leads to less compact, less orderly structure and to decreased density and decreased strength.

Polypropylene, prepared by free-radical polymerization processes, consists of a chain of —$CH$—$CH_2$— units in which the methyl groups are oriented randomly,
$$\underset{CH_3}{|}$$
as in the following:

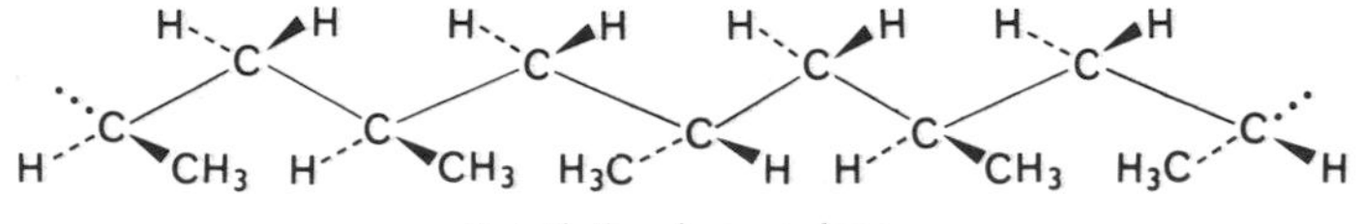

"atatic" polypropylene

The consequence of this random stereochemical orientation is that the polymer molecules cannot assume an orderly, close-packed array, and the polymer lacks hardness and tensile strength. Indeed, polypropylene prepared in this way is usually a soft, often sticky or rubbery material, with quite undesirable properties. This disoriented structure is the result, not of chain branching, but of the presence of the asymmetric centers at the —$CHCH_3$ groups along the chain; any chain branching, of the kind described above for the case of polyethylene, will introduce further randomness and further impair the useful properties of the polypropylene polymer.

Since about 1955, when new polymerization methods were developed through discoveries made in Germany (Karl Ziegler) and in Italy (Giulio Natta), both polyethylene and polypropylene of greatly improved physical properties have been produced, and are today valuable and widely used industrial materials. The Ziegler process involves the use of catalysts of which a combination of triethyl-aluminum and titanium tetrachloride is typical, and involves a polymerization re-action that is ionic and not free-radical in nature. The polymerization occurs at ordinary temperatures and low pressures and leads to polymers of regular struc-ture, free of branching. A simplified representation of the essential feature of the reaction involved in the polymerization of ethylene is the following:

$$>AlCH_2CH_3 \xrightarrow{CH_2=CH_2} >AlCH_2CH_2CH_2CH_3 \xrightarrow{CH_2=CH_2}$$

$$>AlCH_2CH_2CH_2CH_2CH_2CH_3 \xrightarrow{n\ CH_2=CH_2} >Al(CH_2CH_2)_nCH_2CH_2CH_2CH_2CH_2CH_3$$

The catalysts used are so prepared that polymerization occurs on a solid surface, with the result that a regular and stereochemically selective polymerization occurs, with the formation of a polymer chain containing uniformly oriented monomer units. Polypropylene produced by the Ziegler-Natta procedure can be represented as follows:

"isotatic" polypropylene

The isotatic polymer chain adopts a helical conformation in which the methyl groups are arranged in such a way as to minimize the interactions between them; the order and regularity of the resulting polymer chain permits the molecules to pack compactly and to produce a solid that is "crystalline" in character. Poly-propylene prepared in this way is a hard, strong, high-melting material which, because of the chemical inertness of the fully saturated compound, is adaptable to many uses. Polypropylene is a thermoplastic polymer: it can be softened by heating and extruded through molds to yield a variety of shaped forms, such as plastic piping, rods, etc.

Polyethylene produced by the Ziegler process is superior to that produced by the free-radical process, but because of the inherent regularity of the unbranched chain of —$CH_2CH_2$— units the two types of polyethylene do not differ as greatly as do the two types of polypropylene.

The polymerization of isoprene by free-radical processes does not produce polymers with the characteristic properties of natural rubber. However, when iso-prene is polymerized with the use of metal alkyl catalysts (metallic lithium can also be used), the stereoselectivity of the reaction results in the formation of a polymer that possesses the *cis* configuration characteristic of natural rubber, and yields a product that is nearly identical in physical properties with the natural material:

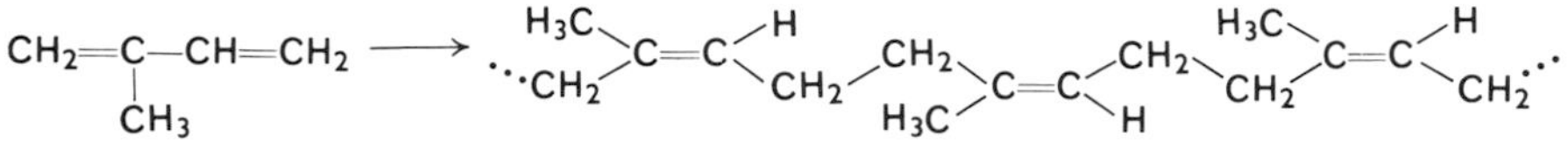

$$CH_2{=}C{-}CH{=}CH_2 \longrightarrow$$

*cis*-polyisoprene

Synthetic rubber (polyisoprene) has not displaced natural rubber from industrial use, chiefly for economic reasons; most of the rubber produced at the present time comes from the natural source, the rubber tree.

## 34.14  Silicones

An interesting and unique class of polymers is found in a group of substances that contain silicon. Silicon is able to form a group of organo-silicon compounds represented by the general formulas $R_3SiOH$, $R_2Si(OH)_2$, and $RSi(OH)_3$. The diol can undergo self-condensation to form rings or linear, polymeric chains:

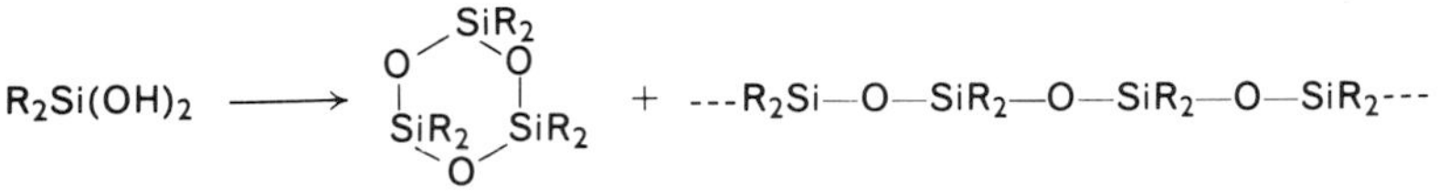

$$R_2Si(OH)_2 \longrightarrow \quad + \quad {-}{-}{-}R_2Si{-}O{-}SiR_2{-}O{-}SiR_2{-}O{-}SiR_2{-}{-}{-}$$

Alkylsilanetriols can give three-dimensional polymers, in which the third —OH group can participate in cross-linking to produce a rigid, complex structure that, it will be seen, bears some resemblance to silica. These polymers, known as silicones, are very stable substances. Because they have a relatively low carbon content they are resistant to oxygen, and their great stability adapts them to special uses, such as for electrical insulation.

By the copolymerization of mixtures of dialkylsilanediols and alkylsilanetriols, polymers with varying degrees of cross-linking can be produced, and substances with a wide range of useful properties can be prepared. The following is a partial representation of a silicone polymer:

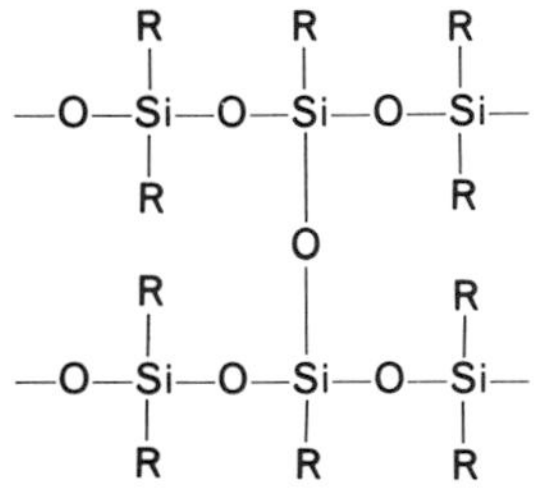

The necessary monomers are prepared by a process that is shown in the following summary equation:

$$Si + RCl \xrightarrow[\text{heat}]{\text{catalyst}} RSiCl_3 + R_2SiCl_2, \text{ etc.}$$

The silanols are prepared from the halides by hydrolysis.

# Chapter thirty-five / The Establishment of Structure of Organic Compounds

## 35.1 Identification and structure proof

An organic compound that is presented to an organic chemist for study may be *(a)* a compound that is already known and is described in the chemical literature; or *(b)* a new compound that has been discovered for the first time. In the first case, the task of the organic chemist is to *identify* the compound by showing that it is indeed identical with the one that has been described earlier. He does this by showing that the physical and chemical properties of the compound agree in all respects with those that have been reported in the literature. In the second case, his task is to establish the structure of the compound and to describe its properties and behavior in such a way that it can later be recognized and identified by others. *Structure proof* usually involves a systematic examination of the compound, starting with purification and elemental analysis and proceeding to the study of its ultraviolet, infrared, and nuclear magnetic resonance spectra and the application of tests that elicit responses which are characteristic of various functional groups. Degradation, by oxidation, cleavage with alkali, or other means, may yield fragments that are themselves identifiable. Specific analytical procedures are often applied to establish the presence and number of such functional groups as alkoxyl, carbon-linked methyl, acetyl, carbon-carbon double bonds, carboxylic acid groups, and so on. When all of this information is assembled it is usually possible to write a structural formula which can then be subjected to confirmatory studies. Finally, when a consideration of all of the experimental evidence leads to the assignment of a structure, the synthesis of a compound with this structure, by unequivocal methods, provides a final confirmation, and a *structure proof* has been accomplished.

If a compound contains asymmetric centers and is optically active, it is necessary to establish the configurations at these centers in order that the complete stereochemistry can be defined. The establishment of the stereochemistry of a compound may consist in the definition of the *relative* orientation of the groups with respect to each other; or it can be and usually is carried further to the complete elucidation of the *absolute* stereochemistry.

In the following sections are described some representative examples of the establishment of the structures of organic compounds by the application of diagnostic and degradative methods, and by total synthesis.*

### 35.2   The structure of a sesquiterpenoid ketone: α-cyperone

α-Cyperone (1), $C_{15}H_{22}O$, is a sesquiterpene found in the essential oil of the plant *Cyperus rotundus*. It is a ketone, forming the usual derivatives (2,4-dinitrophenylhydrazone, semicarbazone), and shows an ultraviolet absorption maximum at 251 mμ ($\epsilon = 19,000$) (Figure 35-1). This shows that it is an α,β-unsaturated ketone that is α,β,β-trisubstituted. Ozonolysis of α-cyperone yields formaldehyde, $C_{15}H_{26}O$. The tetrahydro compound, $C_{15}H_{26}O$ (2), is a saturated ketone; thus, α-cyperone is bicyclic.

Reduction of α-cyperone with sodium and alcohol gives dihydro-α-cyperol (3) [Note *(A)*], dehydrogenation of which yields eudalene (4):†

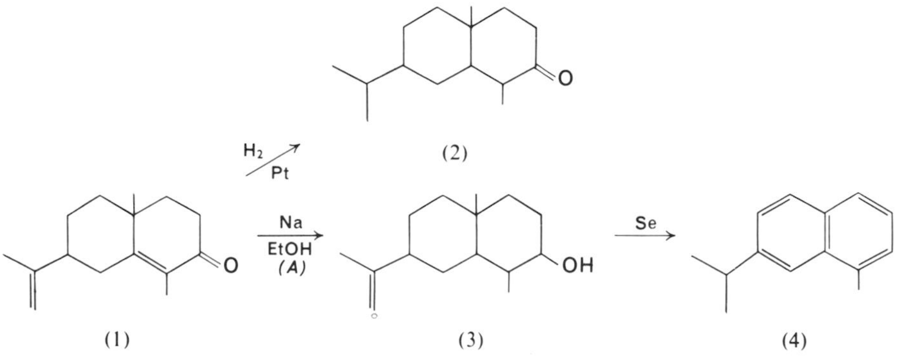

---

\* In the discussions to follow, most of the reactions described are those that have been discussed in relevant sections in earlier chapters. The student is urged to review (with the aid of the Index) the reaction types used in the degradative and synthetic transformations described in what follows.

† The letters [*(A), (B), (C)*] refer to explanatory notes which follow on page 831. Reactions referred to may be found in the Index.

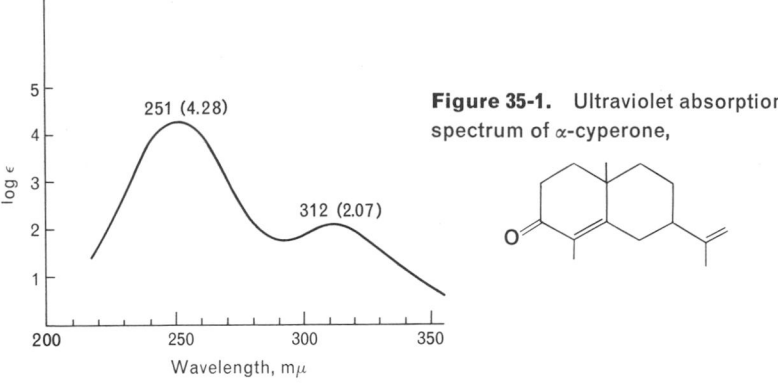

**Figure 35-1.** Ultraviolet absorption spectrum of α-cyperone,

The position of the carbonyl group was established in the following way. Treatment of tetrahydro-α-cyperone (2) with methylmagnesium iodide gave a carbinol (5) which upon dehydration and dehydrogenation gave 1,2-dimethyl-7-isopropylnaphthalene (6):

(2)    CH₃MgI →

(5)    (1) – H₂O
       (2) Se →    (6)

These experimental observations, coupled with an assumption as to the position of the methyl group that is lost on dehydrogenation, lead to the structure (1) for α-cyperone. This structure is "normal" with respect to the isoprene rule, and

(−)-dihydrocarvone    (7)    NaOEt
                             (B) →    (8)    – H₂O
                                            (C) →

α-cyperone (9)

corresponds in its carbon skeleton with eudesmol, selinene, carissone, and numerous other eudalene-yielding sesquiterpenes.

α-Cyperone has been synthesized in a way that confirms the stereochemistry shown in (9) at foot of p. 830. The reaction of the Mannich base [as the methiodide (7)] obtained from methyl ethyl ketone [Note *(B)*], with (−)-dihydrocarvone yields first the keto alcohol (8) which on dehydration gives α-cyperone [Note *(C)*]. This stereochemistry has been confirmed in several ways, one of which is the synthesis of carissone (11) from eudesmol (10), the latter having the stereochemistry shown:

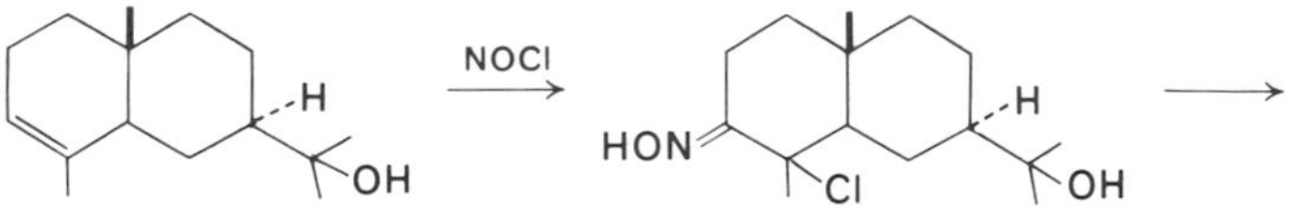

Eudesmol (10)

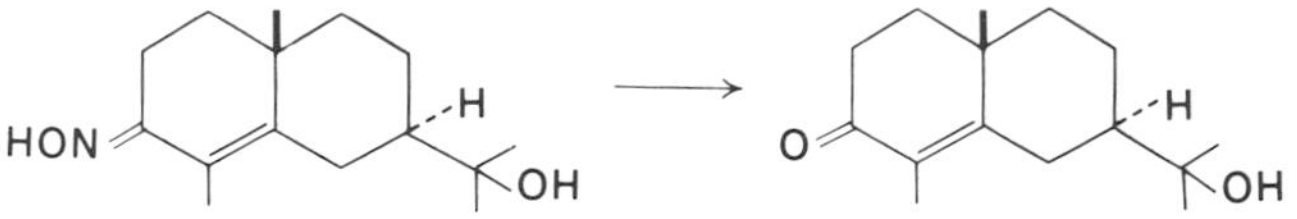

Carissone (11)

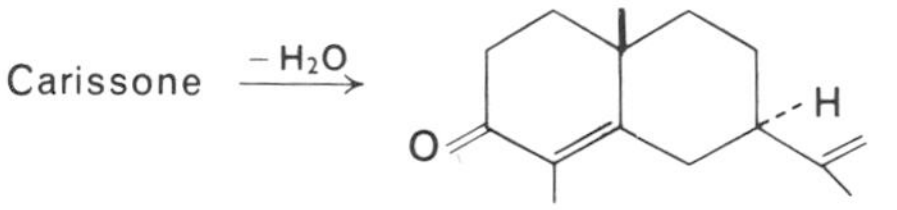

α-Cyperone

*(A)* The reduction with sodium and ethanol does not affect the isolated carbon-carbon double bond in the isopropenyl side chain, but does reduce the α,β-unsaturated ketone to the saturated ketone (first product; not isolated) and then to the secondary alcohol.

*(B)* The Mannich base is prepared by the reaction of methyl ethyl ketone with diethylamine hydrochloride and formaldehyde (Chapter 29):

$$CH_3CH_2COCH_3 \xrightarrow[\text{HCHO}]{\text{Et}_2\text{NH}\cdot\text{HCl}} CH_3CH_2COCH_2CH_2NEt_2$$

Reaction of the tertiary amine with methyl iodide gives the quaternary salt.

When a compound such as 1-diethylamino-3-pentanone methiodide is allowed to react in the presence of the strong base, sodium ethoxide, the first reaction is the elimination of methyldiethylamine, with the formation of the vinyl ketone:

$$CH_3CH_2COCH_2CH_2\overset{+}{N}R_3 \xrightleftharpoons[]{\text{OEt}^-} CH_3CH_2CO\overset{..}{C}H{-}CH_2{-}\overset{+}{N}R_3 \longrightarrow$$

$$CH_3CH_2COCH{=}CH_2 + NR_3$$

The addition of the vinyl ketone to the cyclohexanone ring (dihydrocarvone) is an example of the *Michael reaction*:

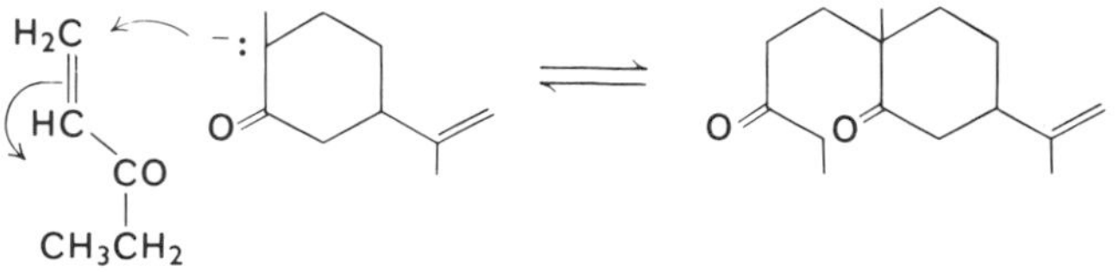

and the final ring closure is an unexceptional aldol condensation:

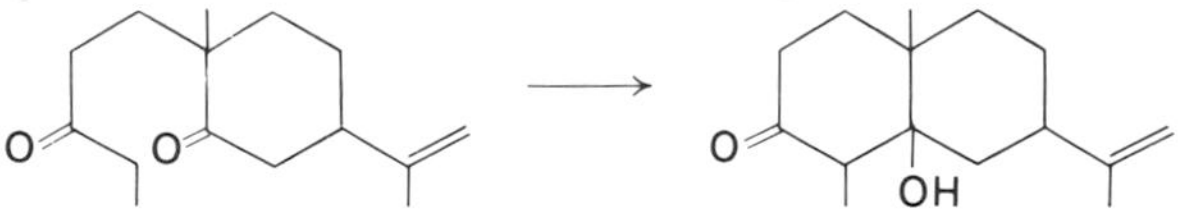

(8)

*(C)* Since the aldol (8) is a β-hydroxy ketone, the loss of water occurs with ease, to give the α,β-unsaturated ketone. Compare this with the dehydration of diacetone alcohol to give mesityl oxide.

## 35.3 The structure of perezone

Perezone is an orange, crystalline compound found in the roots of the plant *Trixis pipitzahuac*, a Mexican composite. The compound has a melting point of 102–103°; it dissolves in alkali with a purple color; and when treated with aniline in hot, ethanolic solution it yields a monoanilino derivative. Perezone can be reduced to a colorless dihydro derivative which is readily reoxidized to the original orange compound. These properties suggest that perezone is a quinone. Perezone has the elementary composition $C_{15}H_{20}O_3$. Ozonolysis of perezone yields acetone, an indication that the structural grouping $=C(CH_3)_2$ is present in the molecule. These observations, in addition to the knowledge that perezone is a $C_{15}$ compound and that upon chromic acid oxidation somewhat more than two moles of acetic acid per mole of compound are produced, showing the presence of (probably) three carbon-linked methyl groups, suggest that perezone is terpenoid in structure.

When oxidized with alkaline hydrogen peroxide, perezone yields acetic acid and an unsaturated acid, $C_9H_{16}O_2$, which can be catalytically hydrogenated to the saturated acid, $C_9H_{18}O_2$. The latter acid was found to be 2,6-dimethylheptanoic acid, which can be synthesized as follows:

$$(CH_3)_2CHCH_2CH_2CH_2Br + CH_3CH(COOEt)_2 \xrightarrow{NaOEt}$$

$$(CH_3)_2CHCH_2CH_2CH_2\overset{\displaystyle CH_3}{\underset{\displaystyle COOEt}{C}}{-}COOEt \xrightarrow[\text{(2) decarboxylate}]{\text{(1) saponify}} (CH_3)_2CHCH_2CH_2CH_2\overset{\displaystyle CH_3}{C}HCOOH$$

These results show that perezone can be partially represented as

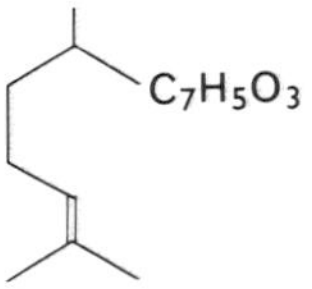

The $C_7$-fragment shows the properties of a benzoquinone with but one unoccupied position, since the addition of aniline yields a monoanilino derivative:

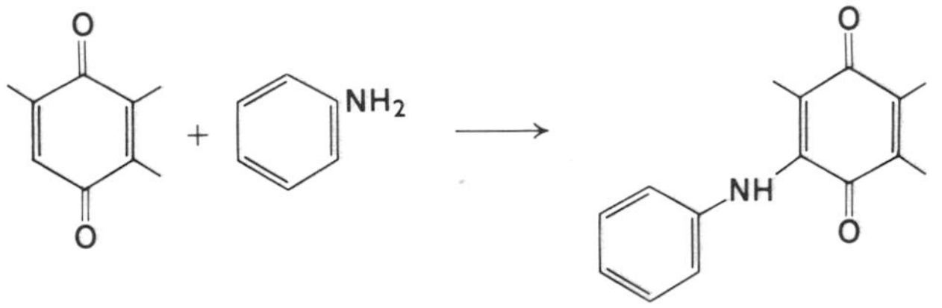

If the quinone had two unoccupied positions in the ring, a dianilino compound would have formed. Since perezone has three carbon-linked methyl groups that yield acetic acid upon oxidation, two of which are in the aliphatic side-chain that appears as the nonenoic acid, the other must be on the quinone nucleus.

When perezone is distilled with zinc dust (a reaction that leads to a deep-seated reduction of aromatic and near-aromatic nuclei to the skeletal structures), a product (uncharacterized) is formed that yields terephthalic acid upon oxidation with potassium permanganate. This shows that the side-chain and the methyl group are in the 1,4-positions with respect to one another.

With all of this information, we can now write two possible structures for perezone:

The first of these (I) was at length shown to be the correct structure for perezone.* The most useful information leading to the establishment of the structure was obtained upon examination of the nuclear magnetic resonance spectrum. The n.m.r. spectrum of perezone itself is shown in Figure 35-2.

---

* Structure II was first proposed for perezone in 1935, and was regarded as correct until the application of n.m.r. studies led to the conclusion that the correct structure was indeed I. The revised structure (I) was published in 1965.

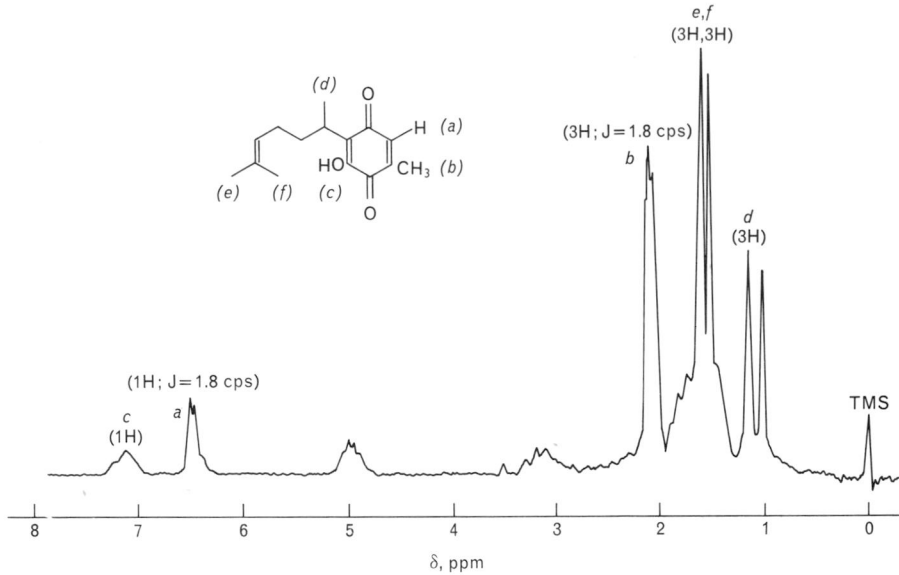

**Figure 35–2.** N.m.r. spectrum of perezone.

The most significant features of this spectrum are the signals for the methyl group (on the quinone ring) and for the hydrogen atom in the adjacent position. These signals are both doublets with a small coupling constant (1.8 cps). This shows that they are coupled in a manner that is quite in accord with their location in adjacent positions (as in I); but were the structure that shown in II, with the methyl group and hydrogen atom in the distant positions on opposite sides of the quinone ring, no such coupling would be expected. The n.m.r. spectra of other derivatives of perezone, which will not be detailed here, are in agreement with this conclusion.

*(A)* The formation of the lithium derivative of 3,5-dimethoxytoluene is the same reaction that is described in Section 27.5, where the preparation of 2,6-dihydroxybenzoic acid is described.

*(B)* The reduction with sodium in liquid ammonia affects only the double bond conjugated with the aromatic ring. The isolated isopropylidene double bond is not affected.

*(C)* The yield in this oxidation is quite low; yet the formation of a methoxyquinone involves no ambiguity in its structure, because the dimethoxytolyl group is symmetrical with respect to the point of attachment of the side chain.

*(D)* The acid-catalyzed hydrolysis of the methoxyquinone occurs with ease, because the compound is a vinyl ether.

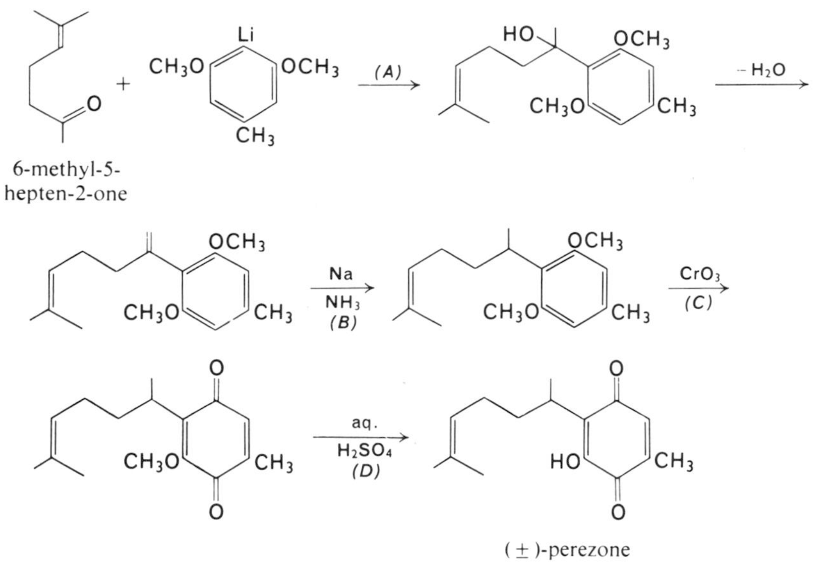

6-methyl-5-
hepten-2-one

+    CH₃O⎯⎯OCH₃ (with Li, CH₃)    $\xrightarrow{(A)}$    HO, OCH₃, CH₃O, CH₃    $\xrightarrow{-H_2O}$

OCH₃ (CH₃O, CH₃)    $\xrightarrow[NH_3 (B)]{Na}$    OCH₃ (CH₃O, CH₃)    $\xrightarrow[(C)]{CrO_3}$

O (CH₃O, CH₃, O)    $\xrightarrow[H_2SO_4 (D)]{aq.}$    O (HO, CH₃, O)

(±)-perezone

## 35.4 The synthesis of equilenin

The first total synthesis of a natural estrogenic hormone (Section 32.8) was carried out by W. E. Bachmann, W. Cole, and A. L. Wilds (1939). Equilenin is a partially aromatic steroid hormone that was first isolated from the urine of pregnant mares. Its synthesis is outlined in the series of reactions shown on the following page, some of which are described in the notes that follow:

(A) The decarbomonoxylation reaction shown here is a method for converting the —COCOOEt grouping into the —COOEt grouping. It involves a loss of the elements of CO, and is brought about by direct heating of the oxalyl ester, usually on a powdered glass surface, upon which the reaction takes place. Since oxalylation with oxalic ester usually gives better yields than carbethoxylation with ethyl carbonate, the introduction of the —COOEt group in the position *alpha* to a carbonyl group is usually carried out by this series of steps.

(B) The alkylation (with methyl iodide) of the β-keto ester requires no comment.

(C) The Reformatsky reaction is now used to introduce the acetic acid grouping as a side chain (see below).

(D) After dehydration (by replacement of —OH by —Cl and removal of the elements of HCl with the base), reduction gives the saturated acid. The reduction of α,β-unsaturated acids with sodium amalgam is a general reaction.

(E) Esterification with diazomethane is not subject to steric interference, thus the diester is formed. Saponification of the diester is, however, strongly influenced by steric factors, and the ester grouping attached to the tertiary carbon atom is relatively difficult to saponify. Hence, preferential saponification of the unhindered primary ester occurs to give the ester acid.

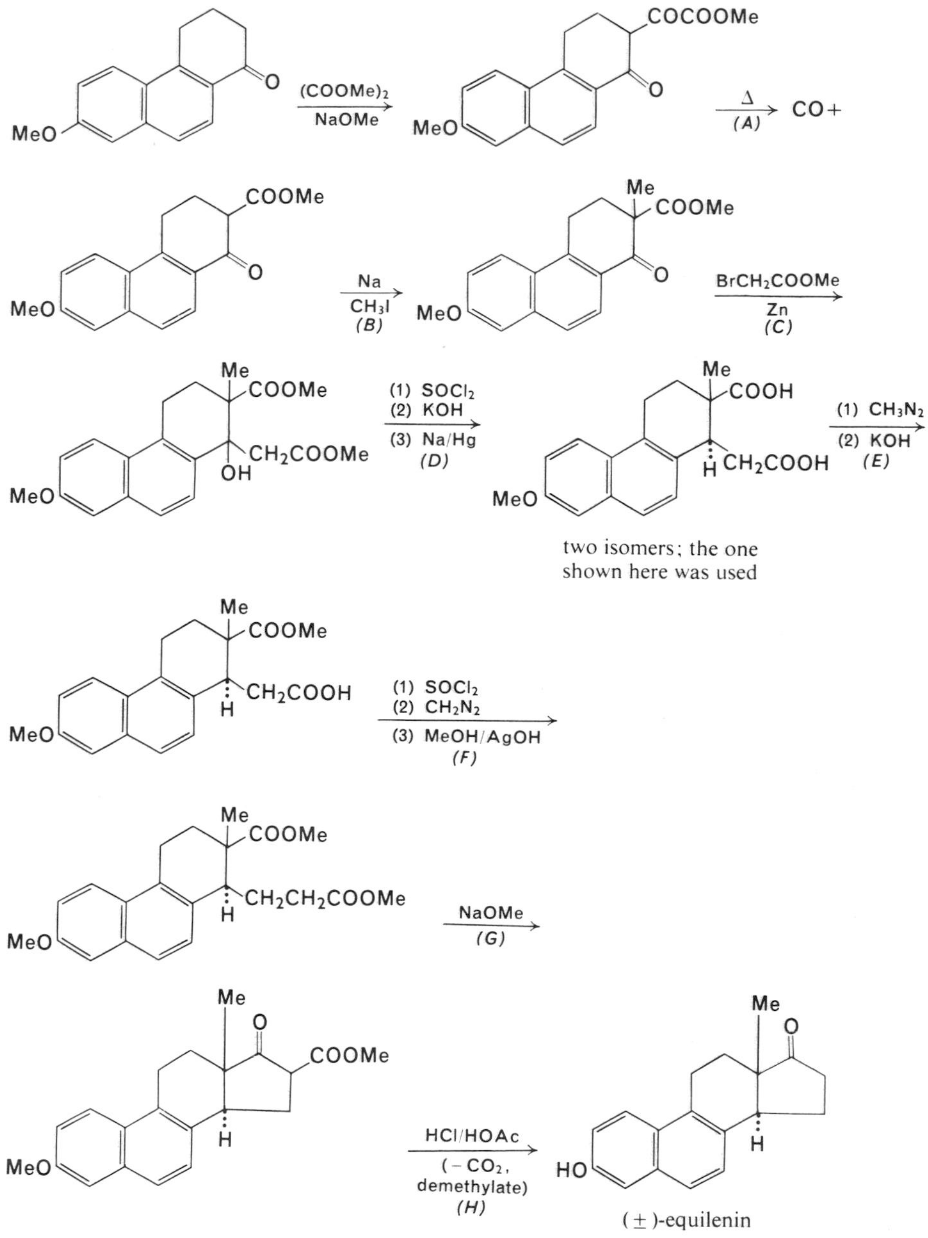

*(F)* This is the Arndt-Eistert reaction, which results in the overall transformation of —CH₂COOMe into —CH₂CH₂COOMe (Section 28.5).

*(G)* The Dieckmann condensation, with closure of the five-membered ring (Section 17.7).

*(H)* Hydrolysis of the keto ester, followed by decarboxylation, and de-methylation, yields equilenin. The natural isomer (but not the optically-active form) is obtained because selection of the proper isomer was made at the stage of the first dicarboxylic acid [following *(D)*]. The preparation of natural, $(+)$-equilenin requires resolution of the racemic product as a final step.

***The Reformatsky reaction.***   Step *(C)* in the above synthetic sequence involves the use of the Reformatsky reaction, which has not been described in the previous chapters. The Reformatsky reaction resembles the Grignard reaction: it involves the addition of an organometallic compound to a carbonyl group, with the formation, after hydrolysis of the intermediate addition compound, of an alcohol. It differs from the Grignard reaction in several respects. The Grignard reagent is prepared by the reaction of an alkyl halide with magnesium in ether solution, and may be preserved; the organometallic compound which is the reagent in the Reformatsky reaction is prepared in the presence of the carbonyl compound with which it is to react, and undergoes reaction as it is formed.

The Reformatsky reaction is carried out with the use of an $\alpha$-bromo ester, and as its use in the equilenin synthesis shows, results in the formation of a $\beta$-hydroxy ester. The reactive component of the Reformatsky reaction mixture is an organozinc compound, and the success of the reaction is due to the fact that this organometallic derivative is not sufficiently reactive to add to the —COOR grouping, but is sufficiently reactive to add to a carbonyl group, such as that present in an aldehyde or a simple ketone.

A representative example of the Reformatsky reaction is that between cyclohexanone and ethyl bromoacetate in the presence of metallic zinc:

The use of the Reformatsky reaction provides a general method of synthesis of substituted cinnamic acids; e.g.,

## 35.5   The synthesis of a carotenoid: crocetin

Crocetin, one of the simpler carotenoid compounds (Section 32.12), occurs naturally as a diester with the disaccharide gentiobiose (Section 18.19) in the flowers of the crocus family (which are found in commerce as an orange-yellow coloring matter called saffron). Hydrolysis of the diester, called crocin, yields two moles of gentiobiose and the dibasic acid crocetin. Crocetin is a deep orange-red crystalline compound of the composition $C_{20}H_{24}O_4$. It contains four carbon-linked methyl groups (by estimation of the acetic acid produced on chromic acid oxidation), and upon catalytic hydrogenation absorbs seven moles of hydrogen to give the completely saturated perhydrocrocetin, $C_{20}H_{38}O_4$. An early synthesis of perhydrocrocetin led to the suggestion that crocetin has the structure

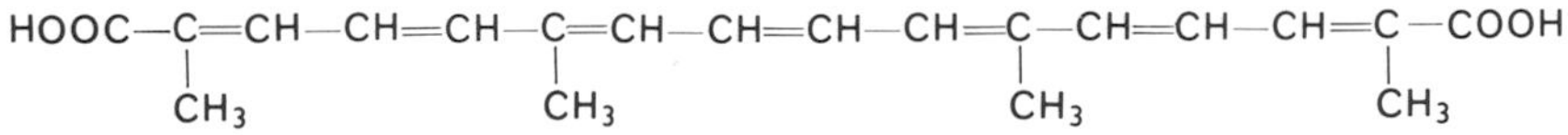

$$HOOC-\underset{\underset{CH_3}{|}}{C}=CH-CH=CH-\underset{\underset{CH_3}{|}}{C}=CH-CH=CH-CH=\underset{\underset{CH_3}{|}}{C}-CH=CH-CH=\underset{\underset{CH_3}{|}}{C}-COOH$$

crocetin (all trans)

The confirmation of this structure by a total synthesis was accomplished in 1953 by a group of Swiss chemists headed by H. H. Inhoffen:

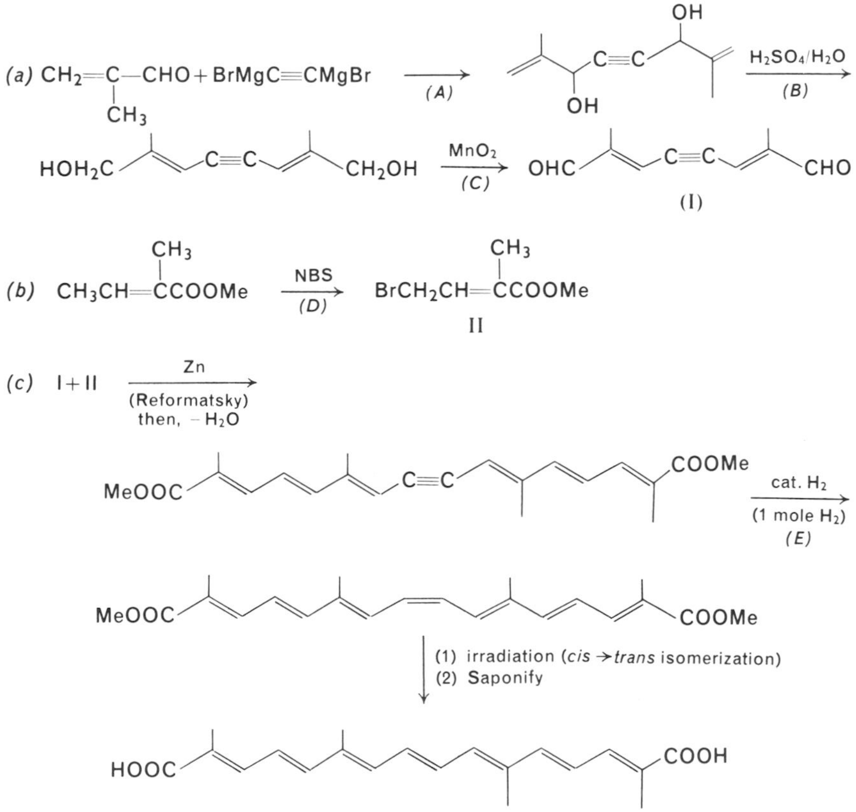

(a) $CH_2{=}\underset{\underset{CH_3}{|}}{C}{-}CHO + BrMgC{\equiv}CMgBr \xrightarrow{(A)}$ [structure] $\xrightarrow[(B)]{H_2SO_4/H_2O}$

$HOH_2C$ [structure with $C{\equiv}C$] $CH_2OH \xrightarrow[(C)]{MnO_2}$ $OHC$ [structure with $C{\equiv}C$] $CHO$  (I)

(b) $CH_3CH{=}\underset{\underset{CH_3}{|}}{C}COOMe \xrightarrow[(D)]{NBS}$ $BrCH_2CH{=}\underset{\underset{CH_3}{|}}{C}COOMe$  II

(c) $I + II \xrightarrow[\substack{(Reformatsky) \\ then, -H_2O}]{Zn}$

$MeOOC$ [polyene structure with $C{\equiv}C$] $COOMe \xrightarrow[\substack{(1\ mole\ H_2) \\ (E)}]{cat.\ H_2}$

$MeOOC$ [polyene structure] $COOMe$

$\downarrow$ (1) irradiation (*cis* →*trans* isomerization)
(2) Saponify

$HOOC$ [polyene structure] $COOH$

crocetin

The identity of the synthetic and natural crocetin was established by their melting points (285°C), and by the distinctive absorption spectrum. Crocetin is a conjugated polyene with seven carbon-carbon double bonds flanked at each end of the chain by carboxyl groups. It is a brick-red crystalline compound; its absorption spectrum (in carbon disulfide solution) shows maxima at 482, 453, and 426 m$\mu$.

*(A)* The formation of the organomagnesium compound of acetylene is carried out by passing acetylene into a convenient Grignard reagent, such as ethylmagnesium bromide (See Section 9.24.)

*(B)* This is an allylic rearrangement of the general form

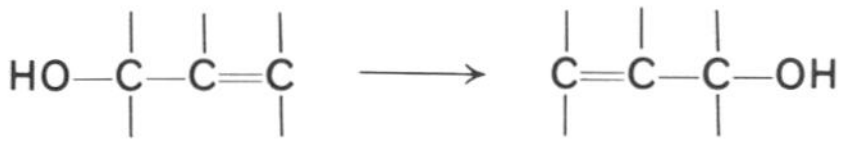

$$\text{HO}-\overset{|}{\text{C}}-\overset{|}{\text{C}}=\overset{|}{\text{C}} \longrightarrow \overset{|}{\text{C}}=\overset{|}{\text{C}}-\overset{|}{\text{C}}-\text{OH}$$

and is catalyzed by acid, probably by way of the carbonium ion formed from the protonated alcohol.

*(C)* The $MnO_2$ oxidation of allylic alcohols has been described in Section 8.24.

*(D)* Bromination of a methyl group attached to a carbon-carbon double bond can be effected by the reagent N-bromosuccinimide (NBS),

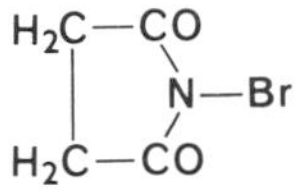

$$\begin{array}{c} \text{H}_2\text{C}-\text{CO} \\ | \qquad\quad \diagdown \\ \qquad\qquad \text{N}-\text{Br} \\ | \qquad\quad \diagup \\ \text{H}_2\text{C}-\text{CO} \end{array}$$

*(E)* The catalytic hydrogenation of the triple bond can be carried out without affecting the remaining double bonds, the hydrogenation being interrupted after one mole has been absorbed. It will be noted that the double bond that is produced is in the *cis* configuration. Isomerization of the *cis* double bond to the more stable *trans* configuration was carried out by irradiation with short-wavelength light in the presence of a trace of iodine as a catalyst (see Section 16.6).

## 35.6 The structure proof and synthesis of an alkaloid: coclaurine

Coclaurine is of special interest and importance among the alkaloids (Chapter 29) because it possesses the hydroxylation pattern found in a number of other benzylisoquinoline alkaloids. Some of these, the bisbenzylisoquinoline alkaloids, have structures that suggest their origin by an oxidative coupling of two molecules of coclaurine or two molecules of a precursor very similar to coclaurine in structure. An example is oxyacanthine; the structure shown below is written in such a way as to indicate its possible genesis from coclaurine:

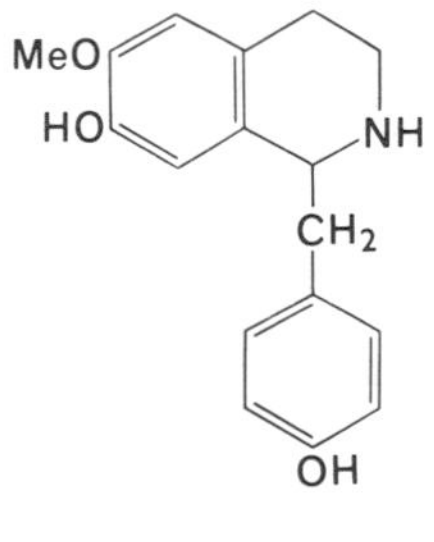

coclaurine 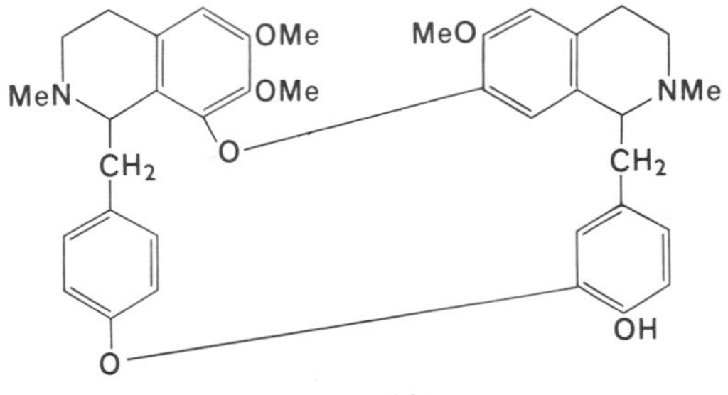                                  oxyacanthine

The structure of coclaurine ($C_{17}H_{19}O_3N$), an alkaloid found in the plant *Cocculus laurifolius*, was first deduced from the following evidence. It is a secondary amine which was found, by the preparation and analysis of the N- and O-methyl derivatives, to contain one —OMe group, two —OH groups, and one $>$NH group. Upon zinc-dust distillation there was obtained *p*-cresol, indicating that the structural element HO—⟨benzene ring⟩—C— was present in the alkaloid.

Complete methylation (to the N-methyl methiodide of the O-dimethyl derivative), followed by the Hofmann degradation, led to a methine base (see the formulation below) that could be further degraded in the same way to a nitrogen-free methine. This compound gave 4,5-dimethoxyphthalic acid (*m*-hemipinic acid) and anisic acid (*p*-methoxybenzoic acid) on oxidation with potassium permanganate. The analogous coclaurine O-diethyl ether gave, on a similar degradation, 3-ethoxy-4-methoxy-6-ethylbenzoic acid and *p*-ethoxybenzoic acid.

The Hofmann degradation, of great importance in the study of nitrogen-containing compounds, and of particular usefulness in the study of alkaloids, is an elimination reaction, related to the dehydrohalogenation of alkyl halides, in which a quaternary ammonium grouping is displaced by internal attack and formation of a double bond:

$$HO^- + H-\overset{|}{\underset{|}{C}}-\overset{|}{\underset{|}{C}}-\overset{+}{N}Me_3 \longrightarrow HOH + {>}C{=}C{<} + NMe_3$$

One method of carrying out the degradation is as follows: the quaternary iodide is treated with a suspension of silver oxide in water, with the formation of silver iodide and the quaternary hydroxide. Concentration of the filtered aqueous solution, followed by heating, leads to the reaction formulated above. In some cases it is sufficient to heat the quaternary iodide in an aqueous or alcoholic solution of alkali.

The degradation of fully methylated coclaurine outlined above can be formulated as follows:

In the degradation of the O-ethyl derivative, hydrogenation of the methine base, with conversion of the vinyl group to an ethyl group, was carried out with the result that 3-ethoxy-4-methoxy-6-ethylbenzoic acid was obtained upon oxidation of the methine.

These observations can be accommodated in the structure written above for coclaurine. The final proof of the structure was established (J. Finkelstein, 1951) by the total synthesis of the alkaloid. This synthesis illustrates the use of methods for "protecting" phenolic hydroxyl groups during a synthetic sequence, and for their eventual regeneration by the removal of the "protecting" groups.

The synthesis consists of three parts: (1) the synthesis of the requisite phenethylamine; (2) the synthesis of the fragment that will constitute the final benzyl

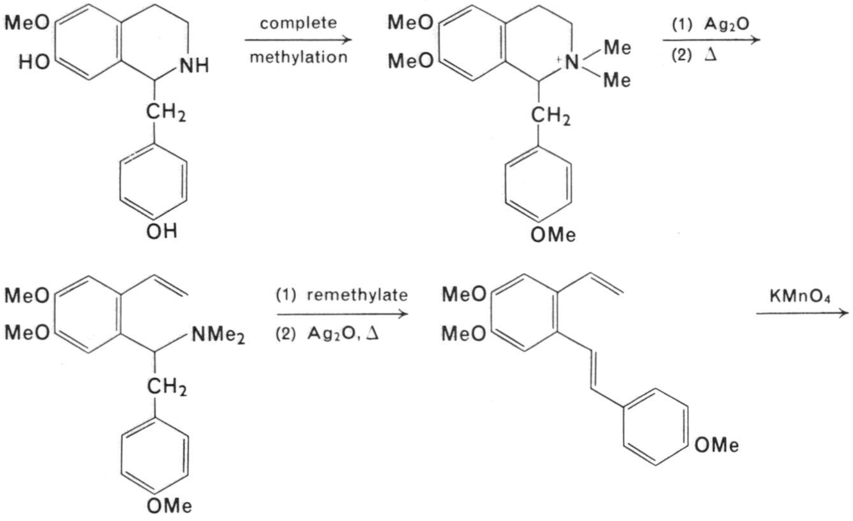


the methine base

the methine

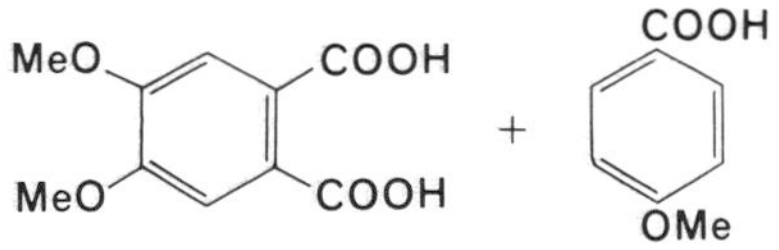

grouping of the alkaloid; and (3) the combining of these two portions in an isoquinoline synthesis to give the final alkaloid:

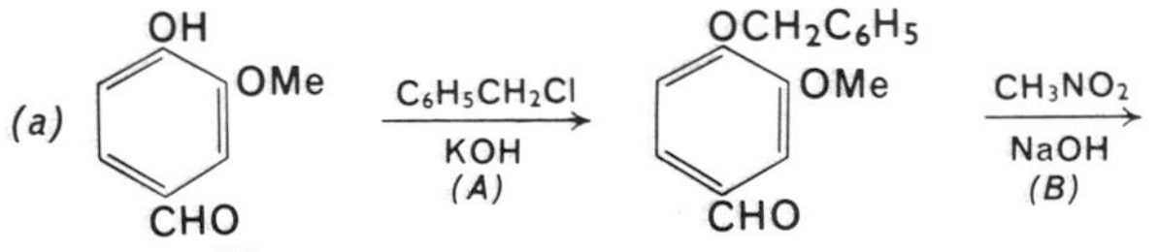


vanillin

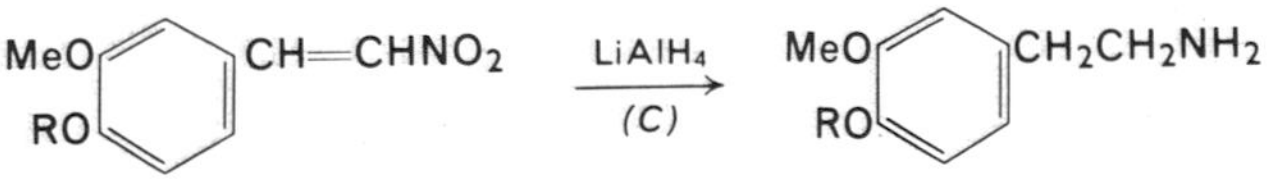


(R = benzyl)

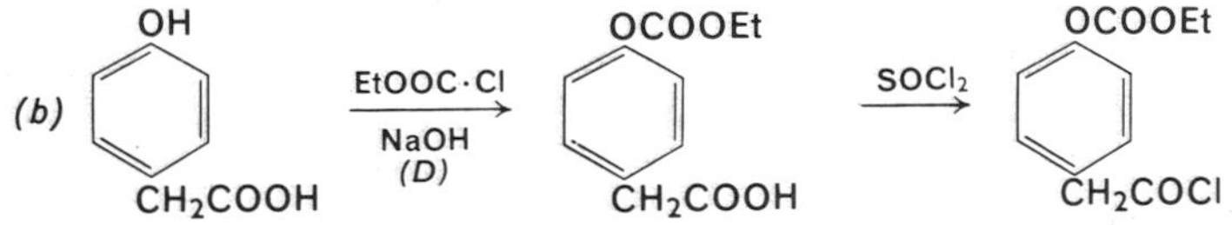


*(A)* This is a straightforward alkylation of the phenolic hydroxyl group and needs no comment.

*(B)* The aldol-like condensation of nitromethane with an aromatic aldehyde was described in Section 13.14.

*(C)* Reduction of β-nitrostyrenes to β-phenylethylamines with lithium aluminum hydride is a useful and general method for the preparation of amines of this class (Section 13.14).

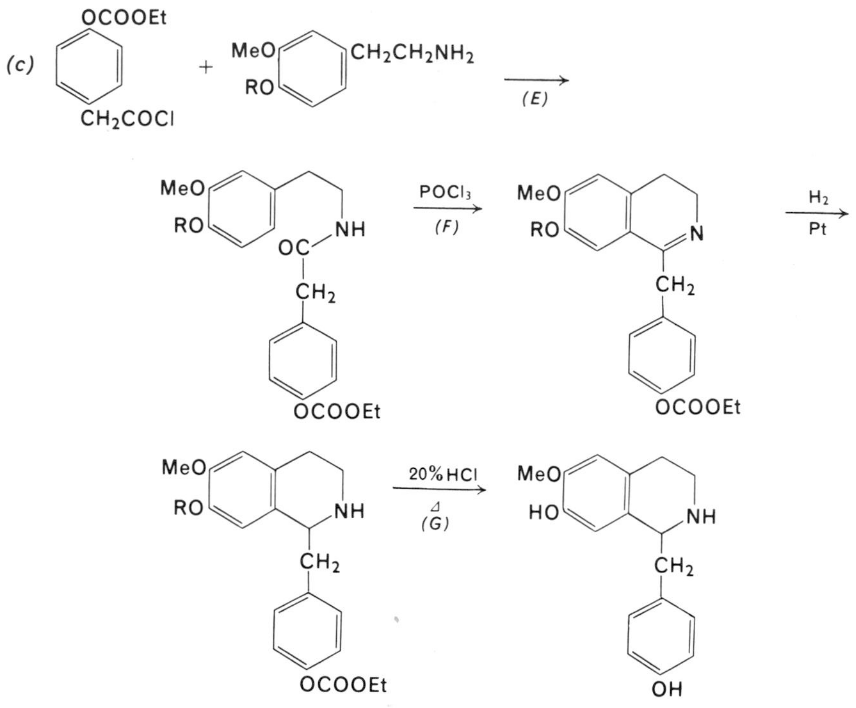

(±) coclaurine

*(D)* Ethyl chloroformate (the half-ester, half-acid chloride of carbonic acid, prepared from phosgene, $COCl_2$, and ethanol) is often used to acylate phenolic hydroxyl groups in this way in order to "protect" such hydroxyl groups. Eventual hydrolysis of the Ar—OCOOEt grouping, with regeneration of Ar—OH, is readily accomplished at the end of the synthetic sequence.

*(E)* This acylation of a primary amine by an acid chloride is unexceptional. The formation of the amide in this way is the first step in the Bischler-Napieralski synthesis of the isoquinoline ring system (Section 29.19).

*(F)* This is the ring-closure step of the Bischler-Napieralski isoquinoline synthesis. It is clearly an electrophilic substitution reaction, and may be written in the (formal) manner shown in the scheme on the opposite page. This is to be regarded as illustrative only, for the exact identity of the phosphorylated intermediate, and of the phosphorus-containing displaced fragment, is only conjectural.

*(G)* The final step involves the acid hydrolysis of the —OCOOEt and benzyl groupings with regeneration of the phenolic hydroxyl groups. The hydrolysis of

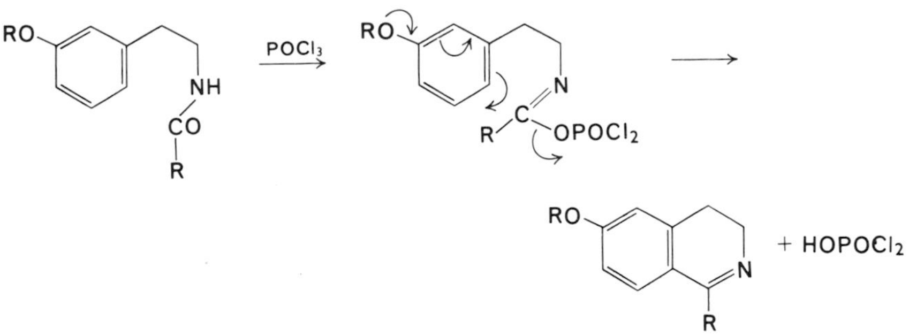

the diester, Ar—O—COOEt, is straightforward and requires no comment. Cleavage of the benzyl ether is an example of aryl ether cleavage similar to that described in Section 27.5. The difference lies in the much greater ease of cleavage of the benzyl, as compared with the methyl, ether. The much greater reactivity of benzyl derivatives in displacement reactions occurring at the benzyl carbon atom accounts for the fact that the cleavage of a benzyl ether can be accomplished with 20% hydrochloric acid, whereas either concentrated HBr or HI at somewhat higher temperatures is needed for the cleavage of methyl ethers. The cleavage reaction can be formulated as follows (compare with Section 27.5):

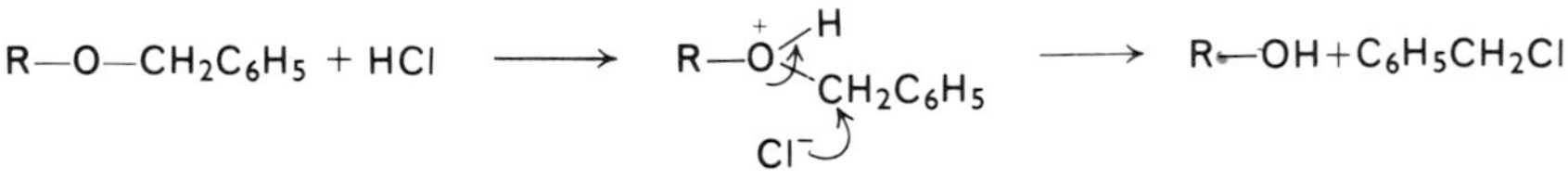

## Chapter One

**5** If the salt is soluble in water, precipitate silver as AgCl by adding HCl. If the salt is not water soluble dissolve in nitric acid, and precipitate silver as AgCl. Dry and weigh AgCl, and calculate silver content from

$$\frac{\text{wt. of AgCl} \times 107.87}{(107.87 + 35.45)} \times \frac{1}{\text{wt. of sample}} \times 100 = \%\,\text{Ag}$$

**6** If the compound contains only C, H, and O the total of the analytical figures for these will add up to 100%. For the analysis C, 48.10%; H, 12.05%; O, 21.40% the total is 81.55%. This indicates that another element is present. Qualitative analysis for nitrogen, halogen, or sulfur can be carried out, followed by quantitative analysis for the other element. [Note: The formula $C_3H_9ON$ will fit this analysis.]

**7** To remove HCN, which otherwise would produce a precipitate with silver nitrate.

**8** Three of these are

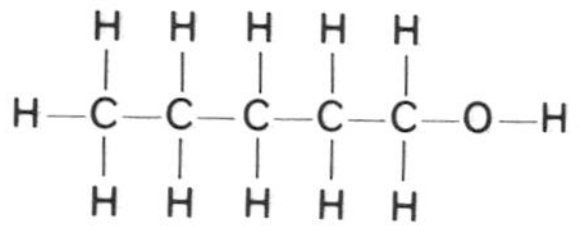

$CH_3CH_2CH_2CH_2CH_2OH$

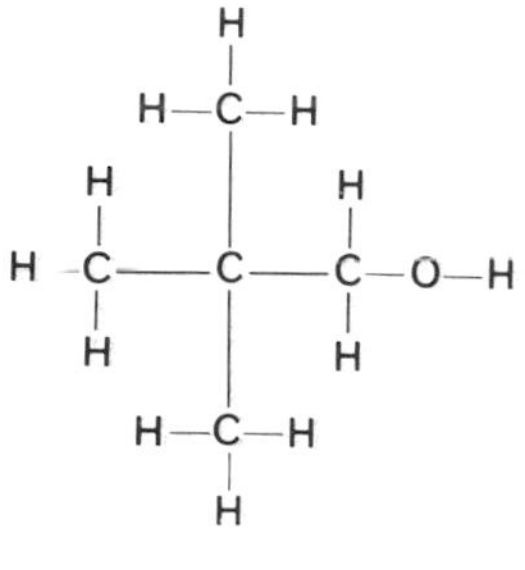

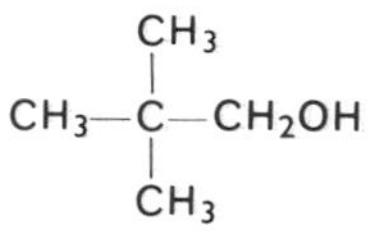

$CH_3 - \overset{\displaystyle CH_3}{\underset{\displaystyle CH_3}{\overset{|}{\underset{|}{C}}}} - CH_2OH$

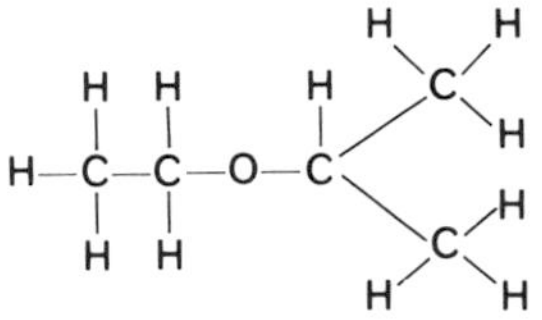

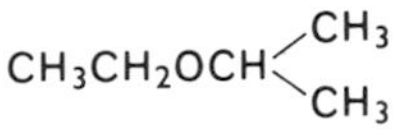

$$CH_3CH_2OCH\begin{array}{c}CH_3\\CH_3\end{array}$$

There are 14 in all (8 alcohols, 6 ethers).

**9** *(a)* only one

*(b)* $CH_3CH_2CH_2Cl$, $CH_3\overset{Cl}{\underset{|}{C}}HCH_3$

*(c)* $ClCH_2CH_2Cl$, $CH_3CHCl_2$

*(d)* $CH_3CH_2NH_2$, $CH_3NHCH_3$

*(e)* $CH_3CH_2CH_2NH_2$, $CH_3\overset{NH_2}{\underset{|}{C}}HCH_3$, $CH_3NHCH_2CH_3$, $CH_3\overset{CH_3}{\underset{|}{N}}CH_3$

**10** 4.52 mg $(CH_3)_2S$ (m.w.$=62)=0.073$ millimole. This will give 0.073 m$M$ $BaSO_4$ (m.w.$=233.3)=17.03$ mg.

**11** 4.40 mg $CO_2=0.1$ millimole. Thus, the amount of each compound that must be burned is *(a)* 0.1 m$M$; *(b)* 0.0333 m$M$; *(c)* 0.1 m$M$; *(d)* 0.050 m$M$; *(e)* 0.0167 m$M$. [Note: m$M\times$ m.w. $=$ mg.]

**12** *(a)* AgBr; $\%Br=42.6$; *(b)* $C_2H_5Br$, $\%Br=73.5$; *(c)* $BrCH_2CH_2Br$, $\%Br=85.1$; *(d)* $BrCH_2CONH_2$, $\%Br=58.0$; *(e)* $CH_2ClBr$; $\%Br=69.2$.

**13** $80/$m.w.$=0.3156$; m.w.$=253$.

**14** 9.17 mg $CO_2=2.50$ mg C;

3.78 mg $CO_2=0.422$ mg H;

$\%C=(2.50/6.23)\times100=40.10$;

$\%H=(0.422/6.23)\times100=6.78$;

mol. wt. : 357 ml $(27°, 750$ mm$)=330$ ml STP;

22.4 l.$=60$ g; m.w.$=60$;

mol. formula$=C_2H_4O_2$ (calc., $\%C=40.00$; $\%H=6.70$).

This formula is obtained as follows:

$$\%C = \frac{\text{grams carbon per mole}}{\text{mol. wt.}}\times 100 = 40.1;$$

grams carbon per mole $= 60\times0.401 = 24$;

$$\%H = \frac{\text{grams H per mole}}{\text{mol. wt.}}\times 100 = 6.78;$$

grams H per mole $= 60\times0.0678 = 4$

The remainder is $60-(24+4)=32$. Thus, the formula $C_2H_4O_2$ will fit the data. It should be noted that the formula $C_2H_4S$ will also give the same carbon-hydrogen analysis.

**15** $100-(69.95+11.70)=18.35$. If this represents only N, the formula $C_9H_{18}N_2$ will fit (calc., C, 70.10; H, 11.70). If both O and N are present, the minimum will be $16+14=30$ per mole. For a residual combined percentage of 18.35, the molecular weight must be 163 $[(30/163)\times100=18.4]$. For this molecular weight,

$$\%C = \frac{\text{grams C per mole}}{163}\times 100 = 69.95;$$

grams C per mole $= 163\times0.6995 = 114$

This cannot represent an integral number of C atoms. For any other combinations of O and N ($O_2N$, $ON_2$, and so on), the molecular weight will be above 170. For example for $ON_2$:

$$\% \text{ combined O and N} = 18.35 = \frac{16+28}{\text{mol. wt.}} \times 100$$

$$\text{mol. wt.} = \frac{44}{0.1845} = 238$$

**16** $C_3H_5O_3Ag$.

**17** The formation of acetic acid shows that the grouping $CH_3$—C must be present. Since the silver salt of an acid contains —COOAg, lactic acid must have a structure

$CH_3(CH_2O)COOAg$. This can only be represented by $CH_3$—$\overset{\overset{\displaystyle H}{|}}{\underset{\underset{\displaystyle OH}{|}}{C}}$—COOAg.

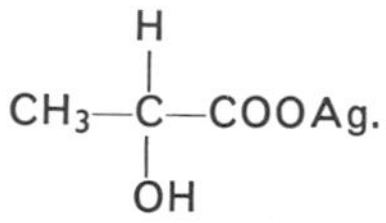

**18** For $C_{27}H_{46}O$, $\%C=84.20$ and $\%H=12.00$. For $C_{28}H_{48}O$; $\%C=84.38$; $\%H=11.74$. Were an analysis of C, 84.30; H, 11.85 to be obtained, it would not distinguish between these formulas, since it would be "acceptable" for either.

**19** $C_{21}H_{22}N_2O_2$.

**20** Composition $C_7H_{12}O$. This is 4H less than the saturated, acyclic compound of seven carbon atoms; thus, there are two double bonds or rings (or one triple bond). If one carbonyl group $\left( {\Large{>}}C{=}O \right)$ is present, the compound could be a cyclic ketone or aldehyde, or an unsaturated ketone or aldehyde; for example,

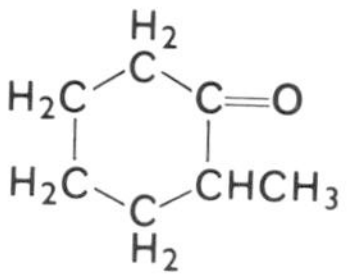

or   $CH_3CH{=}CHCH_2CH_2\overset{\overset{\displaystyle O}{\|}}{C}CH_3$

**21** A possible structure is $CH_3CH{=}CHCH_2$—O—$CH_2CH{=}CH_2$; another is $CH_3CH{=}CH$—$CH{=}CHCH_2OH$.

**23** $CH_3C{\equiv}CCH_2CH{=}N$—$CH_3$ ;

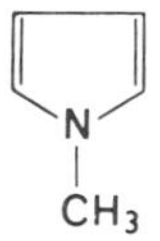

## Chapter Two

**4** Many structures can be written for each of these. Some examples are these:

(a) $CH_3CH{=}CHCH_2C{\Large{\diagdown}}{\overset{\displaystyle O}{\underset{\displaystyle OCH_2CH_2CH_3}{}}}$

(d) $CH_3CH{=}CHCH_2C{\Large{\diagdown}}{\overset{\displaystyle O}{\underset{\displaystyle H}{}}}$ :   (f)

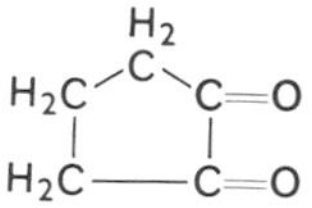

**5** Look these up in a good unabridged dictionary.

## Chapter Three

**10** To form $H_3N \rightarrow Be(CH_3)_2$ and $H_3N \rightarrow Be \leftarrow NH_3$.

$$\overset{\displaystyle CH_3}{\underset{\displaystyle CH_3}{|}}$$

**12** Electron pair in $H:^-$ is close to the nucleus, whereas the external unshared pair in $Na:^-$ is not.

**15** $Et_2O \rightarrow BCl_3$.

**16** The $(CH_3)_3B$ has no unshared electrons; $(CH_3)_3N:$ has a pair capable of forming the $N-O$ bond.

**17** About $90°$.

**18** (a) 0; (b) +1; (c) $H-\ddot{N}=\overset{+}{N}=\ddot{N}:^-$; (d) +1; (e) 0; (f) 0 on each; (g) −1.

**19** Would change from $120°$ to approach tetrahedral angle of $109.5°$.

**20** (a), (b), (c) tetrahedral; (d) about $90°$; (e) about $104°$; (f) $180°$ (linear); (g) about like those in $NH_3$.

**21** Interference between $CH_3$ and *t*-butyl groups; crowding of *t*-butyl groups to rear of N atom.

**22** Inductive effect of Cl atoms; possible participation of Cl in distributing the − charge in $Cl_3C:^-$ (see Chapter 10).

## Chapter Four

**7** (c) $HCO_3^- + HCl \rightleftharpoons H_2CO_3 + Cl^-$

(d) $NH_2NH_2 + HCl \rightleftharpoons NH_2NH_3^+ + Cl^-$

(f) $NH_2Cl + HCl \rightleftharpoons NH_3Cl^+ + Cl^-$

(g) $CH_3CO_2^- + HCl \rightleftharpoons CH_3COOH + Cl^-$

(h) $(CH_3)_2O + HCl \rightleftharpoons (CH_3)_2OH^+ + Cl^-$

**8** Hydronium perchlorate.

**9** (a) $H_3O^+$; (b) $NH_4^+$; (c) $CH_3COOH$, $H_3O^+$; (d) $CH_3COOH$; (e) $NH_4^+$, $H_3O^+$; (f) $HSO_4^-$, $H_2SO_4$, $CH_3COOH_2^+$.

**10** (a) $OH^-$; (b) $H_2O$; (c) $NH_2^-$; (d) $CH_3O^-$; (e) $NH_3$; (f) $CH_3CO_2^-$; (g) $HCO_3^-$; (h) $CH_3NH^-$; (i) $Br^-$; (j) $NH_2NH_2$; (k) $SO_4^=$.

**11** (d) $CH_3OH_2^+ + CH_3COO^- \rightleftharpoons CH_3OH + CH_3COOH$

(g) $H_3PO_4 + CH_3COO^- \rightleftharpoons H_2PO_4^- + CH_3COOH$

**12** (a) $CH_3CH_2-\ddot{O}-H$; $CH_3CH_2-\overset{+}{\ddot{O}}\overset{\displaystyle H}{\underset{\displaystyle H}{\big\langle}}$

(d) $CH_3-\ddot{S}-CH_3$; $H_3C\overset{\displaystyle }{\underset{\displaystyle \underset{\displaystyle H}{|}}{\underset{\displaystyle \overset{+}{S}}{\diagdown}}}CH_3$

**13** (a) $NH_3 + H_2SO_4 \longrightarrow NH_4^+ + HSO_4^-$  (2)

(b) $CH_3OCH_3 + H_2SO_4 \longrightarrow (CH_3)_2OH^+ + HSO_4^-$  (2)

(c) $NH_2NH_3^+Br^- + H_2SO_4 \longrightarrow NH_3NH_3^{++} + Br^- + HSO_4^-$  (3)

(d) $CH_3OH + H_2SO_4 \longrightarrow CH_3OH_2^+ + HSO_4^-$

$CH_3OH_2^+ + HSO_4^- \longrightarrow CH_3OSO_3H + H_2O$

$H_2O + H_2SO_4 \longrightarrow H_3O^+ + HSO_4^-$  (3)

**14** $i>e>f>b>d>c>j>a>h>g>k$.

**15** $pH$ 8.88.

**16** *(a)*

$$\begin{array}{c} H_3C \\ \phantom{H_3C}\diagdown \\ \phantom{H_3C}O \rightarrow BCl_3 \\ \phantom{H_3C}\diagup \\ H_3C \end{array}$$

*(e)*

$$\begin{array}{c} H_3C \\ \diagdown \\ O \rightarrow MgCl_2 \\ \diagup \\ H_3C \end{array} \quad \text{and} \quad \begin{array}{c} \phantom{xx}Cl \\ \phantom{xx}| \\ H_3C\diagdown \phantom{xx} \diagup CH_3 \\ \phantom{xx}O \rightarrow Mg \leftarrow O \\ H_3C\diagup \phantom{xx} \diagdown CH_3 \\ \phantom{xx}| \\ \phantom{xx}Cl \end{array}$$

*(g)*

$$\begin{array}{c} H_3C\diagdown \phantom{x}+ \\ \phantom{xx}O-CH_3 \\ H_3C\diagup \end{array}$$

**17** Stronger; the $\alpha$-Br atom would have a strong influence compared with the more distant $\beta$-Br atoms of $Br_2CHCH_2COOH$.

**19** Stronger; inductive effect of Cl.

**20** HBr in $SO_2$ is undissociated. When water is added, $HBr + H_2O \longrightarrow H_3O^+ + Br^-$. Excess water would not react to produce further ionic species.

## Chapter Six

**4** *a–f, b–c, c–a, d–d, e–g, f–e, g–b*.

**5** *(c)* Diethylacetic acid; *(d)* trimethylacetic acid.

**6** Partial answer given with problem.

**7** *(a)*

$$\begin{array}{c} CH_3-CH-CH_3 \\ | \\ CH_3-CH-\underset{\underset{\underset{\underset{CH_3}{|}}{CH_3}}{|}}{\overset{}{C}}-CH\diagup^{CH_3}_{\diagdown CH_3} \end{array}$$

*(c)* $CH_2{=}CH-CH-CH{=}CH_2$
  with $OH$ below the middle $CH$

*(d)* $CH_2{=}CHCH_2Cl$

*(g)* $CH_2{=}CHCH_2CH_2OH$

**8** *(a)*

$$CH_3-\underset{\underset{CH_3}{|}}{\overset{\overset{CH_3}{|}}{C}}-CH_3$$

*(b)*

$$CH_3CH-\underset{\underset{CH_2CH_3}{|}}{\overset{\overset{CH_3}{|}}{C}}-CH_2-\underset{\underset{CH_3}{|}}{\overset{\overset{CH_3}{|}}{C}}-CH_2CH_3$$

*(c)*

$$CH_3-\underset{\underset{CH_3}{|}}{\overset{\overset{CH_3}{|}}{C}}-CH_2CH_2-\underset{\underset{CH_3}{|}}{\overset{\overset{CH_3}{|}}{C}}-CH_3$$

*(d)* $HOCH_2CHCH_2OH$
  with $OH$ below the middle $CH$

*(e)* $Br_3C-CBr_3$

**9** *(a)*

$$CH_3\underset{\underset{CH_3}{|}}{CH}CH_2\underset{\underset{CH_3}{|}}{CH}CH_3; \text{ 2,4 dimethylpentane}$$

*(d)*

$$CH_3CH_2\underset{\underset{CH_3}{|}}{CH}CH_3; \text{ 2 methylbutane}$$

*(f)*

$$CH_3\underset{\underset{CH_3}{|}}{CH}CH_2CH_2CH_2CH_3; \text{ 2 methylhexane}$$

**10** Two examples are

CH$_3$CH$_2$CH$_2$CH$_2$CH$_3$; pentane, diethylmethane

and

CH$_3$CHCH$_2$CH$_3$; 2=methylbutane, dimethylethylmethane
|
CH$_3$

**11** *(a)* dimethylethylmethane; methylisopropylmethane

    *(b)* trimethylethylmethane; methyl-*t*-butylmethane

    *(c)* diisopropylmethane; dimethylisobutylmethane


## Chapter Seven

**4** CH$_3$CH$_2$CH$_2$OH + H$_3$O$^+$ ⇌ CH$_3$CH$_2$CH$_2$OH$_2^+$ + H$_2$O

  CH$_3$CH$_2$CH$_2$OH$_2^+$ + Br$^-$ ⇌ CH$_3$CH$_2$CH$_2$Br + H$_2$O

  Side reactions: formation of di-*n*-propyl ether by attack of 1-propanol upon protonated 1-propanol; formation of olefin (propylene) by loss of H$_2$O and H$^+$ from protonated 1-propanol.

**5** ClCH$_2$OCH$_2$CH$_3$ is an α-chloro ether; it is very reactive towards nucleophilic reagents because of the stability of $^+$CH$_2$OCH$_2$CH$_3$ in the transition state for displacement. On the other hand, CH$_3$OCH$_2$CH$_2$Cl, a β-chloro ether, would be more nearly comparable to a primary alkyl chloride.

**6** Charge delocalizations (CH$_3$CH=CHCH$_2^+$ ⟷ CH$_3$$\overset{+}{C}$H—CH=CH$_2$) which contributes to stability (lower energy) of transition state for halogen displacement.

**7** Most of these have close counterparts in the text. They are all displacement reactions of the type exemplified by these examples:

  *(c)* CH$_3$CH$_2$CH$_2$CH$_2$CH$_2$Br + CN$^-$ ⟶ CH$_3$CH$_2$CH$_2$CH$_2$CH$_2$CN + Br$^-$

  *(f)* CH$_3$CH$_2$OSO$_2$CH$_3$ + (C$_2$H$_5$)$_2$NH ⟶ CH$_3$CH$_2$$\overset{+}{N}$(C$_2$H$_5$)$_2$ + CH$_3$SO$_2$O$^-$
                                          |
                                          H


## Chapter Eight

**16** *(a)* CH$_3$OCH$_2$CH$_3$; *(b)* CH$_2$=C(CH$_3$)$_2$; *(c)* CH$_3$CH=C(CH$_3$)$_2$; *(d)* CH$_3$OCH$_2$CH$_3$;

    *(e)* both ether and olefin.

**17** *(a)* CH$_3$CH$_2$$\overset{.OH}{C}$HCH$_2$CH$_3$; from CH$_3$CH$_2$MgBr and CH$_3$CH$_2$CHO

    *(b)* (CH$_3$)$_2$CHCH$_2$CH$_2$OH; from (CH$_3$)$_2$CHMgBr and HCHO to give isobutyl alcohol; then isobutylmagnesium bromide (via isobutyl bromide) and HCHO

    *(c)* (CH$_3$CH$_2$)$_3$COH; in reverse order of steps:

      (1) CH$_3$CH$_2$COCH$_2$CH$_3$ and CH$_3$CH$_2$MgBr;

      (2) CH$_3$CH$_2$COCH$_2$CH$_3$ by oxidation of diethylcarbinol [see part *(a)*]

    *(d)* CH$_3$$\overset{OH}{C}$CH$_2$CH$_3$; acetone and ethylmagnesium bromide
         |
         CH$_3$

**18** $CH_3C\begin{smallmatrix}O\\OEt\end{smallmatrix}$ + HA $\rightleftharpoons$ $CH_3C\begin{smallmatrix}\overset{+}{O}H\\OEt\end{smallmatrix}$

$CH_3C\begin{smallmatrix}\overset{+}{O}H\\OEt\end{smallmatrix}$ + PrOH $\rightleftharpoons$ $CH_3-\overset{\overset{\overset{H}{\underset{+}{\overset{\diagup}{O}}}\diagdown Pr}{|}}{\underset{OEt}{C}}-OH$ $\rightleftharpoons$ $CH_3-\overset{\overset{OPr}{|}}{\underset{\overset{O}{\overset{+}{H}}\diagdown Et}{C}}-OH$ $\rightleftharpoons$

$CH_2C\begin{smallmatrix}OPr\\OH\end{smallmatrix}$ + EtOH

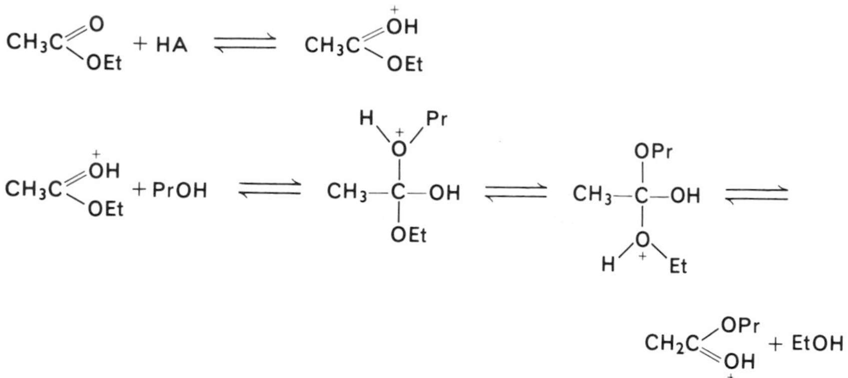

**19** Starts by attack of BuO⁻ on $CH_3COOEt$ to give $CH_3-\overset{\overset{OBu}{|}}{\underset{OEt}{C}}-O^-$

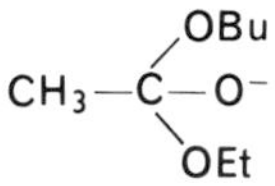

**20** *(a)* $CH_3{=}CHCH_3$ *(b)* $CH_2{=}C(CH_3)_2$ *(c)* $CH_3CH{=}CHCH_3$

*(d)* $(CH_3)_2C{=}CHCH_3$ *(e)*

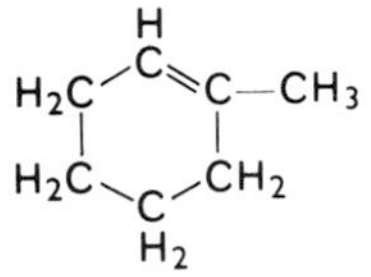

$\begin{smallmatrix} & & H & \\ & H_2C\diagup\overset{C}{\diagdown} & \\ & | & \overset{\|}{C}-CH_3 \\ & H_2C\diagdown\underset{C}{\diagup} & CH_2 \\ & & H_2 & \end{smallmatrix}$

**21** Products are *(a)* $CH_3CH_2OH$; *(b)* $CH_3CH_2CH_2OH+(CH_3)_2CHOH$; *(c)* 2=butanol; *(d)* ethanol + methanol.

**22** *(a)* $CH_3CH_2CH_2OH \xrightarrow{-H_2O} CH_3CH{=}CH_2$

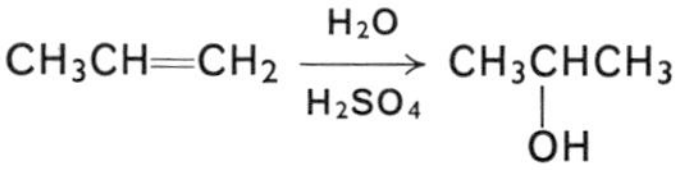

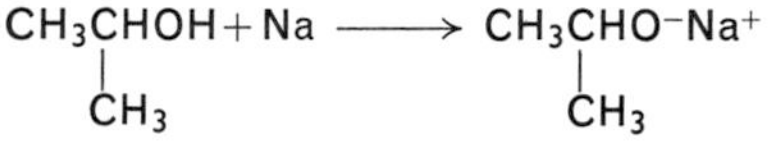

$CH_3CH{=}CH_2 \xrightarrow[H_2SO_4]{H_2O} CH_3\overset{}{\underset{OH}{C}}HCH_3$

$CH_3\overset{}{\underset{CH_3}{C}}HOH + Na \longrightarrow CH_3\overset{}{\underset{CH_3}{C}}HO^-Na^+$

$(CH_3)_2CHO^-Na^+ + CH_3CH_2CH_2Br \longrightarrow$ *i*-Pr—O—*n*-Pr

*(c)* 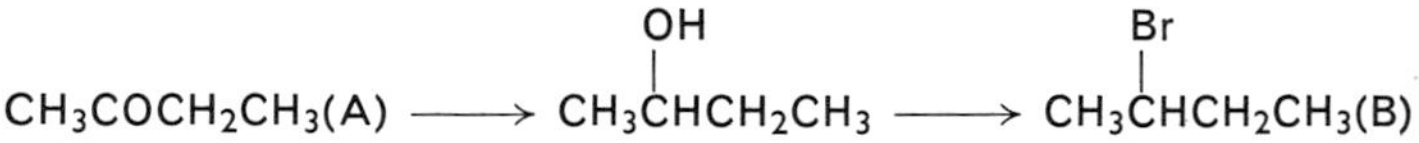 $CH_3COCH_2CH_3(A) \longrightarrow CH_3\overset{\overset{OH}{|}}{C}HCH_2CH_3 \longrightarrow CH_3\overset{\overset{Br}{|}}{C}HCH_2CH_3(B)$

Grignard reagent of (B) added to (A).

## Chapter Nine

**10** *(a)* Ethene; *(b)* propene; *(c)* 3-hexene; *(d)* 2-methylpropene; *(e)* 4-bromo-1-butene; *(f)* 1,3-pentadiene; *(g)* 1,2,3-hexatriene; *(h)* 2,3-dimethyl-2-butene; *(i)* 2,2,5,5-tetramethyl-3-hexene.

**11** *(b)* 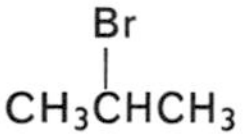 $CH_3\overset{\overset{Br}{|}}{C}HCH_3$ *(e)* 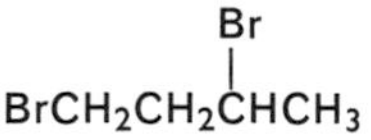 $BrCH_2CH_2\overset{\overset{Br}{|}}{C}HCH_3$

*(f)* largely $CH_3\overset{\overset{Br}{|}}{C}HCH{=}CHCH_3$ *(g)* (complex; omit)

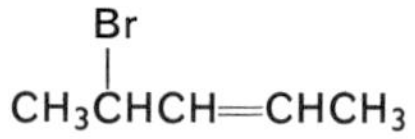

$$\text{Br}$$
*(i)* 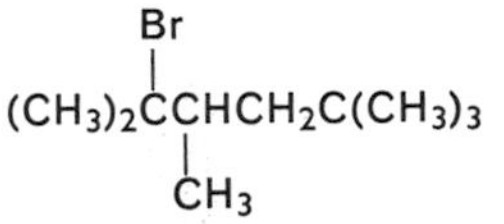 $(CH_3)_2\overset{|}{\underset{|}{C}}CHCH_2C(CH_3)_3$
$$\text{CH}_3$$

**12** *(a)* Ethanol; *(b)* *t*-butyl alcohol; *(c)* 5,5-dimethyl-2-hexanol, *(d)* dimethyl-*n*-propyl-carbinol; *(e)* methyldiethylcarbinol; *(f)* 3-hexanol.

$$\text{Br}$$
**13** *(a)* 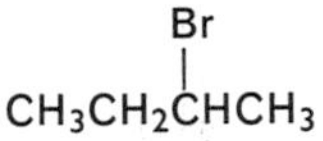 $CH_3CH_2\overset{|}{C}HCH_3$

$$\qquad\qquad\text{Br}\qquad\qquad\qquad\qquad\text{Br}$$
*(b)* $CH_3\overset{|}{C}HCH_2CH_2CH_3$ and $CH_3CH_2\overset{|}{C}HCH_2CH_3$

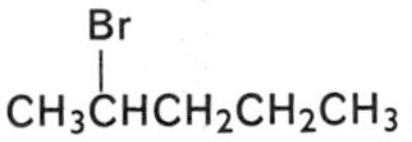

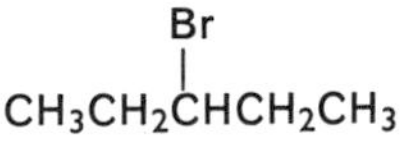

$$\text{Cl Cl}$$
*(c)* $CH_3\overset{|}{C}H\overset{|}{C}HCH_2CH_2CH_3$ as the following ($\pm$) mixture:

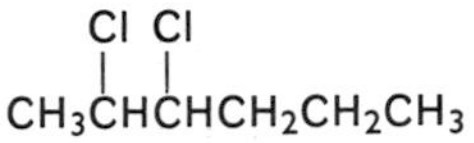

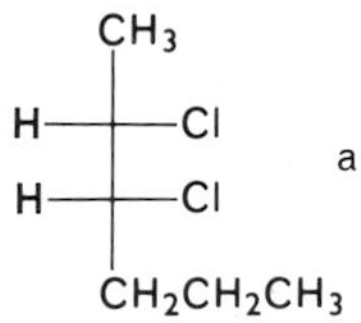

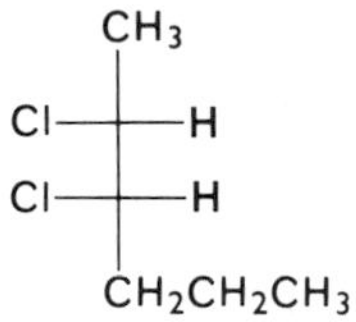

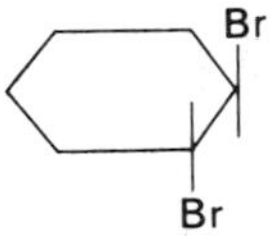

| | | | |
|---|---|---|---|
| | CH$_3$ | | CH$_3$ |
| H— | —Cl | Cl— | —H |
| | | and | |
| H— | —Cl | Cl— | —H |
| | CH$_2$CH$_2$CH$_3$ | | CH$_2$CH$_2$CH$_3$ |

*(d)* [($\pm$)mixture] with Br and Br substituents on cyclohexane ring

*(e)* *cis*-1,2-dimethylcyclohexane

*(f)* *n*-butane   *(g)* ($\pm$)1,2-dibromopropane

*(h)* *t*-butyl bromide

**14** *(a)* propylene will be oxidized readily by aq. KMnO$_4$, and will take up H$_2$ in the presence of a catalyst; cyclopropane will not

*(b)* the cyclopentane derivative will give formaldehyde on ozonization

*(c)* 1-octene will give HCHO with ozone; 4-octene will not

*(d)* 2,4-heptadiene will take up 2 moles of hydrogen on catalytic hydrogenation; 1-methylcyclohexene will take up one mole

*(e)* dicyclohexyl has no double bonds; 1,11-dodecadiene will absorb bromine, can be catalytically hydrogenated, and will react with ozone

**15** *(a)* ethylene   *(b)* 2-butene

*(c)* $(CH_3)_2C{=}CH_2$   *(d)* 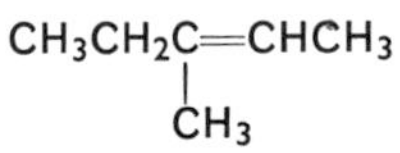 $CH_3CH_2C{=}CHCH_3$
$$\qquad\qquad\qquad\qquad\qquad\qquad\text{CH}_3$$

*(e)* 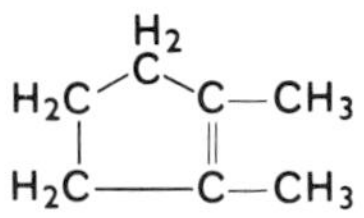 
$$\begin{array}{c} \text{H}_2 \\ \text{H}_2\text{C} \overset{C}{\diagup} \diagdown \text{C}{-}\text{CH}_3 \\ | \quad\quad \| \\ \text{H}_2\text{C}{-}\!\!-\!\!-\text{C}{-}\text{CH}_3 \end{array}$$

*(f)* 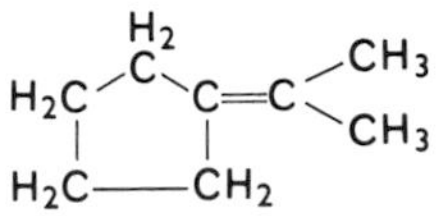 
$$\begin{array}{c} \text{H}_2 \\ \text{H}_2\text{C} \overset{C}{\diagup} \diagdown \text{C}{=}\text{C} \diagup^{\text{CH}_3}_{\diagdown \text{CH}_3} \\ | \quad\quad | \\ \text{H}_2\text{C}{-}\!\!-\!\!-\text{CH}_2 \end{array}$$

*(g)* $CH_3CH{=}C{=}CHCH_3$

*(h)* 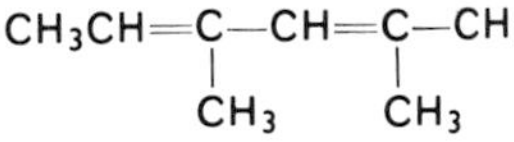 $CH_3CH{=}\underset{|}{C}{-}CH{=}\underset{|}{C}{-}CH_3$
$$\qquad\qquad\quad \text{CH}_3 \qquad \text{CH}_3$$

**16**  A=$CH_3\overset{\overset{O}{\|}}{C}CH\overset{\displaystyle CH_3}{\underset{\displaystyle CH_3}{\diagup}}$ ;  D=$CH_3CH_2\overset{\overset{OH}{|}}{C}\overset{\displaystyle CH_3}{\underset{\displaystyle CH_3}{\diagup}}$

**18**  $CH_3CH_2CH_2CH_2OH \longrightarrow CH_3CH_2CH{=}CH_2 \longrightarrow CH_3CH_2\overset{\overset{OH}{|}}{C}HCH_3.$

**19** Pass mixture through bromine (in an inert solvent); ethylene will be converted to 1,2-dibromoethane, and ethane will pass through. Isolate the dibromo compound, and convert it back into ethylene with zinc in alcohol.

**20** Add bromine to mixture, then distill cyclohexane from higher-boiling 1,2-dibromo-cyclohexane.

**21** One ring; five double bonds.

**22**  $CH_2{=}\overset{\overset{CH_3}{|}}{C}CH_2CH_3+HA \rightleftharpoons \left. CH_3{-}\overset{\overset{CH_3}{|}}{\underset{+}{C}}{-}CH_2CH_3 \right\} A^- \xrightarrow{\overset{CH_3}{|}\atop CH_2{=}CCH_2CH_3}$

$CH_3CH_2{-}\overset{\overset{CH_3}{|}}{\underset{+}{C}}{-}CH_2{-}\overset{\overset{CH_3}{|}}{\underset{\underset{CH_3}{|}}{C}}{-}CH_2CH_3$  (dimer)  $\xrightarrow[\text{before}]{\text{as}}$  (trimer, with loss of H+).

**23** 4.48 ml hydrogen=0.2 millimole. For one mole hydrogen/mole compound, mol. wt. is 19.6/0.2=98. For example,

$C_7H_{14} + H_2 \longrightarrow C_7H_{16}$
0.2 m$M$     0.2 m$M$
(19.6 mg)  (4.48 ml)

If the compound has two double bonds, 0.1 m$M$ compound=19.6 mg; mol. wt.=196.

**24**  $(CH_3)_3CCH{=}CHC(CH_3)_3+HBr \rightleftharpoons \{(CH_3)_3C\overset{+}{C}H{-}CH_2C(CH_3)_3\}Br^- \xrightarrow{\text{rearrange}}$

$(CH_3)_2\overset{+}{\underset{\underset{CH_3}{|}}{C}}{-}CHCH_2(CH_3)_3 \xrightarrow{Br^-} (CH_3)_3C{-}\overset{\overset{Br}{|}}{\underset{\underset{CH_3}{|}}{C}}HCH_2C(CH_3)_3$

**25** (a)  $CH_3C{\equiv}CH+NH_2^- \longrightarrow CH_3C{\equiv}C:^-$
$CH_3C{\equiv}C:^-+CH_3CH_2CH_2Br \longrightarrow CH_3C{\equiv}CCH_2CH_2CH_3$
$CH_3C{\equiv}CCH_2CH_2CH_3+H(Pd) \dashrightarrow cis$-2-hexene

(b) [cyclopentene] + HOCl $\longrightarrow$ [cyclopentane with OH and Cl] $\xrightarrow{CrO_3}$ [cyclopentanone with Cl]

(c) 1. Prepare cyclic ethylene ketal from the keto ester and ethylene glycol.
 2. Reduce —COOEt group with LiAlH$_4$.
 3. Remove —OCH$_2$CH$_2$O— group by acid hydrolysis.

**Chapter Ten**

**11** *(b)* 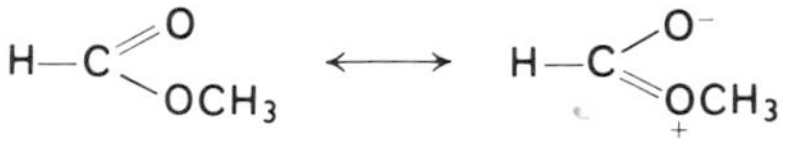 H—C$\overset{O}{\underset{OCH_3}{\diagdown}}$ ⟷ H—C$\overset{O^-}{\underset{OCH_3}{\diagdown}}$

*(c)* BBr$_3$ ⟷ $\overset{+}{Br}$=$\overset{-}{B}$Br$_2$

*(d)*  CH$_3$C$\overset{O}{\underset{O^-}{\diagdown}}$ ⟷ CH$_3$C$\overset{O^-}{\underset{O}{\diagdown}}$

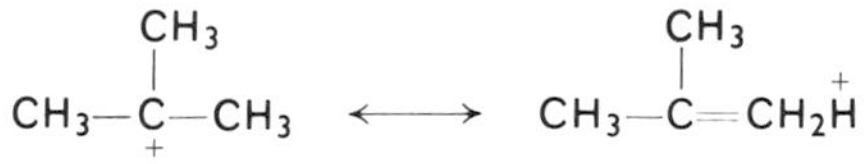

*(e)*  CH$_3$C$\overset{O}{\underset{NH_2}{\diagdown}}$ ⟷ CH$_3$C$\overset{O^-}{\underset{NH_2^+}{\diagdown}}$

*(f)*  CH$_2$=CH—C$\overset{H}{\underset{O}{\diagup}}$ ⟷ $\overset{+}{C}$H$_2$—CH=C$\overset{H}{\underset{O^-}{\diagdown}}$

*(g)* CH$_3$—$\overset{CH_3}{\underset{+}{\underset{|}{C}}}$—CH$_3$ ⟷ CH$_3$—$\overset{CH_3}{\underset{|}{C}}$=CH$_2$$\overset{+}{H}$

*(h)* O—$\overset{+}{N}$$\overset{O^-}{\underset{O}{\diagdown}}$ ⟷ O=$\overset{+}{N}$$\overset{O^-}{\underset{O^-}{\diagdown}}$ ⟷ $^-$O—$\overset{+}{N}$$\overset{O}{\underset{O^-}{\diagdown}}$

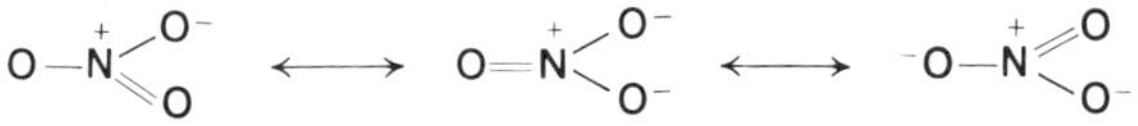

*(i)* H—O—N=O ⟷ H—$\overset{+}{O}$=N—O$^-$

*(j)*  CH$_3$C$\overset{H}{\underset{O}{\diagdown}}$ ⟵ CH$_3$—$\overset{+}{C}$$\overset{H}{\underset{O^-}{\diagdown}}$

**12** H$_2$$\overset{..}{N}$—$\overset{NH_2^+}{\underset{||}{C}}$—$\overset{..}{N}$H$_2$ ⟷ H$_2$$\overset{..}{N}$—$\overset{\overset{..}{N}H_2}{\underset{|}{C}}$=NH$_2^+$ ⟷ H$_2$$\overset{+}{N}$=$\overset{\overset{..}{N}H_2}{\underset{|}{C}}$—$\overset{..}{N}$H$_2$

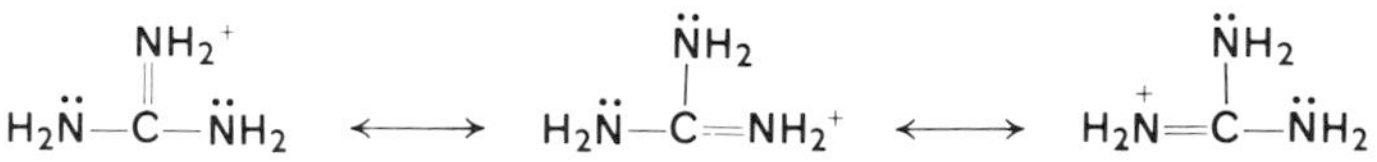

**13** *(b)* 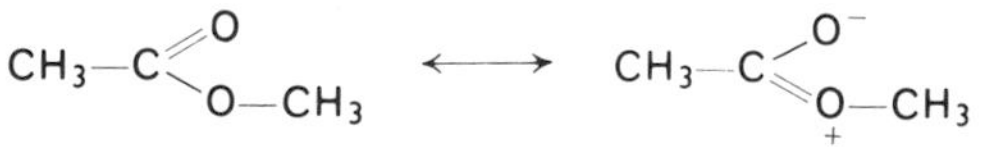 CH$_3$—C$\overset{O}{\underset{O—CH_3}{\diagdown}}$ ⟷ CH$_3$—C$\overset{O^-}{\underset{O—CH_3}{\diagdown}}$

thus, C—O bond "b" normal, C—O bond "a" shorter than normal (i.e., than in an ether)

*(c)* CH$_3$CN ⟷ H$\overset{+}{C}$N$_2$=C=$\overset{-}{N}$; thus, C—C shortened over saturated C—C bond distance

*(d)* CH$_2$=CH—CHO ⟷ $\overset{+}{C}$H$_2$—CH=CHO$^-$; thus, C—C bond "a" somewhat longer than normal C=C; C—C bond "b" somewhat shortened

**14** In CH$_3$—$\overset{OH}{\underset{|}{C}}$=CHCOOEt, ionization of the proton from OH gives the ion

CH$_3$—$\overset{O^-}{\underset{|}{C}}$=CHCOOEt, in which the negative charge is delocalized over the

C=C—C=O system:

*(c)* $(CH_2=CH\ddot{C}HCH=O \longleftrightarrow CH_2=CH-CH=CH-\ddot{\underset{..}{O}}: \longleftrightarrow$

$\qquad\qquad\qquad\qquad\qquad\qquad\qquad :CH_2-CH=CH-CH=O)^-$

*(d)* $(CH_2=CH\ddot{C}HCH=CH_2 \longleftrightarrow :CH_2-CH=CH-CH=CH_2 \longleftrightarrow$

$\qquad\qquad\qquad\qquad\qquad\qquad\quad CH_2=CH-CH=CH-CH_2:)^-$

*(e)* $(CH_2=CH-CH=\overset{+}{O}H \longleftrightarrow \overset{+}{C}H_2-CH=CH-OH)$

**13** *(b)* $CH_3\overset{OH}{\underset{|}{C}}HCH_2COCH_3$
   *(d)* $(CH_3)_2\overset{CH_2OH}{\underset{|}{C}}CHO$

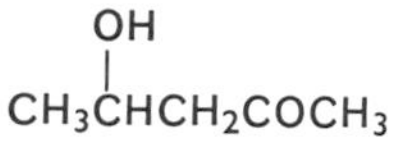

*(e)* $CH_3CO\overset{HOCHCH_2CH_3}{\underset{|}{C}}HCOCH_3$ (followed by dehydration)

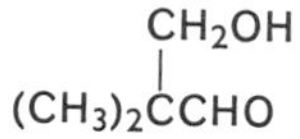

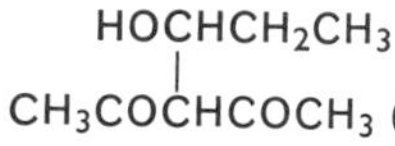

*(f)* $CH_3\underset{\underset{OH}{|}}{C}HCH_2NO_2$

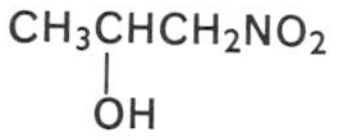

**14** $CH_2=CHCH_2CHO+B:^- \rightleftharpoons (CH_2=CH\ddot{C}HCHO \longleftrightarrow :CH_2CH=CHCHO)^-$
Condensation now occurs at $\gamma$-carbon atom, which attacks acetaldehyde.

**15** $CH_3CO\overset{CH_3}{\underset{\underset{CH_3}{|}}{C}}COCH_3 + OEt^- \rightleftharpoons CH_3-\overset{O^-}{\underset{\underset{OEt}{|}}{C}}-\overset{CH_3}{\underset{\underset{CH_3}{|}}{C}}-COCH_3 \longrightarrow$

$\qquad\qquad\qquad\qquad\qquad\qquad CH_3-C{\overset{O}{\underset{OEt}{\diagup}}} + \left(\overset{CH_3}{\underset{\underset{CH_3}{|}}{\underset{|}{:C}COCH_3}}\right)^-$

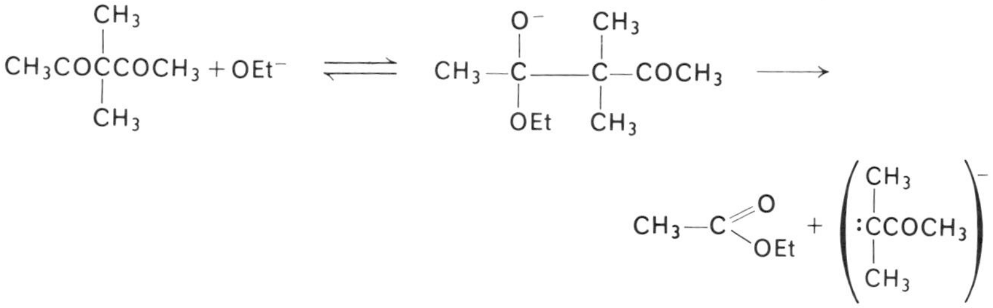

**16** *(a)* $CH_3COCH=CHR$
   *(b)* 

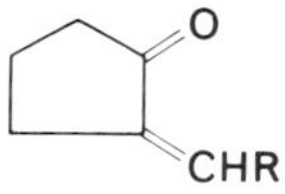

*(c)* $CH_3\overset{CHR}{\overset{\|}{C}}COCH_2CH_3$ or $CH_3\overset{RCH}{\overset{\|}{C}}CO\overset{CHR}{\overset{\|}{C}}CH_3$

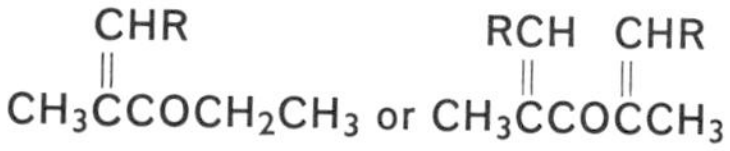

*(d)* $(CH_3)_3CCOCH=CHR$; in these $R=$

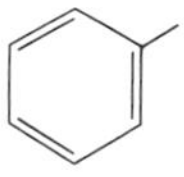

**Chapter Fourteen**

**8** *(a)* $RCOOH+H_2O \rightleftharpoons RCOO^-+H_3O^+$
*(b)* $RCOOH+NH_3 \rightleftharpoons RCOO^-+NH_4^+$
*(c)* $RCOOH+EtNH_2 \rightleftharpoons RCOO^-+EtNH_3^+$
*(d)* $RCOOH+EtO^- \rightleftharpoons RCOOH+EtOH$
*(e)* $RCOOH+EtOH(H_2SO_4) \rightleftharpoons RCOOEt+H_2O$
*(f)* $RCOOH+Me_3N \rightleftharpoons RCOO^-+Me_3NH^+$   $(R=CH_3CH_2-)$
**9** Refer to text.
**10** Refer to text.

**11** *(a)* RCOCl + EtOH $\longrightarrow$ RCOOEt + HCl

*(b)* RCOCl + $(CH_3)_2CHCH_2OH \longrightarrow$ RCOOCH$_2$CH(CH$_3$)$_2$ + HCl

*(c)* RCOCl + CH$_3$COO$^-$Na$^+$ $\longrightarrow$ RCOOCOCH$_3$ + Cl$^-$ + Na$^+$

*(d)* RCOCl + H$_2$O $\longrightarrow$ RCOOH + HCl

*(e)* RCOCl + NH$_2$NH$_2$ $\longrightarrow$ RCONHNHCOR + 2HCl

*(f)* RCOCl + HONH$_2$ $\longrightarrow$ RCONHOH + HCl

*(g)* RCOCl + MeNHEt $\longrightarrow$ RCON$\begin{smallmatrix} \diagup Me \\ \diagdown Et \end{smallmatrix}$ + HCl

[Note: In *e*, *f*, and *g* the HCl formed would react with the basic amine to give the corresponding salt.]

**12** *(a)* nBuBr + KCN $\longrightarrow$ nBuCN; then hydrolysis; *(b)* CH$_3$COCl + sodium acetate (see 11c); *(c)* See 11g; then reduce CH$_3$CH$_2$CONH$_2$ with LiAlH$_4$; *(d)* BrCH$_2$CH$_2$Br + KCN see 12a; *(e)* Saponify with NaOH.

**13** *a*, *b*, and *d*.

**14** *(a)* Remove the n-butyric acid with NaHCO$_3$; remove pentanal with sodium bisulfite; remove di-n-butyl ether with H$_3$PO$_4$; octane remains

*(c)* 2-hexanone will form a bisulfite addition compound, 3-hexanone will not

**15** Acetone + CH$_3$MgBr $\longrightarrow$ t-butyl alcohol $\longrightarrow$ t-butyl chloride $\longrightarrow$ Grignard reagent $\xrightarrow{\text{CO}_2}$ trimethylacetic acid.

## Chapter Fifteen

**6** Products are

*(a)* CH$_3$COCHCOOEt, with $\overset{\text{CH}_3}{\underset{|}{}}$ on the CH

*(b)* CH$_3$CH$_2$CHCH(COOEt)$_2$, with $\overset{\text{CH}_3}{\underset{|}{}}$

*(c)* CH$_3$COCH$_2$CH(COOEt)$_2$

*(d)* cyclohexanone with =O, Et, and COOEt substituents

*(e)* CH$_3$COCHCOOMe with CH$_2$COOEt branch
      $\quad\quad\quad\quad\quad|$
      $\quad\quad\quad$CH$_2$COOEt

**7** *(a)* via n-butyl bromide and acetoacetic ester

*(b)* via 5c

*(c)* via diethyl ethylmalonate and BrCH$_2$COOEt

*(d)* alkylate [H$_3$C-substituted cyclohexanone with =O and COOEt] with EtBr

*(e)* via LiAlH$_4$ reduction of H$_2$C$\overset{\overset{\text{H}_2}{\text{C}}}{\diagup\diagdown}$CHCOOH, which can be prepared via BrCH$_2$CH$_2$Br and malonic ester

*(f)* see 6c

*(g)* HOCH$_2$—C—CH$_2$OH, by reduction of (CH$_3$)$_2$C(COOEt)$_2$.
       $\quad\quad\;\;|$
       $\quad\quad$ with CH$_3$ above and CH$_3$ below the central C

(How could this diol be prepared with the use of an aldol condensation?)

**8** *(a)* $(\ddot{C}H_2CHO)^-$

*(b)* $(CH_3CO\ddot{C}H_2)^-$

*(c)* $[(CH_3)_3CCO\ddot{C}H_2]^-$

*(d)* $(CH_3CO\ddot{C}HCOCH_3)^-$

*(e)* $[(CH_3CO)_3\ddot{C}]^-$

*(f)* $[\ddot{C}H(COOMe)_2]^-$

*(g)* $[CH_3\ddot{C}(COOMe)_2]^-$

*(h)* $[\ddot{C}(COOEt)_3]^-$

*(i)*

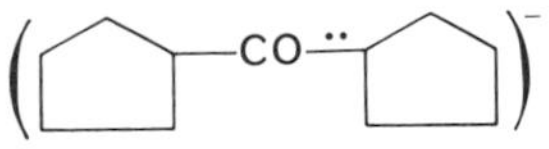

**9** Refer to analogous example in text.

**10**

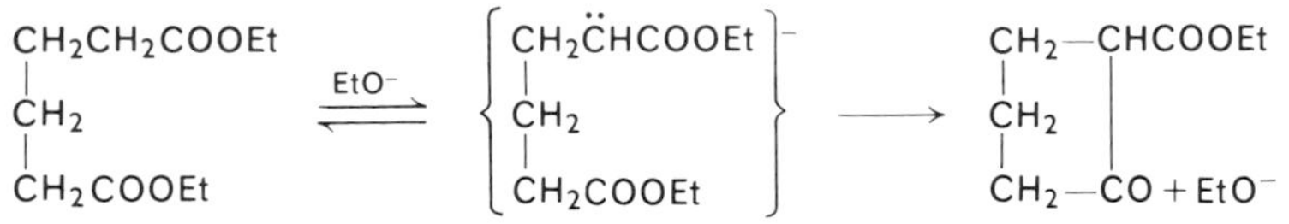

$$\begin{array}{c} CH_2CH_2COOEt \\ | \\ CH_2 \\ | \\ CH_2COOEt \end{array} \underset{EtO^-}{\rightleftharpoons} \left\{ \begin{array}{c} CH_2\ddot{C}HCOOEt \\ | \\ CH_2 \\ | \\ CH_2COOEt \end{array} \right\}^- \longrightarrow \begin{array}{c} CH_2-CHCOOEt \\ | \quad\quad | \\ CH_2 \\ | \\ CH_2-CO+EtO^- \end{array}$$

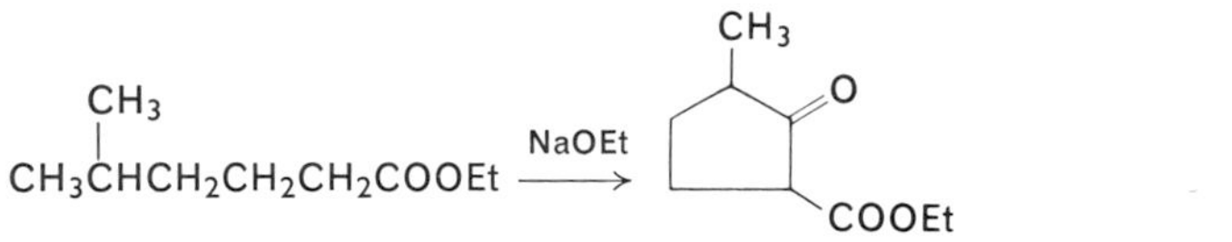

**11** $\underset{\quad\ \ |\ \ \quad}{CH_3\overset{CH_3}{CH}CH_2CH_2CH_2COOEt} \xrightarrow{NaOEt}$

**12** $\underset{\quad\ |}{CH_3CH=CCH_2CH_3} + CH_2=CHCOOMe \longrightarrow CH_3CHCOCH_2CH_3$
$\quad\quad NR_2 \qquad\qquad\qquad\qquad\qquad\qquad\quad\ \ CH_2CH_2COOMe$
$\qquad\qquad\qquad\qquad\qquad\qquad\qquad\qquad\qquad\quad\ \text{(after final hydrolysis)}$

## Chapter Sixteen

**22** Methylcyclohexane has a plane of symmetry.

**23** Two products, diastereomers: same configuration at

$$\text{C-5} \left( \begin{array}{c} CH_3 \\ | \\ -CH- \end{array} \right)$$

both configurations at C-2 ($-CHBr-$).

**24** *(a)* Glycerol (nonresolvable); *(b)* can give both *cis-* and *trans*-1,4-cyclohexanediol (nonresolvable).

**25** *(a)*

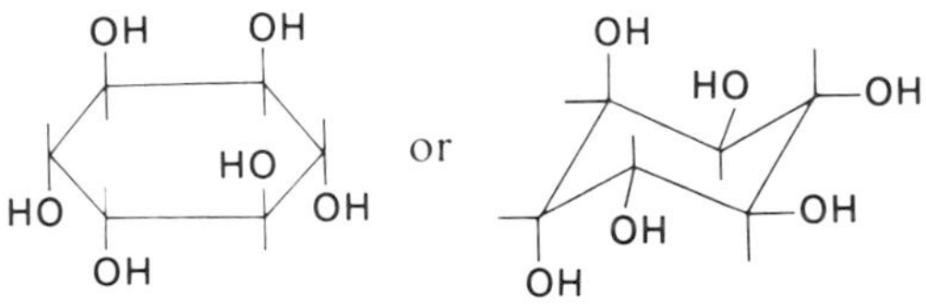

for one of the active isomers.

**26** *(a)* cyclohexanone $\longrightarrow$ cyclohexanol $\longrightarrow$ cyclohexene; oxidize with KMnO$_4$ or OsO$_4$

*(b)* fumaric acid $\longrightarrow$ *trans*-2-butene-1,4-diol, oxidize via the epoxide; *(c)* same, except hydroxylate with OsO$_4$ (see Chapter 9)

**27** 2-Butanol $\longrightarrow$ 2-bromobutane $\longrightarrow$ *sec*-butyl MgBr $\longrightarrow$ carbonation to 2-methylbutanoic acid. Resolve the acid with the active amine, then reduce with LiAlH$_4$.

**28** *(a)* via the ester; *(b)* oxidize the active alcohol to the acid, resolve via the salt; *(c)* reaction of the anhydride with the alcohol gives two diastereometric half ester-acids; separate, saponify each.

**29** Cholesterol has eight asymmetric carbon atoms.

**30** *b* and *e*. Others have a plane of symmetry. Compound *e* is resolvable by virtue of the asymmetric carbon atom, *not* as an allene.

## Chapter Seventeen

**6** *(a)* Starting with $HOOC(CH_2)_5COOH$, via Dieckmann reaction on diester; *(b)* via malonic ester + $BrCH_2CH_2CH_2CH_2CH_2Br$; *(c)* by Diels-Alder reaction; *(d)* by aldol condensation; for example, of 5-ketohexanal.

**8** *(a)* Chlorine equatorial; *(b)* both —$CH_3$ equatorial; *(c)* see text; *(d)* bromine atoms equatorial.

**9**

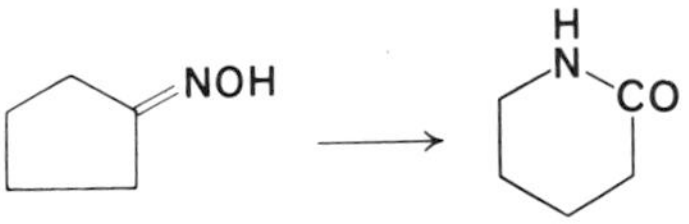

**10**

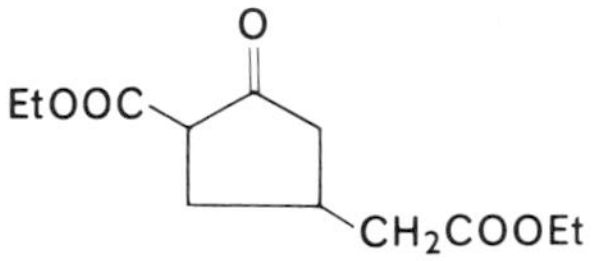

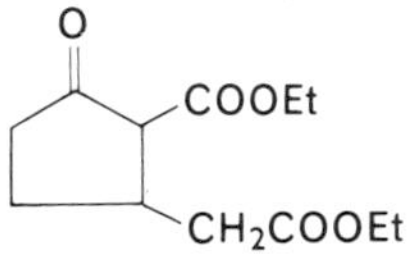

Both would probably be formed.

**11** Gives the lactone of 4-hydroxypentanoic acid.

**12** *Trans*-acid gives *cis*-anhydride.

**13**

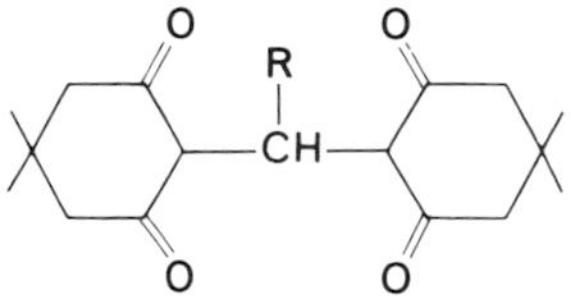

:  *(a)* $R=CH_3$;  *(b)* $R=(CH_3)_2CH$;  *(c)* $R=H$.

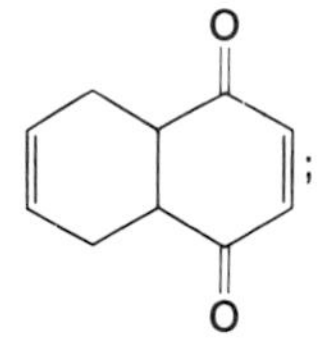

**14**

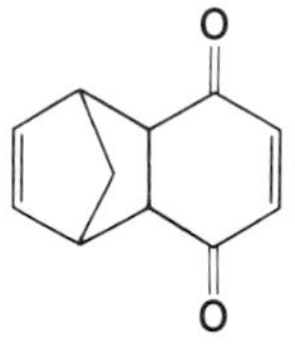

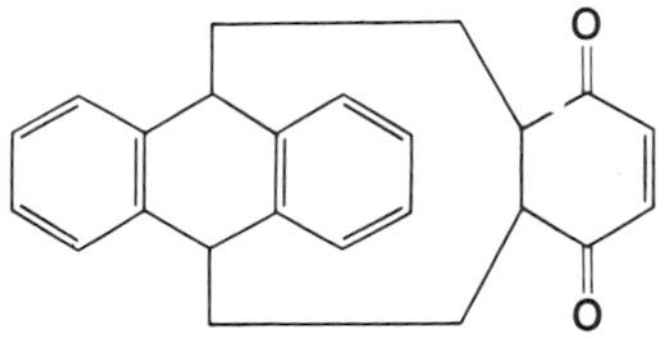

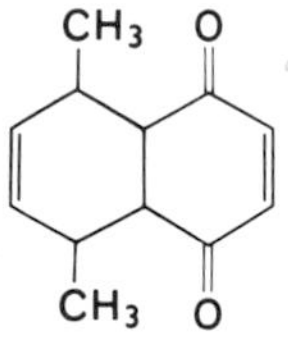

### Chapter Eighteen

**10** *(a)* One; *(b)* one; *(c)* two; *(d)* five; *(e)* two; *(f)* one; *(g)* two; *(h)* none. Products from *(e)* are 1 formaldehyde, 5 formic acid; from *(f)*, EtOOCCHO; from *(g)*, succinic dialdehyde.

**11** *(a)* Same; *(b)* one would, one would not; *(c)* yes; *(d)* D-ribose.

**13** The OH in the 3-position (carbonyl numbered 1).

**14** L-Rhamnose is

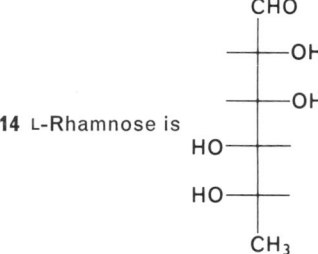

**15** The glucoside is $\alpha$- (or $\beta$-) methyl D-glucofuranoside.

### Chapter Twenty

**4** *(a)* 2, 3, and 4 positions; *(b)* 3 and 4 positions (Br, Br=1, 2); *(c)* Four possible (all positions different); *(d)* only one; *(e)* two possible: 2 and 4 (Cl, Cl=1, 3).

**5** 1, 2, 3, 4; 1, 2, 4, 5; 1, 2, 3, 5. There are three tetramethylbromobenzenes (one from each of the three hydrocarbons).

**6** Hint: *(a)* Disregard conformational isomers; *(b)* since the electrons are not delocalized in a planar, symmetrical orbital (as in benzene), two 1,2-dibromo compounds are theoretically possible.

**8** 1,3-dimethyl-4-bromobenzene.

**9** *p*-Xylene can give only one mononitro compound; ethylbenzene can give three.

**10** *(a)* Substituent enters at position 2; *(b)* substituent can enter at two positions, *ortho* to $CH_3$, *ortho* to $CH_2CH_3$.

**11** *(a)* Ethylbenzene; *(b)* *sec*-butylbenzene; *(c)* *t*-butylbenzene; *(d)* allylbenzene; *(e)* *t*-butylbenzene; *(f)* *sec*-butylbenzene.

### Chapter Twenty-one

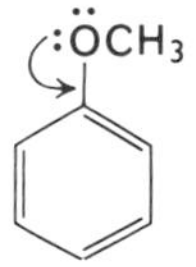

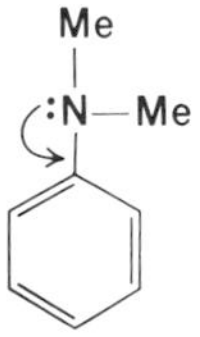

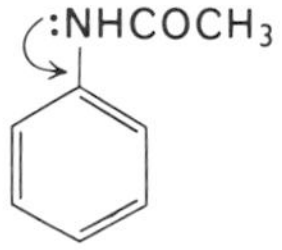

**3**

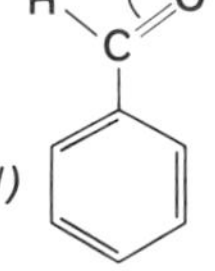

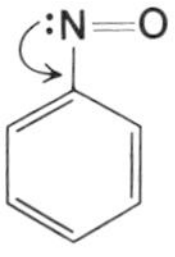

**4** The contribution of the form ⁻O⁻N⁺=⟨benzene ring⟩ to the structure of nitrobenzene is not

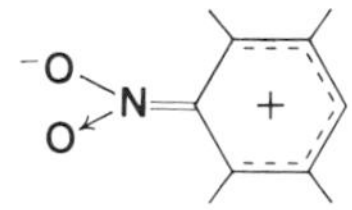

as large as the contribution of the dipolar form in the case of *p*-nitrodimethylaniline; thus steric inhibition of its contribution does not cause so large a change in dipole moment.

**5** There is no steric interference with the contribution Br⁺=⟨ring with H₃C CH₃ groups, (−) ⟩.

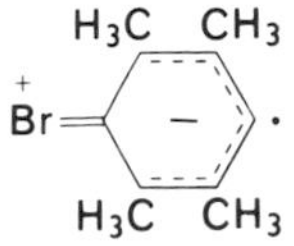

**6** *(a) Meta; (b) ortho* to NHAc; *(c) ortho* to OMe; *(d) ortho* to $CH_2CH_2NO_2$; *(e) ortho* to OMe; *(f) ortho* to $CH_3$; *(g) ortho* to OMe; *(h) ortho* to NHAc; *(i) ortho* to CN, *(j) ortho* to $NH_2$.

**7** *b, c, d, f.*

## Chapter Twenty-two

**2** *(a)* nitrate bromobenzene, reduce $NO_2$ to $NH_2$
*(b)* nitrate acetanilide, hydrolyze $NHCOCH_3$ to $NH_2$
*(c)* nitrate phenetole
*(d)* nitrate phenyl acetate, reduce $NO_2$ to $NH_2$ (catalytically)
*(e)* nitrate anisole, reduce $NO_2$ to $NH_2$
*(f)* dinitrate anisole; better: treat 2,4-dinitrochlorobenzene with NaOMe to replace Cl by OMe

**3**

(a) 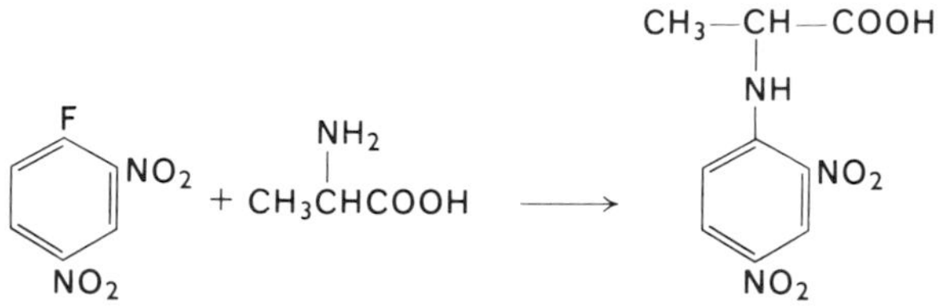 [F, NO₂, NO₂ substituted benzene] $+ CH_3\overset{NH_2}{\underset{}{CH}}COOH \longrightarrow$ [benzene with $NH-CH(CH_3)-COOH$, two $NO_2$ groups]

(c) 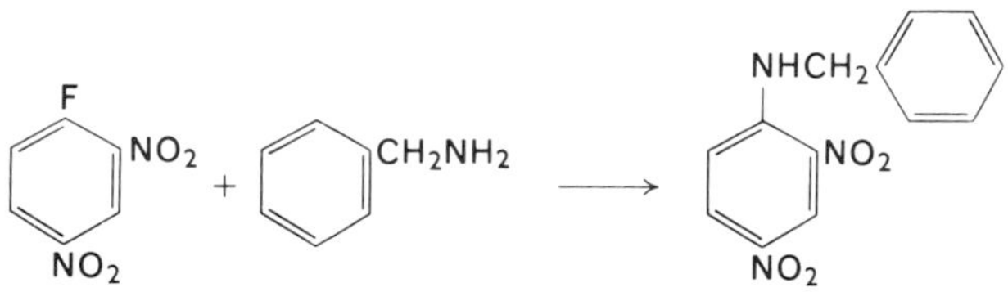 [F, NO₂, NO₂ substituted benzene] $+$ [benzene with $CH_2NH_2$] $\longrightarrow$ [benzene with $NHCH_2$-phenyl, two $NO_2$ groups]

**4** Products are *(a)* 2,6-dimethyl-4-nitro-N-methylaniline; *(b)* 2-methoxy-3-nitrotoluene; *(c)* *p*-nitrophenylhydrazine; *(d)* *p*-nitrosoaniline; *(e)* 2-bromo-4-nitrodiphenylamine.

**5** Greater stability of $C_6H_5CH_2^+$ than $C_6H_{13}CH_2^+$; *p*-methoxybenzyl chloride would be more reactive because *p*-$OCH_3$ participates in delocalization of + charge.

**6** Prepare 4-chlorobutyl benzenesulfonate, treat this with benzylmagnesium chloride.

**7** *(a)* benzene ⟶ bromobenzene ⟶ phenylmagnesium bromide ⟶ benzoic acid ⟶ cyanobenzene (benzonitrile) ⟶ benzylamine *(b)* toluene ⟶ benzyl chloride ⟶ benzyl cyanide ⟶ phenylacetic acid.

*(c)* toluene ⟶ *p*-methylacetophenone ⟶ *p*-toluic acid ⟶ benzene-1,4-dicarboxylic acid (terephthalic acid)

*(d)* benzene ⟶ bromobenzene ⟶ phenylmagnesium bromide; allow to react with diethyl carbonate

*(e)* bromobenzene ⟶ toluene (via Wurtz-Fittig) ⟶ *p*-bromotoluene ⟶ *p*-xylene (Wurtz-Fittig)

**8** Via Ullmann reactions with appropriate bromo or iodo compounds.

[Note: Syntheses can nearly always be accomplished by more than one route. The above are suggested methods, but others can be devised.]

## Chapter Twenty-three

**8** *(a)* by direct Friedel-Crafts acetylation

*(b)* see Chapter 22, problem 7

*(c)* reduce to ethylbenzene, then ⟶ *p*-ethylacetophenone, reduce

*(d)* benzoic acid ⟶ benzyl chloride (2 steps) ⟶ 3-phenylpropanoic acid (via malonic ester) ⟶ ring closure by intramolecular Friedel-Crafts

*(e)* *n*-BuMgBr addition, dehydration, reduction (catalytic)

**9**

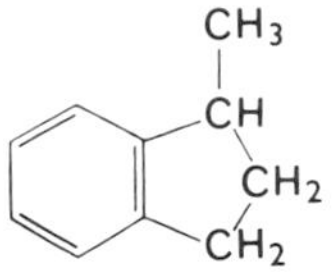

$CH_3$
$CH$
$CH_2$
$CH_2$

**10 (A)** can be

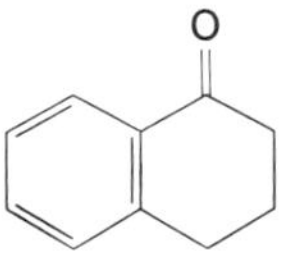

$O$

**11** Toluene ⟶ benzyl chloride ⟶ phenylacetonitrile ⟶ phenylacetic acid ⟶ benzoic acid.

**12** *(a)* the olefin will absorb bromine, react with cold $KMnO_4$

*(b)* benzyl bromide has a very reactive bromine ($AgNO_3$)

*(c)* phenylacetone will give the iodoform test

*(d)* Oxidation ($KMnO_4$); *p*-bromobenzoic acid from the monoalkyl compound

*(e)* iodoform test; carbonyl reagents

[Note: Other tests may be devised, but they should be direct, not requiring elaborate structure proof procedures.]

## Chapter Twenty-four

**6** Entering nitro group goes *(a) para; (b) ortho* to either methyl; *(c) ortho* to $CH_3$; *(d) para*; *(e) para; (f) ortho* to —$NHCOCH_3$; *(g)* position 4; *(h) ortho* to $CH_3$; *(i) meta; (j) ortho* to $CH_3$; *(k)* position 4.

**7** *(a)* aniline ⟶ acetanilide ⟶ *p*-nitroacetanilide; hydrolyze

*(b)* acetanilide ⟶ *p*-bromoacetanilide ⟶ 4-bromo-2-nitroacetanilide ⟶ 4-bromo-2-nitroaniline; oxidize —$NH_2$ to —$NO_2$ (see text)

*(c)* toluene $\longrightarrow$ *o*-nitrotoluene (separated from *p*-compound) $\longrightarrow$ *o*-N-acetyl toluidine $\longrightarrow$ 2-acetylamino-5-nitrotoluene $\longrightarrow$ hydrolyze to —$NH_2$, oxidize —$NH_2$ to —$NO_2$

*(d)* ammonolysis (with alcoholic $NH_3$) of 2,4-dinitrochlorobenzene

**8** Displacement of $Br^-$ by nucleophilic attack of $CH_3CH_2COO^-$.

**9** Displacement of $Br^-$ from Aryl—$COCH_2Br$.

**10** *(a)* Prepare 2,4-dinitrotoluene; *(b)* prepare *p*-nitrosodimethylaniline; *(c)* condense these as in Section 24.4; *(d)* hydrolyze Schiff's base.

**11** Condense with Aryl—NO to give Aryl—N=C(COOEt)$_2$; hydrolyze.

**14** Condensation of aniline with *p*-nitrosoanisole.

## Chapter Twenty-five

**5** Products are *(a)* aniline hydrobromide (anilinium bromide); *(b)* salt (if not heated): N-methylanilinium acetate; *(c)* aniline ($+ NaCl + MeOH$); *(d)* aniline ($+$ sodium acetate); *(e)* *p*-chloroacetanilide.

**6** It is an amide (of nitrous acid).

**7** By reductive alkylation (aniline + propionaldehyde in the presence of $H_2/Pt$).

**8**
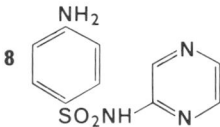

**9** Via Hofmann or Curtius reaction (Chapter 28); or, via —COOH $\longrightarrow$ COCl $\longrightarrow$ —COCH$_3$ (with (CH$_3$)$_2$Cd) $\longrightarrow$ oxime $\longrightarrow$ —NHCOCH$_3$ (Beckmann rearrangement) $\longrightarrow$ —NH$_2$.

**10** R—NHSO$_2$R + OH$^-$ $\rightleftharpoons$ (RṄSO$_2$R)$^-$ + H$_2$O. This can now be alkylated by reaction with an alkyl halide:

$$(\text{R\"NSO}_2\text{R})^- + \text{CH}_3\text{I} \longrightarrow \overset{\overset{\textstyle CH_3}{\textstyle |}}{\text{RNSO}_2\text{R}} + \text{I}^-$$

Hydrolysis of the N-alkylated sulfonamide gives the N-alkyl amine.

**11** Compound A possesses a carbonyl group of the alkanone type, such as that found in acetone; since the phenyl group is attached to carbon (benzoic acid formed on oxidation), one possible structure for the compound is $C_6H_5CH_2COCH_2CH_2NH_2$. Compound B has a carbonyl group whose low C=O stretching frequency is that of an amide, which is in agreement with the fact that the compound is insoluble in dilute acid. Thus, a structure for compound B is $C_6H_5CH_2CH_2CH_2CONH_2$.

## Chapter Twenty-six

**14** Products are *(a)* N-nitroso-N-methylaniline; *(b)* N-nitroso-di-*n*-butylamine; *(c)* no new product; *(d)* *p*-chlorobenzenediazonium salt; *(e)* diazonium salt; *(f)* *p*-nitroso-N,N-diethylaniline.

**15** *(a)* anisole $\longrightarrow$ *p*-nitroanisole $\longrightarrow$ amine; diazotize; replace with —CN

*(b)* nitrobenzene $\longrightarrow$ *m*-bromonitrobenzene $\longrightarrow$ amine; diazotize; replace with Br

*(c)* aniline $\longrightarrow$ 2,4,6-tribromoaniline $\longrightarrow$ remove —NH$_2$ via diazotization, H$_3$PO$_2$

*(d)* nitrobenzene ⟶ *m*-bromonitrobenzene ⟶ *m*-bromoaniline ⟶ *m*-bromoacetanilide; nitrate; reduce

*(e)* toluene ⟶ *o*-nitrotoluene; reduce, diazotize, replace by —Cl

*(f)* *m*-dinitrobenzene ⟶ *m*-nitroaniline; diazotize, replace by OH

*(g)* anisole ⟶ *p*-bromoanisole ⟶ Grignard reagent ⟶ *p*-methoxybenzoic acid; demethylate with HBr

*(h)* *m*-dibromobenzene (as in *b*) ⟶ 2,4-dibromonitrobenzene; reduce

*(i)* *p*-chloronitrobenzene ⟶ *p*-chloroaniline ⟶ *p*-chloroacetanilide; nitrate; reduce; hydrolyze

**16** *c, d, f, g, j.*

**17** *(a)* from aniline via benzonitrile (replacement of $N_2^+$ by CN)

  *(b)* replacement of $N_2^+$ by OH; methylation

  *(c)* coupling of diazotized aniline with phenol

  *(d)* from *p*-bromoaniline, replacement of $N_2^+$ by Br

  *(e)* From *(c)* followed by reduction with $Na_2S_2O_4$

**19** *(a)* N-methyl-*o*-toluidine + HONO ⟶ N-nitroso compound (neutral)

  *(b)* *m*-toluidine ⟶ diazonium salt (forms dye with a phenol)

  *(c)* the amino compound is soluble in dilute aqueous acid

  *(d)* the salt has ionic bromine; detect with $AgNO_3$

  *(e)* phenol is acidic, soluble in dilute alkali

  *(f)* the phenol is acidic

  *(g)* the benzyl bromide has an active halogen ($AgNO_3$)

**20** By selective solubility in dilute aqueous acid and alkali.

**21** Convert to isophthalic (1,3-) acid. This is a known reference compound.

**22** See text.

**22** See text.

**23** *p*-Hydroxybenzyl alcohol (or products derived from it).

**24** The two *ortho*-methyl groups provide a steric inhibition of the resonance delocalization that, in N,N-dimethylaniline, permits participation of the amino nitrogen atom in charge delocalization in the transition state of the electrophilic attack upon the ring.

## Chapter Twenty-seven

**7** *h > f > b > g > d > c > a > e > i.*

**8** *(a)* via *p*-bromoaniline

  *(b)* via sulfonation of benzene

  *(c)* via Kolbe synthesis of salicylic acid

  *(d)* as in *(c)*

  *(e)* via *p*-nitroanisole

  *(f)* via *p*-nitro-*n*-butylbenzene; or via *p*-hydroxybutyrophenone

  *(g)* Reimer-Tiemann on *p*-cresol

  *(h)* via 3,4-dimethoxyacetophenone [Friedel-Crafts on catechol dimethyl ether (veratrole ]

**10** Compound A can be *p*-hydroxybenzyl alcohol; compound B would then be *p*-methoxybenzyl alcohol, and the acid would be *p*-methoxybenzoic acid.

**12** The glucoside is arbutin; that is, the monoglucoside of hydroquinone; the phenol D is *p*-methoxyphenol; E is *p*-dimethoxybenzene.

**13** Picein is the glucoside of *p*-hydroxyacetophenone.

## Chapter Twenty-eight

**5** Products are

(a) 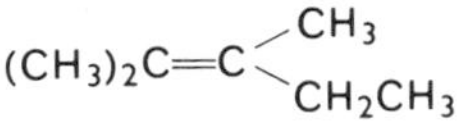 $(CH_3)_2C=C\underset{CH_2CH_3}{\overset{CH_3}{<}}$

(b) mixture of $CH_3CH_2\underset{\overset{|}{CH_3}}{C}=CH_2$, $CH_3CH_2CH=CHCH_3$

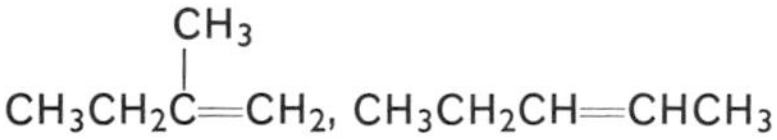

(c) 1,2-dimethylcyclopentene

(d) 1-methyl-2-ethylcyclohexene, perhaps some

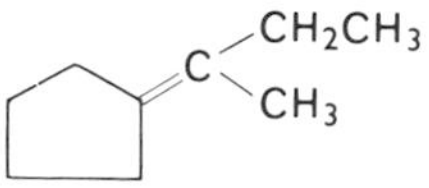

(e) 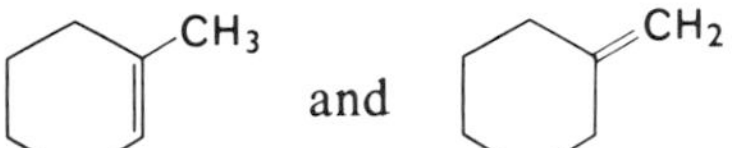 and

**6** Methanol, dimethyl ether; methyl ethyl ether.

**7** Products are

(a) 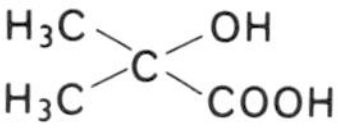 $H_3C\underset{H_3C}{\overset{}{>}}C\underset{COOH}{\overset{OH}{<}}$

(b)

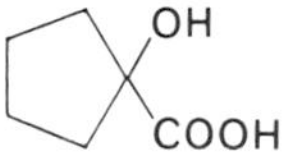

(c)

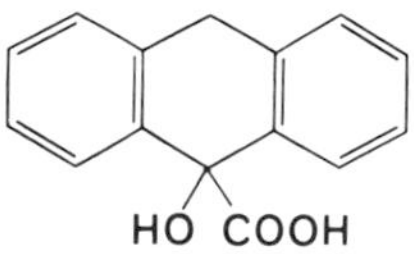

(d)

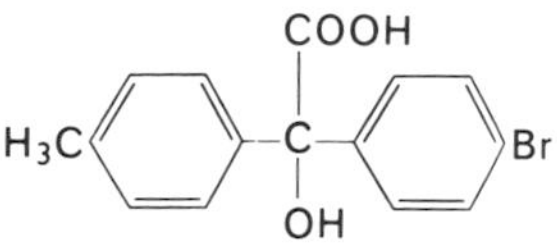

**8** 2,6-Dimethyl-1,4-dihydroxybenzene.

**9** By oxidation with a peracid; $ArCOCH_3 \longrightarrow ArOCOCH_3$.

**10** Succinic acid $\longrightarrow$ 1,4-butane-diol $\longrightarrow$ adipic acid (via dibromide, dinitrile) $\longrightarrow$ cyclopentanone $\longrightarrow$ lactone of 5-hydroxypentanoic acid (via peracid oxidation).

**11** (a) Can introduce $-NH_2$ by first preparing the acid (i.e., benzoic acid $\longrightarrow$ aniline).

**12** (a) via Beckmann rearrangement

   (b) Beckmann rearrangement; $LiAlH_4$ reduction.

   (c) adipic acid $\longrightarrow$ cyclopentanone $\longrightarrow$ 2-piperidone $\longrightarrow$ piperidine by $LiAlH_4$ reduction

   (d) benzoic acid $\longrightarrow$ phenylacetic acid $\longrightarrow$ hydrocinnamic acid (Arndt-Eistert reaction)

**13** By (1) Schmidt reaction, leading to cyclic lactam of 6-aminohexanoic acid; (2) reaction of $-NH-CO-$ grouping with $HN_3$.

**14** Reaction starts by following step:

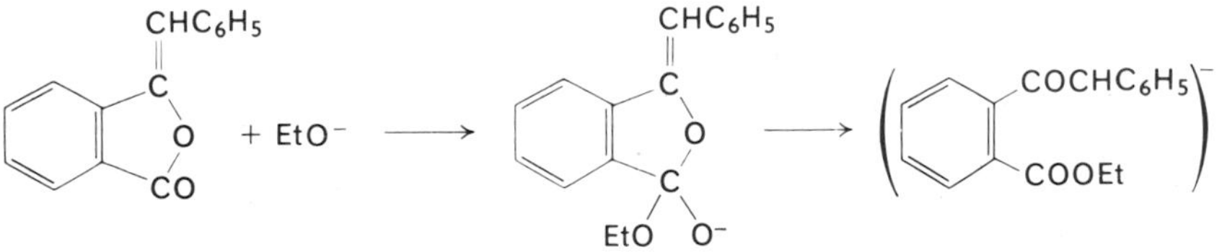

## Chapter Twenty-nine

**5** Product is

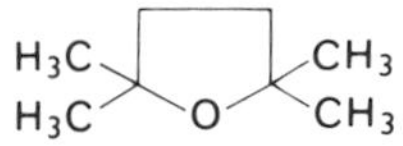

**6** Product is pinacolone.

**7** N,N,N′,N′-tetramethylethylenediamine.

**8** Proceeds via cyclic imonium ion :

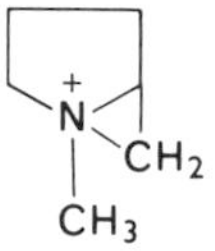

**9** Products are *(a)* 2-ethoxypyridine; *(b)* 2-hydrazinopyridine; or 1,2-dipyridylhydrazine; *(c)* 2-methoxy-3-bromopyridine; *(d)* N-ethyl-2-pyridone; *(e)* 2-aminoquinoline.

**10** Proceeds via 3,4-dimethoxyphenethylamine (by reduction of the nitrile), then Bischler-Napieralski reaction; finally, dehydrogenation (heat with Pd black).

**11** Prepare gramine from indole by means of the Mannich reaction. Gramine can be used as an alkylating agent; for example, it reacts with the sodium derivative of ethyl malonate to give the following indole derivative:

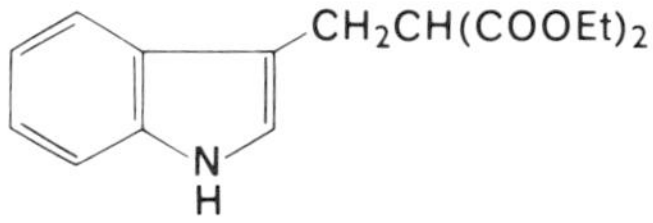

## Chapter Thirty-two

**4** *(a)* Formaldehyde+

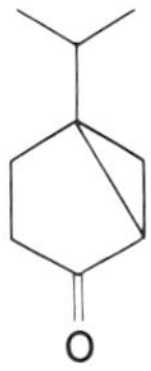

*(b)*

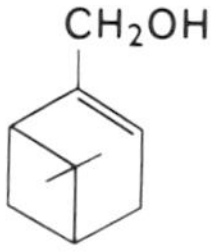

*(c)* O-benzoyl derivative

*(d)* saturated $C_{15}$ alcohol

*(e)* addition of bromine to C=C

*(f)* coupling *ortho* to OH

*(g)* the corresponding aldehyde is formed

*(h)* gives the 4,5-unsaturated-3-ketone

**5** $\gamma$-Carotene would give acetone as a product.

**6** All *trans-*, all *cis-*, and those in which one, two, and three of the double bonds are *cis*, the others *trans*.

**7** via

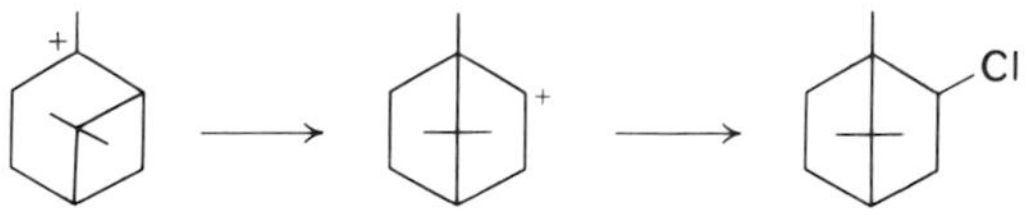

**8** via   +CH₂

$+CH_2$

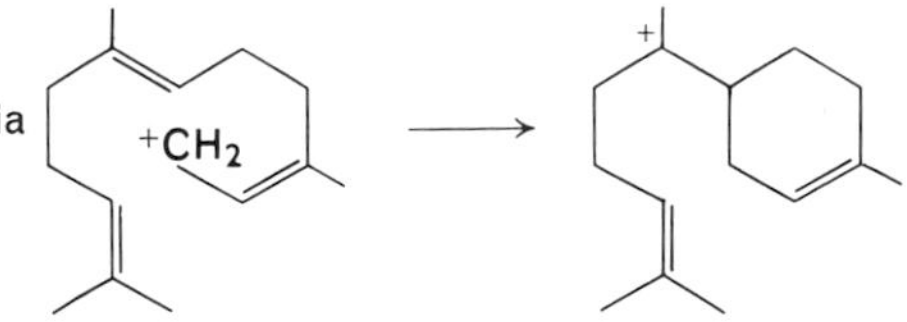

**9** Acetic acid  $\overset{\bullet}{C}H_3COOH$  ⟶   $\overset{\bullet C}{\underset{\bullet C}{>}}C-\overset{\bullet}{C}-C$  ⟶

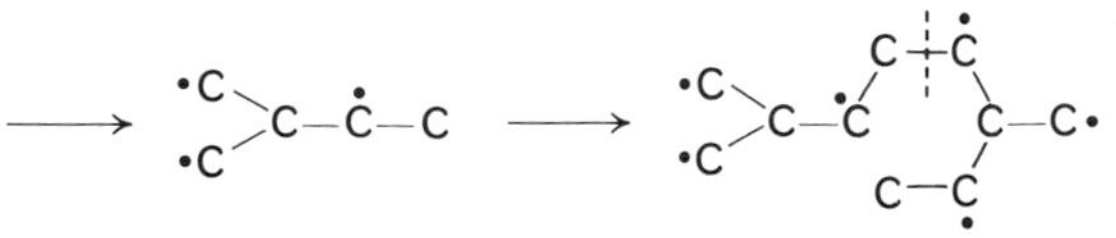

**10** Hint: Starting with the coenzyme A derivative of oleic acid, condense with four malonyl CoA units. The compound shown in Exercise 10 is a cyclic aldol formed by addition of a terminal —COCH₃ group to a carbonyl group in the chain.

# Index